INFRARED DETECTORS

SECOND EDITION

INFRARED DETECTORS

SECOND EDITION

ANTONI ROGALSKI

CRC Press
Taylor & Francis Group
Boca Raton   London   New York

CRC Press is an imprint of the
Taylor & Francis Group, an **informa** business

CRC Press
Taylor & Francis Group
6000 Broken Sound Parkway NW, Suite 300
Boca Raton, FL 33487-2742

First issued in paperback 2020

© 2011 by Taylor and Francis Group, LLC
CRC Press is an imprint of Taylor & Francis Group, an Informa business

No claim to original U.S. Government works

ISBN-13: 978-0-367-57709-4 (pbk)
ISBN-13: 978-1-4200-7671-4 (hbk)

Library of Congress Cataloging-in-Publication Data

Rogalski, Antoni.
    Infrared detectors / Antoni Rogalski. -- 2nd ed.
        p. cm.
    Includes bibliographical references and index.
    ISBN 978-1-4200-7671-4 (hardcover : alk. paper)
    1. Infrared detectors. I. Title.

TA1570.R63 2011
621.36'2--dc22
                                                  2010029532

Visit the Taylor & Francis Web site at
http://www.taylorandfrancis.com

and the CRC Press Web site at
http://www.crcpress.com

In memory of my daughter Marta

Table of Contents

Preface

Progress in infrared (IR) detector technology has been mainly connected to semiconductor IR detectors, which are included in the class of photon detectors. They exhibit both perfect signal-to-noise performance and a very fast response. But to achieve this, the photon detectors require cryogenic cooling. Cooling requirements are the main obstacle to the more widespread use of IR systems based on semiconductor photodetectors making them bulky, heavy, expensive, and inconvenient to use.

Until the 1990s, despite numerous research initiatives and the appeal of room temperature operation and low cost potential, thermal detectors have enjoyed limited success compared with cooled photon detectors for thermal imaging applications. Only the pyroelectric vidicon received much attention with the hope that it could be made practical for some applications. Throughout the 1980s and early 1990s, many companies in the United States (especially Texas Instruments and Honeywell's Research Laboratory) developed devices based on various thermal detection principles. In the mid-1990s, this success caused DARPA (Defense Advanced Research Projects Agency) to reduce support for HgCdTe and attempt a major leap with uncooled technology. The desire was to have producible arrays with useful performance, without the burden of fast ($f/1$) long-wavelength infrared optics.

In order to access these new changes in infrared detector technology, there was need for a comprehensive introductory account of IR detector physics and operational principles, together with important references. In 2000, the first edition of *Infrared Detectors* was published with the intention of meeting this need. The last decade has seen considerable changes with numerous breakthroughs in detector concepts and performance. It became clear that the book needed substantial revision to continue to serve its purpose.

In this second edition of *Infrared Detectors*, about 70% of the contents have been revised and updated, and much of the materials have been reorganized. The book is divided into four parts: fundaments of infrared detection, infrared thermal detectors, infrared photon detectors, and focal plane arrays. The first part provides a tutorial introduction to the technical topics that are fundamental to a thorough understanding of different types of IR detectors and systems. The second part presents theory and technology of different types of thermal detectors. The third part covers theory and technology of photon detectors. The last part concerns IR focal plane arrays (FPAs) where relations between the performance of detector arrays and infrared system quality are considered.

The short description below mainly concerns differences between the original edition and this revision. I have added a discussion of radiometry and flux-transfer issues needed for IR detector and system analysis in the first part. In the next two parts, in addition to updating traditional issues described in the previous book, I have included new achievements and trends in the development of IR detectors, most notably:

- novel uncooled detectors (e.g., cantilever detectors, antenna and optically coupled detectors);
- type II superlattice detectors; and
- quantum dot infrared detectors.

In addition, I have highlighted new approaches to terahertz (THz) arrays and a new generation of infrared detectors—so-called third-generation detectors. THz technologies are now receiving increasing attention, and devices exploiting this wavelength band are set to become increasingly important in a diverse range of human activity applications (e.g., security, biological, drugs and explosion detection, gases fingerprints, imaging, etc.). Today, researchers are developing third-generation systems that provide enhanced capabilities such as a larger number of pixels, higher frame rates, better thermal resolution, multicolor functionality, and other on-chip functions.

This book is written for those who desire a comprehensive analysis of the latest developments in infrared detector technology and basic insight into fundamental processes important to evolving detection techniques. Special attention has been given to the physical limits of detector performance and comparisons of performance in different types of detectors. The reader should gain a good understanding of the similarities and contrasts, the strengths and weaknesses of a multitude of approaches that have been developed over a century to improve our ability to sense IR radiation.

The level of presentation is suitable for graduate students in physics and engineering who have received standard preparation in modern solid-state physics and electronic circuits. This book is also of interest to individuals working with aerospace sensors and systems, remote sensing, thermal imaging, military imaging, optical telecommunications, infrared spectroscopy, and light detection and ranging (LIDAR). To satisfy the needs of the first group, many chapters discuss the principles underlying each topic and some historical background before bringing the reader the most recent information available. For those currently in the field, the book can be used as a collection of useful data, as a guide to the literature, and as an overview of topics covering a wide range of applications. The book could also be used as a reference for participants of relevant workshops and short courses.

This new edition of *Infrared Detectors* gives a comprehensive analysis of the latest developments in IR detector technology and basic insight into the fundamental processes important to evolving detection techniques. The book covers a broad spectrum of IR detectors, including theory, types of materials and their physical properties, and detector fabrication.

Antoni Rogalski

Acknowledgments

In the course of this writing, many people have assisted me and offered their support. I would like, first, to express my appreciation to the management of the Institute of Applied Physics, Military University of Technology, Warsaw, for providing the environment in which I worked on the book. The writing of the book has been partially done under financial support of the Polish Ministry of Sciences and Higher Education, Key Project POIG.01.03.01-14-016/08 "New Photonic Materials and Their Advanced Application."

The author has benefited from the kind cooperation of many scientists who are actively working in infrared detector technologies. The preparation of this book was aided by many informative and stimulating discussions with the author's colleagues at the Institute of Applied Physics, Military University of Technology in Warsaw. The author thanks the following individuals for providing preprints, unpublished information, and in some cases original figures, which were used in preparing the book: Drs. L. Faraone and J. Antoszewski (University of Western Australia, Perth), Dr. J. L. Tissot (Ulis, Voroize, France), Dr. S. D. Gunapala (California Institute of Technology, Pasadena), Dr. M. Kimata (Ritsumeikan University, Shiga, Japan), Dr. M. Razeghi (Northwestern University, Evanston, Illinois), Drs. M. Z. Tidrow and P. Norton (U.S. Army RDECOM CERDEC NVESD, Fort Belvoir, Virginia), Dr. S. Krishna (University of New Mexico, Albuquerque), Dr. H. C. Liu (National Research Council, Ottawa, Canada), G. U. Perera (Georgia State University, Atlanta), Professor J. Piotrowski (Vigo System Ltd., Ożarów Mazowiecki, Poland), Dr. M. Reine (Lockheed Martin IR Imaging Systems, Lexington, Massachusetts), Dr. F. F. Sizov (Institute of Semiconductor Physics, Kiev, Ukraine), and Dr. H. Zogg (AFIF at Swiss Federal Institute of Technology, Zürich). Thanks also to CRC Press, especially Luna Han, who encouraged me to undertake this new edition and for her cooperation and care in publishing this second edition.

Ultimately, it is the encouragement, understanding, and support of my family that provided me the courage to embark on this project and see it to its conclusion.

About the Author

Antoni Rogalski is a professor at the Institute of Applied Physics, Military University of Technology in Warsaw, Poland. He is a leading researcher in the field of IR optoelectronics. During the course of his scientific career, he has made pioneering contributions in the areas of theory, design, and technology of different types of IR detectors. In 1997, he received an award from the Foundation for Polish Science (the most prestigious scientific award in Poland) for achievements in the study of ternary alloy systems for infrared detectors—mainly an alternative to HgCdTe new ternary alloy detectors such as lead salts, InAsSb, HgZnTe, and HgMnTe. In 2004, he was elected as a corresponding member of the Polish Academy of Sciences.

Professor Rogalski's scientific achievements include determining the fundamental physical parameters of InAsSb, HgZnTe, HgMnTe, and lead salts; estimating the ultimate performance of ternary alloy detectors; elaborating on studies of high-quality PbSnTe, HgZnTe, and HgCdTe photodiodes operated in 3–5 μm and 8–12 μm spectral ranges; and conducting comparative studies of the performance limitation of HgCdTe photodiodes versus other types of photon detectors (especially QWIP and QDIP IR detectors).

Professor Rogalski has given about 50 invited plenary talks at international conferences. He is author and co-author of over 200 scientific papers, 11 books, and 13 book chapters. He is a fellow of the International Society for Optical Engineering (SPIE), vice president of the Polish Optoelectronic Committee, vice president of the Electronic and Telecommunication Division at the Polish Academy of Science, editor-in-chief of the journal *Opto-Electronics Review*, deputy editor-in-chief of the *Bulletin of the Polish Academy of Sciences: Technical Sciences*, and a member of the editorial boards of *Journal of Infrared and Millimeter Waves* and *International Review of Physics*.

Professor Rogalski is an active member of the international technical community. He is a chair and co-chair, organizer and member of scientific committees of many national and international conferences on optoelectronic devices and material sciences.

PART I
FUNDAMENTS OF INFRARED DETECTION

1 Radiometry

Traditionally, infrared (IR) technologies are connected with controlling functions and night-vision problems with earlier applications connected simply with detection of IR radiation, and later by forming IR images from temperature and emissive differences (systems for recognition and surveillance, tank sight systems, anti-tank missiles, air–air missiles, etc.). Most of the funding has been provided to fulfill military needs, but peaceful applications have increased continuously, especially since the last decade of the twentieth century (see Figure 1.1). It is predicted currently that the commercial market is about 70% in volume and 40% in value, largely connected with volume production of uncooled imagers [1]. These include medical, industry, earth resources, and energy conservation applications. Medical applications include thermography in which IR scans of the body detect cancers or other trauma, which raise the body surface temperature. Earth resource determinations are done by using IR images from satellites in conjunction with field observation for calibration (in this manner, e.g., the area and content of fields and forests can be determined). In some cases even the health state of a crop is determined from space. Energy conservation in homes and industry has been aided by the use of IR scans to determine the points of maximum heat loss. Demands for these technologies are quickly growing due to their effective applications, for example, in global monitoring of environmental pollution and climate changes, longtime prognoses of agriculture crop yield, chemical process monitoring, Fourier transform IR spectroscopy, IR astronomy, car driving, IR imaging in medical diagnostics, and others.

The infrared range covers all electromagnetic radiation longer than the visible, but shorter than millimeter waves (Figure 1.2). The divisions between these categories are based on the different source and detector technologies used in each region. Many proposals in the division of IR range have been published and these shown in Table 1.1 are based on limits of spectral bands of commonly used IR detectors. Wavelength 1 μm is a sensitivity limit of popular Si detectors. Similarly, wavelength 3 μm is a long wavelength sensitivity of PbS and InGaAs detectors; wavelength 6 μm is a sensitivity limit of InSb, PbSe, PtSi detectors, and HgCdTe detectors optimized for a 3–5 μm atmospheric window; and finally wavelength 15 μm is a long wavelength sensitivity limit of HgCdTe detectors optimized for an 8–14 μm atmospheric window.

The IR devices cannot be designed without an understanding of the amount of radiation power that impinges on the detector from the target, and the radiation of the target cannot be understood without a radiometric measurement. This issue is critical to the overall signal-to-noise ratio achieved by the IR system.

Our discussion in this chapter is simplified due to certain provisions and approximations. We specifically consider the radiometry of incoherent sources and ignore the effects of diffraction. In general, we make small-angle assumptions similar to those made for paraxial optics. The sine of an angle is approximated by the angle itself in radians.

This chapter provides some guidance in radiometry. For further details, see [2–7].

1.1 RADIOMETRIC AND PHOTOMETRIC QUANTITIES AND UNITS

Radiometry is the branch of optical physics that deals with the measurement of electromagnetic radiation in the frequency range between 3×10^{13} and 3×10^{16} Hz. This range corresponds to wavelengths between 10 nm and 10 μm and includes the regions commonly called the ultraviolet, the visible, and the infrared. Radiometry deals with the actual energy content of the light rather than its perception through a human visual system. Typical radiometric units include watt (radiant flux), watt per steradian (radiant intensity), watt per square meter (irradiance), and watt per square meter per steradian (radiance).

Historically, the power of a light source was obtained by observing brightness of the source. It turns out that brightness perceived by the human eye depends upon wavelength; that is, color of the light and differs from the actual energy contained in the light. The eye is most sensitive to the yellow–green light and less sensitive to red and blue lights of the spectrum. To take the difference into account, a new set of physical measures of light is defined for the visible light that parallel the quantities of radiometry, where the power is weighted according to the human response by multiplying the corresponding quantity by a spectral function, called the $V(\lambda)$ function or the spectral luminous efficiency for photopic vision, defined in the domain from 360 to 830 nm, and is normalized to one at its leak, 555 nm (Figure 1.3). The $V(\lambda)$ function tells us the appropriate response of the human eye to various wavelengths. This function was first defined by the Commission Internationale de l'Éclairage (CIE) in 1924 [8] and is an average response

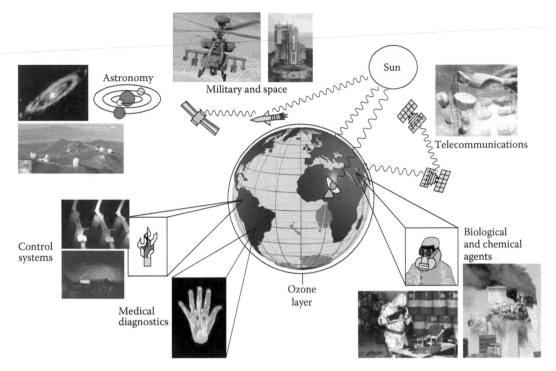

Figure 1.1 Applications of infrared detectors.

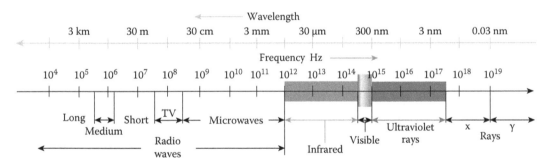

Figure 1.2 Electromagnetic spectrum.

Table 1.1: Division of Infrared Radiation

Region (Abbreviation)	Wavelength Range (µm)
Near infrared (NIR)	0.78–1
Short wavelength IR (SWIR)	1–3
Medium wavelength IR (MWIR)	3–6
Long wavelength IR (LWIR)	6–15
Very long wavelength IR (VLWIR)	15–30
Far infrared (FIR)	30–100
Submillimeter (SubMM)	100–1000

of a population of people in a wide range of ages. It should be noted that the $V(\lambda)$ function was defined assuming additivity of sensation and a 2° field of view at relatively high luminance levels (>1 cd/m²). The spectral responsivity of human eyes deviates significantly at very low levels of luminescence (<10^{-3} cd/m²) when the rods in the eyes are the dominant receptors. This type of vision is called scotopic vision.

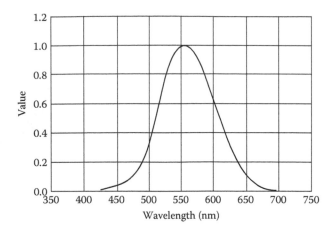

Figure 1.3 CIE $V(\lambda)$ function.

Table 1.2: Radiometric and Photometric Quantities and Units

Photometric Quantity	Unit	Radiometric Quantity	Symbol	Unit	Unit Conversion
Luminous flux	lm (lumen)	Radiant flux	ϕ	W (Watt)	1 W = 683 lm
Luminous intensity	cd (candela) = lm/sr	Radiant intensity	I	W/sr	1 W/sr = 683 cd
Illuminance	lx (lux) = lm/m²	Irradiance	E	W/m²	1 W/m² = 683 lx
Luminance	cd/m² = lm/(sr m²)	Radiance	L	W/(sr m²)	1 W/(sr m²) = 683 cd/m²
Luminous exitance	lm/m²	Radiant exitance	M	W/m²	
Luminous exposure	lx s	Radiant exposure		W/(m² s)	
Luminous energy	lm s	Radiant energy	Q	J (Joule)	1 J = 683 lm s

Photometry is the measurement of light, which is defined as radiation detectable by the human eye. It is restricted to the visible region and all quantities are weighted by the spectral response of the eye. Typical photometric units include lumen (luminous flux), candela (luminous intensity), lux (illuminance), and candela per square meter (luminance).

The units as well as the names of similar properties in photometry differ from those in radiometry. For instance, power is simply called power in radiometry or radiant flux, but it is called the luminous flux in photometry. While the unit of power in radiometry is watt, in photometry it is lumen. A lumen is defined in terms of a superfluous fundamental unit, called candela, which is one of the seven independent quantities of the SI system of units (meter, kilogram, second, ampere, Kelvin, mole, and candela). Candela is the SI unit of the photometric quantity called luminous intensity or luminositry that corresponds to the radiant intensity in radiometry. Table 1.2 lists the radiometric and photometric quantities and units along with translation between both groups of units.

Radiometry is plagued by a confusion of terminology, symbols, definitions, and units. The origin of this confusion is largely because of the parallel or duplicate development of the fundamental radiometric practices by researchers in different disciplines. Consequently, considerable care should be exercised when reading publications. The terminology used in this chapter follows international standards and recommendations [7,9].

1.2 DEFINITIONS OF RADIOMETRIC QUANTITIES

Radiant flux, also called radiant power, is the energy Q (in joules) radiated by a source per unit of time and is defined by

$$\Phi = \frac{dQ}{dt}. \tag{1.1}$$

The unit of radiant flux is the Watt (W = J/s).

Radiant intensity is the radiant flux from a point source emitted per unit solid angle in a given direction, and is expressed as

$$I = \frac{d\Phi}{d\Omega} = \frac{\partial^2 Q}{\partial t \partial \Omega},$$

(1.2)

where $d\Phi$ is the radiant flux leaving the source and propagating in an element of solid angle $d\Omega$ containing the given direction (see Figure 1.4). The unit of radiant intensity is W/sr.

The solid angle may be expressed in differential form as

$$d\Omega = \frac{dA}{r^2}.$$

(1.3)

The unit of solid angle is steradian (sr).

If we use the spherical coordinate system seen in Figure 1.5, and use $dA = r^2 \sin\theta d\theta d\varphi$, we can write an expression for the solid-angle subtense of a flat disc of planar half-angle θ_{max} as

$$\Omega = \int d\Omega = \int_0^{2\pi} d\varphi \int_0^{\theta_{max}} \sin\theta d\theta = 2\pi\left(1 - \cos\theta_{max}\right).$$

(1.4)

Irradiance is the density of incident radiant flux at a point of surface and is defined as radiant flux per unit area [see Figure 1.6a], as follows

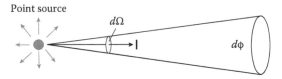

Figure 1.4   Radiant intensity.

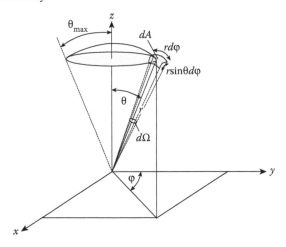

Figure 1.5   Relationship of solid angle to planar angle.

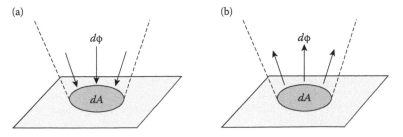

Figure 1.6   (a) Irradiance and (b) radiation exitance.

$$E = \frac{\partial \Phi}{\partial A} = \frac{\partial^2 Q}{\partial t \partial A},$$
(1.5)

where $\partial \Phi$ is the radiant flux incident on an element ∂A of the surface containing the point. The unit of irradiance is W/m^2.

For strictly point source, a common rule of thumb is the "one-over-r-squared falloff." We consider a receiver of area A, which is placed at various distances from a point source having a uniform radiant intensity I (Figure 1.7). Using Equation 1.2 we get

$$\Phi = I \frac{A}{r^2},$$
(1.6)

and next

$$E = \frac{\Phi}{A} = \frac{I}{r^2}.$$
(1.7)

Because the solid angle subtended by the detector falls off as $1/r^2$, the collected flux and the irradiance also decreases proportionally. Note that in the case of extended source, sufficient distance relative to the source is required to assume the relationship (Equation 1.7).

Radiance is the radiant flux per unit solid angle emitted from a surface element in a given direction, per unit projected area of the surface element perpendicular to the direction (see Figure 1.8). This quantity is defined by

$$L = \frac{\partial^2 \Phi}{\partial \Omega \partial A \cos \theta},$$
(1.8)

where $\partial \Phi$ is the radiant flux emitted from the surface element and propagating in the solid angle $\partial \Omega$ containing the given direction, ∂A is the area of the surface element, and θ is the angle between the normal to the surface element and the direction of the beam. The unit of radiance is $W/(sr m^2)$.

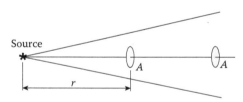

Figure 1.7 Irradiance falloff as a function of r from source.

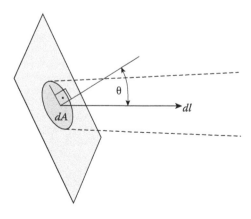

Figure 1.8 Radiance.

The term $\partial A \cos\theta$ gives the projected area of the surface element perpendicular to the direction of measurement.

Radiant exitance is the density radiant flux leaving a surface at a point (see Figure 1.6b) and is defined by

$$M = \frac{\partial \Phi}{\partial A} = \frac{\partial^2 Q}{\partial t \partial A},$$

(1.9)

where $\partial \Phi$ is the radiant flux leaving the surface element. The unit of radiant exitance is W/m².

Irradiance and radiant exitance have the same units, but have different interpretations. Irradiance is the amount of power with respect to unit area that falls on a surface, while radiant exitance is the amount of power per unit area that leaves a surface. Exitance thus characterizes a self-luminous source that is producing energy, while irradiance characterizes a passive receiver surface.

1.3 RADIANCE

Radiance is used to characterize an extended source (see Figure 1.9). Equation 1.8 indicates that the power received by the detector is differential with respect to both incremental projected area of the source and incremental solid angle of detector. Rearranging this equation, we have

$$\partial^2 \Phi = L \partial A_s \cos\theta_s \partial\Omega_d,$$

(1.10)

and integrating once with respect to source, we obtain intensity

$$I = \frac{\partial \Phi}{\partial\Omega_d} = \int_{A_s} L \cos\theta_s dA_s.$$

(1.11)

Similarly, integrating once with respect to detector solid angle, we obtain radiation exitance

$$M = \frac{\partial \Phi}{\partial A_s} = \int_{\Omega_d} L \cos\theta_s d\Omega_d.$$

(1.12)

A Lambertian radiator has a constant radiance that is independent of viewing direction. This type of reflector is also referred to as an ideal diffuse radiator (emitter or reflector); see Figure 1.10. In practice there are no true Lambertian surfaces. Most matte surfaces approximate an ideal diffuse

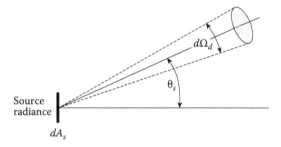

Figure 1.9 Radiance of an extended source.

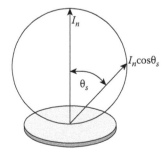

Figure 1.10 Radiant intensity as a function of θ_s for a Lambertian source.

reflector, but typically exhibit semispecular reflection characteristics at oblique viewing angles. An ideal thermal source (blackbody) is perfectly Lambertian, while certain special diffusers also closely approximate the condition. An actual source is typically approximately Lambertian within a range of view angles θ_s that is less than 20°.

Even for a Lambertian source, the intensity depends on θ_s. Making the assumption of L independent of source position, from Equation 1.11 follows

$$I = \frac{\partial \Phi}{\partial \Omega_d} = \int_{A_s} L \cos\theta_s dA_s = LA_s \cos\theta_s = I_n \cos\theta_s. \tag{1.13}$$

It is Lambert's cosine law, where I_n is the intensity of the ray leaving in a direction perpendicular to the surface. For non-Lambertian surfaces, the radiance L is a function of angle itself, and the falloff of I with θ_s is faster than $\cos\theta_s$.

To receive relationship between radiation exitance and radiance for a planar Lambertian source, we return to Equation 1.12 and integrate

$$M = \frac{\partial \Phi}{\partial A_s} = \int_{\Omega_d} L \cos\theta_s d\Omega_d = \int_0^{2\pi} d\varphi \int_0^{\pi/2} L \cos\theta_s \sin\theta d\theta = 2\pi L \frac{1}{2} = \pi L, \tag{1.14}$$

where the Lambertian-source assumption has been used to pull L outside of the angular integrals. For a non-Lambertian source, the integration yields a proportionality constant different from π.

Let us simplify further considerations assuming $\theta_s = 0$. Then, for the geometrical configuration shown in Figure 1.11, the radiant power on the detector can be obtained by multiplying the detector's solid angle by the area of the source and the radiance of the source [10]

$$\Phi_d = LA_s \Omega_d = \frac{LA_s A_d}{r^2} = L\Omega_s A_d. \tag{1.15}$$

From this equation results that the flux on the detector is expressed as the radiance of the source multiplied by an area × solid angle ($A\Omega$) product. To fulfill Equation 1.15, two provisions are required: a small-angle assumption for the approximation of the solid angle of a flat surface by A/r^2 and the flux transfer is unaffected by absorption losses in the system.

Another situation occurs for a tilted receiver shown in Figure 1.12. The source normal is along the line of centers, so $\theta_s = 0$ in this case. The angle θ_d is the angle between the line of centers and the normal to the detector surface. In this situation

$$\Phi_d = LA_s \Omega_d. \tag{1.16}$$

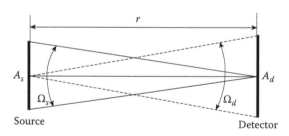

Figure 1.11 Radiant power transfer from source to detector.

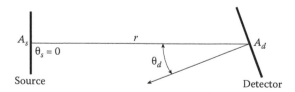

Figure 1.12 Radiant power transfer from source to a titled detector.

Assuming that for a titled surface

$$\Omega_d = \frac{A_d \cos\theta_d}{r^2},$$
(1.17)

we find

$$\Phi_d = LA_s \frac{A_d \cos\theta_d}{r^2}.$$
(1.18)

Thus the flux collected and the irradiance (Φ/A_d) are decreased by a factor of $\cos\theta_d$.

We now proceed to calculate the flux on the detector when both θ_s and θ_d are nonzero assuming a flat Lambertian source (Figure 1.13). A cosine falloff factor arises at both the source and the receiver. In this case

$$\Phi_d = LA_s \cos\theta_s \frac{A_d \cos\theta_d}{\left(r/\cos\theta_s\right)^2}.$$
(1.19)

Assuming that the surfaces of source and detector are parallel and $\theta_s = \theta_d = \theta$, the radiant intensity is proportional to $\cos^4\theta$. On account of this, Equation 1.19 is so-called cosine to the fourth law.

Finally, we consider the flux transfer in image-forming systems assuming the limitations of paraxial optics (small angles) as is shown in Figure 1.14. Only a certain amount of flux Φ is collected by the optical system, which can be calculated letting the lens aperture (A_{lens}) act as an intermediate receiver. It should be noted that in a more complex system, the entrance pupil is this intermediate receiver and the $A_{lens}\Omega_{obj}$ product is the area-solid-angle product of the optical system. At these conditions, the collected radiant flux equal

$$\Phi = LA_{obj}\Omega_{lens} = LA_{lens}\Omega_{obj},$$
(1.20)

and is reformatted by lens and forms an image of the original object at an appropriate magnification.

The image irradiance can be found simply by dividing the flux collected in Equation 1.20 by the image area:

$$\Phi = LA_{lens}\Omega_{lens} = LA_{lens}\Omega_{obj} = L\frac{A_{lens}A_{obj}}{p^2} = L\frac{A_{lens}A_{img}}{q^2}.$$
(1.21)

The last equality was obtained using $A_{img} = A_{obj}(q/p)^2$.

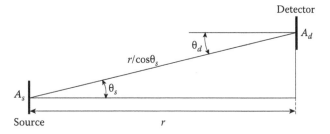

Figure 1.13 Cosine-to-the-fourth ($\cos^4$) law.

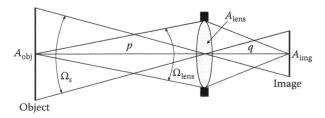

Figure 1.14 Radiant power collected by an optical system. (From Dereniak, E. L., and Boreman, G. D., *Infrared Detectors and Systems*, Wiley, New York, 1996. With permission.)

From Equation 1.21 results, the image irradiance is equal

$$E_{\text{img}} = \frac{\Phi}{A_{\text{imag}}} = L\frac{A_{\text{lens}}}{q^2}.$$

(1.22)

1.4 BLACKBODY RADIATION

All objects are composed of continually vibrating atoms, with higher energy atoms vibrating more frequently. The vibration of all charged particles, including these atoms, generates electromagnetic waves. The higher the temperature of an object, the faster the vibration, and thus the higher the spectral radiant energy. As a result, all objects are continually emitting radiation at a rate with a wavelength distribution that depends upon the temperature of the object and its spectral emissivity, $\varepsilon(\lambda)$.

Radiant emission is usually treated in terms of the concept of a blackbody [5]. A blackbody is an object that absorbs all incident radiation and, conversely according to the Kirchhoff's law, is a perfect radiator. The energy emitted by a blackbody is the maximum theoretically possible for a given temperature. A device of this type is a very useful standard source for the calibration and testing of radiometric instruments. Further, most sources of thermal radiation radiate energy in a manner that can be readily described in terms of a blackbody emitting through a filter, making it possible to use the blackbody radiation laws as a starting point for many radiometric calculations.

The blackbody or Planck equation was one of the milestones of physics. The Planck's law describes the spectral radiance (spectral radiant exitance) of a perfect blackbody as a function of its temperature and the wavelength of the emitted radiation, in the forms

$$L(\lambda, T) = \frac{2hc^2}{\lambda^5}\left[\exp\left(\frac{hc}{\lambda kT}\right) - 1\right]^{-1} \text{W}/(\text{cm}^2 \text{ sr } \mu\text{m}),$$

(1.23)

$$M(\lambda, T) = \frac{2\pi hc^2}{\lambda^5}\left[\exp\left(\frac{hc}{\lambda kT}\right) - 1\right]^{-1} \text{W}/(\text{cm}^2 \mu\text{m}),$$

(1.24)

where λ is the wavelength, T is the temperature, h is the Planck's constant, c is the velocity of light, and k is the Boltzmann's constant. The corresponding equations for spectral radiant exitance, $M(\lambda,T)$, and spectral radiance, $L(\lambda,T)$, are related by $M = \pi L$.

The units listed in Table 1.2 are based on joule as the fundamental quantity. An analogous set of quantities can be based on number of photons. A conversion between two sets of units is easily accomplished using the relationship for the amount of energy carried per photon: $\varepsilon = hc/\lambda$. For example

$$\phi(\text{joule}/s) = \phi(\text{photon}/s) \times \varepsilon(\text{joule}/\text{photon}).$$

(1.25)

In a similar way, the Equations 1.23 and 1.24 may be transformed to the forms

$$L(\lambda, T) = \frac{2c}{\lambda^4}\left[\exp\left(\frac{hc}{\lambda kT}\right) - 1\right]^{-1} \text{photon}/(\text{s cm}^2 \text{ sr } \mu\text{m}),$$

(1.26)

$$M(\lambda, T) = \frac{2\pi c}{\lambda^4}\left[\exp\left(\frac{hc}{\lambda kT}\right) - 1\right]^{-1} \text{photon}/(\text{s cm}^2 \mu\text{m}).$$

(1.27)

Figure 1.15 shows a plot of these curves for a number of blackbody temperatures. As the temperature increases, the amount of energy emitted at any wavelength increases too, and the wavelength of peak emission decreases. The latter is given by the Wien's displacement law [11]:

$$\lambda_{mw}T = 2898\,\mu\text{mK} \quad \text{for maximum watts,}$$

(1.28)

$$\lambda_{mp}T = 3670\,\mu\text{mK} \quad \text{for maximum photons,}$$

(1.29)

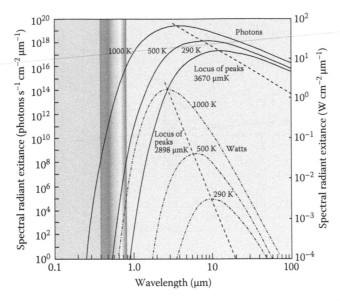

Figure 1.15 Planck's law for spectral radiant exitance. (From Burnay, S. G., Williams, T. L., and Jones, C. H., *Applications of Thermal Imaging*, Adam Hilger, Bristol, England, 1988.)

which is derived from the condition for the peak of the exitance function by setting the derivative equal to zero

$$\frac{dM(\lambda,T)}{d\lambda} = 0 \tag{1.30}$$

and solving for wavelength at maximum exitance.

The loci of these maxima are shown in Figure 1.15. Note that for an object at an ambient temperature of 259 K, λ_{mw} and λ_{mp} occur at 10.0 μm and 12.7 μm, respectively. We need detectors operating near 10 μm if we expect to "see" room temperature objects such as people, trees, and vehicles without the aid of reflected light. For hotter objects such as engines, maximum emission occurs at shorter wavelengths. Thus, the waveband 2–15 μm in infrared or thermal region of the electromagnetic spectrum contains the maximum radiative emission for thermal imaging purposes. It is interesting to note that the λ_{mw} for sun is near 0.5 μm, very close to the peak of sensitivity of the human eye.

Total radiant exitance from a blackbody at temperature T is the integral of spectral exitance over all wavelengths

$$M(T) = \int_0^\infty M(\lambda,T)d\lambda = \int_0^\infty \frac{2\pi h c^2}{\lambda^5 \left[\exp\left(\frac{hc}{\lambda kT} - 1 \right) \right]}\, d\lambda = \frac{2\pi^5 k^4}{15 c^2 h^3}T^4 = \sigma T^4, \tag{1.31}$$

where $\sigma = 2\pi^5 k^4 / 15 c^2 h^3$ is called the Stefan–Boltzmann constant and has an approximate value of 5.67×10^{-12} W/cm²K⁴.

The relation determined by Equation 1.31 between the total radiant exitance of a blackbody and its temperature is called the Stefan–Boltzmann law. The total exitance can be interpreted as the area under the spectral exitance curve for a given temperature, as is shown in Figure 1.16.

The radiant exitance of blackbody between λ_a and λ_b is obtained by integrating the Planck's law over the integral $[\lambda_a, \lambda_b]$ as shown in Figure 1.16:

$$M_{\Delta\lambda}(T) = \int_{\lambda_a}^{\lambda_b} M(\lambda,T)d\lambda = \int_{\lambda_a}^{\lambda_b} \frac{2\pi h c^2}{\lambda^5 \left[\exp\left(\frac{hc}{\lambda kT} - 1 \right) \right]}\, d\lambda. \tag{1.32}$$

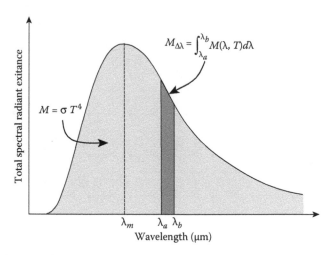

Figure 1.16 Total spectral radiant exitance at temperature T versus wavelength.

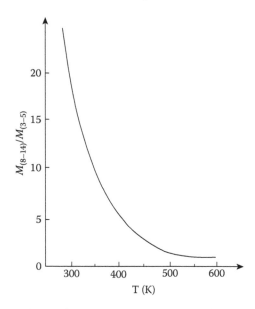

Figure 1.17 Emittance ratio $M_{[8-12]}/M_{[3-5]}$ against temperature. (From Gaussorgues, G., *La Thermographie Infrarouge*, Technique et Documentation, Lavoisier, Paris, 1984.)

The exitance of a human body at 300 K is 500 W/m² and for a skin surface of 2 m², the radiative power is of 10^3 W. But the loss of energy is partially compensated by the radiation absorption.

The exitance at 300 K over the interval 8–14 μm is equal to 1.22×10^2 W/m² and is equivalent to the exitance over the interval 3–5 μm at 410 K. If we reported on a curve (the ratio $M_{[8-12]}/M_{[3-5]}$ shown in Figure 1.17), we notice that the exitance over 8–14 μm is larger than exitance over 3–5 μm up to 600 K [12]. These results show the advantage of using the 8–14 μm region for low temperature targets.

The temperature variation of the spectral radiant exitance is given by the temperature derivation of Equation 1.24; then

$$\frac{\partial M(\lambda, T)}{\partial T} = \frac{(hc/k)\exp(hc/\lambda kT)}{\lambda T^2 [\exp(hc/\lambda kT) - 1]} M(\lambda, T). \tag{1.33}$$

Generally in thermal imaging, the objects are at temperatures near 300 K for $\lambda_{max} \approx 10$ μm or near 700 K for $\lambda_{max} \approx 4$ μm. In these two cases $\lambda \ll hc/kT$ and then

$$\frac{\partial M(\lambda,T)}{\partial T} = \frac{hc}{\lambda kT^2} M(\lambda,T). \tag{1.34}$$

For a system operating within a finite passband ($\Delta\lambda$), it is important at what wavelength the source (target) exitance changes the most with temperature. This question is fundamental for sensitivity of an infrared system. Comparing the second partial derivative to zero

$$\frac{\partial}{\partial\lambda}\left[\frac{\partial M(\lambda,T)}{\partial T}\right] = 0, \tag{1.35}$$

produces a constrain on the wavelength of exitance contrast [10], similar to Wien displacement law

$$\lambda_{\text{max contrast}} = \frac{2410}{T}\mu m. \tag{1.36}$$

For example, at a source temperature of 300 K, the maximum contrast occurs at a wavelength of about 8 µm, which is not the wavelength for maximum exitance.

1.5 EMISSIVITY

As was mentioned previously, the blackbody curve provides the upper limit of the overall spectral exitance of a source for any specific temperature. Most thermal sources are not perfect blackbodies. Many are called graybodies. A graybody is one that emits radiation in exactly the same spectral distribution as a blackbody at the same temperature, but with reduced intensity.

The ratio between the exitance of the actual source and the exitance of a blackbody at the same temperature is defined as emissivity. In general, emissivity depends on λ and T:

$$\varepsilon(\lambda,T) = \frac{M(\lambda,T)_{\text{source}}}{M(\lambda,T)_{\text{blackbody}}}. \tag{1.37}$$

and is a dimensionless number ≤ 1.

For a perfect blackbody $\varepsilon = 1$ for all wavelengths. The emissivity of graybody is independent of λ (see Figure 1.18). A selective source has an emissivity that depends on wavelength.

The total radiant exitance for graybody at all wavelengths is equal

$$M^{gb}(T) = \varepsilon\sigma T^4. \tag{1.38}$$

When radiant energy is incident on a surface, fraction α is absorbed, fraction r is reflected, and fraction t is transmitted. Since energy must be conserved, the following relationship can be written:

$$\alpha + r + t = 1. \tag{1.39}$$

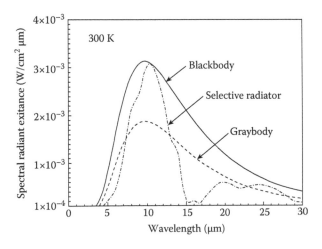

Figure 1.18 Spectral radiant exitance of three different radiators.

Kirchhoff observed that at a given temperature the ratio of the integrated emissivity to the integrated absorptance is a constant for all materials and that it is equal to the radiant exitance of a blackbody at that temperature. Known as Kirchhoff's law, it can be stated as

$$\frac{M(\lambda,T)_{\text{source}}}{\alpha} = M(\lambda,T)_{\text{blackbody}}. \tag{1.40}$$

This law is often paraphrased as "good absorbers are good emitters." Combining Equations 1.31 and 1.37 give

$$\frac{\varepsilon \sigma T^4}{\alpha} = \sigma T^4. \tag{1.41}$$

From this follows that

$$\varepsilon = \alpha. \tag{1.42}$$

Thus the emissivity of any materials at a given temperature is numerically equal to its absorptance at that temperature. Since an opaque material does not transmit energy, $\alpha + r = 1$ and

$$\varepsilon = 1 - r. \tag{1.43}$$

Table 1.3 lists the emissivity of a number of common materials that are frequently a single number, and are seldom given as either a function of λ or T unless it is an essentially well-characterized material [13]. The dependence of emissivity on wavelength results from the fact that many substances (glass, for example) have a negligible absorption and consequent low emissivity at certain wavelengths, while they are almost totally absorbent at other wavelengths. For many materials, emissivity decreases as wavelength increases. For nonmetallic substances, typically $\varepsilon > 0.8$ for room temperature and decreases with increasing temperature. For a metallic substance, the emissivity is very low at room temperature and generally increases proportional to temperature.

Table 1.3: Emissivity of a Number of Materials

Material	Temperature (K)	Emissivity
Tungsten	500	0.05
	1000	0.11
	2000	0.26
	3000	0.33
	3500	0.35
Polished silver	650	0.03
Polished aluminium	300	0.03
	1000	0.07
Polished cooper		0.02–0.15
Polished iron		0.2
Polished brass	4–600	0.03
Oxidized iron		0.8
Black oxidized cooper	500	0.78
Aluminium oxide	80–500	0.75
Water	320	0.94
Ice	273	0.96–0.985
Paper		0.92
Glass	293	0.94
Lampblack	273–373	0.95
Laboratory blackbody cavity		0.98–0.99

Source: Smith, W. J., *Modern Optical Engineering*, McGraw-Hill, New York, 2000.

1.6 INFRARED OPTICS

The optical block in the IR system creates an image of observed objects in a plane of the detector (detectors). In the case of a scanning imager, the optical scanning system creates an image with the number of pixels much greater than the number of elements of the detector. In addition, the optical elements like windows, domes, and filters can be used to protect a system from the environment or to modify detector spectral response.

There is no essential difference in design rules of optical objectives for visible and IR ranges. A designer of IR optics is only more limited because there is significantly fewer materials suitable for IR optical elements, in comparison with those for visible range, particularly for wavelengths over 2.5 µm.

There are two types of IR optical elements: reflective elements and refractive elements. As the names suggest, the role of reflective elements is to reflect incident radiation and the role of refractive elements is to refract and transmit incident radiation.

Mirrors used extensively inside IR systems (especially in scanners) are most often met as reflective elements that serve manifold functions in IR systems. Elsewhere they need a protective coating to prevent them from tarnishing. Spherical or aspherical mirrors are employed as imaging elements. Flat mirrors are widely used to fold an optical path, and reflective prisms are often used in scanning systems.

Four materials are most often used for mirror fabrication: optical crown glass, low-expansion borosilicate glass (LEBG), synthetic fused silica, and Zerodur. Less popular in use are metallic substrates (beryllium, copper) and silicon carbide. Optical crown glass is typically applied in nonimaging systems. It has a relatively high thermal expansion coefficient and is employed when thermal stability is not a critical factor. The LEBG, known by the Corning brand name Pyrex, is well suited for high quality, front-surface mirrors designed for low optical deformation under thermal shock. Synthetic fused silica has a very low thermal expansion coefficient.

Metallic coatings are typically used as reflective coatings of IR mirrors. There are four types of metallic coatings used most often: bare aluminium, protected aluminium, silver, and gold. They offer high reflectivity, over about 95%, in 3–15 µm spectral range. Bare aluminium has a very high reflectance value but oxidizes over time. Protected aluminium is a bare aluminium coating with a dielectric overcoat that arrests the oxidation process. Silver offers better reflectance in near IR than aluminium and high reflectance across a broad spectrum. Gold is a widely used material and consistently offers very high reflectance (about 99%) in the 0.8–50 µm range. However, gold is soft (it cannot be touched to remove dust) and is most often used in laboratory.

The most popular materials used in manufacturing refractive optics of IR systems are: germanium (Ge), silicon (Si), fused silica (SiO_2), glass BK-7, zinc selenide (ZnSe), and zinc sulfide (ZnS). The IR-transmitting materials potentially available for use as windows and lenses are gathered in Table 1.4 and their IR transmission is shown in Figure 1.19 [14].

Germanium is a silvery, metallic-appearing solid of very high refractive index (≈ 4) that enables designing of high resolution optical systems using a minimal number of germanium lenses. Its useful transmission range is from 2 µm to about 15 µm. It is quite brittle and difficult to cut but accept a very good polish. Additionally, due to its very high refractive index, antireflection coatings are essential for any germanium transmitting optical system. Germanium has a low dispersion and is unlikely to need color correcting except in the highest-resolution systems. In spite of the high material price and cost of antireflection coatings, germanium lenses are particularly useful for 8–12 µm band. A significant disadvantage of germanium is serious dependence of its refractive index on temperature, so germanium telescopes and lenses may need to be athermalized.

Physical and chemical properties of silicon are very similar to properties of germanium. It has a high refractive index (≈ 3.45), is brittle, does not cleave, takes an excellent polish, and has large dn/dT. Similarly to germanium, silicon optics must have antireflection coatings. Silicon offers two transmission ranges: 1–7 µm and 25–300 µm. Only the first one is used in typical IR systems. The material is significantly cheaper than germanium. It is used mostly for IR systems operating in 3–5 µm band.

Single crystal material has generally higher transmission than polycrystalline one. Optical-grade germanium used for the highest optical transmission is n-type doped to receive a conductivity of 5–14 Ωcm. Silicon is used in its intrinsic state. At elevated temperatures semiconducting materials become opaque. As a result, germanium is of little use above 100°C. In the 8–14 µm region, semi-insulating GaAs may be used at the temperatures up to 200°C.

Table 1.4: Principal Characteristics of Some Infrared Materials

Material	Waveband (μm)	$n_{4\mu m}, n_{10\mu m}$	dn/dT (10^{-6} K^{-1})	Density (g/cm³)	Other Characteristics
Ge	3–5, 8–12	4.025, 4.004	424 (4 μm) 404 (10 μm)	5.33	Brittle, semiconductor, can be diamond-turned, visibly opaque, hard
Si	3–5	3.425	159 (5 μm)	2.33	Brittle, semiconductor, diamond-turned with difficulty, visibly opaque, hard
GaAs	3–5, 8–12	3.304, 3.274	150	5.32	Brittle, semiconductor, visibly opaque, hard
ZnS	3–5, 8–12	2.252, 2.200	43 (4 μm) 41 (10 μm)	4.09	Yellowish, moderate hardness and strength, can be diamond-turned, scatters short wavelengths
ZnSe	3–5, 8–12	2.433, 2.406	63 (4 μm) 60 (10 μm)	5.26	Yellow-orange, relatively soft and weak, can be diamond-turned, very low internal absorption and scatter
CaF$_2$	3–5	1.410	−8.1 (3.39 μm)	3.18	Visibly clear, can be diamond-turned, mildly hygroscopic
Sapphire	3–5	1.677(n_o) 1.667(n_e)	6 (o) 12 (e)	3.99	Very hard, difficult to polish due to crystal boundaries
AMTIR-1	3–5, 8–12	2.513, 2.497	72 (10 μm)	4.41	Amorphous IR glass, can be slumped to near-net shape
BK7 (Glass)	0.35–2.3		3.4	2.51	Typical optical glass

Source: Couture, M. E., "Challenges in IR optics," *Proceedings of SPIE* 4369, 649–61, 2001. With permission.

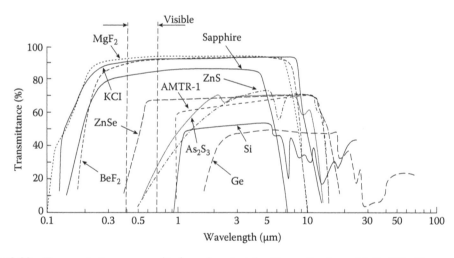

Figure 1.19 Transmission range of infrared materials. (From Couture, M. E., "Challenges in IR optics," *Proceedings of SPIE* 4369, 649–61, 2001. With permission.)

Ordinary glass does not transmit radiation beyond 2.5 μm in IR region. Fused silica is characterized by very low thermal expansion coefficient that makes optical systems particularly useful in changing environmental conditions. It offers transmission range from about 0.3 μm to 3 μm. Because of low reflection losses due to low refractive index (≈1.45), antireflection coatings are not needed. However, an antireflection coating is recommended to avoid ghost images. Fused silica is more expensive than BK-7, but still significantly cheaper than Ge, ZnS, and ZnSe and is a popular material for lenses of IR systems with bands located below 3 μm. The BK-7 glass characteristics are similar to fused silica; the difference is only a bit shorter transmission band up to 2.5 μm.

ZnSe is expensive material comparable to germanium; has a transmission range from 2 to about 20 μm, and a refractive index about 2.4. It is partially translucent when visible and reddish in color.

Due to relatively high refractive index, antireflection coatings are necessary. The chemical resistance of the material is excellent.

ZnS has excellent transmission in the range from 2 μm to 12 μm. It is usually a polycrystalline material that shows only a light yellow color. Recently also colorless, low scatter grades of ZnS are available. Because of a relatively high refractive index of 2.25, antireflection coatings are needed to minimize flux reflection. The hardness and fracture strength are very good. ZnS is brittle, can operate at elevated temperatures, and can be used to color correct high-performance germanium optics.

The alkali halides have excellent IR transmission, however, they are either soft or brittle and many of them are attacked by moisture, making them generally unsuitable for industrial applications.

For more detailed discussion of the IR materials see Smith and Harris [13,15].

1.7 SOME RADIOMETRIC ASPECTS OF INFRARED SYSTEMS

1.7.1 Night-Vision System Concepts

This section concentrates on general aspects of IR systems. A comprehensive compendium devoted to infrared systems was copublished in 1993 by Infrared Information Analysis Center (IRIA) and the International Society for Optical Engineering (SPIE) as *The Infrared and Electro-Optical Systems Handbook* (executive editors: Joseph S. Accetta and David L. Shumaker).

Night-vision systems can be divided into two categories: those depending upon the reception and processing of radiation reflected by an object and those that operate with radiation internally generated by an object.

The human visual perception system is optimized to operate in daytime illumination conditions. The visual spectrum extends from about 420 nm to 700 nm and the region of greatest sensitivity is near the peak wavelength of sunlight at around 550 nm. However, at night fewer visible light photons are available and only large, high contrast objects are visible. It appears that the photon rate in the region from 800 to 900 nm is five to seven times greater than in a visible region around 500 nm. Moreover, the reflectivity of various materials (e.g., green vegetation, because of its chlorophyll content) is higher between 800 nm and 900 nm than at 500 nm. It means that at night more light is available in the NIR than in the visual region and that against certain backgrounds more contrast is available.

A considerable improvement in night-vision capability can be achieved with night viewing equipment that consists of an objective lens, image intensifier, and eyepiece (see Figure 1.20). Improved visibility is obtained by gathering more light from the scene with an objective lens than the unaided eye; by use of a photocathode that has higher photosensitivity and broader spectral response than the eye; and by amplification of photo events for visual sensation.

Thermal imaging is a technique for converting a scene's thermal radiation pattern (invisible to the human eye) into a visible image. In its simplest form, a thermal imager consists of a lens that images the thermal radiation from the scene onto a heat sensitive detector array. The state of

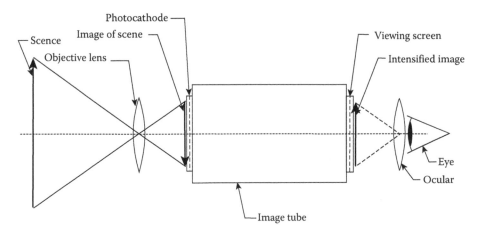

Figure 1.20 Diagram of an image intensifier.

detectors is addressed and displayed so that it is visible to the naked eye. Usefulness of the system is due to the following aspects:

- It is a totally passive technique and allows day and night operation.

- It is ideal for detection of hot or cold spots, or areas of different emissivities within a scene.

- Thermal radiation can penetrate smoke and mist more readily than visible radiation.

- It is a real-time, remote sensing technique.

The thermal image is a pictorial representation of temperature difference. Displayed on a scanned raster, the image resembles a television picture of the scene and can be computer processed to color-code temperature ranges. Originally developed (in the 1960s) to extend the scope of night-vision systems, thermal imagers at first provided an alternative to image intensifiers. As the technology has matured, its range of application has expanded and now extends into the fields that have little or nothing to do with night vision (e.g., stress analysis, medical diagnostics). In most current thermal imagers, an optically focused image is scanned (mechanically or electronically) across detectors (many elements or two-dimensional array), the output of which is converted into a visual image. The optics, mode of scanning, and signal processing electronics are closely interrelated. The number of picture points in the scene is governed by the nature of the detector (its performance) or the size of the detector array. The effective number of picture points or resolution elements in the scene may be increased by an optomechanical scanning device that images different parts of the scene onto the detector sequentially in time.

Figure 1.21 shows representative camera architecture with three distinct hardware pieces [16]: a camera head (which contains optics, including collecting, imaging, zoom, focusing, and spectral filtering assembles), electronics/control processing box, and the display. Electronics and motors to control and drive moving parts must be included. The control electronics usually consist of communication circuits, bias generators, and clocks. Usually a camera's sensor, focal plane array (FPA), needs cooling and therefore some form of cooler is included, along with its closed-loop cooling control electronics. A signal from the FPA is of low voltage and amperage and requires analog preprocessing (including amplification, control, and correction), which is located physically near the FPA and included in the camera head. Often, the A/D is also included here. For user convenience, the camera head often contains the minimum hardware needed to keep volume, weight, and power to a minimum.

Typical costs of cryogenically cooled imagers are around $50,000 and restrict their installation to critical military applications allowing conducting of operations in complete darkness. Moving from a cooled to an uncooled operation (e.g., using silicon microbolometer) reduces the cost of an imager to below $15,000. Less expensive infrared cameras present a major departure from camera architecture presented in Figure 1.21.

1.7.2 Atmospheric Transmission and Infrared Bands

Most of the above mentioned applications require transmission through air, but the radiation is attenuated by the processes of scattering and absorption. Scattering causes a change in the direction of a radiation beam; it is caused by absorption and subsequent reradiation of energy by suspended particles. For larger particles, scattering is independent of wavelength. However, for small particles, compared with the wavelength of the radiation, the process is known as Rayleigh scattering and exhibits a λ^{-4} dependence. Therefore, scattering by gas molecules is negligibly small for wavelengths longer than 2 μm. Also smoke and light mist particles are usually small with respect to IR wavelengths, and IR radiation can therefore penetrate further through smoke and mists than visible radiation. However, rain, fog particles, and aerosols are larger and consequently scatter IR and visible radiation to a similar degree.

Figure 1.22 is a plot of the transmission through 6000 feet of air as a function of wavelength [17]. Specific absorption bands of water, carbon dioxide, and oxygen molecules are indicated, which restricts atmospheric transmission to two windows at 3–5 μm and 8–14 μm. Ozone, nitrous oxide, carbon monoxide, and methane are less important IR absorbing constituents of the atmosphere.

In general, the 8–14 μm band is preferred for high performance thermal imaging because of its higher sensitivity to ambient temperature objects and its better transmission through mist and smoke. However, the 3–5 μm band may be more appropriate for hotter objects, or if sensitivity is less important than contrast. Also additional differences occur; for example, the advantage of

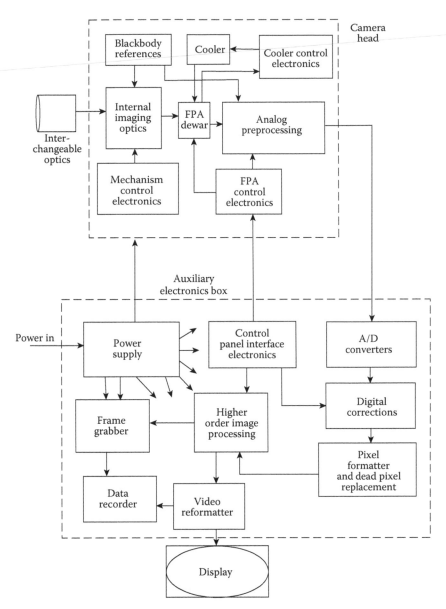

Figure 1.21 Representative infrared camera system architecture. (From Miller, J. L., *Principles of Infrared Technology*, Van Nostrand Reinhold, New York, 1994. With permission.)

MWIR band is a smaller diameter of the optics required to obtain a certain resolution and that some detectors may operate at higher temperatures (thermoelectric cooling) than is usual in the LWIR band where cryogenic cooling is required (about 77 K).

Summarizing, MWIR and LWIR μm spectral bands differ substantially with respect to background flux, scene characteristics, temperature contrast, and atmospheric transmission under diverse weather conditions. Factors which favor MWIR applications are: higher contrast, superior clear-weather performance (favorable weather conditions, e.g., in most countries of Asia and Africa), higher transmittivity in high humidity, and higher resolution due to ~3 times smaller optical diffraction. Factors which favor LWIR applications are: better performance in fog and dust conditions, winter haze (typical weather conditions, e.g., in West Europe, northern United States, Canada), higher immunity to atmospheric turbulence, and reduced sensitivity to solar glints and fire flares. The possibility of achieving higher signal-to-noise (S/N) ratio due to greater radiance levels in LWIR spectral range is not persuasive because the background photon fluxes are higher

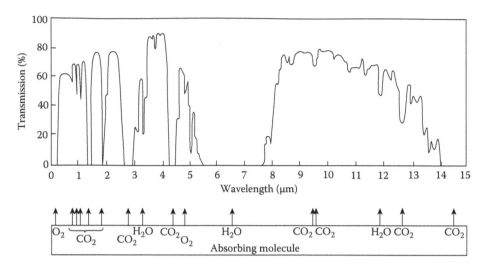

Figure 1.22 Transmission of the atmosphere for a 6000 foot horizontal path at sea level containing 17 mm of precipitate water. (From Hudson, R., *Infrared System Engineering*, Wiley, New York, 1969. With permission.)

Table 1.5: Power Available in Each MWIR and LWIR Imaging Bands

IR region (μm)	Ground-level solar radiation (W/m²)	Emission from 290 K blackbody (W/m²)
3–5	24	4.1
8–13	1.5	127

Source: Burnay, S. G., Williams, T. L., and Jones, C. H., *Applications of Thermal Imaging*, Adam Hilger, Bristol, England, 1988.

to the same extent, and also because of readout limitation possibilities. Theoretically, in staring arrays charge can be integrated for full frame time, but because of restrictions in the charge-handling capacity of the readout cells, it is much less compared to the frame time, especially for LWIR detectors for which background photon flux exceeds the useful signals by orders of magnitude.

1.7.3 Scene Radiation and Contrast

The total radiation received from any object is the sum of the emitted, reflected, and transmitted radiation. Objects that are not blackbodies emit only the fraction $\varepsilon(\lambda)$ of blackbody radiation, and the remaining fraction, $1 - \varepsilon(\lambda)$, is either transmitted or, for opaque objects, reflected. When the scene is composed of objects and backgrounds of similar temperatures, reflected radiation tends to reduce the available contrast. However, reflections of hotter or colder objects have a significant effect on the appearance of a thermal scene. The powers of 290 K blackbody emission and ground-level solar radiation in MWIR and LWIR bands are given in Table 1.5 [11]. We can see that while reflected sunlight have a negligible effect on 8–13 μm imaging, it is important in the 3–5 μm band.

A thermal image arises from temperature variations or differences in emissivity within a scene. When the temperatures of a target and its background are nearly the same, detection becomes very difficult. The thermal contrast is one of the important parameters for IR imaging devices. It is the ratio of the derivative of spectral radiant exitance to the spectral radiant exitance

$$C = \frac{\partial M(\lambda, T)/\partial T}{M(\lambda, T)}. \tag{1.44}$$

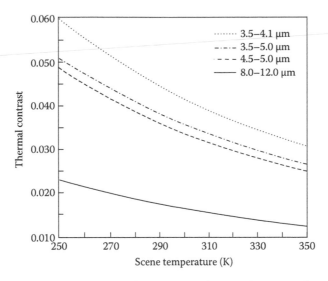

Figure 1.23 Spectral photon contrast in the MWIR and LWIR. (From Kozlowski, L. J., and Kosonocky, W. F., *Handbook of Optics*, McGraw-Hill, New York, 1995.)

Figure 1.23 is a plot of C for several MWIR subbands and the 8–12 LWIR spectral band [18]. The contrast in a thermal image is small when compared with visible image contrast due to differences in reflectivity. We can notice the contrast in the MWIR bands at 300 K is 3.5–4% compared to 1.6% for the LWIR band.

REFERENCES

1. P. R. Norton, "Infrared Detectors in the Next Millennium," *Proceedings of SPIE* 3698, 652–65, 1999.

2. F. Grum and R. J. Becherer, *Optical Radiation Measurements*, Vol. 1. Academic Press, San Diego, CA, 1979.

3. W. L. Wolfe and G. J. Zissis, *The Infrared Handbook*, SPIE Optical Engineering Press, Bellingham, WA, 1990.

4. W. L. Wolfe, "Radiation Theory," in *The Infrared and Electro-Optical Systems Handbook*, ed. G. J. Zissis, SPIE Optical Engineering Press, Bellingham, WA, 1993.

5. W. R. McCluney, *Introduction to Radiometry and Photometry*, Artech House, Boston, MA, 1994.

6. W. L. Wolf, *Introduction to Radiometry*, SPIE Optical Engineering Press, Bellingham, WA, 1998.

7. Y. Ohno, "Basic Concepts in Photometry, Radiometry and Colorimetry," in *Handbook of Optoelectronics*, Vol. 1, eds. J. P. Dakin and R. G. W. Brown, 287–305, Taylor & Francis, New York, 2006.

8. *CIE Compte Rendu*, p. 67, 1924.

9. *Quantities and Units*, ISO Standards Handbook, 3rd ed., 1993.

10. E. L. Dereniak and G. D. Boreman, *Infrared Detectors and Systems*, Wiley, New York, 1996.

11. S. G. Burnay, T. L. Williams, and C. H. Jones, *Applications of Thermal Imaging*, Adam Hilger, Bristol, England, 1988.

12. G. Gaussorgues, *La Thermographie Infrarouge,* Technique et Documentation, Lavoisier, Paris, 1984.

13. W. J. Smith, *Modern Optical Engineering,* McGraw-Hill, New York, 2000.

14. M. E. Couture, "Challenges in IR optics," *Proceedings of SPIE* 4369, 649–61, 2001.

15. D. C. Harris, *Materials for Infrared Windows and Domes,* SPIE Optical Engineering Press, Bellingham, WA, 1999.

16. J. L. Miller, *Principles of Infrared Technology,* Van Nostrand Reinhold, New York, 1994.

17. R. Hudson, *Infrared System Engineering,* Wiley, New York, 1969.

18. L. J. Kozlowski and W. F. Kosonocky, "Infrared Detector Arrays," in *Handbook of Optics,* eds. M. Bass, E. W. Van Stryland, D. R. Williams, and W. L. Wolfe, McGraw-Hill, New York, 1995.

2 Infrared Detector Characterization

Infrared (IR) radiation itself was unknown until 2010 years ago when Herschel's experiment with the thermometer was first reported. The first detector consisted of a liquid in a glass thermometer with a specially blackened bulb to absorb radiation. Herschel built a crude monochromator that used a thermometer as a detector so that he could measure the distribution of energy in sunlight. In April 1800 he wrote [1]:

"Thermometer No. 1 rose 7 degrees in 10 minutes by an exposure to the full red colored rays. I drew back the stand ... thermometer No. 1 rose, in 16 minutes, 8⅜ degrees when its centre was ½ inch out of the visible rays."

The early history of IR was reviewed about 50 years ago in two well-known monographs [2,3]. Many historical information can be also found in a more recently published monograph [4].

The most important steps in the development of IR detectors are the following [5,6]:

- In 1821 Seebeck discovered the thermoelectric effect and soon thereafter demonstrated the first thermocouple.

- In 1829 Nobili constructed the first thermopile by connecting a number of thermocouples in series.

- In 1833 Melloni modified the thermocouple and used bismuth and antimony for its design.

Langley's bolometer appeared in 1880 [7]. Langley used two thin ribbons of platinum foil, connected to form two arms of a Wheatstone bridge. Langley continued to develop his bolometer for the next 20 years (400 times more sensitive than his first efforts). His latest bolometer could detect the heat from a cow at a distance of a quarter of a mile. The beginning of the development of IR detectors was connected with thermal detectors.

The photoconductive effect was discovered by Smith in 1873 when he experimented with selenium as an insulator for submarine cables [8]. This discovery provided a fertile field of investigation for several decades, though most of the effort was of doubtful quality. By 1927, over 1500 articles and 100 patents had been listed on photosensitive selenium [9]. Work on the IR photovoltaic effect in naturally occurring lead sulfide or galena was announced by Bose in 1904 [10], however, this effect was not used in a radiation detector for the next several decades.

The photon detectors were developed in the twentieth century. The first IR photoconductor was developed by Case in 1917 [11]. He discovered that a substance composed of thallium and sulfur exhibited photoconductivity. Later he found that the addition of oxygen greatly enhanced the response [12]. However, instability of resistance in the presence of light or polarizing voltage, loss of responsivity due to overexposure to light, high noise, sluggish response, and lack of reproducibility seemed to be inherent weaknesses.

Since about 1930 the development of IR technology has been dominated by the photon detectors. In about 1930, the appearance of the Cs–O–Ag phototube, with more stable characteristics, to a great extent discouraged further development of photoconductive cells until about 1940. At that time, interest in improved detectors began in Germany [13,14]. In 1933, Kutzscher at the University of Berlin, discovered that lead sulphide (from natural galena found in Sardinia) was photoconductive and had response to about 3 µm. This work was, of course, done under great secrecy and the results were not generally known until after 1945. Lead sulfide was the first practical IR detector deployed in a variety of applications during the war. In 1941, Cashman improved the technology of thallous sulfide detectors, which led to successful production [15]. Cashman, after success with thallous sulfide detectors, concentrated his efforts on lead sulfide and after World War II found that other semiconductors of the lead salt family (PbSe and PbTe) showed promise as IR detectors [15]. Lead sulfide photoconductors were brought to the manufacturing stage of development in Germany in about 1943. They were first produced in the United States at Northwestern University, Evanston, Illinois in 1944 and, in 1945, at the Admiralty Research Laboratory in England [16].

Many materials have been investigated in the IR field. Observing a history of the development of the IR detector technology, a simple theorem, after Norton [17], can be stated: *"All physical phenomena in the range of about 0.1–1 eV can be proposed for IR detectors."* Among these effects are: thermoelectric power (thermocouples), change in electrical conductivity (bolometers), gas expansion (Golay cell), pyroelectricity (pyroelectric detectors), photon drag, Josephson effect (Josephson junctions, SQUIDs), internal emission (PtSi Schottky barriers), fundamental absorption (intrinsic photodetectors),

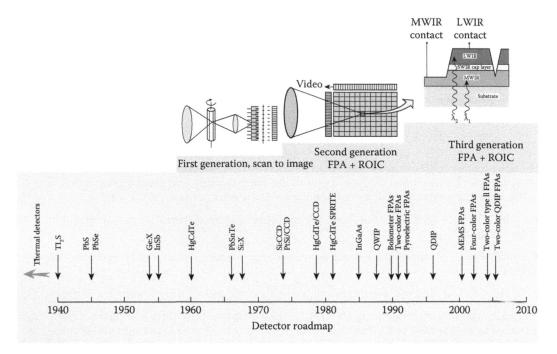

Figure 2.1 History of the development of infrared detectors. Three-generation systems can be considered for principal military and civilian applications: first generation (scanning systems), second generation (staring systems—electronically scanned), and third generation (multicolor functionality and other on-chip functions).

impurity absorption (extrinsic photodetectors), low dimensional solids [superlattice (SL), quantum well (QW) and quantum dot (QD) detectors], different type of phase transitions, and so on.

Figure 2.1 gives approximate dates of significant development efforts for the materials mentioned. The years during World War II saw the origins of modern IR detector technology. Recent success in applying IR technology to remote sensing problems has been made possible by the successful development of high-performance IR detectors over the last six decades. Photon IR technology combined with semiconductor material science, photolithography technology developed for integrated circuits, and the impetus of Cold War military preparedness have propelled extraordinary advances in IR capabilities within a short time period during the last century [18].

2.1 HISTORICAL ASPECTS OF MODERN INFRARED TECHNOLOGY

During the 1950s IR detectors were built using single-element-cooled lead salt detectors, primarily for anti-air-missile seekers. Usually lead salt detectors were polycrystalline and were produced by vacuum evaporation and chemical deposition from a solution, followed by a post-growth sensitization process [16]. The preparation process of lead salt photoconductive detectors was usually not well understood and reproducibility could only be achieved after following well-tried recipes. The first extrinsic photoconductive detectors were reported in the early 1950s [19] after the discovery of the transistor, which stimulated a considerable improvement in the growth and material purification techniques. Since the techniques for controlled impurity introduction became available for germanium at an earlier date, the first high-performance extrinsic detectors were based on germanium. Extrinsic photoconductive response from copper, zinc, and gold impurity levels in germanium gave rise to devices using in the 8 to 14 μm long wavelength IR (LWIR) spectral window and beyond to the 14–30 μm very long wavelength IR (VLWIR) region. The extrinsic photoconductors were widely used at wavelengths beyond 10 μm prior to the development of the intrinsic detectors. They must be operated at lower temperatures to achieve performance similar to that of intrinsic detectors, and a sacrifice in quantum efficiency is required to avoid thick detectors.

In 1967 the first comprehensive extrinsic Si detector-oriented paper was published by Soref [20]. However, the state of extrinsic Si was not changed significantly. Although Si has several advantages over Ge (namely, a lower dielectric constant giving shorter dielectric relaxation time and lower capacitance, higher dopant solubility and a larger photoionization cross section for higher quantum efficiency, and lower refractive index for lower reflectance), these were not sufficient to warrant the necessary development efforts needed to bring it to the level of the highly developed Ge detectors by then. After being dormant for about 10 years, extrinsic Si was reconsidered after the invention of charge-coupled devices (CCDs) by Boyle and Smith [21]. In 1973, Shepherd and Yang [22] proposed the metal-silicide/silicon Schottky-barrier detectors. For the first time it became possible to have much more sophisticated readout schemes—both detection and readout could be implemented in one common silicon chip.

At the same time, rapid advances were being made in narrow bandgap semiconductors that would later prove useful in extending wavelength capabilities and improving sensitivity. The first such material was InSb, a member of the newly discovered III–V compound semiconductor family. The interest in InSb stemmed not only from its small energy gap, but also from the fact that it could be prepared in single crystal form using a conventional technique. The end of the 1950s and the beginning of the 1960s saw the introduction of narrow-gap semiconductor alloys in III–V ($InAs_{1-x}Sb_x$), IV–VI ($Pb_{1-x}Sn_xTe$), and II–VI ($Hg_{1-x}Cd_xTe$) material systems. These alloys allowed the bandgap of the semiconductor and hence the spectral response of the detector to be custom tailored for specific applications. In 1959, research by Lawson and coworkers [23] triggered the development of variable bandgap $Hg_{1-x}Cd_xTe$ (HgCdTe) alloys, providing an unprecedented degree of freedom in IR detector design. This first paper reported both photoconductive and photovoltaic response at the wavelength extending out to 12 μm. Soon thereafter, working under a U.S. Air Force contract with the objective of devising an 8–12 μm background-limited semiconductor IR detector that would operate at temperatures as high as 77 K, the group lead by Kruse at the Honeywell Corporate Research Center in Hopkins, Minnesota developed a modified Bridgman crystal growth technique for HgCdTe. They soon reported both photoconductive and photovoltaic detection in rudimentary HgCdTe devices [24].

The fundamental properties of narrow-gap semiconductors (high optical absorption coefficient, high electron mobility, and low thermal generation rate), together with the capability for bandgap engineering, make these alloy systems almost ideal for a wide range of IR detectors. The difficulties in growing HgCdTe material, significantly due to the high vapor pressure of Hg, encouraged the development of alternative detector technologies over the past 40 years. One of these was PbSnTe, which was vigorously pursued in parallel with HgCdTe in the late 1960s and early 1970s [25–27]. PbSnTe was comparatively easy to grow and good quality LWIR photodiodes were readily demonstrated. However, in the late 1970s two factors led to the abandonment of PbSnTe detector work: high dielectric constant and large temperature coefficient of expansion (TCE) mismatch with Si. Scanned IR imaging systems of the 1970s required relatively fast response times so that the scanned image is not smeared in the scan direction. With the trend today toward staring arrays, this consideration might be less important than it was when first-generation systems were being designed. The second drawback, large TCE, can lead to failure of the indium bonds in hybrid structure (between silicon readout and the detector array) after repeated thermal cycling from room temperature to the cryogenic temperature of operation.

The material technology development was and continues to be primarily for military applications. In the United States, the Vietnam War caused the military services to initiate the development of IR systems that could provide imagery arising from the thermal emission of terrain vehicles, buildings, and people. As photolithography became available in the early 1960s, it was applied to make IR detector arrays. Linear array technology was first applied to PbS, PbSe, and InSb detectors. The discovery in the early 1960s of extrinsic Hg-doped germanium [28] led to the first FLIR systems operating in the LWIR spectral window using linear arrays. Because the detection mechanism was based on an extrinsic excitation, it required a two-stage cooler to operate at 25 K. The cooling requirements of intrinsic narrow bandgap semiconductor detectors are much less stringent. Typically, to obtain the background-limited performance (BLIP), detectors for the 3–5 μm spectral region are operated at 200 K or less, while those for the 8–14 μm are typically operated at liquid nitrogen temperature. In the late 1960s and early 1970s, first-generation linear arrays (in which an electrical contact for each element of a multielement array is brought off the cryogenically cooled focal plane to the outside, where there is one electronic channel at ambient temperature for each detector element) of intrinsic HgCdTe photoconductive detectors were developed. These allowed LWIR FLIR systems to operate with a single-stage cryoengine, making the systems much more compact, lighter, and requiring significantly less power consumption.

Exactly, HgCdTe has inspired the development of the three "generations" of detector devices. The first generation, linear arrays of photoconductive detectors, has been produced in large quantities and is in widespread use today. The second generation, two-dimensional (2-D) arrays of photovoltaic detectors are now in high-rate production. At the present stage of development, staring arrays have about 10^6 elements and are scanned electronically by circuits integrated with the arrays. These 2-D arrays of photodiodes connected with indium bumps to a readout integrated circuit (ROIC) chip as a hybrid structure are often called a sensor chip assembly (SCA). Third-generation devices defined here to encompass the more exotic device structure embodied in two-color detectors and hyperspectral arrays and are now in demonstration programs (see Figure 2.1).

Early assessment of the concept of the second-generation system showed that PtSi Schottky barriers, InSb, and HgCdTe photodiodes or high-impedance photoconductors such as PbSe, PbS, and extrinsic silicon detectors were promising candidates because they have impedances suitable for interfacing with the field-effect transistor (FET) input of readout multiplexes. Photoconductive HgCdTe detectors were not suitable due to their low impedance and high-power dissipation on the focal plane. A novel British invention, the SPRITE detector [29,30], extended conventional photoconductive HgCdTe detector technology by incorporating signal time delay and integration (TDI) within a single elongated detector element. Such a detector replaces a whole row of discrete elements of a conventional serial-scanned detector and external associated amplifiers and time delay circuitry. Although only used in small arrays of about 10 elements, these devices have been produced in the thousands.

In the late 1970s and through the 1980s, HgCdTe technology efforts focused almost exclusively on photovoltaic device development because of the need for low power dissipation and high impedance in large arrays to interface to readout input circuits. This effort is finally paying off with the birth of HgCdTe second-generation IR systems that provide large 2-D arrays in both linear formats with TDI for scanning imagers, and in square and rectangular formats for staring arrays. The 1024 × 1024 hybrid HgCdTe focal plane arrays (FPAs) have been produced recently. However, present HgCdTe FPAs are limited by the yield of arrays, which increases their cost. In such a situation, alternative alloy systems for infrared detectors, such as quantum well infrared photodetectors (QWIPs) and type II superlattices, are investigated.

The trend of increasing pixel numbers is likely to continue in the area of large format arrays. This increasing will be continued using close-butted mosaic of several SCAs. Raytheon manufactured a 4 × 4 mosaic of 2K × 2K HgCdTe SCAs and assisted in assembling it into the final focal-plane configuration to survey the entire sky in the Southern Hemisphere at four IR wavelengths [31]. With 67 million pixels, this is currently the world's largest IR focal plane. Although there are currently limitations to reducing the size of the gaps between active detectors on adjacent SCAs, many of these can be overcome. It is predicted that a focal plane of 100 megapixels and larger will be possible, constrained only by budgets but not by technology [32].

A negative aspect of support by defense agencies has been the associated secrecy requirements that inhibit meaningful collaborations among research teams on a national and especially on an international level. In addition, the primary focus has been on FPA demonstration and much less on establishing the knowledge base. Nevertheless, significant progress has been made over four decades. At present, HgCdTe is the most widely used variable gap semiconductor for IR photodetectors. Over the years it has successfully fought off major challenges from extrinsic silicon and lead–tin telluride devices, but despite that it has more competitors today than ever before. These include Schottky barriers on silicon, SiGe heterojunctions, AlGaAs multiple QWs, GaInSb strain layer superlattices, high temperature superconductors, and especially two types of thermal detectors: pyroelectric detectors and silicon bolometers. However, it is interesting that none of these competitors can compete in terms of fundamental properties. They may promise to be more manufacturable, but never to provide higher performance or, with the exception of thermal detectors, to operate at higher or even comparable temperatures. It should be noticed that from a physics point of view, the type II GaInSb superlattice is an extremely attractive proposition.

As is mentioned above, monolithic extrinsic Si detectors were demonstrated first in the mid-1970s [33–35], but were subsequently set aside because the process of integrated circuit fabrication degraded the detector-quality material properties. Historically, Si:Ga and Si:In were the first mosaic FPA photoconductive materials because early monolithic approaches were compatible with these dopants. The photoconductive material was made in either conventional or impurity band conduction—IBC or blocked impurity band (BIB)—technologies. Extrinsic photoconductors must be made relatively thick because they have much lower photon capture cross section than intrinsic detectors. However, BIB detectors have a unique combination of photoconductive and photovoltaic characteristics, including extremely high impedance, reduced recombination noise, linear photoconductive gain, high uniformity, and superb stability. Megapixel detector arrays with cutoff wavelength to 28 μm are now available [36]. Specially doped IBCs operate as solid-state photomultipliers (SSPMs) and visible light photon converters (VLPC) in which photoexcited carriers allows counting of individual photons at low flux levels. Standard SSPMs respond from 0.4 to 28 μm.

As was mentioned previously, the development of IR technology has been dominated by the photon detectors since about 1930. However, the photon detectors require cryogenic cooling. This is necessary to prevent the thermal generation of charge carriers. The thermal transitions compete with the optical ones, making noncooled devices very noisy. The cooled thermal camera usually uses the Sterling cycle cooler, which is the expensive component in the photon detector IR camera, and the cooler's lifetime is only around 10,000 hours. Cooling requirements are the main obstacle to the widespread use of IR systems based on semiconductor photon detectors making them bulky, heavy, expensive, and inconvenient to use.

The use of thermal detectors for IR imaging has been the subject of research and development for many decades. However, in comparison with photon detectors, thermal detectors have been considerably less exploited in commercial and military systems. The reason for this disparity is that thermal detectors are popularly believed to be rather slow and insensitive in comparison with photon detectors. As a result, the worldwide effort to develop thermal detectors has been extremely small relative to that of the photon detectors.

It must not be inferred from the preceding outline that work on thermal detectors has not also been actively pursued. Indeed, some interesting and important developments have taken place along this line. In 1947, for example, Golay constructed an improved pneumatic infrared detector [37]. This gas thermometer has been used in spectrometers. The thermistor bolometer, originally developed by Bell Telephone Laboratories, has found widespread use in detecting radiation from low temperature sources [38,39]. The superconducting effect has been used to make extremely sensitive bolometers.

Thermal detectors have also been used for infrared imaging. Evaporographs and absorption edge image converters were among the first nonscanned IR imagers. Originally an evaporograph was employed in which the radiation was focused onto a blackened membrane coated with a thin film of oil [40]. The differential rate of evaporation of the oil was proportional to radiation intensity. The film was then illuminated with visible light to produce an interference pattern corresponding to the thermal picture. The second thermal imaging device was the absorption edge image converter [41]. Operation of this device was based upon utilizing the temperature dependence of the absorption edge of semiconductor. The performance of both imaging devices was poor because of the very long time constant and the poor spatial resolution. Despite numerous research initiatives and the attractions of ambient temperature operation and low cost potential, thermal detector technology has enjoyed limited success in competition with cooled photon detectors for thermal imaging applications. A notable exception is the pyroelectric vidicon (PEV) [42] that is widely used by firefighting and emergency service organizations. The PEV tube can be considered analogously to the visible television camera tube except that the photoconductive target is replaced by a pyroelectric detector and germanium faceplate. Compact, rugged PEV imagers have been offered for military applications but suffer the disadvantage of low tube life and fragility, particularly the reticulated vidicon tubes required for enhanced spatial resolution. The advent of the staring FPAs, however, marked the development of devices that would someday make uncooled systems practical for many, especially commercial, applications. The defining effort in this field was undertaken by Texas Instruments with contractual support from the Army Night Vision Laboratory [4]. The goal of this program was to build a staring FPA system based on ferroelectric detectors of barium strontium titanate. Throughout the 1980s and early 1990s, many other companies developed devices based on various thermal detection principles.

The second revolution in thermal imaging began in the end of twentieth century. The development of uncooled IR arrays capable of imaging scenes at room temperature has been an outstanding technical achievement. Much of the technology was developed under classified military contracts in the United States, so the public release of this information in 1992 surprised many in the worldwide IR community [43]. There has been an implicit assumption that only cryogenic photon detectors operating in the 8–12 μm atmospheric window had the necessary sensitivity to image room-temperature objects. Although thermal detectors have been little used in scanned imagers because of their slow response, they are currently of considerable interest for 2-D electronically addressed arrays where the bandwidth is low and the ability of thermal devices to integrate over a frame time is an advantage [44–49]. Much recent research has focused on both hybrid and monolithic uncooled arrays and has yielded significant improvements in the detectivity of both bolometric and pyroelectric detector arrays. Honeywell has licensed bolometer technology to several companies for the development and production of uncooled FPAs for commercial and military systems. At present, the compact 640 × 480 microbolometer cameras are produced by Raytheon, Boeing, and Lockheed-Martin in the United States. The U.S. government allowed these manufacturers to sell their devices to foreign countries, but not to divulge manufacturing technologies. In recent years, several countries, including the United Kingdom, France, Japan, and Korea, have picked up the ball, determined to develop their own uncooled imaging systems. As a result, although the United States has a significant lead, some of the most exciting and promising developments for low-cost uncooled IR systems may come from non-U.S. companies (e.g., microbolometer FPAs with series p-n junction elaborated by Mitsubishi Electric). This approach is unique, based on an all-silicon version of the microbolometer.

2.2 CLASSIFICATION OF INFRARED DETECTORS

Spectral detectivity curves for a number of commercially available IR detectors are shown in Figure 2.2. Interest has centered mainly on the wavelengths of the two atmospheric windows 3–5 μm [middle wavelength IR (MWIR)] and 8–14 μm (LWIR region; atmospheric transmission is the highest in these bands and the emissivity maximum of the objects at T ≈ 300 K is at the wavelength $\lambda \approx 10$ micron), though in recent years there has been increasing interest in longer wavelengths stimulated by space applications. The spectral character of the background is influenced by the transmission of the atmosphere (see Figure 1.22 [50]), which controls the spectral ranges of the infrared for which the detector may be used when operating in the atmosphere.

Progress in IR detector technology is connected with semiconductor IR detectors, which are included in the class of photon detectors. In this class of detectors the radiation is absorbed within the material by interaction with electrons either bound to lattice atoms or to impurity atoms or with free electrons. The observed electrical output signal results from the changed electronic energy distribution. The fundamental optical excitation processes in semiconductors are illustrated in Figure 2.3 [51,52]. The photon detectors show a selective wavelength dependence of response per unit incident radiation power (see Figure 2.4). They exhibit both perfect signal-to-noise performance and a very fast response. But to achieve this, the photon detectors require cryogenic cooling. Photon detectors having long wavelength limits above about 3 μm are generally cooled. This is necessary to prevent the thermal generation of charge carriers. The thermal transitions compete with the optical ones, making noncooled devices very noisy.

Depending on the nature of the interaction, the class of photon detectors is further subdivided into different types as shown in Table 2.1. The most important are: intrinsic detectors, extrinsic detectors, photoemissive (metal silicide Schottky barriers) detectors, and QW detectors. Depending on how the electric or magnetic fields are developed, there are various modes such as photoconductive, photovoltaic, photoelectromagnetic (PEM), and photoemissive ones. Each material system can be used for different modes of operation.

Recently this standard classification has been reconsidered by Kinch [51], as discussed briefly below. Photon detectors can be divided into two broad classes; namely, majority and minority carrier devices. The material systems used are:

1. Direct bandgap semiconductors—minority carriers

   - Binary alloys: InSb, InAs
   - Ternary alloys: HgCdTe, InGaAs
   - Type II, III superlattices: InAs/GaInSb, HgTe/CdTe

2. Extrinsic semiconductors—majority carriers

   - Si:As, Si:Ga, Si:Sb
   - Ge:Hg, Ge:Ga

3. Type I superlattices—majority carriers

   - GaAs/AlGaAs quantum well infrared photodetectors (QWIPs)

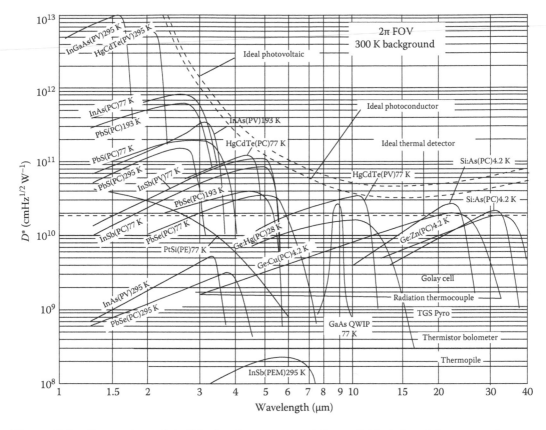

Figure 2.2 Comparison of the D^* of various commercially available infrared detectors when operated at the indicated temperature. Chopping frequency is 1000 Hz for all detectors except the thermopile (10 Hz), thermocouple (10 Hz), thermistor bolometer (10 Hz), Golay cell (10 Hz), and pyroelectric detector (10 Hz). Each detector is assumed to view a hemispherical surrounding at a temperature of 300 K. Theoretical curves for the background-limited D^* (dashed lines) for ideal photovoltaic and photoconductive detectors and thermal detectors are also shown. PC, photoconductive detector; PV, photovoltaic detector; PE, photoemissive; and PEM, photoelectromagnetic detector.

4. Silicon Schottky barriers—majority carriers

- PtSi, IrSi

5. High-temperature superconductors—minority carriers

All of these material systems have been serious players in the IR marketplace with the exception of type III superlattices and high-temperature superconductors.

The second class of IR detectors is composed of thermal detectors. In a thermal detector shown schematically in Figure 3.1, the incident radiation is absorbed to change the material temperature, and the resultant change in some physical property is used to generate an electrical output. The detector is suspended on lags, which are connected to the heat sink. The signal does not depend upon the photonic nature of the incident radiation. Thus, thermal effects are generally wavelength independent (see Figure 2.4); the signal depends upon the radiant power (or its rate of change) but not upon its spectral content. Since the radiation can be absorbed in a black surface coating, the spectral response can be very broad. Attention is directed toward three approaches that have found the greatest utility in infrared technology; namely, bolometers, pyroelectric, and thermoelectric effects. In pyroelectric detectors a change in the internal electrical polarization is measured, whereas in the case of thermistor bolometers a change in the electrical resistance is measured. In contrast to photon detectors, thermal detectors are typically operated at room temperature. They are usually characterized by modest

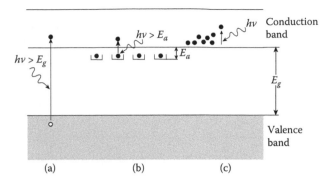

Figure 2.3 Fundamental optical excitation processes in semiconductors: (a) intrinsic absorption, (b) extrinsic absorption, (c) free carrier absorption. (From Elliott, C. T. and Gordon, N. T., *Handbook on Semiconductors,* Vol. 4, 841–936, Elsevier, Amsterdam, 1993. With permission.)

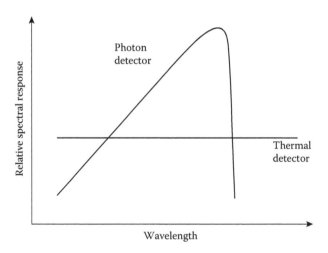

Figure 2.4 Relative spectral response for a photon and thermal detector.

sensitivity and slow response (because heating and cooling of a detector element is a relatively slow process), but they are cheap and easy to use. They have found widespread use in low cost applications, which do not require high performance and speed. Being unselective, they are frequently used in IR spectrometers. A list of thermal effects is included in Table 2.2.

Up until the 1990s, thermal detectors have been considerably less exploited in commercial and military systems in comparison with photon detectors. In the last decade, however, it has been shown that extremely good imagery can be obtained from large thermal detector arrays operating uncooled at TV frame rates. The speed of thermal detectors is quite adequate for nonscanned imagers with 2-D detectors. The moderate sensitivity of thermal detectors can be compensated by a large number of elements in 2-D electronically scanned arrays. With large arrays of thermal detectors the best values of noise equivalent temperature difference (NEDT), below 0.05 K, could be reached because effective noise bandwidths less than 100 Hz can be achieved.

Finally, the third category of infrared detectors, so-called radiation field detectors, which respond directly to the radiation field should be mentioned. This class of detectors was developed after 1970, but has not had significant impact [52]. Recently their importance has increased due to wider application of harmonic mixers in the far-IR (terahertz) and submillimeter regions [53].

Table 2.1: Comparison of Infrared Detectors

Detector Type			Advantages	Disadvantages
Thermal (thermopile, bolometers, pyroelectric)			Light, rugged, reliable, & low cost Room temperature operation	Low detectivity at high frequency Slow response (ms order)
Photon	Intrinsic	IV–VI (PbS, PbSe, PbSnTe)	Easier to prepare More stable materials	Very high thermal expansion coefficient Large permittivity
		II–VI (HgCdTe)	Easy bandgap tailoring Well-developed theory & experience Multicolor detectors	Nonuniformity over large area High cost in growth and processing Surface instability
		III–V (InGaAs, InAs, InSb, InAsSb)	Good material & dopants Advanced technology Possible monolithic integration	Heteroepitaxy with large lattice mismatch Long wavelength cutoff limited to 7 µm (at 77 K)
	Extrinsic (Si:Ga, Si:As, Ge:Cu, Ge:Hg)		Very long wavelength operation Relatively simple technology	High thermal generation Extremely low temperature operation
	Free carriers (PtSi, Pt$_2$Si, IrSi)		Low-cost, high yields Large & close packed 2-D arrays	Low quantum efficiency Low temperature operation
	Quantum wells	Type I (GaAs/AlGaAs, InGaAs/ AlGaAs)	Matured material growth Good uniformity over large area Multicolor detectors	High thermal generation Complicated design and growth
		Type II (InAs/InGaSb, InAs/InAsSb)	Low Auger recombination rate Easy wavelength control Multicolor detectors	Complicated design and growth Sensitive to the interfaces
	Quantum dots	InAs/GaAs, InGaAs/InGaP, Ge/Si	Normal incidence of light Low thermal generation	Complicated design and growth

Table 2.2: Infrared Thermal Detectors

Detector	Method of Operation

Bolometer
    Metal
    Semiconductor
    Superconductor
    Ferroelectric
    Hot electron
	Change in electrical conductivity

Thermocouple/thermopile	Voltage generation, caused by change in temperature of the junction of two dissimilar materials
Pyroelectric	Changes in spontaneous electrical polarization
Golay cell/gas microphone	Thermal expansion of a gas
Absorption edge	Optical transmission of a semiconductor
Pyromagnetic	Changes in magnetic properties
Liquid crystal	Changes of optical properties

2.3 COOLING OF IR DETECTORS

The method of cooling varies according to the operating temperature and the system's logistical requirements [54,55]. Figure 2.5 is a chart depicting infrared operating temperature and wavelength regions spanned by a variety of available infrared detector technologies.

Various types of cooling systems have been developed including dewars with cryogenic liquids or solids, Joule–Thompson open cycle, Stirling closed cycle, and thermoelectric coolers (see Figures 2.5 and 2.6). These systems are discussed briefly below.

2.3.1 Cryogenic Dewars

Most 8–14 μm detectors operate at approximately 77 K and can be cooled by liquid nitrogen. Cryogenic liquid pour-filled dewars are frequently used for detector cooling in laboratories. They are rather bulky and need to be refilled with liquid nitrogen every few hours. For many applications, especially in the field of LN_2, pour-filled dewars are impractical, so many manufacturers are turning to alternative coolers that do not require cryogenic liquids or solids.

2.3.2 Joule–Thompson Coolers

The design of Joule-Thompson coolers is based on the fact that as a high-pressure gas expands upon leaving a throttle valve, it cools and liquefies. The coolers require a high-pressure gas supply from bottles and compressors. Using compressed air, temperatures on the order of 80 K can be achieved in one or two minutes. The gas used must be purified to remove water vapor and carbon dioxide that could freeze and block the throttle valve. Specially designed Joule–Thompson coolers using argon are suitable for ultrafast cooldown (a few seconds cooling time).

2.3.3 Stirling Cycle Coolers

Ever since the late 1970s, military systems have overcome the problem of LN_2 operation by utilizing Stirling closed-cycle refrigerators to generate the cryogenic temperatures necessary for critical IR detector components. These refrigerators were designed to produce operating temperatures of 77 K directly from DC power. Early versions were large and expensive, and suffered from microphonic and electromagnetic interference (EMI) noise problems. Today, smaller and more efficient integral and split Stirling coolers have been developed and refined.

The use of cooling engines, in particular those employing the Stirling cycle, has increased recently due to their efficiency, reliability, and cost reduction. Stirling engines require several minutes of cooldown time; the working fluid is helium. Both Joule–Thompson and engine-cooled detectors are housed in precision-bore dewars into which the cooling device is inserted (see Figure 2.6). The detector, mounted in the vacuum space at the end of the inner wall of the dewar and surrounded by a cooled radiation shield compatible with the convergence angle of the optical system, looks out through an IR window. In some dewars, the electrical leads to detector elements are embedded in the inner wall of the dewar to protect them from damage due to vibration.

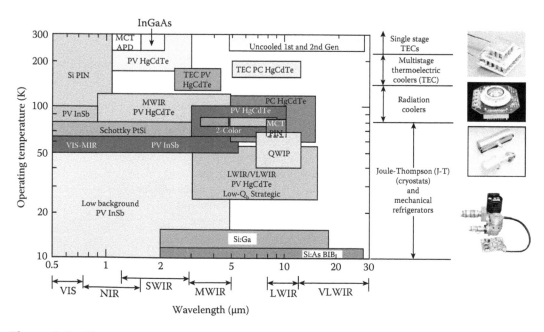

Figure 2.5 The operating temperature and wavelength regions spanned by a variety of available infrared detector technologies.

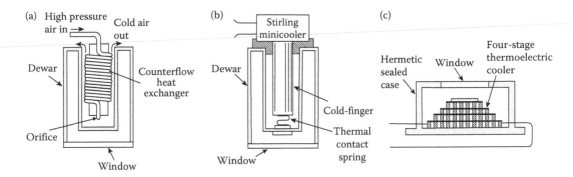

Figure 2.6 Three ways of cooling IR detectors: (a) Joule-Thompson cooler, (b) Stirling-cycle engine, and (c) four-stage thermoelectric cooler (Peltier effect).

Typical Stirling cooler parameters are:

- For integral: mass, 0.3 kg; heat load, 0.15 W; input power, 3.5 W; shelf life, five years; mean time before failure (MTBF) = 2000 hour; size with detector, 9×9 cm; 80 K achievable in four minutes; significant microphonics. After 2000–3000 hours of operation, this cooler will require factory service to maintain performance. Because the dewar and cooler is one integrated unit, the entire unit must be serviced together.

- For split: input power ≈ 20 W, 1.5 W heat load, fewer microphonics, eight-year MTFB. The detector is mounted on the dewar bore, and the cold finger of the cooler is thermally connected to the dewar by a bellows. A fan is necessary to dissipate the heat. The cooler can be easily removed from the detector/dewar for replacement.

Closed-cycle coolers offer the ultimate detector performance without the need for bulk liquid nitrogen. Temperatures as low as 15 K can be obtained. However, the high initial cost, large input power, and limited operating life are not suited for many applications.

2.3.4 Peltier Coolers

Thermoelectric (TE) cooling of detectors is simpler and less costly than closed-cycle cooling. In the case of Peltier coolers, detectors are usually mounted in a hermetic encapsulation with a base designed to make good contact with a heat sink. The TE coolers can achieve temperatures to ≈ 200° K, have ≈ 20-year operating life and low input power (<1 W for a two-stage device and <3 W for a three-stage device), and are small and rugged. The TE coolers used for FPA operation include one-stage (TE1, down to –20°C or 253 K), two-stage (TE2, down to –40°C or 233 K), three-stage (TE3, down to –65°C or 208 K), and four-stage (TE4, down to –80°C or 193 K). Peltier coolers can be used to stabilize the temperature at the required level. There are hopes for much better performance of TE coolers based on superlattice materials [56]. With the TE figure of merit, ZT, almost three times larger than that in Bi_2Te_3, temperatures of <200 K would be achievable with simple two-stage TE coolers based on the new materials.

2.4 DETECTOR FIGURES OF MERIT

It is difficult to measure the performance characteristics of infrared detectors because of the large number of experimental variables involved. A variety of environmental, electrical, and radiometric parameters must be taken into account and carefully controlled. With the advent of large, two-dimensional detector arrays, detector testing has become even more complex and demanding.

This section is intended to serve as an introductory reference for the testing of infrared detectors. Numerous texts and journals cover this issue, including: *Infrared System Engineering* [50], by R. D. Hudson; *The Infrared Handbook* [57], edited by W. L. Wolfe and G. J. Zissis; *The Infrared and Electro-Optical Systems Handbook* [58], edited by W. D. Rogatto; and *Fundamentals of Infrared Detector Operation and Testing* [59], by J. D. Vincent. In this volume we have restricted our consideration to detectors whose output consists of an electrical signal that is proportional to the radiant signal power [59–62].

The measured data described in this text are sufficient to characterize a detector. However, to provide ease of comparison between detectors, certain figures of merit, computed from the measured data, have been defined in this section.

2.4.1 Responsivity

The responsivity of an infrared detector is defined as the ratio of the root mean square (rms) value of the fundamental component of the electrical output signal of the detector to the rms value of the fundamental component of the input radiation power. The units of responsivity are volts per watt (V/W) or ampers per watt (amp/W).

The voltage (or analogous current) spectral responsivity is given by

$$R_v(\lambda, f) = \frac{V_s}{\Phi_e(\lambda)\Delta\lambda},$$
(2.1)

where V_s is the signal voltage due to Φ_e, and $\Phi_e(\lambda)$ is the spectral radiant incident power (in W/m).

An alternative to the above monochromatic quality, the blackbody responsivity, is defined by the equation

$$R_v(T, f) = \frac{V_s}{\int_0^\infty \Phi_e(\lambda)d\lambda},$$
(2.2)

where the incident radiant power is the integral over all wavelengths of the spectral density of power distribution $\Phi_e(\lambda)$ from a blackbody. The responsivity is usually a function of the bias voltage V_b, the operating electrical frequency f, and the wavelength λ.

2.4.2 Noise Equivalent Power

The noise equivalent power (NEP) is the incident power on the detector generating a signal output equal to the rms noise output. Stated another way, the NEP is the signal level that produce a signal-to-noise ratio (SNR) of 1. It can be written in terms of responsivity:

$$\text{NEP} = \frac{V_n}{R_v} = \frac{I_n}{R_i}.$$
(2.3)

The unit of NEP is watt.

The NEP is also quoted for a fixed reference bandwidth, which is often assumed to be 1 Hz. This "NEP per unit bandwidth" has a unit off watts per square root hertz ($W/Hz^{1/2}$).

2.4.3 Detectivity

The detectivity D is the reciprocal of NEP:

$$D = \frac{1}{\text{NEP}}.$$
(2.4)

It was found by Jones [63] that for many detectors the NEP is proportional to the square root of the detector signal that is proportional to the detector area, A_d. This means that both NEP and detectivity are functions of electrical bandwidth and detector area, so a normalized detectivity D^* (or D-star) suggested by Jones [63,64] is defined as

$$D^* = D(A_d\Delta f)^{1/2} = \frac{(A_d\Delta f)^{1/2}}{\text{NEP}}.$$
(2.5)

The importance of D^* is that this figure of merit permits comparison of detectors of the same type, but having different areas. Either a spectral or blackbody D^* can be defined in terms of a corresponding type of NEP.

Useful equivalent expressions to Equation 2.5 include:

$$D^* = \frac{(A_d \Delta f)^{1/2}}{V_n} R_v = \frac{(A_d \Delta f)^{1/2}}{I_n} R_i = \frac{(A_d \Delta f)^{1/2}}{\Phi_e}(\text{SNR}),$$ (2.6)

where D^* is defined as the rms SNR in a 1 Hz bandwidth per unit rms incident radiant power per square root of detector area. D^* is expressed in unit $\text{cmHz}^{1/2}\text{W}^{-1}$, which recently is called "Jones."

The blackbody $D^*(T, f)$ may be found from spectral detectivity:

$$D^*(T, f) = \frac{\int_0^\infty D^*(\lambda, f)\Phi_e(T, \lambda)d\lambda}{\int_0^\infty \Phi_e(T, \lambda)d\lambda} = \frac{\int_0^\infty D^*(\lambda, f)E_e(T, \lambda)d\lambda}{\int_0^\infty E_e(T, \lambda)d\lambda},$$ (2.7)

where $\Phi_e(T,\lambda) = E_e(T,\lambda)A_d$ is the incident blackbody radiant flux (in W), and $E_e(T,\lambda)$ is the blackbody irradiance (in W/cm²).

2.5 FUNDAMENTAL DETECTIVITY LIMITS

The ultimate performance of infrared detectors is reached when the detector and amplifier noise are low compared to the photon noise. The photon noise is fundamental in the sense that it arises not from any imperfection in the detector or its associated electronics but rather from the detection process itself, as a result of the discrete nature of the radiation field. The radiation falling on the detector is a composite of that from the target and that from the background. The practical operating limit for most infrared detectors is not the signal fluctuation limit but the background fluctuation limit, also known as the background limited infrared photodetector (BLIP) limit.

The expression for shot noise can be used to derive the BLIP detectivity

$$D^*_{\text{BLIP}}(\lambda, f) = \frac{\lambda}{hc}\left(\frac{\eta}{2Q_B}\right)^{1/2}$$ (2.8)

where η is the quantum efficiency (the number of electron-hole pairs generated per incident photon) and Q_B is the total background photon flux density reaching the detector

$$Q_B = \sin^2(\theta/2)\int_0^{\lambda_c} Q(\lambda, T_B)\, d\lambda.$$ (2.9)

Planck's photon emittance (in unit photons $\text{cm}^{-2}\text{s}^{-1}\,\mu\text{m}^{-1}$) at temperature T_B is given by

$$Q(\lambda, T_B) = \frac{2\pi c}{\lambda^4[\exp(hc/\lambda k T_B) - 1]} = \frac{1.885 \times 10^{23}}{\lambda^4[\exp(14.388/\lambda T_B) - 1]}.$$ (2.10)

Figure 2.7 shows the dependence of the integral background flux density on wavelength for different blackbody temperatures and 2π field of view (FOV). The values of the integral Equation 2.9 are given in the tables by Lowan and Blanch [65].

From Equation 2.9 results that

$$\frac{Q_B(\theta)}{Q_B(2\pi)} = \sin^2(\theta/2),$$ (2.11)

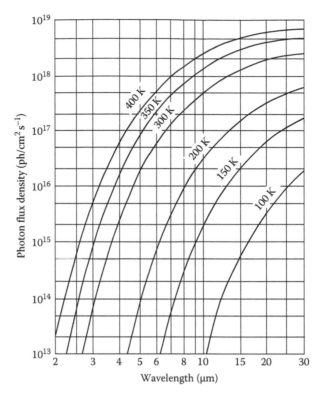

Figure 2.7 Dependence of the integral background flux density on wavelength for different blackbody temperatures and 2π FOV. (From Hudson, R. D., *Infrared System Engineering*, Wiley, New York, 1969. With permission.)

and the background-limited D^* relative to 2π FOV becomes

$$\frac{D^*_{\text{BLIP}}(\theta)}{D^*_{\text{BLIP}}(2\pi)} = \frac{1}{\sin(\theta/2)}. \tag{2.12}$$

The D^* varies with field of view (FOV) as $[\sin(\theta/2)]^{-1}$. Figure 2.8 is a curve showing how ideal D^* is improved as the cone angle θ is reduced for any given background temperature [66].

Equation 2.8 holds for photovoltaic detectors, which are shot-noise limited. Photoconductive detectors that are generation-recombination noise limited have a lower D^*_{BLIP} by a factor of $\sqrt{2}$

$$D^*_{\text{BLIP}}(\lambda, f) = \frac{\lambda}{2hc}\left(\frac{\eta}{Q_B}\right)^{1/2}. \tag{2.13}$$

The photon-noise-limited expressions Equation 2.8 and Equation 2.13 are only for Poisson statistics, when the Bose–Einstein factor $b = [\exp(hc/\lambda kt_B)-1]^{-1}$ is near 1. If the Bose–Einstein factor is included, for example, Equation 2.13 becomes

$$D^*_{\text{BLIP}}(\lambda, f) = \frac{\eta\lambda}{2hc\sin(\theta/2)}\left[\int_0^{\lambda_c}\eta(\lambda)Q(\lambda, T_B)(1+b)d\lambda\right]^{-1/2}. \tag{2.14}$$

The highest performance possible will be obtained by the ideal detector with unity quantum efficiency and ideal spectral responsivity [$R(\lambda)$ increases with wavelength to the cutoff wavelength

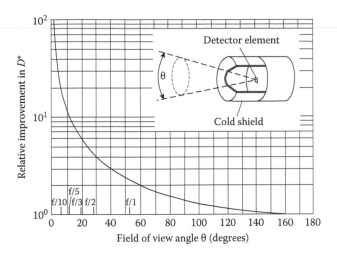

Figure 2.8 Relative improvement factor for detectivity with reduction in the field of view cone angle for a BLIP detector. (From Bratt, P. R., "Impurity Germanium and Silicon Infrared Detectors," in *Semiconductors and Semimetals*, Vol. 12, 39–141, Academic Press, New York, 1977. With permission.)

λ_c at which the responsivity drops to zero]. This limiting performance is of interest for comparison with actual detectors. The detectivity of ideal photoconductors at λ_c as a function of λ_c based on numerical integration is shown as a function of background temperature T_b, for a 2π FOV, in Figure 2.9 [67]. The dashed line for $T_b = 300$ K is the detectivity obtained neglecting the boson factor, which is seen to make a small but increasing effect as the wavelength is extended. As T_b is decreased, the boson factor correction becomes increasingly less significant. Values of D^*_{BLIP} versus λ for various background conditions are given in the literature [3,68–71].

The detectivity of BLIP detectors can be improved by reducing the background photon flux, Φ_b. Practically, there are two ways to do this: a cooled or reflective spectral filter to limit the spectral band or a cooled shield to limit the angular FOV of the detector (as described above). The former eliminates background radiation from spectral regions in which the detector need not respond. The best detectors yield background limited detectivities in quite narrow fields of view.

It can be shown that when the signal source is a blackbody at temperature T_s, and the radiation background is a blackbody at temperature T_b, then the background noise limited blackbody D^*_{BLIP} as a function of the peak spectral D^*_{BLIP} is

$$D^*_{\mathrm{BLIP}}(T_s, f) = D^*_{\mathrm{BLIP}}(\lambda_p, f) \frac{(hc/\lambda_p)}{\sigma T_s^4} \int_0^{\lambda_p} Q(T_s, \lambda) d\lambda, \qquad (2.15)$$

where λ_p is the wavelength of peak detectivity, which is also the cutoff wavelength for an ideal photon detector, and σ is the Stefan–Boltzmann constant. All of the D^*_{BLIP} expressions have assumed a Lambertian source subtending a half-angle of $\pi/2$ radians.

The ratio of the BLIP peak spectral D^* to the BLIP blackbody D^* is

$$K(T, \lambda) = \frac{D^*_{\mathrm{BLIP}}(\lambda_p, f)}{D^*_{\mathrm{BLIP}}(T_s, f)} = \frac{\sigma T_s^4}{\frac{hc}{\lambda_p} \int_0^{\lambda_p} Q(T_s, \lambda) d\lambda}. \qquad (2.16)$$

Figure 2.10 is a plot of $K(\lambda)$ for $T_s = 500$ K and a 2π steradian FOV [67]. The quantity $K(T,\lambda)$ is useful because infrared detector testing yields blackbody D^* values. Peak spectral D^* is then calculated using $K(T,\lambda)$.

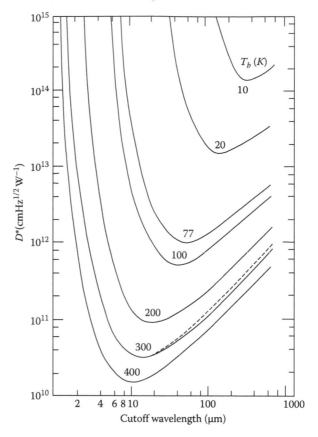

Figure 2.9 Detectivity at λ_c versus λ_c for ideal photoconductive detectors operating at $T_b = 400$, 300, 200, 100, 77, 20, and 10 K for a 2π FOV. The dashed line for 300 K neglects the boson factor. (From Sclar, N., *Progress in Quantum Electronics*, 9, 149–257, 1984. With permission.)

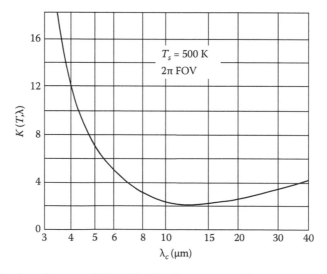

Figure 2.10 Ratio of peak spectral D^* to blackbody D^* versus detector cutoff wavelength for $T_s = 500$ K and 2π FOV. (From Kruse, P. W., *Optical and Infrared Detectors*, 5–69, Springer, Berlin, 1977. With permission.)

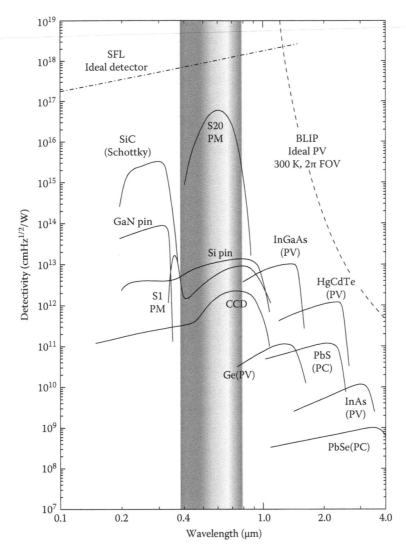

Figure 2.11 Detectivity versus wavelength values of 0.1–4 μm photodetectors. The PC indicates a photoconductive detector; PV, photovoltaic detector; and PM indicates a photomultiplier.

When detectors are operated in conditions where the background flux is less than the signal flux, the ultimate performance of detectors is determined by the signal fluctuation limit (SFL). It is achieved in practice with photomultipliers operating in the visible and ultraviolet region, but it is rarely achieved with solid-state devices, which are normally detector-noise or electronic noise limited. This limit is also applicable to longer wavelength detectors when the background temperature is very low. The NEP and detectivity of detectors operating in this limit have been derived by a number of authors (see e.g., Kruse et al. [3,68]).

The NEP in the signal fluctuation limit is given as [68–72]

$$NEP = \frac{2hc\Delta f}{\eta\lambda},$$ (2.17)

when Poisson statistics are applicable. This threshold value implies a low number of photons per observation interval. A more meaningful parameter is the probability that a photon will be detected during an observation period. Kruse [25] shows that the minimum signal power to achieve 99% probability that a photon will be detected in an observation period t_o is

$$\text{NEP}_{min} = \frac{9.22hc\Delta f}{\eta\lambda}, \tag{2.18}$$

where Δf is assumed to be $1/2t_o$. Note that the detector area does not enter into the expression and that NEP_{min} depends linearly upon the bandwidth, which differs from the case in which the detection limit is set by internal or background noise.

Seib and Aukerman [73] also have derived an expression for the SFL identical to Equation 2.18 except that the multiplicative constant is not 9.22 but $2^{3/2}$ for an ideal photoemissive or photovoltaic detector and $2^{5/2}$ for a photoconductor. This difference in the constant arises from the differing assumptions as to the manner in which the detector is employed and the minimum detectable signal-to-noise ratio.

Assuming Seib and Aukerman approximation for SFL limit of photovoltaic detector, the corresponding detectivity is

$$D^* = \frac{\eta\lambda}{2^{3/2}hc}\sqrt{\frac{A_d}{\Delta f}}. \tag{2.19}$$

It is interesting to determine the composite signal fluctuation and background fluctuation limits. Figure 2.11 illustrates the spectral detectivities over the wavelength range from 0.1 μm to 4 μm assuming a background temperature of 290 K and a 2π steradian FOV (applicable only to the background fluctuation limit). Note that the intersections of curves for signal fluctuation and background fluctuation limits lie about 1.2 μm. At wavelengths below 1.2 μm the SFL dominates; the converse is true above 1.2 μm. Below 1.2 μm the wavelength dependence is small. Above 1.2 μm it is very large, due to steep dependence of detectivity upon wavelength of the short wavelength end of the 290 K background spectral distribution.

It will be seen (Chapter 4) that by employing optical heterodyne detection it is possible to achieve the signal fluctuation limit with infrared detectors even in the presence of ambient background temperature.

REFERENCES

1. W. Herschel, "Experiments on the Refrangibility of the Invisible Rays of the Sun," *Philosophical Transactions of the Royal Society of London* 90, 284, 1800.

2. R. A. Smith, F. E. Jones, and R. P. Chasmar, *The Detection and Measurement of Infrared Radiation*, Clarendon, Oxford, 1958.

3. P. W. Kruse, L. D. McGlauchlin, and R. B. McQuistan, *Elements of Infrared Technology*, Wiley, New York, 1962.

4. L. M. Biberman, ed., *Electro-Optical Imaging: System Performance and Modeling*, SPIE Press, Bellingham, WA, 2000.

5. E. S. Barr, "Historical Survey of the Early Development of the Infrared Spectral Region," *American Journal of Physics* 28, 42–54, 1960.

6. E. S. Barr, "The Infrared Pioneers—II. Macedonio Melloni," *Infrared Physics* 2, 67–73, 1962.

7. E. S. Barr, "The Infrared Pioneers—III. Samuel Pierpont Langley," *Infrared Physics* 3, 195–206, 1963.

8. W. Smith, "Effect of Light on Selenium During the Passage of an Electric Current," *Nature 7*, 303, 1873.

9. M. F. Doty, *Selenium, List of References, 1917–1925*, New York Public Library, New York, 1927.

10. J. C. Bose, U.S. Patent 755840, 1904.

11. T. W. Case, "Notes on the Change of Resistance of Certain Substrates in Light," *Physical Review* **9**, 305–10 (1917).

12. T. W. Case, "The Thalofide Cell: A New Photoelectric Substance," *Physical Review* 15, 289, 1920.

13. R. D. Hudson and J. W. Hudson, *Infrared Detectors,* Dowden, Hutchinson & Ross, Stroudsburg, PA, 1975.

14. E. W. Kutzscher, "Review on Detectors of Infrared Radiation," *Electro-Optical Systems Design* 5, 30, June 1973.

15. D. J. Lovell, "The Development of Lead Salt Detectors," *American Journal of Physics* 37, 467–78, 1969.

16. R. J. Cushman, "Film-Type Infrared Photoconductors," *Proceedings of IRE* 47, 1471–75, 1959.

17. P. R. Norton, "Infrared Detectors in the Next Millennium," *Proceedings of SPIE* 3698, 652–65, 1999.

18. A. Rogalski, *Infrared Detectors*, Gordon and Breach Science Publishers, Amsterdam, 2000.

19. E. Burstein, G. Pines, and N. Sclar, "Optical and Photoconductive Properties of Silicon and Germanium," in *Photoconductivity Conference at Atlantic City*, eds. R. Breckenbridge, B. Russell, and E. Hahn, 353–413, Wiley, New York, 1956.

20. R. A. Soref, "Extrinsic IR Photoconductivity of Si Doped with B, Al, Ga, P, As or Sb," *Journal of Applied Physics* 38, 5201–9, 1967.

21. W. S. Boyle and G. E. Smith, "Charge-Coupled Semiconductor Devices," *Bell Systems Technical Journal* 49, 587–93, 1970.

22. F. Shepherd and A. Yang, "Silicon Schottky Retinas for Infrared Imaging," *IEDM Technical Digest,* 310–13, 1973.

23. W. D. Lawson, S. Nielson, E. H. Putley, and A. S. Young, "Preparation and Properties of HgTe and Mixed Crystals of HgTe-CdTe," *Journal of Physics and Chemistry of Solids* 9, 325–29, 1959.

24. P. W. Kruse, M. D. Blue, J. H. Garfunkel, and W. D. Saur, "Long Wavelength Photoeffects in Mercury Selenide, Mercury Telluride and Mercury Telluride-Cadmium Telluride," *Infrared Physics* 2, 53–60, 1962.

25. J. Melngailis and T. C. Harman, "Single-Crystal Lead-Tin Chalcogenides," in *Semiconductors and Semimetals*, Vol. 5, eds. R. K. Willardson and A. C. Beer, 111–74, Academic Press, New York, 1970.

26. T. C. Harman and J. Melngailis, "Narrow Gap Semiconductors," in *Applied Solid State Science*, Vol. 4, ed. R. Wolfe, 1–94, Academic Press, New York, 1974.

27. A. Rogalski and J. Piotrowski, "Intrinsic Infrared Detectors," *Progress in Quantum Electronics* 12, 87–289, 1988.

28. S. Borrello and H. Levinstein, "Preparation and Properties of Mercury Doped Infrared Detectors," *Journal of Applied Physics* 33, 2947–50, 1962.

29. C. T. Elliott, D. Day, and B. J. Wilson, "An Integrating Detector for Serial Scan Thermal Imaging," *Infrared Physics* 22, 31–42, 1982.

30. A. Blackburn, M. V. Blackman, D. E. Charlton, W. A. E. Dunn, M. D. Jenner, K. J. Oliver, and J. T. M. Wotherspoon, "The Practical Realisation and Performance of SPRITE Detectors," *Infrared Physics* 22, 57–64, 1982.

31. A. Hoffman, "Semiconductor Processing Technology Improves Resolution of Infrared Arrays," *Laser Focus World*, 81–84, February 2006.

32. A. W. Hoffman, P. L. Love, and J. P. Rosbeck, "Mega-Pixel Detector Arrays: Visible to 28 μm," *Proceedings of SPIE* 5167, 194–203, 2004.

33. J. C. Fraser, D. H. Alexander, R. M. Finnila, and S. C. Su, "An Extrinsic Si CCD for Detecting Infrared Radiation," in *Digest of Technical Papers*, 442–445, IEEE, New York, 1974.

34. K. Nummendal, J. C. Fraser, S. C. Su, R. Baron, and R. M. Finnila, "Extrinsic Silicon Monolithic Focal Plane Array Technology and Applications," in *Proceedings of CCD Applications International Conference*, Noval Ocean Systems Center, 19–30, San Diego, CA, 1976.

35. N. Sclar, R. L. Maddox, and R. A. Florence, "Silicon Monolithic Infrared Detector Array," *Applied Optics* 16, 1525–32, 1977.

36. E. Beuville, D. Acton, E. Corrales, J. Drab, A. Levy, M. Merrill, R. Peralta, and W. Ritchie, "High Performance Large Infrared and Visible Astronomy Arrays for Low Background Applications: Instruments Performance Data and Future Developments at Raytheon," *Proceedings of SPIE* 6660, 66600B, 2007.

37. M. J. E. Golay, "A Pneumatic Infrared Detector," *Review of Scientific Instruments* 18, 357–62, 1947.

38. E. M. Wormser, "Properties of Thermistor Infrared Detectors," *Journal of the Optical Society of America* 43, 15–21, 1953.

39. R. W. Astheimer, "Thermistor Infrared Detectors," *Proceedings of SPIE* 443, 95–109, 1983.

40. G. W. McDaniel and D. Z. Robinson, "Thermal Imaging by Means of the Evaporograph," *Applied Optics* 1, 311–24, 1962.

41. C. Hilsum and W. R. Harding, "The Theory of Thermal Imaging, and Its Application to the Absorption-Edge Image Tube," *Infrared Physics* 1, 67–93, 1961.

42. A. J. Goss, "The Pyroelectric Vidicon: A Review," *Proceedings of SPIE* 807, 25–32, 1987.

43. R. A. Wood and N. A. Foss, "Micromachined Bolometer Arrays Achieve Low-Cost Imaging," *Laser Focus World*, 101–6, June, 1993.

44. R. A. Wood, "Monolithic Silicon Microbolometer Arrays," in *Semiconductors and Semimetals*, Vol. 47, eds. P. W. Kruse and D. D. Skatrud, 45–121, Academic Press, San Diego, CA, 1997.

45. C. M. Hanson, "Hybrid Pyroelectric–Ferroelectric Bolometer Arrays," in *Semiconductors and Semimetals*, Vol. 47, eds. P. W. Kruse and D. D. Skatrud, 123–74, Academic Press, San Diego, CA, 1997.

46. P. W. Kruse, "Uncooled IR Focal Plane Arrays," *Opto-Electronics Review* 7, 253–58, 1999.

47. R. A. Wood, "Uncooled Microbolometer Infrared Sensor Arrays," in *Infrared Detectors and Emitters: Materials and Devices*, eds. P. Capper and C. T. Elliott, 149–74, Kluwer Academic Publishers, Boston, MA, 2000.

48. R. W. Whatmore and R. Watton, "Pyroelectric Materials and Devices," in *Infrared Detectors and Emitters: Materials and Devices*, eds. P. Capper and C. T. Elliott, 99–147, Kluwer Academic Publishers, Boston, MA, 2000.

49. P. W. Kruse, *Uncooled Thermal Imaging. Arrays, Systems, and Applications*, SPIE Press, Bellingham, WA, 2001.

50. R. D. Hudson, *Infrared System Engineering*, Wiley, New York, 1969.

51. M. A. Kinch, "Fundamental Physics of Infrared Detector Materials," *Journal of Electronic Materials* 29, 809–17, 2000.

52. C. T. Elliott and N. T. Gordon, "Infrared Detectors," in *Handbook on Semiconductors*, Vol. 4, ed. C. Hilsum, 841–936, Elsevier, Amsterdam, 1993.

53. H.-W. Hübers, "Terahertz Heterodyne Receivers," *IEEE Journal of Selected Topics in Quantum Electronics* 14, 378–91, 2008.

54. J. L. Miller, *Principles of Infrared Technology*, Van Nostrand Reinhold, New York, 1994.

55. P. T. Blotter and J. C. Batty, "Thermal and Mechanical Design of Cryogenic Cooling Systems," in *The Infrared and Electro-Optical Systems Handbook*, Vol. 3, ed. W. D. Rogatto, 343–433, Infrared Information Analysis Center, Ann Arbor, MI, and SPIE Press, Bellingham, WA, 1993.

56. R. J. Radtke and H. E. C. H. Grein, "Multilayer Thermoelectric Refrigeration in $Hg_{1-x}Cd_xTe$ Superlattices," *Journal of Applied Physics* 86, 3195–98, 1999.

57. W. I. Wolfe and G. J. Zissis, eds., *The Infrared Handbook*, Office of Naval Research, Washington, DC, 1985.

58. W. D. Rogatto, ed., *The Infrared and Electro-Optical Systems Handbook*, Infrared Information Analysis Center, Ann Arbor, MI, and SPIE Optical Engineering Press, Bellingham, WA, 1993.

59. J. D. Vincent, *Fundamentals of Infrared Detector Operation and Testing*, Wiley, New York, 1990.

60. W. L. Eisenman, J. D. Merriam, and R. F. Potter, "Operational Characteristics of Infrared Photodetectors," in *Semiconductors and Semimetals*, Vol. 12, eds. R. K. Willardson and A. C. Beer, 1–38, Academic Press, New York, 1977.

61. T. Limperis and J. Mudar, "Detectors," in *The Infrared Handbook*, eds. W. L. Wolfe and G. J. Zissis, 11.1–11.104, Environmental Research Institute of Michigan, Office of Naval Research, Washington, DC, 1989.

62. D. G. Crove, P. R. Norton, T. Limperis, and J. Mudar, "Detectors," in *The Infrared and Electro-Optical Systems Handbook*, Vol. 3, ed. W. D. Rogatto, 175–283, Infrared Information Analysis Center, Ann Arbor, MI, and SPIE Optical Engineering Press, Bellingham, WA, 1993.

63. R. C. Jones, "Performance of Detectors for Visible and Infrared Radiation," in *Advances in Electronics*, Vol. 5, ed. L. Morton, 27–30, Academic Press, New York, 1952.

64. R. C. Jones, "Phenomenological Description of the Response and Detecting Ability of Radiation Detectors," *Proceedings of IRE* 47, 1495–1502, 1959.

65. A. N. Lowan and G. Blanch, "Tables of Planck's Radiation and Photon Functions," *Journal of the Optical Society of America* 30, 70–81, 1940.

66. P. R. Bratt, "Impurity Germanium and Silicon Infrared Detectors," in *Semiconductors and Semimetals*, Vol. 12, eds. R. K. Willardson and A. C. Beer, 39–141, Academic Press, New York, 1977.

67. N. Sclar, "Properties of Doped Silicon and Germanium in Infrared Detectors," *Progress in Quantum Electronics* 9, 149–257, 1984.

68. P. W. Kruse, "The Photon Detection Process," in *Optical and Infrared Detectors*, ed. R. J. Keyes, 5–69, Springer, Berlin, 1977.

69. R. W. Boyd, *Radiometry and the Detection of Optical Radiation*, Wiley, New York, 1983.

70. R. H. Kingston, *Detection of Optical and Infrared Radiation*, Wiley, New York, 1983.

71. E. L. Dereniak and G. D. Boremen, *Infrared Detectors and Systems*, Wiley, New York, 1996.

72. A. Smith, F. E. Jones, and R. P. Chasmar, *The Detection and Measurement of Infrared Radiation*, Clarendon, Oxford, 1968.

73. D. H. Seib and L. W. Aukerman, "Photodetectors for the 0.1 to 1.0 μm Spectral Region," in *Advances in Electronics and Electron Physics*, Vol. 34, ed. L. Morton, 95–221, Academic Press, New York, 1973.

3 Fundamental Performance Limitations of Infrared Detectors

As noted in Chapter 2, infrared detectors fall into two broad categories: photon and thermal detectors. Although thermal detectors have been available commercially in single element form for many decades, their exploitation in imaging arrays started in the last decade of the twentieth century.

This chapter discusses the fundamental limitations to IR detector performance imposed by the statistical nature of the generation, recombination processes, and radiometric considerations. We will try to establish the ultimate theoretical sensitivity limit that can be expected for a detector operating at a given temperature. The models presented here are applicable to any of the detector classes mentioned in Chapter 1. The nonfundamental limitations will be addressed later in this book.

Photon detectors are fundamentally limited by generation–recombination noise arising from photon exchange with radiation background. Thermal detectors are fundamentally limited by temperature fluctuation noise arising from radiant power exchange with a radiating background. Due to fundamentally different types of noise, these two classes of detectors have different dependencies of detectivities on wavelength and temperature. The photon detectors are favored at a long wavelength infrared and lower operating temperatures. The thermal detectors are favored at a very long wavelength spectral range.

In this chapter we first examine fundamental infrared detection processes for both categories of detectors. Next, the comparative studies of thermal and photon detectors are carried out. Different types of thermal as well as photon detectors are discussed in details in Chapters 5–9. However, some elementary detection-process concepts must be understood to fully appreciate the limitation of sensitivity imposed by noise processes within these devices.

3.1 THERMAL DETECTORS

The thermal detectors are classified according to the operating scheme: thermopile scheme, bolometer scheme, and pyroelectric scheme. In the present section, the general principles of thermal detectors are described.

3.1.1 Principle of Operation

The performance of a thermal detector will be calculated in two stages. First, by consideration of the thermal characteristics of the system, the temperature rise produced by the incident radiation is determined. Secondly, this temperature rise is used to determine the change in the property that is being used to indicate the signal. The first stage of the calculations is common to all thermal detectors, but the details of the second stage will differ for the different types of thermal detectors.

Thermal detectors operate on a simple principle; that when heated by incoming IR radiation their temperature increases and the temperature changes are measured by any temperature-dependent mechanism, such as thermoelectric voltage, resistance, or pyroelectric voltage.

The simplest representation of the thermal detector is shown in Figure 3.1. The detector is represented by a thermal capacitance C_{th} coupled via a thermal conductance G_{th} to a heat sink at a constant temperature T. In the absence of a radiation input the average temperature of the detector will also be T, although it will exhibit a fluctuation near this value. When a radiation input is received by the detector, the rise in temperature is found by solving the heat balance equation [1–3]:

$$C_{th} \frac{d\Delta T}{dt} + G_{th} \Delta T = \varepsilon \Phi,$$

(3.1)

where ΔT is the temperature difference due to optical signal Φ, between the detector and its surroundings, and ε is the emissivity of detector. The analogies between thermal and electrical circuits are given in Table 3.1. The thermal circuit (Figure 3.1a) corresponds to an electric circuit shown in Figure 3.1b.

Assuming the radiant power to be a periodic function,

$$\Phi = \Phi_o e^{i\omega t},$$

(3.2)

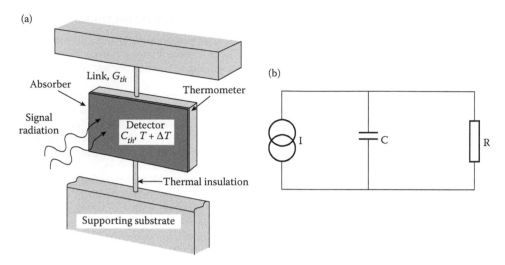

Figure 3.1 (a) Thermal detector and (b) its electrical analogue.

Table 3.1: Thermal–Electric Analogies

Thermal		Electric	
Value	**Unit**	**Value**	**Unit**
Heat energy	J	Charge	C
Heat flow	W	Current	A
Temperature	K	Voltage	V
Thermal impedance	K/W	Impedance	Ω
Heat capacitance	J/K	Capacitance	F

where Φ_o is the amplitude of sinusoidal radiation, the solution of differential heat radiation is

$$\Delta T = \Delta T_o e^{-(G_{th}/C_{th})t} + \frac{\varepsilon \Phi_o e^{i\omega t}}{G_{th} + i\omega C_{th}}. \tag{3.3}$$

The first term is the transient part and as time increases, this term exponentially decreases to zero, so it can be dropped with no loss of generality for the change in temperature. Therefore, the change in temperature of any thermal detector due to incident radiative flux is

$$\Delta T = \frac{\varepsilon \Phi_o}{\left(G_{th}^2 + \omega^2 C_{th}^2\right)^{1/2}}. \tag{3.4}$$

Equation 3.4 illustrates several features of thermal detector. Clearly it is advantageous to make ΔT as large as possible. To do this, the thermal capacity of the detector (C_{th}) and its thermal coupling to its surroundings (G_{th}) want to be as small as possible. The interaction of the thermal detector with the incident radiation need to be optimized while reducing as far as possible all other thermal contacts with its surroundings. This means that a small detector mass and fine connecting wires to the heat sink are desirable.

Equation 3.4 shows that as ω is increased, the term $\omega^2 C_{th}^2$ will eventually exceed G_{th}^2 and then ΔT will fall inversely as ω. A characteristic thermal response time for the detector can therefore be defined as

$$\tau_{th} = \frac{C_{th}}{G_{th}} = C_{th} R_{th}, \tag{3.5}$$

where $R_{th} = 1/G_{th}$ is the thermal resistance. Then Equation 3.4 can be written as

$$\Delta T = \frac{\varepsilon \Phi_o R_{th}}{\left(1 + \omega^2 \tau_{th}^2\right)^{1/2}}.$$ (3.6)

Typical value of thermal time constant is in the millisecond range. This is much longer than the typical time of a photon detector. There is a trade-off between sensitivity, ΔT, and frequency response. If one wants a high sensitivity, then a low frequency response is forced upon the detector.

For further discussion we introduce the coefficient K, which reflects how good the temperature changes translates into the electrical output voltage of detector [4]

$$K = \frac{\Delta V}{\Delta T}.$$ (3.7)

Then, the corresponding rms voltage signal due to temperature changes ΔT is

$$\Delta V = K \Delta T = \frac{K \varepsilon \Phi_o R_{th}}{\left(1 + \omega^2 \tau_{th}^2\right)^{1/2}}.$$ (3.8)

The voltage responsivity R_v of the detector is the ratio of the output signal voltage ΔV to the input radiation power and is given by

$$R_v = \frac{K \varepsilon R_{th}}{\left(1 + \omega^2 \tau_{th}^2\right)^{1/2}}.$$ (3.9)

As the last expression shows, the low frequency voltage responsivity ($\omega \ll 1/\tau_{th}$) is proportional to the thermal resistance and does not depend on the heat capacitance. The opposite is true for high frequencies ($\omega \gg 1/\tau_{th}$). In this case R_v is not dependent on R_{th} and is inversely proportional to the heat capacitance.

As stated previously, the thermal conductance (thermal resistance) from the detector to the outside world should be small (high). The smallest possible thermal conductance would occur when the detector is completely isolated from the environment under vacuum with only radiative heat exchange between it and its heat-sink enclosure. Such an ideal model can give us the ultimate performance limit of a thermal detector. This limiting value can be estimated from the Stefan–Boltzmann total radiation law.

If the thermal detector has a receiving area A of emissivity ε, when it is in thermal equilibrium with its surroundings it will radiate a total flux $A\varepsilon\sigma T^4$ where σ is the Stefan–Boltzmann constant. Now if the temperature of the detector is increased by a small amount dT the flux radiated is increased by $4A\varepsilon\sigma T^3 dT$. Hence, the radiative component of the thermal conductance is

$$G_R = \frac{1}{\left(R_{th}\right)_R} = \frac{d}{dT}\left(A\varepsilon\sigma T^4\right) = 4A\varepsilon\sigma T^3.$$ (3.10)

In this case

$$R_v = \frac{K}{4\sigma T^3 A\left(1 + \omega^2 \tau_{th}^2\right)^{1/2}}.$$ (3.11)

When the detector is in thermal equilibrium with the heat sink, the fluctuation in the power flowing through the thermal conductance into the detector is [5,6]

$$\Delta P_{th} = \left(4kT^2 G\right)^{1/2},$$ (3.12)

which will be the smallest when G assumes its minimum value (i.e., G_R). Then ΔP_{th} will be a minimum and its value gives the minimum detectable power for an ideal thermal detector.

The minimum detectable signal power—or noise equivalent power (NEP)—is defined as the rms signal power incident upon the detector required to equal the rms thermal noise power. Hence if the temperature fluctuation associated with G_R is the only source of noise,

$$\varepsilon NEP = \Delta P_{th} = \left(16A\varepsilon\sigma kT^5\right)^{1/2}, \tag{3.13}$$

or

$$NEP = \left(\frac{16A\sigma kT^5}{\varepsilon}\right)^{1/2}. \tag{3.14}$$

If all the incident radiation is absorbed by the detector, $\varepsilon = 1$, and then

$$NEP = \left(16A\sigma kT^5\right)^{1/2} = 5.0 \times 10^{-11} \text{ W} \tag{3.15}$$

for $A = 1$ cm^2, $T = 290$ K, and $\Delta f = 1$ Hz.

3.1.2 Noise Mechanisms

In order to determine the NEP and detectivity (D^*) of a detector, it is necessary to define a noise mechanism. For any detector there are a number of noise sources that impose fundamental limits to the detection sensitivity.

One major noise is the Johnson noise. This noise in a Δf bandwidth for a resistor of resistance R is

$$V_J^2 = 4kTR\Delta f, \tag{3.16}$$

where k is the Boltzmann constant and Δf is the frequency band. This noise has white character.

Two other fundamental noise sources are important for assessing the ultimate performance of a detector: thermal fluctuation noise and background fluctuation noise.

Thermal fluctuation noise arises from temperature fluctuations in the detector. These fluctuations are caused by heat conductance variations between the detector and the surrounding substrate with which the detector element is in thermal contact.

The variance in temperature ("temperature" noise) can be shown to be [2,5,6]

$$\overline{\Delta T^2} = \frac{4kT^2\Delta f}{1+\omega^2\tau_{th}^2}R_{th}. \tag{3.17}$$

From this equation results that thermal conductance, $G_{th} = 1/R_{th}$, as the principal heat loss mechanism, is the key design parameter that affects the temperature fluctuation noise. Figure 3.2 shows exemplary temperature fluctuation noise (rms value of temperature fluctuation) for a typical IR sensitive micromechanical detector [7]. Note that the signal follows the same roll-off at higher frequencies as the temperature fluctuation noise does.

The spectral noise voltage due to temperature fluctuations is

$$V_{th}^2 = K^2\overline{\Delta T^2} = \frac{4kT^2\Delta f}{1+\omega^2\tau_{th}^2}K^2 R_{th}. \tag{3.18}$$

A third noise source is background noise resulting from radiative heat exchange between the detector at temperature T_d and the surrounding environment at temperature T_b that is being observed. It is the ultimate limit of a detector's performance capability and is given for a 2π field of view (FOV) by [2,5,6]

$$V_b^2 = \frac{8k\varepsilon\sigma A\left(T_d^2 + T_b^2\right)}{1+\omega^2\tau_{th}^2}K^2 R_{th}^2, \tag{3.19}$$

where σ is the Stefan–Boltzmann constant.

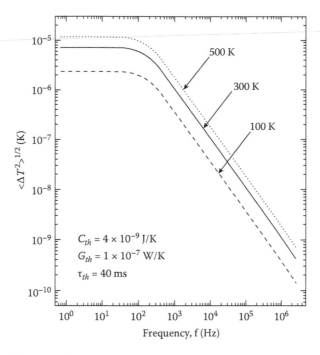

Figure 3.2 Spectral density of temperature fluctuation noise calculated for a typical thermal infrared detector. (From Datskos, P. G., *Encyclopedia of Optical Engineering*, Marcel Dekker, New York, 349–57, 2003. With permission.)

In addition to the fundamental noise sources mentioned above, $1/f$ is an additional noise source that is often found in the thermal detector and can affect detector performance. It can be described by the empirical form

$$V_{1/f}^2 = k_{1/f} \frac{I^\delta}{f^\beta} \Delta f, \tag{3.20}$$

where the coefficient $k_{1/f}$ is a proportionality factor, δ and β are coefficients whose values are about one. The $1/f$ power law noise is difficult to characterize analytically because the parameters $k_{1/f}$, δ and β are very much dependent upon material preparation and processing, including contacts and surfaces.

The square of total noise voltage is

$$V_n^2 = V_{th}^2 + V_b^2 + V_{1/f}^2. \tag{3.21}$$

3.1.3 Detectivity and Fundamental Limits

According to Equations 2.6, 3.16, 3.18, and 3.21, the detectivity of a thermal detector is given by

$$D^* = \frac{K\varepsilon R_{th} A^{1/2}}{\left(1 + \omega^2 \tau_{th}^2\right)^{1/2} \left(\dfrac{4kT_d^2 K^2 R_{th}}{1 + \omega \tau_{th}^2} + 4kTR + V_{1/f}^2\right)^{1/2}}. \tag{3.22}$$

In the case of a typical operation condition of the thermal detector, while it operates in a vacuum or a gas environment at reduced pressures, heat conduction through the supporting microstructure of the device is dominant heat loss mechanism. However, in the case of an extremely good thermal isolation, the principal heat loss mechanism can be reduced to only radiative heat exchange between the detector and its surroundings. In the atmospheric environment, heat conduction through air is likely to be dominant heat dissipation mechanism. Thermal conductivity of air (2.4×10^{-2} Wm^{-1}K^{-1}) is larger than the thermal conductance through supporting beams of a typical micromechanical detector.

The fundamental limit to the sensitivity of any thermal detector is set by temperature fluctuation noise. Under this condition at low frequencies ($\omega \ll 1/\tau_{th}$), from Equation 3.22 results

$$D_{th}^* = \left(\frac{\varepsilon^2 A}{4kT_d^2 G_{th}} \right)^{1/2}. \tag{3.23}$$

It is assumed here that ε is independent of wavelength, so that the spectral D_λ^* and blackbody $D^*(T)$ values are identical.

Figure 3.3 shows the dependence of detectivity on temperature and thermal conductance plotted for different detector active areas. It is clearly shown that improved performance of thermal detectors can be achieved by increasing thermal isolation between the detector and its surrounding.

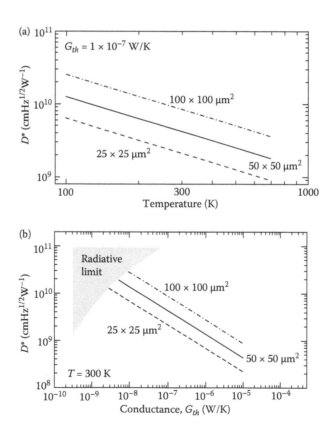

Figure 3.3 Temperature fluctuation noise limited detectivity for thermal infrared detectors of different areas plotted (a) as a function of the detector temperature and (b) as a function of the total thermal conductance between the detector and its surroundings. (From Datskos, P. G., *Encyclopedia of Optical Engineering*, Marcel Dekker, New York, 349–57, 2003. With permission.)

If radiant power exchange is the dominant heat exchange mechanism, then G is the first derivative with respect to temperature of the Stefan–Boltzmann function. In that case, known as the background fluctuation noise limit, from Equations 2.6 and 3.19 we have

$$D_b^* = \left[\frac{\varepsilon}{8k\sigma\left(T_d^5 + T_b^5\right)}\right]^{1/2}.$$

(3.24)

Note that D_b^* is independent of A, as is to be expected.

In many practical instances the temperature of the background, T_b, is room temperature, 290 K. Figure 3.4 shows the photon noise limited detectivity for an ideal thermal detector having an emissivity of unity, operated at 290 K and lower, as a function of background temperature [6].

Equations 3.23 and 3.24 and Figure 3.4 assume that background radiation falls upon the detector from all directions when the detector and background temperature are equal, and from the forward hemisphere only when the detector is at cryogenic temperatures. We see that the highest possible D^* to be expected for a thermal detector operated at room temperature and viewing a background at room temperature is 1.98×10^{10} cmHz$^{1/2}$W^{-1}. Even if the detector or background (not both) were cooled to absolute zero, the detectivity would improve only by the square root of two. This is the basic limitation of all thermal detectors. The background noise limited photon detectors have higher detectivities as a result of their limited spectral responses (what is shown in Figure 2.2).

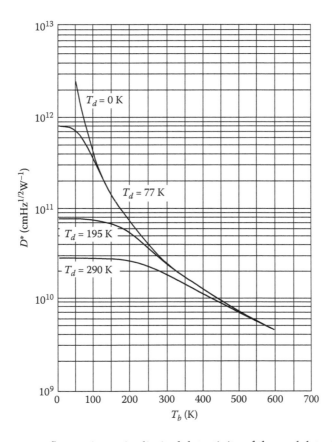

Figure 3.4 Temperature fluctuation noise limited detectivity of thermal detectors as a function of detector temperatures T_d and background temperature T_b for 2π FOV and $\varepsilon = 1$. (From Kruse, W., McGlauchlin, L. D., and McQuistan, R. B., *Elements of Infrared Technology*, Wiley, New York, 1962. With permission.)

Up to now we have only considered a flat spectral response of thermal detector. In practice, it is sometimes necessary to limit the spectral responsivity of a detector by means of cooled filters. Assuming ideal filters, we can calculate the variation of detectivity as a function of both short and long wavelength cutoffs, λ_{c1} and λ_{c2}, respectively. If detector emissivity $\varepsilon = 0$ for wavelengths except between λ_{c1} and λ_{c2}, and if ε is independent of wavelength between λ_{c1} and λ_{c2}, then Equation 3.24 is replaced by [8]

$$D_b^* = \left[\frac{\varepsilon}{8k\sigma T_d^5 + F\left(\lambda_{c1}, \lambda_{c2}\right)} \right]^{1/2}, \tag{3.25}$$

where

$$F\left(\lambda_{c1}, \lambda_{c2}\right) = 2 \int_{\lambda_{c1}}^{\lambda_{c2}} \frac{h^2 c^3}{\lambda^6} \frac{exp\left(hc/\lambda k T_b\right)}{\left[exp\left(hc/\lambda k T_b\right) - 1\right]^2} d\lambda. \tag{3.26}$$

Figure 3.5 illustrates Equation 3.25 as a function of λ for the case of a long wavelength cutoff λ_{c2} (i.e., $\varepsilon = 1$ for $\lambda < \lambda_{c2}$ and $\varepsilon = 0$ for $\lambda > \lambda_{c2}$), and for the case of a short wavelength cutoff λ_{c2} (i.e., $\varepsilon = 0$ for $\lambda < \lambda_{c1}$ and $\varepsilon = 1$ for $\lambda > \lambda_{c1}$). The background temperature is 300 K.

The performance achieved by any real detector will be inferior to that predicted by Equation 3.23. Even in the absence of other sources of noise, the performance of a radiation-noise limited detector will be worse than that of an ideal detector by the factor $\varepsilon^{1/2}$ (see Equation 3.24). Further degradation of performance will arise from:

- Encapsulation of detector (reflection and absorption losses at the window)

- The effects of excess thermal conductance (influence of electrical contacts, conduction through the supports, influence of any gas—conduction and convection)

- The additional noise sources

Figure 3.6 shows the performance of a number of thermal detectors operating at room temperature [9]. Typical values of detectivities of thermal detectors at 10 Hz change in the range between 10^8 to 10^9 cmHz$^{1/2}$W^{-1}.

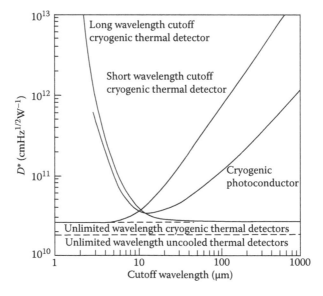

Figure 3.5 Dependence of detectivity upon long wavelength limit for thermal and photon detectors. (From Low, F. J. and Hoffman, A. R., *Applied Optics*, 2, 649–50, 1963. With permission.)

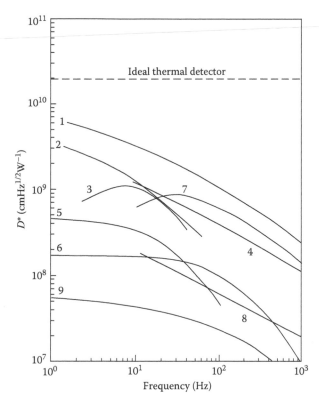

Figure 3.6 Performance of uncooled thermal detectors: 1: alaine doped TGS pyroelectric detector (A = 1.5 × 1.5 mm²); 2: spectroscopic thermopile (A = 0.4 mm², τ_{th} = 40 ms); 3: Golay cell; 4: TGS pyroelectric detector in ruggedized encapsulation (0.5 × 0.5 mm²); 5: Sb–Bi evaporated film thermopile (A = 0.12 × 0.12 mm², τ_{th} = 13 ms); 6: immersed thermistor (A = 0.1 × 0.1 mm², τ_{th} = 2 ms); 7: LiTaO₃ pyroelectric detector; 8: Plessey PZT ceramic pyroelectric detector; and 9: thin film bolometer. (From Putley, E. H., *Optical and Infrared Detectors*, Springer, Berlin, 71–100, 1977. With permission.)

3.2 PHOTON DETECTORS

3.2.1 Photon Detection Process

Photon detectors are based on photon absorption in semiconductor materials. A signal whose photon energy is sufficient to generate photocarriers will continuously lose energy as the optical field propagates through the semiconductor (see Figure 3.7). Inside the semiconductor, the field decays exponentially as energy is transferred to the photocarriers. The material can be characterized by an absorption length, α, and a penetration depth, $1/\alpha$. Penetration depth is the point at which $1/e$ of the optical signal power remains.

The power absorbed in the semiconductor as a function of position within the material is then

$$P_a = P_i(1-r)(1-e^{-\alpha x}). \tag{3.27}$$

The number of photons absorbed is the power (in watts) divided by the photon energy ($E = h\nu$). If each absorbed photon generates a photocarrier, the number of photocarriers generated per number of incident photons for a specific semiconductor with reflectivity r is given by

$$\eta(x) = (1-r)(1-e^{-\alpha x}) \tag{3.28}$$

where $1 \leq \eta \leq 1$ is a definition for the detector's quantum efficiency.

Figure 3.8 shows the measured intrinsic absorption coefficients for various narrow gap photodetector materials. The absorption coefficient and corresponding penetration depth vary among

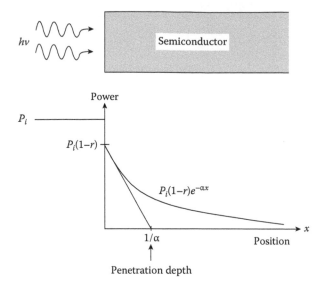

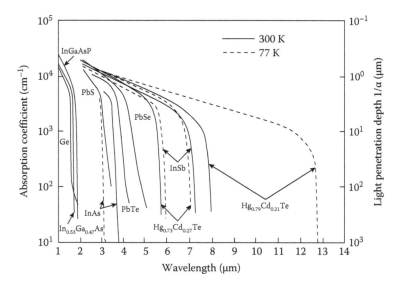

Figure 3.7 Optical absorption in a semiconductor.

Figure 3.8 Absorption coefficient for various photodetector materials in spectral range of 1–14 μm.

the different materials. It is well known that the absorption curve for direct transitions between parabolic bands at photon energy greater than energy gap, E_g, obeys a square-root law

$$\alpha(h\nu) = \beta(h\nu - E_g)^{1/2}, \tag{3.29}$$

where β is a constant. As can be readily seen in Figure 3.8, in a MWIR spectral region, the absorption edge value changes between 2×10^3 cm^{-1} and 3×10^3 cm^{-1}; in the LWIR region it is about 10^3 cm^{-1}.

Since α is a strong function of the wavelength, for a given semiconductor the wavelength range in which appreciate photocurrent can be generated is limited. Near the material's bandgap, there is tremendous variation causing a three orders of magnitude variation in absorption. In the region of the material's maximum usable wavelength, the absorption efficiency drops dramatically. For wavelengths longer than cutoff wavelength, the values of α are too small to give appreciable absorption.

The absorption coefficient, α, for extrinsic semiconductors is given by

$$\alpha = \sigma_p N_i. \qquad (3.30)$$

This is the product of the photoionization cross-section, σ_p, and the neutral impurity concentration, N_i. It is desirable to make α as large as possible. The upper limit of N_i is set by either "hopping" or "impurity band" conduction. Practical values of α for optimized doped photoconductors are in the range 1–10 cm^{-1} for Ge and 10–50 cm^{-1} for Si. Thus, to maximize quantum efficiency, the thickness of the detector crystal should not be less than about 0.5 cm for doped Ge and about 0.1 cm for doped Si. Fortunately, for the most extrinsic detectors, the drift length of photocarriers is sufficiently long that quantum efficiencies approaching 50% can be obtained.

The absorption coefficient is considerably modified for low-dimensional solids. Figure 3.9 shows the infrared absorption spectra for different n-doped, 50 period GaAs/Al$_x$Ga$_{1-x}$As quantum well infrared photodetector (QWIP) structures measured at room temperatures using a 45° multipass waveguide geometry [10]. The spectra of the bound-to-bound continuum (B-C) QWIP (samples A, B, and C) are much broader than the bound-to-bound (B-B; sample F) or bound-to-quasibound (B-QB) QWIP (sample F). Correspondingly, the value of the absorption coefficient for the B-C QWIP is significantly lower than that of the B-B QWIP, due to conservation of oscillator strength. It appears, that the low-temperature absorption coefficient $\alpha_p(77\ K) \approx 1.3\alpha_p(300\ K)$ and $\alpha_p(\Delta\lambda/\lambda)/N_D$ is a constant ($\Delta\lambda$ is the full width at half-α_p, N_D is the well's doping) [10]. Typical value of absorption coefficient in 77 K in LWIR region is between 600 and 800 cm^{-1}. Comparing Figures 3.8 and 3.9 we can notice that the absorption coefficients for direct band-to-band absorption is higher than that for intersubband transitions.

For an ensemble of quantum dots (QDs), the absorption spectra can be modeled using a Gaussian line shape in the form [11]

$$\alpha(E) = \alpha_o \frac{n_1}{\delta} \frac{\sigma_{QD}}{\sigma_{ens}} exp\left[-\frac{(E-E_g)^2}{\sigma_{ens}^2}\right], \qquad (3.31)$$

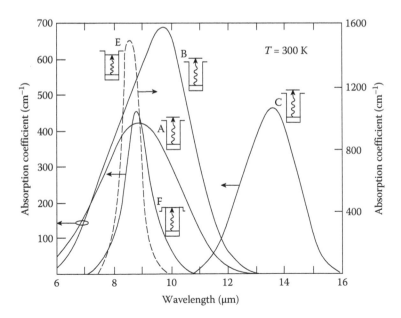

Figure 3.9 Absorption coefficient spectra measured at $T = 300$ K for different QWIP samples. (From Levine, B. F., *Journal of Applied Physics*, 74, R1–R81, 1993. With permission.)

where α_o is the maximum absorption coefficient, n_1 is the areal density of electrons in the quantum dot ground state, δ is the quantum dot density, and $E_g = E_2 - E_1$ is the energy of the optical transition between ground and excited states in the QDs. The expressions σ_{QD} and σ_{ens} are the standard deviations in the Gaussian line shape for intraband absorption in a single quantum dot and for the distribution in energies for the QD ensemble, respectively. Thus, the terms n_1/δ and σ_{QD}/σ_{ens} describe a decrease in absorption due to the absence of available electrons in the QD ground state and inhomogeneous broadening, respectively.

The optical absorption between the ground and excited levels is found to be have a value [12]

$$\alpha_o \approx \frac{3.5 \times 10^5}{\sigma}, \qquad \text{in cm}^{-1} \qquad (3.32)$$

where σ is the linewidth of the transition in meV. Equation 3.32 indicates the trade-off between the absorption coefficient and the absorption linewidth, σ. For very uniform QDs, the theoretically predicted absorption coefficient by Equation 3.31 can be considerably higher in comparison with those measured for narrow gap intrinsic materials.

The spectral dependence of the absorption coefficient has decisive influence on quantum efficiency [13–15]. Figure 3.10 shows the quantum efficiency of some of the detector materials used to fabricate ultraviolet (UV), visible, and infrared detectors [15]. The AlGaN detectors are being developed in the UV region. Silicon p-i-n diodes are shown with and without antireflection coating. Lead salts (PbS and PbSe) have intermediate quantum efficiencies, while PtSi Schottky-barrier types and quantum well infrared photodetectors (QWIPs) have low values. The InSb can respond from the near UV out to 5.5 μm at 80 K. A suitable detector material for near-IR (1.0–1.7 μm) spectral range is InGaAs lattice matched to the InP. Various HgCdTe alloys, in both photovoltaic and photoconductive configurations, cover from 0.7 μm to over 20 μm. Impurity-doped (Sb, As, and Ga) silicon impurity-blocked conduction (IBC) detectors operating at 10 K have a spectral response cutoff in the range of 16–30 μm. Impurity-doped Ge detectors can extend the response out to 100–200 μm.

3.2.2 Theoretical Model of Photon Detectors

Let us consider a generalized model of a photodetector, which by its optical area A_o is coupled to a beam of IR radiation [16–19]. The detector is a slab of homogeneous semiconductor with actual "electrical" area, A_e, and thickness t (see Figure 3.11). Usually, the optical and electrical areas of the

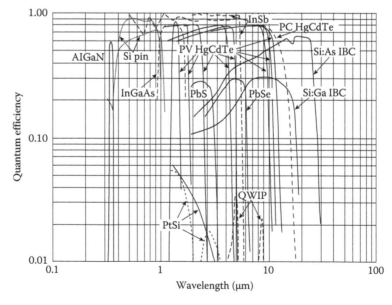

Figure 3.10 Quantum efficiency of ultraviolet, visible, and infrared photon detectors. (From Norton, P., *Encyclopedia of Optical Engineering*, Marcel Dekker Inc., New York, 320–48, 2003. With permission.)

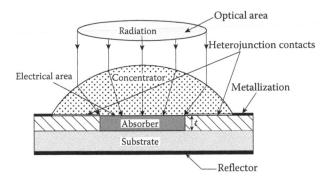

Figure 3.11 Model of a photodetector.

device are the same or similar. However, the use of some kind of optical concentrator can increase the A_o/A_e ratio by a large factor.

The current responsivity of the photodetector is determined by the quantum efficiency, η, and by the photoelectric gain, g. The quantum efficiency describes how well the detector is coupled to the impinging radiation. It is defined here as the number of electron-hole pairs generated per incident photon in an intrinsic detector, the number of generated free unipolar charge carriers in an extrinsic detector, or the number of charge carriers with energy sufficient to cross the potential barrier in a photoemissive detector. The photoelectric gain is the number of carriers passing contacts per one generated pair in an intrinsic detector, or the number of charge carriers in other types of detectors. This value shows how well the generated charge carriers are used to generate current response of a photodetector. Both values are assumed here as constant over the volume of the device.

The spectral current responsivity is equal to

$$R_i = \frac{\lambda \eta}{hc} qg, \tag{3.33}$$

where λ is the wavelength, h is the Planck's constant, c is the velocity of light, q is the electron charge, and g is the photoelectric current gain. The current that flows through the contacts of the device is noisy due to the statistical nature of the generation and recombination processes: fluctuation of optical generation, thermal generation, and radiative and nonradiative recombination rates. Assuming that the current gain for the photocurrent and the noise current are the same, the noise current is

$$I_n^2 = 2q^2 g^2 (G_{op} + G_{th} + R)\Delta f, \tag{3.34}$$

where G_{op} is the optical generation rate, G_{th} is the thermal generation rate, R is the resulting recombination rate, and Δf is the frequency band.

It should be noted that the effects of a fluctuating recombination frequently can be avoided by arranging for the recombination process to take place in a region of the device where it has little effect due to a low photoelectric gain, for example, at the contacts in sweep-out photoconductors, at the backside surface of a photoelectromagnetic detector, or in the neutral regions of the diodes. The generation processes with their associated fluctuations, however, cannot be avoided by any means [20,21].

Detectivity, D^*, is the main parameter to characterize normalized signal-to-noise performance of detectors, and can be defined as

$$D^* = \frac{R_i \left(A_o \Delta f\right)^{1/2}}{I_n}. \tag{3.35}$$

3.2.2.1 *Optical Generation Noise*

Optical generation noise is photon noise due to fluctuation of the incident flux. The optical generation of the charge carriers may result from three different sources:

- Signal radiation generation
- Background radiation generation
- Thermal self-radiation of the detector itself at a finite temperature.

The optical signal generation rate (photons/s) is

$$G_{op} = \Phi_s A_o \eta, \tag{3.36}$$

where Φ_s is the signal photon flux density.

If recombination does not contribute to the noise,

$$I_n^2 = 2\Phi_s A_o \eta q^2 g^2 \Delta f, \tag{3.37}$$

and

$$D^* = \frac{\lambda}{hc}\left(\frac{\eta}{2\Phi_s}\right)^{1/2}. \tag{3.38}$$

This is the ideal situation, when the noise of the detector is determined entirely by the noise of the signal photons. Usually, the noise due to optical signal flux is small compared to the contributions from background radiation or thermal generation–recombination processes. An exception is heterodyne detection, when the noise due to the powerful local oscillator radiation may dominate.

Background radiation frequently is the main source of noise in a detector. Assuming no contribution due to recombination,

$$I_n^2 = 2\Phi_B A_o \eta q^2 g^2 \Delta f, \tag{3.39}$$

where Φ_B is the background photon flux density. Therefore,

$$D^*_{BLIP} = \frac{\lambda}{hc}\left(\frac{\eta}{2\Phi_B}\right)^{1/2} \tag{3.40}$$

Once background-limited performance is reached, quantum efficiency, η, is the only detector parameter that can influence a detector's performance.

Figure 3.12 shows the peak spectral of a photon counter versus the cutoff wavelength plot calculated for 300 K background radiation and hemispherical FOV ($\theta = 90°$). The minimum D^*_{BLIP} (300 K) occurs at 14 μm and is equal to 4.6×10^{10} cmHz$^{1/2}$/W. For some photodetectors that operate at near equilibrium conditions, such as nonsweep-out photoconductors, the recombination rate is equal to the generation rate. For these detectors the contribution of recombination to the noise will reduce D^*_{BLIP} by a factor of $2^{1/2}$. Note that D^*_{BLIP} does not depend on area and the A_o/A_e ratio. As a consequence, the background limited and signal-limited performances cannot be improved by making A_o/A_e large.

In contrast to the signal and background related processes, optical generation is connected with the detector itself and may be of importance for detectors operating at near room temperatures. The related ultimate performance is usually calculated, assuming blackbody radiation and taking into account the reduced speed of light and wavelength due to a greater than one refractive index of the detector material and full absorption of photons with energy larger than the bandgap [23]. The carrier generation rate per unity area is

$$g_a = 8\pi c n^2 \int_0^\infty \frac{d\lambda}{\lambda^4\left(e^{hc/\lambda T} - 1\right)}, \tag{3.41}$$

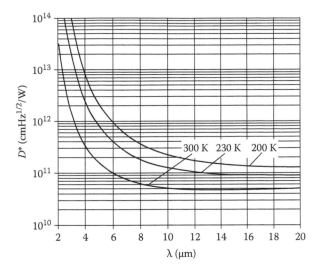

Figure 3.12 Calculated peak spectral detectivities of a photon counter limited by the hemispherical FOV background radiation as a function of the peak wavelength and temperature. (From Piotrowski J. and Rogalski, A., *High-Operating Temperature Infrared Photodetectors*, SPIE Press, Bellingham, WA, 2007. With permission.)

where n is the refractive index. Note that the generation rate is a factor of $4n^2$ larger compared to the 180° FOV background generation for $\eta = 1$. Therefore, the resulting detectivity

$$D^* = \frac{\lambda \eta A_o}{hc\left(2g_a A_e\right)^{1/2}}.$$ (3.42)

will be a factor of $2n$ lower when compared to BLIP detectivity for a background at detector temperature ($A_e = A_o$). The internal thermal radiation limited D^* can be improved by making A_o/A_e large, in contrast to D^*_{BLIP}.

When Humpreys reexamined the existing theories of radiative recombination and internal optical generation [24,25], he indicated that most photons emitted as a result of radiative recombination are immediately reabsorbed inside the detector, generating charge carriers. Due to reabsorption, the radiative lifetime is highly extended, which means that the internal optical generation–recombination processes could be practically noiseless in optimized devices. Therefore, the ultimate limit of performance is set by the signal or background photon noise.

3.2.2.2 Thermal Generation and Recombination Noise

Infrared photodetectors operating at near room temperature and low-temperature devices operated at low background irradiances are generally limited by thermal generation and recombination mechanisms rather than by photon noise. For effective absorption of IR radiation in a semiconductor, we must use material with a low energy of optical transitions compared to the energy of photons to be detected, for example, semiconductors with a narrower bandgap. A direct consequence of this fact is that at near room temperatures, the thermal energy of charge carriers, kT, becomes comparable to the transition energy. This enables thermal transitions, making the thermal generation rate very high. As a result, the long-wavelength detector is very noisy when operated at near room temperature.

For uniform volume generation and recombination rates G and R (in $m^{-6}s^{-1}$), the noise current is

$$I_n^2 = 2\left(G+R\right)A_e t \Delta f q^2 g^2,$$ (3.43)

therefore

$$D^* = \frac{\lambda}{2^{1/2} hc\left(G+R\right)^{1/2}}\left(\frac{A_o}{A_e}\right)^{1/2}\frac{\eta}{t^{1/2}}.$$ (3.44)

At equilibrium, the generation and recombination rates are equal. In this case

$$D^* = \frac{\lambda \eta}{2hc(Gt)^{1/2}} \left(\frac{A_o}{A_e}\right)^{1/2}.$$

(3.45)

3.2.3 Optimum Thickness of Photodetector

For a given wavelength and operating temperature, the highest performance can be obtained by maximizing $\eta/[(G+R)t]^{1/2}$. This is the condition for the highest ratio of the quantum efficiency to the square root of the sum of the sheet thermal generation and recombination rates. This means that high quantum efficiency must be obtained with a thin device.

In further calculations we will assume: $A_e = A_o$, perpendicular incidence of radiation, and negligible front and backside reflection coefficients. In this case

$$\eta = 1 - e^{-\alpha t}.$$

(3.46)

where α is the absorption coefficient. Then

$$D^* = \frac{\lambda}{2^{1/2}hc} \left(\frac{\alpha}{G+R}\right)^{1/2} F(\alpha t),$$

(3.47)

where

$$F(\alpha t) = \frac{1 - e^{-\alpha t}}{(\alpha t)^{1/2}}.$$

(3.48)

Function $F(\alpha t)$ achieves its maximum 0.638 for $t = 1.26/\alpha$. In this case $\eta = 0.716$ and the highest detectivity is

$$D^* = 0.45 \frac{\lambda}{hc} \left(\frac{\alpha}{G+R}\right)^{1/2}.$$

(3.49)

The detectivity can also be increased by a factor of $2^{1/2}$ for double pass of radiation. This can be achieved by the use of a backside reflector. Simple calculation shows that the optimum thickness in this case is half that of the single pass case, while the quantum efficiency remains equal to 0.716.

At equilibrium, the generation and recombination rates are equal. Therefore

$$D^* = \frac{\lambda}{2hc} \eta (Gt)^{-1/2}.$$

(3.50)

If the recombination process is uncorrelated with the generation process that contributes the detector noise

$$D^* = \frac{\lambda}{2^{1/2}hc} \eta (Gt)^{-1/2}.$$

(3.51)

3.2.4 Detector Material Figure of Merit

To summarize the discussion above, the detectivity of an optimized infrared photodetector of any type can be expressed as

$$D^* = 0.31 \frac{\lambda}{hc} k \left(\frac{\alpha}{G}\right)^{-1/2},$$

(3.52)

where $1 \leq k \leq 2$ and dependent on the contribution of recombination and backside reflection as shown in Table 3.2.

As we can see, the ratio of the absorption coefficient to the thermal generation rates is the main figure of merit of any materials for infrared detectors. This figure of merit proposed for the first time by Piotrowski can be utilized to predict ultimate performance of any infrared detector and to select possible material candidates for use as detectors [17].

The α/G ratio versus temperature for different types of tunable materials with a hypothetical energy gap equal to 0.25 eV ($\lambda = 5$ μm) and 0.124 eV ($\lambda = 10$ μm) is shown in Figure 3.13 [26]. Procedures used in calculations of α/G for different material systems are given in Rogalski [27]. It is apparent that HgCdTe is by far the most efficient detector of IR radiation. One may also notice that QWIP is a better material than extrinsic silicon. The above figures are completed by Figure 3.14, where the α/G ratio in dependence on wavelength is presented for different materials at 77 K.

The calculation of the figure of merit involves determination of the absorption coefficient and thermal generation rate taking into account various processes of fundamental and less fundamental nature.

It should be noted that the importance of thermal generation rate as a material figure of merit was recognized for the first time by Long [28]. It was used in many papers of English workers [29,30] related to high operating temperature (HOT) detectors. More recently Kinch [31] introduced the thermal generation rate within $1/\alpha$ depth per unit of area as the figure of merit. This is actually the inversed α/G figure of merit originally proposed by Piotrowski [17].

Table 3.2: Dependence of the Factor k in Equation 3.52, Optimum Thickness and Quantum Efficiency on Contribution of Recombination and Presence of Backside Reflection

Backside Reflection	Contribution of Recombination	Optimum Thickness	Quantum Efficiency	k
0	$R = G$	$1.26/\alpha$	0.716	1
1	$R = G$	$0.63/\alpha$	0.716	$2^{1/2}$
0	No	$1.26/\alpha$	0.716	$2^{1/2}$
1	No	$0.63/\alpha$	0.716	2

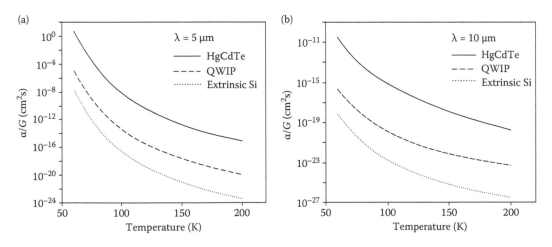

Figure 3.13 α/G ratio versus temperature for (a) MWIR – $\lambda = 5$ μm and (b) LWIR– $\lambda = 10$ μm photon detectors. (From Rogalski, A., *Reports on Progress in Physics*, 68, 2267–2336, 2005. With permission.)

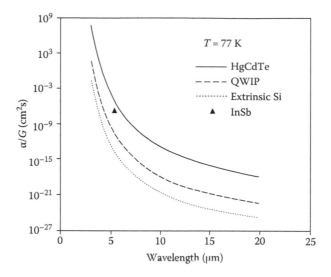

Figure 3.14 α/G ratio versus wavelength for different types of photon detectors operated at 77 K. (From Rogalski, A., *Reports on Progress in Physics*, 68, 2267–2336, 2005. With permission.)

In the Kinch criterion, the BLIP condition can be described as [31]

$$\frac{\eta\Phi_B\tau}{t} > n_{th}, \tag{3.53}$$

where n_{th} is the density of thermal carriers at the temperature T, τ is the carrier lifetime, Φ_B is the total background photon flux density (unit $cm^{-2}s^{-1}$) reaching the detector, and t is the detector's thickness. Rearranging, for the BLIP requirements we have

$$\frac{\eta\Phi_B}{t} > \frac{n_{th}}{\tau}; \tag{3.54}$$

that is, the photon generation rate per unit volume needs to be greater than the thermal generation rate per unit volume. The carriers can be either majority or minority in nature. Using $\eta = \alpha t$, where α is the absorption coefficient in the material, we obtain

$$\Phi_B > \frac{n_{th}}{\alpha\tau} = G_{th}. \tag{3.55}$$

The normalized thermal generation, $G_{th} = n_{th}/(\alpha\tau)$, predicts the ultimate performance of any infrared material and can be used to compare the relative performance of different materials as a function of temperature and energy gap (cutoff wavelength).

3.2.5 Reducing Device Volume to Enhance Performance

One possible way to improve the performance of IR photodetectors is to reduce the total amount of thermal generation within the active element of detector by reducing the detector volume, which is the product of its thickness and physical area. This must be done:

- Without sacrificing quantum efficiency
- Preserving the required optical area
- Keeping the acceptance angle of a detector large enough to intercept radiation from the main optics of an IR system

Interestingly, the performance of thermal detectors is related in a similar way to its physical area when the thermal conductance to the surroundings per area is constant.

Enhancing absorption makes it possible to reduce a detector's thickness without losing quantum efficiency. The quantum efficiency in thin devices can be significantly enhanced using interference phenomena to set up a resonant cavity within the photodetector [32–34]. Various optical resonator structures are shown in Figure 3.15. In the simplest method, interference occurs between the waves reflected at the rear, the highly reflective surface, and at the front surface of a semiconductor. The thickness of the semiconductor is selected to set up the standing waves in the structure with peaks at the front and nodes at the back surface. The quantum efficiency oscillates with thickness of the structure, with the peaks at a thickness corresponding to an odd multiple of $\lambda/4n$, where n is the refractive index of the semiconductor. The gain in quantum efficiency increases with n. Higher gain can be obtained in the structures shown in Figure 3.15b. Even higher improvement is possible in structures with multiple dielectric layers, such as shown in Figure 3.15c.

Enhanced absorption due to interference effects seems to be particularly important for devices where operation is dependent on gradient of minority carriers such as Dember and electromagnetic effect detectors [16,35,36].

It should be noted that the interference effects strongly modify the spectral response of the device, and the gain due to the optical cavity can be achieved only in narrow spectral regions. This may be an important limitation for the applications, which require wide spectral band sensitivity. In practice, infrared systems usually operate in a spectral band (e.g., atmospheric windows) and the use of resonant cavities may yield significant gains. Another limitation comes from the fact that efficient optical resonance occurs only for perpendicular incidence and is less effective for oblique incidence. This limits the use of the devices with fast optics, and especially, the optical immersion [34].

Another possible way to improve the performance of the IR photodetector is an increase in the apparent "optical" size of the detector in comparison with its actual physical size using a suitable concentrator that compresses impinging IR radiation. This must be achieved without reduction of acceptance angle, or at least, with limited reduction to angles required for fast optics of IR systems. Various types of suitable optical concentrators can be used including optical cones, conical fibers, and other types of reflective, diffractive, and refractive optical concentrators.

An efficient way to achieve an effective concentration of radiation is to immerse the photodetector to the hemispherical or hyperhemispherical lenses [37]. Larger gain can be obtained for a hyperhemisphere used as an aplanatic lens [22].

The principle of operation of a hemispherical immersion lens is shown in Figure 3.16. The detector is located at the center of curvature of the immersion lens. The lens produces an image of the detector. No spherical or coma aberration exists (aplanatic imaging). Due to immersion the apparent linear size of detector increases by a factor of n. The image is located at the detector plane. The use of a hemispheric immersion lens in combination with an objective lens of an imaging optical

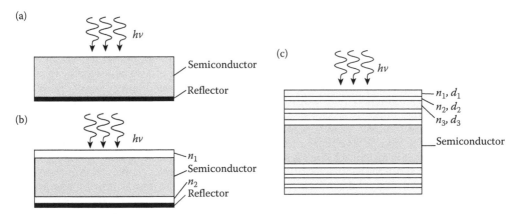

Figure 3.15 Schematic structure of interference enhanced quantum efficiency photodetector: (a) the simplest structure, (b) structure immersed between two dielectric layers and supplied with backside reflector, and (c) structure inserted in a Bragg microcavity. (From Piotrowski J. and Rogalski, A., *High-Operating Temperature Infrared Photodetectors*, SPIE Press, Bellingham, WA, 2007. With permission.)

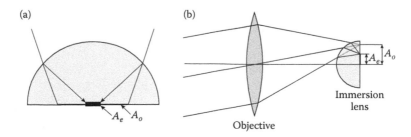

Figure 3.16 (a) Principle of optical immersion and (b) ray tracing for combination of objective lens and immersion lens optical systems. (From Piotrowski, J. and Rogalski, A., *High-Operating Temperature Infrared Photodetectors*, SPIE Press, Bellingham, WA, 2007. With permission.)

Table 3.3: Influence of Optical Immersion on Properties of Photodetectors with CdZnTe Lens

Value	Hemisphere	Hyperhemisphere	Type of Detector
Linear size	n ($\approx$2.7)	n^2 ($\approx$7)	Any
Area	n^2 ($\approx$7)	n^4 ($\approx$50)	Any
Voltage responsivity	n ($\approx$2.7)	n^2 ($\approx$7)	PC, PEM
Voltage responsivity	n^2 ($\approx$7)	n^4 ($\approx$50)	PV, Dember
Detectivity	n ($\approx$2.7)	n^2 ($\approx$7)	Any
Bias power	n^2 ($\approx$7)	n ($\approx$50)	PC, PV
Capacitance	n^2 ($\approx$7)	n^4 ($\approx$50)	PV
Acceptance angle (deg)	180	42	Any
Thickness/radius	1	1.3	Any
Tolerances	Not severe	Stringent	Any

Notes: The numbers show the relative change of a given value compared to a nonimmersed detector of the same optical size. (From Piotrowski, J. and Rogalski, A., *High-Operating Temperature Infrared Photodetectors*, SPIE Press, Bellingham, WA, 2007. With permission.) PC and PV means a photoconductive and photovoltaic detector, respectively.

system is shown in Figure 3.16b. The immersion lens plays the role of a field lens, which increases the FOV of the optical system.

The practical use of immersion technology has been limited due to problems in mechanical matching of detector and lens material, and severe transmission and reflection losses. Another limitation was due to the limited acceptance angle of the devices as a result of the total reflection at the lens-glue interface.

Germanium ($n = 4$) is the most frequently used for immersion lenses. Arsenic doped amorphous selenium or Mylar have been used for the adhesive [3]. Silicon is another important material. The problems of matching the detectors to immersion lenses can also be solved by the use of monolithic technology. For example, VIGO System technology is based on epitaxy of HgCdTe on transparent and high refraction index CdZnTe substrates [38]. The HgCdTe serves as the sensitive element, while the immersion lens is formed directly in the substrate. This technology has been used also for InGaAs and InGaAsSb devices on GaAs substrates [39,40].

As Table 3.3 shows, monolithic optical immersion results in significant improvement of detector parameters. The gain factors achieved with hyperhemispherical immersion are substantially higher compared to those for hemispherical immersion. The hyperhemispherical immersion may restrict the acceptance angle of the detector and require more severe manufacturing tolerances. These restrictions depend on the refraction coefficient of the lens. For CdZnTe they are not so severe as for germanium lenses, however, and have no practical importance in many cases. For example, the minimum f-number of the main optical system is limited to about 1.4 by the CdZnTe immersion lens.

An alternative approach is the use of a Winston cone. QinetiQ have developed a micromachining technique involving dry etching to fabricate the cone concentrators for detector and luminescent devices [41–43].

3.3 COMPARISON OF FUNDAMENTAL LIMITS OF PHOTON AND THERMAL DETECTORS

In further considerations, we follow Kinch [31] assuming that the thermal generation rate of the IR material is the key parameters that enable comparison of different material systems.

The normalized dark current

$$J_{dark} = G_{th}q \qquad (3.56)$$

directly determines thermal detectivity (see Equation 3.51)

$$D^* = \frac{\eta\lambda}{hc}(2G_{th})^{-1/2}. \qquad (3.57)$$

The normalized dark current densities for the various materials used in infrared detector technologies in LWIR spectral region ($E_g = 0.124$ eV, $\lambda_c = 10$ μm) are shown in Figure 3.17 [44]. In addition, the $f/2$ background flux current density is also shown. The extrinsic silicon, the high temperature superconductors (HTSC) and the photoemissive (silicon Schottky barrier) detectors are hypothetical, but are included for comparison. In the calculations, carried out for different material systems we have followed the procedures used in Kinch's paper [31], except quantum dot infrared detectors (QDIPs) where the Phillips' model is used [11]. The parameters representative for self-assembled InAs/GaAs quantum dots reported in the literature are as follows [11,45]: $\alpha_o = 5 \times 10^4$ cm^{-2}, $V = 5.3 \times 10^{-19}$ cm^{-3}, $\delta = 5 \times 10^{10}$ cm^{-2}, $\tau = 1$ ns, $N_d = 1 \times 10^{11}$ cm^{-2} and the detector thickness $t = 1/\alpha_o$. The calculations are described in details in Martyniuk and Rogalski [44].

In the Phillips's paper an ideal QD structure is assumed with two electron energy levels (the excited state coincides with the conduction band minimum of the barrier material). Also inhomogeneous broadening of the dot ensemble is neglected ($\sigma_{QD}/\sigma_{ens} = 1$; see Equation 3.31). Above assumptions determine high performance of QDIPs.

In the MWIR and LWIR regions, the dominant position have HgCdTe photodiodes. Quantum well infrared photodetectors (QWIPs) are mainly used in LWIR tactical systems operating at lower temperature, typically 65–70 K, where cooling is not an issue. Large detector arrays with more than one million detector elements are fabricated by several manufacturers using these material systems. Beyond 15 μm, good performance is achieved using extrinsic silicon detectors. These detectors are termed impurity band conduction (IBC) detectors and found a niche market for the

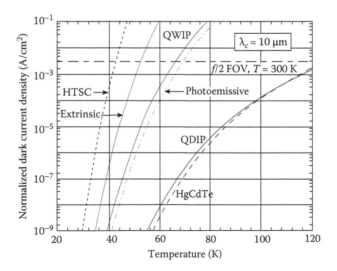

Figure 3.17 Temperature dependence of the normalized dark current of various LWIR ($\lambda_c = 10$ μm) material technologies. The $f/2$ background flux current density is also shown. (From Martyniuk, P. and Rogalski, A., *Progress in Quantum Electronics*, 32, 89–120, 2008. With permission.)

astronomy and civil space communities because HgCdTe has not yet realized its potential at low temperatures and reduced background.

Figure 3.17 displays that tunable bandgap alloy, HgCdTe, demonstrates the highest performance (the lowest dark current/thermal generation and the highest BLIP operating temperature). These estimations are confirmed by experimental data [46,47]. For very uniform QD ensembles, the QDIP performance can be close to the HgCdTe one and potentially can exceed that of HgCdTe in the region of high operation temperatures.

Figure 3.18 compares the thermal detectivities of various photodetectors with cutoff wavelength in MWIR ($\lambda_c = 5$ µm) and LWIR ($\lambda_c = 10$ µm) regions. The assumed typical quantum efficiencies are indicated in the figure. Theoretical estimations for QDIPs are carried out assuming low quantum efficiency ≈ 2% (often measured in practice) and 67%. The last value is typical for HgCdTe photodiodes (without antireflection coating). It should be noticed, however, that rapid progress has recently been made in the performance of QDIP devices, especially at near room temperature. Lim et al. have announced a quantum efficiency of 35% for detectors with peak detection wavelength around 4.1 µm [48].

Estimation of detectivity for InAs/GaInSb strained layer superlattices (SLSs) are based on several theoretical papers [49–51]. Early calculations showed that a LWIR type-II InAs/GaInSb SLS should have an absorption coefficient comparable to an HgCdTe alloy with the same cutoff wavelength [49]. Figure 3.18b predicts that type-II superlattices are the most efficient detector of IR radiation in long wavelength region. It is even better material than HgCdTe; it is characterized by high absorption coefficient and relatively low thermal generation rate. However, hitherto, this theoretical prediction has been not confirmed by experimental data. The main reason of that is influence of the Schockley-Read generation–recombination mechanism, which causes lower carrier lifetime (higher thermal generation rate). It is clear from this analysis that the fundamental performance limitation of QWIPs is unlikely to rival HgCdTe photodetectors. However, the performance of very uniform QDIP [when $\sigma_{QD}/\sigma_{ens} = 1$] is predicted to rival HgCdTe. We can also notice from Figure 3.18 that AlGaAs/GaAs quantum well infrared photoconductor (QWIP) is better material than extrinsic silicon.

The BLIP temperature is defined that the device is operating at a temperature at which the dark current is equal to the background photocurrent, given a FOV and a background temperature.

In Figure 3.19, plots of the calculated temperature required for background limited (BLIP) operation in $f/2$ FOV are shown as a function of cutoff wavelength for various types of detectors. We can see that the operating temperature of ideal QDIPs is comparable with HgCdTe photodiodes. HgCdTe detectors with background limited performance operate in practice with thermoelectric coolers in the MWIR range, but the LWIR detectors ($8 \leq \lambda_c \leq 12$ mm) operate at ≈100 K. HgCdTe photodiodes exhibit higher operating temperature compared to extrinsic detectors, silicide Schottky barriers, QWIPs and HTSCs. Type II SLSs are omitted in our considerations. The cooling

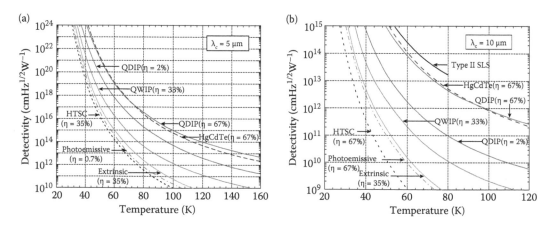

Figure 3.18 The predicted thermal detectivity versus temperature for various (a) MWIR ($\lambda_c = 5$ µm) and (b) LWIR ($\lambda_c = 10$ µm) photodetectors. The assumed quantum efficiencies are indicated. (From Martyniuk, P. and Rogalski, A., *Progress in Quantum Electronics*, 32, 89–120, 2008. With permission.)

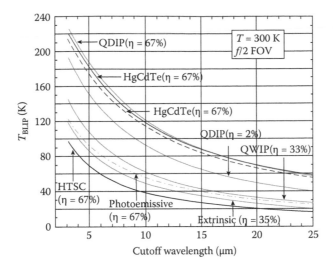

Figure 3.19 Estimation of the temperature required for background limited operation of various types of photodetectors. (From Martyniuk, P. and Rogalski, A., *Progress in Quantum Electronics,* 32, 89–120, 2008. With permission.)

requirements for QWIPs with cutoff wavelengths below 10 μm are less stringent in comparison with extrinsic detectors, Schottky-barrier devices, and HTSCs.

It has been shown by Phillips [45] that the QD detector performance may be degraded by orders of magnitude for the values of $\sigma_{ens}/\sigma_{QD} = 100$, which are indicative of the current state of QD fabrication technology. It is well known that reduced optical absorption in QDs due to size nonuniformity results in an increase in the normalized dark current and a reduction in detectivity. The nonuniformity also has strong influence on BLIP temperature. Increase of σ_{ens}/σ_{QD} ratio from 1 to 100 causes decrease of T_{BLIP} by several tens of degrees [44].

Due to fundamentally different types of noise, thermal and photon detectors have different dependencies of detectivities on wavelength and temperature. The temperature dependence of the fundamental limits of D^* of photon and thermal detectors for different levels of background are shown in Figures 3.20 and 3.21 [52]. In comparison with Kruse's paper [53] these studies are reexamined taking into account updated theories of different types of detectors.

It results from Figure 3.20 that in LWIR spectral range, the performance of intrinsic IR detectors (HgCdTe photodiodes) is higher than for other types of photon detectors. HgCdTe photodiodes with background limited performance operate at temperatures below ≈80 K. HgCdTe is characterized by high optical absorption coefficient and quantum efficiency and relatively low thermal generation rate compared to extrinsic detectors and QWIPs. The extrinsic photon detectors require more cooling than intrinsic photon detectors having the same long wavelength limit.

The theoretical detectivity value for the thermal detectors is much less temperature dependent than for the photon detectors. At temperatures below 50 K and zero background, LWIR thermal detectors are characterized by D^* values lower than those of LWIR photon detectors. However, at temperatures above 60 K, the limits favor the thermal detectors. At room temperature, the performance of thermal detectors is much better than LWIR photon detectors. The above relations are modified by influence of background; this is shown in Figure 3.20 for a background of 10^{17} photons/cm²s. It is interesting to notice that the theoretical curves of D^* for photon and thermal detectors show similar fundamental limits at low temperatures.

Similar considerations have been carried out for VLWIR detectors operated in the 14–50 μm spectral range. The calculation results are presented in Figure 3.21. Detectors operating within this range are cryogenic Si and Ge extrinsic photoconductors and cryogenic thermal detectors, usually bolometers. Nevertheless, in Figure 3.21, theoretical prediction for intrinsic detectors (HgCdTe photodiodes) is also included. Figure 3.21 shows that the theoretical performance limit of VLWIR thermal detectors at zero and high backgrounds in a wide range of temperatures equal or exceed that of photon detectors.

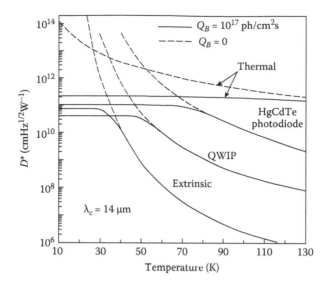

Figure 3.20 Theoretical performance limits of LWIR photon and thermal detectors at wavelength 14 μm, zero background and background of 10^{17} photons/cm²s, as a function of detector temperature. (From Ciupa, R. and Rogalski, A., *Opto-Electronics Review*, 5, 257–66, 1997. With permission.)

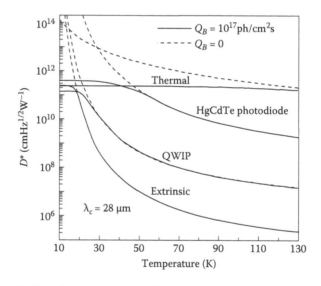

Figure 3.21 Theoretical performance limits of VLWIR photon and thermal detectors at wavelength 28 μm, zero background and background of 10^{17} photons/cm²s, as a function of detector temperature. (From Ciupa, R. and Rogalski, A., *Opto-Electronics Review*, 5, 257–66, 1997. With permission.)

The comparison of both types of detectors indicates that theoretical performance limits for thermal detectors are more favorable as wavelength of operation moves from the LWIR to the VLWIR. It is due to the influence of fundamentally different types of noise (generation–recombination noise in photon detectors and temperature fluctuation noise in thermal detectors); these two classes of detectors have different dependencies of detectivities on wavelength and temperature. The photon detectors are favored at long wavelength infrared and lower operating temperatures. The thermal detectors are favored at very long wavelength spectral range. The temperature requirements to attain background fluctuation noise performance, in general, favor thermal detectors at the higher cryogenic temperatures and photon detectors at the lower cryogenic temperatures.

3.4 MODELING OF PHOTODETECTORS

Traditionally IR photodetectors have been called either photoconductive or photovoltaic detectors based on the principle by which optically generated carriers can be detected as a change in voltage or current across the element. The simple design photoconductors is based on a flake of semiconductor supplied with ohmic contacts, while photovoltaic detectors are p-n junction devices. Dember and photoelectromagnetic effect detectors are less common photovoltaic devices that require no p-n junction.

However, recent advances in heterostructure devices such as the development of the heterojunction photoconductors, double layer heterostructure photodiodes, and introduction of nonequilibrium mode of operation make this distinction unclear. Moreover, the photovoltaic structures are frequently biased, exhibiting signal due to both photovoltages at junctions and from the photoconductivity of some regions.

An optimized photodetector (see Figure 3.11) of any type may be a 3-D monolithic heterostructure that consists of the following regions [18,22,54]:

- Concentrator of IR radiation that directs incident radiation onto absorber (an example is an immersion lens made of a wide gap semiconductor).

- Absorber of IR radiation where the generation of free carriers occurs (this is a narrow gap semiconductor with bandgap, doping and geometry selected for the highest ratio of the optical-to-thermal generation rates).

- Contacts to the absorber, which sense optically generated charge carriers (contacts should not contribute to the dark current of the device; examples are wide gap heterojunction contacts used in the modern devices).

- Passivation of the absorber (the surfaces of the absorber must be insulated from the ambient by a material that also doesn't contribute to carrier generation); in addition, passivation repels the carriers optically generated in absorber, keeping them away from surfaces where recombination can reduce the quantum efficiency.

- Retroreflector to enhance absorption (examples are metal or dielectric mirrors; optical resonant cavity structures can also be used).

The above conditions can be fulfilled using heterojunctions like N+-p-p+ and P+-n-n+ with heavily doped contact regions (symbol "+" denotes strong doping, capital letter denotes wider gap). Homojunction devices (like n-p, n+-p, p+-n) suffer from surface problems; excess thermal generation results in increased dark current and recombination, which reduces photocurrent.

Modeling of the photodetectors is a strategically important task necessary to understand photodetector properties and optimize their design. Analytical models were developed for specific types of IR devices based on idealized structures, operating in equilibrium and nonequilibrium mode. These models make some features of the device's operation easy to understand and analyze.

In general, however, the operation of the advanced devices can no longer be described by analytical models. Omitting specific features of the narrow gap materials, such as degeneracy and nonparabolic conduction band, may result in enormous errors. The nonequilibrium mode of operation of infrared photodetectors brings further complications. The devices are based on an absorber that is near intrinsic or just extrinsic at the operating temperature. The properties of the device differ from those with extrinsic absorber [55]. First, drift and diffusion are dominated by ambipolar effects due to space charge coupling between electrons and holes. Second, the concentration of charge carriers in near intrinsic materials can be driven to levels considerably below intrinsic concentrations. As a result, the perturbation can be described only in terms of large signal theory. Third, carrier concentration in low bandgap materials are dominated by Auger generation and recombination.

An accurate description of more and more complex device architectures including doping and bandgap grading, heterojunctions, 2-D and 3-D effects, ambipolar effects, nonequilibrium operation, and surface, interface, and contact effects can be achieved only with a solution of the fundamental equations that describe the electrical behavior of semiconductor devices. These partial differential equations include continuity equations for electron and holes and Poisson's equation

$$\frac{\partial n}{\partial t} = \frac{1}{q}\vec{\nabla} \times \vec{J}_n + G_n - R_n, \tag{3.58}$$

$$\frac{\partial p}{\partial t} = \frac{1}{q}\vec{\nabla} \times \vec{J}_p + G_p - R_p, \tag{3.59}$$

$$\varepsilon_o \varepsilon_r \nabla^2 \Psi = -q\left(N_d^+ - N_a^- + p - n\right) - \rho_s, \tag{3.60}$$

where Ψ is the electrostatic potential defined as the intrinsic Fermi potential, ρ_s is the surface charge density, N_d^+ and N_a^- are concentrations of ionized donors and acceptors.

The solution of the Equations set 3.58 through 3.60 makes it possible to analyze stationary and transient phenomena in semiconductor devices. The main problem with the solution of these equations is their nonlinearity and complex dependences of their parameters. In many cases, some simplifications are possible. From Boltzmann transport theory, the current densities $\vec{J}_n$ and $\vec{J}_p$ can be written as functions of the carrier concentrations and the quasi-Fermi potentials for electrons and holes, Φ_n and Φ_p

$$\vec{J}_n = -q\mu_n n \vec{\nabla}\Phi_n, \tag{3.61}$$

$$\vec{J}_p = -q\mu_p p \vec{\nabla}\Phi_p. \tag{3.62}$$

Alternatively, $\vec{J}_n$ and $\vec{J}_p$ can be written as functions of Ψ, n, and p, consisting of drift and diffusion components

$$\vec{J}_n = q\mu_n \vec{E}_e + qD_n\vec{\nabla}n, \tag{3.63}$$

and

$$\vec{J}_p = q\mu_p \vec{E}_h - qD_p\vec{\nabla}p, \tag{3.64}$$

where D_n and D_p are the electron and hole diffusion coefficients.

If the effects of bandgap narrowing are neglected and Boltzmann carrier statistic is assumed

$$\vec{E}_n = \vec{E}_p = \vec{E} = -\vec{\nabla}\Psi. \tag{3.65}$$

The steady-state behavior of 1-D devices can be described by the set of five differential equations with suitable boundary conditions: two transport equations for electrons and holes, two continuity equations for electrons and holes and the Poisson equation, which are all related to the Van Roosbroeck [56]:

$$J_n = qD_n\frac{dn}{dx} - q\mu_n n\frac{d\Psi}{dx}, \qquad \text{current transport for electrons} \tag{3.66}$$

$$J_p = qD_p\frac{dp}{dx} - q\mu_p p\frac{d\Psi}{dx}, \qquad \text{current transport for holes} \tag{3.67}$$

$$\frac{1}{q}\frac{dJ_n}{dx} + (G - R) = 0, \qquad \text{continuity equation for electrons} \tag{3.68}$$

$$\frac{1}{q}\frac{dJ_p}{dx} - (G - R) = 0, \qquad \text{continuity equation for holes} \tag{3.69}$$

$$\frac{d^2\Psi}{dx^2} = -\frac{q}{\varepsilon_o \varepsilon_r}\left(N_d^+ - N_a^- + p - n\right) \qquad \text{Poisson's equation.} \tag{3.70}$$

Many papers devoted to the solution of these equations have been published, from papers of Gummel [57] and de Mari [58] to recent commercially available numerical programs. The fundamental equations cannot be solved analytically without the approximations, even for the 1-D steady-state case. Therefore, the numerical solutions must be applied. The numerical solution is composed of three steps: (1) grid generation step, (2) discretization to transform the differential equations into the linear algebraic equations, and (3) the solution. The Newton direct method is usually used to solve the matrix equation [59]. Other methods are also used to improve the convergence speed and reduce number of iterations [60].

Since experiments with complex device structures are complicated, costly, and time consuming, numerical simulations have become a critical tool to develop advanced detectors [61]. Some laboratories have developed suitable software, for example: Stanford University (USA), Military Technical University (Poland) [62], Honyang University (Korea) [60], and others. Commercial simulators are available from several sources, including: Medici (Technology Modeling Associates), Semicad (Dawn Technologies), Atlas/Blaze/Luminouse (Silvaco International, Inc.), APSYS (Crosslight Software, Inc.), and others. The APSYS, for example, is a full 2-D/3-D simulator that solves not only the Poisson's equation and the current continuity equations (including such features as field dependent mobilities and avalanche multiplication), but also the scalar wave equation for photonic waveguiding devices (such as waveguide photodetectors) and heat transfer equations with flexible thermal boundary conditions and arbitrary temperature-dependent parameters.

Although existing simulators still do not fully account for all semiconductor properties important for photodetectors, they are already invaluable tools for analysis and development of improved infrared IR photodetectors. In addition to device simulators, process simulators are being developed that facilitate advanced device growth technologies development [63,64].

REFERENCES

1. A. Rogalski, *Infrared Detectors*, Gordon and Breach, Amsterdam, 2000.

2. J. T. Houghton and S. D. Smith, *Infra-Red Physics*, Oxford University Press, Oxford, 1966.

3. E. L. Dereniak and G. D. Boreman, *Infrared Detectors and Systems*, Wiley, New York, 1996.

4. J. Piotrowski, "Breakthrough in Infrared Technology; The Micromachined Thermal Detector Arrays," *Opto-Electronics Review* 3, 3–8, 1995.

5. A. Smith, F. E. Jones, and R. P. Chasmar, *The Detection and Measurement of Infrared Radiation*, Clarendon, Oxford, 1968.

6. W. Kruse, L. D. McGlauchlin, and R. B. McQuistan, *Elements of Infrared Technology*, Wiley, New York, 1962.

7. P. G. Datskos, "Detectors: Figures of Merit," in *Encyclopedia of Optical Engineering*, ed. R. Driggers, 349–57, Marcel Dekker, New York, 2003.

8. F. J. Low and A. R. Hoffman, "The Detectivity of Cryogenic Bolometers," *Applied Optics* 2, 649–50, 1963.

9. E. H. Putley, "Thermal Detectors," in *Optical and Infrared Detectors*, ed. R. J. Keyes, 71–100, Springer, Berlin, 1977.

10. B. F. Levine, "Quantum-Well Infrared Photodetectors," *Journal of Applied Physics* 74, R1–R81, 1993.

11. J. Phillips, "Evaluation of the Fundamental Properties of Quantum Dot Infrared Detectors," *Journal of Applied Physics* 91, 4590–94, 2002.

12. J. Singh, *Electronic and Optoelectronic Properties of Semiconductor Structures*, Cambridge University Press, New York, 2003.

13. A. Rogalski, "Infrared Detectors: Status and Trends," *Progress in Quantum Electronics* 27, 59–210, 2003.

14. A. Rogalski, "Photon Detectors," in *Encyclopedia of Optical Engineering*, ed. R. Driggers, 1985–2036, Marcel Dekker Inc., New York, 2003.

15. P. Norton, "Detector Focal Plane Array Technology," in *Encyclopedia of Optical Engineering*, ed. R. Driggers, 320–48, Marcel Dekker Inc., New York, 2003.

16. J. Piotrowski, "$Hg_{1-x}Cd_xTe$ Infrared Photodetectors," in *Infrared Photon Detectors*, Vol. PM20, 391–494, SPIE Press, Bellingham, WA, 1995.

17. J. Piotrowski and W. Gawron, "Ultimate Performance of Infrared Photodetectors and Figure of Merit of Detector Material," *Infrared Physics & Technology* 38, 63–68, 1997.

18. J. Piotrowski and A. Rogalski, "New Generation of Infrared Photodetectors," *Sensors and Actuators* A67, 146–52, 1998.

19. J. Piotrowski, "Uncooled Operation of IR Photodetectors," *Opto-Electronics Review* 12, 111–22, 2004.

20. T. Ashley and C. T. Elliott, "Non-Equilibrium Mode of Operation for Infrared Detection," *Electronics Letters* 21, 451–52, 1985.

21. T. Ashley, T. C. Elliott, and A. M. White, "Non-Equilibrium Devices for Infrared Detection," *Proceedings of SPIE* 572, 123–32, 1985.

22. J. Piotrowski and A. Rogalski, *High-Operating Temperature Infrared Photodetectors*, SPIE Press, Bellingham, WA, 2007.

23. S. Jensen, "Temperature Limitations to Infrared Detectors," *Proceedings of SPIE* 1308, 284–92, 1990.

24. R. G. Humpreys, "Radiative Lifetime in Semiconductors for Infrared Detectors," *Infrared Physics* 23, 171–75, 1983.

25. R. G. Humpreys, "Radiative Lifetime in Semiconductors for Infrared Detectors," *Infrared Physics* 26, 337–42, 1986.

26. A. Rogalski, "HgCdTe Infrared Detector Material: History, Status, and Outlook," *Reports on Progress in Physics* 68, 2267–336, 2005.

27. A. Rogalski, "Quantum Well Photoconductors in Infrared Detectors Technology," *Journal of Applied Physics* 93, 4355–91, 2003.

28. D. Long, "Photovoltaic and Photoconductive Infrared Detectors," in *Optical and Infrared Detectors*, ed. R. J. Keyes, 101–47, Springer-Verlag, Berlin, 1977.

29. C. T. Elliott and N. T. Gordon, "Infrared Detectors," in *Handbook on Semiconductors*, Vol. 4, ed. C. Hilsum, 841–936, North-Holland, Amsterdam, 1993.

30. C. T. Elliott, "Photoconductive and Non-Equilibrium Devices in HgCdTe and Related Alloys," in *Infrared Detectors and Emitters: Materials and Devices*, ed. P. Capper and C. T. Elliott, 279–312, Kluwer Academic Publishers, Boston, MA, 2001.

31. M. A. Kinch, "Fundamental Physics of Infrared Detector Materials," *Journal of Electronic Materials* 29, 809–17, 2000.

32. M. S. Ünlü and S. Strite, "Resonant Cavity Enhanced Photonic Devices," *Journal of Applied Physics* 78, 607–39, 1995.

33. E. Rosencher and R. Haidar, "Theory of Resonant Cavity-Enhanced Detection Applied to Thermal Infrared Light," *IEEE Journal of Quantum Electronics* 43, 572–79, 2007.

34. J. Kaniewski, J. Muszalski, and J. Piotrowski, "Resonant Microcavity Enhanced Infrared Photodetectors," *Optica Applicata* 37, 405–13, 2007.

35. J. Piotrowski, W. Galus, and M. Grudzień, "Near Room-Temperature IR Photodetectors," *Infrared Physics* 31, 1–48, 1990.

36. A. Rogalski and J. Piotrowski, "Intrinsic Infrared Detectors," *Progress in Quantum Electronics* 12, 87–289, 1988.

37. R. C. Jones, "Immersed Radiation Detectors," *Applied Optics* 1, 607–13, 1962.

38. M. Grudzień and J. Piotrowski, "Monolithic Optically Immersed HgCdTe IR Detectors," *Infrared Physics* 29, 251–53, 1989.

39. J. Piotrowski, H. Mucha, Z. Orman, J. Pawluczyk, J. Ratajczak, and J. Kaniewski, "Refractive GaAs Microlenses Monolithically Integrated with InGaAs and HgCdTe Photodetectors," *Proceedings of SPIE* 5074, 918–25, 2003.

40. T. T. Piotrowski, A. Piotrowska, E. Kaminska, M. Piskorski, E. Papis, K. Gołaszewska, J. Kątcki, et al., "Design and Fabrication of GaSb/InGaAsSb/AlGaAsSb Mid-Infrared Photodetectors," *Opto-Electronics Review* 9, 188–94, 2001.

41. T Ashley, D. T. Dutton, C. T. Elliott, N. T. Gordon, and T. J. Phillips, "Optical Concentrators for Light Emitting Diodes," *Proceedings of SPIE* 3289, 43–50, 1998.

42. G. R. Nash, N. T. Gordon, D. J. Hall, M. K. Ashby, J. C. Little, G. Masterton, J. E. Hails, et al., "Infrared Negative Luminescent Devices and Higher Operating Temperature Detectors," *Physica E* 20, 540–47, 2004.

43. M. K. Haigh, G. R. Nash, N. T. Gordon, J. Edwards, A. J. Hydes, D. J. Hall, A. Graham, et al., "Progress in Negative Luminescent $Hg_{1-x}Cd_xTe$ Diode Arrays," *Proceedings of SPIE* 5783, 376–83, 2005.

44. P. Martyniuk and A. Rogalski, "Quantum-Dot Infrared Photodetectors: Status and Outlook," *Progress in Quantum Electronics* 32, 89–120, 2008.

45. P. Martyniuk and A. Rogalski, "Insight into Performance of Quantum Dot Infrared Photodetectors," *Bulletin of the Polish Academy of Sciences: Technical Sciences* 57, 103–16, 2009.

46. A. Rogalski, K. Adamiec, and J. Rutkowski, *Narrow-Gap Semiconductor Photodiodes*, SPIE Press, Bellingham, WA, 2000.

47. M. A. Kinch, *Fundamentals of Infrared Detector Materials*, SPIE Press, Bellingham, WA, 2007.

48. H. Lim, S. Tsao, W. Zhang, and M. Razeghi, "High-Performance InAs Quantum-Dot Infrared Photoconductors Grown on InP Substrate Operating at Room Temperature," *Applied Physics Letters* 90, 131112, 2007.

49. D. L. Smith and C. Mailhiot, "Proposal for Strained Type II Superlattice Infrared Detectors," *Journal of Applied Physics* 62, 2545–48, 1987.

50. C. H. Grein, H. Cruz, M. E. Flatte, and H. Ehrenreich, "Theoretical Performance of Very Long Wavelength $InAs/In_xGa_{1-x}Sb$ Superlattice Based Infrared Detectors," *Applied Physics Letters* 65, 2530–32, 1994.

51. C. H. Grein, P. M. Young, M. E. Flatté, and H. Ehrenreich, "Long Wavelength InAs/InGaSb Infrared Detectors: Optimization of Carrier Lifetimes," *Journal of Applied Physics* 78, 7143–52, 1995.

52. R. Ciupa and A. Rogalski, "Performance Limitations of Photon and Thermal Infrared Detectors," *Opto-Electronics Review* 5, 257–66, 1997.

53. P. W. Kruse, "A Comparison of the Limits to the Performance of Thermal and Photon Detector Imaging Arrays," *Infrared Physics & Technology* 36, 869–82, 1995.

54. J. Piotrowski and A. Rogalski, "Uncooled Long Wavelength Infrared Photon Detectors," *Infrared Physics & Technology* 46, 115–131, 2004.

55. M. White, "Auger Suppression and Negative Resistance in Low Gap Diode Structures," *Infrared Physics* 26, 317–24, 1986.

56. W. Van Roosbroeck, "Theory of the Electrons and Holes in Germanium and Other Semiconductors," *Bell Systems Technical Journal* 29, 560–607, 1950.

57. H. K. Gummel, "A Self-Consistent Iterative Scheme for One-Dimensional Steady State Transistor Calculations," *IEEE Transactions on Electron Devices* ED 11, 455–65, 1964.

58. A. De Mari, "An Accurate Numerical Steady-State One-Dimensional Solution of the p-n Junction," *Solid State Electronics* 11, 33–58, 1968.

59. M. Kurata, *Numerical Analysis of Semiconductor Devices*, Lexington Books, DC Heath, 1982.

60. S. D. Yoo, N. H. Jo, B. G. Ko, J. Chang, J. G. Park, and K. D. Kwack, "Numerical Simulations for HgCdTe Related Detectors," *Opto-Electronics Review* 7, 347–56, 1999.

61. K. Kosai, "Status and Application of HgCdTe Device Modeling," *J Electronic Materials* 24, 635–40, 1995.

62. K. Jóźwikowski, "Numerical Modeling of Fluctuation Phenomena in Semiconductor Devices," *Journal of Applied Physics* 90, 1318–27, 2001.

63. J. L. Meléndez and C.R. Helms, "Process Modeling and Simulation of $Hg_{1-x}Cd_xTe$. Part I: Status of Stanford University Mercury Cadmium Telluride Process Simulator," *Journal of Electronic Materials* 24, 565–71, 1995.

64. J. L. Meléndez and C. R. Helms, "Process Modeling and Simulation for $Hg_{1-x}Cd_xTe$. Part II: Self-Diffusion, Interdiffusion, and Fundamental Mechanisms of Point-Defect Interactions in $Hg_{1-x}Cd_xTe$," *Journal of Electronic Materials* 24, 573–79, 1995.

4 Heterodyne Detection

Most current IR detectors that have been considered in the previous chapters are used in the direct detector mode. In direct detection, the output electrical signal is linearly proportional to the signal power. But in terms of the amplitude of the electric field carried by radiation, photodetectors are quadratic. As such, no information regarding the phase of the signal is retained in the electrical output. When the radiation field at the detector is only that of the signal, we talk of direct detection (Figure 4.1a), by far the most usual case in applications.

A heterodyne detection, in contrast, produces an output signal proportional to the electric field strength of the signal and thus the phase of the optical field is preserved in the phase of the electrical signal. In the 1960s the laser made it possible to generate intense coherent beams of light for the first time, similar to those developed at radio frequencies many years earlier. It was first demonstrated that the heterodyne technique worked in the optical region as it did in the radio region. The main virtues of this method detection are higher sensitivity, higher and more easily obtained selectivity, plus the possibility of detection of all types of modulation and easier tuning over wide range. Very weak signals can be detected because mixing the local field with the incoming photons allows the signal to be amplified. However, the main advantage of the technique is that the signals are downconverted to frequencies where extremely low noise electronics can be used to amplify them. Heterodyne receivers are the only detection systems that can offer high spectral resolution ($v/\Delta v > 10^6$, where v is the frequency) combined with high sensitivity. Coherent optical detection has been developed since 1962, but compact and stable production of this system is more difficult and the system is more expensive and troublesome than its radio-technique equivalent. At present, coherent receivers monopolize radio applications; they are not used as widely at infrared or optical frequencies because of their narrow spectral bandwidths, small fields of view and inability to be constructed in simple large-format arrays.

Infrared heterodyne detection has been commonly used for many years in commercial and domestic radio receivers and also for the microwave range of an electromagnetic spectrum. This technique is used for construction of Doppler velocimeters, laser range finders, spectroscopy (particular LIDAR systems), and telecommunication systems. The past 20 years have seen a revolution in terahertz (THz) systems, as advanced materials research provided new and high-power sources, and the potential of THz for advanced physics research and commercial applications was demonstrated. High-resolution heterodyne spectroscopy in the THz region is an important technique for the investigation of the chemical composition, the evolution, and the dynamic behavior of astronomical objects, and in the investigation of the earth's atmosphere. More recently, THz heterodyne receivers have been applied to imaging in biomedicine and security. The THz spectral range is taken to be 0.3–10 THz, which roughly coincides with the somewhat older definitions of the sub-millimeter and far-infrared spectral range.

Infrared heterodyne detection is analogous to millimeter wave techniques. In heterodyne detection, a coherent optical signal beam of photon flux Φ_s is mixed with a laser local oscillator beam of photon flux Φ_{LO} at the input of a detector as shown schematically in Figure 4.1b. Two optical beams are well collimated and are aligned so that their wave fronts are parallel. Thus the weaker coherent signal, which is collinear with the local oscillator and absorbed within the detector is mixed with the local oscillator, producing the photocurrent generated at the intermediate or difference frequency $\omega_{if} = |\omega_{LO} - \omega_s|$ [1–4]

$$I_{ph} = I_{LO} + I_s + 2\eta(\omega_{if})q(\Phi_{LO}\Phi_s)^{1/2} A\cos(\omega_{if}t), \tag{4.1}$$

where I_{LO} and I_s are the DC photocurrents due to Φ_{LO} and Φ_s, respectively. For photodiode:

$$I_{LO} = \eta(0)q\Phi_{LO}A$$
$$I_S = \eta(0)q\Phi_S A$$

The DC quantum efficiency $\eta(0)$ and the AC quantum efficiency $\eta(\omega_{if})$ govern the DC photocurrent ($I_{LO} + I_s$) and the modulated photocurrent, respectively. The values of $\eta(0)$ and $\eta(\omega_{if})$ are different when the frequency response is limited by carrier diffusion to the junction [4].

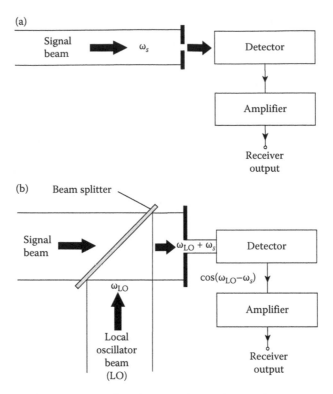

Figure 4.1   The generalized system of (a) direct and (b) coherent detection.

The rms value of the heterodyne signal current $I_H(\omega_{if})$ obtained from Equation 4.1 is equal to

$$I_H(\omega_{if}) = \eta(\omega_{if})q(2\Phi_{LO}\Phi_S)^{1/2}A = \frac{\eta(\omega_{if})q}{h\upsilon}(2P_{LO}P_S)^{1/2} = R_i\left(\frac{2P_{LO}}{P_S}\right)^{1/2},\qquad(4.2)$$

where P_{LO} is the local-oscillator radiation power, P_s is the signal radiation power, and R_i is the current responsivity as measured for direct detection. The above Equation 4.2 indicates that the heterodyne signal is $(2P_{LO}/P_s)^{1/2}$ times greater than in the case of direct detection. Much lower power levels can be detected in the heterodyne mode than in the direct detection mode, if $P_{LO}/P_s \gg 1$. It should be noticed, however, that in heterodyne detection the detector responds only at wavelengths very close to the local-oscillator wavelength (typically $\lambda_{LO} \pm < 0.003\ \mu m$) [5]. This makes them relatively insensitive to background radiation and interference effects.

The modulated photocurrent occurs if the intermediate frequency ω_{if} lies within the range covered by the frequency response of the photomixer. Typically, this frequency response is limited to frequencies less than approximately 1 THz (the shortest response times of photon detectors are in the picosecond range). This implies that the wavelengths of the two coherent sources in indirect detection are nearly equal.

The figure of merit that characterizes the heterodyne detector sensitivity is the heterodyne noise equivalent power NEP_H. It is defined as the signal power P_s necessary to produce the signal-to-noise power ratio $(S/P)_P$ of unity. The noise current arises in the same manner as it did in the direct case. For sufficient LO power ($I_{LO} > I_s$) and reverse-biased photodiode $I_n^2 = 2qI_{LO}\Delta f$ where Δf, is the bandwidth of the intermediate frequency channel following the detector. This is the case of a well-designed heterodyne receiver in which the noise is dominated by local-oscillator induced shot or generation–recombination noise. In this case a final S/P_P is given by

$$\left(\frac{S}{N}\right)_P = \frac{I_H^2}{I_n^2} = \frac{\eta^2(\omega_{if})}{\eta(0)}\frac{P_S}{h\upsilon\Delta f}$$

and then

$$\text{NEP}_H = \frac{\eta(0)}{\eta^2(\omega_{if})} h\nu\Delta f. \tag{4.3}$$

We can see that in heterodyne detection, sensitivity of the photodiode is reduced by a factor $\eta(\omega_{if})/\eta(0)$ from the value $\eta(\omega_{if})P_s/h\nu\Delta f$ obtained when carrier diffusion to the junction is ignored. It should be noticed that the sensitivity degradation factor $\eta(\omega_{if})/\eta(0)$ is common to both heterodyne detection and BLIP direct detection [4].

For the case of a photoconductor heterodyne receiver, the results described above are not directly applicable. In the limit of large local-oscillator power, photoconductors display generation–recombination noise induced by local-oscillator signal fluctuation. For the photoconductor detectors [1]

$$\text{NEP}_H = \frac{2h\nu}{\eta(\omega_{if})} \Delta f. \tag{4.4}$$

These devices are less sensitive by a factor of $2/\eta$ than the perfect quantum counter.

A simple and more general expression for the NEP_H can be obtained assuming that the responsivity and internal noise of the detector are unaffected by the presence of the local oscillator. Then

$$\text{NEP}_H = \left[\frac{(\text{NEP}_D)^2}{2P_{LO}} + \frac{h\nu}{\eta(\omega_{if})}\right]\Delta f, \tag{4.5}$$

where NEP_D is the noise equivalent power for direct detection.

Practical application of heterodyne detection is most successful on the infrared region due to such factors as: sensitivity improvement due to reduced quantum noise, easier alignment of signal and local oscillator, and greater diffraction-limited acceptance angle for a given aperture size [1,5,6]. In this wavelength region strong and stable 10.6 μm line local oscillators exist. Moreover, in contrast to thermal imaging, extremely low dark detector current is not needed for good coherent detection. The main requirement for good heterodyne detection sensitivity is high quantum efficiency at IF and at the LO power level needed for short-noise-limited operation. Achievement of short-noise-limited operation and high frequency at either terahertz frequencies or at high operating temperatures have been the challenges for high-performance photomixers.

The basic block diagram of heterodyne optical receiver is shown in Figure 4.2. Signal radiation containing information is, after being passed through an input optical filter and beam splitter, arranged to coherently combine or "mix" with a light beam of a local oscillator at the detector

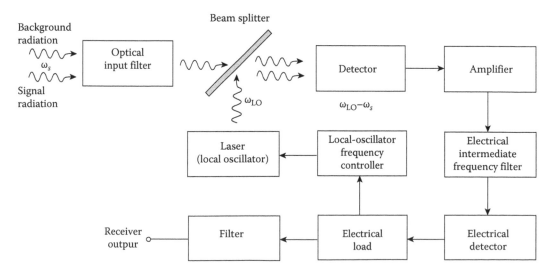

Figure 4.2 Block diagram of heterodyne detection optical receiver.

surface. A beam-splitter can be made in many ways, the simplest being a glass plate with adequate refraction coefficient. In a general case, a device fulfilling such a role is called a direction coupler, as an analogy to microwave or radio devices. A detector used for signal mixing has to have a square-law characteristic for detecting the electronic field of the light, but this is conveniently typical of most optical detectors (photodiode, photoconductor, photomultiplier, avalanche photodiode, etc.). This signal is next amplified. An electrical filter of IF extracts the desired difference component of the signal which next undergoes a demodulation process. The design and operation principle of the subsequent electrical detector depends on the nature of the modulation of the signal. The signal from a load resistance passes through an output filter to a receiver output and by means of a local-oscillator frequency controller it controls a laser. A frequency control loop is used for the local-oscillator laser to maintain a constant frequency difference $\omega_{LO} - \omega_s = \omega_{if}$ with the input signal. An indispensable condition for efficient coherent detection is to match the polarization, and to the shape of both waveforms of both beams to match the profile of the detector surface.

Early work on 10.6 μm photomixers centered on the liquid-helium-cooled copper-doped Ge photoconductors, HgCdTe, and lead salt photodiodes [1,5–9]. However, as HgCdTe photodiodes were developed they quickly replaced the Ge photoconductor, which have very high LO power requirements and a factor-of-2 lower sensitivity, and lead salt photodiodes, which are slow due to a very large dielectric constant of the material [10,11].

Heterodyne detection systems with 1 GHz base-bandwidth and sensitivities that approach the ideal limits were reported during the 1960s [1,7]. Figure 4.3 shows the experimental results of the heterodyne signal-to-noise ratio for Ge:Cu doped operated 4.2 K [2]. The field circles present the observed signal-to-noise power ratio data points, $(S/N)_P$ as a function of the signal beam radiation power P_s. Only noise arising from the presence of the LO beam (which was the dominant contribution to the noise) is considered. A plot of the theoretically expected result $(S/N)_P = \eta P_s / 2h\nu\Delta f$ is also shown. Using an estimated quantum efficiency $\eta = 0.5$, it is seen to be in good agreement with the experimental data. With a heterodyne signal centered at about 70 kHz, and an amplifier bandwidth of 270 kHz, the experimentally observed NEP_H was seen to be 7×10^{-20} W, which is to be compared with the expected value $(2/\eta)h\nu\Delta f \approx 7.6 \times 10^{-20}$ W. Usually detectors have higher noise and require more local-oscillator power that can overpower the cooler or heat the detector, resulting in a NEP_H less than theoretical.

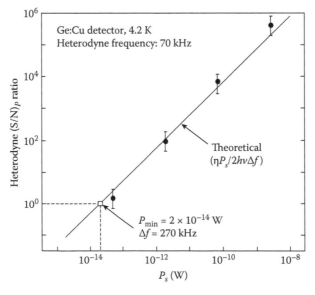

Figure 4.3 The heterodyne signal-to-noise ratio for Ge:Cu doped photoconductor at 4.2 K. The filled circles represent the observed signal-to-noise power ratio data points, the solid line presents the theoretical result. (From Teich, M. C., *Proceedings of IEEE* 56, 37–46, 1968. With permission.)

The frequency spectrum of the shot noise reveals the rolloff into the external circuit (e.g., RC) but is not a good indicator of the bandwidth of the photodiode, because in addition to circuit effects the frequency response of a photodiode depends on the time required for carriers to diffuse and transit across the space-charge region. It is easy to design an optimum heterostructure with transparent wider-energy-gap layers to minimize carrier diffusion, but the p-i-n diode structure needed for low capacitance has been feasible with HgCdTe epitaxial layer. Most 10 μm HgCdTe photomixers have been the n-n⁻-p homojunction photodiodes illuminated from the n-type side. Since the active area of a photomixer is defined by the incident LO pattern, the detector area involves a trade-off between accommodating the LO and junction capacitance. Low-capacitance n-n⁻-p etched mesa and planar HgCdTe photodiodes have been fabricated by Hg diffusion into p-type material. Impurity diffusion and ion implantation have not been as successful in making gigahertz-bandwidth photomixers. Wideband HgCdTe photomixers are operated at reverse biases in the range 0.5–2 V. Operation in this mode results in lower junction capacitance, which facilitate good impedance matching, and good coupling to the first stage preamplifier (preferably cooled together with the detector). The diffused junction with diameters of 100–200 μm typically have capacitances in the range of 1–5 pF and RC rolloff frequencies of 0.5–3 GHz [8,11]. The largest wide band arrays developed for target tracking were 12-element arrays of 1.5-GHz photomixers in the configuration of a central quadrant array surrounded by eight additional detectors.

Figure 4.4 shows the dependence of heterodyne NEP as a function of local-oscillator power [12]. The solid lines are calculated from Equation 4.2 for two different values of NEP_D and quantum efficiency η = 50%. It can be seen that a small value of NEP_D allows use of a low P_{LO} to approach the quantum limit. With a detector exhibiting poor NEP_D, good performance can still be achieved provided the necessary increase in P_{LO} does not affect the internal responsivity and noise processes through increased carrier density or heating effects.

Figure 4.5 shows the best values of NEP_H at 10.6 μm obtained at Lincoln Laboratory and at Honeywell as a function of frequency for HgCdTe photodiodes at 77 K [13]. At 1 GHz, NEP_H is only

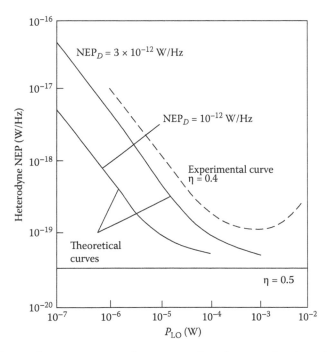

Figure 4.4 The heterodyne noise equivalent power plotted against local-oscillator power. The solid lines are calculated from Equation 9.2 for two different values of noise equivalent power in direct detection and quantum efficiency η = 0.5. The dashed line is a typical curve for experimental performance of HgCdTe performance. (From Wilson, D. J., Constant, G. D. J., Foord, R., and Vaughan, J. M., *Infrared Physics*, 31, 109–15, 1991. With permission.)

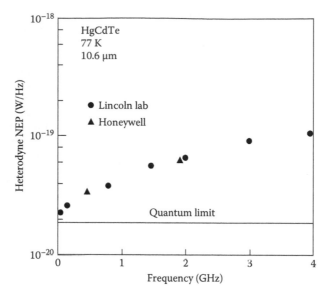

Figure 4.5 Heterodyne NEP$_H$ as a function of *if* frequency. (From Spears, D. L., *Optical and Laser Remote Sensing*, Springer-Verlag, Berlin, 278–86, 1983. With permission.)

Table 4.1: Examples of HgCdTe Heterodyne Photodiodes

x	A $(10^{-4}$ cm^2)	λ_c (μm)	λ_{LO} (μm)	$\eta(0)$ (%)	V (mV)	T (K)	P_{LO} (mW)	Δf (GHz)	NEP $(10^{-19}$ W/Hz)
0.19	1	12.5–14.5		40–60	–500		0.5	1.4	
0.2		10.7–12.5	10.6			77		> 2.0	0.43 (1GHz)
									0.62 (1.8GHz)
									1.1 (4GHz)
0.2	0.12	12		70	–1100	77		3–4	
	1.8		10.6	21		170	0.5–1	0.023	8.0 (10GHz)
			10.6		–800	77		0.85	1.0 (20MHz)
									1.65 (1.5GHz)
									3.0 (1.5GHz, 130K)

Source: Galeczki, G., *Properties of Narrow Gap Cadmium-Based Compounds*, INSPEC, London, 347–58, 1994.

a factor 2 above the theoretical quantum limit that for $\lambda = 10.6$ μm has the value of 1.9×10^{-20} W/Hz. The average sensitivities of many 12-element photodiode arrays, 4.3×10^{-20} W/Hz, have been close to this value [13].

Table 4.1 gives a summary of some of the heterodyne HgCdTe photodiodes detailed in the literature [14].

Most high frequency, 10 μm HgCdTe photomixers were operated at about 77 K, although they can work well at somewhat higher temperatures. At elevated temperatures, p-type photoconductors have higher sensitivities than the more common n-type photoconductors in both heterodyne and direct detection [12–17]. Moreover, for wide bandwidth photoconductors the LO power requirement can be considerably larger than that of a photodiode, due to low photoconductive gain. The dependence of HgCdTe photoconductor NEP on LO power (and bias power) has been successfully modeled by taking into account the effect of heating and bandfilling on carrier lifetime and optical absorption [13]. The best performance is obtained with small (50–100 μm square) devices, because the LO power requirement is proportional to the volume of the photoconductor, and heat sinking improves with reduced size. Figure 4.6 shows the calculated and measured NEP$_H$ as a function of LO power for a 100×100 μm^2 p-type photomixer at 77 and 195 K [13].

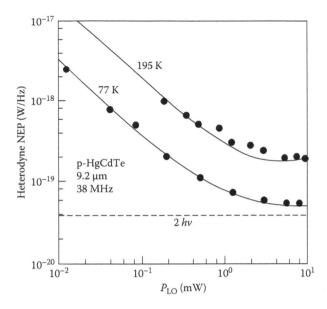

Figure 4.6 Calculated and measured NEP$_H$ versus LO power for p-type HgCdTe photoconductor at 77 and 195 K. (From Spears, D. L., *Optical and Laser Remote Sensing*, Springer-Verlag, Berlin, 278–86, 1983. With permission.)

Table 4.2: HgCdTe Photoconductive Heterodyne Mixer

Material	$x = 0.18$–0.19
Type	p-type, $N_a \approx 2 \times 10^{17}$ cm^{-3}
Surface	Native oxide passivated, ZnS AR coated
Detector temperature	193 K
Sensitive area	100×100 μm^2; $A_{opt} = 10^{-4}$ cm^2
Substrate	Sapphire heat sink
Response time	Down to a few ns
Bandwidths	Up to 100 MHz
Responsivity at 20 kHz	67 V/W
Detectivity at 20 kHz	2.7×10^8 cmHz$^{1/2}$W^{-1}
Minimum NEP at 193 K	2×10^{-19} W/Hz
Theoretical limit	4×10^{-20} W/Hz
P_{LO}	7 mW
λ_{LO}	10.6 μm

Source: Galeczki, G., *Properties of Narrow Gap Cadmium-Based Compounds*, INSPEC, London, 347–58, 1994.

At 77 K, LO-noise-limited operation is easily achieved, but at 195 K, maximum LO-induced noise is comparable to or less than amplifier and dark current g-r noise. At this high LO power level the quantum efficiency has dropped considerably below the low power value of over 70%. Table 4.2 summarizes the relevant information concerning HgCdTe photoconductive heterodyne detector operated at 193 K [14].

Infrared photomixers can be further optimized through modifications of the standard photoconductor geometry using an interdigitated-electrode structure or an immersion lens [17,18].

It should be noticed that wideband coherent detection at 10 μm has been reported for GaAs/AlGaAs quantum well IR photodetectors (QWIPs) [19,20]. In comparison with HgCdTe-based technology, GaAs/AlGaAs QWIPs offers several advantages: the higher electrical bandwidth, more robust and tolerant to high levels of LO power, and monolithically compatible with GaAs HEMT amplifiers. The heterodyne detection up to an IF of 82 GHz has been demonstrated [20]. Recently, THz photon detectors have been demonstrated using intersubband transitions in semiconductor quantum structure. The potential of the very fast time response of these detectors make them attractive for applications in THz heterodyne detection [21].

The most sensitive receivers at microwave, millimeterwave, and THz frequencies are based on the heterodyne principle. These mixer receivers can operate in different modes, depending on the configuration of the receiver and the nature of the measurement. The signal and image frequencies may be separated in the correlator, or the image may be removed by appropriate phase switching of pairs of local oscillators. The function of separating or dumping the image in the receiver is to remove some of the uncorrelated noise to improve the system sensitivity.

In single-sideband (SSB) operation, the receiver is configured so that, at the image sideband, the mixer is connected to a termination within the receiver. There is no external connection to the image frequency, and the complete receiver is functionally equivalent to an amplifier followed by a frequency converter.

In double-sideband (DSB) operation, on the other hand, the mixer is connected to the same input port at both upper and lower sidebands. The DSB receivers can be operated in two modes [22,23]:

- In SSB operation to measure narrow-band signals contained entirely within one sideband—for detection of such narrow-band signals, power collected in the image band of a DSB receiver degrades the measurement sensitivity.

- In DSB operation to measure broadband (or continuum) sources whose spectrum covers both sidebands—for continuum radiometry, the additional signal power collected in the image band of a DSB receiver improves the measurement sensitivity.

Heterodyne receivers can be described by a series of parameters, but the most encountered one is the receiver noise temperature

$$T = T_{\text{mixer}} + LT_{if}. \tag{4.6}$$

Here the insides indicate the noise contribution of the mixer and the intermediate first amplifier stage, respectively, and L is the mixer conversion loss. More information about noise and its measurement in radio frequency equipment can be found in Hewlett Packard [22].

Generally, in terahertz receivers, the noise of a mixer is quoted in terms of a single-sideband (SSB), T^{SSB}, or double-sideband (DSB), T^{DSB}, mixer noise temperature. The quantum-noise limit for a SSB system noise temperature is [23,24]

$$T^{\text{SSB}} = \frac{h\nu}{k}. \tag{4.7}$$

This is the system noise temperature of a broad mixer receiver when performing narrowband measurements (within a single sideband). If we instead perform broadband (continuum) measurements, the desired signal will be twice as large and the ideal system noise temperature will be [23,24]

$$T^{\text{DSB}} = \frac{h\nu}{2k}. \tag{4.8}$$

The noise for a direct detector is normally quoted in terms of a noise equivalent power, or NEP. To convert between an NEP and a T^{SSB}, one can use the relationship [25]

$$T^{\text{SSB}} = \frac{\text{NEP}^2}{2\alpha k P_{\text{LO}}}, \tag{4.9}$$

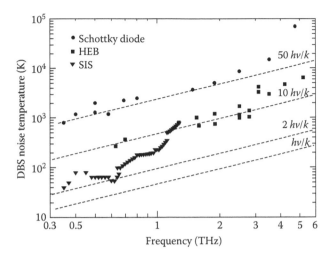

Figure 4.7 DSB noise temperature of Schottky diode mixers, SIS mixers, and HEB mixers operated in terahertz spectral band. (From Hübers, H.-W., "Terahertz Heterodyne Receivers," *IEEE Journal of Selected Topics in Quantum Electronics* 14, 378–91, 2008. With permission.)

where P_{LO} is the incident LO power and α is the coupling factor of the radiation to the mixer. A similar relationship can be found for T^{DSB}

$$T^{DSB} = \frac{NEP^2}{4\alpha k P_{LO}}. \tag{4.10}$$

The technology that has traditionally been available for terahertz receivers utilizes Schottky-barrier diode mixers pumped by gas laser local oscillators. The DSB noise temperatures achieved with Schottky diode mixers are presented in Figure 4.7 [26]. The noise temperature of such receivers has essentially reached a limit of about 50 $h\nu/k$ in frequency range below 3 THz. Above 3 THz, there occurs a steep increase, mainly due to increasing losses of the antenna and reduced performance of the diode itself.

In the last two decades an impressive improvement in receiver sensitivities has been achieved using superconducting mixers, with both superconductor–insulator–superconductor (SIS) and hot electron bolometer (HEB) mixers. In Figure 4.7 selected receiver noise temperatures are plotted. Nb-based SIS mixers yield almost quantum limited performance up to gap frequency of 0.7 THz.

Unlike Schottky diodes and SIS mixers, the HEB mixer is a thermal detector. The overall time constant of the process involved in the mixing is about few tens of picoseconds maximum, so bolometric process is fast to follow the IF, but slow to respond directly to the incident LO or signal field. Figure 4.7 shows that DSB noise temperatures achieved with HEB mixers range from 400 K at 600 GHz up to 6800 K at 5.2 THz. Up to 2.5 THz, the noise temperature follows the 10 $h\nu/k$ line closely. In comparison with Schottky-barrier technology, HEB mixers require three to four orders of magnitude less LO power.

REFERENCES

1. M. C. Teich, "Coherent Detection in the Infrared," in *Semiconductors and Semimetals*, Vol. 5, ed. R. K. Willardson and A. C. Beer, 361–407, Academic Press, New York, 1970.

2. M. C. Teich, "Infrared Heterodyne Detection," *Proceedings of IEEE* 56, 37–46, 1968.

3. R. H. Kingston, *Detection of Optical and Infrared Radiation*, Springer-Verlag, Berlin, 1979.

4. D. L. Spears and R. H. Kingston, "Anomalous Noise Behavior in Wide-Bandwidth Photodiodes in Heterodyne and Background-Limited Operation," *Applied Physics Letters* 34, 589–90, 1979.

5. R. J. Keyes and T. M. Quist, "Low-Level Coherent and Incoherent Detection in the Infrared," in *Semiconductors and Semimetals,* Vol. 5, ed. R. K. Willardson and A. C. Beer, 321–59, Academic Press, New York, 1970.

6. F. R. Arams, E. W. Sard, B. J. Peyton, and F. P. Pace, "Infrared Heterodyne Detection with Gigahertz IF Response," in *Semiconductors and Semimetals,* Vol. 5, eds. R. K. Willardson and A. C. Beer, 409–34, Academic Press, New York, 1970.

7. F. R. Arams, E. W. Sard, B. J. Peyton, and F. P. Pace, "Infrared 10.6-Micron Heterodyne Detection with Gigahertz IF Capability," *IEEE Journal of Quantum Electronics* QE-3, 484–92, 1967.

8. C. Verie and M. Sirieix, "Gigahertz Cutoff Frequency Capabilities of CdHgTe Photovoltaic Detectors at 10.6 μm," *IEEE Journal of Quantum Electronics* QE-8, 180–91, 1972.

9. A. M. Andrews, J. A. Higgins, J. T. Longo, E. R. Gertner, and J. G. Pasko, "High-Speed $Pb_{1-x}Sn_xTe$ Photodiodes," *Applied Physics Letters* 21, 285–87, 1972.

10. D. J. Wilson, R. Foord, and G. D. J. Constant, "Operation of an Intermediate Temperature Detector in a 10.6 μm Heterodyne Rangefinder," *Proceedings of SPIE* 663, 155–58, 1986.

11. I. Melngailis, W. E. Keicher, C. Freed, S. Marcus, B. E. Edwards, A. Sanchez, T. Yee, and D. L. Spears, "Laser Radar Component Technology," *Proceedings of IEEE* 84, 227–67, 1996.

12. D. J. Wilson, G. D. J. Constant, R. Foord, and J. M. Vaughan, "Detector Performance Studies for CO_2 Laser Heterodyne Systems," *Infrared Physics* 31, 109–15, 1991.

13. D. L. Spears, "IR Detectors: Heterodyne and Direct," in *Optical and Laser Remote Sensing,* eds. D. K. Killinger and A. Mooradian, 278–86, Springer-Verlag, Berlin, 1983.

14. G. Galeczki, "Heterodyne Detectors in HgCdTe," in *Properties of Narrow Gap Cadmium-Based Compounds,* ed. P. Capper, 347–58, INSPEC, London, 1994.

15. D. L. Spears, "Theory and Status of High Performance Heterodyne Detectors," *Proceedings of SPIE* 300, 174, 1981.

16. W. Galus and F. S. Perry, "High-Speed Room-Temperature HgCdTe CO_2-Laser Detectors," *Laser Focus/Electro-Optics* 11, 76–79, 1984.

17. J. Piotrowski, W. Galus, and M. Grudzień, "Near Room-Temperature IR Photo-Detectors," *Infrared Physics* 31, 1–48, 1991.

18. T. Kostiuk and D. L. Spears, "30 μm Heterodyne Receiver," *International Journal of Infrared Millimeter Waves,* 8, 1269–79, 1987.

19. E. R. Brown, K. A. McIntosh, F. W. Smith, and M. J. Manfra, "Coherent Detection with a GaAs/AlGaAs Multiple Quantum Well Structure," *Applied Physics Letters* 62, 1513–15, 1993.

20. H. C. Liu, J. Li, E. R. Brown, K. A. McIntosh, K. B. Nichols, and M. J. Manfra, "Quantum Well Intersubband Heterodyne Infrared Detection Up to 82 GHz," *Applied Physics Letters* 67, 1594–96, 1995.

21. H. C. Liu, H. Luo, C. Song, Z. R. Wasilewski, A. J. SpringThorpe, and J. C. Cao, "Terahertz Quantum Well Photodetectors," *IEEE Journal of Selected Topics in Quantum Electronics* 14, 374–77, 2008.

22. *Fundamentals of RF and Microwave Noise Figure Measurements,* Application Note 57-1, Hewlett Packard, July 1983.

23. A. R. Kerr, M. J. Feldman, and S.-K. Pan, "Receiver Noise Temperature, the Quantum Noise Limit, and the Role of the Zero-Point Fluctuations," in *Proceedings of the 8th International Space Terahertz Technology Symposium* March 25–27, pp. 101–11, 1997. http://colobus.aoc.nrao.edu/memos, as MMA Memo 161.

24. E. L. Kollberg and K. S. Yngvesson, "Quantum-Noise Theory for Terahertz Hot Electron Bolometer Mixers," *IEEE Transactions of Microwave Theory Technology* 54, 2077–89, 2006.

25. B. S. Karasik and A. I. Elantiev, "Noise Temperature Limit of a Superconducting Hot-Electron Bolometer Mixer," *Applied Physics Letters* 68, 853–55, 1996.

26. H.-W. Hübers, "Terahertz Heterodyne Receivers," *IEEE Journal of Selected Topics in Quantum Electronics* 14, 378–91, 2008.

PART II
INFRARED THERMAL DETECTORS

5 Thermopiles

The thermocouple was discovered in 1821 by the Russian-born German physicist, J. Seebeck [1]. He discovered that at the junction of two dissimilar conductors a voltage could be generated by a change in temperature (see Figure 5.1). Using this effect, Melloni produced the first bismuth-cooper thermocouple detector in 1833 [2], to investigate the infrared spectrum. The small output voltage of thermocouples, of the order of some μV/K for metal thermocouples, prevented the measurements of very small temperature differences. Connecting several thermocouples in series, for the first time by Nobili in 1829, generates a higher and therefore measurable voltage.

The thermopile is one of the oldest IR detectors, and is a collection of thermocouples connected in series in order to achieve better temperature sensitivity. For a long time, thermopiles were slow, insensitive, bulky, and costly devices. But with developments in semiconductor technology, thermopiles can be optimized for specific applications. Recently, thanks to conventional complementary metal-oxide-semiconductor (CMOS) processes, the thermopile's on-chip circuitry technology has opened the door to mass production. Although thermopiles are not as sensitive as bolometers and pyroelectric detectors, they will replace these in many applications due to their reliable characteristics and good cost/performance ratio.

5.1 BASIC PRINCIPLE AND OPERATION OF THERMOPILES

The internal voltage responsible for current flow in a thermocouple is directly proportional to the temperature difference between the two junctions

$$\Delta V = \alpha_s \Delta T, \tag{5.1}$$

where α_s is the Seebeck coefficient commonly expressed in μV/K.

The coefficient α_s is the effective or relative Seebeck coefficient of the thermocouple composed of two dissimilar conductors "a" and "b" by electrically joining one set of their ends. Consequently, a thermovoltage is equal

$$\Delta V = \alpha_s \Delta T = (\alpha_a - \alpha_b)\Delta T, \tag{5.2}$$

where α_a and α_a are the absolute Seebeck coefficients of the material a and b. It should be noticed that both relative and absolute Seebeck coefficients are temperature dependent and the proportionality between generated potential difference and the temperature gradient is valid only within the limit of a small temperature difference.

The output voltage of a single thermocouple is usually not sufficient; therefore, a number of thermocouples are connected in series to form a so-cold thermopile. Figure 5.2 shows a thermopile that is constructed by a series connection of three thermocouples. The thermopile can be used as an infrared detector if the thermocouples are placed on a suspended dielectric layer and if an absorber layer is placed close to or on the top hot contacts of the thermopile. An important factor for obtaining a large output voltage from a thermopile is to obtain a high thermal isolation in order to maximize the temperature difference between hot cold junctions, ΔT, for a specific absorber power. Corresponding to Equation 5.2, the totalized output voltage will be N times as high as for a single element:

$$\Delta V = N(\alpha_a - \alpha_b)\Delta T, \tag{5.3}$$

with N as the number of joined thermocouples.

Apart from the Seebeck effect, yet another important thermoelectric offsets exist: the Peltier or Thomson effects. The latter two effects are present only when current flows in a closed thermoelectric circuit. The Peltier effect may give rise to considerable asymmetries in thermoelectric effect and special care due to this effect should be taken when designing thermal devices using heating resistors. This effect is reversible, since heat is absorbed or released depending on the direction of the current. The Thompson effect deals with similar heat exchange from a wire instead of a junction; when a wire has a temperature gradient along its length, a Thompson electromotive force is developed.

The Peltier coefficient Π (in V) that quantifies the ratio of heat absorption to electrical current, is equal to the Seebeck coefficient times the absolute temperature, called the first Kelvin relation

$$\Pi = \alpha_s T. \tag{5.4}$$

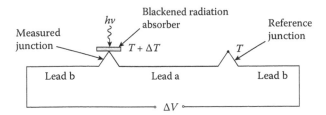

Figure 5.1 Two dissimilar leads connected in series.

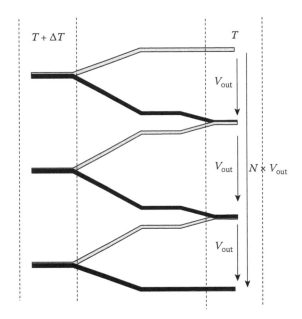

Figure 5.2 Schematic drawing of a three thermocouple thermopile.

A common value for the Peltier coefficient is 100–300 mV, which is significant if heating voltage, for instance, 1–3 V are used. In such cases a Peltier heat flow of 10% of the generated (irreversible) Joule heat will flow from one contact to the other.

   The literature describes a number of theoretical expressions for the Seebeck coefficient. Generally, and without going into thermodynamics, Pollock [3] states that the Seebeck effect does not arise as a result of the junction of dissimilar materials, nor is it directly affected by the Thomson or Peltier effects. Ashcroft and Mermin describe the Seebeck effect as a consequence of the mean electron velocity [4]. This assumption finally implicates a gradient of the electrochemical potential $\Phi(T)$ [in V] with respect of the temperature. Based on this, the output voltage of a thermocouple can be expressed as [5]

$$\Delta V = \left[\frac{d\Phi_a(T)}{dT} - \frac{d\Phi_b(T)}{dT} \right] \Delta T. \tag{5.5}$$

   A more descriptive explanation of the absolute Seebeck effect in extrinsic nondegenerate semiconductor has been given by van Herwaarden [6]. He identifies two contributions to the Seebeck effect:

■ Due to a change in the Fermi–Dirac distribution as the result of temperature gradient.

■ Due to change of the absolute value of the band edges being the result of an electric field induced by a net diffusion current and by phonon-drag currents.

A simplified expression for heavily doped silicon ($>10^{19}$ cm^{-3}) at room temperature is [7]

$$\alpha_n = -\frac{k}{q}\left(\ln\frac{N_c}{N_d}+4\right) \quad \text{for n-type silicon,}$$

(5.6)

$$\alpha_p = \frac{k}{q}\left(\ln\frac{N_v}{N_a}+4\right) \quad \text{for p-type silicon.}$$

(5.7)

where N_c and N_v are the density of states in the conduction and valence bands, respectively, N_d is the donor concentration in n-type silicon, N_a is the acceptor concentration in p-type silicon. Here, α_s has a positive sign for p-type silicon, whereas a negative sign is selected for n-type silicon. Further considerations of the Seebeck effect in semiconductors can be found in Graf and colleagues [8].

The Seebeck effect as it has been treated here is valid for bulk materials. Influence of other effects, such as grain size and grain boundary, can be found in thin film structure. A description of effects in very thin films has been given by Salvadori et al. [9].

Table 5.1 lists the parameters for selected thermoelectric materials [8,10]. The bismuth/antimony (Bi/Sb) thermocouple is the most classical material pair in conventional thermocouples and not only from a historical point of view [11]. The Bi/Sb also has the highest Seebeck coefficient and

Table 5.1: Parameters of Selected Thermoelectric Materials at Near-Room Temperature

Sample	α_a (μV/K)	Reference Electrode	ρ ($\mu\Omega$m)	G_{th} (W/mK)
p-Si	100–1000		10–500	$\approx$150
p-poly-Si	100–500		10–1000	$\approx$20–30
p-Ge	420	Pt		
Sb	48.9	Pt	18.5	0.39
Cr	21.8			
Fe	15		0.086	72.4
Ca	10.3			
Mo	5.6			
Au	1.94		0.023	314
Cu	1.83		0.0172	398
In	1.68			
Ag	1.51		0.016	418
W	0.9			
Pb	−1.0			
Al	−1.66		0.028	238
Pt	−5.28		0.0981	71
Pd	−10.7			
K	−13.7			
Co	−13.3	Pt	0.0557	69
Ni	−19.5		0.0614	60.5
Constantan	−37.25	Pt		
Bi	−73.4	Pt	1.1	8.1
n-Si	−450	Pt	10–500	$\approx$150
n-poly-Si	−100 to −500		10–1000	$\approx$20–30
n-Ge	−548	Pt		

Source: Graf, A., Arndt, M., Sauer, M., and Gerlach, G., *Measurement Science Technology*, 18, R59–R75, 2007 and Schieferdecker, J., Quad, R., Holzenkämpfer, E., and Schulze, M., *Sensors and Actuators A*, 46–47, 422–27, 1995. With permission.

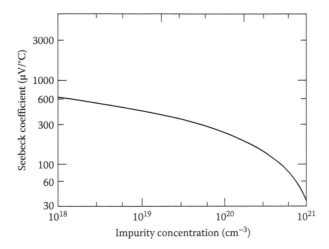

Figure 5.3 Seebeck coefficient for polysilicon. (From Kanno, T., Saga, M., Matsumoto, S., Uchida, M., Tsukamoto, N., Tanaka, A., Itoh, S., et al., "Uncooled Infrared Focal Plane Array Having 128 × 128 Thermopile Detector Elements," *Proceedings of SPIE* 2269, 450–59, 1994. With permission.)

the lowest thermal conductivity of all metal thermocouples. Despite the long tradition of metal thermocouples, new advantages can be found by using semiconducting materials, such as silicon (crystalline, polycrystalline) for thermoelectric materials due to the possibility of using standard integrated circuit processes. The Seebeck coefficient of semiconductor materials depends on the variation of the Fermi level of the semiconductor with respect to temperature; therefore, for semiconductor thermopiles, the magnitude and sign of the Seebeck coefficient and resistivity can be adjusted in the doping type and doping level.

For practical silicon sensor design-purposes it is very convenient to approximate the Seebeck coefficient as a function of electrical resistivity [12]:

$$\alpha_s = \frac{mk}{q} \ln\left(\frac{\rho}{\rho_o}\right) \qquad (5.8)$$

with $\rho_o \cong 5 \times 10^{-6}$ Ωm and $m \cong 2.6$ as constants. Typical values of the Seebeck coefficient of silicon are 500–700 µV/K for the optimum compromise between low resistance and high Seebeck coefficient. Equation 5.8 suggests that the Seebeck coefficient of a semiconductor increases in magnitude with increased resistivity, and therefore, with decreasing doping level. However, a thermopile material with very low electrical resistivity is not necessarily the best choice for a particular infrared detector, as the Seebeck coefficient is only one of the parameters influencing its overall performance.

For a wide variety of surface micromachined devices, polysilicon has rapidly become the most important material. Polysilicon's popularity in this area is a direct result of its mechanical properties and its relatively well-developed deposition and processing technologies. These characteristics along with the capability of utilizing established IC processing techniques make it a natural selection.

The measured Seebeck coefficient for polysilicon is shown in Figure 5.3 [13]. The Seebeck coefficient for p-type polysilicon and n-type polysilicon are almost the same, but the signs are opposite. The value of this coefficient greatly depends on impurity concentration. The used impurity concentrations are between 10^{19} and 10^{20} cm^{-3}.

5.2 FIGURES OF MERIT

In further discussion we follow considerations carried out in Chapter 3. Taking into account Equations 3.7 and 5.1 we notice that $K = \alpha_s$.

A consequence of Equation 3.9 is that the voltage responsivity is equal

$$R_v = \frac{\alpha R_{th} \varepsilon}{(1 + \omega^2 \tau_{th}^2)^{1/2}}. \qquad (5.9)$$

At very low frequencies, $\omega^2 \tau_{th}^2 \ll 1$, and then

$$R_v = \frac{\alpha \varepsilon}{G_{th}}. \tag{5.10}$$

Usually the thermocouple rms noise voltage is dominated in the frequency range 0.1–1000 Hz by the thermal noise of the thermocouple resistance R. Then, according to Equation 3.22

$$D^* = \frac{\alpha_s \varepsilon A_d^{1/2}}{G_{th}(4kTR)^{1/2}}. \tag{5.11}$$

If N thermocouples are placed in series, the responsivity is increased by N:

$$R_v = \frac{N\alpha \varepsilon}{G_{rh}(1 + \omega^2 \tau_{th}^2)^{1/2}}, \tag{5.12}$$

and then for the thermopile with dominant contribution of thermal noise, the detectivity can be expressed as

$$D^* = \frac{\alpha_s \varepsilon (NA_d)^{1/2}}{G_{th}(4kTR_e)^{1/2}}, \tag{5.13}$$

where R_e is the electrical resistance of each thermocouple in the thermopile.

To produce an efficient device, the junction thermal capacity C_{th} must be minimized, to give as short a response as possible (see Equation 3.5), and the absorption coefficient optimized, which is often achieved by blackening the sensor. By careful design it is possible for a thermocouple to be 99% efficient with a spectrally flat response from visible to beyond 40 µm. For further discussion of black absorbers see Blevin and Geist [14]. The spectral response is also determined by the material of encapsulation window.

The junction should be fabricated from two materials with:

- A large Seebeck coefficient α_s

- Low thermal conductivity G_{th} (to minimize the heat transfer between the hot and cold junctions)

- Low volume resistivity (to reduce the noise and heat developed by the flow of current)

Keep in mind that by scaling down a device by a certain percentage the surface area only decreases with the square root (see Equation 5.11), the miniaturization of the thermopile is an appropriate way to increase the overall detectivity.

Unfortunately, these requirements are incompatible in view of the Wiedemann–Franz law [15] relating the thermal conductivity G_{th} and the electrical resistivity ρ:

$$\frac{G_{th}\rho}{T} = L. \tag{5.14}$$

The L is known as the Lorentz number and has very nearly a constant value for most materials, especially metals, except at very low temperatures. This leads naturally to the well-known criterion of figure of merit for thermoelectric materials in which a maximum value of

$$Z = \frac{\alpha_s^2}{\rho G_{th}} \tag{5.15}$$

is sought [16]. It is important to note that this thermoelectric figure of merit is defined in terms of output power delivered into an optimum load resistance, rather than the open-circuit voltage that enters directly into the definition of the responsivity, Equation 5.9.

Thermoelectric materials criteria have been considered in detail in numerous papers and publications. Ioffe's development was based upon the influence of carrier concentration [17], and did not consider the influence of change in effective mass or mobility. These aspects are reviewed and summarized by Egli [18], Cadoff and Miller [16], and in many recently published papers [19–22].

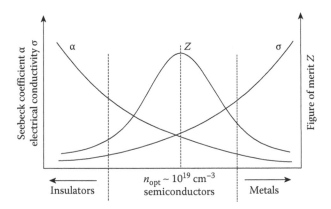

Figure 5.4 Thermoelectric properties of metals, semiconductors, and insulators. (From Voelklein, F., *Sensors and Materials*, 8, 389–408, 1996. With permission.)

Equation 5.11 indicates that the thermocouple materials should be chosen from low-resistivity materials. However, a lower ρ value also gives a lower Seebeck coefficient. Therefore, an optimum point needs to be determined considering all these parameters based on the figure of merit. Figure 5.4 shows a graph for the visual interpretation of the thermoelectric properties of metals, semiconductors, and insulators [20]. An optimum value for the figure of merit for semiconductors is achieved at doping value about 10^{19} cm^{-3} for single crystal silicon and polysilicon materials. A similar conclusion was predicted by Ioffe more than 50 years ago [17].

It should be noticed that when the number of thermocouples is increased to obtain a high output voltage, it also increases the thermal conduction between the hot and cold junctions and the series electrical resistance and thermal noise. This means that care should be taken to optimize the number of thermocouples for a thermopile. Increasing the number of thermocouples does not necessarily increase the performance.

Equation 5.15 is valid for a single thermocouple material. For a thermocouple constructed with two different materials, a and b, the figure of merit is defined as follows [16]

$$Z = \frac{\left(\alpha_a - \alpha_b\right)^2}{\left(\sqrt{\rho_a G_a} + \sqrt{\rho_b G_b}\right)^2},$$ (5.16)

where ρ_a and ρ_b are electrical resistivity of the materials, and G_a and G_b are the thermal conductivity of the materials, respectively.

The total thermal conductivity G_{th} contains all contributions of thermal conductivity between absorber and heat sink (including thermal conductivity of the surrounding gas G_g, the support and thermoelectric conductors G_s as well as the radiation losses G_R):

$$G_{th} = G_g + G_s + G_R + N\left(G_a + G_b\right).$$ (5.17)

The Z values of selected thermoelectric material pairs are listed in Table 5.2 [23].

5.3 THERMOELECTRIC MATERIALS

The first thermopiles were constructed from fine metallic wires, the most popular combinations being bismuth–silver, cooper–constantan, and bismuth–bismuth/tin alloy [24]. The two wires are joined to form the thermoelectric junction and a blackened receiver, usually a thin gold foil that defines the sensitive area is attached directly to the junction.

The development of semiconductors produced materials with much larger Seebeck coefficients, and hence the possibility of constructing thermopiles with increased sensitivities. However, the production of fine wire was impracticable. To make contacts, a new technique was developed in which the gold foil receiver was used as a constructing link between two active elements. The alloys recommended by Schwartz for this construction were (33% Te, 32% Ag, 27% Cu, 7% Se, 1% S) for the positive electrode and (50% Ag$_2$Se, 50% Ag$_2$S) for the negative material [15]. The responsivity

Table 5.2: Z Values of Thermoelectric Junction Pairs at Room Temperature

Junction Pair	Z (K^{-1})
Chromel/Constantan	1.0×10^{-4}
Al/n-polySi or p-polySi	1.1×10^{-5}
n-polySi/p-polySi	1.4×10^{-5}
Bi/Sb	1.8×10^{-4}
$Bi_{0.87}Sb_{0.13}$/Sb	7×10^{-4}
n-PbTe/p-PbTe	1.3×10^{-3}
n-Bi_2Te_3/p-Bi_2Te_3	2×10^{-3}

Source: Kruse, P. W., *Uncooled Thermal Imaging. Arrays, Systems, and Applications,* SPIE Press, Bellingham, WA, 2001. With permission.

of these devices was increased by about an order of magnitude (3×10^9 cm Hz$^{1/2}$W^{-1} had been achieved) if they were mounted in a vacuum filled with a gas that had a low thermal conductivity, for example, xenon. The response time of these devices, usually about 30 minutes, was reduced when the thickness of the deposited films was reduced. However, due to the resistance increase of the device, the Johnson noise increased.

Although the sensitivity of the older thermopiles using metals is much lower than those using semiconductor elements, the metal elements can be made more robust and stable so that they are still widely used where a high degree of reliability and of long-term stability is required. They have been successfully used in a number of space instruments, ground-based meteorological instruments, and in industrial radiation pyrometers [25,26].

Better quality thermopile infrared detectors have been generally realized using vacuum evaporation (bismuth and antimony) and shadow masking of the thermocouple materials on thin plastic or alumina substrates [15,24]. This approach resulted in relatively large structures that lack the batch fabrication and process flexibility typical of devices employing the highly developed silicon IC technology. In order to profit from this technology, thermopile detectors that did use silicon, but only as a supporting structure, were realized [27,28].

Table 5.1 shows the parameters for selected thermoelectric materials. Good electrical conductors (e.g., gold, copper, and silver) have very poor thermoelectric power. Metals with higher resistivity (especially antimony and bismuth) however, posses high thermoelectric power in combination with low thermal conductivity—they became the "classical" thermoelectrical materials. By doping these materials in combination with Se or Te, the thermoelectric coefficient has been improved up to 230 μV/K [11]. Fote and coworkers have improved the performance of thermopile linear arrays by combining Bi–Te and Bi–Sb–Te thermoelectric materials [29,30]. Compared with most other thermoelectric materials, their $D*$ values are highest (what is shown in Figure 5.5). However, Bi–Sb–Te materials are not readily available in a CMOS technology.

Silicon as a promising material for thermoelectric devices due to its high Seebeck coefficient was recognized as early as the 1950s [31,32]. However, early attempts to implement silicon into practical thermopile devices lacked definitive success primarily because the large number of couples, required to generate a meaningful output, made the devices excessively large. Silicon is a very good heat conductor, and the silicon substrate spoils the sensitivity with a thermal short-circuit. It turns out to be very rewarding to remove this silicon using planar technology [27] with its batch fabrication features and micromachining.

The last 20 years have seen significant advances in the development of integrated silicon thermopiles. The introduction of microfabrication facilitated practical solutions to miniaturization and consequently, a number of microfabricated thermopile embodiments have been presented in the literature with varying degrees of success. Early efforts included silicon in their design merely as a substrate, using thin film metals for the thermoelectric materials. For example, Lahiji and Wise fabricated a thermopile in which Bi/Sb pairs were evaporated onto a thin silicon membrane insulated with a chemical vapor deposition (CVD) oxide [28]. Later designs incorporated silicon in the active thermopile junctions. The influence of aluminum interconnections on chip commonly used in ICs is negligible compared to the Seebeck coefficient for silicon.

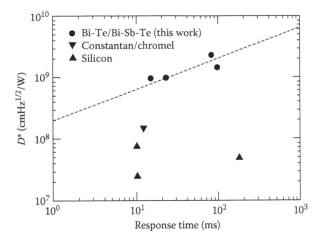

Figure 5.5 Representative data from literature showing reported D^* values as a function of response time for thin film thermopile linear arrays. The dashed line represents the Fote and Jones results. Its slope indicates D^* proportional to the square root of response time, which is typical for thermopiles or bolometers with different geometries and the same material system. (From Fote, M. C. and Jones, E. W., "High Performance Micromachined Thermopile Linear Arrays," *Proceedings of SPIE* 3379, 192–97, 1998. With permission.)

The advantages of semiconductors in thermopile fabrication result from following reasons:

- Semiconductors offer significantly higher Seebeck coefficients than metals.

- Semiconductor micromachining offers considerable miniaturization of devices that effectively reduce their thermal capacity.

- Production of high performance thermopiles is compatible with standard IC processes like CMOS.

The significant progress made in recent years with microsensors is due primarily to micromachining technology. Micromachining is the fabrication of small, robust structures with submicron precision using a combination of photolithography and selective etching. Micromachining is possible in many materials, but silicon is favored because many micromachining techniques are similar to silicon-processing techniques. Silicon also enables the incorporation of electronics monolithically linked with the microstructures. The most widely used approach is n-polySi/Al thermopiles. Although Al has a very low Seebeck coefficient, this approach is used widely, as it is easy to implement with post-CMOS processes. Very attractive is the p-polySi/n-polySi approach, as it provides relatively high Seebeck coefficient. The basic concepts of bulk and surface micromachining technology are discussed, for example, in Ristic's monograph [33]. For a wide variety of surface micromachined devices, polysilicon has rapidly become the most important material. Polysilicon's popularity in this area is a direct result of its mechanical properties and its relatively well-developed deposition and processing technologies. These characteristics along with the capability of utilizing established IC processing techniques make it a natural selection.

Table 5.3 presents some representative parameters of different micromachined CMOS thermopiles [8]. It should be mentioned that different alternative materials not included in Table 5.3 have been proposed. Dehe et al. [34] proposed AlGaAs/GaAs thermopile with high Seebeck coefficient in GaAs of about –670 µV/K, which results from the comparably high charge carrier mobility of 470 cm²/Vs (for comparison, polysilicon has a carrier mobility of 24 cm²/Vs). However, high thermal conductivity avoids the common usage of AlGaAs thermopiles. Also a concept for the realization of InGaAs/InP micromachined thermoelectric sensors has been presented [35]. Key futures of this material system are high thermal resistivity (0.09 Km/W) and high carrier mobility. This is combined with high Seebeck coefficient of 790 µV/K for p-type InGaAs and –450 µV/K for n-InGaAs.

Table 5.3: Characteristics Data of Different Micromachined Thermopiles

Sort	Area (mm²)	D^* (10⁷ Jones)	R (µV/W)	Material System	τ (ms)	α_t (µV/K)	Couples	Atm.	Data
CB	0.013	0.68	10	Al/poly		58	20		Sim
CB	0.77	1.5	25	Al/poly		58	200		Sim
CB	15.2	5		p-Si/Al	300	700	44	Air	Meas
MB	15.2	10	>10	p-Si/Al		700	44	Vac	Meas
MB	0.12	1.7	12	Al/poly	10	−63	4 × 10		Meas
MB	0.3	2	44	n,p poly	18	200	4 × 12		Meas
MB	0.15	2.4	72	n,p poly	10	200	4 × 12		Meas
MB	0.15	2.4	150	n,p poly	22	200	4 × 12	Kr	Meas
MB	0.12	1.74	12	Al/poly AMS	10	65	10	Air	Meas
MB	0.12	1.78	28	Al/poly AMS	20	65	2 × 24	Air	Meas
MB	0.42	4.4	11	InGaAs/InP				Air	Meas
M	0.42	71	184	InGaAs/InP				Vac	Meas
M	4	6	6	Bi/Sb	15	100	60		Meas
M	4	3.5	7	n-poly/Au	15		60		Sim
M	4	4.8	9.6	p-poly/Au	15		60		Sim
M	0.25	9.3	48	p-poly/Al	20		40		
M	3.28	13	12	p-poly/Al	50		68		
M	0.2	55	180	Bi/Sb	19	100	72	Air	Meas
M	0.2	88	290	Bi/Sb	35	100	72	Kr	Meas
M	0.2	52	340	$Bi_{0.50}Sb_{0.15}Te_{0.35}/Bi_{0.87}Sb_{0.13}$	25	330	72	Air	Meas
C	0.2	77	500	$Bi_{0.50}Sb_{0.15}Te_{0.35}/Bi_{0.87}Sb_{0.13}$	44	330	72	Kr	Meas
C	9	26	14.8	Bi/Sb	100		72	Ar	Meas
C	0.785	29	23.5	Bi/Sb	32		15	Ar	Meas
C	0.06	25	194	Si	12		20	Ar	Meas
C	0.37	5.6	36	CMOS	< 6				Meas
C	1.44	8.7	27	CMOS	< 6				Meas
C	0.37	5.6	36	CMOS	< 6				Meas
C	1.44	4.6	12	CMOS	30				Meas
C	0.2	45	200		20		72	N₂	Meas
C	1.44	35	100		30		200	N₂	Meas
C	0.49	21	110	BiSb/NiCr	40		100		Meas
C	0.49	6	35	CMOS	25				Meas
C	1.44	8	20	CMOS	35				Meas
C	0.6	24	80	CMOS	<40			Vac	Meas

Source: Graf, A., Arndt, M., Sauer, M., and Gerlach, G., *Measurement Science Technology* 18, R59–R75, 2007. With permission.

Note: CB: cantilever beam thermopile, MB: micro bridge thermopile, M: membrane thermopile, C: commercial thermopile, Sim: simulated data, Meas: measured data.

5.4 MICROMACHINED THERMOPILES

Thermopile infrared detectors can easily be integrated on standard CMOS chips in several structures by adaptational post-CMOS surface or the bulk micromachining process. The increase in production capacity in recent years is due to a very good compromise between cost-efficiency and performance. Baltes et al. [36] have made a systematic assessment of the compatibility of thermopile fabrication with the CMOS technology, including the issue of materials the thermopile absorbers.

A schematic drawing of a modern thermopile structure is shown in Figure 5.6a [8]. It consists of series thermocouples supported by a micromachined isolating membrane. From a technological and economical point of view, the common industrial device today is the CMOS thermopile, which consists of Al/Si thermocouples. The hot junctions of the thermopile located on a membrane is

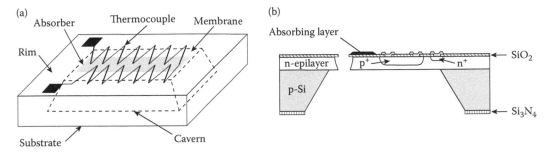

Figure 5.6 Micromachined thermopile: (a) general drawing and (b) schematic cross section.

covered with an absorbing layer. In contrast, the cold junctions lie on the substrate rim, which acts as heat sink. Usually the link between the hot and cold region is formed by a very thin (several micron thick) silicon membrane (Figure 5.6b) or cantilever beam, which contains the thermopile. The heat generated in the hot region flows through the silicon membrane (or cantilever beam) to the cold region, usually at ambient temperature, causing a temperature difference across the thermal resistance of the membrane (or cantilever beam).

5.4.1 Design Optimization

An accurate analysis of micromachined thermopile is complicated due to complex device architecture and influence of 3-D heat flow effects on device performance. For example, to achieve a maximum of detectivity for a thermopile on a membrane, two conditions should be fulfilled [23]. The first is

$$\frac{A_a}{A_b} = \left(\frac{G_{tha}\rho_a}{G_{thb}\rho_b}\right)^{1/2},$$

(5.18)

where A_a is the cross-section area of material a, which is deposited upon a leg connecting the membrane to the substrate, and A_b is the corresponding value for material b. The G_{tha} and G_{thb} are the thermal conductivities of material a and b, respectively.

The second condition is

$$N\left(G_{tha}A_a + G_{thb}A_b\right) = 2G_{thm}tw,$$

(5.19)

where G_{thm} is the thermal conductivity of the membrane material, and t and w are the thickness and width of the legs; it is assumed that there are two identical legs. Equation 5.19 states that the thermal conductance to the substrate through the thermoelectric lines is equal to that through the membrane material.

Assuming the above conditions, detectivity reaches its maximum value given by [23]

$$D^* = \frac{Z^{1/2}\tau_{th}^{1/2}\beta\kappa}{2^{3/2}\left(C_m t + C_{abs}\right)^{1/2}\left(kT\right)^{1/2}},$$

(5.20)

where C_m is the heat capacity per unit volume of the membrane material, and C_{abs} is the heat capacity per unit surface area of the absorbing material. Also, β is a detector fill factor, and κ is an optical absorption coefficient defined as the fraction of the radiant power falling on the sensitive area, which is absorbed by that area.

The thermal response time of the optimized configuration becomes

$$\tau_{th} = \frac{\left(C_m t + C_{abs}\right)Al}{4G_{thm}tw}.$$

(5.21)

Here l is the length of a leg.

Equation 5.20 indicates that thermopile optimization concerns optimization through choice of thermoelectric material and optimization through design of structure. In practice however, silicon compatibility technological steps in fabrication monolithic arrays considerably limit the choices.

5.4.2 Thermopile Configurations

Generally, two thermopile configurations are reported in the literature [8]; deposited either in a one-layer [12,37,38] or a multilayer configuration [11,39].

The most common thermopile structure is the one-layer thermopile where both thermocouple leads are deposited using conventional lithography. Thermocouple leads lie in a single plane one beside the other. The one-layer structures allow simple and fast processing and provide good thermal isolation due to its flatness. An example of this structure is shown in Figure 5.7 [40]. The structure consists of p-type silicon strips in an n-type epilayer or well, interconnected by aluminum strips. The device contains a 10-µm thick cantilever beam with one half covered by an absorbing layer and the other half a 44-strip thermopile. Shallow p- and n-type diffusions were carried out successively. During the n-type diffusion, n+ islands were created to achieve a good contact to the epilayer for the electrochemically controlled etching step. The doping of the strips is optimized so that the Seebeck coefficient is high (700 µV/K). A 750 Å LPCVD Si_3N_4 layer was grown on the back side of the wafer. Next, a two-stage etch process was undertaken: first, electrochemically controlled etching was used to stop etching at the epilayer/substrate junction and to form a membrane; second, reactive plasma etching in $CF_4 + 6\% O_2$ was used to form the cantilever beam structure. Lastly, the absorbing area of each beam was coated with a layer of an absorbent material. The final structure is similar to that shown in Figure 5.6b. The detectors had a responsivity in air of approximately 6 V/W and detectivity measured in air for a 500 K blackbody source of approximately $5 \times 10^7 \ cmHz^{1/2}W^{-1}$.

In the case of multilayer configuration, one thermocouple lead lies upon the other separated by a thin insulating layer (e.g., photoresist). The insulating layer is removed only at the hot and cold ends of the first patterned thermoelectric film, forming a small contact window [11]. However, designing a multilayer thermopile requires a careful balance between higher integration density, higher internal electrical resistance, and lower thermal resistance (the last two effects are due to the thicker layer stack of a multilayer thermopile).

5.4.3 Micromachined Thermopile Technology

From the point of view of micromachining techniques, thermopiles can be generally devoted into bulk-micromachined and surface-bulk micromachined devices [36]. The first type of structure is made by accurate micromachining of the silicon substrate—the membrane etch process is done from the back side to remove the entire bulk silicon under the membrane. In the contrary, in surface-bulk micromachined thermopile the micromachined process is provided from the front side through windows in the deposited thin film stacks. Then, only a part of the silicon substrate is removed, forming a cavity under the membrane.

The fabrication process for bulk-micromachined thermopiles utilizing microjoinery is shown in Figure 5.8. A comprehensive description of this process is given in Allison and colleagues [41]. The strips of alternating p- and n-type silicon are fabricated on separate n- and p-type, single crystal

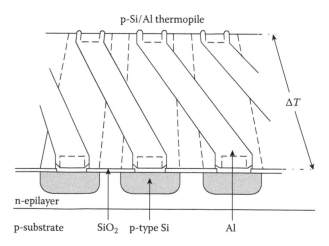

Figure 5.7 Schematic representation of the integrated Al/p-Si thermopile. (From Sarro, P. M. and van Herwaarden, A. W., *Proceedings of SPIE*, 807, 113–18, 1987. With permission.)

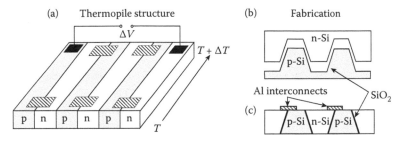

Figure 5.8 Bulk-micromachined silicon thermopile: (a) diagram of thermopile design, (b) fabrication sequence of joints (grooves) formed in both p- and n-type Si by KOH etching, and (c) final device structure (annealed, diced, and packaged).

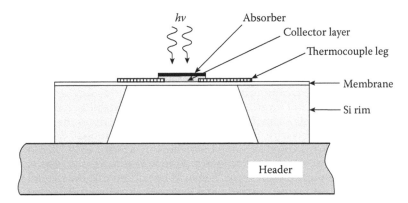

Figure 5.9 Cross section of a bulk-micromachined thermopile.

wafers, bonded together, and electrically connected in series to form the p/n couple junctions, as shown in the cross-section diagram of Figure 5.8c. The basic n- and p-type elements with mating grooves commonly fabricated by anisotropic KOH etching are bonding using either organic adhesive, frit glass, or high temperature bonding. The etching process terminates on (111) crystal planes that provide flat mating surfaces for bonding. Before the wafers can be bonded, an electrically insulating thermal SiO_2 oxide film is grown on the surface of the grooves. Finally, the series connection is carried out by liftoff patterning of e-beam evaporated aluminum. In this way the thermopile as a linear array of thermocouples with a chip size of 0.5×3.5 cm^2 and approximately 100 µm thick is fabricated.

In spite of the simple fabrication of bulk-micromachined thermopile, this kind of device is characterized by high thermal conductivity between cold and hot junctions and what results in poor performance. To improve the thermal isolation, different free standing micromachined structures have been developed. One such solution is a closed micromachined structure shown in Figure 5.9 in which the heat from cold contacts is isolated. The cold region containing cold junctions is formed by a wafer thick rim around the etched membrane. The rim serves both as a heat sink as well as suspension of the etched structure and mechanical protection [42].

The closed-membranes micromachining process has been reported in Lahiji and Wise [28] and Elbel [37]. Lahij and Wise, to develop a fabrication process for a monolithic Bi/Sb detector, began using (100)-oriented Si wafers [28]. The process starts with thermal oxidation of the wafer to a thickness of about 0.8 µm. Next the desired patterns are defined on the front and back side of the wafer. After that, a shallow boron diffusion layer over the membrane area is used as an etch stop for the anisotropic etch process. Because of high electrical conductivity of this layer, a thin dielectric layer is next deposited on the front side isolating the thermopile. The best results have been achieved using a very thin thermal oxide layer followed by thin CVD layers of silicon dioxide and silicon nitride. Finally, the thermocouple materials and interconnecting metallization are deposited and patterned using conventional lithography.

A more simple fabrication process that is widely used today has been proposed by Elbel [37]. In the first step, on the front of a (100)Si wafer, a thin stress-optimized CVD $Si_3N_4/SiO_2/Si_3N_4$ stack is

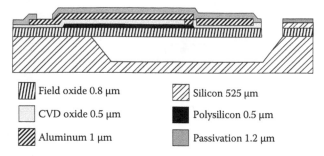

▥ Field oxide 0.8 μm	▨ Silicon 525 μm
▢ CVD oxide 0.5 μm	■ Polysilicon 0.5 μm
▨ Aluminum 1 μm	▨ Passivation 1.2 μm

Figure 5.10 Structure of CMOS process compatible thermopile detector fabricated by bulk micromachining process. The thermopile consists of n^+-polySi/Al junctions. (From Lenggenhager, R., Baltes, H., Peer, J., and Forster, M., *IEEE Electron Devices* 13, 454–56, 1992. With permission.)

deposited. As a pattern mask for the etch process on the back side of the wafer, a single Si_3N_4 layer is deposited. The membrane thickness is defined by the etch-stop bottom layer of the dielectric stack on the front. Using the above describing process, a 1 μm thick membrane with low thermal conductivity (2.4 W/mK) compared to bulk Si (140 W/mK) can be realized.

A further increase of the thermal resistance can be found by reducing the contact between membrane and silicon rim using cantilever beam membranes. In this way sensitivity of the device considerably increases, but unfortunately, the mechanical stability reduces. The thermal beam resistance can be determined by the thermal sheet resistance times the length to width (l/w) ratio of the beam. In comparison with circular membranes, the l/w ratio of cantilever thermocouple is of the order 5, thus increasing the thermal resistance by a factor of 10 [8]. Moreover, due to increasing the absorbing area without reducing the thermal resistance, the responsivity increases considerably compared to circular membrane. It should be noticed that the overall performance reduces considering increased time constant and a mechanically more fragile structure.

Usually, bulk-micromachined beam thermopiles are fabricated by a anisotropic etch process, which consists of two phases [43,44]. In the first, the entire bulk silicon under the membrane is removed from the back side using electrochemically controlled etching to stop the etching at the epilayer/substrate junction. Next, the front side of the wafer is prepared for a plasma etch step, realizing the membrane on three sides. Figure 5.10 shows an example of the pixel structure on n^+-polysilicon/Al thermopile fabricated with a silicon bulk micromachining technology [45].

Surface-bulk micromachining is a combination of bulk and surface micromachining [46]. Fabrication of structures in surface micromachining occurs from surface-deposited thin films being released by etching away sacrificial layers beneath functional layers. However, in contrast to conventional surface micromachined devices, bulk silicon is removed through small openings in the surface layers, which reduces the complexity of the production process. From a technological point of view, a surface-bulk micromachined CMOS thermopile with closed membrane and cap wafer is the best device configuration today [8].

An example of a closed CMOS surface-bulk micromachined thermopile is shown in Figure 5.11. In the first step of fabrication of a Bosch thermopile chip (Figure 5.11a), multiple oxide and epilayers forming membrane and thermocouples on the substrate surface have been provided. An anisotropic etching was applied to perforate surface layers in order to access the substrate and the substrate under the surface was removed through these etch windows. In the final step, the thermopile was hermetically sealed. This thermopile with a closed membrane is mechanically more stable than that with a beam-type membrane.

To improve mechanical stability different kinds of bridge-type structures have been proposed. For example, Figure 5.11b shows a sensor configuration with a 4-bridge structure and absorber on the central membrane [47].

To achieve high sensitivity, the detector's absorption efficiency must be high, but the thermal mass of the absorption layer must be small. In general, absorbers can be divided into three groups: metallic films, porous metal blacks, and thin film stacks.

The metallic films are characterized by narrow spectral bandwidth and their absorption strongly depends on film thickness. A maximum absorption of 50% for a 17 μm thick Au layer has been reported [48]. Porous metal blacks are deposited by electroplating (e.g., platinum) and evaporation (e.g., gold). Another way to enhance absorption is implementation of interference

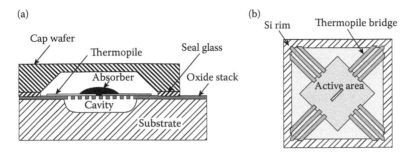

Figure 5.11 Surface-bulk-micromachined thermopile detectors: (a) the Bosch thermopile chip. (From Graf, A., Arndt, M., Sauer, M., and Gerlach, G., *Measurement Science Technology* 18, R59–R75, 2007. With permission.) (b) Bridge thermopile structure.

phenomena to set up a resonant cavity within the detector. With this kind of absorber, 90% of the radiation in the 8–14 μm wavelength range can be absorbed. More information about absorbers can be found in References 48–51.

REFERENCES

1. J. T. Seebeck, "Magnetische Polarisation der Metalle und Erze durch Temperatur-Differenz," *Abhandlung der deutschen Akademie der Wissenschaften zu Berlin*, 265–373, 1822.

2. M. Melloni, "Ueber den Durchgang der Wärmestrahlen durch verschiedene Körper," *Annals of Physics* 28, 371–78, 1833.

3. D. D. Pollock, "Thermoelectric phenomena," in *CRC Handbook of Thermoelectrics*, ed. D. M. Rowe, 7–17, CRC Press, Boca Raton, FL, 1995.

4. N. W. Ashcroft and N. D. Mermin, *Solid State Physics*, Saunders College, Philadelphia, PA, 1976.

5. S. M. Sze, *Semiconductor Devices*, Wiley, New York, 2002.

6. A. W. van Herwaarden, "The Seebeck Effect in Silicon ICs," *Sensors Actuators* 6, 245–54, 1984.

7. J. H. Kiely, D. V. Morgan, and D. M. Rowe, "The Design and Fabrication of a Miniature Thermoelectric Generator Using MOS Process Techniques," *Measurement Science and Technology* 5, 182–89, 1994.

8. A. Graf, M. Arndt, M. Sauer, and G. Gerlach, "Review of Micromachined Thermopiles for Infrared Detection," *Measurement Science Technology* 18, R59–R75, 2007.

9. M. C. Salvadori, A. R. Vaz, F. S. Teixeira, M. Cattani, and I. G. Brown, "Thermoelectric Effect in Very Thin Film Pt/Au Thermocouples," *Applied Physics Letters* 88, 133106, 2006.

10. J. Schieferdecker, R. Quad, E. Holzenkämpfer, and M. Schulze, "Infrared Thermopile Sensors with High Sensitivity and Very Low Temperature Coefficient," *Sensors and Actuators A* 46–47, 422–27, 1995.

11. F. Völklein, A. Wiegand, and V. Baier, "High-Sensitive Radiation Thermopiles Made of Bi-Sb-Te Films," *Sensors and Actuators* 29, 87–91, 1991.

12. A. W. van Herwaarden and P. M. Sarro, "Thermal Sensors Based on the Seebeck Effect," *Sensors and Actuators* 10, 321–46, 1986.

13. T. Kanno, M. Saga, S. Matsumoto, M. Uchida, N. Tsukamoto, A. Tanaka, S. Itoh, et al., "Uncooled Infrared Focal Plane Array Having 128 × 128 Thermopile Detector Elements," *Proceedings of SPIE* 2269, 450–59, 1994.

14. W. R. Blevin and J. Geist, "Influence of Black Coatings on Pyroelectric Detectors," *Applied Optics* 13, 1171–78, 1974.

15. A. Smith, F. E. Jones, and R. P. Chasmar, *The Detection and Measurement of Infrared Radiation,* Clarendon, Oxford, 1968.

16. I. B. Cadoff and E. Miller, *Thermoelectric Materials and Devices,* Reinhold, New York, 1960.

17. A. F. Ioffe, *Semiconductor Thermoelements and Thermoelectric Cooling,* Infosearch Ltd., London, 1957.

18. P. H. Egli, *Thermoelectricity,* Wiley, New York, 1958.

19. H. J. Goldsmid, "Conversion Efficiency and Figure-of-Merit," in *CRC Handbook of Thermoelectric*s, ed. D. M. Rowe, 19–26, CRC Press, Boca Raton, FL, 1995.

20. F. Voelklein, "Review of the Thermoelectric Efficiency of Bulk and Thin-Film Materials," *Sensors and Materials* 8, 389–408, 1996.

21. D. M. Rowe, G. Min, V. Kuznietsov, and A. Kaliazin, "Effect of a Limit to the Figure-of-Merit on Thermoelectric Generation," *Energy Conversion Engineering Conference and Exhibition,* (IECEC), 123–34, 35th Intersociety, Las Vegas, NV, 2000.

22. T. Akin, "CMOS-Based Thermal Sensors," in *Advanced Micro and Nanosystems,* Vol. 2, eds. H. Baltes, O. Brand, G. K. Fedder, C. Hierold, J. Korvink, and O. Tabata, 479–511, Wiley, Weinheim, Germany, 2005.

23. P. W. Kruse, *Uncooled Thermal Imaging. Arrays, Systems, and Applications,* SPIE Press, Bellingham, WA, 2001.

24. B. Stevens, "Radiation Thermopiles," in *Semiconductors and Semimetals,* Vol. 5, eds. R. K. Willardson and A. C. Beer, 287–317, Academic Press, New York, 1970.

25. A. J. Drummond, "Precision Radiometry and Its Significance in Atmospheric and Space Physics," in *Advances in Geophysics,* Vol. 14, 1–52, Academic Press, New York, 1970.

26. R. W. Astheimer and S. Weiner, "Solid-Backed Evaporated Thermopile Radiation Detectors," *Applied Optics* 3, 493–500, 1964.

27. C. Shibata, C. Kimura, and K. Mikami, "Far Infrared Sensor with Thermopile Structure," *Proceedings of the 1st Sensor Symposium* 221–25, Japan, 1981.

28. G. R. Lahiji and K. D. Wise, "A Batch-Fabricated Silicon Thermopile Infrared Detector," *IEEE Transactions on Electron Devices* ED-29, 14–22, 1982.

29. M. C. Fote, E. W. Jones, and T. Caillat, "Uncooled Thermopile Infrared Detector Linear Arrays with Detectivity Greater than 10^9 cmHz$^{1/2}$/W," *IEEE Transactions on Electron Devices* 45, 1896–1902, 1998.

30. M. C. Fote and E. W. Jones, "High Performance Micromachined Thermopile Linear Arrays," *Proceedings of SPIE* 3379, 192–97, 1998.

31. C. Herring, "Theory of the Thermoelectric Power of Semiconductors," *Physical Review* 96, 1163–87, 1954.

32. T. H. Geballe and G. W. Hull, "Seebeck Effect in Silicon," *Physical Review* 98, 940–47, 1955.

33. L. Ristic, ed., *Sensor Technology and Devices,* Artech House, Boston, MA, 1994.

34. A. Dehé, K. Fricke, and H. L. Hartnagel, "Infrared Thermopile Sensor Based on AlGaAs-GaAs Micromachining," *Sensors and Actuators* A 46–47, 432–36, 1995.

35. A. Dehé, D. Pavlidids, K. Hong, and H. L. Hartnagel, "InGaAs/InP Thermoelectric Infrared Sensors Utilizing Surface Bulk Micromachining Technology," *IEEE Transactions on Electron Devices* 44, 1052–58, 1997.

36. H. Baltes, O. Paul, and O. Brand, "Micromachined Thermally Based CMOS Microsensors," *Proceedings of IEEE* 86, 1660–78, 1998.

37. T. Elbel, "Miniaturized Thermoelectric Radiation Sensor," *Sensors and Materials* A3, 97–109, 1991.

38. A. Mzerd, F. Tchelibou, A. Sackda, and A. Boyer, "Improvement of Thermal Sensors Based on Bi_2Te_3, Sb_2Te_3, and $Bi_{0.1}Sb_{1.9}Te_3$," *Sensors and Actuators* A47, 387–90, 1995.

39. T. Elbel, S. Poser, and H. Fischer, "Thermoelectric Radiation Microsensors," *Sensors and Actuators* A42, 493–96, 1994.

40. P. M. Sarro and A. W. van Herwaarden, "Infrared Detector Based on an Integrated Silicon Thermopile," *Proceedings of SPIE* 807, 113–18, 1987.

41. S. C. Allison, R. L. Smith, D. W. Howard, C. Gonzalez, and S. D. Collins, "A Bulk Micromachined Silicon Thermopile with High Sensitivity," *Sensors and Actuators* A102, 32–39, 2003.

42. I. Simon and M. Arndt, "Thermal and Gas Sensing Properties of a Micromachined Thermal Conductivity Sensor for the Detection of Hydrogen in Automotive Applications," *Sensors and Actuators* A98–98, 104–8, 2002.

43. A. W. van Herwaarden, P. M. Sarro, and H. C. Meijer, "Integrated Vacuum Sensor," *Sensors and Actuators* 8, 187–96, 1985.

44. P. M. Sarro and A. W. van Herwaarden, "Silicon Cantilever Beams Fabricated by Electrochemically Controlled Etching for Sensor Applications," *Journal of the Electrochemical Society* 133, 1724–29, 1986.

45. R. Lenggenhager, H. Baltes, J. Peer, and M. Forster, "Thermoelectric Infrared Sensors by CMOS Technology," *IEEE Electron Devices* 13, 454–56, 1992.

46. J. M. Bustillo, R. T. Howe, and R. S. Muller, "Surface Micromachining for Microelectromechanical Systems," *Proceedings of IEEE* 86, 1552–74, 1998.

47. C.-H. Du and C. Lee, "Investigation of Thermopile Using CMOS Compatible Process and Front-Side Si Bulk Etching," *Proceedings of SPIE* 4176, 168–78, 2000.

48. W. Lang, K. Kühl, and H. Sandmaier, "Absorbing Layers for Thermal Infrared Detectors," *Sensors and Actuators* A34, 243–48, 1992.

49. N. Nelms and J. Dowson, "Goldblack Coating for Thermal Infrared Detectors," *Sensors and Actuators* A120, 403–7, 2005.

50. A. Hadni and X. Gerbaux, "Infrared and Millimeter Wale Absorber Structured for Thermal Detectors," *Infrared Physics* 30, 465–78, 1990.

51. A. D. Parsons and D. J. Fedder, "Thin-Film Infrared Absorber Structures for Advanced Thermal Detectors," *Journal of Vacuum Science and Technology* A6, 1686–89, 1988.

6 Bolometers

Another widely used thermal detector is the bolometer. The bolometer is a resistive element constructed from a material with a very small thermal capacity and large temperature coefficient so that the absorbed radiation produces a large change in resistance. In contrast to the thermocouple, the device is operated by passing an accurately controlled bias current through the detector and monitoring the output voltage. The change in resistance is like to the photoconductor, however, the basic detection mechanisms are different. In the case of a bolometer, radiant power produces heat within the material, which in turn produces the resistance change. There is no direct photon-electron interaction.

The first bolometer designed in 1880 by American astronomer S. P. Langley for solar observations [1] used a blackened platinum absorber element and a simple Wheatstone-bridge sensing circuit. Langley was able to make bolometers that were more sensitive than the thermocouples available at that time. Although other thermal devices have been developed since that time, the bolometer remains one of the most used infrared detectors.

Modern bolometer technology development started in the early 1980s with the work of Honeywell on vanadium oxide (VO_x) and Texas Instruments on amorphous silicon (a-Si). Much of the technology was development under classified military contacts in the United States, so the public release of this information in 1992 surprised many in the worldwide infrared community. Figure 6.1 shows the cross section of a thin film bolometer prepared by silicon micromachining compatible with integrated circuit processing technology that enables the development of very large, low-cost, monolithic two-dimensional arrays.

Development efforts in uncooled FPAs are now basically going in two directions:

■ Arrays for military and high-end commercial applications with the highest possible performance

■ Arrays for commercial applications with the lowest possible cost

The key factor is to find a high performance sensor together with high thermal isolation in the smallest possible area.

6.1 BASIC PRINCIPLE AND OPERATION OF BOLOMETERS

The relative temperature coefficient of resistance (TCR) is defined as

$$\alpha = \frac{1}{R}\frac{dR}{dT}.$$

(6.1)

The change of voltage of a constant current-biased bolometer is

$$\Delta V = I\Delta R = IR\alpha\Delta T.$$

So, in this case $K = IR\alpha$ (see Equation 3.7), and according to Equation 3.9 the voltage responsivity is

$$R_v = \frac{IR\alpha R_{th}\varepsilon}{\left(1+\omega^2\tau_{th}^2\right)^{1/2}}.$$

(6.2)

The expressions for voltage responsivity of a bolometer and a thermocouple are similar, with $n\alpha$ replaced by $IR\alpha_s$. The responsivity is inversely proportional to the thermal conductance ($G_{th} = 1/R_{th}$), which is also true to thermocouple.

The maximum bias current is limited by the maximum allowed element temperature T_{max}. Therefore

$$I^2R = G_{th}\left(T_{max} - T\right),$$

(6.3)

and

$$R_v = \alpha\varepsilon\left[\frac{RR_{th}\left(T_{max} - T\right)}{1+\omega^2\tau_{th}^2}\right]^{1/2}.$$

(6.4)

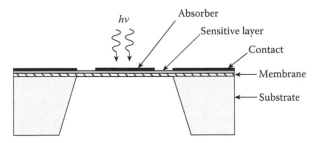

Figure 6.1 Schematic cross section of a thin film bolometer.

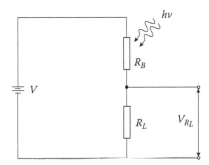

Figure 6.2 Bolometer circuit.

The value of R_v is controlled in part by $R_{th} = 1/G_{th}$; bolometers with a high thermal conductance are fast (see Equation 3.5), but their responsivity is low. The key to developing highly sensitive bolometers is having a high temperature coefficient α, a very low thermal mass C_{th}, and excellent thermal isolation (low thermal conductance G_{th}).

The above considerations are widely used for the simplest model, which omits the Joulean heating of the bias current and assumes a constant electrical bias. An accurate representation of a bolometer is a complex and difficult task. It is analyzed in details (e.g., [2–4]).

More complexity is introduced when the heat flow equation (like Equation 3.1) includes the Joulean heating due to electrical bias and a load resistor, R_L, is introduced into a bolometer circuit in order to distinguish between voltage source operation ($R_L >> R_B$, where R_B is the bolometer resistance) and current source operation ($R_L << R_B$) (see Figure 6.2). If the circuit is opened and if no signal radiation is present, the bolometer is at ambient temperature T_0. Closing the circuit causes current flow and the Joulean heating in the bolometer resistance, R_B. As a result, its temperature increases to T_1. If radiation now falls upon the bolometer, its temperature changes by ΔT to the new value T. This results in a resistance change in the bolometer, causing a change in the voltage across R_L.

Kruse carried out the analysis of bolometer operation including the Joulean heating and constant electrical bias [4]. It appears that the behavior of the detector seriously depends on temperature dependence of bolometer resistance upon temperature.

The resistance of a piece of semiconductor can be shown to be the form

$$R = R_o T^{-3/2} \exp\left(\frac{b}{T}\right),$$ (6.5)

where R_o and b are constants. For a semiconductor at room temperature

$$\alpha = -\frac{b}{T^2}.$$ (6.6)

105

For a metal that has a linear dependence of resistance on temperature; that is

$$R = R[1 + \gamma(T - T_o)],$$ (6.7)

and thus

$$\alpha = \frac{\gamma}{1 + \gamma(T - T_o)}.$$ (6.8)

Here γ is the temperature coefficient of the detector material.

In the case of bolometer operation under constant electrical bias and Joulean heating, the solution of a heat balance equation is similar to that described by Equation 3.3 [4] and has a form

$$\Delta T = \Delta T_o e^{-(G_e/C_{th})t} + \frac{\varepsilon \Phi_o e^{i\omega t}}{G_e + i\omega C_{th}}.$$ (6.9)

The first term of Equation 6.9 represents a transient, whereas the second term is a periodic function. Here G_e is the "effective" thermal conductance defined as

$$G_e = G - G_o(T_1 - T_o)\alpha\left(\frac{R_L - R_B}{R_L + R_B}\right),$$ (6.10)

where G_o is the average thermal conductance through a detector medium in temperature range between T_1 and T_o, and G is the thermal conductance when the bolometer is at temperature T. Equation 6.10 indicates that G_e is the difference in two terms. The G_e is positive if

$$G > G_o(T_1 - T_o)\alpha\left(\frac{R_L - R_B}{R_L + R_B}\right),$$ (6.11)

and then the transient term goes to zero with time and only the periodic function remains. However, in the case when

$$G < G_o(T_1 - T_o)\alpha\left(\frac{R_L - R_B}{R_L + R_B}\right),$$ (6.12)

where G_e is negative that means the bolometer temperature will increase exponentially with time (see Equation 6.9) reaching burnout. This can happen with semiconductors but not with metals [2].

Assuming that $R_L >> R_B$, it can be shown that the responsivity is given by

$$R_v = \frac{\alpha I_b R_B \varepsilon}{G_e(1 + \omega \tau_e)^{1/2}},$$ (6.13)

where τ_e is defined as

$$\tau_e = \frac{C_{th}}{G_e}.$$ (6.14)

Here, τ_e is known as the "effective thermal response time." The dependence of thermal capacity and τ on temperature due to bias current heating is termed the "electrothermal effect."

Usually when large focal plane arrays are implied, the electrical bias is pulsed rather than continuous and heating is due to the electrical bias (Joulean) and the incident absorbed radiant flux. In this situation the heat transfer equation is nonlinear and numerical solutions must be obtained [5].

In addition to radiation noise and temperature noise associated with the thermal impedance of the element, Johnson noise associated with resistance R is one of the most important noise sources. With room temperature bolometers, amplifier noise should not be important but with cryogenic devices it is usually the dominant noise source. With some types of bolometers, low frequency current noise is important and is the principal factor limiting current.

6.2 TYPES OF BOLOMETERS

Bolometers may be divided into several types. The most commonly used are the metal, the thermistor, and the semiconductor bolometers. A fourth type is the superconducting bolometer. This bolometer operates on a conductivity transition in which the resistance changes dramatically over the transition temperature range.

6.2.1 Metal Bolometers

Typical materials used for metal bolometers are nickel, bismuth, platinum, or antimony. These metals are in use where their high, long-term stability is an essential requirement. Being made from metal, these bolometers need to be small so that the heat capacity is low enough to allow reasonable sensitivity. Most metal bolometers are formed as film strips, about 100–500 Å thick, via vacuum evaporation or sputtering. They are often coated with a black absorber such as evaporated gold or platinum black.

A typical value of temperature coefficient is positive for metals and is equal about 0.3%/°C.

The metal bolometers operate at room temperature and have detectivities of the order 1×10^8 cmHz$^{1/2}$W^{-1}, with response times of approximately 10 msec. Unfortunately they are generally rather fragile, thus limiting their use in certain applications. Nevertheless, metal film bolometer arrays have been fabricated in various linear and 2-D formats, and have been successfully employed in nonimaging IR sensors designed for remote surveillance applications [6]. The technology of these devices is generally limited to small arrays due to power consumption and amplifier design constraints associated with matching to the low detector impedance.

Between metals, titanium films are more frequently used in the bolometers due to the following reasons: titanium can be used in a standard silicon process line, low thermal conductivity (0.22 W/Kcm in bulk material—far lower than that of most other metals), and low $1/f$ noise [7,8]. However, metal in thin film form has a temperature coefficient resistance of 0.004%/K, considerably lower than for competitor materials, which causes it to be of little use in uncooled bolometer arrays.

Room temperature antenna-coupled metal microbolometers (usually with bismuth or niobium) are also operated in the very long wavelength IR region (10–100 µm). The noise equivalent power (NEP) values about 10^{-12} W/Hz$^{1/2}$ can be obtained with suspended microbridges or low thermal conductance buffer layers with silicon substrate. See Figure 6.18 for some experimental data for this type of detectors.

6.2.2 Thermistors

Thermistor materials were first developed at Bell Laboratories during World War II. Single pixel thermistor bolometers have been commercially available for about 60 years. They have found wide applications, ranging from burglar alarms, to fire detection systems, industrial temperature measurement, space-borne horizon sensors, and radiometers. They are also useful in radiometric applications where a uniform spectral response is desired. Thermistors have long life, good stability when biased properly, and are highly resistant to nuclear radiation.

Thermistor bolometers are constructed from a sintered mixture of various semiconducting oxides that have a higher TCR than metals (2–4%/°C), and are generally much more rugged. They crystallize in the spinel structure. In their final form they are semiconductor flakes 10 µm thick. The negative temperature coefficient depends on the band gap, the impurity states, and the dominant conduction mechanism. This coefficient is not constant but varies as T^{-2}, which is a result of exponential dependence of semiconductor resistivity [9]. The sensitive material in a thermistor bolometer is typically made of wafers of manganese, cobalt, and nickel oxides sintered together and mounted on electrically insulating but thermally conducting material such as sapphire [10,11]. The sapphire is then mounted on a metallic heat sink, to control the time constant of the device. The sensitive area is blackened to improve its radiation absorption characteristics. Typical room temperature resistivity of a thermistor changes between 250 and 2500 Ωcm. They can be made in sizes ranging from 0.05 to 5 mm square. The principal material investigated is (MnNiCO)$_3$O$_4$, which has a negative TCR of about 0.04/°C at room temperature.

The usual construction uses a matched pair of devices for a single unit (Figure 6.3). One of the pair of thermistors is shielded from radiation and fitted into a bridge such that it acts as the load resistor. This arrangement allows one to optimize the signal from the active element by compensating for any ambient temperature changes. The result can be a dynamic range of a million to

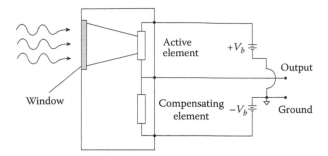

Figure 6.3 Typical bias circuit of a thermistor bolometer. The compensating element is shielded from incident radiation.

one. The device sensitivity and response time cannot both be optimized as improved heat sinking, to reduce the time constant, prevents the detector from reaching its maximum temperature, thus reducing the responsivity. Considerations carried out by Astheimer [12] indicate that Johnson noise limited detectivity of this type of bolometer at room temperature can be described by the equation

$$D^* = 3 \times 10^9 \tau^{1/2} \quad \text{cmHz}^{1/2}\text{W}^{-1} \tag{6.15}$$

with τ in seconds. The time constant changes typically from 1 to 10 msec. Their sensitivity closely approaches that of the thermopile for frequencies higher than 25 Hz. At lower frequencies there may be excess or $1/f$ noise.

Since the thermistor is Johnson noise limited, an improvement in detectivity can be realized by placing a hemispherical or hyperhemispherical lens over its surface [13]. This procedure would not improve the signal-to-noise ratio if the detector is photon noise limited. The detector must be optically coupled with the lens, which can be accomplished by depositing the detector directly on the plane of the lens (see Figure 3.16). The rays directed to the edge of the detector are refracted by the lens, giving the detector an appearance of being larger by a factor of n (or n^2 in the case of hyperhemispherical lenses), where n is the index of refraction of the lens. Since the detector is a two-dimensional device, the virtual area increase is n^2. As a result, the signal-to-noise ratio is increased by n^2 (or n^4). Immersion also reduces the bias power dissipation by factors n^2 (or n^4), thereby reducing the heat load of coolers and enabling higher bias power dissipation densities to be achieved. The requirements on the lens material are that it should have as large an index of refraction as possible, and should be electrically insulating so as not to short the thermistor film. Germanium, silicon, and arsenic triselenide are among the most useful materials. Thermistor detectors are particularly adaptable to immersion because the low thermal impedance inherent in immersion permits higher bias voltage to be applied.

The thermistor detectors are mainly fabricated by bonding thin flakes of the thermistor material to a substrate. Device performance parameters such as responsivity, noise, and response time are very dependent on operator skill and experience, and processing conditions. To overcome these drawbacks and reduce the cost of thermistors the possibility of evaporated films has been investigated [14,15]. The most popular are sputtering techniques.

The thermistor spectral responsivity is essentially flat with an upper wavelength determined by the transmission of the window that encapsulates the chip. Because they exhibit a large amount of $1/f$ noise arising at grain boundaries, thermistor thin films have not been found to be useful in uncooled bolometer thermal imaging arrays.

6.2.3 Semiconductor Bolometers

A significant improvement in the performance of bolometer detectors is achieved if the devices are cooled as the resistance changes are much greater than at room temperature, and thicker devices can be fabricated to improve IR absorption without increasing the thermal capacity due to the reduced specific heat in the cooled material. The ultimate sensitivity can be orders of magnitude higher than that for a room temperature device. In practice, for most applications, it is necessary to have an aperture in the enclosure admitting some room temperature background radiation.

Semiconductor bolometers, are the most highly developed form of thermal detectors for low light levels and are the detectors of choice for many applications, especially in the infrared and submilimeter spectral range. They must be constructed carefully to ensure that they are isolated from the thermal surroundings, and the techniques typically used to construct them do not lend themselves to efficient development of large arrays.

The temperature dependence of semiconductor resistance is described by Equation 6.5. However, at very low temperatures (<10 K), the semiconductor material must be doped more heavily than assumed in Equation 6.5 so that the dominant conductivity mode is hopping. This mechanism freezes out relatively slowly; the resistance is given by an empirical expression of the form:

$$R = R_o \left(\frac{T}{T_o} \right)^{-a}, \qquad (6.16)$$

where typically $a \approx 4$. We then have

$$\alpha(T) = -\frac{a}{T}. \qquad (6.17)$$

Note that $\alpha < 0$ and has strong temperature dependence.

Typical bias circuit of the bolometer is similar to that shown in Figure 6.2. A large value of load resistance is almost always used to minimize the Johnson noise. The general noise considerations can be found in Mather [16].

The modern history of infrared bolometers begins with the introduction of the carbon resistance bolometer by Boyle and Rogers [17]. At this time, carbon radio resistors were widely used by low temperature physicists as thermometers at liquid helium temperatures. The next important step in bolometer development with superior performance was the invention of a low temperature bolometer based on heavily gallium-doped and compensated germanium with sensitivities close to the theoretical limits (see Figure 6.4) over the wavelength range from 5 to 100 μm [18,19]. Its mode of operation has been discussed in detail by Zwerdling et al. [20]. In correctly doped germanium (typical concentrations of gallium are about 10^{16} cm^{-3} with about 10^{15} cm^{-3} in giving p-type conductivity with a compensation ratio of about 0.1), the absorbed energy is transferred rapidly to the lattice, raising the temperature of the sample rather than of the existing free carriers as in a photoconductor. The absorption efficiency can be increased by mounting the device in an integrating cavity, which is discussed by Putley [18]. For small apertures at 4.2 K the inherent Ge bolometer noise was divided equally between Johnson noise and photon noise [21]. At larger apertures the photon noise may predominate over the inherent detector noise, depending on the performance of the background. The advantages of a well-known material with reproducible properties, high stability, and low noise led to its application to infrared astronomy and medium and long wavelengths as well as to laboratory infrared spectroscopy. Over most of the far infrared (FIR) spectrum, the performance of a germanium bolometer is comparable to that of the best photon detectors with the added advantage of being a broadband device. Typical values of Ge and C bolometer parameters are gathered in Table 6.1. Further improvement in device performance was possible. Draine and Sievers have obtained an NEP of 3×10^{-16} WHz$^{-1/2}$ for a device operating at 0.5 K [22]. However, this very low value could probably be exploited in an experiment conducted entirely at cryogenic temperature and requires a cryogenically operated low noise amplifier.

More attention has been recently given to the use of Si as an alternative to Ge. In comparison with Ge, Si has a lower specific heat (by a factor of 5), easier materials preparation, and more advanced device fabrication technology. Silicon bolometers with NEPs of 2.5×10^{-14} WHz$^{1/2}$ that compare favorably with germanium have been reported by Kinch [23].

Details of the fabrication and performance of modern bolometers are described in review papers [24–26]. It appears that Ge and Si submillimeter bolometers with improved performance can be fabricated using more uniform materials obtained by neutron transmutation doping and by ion implantation.

The chip bolometers described above combine the functions of radiation absorption and thermometry. These two functions are difficult to comply especially for chip bolometers designed for millimeter and submillimeter wavelengths. The bulk absorption coefficient of Ge and Si material with a useful resistivity decreases at low frequencies. Consequently, bolometers must typically be one or more millimeters thick and the resulting heat capacity is a significant limitation.

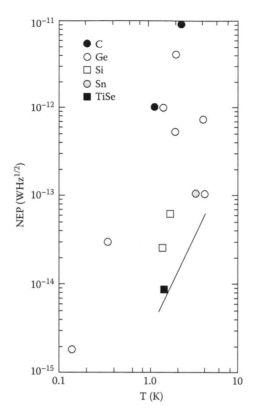

Figure 6.4 Low-temperature dependence of NEP for cooled bolometers of carbon, germanium, silicon, thin and thallium selenide. (From Putley, E. H., *Optical and Infrared Detectors,* Springer, Berlin, 71–100, 1977.) The continuous line is the estimate by Coron of the best attainable performance of a Ge bolometer in the absence of higher temperature background radiation. (From Coron, N. *Infrared Physics,* 16, 411–19, 1976. With permission.)

Table 6.1: Cryogenic Bolometer Characteristics

	Ge Bolometer	C Bolometer
T_{sink} (K)	2.15	2.1
A_d (cm²)	0.15	0.20
Thickness (cm)	0.012	0.0076
R_L (Ω)	5.0×10^5	3.2×10^6
R_d (Ω)	1.2×10^4	1.2×10^6
R_v (V/W)	4.5×10^3	2.1×10^4
τ (μs)	400	10^4
f (Hz)	200	13
G_{th} (μW/K)	183	36
NEP (W)	5×10^{-13}	1×10^{-11}
D^* (cmHz$^{1/2}$/W)	8×10^{11}	4.5×10^{10}

Source: Boyle, W. S., and Rogers, Jr., K. F., *Journal of the Optical Society of America,* 49, 66–69, 1959; Low, F. J., *Journal of the Optical Society of America,* 51, 1300–1304, 1961.

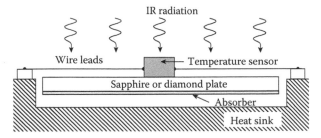

Figure 6.5 Composite bolometer. (From Dereniak, E. L. and Crowe, D. G., *Optical Radiation Detectors*, Wiley, New York, 1984. With permission.)

To overcome these limitations, composite bolometers are fabricated to lower heat capacity and consistently to reduce the time constant [25,27]. They look promising for low-noise and low-temperature detection.

The composite bolometer is made up of three parts: the radiation-absorbing material; the substrate that determines its active area, and the temperature sensor as shown in Figure 6.5 [28]. The absorber is made of a thin film whose thickness and composition are adjusted so that the emissivity is very high—into the hundreds of microns wavelength region. Usually black bismuth and nichrome absorbers have been used. The temperature sensor (e.g., germanium) is bonded both mechanically and thermally to the substrate via an epoxy or varnish. Thus the substrate and film combination act as an efficient absorbing element of a large effective area and low heat capacity for a temperature sensor that can be made very small.

Early composite bolometers used sapphire substrates that have about 1/60th the heat capacity of germanium and have negligible absorptivity up to ~300 cm^{-1} at low temperatures. This means that a larger detector area can be made with no corresponding frequency response loss. Diamond is transparent well beyond 1000 cm^{-1} and has a heat capacity of approximately 1/600th that of germanium, thus much larger active areas can be achieved. It is now widely used for low background bolometers at T ~ 1 K. Silicon substrates have larger lattice specific heat, but are useful for temperature << 1 K because the lattice heat capacity is small enough and they have smaller impurity heat capacity than the diamond. Examples of composite bolometer fabrications are described by Richards [25].

Modern well-developed semiconductor bolometer technologies exist to produce arrays of hundreds of pixels that are operated at temperatures of 100–300 mK (see Section 22.4). They are typically fabricated by lithography and micromachining techniques of Si or Si$_3$N$_4$ and use thermistors of ion-implanted silicon or neutron-transmutation-doped Ge.

6.2.4 Micromachined Room Temperature Silicon Bolometers

A new generation of monolithic Si bolometers has been introduced by Downey et al. [29]. They presented a bolometer concept in which a thin Si substrate supported by narrow Si legs is micromachined from a Si wafer using the techniques of optical lithography. A conventional Bi film absorber was used on the back of the substrate, but the thermometer was created directly in the Si substrate by implanting P and B ions to achieve a suitable donor density and compensation ratio. However, the performance of the bolometer was not good. Further progress in monolithic Si bolometer technology is fascinating. One of the most widely used approaches is microbolometers implemented using surface micromachined bridges on complementary metal-oxide-semiconductor (CMOS)-processed wafers. The surface micromachining technique allows the deposition of temperature-sensitive layers with very small thickness, very low mass, and very good thermal isolation on top of bridges over the readout circuit chips. After removal of the sacrificial layers between the bridge structures and readout circuit chips, thermally isolated and suspended detector structures packaged in vacuum are obtained. Honeywell Sensor and System Development Center in Minneapolis, Minnesota began developing silicon micromachined IR sensors in the early 1980s. The goal of the work that continued under classified programs sponsored by DARPA and U.S. Army Night Vision and Electronic Sensor Directorate was aimed at producing low-cost night vision systems amenable for use throughout the military with noise equivalent temperature difference (NEDT) of 0.1°C using *f*/1 optics. Both Si bolometer arrays and pyroelectric arrays of Texas Instruments have exceeded that goal [30,31].

From considerations carried out in Chapter 20 suggests that focal plane arrays (FPAs) of thermal detector can resolve very small temperature differences, comparable to those attained with the current scanned cryogenic imagers.

Two options for the bolometers structure are used: microbridge and pellicle-supported designs (see Figure 6.6). The former comprises detector elements that are supported on legs above the plane of the microcircuit. The legs are designed to have a high thermal resistance, and carry electrical conductors from the detector to the microcircuit. This approach is implied in the Honeywell microbolometer design [31–33]. The second concept consists of detector elements deposited onto a thin dielectric pellicle that is coplanar with the surface of the wafer, and is the basis of original Australian monolithic detector technology [34,35].

The basic fabrication process used to fabricate Honeywell silicon microbolometers together with a brief explanation of the micromachining process steps is shown in Figure 6.7 [36]. This process consists of depositing and patterning a sacrificial layer on a substrate containing the electronics. The detecting area is defined by a thin membrane upon which a thin film of the detecting material is deposited. The final microbolometer pixel structure is shown in Figure 6.8 [31]. The microbolometer consists of a 0.5 μm thick bridge of Si_3N_4 suspended about 2 μm above the underlying silicon substrate. The bridge is supported by two narrow legs of Si_3N_4. The Si_3N_4 legs provide the thermal

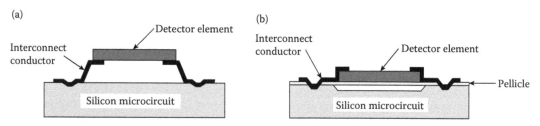

Figure 6.6 Thermal detector element design: (a) microbridge detector element, and (b) pellicle-supported detector element. (From Liddiard, K. C., "Thin Film Monolithic Arrays for Uncooled Thermal Imaging," *Proceedings of SPIE* 1969, 206–16, 1993. With permission.)

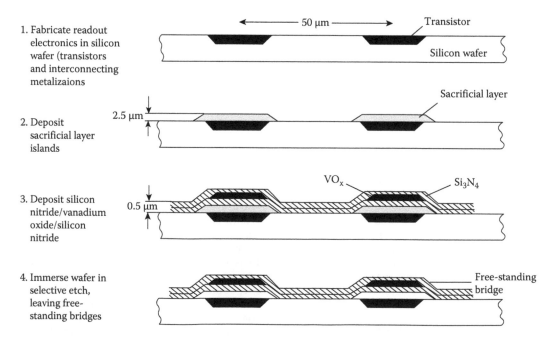

Figure 6.7 Simplified process of two-level Honeywell microbolometer fabrication. (From Wood, R. A., *Electron Devices Meeting, IEDM'93, Technical Digest, International*, 175–77, 1993. With permission.)

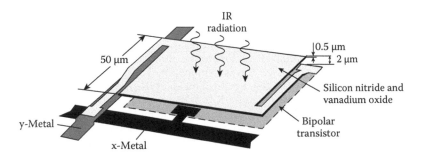

Figure 6.8 Bridge structure of the Honeywell microbolometer. (From Wood, R. A., Han, C. J., and Kruse, P. W., "Integrated Uncooled IR Detector Imaging Arrays," *Proceedings of IEEE Solid State Sensor and Actuator Workshop*, Hilton Head Island, SC, 132–35, June 1992. With permission.)

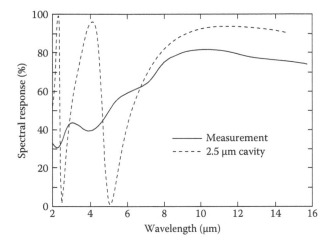

Figure 6.9 Quarter-wave cavity spectrum of the Ulis bolometer. (From Fieque, B., Tissot, J. L., Trouilleanu, C., Crates, A., and Legras, O., *Infrared Physics and Technology*, 49, 187–91, 2007. With permission.)

isolation between the microbolometer and the heat-sink readout substrate. A bipolar input amplifier is normally required, and this can be obtained with biCMOS processing technology. The legs contain a thin metal layer to provide electrical contact between the bolometer material and the readout electronics.

A reflective layer on the substrate below the membrane (typically aluminum) causes incidental infrared radiation that is completely absorbed by the detecting material to be reflected back through the material, thereby increasing the amount absorbed. This is most effective when the spacing between the absorbing layer and the reflecting layer is one-fourth of the wavelength of the incident radiation. The peak absorption wavelength λ_p dictated by the optical cavity results from the following equation

$$nt = \frac{(2k+1)\lambda_p}{4}, \tag{6.18}$$

where n is the index of refraction of the transmission media in the cavity (vacuum), t is the cavity thickness, and k is the resonant order. For $n = 1$ (vacuum), $t = 2.5$ μm, and $k = 0$, $\lambda_p = 10$ μm that corresponds to the standard microbolometer array designated to be efficient in the long wavelength IR (LWIR) region. However, for the next order ($k = 1$), a large absorption in 3–5 μm wavelength band ($\lambda_p = 3.33$ μm) is predicted. For example, Figure 6.9 compares the spectral response of the Ulis microbolometer to the theoretical quarter wave spectrum [37]. It is the most commonly used resonant cavity design. In the second type of design, the resonant optical cavity is part of the bolometer membrane. This solution is rarely used.

The Si_3N_4 is used because of its excellent processing characteristics. This allowed microbolometers to be fabricated with thermal isolation close to the attainable physical limit, which is about 1×10^8 K/W for a 50-μm-square detector. It was demonstrated that, with a microbolometer having a thermal isolation of 1×10^7 K/W [38], a typical incident IR signal of 10 nW was sufficient to change the microbolometer temperature by 0.1 K [32]. The measured thermal capacity was about 10^{-9} J/K, which corresponds to a thermal time constant of 10 ms. Honeywell has determined that the microbridges are robust structures that can tolerate shocks of several thousand g-forces. Encapsulated in the center of the Si_3N_4 bridge is a thin layer (500 Å) of polycrystalline VO_x as the active bolometer's material. The VO_x assures good combination of high TCR, electrical resistivity, and fabrication capability, which has resulted in pixels with responsivity of 250,000 V/W in response to 300 K blackbody radiation [33].

Conventional bolometers are operated in a vacuum package to minimize the thermal conduction between the bolometers and their surroundings through surrounding gas. The vacuum atmosphere in which the bolometers are operated is typically on the order of 0.01 mbr.

6.2.4.1 Bolometer Sensing Materials

Today, the most common bolometer temperature sensing materials are VO_x, amorphous silicon (a-Si), and silicon diodes. The performance of bolometers is typically limited by the $1/f$ noise [38,39] that can vary several orders of magnitude for different materials and even small variations of the material composition can dramatically change the $1/f$ noise value. Monocrystalline materials can have significantly lower $1/f$ noise constant as compared to amorphous or polycrystalline materials [38,40]. However, for most materials $1/f$ noise constant is not well documented in literature.

6.2.4.2 Vanadium Oxide

The most popular thermistor material used in fabrication of the micromachined silicon bolometers is vanadium oxide, VO_x. A thin film of the mixed oxides sputtered on a Si_3N_4 microbridge substrate was originally developed at Honeywell. Vanadium is a metal with a variable valence forming a large number of oxides. Preparation of these materials in both bulk and thin film forms is very difficult given the narrowness of the stability range of any oxide. Some of the vanadium oxides, among them the best known being VO_2, V_2O_3, and V_2O_5, show a temperature-induced crystallographic transformation that is accomplished by reversible semiconductor (low-temperature phase) to metal (high-temperature phase) phase transition with a significant change in electrical and optical properties (Figure 6.10) [41–44]. Vanadium dioxide undergoes their transition in the temperature range from about 50 to 70°C. The V_2O_5 can be formed by the ion beam from a vanadium metal target in high O_2 partial pressure, but its resistance at room temperature is very high. The V_2O_3 has low formation energy and undergoes a transition from semiconductor to metal phase at low temperatures, so its resistance is very low at room temperature. This oxide is important to the fabrication of low-noise microbolometers.

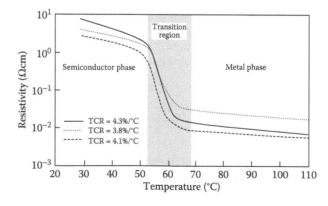

Figure 6.10 Resistivity versus temperature characteristics of three VO_2 films. (From Jerominek, H., Picard, F., and Vincent, D., *Optical Engineering*, 32, 2092–99, 1993. With permission.)

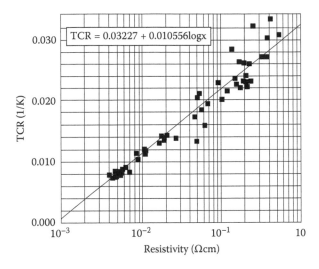

$$TCR = 0.03227 + 0.010556 \log x$$

Figure 6.11 Temperature coefficient of resistance versus resistivity for thin films of mixed vanadium oxides. (From Wood, R. A., *Uncooled Infrared Imaging Arrays and Systems*, Academic Press, San Diego, CA, 43–121, 1997. With permission.)

VO_x films are prepared by a variety of techniques including reactive radio frequency (RF) sputtering, pulsed laser deposition, annealing, and oxidation of evaporated vanadium under controlled conditions [45]. Several groups are looking to find fabrication procedure to obtain high TCR in combination with sufficient low sheet resistance (e.g., [46,47]).

Figure 6.11 shows the TCR in dependence on resistivity for thin films of mixed vanadium oxides [3]. From the point of view of infrared imaging application, the most important property of VO_x is its high negative TCR at ambient temperature, which exceeds 3% per degree. However, there are two reasons for not using VO_x material with higher x-values and substantially higher TCR. In the region of higher x-value, the reproducibility of oxide property suffers due to scatter of experimental data. In addition, a Joulean heating becomes a problem with high resistivity films. This aggravates the nonlinear temperature versus time problem during the pulse duration [3].

6.2.4.3 Amorphous Silicon

Amorphous silicon (a-Si) is extensively used as the active layer in thin-film transistors for liquid crystal displays, small area photovoltaic devices for consumer products, and solar cells. Syllaios et al. have described technology of a-Si microbolometers [48]. The bolometer sensing material has been investigated for at least 20 years. Room temperature TCR values ranging from $-0.025°C^{-1}$ for doped, low resistivity films to $0.06°C^{-1}$ for high resistivity materials have been reported [49]. However, high-resistivity a-Si is characterized by unacceptable levels of $1/f$ noise [50]. Figure 6.12 illustrates the relationship between the TCR and the resistivity of a-Si [51].

Properties of the films depend upon the method of preparation and the type of dopant. Amorphous hydrogenated silicon (a-Si:H) has a metastable state caused by defects arising from prolonged illumination (Staebler and Wronski effect). This is an undesirable feature that requires a specific annealing cycle during preparation (methodology for reliability enhancement is described in [52]). If not removed, it adversely affects long-term reliability. Hydrogenated a-Si can only be produced by a nonequilibrium process, such as plasma-enhanced chemical vapor deposition (PECVD) or sputtering. The TCR and sheet resistance are therefore a direct consequence of the deposition procedure: doping concentration, deposition temperature, and annealing [53]. The a-Si can be deposited at a very low temperature, as low as 75°C.

Because the typical resistivity of a-Si films is several orders of magnitude higher than that of VO_x, a-Si finds application in uncooled arrays in which the bias is continuous rather than pulsed. This choice is dictated by the fact that the bolometer signal depends on $I_b \alpha R$, whereas the power dissipation, which causes the raise in detector temperature, depends on $I_b^2 R$ (here I_b is the bias current).

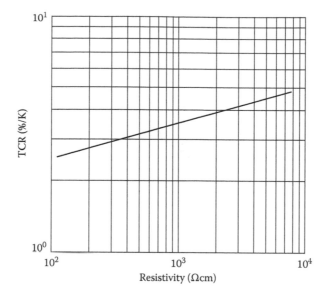

Figure 6.12 TCR as a function of electrical resistivity of a-Si. (From Tissot, J. L., Rothan, F., Vedel, C., Vilain, M., and Yon, J.-J., "LETI/LIR's Amorphous Silicon Uncooled IR Systems," *Proceedings of SPIE* 3379, 139–44, 1998. With permission.)

6.2.4.4 Silicon Diodes

For temperature sensing, the forward biased p-n junction [54–57] or Schottky-barrier junction [58] can be also used.

The voltage temperature dependence of a forward-biased diode is [59]

$$\frac{dV}{dT}\bigg|_{I=\text{const}} = \frac{V}{T} - \frac{q}{kT}\left(\frac{1}{I_s}\frac{dI_s}{dT}\right) = \frac{qV - E_g}{qT}, \tag{6.19}$$

where I_s is the reverse saturation current and E_g is the energy gap. In a silicon diode where $E_g = 1.12$ eV and operating voltage is typically 0.6–0.7 V at room temperature, the current approximately doubles every 10°C, and the voltage decreases linearly with temperature with a coefficient of approximately −2 mV/°C and gives a temperature coefficient of about 0.2%/°C, an order of magnitude lower than the best resistive devices. For example, the diode forward voltage of OMEGA Engineering's temperature sensing diode biased with an accurate 10 μA current in temperature range above 100 K decreases nearly linearly with temperature as indicated by the equation $V = 1.245 - 0.0024(V/kT)T$, where V is expressed in volts [60].

An advantage of diodes as temperature sensing materials is the fabrication possibility of smaller pixel size in comparison to resistors. The diodes have low $1/f$ power law noise and can potentially be manufactured in a standard CMOS technique.

More recently, the first realization and characterization of monolithic uncooled infrared sensor arrays, based on a-Si thin-film transistors have been demonstrated [61,62].

6.2.4.5 Other Materials

At present several research programs are focused toward enhancement of performance levels in excess of 10^9 cmHz$^{1/2}$W^{-1}. It is anticipated that new materials (e.g., SiGe, SiGeO, SiC) will be the basis of the next generation of semiconductor film bolometers [63].

Both amorphous as well as polycrystalline compound materials are studied. Amorphous GeSiO compounds with the Ge content in the order of 85% were grown by reactive sputtering in an Ar or Ar:O$_2$ environment [64–66], or by PECVD [67]. The TCR value up to 5.1%/°C has been reported, however the relatively high $1/f$ noise lowers the potential bolometer performance.

Different techniques have been used in fabrication of SiGe polycrystalline films including reduced-pressure CVD [68], MBE [69], or vapor deposition [70]. Up to now however, the detector performance is lower than VO$_x$ bolometers due to the higher $1/f$ noise of the polycrystalline materials. The crystalline materials, like quantum well Si/GeSi and AlGaAs/GaAs thermistors, are characterized by high TCR (up to 4.5%/°C for AlGaAs/GaAs) and simultaneously very low $1/f$ noise characteristics [71]. The noise level for both materials is being several orders of magnitude lower than that of a-Si and VO_x. These uncooled thermistor materials are hybridized with readout circuits by using conventional flip-chip assembly or wafer level adhesion bonding [45].

Another type of thermistor material is $YBa_2Cu_3O_{6+x}$ (YBaCuO), which belongs to a class of cooper oxides and is best known as a high-temperature superconductor (HTSC). Its conduction properties can be changed from metallic ($0.5 \leq x \leq 1$) to insulating ($0 \leq x \leq 0.5$) by suitably decreasing the oxygen content. For $x \approx 1$, YBaCuO possesses an orthorhombic crystal structure, exhibits metallic conductivity, and becomes superconductive upon cooling below its critical temperature. As x decreased to 0.5, the crystal structure undergoes a phase transition to a tetragonal structure and it exhibits semiconducting conductivity characteristics because it exists in a Fermi glass state. As x is decreased further below 0.3, YBaCuO becomes a Hubbard insulator with a well-defined energy gap on the order of 1.5 eV [72].

In the semiconducting state, YBaCuO exhibits a relatively large TCR (3 to 4%/°C) over a 60 K temperature range near room temperature. The large TCR, combined with the ease of thin film fabrication that is compatible with CMOS processing, makes YBaCuO attractive to microbolometer applications [73–77]. These devices have demonstrated voltage responsivities over 10^3 V/W, detectivities above 10^8 cmHz$^{1/2}$/W, and thermal time constant less than 15 msec using typical air-gap microbolometer structures [77].

It should be mentioned that perovskite metal-oxide manganites with colossal magnetoresistance effect and high room temperature TCR of 4.4%/°C [78] and thin-film carbon [79] have been also proposed for thermal imaging.

6.2.5 Superconducting Bolometers

The concept of a superconducting transition edge bolometer is not new [80–86]. Conventional-type superconducting bolometers are large area structures where the detector film acts as the radiation absorber. These have been built using tin [85] and niobium nitride [82]. The first bolometer to actually utilize the superconducting transition was a composite structure using a blackened aluminum foil absorber in conjunction with tantalum temperature sensor [82]. Another such composite structure makes use of aluminum as the temperature sensor and bismuth as the absorber [86].

Below a certain temperature, defined as the critical temperature T_c, the electrons at the Fermi energy condense into a coherent state made up of Cooper pairs, as shown in Figure 6.13. The parameter 2Δ is the superconducting energy gap. According to Bardeen–Cooper–Schrieffer theory, the value of 2Δ is given by $3.53kT_c$, which is in reasonably good agreement with the measured value of $(3.2–4.6)kT_c$ for superconducting elements [87]. For a transition temperature of 90 K (typical for YBaCuO), the energy gap predicted from this relationship is 27 meV. Often given the scatter in the experimental data for different materials, the precise value of the energy gap cannot be determined. As in an ordinary bolometer, there is not a direct photon–lattice interaction so response is slow, but independent of wavelength.

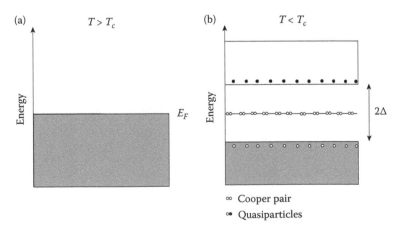

Figure 6.13 Energy diagram (a) above and (b) below the superconducting transition.

The theory, construction principles, and performance of superconductor bolometers are considered in several reviews [25,86–95]. General theory of bolometers described in Section 6.1 is also relevant for superconducting bolometers.

Figure 6.14 illustrates the operation of a bolometer in which the resistive element is a superconductor maintained near the midpoint of the superconducting-to-normal transition edge. This detector is usually called a transition edge sensor (TES). Because of the steepness of the edge near the critical temperature T_c, a small change in temperature causes a large change in resistance; so dR/dT is large and the width can be < 0.001 K. It was shown [89] that the value of D^* depends on the time constant of the detector: $D^* = const \times \tau^{1/2}$ (for comparison, see Equation 6.15). For bolometers especially operated in FIR regions called impulse detectivity as a figure of merit have been introduced. It is defined as the ratio between the detectivity and the square root of the response time, $D^*/\tau^{1/2}$, in units cm/J [90].

The transition between the superconducting and the normal stage can be used as an extremely sensitive detector. For example, Figure 6.15 shows a superconducting transition of a bilayer with 400Å of molybdenum and 750Å of gold, yielding a normal resistance of 330 mΩ [96]. Near its transition temperature of 440 mK, the sensitivity $\alpha \equiv dlogR/dlogT$ reaches 1100. Because the transition region is narrow (≈1 mK) compared to the temperature above the heat sink, the TES is nearly

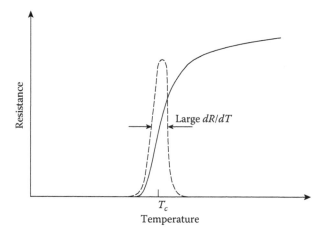

Figure 6.14 Typical resistance versus temperature plot with the derivative dR/dT superimposed on the transition edge.

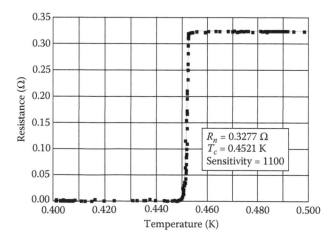

Figure 6.15 Resistance versus temperature for a high-sensitivity TES Mo/Au bilayer with superconducting transition at 444 mK. (From Benford D. J., and Moseley, S. H., "Superconducting Transition Edge Sensor Bolometer Arrays for Submillimeter Astronomy," *Proceedings of the International Symposium on Space and THz Technology.*)

isothermal across the transition. By varying the relative thicknesses of the normal metal (Au or Cu) and superconducting metal (Mo) in the bilayer process, the transition temperature can be tuned. In this manner, detectors optimized for performance in a variety of different optical loads and operating temperatures can be produced.

The experimentally achieved $D*$ values of bolometers by various authors are given in Table 6.2 [90,92]. The most sensitive and slowest bolometers are constructed such that the substrate makes a poor contact with base. This is accomplished either by special thin nylon fibers (Figure 6.16a [90]; samples 1 and 2 in Table 6.2), or the heat is drawn off through the thin substrate, the ends of which are connected to the base (Figure 6.16b; samples 3–5 in Table 6.2). Faster response is realized in bolometers constructions on a solid substrate (see Figure 6.16c). In some cases the superconducting film makes a direct contact with the massive substrate made from material having a high heat conductivity, such as sapphire (sample 7 in Table 6.2), makes contact with liquid helium (samples 8 and 9 in Table 6.2), or sandwiches a heat-insulating layer between the base and the superconducting sensitive element (sample 10 in Table 6.2). The last design has made it possible to build bolometers for the microsecond region having a threshold flux close to the limiting value set by fluctuations of the background radiation power.

A major difficulty to couple incident radiation to a superconducting bolometer resides in the high reflectivity of superconducting materials, especially the long wavelength and FIR. For example, YBaCuO exhibits >98% reflectivity at $\lambda > 20$ µm [97]. To overcome this problem, porous and granular black metals (usually silver and gold) offer both a large absorption and a low specific heat to give a reasonable compromise between the absorbing layer and satisfactory time response. However, this type of device is characterized by rather slow response time. The impulse detectivity ranges in the 10^{10} to 10^{11} cm/J interval and is one or two orders of magnitude lower than the realizable values with antenna-coupled devices.

The antenna-coupled design (Figure 6.16d) gives an effective way of increasing the sensitivity of thermal radiation detector while retaining a fast response. In this case, a thin-film antenna deposited onto a substrate receives the radiation, which induces displacement currents in it having a frequency corresponding to the radiation wavelength. The high-frequency currents heat the thin-film bolometer, which fulfills the role of converting thermal power into an electrical signal.

In 1977 Schwarz and Ulrich had the first paper devoted to room temperature antenna coupled metallic film infrared detectors [98]. In contrast to absorbing layer coupling, antenna coupling gives selective responsivity to both the spatial mode and the polarization of the incoming radiation. Since the effective detector area is an order of λ^2, this leads to a large absorbing area (also to a large thermal mass and slow response time) in the case of FIR radiation. However, fabrication of microbolometers (see Figure 6.16e), using conventional lithography and micromachining techniques, reaches time constants in the µs range and good detectivity. In this case, an antenna with rather large effective area can feed a superconducting microbridge of a much smaller area (a few µm^2).

Two frequency-independent antenna families, shown in Figure 6.17, can be built. In the first one, the antenna geometry is defined by angles (not by geometrical lengths) [92]; bow-tie and spiral antennas fall into so-called equiangular antennas. In the second family (see Figure 6.17c), the antenna is built up from coupled elements (e.g., dipoles). The finite dimensions of the structures restrict the antenna bandwidth roughly from $2r_{min}$ to $2r_{max}$ in terms of wavelength. Perturbations in the radiation bandwidth limits are overcome with the log-periodic structure.

Another class of antenna structures belongs to end-fire antennas, which are derived from a long wire traveling wave antennas. As opposed to previously described structures whose radiation direction lies in a plane orthogonal to the antenna plane, it lies along the antenna plane for end-fire geometries. These structures offer the possibility to make compact detector arrays with pixels containing V-antennas like those shown in Figure 6.16d.

Table 6.2 contains performance characteristics and Figure 6.18 presents detectivity as a function of response time for antenna coupled superconductor bolometers operated in the FIR spectral region. For the best detectors, NEP values close to the 2×10^{-12} W/Hz$^{1/2}$ phonon noise limited prediction have been obtained [92]. For liquid nitrogen (LN) cooled bolometers based on YBaCuO bolometers, a simple relation between detectivity and response time is observed: $D* = 2 \times 10^{12}\tau^{1/2}$, where $D*$ in cmHz$^{1/2}$/W and τ in sec.

The experimental data of room temperature detectors presented in Figure 6.18 concern bismuth or niobium microbridge sensors operated in the 10–100 µm wavelength range. The best NEP values have been obtained with suspended microbridges or low thermal conductance buffer layers with silicon substrates.

119

Table 6.2: Parameters of Superconductor Thermal Radiation Detectors

Material	Element Size (mm^2)	Temperature (K)	Sensitivity (V/W)	Time Constant (s)	D^* $(cmHz^{1/2}W^{-1})$	NEP $(W/Hz^{1/2})$	Remarks (Substrate/Antenna)
1. Sn	3×2	3.05	850	10^{-2}	3.6×10^{11}	7×10^{-13}	
2. Al	4×4	1.27	3.5×10^4	8×10^{-2}	1.2×10^{14}	7×10^{-13}	
3. Ni + Sn	1×1	0.4	2.2×10^6	10^{-3}	2.2×10^{13}	3.4×10^{-15}	
4. Pb + Sn	—	4.8	10^4	6×10^{-3}	—	4.5×10^{-15}	
5. NbN	0.1×0.1	6.5	5×10^5	10^{-4}	—	—	
6. Sn	0.15×0.15	3.7	10^4	6×10^{-3}	10^{10}	1.6×10^{-12}	
7. Pb + Sn	1×1	3.9	24	7×10^{-9}	1.2×10^9	8.4×10^{-11}	
8. Sn	10×10	3.63	1	2×10^{-8}	10^9	10^{-9}	
9. Ag + Sn	2.3×2.3	2.1	2.2	5×10^{-9}	2.6×10^9	9×10^{-10}	
10. Sn	1×1	3.3	4200	2×10^{-6}	5×10^{10}	2×10^{-12}	
11. Pb + Sn	0.02×0.00225	4.7	5700	2×10^{-8}	—	3×10^{-13}	
12. Mo:Ge	—	0.1	10^9	10^{-6}	1×10^{16}	1×10^{-18}	
13. Pb	—	3.7	10^5	10^{-8}	5×10^{11}	2×10^{-14}	Sapphire
14. Au + Pb + Sn	—	3.7	6000	2×10^{-8}	2×10^{11}	5×10^{-14}	Quartz/V antenna
15. YBaCuO	1×1	20	0.1	4×10^{-7}	2.5×10^6	4×10^{-13}	
16. YBaCuO	1×1	86	40	1.3×10^{-2}	6.7×10^7	1.5×10^{-9}	
17. YBaCuO	0.01×0.09	40	4×10^3	10^{-3}	10^8	2.5×10^{-11}	
18. YBaCuO	0.1×0.1	86	15	1.6×10^{-4}	3.3×10^7	3×10^{-10}	
19. YBaCuO	0.1×0.1	80	10^3 (A/W)	6×10^{-2}	3×10^8	10^{-10}	
20. YBaCuO	2.5×4	85	5.2	32	—	5.7×10^{-8}	
21. YBaCuO	—	90	2000	10^{-6}	2×10^9	5×10^{-12}	YSZ
22. YBaCuO	—	91	480	2×10^{-5}	2.2×10^9	4.5×10^{-12}	YSZ/Log-periodic
23. YBaCuO	—	90	4000	2×10^{-7}	4×10^9	2.5×10^{-12}	Si_3N_4/Suspended bridge
24. YBaCuO	—	88	2180	1×10^{-5}	1.1×10^9	9×10^{-12}	Si_3N_4/Log-periodic
25. YBaCuO	—	85	240	3×10^{-7}	8.3×10^8	1.2×10^{-11}	$NdGaO_3$/Bow-tie

Source: Khrebtov, I. A., *Soviet Journal of Optical Technology*, 58, 261–70, 1991; Kreisler, A. J., and Gaugue, A. "Recent Progress in HTSC Bolometric Detectors at Terahertz Frequencies," *Proceedings of SPIE 3481*, 457–68, 1998. With permission.

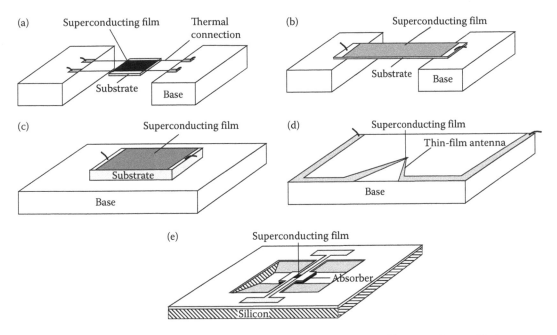

Figure 6.16 Superconductor bolometers: (a) isothermal bolometer, (b) nonisothermal bolometer, (c) bolometer on solid substrate, (d) antenna-coupled bolometer, and (e) micromachined bolometer. (From Khrebtov, I. A., *Soviet Journal of Optical Technology*, 58, 261–70, 1991.)

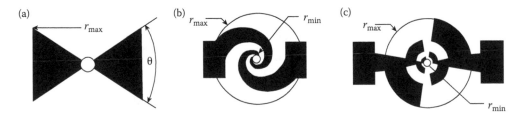

Figure 6.17 Frequency-independent planar antenna geometries: (a) bow-tie equiangular, (b) spiral equiangular, and (c) circular log-periodic. The sensing element is located at the center of the structure. (From Kreisler, A. J., and Gaugue, A., "Recent Progress in HTSC Bolometric Detectors at Terahertz Frequencies," *Proceedings of SPIE* 3481, 457–68, 1998. With permission.)

At present, conventional low critical temperature superconductors are used rarely. They offer unsurpassed performance in terms of both voltage responsivity and NEP. Their low operating temperature leads to high impulse detectivity, order of 10^{15} cm/J, due to lower specific heat. It should be also noted that very low resistance of these sensor causes impedance matching between antenna and sensor.

6.2.6 High-Temperature Superconducting Bolometers

The important discovery by Müller and Bednorz of a new class of superconducting materials [99,100], so called HTSCs, is undoubtedly one of the major breakthroughs in material science at the end of the twentieth century. Figure 6.19 shows the progression of the superconducting transition temperatures from the discovery of the phenomenon in mercury by Onnes in 1911. Between 1911 and 1974, the critical temperatures of metallic superconductors steadily increased from 4.2 K in mercury up to 23.2 K in sputtered Nb_3Ge films. Nb_3Ge held the record for the critical temperature in metallic superconductors until the unexpected discovery of superconductivity at 39 K in the intermetallic MgB_2 [101]. The first superconducting oxide $SrTiO_3$ characterized by transition temperature as low as 0.25 K was discovered in 1964. Discovery of high-temperature

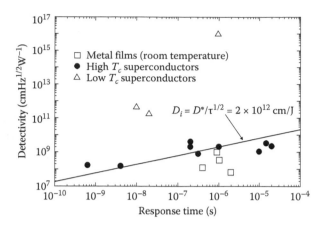

Figure 6.18 Detectivity as a function of response time for antenna coupled far infrared bolometers. (From Kreisler, A. J., and Gaugue, A., "Recent Progress in HTSC Bolometric Detectors at Terahertz Frequencies," *Proceedings of SPIE* 3481, 457–68, 1998. With permission.)

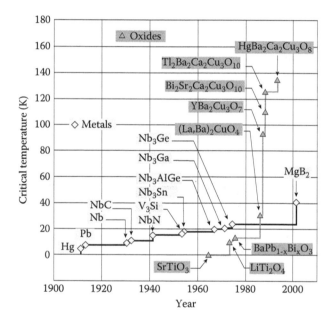

Figure 6.19 Evolution of the superconductive transition temperature subsequent to the discovery of the phenomenon.

superconductivity in the cuprate $(La,Ba)_2CuO_4$ ($T_c \approx 30$ K) opened a new field of research. Within less than a year a critical temperature well above 77 K could be achieved in $YBa_2Cu_3O_{7-x}$ (YBaCuO). So far, the highest transition temperature of 135 K has been found in $HgBa_2Ca_2Cu_3O_{8+x}$ [102].

Prior to a breakthrough in February 1987, superconductivity had been intensively studied, but the development of applications involving superconductors suffered from their extremely low working temperatures. In spite of tremendous progress accomplished in cryogenic technology, superconductivity has been used either when no classical alternative was available or when the required performances were impossible to achieve using a nonsuperconducting solution. Moreover, in the case of superconducting infrared detectors their usefulness was limited due to the stringent temperature control required, their poor radiation absorption characteristics due to being thin, and their fragility. The performance was generally limited by amplifier noise rather than radiation fluctuations.

All HTSC materials are so-called oxygen-deficient perovskites and their basic crystalline structure is similar to that of $CaTiO_3$ (parent mineral for the perovskite family). Although several HTSCs are known, attention is directed toward YBaCuO since it has received far more attention. Enomoto and Murakami made the earliest photoresponse measurements using granular $BaPb_{0.7}Bi_{0.3}O_3$ and reported encouraging results [103]. The parameters of YBaCuO appropriate to the device design are reviewed by Kruse [104]. The interest in using BiSrCaCuO has decreased because of the difficulties to establish one superconducting phase. There have been doubts using TlBaCaCuO because of the poisonous and volatile element Tl, even though the critical temperature is 125 K [105].

For HTSC, several authors have theoretically considered YBaCuO films from a theoretical viewpoint [25,104–108]. The performance of cryogenic superconducting detectors should be one to two orders of magnitude better than uncooled thermal detectors. As the transition temperature of good YBaCuO films is about 90 K, LN is a convenient cryogen for YBaCuO films. Richards and coworkers estimated that NEP in the range $(1–20) \times 10^{-12}$ W/Hz$^{1/2}$ [107], depending on a substrate, should be available. Such performance would exceed that of any other detectors operating at or above LN temperatures for wavelengths greater than 20 μm.

For a properly designed microbolometer detector, the value of the thermal conductance G_{th} is dominated by the support structure, not by YBaCuO. Values as small as 2×10^{-7} W/K are possible [104]. Since the density of YBaCuO is 6.3 g/cm^3, the specific heat at 90 K is 195 mJg^{-1}K^{-1}, a 75×75 μm^2 pixel 0.30 μm thick has a thermal capacity C_{th} of 2.1×10^{-9} J/K, thus the thermal time constant given by Equation 3.5 is 1.0×10^{-2} s. Assuming an absorptance ε of 0.8, the value of the temperature fluctuation-noise-limited detectivity given by Equation 3.23 is 2.1×10^{10} cmHz$^{1/2}$W^{-1}. In order to attain this value, the bias current must equal or exceed 3.5 μA. At lower values of bias current the detector will be Johnson-noise-limited. As the bias is raised to 3.5 μA, the microbolometer will become temperature-fluctuation-noise-limited. Assuming that the TCR is 0.33 K^{-1} the low-frequency responsivity (see Equation 6.2) at bias current of 3.5 μm will be 6.1×10^3 V/W.

The performance of YBaCuO HTSC bolometers in comparison with typical photon detectors in 2D FPAs operated at 77 K has been presented by Verghese et al. [108], and is shown in Figure 6.20. The calculations are carried out using values of measured properties of YBaCuO films on yttria-stabilized zirconia (YSZ) buffer layers on Si and Si$_3$O$_4$ (on Si) with τ = 10 ms, diffraction limited throughput and f/6 optics. The D* can be as high as 3×10^{10} cmHz$^{1/2}$W^{-1}. Also shown, for

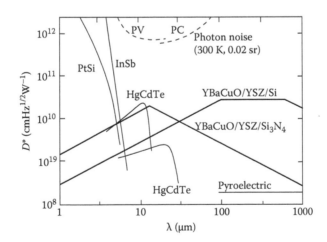

Figure 6.20 Detectivity as a function of wavelength for diffraction-limited pixels with FOV = 0.02 sr (f/6 optics) and τ = 10 ms. The thick lines show the predicted D* for HTSC bolometers on silicon and Si$_3$N$_4$ membranes using YBaCuO films. These lines were calculated using estimates for the minimum achievable heat capacity and thermal conductance and using measurements of voltage noise in HTSC bolometers. Typical values of D* for InSb, PtSi, and HgCdTe detectors in 2D FPAs operated at 77 K are shown for comparison. Also are shown the photon noise limits for photovoltaic and photoconductive detectors which view 300 K radiation in a 0.02 sr FOV. (From Verghese, S., Richards, P. L., Char, K., Fork, D. K., and Geballe, T. H., *Journal of Applied Physics* 71, 2491–98, 1992. With permission.)

comparison, are the photon noise limited D^* and estimates of the performance of detectors now used in large format imaging arrays. We can see that D^* falls at short wavelengths because the membrane technology is not able to provide a small enough G_{th} for a small detector area, and τ becomes shorter than 10 ms. It also falls at long wavelengths because the resistance fluctuation noise becomes important in large area detectors. The D^* of a Si_3O_4 membrane bolometer has a potentially useful peak near 10 μm, which is a very important wavelength for thermal imaging. Because of higher bulk thermal conductivity of silicon, the region of photon-noise-limited D^* appears at longer wavelengths than for Si_3N_4 membrane bolometer. The limit to D^* from bolometer noise is higher than that of bolometers on Si_3O_4 because of the lower NEP of YBaCuO on silicon.

The above analysis clearly indicates that HTSC bolometers are especially useful in the FIR ($\lambda > 20$ μm), where it is difficult to find sensitive detectors that operate at relatively high temperatures (e.g., T > 77 K). Moreover, it is estimated that the pixel production cost of HTSC bolometers is potentially several orders of magnitude less than that of HgCdTe and InSb.

A review of substrate materials and deposition techniques suitable for fabrication of high-quality HTSC films for electronic and optoelectronic applications is given by Sobolewski [105]. The SEM micrographs of HTSC thin films reveal two types of structures: random or granual structure, oriented or epitaxial structure.

The random structure consists of many small (≈ 1 μm) superconducting grains embedded in a nonsuperconducting matrix. Point contacts between grains may act as Josephson junctions. Films made of this type of structure have broad transitions and relatively low T_c and the critical current I_c. The grain boundaries result also in excessive $1/f$ noise.

The oriented structure is a crystal growth with c-axis perpendicular to the plane of the substrate. Films made of this structure have sharp transitions and relatively high T_c and I_c. For detector applications, the film thickness must be less than the optical penetration depth of the material, ≈ 0.15 μm. For granual structures, the film should be patterned into a microbridge to reduce the number of Josephson junctions biased. Ideally we would like a linear chain of junctions so that there is only a single conduction path. For epitaxial structures, we must artificially create weak links in the film to form the Josephson junction.

It is generally accepted that good-quality HTSC films require high-quality dielectric substrates, which combine desired dielectric properties with a good lattice match, enabling epitaxial growth of the films. Except for diamond, most suitable substrate materials have similar volume specific heat at 77–90 K. In all cases, it is very much larger than is seen at liquid helium temperatures. Consequently, the thermal time tends to be long. Therefore, one important requirement for substrate material is strength, so that it can be made very thin. Some substrates that are favorable for film growth, such as $SrTiO_3$ and $LaAlO_3$ are too weak to produce thin layers of millimeter dimensions. However, good quality bolometers have been fabricated using these substrates [25,26,90–93,107–112]. Also such substrates as silicon, sapphire, ZrO_2, or SiN are used [91–93]. Much attention has been given to silicon substrates because of their compatibility with on-chip electronics implementation in semiconductor technology.

In general, however, the substrate should fulfill additional issues. In order to minimize the phonon escape time [e.g., for phonon-cooled, hot-electron bolometers (HEB)], the substrate should have a high thermal conductivity and offer a low thermal interface resistance R_b to the superconducting film. Secondly, it should have good properties to propagate the radiation signal when the readout circuitry is implemented (i.e., in the GHz range); within this respect, the dielectric loss tangent should be low and dielectric constant should be fitted to the propagation line and also comparable with the antenna size. Finally, the substrate material should be transparent to signal radiation; for example, in FIR sensors the receiving antenna is usually illuminated from the substrate back side by means of a focusing lens. See Table 6.3 for gathered substrate parameters [113].

Recently, the main effort in HTSC bolometers technology is directed toward improving the performance of microbolometer FPAs fabricated by micromachining on silicon substrates. At the beginning, the YBaCuO films in these devices were sandwiched between two layers of silicon nitride with thin YSZ layers to buffer YBaCuO from the silicon nitride [114,115]. These 125×125 μm² devices were estimated to have a NEP of 1.1×10^{-12} W/Hz$^{1/2}$ near 5 Hz with a 5 μA bias (neglecting contact noise). A drawback of this design was that the YBaCuO was grown on an a-Si nitride underlayer, which precludes the possibility of epitaxial YBaCuO growth. The YBaCuO, therefore, was polycrystalline with a broad resistance transition, which limits the bolometer responsivity and the grain boundaries result in excessive $1/f$ noise.

Table 6.3: Thermal and Dielectric Characteristics of Some Substrate Materials

Material	MgO	Al_2O_3	$LaAlO_3$	$YAlO_3$	YSZ
Thermal conductivity at 90 K(W/Kcm)	3.4	6.4	0.35	0.3	0.015
R_b with YBaCuO at 90 K ($KW^{-1}cm^2$)	5×10^{-4}	10^{-3}	10^{-3}	—	10^{-3}
$\tan\delta$ at 10 GHz, 77 K	7×10^{-6}	8×10^{-6}	5×10^{-6}	10^{-5}	4×10^{-4}
ε_r at 10 GHz, 77 K	10	10	23	16	32

Source: Burns, M. J., Kleinsasser, A. W., Delin, K. A., Vasquez, R. P., Karasik, B. S., McGraph, W. R., and Gaidis, M. C., *IEEE Transactions on Applied Superconductivity* 7, 3564–67, 1997. With permission.

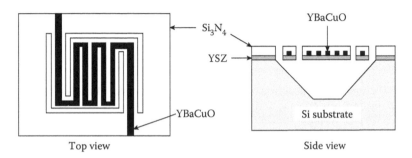

Figure 6.21 Schematic diagram of YBaCuO microbolometer using epitaxial YBaCuO on an epitaxial YSZ buffer layer on a silicon substrate. (From Foote, M. C., Johnson, B. R., and Hunt, B. D., "Transition Edge $YBa_2Cu_3O_{7-x}$ Microbolometers for Infrared Staring Arrays," *Proceedings of SPIE* 2159, 2–9, 1994. With permission.)

Incorporating epitaxial YBaCuO films improve the performance of the bolometers [116–118]. Figure 6.21 shows a schematic diagram of the microbolometer design using epitaxial YBaCuO films. The fabrication process of these devices has been described by Johnson and colleagues [117]. The superconductor film was deposited by pulsed laser deposition onto an epitaxial YSZ buffer layer that had been deposited onto a bare, unoxidized 3-inch silicon wafer. Gold contact metal was deposited onto YBaCuO film by RF sputtering. The YSZ, YBaCuO, and gold were deposited in the same deposition chamber, without breaking vacuum between depositions. The gold and YBaCuO meander line were patterned by conventional photolithography, the YBaCuO was passivated with YSZ and silicon nitride, and silicon etch pits were formed by anisotropic etching to thermally isolate the microbolometers. As is shown in Figure 6.21, the membrane is suspended over an etch pit on a silicon wafer, and supported only by lateral silicon nitride legs approximately 8 μm wide.

For the above devices with the size 140×105 μm², detectivity of $(8 \pm 2) \times 10^9$ cmHz$^{1/2}$W^{-1} has been measured on a single element at a bias current of 2 μA. It is one of the highest D^* reported on any semiconducting microbolometer operating at temperatures higher than 70 K. The NEP was 1.5×10^{-12} W/Hz$^{1/2}$ at 2 Hz, at a temperature of 80.7 K. The thermal constant was 105 ms. The noise power spectral density was found to scale with frequency as $1/f^{3/2}$. Linear arrays of microbolometers have also been fabricated. The measured responsivity of detectors varied by less than 20% over the 6 mm length of the 64-element linear array.

The progress attained toward realizing HTSC photon detectors is described in many papers (e.g., see [90–93]). Since the appearance 20 years ago of the first reports demonstrating a thermal detection action with HTSC sensors, progress in their technology has benefited mainly from newly developed superconducting nanostructures, which are especially promising candidates for FIR detection [91,93]. In the mid-infrared range their detectivities are similar to those of LN cooled photon detectors. The latter, however, are still leading in terms of response time. Figure 6.22 compares the performance of HTSC bolometers, where a plot of detectivity as a function of response time is shown. The performance of a typical cooled photoconductive HgCdTe ($\lambda = 3$–20 μm) is also included in Figure 6.22. As pointed previously, an impulse detectivity $D_i = D^*\tau^{-1/2}$ [cm/J] with D_i value 2×10^{11} cm/J corresponds to state-of-the-art results for the absorber-coupled HTSC bolometers. However, this value is an order of magnitude lower than the average D_i observed for antenna-coupled HTSC bolometers in the FIR region [93].

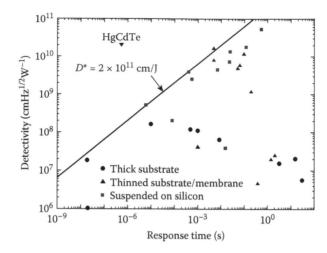

Figure 6.22 The detectivity as a function of the response time for thin-film HTSC bolometers ($\lambda = 0.8$–20 μm). (From Kreisler, A. J., and Gaugue, A., *Superconductor Science and Technology*, 13, 1235–45, 2000. With permission.)

It should be noted that superconducting infrared detectors are also classified as photon detectors. More information on this topic is included in Section 22.4.3.

6.3 HOT ELECTRON BOLOMETERS

In principle, the HEB is quite similar to the transition-edge bolometer described in Section 6.2.5 where small temperature changes caused by the absorption of incident radiation strongly influence resistance of biased sensor near its superconducting transition. The main difference between HEBs and ordinary bolometers is the speed of their response. High speed is achieved by allowing the radiation power to be directly absorbed by the electrons in the superconductor, rather than using a separate radiation absorber and allowing the energy to flow to the superconducting TES via phonons, as ordinary bolometers do. After photon absorption, a single electron initially receives the energy $h\nu$, which is rapidly shared with other electrons, producing a slight increase in the electron temperature. In the next step, the electron temperature subsequently relaxes to the bath temperature through emission of phonons.

In comparison with TES, the thermal relaxation time of the HEB's electrons can be made fast by choosing a material with a large electron-phonon interaction. The development of superconducting HEB mixers has lead to the most sensitive systems at frequencies in the terahertz region, where the overall time constant has to be a few tens of picoseconds. These requirements can be realized with a superconducting microbridge made from NbN, NbTiN, or Nb on a dielectric substrate [94].

The HEBs can work according to two mechanisms that allow electrons to exchange their energy faster than they heat the phonons:

- The phonon-cooled HEB principle was suggested by Gershenzon et al. [119] and the first realized by Karasik et al. [120]

- The diffusion-cooled HEB principle was suggested by Prober [121] and the first realization reported by Skalare et. al. [122]

A synthetic presentation of both mechanisms has been given by McGrath [123].

Figure 6.23a shows the basic operation of the phonon-cooled bolometer. In this type of device, hot electrons transfer their energy to the phonons with the time τ_{eph}. In the next step, the excess of phonon energy escapes towards the substrate with the time τ_{esc}. Several conditions should be fulfilled to make phonon-cooled mechanism effective: (i) the electron–electron interaction time (τ_{ee}) must be much shorter than τ_{eph}, (ii) the superconducting film must be very thin (a few nm) and the film to substrate thermal conductance must be very high ($\tau_{esc} \ll \tau_{eph}$) to obtain an efficient phonon escape from superconductor to substrate, and (iii) the substrate thermal conductivity

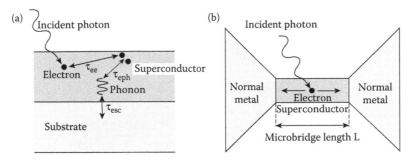

Figure 6.23 Hot-electron bolometer mechanisms: (a) phonon-cooled and (b) diffusion-cooled principles.

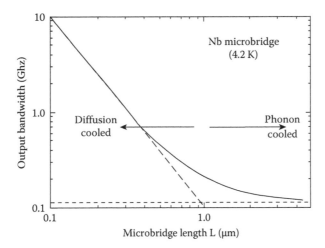

Figure 6.24 Output bandwidth as a function of microbridge length for niobium HEB mixers. For L shorter than 1 µm, the cooling mechanism is by electron diffusion to a normal metal; for larger L, the phonon-cooling mechanism is the dominant one. (From Burke, P. J., Schoelkopf, R. J., Prober, D. E., Skalare, A., McGrath, W. R., Bumble, B., and LeDuc, H. G., *Applied Physics Letters* 68, 3344–46, 1996. With permission.)

must be very high and very good thermal contact between the substrate and a cold finger must be insured.

In diffusion-cooled bolometer, the principle of mechanism is shown in Figure 6.23b, hot electrons transfer their energy by diffusion to a normal metal, which forms the electrical contacts to the external detector readout circuitry and/or the arms of a planar antenna. In this case, the length of the superconducting microbridge must be very short, with maximum value $L_{max} = 2(D_e\tau_{ee})^{1/2}$, where D_e is the electron diffusivity. As was shown by Burke et al. [124], the bolometer bandwidth is inversely proportional to the squared microbridge length, which lies in the submicronic range (see Figure 6.24). The bandwidth of diffusion-cooled bolometers in not limited by such parameter as τ_{eph}. As a result, in comparison with phonon-cooled bolometers, larger intermediate frequency values are obtained. For diffusion cooling, the interface between the contact pads and the superconducting film is crucial, while for phonon cooling, the interface between the film and the substrate is crucial. It should be pointed out that the distinction is, to some extent, arbitrary since phonon cooling also exists in diffusion cooled bolometers and vice versa.

Typically, phonon-cooled HEBs are made from ultrathin films of NbN, whereas diffusion-cooled devices use Nb or Al. Current state-of-the-art NbN technology is capable of routinely delivering 3 nm thick devices that are 500 nm² in size with transition temperature T_c above 9 K and the transition width of 0.5 K. NbN films are deposited on a dielectric (typically high resistivity >10 kΩcm silicon) by dc reactive magnetron sputtering The superconducting bridge is defined by means of electron beam lithography. Its length varies between 0.1 and 0.4 µm and the width between 1 and

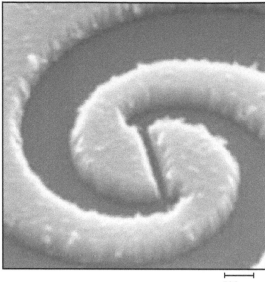

500 nm

Figure 6.25 Central part of a planar logarithmic spiral antenna with the NbN hot-electron microbridge. (From Semenov, A. D., Gol'tsman, G. N., and Sobolewski, R., *Superconductor Science and Technology*, 15, R1–R16, 2002. With permission.)

4 μm. For example, Figure 6.25 shows a micrograph of a central part for a planar logarithmic spiral antenna with the NbN hot-electron microbridge [125].

Theory of HEBs is still under development and typically invokes a hot-spot resistive region in the center, whose size responds to changes in the applied power. This model was originally developed by Skocpol et al. [126], and was later applied to superconducting HEB mixers [125,127,128]. The region where the actual temperature exceeds the critical temperature and switches into the normal state is called the hot spot. When the radiation is absorbed, the length of the hot spot increases and its boundaries begin to move toward the electric contacts until the hot spot reaches thermal equilibrium. The speed of boundaries determines the response time, but other effects such as the interaction of radiation with magnetic vortices play a role. When comparing diffusion-cooled and phonon-cooled HEBs, the latter provides a smaller noise temperature and are therefore preferred.

Superconducting HEB detectors are finding a key role in FIR and terahertz wavelengths. Because the detector is much smaller than the wavelength being received, an antenna and associated coupling circuitry are needed to bring the radiation to the detector. HEB mixers can be made either in a waveguide configuration with a horn antenna or as quasioptical mixers. The more traditional approach is waveguide coupling, in which radiation is first collected by a horn into a single-mode waveguide (typically a rectangular guide), and then, a transition (probe) couples radiation from the waveguide onto a lithographed thin-film transmission line on the detector chip. One major complication of the waveguide approach is that the mixer chip must be very narrow and must be fabricated on an ultrathin substrate. These requirements are helpful using modern micromachining techniques (see Figure 6.26 [129]).

Above ~1 THz, the quasioptical coupling is more common. The quasioptical coupling approach omits the intermediate step of collecting the radiation into a waveguide, and instead uses a lithographed antenna (e.g., twin-slot or a logarithmic spiral antenna) on the detector chip itself. Such mixers are substantially simpler to fabricate and may be produced using thick substrate (see Figure 6.27 [130]). The substrate with the feed antenna and microbridge is mounted to the reserve side of a hyperhemispherical or elliptical lens. The reflection loss at the lens surface can be minimized with a quarter-wavelength antireflection coating.

The performance of selected HEBs is shown in Figure 4.7. At all frequencies, phonon-cooled HEBs have a lower noise temperature than that of diffusion-cooled devices, although improvements to the latter type may be possible [94]. More information about hot-electron bolometric mixers is given in Section 22.4.2.

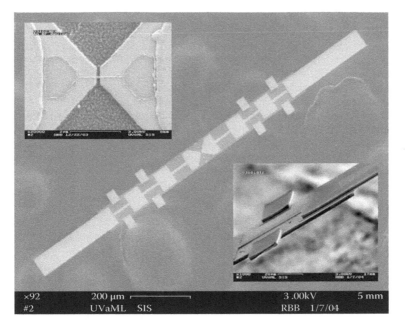

Figure 6.26 Images of a 585-GHz diffusion-cooled HEB mixer chip for a waveguide mount, fabricated using an ultrathin silicon substrate. The dimensions of the HEB bridge are 150 nm long by 75 nm wide; the chip itself is 800 µm long, and 3 µm thick. Protruding from the sides and ends of the chip are 2 µm thick gold leads, which provide electrical and thermal contact to the waveguide block, as well as mechanical support for the chip. (From Bass, R. B., "Hot Electron Bolometers on Ultra-Thin Silicon Chips with Beam Leads for a 585 GHz Receiver, PhD dissertation.)

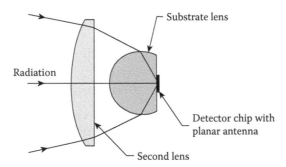

Figure 6.27 Schematic diagram of the "reverse-microscope" quasioptical coupling approach. (From Rutledge, D., and Muha, M., "Imaging Antenna Arrays," *IEEE Transactions on Antennas and Propagation* AP-30, 535–40, 1982. With permission.)

It should be mentioned, in this brief discussion, about other HEB materials such as a normal metal (usually copper), high T_c superconductors, and n-type InSb. HTSCs exhibit a very short electron-phonon interaction time (typically between 1 and 2 ps at 80–90 K in YBaCuO), so only phonon-cooled devices have been realized to date (electron diffusion mechanism is negligible). Moreover, the analysis of HTSC HEBs is rather different from low temperature counterparts due to the high operating temperature [131]. Limited results have been obtained with YBaCuO HEBs [92,93].

InSb HEBs have found practical applications. Since their bandwidth is about 4 MHz, their use is limited. The parameters of these bolometers are largely set by the InSb [132,133]. Typical voltage responsivities are 100 to 1000 V/W, the thermal conductance is about 5×10^{-5} W/K, and the thermal capacity of the electron sea $C_{th} \approx (3/2)nkV$, where n is the carrier concentration and V is the volume of the detector. Assuming detector volume as 10^{-2} cm^{-3} and $n \approx 5 \times 10^{13}$ cm^{-3}; we can estimate thermal capacity as $C_{th} \approx 10^{-11}$ J/K and the detector time constant as 2×10^{-7} s (see Equation 3.5). For this n-type InSb sample operating at 4 K, the thermally limited NEP equal 2×10^{-13} W/Hz$^{1/2}$ has been estimated.

The quantum efficiency of the IR detector depends on the absorption coefficient. Since the free carrier absorption coefficient increases as λ^2, the performance of a device utilizing this effect should improve as the wavelength increases. The value for $\alpha \approx 22$ cm^{-1} at 1 mm wavelength is comparable with that found in extrinsic germanium photoconductive detectors [133], but the value found at 100 μm is to small ($\alpha \approx 0.30$ cm^{-1}) to fabricate an efficient detector at this wavelength. From these estimation results that n-type InSb HEBs are useful at wavelengths of 1 mm or somewhat less, but these devices become ineffective at a wavelength shorter than about 300 μm.

REFERENCES

1. S. P. Langley, "The Bolometer," *Nature* 25, 14–16, 1881.

2. P. W. Kruse, L. D. McGlauchlin, and R. B. McQuistan, *Elements of Infrared Technology: Generation, Transmission, and Detection*, Wiley, New York, 1962.

3. R. A. Wood, "Monolithic Silicon Microbolometer Arrays," in *Uncooled Infrared Imaging Arrays and Systems*, eds. P. W. Kruse and D. D. Skatrud, 43–121, Academic Press, San Diego, CA, 1997.

4. P. W. Kruse, *Uncooled Thermal Imaging. Arrays, Systems, and Applications*, SPIE Press, Bellingham, WA, 2001.

5. C. Jansson, U. Ringh, and K. Liddiard, "Theoretical Analysis of Pulse Bias Heating of Resistance Bolometer Infrared Detectors and Effectiveness of Bias Compensation," *Proceedings of SPIE* 2552, 644–52, 1995.

6. K. C. Liddiard, "Thin-Film Resistance Bolometer IR Detectors," *Infrared Physics* 24, 57–64, 1984.

7. A. Tanaka, S. Matsumoto, N. Tsukamoto, S. Itoh, K. Chiba, T. Endoh, A. Nakazato, et al., "Infrared Focal Plane Array Incorporating Silicon IC Process Compatible Bolometer," *IEEE Transactions on Electron Devices* 43, 1844–80, 1996.

8. S.-B. Ju, Y.-J. Yong, and S.-G. Kim, "Design and Fabrication of High Fill-Factor Microbolometer Using Double Sacrificial Layers," *Proceedings of SPIE* 3698, 180–89, 1999.

9. J. M. Shive, *Semiconductor Detectors*, Van Nostrand, New York, 1959.

10. E. M. Wormser, "Properties of Thermistor Infrared Detectors," *Journal of the Optical Society of America* 43, 15–21, 1953.

11. R. De Waars and E. M. Wormser, "Description and Properties of Various Thermal Detectors," *Proceedings of IRE* 47, 1508–13, 1959.

12. R. W. Astheimer, "Thermistor Infrared Detectors," *Proceedings of SPIE* 443, 95–109, 1984.

13. R. C. Jones, "Immersed Radiation Detectors," *Applied Optics* 1, 607, 1962.

14. S. G. Bishop and W. J. Moore, "Chalcogenide Glass Bolometers," *Applied Optics* 12, 80–83, 1973.

15. S. Baliga, A. Doctor, and M. Rost, "Sputtered Film Thermistor IR Detectors," *Proceedings of SPIE* 2225, 72–78, 1994.

16. J. C. Mather, "Bolometer Noise: Nonequilibrium Theory," *Applied Optics* 21, 1125–29, 1982.

17. W. S. Boyle and K. F. Rogers, Jr., "Performance Characteristics of a New Low-Temperature Bolometer," *Journal of the Optical Society of America* 49, 66–69, 1959.

18. E. H. Putley, "Thermal Detectors," in *Optical and Infrared Detectors,* ed. R. J. Keyes, 71–100, Springer, Berlin, 1977.

19. N. Coron, "Infrared Helium Cooled Bolometers in the Presence of Background Radiation: Optimal Parameters and Ultimate Performances," *Infrared Physics* 16, 411–19, 1976.

20. S. Zwerdling, R. A. Smith, and J. P. Theriault, "A Fast, High-Responsivity Bolometer Detector for the Very Far Infrared," *Infrared Physics* 8, 271–336, 1968.

21. F. J. Low, "Low-Temperature Germanium Bolometer," *Journal of the Optical Society of America* 51, 1300–04, 1961.

22. B. T. Draine and A. J. Sievers, "A High Responsivity, Low-Noise Germanium Bolometer for the Far Infrared," *Optics Communications* 16, 425–28, 1976.

23. M. A. Kinch, "Compensated Silicon-Impurity Conduction Bolometer," *Journal of Applied Physics* 42, 5861–63, 1971.

24. E. E. Haller, "Physics and Design of Advanced IR Bolometers and Photoconductors," *Infrared Physics* 25, 257–66, 1985.

25. P. L. Richards, "Bolometers for Infrared and Millimeter Waves," *Journal of Applied Physics* 76, 1–24, 1994.

26. E. E. Haller, "Advanced Far-Infrared Detectors," *Infrared Physics & Technology* 35, 127–46, 1994.

27. N. S. Nishioka, P. L. Richards, and D. P. Woody, "Composite Bolometers for Submillimeter Wavelengths," *Applied Optics* 17, 1562–67, 1978.

28. E. L. Dereniak and D. G. Crowe, *Optical Radiation Detectors,* Wiley, New York, 1984.

29. P. M. Downey, A. D. Jeffries, S. S. Meyer, R. Weiss, F. J. Bachner, J. P. Donnelly, W. T. Lindley, R. W. Mountain, and D. J. S. Silversmith, "Monolithic Silicon Bolometers," *Applied Optics* 23, 910–14, 1984.

30. R. E. Flannery and J. E. Miller, "Status of Uncooled Infrared Imagers," *Proceedings of SPIE* 1689, 379–95, 1992.

31. R. A. Wood, C. J. Han, and P. W. Kruse, "Integrated Uncooled IR Detector Imaging Arrays," *Proceedings of IEEE Solid State Sensor and Actuator Workshop,* 132–35, Hilton Head Island, SC, June 1992.

32. R. A. Wood, "Micromachined Bolometer Arrays Achieve Low-Cost Imaging," *Laser Focus World,* 101–6, June 1993.

33. R. A. Wood, "Uncooled Thermal Imaging with Monolithic Silicon Focal Planes," *Proceedings of SPIE* 2020, 322–29, 1993.

34. K. C. Liddiard, "Thin Film Monolithic Arrays for Uncooled Thermal Imaging," *Proceedings of SPIE* 1969, 206–16, 1993.

35. M. H. Unewisse, S. J. Passmore, K. C. Liddiard, and R. J. Watson, "Performance of Uncooled Semiconductor Film Bolometer Infrared Detectors," *Proceedings of SPIE* 2269, 43–52, 1994.

36. R. A. Wood, "High-Performance Infrared Thermal Imaging with Monolithic Silicon Focal Planes Operating at Room Temperature," *Electron Devices Meeting, IEDM'93, Technical Digest, International,* 175–77, 1993.

37. B. Fieque, J. L. Tissot, C. Trouilleanu, A. Crates, and O. Legras, "Uncooled Microbolometer Detector: Recent Developments at Ulis," *Infrared Physics and Technology* 49, 187–91, 2007.

38. M. Kohin and N. Buttler, "Performance Limits of Uncooled VO_x Microbolometer Focal Plane Arrays," *Proceedings of SPIE* 5406, 447–53, 2004.

39. P. W. Kruse, "Can the 300 K Radiating Background Noise Limit be Attained by Uncooled Thermal Imagers?" *Proceedings of SPIE* 5406, 437–46, 2004.

40. F. Niklaus, C. Jansson, A. Decharat, J.-E. Källhammer, H. Pettersson, and G. Stemme, "Uncooled Infrared Bolometer Arrays Operating in a Low to Medium Vacuum Atmosphere: Performance Model and Tradeoffs," *Proceedings of SPIE* 6542, 65421M, 2007.

41. H. Jerominek, F. Picard, and D. Vincent, "Vanadium Oxide Films for Optical Switching and Detection," *Optical Engineering* 32, 2092–99, 1993.

42. E. Kuźma, "Contribution to the Technology of Critical Temperature Resistors," *Electron Technology* 26(2/3), 129–42, 1993.

43. H. Jerominek, T. D. Pope, M. Renaud, N. R. Swart, F. Picard, M. Lehoux, S. Savard, et al., "64×64, 128×128 and 240×320 Pixel Uncooled IR Bolometric Detector Arrays," *Proceedings of SPIE* 3061, 236–47, 1997.

44. C. Chen, X. Yi, J. Zhang, and B. Xiong, "Micromachined Uncooled IR Bolometer Linear Array Using VO_2 Thin Films," *International Journal of Infrared Millimeter Waves* 22, 53–58, 2001.

45. F. Niklaus, C. Vieider, and H. Jakobsen, "MEMS-Based Uncooled Infrared Bolometer Arrays: A Review," *Proceedings of SPIE* 6836, 68360D, 2007.

46. M. Soltani, M. Chaker, E. Haddad, R. V. Kruzelecky, and J. Margot, "Effects of Ti-W Codoping on the Optical and Electrical Switching of Vanadium Dioxide Thin Films Grown by a Reactive Pulsed Laser Deposition," *Applied Physics Letters* 85, 1958–60, 2004.

47. Y.-H. Han, K.-T. Kim, H.-J. Shin, and S. Moon, "Enhanced Characteristics of an Uncooled Microbolometer Using Vanadium-Tungsten Oxide as a Thermoelectric Material," *Applied Physics Letters* 86, 254101-3, 2005.

48. A. J. Syllaios, T. R. Schimert, R. W. Gooch, W. L. McCardel, B. A. Ritchey, and J. H. Tregilgas, "Amorphous Silicon Microbolometer Technology," *MRS Proceedings* 609, A14.4 1–6, 2000.

49. K. C. Liddiard, U. Ringh, C. Jansson, and O. Reinhold, "Progress of Swedish-Australian Research Collaboration on Uncooled Smart IR Sensors," *Proceedings of SPIE* 3436, 578–84, 1998.

50. B. I. Craig, R. J. Watson, and M. H. Unewisse, "Anisotropic Excess Noise Within a-Si:H," *Solid-State Electronics* 39, 807–12, 1996.

51. J. L. Tissot, F. Rothan, C. Vedel, M. Vilain, and J.-J. Yon, "LETI/LIR's Amorphous Silicon Uncooled IR Systems," *Proceedings of SPIE* 3379, 139–44, 1998.

52. J. L. Tissot, J. L. Martin, E. Mottin, M. Vilain, J. J. Yon, and J. P. Chatard, "320×240 Microbolometer Uncooled IRFPA Development," *Proceedings of SPIE* 4130, 473–79, 2000.

53. J. F. Brady, T. S. Schimert, D. D. Ratcliff, R. W. Gooch, B. Ritchey, P. McCardel, K. Rachels, et al., "Advances in Amorphous Silicon Uncooled IR Systems," *Proceedings of SPIE* 3698, 161–67, 1999.

54. A. Tanaka, M. Suzuki, R. Asahi, O. Tabata, and S. Sugiyama, "Infrared Linear Image Sensor Using a Poly-Si pn Junction Diode Array," *Infrared Physics* 33, 229–36, 1992.

55. Y. P. Xu, R. S. Huang, and G. A. Rigby, "A Silicon-Diode-Based Infrared Thermal Detector Array," *Sensors and Actuators A* 37–38, 226–30, 1993.

56. M. Ueno, O. Kaneda, T. Ishikawa, K. Yamada, A. Yamada, M. Kimata, and M. Nunoshita, "Monolithic Uncooled Infrared Image Sensor with 160×120 Pixels," *Proceedings of SPIE* 2552, 636–43, 1995.

57. J.-K. Kim and C.-H. Han, "A New Uncooled Thermal Infrared Detector Using Silicon Diode," *Sensors and Actuators* A89, 22–27, 2001.

58. J. E. Murguia, P. K. Tedrow, F. D. Shepherd, D. Leahy, and M. M. Weeks, "Performance Analysis of a Thermionic Thermal Detector at 400K, 300K, and 200K," *Proceedings of SPIE* 3698, 361–75, 1999.

59. E. S. Young, *Fundamentals of Semiconductor Devices*, McGraw-Hill, New York, 1978.

60. *Omega Complete Temperature Measurement Handbook and Encyclopedia*, Vol. 26, Omega Engineering Inc., Stanford, CT, U-1–24, 1989.

61. L. Dong, R. F. Yue, and L. T. Liu, "A High Performance Single-Chip Uncooled a-Si TFT Infrared Sensor," *Proceedings of Transducers 2003*, Vol. 1, 312–15, 2003.

62. L. Dong, R. Yue, and L. Liu, "Fabrication and Characterization of Integrated Uncooled Infrared Sensor Arrays Using a-Si Thin-Film Transistors as Active Elements," *Journal of Microelectromechanical Systems* 14, 1667–77, 2005.

63. M. H. Unewisse, B. I. Craig, R. J. Watson, O. Reinhold, and K. C. Liddiard, "The Growth and Properties of Semiconductor Bolometers for Infrared Detection," *Proceedings of SPIE* 2554, 43–54, 1995.

64. E. Iborra, M. Clement, and L. Herrero, "Sangrador IR Uncooled Bolometers Based on Amorphous GeSiO on Silicon Micromachined Structures," *Journal of Microelectromechanical Systems* 11, 322–29, 2002.

65. D. Butler and M. Rana, "Radio frequency Sputtered SiGe and SiGeO Thin Films for Uncooled Infrared Detectors," *Thin Solid Films* 514, 355–60, 2006.

66. A. Ahmed and R. Tait, "Noise Behavior of Amorphous GeSiO for Microbolometer Applications," *Infrared Physics and Technology* 46, 468–72, 2005.

67. M. Moreno, A. Kosarev, A. Torres, and R. Ambrosio, "Fabrication and Performance Comparison of Planar and Sandwich Structures of Micro-Bolometers with Ge Thermo-Sensing Layer," *Thin Solid Films* 515, 7607–10, 2007.

68. V. Leonov, N. Perova, P. De Moor, B. Du Bois, C. Goessens, B. Grietens, A. Verbist, C. Van Hoof, and J. Vermeiren, "Micromachined Poly-SiGe Bolometer Arrays for Infrared Imaging and Spectroscopy," *Proceedings of SPIE* 4945, 54–63, 2003.

69. I. Chistokhin, I. Michailovsky, B. Fomin, and E. Cherepov, "Polycrystalline Layers of Silicon-Germanium Alloy for Uncooled IR Bolometers," *Proceedings of SPIE* 5126, 407–14, 2003.

70. R. Yue, L. Dong, and L. Liu, "Monolithic Uncooled 8×8 Bolometer Arrays Based on Poly-SiGe Thermistor," *International Journal of Infrared and Millimeter Waves* 27, 995–1003, 2006.

71. S. Wissmar, L. Hoglund, J. Andersson, C. Vieider, S. Susan, and P. Ericsson, "High Signal to Noise Ratio Quantum Well Bolometer Material," *Proceedings of SPIE* 6401, 64010N, 2006.

72. G. Yu and A. J. Heeger, "Photoinduced Charge Carriers in Insulating Cuprates: Fermi Glass Insulator, Metal–Insulator Transition and Superconductivity," *International Journal of Modern Physics* B7, 3751, 1993.

73. P. C. Shan, Z. Celik-Butler, D. P. Butler, A. Jahanzeb, C. M. Travers, W. Kula, and R. Sobolewski, "Investigation of Semiconducting YBaCuO Thin Films: A New Room Temperature Bolometer," *Journal of Applied Physics* 80, 7118–23, 1996.

74. L. Mechin, J. C. Villegier, and D. Bloyet, "Suspended Epitaxial YbaCuO Microbolometers Fabricated by Silicon Micromachining: Modeling and Measurements," *Journal of Applied Physics* 81, 7039–47, 1997.

75. A. Jahanzeb, C. M. Travers, Z. Celik-Butler, D. P. Butler, and S. G. Tan, "A Semiconductor YBaCuO Microbolometer for Room Temperature IR Imaging," *IEEE Transactions on Electron Devices* 44, 1795–801, 1997.

76. L. Phong and S. Qiu, "Room Temperature YBaCuO Microbolometers," *Journal of Vacuum Science and Technology* A18, 635–38, 2000.

77. M. Almasri, Z. Çelik-Butler, D. P. Butler, A. Yarafanakul, and A. Yildiz, "Semiconducting YBaCuO Microbolometers for Uncooled Broad-Band IR Sensing," *Proceedings of SPIE* 4369, 264–73, 2001.

78. J. Kim and A. Grishin, "Free-Standing Epitaxial La$_{1-x}$(Sr,Ca)$_x$MnO$_3$ Membrane on Si for Uncooled Infrared Microbolometer," *Applied Physics Letters* 87, 033502, 2005.

79. M. Liger and Y.-C. Tal, "A 32 × 32 Parylene-Pyrolyzed Carbon Bolometer Imager," *Proceedings of MEMS 2006*, 106–9, 2006.

80. D. H. Andrews, W. F. Brucksch, Jr., W. T. Ziegler, and E. R. Blanchard, "Attenuated Superconductors: I. For Measuring Infra-Red Radiation," *Review of Scientific Instruments* 13, 281–92, 1942.

81. R. M. Milton, "A Superconducting Bolometer for Infrared Measurements," *Chemical Reviews* 39, 419–22, 1946.

82. D. H. Andrews, R. M. Milton, and W. DeSorbo, "A Fast Superconducting Bolometer," *Journal of the Optical Society of America* 36, 518–24, 1946.

83. N. Fuson, "The Infra-Red Sensitivity of Superconducting Bolometers," *Journal of the Optical Society of America* 38, 845–53, 1948.

84. H. D. Martin and D. Bloor, "The Applications of Superconductivity to the Detection of Radiant Energy," *Cryogenics* 1, 159, 1961.

85. C. L. Bertin and K. Rose, "Radiant-Energy Detection by Superconducting Films," *Journal of Applied Physics* 39, 2561–68, 1968.

86. J. Clarke, G. I. Hoffer, P. L. Richards, and N. H. Yeh, "Superconductive Bolometers for Submillimeter Wavelengths," *Journal of Applied Physics* 48, 4865–79, 1977.

87. C. P Poole, H. A. Farach, and R. J. Creswick, *Superconductivity*, Academic Press, San Diego, CA, 1995.

88. K. Rose, C. L. Bertin, and R. M. Katz, "Radiation Detectors," in *Applied Superconductivity*, Vol. 1, ed. V. L. Newhouse, 268–308, Academic Press, New York, 1975.

89. K. Rose, "Superconductive FIR Detectors," *IEEE Transactions on Electron Devices* ED-27, 118–25, 1980.

90. I. A. Khrebtov, "Superconductor Infrared and Submillimeter Radiation Receivers," *Soviet Journal of Optical Technology* 58, 261–70, 1991.

91. H. Kraus, "Superconductive Bolometers and Calorimeters," *Superconductor Science and Technology* 9, 827–42, 1996.

92. A. J. Kreisler and A. Gaugue, "Recent Progress in HTSC Bolometric Detectors at Terahertz Frequencies," *Proceedings of SPIE* 3481, 457–68, 1998.

93. A. J. Kreisler and A. Gaugue, "Recent Progress in High-Temperature Superconductor Bolometric Detectors: From the Mid-Infrared to the Far-Infrared (THz) Range," *Superconductor Science and Technology* 13, 1235–45, 2000.

94. J. Zmuidzinas and P. L. Richards, "Superconducting Detectors and Mixers for Millimeter and Submillimeter Astrophysics," *Proceedings of IEEE* 92, 1597–616, 2004.

95. G. H. Rieke, "Infrared Detector Arrays for Astronomy," *Annual Review of Astronomy and Astrophysics* 45, 77–115, 2007.

96. D. J. Benford and S. H. Moseley, "Superconducting Transition Edge Sensor Bolometer Arrays for Submillimeter Astronomy," *Proceedings of the International Symposium on Space and THz Technology.* Available: http://www.eecs.umich.edu/ ~jeast/benford_2000_4_1.pdf

97. Z. Zhang, T. Le, M. Flik, and E. Carvalho, "Infrared Optical-Constant of the HTc Superconductor $YBa_2Cu_3O_7$," *Journal of Heat Transfer* 116, 253–56, 1994.

98. S. E. Schwarz and B. T. Ulrich, "Antenna Coupled Thermal Detectors," *Journal of Applied Physics* 85, 1870–73, 1977.

99. K. A. Müller and J. G. Bednorz, "The Discovery of a Class of High-Temperature Superconductors," *Science* 237, 1133–39, 1987.

100. J. G. Bednorz and K. A. Müller, "Possible High T_c Superconductivity in the Ba-La-Cu-O System," *Zeitschrift fur Physik B-Condensed Matter* 64, 189–93, 1986; "Perovskite-type Oxides-The New Approach to High-T_c Superconductivity," *Reviews of Modern Physics* 60, 585–600, 1988.

101. J. Nagamatsu, N. Nakagawa, T. Muranaka, Y. Zenitani, and J. Akimitsu, "Superconductivity at 39°K in Magnesium Diboride," *Nature* 410, 63–64, 2001.

102. A. Schilling, M. Cantoni, J. D. Guo, and H. R. Ott, "Superconductivity Above 130 K in the Hg-Ba-Ca-Cu-O System," *Nature* 363, 56–58, 1993.

103. Y. Enomoto and T. Murakami, "Optical Detector Using Superconducting $BaPb_{0.7}Bi_{0.3}O_3$," *Journal of Applied Physics* 59, 3807–14, 1986.

104. P. W. Kruse, "Physics and Applications of High-T_c Superconductors for Infrared Detectors," *Semiconductor Science and Technology* 5, S229–S329, 1990.

105. R. Sobolewski, "Applications of High-T_c Superconductors in Optoelectronics," *Proceedings of SPIE* 1512, 14–27, 1991.

106. Q. Hu and P. L. Richards, "Design Analysis of High T_c Superconducting Microbolometer," *Applied Physics Letters* 55, 2444–46, 1989.

107. P. L. Richards, J. Clarke, R. Leoni, Ph. Lerch, S. Verghese, M. B. Beasley, T. H. Geballe, R. H. Hammond, P. Rosenthal, and S. R. Spielman, "Feasibility of the High T_c Superconducting Bolometer," *Applied Physics Letters* 54, 283–85, 1989.

108. S. Verghese, P. L. Richards, K. Char, D. K. Fork, and T. H. Geballe, "Feasibility of Infrared Imaging Arrays Using High-T_c Superconducting Bolometers," *Journal of Applied Physics* 71, 2491–98, 1992.

109. J. C. Brasunas, S. H. Moseley, B. Lakew, R. H. Ono, D. G. McDonald, J. A. Beall, and J. E. Sauvageau, "Construction and Performance of a High-Temperature-Superconductor Composite Bolometer," *Journal of Applied Physics* 66, 4551–54, 1989.

110. J. Brasunas and B. Lakew, "High T_c Bolometer Developments for Planetary Missions," *Proceedings of SPIE* 1477, 166–73, 1991.

111. S. Verghese, P. L. Richards, S. A. Sachtjen, and K. Char, "Sensitive Bolometers Using High-T_c Superconducting Thermometers for Wavelengths 20–300 µm," *Journal of Applied Physics* 24, 4251–53, 1993.

112. J. Brasunas and B. Lakew, "High T_c Bolometer with Record Performance," *Applied Physics Letters* 64, 777–78, 1994.

113. M. J. Burns, A. W. Kleinsasser, K. A. Delin, R. P. Vasquez, B. S. Karasik, W. R. McGraph, and M. C. Gaidis, "Fabrication of High-Tc Hot-Electron Bolometric Mixers for Terahertz Applications," *IEEE Transactions on Applied Superconductivity* 7, 3564–67, 1997.

114. B. R. Johnson, T. Ohnstein, H. Marsh, S. B. Dunham, and P. W. Kruse, "YBa$_2$Cu$_3$O$_7$ Superconducting Microbolometer Linear Arrays," *Proceedings of SPIE* 1685, 139–45, 1992.

115. B. R. Johnson and P. W. Kruse, "Silicon Microstructure Superconducting Microbolometer Infrared Arrays," *Proceedings of SPIE* 2020, 2–7, 1993.

116. M. C. Foote, B. R. Johnson, and B. D. Hunt, "Transition Edge Yba$_2$Cu$_3$O$_{7-x}$ Microbolometers for Infrared Staring Arrays," *Proceedings of SPIE* 2159, 2–9, 1994.

117. B. R. Johnson, M. C. Foote, H. A. Marsh, and B. D. Hunt, "Epitaxial YBa$_2$Cu$_3$O$_7$ Superconducting Infrared Microbolometers on Silicon," *Proceedings of SPIE* 2267, 24–30, 1994.

118. B. R. Johnson, M. C. Foote, and H. A. Marsh, "High Performance Linear Arrays of YBa$_2$Cu$_3$O$_7$ Superconducting Infrared Microbolometers on Silicon," *Proceedings of SPIE* 2475, 56–61, 1995.

119. E. M. Gershenzon, G. N. Gol'tsman, I. G. Gogidze, Y. P. Gusev, A. J. Elant'ev, B. S. Karasik, and A. D. Semenov, "Millimeter and Submillimeter Range Mixer Based on Electronic Heating of Superconducting Films in the Resistive State," *Superconductivity* 3, 1582–97, 1990.

120. B. Karasik, G. N. Gol'tsman, B. M. Voronov, S. I. Svechnikov, E. M. Gershenzon, H. Ekström, S. Jacobsson, E. Kollberg, and K. S. Yngvesson, "Hot Electron Quasioptical NbN Superconducting Mixer," *IEEE Transactions on Applied Superconductivity* 5, 2232–35, 1995.

121. D. E. Prober, "Superconducting Terahertz Mixer Using a Transition-Edge Microbolometer," *Applied Physics Letters* 62, 2119–21, 1993.

122. A. Skalare, W. R. McGrath, B. Bumble, H. G. LeDuc, P. J. Burke, A. A. Vereijen, R. J. Schoelkopf, and D. E. Prober, "Large Bandwidth and Low Noise in a Diffusion-Cooled Hot-Electron Bolometer Mixer," *Applied Physics Letters* 68, 1558–60, 1996.

123. W. R. McGrath, "Novel Hot-Electron Bolometer Mixers for Submillimeter Applications: An Overview of Recent Developments," *Proceedings of URSI International Symposium on Signals, Systems, and Electronics*, 147–52, 1995.

124. P. J. Burke, R. J. Schoelkopf, D. E. Prober, A. Skalare, W. R. McGrath, B. Bumble, and H. G. LeDuc, "Length Scaling of Bandwidth and Noise in Hot-Electron Superconducting Mixers," *Applied Physics Letters* 68, 3344–46, 1996.

125. A. D. Semenov, G. N. Gol'tsman, and R. Sobolewski, "Hot-Electron Effect in Superconductors and Its Applications for Radiation Sensors," *Superconductor Science and Technology* 15, R1–R16, 2002.

126. W. J. Skocpol, M. R. Beasly, and M. Tinkham, "Self-Heating Hotspots in Superconducting Thin-Film Microbridges," *Journal of Applied Physics* 45, 4054–66, 1974.

127. A. D. Semenov and H.-W. Hübers, "Frequency Bandwidth of a Hot-Electron Mixer According to the Hot-Spot Model," *IEEE Transactions on Applied Superconductivity* 11, 196–99, 2001.

128. H.-W. Hübers, "Terahertz Heterodyne Receivers," *IEEE Journal of Selected Topics in Quantum Electronics* 14, 378–91, 2008.

129. R. B. Bass, "Hot Electron Bolometers on Ultra-Thin Silicon Chips with Beam Leads for a 585 GHz Receiver. PhD dissertation. Available: http://www.ece.virginia.edu/uvml/sis/Papers/rbbpapers/eucas03OI.pdf

130. D. Rutledge and M. Muha, "Imaging Antenna Arrays," *IEEE Transactions on Antennas and Propagation* AP-30, 535–40, 1982.

131. B. S. Karasik, W. R. McGrath, and M. C. Gaidis, "Analysis of a High-Tc Hot-Electron Superconducting Mixer for Terahertz Applications," *Journal of Applied Physics* 83, 1581–89, 1997.

132. M. A. Kinch and B. V. Rollin, "Detection of Millimetre and Submillimetre Wave Radiation by Free Carrier Absorption in a Semiconductor," *British Journal of Applied Physics* 14, 672–76, 1963.

133. E. H. Putley, "InSb Submillimeter Photoconductive Devices," in *Semiconductors and Semimetals*, Vol. 12, eds. R. K. Willardson and A. C. Beer, 143–68, Academic Press, New York, 1977.

7 Pyroelectric Detectors

Whenever a pyroelectric crystal undergoes a change of temperature, surface charge is produced in a particular direction as a result of the change in its spontaneous polarization with temperature. This effect has been known as a physically observable phenomenon for many centuries, being described by Theophrastus in 315 BC [1]. Its name "pyroelectricity" was introduced by Brewster [2]. The concept of using the pyroelectric effect for detecting radiation was proposed very early by Ta [3], however in practice, a little progress was made due to the lack of suitable materials. The importance of the pyroelectric effect in infrared detection was becoming obvious about 50 years ago, due to scientific activity from such authors as Chynoweth [4], Cooper [5,6], Hadni et al. [7], and others [8–13]. A widely acclaimed review of work up to 1969 has been published by Putley [14], and further developments have been reported by Baker et al. [15], Putley [16], Liu and Long [17], Marshall [18], Porter [19], Joshi and Dawar [20], Whatmore [21,22], Ravich [23], Watton [24], and Robin et al. [25]. More recently published papers have been shown that the pyroelectric micromachining version of uncooled thermal detectors reach fundamental limits [26–33].

7.1 BASIC PRINCIPLE AND OPERATION OF PYROELECTRIC DETECTORS

Pyroelectricity has been known for the last twenty-four centuries, but a while ago in 1938 a chemist, Ta, at the Sorbonne in Paris, proposed that tourmaline crystals could be used as IR sensors in spectroscopy [3,32]. Some research on pyroelectric detectors was conducted in the next decade in the United Kingdom, United States, and Germany, but the results appeared only in classified documents. In 1962, Cooper presented the first theory of the pyroelectric detector and conducted experiments using barium titanate [5,6]. Also that year, Lang proposed the use of pyroelectric devices for measuring temperature changes as small as 0.2 μK. Later on, an explosive growth of papers in pyroelectric infrared detectors had begun [32].

Pyroelectric materials are those with a temperature dependent spontaneous electrical polarization. There are 32 known crystal classes. Twenty-one are noncentrosymmetric and 10 of these exhibit temperature-dependent spontaneous polarization. Under equilibrium conditions the electrical asymmetry is compensated by the presence of free charges. If, however, the temperature of the material is changed at a rate faster than these compensating charges can redistribute themselves, an electrical signal can be observed. This means that the pyroelectric detector is an AC device, unlike other thermal detectors that detect temperature levels rather than temperature changes. This generally limits the low frequency operation, and for a maximum output signal the rate of charge of the input radiation should be comparable to the electrical time constant of the element.

Most pyroelectric are also ferroelectric, which means that the direction of their polarization can be reversed at the application of a suitable electric field, and the polarization reduces to zero at some temperature known as the Curie temperature T_C. Generally, the pyroelectric materials considered for thermal detector arrays are the lead-based perovskite oxides such as lead titanate [$PbTiO_3$: PT]. These materials have structural similarities with the mineral perovskite ($CaTiO_3$). The basic formula is ABO_3; where A is lead, O is oxygen, and B may be one, or a mixture, of cations, for example, lead zirconate titanate [$Pb(ZrTi)O_3$: PZT], barium strontium titanate [$BaSrTiO_3$: BST], lead scandium tantalate [$Pb(Sc_{0.5}Ta_{0.5})O_3$: PST], and lead magnesium niobate [$Pb(Mg_{1/3}Nb_{2/3})O_3$: PMN]. Often dopants are added to these basic formulations to enhance or tune the material properties. Above Curie temperature, T_C, these materials form a symmetric nonpolar, cubic structure (Figure 7.1). Above the Curie temperature the material is paraelectric and it has no pyroelectric activity. On cooling they undergo a structural phase transition to form a polar, ferroelectric phase.

The above materials can be further subdivided into two groups. The conventional pyroelectric materials, such as PT and PZT, operate at room temperature well below their Curie temperature without the need for an applied field. Requirements for detector temperature stabilization is minimal or can be eliminated since there is little variation in detector performance over quite a large temperature range.

It is, however, possible to operate ferroelectrics at or above T_C, with an applied bias field, in the mode of a dielectric bolometer. Its operation is associated with the change of permittivity with a temperature in the region of the transition. The permittivity is strongly temperature dependent, but less so with an applied field (see dashed line in Figure 7.1). The applied bias charges the element, and the heating due to the incident radiation results in an increment of permittivity and

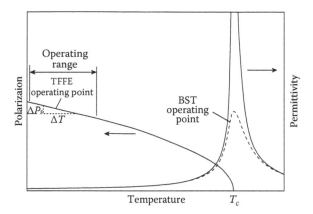

Figure 7.1 Thermal behavior of ferroelectric material. Dashed line shows the effect of an applied field on permittivity.

hence a signal voltage. This second group of materials (including BST, PST, and PMN) has T_C slightly below the detector operating temperature, resulting in minimal pyroelectricity.

In general, the case for ferroelectric materials, the electrical displacement, D, is a sum of contribution from the spontaneous (zero field) polarization, P_s, and the field-induced polarization (i.e., $\varepsilon_o \varepsilon_r E$). It is also important to realize that the permittivity around the transition is nonlinear, hence an integral is required

$$D = P_s(T) + \varepsilon_o \int_0^E \varepsilon_r(E', T) dE', \tag{7.1}$$

where ε_o is the permittivity of free space and ε_r is the relative permittivity of the pyroelectric material.

The pyroelectric coefficient is the change in displacement with temperature

$$p = \frac{dD}{dT}\Big|_E = \frac{dP_s}{dT} + \varepsilon_o \int_0^E \frac{d\varepsilon_r}{dT} dE'. \tag{7.2}$$

To obtain high pyroelectric coefficient from dielectric bolometer materials it is desirable to have a large variation in permittivity with temperature and/or high bias fields should be applied. As mentioned previously, the bias field generally reduces the permittivity variation and even introduces positive slopes, hence there is a limit to the benefits gained from simply applying high fields.

7.1.1 Responsivity

The pyroelectric detector can be considered as a small capacitor with two conducting electrodes mounted perpendicularly to the direction of spontaneous polarization, as shown in Figure 7.2 with its equivalent electrical circuit. To orient the sensitive element before use, the material is heated and an electrical field applied. When the detector is operated, the change in polarization will appear as a charge on the capacitor and a current will be generated, the magnitude of which depends on the temperature rise and the pyroelectrical coefficient, p, of the material.

The polarization change due to a change in temperature ΔT is described by

$$P = p\Delta T. \tag{7.3}$$

The pyroelectric charge generated is given by

$$Q = pA\Delta T, \tag{7.4}$$

so the effect of a temperature change on a pyroelectric material is to cause a current $I_{ph} = dQ/dT$ to flow in an external circuit (see Figure 7.2), such that:

$$I_{ph} = Ap\frac{dT}{dt}, \tag{7.5}$$

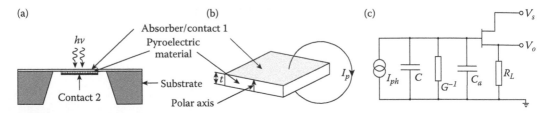

Figure 7.2 Pyroelectric detector: (a) schematic cross section, (b) pyroelectric element, and (c) equivalent electrical circuit.

where A is the detector area, p is the component of the pyroelectric coefficient normal to the electrodes, and dT/dt is the rate of change of temperature with time. Taking into account Equation 3.4, the photocurrent is equal:

$$I_{ph} = \frac{\varepsilon p A \Phi_o \omega}{G_{th}\left(1+\omega^2\tau_{th}^2\right)^{1/2}}.$$ (7.6)

To make a pyroelectric device work it is necessary to modulate the source of energy. This can be achieved by mechanical chopping or by moving the detector relative to the source of radiation. The element with an area A and thickness t gives a thermal capacitance $C_{th} = c_{th}At$ (where c_{th} is the volume specific heat) connected via a thermal conductance G_{th} to a heat sink. This gives a thermal time constant $\tau_{th} = C_{th}/G_{th}$.

The detector possesses an electrical capacitance C and presents an electrical conductance G ($G^{-1} = R$ is a parallel resistance) to a low noise high input impedance buffer amplifier such as a MOSFET, with input capacitance C_a. In practice the amplifier resistance is large compared with the shunt resistor G^{-1}, and can be ignored; but C_a is not always small compared with the detector capacitance C. This produces an electrical time constant $\tau_e = (C_a + C)/G$. The τ_{th} and τ_e are the fundamental parameters that determine the frequency response.

The current responsivity is:

$$R_i = \frac{I_{ph}}{\Phi_o} = \frac{\varepsilon p A \omega}{G_{th}\left(1+\omega^2\tau_{th}^2\right)^{1/2}},$$ (7.7)

and for low frequencies ($\omega \ll 1/\tau_{th}$) the response is proportional to ω. At frequencies greater than this value, the response is constant being:

$$R_i = \frac{\varepsilon p}{c_{th}t}.$$ (7.8)

If the detector is connected to a high impedance amplifier, such as that shown in Figure 7.2c, then the observed signal is equal to the voltage produced by the charge Q. The detector may be represented as a capacitor C, a current generator I_{ph}, and a shunt conductance G, as shown in Figure 7.2c. The voltage generated is therefore given by:

$$V = \frac{I_{ph}}{\left(G^2+\omega^2C^2\right)^{1/2}},$$ (7.9)

and then the voltage responsivity is equal

$$R_v = \frac{V}{\Phi_o} = \frac{R\varepsilon p A \omega}{G_{th}\left(1+\omega^2\tau_{th}^2\right)^{1/2}\left(1+\omega^2\tau_e^2\right)^{1/2}},$$ (7.10)

where $\tau_e = C/G$ is the electrical time constant. The last equations simplifies at frequencies that are high compared with $(\tau_{th})^{-1}$ and $(\tau_e)^{-1}$ to give

$$R_v = \frac{\varepsilon p}{\varepsilon_o\varepsilon_r c_{th}A\omega}.$$ (7.11)

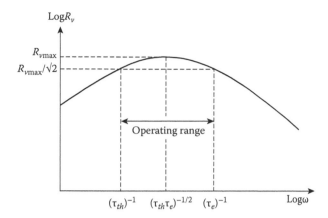

Figure 7.3 Frequency dependence of voltage responsivity of a pyroelectric detector.

Equation 7.11 shows that, at high frequencies, the voltage responsivity of a pyroelectric detector is inversely proportional to frequency. At low frequencies this is modified by the electrical and thermal time constants, as in Equation 7.10, so that the true frequency response is of the form shown in Figure 7.3. The maximum value occurs at a frequency of $(\tau_e\tau_{th})^{-1/2}$ with a value of:

$$R_{v\text{max}} = \frac{\varepsilon p AR}{G_{th}(\tau_e + \tau_{th})}. \tag{7.12}$$

From Equation 7.12 it is easy to show that the responsivity is maximized by minimizing G_{th}. The thermal capacity should also be reduced, within the constraint of maintaining an appropriate thermal time constant τ_{th}.

At the frequencies $\omega = (\tau_e)^{-1}$ and $\omega = (\tau_{th})^{-1}$:

$$R_v = \frac{R_{v\text{max}}}{\sqrt{2}}. \tag{7.13}$$

It is not possible to distinguish between τ_e and τ_{th} from responsivity measurements alone. Putley has discussed in detail the analysis of the performance from responsivity and noise measurements together [34].

The selection of τ_e and τ_{th} is determined by a number of factors. For a low-frequency, high-sensitivity operation, the device is mounted with freely suspended active element to minimize conduction of heat to the surroundings. The thermal capacity of the element is adjusted to maximize the response at the frequency of interest. To realize this device, a thin, low thermal capacity, high electrical capacity element can be used. Typically τ_{th} is within the range 0.01–10 s. The τ_e however, can be anywhere between 10^{-12} s and 100 s, depending on the sizes of the detector capacitance and the shunt resistor.

For high frequency operation, one of the time constants (usually τ_e) is reduced so that its inverse is greater than the maximum frequency of interest. This can be done by minimizing the element's electrical capacitance (using an edge-electroded structure) and feeding the output into a 50 Ω line. Because the speed of the pyroelectric response is limited only by the frequency of the vibrational polarization of the crystal lattice (about 10^{12} Hz), these detectors have the potential to be extremely fast. Austan and Glass have experimentally verified response time of 9 ns [35] while Roundy et al. have demonstrated practical detectors with 170 ps response time [36].

The above considerations of detector response does not take into account the input resistance of the amplifier (R_a) that will appear in parallel with the resistor R. For low-frequency detectors, $R_a \gg R$ and R_a can be ignored in this case. For fast detectors, $R_a \ll R$ and thus R_a determines the electrical time constant and the device responsivity.

More rigorous analyses of pyroelectric detectors have been performed by many authors, taking into account the effects of mounting techniques and black coatings [35–41]. The above treatment, however, is adequate for the majority of applications.

Generally, in bulk material devices $\tau_e < \tau_{th}$. However, considerations carried out by Putley [16] and Porter [19] pointed out that τ_e can also be larger than τ_{th}, depending on materials and electrical elements. This situation is in typical thin film structures. The major consequences of thickness decreasing of the pyroelectric material, t, are an increase of the electrical capacity and a decrease of the heat capacity. In addition, it is difficult to improve thermal insulation to the surroundings by the same amount as the thickness, because bulk pyroelectric materials are good thermal insulators [28]. Since the ratio τ_e/τ_{th} scales roughly as

$$\frac{RC(t)}{C_{th}(t)/G_{th}} \propto \frac{1}{t^2},$$

thus the ratio switch from less than one to larger than one when scaling down from a single crystal to a thin film. The frequency behavior for thin films is similar to those shown in Figure 7.3; however, the time constants have the opposite order (i.e., $\tau_e > \tau_{th}$). As a consequence, the voltage responsivity in intermediate frequency region (operating region; see Figure 7.3) is determined by other parameters:

■ For bulk devices

$$R_v \cong \frac{\varepsilon p A R}{C_{th}} \tag{7.14}$$

■ For thin film devices

$$R_v \cong \frac{\varepsilon p A}{C G_{th}}. \tag{7.15}$$

In bulk devices, a value of parallel resistance of 10 GΩ is typically applied (R should not exceed the gate impedance of the amplifier). The last equation indicates that in thin film detectors the parallel resistance is not directly involved in the voltage responsivity. It may be avoided, because thin film capacitors exhibit larger currents than bulk capacitors.

Assuming that $C = \varepsilon_o \varepsilon_r A/t$, Equation 7.15 can be further modified to

$$R_v \cong \frac{\varepsilon p t}{\varepsilon_o \varepsilon_r G_{th}}, \tag{7.16}$$

which shows that the responsivity is independent of the detector area A.

7.1.2 Noise and Detectivity

There are three major noise sources in a pyroelectric detector with a shunt resistor [14,17,19,21,28]:

■ Thermal fluctuation noise

■ Johnson noise

■ Amplifier noise

The first two types of noises are described in Section 3.1. The Johnson noise connected with the shunt resistor R is described by Equation 3.16. However, in most devices at moderate frequencies of operation (1 Hz–1 kHz), the noise is dominated by the AC electrical conductance of the detector element. The AC conductance of the device has two components: a frequency-independent component R^{-1} and a frequency-dependent component G_d:

$$G_d = \omega C \tan\delta, \tag{7.17}$$

where $\tan\delta$ is the loss tangent of the detector material. For frequencies much less than $\omega = (RC\tan\delta)^{-1}$, the Johnson noise is simply given by:

$$V_{Jr}^2 = \frac{4kTR\Delta f}{1 + \omega^2 \tau_e^2}, \tag{7.18}$$

which leads to an ω^{-1} dependence at frequencies $\omega \gg \tau_e^{-1}$.

For frequencies much greater than $\omega = (RC\tan\delta)^{-1}$, the noise generated by the AC conductance of the detector element will dominate so that:

$$V_{Jd}^2 = 4kT\Delta f \frac{\tan\delta}{C} \frac{1}{\omega} \quad \text{for} \quad C \gg C_a. \tag{7.19}$$

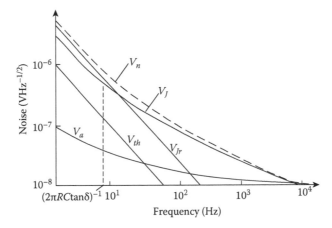

Figure 7.4 Relative magnitudes of noise voltages in a typical pyroelectric detector. (From Whatmore, R. W., "Pyroelectric Devices and Materials," *Reports on Progress in Physics* 49, 1335–86, 1986. With permission.)

This type of noise, also called dielectric noise, dominates at high frequencies.

It is interesting to compare relative magnitudes of these various noise sources for a typical detector. These are shown plotted as a function of frequency in Figure 7.4 [21]. It has been assumed that both the thermal and electrical time constants are longer than one second. In nearly all practical detectors the thermal noise is insignificant and is often ignored in calculations. It can be seen that the loss-controlled Johnson noise dominates above 20 Hz, while below this frequency the resistor-controlled Johnson noise and the amplifier current noise (V_{ai}) contribute almost equally significantly to the total noise. At very high frequencies the amplifier voltage noise (V_{av}) dominates.

At high frequencies [greater than τ_e^{-1} and $(RC\tan\delta)^{-1}$] the detectivity (from Equations 2.6, 7.11, and 7.19) is given by equation

$$D^* = \frac{\varepsilon t}{\left(4kT\right)^{1/2}} \frac{p}{c_{th}\left(\varepsilon_o\varepsilon_r\tan\delta\right)^{1/2}} \frac{1}{\omega^{1/2}}. \tag{7.20}$$

The fall in detectivity with a frequency of $\omega^{-1/2}$ means that the D^* will be at a maximum of a rather higher frequency than R_v (see Equation 7.11) and falls more slowly (as $\omega^{-1/2}$) than R_v (as ω^{-1}) above this maximum. For most detectors, D^* maximizes in the 1–100 Hz range and a reasonably flat D^* can be achieved in the range of a few Hz to several hundred Hz.

There are a number of other sources of unwanted signals in pyroelectric detectors, mostly associated with the environment. Environmental temperature fluctuations can give rise to spurious signals at low frequencies, or, if the rate of external temperature change is very large, the detector amplifier can be saturated.

A major limitation to the usefulness of pyroelectric detectors is that they are microphonic; electrical outputs are produced by mechanical vibration or acoustic noise. This microphonic signal may dominate all other noise sources if the detector is in a high vibration environment. The basic course of microphony is the piezoelectric nature of pyroelectric materials, meaning that a change in polarization is produced by a mechanical strain as well as by a change in temperature. In general lower microphony is obtained by making the mounting of the pyroelectric less rigid. Further reduction in microphony has been achieved by using compensated detectors or by selection of a material with low piezoelectric coupling to the dominant strain components. Shorrocks et al. have discussed methods by which the microphony of pyroelectrical arrays may be reduced to a very low level [42].

The compensating element (see Figure 7.5 [21]) is connected either in series or parallel opposition with the sensitive element, but coated with a reflecting electrode and/or mechanically screened so that it is not subject to the input radiation flux. The compensating element should be placed in a position that makes it thermally and mechanically similar to the detector element so that a signal due to temperature changes or mechanical stresses is canceled.

Two other sources of environment noise affect pyroelectric detector operation. If a pyroelectric detector is subjected to changes in ambient temperature fast pulses are sometimes observed,

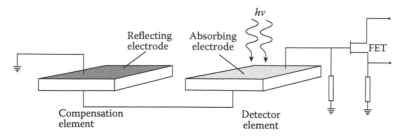

Figure 7.5 Compensated pyroelectric detectors. (From Whatmore, R. W., "Pyroelectric Devices and Materials," *Reports on Progress in Physics* 49, 1335–86, 1986. With permission.)

superimposed on the normal pyroelectric response. These pulses occur in a random fashion, but their number and amplitude increase with a rate of increasing temperature. It is suggested that these spurious noise signals are due to ferroelectric domain wall movements. These can be minimized by good materials selection and appear to be lower in ceramics than some single crystal materials such as LiTaO₃. Finally, electromagnetic interference is a source of unwanted signals. Detectors operated at low frequencies, having a very high input impedance preamplifier, require careful screening. This is generally achieved by using electrically conducting windows of germanium or silicon that are connected to the earthed metal can.

7.2 PYROELECTRIC MATERIAL SELECTION

Many pyroelectric materials have been investigated for detector applications. However the choice is not an obvious one as it will depend on many factors including the size of the detector required, the operating temperature, and the frequency of operation.

It is possible to formulate a number of figures-of-merit (FoM) that describe the contribution of the physical properties of a material to the performance of a device. For example, the current responsivity (see Equation 7.8) is proportional to

$$F_i = \frac{p}{c_{th}},$$
(7.21)

instead the voltage responsivity (see Equation 7.11) is proportional to

$$F_v = \frac{p}{\varepsilon_o \varepsilon_r c_{th}}.$$
(7.22)

For thin film pyroelectric detectors, a voltage responsivity figure-of-merit can be introduced as (see Equation 7.16)

$$F_v^* = \frac{p}{\varepsilon_o \varepsilon_r}.$$
(7.23)

The conventional sensitivity figure-of-merit, detectivity D^*, is of little practical use because of frequency dependencies and filter factors. However, its analytical expression is useful for examining the relative importance of various parameters. In the case of a detector dominated by the AC Johnson noise (see Equation 7.20), the detectivity is proportional to

$$F_d = \frac{p}{c_{th} \left(\varepsilon_o \varepsilon_r \tan\delta\right)^{1/2}},$$
(7.24)

which forms the figure-of-merit for pyroelectric detectors.

A useful figure of merit that includes the effect of input capacitance of the circuit with which the detector is used is

$$F = \frac{1}{C_d + C_L} \frac{p}{c_{th}}.$$
(7.25)

This equation reduces to F_i or F_v when C_L is comparatively small or large, respectively.

The relevant figure-of-merit for the materials used in pyroelectric vidicons is F_{vid}:

$$F_{vid} = \frac{F_v}{G_{th}}, \tag{7.26}$$

where G_{th} is the thermal conductivity of the pyroelectric. The dependence of F_{vid} on G_{th} can be eliminated by dicing a thermal imaging target into individual islands (using the reticulation process).

A responsivity FoM is valuable in selecting material with responsivity sufficiently high that preamplifier noise is small compared to temperature fluctuation noise. A Johnson noise sensitivity FoM is valuable in selecting a material whose Johnson noise is small compared with the temperature fluctuation noise. Thus, both FoM must be large to ensure temperature fluctuation noise-limited performance.

An ideal material should have large pyroelectric coefficient, low dielectric constant, low dielectric loss and low volume specific heat. The possibility of satisfying these requirements in a single material is not promising. While it is generally true that a large pyroelectric coefficient and a small dielectric constant are desirable, it is also true that these two parameters are not independently adjustable. Thus, we find that materials having a high pyroelectric coefficient also have a high dielectric constant, and materials having a low dielectric constant also have a low pyroelectric coefficient. This means that different detector-preamplifier sizes and configurations will be optimized with different materials [21]. Thus, Equation 7.24 is a better responsivity figure-of-merit, assuming one knows the pixel geometry and the circuit with which the detector material will be used. Table 7.1 shows the parameter values and traditional FoM for typical materials. The traditional FoM indicate, for example, that TGS (triglycine sulfate) and LiTaO$_3$ (lithium tantalate) should be much better than BST and PST; however, sensor system results indicate the contrary.

The state of the art in pyroelectric materials and assessments of their relative merits for different applications have been reviewed by Whatmore [21,22], Watton [24], Muralt [28], and others [32,43,44]. Characteristics of pyroelectric detector materials are given in Tables 7.1 through 7.3 [28,31,45].

The pyroelectric materials can broadly be classified into three categories: single crystals, ceramics (polycrystalline), and polymers.

7.2.1 Single Crystals

Among the single crystals, the most notable success has been achieved with TGS [triglycine sulphate, $(NH_2CH_2COOH)_3H_2SO_4$]. It possesses attractive properties, a high pyroelectric coefficient, a reasonably low dielectric constant and thermal conductivity (high value of F_v). However, this material is rather hygroscopic, difficult to handle, and show poor long-term stability, both chemically and electrically. Its low Curie temperature is a major disadvantage, particularly for detectors that are required to meet military specifications. In spite of these problems, TGS is frequently used for high performance single element detectors and it is preferred material for vidicon targets. Several variants on pure TGS have been developed to overcome the problem of the low Curie temperature. The aliane and arsenic acid doped materials (ATGSAs) are particularly interesting because of their low dielectric losses and high pyroelectric coefficients (see Table 7.1). Detectors with D^* values of 2×10^9 cmHz$^{1/2}$W^{-1} have been obtained at 10 Hz (see Figure 7.6 [23]).

Lithium tantalate, LiTaO$_3$, gives inferior performance to TGS, due to its lower pyroelectric coefficient and slightly higher relative permittivity (lower value of F_v). It has the following advantages: high chemical stability, very low loss (so F_d is favorable), very high Curie temperature, and insolubility in water. The material is widely used for single element detectors, although there can be problems associated with thermally induced transient noise spikes from this material when used in very low frequency devices. It is not particularly favorable to use for thermal imaging arrays because of its low permittivity. Its thermal conductivity is quite high so that it is not a good material for pyroelectric vidicons. Good single crystals of LiTaO$_3$ can be produced by the Czochralski technique. It is readily available commercially.

Strontium Barium Niobate (SBN) is the next single crystal pyroelectric material. In fact, it is a family name for a range of solid solutions defined by the formula $Sr_{1-x}Ba_xNb_2O_6$, in which x can be varied from 0.25 to 0.75. SBN-50 ($x = 0.50$) has a favorable F_d figure-of-merit. Depending on composition, the ferroelectric transition can be tuned between 40 and 200°C. A high field-induced effect has been applied in uncooled thermal imaging, based on ferroelectric materials with a near room

Table 7.1: Properties of Bulk and Polymer Pyroelectric Materials

Material	Structure	p (μCm^{-2}K^{-1})	ε	tanδ	c_{th} (10^6Jm^{-3}K^{-1})	F_v^* (kVm^{-1}K^{-1})	F_v (m^{2}C^{-1})	F_d (10^{-5}Pa$^{-1/2}$)	T_C (°C)
NaNO$_2$	Single crystal	40	4	0.02	3.2	1130	0.17		164
LiTaO$_3$	Single crystal	230	47	<0.01	2.3	553	0.36	5–35	620
TGS	Single crystal	280	38	0.01	2.6	832	0.53	6.6	49
DTGS	Single crystal	550	43	0.02	2.6		0.99	8.3	61
ATGSAs	Single crystal	70	32	<0.01	2.3		0.07	>16	51
SBN-50	Single crystal	550	400	0.003	2.0	155	0.22	7.3	121
(Pb,Ba)$_5$Ge$_3$O$_{11}$	Single crystal	320	81	0.001	2.5	446	0.06	18.9	70
PbZrTiO$_3$ PZFNTU	Ceramic	380	290	0.003	3.0	148	0.04	5.5	230
PbTiO$_3$	Ceramic	180	190	0.01	2.5	107	0.08	1.5	490
PbTiO$_3$ PCWT 4-24	Ceramic	380	220	0.01	2.6	195		3.4	255
BaSrTiO$_3$67/33	Ceramic, field-induced	1500	8800	0.004	2.7			12.4	25
PbSc$_{0.5}$Ta$_{0.5}$O$_3$	Ceramic, field-induced	3000–6000	Up to 15,000		2.3	251	0.11	14–16	25
P(VDF/TrFE)50/50	Copolymer film	40	18	0.03	2.3	500	0.22	0.8	49
P(VDF/TrFE)80/20	Copolymer film	31	7	0.015				1.4	135

Source: Muralt, P., "Micromachined Infrared Detectors Based on Pyroelectric Thin Films," *Reports on Progress in Physics* 64 1339–88, 2001; Whatmore, R. W., *Journal of Electroceramics* 13, 139–47, 2004.

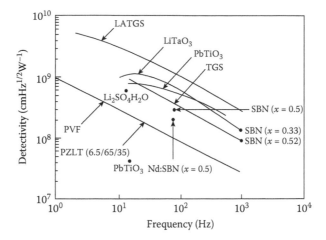

Figure 7.6 Discrete pyroelectric detector performance. (From Ravich, L. E., *Laser Focus/Electro-Optics*, 104–15, July 1986.)

temperature phase transition. Its high dielectric constant makes it a good candidate for thermal imaging arrays. The SBN is produced by the Czochralski technique, but good quality large single crystals are relatively difficult to grow.

7.2.2 Pyroelectric Polymers

Ferroelectric polymers based on polyvinylidene fluoride (PVDF) and on copolymers with trifluoroethylene (PVDF-TrFE) possess relatively low pyroelectric coefficient and low dielectric constants with high losses, so their FoM are inferior to the other materials. This class of materials is attractive for use in pyroelectric vidicon primarily because of their superior mechanical properties, in the case of fabrication in their samples (<6 μm), low permittivities, and low thermal conductivity. Their low permittivities make them well suited to large area detectors, but they are rather poorly suited for large area arrays. They are, however, obvious candidates for very low-cost detectors since they are readily available in large thin sheets that do not require expensive lapping and polishing processes necessary for other materials [46,47]. Their low heat conductivity and dielectric constant reduce the cross-talk between neighboring elements in multielement detectors. The performance of PVDF detectors is inferior to the other categories of materials, except for very large detectors operating at high frequencies. Their low glass temperatures are severe obstacles for many applications.

The PVDF is commercially available as poled and electroded polymer sheets of varying thicknesses and needs to be mechanically stretched before poling to develop its ferroelectric properties. However, the PVDF-TrFE copolymers can be cast from the melt or methyl ethyl ketone solution directly into the ferroelectric phase and, therefore, they are particularly interesting for direct deposition onto a silicon substrate for making arrays.

7.2.3 Pyroelectric Ceramics

Another class of materials, polycrystalline ferroelectric ceramics show promise for use in pyroelectric detectors. They offer a number of advantages over the materials listed above; they:

- Are relatively cheap to manufacture in large areas using standard mixed-oxide processes
- Are both mechanically and chemically robust (they can be processed into thin wafers)
- Possess high Curie temperature
- Do not suffer from thermally induced noise spikes
- Can be modified by the inclusion of selected dopant elements into the lattice to control such parameters as: p, ε_r, $\tan\delta$, Curie temperature, electrical impedance, and mechanical properties (controlling grain size of material)

There is a vast range of ceramic materials that consist of solid solutions of PZ (lead zirconate, $PbZrO_3$) and PT (lead titanate, $PbTiO_3$), and very similar oxides. These have been developed over a period of many years to satisfy a variety of ferroelectric, piezoelectric, electro-optic and pyro-electric requirements. An example of modification of the electrical properties of a pyroelectric ceramic has been given by Whatmore [21]. The ceramic can be poled in any desired direction by the application of a suitable electric field. The morphotropic phase boundary compositions of the PZT system are generally avoided for pyroelectric applications because these have high permittivities, which are detrimental to the FoM.

The conventional pyroelectric ceramics are still favored for most practical applications because of the stability of their properties over the normal operating temperature range (the Currie temperature is typically above 200°C) and because they do not need an applied DC bias field to operate them.

There have been various experimental studies to improve the FoM of modified PZ compositions [45]. One prospect is through exploitation of the step in the spontaneous polarization at phase transition [48].

Ceramic devices have D^* values in the 10^8 cmHz$^{1/2}$W^{-1} range, the performance of these devices is comparable or better than lithium tantalate, except in the case of large detectors.

Resistivities of modified ceramics cover the range of 10^9–10^{11} Ωcm^2. This means that the gate bias resistor in Figure 7.2c, which is generally around 10^{11}–10^{12} Ω, an expensive item, can be eliminated as a separate component by adjusting the resistivity of the material to suit the electrical time constant required. This is particularly important where large numbers of elements are involved in an array.

Using bulk pyroelectrics in fabrication of infrared detectors leads to several drawbacks; the material must be cut, lapped, and polished to make a thin, well-insulated and sensitive layer. In addition, the array fabrication requires metallization on both faces and bonding to a silicon readout circuit to yield a complete hybrid array. On account of this, in the last decade there has been a growth of interest in integration of pyroelectric thin films directly onto silicon substrates as a means for both reducing array fabrication costs and increasing performance through reduced thermal mass and improve thermal isolation [31].

Properties of thin-film materials differ from those of bulk materials in as much as microstructure and substrate influence are of importance [28]. In contrast to bulk ceramics, thin films can be grown textured or even completely oriented in the case of epitaxy (see Table 7.2). The performance similar to the single crystal materials is obtained for the optimal texture when the polar axis stays perpendicular to the electrodes everywhere in the film. Also considerable improvement of thin-film properties is possible in the case of materials that only exist as polycrystalline ceramics in bulk form (e.g., PZT, PLT). For example, a good demonstration of this case is epitaxial PbTiO$_3$, whose FoM F_v^* was measured as 291 kV/mK for thin films, whereas only 107 kV/mK is reached in bulk ceramics [28].

There is a trade-off between temperature stability of materials and size of the pyroelectric effect. Materials with high critical temperatures such as LiTaO$_3$ and PbTiO$_3$ are more adequate for simple and reliable devices. The relevant properties of materials gathered in Table 7.2 indicate that PbTiO$_3$-derived compounds with PZT (15–30%) Zr films are favorite materials, but they can be replaced by PLT or PCT. Pure PbTiO$_3$ has been mostly abandoned because of too high dielectric losses and difficulties to pole. We also notice that LiTaO$_3$ thin film pyroelectric applications are far from being as advanced as its applications in bulk detectors.

The oxide materials (modified lead zirconate titanates or the dielectric bolometer materials) posses the right properties (high ε and high F_d) as ceramics sintered around 1200°C. However, for integrating ferroelectric thin films directly on silicon places, a very important constraint on the temperature at which the ferroelectric can be grown. The interconnect metallization on the chips should not be taken above 500°C for any length of time and this places an upper limit on the ferroelectric layer process temperature. Fortunately, many techniques have been researched for ferroelectric thin film deposition. These include chemical solution deposition (CSD)—particularly sol-gel or metalorganic deposition (MOD) and metalorganic chemical vapor deposition (MOCVD) [28,31]. It appears that sol-gel deposition provides an excellent technique for thin film growth of Mn-doped PZT films at 560°C with a FoM F_d exceeding those of many bulk materials (p of 3.52×10^{-4} C/Km^2 and F_d of 3.85×10^{-5} Pa$^{-1/2}$).

7.2.4 Dielectric Bolometers

The conventional materials discussed above are ferroelectrics operated well below T_C, where the polarization is not permanently affected by changes in ambient temperature. It is, however, possible to operate ferroelectrics at or above T_C, with an applied bias field, in the mode of a dielectric

Table 7.2: Properties of Thin Film Pyroelectric Materials on Silicon Substrates

Material/Texture/Electrode	Deposition Method/Substrate	p ($\mu Cm^{-2}K^{-1}$)	ε	$\tan\delta$	c_{th} ($10^6 Jm^{-3}K^{-1}$)	F_v^* ($kVm^{-1}K^{-1}$)	F_v (m^2C^{-1})	F_d ($10^{-5}Pa^{-1/2}$)
PbTiO$_3$/(001) + (100)Pt	Sol-gel and sputter	130–145	180–260	0.014–0.035	2.7	57–88	0.02–0.03	0.7–1.1
PZT15/85/(111)Pt	Sol-gel	160–220	200–230	0.01–0.015	2.7	78–113	0.03–0.04	1.3–1.5
PZT25/75(111)Pt	Sputter	200	300	0.01	2.7	75	0.028	1.4
PZT30/70(111)Pt	Sol-gel	200	340	0.011	2.7	66	0.025	1.3
Mod. PZT (Mn-doped)	Ceramic	356	218	0.007	2.6		0.07	5.1
PTL10–20Pt and Si	Ion beam, sputter, sol-gel, MOD	200–576	153–550	0.01–0.024	2.7	41–425	0.02–0.15	0.7–4.1
Porous PCT15/(11)Pt	Sol-gel	220	90	0.01	2.0	276	0.14	3.9
LiNbO$_3$/(006)Pt	Sputter	71	30	0.01	3.2	267	0.08	1.4
YBaCuO/Nb	Sputter	4000						3.2
Epitaxial films								
PbTiO$_3$/(001)Pt	Sputter/MgO,	250	97	0.006	3.2	291	0.09	3.4
PZT45/55(001)Pt	Sputter/MgO	420	400	0.013	3.1	119	0.04	2.0
PZT52/48(100)	YBaCuO/LaAlO$_3$ PLD	500	100	0.02	3.1	57	0.02	1.2
PZT90/10(111)Pt	Sapphire/sputter	450	350	0.02	3.2	145	0.05	1.7
PLT5–15/(001)Pt	Sputter/MgO	400–1300	100–350	0.006–0.01	3.2	196–565	0.06–0.17	2.6–8.9
PLZT7.5/8/92–20/80/(001)Pt	Sputter/MgO	360–820	193–260	0.013–0.017	2.6	160–480	0.06–0.18	2.2–6.7
PCT30/(001)Pt	Sputter/MgO	520	290	0.02	3.0	202	0.06	2.4

Source: Muralt, P., "Micromachined Infrared Detectors Based on Pyroelectric Thin Films," *Reports on Progress in Physics* 64, 1339–88, 2001; Whatmore, R. W., *Journal of Electroceramics* 13, 139–47, 2004.

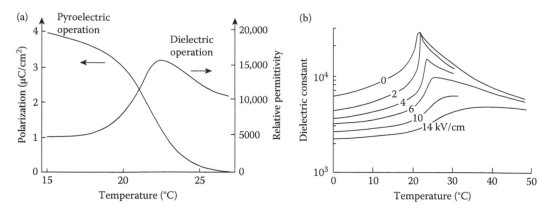

Figure 7.7 Barium strontium titanate ceramic: operating mode for (a) ferroelectric ceramic and (b) dielectric constant. (From Betatan, H., Hanson, C., and Meissner, E. G., "Low Cost Uncooled Ferroelectric Detector," *Proceedings of SPIE* 2274, 147–56, 1994. With permission.)

bolometer [49]. Current developments in the area of pyroelectric materials include the use of dielectric bolometers.

With the application of an external electric field, the total polarization is described by Equation 7.1. Below T_C, P_s is large compared with the second term, thus D and P_s are often used interchangeably. However, it is clear from Figure 7.7a that the maximum pyroelectric effect (i.e., the maximum slope of P versus T) occurs near T_C, and therefore it seems desirable to operate there.

The field-enhanced pyroelectric coefficient is described by Equation 7.2. From this equation, results that the induced part of the pyroelectric coefficient depends not only upon the temperature rate of change of the permittivity, but also upon the field dependence of the rate change. Because of that it is not a simple matter to calculate the field effect. At all temperatures, the dielectric behavior is nonlinear; that is, the gradient of permittivity varies with the applied field and the dielectric peak and $d\varepsilon/dT$ are both depressed with increasing field (see Figure 7.7b). Note that the pyroelectric coefficient maximum is somewhat lower in temperature than the peak capacitance value (see Figure 7.7b). The capacitance data represents a biased sample, and both the dielectric constant and pyroelectric coefficient maxima occur at temperatures above the Curie temperature. As the operating point continues to diverge from the Curie temperature, dielectric contributions to polarization become dominant.

Thus, the application of an electric field gives several benefits to the detector performance [50]:

- It adds an induced polarization to the spontaneous polarization

- It suppresses the dielectric permittivity, especially as it peaks near the transition

- It broadens the response peak, easing temperature control limits

- It suppresses dielectric loss, reducing noise

- It stabilizes polarization near the transitions, providing predictable performance

Several materials have been examined in dielectric bolometer mode, including potassium thallium niobate, $KTa_xNb_{1-x}O_3$ (KTN), lead zinc niobate, $Pb(Zn_{1/3}Nb_{2/3})O_3$ (PZN), barium strontium titanate, $Ba_{1-x}Sr_xTiO_3$ (BST), lead magnesium niobate, $Pb(Mg_{1/3}Nb_{2/3})O_3$, (PMN), and more recently lead scandium tantalate, $Pb(Sc_{1/2}Ta_{1/2})O_3$ (PST) [22]. Dielectric bolometers require stringent bias and temperature stabilization. Properties of pyroelectric materials with transitions near ambient and operating with an electric field are shown in Tables 7.1 and 7.3.

The BST ceramic is a relatively well-behaved material with a very high permittivity. When Sc moves from 40 to 0% in the compound, T_C moves from 0% to 120°C. Typical values of dielectric constant higher than 30,000 was noticed in the BST 67/33 material used in the High-Density Array Development (HIDAD) program [50]; the 17 μm thickness of the BST ceramic reached with difficulty appears as a lowest limit. The peak F_d (10.5×10^{-5} Pa$^{-1/2}$) achieved by BST65/35 is over twice that for the modified PZ or PT ceramics [22].

All the oxide materials for dielectric bolometers have very high dielectric constants (>1000) under the operational temperatures and fields coupled with very high pyroelectric coefficients.

Table 7.3: Properties of Pyroelectric Thin Film Suitable for Induced Pyroelectricity

Material/Texture/Electrode	Deposition Method/ Substrate	Induced p ($\mu Cm^{-2}K^{-1}$)	ε	$\tan\delta$	c_{th} ($10^6 Jm^{-3}K^{-1}$)	F_v^* ($kVm^{-1}K^{-1}$)	F_v (m^2C^{-1})	F_d ($10^{-5}Pa^{-1/2}$)
$PbSc_{0.5}Ta_{0.5}O_3$/sapphire	RF-sputter, 900°C	6000 (25–30°C)	6500	0.03	2.5	104	0.04	6–9
$PbSc_{0.5}Ta_{0.5}O_3$/CdGa-garner	Sol-gel, 900°C	3800	9000	0.002	2.7	50	0.02	11
$PbSc_{0.5}Ta_{0.5}O_3$/Si/Pt	Sol-gel, 700°C	200–450	900	0.02	2.7	25–57	0.02	0.6–1.3
$PbSc_{0.5}Ta_{0.5}O_3$/Si/Pt	Sol-gel, 630°C	490	700	0.008	2.7	60	0.02	2.6
PbMgZn-NbO/(100)Pt/MgO	Sol-gel, 900°C	14,000 (15°C)	1600	0.004	2.85	989	0.34	20–40
$K_{0.89}Na_{0.11}Ta_{0.55}Nb_{0.45}O_3$/KTO$_3$	LPE, 930°C	5200 (66°C)	1200 (66°C)	0.02	2.9	50	0.02	3.9

Source: Muralt, P., "Micromachined Infrared Detectors Based on Pyroelectric Thin Films," *Reports on Progress in Physics* 64, 1339–88, 2001.

These make them well suited for small area detectors in general, and very large arrays of small elements in particular.

Operating in the dielectric mode, single crystal BST should present little if any advantage over ceramic BST. The pyroelectric coefficient of ceramic BST more than doubles that of single crystal material of ostensibly the same composition measured under similar conditions. Likewise, the ceramic dielectric constant exceeds the single crystal value. Other attributes that make the ceramic BST more desirable than the single crystal are: ease and cost of fabrication, material uniformity, superior performance, electrical resistance, resistance to aging, and amenable to doping.

The present BST technology is a cumbersome bulk ceramic technology that requires grounding and polishing of ceramic wafers sliced from a boule, laser reticulation of pixels, multiple thinning, and planarization steps. The arrays are connected to the silicon readout circuit by compression bonds. The process suffers from a thermal isolation problem due to the thick mesa structure and BST surface degradation due to the thinning procedure.

Next generation uncooled pyroelectric detectors are required to operate in normal pyroelectric mode without bias and temperature stabilization. In addition, it is desired to use a thin-film pyroelectric detector technology to utilize the state-of-the-art micromachining technology for fabrication focal plane arrays (FPAs).

7.2.5 Choice of Material

To obtain a direct comparison between the various pyroelectric materials is very difficult as the detector area and operating frequency will affect the performance, and account must also be taken of the environmental operating conditions. Porter has compared devices operating under different conditions, for detector areas ranging from 100 to 0.01 mm^2 [19]. If a maximum detectivity is required then, for a given field-effect transistor (FET), the element area is an important consideration as it will affect the matching between the element capacitance and the amplifier capacitance. For large-area elements the low dielectric constant materials dominate; TGA and lithium tantalate appear to be the best devices for all frequencies, except at very high frequencies (>10 kHz) when the polymer film devices begin to dominate. A more complicated situation is if the element area is reduced to 1 mm^2, the most commonly used order of magnitude. No one material is best at all frequencies. However for small devices, the high dielectric constant materials are better (e.g., SBN because of the better capacitive match between the element and the amplifier). For intermediate area devices the performance of all devices is comparable.

It should be emphasized that the above discussion only shows the trends as varying the detector parameters or the FET amplifier could alter this situation. Other factors such as environmental stability, availability, cost, and manufacturing considerations are also very important.

Stringent requirements are put on the materials used in fabrication of FPAs, where very thin ferroelectric films are used [27,29,51–54]. Most ferroelectrics tend to lose their interesting properties as the thickness is reduced. However, some ferroelectric materials seem to maintain their properties better than others. This seems particularly true for PT and related materials, whereas BST, the material does not hold its properties well in thin-film form.

From Equations 7.22 through 7.24 we can see that for pyroelectric materials, a large pyroelectric coefficient p and a small dielectric constant are desirable. However, these two parameters are generally not independently adjustable. Materials having a higher p usually also have a high dielectric constant., and vice versa, although, it is possible to lower the dielectric constant while maintaining p by doping the material.

7.3 PYROELECTRIC VIDICON

Pyroelectric devices are used in a variety of applications where the characteristics of pyroelectric detectors are of particular significance. Firstly, they only respond to changes in incident radiation and are ideally suited for the detection of very small changes in flux whilst operating with a large background level of incident energy. The second characteristic of note is their broad spectral response from microwaves to X-rays.

Pyroelectric detectors have found a wide variety of applications, including spectrometry, radiometry, remote temperature measurement, direction sensing, laser diagnostics, pollution sensing, and imaging.

The most important application of pyroelectric detectors is thermal imaging. The concept of the pyroelectric vidicon tube was first proposed by Hadni [55], and commercial devices were demonstrated as early as 1970 [56]. A schematic of a tube is shown in Figure 7.8. This device can be

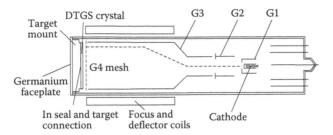

Figure 7.8 Schematic representation of a pyroelectric vidicon tube.

considered analogous to the visible television camera tube except that the photoconductive target is replaced by the pyroelectric detector and germanium faceplate. The target consists of a disc of pyroelectric material (20 μm thick, 2 cm diameter), with a transparent electrode on the front surface. An infrared lens produces a thermal image on this target and the resulting charge distribution is read off the back surface by the scanning electron beam. The original tubes were fabricated with TGS but better results have been achieved with deuterated TGS and TGFB material [57,58].

The major factor limiting the resolution achievable with pyroelectric vidicons is thermal diffusion within the target. This causes the thermal resolution to degrade rapidly as the spatial frequency increases. For this reason reticulated targets are being developed [59]. The vidicon tubes achieve 0.2°C resolution in an image consisting of 100 TV lines and with reticulation the spatial resolution increase to 400 TV lines or more [60]. Of the tube programmers in the United States, United Kingdom, and France, only the UK technology at EEV Co. Ltd. went on to manufacture, with applications in fire service cameras and industrial maintenance.

Although good quality images have been obtained from the pyroelectric vidicon, recent work is mainly directed toward fabrication of large 2-D FPAs. This allows improvement on the temperature resolution of the systems and produces more robust and lighter imagers.

REFERENCES

1. S. B. Lang, "Pyroelectricity: A 2300-year history," *Ferroelectrics* 7, 231–34, 1974.

2. D. Brewster, "Observation of Pyroelectricity of Minerals," *Edinburg Journal of Science* 1, 208–14, 1824.

3. Y. Ta, "Action of Radiations on Pyroelectric Crystals," *Comptes Rendus* 207, 1042–44, 1938.

4. A. G. Chynoweth, "Dynamic Method of Measuring the Pyroelectric Effect with Special Reference to Barium Titanate," *Journal of Applied Physics* 27, 78–84, 1956.

5. J. Cooper, "A Fast-Response Pyroelectric Thermal Detector," *Journal of Scientific Instruments* 39, 467–72, 1962.

6. J. Cooper, "Minimum Detectable Power of a Pyroelectric Thermal Receiver," *Review of Scientific Instruments* 33, 92–95, 1962.

7. A. Hadni, Y. Henninger, R, Thomas, P. Vergnat, and B. Wyncke, "Investigation of Pyroelectric Properties of Certain Crystals and Their Utilization for Detection of Radiation," *Comptes Rendus* 260, 4186, 1965.

8. G. A. Burdick and R. T. Arnold, "Theoretical Expression for the Noise Equivalent Power of Pyroelectric Detectors," *Journal of Applied Physics* 37, 3223–26, 1966.

9. J. H. Ludlow, W. H. Mitchell, E. H. Putley, and N. Shaw, "Infrared Radiation Detection by Pyroelectric Effect," *Journal of Scientific Instruments* 44, 694–96, 1967.

10. H. P. Beerman, "Pyroelectric Infrared Radiation Detector," *American Ceramic Society Bulletin* 46, 737, 1967.

11. A. M. Glass, "Ferroelectric Strontium-Barium-Niobate as a Fast and Sensitive Detector of Infrared Radiation," *Applied Physics Letters* 13, 147–49, 1968.

12. R. W. Astheimer and F. Schwarz, "Thermal Imaging Using Pyroelectric Detectors: Mylar Supported TGS," *Applied Optics* 7, 1687–95, 1968.

13. R. J. Phelan, Jr., R. J. Mahler, and A. R. Cook, "High D* Pyroelectric Polyvinylfluoride Detectors," *Applied Physics Letters* 19, 337–38, 1971.

14. E. H. Putley, "The Pyroelectric Detector," in *Semiconductors and Semimetals*, Vol. 5, eds. R. K. Willardson and A. C. Beer, 259–85, Academic Press, New York, 1970.

15. G. Baker, D. E. Charlton, and P. J. Lock, "High Performance Pyroelectric Detectors," *Radio Electronic Engineers* 42, 260–64, 1972.

16. E. H. Putley, "Thermal Detectors," in *Optical and Infrared Detectors,* ed. R. J. Keyes, 71–100, Springer, Berlin, 1977.

17. S. T. Liu and D. Long, "Pyroelectric Detectors and Materials," *Proceedings of IEEE* 66, 14–26, 1978.

18. D. E. Marshall, "A Review of Pyroelectric Detector Technology," *Proceedings of SPIE* 132, 110–17, 1978.

19. S. G. Porter, "A Brief Guide to Pyroelectric Detectors," *Ferroelectrics* 33, 193–206, 1981.

20. J. C. Joshi and A. L. Dawar, "Pyroelectric Materials, Their Properties and Applications," *Physica Status Solidi A-Applied Research* 70, 353–69, 1982.

21. R. W. Whatmore, "Pyroelectric Devices and Materials," *Reports on Progress in Physics* 49, 1335–86, 1986.

22. R. W. Whatmore, "Pyroelectric Ceramics and Devices for Thermal Infra-Red Detection and Imaging," *Ferroelectrics* 118, 241–59, 1991.

23. L. E. Ravich, "Pyroelectric Detectors and Imaging," *Laser Focus/Electro-Optics* 104–15, July 1986.

24. R. Watton, "Ferroelectric Materials and Design in Infrared Detection and Imaging," *Ferroelectrics* 91, 87–108, 1989.

25. P. Robin, H. Facoetti, D. Broussoux, G. Vieux, and J. L. Ricaud, "Performances of Advanced Infrared Pyroelectric Detectors," *Revue Technique Thompson-CSF* 22(1), 143–86, 1990.

26. H. Betatan, C. Hanson, and E. G. Meissner, "Low Cost Uncooled Ferroelectric Detector," *Proceedings of SPIE* 2274, 147–56, 1994.

27. M. A. Todd, P. A. Manning, O. D. Donohue, A. G. Brown, and R. Watton, "Thin Film Ferroelectric Materials for Microbolometer Arrays," *Proceedings of SPIE* 4130, 128–39, 2000.

28. P. Muralt, "Micromachined Infrared Detectors Based on Pyroelectric Thin Films," *Reports on Progress in Physics* 64, 1339–88, 2001.

29. C. M. Hanson, H. R. Beratan, and J. F. Belcher, "Uncooled Infrared Imaging Using Thin-Film Ferroelectrics," *Proceedings of SPIE* 4288, 298–303, 2001.

30. P. W. Kruse, *Uncooled Thermal Imaging. Arrays, Systems, and Applications,* SPIE Press, Bellingham, WA, 2001.

31. R. W. Whatmore and R. Watton, "Pyroelectric Materials and Devices," in *Infrared Detectors and Emitters: Materials and Devices*, eds. P. Capper and C. T. Elliott, 99–147, Kluwer Academic Publishers, Boston, MA, 2000.

32. S. B. Lang, "Pyroelectricity: From Ancient Curiosity to Modern Imaging Tool," *Physics Today,* 31–36, August 2005.

33. R. W. Whatmore, Q. Zhang, C. P. Shaw, R. A. Dorey, and J. R. Alock, "Pyroelectric Ceramics and Thin Films for Applications in Uncooled Infra-Red Sensor Arrays," *Physica Scripta* T 129, 6–11, 2007.

34. E. H. Putley, "A Method for Evaluating the Performance of Pyroelectric Detectors," *Infrared Physics* 20, 139–47, 1980.

35. D. H. Austan and A. M. Glass, "Optical Generation of Intense Picosecond Electrical Pulses," *Applied Physics Letters* 20, 398–99, 1972.

36. C. B. Roundy, R. L. Byer, D. W. Phillion, and D. J. Kuizenga, "A 170 psec Pyroelectric Detector," *Optics Communications* 10, 374–77, 1974.

37. W. R. Blevin and J. Geist, "Influence of Black Coatings on Pyroelectric Detectors," *Applied Optics* 13, 1171–78, 1974.

38. A. van der Ziel, "Pyroelectric Response and D* of Thin Pyroelectric Films on a Substrate," *Journal of Applied Physics* 44, 546–49, 1973.

39. R. M. Logan and K. More, "Calculation of Temperature Distribution and Temperature Noise in a Pyroelectric Detector: I. Gas-Filled Tube, *Infrared Physics* 13, 37–47, 1973.

40. R. M. Logan, "Calculation of Temperature Distribution and Temperature Noise in a Pyroelectrical Detector: II. Evacuated Tube," *Infrared Physics* 13, 91–98, 1973.

41. R. L. Peterson, G. W. Day, P. M. Gruzensky, and R. J. Phelan, Jr., "Analysis of Response of Pyroelectric Optical Detectors," *Journal of Applied Physics* 45, 3296–303, 1974.

42. N. M. Shorrocks, R. W. Whatmore, M. K. Robinson, and S. G. Parker, "Low Microphony Pyroelectric Arrays," *Proceedings of SPIE* 588, 44–51, 1985.

43. A. Mansingh and A. K. Arora, "Pyroelectric Films for Infrared Applications," *Indian Journal of Pure & Applied Physics* 29, 657–64, 1991.

44. A. Sosnin, "Image Infrared Converters Based on Ferroelectric-Semiconductor Thin-Layer Systems," *Semiconductor Physics, Quantum Electronics and Optoelectronics* 3, 489–95, 2000.

45. R. W. Whatmore, "Pyroelectric Arrays: Ceramics and Thin Films," *Journal of Electroceramics* 13, 139–47, 2004.

46. S. B. Lang and S. Muensit, "Review of Some Lesser-Known Applications of Piezoelectric and Pyroelectric Polymers," *Applied Physics A* 85, 125–34, 2006.

47. J. L. Coutures, R. Lemaitre, E. Pourquier, G. Boucharlat, and P. Tribolet, "Uncooled Infrared Monolithic Imaging Sensor Using Pyroelectric Polymer," *Proceedings of SPIE* 2552, 748–54, 1995.

48. R. Clarke, A. M. Glazer, F. W. Ainger, D. Appleby, N. J. Poole, and S. G. Porter, "Phase Transitions in Lead Zirconate-Titanate and Their Applications in Thermal Detectors," *Ferroelectrics* 11, 359–64, 1976.

49. R. A. Hanel, "Dielectric Bolometer: A New Type of Thermal Radiation Detector," *Journal of the Optical Society of America* 51, 220–25, 1961.

50. C. Hanson, H. Beratan, R. Owen, M. Corbin, and S. McKenney, "Uncooled Thermal Imaging at Texas Instruments," *Proceedings of SPIE* 1735, 17–26, 1992.

51. R. Watton, "IR Bolometers and Thermal Imaging: The Role of Ferroelectric Materials," *Ferroelectrics* 133, 5–10, 1992.

52. R. Watton and P. Manning, "Ferroelectrics in Uncooled Thermal Imaging," *Proceedings of SPIE* 3436, 541–54, 1998.

53. R. K. McEwen and P. A. Manning, "European Uncooled Thermal Imaging Sensors," *Proceedings of SPIE* 3698, 322–37, 1999.

54. C. M. Hanson, H. R. Beratan, and D. L. Arbuthnot, "Uncooled Thermal Imaging with Thin-Film Ferroelectric Detectors," *Proceedings of SPIE* 6940, 694025, 2008.

55. A. Hadni, "Possibilities actualles de detection du rayonnement infrarouge," *Journal of Physics* 24, 694–702, 1963.

56. E. H. Putley, R. Watton, W. M. Wreathall, and S. D. Savage, "Thermal Imaging with Pyroelectric Television Tubes," *Advances in Electronics and Electron Physics* 33A, 285–292, 1972.

57. R. Watton, "Pyroelectric Materials: Operation and Performance in Thermal Imaging Camera Tubes and Detector Arrays," *Ferroelectrics* 10, 91–98, 1976.

58. E. H. Stupp, "Pyroelectric Vidicon Thermal Imager," *Proceedings of SPIE* 78, 23–27, 1976.

59. S. E. Stokowski, J. D. Venables, N. E. Byer, and T. C. Ensign, "Ion-Beam Milled, High-Detectivity Pyroelectric Detectors," *Infrared Physics* 16, 331–34, 1976.

60. A. J. Goss, "The Pyroelectric Vidicon: A Review," *Proceedings of SPIE* 807, 25–32, 1987.

8 Novel Thermal Detectors

At present, vanadium oxide and amorphous silicon (a-Si) microbolometers are technologies of choice for uncooled thermal imaging. However, their sensitivity limitations [1] and the still significant prices encouraged many research teams to explore other IR sensing techniques with the potential for improved performance with reduced detector costs. Recently, thermal-imaging modules for less than $1000 are produced [2]. It means a 10-fold reduction in costs, compared with the approximate price for current IR imaging systems (see Table 8.1) [3].

The Golay cell, invented by Marcel Golay in the late 1940s offers the best performance among thermal infrared detectors [4,5]. Despite some disadvantages such as high cost (more than $5000) and relatively large size, the Golay cell is still commercially available and is used where high performance is essential. New versions of miniaturized micromachined Golay cells that utilize capacitive as well as tunneling displacement transducers have been developed.

One of the most promising IR sensing methods is the use of thermally actuated microelectro-machined structures (MEMS) with a reported detectivity of 10^8 cmHz$^{1/2}$/W. These novel uncooled detectors are also considered in this chapter. It is expected that novel uncooled detector arrays could become very attractive for a number of applications due to their inherent simplicity, high sensitivity, and rapid response to radiation.

8.1 GOLAY CELL

The Golay cell (Figure 8.1) is a thermal detector consisting of a hermetically sealed container filled with gas (usually xenon for its low thermal conductivity) and arranged so that expansion of the gas under heating by a photon signal distorts a flexible membrane on which a mirror is mounted. The movement of the mirror is used to deflect a beam of light shining on a photocell and so producing a change in the photocell current as the output. In modern Golay cells the photocell is replaced by a solid state photodiode and light emitting diode is used for illumination [6]. The reliability and stability of this arrangement is significantly better than that of the earlier Golay cells that used a tungsten filament lamp and a vacuum photocell.

The performance of the Golay cell is only limited by the temperature noise associated with the thermal exchange between the absorbing film and the detector gas, consequently the detector can be extremely sensitive with $D^* \approx 3 \times 10^9$ cmHz$^{1/2}$W^{-1}, and responsivities of 10^5–10^6 V/W. The response time is quite long, typically 15 msec. The detector is fragile and very susceptible to vibration and is only suitable for use in a controlled environment such as a laboratory.

New miniature Golay cells fabricated by silicon micromachining techniques have been described utilizing capacitive [7] as well as tunneling displacement transducers [8,9]. Initial work on micromachined tunneling Golay cells were made at the Jet Propulsion Laboratory (Pasadena, California). The devices were made in a low yield process that involved the hand assembly and gluing together of the sensor parts. This mode of fabrication produced devices with large variations in key operating parameters. Prototype devices have been operated with noise equivalent power (NEP) better than 3×10^{-10} WHz$^{-1/2}$ at 25 Hz [9].

Ajakaiye et al. [10] have described an 80% yield wafer-scale process for fabrication of tunneling displacement transducers. A cross-section view of the sensor is shown in Figure 8.2. The top two parts form gas cells with a square radiation absorbing area of 2 mm on the side and a height of 0.85 μm. Infrared radiation is absorbed by a 50-Å thick platinum film evaporated on the inner side of the 1-μm thick upper nitride membrane. A deflection electrode surrounds a 7 μm high tunneling tip etched on the bottom wafer. The deflection electrode located above the tunneling tip is a flexible 0.5-μm thick membrane. The tip, deflection electrode, and nitride membrane are all covered with gold and make connections to three pads located next to the vent. The performance of this device is comparable to the best commercially available uncooled broadband IR detectors.

8.2 NOVEL UNCOOLED DETECTORS

Despite successful commercialization of uncooled microbolometers suitable for thermal imaging, the community is still searching for a platform for thermal imagers that combines affordability, convenience of operation, and excellent performance. Recent advances in MEMS systems have lead to the development of uncooled IR detectors operating as micromechanical thermal detectors as well as micromechanical photon detectors. Between them the most important are bimaterial microcantilevers that mechanically respond to the absorption of the radiation. These sensing structures were originally invented at the Oak National Laboratory (ORNL) in the mid-1990s [11–15], and subsequently developed by ONRL [16–20], the Sarnoff Corporation [21,22], Sarcon

157

Table 8.1: Performance and Cost Comparison

Features	Microbolometers	Pyrometers	Microcantilevers
Ultimate sensitivity (mK)	20	40	3
Response time (ms)	15–20	15–20	5–10
Dynamic range	10^4	10^3	$>10^5$
Optics	Large, expensive	Large, expensive	Small, cheap
Power requirements	Low	Low	Low
Ease of fabrication	Difficult	Difficult	Standard IC fabrication
Size	Moderately small	Moderately small	Small
Cost of camera	$20–50K	$7–25K	$5–15K

Source: "MEMS Transform Infrared Imaging," *Opto&Laser Europe,* June 2003.

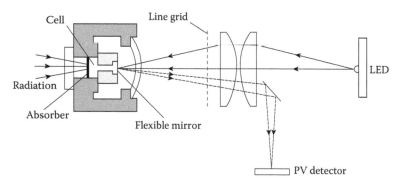

Figure 8.1 Golay cell.

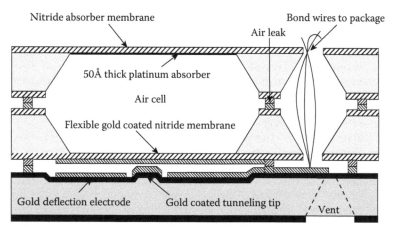

Figure 8.2 Cross-section view of tunneling displacement infrared detector. (From Kenny, T. W., *Uncooled Infrared Imaging Arrays and Systems,* Academic Press, San Diego, CA, 227–67, 1997.)

Microsystems [23–25], and others for imaging [26–33] and photo spectroscopic applications [34,35].

The thermomechanical detector approach was pioneered by Barnes et al. when they coated microcantilevers with a metal as the sensing active layer to form the bimaterial [36]. Figure 8.3 shows a schematic diagram of the capacitively sensed microcantilever structure. The microcantilever is attached mechanically and electrically to the substrate at the end by an anchor, and the second end is free to bend under the influence of any changes in stress along the arm. The IR radiation is absorbed by the microcantilever paddle materials along with a tuned

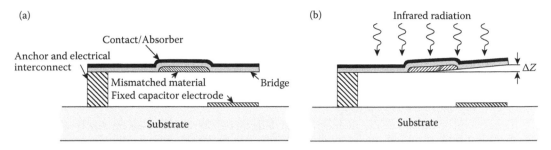

Figure 8.3 Schematic diagram of the operating principle of the bimaterial microcantilever IR detector: (a) nominal position, and (b) with increased radiation.

Table 8.2: Properties of Several Materials Used in Design of Cantilevers

	Young's Module (GPa)	Thermal Conductive Coefficient (W/mK)	Thermal Expansion Coefficient ($10^{-6}K^{-1}$)	Heat Capacity (J/kgK)	Density (10^3 kg/m³)	Emissivity (8–14 μm)
SiN_x	180	5.5 ± 0.5	0.8	691	2.40	0.8
Au	73	296.0	14.2	129	19.3	
SiO_2	46–92	1.1	0.05–12.3	—	2.20	
Al	80	237.0	23.6	908	2.70	0.01
Si	100	135	2.6	700		

resonant absorption cavity. The absorbed radiation is converted to heat in the microcantilever structure thermally isolated from the substrate by thermal isolation arms like those used in bolometers.

The cantilever structure contains a bimaterial region that is fabricated from two layers that possess significantly different thermal expansion coefficients; for example, a low thermal expansion coefficient SiO_2 substrate layer ($\alpha = 0.5 \times 10^{-6}$ K^{-1}), overlying an Al layer with a thermal expansion of $\alpha = 23 \times 10^{-6}$ K^{-1} [24]. When the incident radiation heats the structure, the bimaterial region of the cantilever bends up out of the plane due to the dissimilar thermal expansion of the bimaterials (approximately 0.1 μm for every degree of temperature change). Table 8.2 shows the generally used bimaterial combination. Usually, SiN_x and SiO_2 are used as IR absorbing materials, whereas Au and Al as contacts and reflectors.

The primary fundamental limits to the performance of microcantilevers are related to the properties of the thermal detectors themselves; the background fluctuation limit and the temperature fluctuation limit (see Chapter 3).

In the case of thermomechanical IR detectors, there is an additional fundamental limitation that is related to spontaneous microscopic mechanical motion (oscillation) of any suspended microstructure due to its thermal energy. It appears that for the majority of the readout means, these oscillations are indistinguishable from temperature-induced bending and, therefore, directly contribute to the detector noise [19,37]. According to Sarid [38], thermomechanical noise is frequency independent at low frequencies and the cantilever tip displacement noise is equal

$$\left\langle \delta z_{TM}^2 \right\rangle^{1/2} = \left(\frac{4kT\Delta f}{Q k_s \omega_o} \right)^{1/2},$$

(8.1)

where Q (quality factor) is the ratio of the resonance frequency, ω_o, to the resonance peak width, and k_s is the spring constant, defined as the ratio of the force applied to the microcantilever divided by the displacement of the tip. An alternative model predicts that when the damping is due to intrinsic friction process rather than due to viscous damping of the medium, the density of thermomechanical noise follows a $1/f^{1/2}$ trend below the mechanical resonance [39].

Using Equation 8.1, the predicted limits to NEP and detectivity due to thermomechanical noise are:

$$\text{NEP} = \frac{1}{R_z}\left(\frac{4kT\Delta f}{Qk_s\omega_o}\right)^{1/2} \quad \text{and} \quad D^* = \frac{1}{R_z}\left(\frac{4kT}{AQk_s\omega_o}\right)^{1/2}, \tag{8.2}$$

where R_z is the detector's responsivity.

An important advantage of thermomechanical detectors is that they are essentially free of intrinsic electronic noise and can be combined with a number of different readout techniques with high sensitivity. Depending on readout techniques, the novel uncooled detectors can be devoted to:

- Capacitative [19,21–25]

- Optical [13,15,17,19,20,26–28,30–33]

- Piezoresistive [11,12]

- Electron tunneling [8]

8.2.1 Electrically Coupled Cantilevers

In electrically coupled thermal transducers, bending of the cantilever causes changing its capacitance. This capacitance change is converted into an electrical signal that is proportional to the amount of absorbed IR light. All unwanted external vibrations are damped using an actively tuned resonant RC-circuit. For example, Figure 8.4 shows a schematic diagram of the operation of a capacitive-coupled detector, with the variable plate microcantilever capacitive sensor forming one arm of the bridge circuit [24]. Symmetric and oppositely phased voltage pulses, $\pm V_s$, are applied to the cantilever and bridge reference capacitors, C_s and C_R, respectively, around a reference voltage, V_{Ref}. If the C_s and C_R are the same, the voltage value at the common node between the capacitors is zero. When IR radiation falls on microcantilever, the paddle moves up, increasing the capacitor gap, thereby decreasing the detector capacitance and generating an offset voltage, V_g, at the input to the gain and integrator circuits.

When the detector temperature increases from T_o to T, the deflection of the microcantilever tip (see Figure 8.3b) is given by [24]:

$$\Delta Z = \frac{3L_p^2}{8t_{bi}}(\alpha_{bi} - \alpha_{subs})(T - T_o)K_o, \tag{8.3}$$

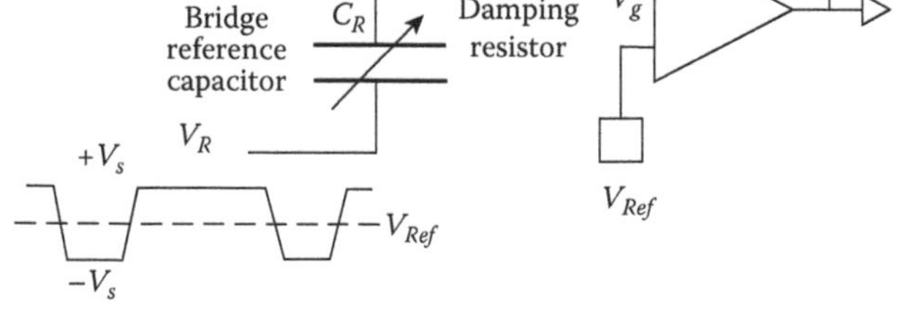

Figure 8.4 Circuit diagram showing the microcantilever bridge circuit, damping resistor, and signal gain amplifier. (From Hunter, S. R., Maurer, G., Jiang, L., and Simelgor, G., "High Sensitivity Uncooled Microcantilever Infrared Imaging Arrays," *Proceedings of SPIE* 6206, 62061J, 2006. With permission.)

where L_p is the length of the bimaterial section of the microcantilever detector, α_{bi} and α_{subs} are the bimaterial and substrate material thermal coefficients of expansion (TCE), respectively, t_{bi} is the thickness of the high TCE bimaterial, and the constant K_o is given by:

$$K_o = \frac{8(1+x)}{4+6x+4x^2+nx^3+1/nx}, \tag{8.4}$$

where $x = t_{subs}/t_{bi}$ is the ratio of the substrate to bimaterial thicknesses and $n = E_{subs}/E_{bi}$ is the ratio of the Young's moduli of the substrate and bimaterial. The last two equations indicate that the maximum microcantilever banding can be obtained using bimaterials with large differences in their thermal expansion of coefficients and optimizing the cantilever geometry.

The voltage responsivity (in V/K) of the detector is given by

$$R_v = \frac{V_s C_s}{C_T Z_{gap}} \frac{\Delta Z}{\Delta T}, \tag{8.5}$$

where Z_{gap} is the effective vacuum gap in the sensor and C_T is the sum of total capacitances appearing at the input to the operational amplifier.

For the electrically coupled cantilevers, a parameter called the temperature coefficient of capacitance (TCC) is defined in a way analogous to the bolometer's temperature coefficient of resistance:

$$TCC = \frac{1}{C_s} \frac{\Delta C}{\Delta T} = \frac{1}{Z_{gap}} \frac{\Delta Z}{\Delta T}. \tag{8.6}$$

For capacitive sensors the TCC > 30%/K has been measured [25]. The modeled performance can be much greater, up to 100%/K depending on the required dynamic range. Thermomechanical noise is comparable to or less than the background thermal conductance noise for a property tuned and damped sensor array. The major noise contribution for the present devices is in the readout integrated circuit (ROIC) (kT/C noise, preamplifier, and switching noise). Table 8.3 summarizes the noise equivalent difference temperature (NEDT) values modeled for various noise sources with different pixel structures [40].

A typical difference in thermal expansion coefficients of metal-ceramic bimaterial designs is inherently limited to $\Delta a < 20 \times 10^{-6}$ K^{-1}. It has recently been suggested that the polymer-ceramic bimaterial cantilevers dramatically enhance thermally induced bending due to much more efficient actuation of readily expandable polymer nanolayers with $\Delta a < 200 \times 10^{-6}$ K^{-1}, combined with low thermal conductivity [41]. These new composite structures have been introduced with

Table 8.3: Modeled NEDT Noise Values (mK) for Different Microcantilever Pixel Structures

Pixel Dimensions	50 μm	25 μm	17 μm
Background thermal noise	1.2	2.1	3.5
Temperature fluctuation noise	5.2	7.3	10.4
Thermomechanical noise	0.7	0.7	1.0
ROIC noise sources			
$1/f$ + white noise	9.7	7.1	7.4
kT/C noise	7.0	8.7	15.1
Switching and other correlated noise	Small?	Small?	Small?
Total NEDT	**13.1**	**13.7**	**19.8**

Source: Hunter, S. R., Maurer, G. S., Simelgor, G., Radhakrishnan, S., and Gray, J., "High Sensitivity 25μm and 50μm Pitch Microcantilever IR Imaging Arrays," *Proceedings of SPIE 6542*, 65421F, 2007. With permission.

a combination of polymer brush layer, silver nanoparticles, and carbon nanotubes to enhance IR absorption and reinforce nanocomposite coating. This new cantilever design allows achievement of nearly fourfold improvement in thermal sensitivity compared to the metal coated counterparts. The serious drawback of the polymer-ceramic cantilevers at the present stage of development is their noncompatibility with traditional microfabrication technology.

Microcantilever IR detectors can have NEDT as small as 5 mK for 50 μm square pixels using silicon nitride for thermal isolation [21]. However, several important issues need to be addressed before the full potential can be realized. Between them we can distinguish: (i) mechanical noise inherent in micromechanical systems, (ii) nonuniformity of microcantilevers in large arrays, and (iii) high sensitivity of thermal IR detectors to the ambient temperature changes.

The effects of mechanical noise can be largely eliminated, by tailoring the resonance frequencies and the stiffness of microcantilevers. Figure 8.5b shows the ribbed single pixel structure to enhance the stiffness of the paddle and thermal isolation arms. The transverse corrugated structure on the bimaterial arm is used to reduce delamination issues and increase the bimaterial responsivity. This figure also shows a design of pixel that is immune to ambient temperature changes and other sources of interfering mechanical stress [42]. The sensor contains a second bimaterial and thermal isolation structure (see also Figure 8.5a). The operation can be described as follows:

- When the ambient temperature rises, Segment A cantilever bends upward

- Segment B cantilever support point follows movement of Segment A cantilever

- Segment B cantilever tracks lower temperature (except for radiation loading)

- Segment B cantilever bends and keeps capacitor plate at same distance from substrate

Both the IR sensing (Segment A) and additional (Segment B) structures respond identically but in opposite sense to the changes in ambient temperature, thus nulling out any substrate temperature induced motion in the cantilever paddle. In practice, leg Segment B is behind Segment A, rather than above it (see Figure 8.5b [40]). The compensation structure has the additional advantage since it also nulls out any mechanical bending of the bimaterial and isolation arms due to residual stresses created during the detector fabrication process.

Detailed modeling of the thermomechanical response of the cantilever pixels give thermal response times in the 5–10 msec range. This theoretical prediction has been confirmed by experimental data [40]. These thermal detectors have slow response times in comparison with photon detectors. However, the micromechanical structures can also be used as photon detectors with faster response times and higher performance than that of micromechanical thermal detectors [43–47].

The absorption of photons by a solid result in temperature changes and thermal expansion that in turn gives rise to acoustic waves at frequencies corresponding to the amplitude modulation of the incident photon radiation. When a silicon microcantilever is exposed to photons, the excess charge carriers generated induce an electronic stress that causes the semiconductor microcantilever to deflect (see Figure 8.6 [44]). Generation of electrons and holes in semiconductors results in

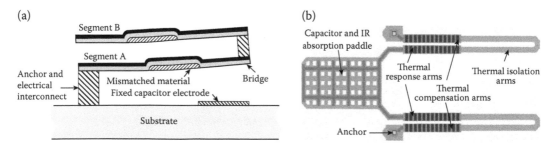

Figure 8.5 Ambient temperature compensation: (a) schematic diagram showing the operating principle, and (b) thermally compensated 25 μm pixel pitch structure elaborated at Multispectral Imaging. (From Hunter, S. R., Maurer, G. S., Simelgor, G., Radhakrishnan, S., and Gray, J., "High Sensitivity 25μm and 50μm Pitch Microcantilever IR Imaging Arrays," *Proceedings of SPIE* 6542, 65421F, 2007. With permission.)

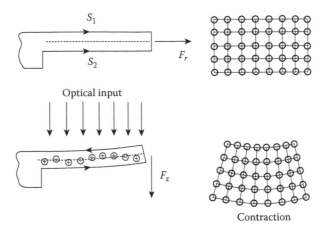

S_1

S_2

F_r

Optical input

F_z

Contraction

Figure 8.6 Schematic diagram showing the bending process of a semiconductor microcantilever exposed to radiation. Surface stresses S_1 and S_2 are balanced at equilibrium. Also depicted is the accompanied contraction of the silicon lattice following the generation of electron-hole pairs. (From Datskos, P. D., Rajic, S., Datskos, I., and Eger, C. M., "Novel Photon Detection Based Electronically-Induced Stress in Silicon," *Proceedings of SPIE* 3379, 173–81, 1998. With permission.)

development of a local mechanical strain. The surface stresses S_1 and S_2 are balanced at equilibrium, generating the radial force F_r along the medial plane of the microcantilevers. These stresses become unequal upon exposure to photons, producing the bending force, F_z, which displaces the tip of the microcantilever. The extent of bending is proportional to the radiation intensity.

Results of works published by Datskos et al. [45–47] demonstrated that microstructures represent an important development in MEMS photon detector technology and can be expected to provide the basis for further development. Up till now however, progress in their development is small.

8.2.2 Optically Coupled Cantilevers

The infrared radiation detection and subsequent reconstruction of an image can be also based on the deflection of individual microcantilever pixels using an optical technique, which was adapted from standard AFM imaging systems [13]. With this approach the array does not require metallization to individually address each pixel. In comparison electrically coupled cantilevers, the optical readout has a number of important advantages [30]:

■ The array is simpler to fabricate enabling reduced cost.

■ The need for an integrated ROIC is eliminated.

■ The layout complexity of matrix addressing is not required.

■ Parasitic heat from ROIC is eliminated.

■ Absence of electrical contacts between pixels and substrate eliminates a thermal leakage path.

The most important practical implication of the above approach is, however, related to their straightforward scalability to much larger (>2000 ×2000) arrays [20].

The responsivities of individual pixels for a particular array of microcantilevers can have slight variations and in addition, part of pixels can be slightly stressed. As a consequence, the deformations of some of them will not be detectable by readout. Fortunately, recently developed computational algorithms restore images or video that contain missing or degraded pixel information [20].

Figure 8.7 demonstrates a schematic diagram and components of the optomechanical IR imaging system [48]. It consists of an IR imaging lens, a microcantilever focal plane array (FPA), and an optical readout. Visible light that comes from the LED becomes parallel via collimating lens. Subsequently the parallel light is reflected by the pixels of the FPA and then passes through a transforming lens. The reflected diffracting rays synthesize the spectra of the cantilever array on the rear focal plane of the transforming lens. When the incident IR flux is absorbed by the pixels, their temperature rises, and then causes a small deflection of the cantilevers. Consequently, the changes in the reflected distribution of visible light are collected and analyzed by a conventional

(a)

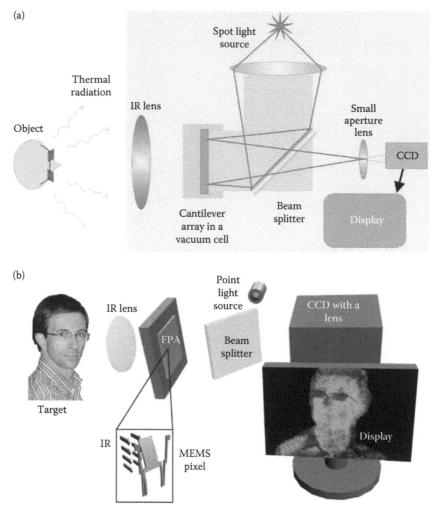

(b)

Figure 8.7 Uncooled optical-readable IR imaging system: (a) schematic diagram, and (b) components of the thermal imager. (From Datskos, P., and Lavrik, N., "Simple Thermal Imagers Use Scalable Micromechanical Arrays," *SPIE Newsroom* 10.1117/2.1200608.036, 2006. With permission.)

charge-coupled-device (CCD) or complementary metal oxide semiconductor (CMOS) camera. The small aperture of the lens mounted on a camera makes it possible to achieve the required angle-to-intensity conversion. This simple optical readout uses 1 mW power of the light beam while the power per FPA pixel is a few nanowatts. The dynamic range, intrinsic noise, and resolution of the camera largely determine the performance of the system.

To minimize fabrication complexity, Datskos and coworkers [20,48] have devised a process flow for the FPA fabrication that involves only three photolithographic steps and relies on well-established methods of surface micromachining (see Figure 8.8). The process starts with double sided polished Si wafers. To pattern 5-μm tall anchoring posts for suspending of the structures, a photoresist mask is etched using SF_6 reactive ion etching. Then, 6.5-μm thick sacrificial layer of silicon oxide is deposited on the Si surface with posts using plasma enhanced chemical vapor deposition (PECVD) at 250°C. This step is followed by chemo-mechanical polishing for surface planarization until a 4.5-μm thick oxide layer flush with the posts remain. This sacrificial layer thickness is chosen to form an optimal resonant cavity between the Si substrate and the pixel layer. Next, a 600-nm thick SiN_x layer is deposited on the planarized oxide layer. After this step, the Au metallization is e-beam evaporated on previously deposited 5-nm thick Cr adhesion layer. The second photolithography involves lift off patterning of a 120-nm Au layer evaporated on SiN_x, what corresponds to the superposition of the bimaterial leg sections and reflective regions of the pixel heads. Definition

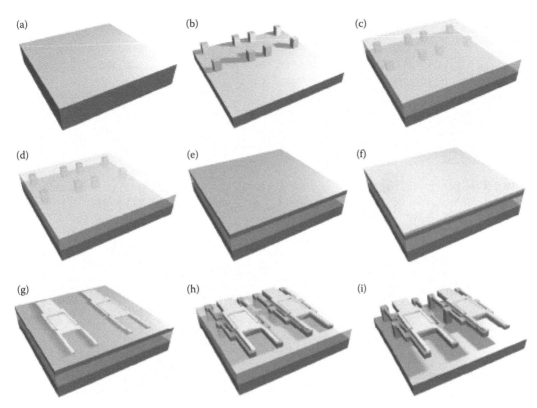

Figure 8.8 FPA fabrication: (a) a double side polished Si wafer, (b) reactive ion etching to create posts on the Si wafer, (c) SiO$_2$ sacrificial layer deposition, (d) chemo-mechanical polishing, (e) low-stress SiN$_x$ layer deposition, (f) lift-off metallization, (g) patterning of the metallized region, (h) patterning of the pixel geometry using RIE, and (i) wet etching of the sacrificial SiO$_2$ layer resulting in a released structure. (From Lavrik, N., Archibald, R., Grbovic, D., Rajic, S., and Datskos, P., "Uncooled MEMS IR Imagers with Optical Readout and Image Processing" *Proceedings of SPIE* 6542, 65421E, 2007. With permission.)

of the detector geometry in the SiN$_x$ layer is carried out in the third photolithography. Finally, a sacrificial layer is removed by wet etching in HF followed by rinsing and CO$_2$ critical point drying.

Cantilever-based IR imaging devices demonstrated to date are less sensitive than the theoretical predictions. There are many ways to improve the sensitivity including the design and improving the processes and readout system. The theoretical prediction indicates that the sensitivity of the microcantilevers is inversely proportional to the gap distance between the cantilevers and substrate [49]. Cantilevers are usually anchored to a silicon substrate with a 2–3 µm spacing between them. Small gaps result in high performance, however, small gaps also led to problems caused by stiction and the sacrificial layer remaining in the released structure. Moreover, the IR flux must transmit through the silicon substrate and only 54% of the incident radiation can reach the cantilevers. As a result of this, new kinds of design structures are explored.

One of the novel designs is substrate-free uncooled IR detector based on an optical-readable method [33]. The detector is composed of a bimaterial cantilever array, without a silicon substrate, which is eliminated in the fabrication process. An example of this structure is shown in Figure 8.9. The cantilever with a 1-µm thick SiN$_x$ main structure layer incorporates an IR absorber/reflector, two bimaterial arms, and two thermal isolation arms. A thin Au reflection layer and thick Au bimaterial layers were deposited on IR absorber and the bimaterial arms, respectively. A bulk silicon process that includes Si-glass anodic bonding and deep reactive ion etching was developed to remove the substrate silicon and form frames for every FPA pixel. Compared with the generally used sacrificial layer cantilever, the loss of incident IR energy caused by the reflection from and absorption by the silicon substrate is eliminated completely

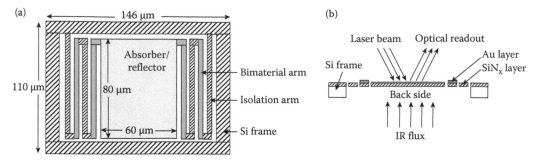

Figure 8.9 Schematic diagram of a cantilever pixel: (a) the top view, and (b) the cross section. (From Yu, X., Yi, Y., Ma, S., Liu, M., Liu, X., Dong, L., and Zhao, L., *Journal of Micromechanics and Microengineering*, 18, 057001, 2008. With permission.)

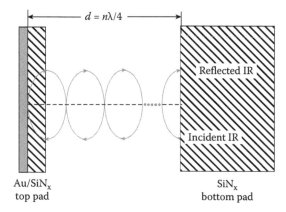

Figure 8.10 Optical model of resonate cavity in the structure without substrate. (From Shi, S., Chen, D., Jiao, B., Li, C., Qu, Y., Jing, Y., Ye, T., et al., *IEEE Sensor Journal*, 7, 1703–10, 2007. With permission.)

in the substrate-free structure. The thermomechanical sensitivity of the cantilever pixel was measured as 0.11 μm/K.

Another modification of substrate-free optical-readable FPA is the introduction of double bimaterial-layer cantilever pixels [50]. The top layer of the cantilever pixels consists of two materials with large mismatching TCE: SiN_x and Au, which convert IR radiation into mechanical deflection (see Figure 8.10). The bottom layer is also SiN_x cantilever, which partially serves as the thermal isolation legs. Such geometry forms the resonant cavity, which together with substrate-free design considerably enhance the absorption of incident IR radiation (IR radiation passes through the bottom SiN_x absorption pad). The theoretical analysis suggests that the temperature resolution of the imaging system can reach 7 mK.

Recently, a multifold interval metallized leg configuration has been proposed to increase sensitivity of optical-readable bimaterial microcantilever array [32]. This multifold configuration consists of alternatively connected unmetallized and metallized legs. However, the measured sensitivity (NEDT about 400 mK for 160 × 160 elements with 120-μm pitch) has been much lower than the theoretical value (28.1 mK). Therefore, better performance could be expected with the improvement of fabrication techniques and using less noise optical readout.

8.2.3 Pyro-Optical Transducers

Among the first nonscanned IR imagers were evaporographs and absorption edge image converters. In an evaporograph the radiation was focused onto a blackened membrane coated with a thin film of oil [51,52]. The differential rate of evaporation of the oil was proportional to radiation intensity. Then, illumination of the film with visible light has produced an interference pattern corresponding to the thermal picture.

The second thermal imaging device was the absorption edge image converter [53]. Operation of the device was based upon utilizing the temperature dependence of the absorption edge of a semiconductor. When the sample of a suitable material is viewed by transmitted monochromatic light at a wavelength near the threshold, any variations in temperature appear as differences in the transmitted intensity. The image converter built by Harding, Hilsum, and Northrop used amorphous selenium as the semiconductor, with absorption edges of 580–660 nm [53]. The absorption edge shifted by 0.27 nm/°C, equivalent to a change in the energy gap of 9.7×10^{-4} eV/°C [54]. The theory and applications of the absorption edge image converter were discussed by Hilsum and Harding [55]. They showed that bodies 10°C above ambient can be imaged.

The performance of both imaging devices was poor because of the very long time constant (up to several seconds) and the poor spatial resolution.

A new generation of solid-state thermal devices with optical readout has been proposed recently. Carr and Setiadi developed pyro-optical pixels with phase-change materials whose absorption changed with temperature [56]. Secundo, Lubianiker, and Granat proposed waveguide-based sensitive pixel arrays based on thermally sensitive electro-optic birefringent crystals [57]. Another approach based on interference on a polymer membrane with high TCE has been described by Flusberg et al. [58,59].

Figure 8.11 shows basic concept of pyro-optical IR-visible transducer. IR radiation from the scene is imaged on IR-visible transducer. Each pixel of thermally tunable transducer acts as a wavelength translator, converting IR radiation into visible signals that can be detected by visible sensor (eye, CCD, or CMOS image cameras). Conversion of IR to visible radiation is received by probing the temperature change of the pixels by visible light. To combine both IR and visible radiation, a high-transmitted splitter is used. The transducer shown in Figure 8.11 as reflective in the visible may instead be transmissive. It is analogous to a bolometer but with optical bias.

An example of a newly designed, optically readable pyro-optical direct-view imager is the RedShift's Thermal Light Valve™ (TLV) [60,61]. The TLV chip is based on a class of optical active thin films developed by Aegis Semiconductor Inc. for telecom applications [62] and is composed of a thermally tunable Fabry-Perot bandpass filters (pixels) standing on thermally resistive posts on an optically reflective and thermally conductive substrate (see Figure 8.12a). A thermal image is obtained by measuring the pixel-to-pixel variation in reflection of the near infrared (NIR) probe signal (an 850 nm vertical cavity surface emitting laser) using CMOS imagers. Because the NIR probe signal reflected from the TLV depends on the incident IR radiation, the intensity of light received by the CMOS imager is effectively modulated by the IR signature of the observed scene.

The Fabry-Perot structure is based on high-index a-Si and low-index SiN_x thin films, which have been used extensively for many years in solar cells and flat panel displays. These materials are deposited using PECVD. This technique produces uniform, dense materials in high volume manufacturing environments. Spectral tunability of the filter can be achieved by changing its cavity—a product of thickness and index of refraction. RedShift achieves tunability by changing the index of refraction. The a-Si layers have a refractive index that changes with temperature at a rate of 6×10^{-5}/K at 300 K (normalized by index).

Figure 8.13 shows the design of TLV and scanning electron microscopy image of tunable 2-D filter array [61]. The array consists of sensor filters that absorb long wavelength IR (LWIR) radiation from the scene, and reference filters, both connected to the substrate through thermal isolating anchors. In its base state, the filter array forms a mirror that reflects the NIR probe beam. Due to IR

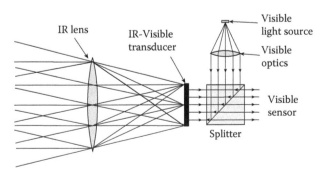

Figure 8.11 Concept of the pyro-optical transducer.

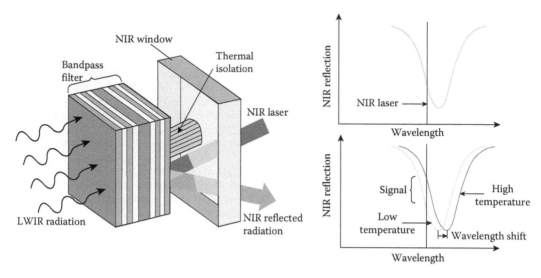

Figure 8.12   RedShift's TLV pixel: (a) filter pixel standing on a post on a substrate and the light paths of IR from scene and NIR probing light, and (b) the "wavelength converter" principle.

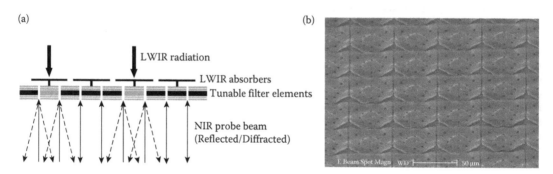

Figure 8.13   (a) Design of TLV, and (b) scanning electron microscopy image of tunable 2-D filter array. (From Wagner, M., "Solid State Optical Thermal Imaging: Performance Update," *Proceedings of SPIE* 6940, 694016, 2008. With permission.)

radiation heating, a sensor filter's temperature diverges from that of the substrate and local reference filters. Then, the phase of the reflected NIR probe beam is shifted and reflection amplitude over the NIR readout wavelength range increases.

Using the above technology, RedShift built a 160 × 120 pixel thermal imager with the spectral sensitivity of 150 mK in the 8–12 μm LWIR region. Because of the low-cost in comparison with microbolometer cameras (about 5 times difference in price/performance), the imagers can be used in large-volume applications like video security and automotive active safety applications.

8.2.4 Antenna-Coupled Microbolometers

A major limitation of thermal detectors seems to be unsuitable for the next generation of IR thermal imaging systems, which are moving toward multispectral operation [63]. Third generation systems enable rapid, efficient, and multidimensional scene interpretation that is especially beneficial for early threat warning and target recognition applications.

Recently, Neikirk et al. [64–66] have proposed a novel design for a monolithic micromachined array of bolometers capable of multispectral imaging in LWIR region. The main ingredient of this approach is to employ a planar multimode antenna structure to couple incident radiation to a microbolometer that is much smaller than the IR wavelength (the effective detector area is an order of λ^2). In contrast to absorbing layer coupling, antenna coupling gives selective responsivity to both the spatial mode and the polarization of the incoming radiation [67].

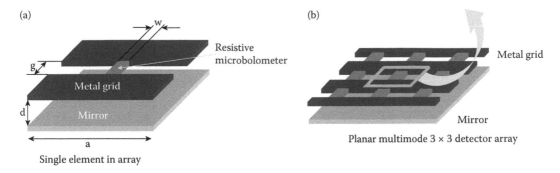

Figure 8.14 Configuration of a planar multimode detector: (a) single element, and (b) multimode array pixel. (From Weling, A. S., Henning, P. F., Neikirk, D. P., and Han, S., "Antenna-Coupled Microbolometers for Multi-Spectral Infrared Imaging," *Proceedings of SPIE* 6206, 62061F, 2006. With permission.)

Figure 8.14a illustrates a simple realization of a multimode detector pixel that consists of an interwoven grid of metallic and resistive "wires" [67]. The pixel shown in Figure 8.14b is composed of a 3×3 array of microbolometric sensors with effective size the same as current FPA pixels (~50 μm). In this way there is a possibility of constructing a conventional thermal detector in which the detection is performed by much smaller devices. The microbolometric sensors are placed one-quarter wavelength above the ROIC substrate, which serves largely as a conducting mirror layer. In general, if n such microbolometer elements are combined together in a single pixel, the antenna-coupled detector is termed "multimode" with effective area $\sim n\lambda^2$ [68]. Its output is an inherent superposition of the individual response of nine sensing elements and the performance is still characterized by a detectivity of a conventional single-mode microbolometer.

The electrical response of a planar multimode bolometric detector is a series and parallel combination of multiple single-mode antenna coupled microbolometers. The wavelength sensitivity can be tuned by optimizing its multiple geometric parameters exploiting resonant antenna coupling and multilayer interference effects in a conventional microbolometer. The grid response depends on the array period a, gap width g, post width w, distance to mirror d, and sheet resistance R_s of microbolometer material.

The electromagnetic behavior of planar microbolometers can be analyzed using transmission line theory [67]. For example [66], Figure 8.15 shows the design and spectral response of three separate pixels, each with peak spectral response at a different wavelength, each located at the identical distance d between the multimode array pixels and the mirror (without mechanical tuning), changing the wavelength selectivity of each pixel by changing only the lithographically drawn parameters: a, g, w. The bolometer sheet resistance, R_s, is assumed to be constant for all pixels, allowing the use of a single bolometric material. In this way, it is possible to accomplish three-color LWIR imaging with no moving parts in the pixels. Moreover, due to greatly reduced thermal mass, the significant enhancement in speed of response of this spectrally selective microbolometer can be realized.

Fabrication of planar antenna-coupled detector array requires adequate thermal impedance to exploit performance comparable to that of the conventional air-bridge microbolometer. A less complex fabrication process can be provided by changing only lithographically parameters with one sacrificial layer thickness.

Finally, it should be mentioned about two types of antenna-coupled IR pixels proposed by Gonzales, Porter, and Boreman [69]. The first type of pixel is serially connected N microbolometers (with areas in the order of 10 μm^2) for which the signal-to-noise ratio of a single detector can increase by a factor of N for an $N \times N$ array. The second IR pixel was fabricated by using a Fresnel lens to collect and focus energy to a single element detector and 2× increase in detectivity was observed for 200 μm detector in diameter.

8.3 COMPARISON OF THERMAL DETECTORS

Table 8.4 summarizes the basic features of thermal detectors [37,70,71]. In thermopiles the signal form is electromotive voltage, ΔV. Bolometers detect the temperature itself by a carrier density change and a mobility change. The signal form is resistance change, ΔR. The pyroelectric scheme detects temperature change by the dielectric constant change and the spontaneous polarization

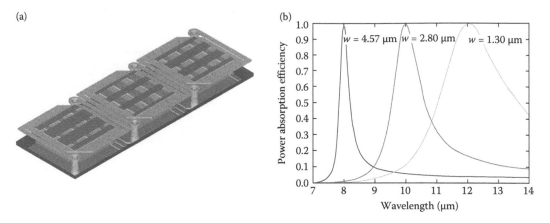

Figure 8.15 (a) Design and (b) spectral response of three separate pixels, each with peak spectral response at a different wavelength, and each located at identical distance d between the multimode array pixels and the mirror ($d_1 = d_2 = d_3$, without mechanical tuning). Power absorptions are almost unity at the designed wavelength. a, g, d, and R_s are 6.80 μm, 0.20 μm, 3.14 μm, and 56.6 Ω/square, respectively. (From Han, S. W., and Neikirk, D. P., "Design of Infrared Wavelength-Selective Microbolometers Using Planar Multimode Detectors," *Proceedings of SPIE* 5836, 549–57, 2005. With permission.)

Table 8.4: Basic Parameters of Thermal Detectors

Type of Detector	Signal Versus Temperature	Characteristic Parameter	K	Electrical Noise Power Density	Bias Power
Thermocouple	$\propto \Delta T$	$\alpha_s = \dfrac{dV}{dT}$	$\alpha_s = \dfrac{dV}{dT}$	$4kTR$	no
Bolometer	$\propto T$	$\alpha = \dfrac{1}{R}\dfrac{dR}{dT}$	IRa	$4kTR$	I^2R
Pyroelectric	$\sim \propto \dfrac{dT}{dt}$	$p = \dfrac{dP}{dT}$	$\dfrac{pA\omega R}{(1+\omega^2\tau^2)^{1/2}}$	$\dfrac{4kTR}{1+\omega^2\tau^2}$	no
Forward-bias diode	$\propto T$	$\alpha = \dfrac{dV}{dT}$	α	$\dfrac{4(kT)^2}{2I}$	IV
Microcantilever	$\propto T$	$TCC = \dfrac{1}{Z_{gap}}\dfrac{\Delta Z}{\Delta T}$			

change. The signal form is polarization change, ΔQ. In the case of thermo-mechanical IR detectors, such as microcantilevers, the intrinsic responsivity should be defined in terms of the mechanical response of the device (i.e., displacement, ΔZ, per absorbed power) in units of meters per watts.

In practice, thermopiles are widely used for low frequency applications including DC operation. They experience serious competition from pyroelectric detectors and bolometers that offer better performance at high frequencies. Bolometers can be used in conjunction with optical immersion, which enables very good performance and response time of ≈1 ms. Since thermopiles detect the temperature difference between the hot junction and the cold junctions, and since the cold junctions are located on the heat reservoir, the cold junction plays an important role of temperature reference. Therefore, thermopiles do not need an operation stabilizer, while the bolometer does. Since the temperature change at the IR absorber by the incident IR is much smaller than the operation temperature change, and since it is difficult for the preamplifier to sense the resistance change according to the range for the whole operation temperature change, operation temperature stabilization is often needed for the bolometer. To increase the temperature dependence of resistance and dielectric constant and to realize large responsivity, bolometer and pyroelectric detectors often use a thermoelectric material with a transition point, and the operation temperature should be set near the transition temperature. In this case, operation temperature control is necessary. However, it should be marked that the improvements of readout circuits can eliminate the bolometer's thermo-electric stabilization.

The thermopile detector has a temperature reference inside, so that chopping is not needed. The bolometer also does not need a chopper, because it detects the temperature itself. On the other hand, the pyroelectric detector needs a chopper, because it detects the temperature change. Moreover, the pyroelectric detector cannot be used under a circumstance where vibration is large, because it is adversely affected by the microphonic noise.

REFERENCES

1. C. M. Hanson, "Barriers to Background-Limited Performance for Uncooled IR Sensors," *Proceedings of SPIE* 5406, 454–64, 2004.

2. *UKTA News*, Issue 27, December 2006.

3. "MEMS Transform Infrared Imaging," *Opto&Laser Europe*, June 2003.

4. M. J. E. Golay, "A Pneumatic Infra-Red Detectors," *Review of Scientific Instruments* 18, 357–62, 1947.

5. M. J. E. Golay, "The Theoretical and Practical Sensitivity of the Pneumatic Infrared Detector," *Review of Scientific Instruments* 20, 816, 1949.

6. J. R. Hickley and D. B. Daniels, "Modified Optical System for the Golay Detector," *Review of Scientific Instruments* 40, 732–33, 1969.

7. J. B. Chevrier, K. Baert, T. Slater, and A. Verbist, "Micromachined Infrared Pneumatic Detector for Gas Sensors," *Microsystem Technologies* 1, 71–74, 1995.

8. T. W. Kenny, J. K. Reynolds, J. A. Podosek, E. C. Vote, L. M. Miller, H. K. Rockstad, and W. J. Kaiser, "Micromachined Infrared Sensors Using Tunneling Displacement Transducers," *Review of Scientific Instruments* 67, 112–28, 1996.

9. T. W. Kenny, "Tunneling Infrared Sensors," in *Uncooled Infrared Imaging Arrays and Systems*, eds. P. W. Kruse and D. D. Skatrud, 227–67, Academic Press, San Diego, CA, 1997.

10. O. Ajakaiye, J. Grade, C. Shin, and T. Kenny, "Wafer-Scale Fabrication of Infrared Detectors Based on Tunneling Displacement Transducers," *Sensors & Actuators* A134, 575–81, 2007.

11. P. I. Oden, P. G. Datskos, T. Thundat, and R. J. Warmack, "Uncooled Thermal Imaging Using a Piezoresistive Microcantilevers," *Applied Physics Letters* 69, 3277–79, 1996.

12. P. G. Datskos, P. I. Oden, T. Thundat, E. A. Wachter, R. J. Warmack, and S. R. Hunter, "Remote Infrared Detection Using Piezoresistive Microcantilevers," *Applied Physics Letters* 69, 2986–88, 1996.

13. E. A. Wachter, T. Thundat, P. I. Oden, R. J. Warmack, P. D. Datskos, and S. L. Sharp, "Remote Optical Detection Using Microcantilevers," *Review of Scientific Instruments* 67, 3434–39, 1996.

14. P. I. Oden, E. A. Wachter, P. G. Datskos, T. Thundat, and R. J. Warmack, "Optical and Infrared Detection Using Microcantilevers," *Proceedings of SPIE* 2744, 345–54, 1996.

15. P. G. Datskos, S. Rajic, and I. Datskou, "Photoinduced and Thermal Stress in Silicon Microcantilevers," *Applied Physics Letters* 73, 2319–21, 1998.

16. P. G. Datskos, S. Rajic, and I. Datskou, "Detection of Infrared Photons Using the Electronic Stress in Metal/Semiconductor Cantilever Interfaces," *Ultramicroscopy* 82, 49–56, 2000.

17. L. R. Senesac, J. L. Corbeil, S. Rajic, N. V. Lavrik, and P. G. Datskos, "IR Imaging Using Uncooled Microcantilever Detectors," *Ultramicroscopy* 97, 451–58, 2003.

18. P. G. Datskos, N. V. Lavrik, and S. Rajic, "Performance of Uncooled Microcantilever Thermal Detectors," *Review of Scientific Instruments* 75, 1134–48, 2004.

19. P. Datskos and N. Lavrik, "Uncooled Infrared MEMS Detectors," in *Smart Sensors and MEMS*, eds. S. Y. Yurish and M. T. Gomes, 381–419, Kluwer Academic, Dordrecht, 2005.

20. N. Lavrik, R. Archibald, D. Grbovic, S. Rajic, and P. Datskos, "Uncooled MEMS IR Imagers with Optical Readout and Image Processing" *Proceedings of SPIE* 6542, 65421E, 2007.

21. R. Amantea, C. M. Knoedler, F. P. Pantuso, V. K. Patel, D. J. Sauer, and J. R. Tower, "An Uncooled IR Imager with 5 mK NEDT," *Proceedings of SPIE* 3061, 210–22, 1997.

22. R. Amantea, L. A. Goodman, F. Pantuso, D. J. Sauer, M. Varghese, T. S. Villani, and L. K. White, "Progress Towards an Uncooled IR Imager with 5 mK NEDT," *Proceedings of SPIE* 3436, 647–59, 1998.

23. S. R. Hunter, R. A. Amantea, L. A. Goodman, D. B. Kharas, S. Gershtein, J. R. Matey, S. N. Perna, Y. Yu, N. Maley, and L. K. White, "High Sensitivity Uncooled Microcantilever Infrared Imaging Arrays," *Proceedings of SPIE* 5074, 469–80, 2003.

24. S. R. Hunter, G. Maurer, L. Jiang, and G. Simelgor, "High Sensitivity Uncooled Microcantilever Infrared Imaging Arrays," *Proceedings of SPIE* 6206, 62061J, 2006.

25. S. R. Hunter, G. Maurer, G. Simelgor, S. Radhakrishnan, J. Gray, K. Bachir, T. Pennell, M. Bauer, and U. Jagadish, "Development and Optimization of Microcantilever Based IR Imaging Arrays," *Proceedings of SPIE* 6940, 694013, 2008.

26. T. Ishizuya, J. Suzuki, K. Akagawa, and T. Kazama, "Optically Readable Bi-Material Infrared Detector," *Proceedings of SPIE* 4369, 342–49, 1998.

27. T. Perazzo, M. Mao, O. Kwon, A. Majumdar, J. B. Varesi, and P. Norton, "Infrared Vision Using Uncooled Micro-Optomechanical Camera," *Applied Physics Letters* 74, 3567–69, 1999.

28. P. Norton, M. Mao, T. Perazzo, Y. Zhao, O. Kwon, A. Majumdar, and J. Varesi, "Micro-Optomechanical Infrared Receiver with Optical Readout—MIRROR," *Proceedings of SPIE* 4028, 72–78, 2000.

29. J. E. Choi, "Design and Control of a Thermal Stabilizing System for a MEMS Optomechanical Uncooled Infrared Imaging Camera," *Sensors & Actuators* A104, 132–42, 2003.

30. J. Zhao, "High Sensitivity Photomechanical MW-LWIR Imaging Using an Uncooled MEMS Microcantilever Array and Optical Readout," *Proceedings of SPIE* 5783, 506–13, 2005.

31. B. Jiao, C. Li, D. Chen, T. Ye, S. Shi, Y. Qu, L. Dong, et al., "A Novel Opto-Mechanical Uncooled Infrared Detector," *Infrared Physics & Technology* 51, 66–72, 2007.

32. F. Dong, Q. Zhang, D. Chen, Z. Miao, Z. Xiong, Z. Guo, C. Li, B. Jiao, and X. Wu, "Uncooled Infrared Imaging Device Based on Optimized Optomechanical Micro-Cantilever Array," *Ultramicroscopy* 108, 579–88, 2008.

33. X. Yu, Y. Yi, S. Ma, M. Liu, X. Liu, L. Dong, and Y. Zhao, "Design and Fabrication of a High Sensitivity Focal Plane Array for Uncooled IR Imaging," *Journal of Micromechanics and Microengineering* 18, 057001, 2008.

34. J. R. Barnes, R. J. Stephenson, C. N. Woodburn, S. J. O'Shea, M. E. Welland, J. R. Barnes, R. J. Stephenson, et al., "A Femtojoule Calorimeter Using Micromechanical Sensors," *Review of Scientific Instruments* 65, 3793–98, 1994.

35. J. Varesi, J. Lai, T. Perazzo, Z. Shi, and A. Majumdar, "Photothermal Measurements at Picowatt Resolution Using Uncooled Micro-Optomechanical Sensors," *Applied Physics Letters* 71, 306–8, 1997.

36. J. R. Barnes, R. J. Stephenson, C. N. Woodburn, S. J. O'Shea, M. E. Welland, T. Rayment, J. K. Gimzewski, et al., "A Femtojoule Calorimeter Using Micromechanical Sensors," *Review of Scientific Instruments* 65, 3793–98, 1994.

37. P. G. Datskos, "Detectors: Figures of Merit," in *Encyclopedia of Optical Engineering*, ed. R. Driggers, 349–57, Marcel Dekker, New York, 2003.

38. D. Sarid, *Scanning Force Microscopy*, Oxford University Press, New York, 1991.

39. E. Majorana and Y. Ogawa, "Mechanical Noise in Coupled Oscillators," *Physics Letters* A233, 162–68, 1997.

40. S. R. Hunter, G. S. Maurer, G. Simelgor, S. Radhakrishnan, and J. Gray, "High Sensitivity 25μm and 50μm Pitch Microcantilever IR Imaging Arrays," *Proceedings of SPIE* 6542, 65421F, 2007.

41. Y. H. Lin, M. E. McConney, M. C. LeMieux, S. Peleshanko, C. Jiang, S. Singamaneni, and V. V. Tsukurk, "Trilayered Ceramic-Metal-Polymer Microcantilevers with Dramatically Enhanced Thermal Sensitivity," *Advanced Materials* 18, 1157–61, 2006.

42. J. L. Corbeil, N. V. Lavrik, S. Rajic, and P. G. Datskos, " 'Self-leveling' Uncooled Microcantilever Thermal Detector," *Applied Physics Letters* 81, 1306–8, 2002.

43. P. G. Datskos, S. Rajic, and I. Datskou, "Photo-Induced Stress in Silicon Microcantilevers," *Applied Physics Letters* 73, 2319–21, 1998.

44. P. D. Datskos, S. Rajic, I. Datskos, and C. M. Eger, "Novel Photon Detection Based Electronically-Induced Stress in Silicon," *Proceedings of SPIE* 3379, 173–81, 1998.

45. P. G. Datskos, "Micromechanical Uncooled Photon Detectors," *Proceedings of SPIE* 3948, 80–93, 2000.

46. P. G. Datskos, S. Rajic, and I. Datskou, "Detection of Infrared Photons Using the Electronic Stress in Metal-Semiconductor Cantilever Interfaces," *Ultramicroscopy* 82, 49–56, 2000.

47. P. G. Datskos, S. Rajic, L. R. Senesac, and I. Datskou, "Fabrication of Quantum Well Microcantilever Photon Detectors," *Ultramicroscopy* 86, 191–206, 2001.

48. P. Datskos and N. Lavrik, "Simple Thermal Imagers Use Scalable Micromechanical Arrays," *SPIE Newsroom* 10.1117/2.1200608.036, 2006.

49. B. Li, "Design and Simulation of an Uncooled Double-Cantilever Microbolometer with the Potential for ~mK NETD," *Sensors and Actuators* A112, 351–59, 2004.

50. S. Shi, D. Chen, B. Jiao, C. Li, Y. Qu, Y. Jing, T. Ye, et al., "Design of a Novel Substrate-Free Double-Layer-Cantilever FPA Applied for Uncooled Optical-Readable Infrared Imaging System," *IEEE Sensor Journal* 7, 1703–10, 2007.

51. P. W. Kruse, L. D. McGlauchlin, and R. B. McQuistan, *Elements of Infrared Technology*, Wiley, New York, 1962.

52. G. W. McDaniel and D. Z. Robinson, "Thermal Imaging by Means of the Evaporograph," *Applied Optics* 1, 311–24, 1962.

53. W. R. Harding, C. Hilsum, and D. C. Northrop, "A New Thermal Image-Converter," *Nature* 181, 691–92, 1958.

54. C. Hilsum, "The Absorption Edge of Amorphous Selenium and Its Change with Temperature," *Proceedings of the Physical Society* B69, 506–12, 1956.

55. C. Hilsum and W. R. Harding, "The Theory of Thermal Imaging, and Its Application to the Absorption-Edge Image Tube," *Infrared Physics* 1, 67–93, 1961.

56. W. Carr and D. Setiadi, "Micromachined Pyro-Optical Structure," U.S. Patent No. 6,770,882.

57. L. Secundo, Y. Lubianiker, and A. J. Granat, "Uncooled FPA with Optical Reading: Reaching the Theoretical Limit," *Proceedings of SPIE* 5783, 483–95, 2005.

58. A. Flusberg and S. Deliwala, "Highly Sensitive Infrared Imager with Direct Optical Readout," *Proceedings of SPIE* 6206, 62061E, 2006.

59. A. Flusberg, S. Swartz, M. Huff, and S. Gross, "Thermal-to-Visible Transducer (TVT) for Thermal-IR Imaging, *Proceedings of SPIE* 6940, 694015, 2008.

60. M. Wagner, E. Ma, J. Heanue, and S. Wu, "Solid State Optical Thermal Imagers," *Proceedings of SPIE* 6542, 65421P, 2007.

61. M. Wagner, "Solid State Optical Thermal Imaging: Performance Update," *Proceedings of SPIE* 6940, 694016, 2008.

62. M. Wu, J. Cook, R. DeVito, J. Li, E. Ma, R. Murano, N. Nemchuk, M. Tabasky, and M. Wagner, "Novel Low-Cost Uncooled Infrared Camera," *Proceedings of SPIE* 5783, 496–505, 2005.

63. A. Rogalski, J. Antoszewski, and L. Faraone, "Third Generation Infrared Photodetector Arrays," *Journal of Applied Physics* 105, 091101, 2009.

64. S. W. Han, J. W. Kim, Y. S. Sohn, and D. P. Neikirk, "Design of Infrared Wavelength-Selective Microbolometers Using Planar Multimode Detectors," *Electronics Letters* 40, 1410–11, 2004.

65. S. W. Han and D. P. Neikirk, "Design of Infrared Wavelength-Selective Microbolometers Using Planar Multimode Detectors," *Proceedings of SPIE* 5836, 549–57, 2005.

66. A. S. Weling, P. F. Henning, D. P. Neikirk, and S. Han, "Antenna-Coupled Microbolometers for Multi-Spectral Infrared Imaging," *Proceedings of SPIE* 6206, 62061F, 2006.

67. S. E. Schwarz and B. T. Ulrich, "Antenna Coupled Thermal Detectors," *Journal of Applied Physics* 85, 1870–3, 1977.

68. D. B. Rutledge and S. E. Schwarz, "Planar Multi-Mode Detector Arrays for Infrared and Millimetre-Wave Applications," *IEEE Journal of Quantum Electronics* QE-17, 407–14, 1981.

69. F. J. Gonzales, J. L. Porter, and G. D. Boreman, "Antenna-Coupled Infrared Detectors," *Proceedings of SPIE* 5406, 863–71, 2004.

70. J. T. Houghton and S. D. Smith, *Infra-Red Physics*, Oxford University Press, Oxford, 1966.

71. J. Piotrowski, "Breakthrough in Infrared Technology—The Micromachined Thermal Detector Arrays," *Opto-Electronics Review* 3, 3–8, 1995.

PART III

INFRARED PHOTON DETECTORS

9 Theory of Photon Detectors

The interaction of infrared radiation with electrons results in several photoeffects such as: photoconductive, photovoltaic, photoelectromagnetic, Dember, and photon drag. Based on these photoeffects, different types of detectors have been of interest, but only photoconductive and photovoltaic (p-n junction and Schottky barrier) detectors have been widely exploited.

Photoeffects, which occur in structures with built-in potential barriers, are essentially photovoltaic and result when excess carriers are injected optically into the vicinity of such barriers. The role of the built-in electric field is to cause the charge carriers of opposite sign to move in opposite directions depending upon the external circuit. Several structures are possible to observe the photovoltaic effect. These include p-n junctions, heterojunctions, Schottky barriers, and metal-insulator-semiconductor (MIS) photocapacitors. Each of these different types of devices has certain advantages for infrared (IR) detection, depending on the particular applications. Recently, more interest has been focused on p-n junction photodiodes for use with silicon hybrid focal plane arrays for direct detection in the 3–5 and 8–14 μm spectral regions. In this application, photodiodes are preferred over photoconductors because of their relatively high impedance, matched directly into the input stage of a silicon readout, and lower power dissipation. Furthermore, the photodiodes have a faster response than photoconductors because the strong field in the depletion region imparts a large velocity to the photogenerated carriers. Also, photodiodes are not affected by many of the trapping effects associated with photoconductors.

The basic theory of different types of photon detectors will be presented in this chapter in a uniform structure convenient for the various detector materials.

9.1 PHOTOCONDUCTIVE DETECTORS

A number of excellent treatises and papers have been published on photoconductive detectors [1–13]. Many of them considered HgCdTe photoconductors, because in the last four decades the work in this area has been devoted almost exclusively to these detectors. Our purpose is to present an up-to-date description of the theory and principles of photoconductors in a form most suitable for design and applications.

9.1.1 Intrinsic Photoconductivity Theory

The photoconductive detector is essentially a radiation-sensitive resistor. The operation of a photoconductor is shown in Figure 9.1. A photon of energy $h\nu$ greater than the band-gap energy E_g is absorbed to produce electron-hole pairs, thereby changing the electrical conductivity of the semiconductor. For direct narrow gap semiconductors, the optical absorption is very much higher than in extrinsic detectors.

In almost all cases the change in conductivity is measured by means of electrodes attached to the sample. For low resistance material, where the sample resistance is typically 100 Ω, the photoconductor is usually operated in a constant current circuit as shown in Figure 9.1. The series load resistance is large compared to the sample resistance, and the signal is detected as a change in voltage developed across the sample. For high resistance photoconductors, a constant voltage circuit is preferred and the signal is detected as a change in current in the bias circuit.

We assume that the signal photon flux density $\Phi_s(\lambda)$ is incident on the detector area $A = wl$ and that the detector is operated under constant current conditions (i.e., $R_L \gg R$). We suppose further that the illumination and the bias field are weak, and the excess carrier lifetime τ is the same for majority and minority carriers. To derive an expression for voltage responsivity, we take a one-dimensional approach for simplicity. This is justified for a detector thickness t that is small with respect to minority carrier diffusion length. We also neglect the effect of recombination at front and rear surfaces. Initially, we will consider simple photoconductivity effects due to the influence of bulk material properties.

The basic expression describing either intrinsic or extrinsic photoconductivity in semiconductors under equilibrium excitation (i.e., steady state) is

$$I_{ph} = q\eta A\Phi_s g, \tag{9.1}$$

where I_{ph} is the short circuit photocurrent at zero frequency (DC); that is, the increase in current above the dark current accompanying irradiation. The photoconductive gain, g, is determined by the properties of the detector (i.e., by which detection effect is used and the material and configuration of the detector).

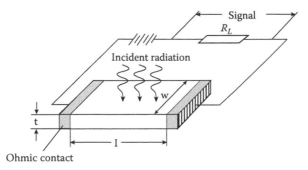

Figure 9.1 Geometry and bias of a photoconductor.

In general, photoconductivity is a two-carrier phenomenon and the total photocurrent of electrons and holes is

$$I_{ph} = \frac{qwt\left(\Delta n\mu_e + \Delta p\mu_h\right)V_b}{l},$$ (9.2)

where μ_e is the electron mobility, μ_h is the hole mobility; V_b is the bias voltage and

$$n = n_o + \Delta n; \quad p = p_o + \Delta p,$$ (9.3)

n_o and p_o are the average thermal equilibrium carrier densities, and Δn and Δp are the excess carrier concentrations.

Taking the conductivity to be dominated by electrons (in all known high sensitivity photoconductors this is found to be the case) and assuming uniform and complete absorption of the light in the detector, the rate equation for the excess electron concentration in the sample is [14]

$$\frac{d\Delta n}{dt} = \frac{\Phi_s \eta}{t} - \frac{\Delta n}{\tau},$$ (9.4)

where τ is the excess carrier lifetime. In the steady condition, the excess carrier lifetime is given by the equation

$$\tau = \frac{\Delta n t}{\eta \Phi_s}.$$ (9.5)

Equating Equation 9.1 to Equation 9.2 gives

$$g = \frac{t V_b \mu_e \Delta n}{l^2 \eta \Phi_s},$$ (9.6)

and invoking Equation 9.5 we get for the photoconductive gain

$$g = \frac{\tau \mu_e V_b}{l^2} = \frac{\tau}{l^2/\mu_e V_b}.$$ (9.7)

So, the photoconductive gain can be defined as

$$g = \frac{\tau}{t_t},$$ (9.8)

where t_t is the transit time of electrons between ohmic contacts. This means that the photoconductive gain is given by the ratio of free carrier lifetime, τ, to transit time, t_t, between the sample electrodes. The photoconductive gain can be less than or greater than unity depending upon whether the drift length, $L_d = v_d \tau$, is less than or greater than interelectrode spacing, l. The value of $L_d > l$ implies that a free charge carrier swept out at one electrode is immediately replaced by injection of an equivalent free charge carrier at the opposite electrode. Thus, a free charge carrier will continue to circulate until recombination takes place.

When $R_L \gg R$, a signal voltage across the load resistor is essentially the open circuit voltage

$$V_s = I_{ph} R_d = I_{ph} \frac{l}{qwtn\mu_e},$$
(9.9)

where R_d is the detector resistance. Assuming that the change in conductivity upon irradiation is small compared to the dark conductivity, the voltage responsivity is expressed as

$$R_v = \frac{V_s}{P_\lambda} = \frac{\eta}{lwt} \frac{\lambda\tau}{hc} \frac{V_b}{n_o},$$
(9.10)

where the absorbed monochromatic power $P_\lambda = \Phi_s A h\nu$.

The expression (Equation 9.10) shows clearly the basic requirements for high photoconductive responsivity at a given wavelength λ: one must have high quantum efficiency η, long excess-carrier lifetime τ, the smallest possible piece of crystal, low thermal equilibrium carrier concentrations n_o, and the highest possible bias voltage V_b.

The frequency dependent responsivity can be determined by the equation

$$R_v = \frac{\eta}{lwt} \frac{\lambda\tau_{ef}}{hc} \frac{V_b}{n_o} \frac{1}{\left(1+\omega^2\tau_{ef}^2\right)^{1/2}},$$
(9.11)

where τ_{ef} is the effective carrier lifetime.

The above simple model takes no account of additional limitations related to the practical conditions of photoconductor operation such as sweep-out effects or surface recombination. These are specified below.

9.1.1.1 Sweep-Out Effects

Equation 9.11 shows that voltage responsivity increases monotonically with the increase of bias voltage. However, there are two limits on applied bias voltage; namely, thermal conditions (Joule heating of the detector element) and sweep-out of minority carriers. The thermal conductance of the detector depends on the device fabrication procedure. The trend to smaller element dimensions (typically, e.g., 50×50 μm^2) is conditioned by the extension of photoconductor technology to two-dimensional close-packed arrays for thermal imaging. If the excess carrier lifetime is long (usually exceeds 1 μs in 8–14 μm devices at 77 K and 10 μs in 3–5 μm devices at higher temperatures) we cannot ignore the effects of contacts and of drift and diffusion on the device performance. At moderate bias fields minority carriers can drift to the ohmic contacts in a short time compared to the recombination time in the material. Removal of carriers at an ohmic contact in this way is referred to as sweep-out [15,16]. Minority carrier sweep-out limits the maximum applied voltage of V_b. The effective carrier lifetime can be reduced considerably in detectors where the minority carrier diffusion length exceeds the detector length (even at very low bias voltages) [17–21]. At low bias the average drift length of the minority carriers is very much less than the detector length l, and the minority carrier lifetime is determined by the bulk recombination modified by diffusion to surface and contacts. The carrier densities are uniform along the length of the detector. At higher values of the applied field, the drift length of the minority carriers is comparable to or greater than l. Some of the excess minority carriers are lost at an electrode, and to maintain space charge equilibrium, a drop in excess majority carrier density is necessary. This way the majority carrier lifetime is reduced. It should be pointed out that the loss of the majority carriers at one ohmic contact is replenished by injection at the other, but minority carriers are not replaced. At high bias the excess carrier density is nonuniformly distributed along the length of the sample.

We follow Rittner and derive the optically generated excess minority carrier concentration under sweep-out conditions [15]. The excess carrier concentration $\Delta p(x,t) = p(x,t) - p_o$ within the semiconductor is governed by ambipolar transport. The ambipolar continuity equation for a one-dimensional case under steady state and electrical neutrality conditions may be written as

$$\frac{\partial^2(\Delta p)}{\partial x^2} + \frac{L_d}{L_D^2} \frac{\partial(\Delta p)}{\partial x} + \frac{\Delta p}{L_D^2} + G_s = 0,$$
(9.12)

where

$$L_d = \tau\,\mu_a E \qquad\qquad \text{drift length,}$$

$$L_D = \left(D_D\tau\right)^{1/2} \qquad\qquad \text{diffusion length,}$$

$$\mu_a = \frac{\left(n_o - p_o\right)\mu_e\mu_h}{n_o\mu_e + p_o\mu_h} \qquad\qquad \text{ambipolar drift mobility,}$$

$$D_D = \frac{D_e p_o\mu_h + D_h n_o\mu_e}{n_o\mu_e + p_o\mu_h} \qquad \text{ambipolar diffusion coefficient.}$$

Other marks have their usual meanings: $D_{e,h} = (kT/q)\mu_{e,h}$ are the respective carrier diffusion coefficients, G_s is the signal generation rate, $E = V_b/l$ is the bias electrical field, and k is the Boltzmann constant.

The major assumption in the Rittner model relates to the boundary conditions at the metal-semiconductor interface, at $x = 0$ and $x = l$. In this model one assumes that this interface is characterized by infinite recombination velocity, which means that the photoconductor contacts are completely ohmic. The appropriate boundary conditions are

$$\Delta p(0) = \Delta p(l) = 0. \tag{9.13}$$

The solution to Equation 9.12 is

$$\Delta p = G_s\tau\left[1 - C_1\exp\left(\alpha_1 x\right) + C_2\exp\left(\alpha_2 x\right)\right], \tag{9.14}$$

where

$$\alpha_{1,2} = \frac{1}{2L_D^2}\left[-L_d \pm \left(L_d^2 + 4L_D^2\right)^{1/2}\right]. \tag{9.15}$$

Taking into account the boundary conditions (Equation 9.13) we have

$$C_1 = \frac{1 - \exp\left(\alpha_2 l\right)}{\exp\left(\alpha_2 l\right) - \exp\left(\alpha_1 l\right)}; \quad C_2 = \frac{1 - \exp\left(\alpha_1 l\right)}{\exp\left(\alpha_2 l\right) - \exp\left(\alpha_1 l\right)}. \tag{9.16}$$

The total number of carriers contributing to photoconductivity is obtained by integrating Equation 9.14 over the length of the samples:

$$\Delta P = wt\int_0^l \Delta p(x)dx.$$

Note that the signal generation rate G_s is related to the total signal flux Φ_s and the quantum efficiency η, by the expression $G_s = \eta\Phi_s/t$. Then

$$\Delta P = \eta\Phi_s\tau w\int_0^l \left[1 + C_1\exp\left(\alpha_1 x\right) + C_2\exp\left(\alpha_2 x\right)\right]dx. \tag{9.17}$$

Alternatively we can write

$$\Delta P = \eta\Phi_s\tau_{ef} \qquad\qquad \text{where} \qquad \tau_{ef} = \gamma\tau. \tag{9.18}$$

It can be proved that

$$\gamma = 1 + \frac{\left(\alpha_1 - \alpha_2\right)th\left(\alpha_1 l/2\right)th\left(\alpha_2/2\right)}{\alpha_1\alpha_2\left(l/2\right)\left[th\left(\alpha_2 l/2\right) - th\left(\alpha_1 l/2\right)\right]}. \tag{9.19}$$

In this situation, the voltage responsivity is [3]

$$R_v = \frac{\eta}{lwt} \frac{\lambda \tau_{ef}}{hc} \frac{V_b(b+1)}{bn+p} \frac{1}{\left(1+\omega^2 \tau_{ef}^2\right)^{1/2}} , \tag{9.20}$$

were $b = \mu_e/\mu_h$, and at the low frequency modulation ($\omega \tau_{ef} \ll 1$)

$$R_v = \frac{\eta}{lwt} \frac{\lambda \tau_{ef}}{hc} \frac{V_b(b+1)}{bn+p}. \tag{9.21}$$

We obtain similar formulas as previously seen (Equation 9.11), except that the carrier lifetime τ is replaced by τ_{ef}. Because $\gamma \le 1$, so always $\tau_{ef} \le \tau$. The lifetime degradation problem associated with ohmic contacts can be eliminated by the use of an overlap structure [17,18], heterojunction contact [19,22,23], or highly doped contact [18,19,23].

Practically, the contacts are characterized by a recombination velocity that can be varied from infinity (ohmic contacts) to zero (perfectly blocking contacts). In the latter case, a more intensely doped region at the contact (e.g., n+ for n-type devices) causes a built-in electric field that repels minority carriers, thereby reducing recombination and increasing the effective lifetime and the responsivity. More sophisticated blocking contacts and their influence on the performance of intrinsic photoconductors have been considered by Kumar et al. [24,25]. The experimental results show that contact recombination velocities as low as a few hundred cm/s can be achieved [19,23,26].

In general, the electric field distribution in photoconductors is not homogeneous. In this instance, these structures cannot be adequately described by analytical methods and require a numerical solution. Numerical techniques have been used to solve the carrier transport equations for several device configurations [21]. Usually, the Van Roosbroeck model is used [27] (see Section 3.4).

Analysis of the influence of the sweep-out effect on the photoconductor performance has been carried out by Elliott et al. [13,20]. It appears that the formulas (Equation 9.20 and 9.21) given above are quite generally applicable provided that n and p are replaced by somewhat different values n' and p' under high bias conditions. The values of n' and p' depend on the source of the minority carriers and on the nature of the minority carrier injecting contact. For general case the responsivity can be written as

$$R_v = \frac{\eta}{lwt} \frac{\lambda \tau_{ef}}{hc} \frac{V_b(b+1)}{bn'+p'} , \tag{9.22}$$

where

$$\tau_{ef} = \tau \left[1 - \frac{\tau}{\tau_a} \left\{ 1 - \exp\left(-\frac{\tau_a}{\tau}\right) \right\} \right],$$

and $\tau_a = 1/\mu_a E$, the time for a minority carrier to drift the sample length.

Under very high bias conditions, the voltage responsivity saturates to the value [20]

$$R_v = \frac{\eta q \lambda}{2hc}(1+b)\frac{\mu_h}{\mu_a} R', \tag{9.23}$$

where R' is the device resistance; $\mu_a = \mu_h$ in the n-type material and $\mu_a = \mu_e$ in p-type material.

9.1.1.2 Noise Mechanisms in Photoconductors

All detectors are limited in the minimum radiant power that they can detect by some form of noise that may arise in the detector itself, in the radiant energy to which the detector responds, or in the electronic system following the detector. Careful electronic design including that of low noise amplification can reduce system noise below that in the output of the detector. That topic will not be treated here.

We can distinguish two groups of noise; the radiation noise and the noise internal to the detector. The radiation noise includes signal fluctuation noise and background fluctuation noise [28,29]. Under most operating conditions the background fluctuation limit discussed in Section 3.2 is operative for infrared detectors, whereas the signal fluctuation limit is operative for ultraviolet and visible detectors.

The random processes occurring in semiconductors give rise to internal noise in detectors even in the absence of illumination. There are two fundamental processes responsible for the noise: fluctuations in the velocities of free carriers due to their random thermal motion, and fluctuations in the densities of free carriers due to randomness in the rates of thermal generation and recombination [30].

The photon noise voltage can be calculated according to the theory presented in Van der Ziel's monograph [30]:

$$V_{ph} = \frac{2\pi^{1/2}V_b}{(lw)^{1/2}t} \frac{1+b}{bn+p} \int_0^\infty \frac{\eta(v)v^2\exp(hv/kT_b)dv}{c^2\left[\exp(hv/kT_b)-1\right]^2} \frac{\tau(\Delta f)^{1/2}}{(1+\omega^2\tau^2)^{1/2}}, \tag{9.24}$$

where T_b is the temperature of the background and v_o is the frequency corresponding to the detector long-wavelength limit λ_c.

A number of internal noise sources are usually operative in photoconductive detectors. The fundamental types are: Johnson–Nyquist (sometime called thermal) noise and generation–recombination (g–r) noise. The third form of noise, not amenable to exact analysis, is called $1/f$ noise because it exhibits a $1/f$ power law spectrum to a close approximation.

The total noise voltage of a photoconductor is

$$V_n^2 = V_{gr}^2 + V_f^2 + V_{1/f}^2. \tag{9.25}$$

Johnson–Nyquist noise is associated with the finite resistance R of the device. This type of noise is due to the random thermal motion of charge carriers in the crystal and not due to fluctuations in the total number of these charge carriers. It occurs in the absence of external bias as a fluctuating voltage or current depending upon the method of measurement. Small changes in the voltage or current at the terminals of the device are due to the random arrival of charge at the terminals. The root mean square of Johnson–Nyquist noise voltage in the bandwidth Δf is given by Equation 3.16. This type of noise has a "white" frequency distribution.

At finite bias currents, the carrier density fluctuations cause resistance variations, which are observed as noise exceeding Johnson–Nyquist noise. This type of excess noise in photoconductive detectors is referred to as g–r noise. The g–r noise is due to the random generation of free charge carriers by the crystal vibrations and their subsequent random recombination. Because of the randomness of the generation and recombination processes, it is unlikely that there will be exactly the same number of charge carriers in the free state at succeeding instances of time. This leads to conductivity changes that will be reflected as fluctuations in current flow through the crystal.

The (g–r) noise voltage for equilibrium conditions is equal

$$V_{gr}^2 = 2(G+R)lwt(Rqg)^2\,\Delta f, \tag{9.26}$$

where G and R in the first bracket are the volume generation and recombination rates.

Many forms of g–r noise expression exist, depending upon the internal properties of the semiconductors. The expression for noise in a near intrinsic photoconductor has been given by Long [31]:

$$V_{gr} = \frac{2V_b}{(lwt)^{1/2}} \frac{1+b}{bn+p} \left(\frac{np}{n+p}\right)^{1/2} \left(\frac{\tau\Delta f}{1+\omega^2\tau^2}\right)^{1/2}. \tag{9.27}$$

Generation–recombination noise usually dominates the noise spectrum of photoconductors at intermediate frequencies. It should be noted that in the high bias regime, the expressions for g–r noise are different from those at low bias [20].

The rms g–r noise current for an extrinsic n-type photoconductor with carrier lifetime τ can be written [30]

$$I_{gr}^2 = \frac{4I^2\overline{\Delta N^2}\tau\Delta f}{N^2\left(1+\omega^2\tau^2\right)}, \tag{9.28}$$

where N is the number of carriers in the detector. Usually, in an extrinsic semiconductor there will be some counterdoping (i.e., electrons trapped at deep lying levels). If the number of deep traps is small compared to the number of electrons (electrons being the majority carriers), then the variance ΔN^2 is equal to N [30]. The current flowing in the device is $I = Nqg/\tau$, hence

$$I_{gr}^2 = \frac{4qIg\Delta f}{1+\omega^2\tau^2}. \tag{9.29}$$

The $1/f$ noise is characterized by a spectrum in which the noise power depends approximately inversely upon frequency. Infrared detectors usually exhibit $1/f$ noise at low frequency. At higher frequencies the amplitude drops below that of one of the other types of noise: the generation-recombination noise and Johnson noise.

The general expression for the noise current is

$$I_{1/f} = \left(\frac{KI_b^\alpha \Delta f}{f^\beta}\right)^{1/2}, \tag{9.30}$$

where K is a proportionality factor, I_b is the bias current, α is a constant whose value is about 2, and β is a constant whose value is about unity.

In general, $1/f$ noise appears to be associated with the presence of potential barriers at the contacts, interior, or surface of the semiconductor. Reduction of $1/f$ noise to an acceptable level is an art that depends greatly on the processes employed in preparing the contacts and surfaces. Up until now, no fully satisfactory general theory has been formulated. The two most current models for the explanation of $1/f$ noise were considered [32]: Hooge's model [33], which assumes fluctuations in the mobility of free charge carriers, and McWhorter's model [30], based on the idea that the free carrier density fluctuates.

The low frequency noise voltage described by the Hooge's expression

$$V_{1/f}^2 = \alpha_H \frac{V^2}{Nf}\Delta f, \tag{9.31}$$

where α_H is the Hooge's constant and N is the number of charge carriers. Frequently, the low frequency is characterized by the $1/f$ noise knee frequency $f_{1/f}$

$$V_{1/f}^2 = V_{gr}^2 \frac{f_{1/f}}{f}. \tag{9.32}$$

The value of the Hooge constant and $f_{1/f}$ is usually considered the technology-related property of the device. There are, however, quantum $1/f$ noise theories describing the $1/f$ noise as the fundamental material property [7]. Hooge's constant in the range $5 \times 10^{-3} - 3.4 \times 10^{-5}$ have been measured frequently below the lower limit calculated according to existing theories [34].

9.1.1.3 Quantum Efficiency

In most photoconductor materials the internal quantum efficiency η_o is nearly unity; that is, almost all photons absorbed contribute to the photoconductive phenomenon. For a detector, as a slab of material, shown in Figure 9.1, with surface reflection coefficients r_1 and r_2 (on the top and bottom surfaces, respectively) and absorption coefficient α, the internal photogenerated charge profile in the y-direction is [35]

$$S(y) = \frac{\eta_o(1-r_1)\alpha}{1-r_1r_2\exp(-2\alpha t)}[\exp(-\alpha y)+r_2\exp(-2\alpha t)\exp(-\alpha y)]. \tag{9.33}$$

The external quantum efficiency is simply the integral of this function over the detector thickness:

$$\eta = \int_0^t S(y)dy = \frac{\eta_o(1-r_1)[1+r_2\exp(-\alpha t)][1-\exp(\alpha t)]}{1-r_1r_2\exp(-\alpha t)}. \tag{9.34}$$

When r_1 and $r_2 = r$, the quantum efficiency is reduced to

$$\eta = \frac{\eta_o(1-r)[1-\exp(\alpha t)]}{1-r\exp(-\alpha t)}. \tag{9.35}$$

182

Intrinsic detector materials tend to be highly absorptive; so in a practical well-designed detector assembly only the top surface reflection term is significant, and then

$$\eta \approx \eta_o (1 - r) \approx 1 - r. \tag{9.36}$$

By antireflection coating the front surface of the detector, this quantity can be made greater than 0.9.

9.1.1.4 Ultimate Performance of Photoconductors

Usually intrinsic or lightly doped n-type materials are used for fabrication of infrared photoconductors. However, if band-to-band recombination mechanisms are dominant, ultimate photoconductor performances are expected in lightly doped p-type materials. This situation usually occurs in the case of long-wavelength near room temperature HgCdTe photoconductors [36,37].

The classical long-wavelength near-room-temperature photoconductors operated at weak optical excitation and at steady-state conditions can be satisfactory described by a simple model, in which such phenomena as sweep-out, surface recombination, interference within the device, edge effects and the influence of background radiation are neglected. In the optimum case, when reflection coefficients from front and rear surface are $r_1 = 0$ and $r_2 = 1$, the quantum efficiency is given by

$$\eta = \eta_o [1 - \exp(-2\alpha t)] \approx 1 - \exp(-2\alpha t). \tag{9.37}$$

Under the above conditions, the expression for voltage responsivity specified by Equation 9.21 becomes

$$R_v = \frac{V_b}{hc} \frac{\mu_e + \mu_h}{n_o \mu_e + p_o \mu_h} \frac{\tau [1 - \exp(-2\alpha t)]}{lwt}. \tag{9.38}$$

Taking into account only Johnson–Nyquist and g–r noise (the $1/f$ noise can be minimized by appropriate fabrication techniques and can be neglected), the detectivity is equal

$$D^* = \frac{R_v (lw\Delta f)^{1/2}}{\left(V_f^2 + V_{gr}^2\right)^{1/2}}. \tag{9.39}$$

We can distinguish two cases: the first when V_{gr} saturates at a level above V_J, the g–r noise limited case, and the second when the saturated level of V_{gr} is less than V_J, the Johnson noise/sweep-out limited case.

The g–r noise limited case always applies to background-limited detectors. The g–r noise limited detectivity is obtained from Equations 9.21, 9.27, and 9.39 as

$$D_{gr}^* = \frac{\lambda}{2hc} \frac{\eta}{t^{1/2}} \left(\frac{n+p}{np}\right)^{1/2} \tau^{1/2}. \tag{9.40}$$

This can be written as Equation 3.50, where G is the sum of all the generation processes per unit volume. The $(n + p)\tau/(np)$ can be used as a generalized, doping dependent figure of merit of the semiconductor that determines the ultimate performance of the photoconductor. Since $\alpha \approx 1/t$, the Equation 9.40 can be written as

$$D_{gr}^* = \frac{\lambda\eta}{2hc} \left(\frac{n+p}{n_i}\right)^{1/2} \left(\frac{\alpha\tau}{n_i}\right)^{1/2}, \tag{9.41}$$

where the $\alpha\tau/n_i$ can be treated as material figure of merit for photoconductors [38], which actually is the α/G figure of merit (see Section 3.2).

If, for an idealized detector structure, we ignore the nonfundamental generation processes that occur at surfaces and electrodes, the total generation rate can be expressed as the sum of the rates due to three types of bulk processes: Auger, radiative, and Shockley-Read. The radiative term is due to photons absorbed in the detector that have been emitted from the detector enclose or have been received through a lens from the ambient temperature scene. The fundamental limit to detector performance is reached when the detector is cooled sufficiently for radiative generation to dominate, provided that this term is principally caused by photons from the scene (see below).

In most practical applications, photoconductive detectors are operated at reduced temperatures to eliminate thermally generated transitions and noise due to power dissipation. Joule heating due to bias current produces a rise in detector temperature and as a consequence an interface between the detector and the cooling receiver is necessary. Johnson noise-limited detectivity due to power dissipation is observed in large detectors operating in the short wavelength range under reduced background.

9.1.1.5 Influence of Background

Under a condition of excess carrier density generation by a background radiation flux density Φ_b, the carrier densities are given by

$$n = n_o + \frac{\eta\Phi_b\tau}{t}, \qquad p = p_o + \frac{\eta\Phi_b\tau}{t}.$$

As Φ_b increases, the influence of the background appears initially as an increase of minority carrier density. In normal operation the detector is cooled sufficiently so that the thermally excited minority carriers are negligible compared to the photon excited excess carriers. Then, the g–r noise is due entirely to the background photon flux density. For the photoconductors operating in the background flux density, background limited performance requires that two conditions be satisfied $\eta\Phi_b\tau/t > p_o$ for an n-type sample ($\eta\Phi_b\tau/t > n_o$ for a p-type sample) and $V_{gr}^2 > V_j^2$. The second condition states that the applied bias voltage across the detector must be large enough that the g–r noise dominates the Johnson–Nyquist noise contributions. If the above conditions are satisfied, the detectivity is given by

$$D_b^* = \frac{\lambda}{2hc}\left[\frac{\eta(n+p)}{\Phi_b n}\right]^{1/2}. \tag{9.42}$$

At moderate background influence ($p_o < \Delta n = \Delta p < n_o$) we obtain the detectivity defined by equation

$$D_b^* = \frac{\lambda}{2hc}\left(\frac{\eta}{\Phi_b}\right)^{1/2}. \tag{9.43}$$

However, with high background fluxes and high-purity material, fulfillment of the condition $\Delta n = \Delta p \gg n_o, p_o$ is possible, and "photovoltaic" BLIP detectivity is achieved:

$$D_b^* = \frac{\lambda}{hc}\left(\frac{\eta}{2\Phi_b}\right)^{1/2}. \tag{9.44}$$

It should be noted that the photovoltaic BLIP detectivity is difficult to achieve in practice because it is a function of carrier densities and decreases at high background levels. This type of lifetime behavior has been observed experimentally [39].

9.1.1.6 Influence of Surface Recombination

The photoconductive lifetime in general provides a lower limit to the bulk lifetime, due to the possibility of enhanced recombination at the surface. Surface recombination reduces the total number of steady-state excess carriers by reducing the recombination time. It can be shown that τ_{ef} is related to the bulk lifetime by the expression [40]

$$\frac{\tau_{ef}}{\tau} = \frac{A_1}{\alpha^2 L_D^2 - 1}, \tag{9.45}$$

where

$$A_1 = L_D\alpha\left[\frac{(\alpha D_D + s_1)\{s_2[ch(t/L_D)-1]+(D_D/L_D)sh(t/L_D)\}}{(D_D/L_D)(s_1+s_2)ch(t/L_D)+(D_D^2/L_D^2+s_1s_2)sh(t/L_D)}\right.$$
$$\left. -\frac{(\alpha D_D - s_2)sh\{s_1[ch(t/L_D)-1]+(D_D/L_D)sh(t/L_D)\}\exp(\alpha t)}{(D_D/L_D)(s_1+s_2)ch(t/L_D)+(D_D^2/L_D^2+s_1s_2)sh(t/L_D)}-[1-\exp(-\alpha t)]\right].$$

D_D is the ambipolar diffusion coefficient, s_1 and s_2 are the surface recombination velocities at the front and back surfaces of the photoconductor and $L_D = (D_D\tau)^{1/2}$.

If the absorption coefficient α is large, $\exp(-\alpha t) \approx 0$ and $s_1 \ll \alpha D_D$, Equation 9.45 is simplified to the well-known expression [15,20,28]

$$\frac{\tau_{ef}}{\tau} = \frac{D_D}{L_D} \frac{s_2\left[ch(t/L_D)-1\right]+(D_D/L_D)sh(t/L_D)}{L_D(D_D/L_D)(s_1+s_2)ch(t/L_D)+(D_D^2/L_D^2+s_1+s_2)sh(t/L_D)}. \tag{9.46}$$

Further simplification for $s_1 = s_2 = s$ leads to

$$\frac{1}{\tau_{ef}} = \frac{1}{\tau} + \frac{2s}{t}. \tag{9.47}$$

Considerations carried out by Gopal indicate that for accurate modeling of photoconductors, surface recombination effects should be considered as directly influencing the quantum efficiency rather than the carrier lifetime [40]. According to this in the case $r_1 = r_2 = r$,

$$\eta = \frac{(1-r)A_1}{[1-r\exp(-\alpha t)](\alpha^2 L_D^2 - 1)}. \tag{9.48}$$

If $s_1 = s_2 = 0$, this equation reduces to Equation 9.34.

For low temperatures, the diffusion length is so large that the typical photoconductor is invariably operated in the mode $t/L < 1$, and if $s \ll 1$, $\tau_{ef} = [1/\tau + 2s/t]^{-1} \approx t/2s$, the detectivity becomes [41]

$$D^* = \frac{\eta\lambda}{2hc}\left(\frac{n_o + p_o}{2n_i^2 s}\right)^{1/2}. \tag{9.49}$$

The essential point of the above discussion is that a finite value of surface recombination velocity may have a strong effect on the attainable detectivity.

9.1.2 Extrinsic Photoconductivity Theory

A number of extrinsic photoconductor reviews have previously been published, the first of which, "Optical and Photoconductive Properties of Silicon and Germanium," by Burstein, Picus, and Sclar [42], appeared in 1956. This was followed by "Photoconductivity of Germanium" in 1959 by Newman and Tyler [43], "Far Infrared Photoconductivity" in 1964 by Putley [44], "Impurity Germanium and Silicon Infrared Detectors" by Bratt [5] in 1977, and "Properties of Doped Silicon and Germanium Infrared Detectors" by Sclar [8] in 1984. The last two reviews are still timely and very comprehensive and are ones to which we will make numerous references. In a more recently published review [45], Kocherov et al. have considered certain peculiarities of the operation of extrinsic detectors under a low background.

In the beginning, major emphasis was directed to the Ge detectors. At present, however, there is considerable interest in Si devices because of their potential for the fabrication of very large focal plane arrays (FPAs) for thermal imaging [46,47]. The attraction of extrinsic silicon lies in the highly developed MOS technology and the possibility of integrating the detectors with charge transfer devices (CTD) for the readout and signal processing.

There are two simple configurations used in biasing extrinsic photoconductors: transverse bias and parallel bias. These are illustrated in Figure 9.2 [8]. In the transverse case, the electric field and the resulting current flow are transverse to the incident photon flux; the photocarrier generation profile is independent of distance in the direction of current flow. In the longitudinal case, the electric field is parallel to the photon flux, and the photocarrier generation profile varies exponentially in the direction of current flow. The distinction between bias configurations becomes important for large absorptance ($\alpha l > 1$). Analysis carried out by Nelson indicates (Figure 9.3) that for the optimum condition of longitudinal geometry, the responsivity peak is about 87% of the normalized value at $\alpha l \cong 1.5$ and then declines with further increase in αl [35]. Adherence to the condition $\alpha l \cong 1.5$, accordingly, represents an optimum design criterion for detectors employing longitudinal geometries. The reason for the inferiority of the longitudinal geometry is illustrated in Figure 9.2. For the transverse geometry, the unactivated detector depth merely represents a high resistance shunt that has little influence on signal or noise. For the longitudinal geometry, this unactivated depth is electrically in series with the activated depth. Consequently, it acts to quench the signal

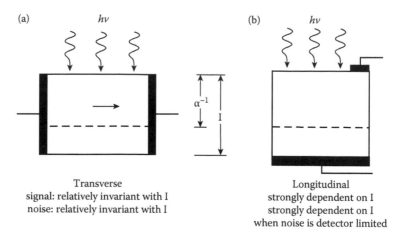

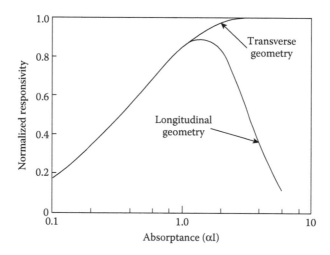

Figure 9.2 Activation geometry comparison for (a) transverse and (b) longitudinal detectors. (From Sclar, N., *Progress in Quantum Electronics*, 9, 149–257, 1984. With permission.)

Figure 9.3 Normalized responsivity for longitudinal and transverse detector geometry versus absorptance for a detector with ideal surface coatings ($r_1 = 0$; $r_2 = 1$) and a photoconductive gain of unity. (From Nelson, R. D., *Optical Engineering*, 16, 275–83, 1977. With permission.)

level and to possibly increase the noise. Transverse bias has historically been used in arrays of discrete detectors while parallel bias is now used in monolithic arrays. The longitudinal geometry detector exhibits a far more uniform sensitivity to a spot scan compared to the transverse detector, which can be very significant in scanned arrays. Because the latter promises better economy and performance the subsequent analysis will assume parallel bias.

For the following discussion, we will analyze the geometrical model of Figure 9.2b and assume a simple energy level model of an n-type extrinsic semiconductor consisting of a photoionizable donor level and a compensating acceptor level; properties of a corresponding p-type model would be analogous. We assume that the photoconductor crystal contains N_d majority shallow donor impurities and N_a minority shallow acceptor impurities (i.e., $N_d > N_a$). At very low temperatures ($kT \ll E_i$, where E_i is a binding energy of the electrons to the donors) and in dark ($N_d - N_a$) donors will bind an electron and are therefore neutral while N_a donors will have given up their electron to the compensating acceptors. The number of electrons in the conduction band will be extremely small, resulting in high resistivity. The semiconductor material is further characterized by the lifetime τ, the mobility μ, and the quantum efficiency η.

When the signal photon flux density Φ_s with $h\nu \geq E_d$ enter the crystal and are absorbed by neutral donors, bound electrons will be excited into the conduction band. The free electrons will travel in the externally applied electric field E with a velocity $v = \mu E$. The photocurrent is then given by Equation 9.1, where the photoconductive gain g is

$$g = \frac{\mu\tau E}{l}. \tag{9.50}$$

Then

$$I_{ph} = \frac{q\eta\mu\tau}{l} E\Phi_s A. \tag{9.51}$$

It is clear from Equation 9.51 that a high photocurrent requires high mobility, long lifetime, and as short a detector as is consistent with a high quantum efficiency, to be discussed later.

The photoconductive gain of extrinsic photoconductors depends on frequency due to carrier sweep-out and dielectric relaxation. Sweep-out effects are more difficult to understand [45,48–50] but not generally as important in practice as in intrinsic photoconductors.

Let us consider a detector subjected to a short light pulse. The pulse will produce n_{op} electrons and an equal density of positively charged donors. The electrons are swept out of the detector in a transit time, leaving behind a uniform distribution of ionized donors. It is assumed here that the drift length $L_d = \mu\tau E$ is larger than the detector length l. The detector relaxes back to its neutral state within the dielectric relaxation time

$$\tau_\rho = \varepsilon\varepsilon_o\rho \tag{9.52}$$

where ε is the dielectric constant, ε_o is the permittivity of space, and ρ is the resistivity of the detector. Assuming that $\rho = (qn_{op}\mu)^{-1}$ and $n_{op} = \eta\Phi_s\tau/l$, the dielectric relaxation frequency is given by

$$f_\rho = \frac{q\eta\mu\tau\Phi_s}{2\pi\varepsilon\varepsilon_o}. \tag{9.53}$$

For typical parameters of Si, $\eta = 0.3$, $\mu = 8 \times 10^3$ cm^2/Vs, $\tau = 10^{-8}$ s, and $l = 0.05$ cm; the last equation gives $f_\rho \cong 1.2 \cong 10^{-11}\,\Phi_s$ Hz. For low background applications, where $\Phi_b \approx 10^{12}$ photons/cm^{2}s, f_ρ is only 12 Hz, while for conventional 300 K terrestrial imaging, f_ρ is only in the low-kHz range.

Dielectric relaxation time effects are observed when holes are swept out of the detector without replenishment from contacts. It means that the photoconductive gain should be frequency dependent. There are several models that describe this frequency dependence. The first model [48], predicts a gain drop at f_ρ, while the second model [49] predicts a corner frequency of $f_\rho/2g_o$, where g_o is the low-frequency gain given by Equation 9.50.

More recently published papers indicate on nonlinear phenomena and anomalous transient response of cooled extrinsic photoconductors (see e.g., [45,51–55]). The commonly observed behavior in these photoconductors is investigated by performing the dynamical response analysis of the space charge to illumination with full account of the regions near the injecting electrical contacts. Detector anomalies in the transient response, spiking and noise are currently attributed to electric field effects at the injecting contacts. A high local electric field value creates, for instance, a hot carrier distribution with changes substantially in the mobility, changes drastically in the capture cross section, the impact ionization coefficient, and consequently the dynamic state of the carriers.

Excess carriers generated in response to an increase in photon illumination, can either drift or diffuse to a contact region, where they recombine. This limits the initial gain of the device. Since changes in injection require local changes in the space charge electric field in the region adjacent to the contact, the charge that is lost to the contact cannot be immediately replaced in the bulk by increased injection. As a result, the transient response consists of a slow and a fast component, with their relative magnitudes dependent on the ratio of diffusion and drift lengths to the device length. The slow transient response is controlled by out-diffusion and sweep-out and the establishment of a counteracting electric field barrier, but the fast component is determined by the carrier lifetime.

Photoconductors made from doped Ge and Si exhibit values of g to 10, but values between 0.1 and 1 are more typical, because of the low lifetime achieved thus far. Hence using a frequency-independent gain is reasonable. However, with material improvements, lifetime improvements can be expected with resultant gain increases, and then the frequency dependence of the gain will need to be considered.

Since photoconductivity from extrinsic detectors arises from the photoionization of impurities, it is necessary that the detector be operated under circumstances that permit the free charge carriers to be trapped at the impurities. The major competing process is that of thermal ionization, which dictates a cooling requirement for the detector to suppress this contribution. In the absence of background, the thermal equilibrium concentration of the electrons, n_{th}, is determined by a balance between the rate of thermal ionization of the neutral impurity centers and the rate of recombination at the ionized centers. The general model is rather complicated (e.g., see the discussion carried out in [5] and [8]). For steady state conditions, at the low temperature of operation of impurity photoconductors (when $kT \ll E_i$ and $n \ll N_d, N_a$), the thermal equilibrium free-charge carrier is equal to

$$n_{th} = \frac{N_c}{\delta}\left(\frac{N_d - N_a}{N_a}\right)\exp\left(-\frac{E_d}{kT}\right). \tag{9.54}$$

Here N_C is the density of states in the conduction band and δ is the degeneracy factor, which is four for p-type and two for n-type impurities. High n_{th} would make the detector useless, and there are two options to reduce n_{th}: reduce the temperature to freeze the electrons, or add compensating acceptors. The former is clearly undesirable, and hence the latter method is used. For example, the effects of residual boron impurities in the Si:In detector are compensated by donor concentrations to achieve moderate cooling requirements (50–60 K). For Czochralski-grown Si, where $N_B = (5-10) \times 10^{13}$ cm^{-3}, it is obviously very difficult to achieve the desired compensation. For the float zone, where the boron concentration is lower by a factor of 10–50, precise compensation is easier to obtain, provided a compensating impurity like phosphorus can be introduced at such low levels. A promising method is the use of neutron transmutation doping where the thermal neutrons in a nuclear reactor interact with the Si lattice transmuting a small fraction of the silicon atoms into a known concentration of phosphorus donors [56]. Figure 9.4 demonstrates the power of

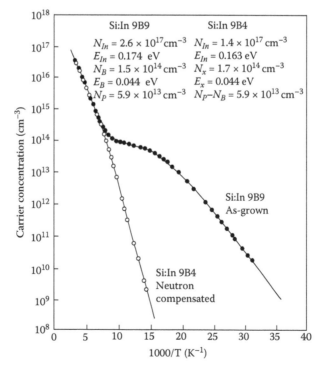

Figure 9.4 Carrier concentration versus reciprocal temperature for an uncompensated as-grown sample (9B9) and a neutron-compensated sample (9B4). The concentrations on the figure refer to In, X (0.11 eV level), B, and P-B (net concentration). (From Thomas, R. N., Braggins, T. T., Hobgood, H. M., and Takei, W. J., *Journal of Applied Physics*, 49, 2811–20, 1978. With permission.)

the neutron transmutation doping technique for very precise compensation [56]. The Czochralski-grown sample had a fairly high N_B of 1.5×10^{14} cm^{-3} with a residual phosphorus concentration of 5.9×10^{13} cm^{-3}. After neutron irradiation $N_P - N_B = 1.9 \times 10^{14}$ cm^{-3}.

Additional charge carriers may be added to the semiconductor by absorption of external radiation or by impact ionization. Theoretical and experimental results indicate that phonon-assisted cascade recombination process is the dominant mechanism for free electron or hole recombination at ionized impurities in Ge and Si [28]. Thus

$$\tau = \frac{1}{B(N_a + n)}. \tag{9.55}$$

In most practical cases, $n \ll N_a$ so that Equation 9.55 becomes

$$\tau = \frac{1}{BN_a}. \tag{9.56}$$

The recombination coefficient B is given by

$$B = \langle v \rangle \sigma_c, \tag{9.57}$$

where $<v> = (8kT/\pi m^*)^{1/2}$ is the average free carrier velocity and σ_c is the capture cross section of the recombination center.

Impact ionization is caused by free carriers gaining sufficient energy from the applied electric field to ionize neutral impurity atoms. This effect is manifested by a sharp increase in current through the crystal at some critical field strength E_c. Impact ionization not only creates additional free carriers, but also produces excessive electrical noise due to the sporadic nature of the breakdown in different regions of the crystal. The critical field increases with an increasing majority impurity concentration, because higher concentrations reduce the carrier mobility through neutral impurity scattering. Figure 9.5 shows representative experimental data [5].

As the concentration is increased and the distance between atoms becomes sufficiently small, carriers can hop from one impurity to another. The probability of hopping is enhanced by compensating impurities, which by ionizing some of the majority impurities make empty sites available for carriers to hop into. For still higher concentrations, the impurity level forms into a band, and conduction takes place by carriers flowing within this band. For both hopping and impurity band conduction, current flows without the need to excite holes into the valence band. Detector

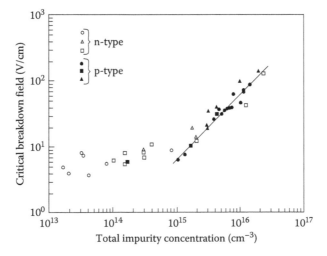

Figure 9.5 Critical impact ionization breakdown field for shallow level impurities in Ge at 4–5 K. (From Bratt, P. R., *Semiconductors and Semimetals*, Academic Press, New York, 12, 39–142, 1977. With permission.)

performance is degraded by reducing the ratio of photoconductive/dark current and by increasing device noise.

Quantum efficiency assumes maximum value when the reflectivity approaches zero at the front surface and unity at the back (see Equation 9.37). It should be noticed, however, that this case can introduce optical cross talk in FPAs by allowing nonabsorbed irradiation to be reflected back into the device.

The absorption coefficient α is given by

$$\alpha = \sigma_p N_i. \tag{9.58}$$

This is the product of the photoionization cross-section σ_p and the neutral impurity concentration. It is desirable to make α as large as possible. The upper limit of N_i is set by either "hopping" or "impurity band" conduction, as discussed earlier and is around 10^{15}–10^{16} cm^{-3} for silicon and somewhat lower for germanium (see Table 9.1) [5,8].

Various attempts have been made to develop theories that can predict the photoionization cross section [8]. Some of these are applicable to deep-lying impurities, while others are better suited to impurities with shallow energy levels. The functional dependence of σ_p on wavelength for Si:In and Si:Ga detector materials is shown in Figure 9.6 [57]. It rises from zero wavelength to a maximum at $\lambda_c/2$ and then decreases. Although it is not constant, it has a rather broad maximum and the absorption coefficient is reasonably constant over a useful wavelength range. The dependence of the maximum value for photoionization cross-section σ_o on E_i for hydrogenic approximation is given for Si as

$$\sigma_o = 2.65 \times 10^{-18} E_i^{-2} \quad \text{in cm}^2(\text{eV})^2, \tag{9.59}$$

which is shown in Figure 9.7 [8] to give a reasonable fit to experimental data. The maximum value varies with energy level of the extrinsic impurity. Note that the shallower the energy level, the larger the photoionization cross section. With some exceptions, the available data indicate that, for a given energy, the donors achieve a higher value for the cross section than the acceptors.

Using typical acceptable impurity concentrations and the photoionization cross sections, it can be seen from Equation 9.58 that the absorption coefficients for extrinsic photodetectors are some three orders of magnitude less than those for direct absorption in intrinsic photoconductors. Practical values of α for optimized photoconductors are in the range 1–10 cm^{-1} for Ge and 10–50 cm^{-1} for Si. Thus, to maximize quantum efficiency, the thickness of the detector crystal should be not less than about 0.5 cm for doped Ge and about 0.1 cm for doped Si. There is a limit in thickness of extrinsic detectors, because photocarriers generated beyond the drift length $L_d = \mu \tau E$ recombine before being collected (photoconductive gain $g = L_d/l$ decreases as l increases). Fortunately, for the

Table 9.1: Photoionization Cross Section of Impurity Atoms in Ge and Si

Impurity	Type	Ge λ_c (µm)	Ge σ_p (cm^{-2})	Si λ_c (µm)	Si σ_p (cm^{-2})
Al	p			18.5	8×10^{-16}
B	p	119	1.0×10^{-14}	28	1.4×10^{-15}
Be	p	52		8.3	5×10^{-18}
Ga	p	115	1.0×10^{-14}	17.2	5×10^{-16}
In	p	111		7.9	3.3×10^{-17}
As	n	98	1.1×10^{-14}	23	2.2×10^{-15}
Cu	p	31	1.0×10^{-15}	5.2	5×10^{-18}
P	n	103	1.5×10^{-14}	27	1.7×10^{-15}
Sb	n	129	1.6×10^{-14}	29	6.2×10^{-15}

Source: Bratt, P. R., *Semiconductors and Semimetals*, Academic Press, New York, 12, 39–142, 1977; Sclar, N., *Progress in Quantum Electronics*, 9, 149–257, 1984. With permission.

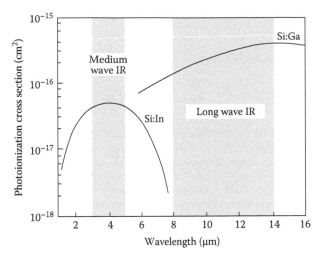

Figure 9.6 Photoionization cross section versus wavelength for Si:In and Si:Ga infrared detector material. (From Hobgood, H. M., Braggins, T. T., Swartz, J. C., and Thomas, R. N., *Neutron Transmutation in Semiconductors*, Plenum Press, New York, 65–90, 1979.)

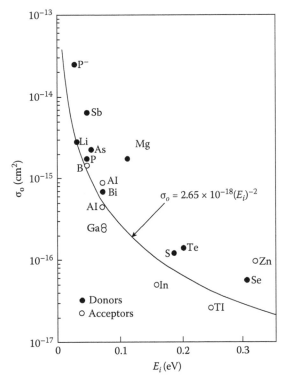

Figure 9.7 Impurity photoionization cross section at peak response versus impurity binding energy in Si. (From Sclar, N., *Progress in Quantum Electronics*, 9, 149–257, 1984. With permission.)

most extrinsic detectors the drift length is sufficiently long that quantum efficiencies approaching 50% can be obtained.

In describing the performance of an IR detector, a quantity of interest is the current (or voltage) responsivity. Analogical to consideration of intrinsic photoconductors (see Section 9.1.1), the short circuit current responsivity is

$$R_i = \frac{I_{ph}}{P_\lambda}, \tag{9.60}$$

which can be converted to equation

$$R_i = \frac{\eta\lambda}{hc} \frac{\tau}{lwt} \frac{I}{n} \frac{1}{\left(1+\omega^2\tau^2\right)^{1/2}}, \tag{9.61}$$

where I is the dark current flow through the detector circuit, and lwt is the volume of the detector element. It can be shown that for $\alpha l < 1$, $R_i \propto \alpha\lambda \propto \sigma_o$, the responsivity is proportional to σ_o.

Figure 9.8 shows the relative spectral response of neutron-compensated Si:In detector at 10 K [56]. The measured response differs only slightly over the 2–8 μm region from the generally accepted theoretical model for deep impurities developed by Lucovsky [58]. Usually values of R_i reach up to 100 A/W for the very best Si photoconductors, while typical values range from 1 to 20 A/W. It has been found that for a given energy level, n-type extrinsic impurities in Si have their peak response at longer wavelengths than do p-type impurities, so n-type detectors are expected to provide superior temperature characteristics for a given wavelength response [8,59,60].

To determine voltage responsivity, the photoconductive detector circuit should be considered. The practical detector circuit is shown in Figure 9.9. The detector is connected in series with a load resistor R_L and a source of direct current such as a battery V_b. The photocurrent produced by incoming signal photons is usually very small compared to the dark current. By AC coupling of the detector circuit to the preamplifier, the large direct current is blocked out and only the fluctuating signal current is measured.

It can be shown that

$$\Delta V = I \frac{\Delta R}{R} \frac{RR_L}{R+R_L}, \tag{9.62}$$

where ΔV is the signal voltage, and where ΔR is a small change in detector resistance R due to signal radiation. This equation is valid for an "ohmic" photoconductor provided that the amplifier input resistance R_a is much greater than the detector resistance. Most impurity detectors are decidedly "non-ohmic," and in this case, Equation 9.62 should be replaced by

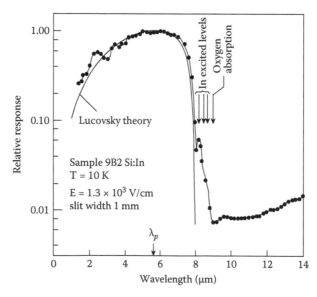

Figure 9.8 Experimental and theoretical relative spectral response for neutron-compensated Czochralski Si:In sample with $N_{In} = 2.5 \times 10^{17}$ cm^{-3}, $N_P - N_B = 1.6 \times 10^{14}$ cm^{-3}, and $N_B = 1.3 \times 10^{14}$ cm^{-3}. (From Thomas, R. N., Braggins, T. T., Hobgood, H. M., and Takei, W. J., *Journal of Applied Physics* 49, 2811–20, 1978. With permission.)

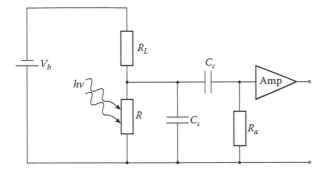

Figure 9.9 Practical detector circuit.

$$\Delta V = I \frac{\Delta R}{R_{dc}} \frac{R_{ac} R_L}{R_{ac} + R_L},$$

(9.63)

R_{ac}, given by dV/dI, is the AC resistance of the detector; R_{dc}, given by V/I, is the DC resistance. Voltage responsivity may now be easily obtained because

$$R_v = R_i \frac{R_{ac} R_L}{R_{ac} + R_L}.$$

(9.64)

The open circuit voltage responsivity is sometimes required, and this can be obtained from Equation 9.64 by letting $R_L \gg R_{ac}$, so that

$$R_{vo} = R_i R_{ac}.$$

(9.65)

The ultimate sensitivity of the detector is determined by the signal-to-noise ratio. Maximum performance can be achieved only when the noise in the device is of the g–r type. For the usual case at low temperatures, $n \ll N_a$, N_d and g–r noise current is given by [8]

$$I_{gr} = 2I \left(\frac{\tau \Delta f}{nlwt} \right)^{1/2} \frac{1}{\left(1 + \omega^2 \tau^2 \right)^{1/2}}.$$

(9.66)

Then, because the detectivity

$$D^* = \frac{R_i (A \Delta f)^{1/2}}{I_n},$$

(9.67)

so inserting the g–r noise (Equation 9.66) and the current responsivity (Equation 9.61) into Equation 9.67 gives

$$D^* = \frac{\eta \lambda}{2hc} \left(\frac{\tau}{nl} \right)^{1/2}.$$

(9.68)

When both thermal and photon generation are important, the free carrier density may be written as a sum of two terms, $n = n_{th} + n_{op}$, with n_{th} given by Equation 9.54 and

$$n_{op} = \frac{\eta \Phi \tau}{l} = \frac{\eta \Phi}{l} \frac{1}{BN_a},$$

(9.69)

where Equation 9.56 has been used to express the lifetime in terms of B, which contains the temperature dependence. Then the detectivity can be written with explicit temperature dependence as

$$D^* = \frac{\eta \lambda}{2hc} \left[\eta \Phi + \frac{lB}{\delta} (N_d - N_a) N_c \exp\left(-\frac{E_i}{kT} \right) \right]^{-1/2}.$$

(9.70)

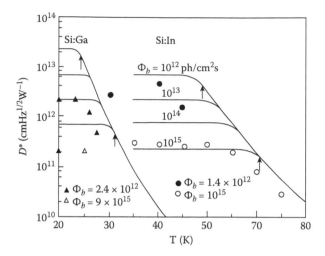

Figure 9.10 Measured and calculated detectivity D^* as a function of temperature for a neutron-compensated Si:In photoconductor and a doped Si:Ga photoconductor. The solid lines are calculated for $\eta = 50\%$ and $l = 0.05$ cm at $\lambda = 12$ μm (Si:Ga) and 4 μm (Si:In). The data points are measured values. (From Schroder, D. K., *Charge-Coupled Devices*, Springer-Verlag, Heidelberg, 57–90, 1980. With permission.)

Maximum D^* is achieved when $n_{th} \ll n_{op}$; that is, the thermal carrier concentration should be reduced for the detector to be dominated by optically generated carriers. This is shown graphically in Figure 9.10 for Si:In and Si:Ga photoconductors [50]. For temperature $T < T_{BLIP} \approx 60$ K, the Si:In detector is background limited for background radiation flux density $\Phi_b = 10^{15}$ photons/cm²s. T_{BLIP} is a function of Φ_b to which the detectors are exposed. We can see from Figure 9.10 that the Si:In data lie 5–10 K below theory, mainly as a result of the 0.11 eV level contaminant, while the Si:Ga data are about 3–5 K below. As expected, the temperature decreases for reduced backgrounds, as shown by the 3 dB arrows.

9.1.3 Operating Temperature of Intrinsic and Extrinsic Infrared Detectors

In this section we compare detectivity, as a function of operating temperature, for intrinsic and extrinsic photoconductors based on previously established relations and measured data. The effects of such parameters as impurity concentration, free-carrier lifetime, and capture cross section are evaluated.

We begin with the expression for D^* at peak when the g–r noise is dominant, as this sets the upper D^* limit independent of bias and detector area. Combining Equations 9.21, 9.27, and 9.39, D^* for intrinsic photoconductors with low excess impurity density can be expressed as follows:

$$D_{in}^* = \frac{\eta\lambda}{2hc}\left[\frac{\tau_{in}}{t_{in}\left(n_{ph}+n_i\right)}\right]^{1/2}, \tag{9.71}$$

where n_{ph} is the optically generated carrier density, and n_i is the intrinsic carrier concentration. Equation 9.71 gives an upper limit of D^* not achievable in measurements because of such effects as trapping centers in the forbidden band, excess impurities, temperature-dependent excess noise, or preamplifier requirements.

In the case of extrinsic photoconductors with the same assumptions of thermal and optical g–r noise dominant, D^* for an n-type detector can be expressed as

$$D_{ex}^* = \frac{\eta\lambda}{2hc}\left[\frac{\tau_{ex}}{t_{ex}\left(n_{ph}+n_{th}\right)}\right]^{1/2}, \tag{9.72}$$

according to Equation 9.68.

Both for intrinsic and extrinsic detectors, the BLIP conditions are fulfilled when $n_{ph} > (n_{th}$ or $n_i)$ and D^* is determined by Equation 9.43. The temperature corresponding to the transition from thermal to background limited noise is found by equating the thermal and background generated free carrier densities. For extrinsic photoconductors by equating Equations 9.54 and 9.69 we can obtain:

$$T_{\text{BLIP}} = \frac{E_d}{k} \left\{ \ln \left[\left(\frac{t N_d}{\eta} \right) \frac{B N_c}{\delta \, \Phi_b} \right] \right\}^{-1}.$$

(9.73)

This equation shows that T_{BLIP} is, for given field of view (FOV), a function of the impurity parameters E_d, σ_c ($B \propto \sigma_c$), σ_p (which determines the absorption coefficient and quantum efficiency), and the background flux Φ_b. The temperature limitation of extrinsic Si detectors has been considered by Bryan [61].

One of the principal factors that determines the stringent cooling requirements for extrinsic detectors is the large value of σ_c associated with the commonly used dopants, since very low values for the recombination time result even when a small fraction of the dopants are thermally ionized. The capture cross-section σ_{ex} of extrinsic photoconductors is larger than the corresponding σ_{in} of intrinsic photoconductors. The shallow-level impurities (B and As) show typically $\sigma_c \approx 10^{-11}$ cm^2, while the deep-level impurities (In, Au, Zn) show $\sigma_c = 10^{-13}$ cm^2 (see Table 9.2) [8]. By comparison, Figure 9.11 shows the σ_c of several intrinsic photoconductors to be $\sigma_{in} = 1.2 \times 10^{-17}$ cm^2 [62].

Milnes indicates that conventional dopants, with an attractive charge for the recapture of photocarriers, have values for σ_c between 10^{-15} and 10^{-12} cm^2; neutral impurities have σ_c about 10^{-17} cm^2 and

Table 9.2: Capture Cross Section of Impurity Atoms in Ge and Si

Impurity Atom	Germanium		Silicon	
	T(K)	σ_c (cm^2)	T(K)	σ_c (cm^2)
B			4.2	8×10^{-12}
Al	10	2×10^{-12}		
In			77	10^{-13}
As	10	10^{-12}	10	10^{-11}
Cu	10	5×10^{-12}		
Au	80	1×10^{-13}	77	10^{-13}
Zn			80–200	10^{-13}
Cd	8	1×10^{-11}		
Hg	20	3.6×10^{-12}		

Source: Sclar, N., *Progress in Quantum Electronics*, 9, 149–257, 1984. With permission.

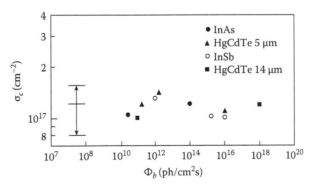

Figure 9.11 σ_c of some intrinsic photoconductors, calculated from response-time data. (From Borrello, S. R., Roberts, C. G., Breazeale, B. H., and Pruett, G. R., *Infrared Physics*, 11, 225–32, 1971. With permission.)

10^{-15} cm^2, and repulsive centers have σ_c less than 10^{-22} cm^2 (see Figure 9.12) [63]. The σ_c of an impurity atom depends on its recombination potential, which is smaller for neutral or repulsive centers than present attractive coulomb centers. Elliott et al. have discussed the possibility of obtaining higher temperature operation using neutral or repulsive centers [64]. It is suggested that this might be achieved by using very deep levels, for example, an acceptor level at the appropriate ionization energy from the conduction band, counterdoped with a shallow donor impurity; compensation with impurities of the opposite type produces recombination sites that are neutral or repulsive, respectively. Although some increase in operating temperature has been obtained from counterdoping, the benefits in operating temperatures are not as large as was predicted. The probable reason for this is that the capture-cross sections for neutral and repulsive centers increase with temperature and the values given by Milnes [63] have generally been measured at very low temperatures.

Figure 9.13 illustrates the direct effect of smaller σ_c on higher operating temperatures for extrinsic Si photoconductors [20]. In calculations the photoionization cross section, σ_p, is assumed to have

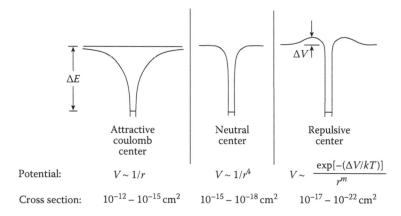

Figure 9.12 Potential distributions and estimated cross sections for recombination for attractive coulomb, neutral, and repulsive charge centers. (From Sclar, N., *Progress in Quantum Electronics*, 9, 149–257, 1984. With permission.)

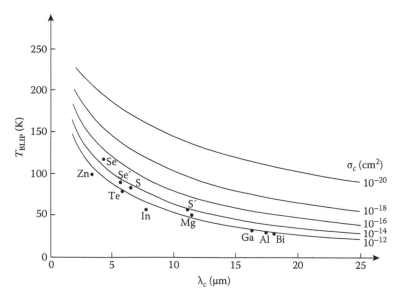

Figure 9.13 T_{BLIP} versus cutoff wavelength for extrinsic Si photoconductors as a function of the thermal capture cross section. (Experimental data are taken from *Progress in Quantum Electronics*, 9, 149–257, 1984.) 295 K background, 30° FOV. (From Elliott, C. T., and Gordon, N. T., *Handbook on Semiconductors*, 4, 841–936, North-Holland, Amsterdam, 1993. With permission.)

the wavelength dependence predicted by Lucovsky [58] and the following numerical values were used for p-type material: refractive index = 3.44, $N_v = 1.7 \times 10^{15} \, T^{3/2}$ cm^{-3}, $\delta = 4$, $v_{th} = 9.5 \times 10^5 \, T^{1/2}$ cm/s, and $E_{ef}/E_o = 2.5$; where E_{ef} and E_o are electric fields as defined in Lucovsky's paper. The scene temperature is 295 K and the FOV is 30°. The experimental results for Si infrared detectors for use at higher background fluxes are taken from Sclar [8]. The plots are consistent with the observed behavior of the most studied impurities assuming $\sigma_c = 10^{-12}$ cm^2. They also indicate that T_{BLIP} greater than 50 K should be achievable in the 8–14 μm band and greater than 80 K in the 3–5 μm band using dopants with a similar value of σ_c.

In Figure 3.19 plots of the calculated temperature required for background limited operation in $f/2$ FOV are shown as a function of cutoff wavelength [65,66]. The calculation for silicon extrinsic detectors has been carried out in terms of a dimensionless parameter $a(\eta)$ [8], which is of order unity, and the value of his parameter $Q = [a(\eta)(m^*/m)^{3/2}/\delta](B/\sigma_p)$ of 10^{10}. We can see that the operating temperature of "bulk" intrinsic IR detectors (HgCdTe) is higher than for other types of photon detectors. Intrinsic materials are characterized by high optical absorption coefficient and quantum efficiency and relatively low thermal generation rate compared to extrinsic detectors, silicide Schottky barriers and quantum well IR photodetectors (QWIPs).

Extrinsic photoconductors are made thicker than intrinsic photoconductors since the absorption cross section of extrinsic photoconductors is smaller than that of intrinsic photoconductors. For Si detectors, typically $t_{ex} = 0.1$ cm, whereas for intrinsic ones $t_{in} = 10^{-3}$ cm. As is pointed out above, t_{ex} is a direct consequence of σ_p, which is a function of wavelength (see Figure 9.6) and is independent of temperature, irradiance, and impurity concentration [67]. Therefore, σ_p is not subject to control and is a fixed parameter for each impurity.

9.2 P-N JUNCTION PHOTODIODES

The most common example of a photovoltaic detector is the abrupt p-n junction prepared in the semiconductor, which is often referred to simply as a photodiode. The operation of the p-n junction photodiode is illustrated in Figure 9.14. Photons with energy greater than the energy gap, incident on the front surface of the device, create electron-hole pairs in the material on both sides of the junction. By diffusion, the electrons and holes generated within a diffusion length from the junction reach the space-charge region. Then electron-hole pairs are separated by the strong electric field; minority carriers are readily accelerated to become majority carriers on the other side. This way a photocurrent is generated that shifts the current-voltage characteristic in the direction of negative or reverse current, as shown in Figure 9.14d.

The equivalent circuit of a photodiode is shown in Figure 9.15. The photodiode has a small series resistance R_s, a total capacitance C_d consisting of junction and packaging capacitances, and a bias (or load) resistor R_L. The amplifier following the photodiode has an input capacitance C_a and a resistance R_a. For practical purposes, R_s is much smaller than the load resistance R_L and can be neglected.

The total current density in the p-n junction is usually written as

$$J(V, \Phi) = J_d(V) - J_{ph}(\Phi), \qquad (9.74)$$

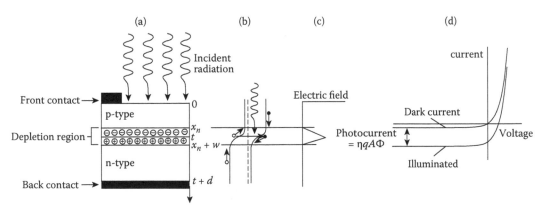

Figure 9.14 p-n junction photodiode: (a) structure of abrupt junction, (b) energy band diagram, (c) electric field, and (d) current-voltage characteristics for the illuminated and nonilluminated photodiode.

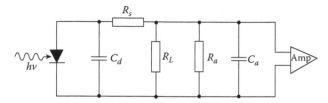

Figure 9.15 Equivalent circuit of an illuminated photodiode (the series resistance includes the contact resistance as well as the bulk p- and n-regions).

where the dark current density, J_d, depends only on V and the photocurrent depends only on the photon flux density Φ.

Generally, the current gain in a simple photovoltaic detector [e.g., not an avalanche photodiode (APD)] is equal to 1, and then according to Equation 9.1, the magnitude of photocurrent is equal

$$I_{ph} = \eta q A \Phi. \tag{9.75}$$

However, the electric field that is responsible for the removal of excess majority carriers created when a photodiode is illuminated induces an additional minority carrier flow to the junction. As a result, for photodiodes where mixed conduction is significant there is a gain associated with the collection photocurrent, which depends on the mobility ratio and that can be increased or decreased by application of bias [68]. The gain applies to excitation close to the junction. This effect can be the cause of anomalously low junction resistance in photodiodes. Below we consider the theory of a conventional photodiode with gain equal to 1.

The dark current and photocurrent are linearly independent (which occurs even when these currents are significant) and the quantum efficiency can be calculated in a straightforward manner [69–72].

If the p-n diode is open-circuited, the accumulation of electrons and holes on the two sides of the junction produces an open-circuit voltage (Figure 9.14d). If a load is connected to the diode, a current will conduct in the circuit. The maximum current is realized when an electrical short is placed across the diode terminals and this is called the short-circuit current.

The open-circuit voltage can be obtained by multiplying the short-circuit current by the incremental diode resistance $R = (\partial I/\partial V)^{-1}$ at $V = V_b$:

$$V_{ph} = \eta q A \Phi R, \tag{9.76}$$

where V_b is the bias voltage and $I = f(V)$ is the current-voltage characteristic of the diode.

In many direct applications the photodiode is operated at zero-bias voltage:

$$R_o = \left(\frac{\partial I}{\partial V} \right)^{-1}_{|V_b=0}. \tag{9.77}$$

A frequently encountered figure of merit for a photodiode is the $R_o A$ product

$$R_o A = \left(\frac{\partial J}{\partial V} \right)^{-1}_{|V_b=0}, \tag{9.78}$$

where $J = I/A$ is the current density.

In the detection of radiation, the photodiode is operated at any point of the I-V characteristic. Reverse bias operation is usually used for very high frequency applications to reduce the RC time constant of the devices.

9.2.1 Ideal Diffusion-Limited p-n Junctions

9.2.1.1 Diffusion Current

Diffusion current is the fundamental current mechanism in a p-n junction photodiode. Figure 9.14a shows a one-dimensional photodiode model with an abrupt junction where the spatial charge of width w surrounds the metallographic junction boundary $x = t$, and two quasineutral regions $(0, x_n)$ and $(x_n + w, t + d)$ are homogeneously doped. The dark current density consists of electrons injected from the n-side over the potential barrier into the p-side and an analogous current due to holes injected from the p-side into the n-side. The current-voltage characteristic for an ideal diffusion-limited diode is given by

$$I_D = AJ_s \left[\exp\left(\frac{qV}{kT}\right) - 1 \right],$$

(9.79)

where [70,71]

$$J_s = \frac{qD_h p_{no}}{L_h} \frac{\gamma_1 ch(x_n/L_h) + sh(x_n/L_h)}{\gamma_1 sh(x_n/L_h) + ch(x_n/L_h)}$$

$$+ \frac{qD_e n_{po}}{L_e} \frac{\gamma_2 ch[(t+d-x_n-w)/L_e] + sh[(t+d-x_n-w)/L_e]}{\gamma_2 sh[(t+d-x_n-w)/L_e] + ch[(t+d-x_n-w)/L_e]},$$

(9.80)

in which case $\gamma_1 = s_1 L_h/D_h$, $\gamma_2 = s_2 L_e/D_e$, p_{no} and n_{po} are the concentrations of minority carriers on both sides of the junction, s_1 and s_2 are the surface recombination velocities at the illuminated (for holes in n-type material) and back photodiode surface (for electrons in p-type material), respectively. The value of the saturation current density, J_s, depends on minority carrier diffusion lengths (L_e, L_h), minority carrier diffusion coefficients (D_e, D_h), surface recombination velocities (s_1, s_2), minority carrier concentrations (p_{no}, n_{po}), and junction design (x_n, t, w, d).

For a junction with thick quasineutral regions $[x_n \gg L_h, (t + d - x_n - w) \gg L_e]$ the saturation current density is equal to

$$J_s = \frac{qD_h p_{no}}{L_h} + \frac{qD_e n_{po}}{L_e},$$

(9.81)

and when the Boltzmann statistic is valid $[n_o p_o = n_i^2, D = (kT/q)\mu, \text{ and } L = (D\tau)^{1/2}]$, then

$$J_s = (kT)^{1/2} n_i^2 q^{1/2} \left[\frac{1}{p_{po}} \left(\frac{\mu_e}{\tau_e}\right)^{1/2} + \frac{1}{n_{no}} \left(\frac{\mu_h}{\tau_h}\right)^{1/2} \right],$$

(9.82)

where p_{po} and n_{no} are the hole and electron majority carrier concentrations, and τ_e and τ_h the electron and hole lifetimes in the p- and n-type regions, respectively. Diffusion current varies with temperature as n_i^2.

The resistance at zero bias can be obtained from Equation 9.79 by differentiation of I-V characteristic

$$R_o = \frac{kT}{qI_s},$$

(9.83)

and then the $R_o A$ product determined by diffusion current is:

$$(R_o A)_D = \left(\frac{dJ_D}{dV}\right)^{-1}_{V_b=0} = \frac{kT}{qJ_s}.$$

(9.84)

If $\gamma_1 = \gamma_2 = 1$ and in the case of a diode with thick regions on both sides of the junction, the $R_o A$ given by equation

$$(R_o A)_D = \frac{(kT)^{1/2}}{q^{3/2} n_i^2} \left[\frac{1}{n_{no}} \left(\frac{\mu_h}{\tau_h}\right)^{1/2} + \frac{1}{p_{po}} \left(\frac{\mu_e}{\tau_e}\right)^{1/2} \right]^{-1}.$$

(9.85)

The photodiodes with thick regions on both sides of the junction are not realized in practice. Analysis of the effect of the structure of a classical photodiode (of thick p-type region) has shown that the $R_o A$ product for junction depths $0 < t < 0.2L_h$ and surface recombination velocities $0 < s_1 < 10^6$ cm/s differs from product $(R_o A)_o$ for a photodiode with thick regions on both sides of the junction by a factor of 0.3–2 [72]. This shows that the $R_o A$ product calculated for photodiodes

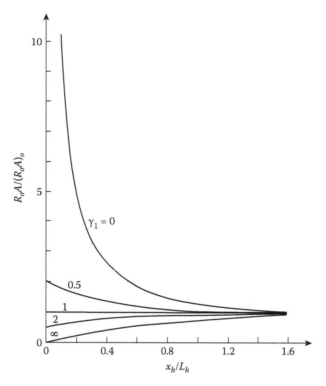

Figure 9.16 The dependence of $R_oA/(R_oA)_o$ on the normalized depth of the n-p$^+$ (n$^+$-p) junction at different surface recombination velocities $\gamma_1 = 0$; 0.5; 1; 2; and ∞. (From Rogalski, A., and Rutkowski, J., *Infrared Physics*, 22, 199–208, 1982. With permission.)

with thick p-type and n-type regions is a good approximation of the product for photodiodes of optimum construction. For n-p$^+$ type junctions the $R_oA/(R_oA)_o$ ratio has the form

$$\frac{R_oA}{(R_oA)_o} = \frac{\gamma_1 ch(x_n/L_h) + sh(x_n/L_h)}{\gamma_1 sh(x_n/L_h) + ch(x_n/L_h)}. \tag{9.86}$$

In Figure 9.16 the dependence of the $R_oA/(R_oA)_o$ on the n-p$^+$ (n$^+$-p) junction depth at various γ_1 values is presented [72]. For $\gamma_1 < 1$ (blocking contact [69]) we get $R_oA > (R_oA)_o$, the increase of R_oA being particularly high for small values of γ_1 and $x_n/L_h \to 0$. Fabrication of such structures may involve great technological difficulties connected with the need to fulfill the condition $s_2 = 0$. In that case it is advantageous to use n-p-p$^+$ (p-n-n$^+$) structures, since the potential barrier between the p- (n-) and p$^+$-type (n$^+$-type) regions limits the flow of minority carriers to the region with more impurities.

For the n$^+$-p diode structure, the junction resistance is limited by diffusion of minority carriers from the p side into the depletion region. In the case of conventional bulk diodes, where $d \gg L_e$:

$$(R_oA)_D = \frac{(kT)^{1/2}}{q^{3/2}n_i^2} N_a \left(\frac{\tau_e}{\mu_e}\right)^{1/2}. \tag{9.87}$$

By thinning the substrate to a thickness smaller than the minority-carrier diffusion length (thus reducing the volume in which diffusion current is generated) the corresponding R_oA product increases, provided that the back surface is properly passivated to reduce surface recombination. In the case of the n$^+$-p junction, if the thickness of the p type region is such that $d \ll L_e$, we obtain:

$$(R_oA)_D = \frac{kT}{q^2} \frac{N_a}{n_i^2} \frac{\tau_e}{d}. \tag{9.88}$$

As result, R_oA can increase by a factor of L_e/d. Of course, analogical formulas can be obtained for p^+-n junctions.

The thickness of the illuminated p^+-n (n^+-p) junction from the p^+ (n^+) side must be small to eliminate absorption of radiation by the free carriers. In n-p^+ (p-n^+) structures illuminated from the $n(p)$-type side, the major contribution to the quantum efficiency comes from the region with less impurities of the $n(p)$-type. That is why the thickness of that region, and hence the depth of the junction, should be greater (i.e., $0.2L_h < t < 0.4L_h$ [72]). On the other hand, at lower t values and $0 < \gamma_1 < 1$ we can obtain a significant increase of the R_oA product (see Figure 9.16). It follows that the optimum depth of the junction is shifted toward smaller t values.

In conclusion it should be noted that the influence of design on the R_oA product is determined by the diffusion component of current density. In the case when the R_oA product is determined by another mechanism, the above considerations are not justifiable. The consideration concerning the quantum efficiency still remains valid.

9.2.1.2 Quantum Efficiency

Three regions contribute to photodiode quantum efficiency: two neutral regions of different types of conductivity and the spatial charge region (see Figure 9.14). Thus [70,71]:

$$\eta = \eta_n + \eta_{DR} + \eta_p, \tag{9.89}$$

where

$$\eta_n = \frac{(1-r)\alpha L_h}{\alpha^2 L_h^2 - 1}\left\{\frac{\alpha L_h + \gamma_1 - e^{-\alpha x_n}\left[\gamma_1 ch\left(x_n/L_h\right) + sh\left(x_n/L_h\right)\right]}{\gamma_1 sh\left(x_n/L_h\right) + ch\left(x_n/L_h\right)} - \alpha L_h e^{-\alpha\,x_n}\right\}, \tag{9.90}$$

$$\eta_p = \frac{(1-r)\alpha L_e}{\alpha^2 L_e^2 - 1}e^{-\alpha(x_n+w)}$$

$$\times \left\{\frac{(\gamma_2 - \alpha L_e)e^{-\alpha(t+d-x_n-w)} - sh\left[(t+d-x_n-w)/L_e\right] - \gamma_2 ch\left[(t+d-x_n-w)/L_e\right]}{ch\left[(t+d-x_n-w)/L_e\right] - \gamma_2 sh\left[(t+d-x_n-w)/L_e\right]} + \alpha L_e\right\}, \tag{9.91}$$

$$\eta_{DR} = (1-r)\left[e^{-\alpha x_n} - e^{-\alpha(x_n+w)}\right]. \tag{9.92}$$

In the following we shall consider the internal quantum efficiency, neglecting the losses due to reflection of the radiation from the illuminated photodiode surface. Obtaining high photodiode quantum efficiency requires that the illuminated region of the junction be sufficiently thin so that the generated carriers may reach the junction potential barrier by diffusion.

In Figure 9.17 the relationship is presented between the components of the photodiode quantum efficiency and the normalized thickness of the junction-illuminated region t/L_h at infinite thickness of the p-type region [72]. The calculation are carried out for the typical absorption of 5×10^3 cm^{-1} at a wavelength close to the intrinsic absorption edge in narrow-gap semiconductors and for $L_e = L_h = 15$ µm, $w = 0.3$ µm. The quantum efficiency of the depletion layer gradually decreases with increasing t, but it is small and plays no major role. The total quantum efficiency attains its maximum at $t \approx 0.2L_h$ for $s_1 = 0$. This maximum is shifted toward smaller t values as the surface recombination velocity s_1 increases. The position of the total quantum efficiency maximum also depends on the absorption coefficient. When the absorption coefficient increases, the depth of the junction at which the total efficiency attains maximum decreases.

The surface recombination velocity significantly affects η in the range of high absorption coefficient values (small wavelengths) so the depth of radiation penetration $1/\alpha$ is very small. The quantum efficiency is constant for all values of the absorption coefficient when the surface recombination velocity is much smaller than a certain characteristic value s_o and then decreases to a smaller but also constant value in the range $s_1 \gg s_o$. In Van De Wiele [71], it was found that the value s_o can be determined from the formula $s_o = (D_h/L_h)cth(x_n/L_h)$ and that it is independent of the absorption coefficient.

Normally, the photodiode is designed so that most of the radiation is absorbed in one side of the junction, for example, in the p-type side in Figure 9.14a. This could be achieved in practice either by making the n-type region very thin or by using a heterojunction in which the band gap in the n-region is larger than the photon energy so that most of the incident radiation can reach the

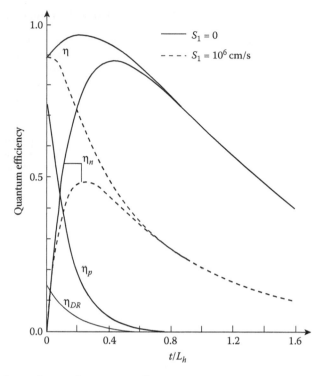

Figure 9.17 The dependence of quantum efficiency on the normalized thickness of the junction illuminated region at $s_1 = 0$ ($\gamma_1 = 0$) and $s_1 = 10^6$ cm/s ($\gamma_1 = 7$). In the calculations it was assumed that $d = \infty$, $w = 0.3$ μm, $r = 0$, and $\alpha = 5 \times 10^3$ cm^{-1}. (From Rogalski, A., and Rutkowski, J., *Infrared Physics*, 22, 199–208, 1982. With permission.)

junction without being absorbed. If the back contact is several minority carrier diffusion lengths, L_e, away from the junction, the quantum efficiency is given by

$$\eta(\lambda) = (1-r)\frac{\alpha(\lambda)L_e}{1+\alpha(\lambda)L_e}. \tag{9.93}$$

If the back contact is less than a diffusion length away from the junction, the quantum efficiency tends to

$$\eta(\lambda) = (1-r)\left[1-e^{-\alpha(\lambda)d}\right], \tag{9.94}$$

where d is the thickness of the p-type region. It has been assumed that the back contact has zero surface recombination velocity and that no radiation is reflected from the back surface. Thus, if the above conditions hold, a high quantum efficiency can be achieved using an antireflection coating to minimize the reflectance of the front surface ensuring that the device is thicker than the absorption length.

It should be noticed that many works (e.g., [73–75]), based on numerical and analytical approaches presented computer solutions for the two-dimensional and three-dimensional cases of photodiodes.

9.2.1.3 Noise

In comparison with photoconductive detectors, the two fundamental processes responsible for thermal noise mechanisms (fluctuations in the velocities of free carriers due to their random motion, and due to randomness in the rates of thermal generation and recombination) are less readily distinguishable in the case of junction devices, giving rise jointly to shot noise on the minority carrier components, which make up the net junction current. The random thermal motion is responsible for fluctuations in the diffusion rates in the neutral regions of junction devices and g–r fluctuations both in the neutral regions and in the depletion region. We will show later that for a junction device at zero bias (i.e., when the net junction current is zero), the resulting noise is identical to Johnson noise associated with the incremental slope of the device.

A general theory of noise in photodiodes that is applicable at arbitrary bias and to all sources of leakage current has not been developed [76]. The intrinsic noise mechanism of a photodiode is shot noise in the current passing through the diode. It is generally accepted that the noise in an ideal diode is given by

$$I_n^2 = \left[2q\left(I_D + 2I_s\right) + 4kT\left(G_J - G_o\right)\right]\Delta f,$$
(9.95)

where $I_D = I_s[\exp(qV/kT) - 1]$, G_J is the conductance of the junction and G_o is the low-frequency value of G_J. In the low-frequency region, the second term on the right-hand side is zero. For a diode in thermal equilibrium (i.e., without applied bias voltage and external photon flux) the mean square noise current is just the Johnson–Nyquist noise of the photodiode zero bias resistance ($R_o^{-1} = qI_s/kT$):

$$I_n^2 = \frac{4kT}{R_o}\Delta f,$$
(9.96)

and

$$V_n = 4kTR_o\Delta f.$$
(9.97)

Note that the mean-square shot noise in reverse bias is half the mean-square Johnson–Nyquist noise at zero bias.

For a diode exposed to background flux density Φ_b, an additional current $I_{ph} = q\eta A\Phi_b$ constitutes a statistically independent contribution to the mean square noise current. Then [3]

$$I_n^2 = 2q\left[q\eta A\Phi_b + \frac{kT}{qR_o}\exp\left(\frac{qV}{\beta kT}\right) + \frac{kT}{qR_o}\right]\Delta f,$$
(9.98)

where

$$R_o = \left(\frac{dI}{dV}\right)_{V=0}^{-1} = \frac{\beta kT}{qI_s},$$
(9.99)

is the dark resistance of the diode at zero bias voltage. In the case of zero bias voltage Equation 9.96 becomes

$$I_n^2 = \frac{4kT\Delta f}{R} + 2q^2\eta A\Phi_b\Delta f,$$
(9.100)

and then

$$V_n = \left(4kT + 2q^2\eta A\Phi_b R_o\right)R_o\Delta f.$$
(9.101)

These forms of the noise equation at zero bias voltage are generally assumed to be applicable independent of the other sources of current.

Equation 9.98 predicts that the shot noise is decreased by operating the diode under a reverse bias. In the absence of a background-generated current, the current noise is equal to the Johnson noise ($4kT\Delta f/R_o$) at zero bias, and it tends to the usual expression for shot noise ($2qI_D\Delta f$) for voltages greater than a few kT in either direction. However, this improved performance under reverse bias is quite difficult to achieve in practice. In real devices, the current noise often increases under a reverse bias, due to presumably $1/f$ noise in the leakage current. In the above analysis $1/f$ noise is ignored.

9.2.1.4 Detectivity

In the case of the photodiode, the photoelectric gain is usually equal to 1, so according to Equation 3.33, the current responsivity is

$$R_i = \frac{q\lambda}{hc}\eta,$$
(9.102)

and detectivity (see Equation 3.35) can be determined as

$$D^* = \frac{\eta\lambda q}{hc}\left[\frac{4kT}{R_o A} + 2q^2\eta\Phi_b\right]^{-1/2},$$

(9.103)

and is obtained from Equations 9.100 and 9.102.

For the last formula we may distinguish two important cases:

- Background-limited performance; if $4kT/R_o A \ll 2q^2\eta\Phi_b$, then we obtain Equation 3.40
- Thermal noise limited performance; if $4kT/R_o A \gg 2q^2\eta\Phi_b$, then

$$D^* = \frac{\eta\lambda q}{2hc}\left(\frac{R_o A}{kT}\right)^{1/2}.$$

(9.104)

Figure 9.18 shows detectivity against $R_o A$ for diodes of 12 μm and 5 μm cutoff with $\eta = 50\%$ operated in 300 K, $f/1$ background. The minimum requirements for near-background-limited operation are 1.0 and 160 Ωcm^2, at 77 and 195 K, respectively, if the criterion $4kT/R_o A = 2q^2\eta\Phi_b$ is used [77].

The detectivity in the absence of background photon flux can be also expressed as:

$$D^* = \frac{\eta\lambda q}{hc}\left[\frac{A}{2q(I_D + 2I_s)}\right]^{1/2}.$$

(9.105)

In reverse biases, I_d tends to $-I_s$ and the expression in brackets tends to I_s.

From the discussion it is shown that the performance of an ideal diffusion-limited photodiode can be optimized by maximizing the quantum efficiency and minimizing the reverse saturation current, I_s. For a diffusion-limited photodiode, the general expression for the saturation current of electron from the p-type side is (see Equation 9.80)

$$I_s^p = A\frac{qD_e n_{po}}{L_e}\frac{sh(d/L_e) + (s_2 L_e/D_e)ch(d/L_e)}{ch(d/L_e) + (s_2 L_e/D_e)sh(d/L_e)}.$$

(9.106)

It is normally possible to minimize the leakage current from the side that does not contribute to the photosignal. The minority carrier generation rate and hence the diffusion current can be greatly reduced, in theory at least, by increasing doping or the band gap on the inactive side of the junction.

If the back contact is several diffusion lengths away from the junction, then Equation 9.106 tends to

$$I_s = \frac{qD_e n_{po}}{L_e}A.$$

(9.107)

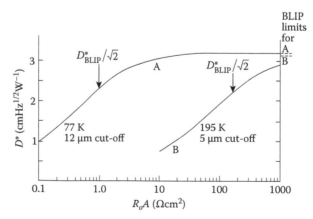

Figure 9.18 Detectivity as a function of $R_o A$ for zero bias photodiodes with $\eta = 50\%$ operated in a 300 K, $f/1$ background. (From Knowles, P., *GEC Journal of Research*, 2, 141–56, 1984.)

As the back contact is brought closer to the junction, the leakage current can either increase or decrease, depending on whether the surface recombination velocity is greater than the diffusion velocity D_e/L_e. In the limiting case where $d \ll L_e$, the saturation current is reduced by a factor d/L_e relative to Equation 9.107 for $s = 0$ and increased by a factor of L_e/d for $s = \infty$. If the surface recombination velocity is small then Equation 9.107 can usually be written in the form

$$I_s = qGV_{diff},$$ (9.108)

where G is the bulk minority carrier generation rate per unit volume and V_{diff} is the effective volume of material from which the minority carriers diffuse to the junction. The effective volume is AL_e for $L_e \ll d$ and tends to Ad for $L_e \gg d$. For p-type material, the generation rate is given by

$$G = \frac{n_{po}}{\tau_e} = \frac{n_i^2}{N_d\tau}.$$ (9.109)

The discussion indicates that the performance of the device is strongly dependent on the properties of the back contact. The most common solution to this problem is to move the back contact many diffusion lengths away to one side and to ensure that all the surfaces are properly passivated. Alternatively, the back contact itself can sometimes be designed to have a low surface recombination velocity by introducing a barrier for minority carriers between the metal contact and the rest of the device. This barrier can be made by increasing the doping or the band gap near the contact, which effectively isolates the minority carriers from the high recombination rate at the contact.

9.2.2 Real p-n Junctions

In the previous section, photodiodes were analyzed in which the dark current was limited by diffusion. However, this behavior is not always observed in practice, especially for wide-gap semiconductor p-n junctions. Several additional excess mechanisms are involved in determining the dark current-voltage characteristics of the photodiode. The dark current is the superposition of current contributions from three diode regions: bulk, depletion region and surface. Between them we can distinguish:

1. Thermally generated current in the bulk and depletion region

   - Diffusion current in the bulk p and n regions
   - Generation–recombination current in the depletion region
   - Band-to-band tunneling
   - Intertrap and trap-to-band tunneling
   - Anomalous avalanche current
   - The ohmic leakage across the depletion region

2. Surface leakage current

   - Surface generation current from surface states
   - Generation current in a field-induced surface depletion region
   - Tunneling induced near the surface
   - The ohmic or nonohmic shunt leakage
   - Avalanche multiplication in a field-induced surface region

3. Space-charge-limited current

Figure 9.19 illustrates schematically some of these mechanisms [20]. Each of the components has its own individual relationship to voltage and temperature. On account of this, many researchers analyzing the I-V characteristics assume that only one mechanism dominates in a specific region of the diode bias voltage. This method of analysis of the diode's I-V curves is not always valid. A better solution is to numerically fit the sum of the current components to experimental data over a range of both applied voltage and temperature.

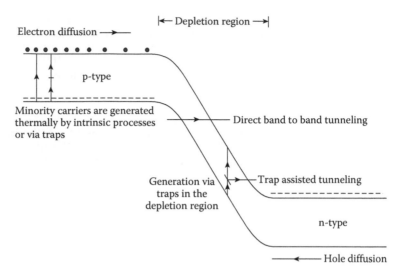

Electron diffusion ⟶

|← Depletion region →|

p-type

Minority carriers are generated
thermally by intrinsic processes
or via traps

Direct band to band tunneling

Generation via
traps in the
depletion region

Trap assisted tunneling

n-type

⟵ Hole diffusion

Figure 9.19 Schematic representation of some of the mechanisms by which dark current is generated in a reverse biased p-n junction. (From Elliott, C. T., and Gordon, N. T., *Handbook on Semiconductors*, North-Holland, Amsterdam, 4, 841–936, 1993. With permission.)

Below, we will be concerned with the current contribution of high-quality photodiodes with high R_oA products limited by:

- Generation–recombination within the depletion region
- Tunneling through the depletion region
- Surface effects
- Impact ionization
- Space-charge limited current

9.2.2.1 Generation–Recombination Current

The importance of this current mechanism was first pointed out by Sah et al. [78], and was later extended by Choo [79]. It appears that one space-charge region g–r current could be more important than its diffusion current, especially at low temperatures, although the width of the space-charge region is much less than the minority carrier diffusion length. The generation rate in the depletion region can be very much greater than in the bulk of the material. In reverse bias, the current can be given by an equation similar to Equation 9.108

$$I = qG_{dep}V_{dep},$$ (9.110)

where G_{dep} is the generation rate and V_{dep} is the volume of the depletion region. In particular, the generation rate from traps in the depletion region is given by the Shockley–Read–Hall formula

$$G_{dep} = \frac{n_i^2}{n_1\tau_{eo} + p_1\tau_{ho}},$$ (9.111)

where n_1 and p_1 are the electron and hole concentrations that would be obtained if the Fermi energy was at the trap energy and τ_{eo} and τ_{ho} are the lifetimes in strong n-type and p-type material. Normally, one of the terms in the denominator of Equation 9.111 will dominate and for the case of a trap at the intrinsic level holds n_i and $p_1 = n_i$, giving

$$G_{dep} = \frac{n_i}{2\tau_o},$$ (9.112)

and then the g–r current of the depletion region is equal

$$J_{GR} = \frac{qwn_i}{2\tau_o}.$$ (9.113)

The comparison of Equations 9.109 and 9.112 indicates that the generation rate in the bulk of the material is proportional to n_i^2, whereas the generation rate in the depletion region is proportional to n_i for a mid-gap state.

The width of the depletion region, and therefore its volume, increases with reverse voltage. For an abrupt junction

$$w = \sqrt{\frac{2\varepsilon_o\varepsilon_r \left(V_{bi} \pm V\right)}{qN_aN_d \left(N_a + N_d\right)}},$$ (9.114)

where N_a and N_d are the acceptor and donor concentrations, respectively; $V_{bi} = (kT/q)\ln(N_aN_d/n_i^2)$ is the built-in voltage and V is the applied voltage. For a linearly graded junction, the width of depletion region depends on $V^{1/3}$.

The g–r current varies roughly as the square root of the applied voltage ($w \sim V^{1/2}$) for the case of an abrupt junction, or as the cube root ($w \sim V^{1/3}$) for a linearly graded junction. This behavior, in which the current increases with reverse bias, can be contrasted with a diffusion-limited diode, where the reverse current is independent of voltage above a few kT/q.

Space-charge region g–r current varies with temperature as n_i; that is, less rapidly than the diffusion current that varies as n_i^2; so that a temperature is finally reached at which the two currents are comparable and below this temperature the g–r current dominates. Usually, at the lowest temperatures current density may saturate due to the existence of a weakly temperature-dependent shunt resistance. The effect of space-charge recombination is to be seen most readily in wide band gap semiconductors.

In the Sah–Noyce–Shockley theory [78], the doping levels were assumed to be the same on the two sides of the junction, and a single recombination center located in the vicinity of the gap was also assumed. The g–r current density under reverse-bias voltage and for forward-bias voltage values that are less than V_{bi} by several kT/q was derived as

$$J_{GR} = \frac{qn_iw}{\left(\tau_{eo}\tau_{ho}\right)^{1/2}} \frac{2sh\left(qV/2kT\right)}{q\left(V_{bi} - V\right)/kT} f(b),$$ (9.115)

where τ_{eo}, τ_{ho} are the carrier lifetimes with the depletion region. The function $f(b)$ is a complicated expression involving the trap level E_t, the two lifetimes and applied voltage V:

$$f(b) = \int_0^\infty \frac{dx}{x^2 + 2bx + 1},$$

$$b = \exp\left(-\frac{qV}{2kT}\right)ch\left[\frac{E_t - E_i}{kT} + \frac{1}{2}\ln\left(\frac{\tau_{eo}}{\tau_{ho}}\right)\right],$$

where E_i is the intrinsic Fermi level. The function $f(b)$ has a maximum value of $\pi/2$, which occurs at small values of b (forward biases $> 2kT/q$); $f(b)$ decreases as b increases. When $E_t = E_i$ and $\tau_{eo} = \tau_{ho} = \tau_o$, the recombination center has its maximum effect and then for symmetrical junction parameters $f(b) \approx 1$, and J_{GR} is determined by Equation 9.113.

For small applied bias, the function $f(b)$ may be taken as independent of V. We can also neglect the bias dependence of the depletion width w for small bias. The zero bias resistance is then found by differentiating Equation 9.115 and setting $V = 0$:

$$\left(R_oA\right)_{GR} = \left(\frac{dJ_{GR}}{dV}\right)_{V=0}^{-1} = \frac{V_b\left(\tau_{eo}\tau_{ho}\right)^{1/2}}{qn_iwf(b)}.$$ (9.116)

For simplicity, we further assume $\tau_{ho} = \tau_{eo} = \tau_o$, $E_t = E_i$ and $f(b) = 1$. Then, Equation 9.116 becomes

$$\left(R_oA\right)_{GR} = \frac{V_b\tau_o}{qn_iw}.$$ (9.117)

In evaluating Equation 9.117, the term of greatest uncertainty is τ_o.

9.2.2.2 *Tunneling Current*

The third type of dark current component that can exist is a tunneling current caused by electrons directly tunneling across the junction from the valence band to the conduction band (direct tunneling) or by electrons indirectly tunneling across the junction by way of intermediate trap sites in the junction region (direct tunneling or trap-assisted tunneling; see Figure 9.19).

The usual direct tunneling calculations assume a particle of constant effective mass incident on a triangular or parabolic potential barrier. For the triangular potential barrier [80]

$$J_T = \frac{q^2 E V_b}{4\pi^2 \hbar^2} \left(\frac{2m^*}{E_g} \right)^{1/2} \exp\left[-\frac{4(2m^*)^{1/2} E_g^{3/2}}{3q\hbar E} \right]. \tag{9.118}$$

For an abrupt p-n junction the electric field can be approximated by

$$E = \left[\frac{2q}{\varepsilon_o \varepsilon_s} \left(\frac{E_g}{q} \pm V_b \right) \frac{np}{n+p} \right]^{1/2}. \tag{9.119}$$

The tunnel current is seen to have an extremely strong dependence on energy gap, applied voltage and an effective doping concentration $N_{ef} = np/(n + p)$. It is relatively insensitive to temperature variation and the shape of the junction barrier. For the parabolic barrier [80]

$$J_T \propto \exp\left[-\frac{(\pi m^*)^{1/2} E_g^{3/2}}{2^{3/2} q\hbar E} \right]. \tag{9.120}$$

Anderson [81], based on the WKB approximation and Kane **k·p** theory, developed expressions for direct interband tunneling in narrow-gap semiconductors for asymmetrically abrupt p-n homojunctions. Anderson's expressions are convenient to use as a first estimation, especially for bias voltage near zero. But due to the extreme sensitivity of tunneling to the electric field, it may lead to orders of magnitudes of error when applied to general device structure. This was demonstrated by Beck and Byer in their tunneling calculations of linearly graded junctions with various slopes [82]. Next, Adar has presented a new technique for direct band-to-band tunneling in narrow gap semiconductors including spatial integration throughout the depletion region at each bias point [83].

Apart from direct band-to-band tunneling, tunneling is possible by means of indirect transitions in which impurities or defects within the space-charge region act as intermediate states [84]. This is a two-step process in which one step is a thermal transition between one of the bands and the trap and the other is tunneling between the trap and the other band. The tunneling process occurs at lower fields than direct band-to-band tunneling because the electrons have a shorter distance to tunnel (see Figure 9.19). This type of tunneling has been reported in HgCdTe p-n junctions at low temperatures [85–93]. Small but finite acceptor activation energies are generally observed in p-type HgCdTe.

Trap-assisted tunneling can occur via:

- A thermal transition with rate $\gamma_p p_1$ to trap center in the band gap located at energy E_t from the valence band [where γ_p is the hole recombination coefficient for the trap centers of density N_t, and $p_t = N_v \exp(-E_t/kT)$]

- A tunnel transition with rate $\omega_v N_v$ (where ω_v represents the carrier tunneling probability between the center and the band)

- Followed by tunnel transition to the conduction band with rate $\omega_c N_c$

The total trap-assisted current is then given by [86]

$$J_T = q N_t w \left(\frac{1}{\gamma_p p_1 + \omega_v N_v} + \frac{1}{\omega_c N_c} \right)^{-1}. \tag{9.121}$$

On account of the low density of states associated with the small value of the conduction band electron mass, thermal transitions from the conduction band are ignored. For the limiting case $\omega_c N_c < \gamma_p p_1$ and $\omega_c N_c \cong \omega_v N_v$, then

$$J_T = q N_t \omega_c N_c w. \tag{9.122}$$

Assuming a parabolic barrier and uniform electric field, the tunneling rate from a neutral center and to the conduction band is given by

$$\omega_c N_c = \frac{\pi^2 q m^* E M^2}{h^3 \left(E_g - E_t\right)} \exp\left[-\frac{\left(m^*/2\right)^{1/2} E_g^{3/2} F(a)}{2qE\hbar}\right],$$

(9.123)

where $a = 2(E_t/E_g) - 1$, $F(a) = (\pi/2) - a(1 - a^{1/2})^{1/2} - (1/\sin a)$. The matrix element M is associated with the trap potential. The experimentally determined value for the quantity $M^2(m^*/m)$ for silicon is found to be 10^{23} Vcm^3 [84]. A similar value is assumed for HgCdTe [86,91,92]. Tunneling increases exponentially with decreasing effective mass, thus the light-hole mass dominates tunnel processes between the valence band and trap centers. In comparison with direct tunneling, indirect tunneling is critically dependent not only on the electric field (doping concentration) but also on the density of recombination centers and their location in the band gap (via the geometrical factor $0 < F(a) < \pi$). Most of the tunnel current will pass through the trap level with the highest transition probability (i.e., the carriers choose the path of least resistance). If the conduction-band and light-hole masses are approximately equal, then the maximum tunneling probability occurs for midgap states for which $\omega_c N_c \cong \omega_v N_v$. Often in the absence of detailed information on the location of trap states in material, the theoretical treatment assumes a single midgap SRH center. Assuming another trap level changes the overall magnitude of the calculated tunnel current, the general behavior of the tunnel current with electric field and temperature will be similar. In general, tunnel current varies exponentially with an electric field and has a relatively weak temperature dependence in comparison with diffusion and depletion currents.

The influence of different junction current components on the R_oA product for various types of junctions fabricated in narrow gap semiconductors has been considered in many papers. For example, Figure 9.20 shows the dependence of the R_oA product components on the dopant concentrations for n^+-p Hg$_{0.78}$Cd$_{0.22}$Te abrupt junctions at 77 K [94]. We can see that the performance

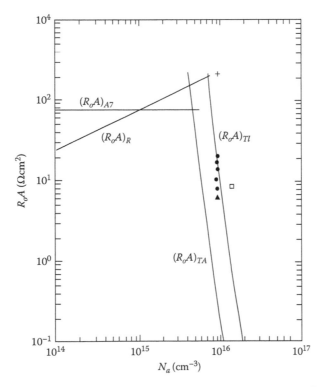

Figure 9.20 The dependence of the R_oA product on acceptor concentration for n^+-p Hg$_{0.78}$ Cd$_{0.22}$Te photodiodes at 77 K. The experimental values (•, +, ▲, □) are taken from different papers. (From Rogalski, A., *Infrared Physics*, 28, 139–53, 1988. With permission.)

of the junctions at zero bias is limited by the tunneling current if the substrate doping is too high. The doping concentration of 10^{16} cm^{-3} or less is required to produce high R_oA products. Below this concentration the R_oA product is determined by diffusion current with minority carrier lifetime conditioned by the Auger 7 process. However, to obtain the highest possible value of the R_oA product and avoid the effect of a fixed insulator charge of a passivation junction layer, the technological process of photodiode preparation should be performed in a manner to obtain a dopant concentration above 10^{15} cm^{-3} [94]. From above, it follows that the optimum concentration range on the lower doping side of the junction is $10^{15} < N_a < 10^{16}$ cm^{-3}. From the comparison of the theoretical curves with the experimental data it may be seen that a satisfactory consistence has not been achieved. The main reason for these discrepancies is probably due to inconsistency of the abrupt junction model with the experimental junction profiles. The Anderson tunneling theory assumes a uniform charge model for the potential barrier and nonparabolicity of the band structure. Smaller values of tunnel current (higher experimental R_oA products for acceptor concentration $10^{16} < N_a < 2 \times 10^{16}$ cm^{-3}; see Figure 9.20) can be associated with a decrease of electric field away from the metallographic junction boundary. Furthermore, the tunnel current calculations are strictly valid only for the approximately empty well condition. As the well fills, the tunnel current tends to decrease due to the collapse of the well.

9.2.2.3 Surface Leakage Current

In a real p-n junction, particularly in wide gap semiconductors and at low temperatures, additional dark current related to the surface occurs. Surface phenomena play an important part in determining photovoltaic detector performance. The surface provides a discontinuity that can result in a large density of interface states. These generate minority carriers by the Shockley-Read-Hall (SRH) mechanism, and can increase both the diffusion and depletion region generated currents. The surface can also have a net charge, which affects the position of the depletion region at the surface.

The surface of actual devices is passivated in order to stabilize the surface against chemical and heat-induced changes as well as to control surface recombination, leakage, and related noise. Native oxides and overlying insulators are commonly employed in p-n junction fabrication that introduce fixed charge states, which then induce accumulation or depletion at the semiconductor-insulator surface. We can distinguish three main types of states on the semiconductor-insulator interface; namely, fixed insulator charge, low surface states, and fast surface states. The fixed charge in the insulator modifies the surface potential of the junction. A positively charged surface pushes the depletion region further into the p-type side and a negatively charged surface pushes the depletion region toward the n-type side. If the depletion region is moved toward the more highly doped side, the field will increase, and tunneling becomes more likely. If it is moved toward the more lightly doped side, the depletion region can extend along the surface, greatly increasing the depletion region generated currents. When sufficient fixed charge is present, accumulated and inverted regions as well as n-type and p-type surface channels are formed (see Figure 9.21) [95]. An ideal surface would be electrically neutral and would have a very low density of surface states. The ideal passivation would be a wide gap insulator grown with no fixed charge at the interface.

Fast interface states, which act as g–r centers, and fixed charge in the insulator cause a variety of surface-related current mechanisms. The kinetics of g–r through fast surface states is identical to that through bulk SRH centers. The current in a surface channel is given by

$$I_{GRS} = \frac{q n_i w_c A_c}{\tau_o},$$
(9.124)

where w_c is the channel width and A_c is the channel area.

Apart from g–r processes occurring at the surface and within surface channels, there are other surface-related current mechanisms, termed surface leakage, with ohmic or breakdown-like current-voltage characteristics. They are nearly temperature independent. Surface breakdown occurs in a region of high electric field (Figure 9.21c, where the depletion layer intersects the surface; Figure 9.21d, very narrow depletion layer).

The zero bias resistance area product for the case of uniformly distributed bulk and surface g–r centers is given by

$$\left(\frac{1}{R_oA}\right)_{GR} = \frac{e n_i w}{V_b}\left(\frac{1}{\tau_o} + \frac{S_o P}{A}\right),$$
(9.125)

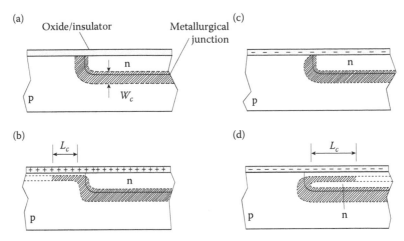

Figure 9.21 Effect of fixed insulator charge on the effective junction space-charge region: (a) flat band condition, (b) positive fixed charge (inversion of the p side, formation of an n-type surface channel), (c) negative fixed channel (accumulation of the p side, field induced junction at the surface), and (d) large negative fixed charge (inversion of the n side, formation of p-type surface channel). (From Reine, M. B., Sood, A. K., and Tredwell, T. J., *Semiconductors and Semimetals*, Academic Press, New York, 18, 201–311, 1981. With permission.)

where V_b is the built-in voltage of the p-n junction. The second term in parentheses is for g–r centers localized at the surface, and the parameter S_o is referred to as the g–r surface recombination velocity, which is proportional to the density of g–r defects. P is the junction perimeter. Note that the bulk and surface components of the g–r R_oA product have different dependencies of junction geometry.

Band bending at the p-n junction surface can be controlled by a gate electrode overlaid around the junction perimeter on an insulating film.

The dark current as a sum of several independent contributions can also be written as follows:

$$I = I_s\left[\exp\frac{q(V-IR_s)}{\beta kT}-1\right]+\frac{V-IR_s}{R_{sh}}+I_T,$$

where R_s is the series resistance and R_{sh} is the shunt resistance of the diode. In the case of predominant diffusion current the β coefficient approaches unity, but when the g–r current is mainly responsible for carrier transport, $\beta = 2$.

9.2.2.3 Space-Charge Limited Current

In the case of wide band-gap p-n junctions, the forward current-voltage characteristics are often described by equation

$$J \propto \exp\left(\frac{qV}{\beta kT}\right),\tag{9.126}$$

where the diode ideality factor $\beta > 2$. This value of β does not fall within the range that result when diffusion current ($\beta = 1$) or depletion layer current ($\beta = 2$) dominate the forward-bias current conduction. This behavior is typical for an insulator with shallow and/or deep traps and thermally generated carriers.

Space-charge-limited (SCL) current flow in solids has been considered in detail by Rose [96], Lampert and Mark [97,98]. They loosely defined materials with $E_g \leq 2$ eV as semiconductors and those with $E_g \geq 2$ eV as insulators.

When a sufficiently large field is applied to an insulator with ohmic contacts, electrons will be injected into the bulk of the material to form a current that is limited by space-charge effects. When trapping centers are present, they capture many of the injected carriers, thus reducing the density of free carriers.

At low voltages where the injection of carriers into the semi-insulating material is negligible, Ohm's law is obeyed and the slope of the J-V characteristics defines the resistivity ρ of the material.

At some applied voltage V_{TH}, the current begins to increase more rapidly than linearly with applied voltages. Here V_{TH} is given by

$$V_{TH} = 4\pi \times 10^{12} q p_t \frac{t^2}{\varepsilon}, \tag{9.127}$$

where t is the thickness of the semi-insulating material, and p_t is the density of trapped holes, if one carrier (holes) space-charge limited current is considered. As the voltage is continuously increased beyond V_{TH}, additional excess holes are injected into the material and the current density is given by

$$J = 10^{-13} \mu_h \varepsilon \theta \frac{V^2}{t^2}, \tag{9.128}$$

where θ is the probability of trap occupation determined as a ratio of the densities of free to trapped holes. Here θ is given by

$$\theta = \frac{p}{p_t} = \frac{N_v}{N_t} \exp\left(-\frac{E_t}{kT}\right), \tag{9.129}$$

where N_v is the effective density of states in the valence band, N_t is the density of traps, and E_t is the depth of traps from the top of the valence band. When the applied voltage further increases, the square-law region of Equation 9.128 will terminate in a steeply rising current that increases till it becomes the trap-free SCL currents given by

$$J = 10^{-13} \mu_h \varepsilon \frac{V^2}{t^2}. \tag{9.130}$$

The density of traps N_t can be determined from the voltage V_{TFL} at which the traps are filled and the currents rise sharply;

$$V_{TFL} = 4\pi \times 10^{12} q N_t \frac{t^2}{\varepsilon}. \tag{9.131}$$

The trap depth E_t can also be calculated from the Equation 9.131 using the value of N_t. But if Equation 9.128) is measured as the function of temperature and a plot of $\ln(\theta T^{-3/2})$ versus $1/T$ is possible, E_t and N_t can be obtained directly from these data without referring to Equation 9.131.

For the purpose of illustration of the above phenomenon in semi-insulating materials, consider the four $\log J$ versus $\log V$ graphs in Figure 9.22 [99]. Figure 9.22a represents an ideal insulator where $I \propto V^2$, indicating SCL current flow. In other words, there are no thermally generated carriers resulting from impurity-band or band-to-band transitions; the conduction is only within the

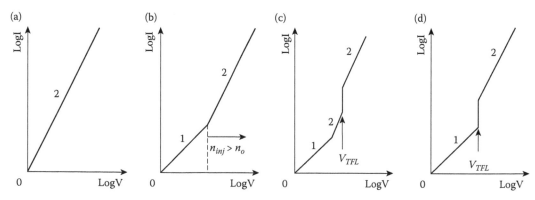

Figure 9.22 Schematic drawings of the logarithmic dependence of current versus voltage for (a) an ideal insulator, (b) a trap insulator with thermally generated free carriers, (c) an insulator with shallow traps and thermal free carriers, and (d) an insulator with deep traps and thermal-free carriers. (From Edmond, J. A., Das, K., and Davis, R. F., *Journal of Applied Physics*, 63, 922–29, 1988. With permission.)

conduction band as a result of carrier injection. As shown in Figure 9.22b, ohmic conduction is obtained in trap-free insulators in the presence of thermally generated free carriers, n_o. When the injected carrier density n_{inj} exceeds n_o ($n_{inj} > n_o$), ideal insulator characteristics are observed ($I \propto V^2$). Shallow traps contribute to an $I \propto V^2$ regime at a lower voltage followed by a sharp transition to an ideal insulator, square-law regime as shown in Figure 9.22c. The sharp transition corresponds to an applied voltage; V_{TFL} is required to fill a discrete set of traps that are initially unoccupied. Figure 9.22d illustrates the case of a material with deep traps that have become filled when n_{inj} becomes comparable to n_o. The voltage at which this occurs is V_{TFL}. Therefore, at $V < V_{TFL}$ ohmic conduction is observed and at $V > V_{TFL}$ SCL current flow dominates.

9.2.3 Response Time

Wide-bandwidth photodiodes are used both for direct and heterodyne detection. The considerable interest in high-frequency photodiodes has been mostly due to their application at 10.6 μm (involving CO_2 laser heterodyne detection) for lidar systems and applications for optical fiber communications.

The upper-frequency response of a photodiode may be determined by basically three effects: the time of carrier diffusion to the junction depletion region, τ_d; the transit time of carrier drift across the depletion region, τ_t; and the RC time constant associated with circuit parameters including the junction capacitance C and the parallel combination of diode resistance and external load (the series resistance is neglected).

The photodiode parameters responsible for these three factors are the absorption coefficient α, the depletion region width w_{dep}, the photodiode junction and package capacitances C_d, the amplifier capacitance C_a, the detector load resistance R_L, the amplifier input resistance R_a, and the photodiode series resistance R_s (see Figure 9.15).

Photodiodes designed for fast response are generally constructed so that the absorption of radiation occurs in the p-type region. This ensures that most of the photocurrent is carried by electrons that are more mobile than holes (whether by diffusion or drift). The frequency response for the diffusion process in a backside illuminated diode has been calculated by Sawyer and Rediker as a function of diode thickness, diffusion constant, absorption depth, minority carrier lifetime, and surface-recombination velocity [100]. Assuming that the diffusion length is greater than both the diode thickness and the absorption depth, the cutoff frequency where the response drops by $\sqrt{2}$, is given by [20,101]

$$f_{diff} = \frac{2.43D}{2\pi t^2},$$ (9.132)

where D is the diffusion constant and t is the diode thickness.

The depletion region transit time is equal to

$$\tau_t = \frac{w_{dep}}{v_s},$$ (9.133)

where w_{dep} is the depletion region width and v_s is the carrier saturation drift velocity in the junction field, which has a value of about 10^7 cm/s. The frequency response of a transit time limited diode has been derived by Gartner [102]. This source of delay is virtually eliminated in properly constructed photodiodes where the depletion region is sufficiently close to the surface. For typical parameters $\mu_e = 10^4$ cm^2/Vs, $v_s = 10^7$ cm/s, $\alpha = 5 \times 10^3$ cm^{-1} and $w_{dep} = 1$ μm, the transit time as well as the diffusion time are about 10^{-11} s.

The diffusion processes are generally slow compared with the drift of carriers in the high-field region. Therefore, to have a high-speed photodiode, the photocarriers should be generated in the depletion region or close to it so that the diffusion times are less than or equal to the carrier drift times.

The effect of long diffusion times can be seen by considering the photodiode response time when the detector is illuminated by a step input of optical radiation (see Figure 19.23 [103]). For fully depleted photodiodes, when $w_{dep} \gg 1/\alpha$, the rise and fall time are generally the same. The rise and fall times of the photodiode follow the input pulse quite well (Figure 9.23b). If the photodiode capacitance is larger, the response time becomes limited by the RC time constant of the load resistor R_L and the photodiode capacitance (Figure 9.23c):

$$\tau_{RC} = \frac{AR_T}{2}\left(\frac{q\varepsilon_o\varepsilon_s N}{V}\right)^{1/2}.$$ (9.134)

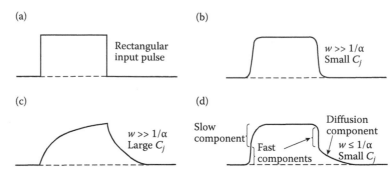

Figure 9.23 Photodiode pulse responses under various detector parameters. (From Kreiser, G., *Optical Fiber Communications*, McGraw-Hill Book Co., Boston, 2000.)

$R_T = R_d(R_s + R_L)/(R_s + R_d + R_L)$ where R_s, R_d and R_L are the series, diode, and load resistances. The detector behaves approximately like a simple RC low-pass filter with a passband given by

$$\Delta f = \frac{1}{2\pi R_T C_T}. \tag{9.135}$$

In a general case, R_T is the combination of the load and amplifier resistance and C_T is the sum of the photodiode and amplifier capacitance (see Figure 9.15).

To reduce the RC time constant, we can decrease the majority carrier concentration adjacent to the junction, increase V by means of a reverse bias, decrease the junction area, or lower either the diode resistance or the load resistance. Except for the application of a reverse bias, all of these changes reduce the detectivity. The trade-off between response time and detectivity is then apparent.

If the depletion layer is too narrow, electron-hole pairs generated in the n and p regions would have to diffuse back to the depletion region before they could be collected. Devices with very thin depletion regions thus tend to show distinct slow- and fast-response components, as shown in Figure 9.23d; the fast component is due to carriers generated in the depletion region, whereas the slow component arises from the diffusion carriers.

Generally, a reasonable compromise between high-frequency response and high quantum efficiency is required and usually thickness of the absorption region is between $1/\alpha$ and $2/\alpha$.

9.3 p-i-n PHOTODIODES

The p-i-n photodiode is a popular alternative to the simple p-n photodiodes especially for ultra-fast photodetection in optical communication, measurement, and sampling systems. In p-i-n photodiodes an undoped i-region (p⁻ or n⁻, depending on the method of junction formation) is sandwiched between p⁺ and n⁺ regions. Figure 9.24 shows schematic representation of a p-i-n diode, an energy band diagram under reverse-bias conditions together with optical absorption characteristics. Because of the very low density of free carriers in the i-region and its high resistivity, any applied bias drops entirely across the i-region, which is fully depleted at zero bias or very low value of reverse bias. Typically, for a doping concentration of $\sim 10^{14}$–10^{15} cm^{-3} in the intrinsic region, a bias voltage of 5–10 V is sufficient to deplete several micrometers, and the electron velocity also reaches the saturation value.

The p-i-n photodiode has a "controlled" depletion layer width, which can be tailored to meet the requirements of photoresponse and bandwidth. A tradeoff is necessary between response speed and quantum efficiency. For high response speed, the depletion layer width should be small but for high quantum efficiency (or responsivity) the width should be large. An external resonant microcavity approach has been proposed to enhance quantum efficiency in such a situation [104,105]. In this approach the absorption region is placed inside a cavity so that a large portion of the photons can be absorbed even with small detection volume.

The response speed of a p-i-n photodiode is ultimately limited either by transit time or by circuit parameters. The transit time of carriers across the i-layer depends on its width and the carrier velocity. Usually, even for moderate reverse biases that carriers drift across the i-layer with saturation velocity. The transit time can be reduced by reducing the i-layer thickness. Fabricating the

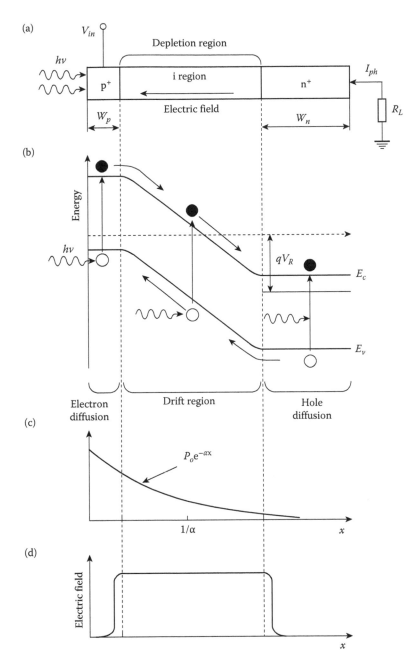

Figure 9.24 p-i-n photodiode: (a) structure, (b) energy band diagram, (c) carrier generation characteristic, and (d) electric field profile.

junction close to the illuminated surface can minimize the effect of diffusion of carrier created outside the i-layer.

The transit time of the p-i-n photodiode is shorter than that obtained in a p-n photodiode even though the depletion region is longer than in the p-n photodiode case due to carriers traveling at near their saturation velocity virtually the entire time they are in the depletion region (in p-n junction the electric field peaked at the p-n interface and then rapidly diminished). For p and n regions less than one diffusion length in thickness, the response time to diffusion alone is typically 1 ns/μm in p-type silicon and about 100 ps/μm in p-type III-V materials. The corresponding values for n-type III-V materials is several nanoseconds per micrometer due to the lower mobility of holes.

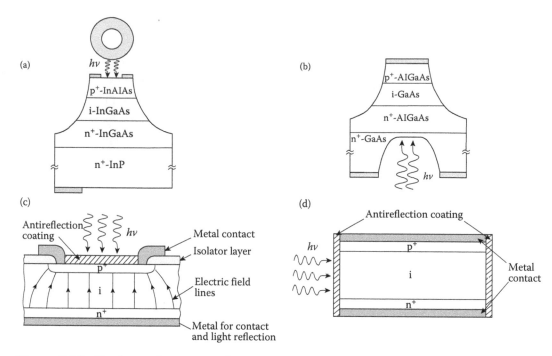

Figure 9.25 Device configurations of p-i-n photodiodes: (a) front-illuminated mesa, (b) back-illuminated mesa, (c) front-illuminated planar, and (d) parallel-illuminated planar.

Two generic p-i-n photodiodes, front-illuminated and back-illuminated are commonly used (see Figure 9.25). For practical applications, photoexcitation is provided through either an etched opening in the top contact, or an etched hole in the substrate. The latter reduces the active area of the diode to the size of the incident light beam. Sidewalls of mesas are covered using passivation materials such as polyimide. A compromise between quantum efficiency and response can be reached if the light is incident from the side, parallel to the junction as shown in Figure 9.25d. The light can also be allowed to strike at an angle that creates multiple reflection inside the device, substantially increasing the effective absorption and quantum efficiency. This solution is often used to couple a detector with one-mode fiber.

For the 1.0–1.55-μm wavelength range suitable for optical communications, Ge and a few III-V compound semiconductor alloys are materials for p-i-n photodiodes, primarily because of their large absorption coefficients (see Figure 3.8). Typical p-i-n photodiode responsivities as a function of wavelength in NIR spectral region are shown in Figure 9.26 [103]. Representative values are 0.65 A/W for silicon at 900 nm and 0.45 A/W for germanium at 1.3 μm. For InGaAs, typical values are 0.9 A/W at 1.3 μm and 1.0 A/W at 1.55 μm. At present, III-V semiconductors have largely replaced germanium as materials for fiber optical compatible detectors. Due to small bandgap, the dark current of Ge photodiodes degrades the signal-to-noise ratio. Also passivation technique for Ge photodiodes is not satisfactory. Table 9.3 compares the performance values for p-i-n photodiodes derived from various vendor data sheets.

The principal source of noise in a p-i-n photodiode is g–r noise; it is larger than the Johnson noise, since the dark current in a reverse-biased junction is very low.

9.4 AVALANCHE PHOTODIODES

When the electric field in a semiconductor is increased above a certain value, the carriers gain enough energy (greater than the band gap) so that they can excite electron-hole pairs by impact ionization. An avalanche photodiode (APD) operates by converting each detected photon into a cascade of moving carrier pairs [80,106–112]. The device is a strongly reverse-biased photodiode in which the junction electric field is large; the charge carriers therefore accelerate, acquiring enough energy to excite new carriers by the process of impact ionization.

In an optimally designed photodiode, the geometry of the APD should maximize photon absorption (for example, by assuming the form of a p-i-n structure) and the multiplication region

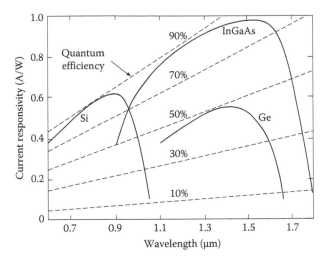

Figure 9.26 Comparison of the current responsivity and quantum efficiency as a function of wavelength for p-i-n NIR photodiodes. (From Kreiser, G., *Optical Fiber Communications*, McGraw-Hill Book Co., Boston, 2000.)

Table 9.3: Summary of Si, Ge, and InGaAs p-i-n Photodiode Characteristics

Parameter	Si	Ge	InGaAs
Wavelength range (nm)	400–1100	800–1650	1100–1700
Peak wavelength (nm)	900	1550	1550
Current responsivity (A/W)	0.4–0.6	0.4–0.5	0.75–0.95
Quantum efficiency (%)	65–90	50–55	60–70
Dark current (nA)	1–10	50–500	0.5–2.0
Rise time (ns)	0.5–1	0.1–0.5	0.05–0.5
Bandwidth (GHz)	0.3–0.7	0.5–3	1–2
Bias voltage (–V)	5	5–10	5
Capacity (pF)	1.2–3	2–5	0.5–2

should be thin to minimize the possibility of localized uncontrolled avalanches (instabilities or microplasmas) being produced by strong electric field. Greater electric-field uniformity can be achieved in a thin region. These two conflicting requirements call for an APD design in which the absorption and multiplication regions are separate. For example, Figure 9.27 shows a reach-through p^+-π-p-n^+ APD structure that accomplishes this. Photon absorption occurs in the wide π-region (very lightly doped p region). Electrons drift through the π-region into a thin p-n^+ junction, where they experience a sufficiently strong electric field to cause avalanching. The reverse bias applied across the device is large enough for the depletion layer to reach through the p and π regions into the p^+ contact layer.

The avalanche multiplication process is illustrated in Figure 9.27d. A photon absorbed at point 1 creates an electron-hole pair. The electron accelerates under the effect of the strong electric field. The acceleration process is constantly interrupted by random collisions with the lattice in which the electron loses some of its acquired energy. These competing processes cause the electron to reach an average saturation velocity. The electron can gain enough kinetic energy that, upon collision with an atom, can break the lattice bonds, creating a second electron-hole pair. This is called impact ionization (at point 2). The newly created electron and hole both acquire kinetic energy from the field and create additional electron-hole pairs (e.g., at point 3). These in turn continue the process, creating other electron-hole pairs. The microscopic manner in which a carrier gains energy from an electric field and undergoes impact ionization depends on the semiconductor band structure and the scattering environment (mainly optical phonons) it finds itself. A good

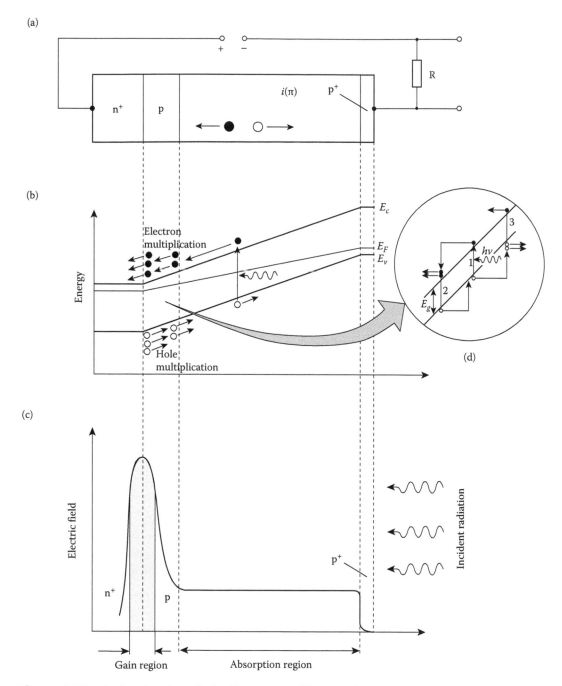

Figure 9.27 Avalanche photodiode: (a) structure, (b) energy band diagram, (c) electric field profile, and (d) schematic representation of the multiplication process.

review of the theory of the impact ionization process in semiconductor at low fields is given by Capasso [107].

The abilities of electrons and holes to impact ionize are characterized by the ionization coefficients α_e and α_h. These quantities represent ionization probabilities per unit length. The ionization coefficients increase with the depletion layer electric field and decrease with increasing device temperature.

Figure 9.28 shows the curves of α_e and α_h as a function of the electric field for several semiconductors important in APDs [113]. As is shown, starting from a few 10^5 V/cm, the ionization

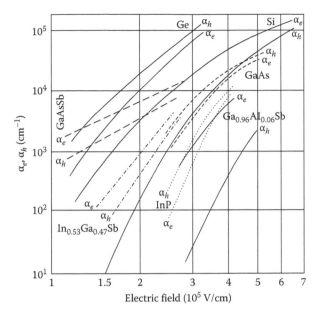

Figure 9.28 Ionization coefficients of electrons (α_e) and holes (α_h) as a function of the electric field for some semiconductors used in avalanche photodiodes. (From Donati, S., *Photodetectors. Devices, Circuits, and Applications*, Prentice Hall, New York, 2000.)

coefficients steeply increase with a small increase of the electric field, while for field $< 10^5$ V/cm ionization is negligible in all of the semiconductor compounds. In some semiconductors electrons ionize more efficiently than holes (Si, GaAsSb, InGaAs, for which $\alpha_e > \alpha_h$), while in others the reverse is true (Ge, GaAs, where $\alpha_h > \alpha_e$).

The ionization coefficients increase with the applied electric field and decrease with increasing device temperature. The increase with the field is due to additional carrier velocity, while the decrease with temperature is due to an increase in nonionizing collisions with thermally excited atoms. For a given temperature, the ionization coefficients are exponentially dependent on the electric field and have a functional form of

$$\alpha = a\exp\left(-\left[\frac{b}{E}\right]^c\right),\tag{9.136}$$

where a, b, c are experimentally determined constants, and E is the magnitude of an electric field.

An important parameter for characterizing the performance of an APD is the ionization ratio $k = \alpha_h/\alpha_e$. When holes do not ionize appreciably (i.e., $\alpha_h \ll \alpha_e$; $k \ll 1$), most of the ionization is achieved by electrons. The avalanching process then proceeds principally from right to left (i.e., from the p side to n side) in Figure 9.27d. It terminates some time later when all the electrons arrive at the n side of the depletion layer. If electrons and holes both ionize appreciably ($k \approx 1$), on the other hand, those holes moving to the right create electrons that move to the left, which, in turn, generate further holes moving to the right, in a possibly unending circulation. Although this feedback process increases the gain of the device (i.e., the total generated charge in the circuit per photocarrier pair), it is nevertheless undesirable for several reasons: it is time consuming and therefore reduces the device bandwidth, it is random and therefore increases the device noise, and it can be unstable, thereby causing avalanche breakdown. It is therefore desirable to fabricate APDs from materials that permit only one type of carrier (either electrons or holes) to impact ionize. If electrons have the higher ionization coefficient, for example, optimal behavior is achieved by injecting the electron of a photocarrier pair at the p edge of the depletion layer and by using a material whose value of k is as low as possible. If holes are injected, the hole of a photocarrier pair should be injected at the n edge of the depletion layer and k should be as large as possible. The ideal case of single-carrier multiplication is achieved when $k = 0$ or ∞.

A comprehensive theory of avalanche noise in APDs was developed by McIntyre [114,115]. The noise of an APD per unit bandwidth can be described by the formula

$$\langle I_n^2 \rangle = 2qI_{ph}\langle M \rangle^2 F, \tag{9.137}$$

where I_{ph} is the unmultiplied photocurrent (signal), $<M>$ is the average avalanche gain and F is the excess noise factor associated with M that arises from the stochastic nature of the ionization process. McIntyre showed that

$$F_e\left(M_e\right) = k\langle M_e \rangle + \left(1-k\right)\left(2 - \frac{1}{\langle M_e \rangle}\right), \tag{9.138}$$

and for the case when holes initiate multiplication

$$F_h\left(M_h\right) = \frac{1}{k}\langle M_h \rangle + \left(1-\frac{1}{k}\right)\left(2 - \frac{1}{\langle M_h \rangle}\right). \tag{9.139}$$

In p-n and p-i-n reverse-biased photodiodes without gain, $<M> = 1$, $F = 1$ and the well-known shot noise formula will indicate the device's noise performance. In the avalanche process, if every injected photocarrier underwent the same gain M, the factor would be unity, and the resulting noise power would only be the input shot noise due to the random arrival of signal photons, multiplied by the gain squared. The avalanche process is, instead, intrinsically statistical, so that individual carriers generally have different avalanche gains characterized by a distribution with an average $<M>$. This causes additional noise called avalanche excess noise, which is conveniently expressed by the F factor in Equation 9.137. As was mentioned above, to achieve a low F, not only must α_e and α_h be as different as possible, but also the avalanche process must be initiated by carriers with the higher ionization coefficient. According to McIntyre's rule, the noise performance of ADP can be improved by more than a factor of 10 when the ionization ratio is increased to 5. Most of III-V semiconductors have $0.4 \leq k \leq 2$.

The gain mechanisms are very temperature-sensitive because of the temperature dependence of the electron and hole ionization rates.

Equations 9.138 and 9.139 have been derived under the condition that the ionization coefficients are in local equilibrium with the electric field, hence, the designation "local field" model. In most semiconductor materials, this local approximation provides an accurate prediction of the excess noise factors for thick avalanche regions (>1 μm). From Equation 9.138 we can see that the lowest excess noise is obtained when k is minimized for electron initiated multiplication, as shown in Figure 9.29. It is well known, however, that the impact ionization is nonlocal and carriers injected into the high-field are "cool" and require a certain distance to attain a sufficient energy to ionize [116]. The distance in which no impact ionization occurs is referred to as the "dead space" $d_{e(h)}$ for electrons (holes). If the multiplication region is thick, the dead space can be neglected and the local field model provides an accurate description of the APD characteristics. Figure 9.30 shows a schematic of the ionization path length probability distribution function (PDF) for electrons, $h_e(x)$,

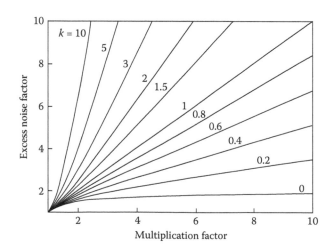

Figure 9.29 Excess noise factor versus multiplication factor for different k.

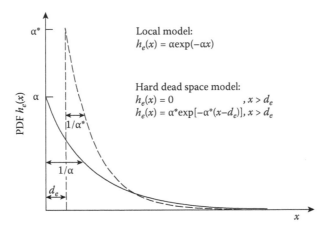

Local model:
$$h_e(x) = \alpha \exp(-\alpha x)$$

Hard dead space model:
$$h_e(x) = 0 \qquad\qquad\quad , x > d_e$$
$$h_e(x) = \alpha^* \exp[-\alpha^*(x-d_e)], x > d_e$$

Figure 9.30 PDFs of ionization path lengths in local and hard dead space models. (From David, J.P.R., and Tan, C.H., *IEEE Journal of Selected Topics in Quantum Electronics* 14, 998–1009, 2008. With permission.)

including both local model and hard dead space model [116]. The value of d for electrons and holes is approximately given by E_{th}/qE, where E_{th} is the threshold energy for ionization and depends on the semiconductor band structure and the electric field, E. The dead-space effect can be significant and a large reduction in the excess noise can occur as a result of a much narrow PDF than the local model. In consequence, APDs with low excess noise can be obtained even with $k \sim 1$ [117–119].

Reducing of the avalanche region length has other unexpected benefits—increased speed. The photodiode gain-bandwidth product results from the time required for the avalanche process to build up or decay; the higher the gain, the higher the associated time constant, thus lower the bandwidth. However, Emmons showed that the bandwidth limitation is removed when either α_e or $\alpha_h = 0$ [120]. For nonzero ionization coefficients, the frequency dependence of electron initiated mean gain is approximately given by

$$M(\omega) = \frac{M_o}{\sqrt{1 + (\omega M_o k \tau)^2}}, \qquad (9.140)$$

where M_o is the DC gain, and τ is approximately (within a factor of ~2) the carrier transit time across the multiplication region.

Two of the most important objectives in APD design are reduction of dark current and enhancement of device speed. In order to obtain the best performance, several structural and materials requirements must be met. First and foremost it is important to ensure uniformity of carrier multiplication over the entire photosensitive area of the diode. The device material in which avalanching occurs must be defect free, and great care must be taken in device fabrication. An essential problem is the excessive leakage current at the junction edges. In Si APDs the common technique used to alleviate this problem is to incorporate a guard ring, which is an n-p junction created by selective-area diffusion around the periphery of the diode. Very careful regulation of the detector bias is required for stable operation of APDs. The commonly available silicon APD structures are optimized to meet specific paradigms [121].

The choice of materials for an APD is dictated by the applications; the most popular are: laser range finding, high speed receivers, and single-photon counting. Silicon APDs are used between 400 and 1100 nm, germanium between 800 and 1550 nm, and InGaAs between 900 and 1700 nm. InGaAs used for fiber optic communication is more expensive than germanium, but provide lower noise and higher frequency response for a given active area. Germanium APDs are recommended for applications in which amplifier noise is high or cost is a prime consideration. Table 9.4 lists the parameters of Si, Ge, and InGaAs APDs. They are given as guidelines for comparison purposes.

Much of recent work on APDs has focused on developing new structures and incorporating alternative materials that will yield lower noise and higher speed while maintaining optimal gain levels [112,118]. For example, low excess noise, due to dead space effect, can be achieved using submicron InAlAs or InP avalanche regions with InGaAs absorption regions. Both InAlAs and InP

Table 9.4: Summary of Si, Ge, and InGaAs Avalanche Photodiode characteristics

Parameter	Si	Ge	InGaAs
Wavelength range (nm)	400–1100	800–1650	1100–1700
Peak wavelength (nm)	830	1300	1550
Current responsivity (A/W)	50–120	2.5–25	—
Quantum efficiency (%)	77	55–75	60–70
Avalanche gain	20–400	50–200	10–40
Dark current (nA)	0.1–1	50–500	10–50 ($M = 10$)
Rise time (ns)	0.1–2	0.5–0.8	0.1–0.5
Gain × Bandwidth (GHz)	100–400	2–10	20–250
Bias voltage (V)	150–400	20–40	20–30
Capacity (pF)	1.3–2	2–5	0.1–0.5

are lattice matched to InGaAs. The α_e/α_h ratio has been found to be significantly larger than α_h/α_e ratio in InP at low electric fields. The excess noise factor at a given gain is significantly lower in InAlAs than that in InP due to a large α_e/α_h ratio in the former and the beneficial effect of the dead space in the latter.

It should be mentioned that the APD can also be biased at voltage larger than the infinity-gain voltage in such a way that the arrival of a single photon precipitates avalanche breakdown, thereby creating a large current pulse that signifies a subsequent photon. This may be carried out either by passive or active means. This working regime is called counting mode or single photon avalanche detector (SPAD), also known as a Geiger-mode avalanche detector, after the work pioneered by Cova and coworkers [122]. SPAD is potentially very sensitive, comparable to that of photomultipliers. However, it has to be noted that once the avalanche at infinite gain is initiated, further photons eventually detected during pulse duration and circuit recovery time is ignored. From this point of view, the SPAD is similar to a Geiger counter than a photomultiplier.

9.5 SCHOTTKY-BARRIER PHOTODIODES

Schottky-barrier photodiodes have been studied quite extensively and have also found application as ultraviolet, visible, and IR detectors [123–132]. These devices reveal some advantages over p-n junction photodiodes: fabrication simplicity, absence of high-temperature diffusion processes, and high speed response.

9.5.1 Schottky–Mott Theory and Its Modifications

According to a simple Schottky–Mott model, the rectifying property of the metal-semiconductor contact arises from the presence of an electrostatic barrier between the metal and the semiconductor, which is due to the difference in work functions ϕ_m and ϕ_s of the metal and semiconductor, respectively. For example, for a metal contact with an n-type semiconductor, ϕ_m should be greater, while for a p-type semiconductor it should be less than ϕ_s. The barrier heights in both these cases, shown in Figure 9.31a and 9.31b, are given by

$$\phi_{bn} = \phi_m - \chi_s \tag{9.141}$$

and

$$\phi_{bp} = \chi_s + E_g - \phi_m, \tag{9.142}$$

respectively, where χ_s is the electron affinity of the semiconductor. The potential barrier between the interior of the semiconductor and the interface, known as band bending, is given by

$$\psi_s = \phi_m - \phi_s \tag{9.143}$$

in both cases. If $\phi_b > E_g$, the layer of the p-type semiconductor adjacent to the surface is inverted in type and we have a p-n junction within the material. However, in practice the built-in barrier does not follow such a simple relationship with ϕ_m and is effectively reduced due to interface states

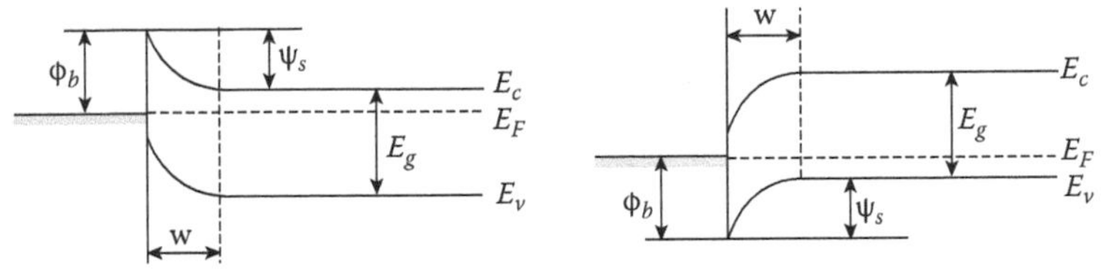

Figure 9.31 Equilibrium energy band diagram of Schottky-barrier junctions: (a) metal-(n-type) semiconductor, and (b) metal-(p-type) semiconductor.

originating either from surface states or from metal-induced gap states and/or due to interface chemical reactions of metal and semiconductor atoms.

In the literature, there are numerous and considerable variations among experimental data on ϕ_m [128,130]. Their analysis indicates an empirical relationship of the type

$$\phi_b = \gamma_1 \phi_m + \gamma_2, \tag{9.144}$$

where γ_1 and γ_2 are constants characteristic of the semiconductors. Two limit cases, namely $\gamma_1 = 0$ and $\gamma_1 = 1$ indicative of Bardeen barrier (when influence of localized surface states is decisive) and ideal Schottky barrier, respectively, can be visualized from such an empirical relation. It has also been pointed out by various workers that the slope parameter $\gamma_1 = \partial \phi_b / \partial \phi_m$ can be used for describing the extent of Fermi level stabilization or pinning for a given semiconductor. The parameters γ_1 and γ_2 have been used by some workers for estimating the interface state density.

It was shown by Cowley and Sze that, according to the Bardeen model, the barrier height in the case of an n-type semiconductor is given approximately by [133]

$$\phi_{bn} = \gamma(\phi_m - \chi_s) + (1 - \gamma)(E_g - \phi_o) - \Delta\phi, \tag{9.145}$$

where $\gamma = \varepsilon_i/(\varepsilon_i + q\delta D_s)$. The ϕ_o is the position of the neutral level of the interface states measured from the top of the valence band, $\Delta\phi$ is the barrier lowering due to image forces, δ is the thickness of the interfacial layer, and ε_i its total permittivity. The surface states are assumed to be uniformly distributed in energy within the bandgap, with a density D_s per electron-volt per unit area. If there are not surface states, $D_s = 0$, and neglecting $\Delta\phi$, Equation 9.145 reduces to Equation 9.141. If the density of states is very high, γ becomes very small and ϕ_{bn} approaches the value $E_g - \phi_o$. This is because a very small deviation of the Fermi level from the neutral level can produce a large dipole moment, which stabilizes the barrier height by a sort of negative feedback effect. The Fermi level is pinned relative to the band edges by the surface states.

A similar analysis for the case of a p-type semiconductor shows that ϕ_{bp} is approximately given by

$$\phi_{bp} = \gamma(E_g - \phi_m + \chi_s) + (1 - \gamma)\phi_o. \tag{9.146}$$

Hence if ϕ_{bn} and ϕ_{bp} refer to the same metal on n- and p-type specimens of the same semiconductor, we should have

$$\phi_{bn} + \phi_{bp} \cong E_g, \tag{9.147}$$

if the semiconductor surface is prepared in the same way in both cases, so that δ, ε_i, D_s, and ϕ_o are the same. This relationship holds quite well in practice. It is usually true that $\phi_{bn} > E_g/2$, and $\phi_{bn} > \phi_{bp}$.

9.5.2 Current Transport Processes

The current transport in metal-semiconductor contacts is due mainly to majority carriers, in contrast to p–n junctions, where current transport is due mainly to minority carriers. The current can be transported in various ways under forward bias conditions as shown in Figure 9.32. The four processes are [129]:

1. Emission of electrons from the semiconductor over the top of the barrier into the metal

2. Quantum mechanical tunneling through the barrier

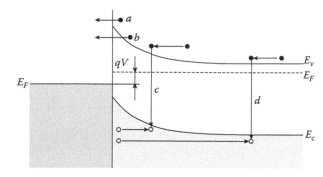

Figure 9.32 Four basic transport processes in forward-biased Schottky barrier on an n-type semiconductor. (From Rhoderick, E. R., "Metal-Semiconductor Contacts," *IEE Proceedings* 129, 1–14, 1982. With permission.)

3. Recombination in the space-charge region

4. Recombination in the neutral region (equivalent hole injection from the metal to the semiconductor)

The inverse processes occur under reverse bias. In addition, we may have edge leakage current due to a high electric field at the contact periphery or interface current due to traps at the metal-semiconductor interface.

The transport of electrons over the potential barrier have been described by various theories, namely: diffusion [134,135], thermionic emission [136], and unified thermionic emission-diffusion [133]. It is now widely accepted that, for high-mobility semiconductors with impurity concentrations of practical interest, the thermionic emission theory appears to explain qualitatively the experimentally observed *I-V* characteristics [137]. Some workers [138–140] have also included in the simple thermionic theory the quantum effects (i.e., quantum-mechanical reflection and tunneling of carriers through the barrier) and have tried to obtain modified analytical expressions for the current-voltage relation. This, however, has essentially led to a lowering of the barrier height and a rounding-off of the top.

The thermionic emission theory by Bethe [136] is derived from the assumptions that the barrier height is much larger than *kT*, thermal equilibrium is established at the plane that determines emission, and the existence of a net current flow does not affect this equilibrium. Bethe's criterion for the slope of the barrier is that the barrier must decrease by more than *kT* over a distance equal to the scattering length. The resulting current flow will depend only on the barrier height and not on the width, and the saturation current is not dependent on the applied bias. Then the current density of majority carriers from the semiconductor over the potential barrier into the metal is expressed as

$$J_{MSt} = J_{st}\left[\exp\left(\frac{qV}{\beta kT}\right) - 1\right],$$ (9.148)

where saturation current density

$$J_{st} = A^* T^2 \exp\left(-\frac{\phi_b}{kT}\right)$$ (9.149)

and $A^* = 4\pi q k^2 m^*/h^3$ equal $120(m^*/m)$ Acm^{-2}K^{-2}, is the Richardson constant, m^* is the effective electron mass, and β is an empirical constant close to unity. Equation 9.148 is similar to the transport equation for p-n junctions. However, the expression for the saturation current densities are quite different.

It appears that the diffusion theory is applicable to low-mobility semiconductors and the current density expressions of the diffusion and thermionic emission theories are basically very similar. However, the saturation current density for the diffusion theory

$$J_{sd} = \frac{q^2 D_e N_c}{kT}\left[\frac{q(V_{bi} - V)2N_d}{\varepsilon_o \varepsilon_s}\right]$$ (9.150)

varies more rapidly with the voltage but is less sensitive to temperature compared with the saturation current density J_{st} of the thermionic emission theory.

A synthesis of the thermionic emission and diffusion approaches described above has been proposed by Crowell and Sze [137]. They assumed Bethe's criterion for the validity of the thermionic emission (the mean free path should exceed the distance within which the barrier falls by kT/q from its maximum value) and also took into account the effects of optical phonon scattering in the region between the top of the barrier and the metal and of the quantum mechanical reflection of electrons that have sufficient energy to surmount the barrier. Their combined effect is to replace the Richardson constant A^* with $A^{**} = f_p f_q q A^*$, where f_p is the probability of an electron reaching the metal without being scattered by an optical phonon after having passed the top of the barrier, and f_q is the average transmission coefficient. The f_p and f_q depend on the maximum electric field in the barrier, the temperature, and the effective mass. Generally speaking, the product $f_p f_q$ is on the order of 0.5.

In respect to the band diagram shown in Figure 9.31, at smaller values of photon energy, such that $q\phi_b < h\nu < E_g$, electrons photoexcited in the metal can surmount the barrier by thermionic emission, transit across the semiconductor, and be collected at the contact. This process extends the spectral response of the diode to photon energies lower than the bandgap. However, since the thermionic emission process can be slow, it is not a very desirable mode of operation. The most efficient mode of operation is when photon energy is greater than the energy gap ($h\nu > E_g$). If the metal layer is semitransparent, photons absorbed in the semiconductor create the electron-hole pairs, which are moving in opposite directions with their respective saturation and are collected.

Knowing J_{st} with Equation 9.149, $R_o A$ can be calculated from

$$(R_o A)_{MS} = \left(\frac{dJ_{MSt}}{dV}\right)^{-1}_{|V=0} = \frac{kT}{qJ_{st}} = \frac{k}{qA^*T} \exp\left(\frac{\phi_b}{kT}\right). \tag{9.151}$$

The current responsivity may be written in the form

$$R_i = \frac{q\lambda}{hc}\eta \tag{9.152}$$

and the voltage responsivity

$$R_v = \frac{q\lambda}{hc}\eta R, \tag{9.153}$$

where $R = (dI/dV)^{-1}$ is the differential resistance of the photodiode.

At this point it is important to discuss some significant differences between photoconductive, p-n junction, and Schottky-barrier detectors.

The photoconductive detectors exhibit the important advantage of the internal photoelectric gain, which relaxes requirements to a low noise preamplifier. The advantages of p-n junction detectors relative to photoconductors are: low or zero bias currents; high impedance, which aids coupling to readout circuits in focal plane arrays; capability for high-frequency operation and the compatibility of the fabrication technology with planar-processing techniques. In comparison with Schottky barriers, the p-n junction photodiodes also indicate some important advantages. The thermionic emission process in a Schottky barrier is much more efficient than the diffusion process and therefore for a given built-in voltage, the saturation current in a Schottky diode is several orders of magnitude higher than in the p-n junction. In addition, the built-in voltage of a Schottky diode is smaller than that of a p-n junction with the same semiconductor. However, the high-frequency operation of p-n junction photodiodes is limited by the minority-carrier storage problem. The Schottky-barrier structures are majority-carrier devices and therefore have inherently fast responses and large operating bandwidths. In other words, the minimum time required to dissipate the carriers injected by the forward bias is dictated by the recombination lifetime. In a Schottky barrier, electrons are injected from the semiconductor into the metal under forward bias if the semiconductor is n-type. Next they thermalize very rapidly ($\approx 10^{-14}$ s) by carrier–carrier collisions, and this time is negligible compared to the minority-carrier recombination lifetime. There are examples of photodiodes with bandwidths in excess of 100 GHz. The diode is usually operated under a reverse bias.

9.5.3 Silicides

Most contacts used in semiconductor devices are subjected to heat treatment. This may be deliberate, to promote adhesion of the metal to the semiconductor, or unavoidable, because high

Table 9.5: Properties of Silicides

Silicide	Resistivity ($\mu\Omega$cm)	Formation Temperature (°C)	Å of Si Per Å of Metal	Å of Silicide Per Å of Metal
$CoSi_2$	18–25	> 550	3.64	3.52
$MoSi_2$	80–250	> 600	2.56	2.59
$NiSi_2$	≈ 50	750	3.65	3.63
Pd_2Si	30–35	> 400	0.68	≈ 1.69
PtSi	28–35	600–800	1.32	1.97
$TaSi_2$	30–45	> 600	2.21	2.40
TiS_2	14–18	> 700	2.27	2.51
WSi_2	30–70	> 600	2.53	2.58

temperatures are needed for other processing stages that occur after the metal is deposited. It is important to avoid the melting of rectifying contacts, because if this happens the interface may become markedly nonplanar, with sharp metallic spikes projecting into the semiconductor that can cause tunneling through the high-field region at the tip of the spike and may severely degrade the electrical characteristics.

The effect of heat on metal-silicon contacts is particularly important if the metal is capable of forming a silicide, which is a stoichiometric compound. Most metals, including all the transition metals, form silicides after appropriate heat treatment (see Table 9.5). These silicides may form as a result of solid-state reactions at temperatures of about one-third to one-half the melting point of the silicide in degrees Kelvin [139]. Studies on a number of transition metal silicides-Si systems have revealed that ϕ_b decreases almost linearly with the eutectic temperature [140]. The vast majority of silicides exhibit metallic conductivity so that, if a metallic silicide is formed as a result of heat treatment of a metal–silicon contact, the silicide–silicon junction behaves like a metal-semiconductor contact and may exhibit rectifying properties. Moreover, because the silicide–silicon interface is formed some distance below the original surface of the silicon it is free from contamination and very stable at room temperature. They also exhibit very good mechanical adhesion. Contacts formed in this way generally show stable electrical characteristics, which are very close to ideal [141,142]. The unique feature of silicide–silicon devices is compatible with the silicon planar processing technology.

In 1973 Shepherd and Yang, of the Rome Air Development Center, Hanscom AFB, Massachusetts, proposed the concept of silicide Schottky-barrier detector FPAs as a much more reproducible alternative to HgCdTe FPAs for infrared imaging [143]. Since then, the development of silicide Schottky-barrier FPA technology has progressed from the demonstration of the initial concepts in the 1970s to the development of high-resolution scanning and staring FPAs that are being considered for many applications for infrared imaging in the 1–3 and 3–5 μm spectral bands [144]. PtSi/ p–Si detectors are well situated for the 3–5 μm wavelength range. Alternative silicides of interest are Pd_2Si and IrSi; Pd_2Si Schottky barriers exhibit a cutoff wavelength $\lambda = 3.7$ μm that does not match to the IR transparent atmospheric window. IrSi Schottky barriers exhibit a low barrier energy and cutoff wavelengths reported in the range from $\lambda = 7.3$ to 10.0 mm [145].

More information about properties and technology of silicide Schottky-barrier detectors and FPAs can be found in Chapter 12.

9.6 METAL-SEMICONDUCTOR–METAL PHOTODIODES

Another form of metal-semiconductor photodiode is the metal-semiconductor-metal or MSM photodiode illustrated in Figure 9.33 [80,110,111]. This structure is physically similar to the interdigitated photoconductor that is illustrated in Figure 9.33b except that the metal-semiconductor and semiconductor-metal junctions are fabricated as Schottky barriers instead of ohmic contacts. Being a planar structure, the MSM photodiode lends itself to monolithic integration and can be fabricated using processing steps nearly identical to those required for making field-effect transistors [146].

The MSM photodiode is essentially a pair of Schottky diodes connected back-to-back. Absorbed photons generate electron-hole pairs in the semiconductor. The holes drift with the applied electric field to the negative contacts while electrons drift to the positive contacts. The quantum efficiency of the MSM photodiode is dependent on the shadowing caused by the metal electrodes.

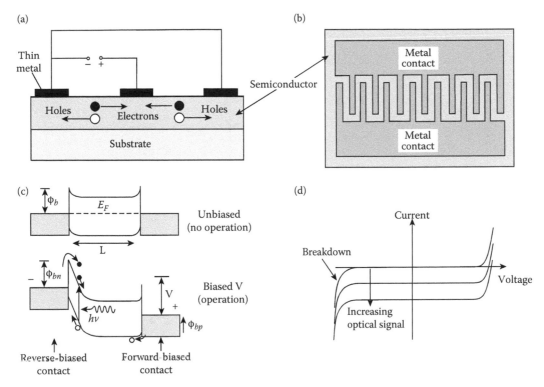

Figure 9.33 Metal-semiconductor-metal photodiode: (a) cross-section structure, (b) top view, (c) unbiased (nonoperational) and biased (operational) energy band diagrams, and (d) current-voltage characteristics.

Dual Schottky barriers provide a lower dark current than a Schottky diode alone, and the capacitance of a MSM diode is smaller than that of a p-i-n photodiode, which enhances the response speed. The MSM diode operates under a bias, which forward-biases one contact while reverse-biasing the other. Diagrams of the biased and unbiased energy band states are shown in Figure 9.33c. At low biases, electron injection at the reverse-biased contact dominates the conduction mechanism. At higher biases, this conduction is supplemented by the injection of holes at the forward-biased contact. Figure 9.33d illustrates the I-V characteristics for an MSM photodiode for three levels of illumination. The depletion region at the reverse-biased contact is much larger than the depleted region near the forward-biased contact, but when they meet the device is said to achieve the "reach through" condition. Although internal gain has been observed for these structures, it has not been fully understood and modeled [110].

For example, Figure 9.34 shows the current-voltage characteristics of a GaAs MSM device with WSi_x contacts [147]. The dark current is of order 1 nA, which is comparable to a p-i-n photodiode. For both polarities the capacitance decreases with bias up to the reach-through voltage, after which it remains nearly unchanged. The increase of the responsivity with bias (see Figure 9.34c) indicates an internal gain, even at low bias values. The latter precludes the avalanche multiplication effect. It is suggested that the gain at low biases can be operative due to traps or surface defects having a long lifetime. The barrier for hole transit can be lowered by electrons accumulated at the conduction band minimum.

9.7 MIS PHOTODIODES

The metal-insulator-semiconductor (MIS) structure is the most useful device in the study of semiconductor surface. This structure has been extensively studied because it is directly related to most planar devices and integrated circuits. Before the 1970s, MIS devices as IR detectors had no importance whatever. In 1970, Boyle and Smith [148] published a paper reporting the charge-coupling principle, which is a simple, extremely powerful concept based upon the transfer of charge packets in a MIS structure. In the area of IR technique, the use of CTD holds the key to

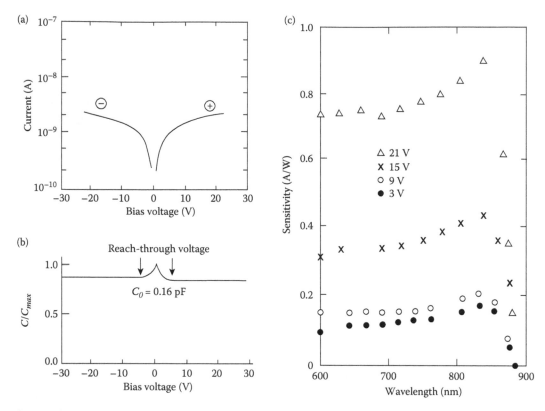

Figure 9.34 Characteristics of GaAs MSM photodiode with tungsten silicide Schottky contacts: (a) dark current characteristics, (b) capacitance-voltage characteristics, and (c) spectral responsivity as a function of bias. (From Ito, M., and Wada, O., *IEEE Journal of Quantum Electronics* QE-62, 1073–77, 1986. With permission.)

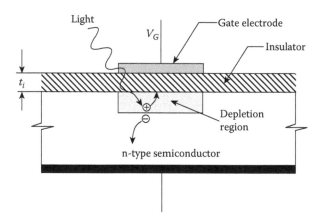

Figure 9.35 MIS structure.

substantial improvements in thermal imaging. The general theory of MIS capacitors is reviewed in many excellent papers (e.g., [86,149–151]).

In this section only p-channel MIS with n-type substrate will be considered, although all of the arguments can be extended to n-channel MIS with appropriate changes.

The simple MIS devices consist of a metal gate separated from a semiconductor surface by an insulator of thickness t_i and dielectric constant ε_i (Figure 9.35). By applying a negative voltage V_b to the metal electrode, electrons are repelled from the I-S interface, creating a depletion region

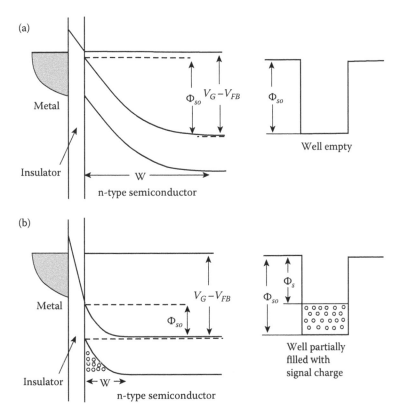

Figure 9.36 Energy band diagram for a p-channel MIS structure: (a) band bending at deep depletion and the empty potential well representation, and (b) band bending with mobile charge at the I-S interface and the partially filled potential well representation.

(Figure 9.36). In effect a potential well for minority carriers (holes) is created. The surface potential Φ_s is related to the gate voltage and other parameters by [152–154]

$$\Phi_s = V_G - V_{FB} + \frac{qN}{C_i} - \frac{qN_d \varepsilon_o \varepsilon_s}{C_i^2} + \frac{1}{C_i}\left[-2q\varepsilon_o \varepsilon_s N_d \left(V_G - V_{FB} + \frac{qN}{C_i} \right) + \left(\frac{qN_d \varepsilon_o \varepsilon_s}{C_i} \right)^2 \right]^{1/2}, \qquad (9.154)$$

where V_{FB} is the flatband voltage, $C_i = \varepsilon_o \varepsilon_i / t$ is the insulator capacitance per unit area, N is the number of mobile electrons in the inversion layer per unit area, $N_d = n_o$ is the substrate doping density and ε_s is the dielectric constant of the semiconductor. From Equation 9.154 we can see that the surface potential can be controlled by the proper choice of gate voltage, doping density, and insulator thickness.

Initially, no charge is present in the potential well ($N = 0$ in Equation 9.154) resulting in a relatively large surface potential Φ_{so}. However, carriers are collected in the potential well as a result of photon absorption, injection from an input diffusion, thermal generation, or tunneling, and the potential across the insulator and the semiconductor will be redistributed as shown in Figure 9.36b. In steady state the potential well is completely filled up and the surface potential assumes its final value [80]

$$\Phi_{sf} \approx 2\Phi_F = -\frac{2kT}{q}\ln\left(\frac{n_o}{n_i} \right) = -\frac{kT}{q}\ln\left(\frac{n_o}{p_o} \right). \qquad (9.155)$$

Φ_F is the potential difference between the bulk Fermi level and the intrinsic Fermi level. It should be noticed that at this surface potential strong inversion begins (Figure 9.37 [86]), and surface concentration of holes becomes greater than the bulk majority carrier concentration.

The thermal current-generation mechanisms for MIS structure are similar to those previously discussed for p-n junction (see Section 9.2). For n-type material [86],

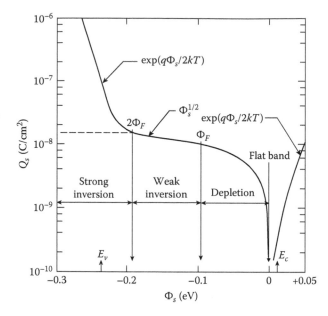

Figure 9.37 Variation of space-charge density $Q_s = qN$ as a function of surface potential Φ_s for 0.25 eV p-channel MIS HgCdTe with $N_d = 2 \times 10^{15}$ cm^{-3}, $T = 77$ K. (From Kinch, M. A., *Semiconductors and Semimetals*, 18, 313–78, Academic Press, New York, 1981. With permission.)

$$J = qn_i \left(\frac{n_i L_h}{N_d \tau_h} + \frac{w_d(t)}{\tau} + \frac{1}{2} S \right) + \eta q \Phi_b + J_t \tag{9.156}$$

where the terms in brackets represent the currents due to thermal generation in the neutral bulk, the depletion region and via interface states. The interface state generation velocity, S, can be greatly reduced by maintaining a bias charge of "flat zero" at the interface at all time. Syllaios and Colombo [155] have found a strong correlation between the dark current and the dislocation density in both MWIR and LWIR HgCdTe MIS devices. The dislocations are thought to act as recombination centers and degrade the lifetime, and these defects limit breakdown bias voltage of HgCdTe MIS devices. The third term in Equation 9.156 is the current due to the background flux, and the condition for background-limited performance is that this should be greater than the dark current. The fourth term is due to tunneling of carriers from the valence band to the conductance band across the bandgap. The considerations carried out by Kinch [86] indicate that the dominant dark current in HgCdTe MIS devices, for reasonable surface recombination velocities ($s < 10^2$ cm/s), is due to generation in the depletion region.

Equation 9.155 can be used to calculate the maximum charge density that can be stored on a MIS capacitor. Solving Equation 9.154 for $N = N_{max}$ gives

$$N_{max} = \frac{C_i}{q} \left(V_G - V_{FB} - 2\Phi_{sf} - V_B \right), \tag{9.157}$$

where $V_B = (4q\varepsilon_o \varepsilon_s \Phi)^{1/2}/C_i$. Typically, V_{FB}, $2\Phi_{sf}$, and V_B are smaller than V_b, therefore $N_{max} \approx C_i V_b / q$. If we assume similar values for C_i and V_b to those achieved in the Si/SiO$_2$ structure then the maximum stored charge is $\approx 10^{12}$ electrons/cm^2. In practice, the maximum storage capacity is taken to about 0.5 N_{max}, to prevent the stored charge from diffusing away and is around 0.4×10^{12} electrons/cm^2 at 4.5 μm and 0.15×10^{12} electrons/cm^2 at 9.5 μm [20].

From Equation 9.157 results that the storage capacity can be increased by increasing the insulator capacitance, or the gate voltage, or by reducing the doping in semiconductor. However, the electric field in the semiconductor must be kept below breakdown. A common breakdown mechanism in narrow bandgap semiconductors is tunneling. Calculations carried out by Anderson for MIS devices of HgCdTe [156], InSb and PbSnTe indicate that tunneling current increases rapidly as the bandgap is reduced and this current impose a severe limitation for cutoff wavelengths greater than 10 µm. Unlike thermal processes, the tunneling current cannot be reduced by cooling. Goodwin et al. [157] have shown that the tunneling currents can be greatly reduced in HgCdTe by growing a heterojunction, which is arranged such that the highest fields occur in the wide band-gap material.

The maximum storage time for an unilluminated device can be approximated by

$$\tau_c = \frac{Q_{max}}{J_{dark}} = \frac{qN_{max}}{J_{dark}} \tag{9.158}$$

and its typical value at 77 K varies from 100 s at 4.5 µm to 100 µs at 10 µm [86]. The storage time is a critical parameter in the operation of a CTD because it establishes the minimum frequency of operation. A long storage time is obtained by: reducing the number of bulk generation centers, reducing the number of surface states, and by reducing the temperature. Serious limitation of the storage time in narrow-gap semiconductors is caused by tunneling.

The total capacitance of the idealized MIS device (with $V_{FB} = 0$) is a series combination of the insulator capacitance C_i and the semiconductor depletion layer capacitance C_d

$$C = \frac{C_i C_d}{C_i + C_d}, \tag{9.159}$$

where

$$C_d = \left(\frac{\varepsilon_o \varepsilon_s q^2}{2kT}\right)^{1/2} \frac{n_{no}\left[\exp(q\Phi_s/kT)-1\right] - p_{no}\left[\exp(-q\Phi_s/kT)-1\right]}{\left\{n_{no}\left[\exp(q\Phi_s/kT) - q\Phi_s/kT - 1\right] - p_{no}\left[\exp(-q\Phi_s/kT) + q\Phi_s/kT - 1\right]\right\}^{1/2}}. \tag{9.160}$$

The theoretical variation of C versus gate voltage V_b for an idealized p-channel MIS HgCdTe with $E_g = 0.25$ eV and $n_o = 10^{15}$ cm^{-3} is shown in Figure 9.38 [86]. At positive gate voltage we have an accumulation of electrons and therefore C_d is large. As the voltage is reduced to the negative value a depletion region is formed near the I–S interface and total capacitance decreases. The capacitance achieves a minimum value and then increases until it again approximates C_i in the strong inversion range. Note that the increase of the capacitance in the negative voltage region depends on the ability of minority carriers to follow the applied AC signal. MIS capacitance-voltage curves measured at higher frequencies do not show the increase of capacitance in this voltage region. Figure 9.38 also shows the capacitance under deep depletion, sufficiently fast pulse conditions, which is directly related to the operation of a CCD.

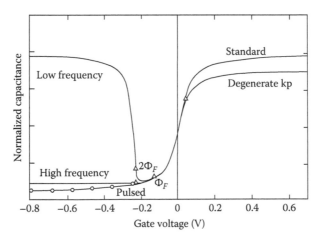

Figure 9.38 Capacitance versus gate voltage for p-channel MIS HgCdTe at 77 K, $E_g = 0.25$ eV, $N_d = 10^{15}$ cm^{-3}, $\Phi_F = 0.092$ eV, $C_i = 2.1 \times 10^{-7}$ F/cm^2. (From Kinch, M. A., *Semiconductors and Semimetals*, 313–78, Academic Press, New York, 1981. With permission.)

For the case of an incident photon flux, surface inversion occurs for lower values of Φ_s than the thermal equilibrium case. The associated depletion width will also narrower for the incident photon flux case. The relationship between p_o and Φ_s is given by Equation 9.155 to be

$$\Delta\Phi_{sf} = \frac{kT}{q}\frac{\Delta p_o}{p_o}.$$ (9.161)

thus a change of Δp_o due to a change in photon flux results in a corresponding change in $\Delta\Phi_{sf}$.

The last expression can be recast into more familiar terms. Consider the case of the MIS device limited by diffusion current, the impedance of the diode region is given by (see Section 9.2.1)

$$RA = \frac{kT}{qJ_D} = \frac{kT\tau_h}{q^2 p_o L_h}.$$ (9.162)

Assuming that Δp_o is due to an incident signal photon flux Φ_s then $\Delta p_o = \eta\Delta\Phi_s\tau_h/L_h$, and substituting in Equations 9.161 and 9.162 we have

$$\Delta\Phi_{sf} = \eta q\Phi_s RA.$$ (9.163)

Thus, the change in surface potential due to an incident photon flux is exactly that expected for an open-circuit photodiode of impedance R (see Equation 9.76).

9.8 NONEQUILIBRIUM PHOTODIODES

The major drawback of IR detectors is the need for cooling to suppress thermal generation of free carriers, which results in noise. Elliott and other British scientists [158,159] have proposed a new approach to reduce the detector cooling requirements, which is based on the nonequilibrium mode of operation. Their concept relies on the suppression of the Auger processes by decreasing the free carrier concentration below its equilibrium values. This can be achieved, for example, in a biased l-h or heterojunction contacts. The above possibilities have been demonstrated in the case of HgCdTe n-type photoconductors [159,160] and photodiodes [161], and InSb photodiodes [162].

The nonequilibrium devices are based on a near-intrinsic, narrow gap, epitaxial layer that is contained between two wider gap layers or between one wider gap layer and one very heavily doped layer. Examples are P-π-N, P-π-N$^+$, P-ν-N$^+$, where the capital letter means wide gap, the + symbol indicates high doping in excess of 10^{17} cm^{-3}, π is near intrinsic p, and ν is near intrinsic n. These devices contain one p-n junction, which is operated in reverse bias producing extraction of minority carriers. The other, isotype junction is excluding to minority carriers preventing their injection into the π or ν layer. For example, let us consider a heterostructure P-π-N shown in Figure 9.39 [163]. Both P and N regions are transparent to photons with energy close to or just above the band gap of the π region, so that both can act as windows. In equilibrium, the electron and hole concentrations in the π region, n_o and p_o are close to the intrinsic value, n_i, which is typically 10^{16}–10^{17} cm^{-3} as illustrated in Figure 9.39c.

Even at zero bias, the device structure shown in Figure 9.39 has two important advantages over homojunction for IR detector:

- The noise generation is confined to the active volume (the wide gap regions have very low thermal generation rates and isolate the active region from carrier generation at the contacts).

- The doping level and type in the active region can be chosen to maximize carrier lifetime and minimize noise.

However, larger improvements of detector performance are expected under reverse bias as a result of the following phenomena:

- Minority carrier exclusion and extraction occur at the P-π and π-P junctions, respectively; as a result both carrier types in active region decrease (minority electron density by several orders, while the majority hole concentration falls to the net doping level) as shown in Figure 9.39c.

- As a consequence of these processes, the thermal generation involving Auger processes falls, so that the saturation current (I_s) is less than would be expected from the zero bias resistance (R_o); that is, $I_s < kT/qR_o$ (see Equation 9.83) and a region of negative conductance is predicted to occur [164].

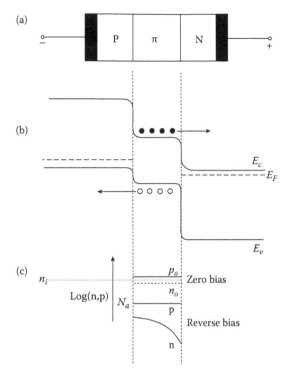

Figure 9.39 Schematic drawing of an extracting P-π-N heterostructure photodiode: (a) multi-layer structure, (b) band edges under reverse bias, and (c) current densities in the π region. (From Elliott, C. T., "Advanced Heterostructures for In$_{1-x}$Al$_x$Sb and Hg$_{1-x}$Cd$_x$Te Detectors and Emiters," *Proceedings of SPIE* 2744, 452–62, 1996. With permission.)

At the present stage of device technology, the Auger suppression nonequilibrium photodiodes suffer from large $1/f$ noise and hence the improvement in the detectivity resulting from the reduced leakage currents can only be realized at high frequencies. However, this is not a problem for heterodyne systems operated at higher intermediate frequencies.

9.9 NBN DETECTOR

Recently, a new concept of infrared detector named the nBn detector has been proposed by Maimon and Wicks [165]. This type of detector can be implemented in different semiconductor materials. Its practical application has been demonstrated in InAs, InAsSb [165], and InAs/GaSb type-II superlattices [166,167].

The nBn structure is shown in Figure 9.40 [165]. It consists of a n-type narrow-bandgap thin contact layer, a 50–100 nm thick wide-bandgap layer with a barrier for electrons and no barrier for holes, and a thick n-type narrow-bandgap absorbing layer. The high barrier layer is thick enough so that there is negligible electron tunneling through it (i.e., 5–100 nm thick and height over 1 eV). As a result, the majority carrier current between two contacts is blocked by the large energy offset, while there is no barrier for photogenerated minority carriers. In fact, this type of device operates as a "minority carrier photoconductor."

Due to a new heterostructure device design and processing, the nBn detectors demonstrate promising results for suppression of surface leakage currents. For example, the insert of Figure 9.41 shows the nBn InAs structure after standard processing [165]. The detector is defined by etching the contact layer with a selective etchant that stops at the barrier. No other layers are etched. Gold contact is deposited on the contact layer and on the substrate and the active layer is covered with the barrier layer. As a result, an additional surface passivation can be eliminated. This is a major advantage compared to InAs-InAsSb-GaSb material system photodiodes, as there is no good passivation.

The InAs device consists of three MBE-grown layers: 3-μm thick InAs ($N_d \sim 2 \times 10^{16}$ cm^{-3}), a 100 nm thick AlAsSb barrier and n-type InAs contact layer ($N_d \sim 1 \times 10^{18}$ cm^{-3}). The growth

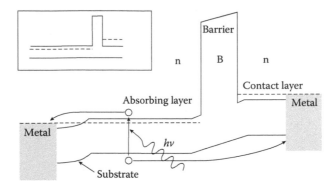

Figure 9.40 Band diagram of an nBn structure (InAs-based), biased under operating conditions. Inset: flatband condition in the barrier junction. (From Maimon, S., and Wicks, G. W., *Applied Physics Letters*, 89, 151109, 2006. With permission.)

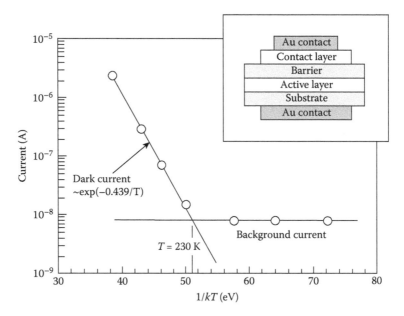

Figure 9.41 Arrhenius plot of the current of an InAs nBn exposed to room temperature background radiation via 2π steradians. Inset: schematic of an nBn device. (From Maimon, S., and Wicks, G. W., *Applied Physics Letters*, 89, 151109, 2006. With permission.)

was made on InAs substrate lattice matched to GaSb. As Figure 9.41 shows, at higher temperatures, the dark device current exhibits a thermal activation energy of 0.439 eV close to InAs band gap. At temperatures below 230 K, the device is operating in BLIP conditions. The BLIP temperature is at least 100 K higher than that of commercial InAs photodiodes [165]. Greatly reduced cooling requirements and higher room temperature performance are the benefits of the nBn detectors.

9.10 PHOTOELECTROMAGNETIC, MAGNETOCONCENTRATION, AND DEMBER DETECTORS

Aside from photoconductive detectors and photodiodes, three other junctionless devices have been used mainly for uncooled IR photodetectors: photoelectromagnetic (or PEM) detectors, magnetoconcentration detectors, and Dember effect detectors. Certainly they are niche devices, but they are still in production and used with important applications, including very fast uncooled detectors with long-wavelength IR radiation. The devices have been reviewed by Piotrowski and Rogalski [38,168].

9.10.1 Photoelectromagnetic Detectors

The first experiments on the PEM effect were performed with Cu_2O by Kikoin and Noskov in 1934 [169]. Nowak's monographic paper summarizes the results of his investigations on the PEM effect [170], which has been replicated worldwide during the last 60 years. For a long time, the PEM effect has been used mostly for InSb room temperature detectors in the middle- and far-IR band [171]. However, the uncooled InSb devices with cutoff wavelength at $\approx 7\ \mu m$ exhibit no response in the 8–14 μm atmospheric window and relatively modest performance in the 3–5 μm window. $Hg_{1-x}Cd_xTe$ and closely related $Hg_{1-x}Zn_xTe$ and $Hg_{1-x}Mn_xTe$ alloys made it possible to optimize performance of PEM detectors at any specific wavelength [172].

9.10.1.1 PEM Effect

The PEM effect is caused by diffusion of photogenerated carriers due to the photo-induced carrier in-depth concentration gradient and by deflection of electron and hole trajectories in opposite directions by the magnetic field (Figure 9.42). If the sample ends are open-circuit in the x-direction, a space charge builds up that gives rise to an electric field along the x axis (open-circuit voltage). If the sample ends are short-circuited in the x-direction, a current flows through the shorting circuit (short-circuit current). In contrast to photoconductors and PV devices, the generation of PEM photovoltage (or photocurrent) requires not simply optical generation, but instead, the formation of an in-depth gradient of photogenerated carriers. Typically, this is accomplished by nonhomogeneous optical generation due to radiation absorption in the near surface region of the device.

Generally, the carrier transport in PEM devices cannot be adequately described by analytical methods and therefore require a numerical solution. Consider the assumptions that make it possible to obtain analytic solutions: homogeneity of the semiconductor, nondegenerate statistics, negligible interface and edge effects, independence of material properties on the magnetic and electric fields, and equal Hall and drift mobilities.

The transport equations for electrons and holes both in x and y take the form:

$$J_{hx} = qp\mu_h E_x + \mu_h B J_{hy}, \tag{9.164}$$

$$J_{hy} = qp\mu_h E_y - \mu_h B J_{hx} - qD_h \frac{dp}{dy}, \tag{9.165}$$

$$J_{ex} = qn\mu_e E_x - \mu_e B J_{ey}, \text{ and} \tag{9.166}$$

$$J_{ey} = qn\mu_e E_y + \mu_e B J_{ex} + qD_e \frac{dn}{dy}, \tag{9.167}$$

where B is the magnetic field in the z direction, E_x and E_y are the x and y components of the electric field, J_{ex} and J_{ey} are the x and y components of the electron current density, and J_{hx} and J_{hy} are the analogous components of the hole current density.

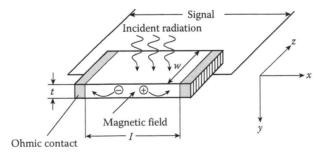

Figure 9.42 Schematic of PEM effect.

The E_y can be eliminated in Equations 9.164 through 9.167 with the condition

$$J_y = J_{ey} + J_{hy} = 0. \tag{9.168}$$

The other equation to be used is the continuity equation for y-direction currents:

$$\frac{dJ_{hy}}{dy} = -\frac{dJ_{ey}}{dy} = q(G - R), \tag{9.169}$$

where G and R denote the carrier generation and recombination rates, respectively.

As a result, a nonlinear second order differential equation for p can be obtained from the set of transport equations for electrons and holes,

$$A_2 \frac{d^2 p}{dy^2} + A_1 \left(\frac{dp}{dy} \right)^2 + A_0 \frac{dp}{dy} - (G - R) = 0, \tag{9.170}$$

where A_2, A_1, and A_0 are coefficients dependent on semiconductor parameters, and electric and magnetic fields. Equation 9.170 with boundary conditions for the front and back side surfaces determines the hole distribution in the y-direction. The electron concentration can be calculated from electric quasineutrality equations. In consequence, x-direction currents and electric fields can be calculated.

9.10.1.2 Lile Solution

Lile reported analytical solution for the small-signal steady-state PEM photovoltage. The voltage responsivity of the PEM detector that can be derived from the Lile solution is [173,174]

$$R_v = \frac{\lambda}{hc} \frac{B}{wt} \frac{\alpha z(b+1)}{n_i(b+z^2)} \frac{Z(1-r_1)}{Y(a^2 + \alpha^2)}, \tag{9.171}$$

where $b = \mu_e/\mu_h$, $z = p/n_i$, w and t are the width and thickness of the detector, r_1 is the front reflectance, and a is the reciprocal diffusion length in the magnetic field equal to

$$a = \left[\frac{(1 - \mu_e^2 B^2)z^2 + b(1 + \mu_h^2 B^2)}{L_e^2(z^2 + 1)} \right]^{1/2}, \tag{9.172}$$

Z and Y are rather complicated functions of semiconductor parameters [174].

The sheet resistivity of the PEM detector is

$$R = \frac{zl}{qn_i\mu_h(b+z^2)wt} \left[1 - \frac{b(b+1)^2 z^2 \mu_h^2 B^2}{a^2 L_e^2(1-z^2)(b+z^2)} \right]^{-1}. \tag{9.173}$$

Since the PEM detector is not biased, the Johnson–Nyquist noise is the only noise of the device

$$V_j = (4kTR\Delta f)^{1/2} \tag{9.174}$$

so the detectivity can be calculated from

$$D^* = \frac{R_v(A\Delta f)^{1/2}}{V_j}. \tag{9.175}$$

The PEM photovoltage is generated along the length of the detector, so the signal linearly increases with the length of the detector and is independent on the device widths for the same photon flux density. This results in a good responsivity for large area devices, in contrast to conventional junction photovoltaic devices.

Analysis of Equation 9.171 indicates that the maximum voltage responsivities can be reached in strong magnetic fields ($B \approx 1/\mu_e$) for samples with high resistance. In the case $\mu_e/\mu_h \gg 1$, the resistance of the detector reaches its maximum value at the point $p/n_i \approx (\mu_e/\mu_h)^{1/2}$ and the highest value for R_v is reached for lightly doped p-type material [173,174]. The acceptor concentration in narrow gap semiconductors is adjusted to a level of about $(2–3) \times 10^{17}$ cm^{-3}. At room temperature the ambipolar diffusion length in narrow gap semiconductors is small (several µm) while the absorption of radiation is relatively weak ($1/\alpha \approx 10$ µm). In such cases the radiation is almost uniformly absorbed within the diffusion length. Thus, a low recombination velocity at the front surface and a high recombination velocity at the back surface is necessary for a good PEM detector response. In such devices the polarity of signal reverses with a change of illumination direction from the low to the high recombination velocity surface, while the responsivity remains almost unchanged.

Figure 9.43 shows the properties of detectivity-optimized room temperature 10.6 µm Hg$_{1-x}$Cd$_x$Te PEM detectors as a function of doping [38,172,174]. The best performance is achieved with p-type material doped to $\approx 2 \times 10^{17}$ cm^{-3}. Due to the high mobility of minority carriers in p-type devices, a magnetic field of ≈ 2 T is sufficient for good performance. The voltage responsivity of the detectivity-optimized device is ≈ 0.6 V/W. The maximum theoretical detectivity of uncooled 10.6 µm device is $\approx 3.4 \times 10^7$ cmHz$^{1/2}$/W.

The response time of the PEM detector may be determined either by the RC time constant or by the decay time of the gradient of excess charge carrier concentration [172]. Typically, the RC time constant of uncooled long wavelength devices is low (<0.1 ns) due to the small capacitance of high frequency optimized devices (≈ 1 pF or less) and low resistance (≈ 50 Ω).

The decay of gradient of carrier's concentration may be caused by volume recombination or ambipolar diffusion. While the first mechanism reduces excess concentration, the second one tends to make the excess concentration uniform. Therefore, the response time can be significantly shorter than the recombination time if the thickness is shorter than the diffusion length. The resulting response time is

$$\frac{1}{\tau_{ef}} = \frac{1}{\tau} + \frac{2D_a}{t^2}.$$ (9.176)

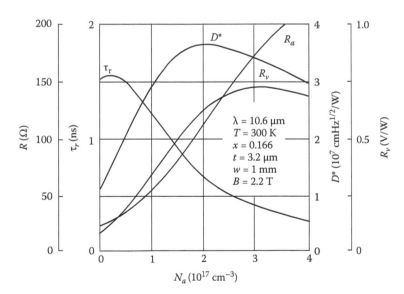

Figure 9.43 Sheet resistivity, voltage responsivity, detectivity, and response time of uncooled 10.6 µm Hg$_{1-x}$Cd$_x$Te PEM detectors as a function of acceptor doping. (From Piotrowski, J., and Rogalski, A., *High-Operating Temperature Infrared Photodetectors*, SPIE Press, Bellingham, WA, 2007. With permission.)

For a p-type HgCdTe device with a thickness t of 2 μm the response time is $\approx 5 \times 10^{-11}$ s. Even shorter response times are achieved with thinner layers, but at the cost of radiation absorption, quantum efficiency, and detectivity.

9.10.1.3 Fabrication and Performance

Only epitaxial devices are manufactured at present [175,176]. The preparation of PEM detectors is essentially very similar to that of photoconductive ones, with the exception of the back surface preparation, which in PEM devices is subjected to a special mechanochemical treatment to achieve a high recombination velocity. This procedure is not mandatory when using graded gap structures. Electrical contacts are usually made by Au/Cr deposition and gold wires are then attached. Figure 9.44 shows the cross section of a sensitive element of a PEM detector.

Figure 9.45 schematically shows the housings of PEM detectors, which are based on standard TO-5 or, for larger elements, TO-8 transistor cans. Frequently, the PEM detectors are accommodated in housing dedicated for high frequency operation. The active elements are mounted in a housing, which incorporates a miniature two-element permanent magnet and pole pieces. Magnetic fields approaching 2 T are achievable with the use of modern rare-earth magnetic materials for the permanent magnet and cobalt steel for the pole pieces.

The best PEM devices exhibit measured voltage responsivity of exceeding 0.15 V/W (width of 1 mm) and detectivities of 1.8×10^7 cmHz$^{1/2}$/W, a factor of ≈ 2 below the predicted ultimate value [174]. The reason for this is a lower (than optimum) magnetic field and faults in detector construction, which are probably nonoptimum surface processing and material composition/doping profile.

The fast response of PEM detectors has been confirmed by observations of CO_2 laser self-mode-locking and free-electron laser experiments [38]. When detecting ordinary or low repetition rate short pulses, signal voltages up to ≈ 1 V are obtained with 1 mm long detectors, being limited by strong optical excitation effects. The maximum signal voltages for chopped CO_2 radiation are much lower due to radiation heating. In good heat dissipation design devices they exceed 30 mV per mm.

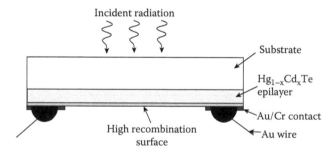

Figure 9.44 Cross section of a back side illuminated sensitive element of PEM detector.

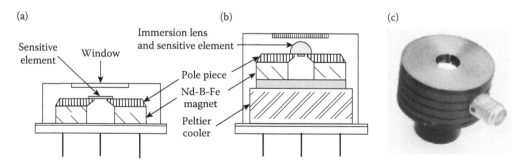

Figure 9.45 Schematic of the (a) housing of an ambient temperature and (b) a thermoelectrically cooled, optically immersed PEM detector, and (c) a picture of high frequency optimized specialized housing. (From Piotrowski, J., and Rogalski, A., *High-Operating Temperature Infrared Photodetectors*, SPIE Press, Bellingham, WA, 2007. With permission.)

Both theoretical and measured performance of PEM detectors are inferior to these of PC detectors. PEM detectors have, however, additional important advantages, which make them useful in many applications. In contrast to photoconductors, they do not require electric bias. The frequency characteristics of PEM detector are flat over a wide frequency range, starting from zero frequency. This is due to the lack of the low-frequency noise and very short response time. The resistance of PEM detectors does not decrease with increasing size, which makes it possible to achieve the same performance for small and large area devices. With the resistance typically close to 50 Ω, the devices are conveniently coupled directly to wideband amplifiers.

PEM detectors have been also fabricated with other HgTe-based ternary alloys; $Hg_{1-x}Zn_xTe$ and $Hg_{1-x}Mn_xTe$ [38]. It seems that the use of $Hg_{1-x}Zn_xTe$ and $Hg_{1-x}Mn_xTe$ offers no advantages when compared to $Hg_{1-x}Cd_xTe$ in terms of performance and speed of response.

9.10.2 Magnetoconcentration Detectors

If a semiconductor plate is electrically biased and placed in a crossed magnetic field, then the spatial distribution of electron-hole pairs along the crystal section deviates from the equilibrium value. It is so called magnetoconcentration effect. Such redistribution is efficient in the plate whose thickness is comparable with ambipolar diffusion length of charge carriers. The Lorentz force deflects electrons and holes drifting in the electric field in the same direction, resulting in an increase of concentration of carriers at one surface and decrease at the opposite one, in dependence on the direction of magnetic and electric field.

The sample shown in Figure 9.46 with small recombination velocity at the front side surface and a high at the back side is especially interesting for IR detector and source application. When Lorentz force moves carriers toward the high recombination velocity surface (depletion mode), the carriers recombine there, resulting in depletion in the volume with exception of the region close to the high surface recombination surface. With opposite direction of the Lorentz force (enrichment mode), the carriers are moved toward the low recombination velocity surface being replenished by generation at the high recombination velocity surface. Changes of the carrier concentration will result in positive or negative luminescence [177–179].

Djuric and Piotrowski proposed to use depletion mode magnetoconcentration effect to suppress the Auger [180–182]. They have performed numerical simulations of the magnetoconcentration devices and reported first practical devices. Theoretical analysis of magnetoconcentration devices can be done only by the numerical solution of the set of transport equations for electrons and holes in the presence of a magnetic field (see Equations 9.164 through 9.167). The steady-state numerical analysis was performed solving the equations using the fourth-order Runge-Kutta method. For more detail see [38].

Practical 10.6 µm uncooled and thermoelectrically cooled magnetoconcentration detectors have been reported [183]. The expected shape of the *I-V* curve with current saturation, negative resistance, and oscillation regions has been observed. The devices exhibited a large low-frequency noise when biased to achieve a sufficient depletion of semiconductor. Anomalously high noise has been generated in the region of negative resistance. Significant reduction of noise and improvement of performance was observed at high frequencies (>100 kHz) with careful selection of bias current. With such properties the devices are suitable only for some broadband applications and further efforts are necessary to make them useful in typical applications.

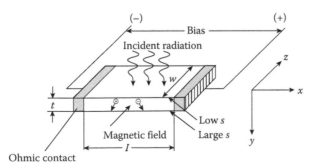

Figure 9.46 Magnetoconcentration effect (depletion mode). Electrons and holes are pushed by action of the Lorentz force toward back side surface with a high recombination velocity.

9.10.3 Dember Detectors

Dember detectors are a type of photovoltaic devices, based on bulk photodiffusion voltage in a simple structure with only one type of semiconductor doping supplied with two contacts [184]. When radiation is incident on the surface of a semiconductor generating electron-hole pairs, a potential difference is usually developed in the direction of the radiation (see Figure 9.47) as a result of difference in diffusion of electrons and holes. The Dember effect electrical field restrains the electrons with higher mobility, while holes are accelerated, thus making both fluxes equal.

The Dember effect device can be analyzed solving the transport and continuity equations assuming zero total currents in x and y directions. The steady-state photovoltage under conditions of weak optical excitation and assuming electroneutrality can be expressed as [185]

$$V_d = \int_0^t E_z(z)dz = \frac{kT}{q}\frac{\mu_e - \mu_h}{n_o\mu_e + p_o\mu_h}\left[\Delta n(0) - \Delta n(t)\right], \tag{9.177}$$

where E_z is electric field in the z-direction and $\Delta n(z)$ is the excess electron concentration. As this expression shows, two conditions are required for generation of the photovoltage: the distribution of photogenerated carriers should be nonuniform and the diffusion coefficients of electrons and holes must be different. The gradient may result from nonuniform optical generation or/and from different recombination velocities at the front and back surfaces of the device.

For boundary conditions at the top and bottom surface

$$D_a\frac{\partial\Delta n}{\partial z} = s_1\Delta n(0) \text{ at } z = 0 \tag{9.178}$$

$$D_a\frac{\partial\Delta n}{\partial z} = s_2\Delta n(t) \text{ at } z = t \tag{9.179}$$

the voltage responsivity can be expressed as [186]

$$\begin{aligned}
R_v A = C\{&-\left[A(\alpha) + r_2 e^{-2\alpha t}A(-\alpha)\right]\sinh(t/L_a)\\
&+\left[B(\alpha) + r_2 e^{-2\alpha t}B(-\alpha)\right]\times\left[1 - \cosh(t/L_a)\right] + \left(1 - e^{-\alpha t}\right)\left(1 - r_2 e^{-2\alpha t}\right)\}\\
&+\left(1 - e^{-\alpha t}\right)\left(1 - r_2 e^{-\alpha t}\right)\},
\end{aligned} \tag{9.180}$$

where

$$C = \frac{\lambda\alpha\tau}{hc}\frac{\mu_e - \mu_h}{n\mu_e + p\mu_h}\frac{(1 - r_1)kT}{q(1 - \alpha^2 L_a^2)(1 - r_1 r_2 e^{-2\alpha t})} \tag{9.181}$$

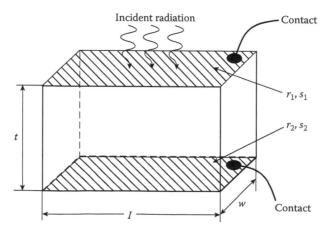

Figure 9.47 Schematic of Dember detector.

$$A(\alpha) = \frac{(\alpha L_a + \Gamma_1)\left[\sinh(t/L_a) + \Gamma_2 \cosh(t/L_a)\right] + \Gamma_1(\alpha L_a - \Gamma_2)e^{-\alpha t}}{(\Gamma_1 + \Gamma_2)\cosh(t/L_a) + (1 + \Gamma_1\Gamma_2)\sinh(t/L_a)}$$ (9.182)

$$B(\alpha) = \frac{(\alpha L_a - \Gamma_2)e^{-\alpha t} - (\Gamma_1 + \alpha L_a)\left[\cosh(t/L_a) + \Gamma_2 \sinh(t/L_a)\right]}{(\Gamma_1 + \Gamma_2)\cosh(t/L_a) + (1 + \Gamma_1\Gamma_2)\sinh(t/L_a)}$$ (9.183)

$$\Gamma_1 = \frac{s_1 L_a}{D_a}$$ (9.184)

$$\Gamma_1 = \frac{s_2 L_a}{D_a}.$$ (9.185)

The analysis of Equation 9.180 shows the maximum response can be achieved for:

- Optimized p-type doping

- A low surface recombination velocity and a low reflection coefficient at illuminated side contact

- A large recombination velocity and a large reflection coefficient at nonilluminated side contact

Since the device is not biased and the noise voltage is determined by the Johnson–Nyquist thermal noise so the detectivity (see Equation 9.175) can be calculated from Equation 9.180 and the expression for the Johnson–Nyquist noise (see Equation 9.174).

The theoretical design of $Hg_{1-x}Cd_xTe$ and practical Dember effect detectors have been reported [172,183,186]. The best performance is achievable for a device thickness slightly larger than the ambipolar diffusion length. More thin devices exhibit low voltage responsivity while more thick have excessive resistance and large related Johnson noise.

Figure 9.48 shows the calculated resistivity, detectivity, and bulk recombination time for the uncooled 10.6 μm Dember detector with detectivity-optimized thickness [172]. As in the case of PEM detectors, the best performance is achieved in p-type material. The calculated detectivity of Dember detectors is comparable to that of photoconductors operated under the same conditions. Detectivities as high as $\approx 2.4 \times 10^8$ cmHz$^{1/2}$/W and of $\approx 2.2 \times 10^9$ cmHz$^{1/2}$/W are predicted for optimized 10.6 μm devices at 300 and 200 K, respectively.

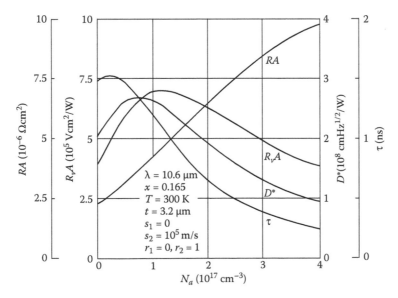

Figure 9.48 The calculated normalized resistivity (RA), normalized responsivity (R_vA), detectivity (D^*) and bulk recombination time (τ) of uncooled 10.6-μm $Hg_{1-x}Cd_xTe$ Dember detector as a function of acceptor concentration. (From Piotrowski, J., and Rogalski, A., *High-Operating Temperature Infrared Photodetectors*, SPIE Press, Bellingham, WA, 2007. With permission.)

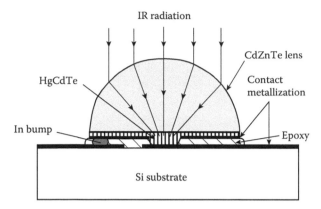

Figure 9.49 Schematic cross section of an experimental monolithic optically immersed Dember detector. (From Piotrowski, J., and Rogalski, A., *High-Operating Temperature Infrared Photodetectors*, SPIE Press, Bellingham, WA, 2007. With permission.)

The interesting feature of the Dember device is a significant photoelectric gain larger at zero bias condition. For the optimum doping the gain is about 1.7 and increases in decreasing thickness. The gain is caused by the ambipolar effects. At zero bias and shortened device, the photo-generated electrons may travel several times between contacts before the holes will recombine or diffuse to the back side contact.

Very low resistances, low voltage responsivities and noise voltages well below the noise level of the best amplifiers pose serious problems in achieving the potential performance, however. For example, the optimized uncooled 10.6 μm devices with only 7×7 μm² size will have a resistance of ≈ 7 Ω, voltage responsivity of ≈ 82 V/W and noise voltage of ≈ 0.35 nV/Hz$^{1/2}$. The low resistance and the low voltage responsivity is the reason why the simple Dember detectors cannot compete with other types of photodetectors.

There are some potential ways to overcome this difficulty. One is the connection of small area Dember detectors in series. The other possibility is to use optical immersion [38].

Due to ambipolar diffusion of excess carriers, the response time of a Dember detector is shorter than the bulk recombination time and can be very short for devices with thickness much less than the diffusion length. The response time of detectivity optimized devices is about 1 ns, shortening with p-type doping, with some expense in performance. The high-frequency optimized Dember detectors should be prepared from heavily doped material using optical resonant cavity.

Figure 9.49 shows a schematic of a small size optically immersed Dember effect detector manufactured by Vigo Systems [38]. The sensitive element has been prepared from a $Hg_{1-x}Cd_xTe$ graded gap epilayers grown by the isothermal vapor phase epitaxy and supplied with hyperhemispherical lens formed directly in the transparent (Cd,Zn)Te substrate. The surface of the sensitive element was subjected to a special treatment to produce a high recombination velocity and covered with reflecting gold. The "electrical" size of the sensitive element is 7×7 μm² while, due to the hyperhemispherical immersion, the apparent optical size is increased to $\approx 50 \times 50$ μm². Multiple cells connected in series Dember detectors have been used for large area devices.

HgCdTe Dember detectors have found an application in high-speed laser beam diagnostics of industrial CO_2 lasers and other applications requiring fast speed of operation [187].

9.11 PHOTON-DRAG DETECTORS

In heavily doped semiconductors, free carrier absorption is the dominant absorption mechanism for wavelengths longer than the absorption edge. An incident stream of photons having wavelengths consistent with free carrier absorption will transfer momentum to free carriers, effectively pushing them in the direction of the Poynting vector. Thus a longitudinal electric field is established within the semiconductor that can be detected through electrodes attached to the sample. The transfer of momentum from photons to free carriers in semiconductors is called photon-drag effect. The radiation pressure builds up a voltage difference between the front and back surfaces. The photon-drag detector structure is shown in Figure 9.50. When the device is operated with a high-impedance amplifier, the net current is zero, an electric field is set-up to oppose the photon-drag force and the potential change along the bar provides the output signal.

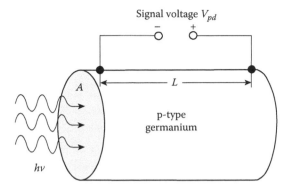

Figure 9.50 Photon-drag detector structure.

In the classical model, when the photon energy $h\nu \ll kT$ and absorption is determined by free carriers, the photon drag voltage can be estimated by considering the momentum p possessed by each photon, $p = E/c$, where E is photon energy and c is the speed of light. The rate of change of momentum in x-direction per unit volume is given by

$$\frac{dp(x)}{dt} = \frac{n_r P \alpha \exp(-\alpha x)}{Ac},$$ (9.186)

where P is the power of the light flux, which falls on the cross sectional A. In open-circuit condition, the average rate of change of momentum to each carrier must be balanced by an electromotive force acting on the carrier (taking n-type as an example)

$$qE(x) = \frac{n_r P \alpha \exp(-\alpha x)}{Acn}.$$ (9.187)

Integrating this longitudinal electric field over the entire length, L, the photon-drag voltage is equal

$$V_{pd} = \frac{n_r P[1 - \exp(-\alpha L)]}{qAcn}.$$ (9.188)

The last equation is found to obey only for wavelengths longer than a few hundred micrometers. For shorter wavelengths, responsivity is wavelength dependent and V_{pd} value is less than that given by Equation 9.188.

The microscopic theory of the classical photon-drag effect was developed by Gurevich and Rumyantsev in 1967 [188]. The theory of the photon drag given by Gibson and Montasser is valid in the wavelength range 1–10 µm [189]. A photon-drag effect on quantum-well structures, arising from intersubband transition rather than interbad transition, was suggested by Grinberg and Luryi [190]. The photon-drag detectors are described in more detail in Gibson and Kimmitt [191].

In general, the mechanism controlling the photon drag effect is complicated especially in the case when light absorption is determined by interband optical transitions. Due to participation of the lattice in the momentum conservation law, the drag of carriers can be opposite to the light propagation direction. As a result, the sign of the photon-drag voltage changes with the wavelength in the range 1–10 µm [192].

The absorption of p-type germanium at $\lambda = 10.6$ µm is due to electron transitions from the light-hole band to vacant states in the heavy-hole band. Photoconductivity due to these intraband transitions was first observed in the heavily doped sample by Feldman and Hergenrother [193]. The photon-drag effect was used for the detection of short CO_2-laser pulses by Gibson et al. [194] and Danishevskii et al. [195]. Gibson et al. showed that a Q-switched CO_2 laser working at 10.6 µm wavelength can transfer sufficient momentum to produce a longitudinal electromagnetic force in a rectangular germanium rod of 4 cm length [194]. Similar results have been achieved using p-type tellurium [196].

The responsivity measured experimentally is very low, typically in the range 1–40 µV/W, but the devices are extremely fast (less than 1 ns) and they operate at room temperature. At the shortest wavelengths, p-GaAs has the best sensitivity (see Figure 9.51) [191]. For 2–11 µm, n-GaP is a good choice. The peak response around 3 µm is due to optical rectification and severe absorption sets extends to around 12 µm. At longer wavelengths (Figure 9.52) p-type silicon appears to be a good available detector. The nature of the direct valence band transitions leads to high photon drag coefficients.

Detectors exploiting the photon-drag effect are useful as laser detectors because of their rapid response and ability to absorb large amounts of power without damage because their absorption is small (the radiation is absorbed over a large volume of material). Signal linearity with power is good up to a power density of ≈ 50 MW/cm². Damage of the detector starts to occur around 100 MW/cm². Their lack of sensitivity makes them of little use as infrared detectors for most applications.

Recently, interest in photon-drag detectors has increased and is stimulated by impressive progress in the development of THz detector technology [197]. For THz detection p-type Ge detectors are not well suited. The reason is that due to direct intersubband transitions in the

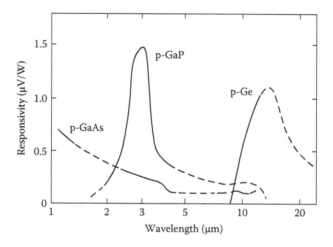

Figure 9.51 Responsivity of 3.2 Ωcm p-Ge, 5 Ωcm n-GaP, and 2.5 Ωcm p-Ge longitudinal detectors. All the detectors have an active area of 4 × 4 mm, a resistance of 50 Ω, and are oriented in a 111 direction. Responsivity of 1 µV/W is equivalent to a NEP of ≈ 10^{-3} W/Hz$^{1/2}$. (From Gibson, A. F., and Kimmitt, M. F., *Infrared and Millimeter Waves*, Academic Press, New York, Vol. 3, 182–219, 1980. With permission.)

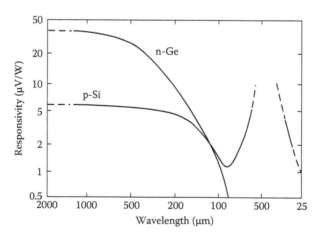

Figure 9.52 Responsivity of 30 Ωcm n-Ge and p-Si, longitudinal detectors oriented in a 100 direction. The active area is 4 × 4 mm and the detector resistance are n-Ge, 250 Ω, and p-Si, 350 Ω. Responsivity of 10 µV/W is equivalent to a NEP of ≈ 2×10^{-4} W/Hz$^{1/2}$ for n-Ge and 2.6×10^{-4} W/Hz$^{1/2}$ for p-Si. (From Gibson, A. F., and Kimmitt, M. F., *Infrared and Millimeter Waves*, Academic Press, New York, Vol. 3, 182–219, 1980. With permission.)

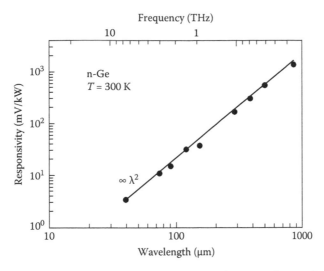

Figure 9.53 Responsivity of an n-type Ge:Sb photon drag detector after 20 dB amplification as a function of wavelength. (From Ganichev, S. D., Terent'ev, Ya., V., and Yaroshetskii, I. D., *Sov. Tech. Phys. Lett.* 11, 20–21, 1985.)

valence band and Drude absorption of free carriers, the spectrum of responsivity is sharply structured with several zeros and sign inversions. In Figure 9.53, the responsivity after 20 dB amplification of an n-type germanium detector is plotted as a function of wavelength [198]. Due to rapid free carrier momentum relaxation at room temperature, the response time is very short and may be less than 1 ps.

REFERENCES

1. A. Rose, *Concepts in Photoconductivity and Allied Problems,* Interscience, New York, 1963.

2. D. Long and J. Schmit, "Mercury-Cadmium Telluride and Closely Related Alloys," in *Semiconductors and Semimetals,* Vol. 5, eds. R. K. Willardson and A. C. Beer, 175–255, Academic Press, New York, 1970.

3. D. Long, "Photovoltaic and Photoconductive Infrared Detectors," in *Optical and Infrared Detectors,* ed. R. J. Keyes, 101–47, Springer-Verlag, Berlin, 1977.

4. W. L. Eisenman, J. D. Merriam, and J. F. Potter, "Operational Characteristics of Infrared Photodetectors," in *Semiconductors and Semimetals,* Vol. 12, eds. R. K. Willardson and A. C. Beer, 1–38, Academic Press, New York, 1977.

5. P. R. Bratt, "Impurity Germanium and Silicon Infrared Detectors," in *Semiconductors and Semimetals,* Vol. 12, eds. R. K. Willardson and A. C. Beer, 39–142, Academic Press, New York, 1977.

6. R. H. Kingston, *Detection of Optical and Infrared Radiation,* Springer-Verlag, Berlin, 1979.

7. R. M. Broudy and V. J. Mazurczyck, "(HgCd)Te Photoconductive Detectors," in *Semiconductors and Semimetals,* Vol. 18, eds. R. K. Willardson and A. C. Beer, 157–99, Academic Press, New York, 1981.

8. N. Sclar, "Properties of Doped Silicon and Germanium Infrared Detectors," *Progress in Quantum Electronics* 9, 149–257, 1984.

9. A. Rogalski and J. Piotrowski, "Intrinsic Infrared Detectors," *Progress in Quantum Electronics* 12, 87–289, 1988.

10. A. Rogalski, "Photoconductive Detectors," in *Infrared Photon Detectors,* ed. A. Rogalski, 13–49, SPIE Optical Engineering Press, Bellingham, WA, 1995.

11. E. L. Dereniak and G. D. Boreman, *Infrared Detectors and Systems*, Wiley, New York, 1996.

12. A. Rogalski, *Infrared Detectors,* Gordon and Breach Science Publishers, Amsterdam, 2000.

13. C. T. Elliott, "Photoconductive and Non-Equilibrium Devices in HgCdTe and Related Alloys," in *Infrared Detectors and Emitters: Materials and Devices,* eds. P. Capper and C. T. Elliott, 279–312, Kluwer Academic Publishers, Boston, 2001.

14. R. A. Smith, *Semiconductors*, Cambridge University Press, Cambridge, 1978.

15. E. S. Rittner, "Electron Processes in Photoconductors," in *Photoconductivity Conference at Atlantic City 1954,* eds. R. G. Breckenridge, B. Russel, and E. Hahn, 215–68, Wiley, New York, 1956.

16. R. L. Williams, "Sensitivity Limits of 0.1 eV Intrinsic Photoconductors," *Infrared Physics* 8, 337–43, 1968.

17. M. A. Kinch, S. R. Borrello, B. H. Breazeale, and A. Simmons, "Geometrical Enhancement of HgCdTe Photoconductive Detectors," *Infrared Physics* 17, 137–45, 1977.

18. J. F. Siliquini, C. A. Musca, B. D. Nener, and L. Faraone, "Performance of Optimized $Hg_{1-x}Cd_xTe$ Long Wavelength Infrared Photoconductors," *Infrared Physics & Technology* 35, 661–71, 1994.

19. Y. J. Shacham-Diamand and I. Kidron, "Contact and Bulk Effects in Intrinsic Photoconductive Infrared Detectors," *Infrared Physics* 21, 105–15, 1981.

20. C. T. Elliott and N. T. Gordon, "Infrared Detectors," in *Handbook on Semiconductors,* Vol. 4, ed. C. Hilsum, 841–936, North-Holland, Amsterdam, 1993.

21. A. Jóźwikowska, K. Jóźwikowski, and A. Rogalski, "Performance of Mercury Cadmium Telluride Photoconductive Detectors," *Infrared Physics* 31, 543–54, 1991.

22. D. L. Smith, D. K. Arch, R. A. Wood, and M. W. Scott, "HgCdTe Heterojunction Contact Photoconductor," *Applied Physics Letters* 45, 83–85, 1984.

23. D. L. Smith, "Effects of Blocking Contacts on Generation-Recombination Noise and Responsivity in Intrinsic Photoconductors," *Journal of Applied Physics* 56, 1663–69, 1984.

24. R. Kumar, S. Gupta, V. Gopal, and K. C. Chhabra, "Dependence of Responsivity on the Structure of a Blocking Contact in an Intrinsic HgCdTe Photoconductor," *Infrared Physics* 31, 101–7, 1991.

25. V. Gopal, R. Kumar, and K. C. Chhabra, "A n^{2+}-n^+-n Blocking Contact Structure for an Intrinsic Photoconductor," *Infrared Physics* 31, 435–40, 1991.

26. T. Ashley and C. T. Elliott, "Accumulation Effects at Contacts to n-Type Cadmium-Mercury-Telluride Photoconductors," *Infrared Physics* 22, 367–76, 1982.

27. W. Van Roosbroeck, "Theory of the Electrons and Holes in Germanium and Other Semiconductors," *Bell Systems Technical Journal* 29, 560–607, 1950.

28. P. W. Kruse, L. D. McGlauchlin, and R. B. McQuistan, *Elements of Infrared Technology,* Wiley, New York, 1962.

29. P. W. Kruse, "The Photon Detection Process," in *Optical and Infrared Detectors,* ed. R. J. Keyes, 5–69, Springer-Verlag, Berlin, 1977.

30. A. Van der Ziel, *Fluctuation Phenomena in Semiconductors*, Butterworths, London, 1959.

31. D. Long, "On Generation-Recombination Noise in Infrared Detector Materials," *Infrared Physics* 7, 169–70, 1967.

32. T. D. Kleinpenning, "1/f Noise in p-n Junction Diodes," *Journal of Vacuum Science and Technology* A3, 176–82, 1985.

33. F. N. Hooge, "1/f Noise Is No Surface Effect," *Physics Letters* 29A, 123–40, 1969.

34. Z. Wei-jiann and Z. Xin-Chen, "Experimental Studies on Low Frequency Noise of Photoconductors," *Infrared Physics* 33, 27–31, 1992.

35. R. D. Nelson, "Infrared Charge Transfer Devices: The Silicon Approach," *Optical Engineering* 16, 275–83, 1977.

36. K. Jóźwikowski and J. Piotrowski, "Ultimate Performance of $Cd_xHg_{1-x}Te$ Photoresistors as a Function of Doping," *Infrared Physics* 25, 723–27, 1985.

37. J. Piotrowski, W. Galus, and M. Grudzień, "Near Room-Temperature IR Photo-Detectors," *Infrared Physics* 31, 1–48, 1991.

38. J. Piotrowski and A. Rogalski, *High-Operating Temperature Infrared Photodetectors*, SPIE Press, Bellingham, WA, 2007.

39. S. Borrello, M. Kinch, and D. LaMont, "Photoconductive HgCdTe Detector Performance with Background Variations," *Infrared Physics* 17, 121–25, 1977.

40. V. Gopal, "Surface Recombination in Photoconductors," *Infrared Physics* 25, 615–18, 1985.

41. A. Kinch and S. R. Borrello, "0.1 eV HgCdTe Photodetectors," *Infrared Physics* 15, 111–24, 1975.

42. E. Burstein, G. Picus, and N. Sclar, "Optical and Photoconductive Properties of Silicon and Germanium," in *Photoconductivity Conference at Atlantic City*, eds. R. Breckenbridge, B. Russell, and E. Hahn, 353–413, Wiley, New York, 1956.

43. R. Newman and W. W. Tyler, "Photoconductivity in Germanium," in *Solid State Physics* 8, eds. F. Steitz and D. Turnbill, 49–107, Academic Press, New York, 1959.

44. E. H. Putley, "Far Infrared Photoconductivity," *Physica Status Solidi* 6, 571–614, 1964.

45. V. F. Kocherov, I. I. Taubkin, and N. B. Zaletaev, "Extrinsic Silicon and Germanium Detectors," in *Infrared Photon Detectors*, ed. A. Rogalski, 189–297, SPIE Optical Engineering Press, Bellingham, WA, 1995.

46. A. W. Hoffman, P. J. Love, and J. P. Rosbeck, "Mega-Pixel Detector Arrays: Visible to 28 μm," *Proceedings of SPIE* 5167, 194–203, 2004.

47. G. H. Rieke, "Infrared Detector Arrays for Astronomy," *Annual Review of Astronomy and Astrophysics* 45, 77–115, 2007.

48. M. M. Blouke, E. E. Harp, C. R. Jeffus, and R. L. Williams, "Gain Saturation in Extrinsic Photoconductors Operating at Low Temperatures," *Journal of Applied Physics* 43, 188–94, 1972.

49. A. F. Milton and M. M. Blouke, "Sweepout and Dielectric Relaxation in Compensated Extrinsic Photoconductors," *Physical Review* 3B, 4312–30, 1971.

50. D. K. Schroder, "Extrinsic Silicon Focal Plane Arrays," in *Charge-Coupled Devices*, ed. D. F. Barbe, 57–90, Springer-Verlag, Heidelberg, 1980.

51. R. W. Westervelt and S. W. Teitsworth, "Nonlinear Transient Response of Extrinsic Ge Far-Infrared Photoconductors," *Journal of Applied Physics* 57, 5457–69, 1985.

52. N. M. Haegel, C. A. Latasa, and A. M. White, "Transient Response of Infrared Photoconductors: The Role of Contacts and Space Charge," *Applied Physics* A56, 15–21, 1993.

53. N. M. Haegel, C. B. Brennan, and A. M. White, "Transport in Extrinsic Photoconductors: A Comprehensive Model for Transient Response," *Journal of Applied Physics* 80, 1510–14, 1996.

54. N. M. Haegel, S. A. Sampei, and A. M. White, "Electric Field and Responsivity Modeling for Far-Infrared Blocked Impurity Band Detectors," *Journal of Applied Physics* 93, 1305–10, 2003.

55. N. M. Haegel, W. R. Schwartz, J. Zinter, A. M. White, and J. W. Beeman, "Origin of the Hook Effect in Extrinsic Photoconductors," *Applied Optics* 40, 5748–754, 2001.

56. R. N. Thomas, T. T. Braggins, H. M. Hobgood, and W. J. Takei, "Compensation of Residual Boron Impurities in Extrinsic Indium-Doped Silicon by Neutron Transmutation of Silicon," *Journal of Applied Physics* 49, 2811–20, 1978.

57. H. M. Hobgood, T. T. Braggins, J. C. Swartz, and R. N. Thomas, "Role of Neutron Transmutation in the Development of High Sensitivity Extrinsic Silicon IR Detector Material," in *Neutron Transmutation in Semiconductors,* ed. J. M. Meese, 65–90, Plenum Press, New York, 1979.

58. G. Lucovsky, "On the Photoionization of Deep Impurity Centers in Semiconductors," *Solid State Communications* 3, 299–302, 1965.

59. N. Sclar, "Extrinsic Silicon Detectors for 3–5 and 8–14 μm," *Infrared Physics* 16, 435–48, 1976.

60. N. Sclar, "Survey of Dopants in Silicon for 2–2.7 and 3–5 μm Infrared Detector Application," *Infrared Physics* 17, 71–82, 1977.

61. E. Bryan, "Operation Temperature of Extrinsic Si Photoconductive Detectors," *Infrared Physics* 23, 341–48, 1983.

62. S. R. Borrello, C. G. Roberts, B. H. Breazeale, and G. R. Pruett, "Cooling Requirements for BLIP Performance of Intrinsic Photoconductors," *Infrared Physics* 11, 225–32, 1971.

63. A. G. Milnes, *Deep Impurities in Semiconductors,* John Wiley, New York, 1973.

64. C. T. Elliott, P. Migliorato, and A. W. Vere, "Counterdoped Extrinsic Silicon Infrared Detectors," *Infrared Physics* 18, 65–71, 1978.

65. A. Rogalski, "Quantum Well Infrared Photodetectors Among the Other Types of Semiconductor Infrared Detectors," *Infrared Physics & Technology* 38, 295–310, 1997.

66. P. Martyniuk and A. Rogalski, "Quantum-Dot Infrared Photodetectors: Status and Outlook," *Progress in Quantum Electronics* 32, 89–120, 2008.

67. N. Sclar, "Temperature Limitation for IR Extrinsic and Intrinsic Photodetectors," *IEEE Transactions on Electron Devices* ED-27, 109–18, 1980.

68. A. P. Davis, C. T. Elliott, and A. M. White, "Current Gain in Photodiode Structures," *Infrared Physics* 31, 575–77, 1991.

69. R. W. Dutton and R. J. Whittier, "Forward Current-Voltage and Switching Characteristics of p⁺-n-n+ (Epitaxial) Diodes," *IEEE Transactions on Electronic Devices* ED-16, 458–67, 1969.

70. H. J. Hovel, *Semiconductors and Semimetals*, Vol. 11, eds. R. K. Willardson and A. C. Berr, Academic Press, New York, 1975.

71. F. Van De Wiele, "Quantum Efficiency of Photodiode," in *Solid State Imaging*, eds. P. G. Jespers, F. Van De Wiele, and M. H. White, 41–76, Noordhoff, Leyden, The Netherlands, 1976.

72. A. Rogalski and J. Rutkowski, "Effect of Structure on the Quantum Efficiency and R_oA Product of Lead-Tin Chalcogenide Photodiodes," *Infrared Physics* 22, 199–208, 1982.

73. Z. Djuric and Z. Jaksic, "Back Side Reflection Influence on Quantum Efficiency of Photovoltaic Devices," *Electronics Letters* 24, 1100–01, 1988.

74. J. Shappir and A. Kolodny, "The Response of Small Photovoltaic Detectors to Uniform Radiation," *IEEE Transactions on Electronic Devices* ED-24, 1093–98, 1977.

75. D. Levy, S. E. Schacham, and I. Kidron, "Three-Dimensional Analytical Simulation of Self- and Cross-Responsivities of Photovoltaic Detector Arrays," *IEEE Transactions on Electronic Devices* ED-34, 2059–70, 1987.

76. M. J. Buckingham and E. A. Faulkner, "The Theory of Inherent Noise in p-n Junction Diodes and Bipolar Transistors," *Radio and Electronic Engineers* 44, 125–40, 1974.

77. P. Knowles, "Mercury Cadmium Telluride Detectors for Thermal Imaging," *GEC Journal of Research* 2, 141–56, 1984.

78. C. T. Sah, R. N. Noyce, and W. Shockley, "Carrier Generation and Recombination in p–n Junctions and p–n Junction Characteristics," *Proceedings of IRE* 45, 1228–43, 1957.

79. S. C. Choo, "Carrier Generation-Recombination in the Space-Charge Region of an Asymmetrical p–n Junction," *Solid-State Electronics* 11, 1069–77, 1968.

80. S. M. Sze, *Physics of Semiconductor Devices*, Wiley, New York, 1981.

81. W. W. Anderson, "Tunnel Contribution to $Hg_{1-x}Cd_xTe$ and $Pb_{1-x}Sn_xTe$ p–n Junction Diode Characteristics," *Infrared Physics* 20, 353–61, 1980.

82. W. A. Beck and N. Y. Byer, "Calculation of Tunneling Currents in (Hg,Cd)Te Photodiodes Using a Two-Sided Junction Potential," *IEEE Transactions on Electronic Devices* ED-31, 292–97, 1984.

83. R. Adar, "Spatial Integration of Direct Band-to-Band Tunneling Currents in General Device Structures," *IEEE Transactions on Electronic Devices* 39, 976–81, 1992.

84. C. T. Sah, "Electronic Processes and Excess Currents in Gold-Doped Narrow Silicon Junctions," *Physical Review* 123, 1594–612, 1961.

85. J. Y. Wong, "Effect of Trap Tunneling on the Performance of Long-Wavelength $Hg_{1-x}Cd_xTe$ Photodiodes," *IEEE Transactions on Electronic Devices* ED-27, 48–57, 1980.

86. M. A. Kinch, "Metal-Insulator-Semiconductor Infrared Detectors," in *Semiconductors and Semimetals*, Vol. 18, eds. R. K. Willardson and A. C. Beer, 313–78, Academic Press, New York, 1981.

87. H. J. Hoffman and W. W. Anderson, "Impurity-to-Band Tunneling in $Hg_{1-x}Cd_xTe$," *Journal of Vacuum Science and Technology* 21, 247–50, 1982.

88. D. K. Blanks, J. D. Beck, M. A. Kinch, and L. Colombo, "Band-to-Band Processes in HgCdTe: Comparison of Experimental and Theoretical Studies," *Journal of Vacuum Science and Technology* A6, 2790–94, 1988.

89. Y. Nemirovsky, D. Rosenfeld, A. Adar, and A. Kornfeld, "Tunneling and Dark Currents in HgCdTe Photodiodes," *Journal of Vacuum Science and Technology* A7, 528–35, 1989.

90. Y. Nemirovsky, R. Fastow, M. Meyassed, and A. Unikovsky, "Trapping Effects in HgCdTe," *Journal of Vacuum Science and Technology* B9, 1829–39, 1991.

91. D. Rosenfeld and G. Bahir, "A Model for the Trap-Assisted Tunneling Mechanism in Diffused n–p and Implanted n^+–p HgCdTe Photodiodes," *IEEE Transactions on Electronic Devices* 39, 1638–45, 1992.

92. Y. Nemirovsky and A. Unikovsky, "Tunneling and $1/f$ Noise Currents in HgCdTe Photodiodes," *Journal of Vacuum Science and Technology* B10, 1602–10, 1992.

93. G. Sarusi, A. Zemel, A. Sher, and D. Eger, "Forward Tunneling Current in HgCdTe Photodiodes," *Journal of Applied Physics* 76, 4420–25, 1994.

94. A. Rogalski, "Analysis of the R_oA Product in n^+-p $Hg_{1-x}Cd_xTe$ Photodiodes," *Infrared Physics* 28, 139–53, 1988.

95. M. B. Reine, A. K. Sood, and T. J. Tredwell, "Photovoltaic Infrared Detectors," in *Semiconductors and Semimetals*, Vol. 18, eds. R. K. Willardson and A. C. Beer, 201–311, Academic Press, New York, 1981.

96. A. Rose, "Space-Charge-Limited Currents in Solids," *Physical Review* 97, 1538–44, 1955.

97. M. A. Lampert, "Simplified Theory of Space-Charge-Limited Currents in an Insulator with Traps," *Physical Review* 103, 1648–56, 1956.

98. M. A. Lampert and P. Mark, *Current Injections in Solids*, eds. H. G. Booker and N. DeClaris, Academic Press, New York, 1970.

99. J. A. Edmond, K. Das, and R. F. Davis, "Electrical Properties of Ion-Implanted p-n Junction Diodes in β-SiC," *Journal of Applied Physics* 63, 922–29, 1988.

100. D. E. Sawyer and R. H. Rediker, "Narrow Base Germanium Photodiodes," *Proceedings of IRE* 46, 1122–30, 1958.

101. G. Lucovsky and R. B. Emmons, "High Frequency Photodiodes," *Applied Optics* 4, 697–702, 1965.

102. W. W. Gartner, "Depletion-Layer Photoeffects in Semiconductors," *Physical Review* 116, 84–87, 1959.

103. G. Kreiser, *Optical Fiber Communications*, McGraw-Hill Book Co., Boston, 2000.

104. A. G. Dentai, R. Kuchibhotla, J. C. Campbell, C. Tasi, and C. Lei, "High Quantum Efficiency, Long-Wavelength InP/InGaAs Microcavity Photodiode," *Electronics Letters* 27, 2125–27, 1991.

105. M. S. Ünlü and M. S. Strite, "Resonant Cavity Enhanced Photonic Devices," *Journal of Applied Physics* 78, 607–39, 1995.

106. G. E. Stillman and C. M. Wolfe, "Avalanche Photodiodes," in *Semiconductors and Semimetals*, Vol. 12, eds. R. K. Willardson and A. C. Beer, 291–393, Academic Press, New York, 1977.

107. F. Capasso, "Physics of Avalanche Diodes," in *Semiconductors and Semimetals*; Vol. 22D, ed. W. T. Tang, 2–172, Academic Press, Orlando, 1985.

108. T. Kaneda, "Silicon and Germanium Avalanche Photodiodes," in *Semiconductors and Semimetals*, Vol. 22D, ed. W. T. Tang, 247–328, Academic Press, Orlando, 1985.

109. T. P. Pearsall and M. A. Pollack, "Compound Semiconductor Photodiodes," in *Semiconductors and Semimetals*, Vol. 22D, ed. W. T. Tang, 173–245, Academic Press, Orlando, 1985.

110. P. Bhattacharya, *Semiconductor Optoelectronics Devices*, Prentice Hall, New Jersey, 1993.

111. S. B. Alexander, *Optical Communication Receiver Design*, SPIE Optical Engineering Press, Bellingham, WA, 1997.

112. J. C. Campbell, S. Demiguel, F. Ma, A. Beck, X. Guo, S. Wang, X. Zheng, et al., "Recent Advances in Avalanche Photodiodes," *IEEE Journal of Selected Topics in Quantum Electronics* 10, 777–87, 2004.

113. S. Donati, *Photodetectors. Devices, Circuits, and Applications*, Prentice Hall, New York, 2000.

114. R. J. McIntryre, "Multiplication Noise in Uniform Avalanche Diodes," *IEEE Transactions on Electronic Devices* ED-13, 164–68, 1966.

115. R. J. McIntyre, "The Distribution of Gains in Uniformly Multiplying Avalanche Photodiodes: Theory," *IEEE Transactions on Electronic Devices* ED-19, 703–13, 1972.

116. Y. Okuto and C. R. Crowell, "Ionization Coefficients in Semiconductors: A Nonlocalized Property," *Physical Review* B10, 4284–96, 1974.

117. J. P. R. David and C. H. Tan, "Material Considerations for Avalanche Photodiodes," *IEEE Journal of Selected Topics in Quantum Electronics* 14, 998–1009, 2008.

118. J. C. Campbell, "Recent Advances in Telecommunications Avalanche Photodiodes," *Journal of Lightwave Technology* 25, 109–21, 2007.

119. G. J. Rees and J. P. R. David, "Why Small Avalanche Photodiodes are Beautiful," *Proceedings of SPIE* 4999, 349–62, 2003.

120. R. B. Emmons, "Avalanche-Photodiode Frequency Response," *Journal of Applied Physics* 38, 3705–14, 1967.

121. S. Melle and A. MacGregor, "How to Choice Avalanche Photodiodes," *Laser Focus World*, 145–56, October 1995.

122. S. Cova, A. Lacaita, M. Ghioni, G. Ripamonti, and T. A. Louis, "20 ps Timing Resolution with Single-Photon Avalanche Diodes," *Review of Scientific Instruments* 60, 1104–10, 1989.

123. H. K. Henish, *Rectifying Semiconductor Contacts*, Clarendon Press, Oxford, 1957.

124. M. M. Atalla, "Metal-Semiconductor Schottky Barriers, Devices and Applications," in *Proceedings of 1966 Microelectronics Symposium* 123–57, Munich-Oldenberg, 1966.

125. F. A. Padovani, "The Voltage-Current Characteristics of Metal-Semiconductor Contacts," in *Semiconductors and Semimetals*, Vol. 7A, eds. R. K. Willardson and A. C. Beer, 75–146, Academic Press, New York, 1971.

126. A. G. Milnes and D. L. Feught, *Heterojunctions and Metal-Semiconductor Junctions*, Academic Press, New York, 1972.

127. V. L. Rideout, "A Review of the Theory, Technology and Applications of Metal-Semiconductor Rectifiers," *Thin Solid Films* 48, 261–291, 1978.

128. E. H. Rhoderick, *Metal-Semiconductor Contacts*, Clarendon Press, Oxford, 1978.

129. E. R. Rhoderick, "Metal-Semiconductor Contacts," *IEE Proceedings* 129, 1–14, 1982.

130. S. C. Gupta and H. Preier, "Schottky Barrier Photodiodes," in *Metal-Semiconductor Schottky Barrier Junctions and Their Applications,* ed. B. L. Sharma, 191–218, Plenum, New York, 1984.

131. W. Monch, "On the Physics of Metal-Semiconductor Interfaces," *Reports on Progress in Physics* 53, 221–78, 1990.

132. R. T. Tung, "Electron Transport at Metal-Semiconductor Interfaces: General Theory," *Physical Review* B45, 13509–23, 1992.

133. A. M. Cowley and S. M. Sze, "Surface States and Barrier Height of Metal-Semiconductor Systems," *Journal of Applied Physics* 36, 3212–20, 1965.

134. W. Schottky and E. Spenke, "Quantitative Treatment of the Space-Charge Boundary-Layer Theory of the Crystal Rectifiers," *Wiss. Veroff. Siemens-Werken.* 18, 225–91, 1939.

135. E. Spenke, *Electronic Semiconductors,* McGraw-Hill, New York, 1958.

136. H. A. Bethe, "Theory of the Boundary Layer of Crystal Rectifiers," *MIT Radiation Laboratory Report* 43-12, 1942.

137. C. R. Crowell and S. M. Sze, "Current Transport in Metal-Semiconductor Barriers," *Solid-State Electronics* 9, 1035–48, 1966.

138. S. Y. Wang and D. M. Bloom, "100 GHz Bandwidth Planar GaAs Schottky Photodiode," *Electronics Letters* 19, 554–55, 1983.

139. J. W. Mayer and K. N. Tu, "Analysis of Thin-Films Structures with Nuclear Backscattering and X-ray Diffraction," *Journal of Vacuum Science and Technology* 11, 86–93, 1974.

140. G. Ottaviani, K. N. Tu, and J. M. Mayer, "Interfacial Reaction and Schottky Barrier in Metal-Silicon Systems," *Physical Review Letters* 44, 284–87, 1980.

141. M. P. Lepselter and S. M. Sze, "Silicon Schottky Barrier Diode with Near-Edeal I–V Characteristics," *Bell Systems Technology Journal* 47, 195–208, 1968.

142. J. M. Andrews and M. P. Lepselter, "Reverse Current-Voltage Characteristics of Metal-Silicide Schottky Diodes," *Solid-State Electronics* 13, 1011–23, 1970.

143. F. D. Shepherd and A. C. Yang, "Silicon Schottky Retinas for Infrared Imaging," *1973 IEDM Technical Digest,* 310–13, Washington, DC, 1973.

144. W. F. Kosonocky, "Review of Infrared Image Sensors with Schottky-Barrier Detectors," *Optoelectronics—Devices and Technologies* 6, 173–203, 1991.

145. W. A. Cabanski and M. J. Schulz, "Electronic and IR-Optical Properties of Silicide/Silicon Interfaces," *Infrared Physics* 32, 29–44, 1991.

146. J. S. Wang, C. G. Shih, W. H. Chang, J. R. Middleton, P. J. Apostolakis, and M. Feng, "11 GHz Bandwidth Optical Integrated Receivers Using GaAs MESFET and MSM Technology," *IEEE Photonics Technology Letters* 5, 316–18, 1993.

147. M. Ito and O. Wada, "Low Dark Current GaAs Metal-Semiconductor-Metal (MSM) Photodiodes using WSi_x Contacts," *IEEE Journal of Quantum Electronics* QE-62, 1073–77, 1986.

148. W. S. Boyle and G. E. Smith, "Charge-Coupled Semiconductor Devices," *Bell Systems Technology Journal* 49, 587–93, 1970.

149. A. S. Grove, *Physics and Technology of Semiconductor Devices*, Wiley, New York, 1967.

150. E. R. Nicollian and J. R. Brews, *MOS Physics and Technology*, Wiley, New York, 1982.

151. D. G. Ong, *Modern MOS Technology: Process, Devices and Design*, McGraw-Hill Book Company, New York, 1984.

152. D. F. Barbe, "Imaging Devices Using the Charge-Coupled Concept," *Proceedings of IEEE* 63, 38–67, 1975.

153. E. S. Young, *Fundamentals of Semiconductor Devices*, McGraw-Hill Book Company, New York, 1978.

154. W. D. Baker, "Intrinsic Focal Plane Arrays," in *Charge-Coupled Devices*, ed. D. F. Barbe, 25–56, Springer, Berlin, 1980.

155. A. J. Syllaios and L. Colombo, "The Influence of Microstructure on the Impedance Characteristics on HgCdTe MIS Devices," *IEDM Technical Digest* 137–48, 1982.

156. W. W. Anderson, "Tunnel Current Limitation of Narrow Bandgap Infrared Charge Coupled Devices," *Infrared Physics* 17, 147–64, 1977.

157. M. W. Goodwin, M. A. Kinch, and R. J. Koestner, "Metal-Insulator-Semiconductor Properties of Molecular-Beam Epitaxy Grown HgCdTe Heterostructure," *Journal of Vacuum Science and Technology* A8, 1226–32, 1990.

158. T. Ashley and C. T. Elliott, "Non-Equilibrium Devices for Infrared Detection," *Electronics Letters* 21, 451–52, 1985.

159. T. Ashley, C. T. Elliott, and A. T. Harker, "Non-Equilibrium Modes of Operation for Infrared Detectors," *Infrared Physics* 26, 303–15, 1986.

160. T. Ashley, C. T. Elliott, and A. M. White, "Non-Equilibrium Devices for Infrared Detection," *Proceedings of SPIE* 572, 123–33, 1985.

161. C. T. Elliott, "Non-Equilibrium Modes of Operation of Narrow-Gap Semiconductor Devices," *Semiconductor Science and Technology* 5, S30–S37, 1990.

162. T. Ashley, A. B. Dean, C. T. Elliott, M. R. Houlton, C. F. McConville, H. A. Tarry, and C. R. Whitehouse, "Multilayer InSb Diodes Grown by Molecular Beam Epitaxy for Near Ambient Temperature Operation," *Proceedings of SPIE* 1361, 238–44, 1990.

163. C. T. Elliott, "Advanced Heterostructures for $In_{1-x}Al_xSb$ and $Hg_{1-x}Cd_xTe$ Detectors and Emiters," *Proceedings of SPIE* 2744, 452–62, 1996.

164. A. M. White, "Auger Suppression and Negative Resistance in Low Gap Pin Diode Structure," *Infrared Physics* 26, 317–24, 1986.

165. S. Maimon and G. W. Wicks, "nBn Detector, an Infrared Detector with Reduced Dark Current and Higher Operating Temperature," *Applied Physics Letters* 89, 151109, 2006.

166. J. B. Rodriguez, E. Plis, G. Bishop, Y. D. Sharma, H. Kim, L. R. Dawson, and S. Krishna, "nBn Structure Based on InAs/GaSb Type-II Strained Layer Superlattices," *Applied Physics Letters* 91, 043514, 2007.

167. G. Bishop, E. Plis, J. B. Rodriguez, Y. D. Sharma, H. S. Kim, L. R. Dawson, and S. Krishna, "nBn Detectors Based on InAs/GaSb type-II Strain Layer Superlattices," *Journal of Vacuum Science and Technology* B26, 1145–48, 2008.

168. J. Piotrowski and A. Rogalski, "Photoelectromagnetic, Magnetoconcentration and Dember Infrared Detectors," in *Narrow-Gap II-VI Compounds for Optoelectronic and Electromagnetic Applications,* ed. P. Capper, 507–25, Chapman & Hall, London, 1997.

169. I. K. Kikoin and M. M. Noskov, "A New Photoelectric Effect in Copper Oxide," *Physik Zeit Der Soviet Union,* 5, 586, 1934.

170. M. Nowak, "Photoelectromagnetic Effect in Semiconductors and Its Applications," *Progress in Quantum Electronics* 11, 205–346, 1987.

171. P. W. Kruse, "Indium Antimonide Photoelectromagnetic Infrared Detector," *Journal of Applied Physics* 30, 770–78, 1959.

172. J. Piotrowski, W. Galus, and M. Grudzień, "Near Room-Temperature IR Photodetectors," *Infrared Physics* 31, 1–48, 1991.

173. D. L. Lile, "Generalized Photoelectromagnetic Effect in Semiconductors," *Physical Review* B8, 4708–22, 1973.

174. D. Genzow, M. Grudzień, and J. Piotrowski, "On the Performance of Noncooled CdHgTe Photoelectromagnetic Detectors for 10,6 µm Radiation," *Infrared Physics* 20, 133–38, 1980.

175. J. Piotrowski, "HgCdTe Detectors" in *Infrared Photon Detectors,* ed. A. Rogalski, 391–493, SPIE Optical Engineering Press, Bellingham, WA, 1995.

176. J. Piotrowski, "Uncooled Operation of IR Photodetectors," *Opto-Electronics Review* 12, 11–122, 2004.

177. P. Berdahl, V. Malutenko, and T. Marimoto, "Negative Luminescence of Semiconductors," *Infrared Physics* 29, 667–72, 1989.

178. V. Malyutenko, A. Pigida, and E. Yablonovsky, "Noncooled Infrared Magnetoinjection Emitters Based on $Hg_{1-x}Cd_xTe$," *Optoelectronics—Devices and Technologies* 7, 321–28, 1992.

179. T. Ashley, C. T. Elliott, N. T. Gordon, R. S. Hall, A. D. Johnson, and G. J. Pryce, "Negative Luminescence from $In_{1-x}Al_xSb$ and $Cd_xHg_{1-x}Te$ Diodes," *Infrared Physics & Technology* 36, 1037–44, 1995.

180. Z. Djuric and J. Piotrowski, "Room Temperature IR Photodetector with Electromagnetic Carrier Depletion," *Electronics Letters* 26, 1689–91, 1990.

181. Z. Djuric and J. Piotrowski, "Infrared Photodetector with Electromagnetic Carrier Depletion," *Optical Engineering* 31, 1955–60, 1992.

182. Z. Djuric, Z. Jaksic, A. Vujanic, and J. Piotrowski, "Auger Generation Suppression in Narrow-Gap Semiconductors Using the Magnetoconcentration Effect," *Journal of Applied Physics* 71, 5706–8, 1992.

183. J. Piotrowski, W. Gawron, and Z. Djuric, "New Generation of Near Room-Temperature Photodetectors," *Optical Engineering* 33, 1413–21, 1994.

184. H. Dember, "Uber die Vorwartsbewegung von Elektronen Durch Licht," *Physik Z.* 32, 554, 856, 1931.

185. J. Auth, D. Genzow, and K. H. Herrmann, *Photoelectrische Erscheinungen,* Akademie Verlag, Berlin, 1977.

186. Z. Djuric and J. Piotrowski, "Dember IR Photodetectors," *Solid-State Electronics* 34, 265–69, 1991.

187. H. Heyn, I. Decker, D. Martinen, and H. Wohlfahrt, "Application of Room-Temperature Infrared Photo Detectors in High-Speed Laser Beam Diagnostics of Industrial CO_2 Lasers," *Proceedings of SPIE* 2375, 142–53, 1995.

188. L. E. Gurevich and A. A. Rumyantsev, "Theory of the Photoelectric Effect in Finite Crystals at High Frequencies and in the Presence of an External Magnetic Field," *Soviet Physics Solid State* 9, 55, 1967.

189. A. F. Gibson and S. Montasser, "A Theoretical Description of the Photon-Drag Spectrum of p-Type Germanium," *Journal of Physics C: Solid State Physics* 8, 3147–57, 1975.

190. A. A. Grinberg and S. Luryi, "Theory of the Photon-Drag Effect in a Two-Dimensional Electron Gas," *Physical Review B* 38, 87, 1987.

191. A. F. Gibson and M. F. Kimmitt, "Photon Drag Detection," in *Infrared and Millimeter Waves*, Vol. 3, ed. K. J. Button, 182–219, Academic Press, New York, 1980.

192. A. F. Gibson and A. C. Walker, "Sign Reversal of the Photon Drag Effect in p Type Germanium," *Journal of Physics C* 4, 2209–19, 1971.

193. J. M. Feldman and K. M. Hergenrother, "Direct Observation of the Excess Light Hole Population in Optically Pumped p-Type Germanium," *Applied Physics Letters* 9, 186, 1966.

194. A. F. Gibson, M. F. Kimmitt, and A. C. Walker, "Photon Drag in Germanium," *Applied Physics Letters* 17, 75–77, 1970.

195. A. M. Danishevskii, A. A. Kastalskii, S. M. Ryvkin, and I. D. Yaroshetskii, "Dragging of Free Carriers by Photons in Direct Interband Transitions," *Soviet Physics JETP* 31, 292, 1970.

196. S. Panyakeow, J. Shirafuji, and Y. Inuishi, "High-Performance Photon Drag Detector for a CO_2 Laser Using p-Type Tellurium," *Applied Physics Letters* 21, 314–16, 1972.

197. S. D. Ganichev and W. Prettl, *Intense Terahertz Excitation of Semiconductors*, Clarendon Press, Oxford, 2005.

198. S. D. Ganichev, Ya. V. Terent'ev, and I. D. Yaroshetskii, "Photon-Drag Photodetectors for the Far-IR and Submillimeter Regions," *Soviet Technical Physics Letters*, 11, 20–21, 1985.

10 Intrinsic Silicon and Germanium Detectors

Silicon is the semiconductor that has dominated the electronic industry for over 40 years. While the first transistor fabricated in Ge and III-V semiconductor material compounds may have higher mobilities, higher saturation velocities, or larger bandgaps, silicon devices account for over 97% of all microelectronics [1]. The main reason is that silicon is the cheapest microelectronic technology for integrated circuits. The reason for the dominance of silicon can be traced to a number of natural properties of silicon but more importantly, two insulators of silicon, SiO_2 and Si_3N_4, allow deposition and selective etching processes to be developed with exceptionally high uniformity and yield.

Photodetectors are perhaps the oldest and best understood silicon photonic devices [2,3]. Recently, the interest in utilizing Si-based optical components to realize a fully monolithic solution for high performance optical interconnects is on the rise [4]. Silicon being an indirect bandgap semiconductor with a centro-symmetric crystalline structure, is not directly suited for optoelectronics. Moreover, a Si bandgap of 1.16 eV prevents its use in the second (1.3 µm) and third (1.55 µm) window of optical fiber communications. Despite these facts, the rather unique success of Si as a semiconductor for electronics has motivated a large amount of research toward the development of silicon-based optoelectronic devices. The cost of silicon integrated circuits has remained constant around 1 U.S. cent per square mm for a number of decades, but the number of their elements (transistors, passive, and other components) have been increasing at an exponential rate with time [1]. In addition, in complementary metal oxide semiconductor (CMOS) architectures now dominated, apart from leakage currents, power is only dissipated when gates are switched. The low leakage currents achievable with silicon insulators and p-n implanted isolation combined with the higher thermal conductivity than many other semiconductors have allowed higher densities in silicon technology than any other technology driving the integrated circuit developments. While microelectronics has dominated the twentieth-century technologies, some of the authors have predicted that silicon photonics will be a major technology in the twenty-first century [2,3]. Currently silicon is becoming an important candidate for optical functionalities.

In this chapter we review the recent achievements of silicon and germanium technologies for the fabrication of near infrared photodetectors. To probe beyond this chapter, readers should consult the excellent monographs [4–7] and reviews [8–10]. Table 10.1 lists the properties of silicon and germanium in room temperatures [4,5].

10.1 SILICON PHOTODIODES

Silicon photodiodes are widely applied in spectral range below 1.1 µm and even are used for X-ray and gamma ray detectors. The main types are as follows:

- p-n junctions generally formed by diffusion (ion implantation is also used),

- p-i-n junctions (because of thicker active region, they have enhanced near-IR spectral response),

- UV- and blue-enhanced photodiodes, and

- avalanche photodiodes.

In the planar photodiode structure (diffused or implanted); cross section is shown in Figure 10.1, the highly doped p$^+$-region is very thin (typically about 1 µm) and is coated with the thin dielectric film (SiO_2 or Si_3N_4) that serves as an antireflection layer. The diffused junction can be formed either by a p-type impurity such as born into a n-type bulk silicon wafer, or the n-type impurity, such as phosphorous, into a p-type bulk silicon wafer. To form an ohmic contact another impurity diffusion (often coupled with implanting technique) into the back side of the wafer is necessary. The contact pads are deposited on the front defined active area, and on the back side, completely covering the surface. An antireflection coating reduces the reflection of the light for specific predefined wavelength. The nonactive on the top is covered with a thick layer of SiO_2. In dependence of the photodiode application, different design structures are used. By controlling the thickness of bulk substrate, both speed response and sensitivity of the photodiode can be controlled (see Section 9.2).

Note that the photodiodes can be operated as unbiased (photovoltaic) or reverse biased (photoconductive) modes (Figure 10.2). The amplifiers function is a simple current to voltage conversion (photodiode operates in a short circuit mode). Mode selection depends upon the speed requirements of the application, and the amount of dark current that is tolerable. The unbiased mode of

Table 10.1: Properties of Si and Ge at 300 K

Properties	Si	Ge
Atoms (cm⁻³)	5.02×10^{22}	4.42×10^{22}
Atomic weight	28.09	72.60
Breakdown field (V/cm)	$\sim 3 \times 10^5$	$\sim 10^5$
Crystal structure	Diamond	Diamond
Density (g/cm³)	2.329	5.3267
Dielectric constant	11.9	16.0
Effective density of states in conduction band (cm⁻³)	2.86×10^{19}	1.04×10^{19}
Effective density of states in valence band (cm⁻³)	2.66×10^{19}	6.0×10^{19}
Effective mass (conductivity)		
Electrons (m_e/m_o)	0.26	0.39
Holes (m_h/m_o)	0.19	0.12
Electron affinity (V)	4.05	4.0
Energy gap (eV)	1.12	0.67
Index of refraction	3.42	4.0
Intrinsic carrier concentration (cm⁻³)	9.65×10^9	2.4×10^{13}
Intrinsic resistivity (Ωcm)	3.3×10^5	
Lattice constant (Å)	5.43102	5.64613
Linear coefficient of thermal expansion (°C⁻¹)	2.59×10^{-6}	5.8×10^{-6}
Melting point (°C)	1412	937
Minority carrier lifetime (μs)		
Electrons (p-type)	800	1000
Holes (n-type)	1000	1000
Mobility (cm²/Vs)		
μ_e (electrons)	1450	3900
μ_h (holes)	505	1900
Optical phonon energy (eV)	0.063	0.037
Specific heat (J/g°C)	0.7	
Thermal conductivity (W/cmK)	1.31	0.31

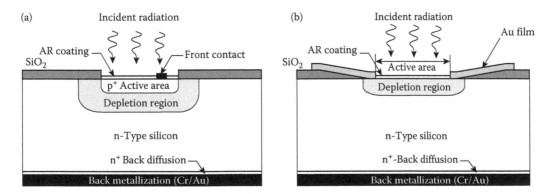

Figure 10.1   Cross section of silicon photodiodes: (a) p-n junction, and (b) Schottky barrier.

operation is preferred when a photodiode is used in low frequency applications (up to 350 kHz) as well as ultra low light applications. Application of a reverse bias can greatly improve the speed of response and linearity of the devices. This is due to an increase in the depletion region width and consequently a decrease in junction capacitance. The drawback of applying a reverse bias is an increase in the dark and noise currents.

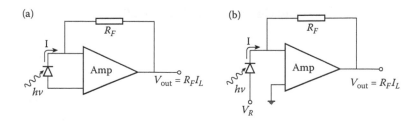

Figure 10.2 Modes of photodiode operation: (a) photovoltaic mode, and (b) photoconductive mode.

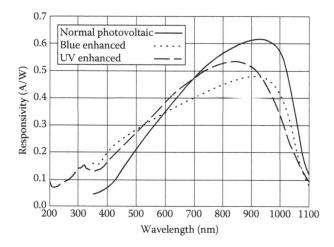

Figure 10.3 Typical current responsivity of several different types of planar diffused silicon photodiodes (After UDT Sensors, Inc., Catalog; http://129.105.69.13/datasheets/optoelectronics)

Typical spectral characteristics of planar diffusion photodiodes are shown in Figure 10.3. The time constant of p-n junction silicon photodiodes is generally limited by RC constant rather than by the inherent speed of the detection mechanism (drift and/or diffusion) and is on the order of microsecond. Detectivity is typically between mid-10^{12} and 10^{13} cmHz$^{1/2}$/W usually amplifier-limited for small area detectors.

The p-i-n detector is faster but is also less sensitive than conventional p-n junction detector and has slightly extended red response. It is a consequence of extension of the depletion layer width, since longer wavelength photons will be absorbed in the active device region. Incorporation of a very lightly doped region between the p and n regions and a modest reverse bias form a depletion region the full thickness of the material (≈ 500 µm for a typical silicon wafer). The higher dark current collected from generation within the wider depletion layer results in lower sensitivity.

High absorption coefficient of silicon in the blue and UV spectral regions causes the generation of carriers within the heavily doped p$^+$ (or n$^+$) contact surface of p-n and p-i-n photodiodes, where the lifetime is short due to the high and/or surface recombination. As a result, the quantum efficiency degrades rapidly in these regions. Blue- and UV-enhanced photodiodes optimize the response at short wavelengths by minimizing near-surface carrier recombination. This is achieved by using very thin and highly graded p$^+$ (or n$^+$ or metal Schottky) contacts, by using lateral collection to minimize the percentage of the surface area that is heavily doped, and/or passivating the surface with a fixed surface charge to repeal minority carriers from the surface.

Impressive progress in the development of hybrid p-i-n Si-CMOS arrays for the large visible and near IR imaging market has been obtained by Raytheon Vision Systems (see Figure 10.4) [11–13]. The large format imagers are loosely defined as detector array that exceeds 2k × 2k elements and are generally over 2 × 2 cm^2 in area. At the same time, the pitch is less than 10 µm. The arrays as large as 4096 × 4096 (H4RG-10) with 10 µm pitch are being produced with > 99.9% operability [12]. The hybrid imagers independently optimize the readout chip and the detector chip. This flexibility decreases cycle time and allows the ROIC to be updated independent of the detector and vice versa.

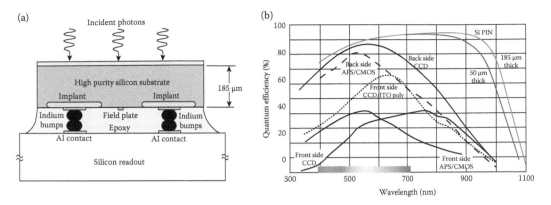

Figure 10.4 (a) Hybrid p-i-n Si-CMOS imager unit cells, and (b) quantum efficiency comparison of various technologies. (From Kilcoyne, S., Malone, N., Harris, M., Vampola, J., and Lindsay, D., "Silicon p-i-n Focal Plane Arrays at Raytheon," *Proceedings of SPIE* 7082, 70820J, 2008. With permission.)

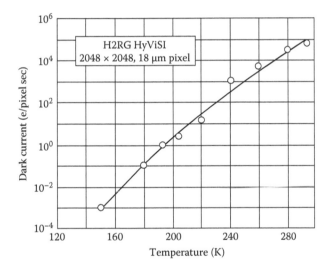

Figure 10.5 Dark current of 18 μm pixel p-i-n Si detector (H2RG-18 FPA fabricated by Teledyne Imaging Sensors). (From Bai, Y., Bajaj, J., Beletic, J. W., and Farris, M. C., "Teledyne Imaging Sensors: Silicon CMOS Imaging Technologies for X-Ray, UV, Visible and Near Infrared," *Proceedings of SPIE* 7021, 702102, 2008. With permission.)

The arrays are characterized by 100% fill factor, high quantum efficiency in 400–900 nm wavelength range and high modulation transfer function (MTF). One of the key advantages is the use of high resistivity silicon starting material, which results in the ability to fully deplete the intrinsic region of detectors up to 200 μm thick. The deep depleted (intrinsic) absorption region is more sensitive to the longer (red) wavelengths of the visible spectrum compared to conventional charge coupled device (CCD) imagers (Figure 10.4b). The plot in Figure 10.4b shows a comparison of spectral characteristics of the various front and back side illuminated imagers offered in the imaging market.

Silicon p-i-n detector dark current is due to thermal generation in the depletion region and at the surface of the detector. At present stage of development, the dark current is 5–10 nA/cm² at room temperature for 18 μm pixel, which corresponds to one electron per pixel per second at ~195 K (see Figure 10.5 [12]). Further reduction, to ~1 nA/cm² is possible. With a thick detector layer, the p-i-n photodiode must have a strong electric field within the detector layer to push the photocharge to the p-n junction to minimize charge diffusion and obtain good point spread function. A bias up to 50 volts is applied to the back surface contact.

The avalanche photodiodes are especially useful where both fast response and high sensitivity are required. An n⁺-p-π-p⁺ epiplanar avalanche Si photodiode configuration developed at

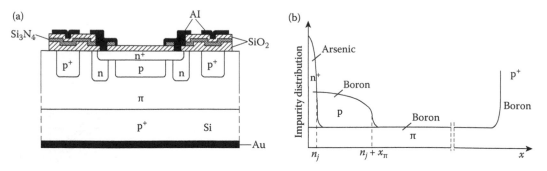

Figure 10.6 n⁺-p-π-p⁺ Si avalanche photodiode: (a) cross section of the structure, and (b) distribution of dopants in active region. (From Węgrzecka, I., and Węgrzecki, M., *Opto-Electronics Review*, 5, 137–46, 1997. With permission.)

Institute of Electron Technology (ITE) is shown in Figure 10.6 [14]. An initial material is Si wafer with π-type epilayer ($\rho_\pi = 200$–300 Ωcm, $x_\pi = 30$–35 μm; see Figure 10.6b) on p⁺ Si(111) substrate. The choice of π-type highly resistive layer ensures higher participation of electrons than holes in the detection process. The n-type guard ring is provided by prediffusion of phosphorous followed by rediffusion that takes place during a thermal treatment of the active region. The p⁺-type channel stopper is made by implanting and then rediffusing boron. The 150 nm thick SiO₂ antireflection layer covers the photodiode active region. The active (photosensitive, avalanche) region constitutes the central region of the n⁺-p abrupt junction obtained by arsenic diffusion from amorphous silicon to the p-type area previously formed by boron implantation followed by boron rediffusion. The basic parameters of the photodiodes developed at ITE are listed in Table 10.2 [14].

Whereas normal photodiodes become Johnson- or thermal-noise-limited when used with low-impedance load resistor for fast response, avalanche photodiodes make use of internal multiplication and keep the detector noise above the Johnson noise level. An optimum gain exists below which the system is limited by receiver noise and above which the shot noise dominates receiver noise and the overall noise increases faster than the signal (see Figure 10.7). Very careful bias control is essential for stable performance. Noise is a function of the detector area and increases as gain increases. Signal-to-noise ratio improvements of one to two orders of magnitude over a nonavalanche detector can be achieved. Typical detectivity is $(3$–$5) \times 10^{14}$ cmHz$^{1/2}$/W.

Table 10.3 lists the characteristics of commonly available silicon avalanche photodiode (APD) structures optimized to meet specific paradigms [15]. The excessive leakage current along the junction edges due to junction curvature effect or high-field concentration is eliminated by using a guard-ring or surface-beveled structure.

The Schottky-barrier diode can be used as an efficient photodiode. Since it is a majority carrier device, minority carrier storage and removal problems do not exist and therefore higher bandwidths can be expected. In comparison with p-i-n photodiode, Schottky photodiode has narrower active regions and hence transit times are very short. This type of device also offers lower parasitic resistance and capacitance and has the capability to operate at frequencies > 100 GHz. However, narrow active regions also cause lower quantum efficiency. Surface traps and recombination cause substantial loss of generated carriers at the surface.

The general structure of silicon Schottky photodiode is shown in Figure 10.1b. The silicon devices are usually fabricated by evaporating a thin Au layer (~150 Å) on high resistivity n-type material. Due to deposited gold layer, the reflection coefficient in wavelength range λ > 800 nm is above 30%. This results in decreasing the responsivity in this spectral range. Response time is in the picosecond region, corresponding to a bandwidth of about 100 GHz.

The absorption in silicon is generally poor because of the indirect bandgap. The absorption length in Si is almost 20 μm compared with 1.1 μm for GaAs and 0.27 μm for Ge. Consequently, it is difficult to design a Si photodetector with high efficiency in silicon CMOS processes.

Due to the long absorption length of silicon, the traditional vertical photodiode would have to be many microns thick to attain reasonable quantum efficiency. In bulk Si detectors, a majority of the electron-hole pairs are generated deep within the substrate, far from the high-field drift region or generated by surface electrodes. Secondly, the operating voltage would have to be high to deplete

Table 10.2: Typical Parameters of Silicon Avalanche Photodiodes Developed at ITR

Parameter	Units	BPYP 52 (0.3 mm)	BPYP 54 (0.5 mm)	BPYP 53 (0.9 mm)	BPYP 58 (1.5 mm)	BPYP 59 (3 mm)	Test Conditions*
Operating voltage, V_R	V	180–220 (min. 130 – max. 280)					$\lambda = 850$ nm
Temperature coefficient of V_R	V/°C						
Responsivity	A/W			50	0.75		$\lambda = 850$ nm
Noise current	pA/Hz$^{1/2}$	0.07	0.12	0.3	0.45	1.5	$P = 0$
Noise equivalent power	fW/Hz$^{1/2}$	1.4	2.4	6	9	30	$\lambda = 850$ nm
Excess noise factor				4			$\lambda = 850$ nm
Dark current	nA	0.7	1.3	2.2	5	12	$P = 0$
Capacitance	pF	1.7	3	7	12	40	$P = 0$

Source: I. Węgrzecka and M. Węgrzecki, *Opto-Electronics Review*, 5, 137–46, 1997. With permission.

*V_R for the gain $M = 100$; $t_{amb} = 22°C$.

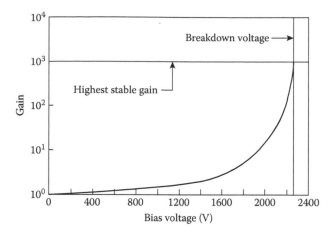

Figure 10.7 Gain as a function of reverse bias can reach 1000. This operating point is very close to breakdown and requires careful bias control (After Advanced Photonix Inc. Avalanche Catalog.) http://www.advancephotonix.com/ap_products/

Table 10.3: Properties of Commonly Available Silicon APD Structures

Structure	Beveled Edge	Epitaxial	Reach-Through
Absorption region	wide	narrow	medium to wide
Multiplication region	wide	narrow	narrow
Typical size (diameter)	up to 15 mm	up to 5 mm	up to 5 mm
Gain	50–1000	1–200	13–300
Excess noise factor	excellent ($k \approx 0.0015$)	good ($k \approx 0.03$)	good to excellent ($k \approx 0.0015$)
Operating voltage	500–2000 V	80–300 V	150–500 V
Response time	slow	fast	fast
Capacitance	low	high	low
Blue response	good	good	poor
Response	excellent	poor	good

Source: S. Melle and A. MacGregor, *Laser Focus World*, 145–56, October 1995.

the absorption region. As a result, a 3-dB bandwidth is seriously limited. Moreover, attempting to incorporate a thick p-i-n structure into the silicon CMOS process is simply impractical.

A photodiode structure that is more compatible with CMOS is the lateral p-i-n structure shown in Figure 10.8a. The design consists of alternating p-type and n-type interdigitated fingers separated by the absorption region, similar to a MSM photodetector layout. This structure features low capacitance per unit area, however, slow drift of the carrier to the electrodes from a deep region severely limits the bandwidth. Therefore, it is beneficial to block the deep carriers at the expense of quantum efficiency. One method of solution involves placing an insulating layer, such as SiO_2, a couple of microns below the surface. The thickness of the oxide is adjusted to maximize the reflectivity at the desired wavelength. An alternative way to block the slow carriers is to use a p-n junction as a screening terminal. The active area is placed inside the n-well surrounded by substrate contacts.

A novel approach to improve bulk detectors was described by Yang et al. [16], who demonstrated a lateral trench detector (LTD) that consisted of a lateral p-i-n detector with 7 μm deep trench electrodes (see Figure 10.8b). These detectors were fabricated on a p-type (100) silicon with

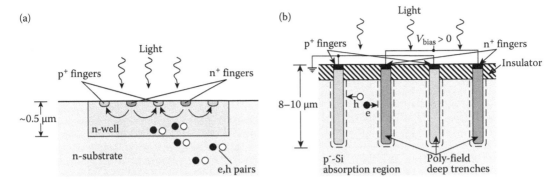

Figure 10.8 Si p-i-n lateral photodiodes: (a) interdigitated, and (b) trenched.

resistivity of 11–16 Ωcm. The trench width at the top surface is approximately 0.35 μm. To fill the trenches with n+ and p+ amorphous silicon (a-Si) sequentially, borosilicate glass was used as a sacrificial material. The a-Si inside the trenches was *in situ* doped with a concentration of 1×10^{20} cm^{-3} for phosphorous and 6×10^{20} cm^{-3} for boron. In the final step, devices were annealed at a high temperature to crystallize a-Si into polysilicon, activate the dopants, and drive the dopants from trenches into the silicon substrate, forming junctions away from the trench sidewalls.

LDTs presented in Yang and colleagues [16] were able to simultaneously show high quantum efficiency (68%) and high-speed operation (3 GHz) at 670 nm. However, at longer wavelengths the devices suffered from degraded bandwidth due to carrier generation and collection below electrodes.

To improve the performance of Si-based photodetectors, the use of silicon-on-insulator (SOI) has been investigated extensively [17]. This technology is particularly attractive when taken into account the widespread acceptance of the SOI technology as a platform for high performance CMOS devices [18]. The main benefit of using SOI technology is the buried insulator that prevents carriers generated in the substrate, below the oxide, from reaching the surface electrodes above the oxide. In addition, the refractive index contrast of the buried oxide causes the reflection of a portion of the incident light back into the absorbing layer, thus improving quantum efficiency.

Liu et al. has demonstrated MSM detectors on thin SOI substrates with bandwidth as high as 140 GHz [19]. However, due to the thin absorbing layer of 200 nm, these devices have very low external quantum efficiency, below 2%. The SOI detectors with thicker absorbing layers were characterized by improved quantum efficiency (to 24% at 840 nm), but also resulted in a reduced bandwidth of 3.4 GHz [20].

The resonance cavity-enhanced (RCE) detector design is another technique that is applied to Si detectors [21,22]. Figure 10.9a shows a schematic representation of a general RCE active layer structure that can be applied for different device configurations, such as Schottky diodes, MSM photodetectors, and APDs. The top and bottom distributed Bragg reflection mirrors consist of alternating layers of nonabsorbing larger band gap materials. The active layer of thickness d is a small bandgap semiconductor positioned between two mirror structures at respective distances l_1 and l_2 from the top and bottom mirrors. The two end mirrors can be formed with quarter-wave ($\lambda/4$) stacks of large bandgap semiconductors. The wavelength dependent quantum efficiency of the detector for a cavity of optical length βl, which an active region of thickness d and absorption coefficient α, is given as [21,22]

$$\eta = \left[\frac{1 + r_2 \exp(-\alpha d)}{1 - 2\sqrt{r_1 r_2} \exp(-\alpha d)\cos(2\beta L) + r_1 r_2 \exp(-\alpha d)} \right](1 - r_1)[1 - \exp(-\alpha d)], \tag{10.1}$$

where $\beta = 2\pi/\lambda$, r_1 and r_2 are the refraction coefficients of materials around the active layers with thicknesses l_1 and l_2, respectively. The quantum efficiency is enhanced periodically at the resonant wavelengths, which occur when $\beta l = m\pi$. It should be mentioned that devices based upon the resonant cavity design could be very sensitive to process-induced variations.

For example, Figure 10.9b shows a RCE interdigitated p-i-n Si photodiode grown by lateral overgrowth [23]. The bottom mirror was formed on top of a p-type (100) substrate by depositing three pairs of quarter-wavelength SiO$_2$ and polysilicon layers. Two 20×160 μm trenches separated

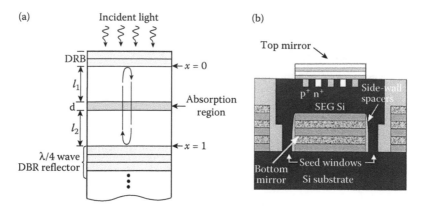

Figure 10.9 Resonant cavity-enhanced photodetector: (a) a general structure, and (b) schematic cross section of interdigitated p-i-n photodiode. (From Schaub, J. D., Li, R., Schow, C. L., Campbell, L. C., Neudeck, G. W., and Denton, J., *IEEE Photonics Technology Letters*, 11, 1647–49, 1999. With permission.)

by 40 µm were etched in the mirror to serve as seed windows for the subsequent selective epitaxial growth. SiO_2 sidewall spacers on the mirror were formed to prevent the nucleation of defects at the edges of the polysilicon during Si epitaxial process. Next, interdigitated p-i-n photodiodes were formed in the epitaxial silicon by sequential As and BF_2 implants and anneals. Following metallization, two dielectric mirror pairs (ZnS–MgF) were evaporated to form the Fabry-Perot cavity.

10.2 GERMANIUM PHOTODIODES

Germanium photodiodes are usually fabricated by diffusion of arsenide into a p-type germanium (gallium doped to a 10^{15} cm^{-3} concentration and resistivity of 0.8 Ωcm). After creation of 1 µm thick n-type region, an oxide passivation film is deposited to reduce surface conductivity in the vicinity of the p-n junction. Finally, an antireflection coating is deposited (germanium has a high refractive index, $n \sim 4$, and a useful transmission range of 2–23 µm).

Germanium does not form a stable oxide. GeO_2 is soluble in water, which leads to two process challenges: device passivation and stability. The lack of a high-quality passivation layer makes it difficult to achieve a low dark current. Interestingly, by scaling the device to smaller dimensions, a higher dark current can be tolerated.

Three germanium photodiode types are available: p-n junction, p-i-n junction, and APD [24]. They are readily made in areas ranging from 0.05 to 3 mm², and capability exists to make the area as small as 10×10 µm² or as large as 500 mm². The upper limit is imposed by raw material uniformity.

The previous discussion on silicon detectors applies in general to germanium ones, with the exception that blue- and UV-enhanced devices are not relevant to germanium detectors. Because of narrower bandgap, the germanium photodiodes have higher leakage currents, compared to silicon detectors. They offer submicrosecond response or high sensitivity from the visible to 1.8 µm. Zero bias is generally used for high sensitivity and large reverse bias for high speed. Peak of detectivity at room temperature is above 2×10^{11} cmHz$^{1/2}$W^{-1}. The performance can be improved significantly with thermoelectric cooling or cooling to liquid nitrogen temperature, what is shown in Figure 10.10 (detector impedance increases about the order of magnitude by cooling 20°C below room temperature).

Germanium, with $n = 4$, is an excellent candidate material for the immersion lens because of its high refractive index. Usually, the performance of germanium photodiodes are Johnson-noise limited, and then we can improve detector performance by immersion in a hemispherical lens. The effective area of the detector increases by n^2, where n is the refractive index of the medium.

At the present stage of development, the research efforts are directed to integrate Ge detectors on silicon (Ge/Si) substrates into a CMOS-compatible process. It is an attractive goal for making arrays of on-chip detectors that can be used in an electronic-photonic chip.

Since the pioneering work of Luryi et al. [25], a number of different approaches have been proposed for growing Ge (or GeSi) films on silicon substrates aimed at both optimizing the electronic quality of the films while preserving compatibility with the standard silicon technology. The

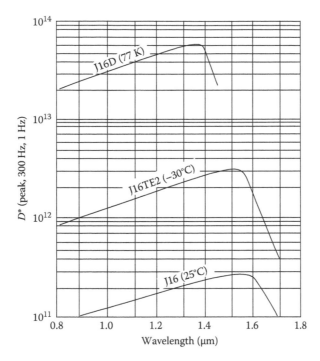

Figure 10.10 Detectivity as a function of wavelength for germanium photodiode at three temperatures (After Judson Catalog, *Infrared Detectors*) http://www.judsontechnologies.com/germanium.html.

lattice constant of Ge is 4.17% larger than that of Si. The strain resulting from lattice mismatch modifies the band structure and causes dislocation defects that increase the leakage current of photodiodes. Strain limits the thickness of Ge layers that can be epitaxially grown on silicon. Taking into account the bandwidth point of view, thin Ge films are preferred since it minimizes the carrier transit time. However, it comes at the expense of reduced absorption and diminished responsivity. Nevertheless, excellent progress is being made toward Ge on silicon detectors.

To overcome the above limitations, the following approaches have been pursued: low Ge content alloys, incorporating of carbon, graded buffers, low temperature thin buffers, and, more recently, the growth of polycrystalline films [26,27]. p-i-n devices with thick graded buffer layers demonstrated excellent quality [28,29], however they suffer from integration difficulties due to nonplanarity with the CMOS devices. For this reason, more recent work on Ge-on-Si detectors has focused on the use of thin buffer layers, or even direct growth of Ge on Si. The recent research activities on Ge photodiodes on Si (Figure 10.11) have shown a dramatic increase in speed to 40 GHz with prospects to 100 GHz and more [30].

Successful use of low temperature buffers for the grown of Ge films on Si substrates was demonstrated for the first time (see Colace and colleagues [31]). The addition of thermal cycles after growth allowed to reduce the dislocation density down to 2×10^7 cm^{-2} [32,33]. Using this technology, high performance p-i-n photodiodes were fabricated with maximum responsivities of 0.89 A/W at 1.3 μm and 0.75 A/W at 1.55 μm, respectively; reverse dark currents of 15 mA/cm^2 at 1 V and response time as short as 180 ps [33].

Figure 10.12 shows I-V and spectral characteristics of a 100×100 μm^2 p-i-n Ge-on-Si photodiode [34]. The insert of Figure 10.12a presents the device structure. Approximately 2-μm thick intrinsic Ge film was deposited on p$^+$-(100)Si substrate by a two-step deposition process (by ultra-high vacuum chemical vapor deposition (CVD) and by low-pressure CVD), and capped with a 0.2 μm n$^+$-polysilicon layer. After Ge growth, standard CMOS processes were used to deposit and pattern a dielectric (SiON) film to open up windows to the Ge surface. Then poly Si was deposited and implanted with phosphorus into the underlying Ge to form a vertical p-i-n junction in germanium. The diode has an ideality factor less than 1.2 at 300 K with a perimeter dominated reverse leakage current of ~40 mA/cm^2 at –1 V bias. The leakage current is believed to be due to surface states

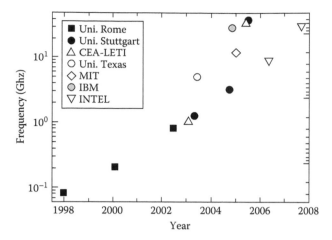

Figure 10.11 Speed evolution of Ge photodetectors. (From Kasper, E., and Oehme, M., *Physic Status Solidi (c)*, 5, 3144–49, 2008. With permission.)

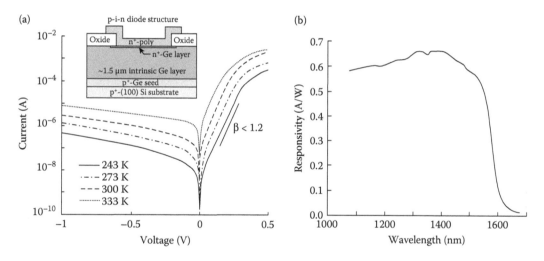

Figure 10.12 Characteristics of a $100 \times 100 \ \mu m^2$ p-i-n Ge-on-Si photodiode: (a) I-V characteristics as a function of temperature, and (b) responsivity versus wavelength (without AR coating). (From Kärtner, F. X., Akiyama, S., Barbastathis, G., Barwicz, T., Byun, H., Danielson, D. T., Gan, F., et al., *Proceedings of SPIE* 6125, 612503, 2006. With permission.)

introduced during the Ge passivation process. The responsivity without AR coating is 0.5 A/W at 1.55 μm.

Recently, NoblePeal Vision developed monolithic germanium imaging arrays in which Ge islands are integrated with the silicon transistors and metal layers of a CMOS process [35]. The key to growing germanium islands is illustrated schematically in Figure 10.13a. Selective germanium epitaxial growth is carried out in an aperture formed in the dielectric layer that overlies the transistors. Dislocations at the Si–Ge interface propagate at an angle of 60° to the interface and terminate on the sidewalls. After formation of the germanium island, the conventional CMOS process is proceeded to connect the germanium photodiode to the circuitry (see Figure 10.13b). An ion implantation was used to form lateral photodiodes. Using this innovative growth technique, a prototype 128×128 imager at a 10-mm pitch was designed in a high volume 0.18 μm CMOS technology.

10.3 SiGe PHOTODIODES

There are several technologies such as InGaAs, PbS, and HgCdTe, which cover near infrared (NIR) spectrum. SiGe offers a low cost alternative approach for NIR sensors that can cover spectral band

(a)

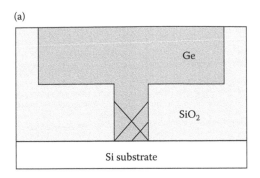

(b)

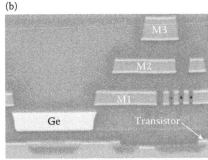

Figure 10.13 Growth of germanium island: (a) dislocation trapping, and (b) SEM of germanium photodiode embedded in CMOS stack. (From Rafferty, C. S., King, C. A., Ackland, B. D., Aberg, I., Sriram, T. S., and O'Neill, J. H., *Proceedings of SPIE* 6940, 69400N, 2008. With permission.)

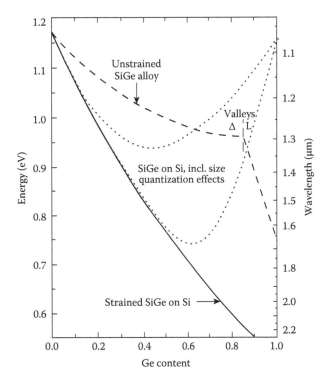

Figure 10.14 Fundamental bandgap of SiGe alloys versus Ge-content. The dashed line corresponds to unstrained SiGe bulk samples, the solid line to strained SiGe with the lateral constant of Si. For the dotted lines the effect of critical layer thickness and size quantization is included. (From Presting, H., *Handbook of Infrared Detection Technologies*, 393–448, Elsevier, Kidlington, UK, 2002. With permission.)

up to 1.6 microns. The attractive features of SiGe-based IR FPAs take advantage of silicon-based technology, which can promise very small feature size and compatibility with the silicon CMOS circuit for signal processing.

The $Si_{1-x}Ge_x$ alloy grown on Si is an ideal material because, due to its 100% complete miscibility the SiGe can be continuously tuned from the Si (1.1 eV) down to the Ge bandgap (0.66 eV). The lattice mismatch between Si and Ge leads to strained hetero-epitaxy that for a given Ge-content x, a $Si_{1-x}Ge_x$ alloy can only be deposited on Si up to a critical thickness without the formation of dislocations. The strain also leads to change in the SiGe bandgap structure (see Figure 10.14 [36]). One can see the big influence of the strain to the bandgap when one compares the unstrained curve

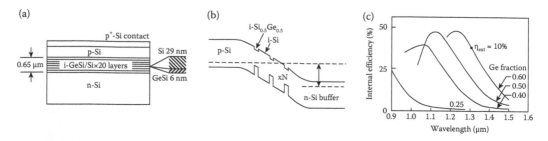

Figure 10.15 p-i-n SiGe superlattice photodiode: (a) schematic diagram of the device structure, (b) energy band diagram, and (c) room-temperature spectral response as a function of Ge content of the active layer. (From Paul, D. J., *Semiconductor Science and Technology,* 19, R75–R108, 2004. With permission.)

(dashed) with the strained line (solid lines). SiGe technology has been introduced to capture the increased carrier mobility advantages of strained-layer epitaxy, which is 46% for n-channel devices and 60–80% for p-channel devices, is decisive in capture SiGe technology into high-speed CMOS technology. The sharp drop of the bandgap for $x > 0.85$ for the unstrained alloy, indicates the transition point to the Ge-like band structure for this high Ge-content where the conduction band minimum changes from the Δ-point to the L-point $< 111 >$ of the Brillouin zone [36,37].

A number of epitaxial systems have been demonstrated including molecular beam epitany (MBE), plasma enhanced CVD, sputtering and laser assisted growth. For production lines used in CMOS and Si bipolar factories, only conventional CVD at high temperatures (>1000°C) has to date demonstrated the quality, uniformity, and through-put required by companies for production.

In one of the first NIR p-i-n Ge on silicon detector, a graded SiGe buffer was used in order to separate the active intrinsic layer of the diode from the highly dislocated Si–Ge interface region [25]. The large dislocation density, however, gave rise to reverse dark currents in excess of 50 mA/ cm² at 1 V. In order to reduce the dark current, an active layer made of a coherently strained GeSi/ Si superlattice (SLS) was introduced by Temkin et al. [38]. The bandgap of SLS is smaller than that of unstrained layers, and therefore, interesting to increase the optical absorption coefficient at a small detector thickness.

The SLS are preferred to SiGe alloys of similar average composition because they allowed the deposition of layers well above the critical thickness (strain symmetrization). When strained SiGe is grown on relaxed Si, the band alignment at the heterointerface of Si and SiGe is of type I, what means that the offset lies predominantly within the valence band (see Figure 10.15b). On the other hand, when strained Si is grown on relaxed SiGe, a staggered band alignment of type II results at the heterointerface. In this case, the conduction band of SiGe is higher than that of Si and the valence band in SiGe is lower than that of Si.

Commonly the photodetector is grown layer by layer on a substrate and the light hits the detector perpendicularly to the surface. To overcome the poor sensitivity due to weak absorption in the thin layers of vertical diodes, so called waveguide photodetectors were built. Here the absorber is shaped as a thin, narrow but very long rectangular waveguide. The light travels parallel to the surface through the detector structure and the absorption length can be up to some millimeters depending on the length of the waveguide.

Figure 10.15 presents p-i-n SLS SiGe waveguide photodiode characteristics [37]. The p-i-n detector consists of n-type and p-type Si layers on either side of an undoped 20-period superlattice active region grown by MBE on n-type Si substrate. Serious change in spectra responsivity occurs for composition $x > 0.40$, which clearly shows efficient response at wavelengths of 1.3 μm. As the Ge fraction is increased from $x = 40$ in 10% increments the photocurrent response peaks at 1.08, 1.12, and 1.23 μm, respectively. Using waveguide geometry, the internal quantum efficiency on the order of 40% at 1.3 μm in SL with the Ge fraction of $x = 0.6$ was measured.

The first avalanche SiGe photodiodes were demonstrated at Bell Labs in 1986 with overall thickness ranging from 0.5 to 2 μm [39]. Among them, the most effective was based on 3.3 nm thick $Ge_{0.6}Si_{0.4}$ layers with 39 nm Si-spacers with a quantum efficiency of about 10% at 1.3 μm and a maximum responsivity of 1.1 A/W. More stable characteristics were obtained at 30 V reverse bias with responsivity of 4 A/W.

After the mid-1990s, a large effort in the investigation of SiGe detectors was devoted to structures able to operate in the third spectral window (1.55 µm) [7]. Due to the vanishing absorption of SiGe at this wavelength, the pure Ge material was used to reduce the dislocation density in the epitaxial layer. For the growth of Ge on Si, in order to minimize the threading dislocation density due to lattice mismatch, a graded $Si_{1-x}Ge_x$ films are used that act as a virtual substrate for Ge growth, or depositing a low-temperature Ge seed layer followed by a thicker film that undergoes annealing [40].

It should be mentioned that short-period SiGe superlattices and SiGe quantum well structures have been grown by MBE epitaxy on Si substrates for NIR, middle and long wavelength infrared detection [41,42]. More information about quantum well infrared detectors can be found in Chapter 16.

REFERENCES

1. D. J. Paul, "Si/SiGe Heterostructures: From Material and Physics to Devices and Circuits," *Semiconductor Science and Technology* 19, R75–R108, 2004.

2. L. Pavesi, "Will Silicon be the Photonic Material of the Third Millennium?" *Journal of Physics: Condensed Matter* 15, R1169–R1196, 2003.

3. B. Jajali and S. Fathpour, "Silicon Photonics," *Journal of Lightwave Technology* 24, 4600–615, 2006.

4. S. M. Sze, *Physics of Semiconductor Devices,* Wiley, New York, 1981.

5. S. Adachi, *Properties of Group-IV, III-V and II-VI Semiconductors*, John Wiley & Sons, Chichester, 2005.

6. D. Wood, *Optoelectronic Semiconductor Devices*, Prentice-Hall, Trowbridge, UK, 1994.

7. H. Zimmermann, *Integrated Silicon Optoelectronics*, Springer, New York, 2000.

8. T. P. Pearsall, "Silicon-Germanium Alloys and Heterostructures: Optical and Electronic Properties," *CRS Critical Reviews in Solid State and Materials Sciences*, 15, 551–600, 1989.

9. J. C. Bean, "Silicon-Based Semiconductor Heterostructures: Column IV Bandgap Engineering," *Proceedings of IEEE* 80, 571–87, 1992.

10. R. A. Soref, "Silicon-Based Optoelectronics," *Proceedings of IEEE* 81, 1687–1706, 1993.

11. T. Chuh, "Recent Developments in Infrared and Visible Imaging for Astronomy, Defense and Homeland Security," *Proceedings of SPIE* 5563, 19–34, 2004.

12. Y. Bai, J. Bajaj, J. W. Beletic, and M. C. Farris, "Teledyne Imaging Sensors: Silicon CMOS Imaging Technologies for X-Ray, UV, Visible and Near Infrared," *Proceedings of SPIE* 7021, 702102, 2008.

13. S. Kilcoyne, N. Malone, M. Harris, J. Vampola, and D. Lindsay, "Silicon p-i-n Focal Plane Arrays at Raytheon," *Proceedings of SPIE* 7082, 70820J, 2008.

14. I. Węgrzecka and M. Węgrzecki, "Silicon Photodetectors: The State of the Art," *Opto-Electronics Review* 5, 137–46, 1997.

15. S. Melle and A. MacGregor, "How to Choice Avalanche Photodiodes," *Laser Focus World*, 145–56, October 1995.

16. M. Yang, K. Rim, D. L. Rogers, J. D. Schaub, J. J. Welser, D. M. Kuchta, D. C. Boyd, et al., "A High-Speed, High-Sensitivity Silicon Lateral Trench Photodetector," *IEEE Electron Device Letters* 23, 395–97, 2002.

17. S. J. Koester, J. D. Schaub, G. Dehlinger, and J. O. Chu, "Germanium-on-SOI Infrared Detectors for Integrated Photonic Applications," *IEEE Journal of Selected Topics in Quantum Electronics* 12, 1489–1502, 2006.

18. G. G. Shahidi, "SOI Technology for the GHz Era," *IBM Journal of Research and Development* 46, 121–31, 2002.

19. M. Y. Liu, E. Chen, and S. Y. Chou, "140-GHz Metal-Semiconductor-Metal Photodetectors on Silicon-on-Insulator Substrate with a Scaled Active Layer," *Applied Physics Letters* 65, 887–88, 1994.

20. C. L. Schow, R. Li, J. D. Schaub, and J. C. Campbell, "Design and Implementation of High-Speed Planar Si Photodiodes Fabricated on SOI Substrates," *IEEE Journal of Quantum Electronics* 35, 1478–82, 1999.

21. M. S. Ünlü and M. S. Strite, "Resonant Cavity Enhanced Photonic Devices," *Journal of Applied Physics* 78, 607–39, 1995.

22. M. S. Ünlü, G. Ulu, and M. Gökkavas, "Resonant Cavity Enhanced Photodetectors," in *Photodetectors and Fiber Optics*, ed. H. S. Nalwa, 97–201, Academic Press, San Diego, CA, 2001.

23. J. D. Schaub, R. Li, C. L. Schow, L. C. Campbell, G. W. Neudeck, and J. Denton, "Resonant-Cavity-Enhanced High-Speed Si Photodiode Grown by Epitaxial Lateral Overgrowth," *IEEE Photonics Technology Letters* 11, 1647–49, 1999.

24. A. Bandyopadhyay and M. J. Deen, "Photodetectors for Optical Fiber Communications," in *Photodetectors and Fiber Optics*, ed. H. S. Nalwa, 307–68, Academic Press, San Diego, CA, 2001.

25. S. Luryi, A. Kastalsky, and J. C. Bean, "New Infrared Detector on a Silicon Chip," *IEEE Transactions on Electron Devices* ED-31, 1135–39, 1984.

26. G. Masini, L. Colace, and G. Assanto, "Poly-Ge Near-Infrared Photodetectors for Silicon Based Optoelectronics," *ICTON* Th.C.4, 207–10, 2003.

27. G. Masini, L. Colace, F. Petulla, G. Assanto, V. Cencelli, and F. DeNotaristefani, *Optical Materials* 27, 1079–83, 2005.

28. S. B. Samavedam, M. T. Currie, T. A. Langdo, and E. A. Fitzgerald, "High-Quality Germanium Photodiodes Integrated on Silicon Substrates Using Optimized Relaxed Buffers," *Applied Physics Letters* 73, 2125–27, 1998.

29. J. Oh, J. C. Campbell, S. G. Thomas, S. Bharatan, R. Thoma, C. Jasper, R. E. Jones, and T. E. Zirkle, "Interdigitated Ge p-i-n Photodetectors Fabricated on a Si Substrate Using Graded SiGe Buffer Layers," *IEEE Journal of Quantum Electronics* 38, 1238–41, 2002.

30. E. Kasper and M. Oehme, "High Speed Germanium Detectors on Si," *Physic Status Solidi (c)* 5, 3144–49, 2008.

31. L. Colace, G. Masini, G. Assanto, G. Capellini, L. Di Gaspare, E. Palange, and F. Evangelisti, "Metal-Semiconductor-Metal Near-Infrared Light Detector Based on Epitaxial Ge/Si," *Applied Physics Letters* 72, 3175, 1998.

32. H.-C. Luan, D. R. Lim, K. K. Lee, K. M. Chen, J. G. Sandland, K. Wada, and L. C. Kimerling, "High-Quality Ge Epilayers on Si with Low Threading-Dislocation Densities," *Applied Physics Letters* 75, 2909–11, 1999.

33. S. Farma, L. Colace, G. Masini, G. Assanto, and H.-C. Luan, "High Performance Germanium-on-Silicon Detectors for Optical Communications," *Applied Physics Letters* 81, 586–88, 2002.

34. F. X. Kärtner, S. Akiyama, G. Barbastathis, T. Barwicz, H. Byun, D. T. Danielson, F. Gan, et al., "Electronic Photonic Integrated Circuits for High Speed, High Resolution, Analog to Digital Conversion," *Proceedings of SPIE* 6125, 612503, 2006.

35. C. S. Rafferty, C. A. King, B. D. Ackland, I. Aberg, T. S. Sriram, and J. H. O'Neill, "Monolithic Germanium SWIR Imaging Array," *Proceedings of SPIE* 6940, 69400N, 2008.

36. H. Presting, "Infrared Silicon/Germanium Detectors," *Handbook of Infrared Detection Technologies*, eds. M. Henini and M. Razeghi, 393–448, Elsevier, Kidlington, UK, 2002.

37. D. J. Paul, "Si/SiGe Heterostructures: From Material and Physics to Devices and Circuits," *Semiconductor Science and Technology* 19, R75–R108, 2004.

38. H. Temkin, T. P. Pearsall, J. C. Bean, R. A. Logan, and S. Luryi, "Ge_xSi_{1-x} Strained-Layer Superlattice Waveguide Photodetectors Operating near 1.3 µm," *Applied Physics Letters* 48, 963–65, 1986.

39. T. P. Pearsall, H. Temkin, J. C. Bean, and S. Luryi, "Avalanche Gain in Ge_xSi_{1-x}/Si Infrared Waveguide," *IEEE Electron Device Letters* 7, 330–32, 1986.

40. N. Izhaky, M. T. Morse, S. Koehl, O. Cohen, D. Rubin, A. Barkai, G. Sarid, R. Cohen, and M. J. Paniccia, "Development of CMOS-Compatible Integrated Silicon Photonics Devices," *IEEE Journal of Selected Topics in Quantum Electronics* 12, 1688–98, 2006.

41. H. Presting, "Near and Mid Infrared Silicon/Germanium Based Photodetection," *Thin Solid Films* 321, 186–95, 1998.

42. H. Presting, J. Konle, M. Hepp, H. Kibbel, K. Thonke, R. Sauer, W. Cabanski, and M. Jaros, "Mid-Infrared Silicon/Germanium Focal Plane Arrays," *Proceedings of SPIE* 3630, 73–89, 1999.

11 Extrinsic Silicon and Germanium Detectors

Historically, an extrinsic photoconductor detector based on germanium was the first extrinsic photodetector. After that, photodetectors based on silicon and other semiconductor materials, such as GaAs or GaP, have appeared.

Extrinsic photodetectors are used in a wide range of the IR spectrum extending from a few μm to approximately 300 μm. They are the principal detectors operating in the range $\lambda > 20$ μm. The spectral range of particular photodetectors is determined by the doping impurity and by the material into which it is introduced. For the most shallow impurities in GaAs, the long wavelength cutoff of photoresponse is around 300 μm. Detectors based on silicon and germanium have found the widest application as compared with extrinsic photodetectors on other materials and will be considered in the present chapter.

Research and development of extrinsic IR photodetectors have been ongoing for more than 50 years [1–3]. In the 1950s and 1960s, germanium could be made purer than silicon; doped Si then needed more compensation than doped Ge and was characterized by shorter carrier lifetimes than extrinsic germanium. Today, the problems with producing pure Si have been largely solved, with the exception of boron contamination. Si has several advantages over Ge; for example, three orders of magnitude higher impurity solubilities are attainable, hence thinner detectors with better spatial resolution can be fabricated from silicon. Si has lower dielectric constant than Ge, and the related device technology of Si has now been more thoroughly developed, including contacting methods, surface passivation, and mature MOS and charge-coupled device (CCD) technologies. Moreover, Si detectors are characterized by superior hardness in nuclear radiation environments.

The availability of a highly developed silicon MOS technology facilitates the integration of large detector arrays with charge-transfer devices (CTD) for readout and signal processing. The well-established technology also helps in the manufacturing of uniform detector arrays and the formation of low-noise contacts. Although the potential of large extrinsic silicon FPAs for terrestrial applications has been examined, interest has declined in favor of HgCdTe and InSb with their more convenient operating temperatures. Strong interest in doped silicon continues for space applications, particularly in low background flux and for wavelengths from 13 to 30 μm, where compositional control is difficult for HgCdTe. The shallower impurity energies in germanium allow detectors with spectral response up to beyond 100 μm wavelength and major interest still exists in extrinsic germanium for wavelengths beyond about 200 μm.

Increasing interest in extrinsic detectors has been observed due to creation of multielement focal plane arrays (FPAs) for application in space and ground-based IR astronomy and on defense space vehicles. Successes in the technology of photodetectors, creation of deep-cooled, low-noise semiconductor preamplifiers and multiplexers, as well as unique designs of photodetector devices and equipment for deep cooling have ensured the achievement of a record-breaking detectivity close to the radiation limit even under exceedingly low space backgrounds, 8–10 orders of magnitude lower than that of the room background [4,5]. This is unattainable for the other types of photodetectors at the present time.

The development and manufacture of extrinsic photodetectors are mainly concentrated in the United States. The programs on the use of off-atmospheric astronomy have spread especially intensively after the outstanding success of the Infrared Astronomical Satellite (IRAS) [6,7], which used 62 discrete photodetectors arranged in the focal plane. A number of NASA- and NSF-supported programs have been initiated by various research centers and universities. The array manufacturers have taken a strong interest and have done far more for astronomers than they might have anticipated [8,9]. They are Raytheon Vision Systems (RVS), DRS Technologies, and Teledyne Imaging Sensors. A deep-cooled space telescope located in vacuum can cover the entire IR range from 1 μm to 1000 μm. Far-infrared astronomy provides key information about the formation and evolution of galaxies, stars, and planets. The low level of background irradiation makes it possible to improve the sensitivity of such systems by increasing the integration time (to hundreds of seconds).

The fundamentals of the classical theory of extrinsic photoconductor detectors have been disclosed earlier in this book (see Sections 9.1.2). In comparison with intrinsic photoconductivity, the extrinsic photoconductivity is far less efficient because of limits in the amount of impurity that can be introduced into a semiconductor without altering the nature of the impurity states. Implicit in the treatment of low background detectors is that the generation of free carriers is dominated by photon absorption, not by thermal excitation. As a result, the lower temperature is required as the long wavelength cutoff of the detector increases and can be approximated as

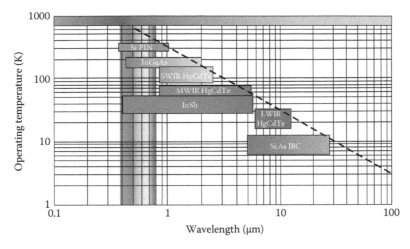

Figure 11.1 Operating temperatures for low-background material systems with their spectral band of greatest sensitivity. The dashed line indicates the trend toward lower operating temperature for longer wavelength detection.

$$T_{max} = \frac{300}{\lambda_c[\mu m]}. \tag{11.1}$$

This general trend is illustrated in Figure 11.1 for five high performance detector materials suitable for low-background applications: Si, InGaAs, InSb, HgCdTe photodiodes, and Si:As impurity band conductor (IBC) detectors also known as the blocked impurity band (BIB) detectors. The longer cutoff wavelength of the detector generally requires a lower operating temperature to achieve the same dark current performance. The operating temperature range is fairly broad due to the wide range of system backgrounds.

This chapter gives a brief survey of extrinsic detector technology, comments on certain peculiarities of the operation of these devices, and describes different types of extrinsic photodetectors that have been developed: blocked impurity band (BIB) devices and solid-state photomultipliers (SSPM). The main versions of extrinsic photodetector FPAs are presented in Section 21.2.

11.1 TECHNOLOGY

Silicon and germanium extrinsic photoconductors are usually made using techniques drawn from the semiconductor industry [4,10]. In the Czochralski method a crystal is pulled out of the melt without touching the walls of the vessel; this technique avoids stresses in the crystal that can develop if it solidifies in contact with the crucible walls and also prevents the escape of impurities from the walls into the crystal. The material can be doped by adding impurities to the melt from which the crystal is grown. Alternatively, the purity of the material can be improved after grown by zone refining the crystal. The dopant impurities can be added in a late stage of the zone refining by placing the impurity in the melted zone. Doping of grown crystals can also be accomplished by exposing a crystal to impurity vapors or by bringing its surface into contact with a sample of the impurity material; in both of these cases, the impurity atoms are distributed into the crystal by diffusion at high temperature. Floating zone growth combines features of Czochralski growth and zone refining. The crystal is grown onto a seed from a melted portion of an ingot that is suspended to avoid contact with the walls of the crucible.

To counteract the problem of unwanted thermal conductivity due to impurities with small values of excitation energy E_i, these dopants are compensated by adding carefully controlled amounts of the opposite kind of impurity, for example, n-type if the unwanted impurity is p-type (see Section 9.1.2). Although some degree of compensation is frequently critical in obtaining good detector performance, heavily compensated material is usually inferior to material that starts as pure as possible and needs to be only lightly compensated. A high concentration of compensating impurities usually results in the rapid capture of photoexcited charge carriers and reduces the photoconductive gain. Problems with control of the compensation process arise because of nonuniformities in the sample or because of imprecision in the exact amount of compensating material. Finally, hopping conductivity may be enhanced because the atoms of the compensating impurity ionize some of the majority impurity atoms, making available empty sites into which carriers can hop. The effect of proper compensation is to fill up the contaminating impurity while only partially filling the desired impurity; the material then freezes out at a temperature determined by the energy level of the desired impurity. In compensating, one raises operating temperature at the expense of decreased carrier lifetime. Undercompensation leads to thermal generation from shallow impurities, which results in excess noise, unless the operating temperature is lowered. Overcompensation degrades carrier lifetime and causes poor detector responsivity, but does not further increase the operating temperature.

Achieving uniformity in both the dopant and the compensating impurities is a paramount importance in creating high quality Si:X (Ge:X) material, especially that used for an array. Factors that influence uniformity include: level of dopant concentration, the segregation coefficient and vapor pressure of the dopant at the solidification of the host crystal, pick-up from the atmosphere or the crucible during crystal growth, pulling and rotation rates during growth, defects in the form of precipitation and dislocations, and possible high temperature induced effects from processing material into an array. High quality Si:X material is produced by using the float zone growth technique. This technique gives material with the lowest density of contamination elements. Uniformity of the major dopant must be grown-in by using high speed rotation rates (6 rpm) and low pulling rates (4 mm/min). In the case of Si:Ga, annealing for 16 days at 1300°C is necessary to achieve maximum uniformity. This annealing process is successful only for crystals low in oxygen and carbon that must necessarily be produced by float zone growth. By comparison, Czochralski grown crystals are full of contaminating impurities including oxygen, which can introduce unwanted donors, and carbon and can alter the impurity activation energy level.

Both phosphorus and boron are found in polycrystalline silicon. Phosphorus can be removed by a succession of vacuum float zone operations, whereas boron cannot. Phosphorus is usually removed, then reintroduced to compensate for the boron. A typical compensated impurity concentration is $N_d - N_a = 10^{13}$ cm^{-3}. A method of compensating the boron in extrinsic Si, called neutron transmuting doping and involves the transmuting of a fraction of the Si atoms into P donor via controlled neutron irradiation.

To make the detector, a thin wafer of material is cut from the crystal with a precision saw and then polished. Two sides of this wafer form the contacts of the detector. To smooth the variations in the electric fields near the contacts, heavy doping is used in the semiconductor where it joins the actual contact. The doping can be applied by implantation of impurity atoms of a similar type to the majority impurity (for example, boron for a p-type material and phosphorus for n-type material) and then annealing the crystal damage by heating [11]. Thermal annealing restores single crystallinity and activates the implanted atoms. Metallization with a thin 200 Å Pd adhesion layer followed by a few thousand Å of Au completes the contact formation process. A transparent contact can be made by adding a second layer of ion-implanted impurities at a lower energy than the first so it does not penetrate so far into the crystal and at a higher impurity density so it has large electrical conductivity. For an opaque contact, metal is evaporated over the ion-implanted layers. Individual detectors or arrays of detectors are cut from the wafer with a precision saw. The surface damage left by the saw can be removed by etching the detectors. Finally, naturally grown silicon oxide (at elevated temperatures) is excellent protective layers for silicon photodetectors. In the case of germanium and many other semiconductors, protective layers must be added by more complex processing steps.

Different additional techniques, including standard photolithography, are used to produce the semiconductor detectors. Epitaxy is a particularly useful technique for growing a thin crystalline layer on a preexisting crystal structure. In comparison with silicon, germanium and many other semiconductors require more complex processing steps especially in fabrication protection or insulating layers.

11.2 PECULIARITIES OF THE OPERATION OF EXTRINSIC PHOTODETECTORS

The studies on the photoconductivity in high-resistivity crystals, to which extrinsic photoconductors belong at low temperatures and backgrounds, began simultaneously with the works or the studies of the extrinsic photoconductivity as a phenomenon since the early 1950s. The most interesting works carried out at that time were: the studies of the mechanism of photocurrent gain, estimation of the space-charge-limited currents (SCLC) [12–14], ascertainment of the peculiarities of the sweep-out of photogenerated carriers in strong electric fields [15–19], screening of electric fields in compensated semiconductors [20], and the study of thermal and generation–recombination noises in photoconductors [16]. Conventional extrinsic photoconductors have only one type of mobile carrier, unlike intrinsic devices. The low impedance of the intrinsic photoconductive devices results in space charge neutrality being relatively easy to maintain, and hence the excess distribution of both electrons and holes moves in one direction under an applied bias. Space charge neutrality can be violated in high-impedance photoconductors, and this can lead to unusual effects as noted later.

The earliest experimental studies have revealed a strong dependence of the parameters of extrinsic photoconductor detectors on the methods of fabrication and characteristics of the contacts especially at the frequencies that are higher than the inverse of the dielectric-relaxation time. The necessity of detailed studies of the mechanisms of this influence including the peculiarities of the behavior of the contacts with compensated semiconductors at low temperatures and backgrounds was emphasized in Sclar's review papers published in the 1980s [4,21–25]. It should be noted that these review papers have been of great importance in systematization of the data on extrinsic photoconductor detectors as well as in the popularization and incorporation of these devices into infrared optoelectronic equipment.

Considerable progress in understanding the operation of extrinsic photoconductors at low temperatures and backgrounds has been achieved in the last three decades [4,8,26–33]. An interesting historical overview of the theory of nonstationary behavior of low background extrinsic detectors, developed in Russia and ignored elsewhere, is given by Fouks [29]. More comprehensive theoretical discussion is presented by Kocherov and colleagues [31]. It has been shown that the heavy-doped contacts with the compensated semiconductors, being ohmic under the influence of the constant or low-frequency-modulated electric fields, become the effective injectors at the frequencies that are higher than the inverse of the trap recharging time. In this case, the injection of charge carriers is controlled by the electric field near the contact and observed with the essentially smaller bias than for SCLC.

The screening length in extrinsic semiconductors is dependent on the frequency. At low frequencies, the screening of electric fields takes place over the distance of the order of the Debye length, which is much smaller than the length of detector. The screening length greatly increases and can exceed the detector length at the frequencies that are higher than the dielectric-relaxation frequency, equal to the inverse of the dielectric-relaxation time. In this case, the sweep-out of the photogenerated carriers and photoconductive gain saturation with increasing the electric field takes place.

The frequency dependence of the impedance and photoresponse, transient behavior on an abrupt change of photosignal or voltage, as well as frequency dependence of the spectral density of noise for an extrinsic photoconductive detector were obtained using the new concepts of contact properties [27,29–35]. In addition, it has been shown that the trap recharging resulting from injection of charge carriers leads to unstable behavior of the monopolar plasma of carriers in a compensated semiconductor even with the linear current–voltage characteristic. The leading role of the excessive concentration of the majority carriers in nonsteady-state recombination processes has been established, and the mechanism of suppression of the shot noise of the current of the injecting contact at the intermediate frequencies has been found.

In the traditional discussion of photoconductivity, the initial rapid response is due to the generation and recombination of charges in the active region of the detector (see Figure 11.2) [36]. For the moderate background, the detector follows the signal reasonably (dashed line). In low background conditions, typical for extrinsic detectors, the necessity to emit a charge carrier to maintain equilibrium is communicated across the detector at roughly the dielectric relaxation-time constant, and the detector space charge adjusts to a new configuration at this relatively slow rate. The result of this slow injection of new carriers is a slow adjustment of the electric field near the injecting contact to a new equilibrium under a new level of illumination. Due to this phenomenon, the response shows multiple components, including fast and slow response, a hook anomaly, voltage spikes, and oscillations (solid line).

The hook response artifact (so named because of the appearance of the electric output waveform) can result from nonuniform illumination of the detector volume in transverse contact detectors; due to shading by the contacts, the portion of light under them is reduced. When illumination is increased, the resistance of the rest of the detector volume is driven down, leaving a high-resistance layer near the contacts for the carriers in most of the electric field and where the recovery to an equilibrium state occurs only over the dielectric-relaxation time. As a result, after the initial fast response the photoconductive gain in the detector area is driven down, and the overall response

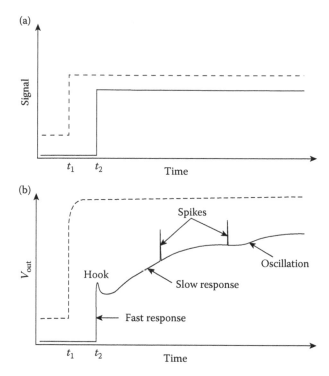

Figure 11.2 Response of an extrinsic photoconductor to two step input signals: (a) one from a moderate background (dashed line) and the other from a nearly zero one (solid line). (b) The response of the detector. (From Rieke, G. H., *Detection of Light: From the Ultraviolet to the Submillimeter*, 2nd ed., Cambridge University Press, Cambridge, 2003.)

decreases slowly. Hook response is associated with the quantum mechanical tunneling of charge carriers through the voltage barrier at the contact. Spikes are produced when charges are acceler-ated in the electric fields near contacts and acquire sufficient energy to impact-ionize impurities in the material, creating a mini-avalanche of charge carriers. Care in the construction of the detec-tors' electrical contacts can minimize hook response and spiking. In transparent Ge detectors, where the entire detector volume including the areas under the contacts is fully illuminated, the hook behavior is reduced or eliminated [35]. Spiking and hook response are also reduced at low photoconductive gain (for example, by operating the detector at reduced bias voltage).

Extrinsic photoconductors are relatively well behaved in high-background condition that pro-duces a steady concentration of free carriers that reduces the dielectric-relaxation time constant and hence allows rapid approach to a new equilibrium appropriate for the new signal level. At low backgrounds however, the signal must be extracted while the detector is still in a nonequilibrium state and the input signal must be deduced from partial response of the detector. A variety of approaches, including empirically fitted corrections and approximate analytical models, has been used [37,38]. Calibration of data under low backgrounds can be challenging.

11.3 PERFORMANCE OF EXTRINSIC PHOTOCONDUCTORS

11.3.1 Silicon-Doped Photoconductors

The spectral response of the detector depends on the energy level of the particular impurity state and the density of states as a function of energy in the band to which the bound charge carrier is excited. A number of other impurities have been investigated. Table 11.1 lists some of the common impurity levels and the corresponding long wavelength cutoff of the extrinsic silicon detector based on them. Note that the exact long wavelength spectral cutoff is a function of the impurity doping density, with higher densities giving slightly longer spectral response. Figure 11.3 illus-trates the spectral response for several extrinsic detectors [39]. The longer spectral response of the BIB Si:As device (see Section 11.4) compared with the bulk Si:As device is due to the higher doping level in the former that reduces the binding energy of an electron.

Table 11.1: Common Impurity Levels Used in Extrinsic Si Infrared Detectors

Impurity	Energy (meV)	Cutoff (μm)	Temperature (K)
Indium	155	8	40–60
Bismuth	69	18	20–30
Gallium	65	19	20–30
Arsenic	54	23	13
Antimony	39	32	10

Note: Operating temperature depends upon background flux level.

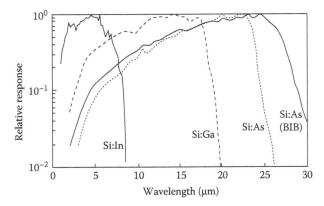

Figure 11.3 Examples of extrinsic silicon detector spectral response. Shown are Si:In, Si:Ga, and Si:As bulk detectors and a Si:As IBC. (From Norton, P. R., *Optical Engineering*, 30, 1649–63, 1991. With permission.)

Extrinsic detectors are frequently cooled with liquid He for applications such as ground- and space-based astronomy. Closed cycle two- and three-stage refrigerators are available for use with these detectors for cooling from 20 to 60 and 10 to 20 K, respectively.

The performance of extrinsic detectors is generally background limited with a quantum efficiency that varies with the specific dopant and dopant concentration, wavelength, and device thickness (see Section 9.2.2). Typical quantum efficiencies are in the range of 10–50% at the response peak.

For the 3–5 μm range, Si doped with the acceptor In (Si:In) offers an ideal choice. The In hole ground state lies at $E_v + 156$ meV (E_v is energy at the valence band top) leading to a photoconductive onset at $\lambda \approx 8$ μm, with peak response at 7.4 μm. Every Si crystal unavoidably contains some residual B acceptors with a hole binding energy of $E_v + 45$ meV. A further level called In-X that has been linked to an In-C center is located at $E_v + 111$ meV. Besides these three acceptor levels, there are always P donor levels present. Alexander et al. have evaluated the influence of the temperature and compensation of the shallow levels on detector responsivity [40]. At a very low temperature operation, when thermal generation from even the shallowest residual acceptor ($E_B = 45$ meV) is negligible, highest responsivity is always obtained by minimizing the concentration of residual donors. Very high responsivities may be obtained when the shallow acceptors are closely compensated because the effective cross section for capture in these centers is small due to high probability of thermal reemission, leading to long free-carrier lifetimes (as high as 200 ns). Values of 100 A/W have been obtained in Si:In. The close to perfect compensation has been achieved by a doping process using the transmutation of ^{30}Si nuclei into ^{31}P nuclei by thermal-neutron capture followed by β-decay [41,42]. The performance achieved with Si:In detectors are shown in Figure 9.10 and are compared with the theoretical curves.

Table 11.2: Performance of Si IR Detectors

Detector	$(E_i)_{op}$ (meV)	$(E_i)_{th}$ (meV)	λ_p (μm)	$\lambda_c(T)$ [μm; (K)]	$\eta(\lambda_p)$ (%)	T_{BLIP} (30 FOV; λ_p) (K; μm)
Si:Zn(p)		316	2.3	3.2 (50–110)	20	103(32)
Si:Tl(p)	246	240	3.5	4.3(78)	>1	
Si:Se(n)	306.7	300	3.5	4.1(78)	24	122
Si:In(p)	156.9	153	5.0	7.4(78)	48	60
Si:Te(n)	198.8	202	4.6	6.3(78)	25	77
Si:S(n)	186.42	174	5.5	6.8(78)	13	78
Si:Se'(n)	205	200	5.5	6.2(78)		85
Si:Ga(p)	74.05	74	15.0	17.8(27)	30(13.5)	32(13.5)
Si:Al(p)	70.18	67	15.0	18.4(29)	6(13.5)	32(13.5)
Si:Bi(n)	70.98	69	17.5	18.7(29)	35(13.5)	32(13.5)
Si:Mg(n)	107.5	108	11.5	12.1–12.4 (29)	2(11)	50(11)
Si:S'(n)	109	102	11.0	12.1(5)	<1	55

Source: N. Sclar, *Progress in Quantum Electronics* 9, 149–257, 1984. With permission.

For the longer wavelength detectors, Ga is chiefly used. The detector material is usually prepared by doping the melt during vertical float zone crystal growth. However, low activation energy (0.074 eV) of Ga is not optimum for use in the 8–14 μm window and unnecessarily contributes to its low operational temperature. Magnesium (Mg) has a more suitable energy level, but according to Sclar appears to have a shallow 0.044 eV level associated with it, requiring low temperatures or extra compensation [21]. If this level can be eliminated, then perhaps Mg is the ideal 8–14 μm level. Of other detectors, Si:Al and Si:Bi reveal various disadvantages [21]. During heat treatments of Si:Al wafers, Al precipitation and the formation of interstitial Al_2C_3 is frequently experienced, which is detrimental for preparing uniform detector arrays. In the case of Si:Bi, growing uniform crystals encompassing suitable doping concentrations by the float zone method has proved to be difficult because of the high vapor pressure of Bi at growth temperatures. Si:Bi grown by the Czochralski method exhibits higher B concentrations that decreases detector performance.

The comprehensive reviews of Si and Ge detectors operated in various spectral ranges are presented in Sclar's papers [4,21–25]. Table 11.2 summarizes the status of Si detectors for use in the atmospheric windows of 2–2.5 μm, 3–5 μm, and 8–14 μm. The next table (Table 11.3) presents a summary of the properties of some doped silicon and doped germanium devices for low-background space applications.

11.3.2 Germanium-Doped Photoconductors

Germanium extrinsic detectors have largely been supplanted by silicon detectors, as discussed previously, for both high and low background applications where comparable spectral response can be obtained, but germanium devices are still of interest for very long wavelengths. Germanium photoconductors have been used in a variety of infrared astronomical experiments, both airborne and space-based. As an example of the space application is ISOPHOT, the photometer for ESA's infrared space observatory (ISO), which uses extrinsic photoconductors at wavelength ranging from 3 to more than 200 μm [43]. The highly successful IRAS mission marks the beginning of modern, far infrared photoconductor research, and development [44]. Very shallow donors, such as Sb, and acceptors, such as B, In, or Ga, provide cutoff wavelengths in the region of 100 μm. Figure 11.4 shows the spectral response of the extrinsic germanium photoconductors doped with Zn, Be, Ga, and of stressed gallium doped germanium [28]. Despite a large amount of effort recently in the development of very sensitive thermal detectors, germanium photoconductors remain the most sensitive detectors for wavelength shorter than 240 μm.

The achievement of low NEP values in the range of a few parts 10^{-17} WHz$^{-1/2}$ (see Table 11.3) was made possible by advances in crystal growth development and control the residual minority impurities down to 10^{10} cm^{-3} in a doped crystal [11,45]. As a result, a high lifetime and mobility value and thus a higher photoconductive gain have been obtained.

Table 11.3: Status Summary of Some Si and Ge IR Detectors for Low-Background Applications

Detector	$(\Delta E)_{opt}$ (meV)	λ_p (μm)	$\lambda_c(T)$ μm(K)	$\eta(\lambda_p)$ (%)	Φ_B (phcm^{-2} s^{-1})	$NEP(\lambda; T; f)$ (WHz$^{-1/2}$)	λ(μm); T(K); f(Hz)
Si:As	53.76	23	24–24.5 (5)	50(T)	9×10^6	0.88×10^{-17}	(19; 6; 1.6)
				20(L)	6.4×10^7	4.0×10^{-17}	(23; 5; 5)
Si:P	45.59	24/26.5	28/29 (5)	~30(T)	2.5×10^8	7.5×10^{-17}	(28; 4.2; 10)
Si:Sb	42.74	28.8	31 (5)	58(T)	1.2×10^8	5.6×10^{-17}	(28.8; 5; 5)
				13(L)	1.2×10^8	5.5×10^{-17}	(28.8; 5; 5)
Si:Ga	74.05	15.0	18.4 (5)	47(T)	6.6×10^8	1.4×10^{-17}	(15; 5; 5)
Si:Bi	70.98	17.5	18.5 (27)	34(L)	$<1.7 \times 10^8$	3×10^{-17}	(13; 11; −)[a]
Ge:Li	9.98	125 (calc)			8×10^8	1.2×10^{-16}	(120; 2; 13)
Ge:Cu	43.21	23	29.5 (4.2)	50	5×10^{10}	1.0×10^{-15}	(12; 4.2; 1)
Ge:Be[b]	24.81	39	50.5 (4.2)	100[b]	1.9×10^{10}	1.8×10^{-16}	(43; 3.8; 20)
Ge:Ga	11.32	94	114 (3)	34	6.1×10^9	5.0×10^{-17}	(94; 3; 150)
Ge:Ga[b]	11.32	94	114 (3)	~100[b]	5.1×10^9	2.4×10^{-17}	(94; 3; 150)
Ge:Ga[b](s)[c]	~6	150	193 (2)	73[b]	2.2×10^{10}	5.7×10^{-17}	(150; 2; 150)

Source: N. Sclar, *Progress in Quantum Electronics*, 9, 149–257, 1984. With permission.
Note: T and L indicate transverse and longitudinal geometry detector.
[a] Signal integrated for 1 s.
[b] Results obtained with an integrating cavity.
[c] (s): stress = 6.6×10^3 kgcm^{-2}.

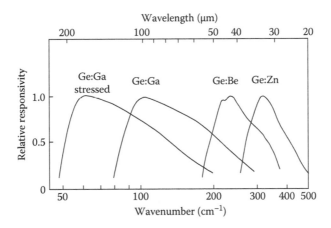

Figure 11.4 Relative spectral response of some germanium extrinsic photoconductors. (From Leotin, J., "Far Infrared Photoconductive Detectors," *Proceedings of SPIE* 666, 81–100, 1986. With permission.)

Ge:Be photoconductors cover the spectral range from ≈30 to 50 μm. Beryllium, a double acceptor in Ge with energy levels at $E_v + 24.5$ meV and $E_v + 58$ meV, poses special doping problems because of its strong oxygen affinity. The Be doping concentrations of 5×10^{14}–1×10^{15} cm^{-3} give significant photon absorption in 0.5–1 mm thick detectors, while at the same time keeping dark currents caused by hopping conduction at levels as low as a few tens of electrons per second. Ge:Be detectors with responsivities >10 A/W at $\lambda = 42$ μm and quantum efficiency 46% have been reported at low background [11].

Ge:Ga photoconductors are the best low background photon detectors for the wavelength range from 40 to 120 μm. Since the absorption coefficient for a material is given by the product of the photoionization cross section and the doping concentration (see Equation 9.58), it is generally desirable to maximize this concentration. The practical limit occurs when the concentration is so high that impurity band conduction results in excessive dark current. For Ge:Ga the onset of impurity banding occurs at approximately 2×10^{14} cm^{-3}, resulting in an absorption coefficient of

Table 11.4: Typical Ge:Ga Detector Parameters

Parameter	Value
Acceptor concentration	2×10^{14} cm^{-3}
Donor concentration	$< 1 \times 10^{11}$ cm^{-3}
Typical bias voltage	50 mV/mm
Operating temperature	< 1.8 K
Responsivity	7 A/W
Quantum efficiency	20%
Dark current	< 180 e/s

Source: Young, E., Stansberry, J., Gordon, K., and Cadien, J., "Properties of Germanium Photoconductor Detectors," in *Proceedings of the Conference of ESA SP-481*, 231–35, VilSpa, 2001.

only 2 cm^{-1} and typical values of quantum efficiency range from 10 to 20% [46]. Consequently, the detectors must either have long physical absorption path lengths or be mounted in an integrating cavity. Table 11.4 gives some characteristic parameters for the Ge:Ga detectors [47].

Application of uniaxial stress along the (100) axis of Ge:Ga crystals reduces the Ga acceptor binding energy, extending the cutoff wavelength to ≈240 μm [48,49]. In making practical use of this effect, it is essential to apply and maintain very uniform and controlled pressure to the detector so that the entire detector volume is placed under stress without exceeding its breaking strength at any point. A number of mechanical stress modules have been developed. The stressed Ge:Ga photoconductor systems have found a wide range of astronomical and astrophysical applications [43,47,50,51].

11.4 BLOCKED IMPURITY BAND DEVICES

From Section 9.1.2 results that in order to maximize the quantum efficiency and detectivity of extrinsic photoconductors, the doping level should be as high as possible. This is particularly important when the devices are required to be radiation hard and are made as thin as possible to minimize the absorbing volume for ionizing radiation. The limit to the useful doping that is possible in conventional extrinsic detectors is set by the onset of impurity banding. This occurs when the doping level is sufficiently high that the wave functions of neighboring impurities overlap and their energy level is broadened to a band that can support hopping conduction. When this occurs it limits the detector resistance and photoconductive gain and also increases the dark current and noise. In Si:As, for example, theses effects become important for doping levels above 7×10^{16} cm^{-3}. To overcome the impurity banding effect and, in addition, to improving radiation hardness and reducing the optical cross-talk between adjacent elements of an array, the BIB device was proposed. The BIB detectors, also called the impurity band conduction (IBC) detectors, have demonstrated another significant advantages, such as freedom from the irregular behavior typical of photoconductive detectors (spiking, anomalous transient response) increased frequency range for constant responsivity and superior uniformity of response over the detector area and from detector to detector.

The BIB devices made from either doped silicon and doped germanium are sensitive to infrared wavelength range located between 2 and 220 μm. They were first conceived at Rockwell International Science Center in 1977 by Petroff and Stapelbroek [52]. At the beginning, most of the BIB detector development centered on arsenic-doped silicon, Si:As [53,54]. The Si:As detector is sensitive to IR radiation only in the 2–30 μm wavelength range—they are widely used on ground-based telescopes and in all Spitzer instruments. Extension of BIB performance to longer wavelengths, awaited the development of suitable materials. Phosphorus is another attractive dopant (wavelength cutoff ~34 μm) because it is widely used in commercial integrated circuits and hence it would be relatively straightforward to fabricate detectors [55]. Si:Sb together with Si:As are installed in Spitzer infrared spectrograph [56,57]. BIB detectors have also been fabricated with Ge:Ga [58,59], Ge:B [60,61], Ge:Sb [62], and GaAs:Te [63,64]. Data on GaAs:Te can extend detector spectral approach to beyond 300 μm without having to apply uniaxial stress close to the breaking limit of detector's pixel. However, Si:As BIB detectors are the only form currently readily available in large format arrays.

The detector arrays are fabricated with an advanced well-established process currently in use in many manufacturers. Polished intrinsic silicon wafers and doped and undoped chemical vapor deposition are usually used in device preparation. In place of one uniform, moderately doped piece of single crystal semiconductor, we deal with a two layer structure consisting of a pure undoped blocking layer that is a few micrometers thick and a tens of micron thick, heavily doped IR absorbing layer. These two sections are contacted by degenerately doped, ohmic layers. Like bulk photoconductors, BIB devices are unipolar (either all n-type or p-type).

A BIB detector structure based on epitaxially grown n-type material is displayed in Figure 11.5. The active layer is sandwiched between a higher doped degenerate substrate electrode and an undoped blocking layer. Doping of an active layer with a thickness value in the 10 μm range is high enough for the onset of an impurity band in order to display a high quantum efficiency for impurity ionization (in the case of Si:As BIB, the active layer is doped to $\approx 5 \times 10^{17}$ cm^{-3}). The device exhibits a diode-like characteristic, except that photoexcitation of electrons takes place between the donor impurity and the conduction band. The heavily doped n-type IR-active layer has a small concentration of negatively charged compensating acceptor impurities ($N_a \approx 10^{13}$ cm^{-3}). In the absence of an applied bias, charge neutrality requires an equal concentration of ionized donors. Whereas the negative charges are fixed at acceptor sites, the positive charges associated

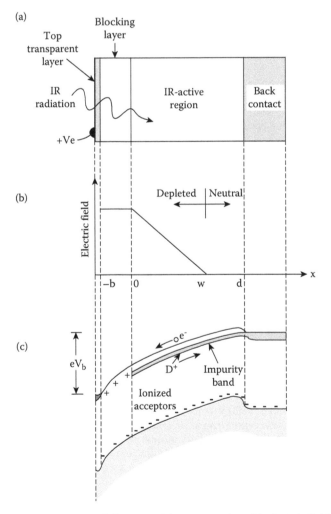

Figure 11.5 Blocked impurity band detector: (a) cross section, (b) electric field, and (c) energy-band diagram of a positively biased detector.

281

with ionized donor sites (D^+ charges) are mobile. The D^+ charges can propagate through the IR-active layer via the mechanism of hopping between occupied (D^0) and vacant (D^+) neighboring sites. Applying a positive bias to the transparent contact creates a field that drives the preexisting D^+ charges toward the substrate, while the undoped blocking layer prevents the injection of new D^+ charges. A region depleted of D^+ charges is therefore created, with a width depending on the applied bias and on the compensating acceptor concentration.

Below, we will focus our attention on n-type silicon; in particular, Si:As. The device consists of two layers deposited on a degenerately doped n-type Si substrate. The first layer is a heavily, but not degenerately, doped IR active layer with thickness d. Because of the heavy doping of the active layer, the impurity band increases in width around its nominal energy level and the energy gap decreases between the conduction band (or valence band for p-type material) and the nearest portion of the impurity band. Consequently, the minimum photon energy required to excite photoconduction is lower for a BIB detector than for a conventional bulk photoconductor with the same dopant, and the spectral response extends to longer wavelengths than that of the conventional detector. This effect is illustrated in Figure 11.3.

The IR active layer is divided into a depletion region of thickness w and a neutral region of thickness $d - w$. The second epilayer called the blocking layer, of thickness b, is intrinsic or, at most, lightly doped. Finally, a shallow n^+ implant transparent contact for incoming IR radiation, is put on top of the blocking layer. In the IR active layer there is a small concentration of acceptors (boron), which are assumed to be totally compensated and hence all ionized, $N_a^- = N_a (N_a \approx 10^{13}$ cm^{-3}). In the absence of an applied bias, charge neutrality requires an equal concentration of ionized donors, $N_d^+ = N_a$. The negative charges associated with the fixed acceptor sites are immobile, whereas the positive charges associated with ionized donor sites are mobile. Because of the heavy doping concentration of donors, there is a high probability of donor charge hopping from one site to another (called the impurity band effect). Electrons associated with the N_d^0 neutral sites hop to the vacant N_d^+ sites, which can be viewed as "holes" in the impurity band moving in the opposite direction. Applying a positive bias to the transparent contact, the N_d^+ charges are swept out through the IR active layer away from the interface with the blocking layer, while the undoped blocking layer prevents the injection of new N_d^+ charges. A region depleted of N_d^+ charges is therefore created. Since the ionized acceptor charges are not mobile, a negative space charge is left in the depletion region. The electric field is largest in the blocking layer and decreases linearly with the distance in the IR-active layer (see Figure 11.5b) in accordance with Poisson's equation. If the blocking layer is taken ideally as intrinsic (is devoid of all impurities), the width, w, of the depletion region is given by

$$w = \left[\frac{2\varepsilon_0 \varepsilon_s}{q N_a} V_b + b^2 \right]^{1/2} - b, \tag{11.2}$$

where V_b is the applied bias voltage. The width of the depletion region defines the active region of the device, since an appropriate electric field exists only in this region.

Assuming $N_a = 10^{13}$ cm^{-3}, $V_b = 4$ V, and $b = 4$ μm, we derive $w = 19.2$ μm. For a typical donor concentration $N_d = 5 \times 10^{17}$ cm^{-3}, it can be see that $\sigma_i N_d w = 2.12$, and the absorptive quantum efficiency will be about 88% for a single pass, or as high as 98% for a double pass in a detector with reflective back surface. The quantum efficiency will increase with increasing V_b until $w \geq d$, where d is the thickness of IR-active layer. Assuming that $w = d$ and typical values of other parameters, the electric field near the blocking layer is large, ≈ 2800 V/cm, far larger than the bulk conventional photoconductors [36].

The mean free path of the electrons in this large region is ≈ 0.2 μm. An electron that is accelerated across this path by a field of 2800 V/cm will acquire an energy of 0.056 eV. The excitation energy for Si:As is 0.054 eV (corresponding λ_c of 23 μm), therefore, 0.054 eV of energy will be adequate to ionize the neutral arsenic impurity atoms, resulting in two conduction band electrons. Repeated collisions can lead to significant gain and, because this is a statistical process, additional noise due to gain dispersion that can be characterized by an excess noise factor, β. This factor exceeds unity at higher bias voltages, and the detector therefore operates with increased noise. The product of the quantum efficiency η and the gain g gives the quantum yield that is proportional to the responsivity $R = (\lambda \eta / hc) qg$ (see Equation 3.33). The detective quantum efficiency is defined as the ratio of the quantum efficiency to the excess noise factor η / β. Under background-limited conditions the signal to noise ratio is proportional to the square root of η/β. For a more detailed analysis of the BIB detector see Szmulowicz and Madarsz [65].

There are several advantages of a BIB detector compared to a bulk photodetector. Because the intrinsic layer blocks the dark current, the doping in the active layer can be two orders of magnitude greater than in a bulk photoconductor. Second, the increase of doping allows a corresponding decrease in detector volume, effectively reducing interference from energetic particles or photons. Third, the electron travels to the positive contact traversing the blocking layer while the positive donor state (not the donor itself) travels to the undepleted region, which is the negative contact. The net result is equivalent to one carrier traveling the full distance between two contacts, independent of the location of the ionization event—there is only a single random event associated with the detection of a single photon. Consequently, the rms g-r noise current is reduced by a factor of $\sqrt{2}$ compared with conventional photoconductors (the recombination in BIB detectors occurs in relatively low resistance material and is not a random process distributed through the high impedance section of the detector).

The BIB detector resembles the photovoltaic detector in its operation. Pairs are created when IR radiation, with photon energy above the photoionization threshold for the chosen donor impurity, is incident on the detector. Those pair created in the depletion region get swept out by its electric field; pairs created in the neutral region recombine. There are also pairs that are produced thermally in the depletion region that give rise to the dark current and to the associated dark current noise. For large electric fields, electrons accelerating through the depletion region will impact ionize the neutral donors producing internal gain amplification. In addition, the heavily doped material under one contact of BIB detectors reduces the possibility of quantum mechanical tunneling and thus hook response, while the intrinsic material under the other contact, with its low impurity concentration, reduces impact ionization and thus spiking relative to bulk photoconductors under low backgrounds.

The details of the arrangement of steps in detector preparation depend on whether the detector is to be front illuminated through the first contact and blocking layer or back illuminated through the second contact. These two geometries are illustrated in Figure 11.6 [36]. Fabrication of a back-illuminated BIB detector begins with an ion-implanted buried electrical contact. The relatively highly doped IR-active layer and undoped blocking layer are grown epitaxially over the buried contact. Detector elements (pixels in array) are patterned and defined by another ion implant to form the second contact. The detector structures are completed by using standard silicon microlithographic processing to delineate metal lines, provide for electrical connection to the buried contact, and form indium bumps. In the front illuminated detector, a transparent contact is implanted into the blocking layer and the second contact is made by growing the detector on an

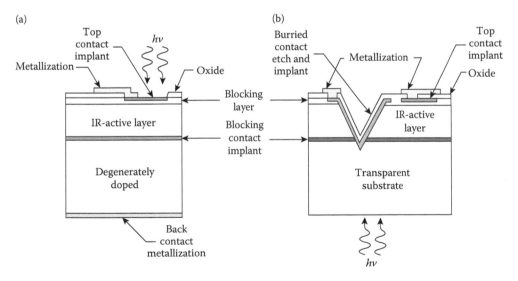

Figure 11.6 (a) Front illuminated BIB detector compared with (b) a back illuminated detector. (From Rieke, G. H., *Detection of Light: From the Ultraviolet to the Submillimeter*, 2nd ed., Cambridge University Press, Cambridge, 2003.)

extremely heavily doped, electrically conducting (degenerate) substrate. In the latter case, a thin degenerate but transparent contact layer is grown underneath the active layer on a high purity, transparent substrate. The bias voltage on the buried contact is established through a V-shaped etched trough, metal-coated to make it conductive, and placed to one side of the detector.

The BIB detectors exhibit similar spectral responses as conventional detector, except that the long wavelength cutoffs of these detectors are more voltage dependent, and extend to slightly longer wavelengths. The great success achieved with Si:As BIB arrays has so far not been carried over successfully to Ge BIB devices. The art and the science of Ge epitaxy was not nearly as far developed as for Si and it was also very different in many aspects. Several technological hurdles exist, including the very limited experience with Ge epitaxy and the nonexistence of high purity Ge epitaxial growth techniques. Structural defects and impurities continue to affect detector performance.

The fabrication of a Ge BIB detector requires the use of a high purity epitaxial technique. A different approach to fabricate these detectors that are based on the use of ultrapure Ge doped with ion implantation has led to functional devices that show low leakage currents but exhibit low responsivity because of the rather thin implanted layer [60]. It appears that liquid phase epitaxy has many advantages over MBE and CVD techniques (lower growth temperature, lower dopand level) for growth of epitaxial Ge layers for BIB detectors [66,67]. It can be expected that continued development efforts together with some new ideas will lead to functional Ge BIB detectors that can be used in low background astronomical and astrophysical observations.

For optimal GaAs BIB operation, unintentional doping of the blocking layer must be below 10^{13} cm^{-3}, but the requirements for the absorbing layer are more challenging. Compared to elemental Ge and Si semiconductors, one encounters difficulties that are due to the nature of GaAs. Any deviation from perfect stoichiometry leads to large concentrations of native defects. These defects are typically charged and form deep centers. In the interest of the highest possible purity, a LPE technique is preferred. A reproducible growth of GaAs LPE layers with residual doping below 10^{12} cm^{-3} has been demonstrated [68].

11.5 SOLID-STATE PHOTOMULTIPLIERS

The solid state photomultiplier (SSMP) is closely related to the BIB detector, and is illustrated schematically in Figure 11.7 [69]. Successful operation of SSPM requires optimization of a number of parameters, including not only the construction of the detector but also its bias voltage and operating temperature. The internal structure of the detector is similar to a Si:As BIB except that a well-defined gain region is grown between the blocking layer and IR-absorbing layer with higher compensating acceptor level ($5 \times 10^{13} - 1 \times 10^{14}$ cm^{-3}) and a thickness of about 4 μm. Typically, the

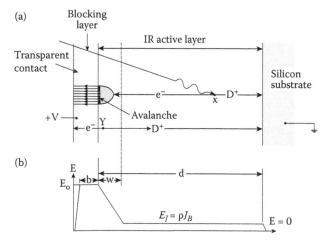

Figure 11.7 (a) Solid state photomultiplier configuration with schematic representation of the generation of bias current and effects following absorption of a photon, and (b) electric-field profile. (From Petroff, M. D., Stapelbroek, M. G., and Kleinhaus, W. A., *Applied Physics Letters*, 51, 406–8, 1987. With permission.)

IR-absorbing region has a lower acceptor concentration. When the detector is properly biased, the gain region goes into depletion and a strong electric field is developed across it as with the BIB detector. A higher electric field falling from about 8000 V/cm in the blocking layer over a depletion width in the IR-active layer. Over part of this depletion layer (4 μm thick) the electric field exceeds the critical field for impact ionization of neutral donors (≈2500 V/cm). A larger field than with a normal BIB detector increases the amount of avalanching. To the right of the depletion region is a uniform-field (≈ 1000 V/cm) drift region, ≈25 μm thick (see Figure 11.7b).

As in the case of BIB detectors, the positive charges associated with ionized donor sites (D^+ charges) are mobile. An electron-D^+ pair created by an IR photon (or by thermal generation) at point x in Figure 11.7a is separated by the electrical field $E_J = \rho J_B$ (where J_B is the bias current, and ρ is the resistivity of IR-active layer). The electron drifts rapidly to the left, with a negligible probability for impact ionization while in low-field region, and the D^+-charge drifts more slowly to the right. An avalanche of well-defined gain ($M \approx 4 \times 10^4$ at V = 7 V and T = 77 K) occurs for each electron entering the steeply increasing electric-field region adjacent to the blocking layer [69].

The SSPM devices are capable of detecting continuously the individual photons in the wavelength range from 0.4 to 28 μm [69], indicating the applicability of these devices to astronomical applications. The output pulses have submicrosecond rise times and amplitudes well above the readout noise. A counting quantum efficiency of over 30% was observed at 20 μm wavelength and over 50% in the visible-light region. Optimum photon-counting performance occurs for temperatures between 6 and 10 K for count rates less than 10^{10} counts/cm²s of detector area.

The performance of the SSPM is described in detail by Hays et al. [70].

REFERENCES

1. B. V. Rollin and E. L. Simmons, "Long Wavelength Infrared Photoconductivity of Silicon at Low Temperatures," *Proceedings of the Physical Society* B65, 995–96, 1952.

2. E. Burstein, J. J. Oberly, and J. W. Davisson, "Infrared Photoconductivity Due to Neutral Impurities in Silicon," *Physical Review* 89(1), 331–32, 1953.

3. B. V. Rollin and E. L. Simmons, "Long Wavelength Infrared Photoconductivity of Silicon at Low Temperatures," *Proceedings of the Physical Society* B66, 162–68, 1953.

4. N. Sclar, "Properties of Doped Silicon and Germanium Infrared Detectors," *Progress in Quantum Electronics* 9, 149–257, 1984.

5. C. R. McCreight, M. E. McKelvey, J. H. Goebel, G. M. Anderson, and J. H. Lee, "Detector Arrays for Low-Background Space Infrared Astronomy," *Laser Focus/Electro-Optics* 22, 128–33, November 1986.

6. F. J. Low, C. A. Beichman, F. C. Gillett, J. R. Houck, G. Neugebauer, D. E. Langford, R. G. Walker, and R. H. White, "Cryogenic Telescope on the Infrared Astronomical Satellite (IRAS)," *Proceedings of SPIE* 430, 288–96, 1983.

7. N. W. Boggess, "NASA Space Programs in Infrared Astronomy," *Laser Focus/Electro-Optics,* 116–27, June 1984.

8. G. H. Rieke, "Infared Detector Arrays for Astronomy," *Annual Review of Astronomy and Astrophysics* 45, 77–115, 2007.

9. T. Sprafke and J. W. Beletic, "High-Performance Infrared Focal Plane Arrays for Space Applications," *OPN* 22–27, June 2008.

10. P. R. Bratt, "Impurity Germanium and Silicon Infrared Detectors," in *Semiconductors and Semimetals,* Vol. 12, eds. R. K. Willardson and A. C. Beer, 39–142, Academic Press, New York, 1977.

11. E. E. Haller, "Advanced Far-Infrared Detectors," *Infrared Physics & Technology* 35, 127–46, 1994.

12. A. Rose, "Space-Charge-Limited Currents in Solids," *Physical Review* 97, 1538–44, 1955.

13. M. A. Lampert, "Simplified Theory of Space-Charge-Limited Currents in an Insulator with Traps," *Physical Review* 103, 1648–56, 1956.

14. M. A. Lampert and P. Mark, *Current Injections in Solids*, eds. H. G. Booker and N. DeClaris, Academic Press, New York, 1970.

15. A. Rose, "Performance of Photoconductors," *Proceedings of IRE* 43, 1850–69, 1955.

16. A. Rose, *Concepts in Photoconductivity and Allied Problems*, Interscience, New York, 1963.

17. R. L. Williams, "Response Characteristics of Extrinsic Photoconductors," *Journal of Applied Physics* 40, 184–90, 1969.

18. A. F. Milton and M. M. Blouke, "Sweepout and Dielectric Relaxation in Compensated Extrinsic Photoconductors," *Physical Review* 3B, 4312–30, 1971.

19. M. M. Blouke, E. E. Harp, C. R. Jeffus, and R. L. Williams, "Gain Saturation in Extrinsic Photoconductors Operating at Low Temperatures," *Journal of Applied Physics* 43, 188–94, 1972.

20. S. M. Ryvkin, *Photoelectric Effects in Semiconductors*, Consultants Bureau, New York, 1964.

21. N. Sclar, "Extrinsic Silicon Detectors for 3–5 and 8–14 μm," *Infrared Physics & Technology* 16, 435–48, 1976.

22. N. Sclar, "Survey of Dopants in Silicon for 2–2.7 and 3–5 μm Infrared Detector Application," *Infrared Physics & Technology* 17, 71–82, 1977.

23. N. Sclar, "Temperature Limitation for IR Extrinsic and Intrinsic Photodetectors," *IEEE Transactions on Electron Devices* ED-27, 109–18, 1980.

24. N. Sclar, "Development Status of Silicon Extrinsic IR Detectors," *Proceedings of SPIE* 409, 53–61, 1983.

25. N. Sclar, "Development Status of Silicon Extrinsic IR Detectors," *Proceedings of SPIE* 443, 11–41, 1984.

26. D. K. Schroder, "Extrinsic Silicon Focal Plane Arrays," in *Charge-Coupled Devices*, ed. D. F. Barbe, 57–90, Springer-Verlag, Heidelberg, 1980.

27. R. W. Westervelt and S. W. Teitsworth, "Nonlinear Transient Response of Extrinsic Ge Far-Infrared Photoconductors," *Journal of Applied Physics* 57, 5457–69, 1985.

28. J. Leotin, "Far Infrared Photoconductive Detectors," *Proceedings of SPIE* 666, 81–100, 1986.

29. B. I. Fouks, "Nonstationary Behaviour of Low Background Photon Detectors," *Proceedings of ESA Symposium on Photon Detectors for Space Instruments* 167–74, 1992.

30. N. M. Haegel, C. A. Latasa, and A. M. White, "Transient Response of Infrared Photoconductors: The Roles of Contacts and Space Charge," *Applied Physics* A 56, 15–21, 1993.

31. V. F. Kocherov, I. I. Taubkin, and N. B. Zaletaev, "Extrinsic Silicon and Germanium Detectors," in *Infrared Photon Detectors*, ed. A. Rogalski, 189–297, SPIE Optical Engineering Press, Bellingham, WA, 1995.

32. N. M. Haegel, C. B. Brennan, and A. M. White, "Transport in Extrinsic Photoconductors: A Comprehensive Model for Transient Response," *Journal of Applied Physics* 80, 1510–14, 1996.

33. N. M. Haegel, J. C. Simoes, A. M. White, and J. W. Beeman, "Transient Behavior of Infrared Photoconductors: Application of a Numerical Model," *Applied Optics* 38, 1910–19, 1999.

34. A. Abergel, M. A. Miville-Deschenes, F. X. Desert, M. Perault, H. Aussel, and M. Sauvage, "The Transient Behaviour of the Long Wavelength Channel of ISOCAM," *Experimental Astronomy* 10, 353–68, 2000.

35. N. M. Haegel, W. R. Schwartz, J. Zinter, A. M. White, and J. W. Beeman, "Origin of the Hook Effect in Extrinsic Photoconductor," *Applied Optics* 34, 5748–54, 2001.

36. G. H. Rieke, *Detection of Light: From the Ultraviolet to the Submillimeter,* 2nd ed., Cambridge University Press, Cambridge, 2003.

37. A. Coulais, B. I. Fouks, J.-F. Giovanelli, A. Abergel, and J. See, "Transient Response of IR Detectors Used in Space Astronomy: What We Have Learned From the ISO Satellite," *Proceedings of SPIE* 4131, 205–17, 2000.

38. J. Schubert, B. I. Fouks, D. Lemke, and J. Wolf, "Transient Response of ISOPHOT Si:Ga Infrared Photodetectors: Experimental Results and Application of the Theory of Nonstationary Processes," *Proceedings of SPIE* 2553, 461–69, 1995.

39. P. R. Norton, "Infrared Image Sensors," *Optical Engineering* 30, 1649–63, 1991.

40. D. H. Alexander, R. Baron, and O. M. Stafsudd, "Temperature Dependence of Responsivity in Closely Compensated Extrinsic Infrared Detectors," *IEEE Transactions on Electron Devices* ED-27, 71–77, 1980.

41. R. N. Thomas, T. T. Braggins, H. M. Hobgood, and W. J. Takei, "Compensation of Residual Boron Impurities in Extrinsic Indium-Doped Silicon by Neutron Transmutation of Silicon," *Journal of Applied Physics* 49, 2811–20, 1978.

42. H. M. Hobgood, T. T. Braggins, J. C. Swartz, and R. N. Thomas, "Role of Neutron Transmutation in the Development of High Sensitivity Extrinsic Silicon IR Detector Material," in *Neutron Transmutation in Semiconductors,* ed. J. M. Meese, 65–90, Plenum Press, New York, 1979.

43. J. Wolf, C. Gabriel, U. Grözinger, I. Heinrichsen, G. Hirth, S. Kirches, D. Lemke, et al., "Calibration Facility and Preflight Characterization of the Photometer in the Infrared Space Observatory," *Optical Engineering* 33, 26–36, 1994.

44. G. H. Rieke, M. W. Werner, R. I. Thompson, E. E. Becklin, W. F. Hoffmann, J. R. Houck, F. J. Low, W. A. Stein, and F. C. Witteborn, "Infrared Astronomy after IRAS," *Science* 231, 807–14, 1986.

45. N. M. Haegel and E. E. Haller, "Extrinsic Germanium Photoconductor Material: Crystal Growth and Characterization," *Proceedings of SPIE* 659, 188–94, 1986.

46. J.-Q. Wang, P. I. Richards, J. W. Beeman, J. W. Haegel, and E. E. Haller, "Optical Efficiency of Far-Infrared Photoconductors," *Applied Optics* 25, 4127–34, 1986.

47. E. Young, J. Stansberry, K. Gordon, and J. Cadien, "Properties of Germanium Photoconductor Detectors," in *Proceedings of the Conference of ESA SP-481,* eds. L. Metcalfe, A. Salama, S. B. Peschke and M. F. Kessler, 231–35, VilSpa, 2001.

48. A. G. Kazanskii, P. L. Richards, and E. E. Haller, "Far-Infrared Photoconductivity of Uniaxially Stressed Germanium," *Applied Physics Letters* 31, 496–97, 1977.

49. E. E. Haller, M. R. Hueschen, and P. L. Richards, "Ge:Ga Photoconductors in Low Infrared Backgrounds," *Applied Physics Letters* 34, 495–97, 1979.

50. N. Hiromoto, M. Fujiwara, H. Shibai, and H. Okuda, "Ge:Ga Far-Infrared Photoconductors for Space Applications," *Japanese Journal of Applied Physics* 35, 1676–80, 1996.

51. Y. Doi, S. Hirooka, A. Sato, M. Kawada, H. Shibai, Y. Okamura, S. Makiuti, T. Nakagawa, N. Hiromoto, and M. Fujiwara, "Large-Format and Compact Stressed Ge:Ga Array for the Astro-F (IRIS) Mission," *Advances in Space Research* 30, 2099–2104, 2002.

52. M. D. Petroff and M. G. Stapelbroek, "Blocked Impurity Band Detectors," *U.S. Patent,* No. 4 568 960, filed October 23, 1980, granted February 4, 1986.

53. S. B. Stetson, D. B. Reynolds, M. G. Stapelbroek, and R. L. Stermer, "Design and Performance of Blocked-Impurity-Band Detector Focal Plane Arrays," *Proceedings of SPIE* 686, 48–65, 1986.

54. D. B. Reynolds, D. H. Seib, S. B. Stetson, T. L. Herter, N. Rowlands, and J. Schoenwald, "Blocked Impurity Band Hybrid Infrared Focal Plane Arrays For Astronomy," *IEEE Transactions on Nuclear Science* 36, 857–62, 1989.

55. H. H. Hogue, M. L. Guptill, D. Reynolds, E. W. Atkins, and M. G. Stapelbroek, "Space Mid-IR Detectors from DRS," *Proceedings of SPIE* 4850, 880–89, 2003.

56. J. E. Huffman, A. G. Crouse, B. L. Halleck, T. V. Downes, and T. L. Herter, "Si:Sb Blocked Impurity Band Detectors for Infrared Astronomy," *Journal of Applied Physics* 72, 273–75, 1992.

57. J. E. van Cleve, T. L. Herter, R. Butturini, G. E. Gull, J. R. Houck, B. Pirger, and J. Schoenwald, "Evaluation of Si:As and Si:Sb Blocked-Impurity-Band Detectors for SIRTF and WIRE," *Proceedings of SPIE* 2553, 502–13, 1995.

58. D. M. Watson and J. E. Huffman, "Germanium Blocked Impurity Band Far Infrared Detectors," *Applied Physics Letters* 52, 1602–4, 1988.

59. D. M. Watson, M. T. Guptill, J. E. Huffman, T. N. Krabach, S. N. Raines, and S. Satyapal, "Germanium Blocked Impurity Band Detector Arrays: Unpassivated Devices with Bulk Substrates," *Journal of Applied Physics* 74, 4199–4206, 1993.

60. I. C. Wu, J. W. Beeman, P. N. Luke, W. L. Hansen, and E. E. Haller, "Ion-Implanted Extrinsic Ge Photodetectors with Extended Cutoff Wavelength," *Applied Physics Letters* 58, 1431–33, 1991.

61. J. W. Beeman, S. Goyal, L. R. Reichetz, and E. E. Haller, "Ion-Implanted Ge:B Far-Infrared Blocked Impurity-Band Detectors," *Infrared Physics & Technology* 51, 60–65, 2007.

62. J. Bandaru, J. W. Beeman, and E. E. Haller, "Growth and Performance of Ge:Sb Blocked Impurity Band (BIB) Detectors," *Proceedings of SPIE* 4486, 193–99, 2002.

63. L. A. Reichertz, J. W. Beeman, B. L. Cardozo, N. M. Haegel, E. E. Haller, G. Jakob, and R. Katterloher, "GaAs BIB Photodetector Development for Far-Infrared Astronomy," *Proceedings of SPIE* 5543, 231–38, 2004.

64. L. A. Reichertz, J. W. Beeman, B. L. Cardozo, G. Jakob, R. Katterloher, N. M. Haegel, and E. E. Haller, "Development of a GaAs-Based BIB Detector for Sub-mm Wavelengths," *Proceedings of SPIE* 6275, 62751S, 2006.

65. F. Szmulowicz and F. L. Madarsz, "Blocked Impurity Band Detectors: An Analytical Model: Figures of Merit," *Journal of Applied Physics* 62, 2533–40, 1987.

66. J. E. Huffman, "Infrared Detectors for 2 to 220 μm Astronomy," *Proceedings of SPIE* 2274, 157–69, 1994.

67. N. M. Haegel, "BIB Detector Development for the Far Infrared: From Ge to GaAs," *Proceedings of SPIE* 4999, 182–94, 2003.

68. E. E. Haller and J. W. Beeman, "Far Infrared Photoconductors: Recent Advances and Future Prospects," *Far-IR Sub-mm&MM Detectors Technology Workshop*, 2-06, Monterey, April, 1–3, 2002.

69. M. D. Petroff, M. G. Stapelbroek, and W. A. Kleinhaus, "Detection of Individual 0.4–28 μm Wavelength Photons via Impurity-Impact Ionization in a Solid-State Photomultiplier," *Applied Physics Letters* 51, 406–8, 1987.

70. K. M. Hays, R. A. La Violette, M. G. Stapelbroek, and M. D. Petroff, "The Solid State Photomultiplier-Status of Photon Counting Beyond the Near-Infrared," in *Proceedings of the Third Infrared Detector Technology Workshop,* ed. C. R. McCreight, 59–80, NASA Technical Memorandum 102209, 1989.

12 Photoemissive Detectors

In 1973, Shepherd and Yang of Rome Air Development Center (Rome, New York) proposed the concept of silicide Schottky-barrier detector FPAs as a much more reproducible alternative to HgCdTe FPAs for infrared thermal imaging [1]. For the first time it became possible to have much more sophisticated readout schemes—both detection and readout could be implemented in one common silicon chip. Since then, the development of the Schottky-barrier technology has progressed continuously and currently offers large IR image sensor formats. Despite lower quantum efficiency (QE) than other types of infrared detectors, the PtSi Schottky-barrier detector technology has exhibited remarkable advances. Such attributes as monolithic construction, uniformity in responsivity and signal to noise (the performance of an infrared system ultimately depends on the ability to compensate for the nonuniformity of an FPA using external electronics and a variety of temperature references), and the absence of discernible $1/f$ noise make Schottky-barrier devices a formidable contender to the mainstream infrared systems and applications [2–9]. While PtSi Schottky-barrier detectors are operated in the short and middle wavelength infrared spectral bands, long wavelength infrared detectors have already been demonstrated with Si-based heterojunction infrared photoemission detectors. The photodetection mechanism in the later detectors is the same as that of the Schottky-barrier detectors.

In this chapter the Si-based photoemissive detectors are reviewed.

12.1 INTERNAL PHOTOEMISSION PROCESS

The original electron photoemission model from metals into vacuum was described by Fowler in the 1930s [10]. In the 1960s Fowler's photoyield model was modified based on studies of internal photoemission of hot electrons from metal films into a semiconductor [11,12]. Cohen et al. modified the Fowler emission theory to account for emission into semiconductors [13].

As illustrated in Figure 12.1, internal photoemission resembles electron emission from metal to vacuum by photon irradiation. The incident photons are absorbed in the metal and generate electron-hole pairs. The excited electrons randomly walk in the metal films until they reach the interface between the metal and the semiconductor. Finally, the electrons surmount the barrier and are emitted into the semiconductor. The internal photoemission process involves three steps:

- Photoabsorption in the electrode that gives rise to a hot carrier gas

- Hot carrier transport in the electrode and in the semiconductor prior to barrier emission

- Emission over the Schottky barrier

Unlike intrinsic detectors, the QE of the Schottky-barrier detector depends on the photon energy because of the strong dependence of the emission probability on the energy of the excited electron.

Assuming that the probability of electron excitation is independent of the energy of the initial and final states and that there is an abrupt transition from filled to empty states at the Fermi level (see Figure 12.1), the total number of possible excited states is given by

$$N_T = \int_{E_F}^{E_F + h\nu} \frac{dN}{dE} dE, \tag{12.1}$$

where dN/dE is the density of states of the metal, E_F is the Fermi energy, $h\nu$ is the incident photon energy, and E is the electron energy with respect to the metal conduction band edge. The photoemission occurs when an electron is excited to a state for which the component of the momentum normal to the interface corresponds to a kinetic energy equal to or greater than the barrier. Therefore the number of states that meet the momentum criterion is

$$N_E = \int_{E_F + \phi_b}^{E_F + h\nu} \frac{dN}{dE} P(E) dE, \tag{12.2}$$

where $P(E)$ is the photoemission probability for the electron with an energy E and ϕ_b is the height of the barrier. If the momentum distribution of the electrons is isotropic, $P(E)$ can be calculated

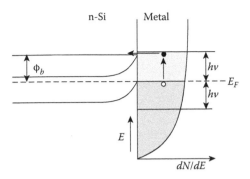

Figure 12.1 Internal photoemission in Schottky-barrier detector.

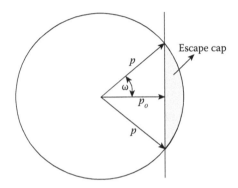

Figure 12.2 Momentum criterion for internal photoemission. p_o is the momentum corresponding to the barrier height. The excited electrons with momentum included within the escape cap are emitted into the semiconductor.

as shown in Figure 12.2. In this figure, p is the momentum of the excited electron and p_o is the momentum corresponding to the barrier height:

$$p = \sqrt{2m^* E}, \tag{12.3}$$

$$p_o = \sqrt{2m^* (E_F + \phi_b)}, \tag{12.4}$$

where m^* is the effective mass of the electron. $P(E)$ is the ratio of the surface area of the sphere included within the escape cap to the total surface area of the sphere and is equal to

$$P(E) = \frac{1}{2}(1 - \cos\omega) = \frac{1}{2}\left(1 - \sqrt{\frac{E_F + \phi_b}{E}}\right). \tag{12.5}$$

For further discussion we assume that dN/dE can be considered to be independent of the energy over the energy range of interest because the Fermi energy is much greater than the photon energy. Then Equations 12.1 and 12.2 become

$$N_T = \frac{dN}{dE}h\nu. \tag{12.6}$$

$$N_E = \frac{dN}{dE}\frac{(h\nu - \phi_b)^2}{8E_F}. \tag{12.7}$$

Here we assume that there are no collisions of electrons or energy losses before the excited electron reaches the interface. Thus, the internal QE, which is the ratio of N_E to N_T, is given by

$$\eta_i = \frac{1}{8E_F} \frac{(h\nu - \phi_b)^2}{h\nu}. \tag{12.8}$$

This simple theory described by Cohen et al. [13] was later extended by Dalal [14], Vickers [15], and Mooney and Silverman [16]. Following theses authors, the general form for the QE for internal photoemission is given by

$$\eta = C_f \frac{(h\nu - \phi_b)^2}{h\nu}, \tag{12.9}$$

where C_f is the Fowler emission coefficient. The Fowler coefficient provides an energy independent measure of the efficiency of the internal photoemission. Its value may be approximated by

$$C_f = \frac{H}{8E_F}, \tag{12.10}$$

where H is a device and a voltage dependent factor.

Equation 12.9 converted to wavelength variables is given by

$$\eta = 1.24C_f \frac{(1 - \lambda/\lambda_c)^2}{\lambda}. \tag{12.11}$$

C_f depends upon the physical and geometric parameters of the Schottky electrode. Values of λ_c and C_f as high as 6 μm and 0.5 (eV)$^{-1}$, respectively, have been obtained in PtSi-Si [3]. Schottky photoemission is independent of such factors as semiconductor doping, minority carrier lifetime, and alloy composition and, as a result of this, has spatial uniformity characteristics that are far superior to those of other detector technologies. Uniformity is limited only by the geometric definition of the detectors.

Figure 12.3 compares the spectral QE of the typical photon detectors. From this figure, it seems reasonable to choose high-QE intrinsic photodetectors. The effective QE of Schottky-barrier detectors in the 3–5 μm atmospheric window is very low, of the order of 1%, but useful sensitivity is obtained by means of near full frame integration in area arrays. An extension of this technology to the long wavelength band is possible using IrSi (see Figure 12.3), but this will require cooling below 77 K [4].

The current responsivity (see Equation 3.33 with $g = 1$) can be expressed as

$$R_i = qC_f \left(1 - \frac{\lambda}{\lambda_c}\right)^2. \tag{12.12}$$

Two specific properties of photoemissive detectors follow from the last two equations. The photoresponse decreases with wavelength and the QE is low, compared to that of bulk detectors. Both of these properties are a direct result of conservation of momentum during carrier emission over the potential barrier. The majority of excited carriers, which do not have enough momentum normal to the barrier, are reflected and not emitted. Figure 12.4 shows typical spectral responses of Pd_2Si, PtSi, and IrSi Schottky-barrier detectors [7].

Figures 9.31 and 12.1, which are often given as the energy band diagram for a Schottky barrier, are misleading because they give the impression that the peak of the Schottky barrier potential occurs at the semiconductor–electrode interface. The electric field near the Schottky barrier has influence on the barrier height. When the carrier is injected into the semiconductor, it feels an

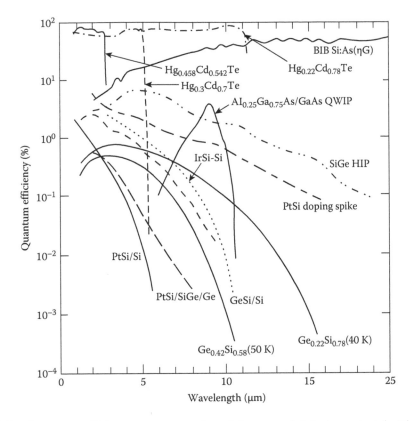

Figure 12.3 Quantum efficiency versus wavelength for several detector materials. The detectors include HgCdTe intrinsic photodiodes, blocked impurity band (BIB) extrinsic detectors, GaAs-based quantum well infrared photodetectors (QWIPs), and Si-based photoemissive detectors (PtSi, IrSi, PtSi/SiGe, PtSi doping spike, and SiGe heterojunction internal photoemission).

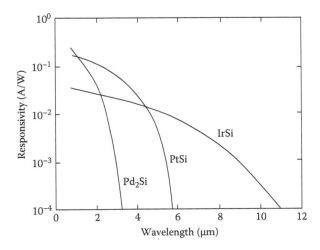

Figure 12.4 Spectral response of Pd$_2$Si, PtSi, and IrSi Schottky-barrier detectors. (From Kimata, M., and Tsubouchi, N., *Infrared Photon Detectors*, SPIE Optical Engineering Press, Bellingham, WA, 299–349, 1995. With permission.)

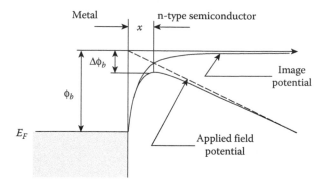

Figure 12.5 Barrier lowering by Schottky effect. The attraction force between the emitted electron and the induced positive charge reduces the barrier height by $\Delta\phi_b$. (From Sze, S. M., *Physics of Semiconductor Devices*, Wiley, New York, 1981. With permission.)

attractive force called the image force. As a result, the effective barrier height is reduced. This lowering is called the Schottky effect. As a result of this effect, the peak potential always occurs in the semiconductor, typically at a depth of 5 to 50 nm (see Figure 12.5 [17]).

The magnitude of the barrier lowering $\Delta\phi_b$ is given by [17]

$$\Delta\phi_b = \sqrt{\frac{qE}{4\pi\varepsilon_o\varepsilon_s}}, \qquad (12.13)$$

where q is the electron charge, E is the electric field near the barrier, ε_o is the permittivity of free space, and ε_s is the dielectric constant of silicon. The electric field is given by

$$E = \sqrt{\frac{2qN_d}{\varepsilon_o\varepsilon_s}\left(V + V_{bi} - \frac{kT}{q}\right)}, \qquad (12.14)$$

where N_d is the impurity concentration of silicon, V is the applied voltage, and V_{bi} is the built-in potential. The last equation indicates that the barrier height can be controlled by the reverse bias voltage and the impurity concentration of the substrate. The distance between the interface and potential maximum becomes shorter for larger electric fields, and this shift of the potential maximum enhances the QE coefficient.

From Equation 12.9 we find that the QE of the Schottky-barrier detector is expressed by two parameters: the barrier height and the Fowler emission coefficient. We can determine those two parameters from the plot of $\sqrt{\eta \times h\nu}$ versus $h\nu$. This type of analysis is known as a Flower plot. The Flower coefficient is determined from the square of the slope and the barrier potential from the intercept of the plot. Figure 12.6 shows the Fowler plot based on spectral responsivity data from a PtSi/p-Si detector fabricated at the Rome Laboratory [18]. Image lowering improves the emission efficiency and extends the spectral response to longer wavelengths with increased voltage.

12.1.1 Scattering Effects

Hot carrier transport in the electrode includes elastic scattering at surfaces and grain boundaries, as well as inelastic scattering with phonons and Fermi electrons [16]. Elastic scattering redirects carriers, thereby increasing the emission probability. Also, the phonon scattering redirects carriers and may improve the emission probability. However, the loss of energy to the lattice during phonon emission lowers the probability for transit of the potential barrier. In the case of multiple phonon scattering events, the carrier energy drops below that of the potential peak and the photoemission does not occur. The carrier is "thermalized" to the Fermi level.

The choice of electrode thickness is a compromise between keeping it thin enough for hot carriers to reach the silicon interface without loss of energy and making it thick enough to absorb the radiation. To optimize emission efficiency, the photoemission electrode must be

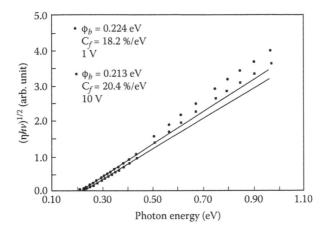

Figure 12.6 Flower photoemission analysis for PtSi/p-Si Schottky diode, with $\lambda_c = 5.5$ μm at 1 V bias and $\lambda_c = 5.8$ μm at 10 V. (From Shepherd, F. D., "Infrared Internal Emission Detectors," *Proceedings of SPIE* 1735, 250–61, 1992. With permission.)

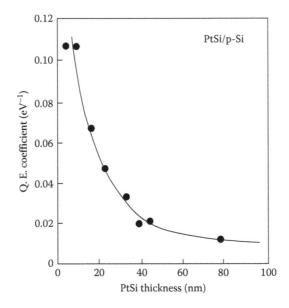

Figure 12.7 Dependence of quantum efficiency coefficient of PtSi Schottky-barrier detector on PtSi thickness. (From Kimata, M., Denda, M., Fukumoto, T., Tsubouchi, N., Uematsu, S., Shibata, H., Higuchi, T., Saeki, T., Tsunoda, R., and Kanno, T., *Japanese Journal of Applied Physics* 21(Suppl. 21-1), 231–35, 1982. With permission.)

designed to maximize the ratio of elastic to inelastic scattering events. Many papers reported a great improvement in the QE of PtSi/p-Si Schottky-barrier detectors by thinning the metal films as shown in Figure 12.7 [19–22]. This observation has been explained assuming that the interface scattering (at the metal–silicon and metal–dielectric interfaces) redirects the momentum without energy loss and the direction of momentum is independent of its previous orientation [13–15]. When the thickness of the metal film becomes much less than the attenuation length of the electron, multiple interface scattering occurs in the metal film and the QE is improved, because the redirection of the momentum by the interface scattering increases the probability of emission.

Mooney et al. [16,23] extended the Vickers model [15] to take into account the effect of carriers removed by emission and the energy loss by electron–phonon scattering. This model explains the fine structure of the spectral response, including roll-off from a linear fit of the modified Fowler plot for higher photon energy and a finite response for photon energies below the intercept of the linear region with the photon energy axis. The roll-off for large photon energy is caused by the reduction in the number of available carriers by prior emission events in the high QE region. The finite response below the energy of the extrapolated barrier height is related to the energy loss by the electron–phonon scattering. While only a few phonon collisions suffice to thermalize the hot carrier for low excitation energy, at higher energy, the carrier is less easily thermalized and more probably redirected into the escape direction and hence phonon scattering tends to increase the QE. This effect makes the apparent extrapolated barrier height higher than the actual barrier height. This is the reason why the barrier height measured by electrical methods is always lower than that measured by optical methods.

12.1.2 Dark Current

The current flowing through the barrier in silicide Schottky-barrier diodes is dominated by thermionic emission current. The thermionic emission theory gives the current-voltage characteristic expressed by Equation 9.148. The effective Richardson constant for holes in silicon, A^*, is about 30 A/cm²K² in a moderate electric field range [17].

The Schottky-barrier diode is operated under reverse bias in the IR focal plane arrays (FPAs). In the reverse biased condition, barrier lowering due to the Schottky effect has to be taken into consideration. For a reverse bias greater than $3kT/q$, from Equation 9.149 we get

$$J_{st} = A^* T^2 \exp\left[-\frac{q(\phi_b - \Delta\phi_b)}{kT}\right],$$ (12.15)

where $\Delta\phi_b$ is the magnitude of the barrier lowering calculated by Equation 12.13. By Equation 12.15, we can determine the effective barrier height at a certain reverse bias from the plot of J_{st}/T^2 versus $1/T$. Figure 12.8 is a Richardson analysis for the PtSi diode of Figure 12.6, biased at 1 V [18]. The Richardson constant is determined with less accuracy from the vertical intercept. Richardson analysis is often linear over more than five orders of magnitude. The presence of any leakage current or excess series resistance causes this plot to saturate. Thus, the Richardson analysis allows assessment of data quality. It is important to note that the barrier height obtained from the electrical measurement (ϕ_{bt}) is lower than that from the optical measurement (ϕ_{bo}), as discussed previously:

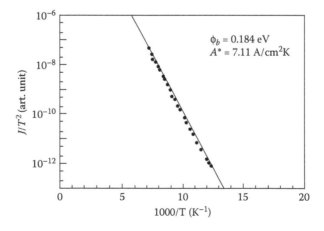

Figure 12.8 Richardson thermionic emission analysis PtSi/p-Si diode, at 1 V bias. Note difference in barrier potential compared to Flower analysis in Figure 12.6. (From Shepherd, F. D., *Proceedings of SPIE* 1735, 250–61, 1992. With permission.)

$$\phi_{bt} = \phi_{bo} - nh\nu, \tag{12.16}$$

where $nh\nu$ is the average energy loss by electron-phonon scattering. The typical measured value of this energy loss is from 20 to 50 meV. The lower values are observed in the most efficient devices where elastic scattering is dominant.

Using Equations 12.11 and 12.15, we can estimate the temperature required for background-limited operation of Schottky-barrier detectors as a function of cutoff wavelength, assuming that a range of barrier heights could be made. The temperature T_{BLIP} is calculated at which J_{st} is equal to the background current. The results are shown in Figure 3.19. The cooling requirements are more stringent in comparison with intrinsic detectors but are comparable with extrinsic silicon devices. For operation in the 8–12 μm band, cooling below 80 K is required—typical temperature is about 45 K.

12.1.3 Metal Electrodes

There are five silicides used for Schottky-barrier IR detectors: palladium silicide (Pd_2Si), platinum silicide (PtSi), iridium silicide (IrSi), cobalt silicide (Co_2Si), and nickel silicide (NiSi).

The solid–solid chemical reaction between the Pt and Si substrate is well defined and controllable. The main steps in the formation of PtSi are illustrated in Figure 12.9 [24]. The initial phase of a Pt_2Si layer begins to form at a temperature of 300°C. The thickness of the Pt_2Si layer grows with the square root of the anneal time, until all of the Pt is consumed, yielding a Pt_2Si–Si interface. Further absence of source elemental Pt leads to a change in the composition of the interface in favor of the formation of an Si-rich phase, namely PtSi. Thickness of PtSi layer is also proportional to the square root of the annealing time. Finally, additional annealing causes consumption of the Pt_2Si and the formation of the PtSi/Si junction interface. Postannealing oxidation forms an SiO_2 layer on the outer surface of the PtSi layer. This procedure is suitable for forming the resonant cavity in the Schottky-barrier structure.

The first thermal image was achieved at the General Electric Company using a high landing velocity vidicon, as described by Spratt and Schwarz [25], and a palladium silicide diode focal plane target, which was fabricated at Rome Laboratory. The vidicon sensitivity, angular resolution, and dynamic range were limited. As a result, tube sensor development was discontinued.

In the same year Kohn et al. [26] at the David Sarnoff Research Center (then RCA Laboratory, Princeton, New Jersey) demonstrated the first solid state IR imaging using a monolithic CCD array with Pd_2Si/p-Si Schottky photodiodes, fabricated using the process developed at Rome Laboratory. The CCD sensor had excellent dynamic range, good sensitivity from 1 μm to 3 μm, and angular resolution equal to the detector pitch. The barrier of the Pd_2Si detector was 0.34 eV and the cutoff wavelength was 3.5 μm. As a result, these devices had limited thermal sensitivity and research efforts were directed toward the development of PtSi, which had potential for photoresponse beyond 5 μm. There was continuing interest in the use of Pd_2Si to measure reflected energy in the spectral band between 1 and 3 μm, for satellite-based Earth resources assessment (e.g., [27–31]).

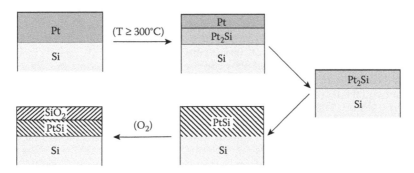

Figure 12.9 Formation of PtSi and Pt_2Si films. Arrows indicate increasing time or temperature. (From Dereniak, E. L., and Boreman, G. D., *Infrared Detectors and Systems*, Wiley, New York, 1996. With permission.)

The operating temperature was around 130 K, which is compatible with the current satellite passive cooling technology.

The most popular Schottky-barrier detector is the PtSi detector on p-type silicon, which can be used for detection in the 3–5 μm spectral range. In 1979, Capone et al. [32] described a 25×50 element PtSi monolithic IR-CCD array with $\lambda_c = 4.6$ μm (barrier height of 0.27 eV), QE coefficient of 0.036 eV^{-1}, 16% fill factor, 2% photoresponse uniformity, and an NEDT of 0.8°C. There have been steady advances in PtSi device processing and focal plane array development ever since. Significant advances were reported by Kosonocky [33] and Sauer [34] and their coworkers at the Sarnoff Center, Gates et al. [35] at Hughes Aircraft, Kimata et al. [36] at Mitsubishi, and Clark et al. [37] at Kodak. The state of the art has advanced to where several vendors in the United States and Japan offer PtSi monolithic arrays with sizes larger than 480×640 and NEDTs below 0.1 K. These devices have been incorporated in high performance IR cameras. In 1991 Yutani and coworkers [38] at Mitsubishi Electric Company reported operation of a 1040×1040 element PtSi array with an NEDT of 0.1 K.

The IrSi Schottky-barrier on p-type silicon is expected to have the lowest barrier. In 1982, Pellegrini et al. [39] measured photoresponse beyond 8 μm in iridium silicide detectors. They also discussed the difficulty of IrSi formation compared to PtSi formation [40]. Unlike PtSi formation, Si is the major diffusing species during the whole reaction process, and the contaminations at the original Si surface remain on the IrSi/Si interface and thus degrade diode characteristics. The other difficulty is related to the phase control. IrSi process repeatability is poor and array uniformities are substantially inferior to that of either Pd$_2$Si or PtSi. It is very difficult to obtain a clean interface with the IrSi detector because the diffusing atom is not Ir but Si. Despite these problems, IrSi arrays have been demonstrated [41].

CoSi$_2$ and NiSi detectors have also been developed for remote sensing applications in short wavelength IR bands. Optical barrier heights of CoSi$_2$ and NiSi detectors on p-type silicon were reported to be 0.44 and 0.40 eV, respectively [42,43].

12.2 CONTROL OF SCHOTTKY-BARRIER DETECTOR CUTOFF WAVELENGTH

The cutoff wavelength or operating temperature of Schottky-barrier detectors limits specific applications. The barrier potential may be raised to reduce dark current, or to allow an increase in operating temperature. Lowering the barrier of the PtSi detector improves cold night thermal imaging performance but requires operation at lower temperature.

Enhancement of the electric field near the barrier reduces the effective barrier height. Tsubouchi et al. have observed the barrier lowering of PtSi Schottky-barrier detectors by the Schottky effect [44]. The effective barrier height is also reduced by narrowing the potential profile to the onset of tunneling. Shannon demonstrated raising and lowering of Ni Schottky barrier potential by very shallow ion implantation [45]. This technique was employed by Pellegrini et al. [46] and by Wei et al. [47] to adjust the cutoff wavelength of PtSi Schottky-barrier detectors. Tsaur et al. applied this technique to an IrSi Schottky-barrier detector, and obtained a cutoff wavelength longer than 12 μm [48].

Fathauer et al. controlled the barrier potential of both p-type and n-type CoSi$_2$/Si Schottky diodes by MBE growth of thin n$^+$ Ga doped Si layer at the metal-semiconductor interface [49]. They reported that the barrier height was reduced from 0.35 to 0.25 eV. Lin et al. demonstrated a PtSi detector with a 22 μm cutoff wavelength by introducing a 1 nm thick doping spike at the interface [50–52]. A thin doping spike is essential to reduce the optical barrier height. In order to make a very thin doping spike with high impurity concentrations, they developed a low-temperature Si-MBE technology using elemental boron as a dopant source.

The barrier height may also be controlled by the use of an alloy of two metals. Tsaur et al. fabricated a Schottky-barrier detector with a combination of PtSi and IrSi [53]. A Pt-Ir silicde Schottky-barrier equal 0.135 eV was obtained by sequentially depositing 1.5 nm Ir and 0.5 nm Pt films onto p-type silicon. They reported that better diode characteristics can be obtained using this technology compared with IrSi detectors alone.

Schottky-barrier detectors are generally fabricated on (100) substrates. However, it appears that the use of a (111) orientation raises the barrier potential nearly 0.1 eV to 0.313 eV. This reduces the detector cutoff wavelength to 4.0 μm and raises the operating temperature above 100 K.

The existence of surface states also affects the barrier height. The barrier height of PtSi barriers decreases with the reduction of the surface states [54].

12.3 OPTIMIZED STRUCTURE AND FABRICATION OF SCHOTTKY-BARRIER DETECTORS

The Schottky-barrier detector is typically operated in back side illumination mode. This mode of operation is possible because silicon is transparent in the IR spectral range of interest. The silicon substrate serves as a refraction index-matching layer for the silicide film, and then the back side illumination mode has lower reflection loss than the front side illumination mode. The difference in the quantum efficiency between the back side and front side illumination modes was reported to be a factor of 3 [55].

To achieve maximum responsivity, the Schottky-barrier detector is usually constructed with an "optical cavity." The optical cavity structure consists of the metal reflector and the dielectric film between the reflector and the metal electrode of the Schottky-barrier diode. According to fundamental optical theory, the effect of the optical cavity depends on the thickness and refractive index of the dielectric films and the wavelength. The conventional 1/4 wavelength design for the optical cavity thickness is a good first approximation for optimizing the responsivity [56]. For example, Figure 12.10 shows the cross section of a PtSi Schottky-barrier detector with optical cavity tuned for maximum response at 4.3 μm [6].

It should be added that the front side illuminated Schottky-barrier detectors were reported with the IR signal introduced on the silicide side of the detector to achieve a broad spectral response including infrared, visible, and ultraviolet [55,57], and in the case of direct Schottky injection for achieving 100% fill factor [58]. Figure 12.11 shows the spectral response of a front side illuminated PtSi Schottky-barrier detector in the wavelength range between 0.4 and 5.2 μm [59]. The quantum efficiency in the visible and near infrared spectral regions is higher than in the infrared spectral range due to electron-hole creation in the silicon substrate for a wavelength range shorter than the fundamental absorption edge of silicon.

The main advantage of the Schottky-barrier detectors is that they can be fabricated as monolithic arrays in a standard silicon VLSI process. Typically, the silicon arrays are completed up to the Al metallizational step. A Schottky-contact mask is used to open the SiO_2 surface to p-type (100) silicon (with resistivity 30–50 Ωcm) at the Schottky-barrier detector location. In the case of PtSi detectors, a very thin layer of Pt (1 to 2 nm) is deposited and sintered (annealed at a temperature in the range of 300–600°C) to form PtSi, and the unreacted Pt on the SiO_2 surfaces is removed by dip-etching in hot aqua regia [60]. The Schottky-barrier structure is then completed by a deposition of a suitable dielectric (usually SiO_2) for forming the resonant cavity, removing this dielectric outside the Schottky-barrier regions, and depositing and defining Al for the detector reflector and the interconnects of the Si readout multiplexer. In the case of the 10 μm IrSi Schottky-barrier detectors, the IrSi was formed by *in situ* vacuum annealing and the unreacted Ir was removed by reactive ion etching [61].

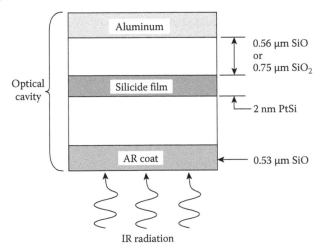

Figure 12.10 Cross-sectional view of PtSi Schottky-barrier detector with optical cavity tuned for maximum response at a wavelength of 4.3 μm. (From Kosonocky, W. F., *Optoelectronics: Devices and Technologies*, 6, 173–203, 1991. With permission.)

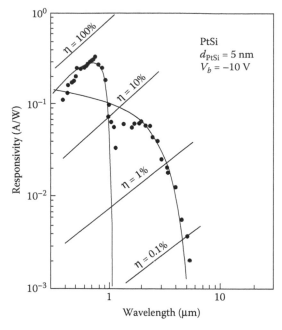

Figure 12.11 Spectral response of a front side illuminated PtSi Schottky-barrier detector. The fitting curves in the infrared and visible spectral regions are calculated based on internal photoemission and intrinsic mechanism, respectively. (From Kimata, M., Denda, M., Iwade, S., Yutani, N., and Tsubouchi, N., *International Journal of Infrared and MM Waves*, 6, 1031–41, 1985. With permission.)

12.4 NOVEL INTERNAL PHOTOEMISSIVE DETECTORS
12.4.1 Heterojunction Internal Photoemissive Detectors

In 1971 Shepherd et al. recommended replacing the metal electrode of Schottky-barrier detector with degenerate semiconductors [62]. The free carrier density of a degenerate semiconductor is at least two orders of magnitude less than that of metal-silicides. Because of this, the carrier absorption coefficient is lower and degenerate semiconductor electrodes must be made thicker, to match metal-silicide absorption efficiencies. In 1970, degenerate electrode devices were fabricated at Rome Laboratory using alloying and liquid phase epitaxy, however due to higher phonon scattering losses during transport in the thicker electrode, the emission efficiencies of these devices were not competitive with those of metal-silicide devices and work was discontinued [62].

The development in the MBE technology made it possible to fabricate high quality Ge_xSi_{1-x} (GeSi) thin films onto silicon substrates, and several works concerning the realization of ideas utilizing the internal photoemission of GeSi/Si heterojunction diodes for infrared detection have been reported [63–74]. Lin et al. demonstrated the first GeSi/Si heterojunction detector in 1990 [63,64]. Figure 12.12 shows the structure and energy band diagram of this detector [68]. An energy barrier ϕ_b is established between GeSi and Si. The band gap energy of the strained SiGe is smaller than that of Si, and by varying Ge composition and the doping concentration of the Ge_xSi_{1-x} layer, the cutoff wavelength of the device can be varied from 2 to 25 μm [65,68]. Figure 12.3 compares the quantum efficiencies of several technologies, between them a conventional PtSi/p-Si detector, IrSi/p-Si detector, PtSi/p-$Si_{0.85}Ge_{0.15}$/p-Si detector, PtSi/p-Si detector with a 1 nm doping spike of 2×10^{20} cm^{-3} boron, and p-$Si_{0.7}Ge_{0.3}$/p-Si heterojunction detector with a boron doping of 5×10^{20} cm^{-3} in the SiGe layer.

Absorption mechanisms in the $Ge_{1-x}Si_x$/Si detector are free carrier absorption and intravalence band transitions. Assuming that the density of state of the valence band is proportional to the square root of the energy, the quantum efficiency can be expressed as [70]:

$$\eta = \frac{A}{8E_F^{1/2}\left(E_F + \phi_b\right)^{1/2}} \frac{\left(h\nu - \phi_b\right)^2}{h\nu}, \tag{12.17}$$

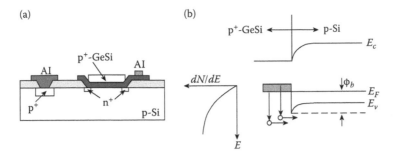

Figure 12.12 Construction (a) and operation (b) of Ge$_x$Si$_{1-x}$ heterojunction photoemissive detector. (From Tsaur, B.-Y., Chen, C. K., and Marino, S. A., "Long-Wavelength Ge$_{1-x}$Si$_x$/Si Heterojunction Infrared Detectors and Focal Plane Arrays," *Proceedings of SPIE* 1540, 580–95, 1991. With permission.)

where A is the absorptance relatively wavelength-independent in the long wavelength region. This expression is similar for silicide-barrier detectors (see Equation 12.9). Since E_F is usually much smaller for semiconductors than for metals, η is expected to be significantly greater for heterojunction detectors than for silicide Schottky-barrier detectors. On the other hand, the absorptance A of Ge$_{1-x}$Si$_x$ is expected to be smaller than that of the silicide Schottky-barrier detectors because of the lower density of free carriers in Ge$_{1-x}$Si$_x$.

The optimum thickness of Ge$_{1-x}$Si$_x$ is reported to be about 20 nm [68], which is an order of magnitude thicker than that of the metal electrode for silicide-barrier detectors. This thickness is a compromise between achieving a possible high absorptance of GeSi layer and an influence of elastic and inelastic scattering (see discussion of scattering effects in Section 12.1.1). Tsaur et al. discussed the improvement of the quantum efficiency of the heterojunction detector with a metal overlayer and double-heterojunction structure [68]. Park et al. proposed a stacked GeSi/Si heterojunction detector, with which both the optical absorption and internal quantum efficiency can be optimized [71–73]. The stacked Si$_{0.7}$Ge$_{0.3}$/Si detector consisted of several periods of degenerated boron doped thin (≤ 5 nm) Si$_{0.7}$Ge$_{0.3}$ layers and undoped thick (≈ 30 nm) Si layers has exhibited photoresponse at wavelengths ranging from 2 to 20 µm with quantum efficiency of about 4% and 1.5% at 10 and 15 µm wavelengths, respectively. The detectors showed near ideal thermionic-emission limited dark current characteristics.

12.4.2 Homojunction Internal Photoemissive Detectors

The concept of homojunction internal photoemissive detector for IR was first realized in 1988 by Tohyama et al. [75]. This detector had useful spectral response from 1 to 7 µm, although the reported quantum efficiency was low. Next, O'Neil reported that the quantum efficiency of the homojunction detector had been improved to a practical level (several percent). His group has developed a 128×128 element array using this technology [76].

More recently, various detector approaches based on a high-low Si and GaAs homojunction interfacial workfunction internal photoemission (HIWIP) junctions have been discussed by Perera et al. [77–84]. The operation of HIWIP detectors is based on the internal photoemission occurring at the interface between a heavily doped absorber/emitter layer and an intrinsic layer, with the cutoff wavelength mainly determined by the interfacial workfunction. The detection mechanism involves far infrared (FIR) absorption in the highly doped thin emitter layers by free carrier absorption followed by the internal photoemission of photoexcited carriers across the junction barrier and then collection.

The basic structure of the HIWP detector consists of a heavily doped layer, which act as the IR absorber region, and an intrinsic (or lightly doped) layer across which most of the bias is dropped. According to the doping concentration level in the heavily doped layer, these detectors can be divided into three types as shown in Figure 12.13 [77].

In the Type I detector, when the doping concentration, N_d, in the n$^+$-layer is high but below the Mott critical value, N_c, an impurity band is formed. At low temperatures, the Fermi energy is located in the impurity band. The incident IR radiation is absorbed due to the impurity photoionization, with a workfunction given by $\Delta = E_c^{n^+} - E_F$, where $E_c^{n^+}$ is the conduction band edge in the n$^+$-layer. An electric field is formed in the i-layer by an external bias to collect photoexcited electrons generated in the n$^+$-layer. Type I detectors are analogous

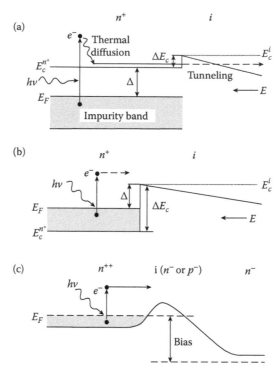

Figure 12.13 Energy band diagram of three different types of heterojunction internal photo-emissive detectors. (a) Type I: $N_d < N_c$ ($E_F < E_c^{n^+}$), (b) Type II: $N_d < N_d < N_o$ ($E_c^{n^+} < E_F < E_c^i$), and (c) Type III: $N_d > N_o$ ($E_F > E_c^i$). Here N_c is the Mott's critical concentration and N_o is the critical concentration corresponding to $\Delta = 0$. In (a) and (b) the conduction band edge of the i-layer is represented by a dotted line for $V_b = 0$ and by solid line for $V_b > V_o$. (From Perera, A. G. U., Yuan, H. X., and Francombe, M. H., *Journal of Applied Physics*, 77, 915–24, 1995. With permission.)

to semiconductor photoemissive detectors in their operation, which can be described by a three-step process:

- Photoexcitation of electrons from filled impurity band states below Fermi level into empty states in conduction band

- Rapid thermalization of photoexcited electrons into the bottom of the conduction band by pho-non relaxation and next their diffusion to the emitting interface

- Tunneling of the electrons through an interfacial barrier, ΔE_c, which is due to the offset of the conduction band edge caused by the bandgap narrowing effect

Type II detectors are analogous to Schottky-barrier IR detectors in their operation. When the doping concentration is above the Mott transition, the impurity band is linked with the conduction band edge, and n+-layer becomes metallic. Even in this case, the Fermi level can still be below the conduction band edge of the i-layer ($E_F < E_c^i$) due to the bandgap narrowing effect, giving rise to a workfunction $\Delta = E_c^i - E_F$, unless N_d exceeds a critical concentration N_o at which $\Delta = 0$. One of unique features of Type II detectors is the fact that there is no restriction on cutoff wavelength, because the workfunction can become arbitrarily small with increasing doping concentration. High perfor-mance Si detectors with $\lambda_c > 40\ \mu m$ can be realized [77].

In the Type III detector the doping concentration is so high that the Fermi level is above the conduction band edge of the i-layer, the n+-layer becomes degenerate, and a barrier associated with a space-charge region is formed at the n+-i interface due to electron diffusion. The barrier height depends on the doping concentration and the applied voltage, giving rise to a electrically tunable λ_c. As the barrier voltage is increased, the barrier height is reduced, the spectral response shifts toward a longer wavelength and the signal increases at a given wavelength. This change in response is caused by image lowering, leveraged by the p-n junction at the interface. Type III

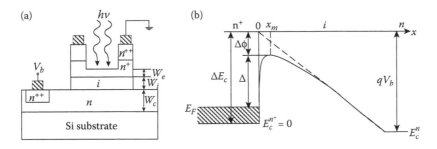

Figure 12.14 (a) Basic structure, and (b) energy band diagram of a front side illuminated n^+-i HIWIP. (From Yuan, H. X., and Perera, A. G. H., *Applied Physics Letters*, 66, 2262–64, 1995. With permission.)

device was first demonstrated by Tohyama et al. [75] using a structure composed of a degenerate n^{++} hot-carrier, a depleted barrier layer (lightly doped p, n or i), and a lightly doped n-type hot-carrier collector. This detector had useful spectral response from 1 to 7 μm, although the reported quantum efficiency was low.

Figure 12.14 shows the basic structure and energy band diagram for a n^+-i HIWIP detector [79]. The structure consists of the emitter, intrinsic, and collector layers, with the respective thickness represented by W_e, W_i, and W_c. The top contact layer is formed as a ring surrounding the active area to minimize absorption loss. The collector layer is moderately doped and has a relatively low resistance due to the impurity band conduction, while it is still transparent in the FIR range when the photon energy is smaller than the impurity ionization energy. The interfacial work function is given by $\Delta = \Delta E_c - E_F - \Delta\phi$, where ΔE_c is the conduction band edge offset due to band-gap narrowing in the heavily doped emitter layer, E_F is the Fermi energy, and $\Delta\phi$ is a lowering of interfacial barrier height due to image force effect. In order to get high internal quantum efficiency, the emitter layer thickness should be thin enough, in spite of the reduction of the photon absorption efficiency. Thus, the optimal thickness is a tradeoff of photon absorption and hot electron scattering.

Si HIWIP FIR detectors could have a performance comparable to that of conventional Ge FIR photoconductors [85] or Ge BIB detectors [86]. In addition to Si, significant bandgap shrinkage has been observed for heavily doped p-GaAs. Better carrier transport properties of GaAs may produce improved performance for this type of device. The p-GaAs HIWIP FIR detectors show great potential in becoming a strong competitor in FIR applications: responsivity of 3.10 ± 0.05 A/W, quantum efficiency of 12.5%, and detectivity of 5.9×10^{10} cmHz$^{1/2}$/W at 4.2 K, for λ_c from 80 to 100 μm.

REFERENCES

1. F. D. Shepherd and A. C. Yang, "Silicon Schottky Retinas for Infrared Imaging," *Technical Digest of IEDM*, 310–13, 1973.

2. W. F. Kosonocky, "Infrared Image Sensors with Schottky-Barrier Detectors," *Proceedings of SPIE* 869, 90–106, 1987.

3. F. D. Shepherd, "Schottky Diode Based Infrared Sensors," *Proceedings of SPIE* 443, 42–49, 1984.

4. F. D. Shepherd, "Silicide Infrared Staring Sensors," *Proceedings of SPIE* 930, 2–10, 1988.

5. W. F. Kosonocky, "Review of Schottky-Barrier Imager Technology," *Proceedings of SPIE* 1308, 2–26, 1990.

6. W. F. Kosonocky, "Review of Infrared Image Sensors with Schottky-Barrier Detectors," *Optoelectronics: Devices and Technologies* 6, 173–203, 1991.

7. M. Kimata and N. Tsubouchi, "Schottky Barrier Photoemissive Detectors," in *Infrared Photon Detectors*, ed. A. Rogalski, 299–349, SPIE Optical Engineering Press, Bellingham, WA, 1995.

8. M. Kimata, "Metal Silicide Schottky Infrared Detector Arrays," in *Infrared Detectors and Emitters: Materials and Devices*, eds. P. Capper and C. T. Elliott, 77–98, Kluwer Academic Publishers, Boston, MA, 2000.

9. M. Kimata, "Silicon Infrared Focal Plane Arrays," in *Handbook of Infrared Detection Technologies*, eds. M. Henini and M. Razeghi, 353–92, Elsevier, 2002.

10. R. H. Fowler, "The Analysis of Photoelectric Sensitivity Curves for Clean Metals at Various Temperatures," *Physical Review* 38, 45–57, 1931.

11. C. R. Crowell, W. G. Spitzer, L. E. Howarth, and E. E. Labate, "Attenuation Length Measurements of Hot Electrons in Metal Films," *Physical Review* 127, 2006, 1962.

12. R. Stuart, F. Wooten, and W. E. Spicer, "Monte-Carlo Calculations Pertaining to the Transport of Hot Electrons in Metals," *Physical Review* 135(2A), 495–504, 1964.

13. J. Cohen, J. Vilms, and R. J. Archer, "Investigation of Semiconductor Schottky Barriers for Optical Detection and Cathodic Emission," Air Force Cambridge Research Labs. Report No. 68-0651 (1968) and No. 69-0287 (1969).

14. V. L. Dalal, "Simple Model for Internal Photoemission," *Journal of Applied Physics* 42, 2274–79, 1971.

15. V. E. Vickers, "Model of Schottky-Barrier Hot Electron Mode Photodetection," *Applied Optics* 10, 2190–92, 1971.

16. J. M. Mooney and J. Silverman, "The Theory of Hot-Electron Photoemission in Schottky Barrier Detectors," *IEEE Transactions on Electron Devices* ED-32, 33–39, 1985.

17. S. M. Sze, *Physics of Semiconductor Devices*, Wiley, New York, 1981.

18. F. D. Shepherd, "Infrared Internal Emission Detectors," *Proceedings of SPIE* 1735, 250–61, 1992.

19. R. Taylor, L. Skolnik, B. Capone, W. Ewing, F. Shepherd, S. Roosild, B. Cochrun, M. Cantella, J. Klein, and W. Kosonocky, "Improved Platinum Silicide IRCCD Focal Plane," *Proceedings of SPIE* 217, 103–10, 1980.

20. H. Elabd and W. F. Kosonocky, "The Photoresponse of Thin-Film PtSi Schottky Barrier Detector with Optical Cavity," *RCA Review* 43, 543–47, 1982.

21. H. Elabd and W. F. Kosonocky, "Theory and Measurements of Photoresponse for Thin Film Pd_2Si and PtSi Infrared Schottky-Barrier Detectors with Optical Cavity," *RCA Review* 43, 569–89, 1982.

22. M. Kimata, M. Denda, T. Fukumoto, N. Tsubouchi, S. Uematsu, H. Shibata, T. Higuchi, T. Saeki, R. Tsunoda, and T. Kanno, "Platinum Silicide Schottky-Barrier IR-CCD Image Sensor," *Japanese Journal of Applied Physics* 21(Suppl. 21-1), 231–35, 1982.

23. J. M. Mooney, J. Silverman, and M. M. Weeks, "PtSi Internal Photoemission; Theory and Experiment," *Proceedings of SPIE* 782, 99–107, 1987.

24. E. L. Dereniak and G. D. Boreman, *Infrared Detectors and Systems*, Wiley, New York, 1996.

25. J. P. Spratt and R. F. Schwarz, "Metal-Silicon Schottky Diode Arrays as Infrared Retinae," *Technical Digest of IEDM*, 306–9, 1973.

26. E. Kohn, S. Roosild, F. Shepherd, and A. Young, "Infrared Imaging with Monolithic CCD-Addressed Schottky-Barrier Detector Arrays," *International Conference on Application of CCD's*, 59–69, 1975.

27. H. Elabd, T. S. Villani, and J. R. Tower, "High Density Schottky-Barrier Infrared Charge-Coupled Device (IRCCD) Sensors for Short Wavelength Infrared (SWIR) Application at Intermediate Temperature," *Proceedings of SPIE* 345, 161–71, 1982.

28. H. Elabd, T. Villani, and W. Kosonocky, "Palladium-Silicide Schottky-Barrier IR-CCD for SWIR Applications at Intermediate Temperatures," *IEEE Electron Device Letters* EDL-3, 89–90, 1982.

29. J. R. Tower, A. D. Cope, L. E. Pellon, B. M. McCarthy, R. T. Strong, K. F. Kinnard, A. G. Moldovan, et al., "Development of Multispectral Detector Technology," *Proceedings of SPIE* 570, 172–83, 1985.

30. J. R. Tower, L. E. Pellon, B. M. McCarthy, H. Elabd, A. G. Moldovan, W. F. Kosonocky, J. E. Kakshoven, and D. Tom, "Shortwave Infrared 512×2 Line Sensor for Earth Resources Applications," *IEEE Transactions on Electron Devices* ED-32, 1574–83, 1985.

31. J. R. Tower, A. D. Cope, L. E. Pellon, B. M. McCarthy, R. T. Strong, K. F. Kinnard, A. G. Moldovan, et al., Design and Performance of 4×5120-Element Visible and 5×2560-Element Shortwave Infrared Multispectral Focal Planes," *RCA Review* 47, 226–55, 1986.

32. B. Capone, L. Skolnik, R. Taylor, F. Shepherd, S. Roosild, and W. Ewing, "Evolution of a Schottky Infrared Charge-Coupled (IRCCD) Staring Mosaic Focal Plane," *Optical Engineering* 18, 535–41, 1979.

33. W. F. Kosonocky, F. V. Shallcross, T. S. Villani, and J. V. Groppe, "160×244 Element PtSi Schottky-Barrier IR-CCD Image Sensor," *IEEE Transactions on Electron Devices* ED-32, 1564–73, 1985.

34. D. J. Sauer, F. V. Shallcross, F. L. Hsueh, G. M. Meray, P. A. Levine, H. R. Gilmartin, T. S. Villani, B. J. Esposito, and J. R. Tower, "640×480 MOS PtSi IR Sensor," *Proceedings of SPIE* 1540, 285–96, 1991.

35. J. L. Gates, W. G. Connelly, T. D. Franklin, R. E. Mills, F. W. Price, and T. Y. Wittwer, "488×640-Element Platinum Silicide Schottky Focal Plane Array," *Proceedings of SPIE* 1540, 262–73, 1991.

36. M. Kimata, M. Denda, N. Yutani, S. Iwade, and N. Tsubouchi, "512×512 Element PtSi Schottky-Barrier Infrared Image Sensor," *IEEE Journal of Solid-State Circuits* SC-22, 1124–29, 1987.

37. D. L. Clark, J. R. Berry, G. L. Compagna, M. A. Cosgrove, G. G. Furman, J. R. Heydweiller, H. Honickman, R. A. Rehberg, P. H. Solie, and E. T. Nelson, "Design and Performance of a 486–640 Pixel Platinum Silicide IR Imaging System," *Proceedings of SPIE* 1540, 303–11, 1991.

38. N. Yutani, H. Yagi, M. Kimata, J. Nakanishi, S. Nagayoshi, and N. Tsubouchi, "1040×1040 Element PtSi Schottky-Barrier IR Image Sensor," *IEDM Technical Digest*, 175–78, 1991.

39. P. W. Pellegrini, A. Golubovic, C. E. Ludington, and M. M. Weeks, "IrSi Schottky Barrier Diodes for Infrared Detection," *IEDM Technical Digest*, 157–60, 1982.

40. P. W. Pellegrini, A. Golubovic, and C. E. Ludington, "A Comparison of Iridium Silicide and Platinum Silicide Photodiodes," *Proceedings of SPIE* 782, 93–98, 1987.

41. B-Y. Tsaur, M. J. McNutt, R. A. Bredthauer, and B. R. Mattson, "128×128-Element IrSi Schottky-Barrier Focal Plane Arrays for Long Wavelength Infrared Imaging," *IEEE Electron Devices Letters* 10, 361–63, 1989.

42. J. Kuriański, J. Vermeiren, C. Claeys, W. Stessens, K. Maex, and R. De Keersmaecker, "Development and Evaluation of $CoSi_2$ Schottky Barrier Infrared Detectors," *Proceedings of SPIE* 1157, 145–52, 1989.

43. J. Kurianski, J. Van Dammer, J. Vermeiren, M. Maex, and C. Claeys, "Nickel Silicide Schottky Barrier Detectors for Short Wavelength Infrared Applications," *Proceedings of SPIE* 1308, 27–34, 1990.

44. N. Tsubouchi, M. Kimata, M. Denda, M. Yamawaki, N. Yutani, and S. Uematsu, "Photoresponse Improvement of PtSi-Si Schottky-Barrier Infrared Detectors by Ion Implantation," *Technical Digest 12th ESSDERC*, 169–71, 1982.

45. J. M. Shannon, "Reducing the Effective Height of a Schottky Barrier Using Low-Energy Ion Implantation," *Applied Physics Letters* 24, 369–71, 1974; "Control of Schottky Barrier Height Using Highly Doped Surface Layers," *Solid-State Electronics* 19, 537–43, 1976.

46. P. Pellegrini, M. Weeks, and C. E. Ludington, "New 6.5 μm Photodiode for Schottky Barrier Array Applications," *Proceedings of SPIE* 311, 24–29, 1981.

47. C-Y. Wei, W. Tantraporn, W. Katz, and G. Smith, "Reduction of Effective Barrier Height in PtSi-p-Si Schottky Diode Using Low-Energy Ion Implantation," *Thin Solid Films* 93, 407–12, 1982.

48. B-Y. Tsaur, C. K. Chen, and B. A. Nechay, "IrSi Schottky-Barrier Infrared Detectors with Wavelength Response Beyond 12 μm," *IEEE Electron Device Letters* 11, 415–17, 1990.

49. R. W. Fathauer, T. L. Lin, P. J. Grunthaner, P. O. Andersson, and J. Maserijian, "Modification of the Schottky Barrier Height of MBE-Grown $CoSi_2$/Si(111) Diodes by the Use of Selective Ga Doping," *Proceedings of the 2nd International Symposium on MBE* (Honolulu), 228–34, 1987.

50. T. L. Lin, J. S. Park, T. George, E. W. Jones, R. W. Fathauer, and J. Maserijian, "Long-Wavelength PtSi Infrared Detectors Fabricated by Incorporating a p^+ Doping Spike Grown by Molecular Beam Epitaxy," *Applied Physics Letters* 62, 3318–20, 1993.

51. T. L. Lin, J. P. Park, T. George, E. W. Jones, R. W. Fathauer, and J. Maserijian, "Long-Wavelength Infrared Doping-Spike PtSi Detectors Fabricated by Molecular Beam Epitaxy," *Proceedings of SPIE* 2020, 30–35, 1993.

52. T. L. Lin, J. S. Park, S. D. Gunapala, E. W. Jones, and H. M. Del Castillo, "Doping-Spike PtSi Schottky Infrared Detectors with Extended Cutoff Wavelengths," *IEEE Transactions on Electron Devices* 42, 1216–20, 1995.

53. B-Y. Tsaur, M. M. Weeks, and P. W. Pellegrini, "Pt-Ir Silicide Schottky-Barrier IR Detectors," *IEEE Electron Device Letters* 9, 100–102, 1988.

54. B-Y. Tsaur, J. P. Mattia, and C. K. Chen, "Hydrogen Annealing of PtSi-Si Schottky Barrier Contacts," *Applied Physics Letters* 57, 1111–13, 1990.

55. B-Y. Tsaur, C. K. Chen, and J. P. Mattia, "PtSi Schottky-Barrier Focal Plane Arrays for Multispectral Imaging in Ultraviolet, Visible, and Infrared Spectral Bands," *IEEE Electron Device Letters* 11, 162–64, 1990.

56. J. M. Kurianski, S. T. Shanahan, U. Theden, M. A. Green, and J. W. V. Storey, "Optimization of the Cavity for Silicide Schottky Infrared Detectors," *Solid-State Electronics* 32, 97–101, 1989.

57. C. K. Chen, B. Nechay, and B.-Y. Tsaur, "Ultraviolet, Visible, and Infrared Response of PtSi Schottky-Barrier Detectors Operated in the Front Illumination Mode," *IEEE Transactions on Electron Devices* 38, 1094–1103, 1991.

58. W. F. Kosonocky, T. S. Villani, F. V. Shallcross, G. M. Meray, and J. J. O'Neil, "A Schottky-Barrier Image Sensor with 100% Fill Factor," *Proceedings of SPIE* 1308, 70–80, 1990.

59. M. Kimata, M. Denda, S. Iwade, N. Yutani, and N. Tsubouchi, "A Wide Spectral Band Detector with PtSi/p-Si Schottky-Barrier," *International Journal of Infrared and MM Waves* 6, 1031–41, 1985.

60. W. F. Kosonocky, F. V. Shallcross, T. S. Villani, and J. V. Groppe, "160 × 244 Element PtSi Schottky-Barrier IR-CCD Image Sensor," *IEEE Transactions on Electron Devices* ED-32, 1564–72, 1995.

61. B.-Y. Tsaur, M. M. Weeks, R. Trubiano, P. W. Pellegrini, and T.-R. Yew, "IrSi Schottky-Barrier Infrared Detectors with 10-μm Cutoff Wavelength," *IEEE Transactions on Electron Device Letters* 9, 650–53, 1988.

62. F. D. Shepherd, V. E. Vickers, and A. C. Yang, "Schottky-Barrier Photodiode with Degenerate Semiconductor Active Region," U.S. Patent No. 3,603,847, September 7, 1971.

63. T. L. Lin, A. Ksendzov, S. M. Dajewski, E. W. Jones, R. W. Fathauer, T. N. Krabach, and J. Maserjian, "A Novel Si-Based LWIR Detector: The SiGe/Si Heterojunction Internal Photoemission Detector," *IEDM Technical Digest*, 641–44, 1990.

64. T. L. Lin and J. Maserjian, "Novel $Si_{1-x}Ge_x$/Si Heterojunction Internal Photoemission Long-Wavelength Infrared Detectors," *Applied Physics Letters* 57, 1422–24, 1990.

65. T. L. Lin, A. Ksendzov, S. M. Dejewski, E. W. Jones, R. W. Fathauer, T. N. Krabach, and J. Maserjian, "SiGe/Si Heterojunction Internal Photoemission Long-Wavelength Infrared Detectors Fabricated by Molecular Beam Epitaxy," *IEEE Transactions on Electron Devices* 38, 1141–44, 1991.

66. T. L. Lin, E. W. Jones, T. George, A. Ksendzov, and M. L. Huberman, "Advanced Si IR Detectors Using Molecular Beam Epitaxy," *Proceedings of SPIE* 1540, 135–39, 1991.

67. B.-Y. Tsaur, C. K. Chen, and S. A. Marino, "Long-Wavelength GeSi/Si Heterojunction Infrared Detectors and 400 × 400-Element Imager Arrays," *IEEE Electron Device Letters* 12, 293–96, 1991.

68. B.-Y. Tsaur, C. K. Chen, and S. A. Marino, "Long-Wavelength $Ge_{1-x}Si_x$/Si Heterojunction Infrared Detectors and Focal Plane Arrays," *Proceedings of SPIE* 1540, 580–95, 1991.

69. B.-Y. Tsaur, C. K. Chen, and S. A. Marino, "Heterojunction $Ge_{1-x}Si_x$/Si Infrared Detectors and Focal Plane Arrays," *Optical Engineering* 33, 72–78, 1994.

70. T. L. Lin, J. S. Park, S. D. Gunapal, E. W. Jones, and H. M. Del Castillo, "Photoresponse Model for $Si_{1-x}Ge_x$/Si Heterojunction Internal Photoemission Infrared Detector," *IEEE Electron Device Letters* 15, 103–5, 1994.

71. T. L. Lin, J. S. Park, S. D. Gunapala, E. W. Jones, and H. M. Del Castilo, "$Si_{1-x}Ge_x$/Si Heterojunction Internal Photoemission Long Wavelength Infrared Detector," *Proceedings of SPIE* 2474, 17–23, 1994.

72. J. S. Park, T. L. Lin, E. W. Jones, H. M. Del Castillo, T. George, and S. D. Gunapala, "Long-Wavelength Stacked $Si_{1-x}Ge_x$/Si Heterojunction Internal Photoemission Infrared Detectors," *Proceedings of SPIE* 2020, 12–21, 1993.

73. J. S. Park, T. L. Lin, E. W. Jones, H. M. Del Castillo, and S. D. Gunapala, "Long-Wavelength Stacked SiGe/Si Heterojunction Internal Photoemission Infrared Detectors Using Multiple SiGe/Si Layers," *Applied Physics Letters* 64, 2370–72, 1994.

74. H. Wada, M. Nagashima, K. Hayashi, J. Nakanishi, M. Kimata, N. Kumada, and S. Ito, "512 × 512 Element GeSi/Si Heterojunction Infrared Focal Plane Array," *Opto-Electronics Review* 7, 305–11, 1999.

75. S. Tohyama, N. Teranishi, K. Kunoma, M. Nishimura, K. Arai, and E. Oda, "A New Concept Silicon Homojunction Infrared Sensor," *IEDM Technical Digest*, 82–85, 1988.

76. W. F. O'Neil, "Nonuniformity Corrections for Spectrally Agile Sensor," *Proceedings of SPIE* 1762, 327–39, 1992.

77. A. G. U. Perera, H. X. Yuan, and M. H. Francombe, "Homojunction Internal Photoemission Far-Infrared Detectors: Photoresponse Performance Analysis," *Journal of Applied Physics* 77, 915–24, 1995.

78. A. G. U. Perera, "Physics and Novel Device Applications of Semiconductor Homojunctions," in *Thin Solid Films*, Vol. 21, eds. M. H. Francombe and J. L. Vossen, 1–75, Academic Press, New York, 1995.

79. H. X. Yuan and A. G. H. Perera, "Dark Current Analysis of Si Homojunction Interfacial Work Function Internal Photoemission Far-Infrared Detectors," *Applied Physics Letters* 66, 2262–64, 1995.

80. A. G. U. Perera, H. X. Yuan, J. W. Choe, and M. H. Francombe, "Novel Homojunction Interfacial Workfunction Internal Photoemission (HIWIP) Tunable Far-Infrared Detectors for Astronomy," *Proceedings of SPIE* 2475, 76–87, 1995.

81. W. Shen, A. G. U. Perera, M. H. Francombe, H. C. Liu, M. Buchanan, and W. J. Schaff, "Effect of Emitter Layer Concentration on the Performance of GaAs p⁺–i Homojunction Far-Infrared Detectors: A Comparison of Theory and Experiment," *IEEE Transaction on Electron Devices* 45, 1671–77, 1998.

82. A. G. U. Perera and W. Z. Shen, "GaAs Homojunction Interfacial Workfunction Internal Photoemission (HIWIP) Far-Infrared Detectors," *Opto-Electronics Review* 7, 153–80, 1999.

83. A. G. H. Perera, "Semiconductor Photoemissive Structures for Far Infrared Detection," in *Handbook of Thin Devices,* Vol. 2, ed. M. H. Francombe, 135–70, Academic Press, San Diego, CA, 2000.

84. A. G. H. Perera, "Silicon and GaAs as Far-Infrared Detector Material," in *Photodetectors and Fiber Optics*, ed. H. S. Nalwa, 203–37, Academic Press, San Diego, CA, 2001.

85. E. E. Haller, "Advanced Far-Infrared Detectors," *Infrared Physics & Technology* 35, 127, 1994.

86. D. W. Watson, M. T. Guptill, J. E. Huffman, T. N. Krabach, S. N. Raines, and S. Satyapal, "Germanium Blocked-Impurity-Band Detector Arrays: Unpassivated Devices with Bulk Substrates," *Journal of Applied Physics* 74, 4199, 1993.

13 III-V Detectors

In the middle and late 1950s it was discovered that InSb had the smallest energy gap of any semiconductor known at that time and its applications as a middle wavelength infrared detector became obvious [1,2]. The energy gap of InSb is less well matched to the 3–5 μm band at higher operating temperatures, and better performance can be obtained from $Hg_{1-x}Cd_xTe$. InAs is a similar compound to InSb, but has a larger energy gap [3], so that the threshold wavelength is 3–4 μm, and both photoconductive and photovoltaic detectors have been fabricated. The photoconductive process in InSb has been studied extensively, and more details can be found in Morten and King [4], Kruse [5], and Elliott and Gordon [6].

Indium antimonide detectors have been extensively used in high-quality detection systems and have found numerous applications in the defense and space industry for more than 40 years. Perhaps the best known (and most successful) of these systems has been the Sidewinder air-to-air antiaircraft missile. Manufacturing techniques for InSb are well established and the invention of CCD and CMOS hybrid devices has increased the interest in this semiconductor.

13.1 SOME PHYSICAL PROPERTIES OF III-V NARROW GAP SEMICONDUCTORS

Development of crystal growth techniques in the early 1950s led to InSb and InAs bulk single crystal detectors. Since then the quality of single crystal growth has improved immensely.

Two methods of obtaining these crystals have been used commercially: Czochralski or horizontal Bridgman techniques. Single crystals can be grown with relatively high purity, low dislocation density, and ingot sizes that permit wafer diameters in the 5–100 mm range, suitable for convenient handling and photolithography. The growth of InSb single crystals is reviewed by Hulme, Mullin, Liang, Micklethwaite, and Johnson [7–11]. A wide spectrum of topics in materials that today's engineers, material scientists, and physicists need is included in a comprehensive treatise on electronic and photonic materials gathered in the *Springer Handbook of Electronic and Photonic Materials* [12].

Unlike InSb, reaction of the elements to form an InAs compound is not a simple matter. To keep the arsenic from disappearing because of its high vapor pressure near the melting point, it is necessary to let the constituents react in a sealed quartz ampoule. Purification of InAs is also more difficult than InSb.

The $In_xGa_{1-x}As$ ternary alloy has been of great interest for the short wavelength infrared (SWIR), low-cost detector applications. The $x = 0.53$ alloy is lattice -matched to InP substrates, has a bandgap of 0.73 eV, and covers the wavelength range from 0.9 to 1.7 μm. High quality InP substrates are available with diameters as large as 100 mm. The $In_xGa_{1-x}As$ with 53% InAs is often called "standard InGaAs" without bothering to note the values of "x" or "1–x". This is mature material, driven by the mass-production of fiber optic receivers at 1.3 μm and 1.55 μm. At present, InGaAs is also becoming the choice for high temperature operation in the 1–3 μm spectrum. By increasing the indium content to $x = 0.82$, the wavelength response of $In_xGa_{1-x}As$ can be extended out to 2.6 μm. Single element InGaAs detectors have been made with up to 2.6 μm cutoffs while linear arrays and cameras have been demonstrated to 2.2 μm.

The energy bandgap of InGaAsP quaternary system range from 0.35 eV (InAs) to 2.25 eV (GaP), with InP (1.29 eV) and GaAs (1.43 eV) falling between [13,14]. Figure 13.1 shows values of energy gap at 300 K versus composition y for $Ga_xIn_{1-x}As_yP_{1-y}$ lattice matched to InP. InGaAsP alloys have been epitaxially grown by: hydride and chloride vapor phase epitaxy (VPE), liquid phase epitaxy (LPE), molecular beam epitaxy (MBE), and metalorganic chemical vapor deposition (MOCVD) [15]. A brief comparison of the four techniques is given by Olsen and Ban [16]. While each of these techniques has certain advantages, hydride VPE is well suited for InGaAsP/InP optoelectronic devices. Epitaxy methods are also used for more sophisticated structures of modern InSb, InAs, InGaAs, $InAs_{1-x}Sb_x$ (InAsSb), and $Ga_xIn_{1-x}Sb$ (GaInSb) devices [12,17–20].

Figure 13.2 shows the composition dependence of the energy gap and the electron effective mass at the Γ-conduction bands of $Ga_xIn_{1-x}As$, $InAs_xSb_{1-x}$, and $Ga_xIn_{1-x}Sb$ ternaries.

InAsSb is an attractive semiconductor material for detectors covering the 3–5 μm and 8–14 μm spectral ranges [17]. However, progress in this ternary system has been limited by crystal synthesis problems. The large separation between the liquidus and solidus (Figure 13.3) and the lattice mismatch (6.9% between InAs and InSb) place stringent demands upon the method of crystal growth [21]. These difficulties are being overcome systematically using MBE and MOCVD.

For long wavelength (8–12 μm) detector technology, the dominant material is HgCdTe. Despite considerable progress in HgCdTe photovoltaic technology over the past three decades [22,23], difficulties still remain—particularly for wavelengths exceeding 10 μm where device performance is

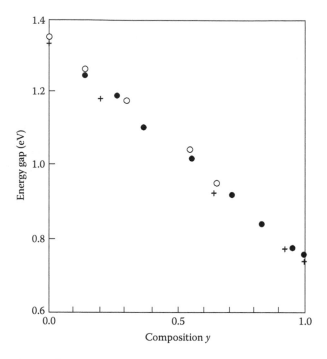

Figure 13.1 Dependence of energy gap at 300 K versus composition y for $Ga_xIn_{1-x}As_yP_{1-y}$ lattice matched to InP. (From Adachi, S., *Physical Properties of III-V Semiconducting Compounds: InP, InAs, GaAs, GaP, InGaAs, and InGaAsP*, Wiley-Interscience, New York, 1992; *Properties of Group-IV, III-V and II-VI Semiconductors*, John Wiley & Sons, Ltd., Chichester, 2005. With permission.)

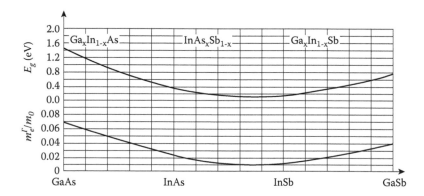

Figure 13.2 Variation of the bandgap energy and electron effective mass at the Γ-conduction bands of $Ga_xIn_{1-x}As$, $InAs_xSb_{1-x}$, and $Ga_xIn_{1-x}Sb$ ternary alloys at room temperature.

limited by large tunneling dark current and a sensitive dependence on precise composition control to accurately determine the energy gap. Given the maturity of III-V growth and processing technology, alternatives to HgCdTe based III–V semiconductors have been investigated and suggested [24]. The approaches can be divided into three categories:

- Use of superlattices such as AlGaAs/GaAs [25–30], InAsSb/InSb [17,26,31], and InGaSb/InAs [20,32,33] (see Chapters 16 and 17)

- Use of quantum dots (see Chapter 18)

- Addition of large group V element, Bi, in InAs, InSb, and InAsSb, or large group III element, Tl, in InAs, InP, and InSb [20,34]

Hitherto, the best results have been obtained using AlGaAs/GaAs and InGaSb/InAs superlattices.

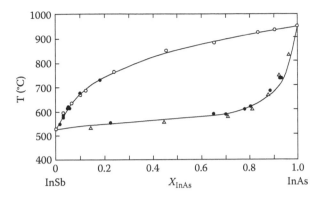

Figure 13.3 Pseudobinary phase diagram for the InAs-InSb system. (From Stringfellow, G. B., and Greene, P. R., *Journal of the Electrochemical Society*, 118, 805–10, 1971.)

The III-V detector materials have a zinc-blende structure and direct energy gap at the Brillouin zone center. The shape of the electron band and the light mass hole band is determined by the **k p** theory. The momentum matrix element varies only slightly for different materials and has an approximate value of 9.0×10^{-8} eVcm. Then, the electron effective masses and conduction band densities of states are similar for materials with the same energy gap.

The III-V materials have a conventional negative temperature coefficient of the energy gap which is well described by the Varshni relation [35]

$$E_g(T) = E_o - \frac{\alpha T^2}{T + \beta},$$

(13.1)

where α and β are fitting parameter characteristics of a given material.

It has been found that the bandgap energy of a ternary compound is generally a square function of the composition. For $InAs_{1-x}Sb_x$ it may be described by the following expression [36]:

$$E_g(x,T) = 0.411 - \frac{3.4 \times 10^{-4} T^2}{210 + T} - 0.876x + 0.70x^2 + 3.4 \times 10^{-4} xT(1-x),$$

(13.2)

which indicates a fairly weak dependence of the band edge on composition in comparison with HgCdTe. The minimum of E_g appears at composition $x \approx 0.65$.

Table 13.1 contains some physical parameters of the InAs, InSb, GaSb, $InAs_{0.35}Sb_{0.65}$, and $In_{0.53}Ga_{0.47}As$ semiconductors [37].

The III-V detector material that has been investigated most broadly is InSb. The temperature independent portions of the Hall curves indicate that most of the electrically active impurity atoms in InSb have shallow activation energies and above 77 K are thermally ionized. The Hall coefficient for p-type samples is positive in the low-temperature extrinsic range, and reverses sign to become negative in the intrinsic range because of the higher mobility of the electrons (the mobility ratio $b = \mu_e/\mu_h$ of the order of 10^2 is observed). The transition temperature for the p-type samples, at which R_H changes sign, depends on the purity. The samples become intrinsic above a certain temperature (above 150 K for pure n-type samples) and below these temperatures (below 100 K for pure n-type samples) there is little variation of Hall coefficients.

There are various carrier scattering mechanisms in semiconductors, as shown in Figure 13.4 for InSb [38]. Reasonably pure n-type and p-type samples exhibit an increase in mobility up to approximately 20–60 K after which the mobility decreases due to polar and electron hole scattering. Carrier mobility systematically increases with a decrease in impurity concentration both in temperature 77 as well as in 300 K.

In alloy semiconductors the charged carriers see potential fluctuations as a result of the composition disorder. This kind of scattering mechanisms, so-called alloy scattering, is important in some III-V ternaries and quaternaries. Let us simply express the total carrier mobility μ_{tot} in alloy $A_xB_{1-x}C$ as [12]

Table 13.1: Physical Properties of Narrow Gap III–V Compounds

	T(K)	InAs	InSb	GaSb	InAs$_{0.35}$Sb$_{0.65}$	In$_{0.53}$Ga$_{0.47}$As
Lattice structure		cub. (ZnS)	cub. (ZnS)	cub. (ZnS)	cub.(ZnS)	cub.(ZnS)
Lattice constant a (nm)	300	0.60584	0.647877	0.6094	0.636	0.58438
Thermal expansion coefficient α (10^{-6}K^{-1})	300	5.02	5.04	6.02		
	80		6.50			
Density ρ (g/cm^3)	300	5.68	5.7751	5.61		5.498
Melting point T_m (K)		1210	803	985		
Energy gap E_g (eV)	4.2	0.42	0.2357	0.822	0.138	0.627
	80	0.414	0.228		0.136	
	300	0.359	0.180	0.725	0.100	0.75
Thermal coefficient of E_g	100–300	-2.8×10^{-4}	-2.8×10^{-4}			-3.0×10^{-4}
Effective masses:						
m_e^*/m	4.2	0.023	0.0145	0.042		0.041
	300	0.022	0.0116		0.0101	
m_{lh}^*/m	4.2	0.026	0.0149			0.0503
m_{hh}^*/m	4.2	0.43	0.41	0.28	0.41	0.60
Momentum matrix element P (eVcm)		9.2×10^{-8}	9.4×10^{-8}			
Mobilities:						
μ_e (cm^2/Vs)	77	8×10^4	10^6		5×10^5	70000
	300	3×10^4	8×10^4	5×10^3	5×10^4	13800
μ_h (cm^2/Vs)	77		1×10^4	2.4×10^3		
	300	500	800	880		
Intrinsic carrier concentration n_i (cm^{-3})	77	6.5×10^3	2.6×10^9		2.0×10^{12}	
	200	7.8×10^{12}	9.1×10^{14}		8.6×10^{15}	
	300	9.3×10^{14}	1.9×10^{16}		4.1×10^{16}	5.4×10^{11}
Refractive index n_r		3.44	3.96	3.8		
Static dielectric constant ε_s		14.5	17.9	15.7		14.6
High frequency dielectric constant e_∞		11.6	16.8	14.4		
Optical phonons:						
LO (cm^{-1})		242	193		≈ 210	
TO (cm^{-1})		220	185		≈ 200	

Source: S. Adachi, *Physical Properties of III-V Semiconducting Compounds: InP, InAs, GaAs, GaP, InGaAs, and InGaAsP*, Wiley-Interscience, New York, 1992; *Properties of Group-IV, III-V and II-VI Semiconductors*, John Wiley & Sons, Ltd., Chichester, 2005; I. Vurgaftman, J. R. Meyer, and L. R. Ram-Mohan, *Journal of Applied Physics*, 89, 5815–75, 2001; A. Rogalski, K. Adamiec, and J. Rutkowski, *Narrow-Gap Semiconductor Photodiodes*, SPIE Press, Bellingham, WA, 2000. With permission.

$$\frac{1}{\mu_{tot}(x)} = \frac{1}{x\mu_{tot}(AC) + (1-x)\mu_{tot}(BC)} + \frac{1}{\mu_{al,0}/[x/(1-x)]}. \tag{13.3}$$

The first term in Equation 13.3 comes from the linear interpolation scheme and the second term accounts for the effects of alloying. For example, Figure 13.5 plots the electron Hall mobility in Ga$_x$In$_{1-x}$P$_y$As$_{1-y}$/InP quaternary [12]. The quaternary is an alloy of the constituents In$_{0.53}$Ga$_{0.47}$As ($y = 0$) and InP ($y = 1.0$) and the values μ_{tot}(In$_{0.53}$Ga$_{0.47}$As) = 13,000 and $\mu_{al,0}$ = 3000 cm^2/Vs have been used. The experimental data correspond to those for relatively pure samples.

Optical properties of InSb have been reviewed by Kruse [5]. Because of the very small effective mass of electrons, the conduction band density of states is small and it is possible to fill the

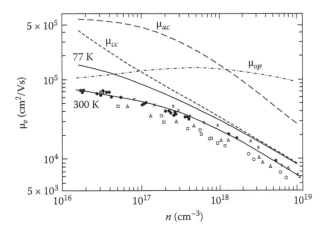

Figure 13.4 Electron mobility in n-type InSb at 300 K and 77 K versus free electron concentration—dashed lines denote the theoretical mobilities at 300 K for charged center, polar optical, and acoustic scattering modes. The experimental data are taken at 300 K. (From Zawadzki, W., *Advances in Physics*, 23, 435–522, 1974. With permission.)

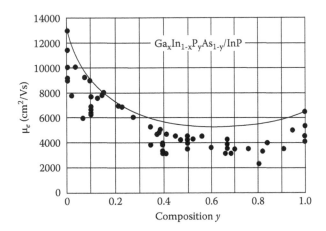

Figure 13.5 Electron Hall mobility in the $Ga_xIn_{1-x}P_yAs_{1-y}$/InP quaternary. The experimental data correspond to those for relatively pure samples. The solid line represents the results calculated using Equation 13.3 with the values of $\mu_{tot}(In_{0.53}Ga_{0.47}As) = 13,000$ and $\mu_{al,0} = 3000$ cm²/Vs. (From Kasap S., and Capper, P., *Springer Handbook of Electronic and Photonic Materials*, Springer, Heidelberg, 2006. With permission.)

available band states by doping, thereby appreciably shifting the absorption edge to shorter wavelengths. This has been referred as the Burstein-Moss effect (Figure 13.6 [39]).

The physical properties of InAs are similar to those of InSb. InAs material with cutoff wavelength of 3.5 μm makes it of limited utility for the middle wavelength band even though it could theoretically be operated near 190 K. Development work has been limited by growth and passivation problems.

Although effort has been devoted to the development of InAsSb as an alternative to HgCdTe for IR applications, there is relatively little information available on its physical properties. A III–V detector technology would benefit from superior bond strengths and material stability (compared to HgCdTe), well-behaved dopants, and high-quality III-V substrates. The properties of InAsSb were first investigated by Woolley and coworkers. They established the InAs-InSb miscibility [40], the pseudobinary phase diagram [41], scattering mechanisms [42] and the dependence of fundamental properties such as bandgap [43], and effective masses on composition [43,44]. All of the above measurements were performed on polycrystalline samples prepared by various freezing and annealing techniques. The review of the development of InAsSb crystal growth techniques, physical properties, and detector fabrication procedures is presented in Rogalski's papers [17,45].

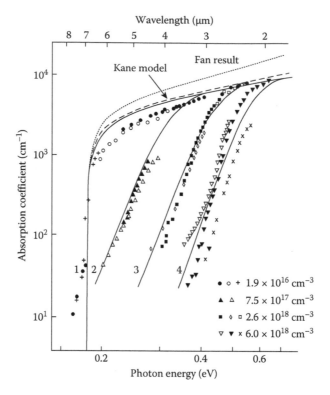

Figure 13.6 Dependence of the optical absorption coefficient of InSb upon photon energy at 300 K. Carrier concentration: 1.9×10^{16} cm^{-3} (1), 7.5×10^{17} cm^{-3} (2), 2.6×10^{18} cm^{-3} (3), and 6.0×10^{18} cm^{-3} (4). (From Kesamanli, F. P., Malcev, Yu. V., Nasledov, D. N., Uhanov, Yu. I., and Filipczenko, A. S., *Fizika Tverdogo Tela*, 8, 1176–81, 1966.)

The intrinsic carrier concentration in InAsSb as a function of composition x for various temperatures can be approximated by the following relation [17]:

$$n_i = \left(1.35 + 8.50x + 4.22 \times 10^{-3}T - 1.53 \times 10^{-3}xT - 6.73x^2\right) \times 10^{14} T^{3/2} E_g^{3/4} \exp\left(-\frac{E_g}{2kT}\right). \quad (13.4)$$

For a given temperature, the maximum of n_i appears at $x \approx 0.65$, which corresponds to the minimum energy gap.

Conventional InAsSb (although it is the smallest bandgap for any bulk III-V materials) does not have a sufficiently small gap at 77 K for operation in the 8–14 μm wavelength range. To build detectors that span the 8–14 μm atmospheric window, Osbourn proposed the use of InAsSb strain layer superlattices [46]. This theoretical prediction reinforced interest in the investigation of InAsSb ternary alloy as a material for infrared detector applications. More information about superlattice InAsSb infrared detectors can be found in Chapter 17.

The ternary alloy Ga$_x$In$_{1-x}$Sb is important material for the fabrication of detectors designed for middle wavelength IR applications. The long wavelength limit of Ga$_x$In$_{1-x}$Sb detectors has been tuned compositionally from 1.52 μm ($x = 1.0$ at 77 K) to 6.8 μm ($x = 0.0$ at room temperature). The band gap energy of In$_x$Ga$_{1-x}$As$_y$Sb$_{1-y}$ at room temperature can be fitted by the relationship [47]:

$$E_g(x) = 0.726 - 0.961x - 0.501y + 0.08xy + 0.451x^2 - 1.2y^2 + 0.021x^2y + 0.62xy^2. \quad (13.5)$$

The lattice matching condition to GaSb imposes the additional constraint that x and y are related as $y = 0.867/(1 - 0.048x)$.

13.2 InGaAs PHOTODIODES

The need for high-speed, low-noise $In_xGa_{1-x}As$ (InGaAs) photodetectors for use in lightwave communication systems operating in the 1–1.7 µm wavelength region is well established [48–56]. Having lower dark current and noise than indirect-bandgap germanium, the competing near-IR material, the material is addressing both entrenched in a variety of thermal-imaging applications not practical with cryogenically cooled detectors [57–69]. The applications now include low-cost industrial thermal imaging, eye-safe surveillance, online process control, and subsurface inspection of fine art [62,70].

The SWIR wavelength band offers unique imaging advantages over visible and thermal bands. Like visible cameras, the images are primarily created by reflected broadband light sources, so SWIR images are easier for viewers to understand. Most materials used to make windows, lenses, and coatings for visible cameras are readily usable for SWIR cameras, keeping costs down. Ordinary glass transmits radiation to about 2.5 µm. SWIR cameras can image many of the same light sources, such as YAG laser wavelengths. Thus, with safety concerns shifting laser operations to the "eye-safe" wavelengths where beams won't focus on the retina (beyond 1.4 µm), SWIR cameras are in a unique position to replace visible cameras for many tasks. Due to the reduced Rayleigh scatter of light at longer wavelengths, particulate in the air, such as dust or fog, SWIR cameras can see through haze better than visible cameras.

The InAs/GaAs ternary system bandgaps span 0.35 eV (3.5 µm) for InAs to 1.43 eV (0.87 µm) for GaAs. By changing the alloy composition of the InGaAs absorption layer, the photodetector responsivity can be maximized at the desired wavelength of the end user to enhance the signal to noise ratio. Figure 13.7 shows the spectral response of three such InGaAs detectors at room temperature, whose cutoff wavelength is optimized at 1.7 µm, 2.2 µm, and 2.5 µm, respectively. The spectral response of an $In_{0.53}Ga_{0.47}As$ focal plane array (FPA) to the night spectrum makes it a better choice for use in a night-vision camera in comparison with the current state-of-art technology for enhancing night vision—GaAs Gen III image-intensifier tubes. Figure 13.7 also marks the key laser wavelengths.

The fundamental device parameters (energy bandgap, absorption coefficient, and background carrier concentration) distinguish InGaAs from germanium [16]. Low background doping level ($n = 1 \times 10^{14}$ cm^{-3}) and high mobilities (11,500 cm^2/Vs) for InGaAs at room temperature were achieved [71].

InGaAs-detector processing technology is similar to that used with silicon, but the detector fabrication is different. The InGaAs detector's active material is deposited onto a substrate using chloride VPE [16,72] or MOCVD [73,74] techniques adjusted for thickness, background doping,

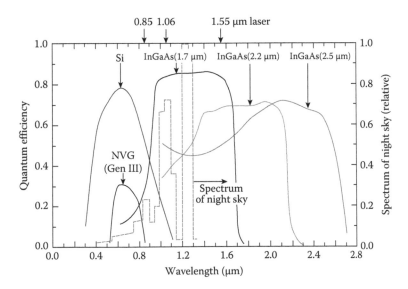

Figure 13.7 Quantum efficiency of silicon, InGaAs, and night vision tube detectors in the visible and SWIR. Key laser wavelengths are also noted. $In_{0.53}Ga_{0.47}As$ photodiode has nearly three times higher quantum efficiency than GaAs Gen III photocathodes; InGaAs also overlaps the illumination spectrum of the night sky more.

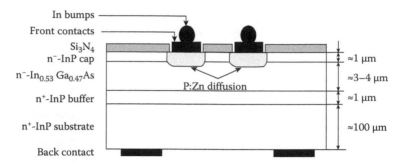

Figure 13.8 Cross section of a planar back side illuminated p-i-n InGaAs photodiode. (From Joshi, A. M., Ban, V. S., Mason, S., Lange, M. J., and Kosonocky, W. F., "512 and 1024 Element Linear InGaAs Detector Arrays for Near-Infrared (1–3 µm) Environmental Sensing," *Proceedings of SPIE* 1735, 287–95, 1992. With permission.)

and other requirements. Planar technology evolved from the older mesa technology and at present is widely used due to its simple structure and processing as well as the high reliability and low cost.

Both p-i-n and avalanche InGaAs photodiode structures with total layer thicknesses of several microns are fabricated [55]. The absorption coefficient of $In_{0.53}Ga_{0.47}As$ at 1.55 µm of 7000 cm^{-1} is more than an order of magnitude larger than that of Ge, so that a thickness of 1.5 µm can absorb > 70% of the incident photons. This enables high-speed p-i-n photodiodes to operate at up to ~10 GHz with good quantum efficiency.

It is well known that the internal gain of APDs provides a higher sensitivity in optical receivers than p-i-n photodiodes [75–78], however, at the cost of more complex epitaxial wafer structures and bias circuits. In comparison to APDs operating in the same wavelength region, p-i-n photodiodes offer the advantages of lower dark current, larger frequency bandwidth, and simpler driving circuitry. Thus, although p-i-n diodes do not have internal gain, an optimal combination of a p-i-n diode with a low-noise, large-bandwidth transistor has led to high sensitivity optical receivers operating up to several Gbit/s.

13.2.1 p-i-n InGaAs Photodiodes

Figure 13.8 shows the structure of a InGaAs back side illuminated (BSI) p-i-n photodiode. The starting substrate is n$^+$-InP on which is deposited approximately 1 µm of n$^+$-InP as a buffer layer. The 3–4 µm of the n$^-$-InGaAs active layer is then deposited followed by a 1 µm n$^-$-InP cap layer. The structure is covered with Si_3N_4. The p-on-n photodiodes are formed by the diffusion of zinc through the InP cap into the active layer. Ohmic contacts were formed by the sintering of a Au/Zn alloy. At this point, the substrate is thinned to approximately 100 µm and a sintered Au/Ge alloy is used as the back ohmic contact. The last step is the deposition of a 20 µm columns of indium on the front contacts.

When the indium content of the alloy is increased, the long wavelength cut off extends to cover the entire traditional near IR band [72–74]. The lattice constant of $In_{0.8}Ga_{0.2}As$ ($\lambda_c \approx 2.5$ µm) is about 2% larger than that of the InP substrate and the $InAs_yP_{1-y}$ gradient layer structure is used to accommodate the difference [57]. Published literature describes a structure having 15 layers with y increasing from 0.0 to 0.68 [79]. By selective removing of various layers, the p-n junctions with different wavelength response could be formed. In such a way, a novel three wavelength InGaAs FPA pixel element for detection at wavelengths from 1.0 to 2.5 µm has been fabricated [74]. However, due to the smaller bandgap and to interface defects resulting from the lattice mismatch, more long wavelength InGaAs photodiodes have considerably higher dark currents than those fabricated from the lattice-matched alloy. The dark current of three types of photodiodes under reverse bias is shown as points in Figure 13.9 [74]. The dark currents of $In_{0.7}Ga_{0.3}As$ and $In_{0.75}Ga_{0.15}As$ detectors are considerably larger than for the $In_{0.53}Ga_{0.47}As$ detector, especially at lower voltages. The mismatch between the absorption layers and the InP substrate provide midgap generation–recombination (g–r) centers, increasing the g–r current. Depending on wavelength and illumination direction, the quantum efficiency between 15 and 95% has been measured.

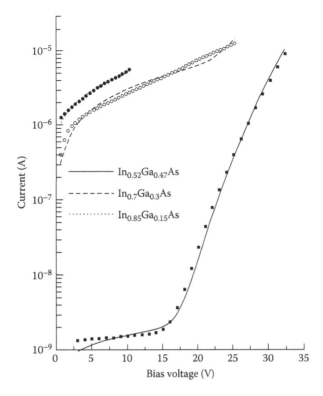

Figure 13.9 Dark current versus reverse bias voltage of the InGaAs photodiodes. The points are measured data and the lines are from theoretical fits. (From Olsen, G. H., Lange, M. J., Cohen, M. J., Kim, D. S., and Forrest, S. R., "Three-Band 1.0–2.5 μm Near-Infrared InGaAs Detector Array," *Proceedings of SPIE* 2235, 151–59, 1994. With permission.)

Joshi et al. have reported "Four Rules" for reducing the leakage currents in the lattice mismatched InGaAs photodiodes [72]:

- Grow epitaxial layers with compositionally abrupt interfaces

- Keep lattice mismatch between the adjacent layers below 0.12%

- Dope the active InGaAs layer to 1 to 5×10^{17} cm^{-3}

- Thermally cycle the wafer after growth

A useful photodiode parameter is the $R_o A$ product. Figure 13.10 shows that the highest quality InGaAs photodiodes have been grown by MOCVD [65,80,81]. Their performance agrees with the radiative limit. Due to similar band structure of InGaAs and HgCdTe ternary alloys, the ultimate fundamental performance of both types of photodiodes are similar in the wavelength range $1.5 < \lambda < 3.7$ μm [82]. Figure 13.10b shows the temperature dependence of zero-bias resistance in the region from –20 to 40°C for 1.7 μm InGaAs 20 μm pixel diode [68]. It can be seen that $R_o A$ is diffusion limited throughout the temperature range tested. At –100 mV, the current is diffusion limited at T ≥ 7°C and becomes g–r limited at lower temperatures. At this bias voltage, mean dark current is about 70 fA at room temperature and about 25 fA at 4°C.

Standard InGaAs photodiodes have detector-limited room temperature detectivity of ~10^{13} cmHz$^{1/2}$W^{-1}. Figure 13.11 shows the spectral responsivities and detectivities of InGaAs photodiodes with different cutoff wavelengths operated at room temperatures. Further insight in device performance gives Figure 13.12 [65], where the mean D^* versus operating temperature and background is shown. The photodiodes fabricated with radiatively limited MOCVD-fabricated material would exhibit D^* of about 8×10^{13} cmHz$^{1/2}$W^{-1} at 295 K. The dependence of detector $R_o A$ product on D^* at the very low short wavelength IR backgrounds is shown in Figure 13.12b. The highest mean D^* exceeding 10^{15} cmHz$^{1/2}$W^{-1} implies $R_o A$ product > 10^{10} Ωcm^2.

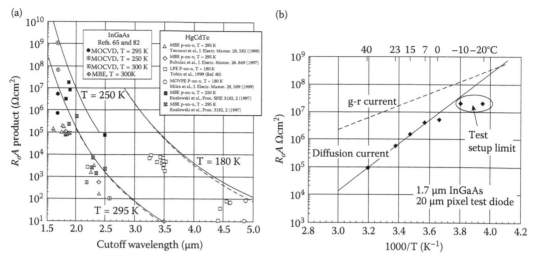

Figure 13.10 R_oA product of InGaAs photodiodes: (a) versus wavelength at 295 K and 250 K. (From Kozlowski, L. J., Vural, K., Arias, J. M., Tennant, W. E., and DeWames, R. E., "Performance of HgCdTe, InGaAs and Quantum Well GaAs/AlGaAs Staring Infrared Focal Plane Arrays," *Proceedings of SPIE* 3182, 2–13, 1997; Rogalski, A., and Ciupa, R., *Journal of Electronic Materials*, 28, 630–36, 1999.) (b) Temperature dependence for 1.7 µm InGaAs 20 µm pixel photodiode. (From Yuan, H., Apgar, G., Kim, J., Laquindanum, J., Nalavade, V., Beer, P., Kimchi, J., and Wong, T., "FPA Development: From InGaAs, InSb, to HgCdTe," *Proceedings of SPIE* 6940, 69403C, 2008. With permission.)

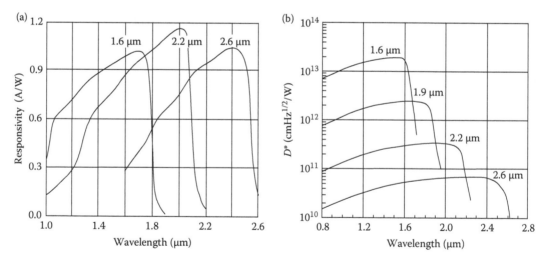

Figure 13.11 Room-temperature spectral response (a) responsivity and (b) detectivity of InGaAs photodiodes with different cutoff wavelengths.

13.2.2 InGaAs Avalanche Photodiodes

The earliest APDs for telecommunication application were based on silicon utilizing a reach-through structure with a relatively thin high electric field multiplication region and a much thicker carrier drift region with lower field [83]. This structure with a relatively low operating voltage gives high quantum efficiency, a large α_e/α_h ratio (typically 10–100), modest speed, high gain, and very low noise. As fiber-based optical communication systems evolved, there was requirement for APDs capable of detecting light at 1.3 µm and 1.55–1.65 µm. Germanium APDs capable of detecting photons in this wavelength range have α_h/α_e ratio of 1.5, so low excess noise could not be

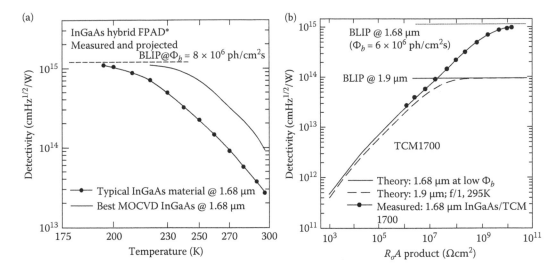

Figure 13.12 Measured and theoretical detectivity of InGaAs photodiodes: (a) versus temperature for 1.7 μm InGaAs, projected trendline shows MOCVD InGaAs capability, and (b) versus R_oA product. (From Kozlowski, L. J., Vural, K., Arias, J. M., Tennant, W. E., and DeWames, R. E., "Performance of HgCdTe, InGaAs and Quantum Well GaAs/AlGaAs Staring Infrared Focal Plane Arrays," *Proceedings of SPIE* 3182, 2–13, 1997. With permission.)

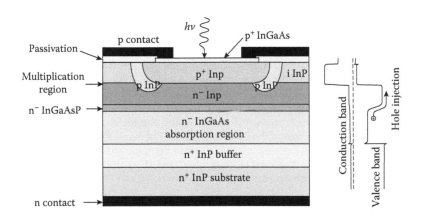

Figure 13.13 Cross section of SAGM-APD InP-based photodiode.

obtained [84–86]. Moreover, the quantum efficiency drops off rapidly beyond 1.5 μm and photodiodes suffer from a high thermal generation rate.

For high-speed receivers, one has to produce a short response time of the photogenerated carriers and a high-bandwidth product. These parameters are limited mainly by the shape of the heterojunction between absorption and multiplication zone and the doping profile throughout the device. In the case of InGaAs attempts to obtain significant avalanche gain by increasing the electric field is not possible due to the onset of a tunneling mechanism that causes very high leakage current. The small electron effective mass results in rapidly increasing the tunneling current at low fields above 150 kV/cm [51,81,87]. These problems were overcome by the combination of an InGaAs region capable of absorbing photons under a low electric field, and a lattice-matched wider bandgap InP region producing avalanche multiplication. The resulting structure is known as a separate absorption multiplication APD (SAM-APD).

An example of SAM-APD is shown in Figure 13.13 together with the band diagram of the entire heterostructure. Light is absorbed in the InGaAs and holes (with a higher impact ionization coefficient than electrons; this ensures low-noise operation) are swept to an InP junction, where the avalanche multiplication take place. This structure combines low leakage, due to the junction being placed in the high bandgap material (InP), which sensitivity at longer wavelength provided by the lower-bandgap InGaAs absorption region. Dark currents of the order pA can be obtained. However, there is a potential problem with the operation of a SAM-ADP. Holes can accumulate at the valence band discontinuity at the InGaAs/InP heterojunction and thereby increase the response time. To alleviate this problem, a graded bandgap InGaAsP layer can be inserted between the InP and InGaAs. This modified structure is known as a separate absorption graded multiplication APD (SAGM-APD). The APD can achieve 5–10 dB better sensitivity than p-i-n, provided that the multiplication noise is low and the gain bandwidth product is sufficiently high.

In practical devices, the 1–2 μm thick absorption region is undoped. The graded layer (0.1–0.3 μm) and the avalanching layer (1–2 μm) are doped to 1×10^{16} cm^{-3}. The p$^+$-layer can be thin and doped to 10^{17}–10^{18} cm^{-3}. The junction is usually fabricated by zinc p$^+$-type diffusion in the InP for multiplication and Cd diffusion (or implantation) for guard-ring into the top InP layer through structures SiO$_2$ masks in closed ampoules. According to Gyuro [71], the device shows a dark current of < 6 nA, a 3 dB bandwidth of 7 GHz and a gain bandwidth product of 74 GHz. The excellent stability of the device was demonstrated by long-term aging measurements (after 30,000 hours at 150°C and a 100 μA reverse current).

The critical part of InGaAs SAM-APD is the thickness and doping in the field control layer between InGaAs absorption layer and InP multiplication layer. The avalanche gain and excess noise are predominantly determined by the ionization coefficients of InP [88]. In InP, the hole ionization coefficient, α_h, is greater than electron ionization coefficient, α_e, and the α_h/α_e ratio varies from 4 at low fields to 1.3 at the highest fields. Then, according to the McIntyre local theory, this gives rise to APDs with noise corresponding to $1/k \sim 0.4$ in 1 μm thick avalanching structures using hole initiated multiplication (see Figure 13.14).

In the past decade, the performance of APDs for optical fiber communication systems has improved as a result of improvements in materials and the development of advanced device structures [56,89,90]. Improvements concern introduction as continual or step grading in the band gap of the absorption/multiplication regions to prevent carrier trapping and attempts to introduce a field control layer. These novel structures are described in more detail in a recent review of telecommunication APDs by Campbell [89]. The width of the avalanching region in these devices has continued to shrink due to a requirement for higher operating speed. McIntyre local model is incapable of predicting noise performance of devices with thin avalanching widths (see Figure 13.15 [90]). An InP diode with an avalanching width of 0.25 μm would exhibit a noise performance equivalent of $1/k \sim 0.25$ for hole initiated multiplication, although the actual ionization coefficient ratio at such electric fields would yield a much larger value of ~0.7 using the McIntyre model [90]. The significant reduction of excess noise is due to the dead space effect observed in submicron avalanche regions. Further improvement in device performance can be achieved by introducing InAlAs, also lattice matched to InGaAs and InP, as a replacement for the InP as the multiplication region. The α_e/α_h ratio has been found to be significant larger than the α_h/α_e ratio in InP at a low electric field.

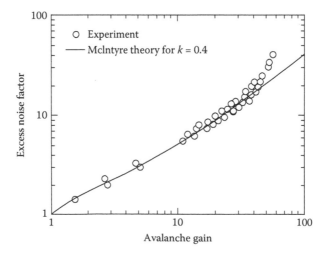

Figure 13.14 Excess noise factor versus avalanche gain for SAGM-APD InP-based photodiode.

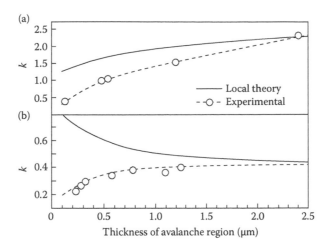

Figure 13.15 The effective ionization coefficient ratio in InP versus avalanche region thickness for (a) electron and (b) hole initiated multiplication. (From David, J. P. R., and Tan, C. H., *IEEE Journal of Selected Topics in Quantum Electronics*, 14, 998–1009, 2008. With permission.)

Also the excess noise factor at a given gain is significantly lower in InAlAs than that in InP due to the large α_e/α_h ratio in the former and the beneficial effect of the dead space in the latter.

13.3 BINARY III-V DETECTORS

13.3.1 InSb Photoconductive Detectors

The basic requirements for a photoconductive detector is material of very high purity into which small and carefully controlled amounts of a particular impurity can be introduced. InSb photoconductors are usually fabricated of p-type material (a dopant such as germanium is often used) with a free carrier concentration of less than 10^{14} cm^{-3} at 77 K, hole mobility of 7000 cm^2/Vs and dislocation density of less than 10^2 cm^{-2}. The higher resistance of p-type photoconductors and the absence of sweep-out allow higher responsivities than are possible in an n-type device [6]. In p–type material at low temperatures the lifetime for the two types of carriers are very different ($\tau_e < 10^{-9}$ s, $\tau_h \approx 10^9/p$) [91]. The effect of the strong minority carrier trapping reduces their drift length in the device, thereby increasing the bias field at which sweep-out onset occurs. It allows achievement of higher D^* values than in small detectors made of trap-free, p-type material. High-quality photoconductors are also made of n-type material with carrier concentration below 10^{14} cm^{-3} and electron mobility of 7.5×10^5 cm^2/Vs at 77 K [92,93].

The photosensitive area of the detectors is made from several tens of μm^2 to several mm^2, depending upon their usage. A small photosensitive area is used for quick response, and a large photosensitive area for high responsivity. For the 300 K operating devices the optimum thickness of elements is 5–10 μm, while for 195 K and 77 K photoconductive element thickness is about 25 μm [5].

The steps in preparing InSb photoconductive detectors are exactly described by Kruse [5]. A wafer that has the desired electrical properties is cut into element size and both surfaces are mirror-lapped using fine alumina powder. Next the surface of the oxidized film and damaged layer are chemically removed by etching in a mixture of equal volumes of undiluted HF, HNO$_3$, and CH$_3$COOH to a final thickness. Elements are then mounted on a sapphire substrate with epoxy resin. Also, other substrates such as Ge and Irtran 2, for which the thermal expansion coefficient is reasonably close to that of InSb, can be used. An alternative approach for removal of the sensitive elements from the slab employs photolithographic techniques. Electrical contacts are applied by soldering, using indium-based contact material.

In order to control and stabilize the behavior of InSb detectors for long periods, a passivation layer on the surface is needed. This passivation alters the characteristics of the surface as well as the behavior of the detector [92]. Sunde has found that degradation of photoconductors is caused by shallow traps in the oxide layer on the InSb surface [94]. A good passivation of the device would imply a surface with an oxide layer free from such traps. A passivation layer approximately 50 nm thick is obtained by anodization, which is usually carried out in a 0.1 N KOH solution.

Additionally, on top of the passivation layer a layer of SiO_x or ZnS about 0.5 μm thick is evaporated to provide a more stable surface. The last mentioned layer is also the antireflection coating.

Detectors for different applications are usually supplied with self-contained, thermally insulated encapsulations complete with window, radiation shield, lead-outputs, and provision for cooling.

The detectivity of p-type InSb photoconductive detectors at 77 K over a wide range of doping levels are g–r noise limited at bias voltages well below the sweep-out onset or the power dissipation limit. The g–r noise expression in the case of excitation and recombination for a single type of center is given by [95]

$$V_{gr}(0) = \frac{2V_b \tau_h^{1/2}}{(lwtp)^{1/2}},$$ (13.6)

and the responsivity [5,6]

$$R_v = \frac{\eta \lambda V_b \tau_h}{hclwtp}.$$ (13.7)

Thus from the above equations and Equation 3.35

$$D^* = \frac{\eta \lambda \tau_h^{1/2}}{2hct^{1/2}p^{1/2}}.$$ (13.8)

A more detailed theory of the photoconductive effect in InSb in which expressions are derived for the spectral responsivity and detectivity for three operating temperatures (77, 195, and 300 K) is given by Kruse [5]

Figure 13.16 illustrates the detectivities of InSb photoconductors as functions of wavelength, with modulation frequency as an independent parameter. The detectivity of available detectors operating at 77 K is usually in the range of 5×10^{10}–10^{11} cmHz$^{1/2}$W^{-1} at a sensitivity peak of 5.3 μm. Devices approaching the background-limited detectivity can be made. Responsivities in excess of 10^5 V/W for a 0.5×0.5 mm^2 element at optimum signal-to-noise ratio are obtained, and the resistance is typically 2 kΩ per square. The noise characteristics do not always vary linearly to the bias current and over a certain critical value, noise increases rapidly. Up to 100–150 Hz the noise is proportional to $1/f$. From 150 Hz to several kHz, g–r noise is predominant. At higher frequency ranges, noise is limited by Johnson noise.

InSb photoconductors are also used without cooling or with thermoelectric cooling in the temperature range from 190 to 300 K. As the operating temperature increases, the long wavelength response increases and the detectivity decreases considerably. At room temperature, detectivities at peak wavelength $\lambda_p \approx 6$ μm of 2.5×10^8 cmHz$^{1/2}$W^{-1} may be achieved. The response of a 1×1 mm^2 detector may be 0.5 V/W. The low time constant, about 0.05 μs, makes these detectors suitable for high speed operation applications. For 77 K photoconductive detectors the time constant is normally in the range of 5–10 μs.

Pines and Stafsudd have reported photoconductive data on high-quality n-type InSb photoconductors [93]. Their studies show that the parameters of surface passivated detectors are limited by bulk material properties. Figure 13.17 shows the responsivity and noise as a function of temperature in n-type photoconductive detector. At higher temperatures, the responsivity and noise depend slightly on the background photon flux density. The roll-off in responsivity and noise at higher temperatures can be correlated with the decrease in the carrier lifetime and the increase in the majority carrier concentrations.

13.3.2 InSb Photoelectromagnetic Detectors

InSb photoelectromagnetic (PEM) detectors operated at room temperature offer excellent performance in the 5–7 μm interval. Difficulties associated with cooling prevent operating the detectors at 195 K or 77 K. Moreover, when cooled they offer no advantage over photoconductive or

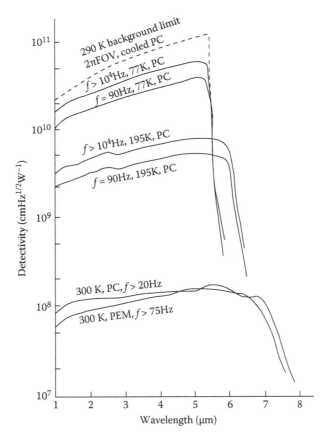

Figure 13.16 Typical spectral variation of detectivity for InSb PC and PEM detectors operating at 77 K, 195 K, and 300 K. (From Kruse, P. W., *Semiconductors and Semimetals,* Academic Press, New York, Vol. 5, 15–83, 1970. With permission.)

photovoltaic detectors. A comprehensive review of the operation, technology, and properties of InSb PEM detectors has been done by Kruse [5].

The preparation of the PEM InSb detectors is the same as for the photoconductive devices. However, whereas photoconductive detectors require a low surface recombination velocity on both the front and the back surface, PEM detectors require a low recombination velocity on the front surface and high recombination velocity on the back (see Section 9.10.1).

The highest theoretically calculated magnitudes of responsivity and detectivity have been found in p-type material having a hole concentration of approximately 7×10^{16} cm^{-3}. For the assumed values of the parameters ($\lambda_c = 6.6$ μm, $B = 0.7$ T, and $t = 20$ μm) the maximum detectivity is about 6×10^8 cmHz$^{1/2}$W^{-1} and responsivity 5 V/W. To obtain slightly p-type material, a residual donor in InSb single crystals is usually compensated with zinc or cadmium. For the 300 K PEM detectors the optimum thickness is about 25 μm. The housing design of the PEM detector incorporates a permanent magnet (see Figure 9.45). The requirement for a high magnetic flux density within the sample restricts the sample width to a value of about 1 mm or less.

Figure 13.16 illustrates the spectral detectivity of the 300 K InSb PEM detector. The detectivity has a peak of approximately 2×10^8 cmHz$^{1/2}$W^{-1} for frequencies greater than 75 Hz. Usually the peak value of D^* occurs at 6.2 μm [96,97]. The response time no greater than 0.2 μs has been found. It is an attractive feature in wideband applications.

The region of maximum responsivity of uncooled InSb PEM detectors (5.5–6.5 μm) is outside the "atmospheric windows." The realization of noncooled detectors for radiation with wavelengths longer than 8 μm is a real problem. Jóźwikowski et al. have indicated that the PEM effect in InAs$_{0.35}$Sb$_{0.65}$ can be used as a basic principle for the long-wavelength, noncooled detectors [98].

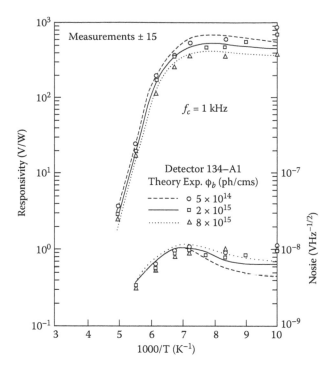

Figure 13.17 Responsivity and noise as a function of temperature in n-type InSb photoconductors. (From Pines, M. Y., and Stafsudd, O. M., *Infrared Physics*, 19, 563–69, 1979. With permission.)

13.3.3 InSb Photodiodes

Recent efforts in the technology of infrared detectors have mostly been on large electronically scanned FPAs. The design of hybrid FPAs is currently based on junction photodiodes, due to reduced electrical power dissipation, high impedance matched directly into the input stage of a silicon readout, and less stringent noise requirements for the readout devices and circuits [99].

The III-V photodiodes are generally fabricated by impurity diffusion [100–110], ion implantation [111–123], LPE [124–132], MBE [20,133–136], and MOCVD [20,136,137].

Initially p-n junctions in InSb were made by diffusing Zn or Cd into n-type substrates with net donor concentration in the range of 10^{14}–10^{15} cm^{-3} at 77 K [100]. To provide a flat, damage-free surface, the substrates were chemically or electrochemically polished. Catagnus et al. [104] made p-n junctions in closed ampoules using the In:Sb:Cd source of composition of 5:45:50 atm%. At any temperature between 250 and 400°C, the variation of junction depth with time may be determined from the formula $x = 40.5$ cm/hr$^{1/2}(t)^{1/2}$exp(–0.80eV/kT). Nishitani et al. [105] used either elemental Zn or a combination of elemental Zn and Sb as the diffusion source over the temperature range of 355–455°C. The diffusion source of $N_{Sb}/N_{Zn} \geq 5$ was recommended as the optimum source condition in the preparation of a p-n junction. A modified technology utilizing a two-temperature-zone method with the Cd source at 380°C and the InSb substrate at 440°C has been also used [109,110].

Figure 13.18 illustrates 10 major processing steps for the diffusion fabrication of InSb mesa photodiodes [138]. The active surface area of the detector is coated with an anodic layer for stabilizing surface states and an antireflection layer of SiO.

It was found, however, by several researchers [113] that Zn and Cd produced a porous surface that is difficult to remove. To circumvent this surface problem and to make high-quality p-n junctions, implantation of the light ions Be and Mg have been used [111–123]. Implantation of sulfur [112] and protons [111] to form n-p diodes and zinc and cadmium to form p-n diodes have been reported. It appears that InSb can be made amorphous by implanting it at room temperature with heavy ions [117]. Zn and Cd may be too heavy to implant in InSb.

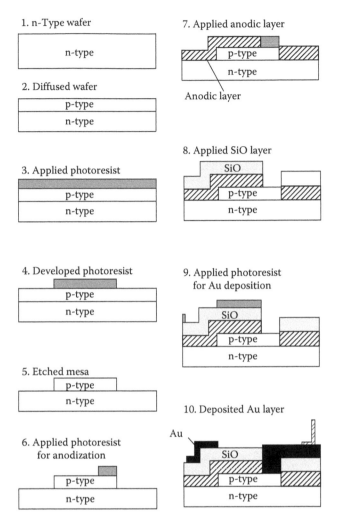

Figure 13.18 Processing steps for the diffusion fabrication of InSb mesa photodiodes. (From Chan, W. S., and Wan, J. T., *Journal of Vacuum Science Technology*, 14, 718–22, 1977. With permission.)

InSb photodiodes produced by Be implantation have been reported by Hurwitz and Donnelly [113]. This developed diode process has been used for monolithic InSb CCD arrays, for flat-zero input structures and simple extraction of charge at the output or in charge-skimming circuits [114]. Beryllium implantation is also used for source and drain formation for InSb MOSFETs. Diodes were fabricated on (100)-oriented slices with donor concentrations of 10^{14}–10^{15} cm^{-3}. Before implantation, each wafer was polished and anodized in a 0.1 N solution of NH$_4$OH at 30 V for 30 s to protect the surface during subsequent processing. The 5 μm thick photoresist was then applied to each sample and an array of 0.5 mm diameter holes opened in the photoresist that formed a mask against ions. The best photodiodes were made with ion implantation energies of 100–200 keV, with doses of $(2–5) \times 10^{14}$ cm^{-2}. Samples were tilted 7° from the normal to reduce channeling. After removal of the photoresist, samples were coated with 100 nm of pyrolytic SiO$_2$ deposited at 350°C for 2–3 minutes and were annealed at 350°C for 15 minutes. This process acts effectively to anneal out the radiation damage caused by the room temperature Be ion implantation. Next, approximately 0.4 μm of InSb was removed together with a surface inversion layer formed during annealing. It was found to be vital for the fabrication of good diodes. Immediately after etching and rinsing the wafers were coated with 150 nm of silicon oxynitride and with 100 nm of SiO$_2$ to provide a stable surface. To permit adjustment of the surface potential for optimum diode performance, a field guard ring was employed. Contact to the p-type region was made by plating of 0.5-μm-thick Au covered by

1.5 μm of In. The substrate was contacted with In. A cross-sectioned wafer of a finished diode is shown in Figure 13.19a [113].

At present in InSb photodiode fabrication, epitaxy is usually not used; instead, the standard manufacturing technique begins with bulk n-type single crystal wafers with donor concentration about 10^{15} cm^{-3}. Relatively large bulk grown crystals with 2 and 3 inch diameters are available on the market. An array hybrid size up to 2048 × 2048 is possible because the InSb detector material is thinned to less than 10 μm (after surface passivation and hybridization to a readout chip) that allows it to accommodate the InSb/silicon thermal mismatch [139]. As is shown in Figure 13.19b, the BSI InSb p-on-n detector is a planar structure with an ion-implanted junction. After hybridization, epoxy is wicked between the detector and the Si ROIC and the detector is thinned to 10 μm or less by diamond-point-turning. One important advantage of a thinned InSb detector is that no substrate is needed; these detectors also respond in the visible portion of the spectrum. Also growing of InSb and related alloys by MBE together with doping of substrate to induce transparency has been demonstrated [140]. In the last case the thinning of the detector material is not required.

The forward I-V characteristics are given by $J = J_o \exp(qV/\beta kT)$ with $\beta \approx 1.7$. The diffusion current component $J_D = J_s[\exp(qV/kT)-1]$ was determined from the forward I-V curves. J_s was 5 to 7 × 10^{-10} A/cm^2 at 77 K and was negligibly small compared to generation–recombination current density in reverse bias. The measured values of R_oA were above 10^6 Ωcm^2 at 77 K. Figure 13.20a shows measured current density versus reverse voltage for planar p$^+$-n junctions in InSb formed by Be-ion implantation [114]. The base material was obtained using either Czochralski or LPE growth techniques. The current in the small bias region (A) is clearly g–r limited. The solid curve labeled J_{gr} is calculated using Equation 9.113 and the parameters $N_d = 6.6 \times 10^{14}$ cm^{-3}, $V_{bi} = 0.209$ V, $\tau_o = 10^{-8}$ s, $E_t = E_i$, and $n_i = 2.7 \times 10^9$ cm^{-3}. As V increases, the photodiodes enter a breakdown region where J increases superlinearly with V. The current in the breakdown region is a strong function of doping and is believed to be due to interband tunneling. Solid curves (B) and (C) are the tunnel currents calculated using an equation similar to Equation 9.118 for $N_d = 6.6 \times 10^{14}$ cm^{-3} and 1.4×10^{14} cm^{-3}. The broken curves (D) and (E) are calculated for liquid-helium temperature assuming an interband tunneling model. We can see that tunneling is dominating for reverse voltage exceeding ≈ 2 V. The current densities of the LPE diodes are about a factor of five lower than for Czochralski material. The longer minority carrier lifetime observed in the epitaxial material translates directly into reduced dark current density in the photodiodes for a given operating temperature [141].

Figure 13.20b compares the dependence of dark current on temperature between InSb and HgCdTe photodiodes used in the current high performance large FPAs [142]. This comparison

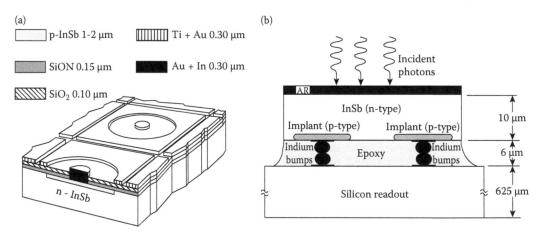

Figure 13.19 Cross-section views of InSb photodiodes: (a) ion-implanted, field-plate-guarded InSb photodiode (the implanted region is 0.5 mm in diameter and the contact 75 μm in diameter, distance between adjacent field plates is 5 μm). (From Hurwitz, C. E., and Donnelly, J. P., "Planar InSb Photodiodes Fabricated by Be and Mg Ion Implantation," *Solid State Electronics* 18, 753–56, 1975.) (b) Architecture of an InSb sensor chip assembly. (From Love, P. J., Ando, K. J., Bornfreund, R. E., Corrales, E., Mills, R. E., Cripe, J. R., Lum, N. A., Rosbeck, J. P., and Smith, M. S., "Large-Format Infrared Arrays for Future Space and Ground-Based Astronomy Applications," *Proceedings of SPIE* 4486, 373–84, 2002. With permission.)

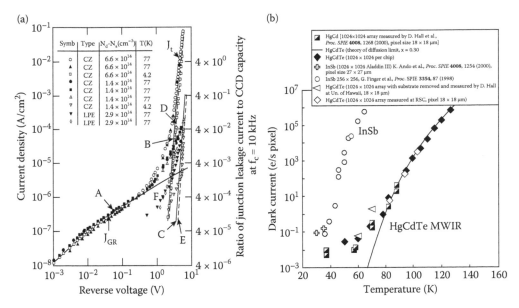

Figure 13.20 Current density versus reverse voltage for InSb photodiodes: (a) planar p⁺-n junctions formed by Be-ion implantation, solid curves are the theoretical fit to the data. (From Thom, R. D., Koch, T. L., Langan, J. D., and Parrish, W. L., *IEEE Transactions on Electron Devices* ED-27, 160–70, 1980. With permission.) (b) The comparison of dependence of dark current on temperature between highest reported value for InSb arrays and MBE-grown HgCdTe MWIR FPAs (with 18 × 18 μm pixels). (From Zandian, M., Garnett, J. D., DeWames, R. E., Carmody, M., Pasko, J. G., Farris, M., Cabelli, C. A., et al., *Journal of Electronic Materials*, 32, 803–9, 2003. With permission.)

suggests that MWIR HgCdTe photodiodes have significant higher performance in the 30–120 K temperature range. The InSb devices are dominated by g–r currents in the 60–120 K temperature range because of a defect center in the energy gap, whereas MWIR HgCdTe detectors do not exhibit g–r currents in this temperature range and are limited by diffusion currents. In addition, wavelength tunability has made HgCdTe the preferred material.

Capacitance–voltage measurements gave a functional dependence of C versus V as C^{-2}, confirming that the Be diodes are one-sided abrupt junctions. A zero-bias RC time constant is about 0.4 ms.

The simplest possible InSb photodiode structures, requiring only an insulating layer and surface metallization for contacts are typically used for large, single element detectors. However, there are several disadvantages of this scheme: the bonding pad capacitance becomes a significant portion of the total device capacitance (especially for small area detectors); in closely spaced, multiple element arrays the bonding pads can no longer be used to completely blind all active area between elements. In multiple element imaging and spectroscopy systems, the resulting exposed "semiactive" area between photodiodes may result in a loss of resolution (the minority carrier diffusion length in the n-region of InSb diodes is on the order of 20–30 μm). To maximize the available resolution and response in InSb photodiodes, the bulk material is thinned to about 10 μm. For high-performance FPAs in the 2–5 μm range, the photodiode parameters must be well defined to design a complete detector/preamplifier package. InSb material is highly uniform and, combined with diffusion or an implanted process in which the device geometry is precisely controlled, the resulting detector array responsivity is excellent.

Bloom and Nemirovsky have presented an improved processing technology of BSI, planar, gate-controlled InSb photodiodes with active n-type region with carrier concentration about 10^{15} cm⁻³ [143,144]. The photodiode design is shown in Figure 13.21. The p⁺ junctions were formed by implanting beryllium on (111) wafers with a dose of 5×10^{14} cm⁻² and energy of 100 keV, followed by annealing at 350°C for half an hour in a nitrogen ambient. For both front and back side surfaces they used improved surface passivation utilizing a modified UV photo-assisted SiO_x deposition (PHOTOX). In this process, the photolysis of the reactant gases (SiH₄ and N₂O) occurs at 50°C by collisions with excited mercury atoms that absorb the ultraviolet

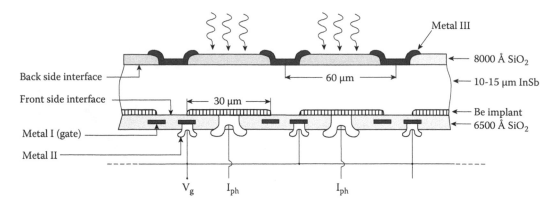

Figure 13.21 Detailed cross section of back side illuminated, gate-controlled InSb matrix photodiodes. (From Bloom, I., and Nemirovsky, Y., *IEEE Transactions on Electron Devices*, 40, 309–14, 1993. With permission.)

radiation at a wavelength of 2537 Å. The strongly accumulated interface with reduced surface recombination velocity was formed on In face. The resultant band-banding produces an electric field that retards the flow of minority carriers to the surface recombination sites, sweeps the generated holes to the junctions, and increases quantum efficiency. The requirements of the surface passivation at the Sb face where the junctions are implanted are more stringent. To obtain the required surface control at the periphery of the junctions, the gate (evaporated titanium) covers the junctions along the periphery and all the gates of the sensing elements of the array are connected with a second metallization of titanium and gold (see Figure 13.21). The Sb face forms a slightly accumulated interface, with a relatively small concentration of fast and slow surface states, as exhibited by near ideal capacitance–voltage characteristics of MIS devices and very small hysteresis. In addition, the electrical performance of photodiodes at optimum gate bias (up to 1 V reverse bias) with small-geometry junctions (30×30 μm^2) is the same as that observed in large junctions indicating that the performance is limited by the properties of the starting bulk material rather than the passivation of the front side (Sb face). The R_oA product is 5×10^4 Ωcm^2 at 77 K.

Wimmers et al. have presented the status of InSb photodiode technology at Cincinnati Electronics (Mason, Ohio) for a wide variety of linear and FPAs [106,108]. Fabrication techniques for InSb photodiodes use gaseous diffusion and a subsequent etch results in a p-type mesa on n-type substrate with donor concentration about 10^{15} cm^{-3}. A highly controlled diffusion process allows p-layer diffusion to occur with little surface damage, eliminating the need for deep diffusion and subsequent etch-back. This permits total "mesa" heights of only a few microns. A grounded "buried-metallization" process, independent of bond-pad metallization, was developed to render the surface of the InSb opaque, with the exception of the active area and the contact area. The accuracy of the photolithography along with the controlled diffusion process provides excellent uniformity of response.

Typical InSb photodiode RA product at 77 K is 2×10^6 Ωcm^2 at zero bias and 5×10^6 Ωcm^2 at slight reverse biases of approximately 100 mV (see Figures 13.22 and 13.23 [106,108]). This characteristic is beneficial when the detector is used in the capacitive discharge mode. As element size decreases below 10^{-4} cm^{-2} the ratio of the circumference to area is increased and some slight degradation in resistance due to surface leakage occurs.

The performance of FPAs depend upon the capacitance of the detector element. InSb photodiode capacitance can be effectively modeled as the sum of a voltage-dependent junction capacitance (which agree well with the abrupt junction model) and a bond pad capacitance [106]. For diodes with a relatively large active area, neither the bond pad capacitance nor the capacitance associated with the contact area represent a significant fraction of the total diode capacitance. However, as the active area decreases the total diode capacitance becomes more strongly dependent on these two terms. Figure 13.24 shows zero-bias capacitance as a function of active area for InSb photodiodes at 77 K [106].

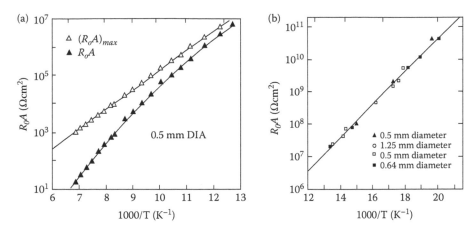

Figure 13.22 R_oA and $(RA)_{max}$ versus temperature for InSb photodiodes: (a) 150 K to 80 K. (From Wimmers, J. T., Davis, R. M., Niblack, C. A., and Smith, D. S., "Indium Antimonide Detector Technology at Cincinnati Electronics Corporation," *Proceedings of SPIE* 930, 125–38, 1988. With permission.) (b) 80 K to 50 K. (From Wimmers, J. T., and Smith, D. S., "Characteristics of InSb Photovoltaic Detectors at 77 K and Below," *Proceedings of SPIE* 364, 123–31, 1983. With permission.)

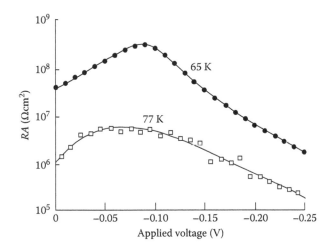

Figure 13.23 Dependence of the RA product on applied reverse voltage for 3.2×3.2 mil InSb photodiode at 77 K and 65 K. (From Wimmers, J. T., Davis, R. M., Niblack, C. A., and Smith, D. S., "Indium Antimonide Detector Technology at Cincinnati Electronics Corporation," *Proceedings of SPIE* 930, 125–38, 1988. With permission.)

The InSb photovoltaic detectors are widely used for ground- and space-based infrared astronomy. For applications in astrophysics, these devices are very often operated at 4–7 K with a resistive or capacitive transimpedance amplifier to achieve the lowest noise performance [145]. At these low temperatures, the InSb photodiode resistance is so high that the detector Johnson noise is negligible, and the dominant noise sources are either the feedback resistance or input amplifier noise. Since the latter scales directly with the combined detector and input circuit capacitance, it becomes important to minimize these. The InSb photodiodes described above, whose performance is optimized for 60–80 K operation, have been shown to lose long wavelength quantum efficiency at lower temperatures, due to a decrease in the minority carrier lifetime in the n-region [146]. Thus, the process of device optimization must be redone for low temperature applications with emphasis on reducing detector capacitance while simultaneously maximizing the quantum efficiency [147]. A reduction in doping density in the n-type region to $\approx 10^{14}$ cm^{-3}, along with other minor process modifications, minimizes decrease of the carrier lifetime, providing an added

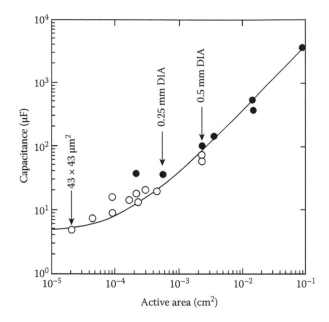

Figure 13.24 Zero-bias capacitance as a function of active area for InSb photodiodes at 77 K. The solid line is theoretically calculated. (From Wimmers, J. T., and Smith, D. S., "Characteristics of InSb Photovoltaic Detectors at 77 K and Below," *Proceedings of SPIE* 364, 123–31, 1983. With permission.)

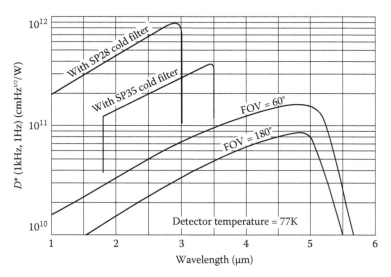

Figure 13.25 Detectivity as a function of wavelength for a InSb photodiode operating at 77 K. (From Judson Catalog, *Infrared Detectors*, http://www.judsontechnologies.com.)

benefit of decreases in capacitance. This approach also reduces the RA product slightly, but the RA product still increases exponentially with decreasing temperature, until detector resistance once again is not a significant noise contribution.

The detectivity versus wavelength for an typical InSb photodiode is shown in Figure 13.25. Detectivity increases with reduced background flux (narrow FOV and/or cold filtering) as illustrated in the figure. InSb photodiodes can also be operated in the temperature range above 77 K. Of course, the RA products degrade in this region. The quantum efficiency in InSb photodiodes optimized for this temperature range remains unaffected up to 160 K.

InSb photodiodes can also be operated in the temperature range above 77 K. Of course, the RA products degrade in this region. At 120 K, RA products of 10^4 Ωcm^2 are still achieved with slight

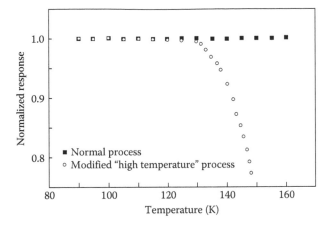

Figure 13.26 Normalized response of InSb photodiodes in elevated temperature range. (From Wimmers, J. T., Davis, R. M., Niblack, C. A., and Smith, D. S., "Indium Antimonide Detector Technology at Cincinnati Electronics Corporation," *Proceedings of SPIE* 930, 125–38, 1988.)

reverse bias, making BLIP operation possible. The quantum efficiency in InSb photodiodes optimized for this temperature range remains unaffected up to 160 K (see Figure 13.26 [108]). Process modifications such as increases in doping density allow responsivity to remain unchanged by increasing temperature.

Some of the design constraints associated with bulk devices with implanted or diffusion junctions can be relaxed using epitaxial methods. Epitaxial, BSI InSb photodiodes have been grown on Te-doped InSb substrates. Degenerate n-type doping of the substrates allows them to be made transparent through the Burstein-Moss effect. A doping of 2×10^{18} cm^{-3} is sufficient to obtain transparency at 80 K over the majority of the 3–5 μm wavelength range [148]. Free carrier absorption requires that the substrates are polished back to a thickness of 100 μm. However, to obtain high quantum efficiency the substrate should be thinned to a value of the order of 10 μm, as in the case for bulk devices. The p$^+$ and n$^+$ regions are doped to 3×10^{18} cm^{-3} and 2×10^{18} cm^{-3}, respectively, and the active region has an n-type doping of approximately 2×10^{15} cm^{-3}. The thicknesses of the p$^+$, n$^+$, and active regions are 1 μm, 4 μm, and 3 μm, respectively. During array fabrication procedures, the wafers are patterned on a 30 μm pitch using standard lithography, then profiled to mesa shapes by a chemical etch down to the n$^+$ region. The final size is approximately 17 μm at the diode junction.

The InSb photodiodes grown heteroepitaxially on Si and GaAs substrates by MBE have been also reported [20,149–154]. Recently, Kuze et al. [153,154] have developed a novel microchip-sized InSb photodiode sensor, on semi-insulating GaAs(100) substrate, operating at detection. The sensor consists of 910 photodiodes connected in series (see Figure 13.27). Each photodiode consists of MBE grown 1 μm thick n$^+$-InSb layer, followed by a 2 μm thick π-InSb absorber layer. To reduce the diffusion of photoexcited electrons, a 20 nm thick p$^+$-Al$_{0.17}$In$_{0.83}$Sb barrier layer was grown on the π-InSb layer. Finally, a 0.5 μm thick π-InSb layer was grown as the top contact. As the n- and p-type dopants Sn and Zn were used, respectively, with concentrations of 7×10^{18} cm^{-3} for the n$^+$-layer, 6×10^{16} cm^{-3} for the π layer, and 2×10^{18} cm^{-3} for the p$^+$ layer. To insulate mesa structures, a 300 nm thick plasma CVD passivation Si$_3$N$_4$ layer was deposited. Finally, after Ti/Au lift-off metallization, a 300 nm thick SiO$_2$ passivation also grown by plasma CVD was made. The length of a single InSb photodiode was 20 μm. The final external dimensions of photovoltaic infrared sensor were $1.9 \times 2.7 \times 0.4$ mm^2. The sensitivity and the noise equivalent difference temperature were 127 μV/K and 1 mK/Hz$^{1/2}$, respectively.

13.3.4 InAs Photodiodes

InAs detectors have been made to operate in the photoconductive, photovoltaic, and PEM modes. Recently however, wider applications (laser warning receivers, process control monitors, temperature sensors, pulsed laser monitors, and infrared spectroscopy) have found InAs photodiodes operated at near room temperature. The photodiodes are mainly fabricated by ion implantation [112,123,124] and the diffusion method [100,102].

(a)

(b)

Ti/Au electrodes   Si$_3$N$_4$ passivation

p$^+$-InSb; 0.5 μm
p$^+$-AlInSb; 20 nm
π-InSb; 2 μm
n$^+$-InSb; 1 μm

Semi-insulating GaA substrate

Incident photons

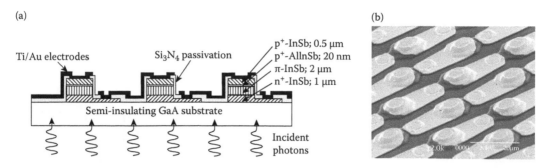

Figure 13.27 InSb photovoltaic infrared sensor: (a) schematic structure of multiple photodiodes connected in series, and (b) SEM photograph. (From Camargo, E. G., Ueno, K., Morishita, T., Sato, M., Endo, H., Kurihara, M., Ishibashi, K., and Kuze, M., *IEEE Sensors Journal*, 7, 1335–39, 2007. With permission.)

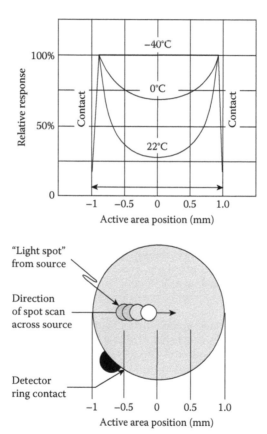

Figure 13.28 InAs photodiodes: (a) response variation across 2 mm active area of InAs photodiodes (After product brochure of Judson Inc., http://www.judsontechnologies.com).

The diode sensitivity, speed of response, impedance, and peak wavelength can be optimized by operation at the proper temperature. At room temperature the shunt resistance of InAs photodiode is comparable with series resistance, which effects the response of photodiodes (see Figure 13.28). This effect is less pronounced in small area detectors, which have higher shunt resistance and less surface area. The effect is also reduced or eliminated by cooling the diode, thereby increasing the junction resistance. InAs photodiodes are sensitive in the 1–3.6 μm wavelength range. A typical range of detectivity for InAs photodiodes is shown in Figure 13.29.

Kuan et al. have presented high performance InAs photodiodes grown by MBE [135,155]. The diode structure was grown on the (100) n-type InAs wafers. After removing the surface oxides

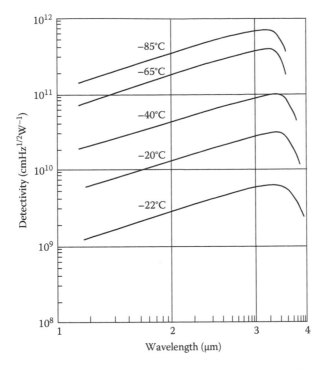

Figure 13.29 Detectivity versus wavelength for InAs photodiodes at different temperatures (After product brochure of Judson Inc., http://www.judsontechnologies.com).

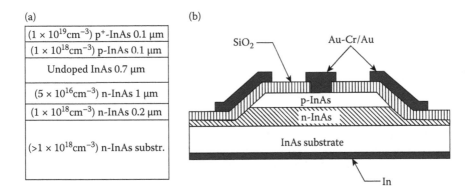

Figure 13.30 InAs gate-controlled photodiode: (a) the schematic diagram, and (b) the device structure. (From Lin, R. M., Tang, S. F., Lee, S. C., Kuan, C. H., Chen, G. S., Sun, T. P., and Wu, J. C., *IEEE Transactions on Electron Devices*, 44, 209–13, 1997. With permission.)

(by slowly heating to 500°C), the InAs epilayers were grown at 500°C under optimized growth conditions. The p-i-n photodiode structure consisted of a 0.2 μm thick n-type buffer layer (Si-doped to 1×10^{18} cm^{-3}), followed by a 1 μm thick n-type InAs active layer (Si-doped to 5×10^{16} cm^{-3}). Then a 0.72 μm thick undoped InAs layer was grown, followed by a 0.1 μm thick p-type InAs layer (Be-doped to 1×10^{18} cm^{-3}), and finally a 0.1 μm thick InAs contact layer (exponentially graded doping from 1×10^{18} cm^{-3} to 1×10^{19} cm^{-3}) was deposited. The same structure was also grown except for the undoped InAs layer for p-n diode. The schematic diagram and device structure of the InAs gate-controlled photodiode are shown in Figure 13.30 [155].

Before fabricating unpassivated and passivated InAs diodes, a special chemical treatment and two-step photolithographic procedures were used [155]. In the case of gate-controlled photodiodes, the epilayers were first mesa etched into circular dots measured 200 μm in diameter and then the

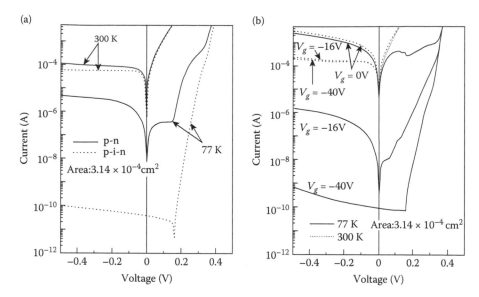

Figure 13.31 The 77 K and room temperature I-V characteristics of the (a) unpassivated, and (b) gate-controlled InAs p-i-n and p-n photodiodes. The gate biases ($V_g = 0$, –16, and –40V) are marked in (b). (From Lin, R. M., Tang, S. F., Lee, S. C., Kuan, C. H., Chen, G. S., Sun, T. P., and Wu, J. C., *IEEE Transactions on Electron Devices*, 44, 209–13, 1997. With permission.)

photo-CVD technique was applied to deposit 300 nm thick SiO_2. A second photolithographic step was used to remove the SiO_2 from the defined area of 10×4 µm² on the p-type layer in order to make electrical contacts. A double layer of 100 nm thick Au-Be and 300 nm thick Au was evaporated sequentially and lifted off to form a p-type ohmic contact. A third photolithographic step was used to define the pad with a dot having a 40 µm diameter and the gate electrode that covered the junction perimeter. Then a double layer of 12 nm of Cr and 300 nm of Au were evaporated and lifted off.

Analysis and comparison of the dark current characteristics between InAs p-n and p-i-n diodes at the temperature range from 30 to 300 K carried out by Kuan and colleagues [135] indicated that incorporation of a thick i-layer (i.e., 720 nm) is the key to success of the detector. The advantage of the p-i-n structure is not only to cut down the tunneling current but also to increase the uniformity.

Figure 13.31a shows the typical 77 K and room temperature I-V characteristics of unpassivated p-i-n and p-n photodiodes, respectively [155]. At 77 K, the dark current of unpassivated p-i-n photodiode is disturbed by the background thermal radiation, proof of which is the existence of a photovoltage. We can also see that the reverse dark current of unpassivated photodiode depends on the diode reverse bias both at 77 K and 300 K indicating the existence of a shunt leakage current. Figure 13.31b shows the typical 77 and 300 K I-V characteristics of gate-controlled p-i-n photodiodes under different gate bias at 0, –16, and –40 V. The strong dependence of the diode reverse dark current on the gate voltage indicates that the reverse leakage current is flowing through the surface region. When the gate bias V_g approaches –40 V, the reverse dark current becomes nearly independent of reverse biases, which indicates that the diode is leakage free. It is obvious that the I-V characteristics of unpassivated p-i-n photodiode is similar to and even better than the p-i-n gate-controlled photodiode under gate bias $V_g = -40$ V. This indicates that passivation of InAs p-i-n photodiode degrades the device performance. The unpassivated p-i-n photodiode exhibits R_oA product of 8.1 Ωcm² at room temperature and 1.3 MΩcm² at 77 K. When illuminated under a 500 K blackbody source, the photodiode detectivity limited by Johnson noise is 1.2×10^{10} cmHz$^{1/2}$W^{-1} at room temperature and 8.1×10^{11} cmHz$^{1/2}$W^{-1} at 77 K.

Alternative substrates in fabrication InAs photodiodes have also been used. Dobbelaere et al. have produced InAs photodiodes grown on GaAs and GaAs-coated Si by MBE [156]. This technique is suitable for fabrication of monolithic near-infrared imagers where a combination of detection with silicon readout electronics is possible.

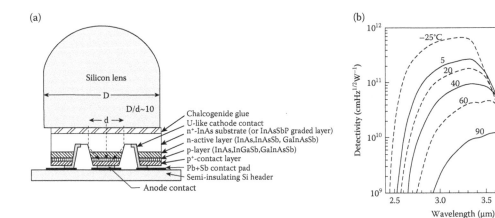

Figure 13.32 InAs heterostructure immersion photodiode: (a) construction of the immersion photodiode, and (b) detectivity spectral at near-room temperatures. (From Remennyy, M. A., Matveev, B. A., Zotova, N. V., Karandashev, S. A., Stus, N. M., and Ilinskaya, N. D., "InAs and InAs(Sb)(P) (3–5 μm) Immersion Lens Photodiodes for Potable Optic Sensors," *Proceedings of SPIE* 6585, 658504, 2007. With permission.)

Recently, the research group at the Ioffe Physico-Technical Institute (St. Petersburg, Russia) has developed InAs immersion-lens photodiodes operated at near-room temperatures [132]. InAs heterostructure photodiodes (see Figure 13.32a) were LPE grown onto n^+-InAs transparent substrates (due to the Burstein–Moss effect) and consisted of ~3 μm thick n-InAs layers and ~3 μm thick p-InAs$_{1-x-y}$Sb$_x$P$_y$ cladding layers lattice matched with InAs substrate ($y \sim 2.2x$). Due to an energy step at the n^+-InAs/n-InAs interface, a beneficial hole confinement for the photodiode operation is expected. Flip-chip mesa devices with the diameter of 280 μm were processed by a multistage wet photolithography process. A cathode as well as anode contacts were formed by sputtering of Cr, Ni, Au(Te), and Cr, Ni, and Au(Zn) metals followed by an electrochemical deposition of a 1–2 μm thick gold layer. Next the substrate was thinned down to 150 μm and chips were soldered onto silicon submounts with Pb–Sn contact pads. Finally, the 3.5 mm wide silicon lens was attached to the substrate side of a chip by a chalcogenide glass with high refractive index ($n = 2.4$). It is obvious that the field of view of immersion photodiodes is considerable lower than for uncoated device (decreased down to 15°).

Figure 13.32b shows the detectivity spectra of InAs heterostructure immersion photodiode. Superior detectivity of these photodiodes (for comparison see Figure 13.29) reflects improvements associated with board mirror contact, asymmetric doping, immersion effect, and radiation collection by inclined mesa walls. The narrow spectral responses are a result of filtering in the substrate and intermediate layers. Peak wavelengths shift to long wavelengths as temperature is increased due to bandgap narrowing at higher temperatures. However, the short wavelength spectra are more sensitive to temperature than the long wavelength ones probably due to progressively poor transparency of n^+-InAs near the absorption edge at elevated temperatures due to elimination of the conduction band electron degeneration [132].

13.3.5 InSb Nonequilibrium Photodiodes

The first nonequilibrium InSb detectors had a p^+-π-n^+ structure, where π represents low doped p-type material that is intrinsic at the temperature of operation [18]. An accurate analysis of the source of current in the diode at room temperature indicates the predominant contribution of Auger seven generation in the p^+ material. At temperatures below 200 K, the performance of photodiodes is determined by Shockley–Read generation in the π region (Figure 13.33). Davies and White investigated the residual currents in Auger suppressed photodiodes [157]. They find that removing electrons from the active region alters the occupancy of the traps leading to an increased generation rate from the Shockley–Read traps in the active region.

Next it was shown that a thin strained layer of InAlSb between the p^+ and π regions produces a barrier in the conduction band that substantially reduces the diffusion of electrons from the p^+ layer to the π region leading to an improvement in room temperature performance [19,158]. This type of p^+-P^+-π-n^+ InSb/In$_{1-x}$Al$_x$Sb structure shown schematically in Figure 13.34 was fabricated

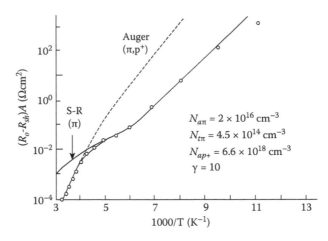

Figure 13.33 Temperature dependence of the R_oA product for p$^+$-π-n+ InSb photodiode. Circles indicate experimental data; the solid lines show calculated values based on both Shockley–Read and Auger generation mechanisms in the π and p$^+$ regions. (From Ashley, T., Dean, A. B., Elliott, C. T., Houlton, M. R., McConville, C. F., Tarry, H. A., and Whitehouse, C. R., "Multilayer InSb Diodes Grown by Molecular Beam Epitaxy for Near Ambient Temperature Operation," *Proceedings of SPIE* 1361, 238–44, 1990. With permission.)

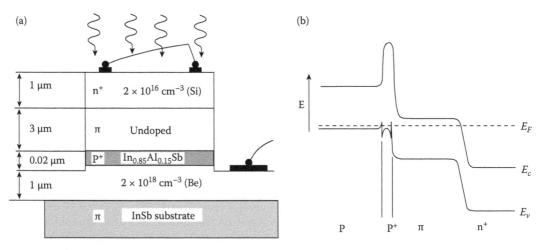

Figure 13.34 Schematic cross section of the (a) p$^+$-P$^+$-π-n$^+$ InSb/In$_{1-x}$Al$_x$Sb heterostructure photodiode, and (b) its energy band diagram. (From Elliott, C. T., "Advanced Heterostructures for In$_{1-x}$Al$_x$Sb and Hg$_{1-x}$Cd$_x$Te Detectors and Emiters," *Proceedings of SPIE* 2744, 452–62, 1996. With permission.)

by MBE at 420°C using silicon-doped (n-type) and beryllium-doped (p-type) InSb layers. Typical doping concentrations and thickness of individual layers are marked in Figure 13.34a. The composition, x, of the In$_{1-x}$Al$_x$Sb layer, is 0.15 giving a conduction band barrier height which is estimated as 0.26 eV. The central region is typically 3 μm thick and not deliberately doped. Circular diodes of 300 μm diameter were fabricated by mesa etching to the p$^+$ region and were passivated with an anodic oxide. Sputtered chromium/gold contacts were applied to the top of each mesa, with an annual geometry having an internal diameter of 180 μm and external diameter of 240 μm and to the p$^+$ region. For the antireflection coating, the 0.7 μm thick thin oxide layer was used.

The P$^+$-P$^+$-π-n$^+$ InSb/In$_{1-x}$Al$_x$Sb unbiased heterostructure photodiodes give detectivity above 2×10^9 cmHz$^{1/2}$W^{-1}, with peak responsivity at 6 μm. This value is an order of magnitude higher than that of typical, commercially available, single-element thermal detectors. Figure 13.35 compares theoretical curves for detectivity, calculated from the zero bias resistance, for a conventional p$^+$-n diode and an epitaxially grown p$^+$-P$^+$-v-n$^+$ structure with a 3 μm thick active region. We see, for example, an increase in operating temperature of about 40 K in the vicinity of 200 K. However, InSb

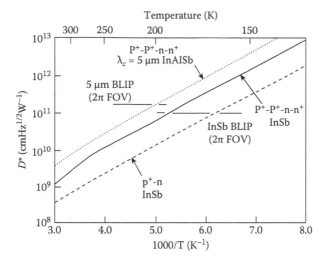

Figure 13.35 Calculated detectivity versus temperature for InSb photodiodes comparing bulk InSb (dashed), epitaxial InSb (solid), and epitaxial InAlSb with 5 μm cutoff wavelength (dotted). (From Elliott, C. T., "Advanced Heterostructures for In$_{1-x}$Al$_x$Sb and Hg$_{1-x}$Cd$_x$Te Detectors and Emiters," *Proceedings of SPIE* 2744, 452–62, 1996. With permission.)

detectors operated near ambient temperature are not well matched to the 3–5 μm atmospheric transmission window. One solution is to form the active region from In$_{1-x}$Al$_x$Sb with a composition such that the cutoff wavelength is decreased to the optimum value. To obtain a 5 μm cutoff would require $x \approx 0.023$ at 200 K and $x \approx 0.039$ at room temperature. The predicted detectivity for material with a constant 5 μm cutoff wavelength is plotted by the dotted line in Figure 13.35 [158]. The total increase in D^* compared with a conventional bulk device is more than a factor of 10. Background limited detectivity in 2π FOV can be achieved at 200 K, which is achievable with Peltier coolers leading to the possibility of high performance, compact, and comparatively inexpensive imaging systems.

Optimum material quality of devices with In$_{1-x}$Al$_x$Sb active regions is only obtained for lattice-matched growth. InGaSb lattice-matched substrates with the (111) orientation have been grown by a double crucible Czochralski technique [148]. This technique provides replenishment of the melt from the outer crucible to ensure that the melt composition in the inner crucible is maintained at the correct value to permit a uniform ingot composition. Photodiodes with Al composition up to 6.7% have been grown. As expected, at room temperature the R_oA product is raised by a factor of 10. However, at temperatures below 200 K this performance gain is not retained, probably due to the introduction of additional Shockley–Read traps through defects associated with imperfect lattice matching and the quality of the InGaSb substrates.

At present, the density of Shockley–Read traps in the InSb material is such that the benefits of nonequilibrium photodiodes is not realized, as the generation rate through the traps tends to rise under reverse bias. Growth on (001) substrates oriented 2° toward (111) has achieved better performance; for the first time the devices display negative differential resistance [148].

13.4 TERNARY AND QUATERNARY III-V DETECTORS

Ternary and quaternary III-V compound materials are suitable for fabricating optoelectronic devices in the near and mid-infrared wavelength range. The availability of binary substrates, such as InAs and GaSb, allows the growth of multilayer homo- and heterostructures, where lattice-matched ternary and quaternary layers could be tailored to detect wavelengths in the range of 0.8–4 μm. The bandgap of Ga$_x$In$_{1-x}$As$_y$Sb$_{1-y}$ can be continuously tuned from about 475 to 730 meV while remaining lattice matched to a GaSb substrate [13,14], as shown in Figure 13.36, and in contrast to leading ternary materials in this range such as InGaAs on InP. Both ternary (InGaSb and InAsSb) and quaternary (InGaAsSb and AlGaAsSb) indicated good performance for wavelength range ≥ 2 μm, but still they are on the research level not being commercially available. The availability of ternary InGaSb virtual substrates has a promising potential for developing high performance detectors, without the influence of the binary substrates usually used for processing the ternary materials [159].

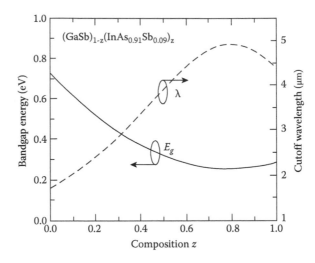

Figure 13.36   The bandgap of $Ga_xIn_{1-x}As_ySb_{1-y}$ with x and y concentrations chosen in the ratio $(GaSb)_{1-z}(InAs_{0.91}Sb_{0.09})_z$ can be tuned continuously from about 475 to 730 meV while remaining lattice matched to a GaSb substrate.

The InAsSb ternary alloy is more stable in comparison with HgCdTe and has a fairly weak dependence of the band edge on composition. The stability of this material is conditioned by the stronger chemical bonds available in the lower atomic number III-Vs family and the larger covalent bonding contribution compared to ionic bonding in HgCdTe. Some other physical properties of InAsSb material are better compared to those of HgCdTe. For instance, its dielectric constant is low (≈ 11.5) and the room temperature self-diffusion coefficient is low ($\approx 5.2 \times 10^{-16}$ cm^2/s).

The three semiconductors InAs, GaSb, and AlSb form an approximately lattice-matched set around 6.1 Å, with (room temperature) energy gaps ranging from 0.36 eV (InAs) to 1.61 eV (AlSb) [160]. Like other semiconductor alloys, they are of interest principally for their heterostructures, especially heterostructures combining InAs with the two antimonides and their alloys. This combination offers band lineups that are drastically different from those of the more widely studied AlGaAs system, and the lineups are one of the principal reasons for interest in the 6.1 Å family. The most exotic lineup is that of InAs/GaSb heterojunctions, for which it was found in 1977 by Sakaki et al. [161] and that they exhibit a broken gap lineup: at the interface the bottom of conduction band of InAs lines up below the top of the valence band of GaSb with a break in the gap of about 150 meV. In such a heterostructure, with partial overlapping of the InAs conduction band with the GaSb-rich solid solution valence band, electrons and holes are spatially separated and localized in self-consistent quantum wells formed on both sides of the heterointerface. This leads to unusual tunneling-assisted radiative recombination transitions and novel transport properties. For example, Figure 13.37 shows the simulation of approximate energy band diagrams for the four types of GaInAsSb/InAs heterojunction (N-n, N-p, P-p, and P-n) [162]. As one can see in the figure all of the rectifying heterostructures (N-n, N-p, and P-p) demonstrate a large space-charge region in the heterojunction. The significant overlap owing to a large bending of the conduction band in GaInAsSb and the valence band in InAs at the heterointerface leads to a strong confinement of the carriers in the self-consistent potential wells on both sides of the heteroboundary. When this overlap disappears, an unconfined movement of carriers on both sides of the junction resulted in the ohmic (metallic) behavior of the P-GaInAsSb/n-InAs structure. For the N-n heterojunction, the barrier height is close to the bandgap value of the GaInAsSb solid solution and for the case of the P-p structure to the InAs bandgap.

Many articles reported different device structures using materials such as InGaAsSb and AlGaAsSb [126–132] including APDs [163–165] and phototransistors [166]. Such devices usually involve complicated structures with difficult material processing. Aside from an APD, a phototransistor can achieve higher gain and better signal-to-noise ratio, without the excess noise effects that makes it attractive for 2 μm applications.

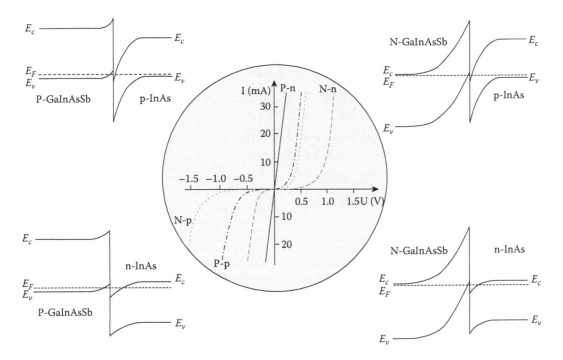

Figure 13.37 Energy band diagrams for four types of single broken-gap GaInAsSb/InAs heterojunctions and their I–V characteristics at 77 K. (From Mikhailova, M. P., Moiseev, K. D., and Yakovlev, Yu. P., *Semiconductor Science and Technology*, 19, R109–R128, 2004. With permission.)

13.4.1 InAsSb Detectors

13.4.1.1 InAsSb Photoconductors

Only a few papers are devoted to some aspects of theory and technology of $InAs_{1-x}Sb_x$ (InAsSb) detectors [167–174]. Bethea et al. [168,169] have fabricated and characterized photoconductive detectors obtained by MBE growth of InAsSb on semi-insulating GaAs substrate. In spite of large lattice mismatch (about 14%) between the epitaxial layer and substrate, good quality $InAs_{0.02}Sb_{0.98}$ photoconductors have been obtained [168]. Such detectors have achieved a detectivity value of 3×10^{10} cmHz$^{1/2}$W^{-1} at $\lambda = 5.4$ μm, which is within an order of magnitude of the best InSb detectors; a high internal quantum efficiency of 47% and a high-speed response of less than 10 ns were achieved.

The performance of photoconductors fabricated from alloys, which are closer to the minimum bandgap composition, are inferior. The voltage responsivity of this detector is 1.5 V/W at 8 μm for a bias voltage 9 V, which corresponds to a detectivity of $D^* = 10^8$ cmHz$^{1/2}$W^{-1} at 77 K. A photoconductive lifetime determined for this detector is low and equals 9 ns. This parameter and the low value of electron mobility indicate a poor material quality. Continued improvement of the material may lead to efficient photoconductive detectors operating in the 8–14 μm spectral region.

More recently, a research group at the Interuniversity Micro-Electronics Center (Leuven, Belgium) has demonstrated the first coplanar technology, which is ideally suited for the monolithic integration of InAsSb-based infrared detectors with Si charge coupled devices [170,171]. The fabrication and the performance of $InAs_{1-x}Sb_x$ photoconductive detectors (with x equal to 0.80 and 0.95) grown in recessed Si wells by MBE have been demonstrated. The InAsSb epilayer morphology was compared for different substrate conditions: Si-well, Si-mesa, and GaAs [171,172]. No degradation in morphology is observed for epilayers grown in the wells, as compared with large-area InAsSb films on GaAs-coated Si or GaAs. The threading defect density of the InAsSb layers is high (about 10^8 cm^{-2}) but decreases with thickness. The lattice mismatch is relaxed in the early stage of growth by a regular array of misfit dislocations.

The drawing in Figure 13.38a shows a schematic diagram of InAsSb detector structure. As substrates, the silicon 3-inch (001) wafers 4° off toward (011) were used. The wafers were

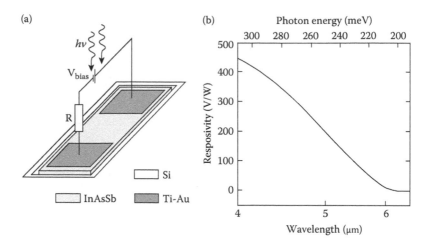

Figure 13.38 InAsSb photoconductive detectors: (a) schematic diagram of processed detectors, and (b) spectral response of InAs$_{0.05}$Sb$_{0.095}$ photoconductor at 77 K and 1.5 V bias. (From Dobbelaere, W., De Boeck, J., Van Hove, M., Deneffe, K., De Raedt, W., Mertens, R., and Borghs, G., *Electronics Letters*, 26, 259–61, 1990. With permission.)

patterned with an SiO$_2$ window mask using standard photolithography and etched using a 69% HNO$_3$:48% HF (19:1) mixture to form 80 × 400 μm rectangular wells about 6 μm deep, parallel to the (110) directions. Next the growth of GaAs on Si was initiated at low temperature (350°C) with a growth rate of 1 μm/h. The substrate temperature and growth rate were then increased to 580°C and 1.3 μm/h, respectively, and a 2 μm thick buffer layer was grown. Then the substrate was cooled to 380°C and a thin InAs$_{0.05}$Sb$_{0.95}$ layer grown. Finally a 3 μm thick InAs$_{0.05}$Sb$_{0.95}$ layer was grown at 410°C and 1 μm/h. The detector structure was then covered with photoresist and the polycrystalline InAsSb grown on top of the SiO$_2$ protective layer was etched away in a 25:1:50 (H$_2$SO$_4$:H$_2$O$_2$:H$_2$O) solution. This etchant stops at the SiO$_2$ layer so that in the case of real integration, the preprocessed electronic devices on Si are protected by a SiO$_2$ layer. Lastly, Ti-Au (40 nm–260 nm) contacts were evaporated on the embedded InAsSb layers to form photoconductive detectors. The gap between the detector and the surrounding substrate assures lateral isolation, but should be filled with a material with good reflow properties to avoid problems with metallization. The spectral response of such a photoconductor is shown in Figure 13.38b. This photoconductor exhibited a voltage responsivity of 420 V/W at 77 K and 4.2 μm wavelength with a load resistor of 100 Ω and a bias voltage of 1.5 V. It means that performance of photoconductors are not high, which can be caused by a very high doping level (of the order of the intrinsic carrier concentration at 300 K) and a large number of defects in the interface region between the InAsSb epilayer and the GaAs buffer.

Coplanar technology can be adapted for vertical, photovoltaic detectors that are expected to be superior to the present photoconductive detectors for two reasons: they allow operation at zero bias conditions and the active region can be placed at some distance from the interface so that bulk-like behavior with high resistivities is expected.

Podlecki et al. [173] have described the properties of InAs$_{0.91}$Sb$_{0.09}$ photoconductors deposited by MOCVD on GaAs substrates. The 2 μm thick detectors with an active area of 200 μm^2 and sputtered Au-Sn contacts have been fabricated. The excess noise led to a rather low 3.65 μm detectivity $D^* = 5.3 \times 10^{10}$ cmHz$^{1/2}$W^{-1} ($V_b = 3.75$ V, $f = 10^5$ Hz, $T = 80$ K). Also Kim et al. [174] have reported the room temperature operating photoconductors based on p-type InAs$_{0.23}$Sb$_{0.77}$ grown on GaAs substrates by MOCVD. The photoconductor structure composed of two epitaxial p-InAs$_{0.23}$Sb$_{0.77}$/ p-InSb layers. InSb layer was used as a buffer layer with 2% lattice mismatch to InAs$_{0.23}$Sb$_{0.77}$ and also as a confinement layer for the electron in the active layer. P-type doping in the InSb-like band structures ensures a better compromise between optical and thermal generation than n-type doping at near room temperatures [175]. A cutoff wavelength around 14 μm at room temperature was observed that indicates a smaller bandgap than the expected value. The decrease may have been caused by structural ordering. Figure 13.39 displays the voltage-dependent responsivity at 10.6 μm [174]. It increases with applied voltage and reaches saturation at around 3 V, which

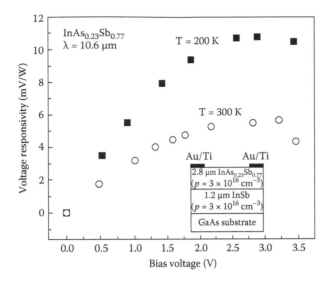

Figure 13.39 Voltage-dependent responsivity of the InAs$_{0.23}$Sb$_{0.77}$ photoconductor at 10.6 μm. (From Kim, J. D., Wu, D., Wojkowski, J., Piotrowski, J., Xu, J., and Razeghi, M., *Applied Physics Letters*, 68, 99–101, 1996. With permission.)

corresponds to the values of 5.8 mV/W at 300 K and 10.8 mV/W at 200 K, respectively. The corresponding Johnson-noise limited detectivity was estimated to be ≈ 3.27 × 10^7 cmHz$^{1/2}$W^{-1} at 300 K. This is below the theoretical limit (≈ 1.5 × 10^8 cmHz$^{1/2}$W^{-1}) set by Auger g–r process.

13.4.1.2 InAsSb Photodiodes

The main effort in InAsSb detector technology has been shifted to the development of photodiodes as useful devices for the second-generation thermal-imaging systems and the next generation very low loss fiber communication systems [17,37,45]. During the last three decades high-quality InAsSb photodiodes for the 3–5 μm spectral region have been developed [132,176–197]. The long wavelength limit of InAs$_{1-x}$Sb$_x$ detectors has been tuned compositionally from 3.1 μm ($x = 0.0$) to 7.0 μm ($x ≈ 0.6$). Potentially the material is capable of operating at the longest cutoff wavelength (≈ 9.0 μm at 77 K) of the entire III–V alloy family [17,37]. To realize IR detectors in all potentially capable operating regions, lattice-matched substrates are necessary. This problem seems to be resolved by using Ga$_{1-x}$In$_x$Sb substrates. In this case the lattice parameter can be tuned between 6.095 Å (GaSb) and 6.479 Å (InSb). Several research groups have succeeded in growing GaInSb single crystals [148,198–200]. One composition worth noting is Ga$_{0.38}$In$_{0.62}$Sb that is lattice matched to InAs$_{0.35}$Sb$_{0.65}$, which has a bandgap minimum, corresponding to ~12 μm at room temperature.

A variety of InAsSb photodiode configurations have been proposed including mesa and planar, n-p, n-p$^+$, p$^+$-n, and p-i-n structures. The techniques used to form p-n junctions have included diffusion of Zn, Be ion implantation, and the creation of p-type layers on n-type material by LPE, MBE, and MOCVD. The photodiode technology relies essentially on n-type material with concentrations generally about 10^{16} cm^{-3}. A summary of works for the fabrication of InAsSb photodiodes is given in Table 13.2.

At first, InAs$_{0.85}$Sb$_{0.15}$ photodiodes grown by a step-graded LPE technique on InAs substrates were reported [176]. A series of InAsSb compositionally step-graded buffer layers were introduced between active layers and the InAs substrate to relieve strain caused by lattice mismatch. The device was operated in the BSI mode. In this case, photons enter through the InAs substrate and through sufficiently thick buffer layers and on a filter layer to absorb most incoming photons with energy greater than the energy bandgap of the filter layer. The cutoff wavelength is mainly determined by the energy bandgap of the active layer and spectral characteristics can be controlled by the Sb composition of the filter layer and active layer. The carrier concentration across the junction is ≈ 10^{15} cm^{-3} for n-type and ≈ 10^{16} cm^{-3} for p-type. Mesa photodiodes fabricated in this way exhibit excellent characteristics as narrow-band BSI infrared detectors. Half-widths of spectral response

Table 13.2: Performance of InAs$_{1-x}$Sb$_x$ Photodiodes

Material	Fabrication	T (K)	R_oA (Ωcm^2)	λ_p (μm)	η_p (%)	D^* (cmHz$^{1/2}$W^{-1})	Comments	Refs.
n-p$^+$ InAs$_{0.85}$Sb$_{0.15}$	mesa, step-graded LPE (n $\approx$ 10^{15} cm^{-3})	77	2 × 10^7	4.2	60		InAs substrate, back side illuminated	176
n-p InAs$_{0.86}$Sb$_{0.14}$	mesa, LPE, (n $\approx$ p $\approx$ 10^{16} cm^{-3})	77	10^9	4.2	65		GaSb substrate, back side illuminated	178
n-p$^+$ InAs$_{0.85}$Sb$_{0.15}$	planar, LPE, (n $\approx$ 10^{16} cm^{-3}), Be implantation	77	10^7		80	4 × 10^{11}	GaSb substrate, CVD SiO$_2$ as a implantation mask, back side illuminated, 10^{15} ph/cm^{2}s	177
n-p InAs$_{0.85}$Sb$_{0.15}$	mesa, LPE, (n $\approx$ p $\approx$ 10^{17} cm^{-3})	77 / 200		3.5 / 3.0	40 / 40	1.5 × 10^{11} / 2.0 × 10^{10}	InAs substrate / 2π FOV	180
n-i-p InAs$_{0.85}$Sb$_{0.15}$	mesa, MBE n $\approx$ p $\approx$ (2–20) × 10^{16}cm^{-3}, i 3 × 10^{15} cm^{-3}	77 / 77 / 77	49 / 25 / 1.9	3.5 / 3.5 / 3.5		1.5 × 10^{11} / 1.6 × 10^{11} / 4.5 × 10^{10}	InAs substrate / GaAs substrate / Si substrate	182
n-p InAs$_{0.88}$Sb$_{0.12}$	mesa, LPE	77 / 200	2	3.8 / 4.3		3.0 × 10^{11} / 5 × 10^9	InAs substrate, front side illumination	184
P-GaSb/i-InAs$_{0.91}$Sb$_{0.09}$/N-GaSb	mesa, MBE i(1 μm) $\approx$ 10^{16} cm^{-3}	250		3.39		2.5 × 10^{10}	N-GaSb substrate, double heterojunction	195
n-InAs$_{0.91}$Sb$_{0.09}$/N-GaSb	mesa, MOCVD n(2.65 μm) $\approx$ 1.06 × 10^{16} cm^{-3}, N $\approx$ 1.1 × 10^{18} cm^{-3}	300 / 180	2–3 / 180	$\approx$ 4		4.9 × 10^9 / 1.3 × 10^{10}	N-GaSb substrate, front side illumination, izotype heterojunction, detectivity measured at reverse bias, bias tuning two-color detection	197
p$^+$-InAs$_{0.91}$Sb$_{0.09}$/P$^+$-InAlAsSb/ n-InAs$_{0.91}$Sb$_{0.09}$/p$^+$-InAs$_{0.91}$Sb$_{0.09}$	mesa, MBE n(2 μm) $\approx$ 2 × 10^{16} cm^{-3}	300 / 230	0.19 / 10.9	4 / 3.7		2.6 × 10^9 / 4.2 × 10^{10}	P-GaSb substrate, front side illumination, 2 μm thick In$_{0.88}$Al$_{0.12}$As$_{0.80}$Sb$_{0.20}$ barrier nearly matches with active layer leads to efficient transport of photogenerated holes	193
N$^+$-InAs/n-InAsSb/P-InAsSbP	mesa, LPE N$^+$ $\approx$ 10^{18} cm^{-3} n(3–8 μm) undoped, $\approx$ 10^{16} cm^{-3}	300	2 × 10^{-2}	4.2		2 × 10^{10}	N$^+$-InAs substrate, back side illumination, optical immersion (Si lens)	132

Structure / growth / layers	T (K)			D^*	Comments	Ref.
N^+-InAs/N^+-InAs$_{0.89}$Sb$_{0.11}$/P-InAs$_{0.55}$Sb$_{0.15}$P$_{0.30}$/ n-InAs$_{0.89}$Sb$_{0.11}$/P-InAs$_{0.55}$Sb$_{0.15}$P$_{0.30}$ mesa, LPE n(5 μm) undoped active region sandwiched between two (3 μm thick) InAsSbP cladding layers	300		4.5	1.26×10^9	N^+-InAs substrate, front side illumination	185
P^+-AlGaAsSb/AlInAsSb/n-InAs$_{0.91}$Sb$_{0.09}$/N^+-GaSb mesa, MBE n(1.6 μm) $\approx 3 \times 10^{16}$ cm^{-3} undoped active region, alkali sulfur passivation	300	3	4.3	1×10^{10}	N-GaSb substrate, AlGaAsSb used as a transparent window, AlInAsSb reduces the hole confinement at the InAsSb/ AlGaAsSb heterointerface, front side illumination	191
p^+-InSb/π-InAs$_{0.15}$Sb$_{0.85}$/n^+-InSb mesa, MOCVD n^+(2 μm) $\approx 3 \times 10^{18}$ cm^{-3} π(3 μm) $\approx 3.6 \times 10^{16}$ cm^{-3} p^+(0.5 μm)) $\approx 3 \times 10^{18}$ cm^{-3}	300	2×10^{-4}	≈ 8	1.5×10^8	GaAs substrate, back side illuminated $\lambda_c \approx 13$ μm	183
p^+-InSb/π-InAs$_{0.15}$Sb$_{0.85}$/n^+-AlInSb mesa, MBE n^+(2 μm) $\approx 3 \times 10^{18}$ cm^{-3} π(3 μm) $\approx 3.6 \times 10^{16}$ cm^{-3} p^+(0.5 μm) $\approx 3 \times 10^{18}$ cm^{-3}	300	0.11	6	$\approx 3 \times 10^8$	GaAs substrate, back side illuminated, incorporating of AlInSb buffer layer blocks carriers from the highly dislocation interface	20

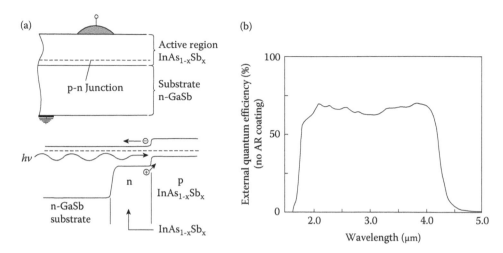

Figure 13.40 Back side illuminated $InAs_{0.86}Sb_{0.14}$/GaSb photodiode: (a) device structure and energy band diagram of the structure, and (b) spectral response at 77 K. (From Bubulac, L. O., Andrews, A. M., Gertner, E. R., and Cheung, D. T., *Applied Physics Letters*, 36, 734–36, 1980. With permission.)

as narrow as 176 nm at 77 K with peak internal quantum efficiency of 70% has been achieved. The zero-bias resistance area R_oA products are in the 10^5 Ωcm^2 range, with the best ones achieving 2×10^7 Ωcm^2.

The best performances of InAsSb photodiodes have been obtained when lattice-matched $InAs_{1-x}Sb_x$/GaSb ($0.09 \leq x \leq 0.15$) device structures were used [178]. Lattice mismatch up to 0.25% for the $InAs_{0.86}Sb_{0.14}$ epitaxial layer can be accommodated in terms of low etch-pit density ($\approx 10^4$ cm^{-2}). The structure of a BSI $InAs_{1-x}Sb_x$/GaSb photodiode is shown in Figure 13.40a [178]. The photons enter through the GaSb transparent substrate and reach the $InAs_{1-x}Sb_x$ active layer where they are absorbed. The GaSb substrate determines the short-wavelength cut-on value, which is 1.7 μm at 77 K; instead the active region establishes the long-wavelength cutoff value (see Figure 13.40b). The p-n junctions were obtained as homojunctions using the LPE technique. The carrier concentrations, both in the undoped n-type layer and in the Zn doped p-type layer, were approximately 10^{16} cm^{-3}. The high quality of $InAs_{0.86}Sb_{0.14}$ photodiodes was demonstrated by a high R_oA product in excess of 10^9 Ωcm^2 at 77 K.

High performance $InAs_{0.89}Sb_{0.11}$ photodiodes have also been obtained by Be ion implantation [177]. The as-grown LPE layers on (100) GaSb substrates were n-type with a typical carrier concentration of 10^{16} cm^{-3}. The implantation mask was formed by 100 nm of CVD SiO_2 deposited at 200°C and next covered with about 5 μm of photoresist or about 700 nm of aluminum. The Be ion implantation was performed using a 100 keV beam and a total dose of 5×10^{15} cm^{-2}. Following the implantation, annealing was carried out at 550°C for about 1 hr. The EBIC analysis of the $InAs_{0.89}Sb_{0.11}$ planar junction and C-V data confirmed the junction formation by the thermodiffusion mechanism.

Attempts to MBE grow $InAs_{0.85}Sb_{0.15}$ p-i-n junctions on lattice mismatched substrates—InAs (lattice mismatch 1%), GaAs (8.4%), and Si (12.8%)—have not given good results [182]. The performance of these photodiodes was inferior in comparison with the ones fabricated using LPE. Their R_oA product was almost three orders lower than photodiodes obtained by LPE [176–178]—below 50 Ωcm^2 at 77 K for diodes on InAs. The diodes exhibited significantly larger reverse leakage currents. The presence of defects reduces the carrier lifetime so that the g–r currents become increasingly important. To decrease the influence of misfit dislocations, different procedures were followed in deposition of interface regions, which are described exactly by Dobbelaere and colleagues [201,202].

Rogalski has performed an analysis of a resistance-area product (R_oA) of n-p+ abrupt $InAs_{0.85}Sb_{0.15}$ junctions at 77 K [45]. The dependence of the ultimate values of the R_oA product on the concentrations of dopants for abrupt n-p+ $InAs_{0.85}Sb_{0.15}$ photodiodes at 77 K is shown in Figure 13.41a. We can see that the R_oA product is determined by the g–r current of the junction depletion layer. The characteristic lifetime in the depletion region τ_o determined from

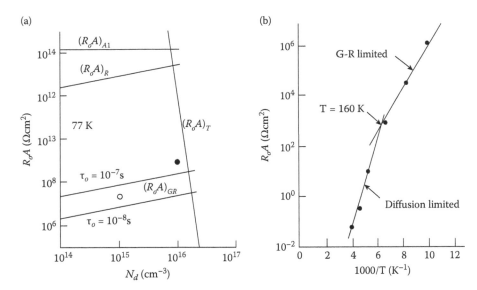

(a)

(b)

Figure 13.41 R_oA product of InAs$_{1-x}$Sb$_x$ photodiodes: (a) dependence on the doping concentration for n-p$^+$ abrupt InAs$_{0.85}$Sb$_{0.15}$ junctions at 77 K, the experimental values are taken from Cheung [176] (o) and Bubulac [178] (•). (From Rogalski, A., *Progress in Quantum Electronics*, 13, 191–231, 1989.) (b) Dependence on the temperature for a typical planar Be implanted InAs$_{0.89}$Sb$_{0.11}$ photodiode. (From Bubulac, L. O., Barrowcliff, E. E., Tennant, W. E., Pasko, J. P., Williams, G., Andrews, A. M., Cheung, D. T., and Gertner, E. R., *Institute of Physics Conference Series No. 45*, 519–29, 1979. With permission.)

the theoretical fit to the g–r model is found to be 0.03–0.5 μs. In the best photodiodes it was determined to be 0.55 μs [178]. The theoretical estimates yield for the radiative $(R_oA)_R$ and Auger recombination $(R_oA)_{A1}$ values of the R_oA product of several orders of magnitude larger. Tunneling current produces an abrupt lowering of R_oA at a concentration a little above 10^{16} cm^{-3}. To obtain a possibly high value of the zero-bias resistance of the junction, the technological process of photodiode preparation should be conducted so that the concentration of dopants is slightly below 10^{16} cm^{-3}.

Each of the current components of the p-n junction has its own individual relationship to voltage and temperature and can be associated with either the bulk or the surface. In Figure 13.41b the R_oA product against $1/T$ for Be ion implantation InAs$_{0.89}$Sb$_{0.11}$ junction is plotted [177]. In a semilog scale representation R_oA varies linearly for the g–r and diffusion model, as $1/n_i$ and $1/n_i^2$, respectively. In the temperature range above 160 K, the R_oA product follows the diffusion model, whereas in the temperature range $80 \leq T \leq 160$ K, R_oA fits a g–r model. At temperatures below 80 K the R_oA product is limited by surface effects. Therefore the operation of planar Be-implanted photodiodes is bulk-limited above 80 K.

Attempts to fabricate p-n junction formation in the miscibility range of InAs$_{1-x}$Sb$_x$ ternary alloy over the composition range $0.4 < x < 0.7$ using MOCVD have not given positive results [203]. The p$^+$-n junctions were formed by Zn diffusion into the undoped n-type epitaxial layer with carrier concentration in the range of 10^{16} cm^{-3}. The forward and the reverse characteristics were affected by the g–r current of the depletion region and by surface leakage current. It is believed that recombination centers in the depletion layer were caused by diffusion-induced damages and by lattice mismatch dislocations between the InAs$_{0.60}$Sb$_{0.40}$ epilayer and the InSb substrate.

Generally, the low-temperature homojunction devices suffer from Shockley–Read generation current and the activation energy decreases to about half the bandgap energy [45,188]. At high temperatures, where the diffusion current mechanism is dominant, these homojunction devices exhibit a large dark current and small R_oA product, below 10^{-2} Ωcm^2, which leads to relatively low detectivity.

In order to improve device performance (lower dark current and higher detectivity) several groups have developed P-i-N heterostructure devices of an unintentionally doped InAsSb active layer sandwiched between P and N layers of larger bandgap materials. As is shown in Section 3.2,

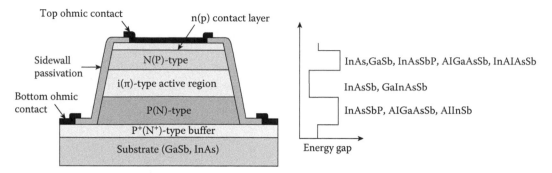

Figure 13.42 Schematic band diagram of the N-i-P double heterostructure antimonide based III-V photodiodes. Different combinations of active and cladding layers are also shown.

the lower minority carrier concentration in the high bandgap layers resulted in a lower diffusion dark and higher R_oA product and detectivity. Figure 13.42 shows a schematic band diagram of the N-i-P double heterostructure antimonide based III-V photodiodes together with the different combinations of active and cladding layers in the device structure. In dependence on contact configurations and transparency of substrates, both back side and front side illumination can be used. Usually p-type GaSb and n-type InAs are used. Despite the relatively low absorption coefficients, substrates required thinning to small thicknesses, even less than 25 μm. InAs is fragile and many fabrication processes are not possible. This obstacle can be overcame by using heavily doped n$^+$-InAs substrates where strong degeneracy of the electrons in the conduction band occurs at relatively low electron concentration (> 10^{17} cm^{-3}). For example, the Burstein-Moss shift in heavily doped n$^+$-InAs ($n = 6 \times 10^{18}$ cm^{-3}) makes the corresponding substrates transparent to the 3.3 μm [132].

Table 13.2 has short information about InAsSb heterostructure photodiode fabrications and their performance. Considerable progress in the development of LPE antimonide based heterostructure photodiodes has been achieved at the Ioffe Physical-Technical Institute in St. Petersburg, Russia. The knowledge about a thermodynamic data of coexisting phases in LPE process has been achieved by the development of the original method based on excess thermodynamic functions and linear combinations of chemical potentials [204]. In this way the phase diagrams for the Ga-In-As-Sb, In-As-Sb-P, Ga-In-As-Pb, and Ga-Al-As-Sb systems have been calculated. An idea was introduced using Pb as a neutral solvent during LPE growth of both GaSb and InAs layers and caused considerable decrease of the structural defect concentration in GaSb solid solutions from 2.8×10^{17} up to 2×10^{15} cm^{-3} [205]. Moreover, the use of Pb introduces undoped GaInAsSb solid solution with low concentration of defects and impurities and with high carrier mobility [206].

The LPE InAsSb heterostructure photodiodes were grown on heavily doped (111) n$^+$-InAs(Sn) substrates with electron concentration of 10^{18} cm^{-3} and the density of each pits of 10^4–10^5 cm^{-2}. They consisted of 5 μm thick Zn doped p-InAsSbP [$E_g = 375$ meV, $p = (2–5) \times 10^{17}$ cm^{-3}] deposited on 3–8 μm thick undoped InAsSb active layers. Figure 13.43a shows the near room temperature detectivity variation of InAsSb photodiodes with immersion Si lens (see Figure 13.32a) [132]. The short wavelength sharp fall of spectral response is related to InAs substrate transparency.

Also InAsSbP immersion photodiodes, which spectral detectivities are shown in Figure 13.43b, have been fabricated by LPE. However in this case 25–60 μm thick InAsSbP graded bandgap epilayers with low dislocation density (10^4 cm^{-2}) have been grown at 650°C –680°C onto 350μm thick (111) n-InAs substrates with carrier density of 10^{16} cm^{-3}. The wide band part of the InAsSbP at the interface was a transparent for the photons whose energy is close to the narrow bandgap value of the InAsSb(P) [207]. The high energy sensitivity decline in the graded bandgap photodiodes and is related to a diffusion mechanism. The carriers optically created near a broad band InAsSbP surface diffuse into the narrow band part of the diode (estimated diffusion length is 11–15 μm and is shorter than the thickness of the graded bandgap InAsSbP layer).

The first InAsSb-based long wavelength (8–14 μm) photodiode operating at room temperature has been described by Kim et al. [183]. The structure, grown by low-pressure MOCVD, is designed for the operation of back side (GaAs substrate side) illumination. Figure 13.44 shows the voltage responsivity at various temperatures and inside—schematic of device structure [186].

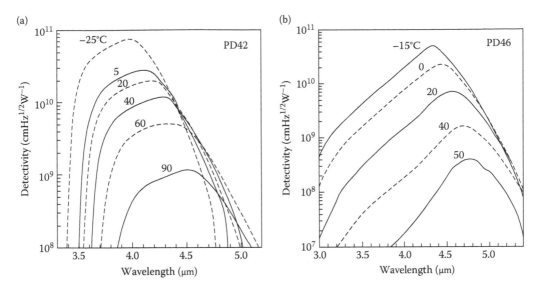

Figure 13.43 Detectivity spectral at (a) near-room temperature InAsSb, and (b) InAsSbP heterostructure photodiodes with optical immersion. (From Remennyy, M. A., Matveev, B. A., Zotova, N. V., Karandashev, S. A., Stus, N. M., and Ilinskaya, N. D., "InAs and InAs(Sb)(P) (3–5 μm) Immersion Lens Photodiodes for Potable Optic Sensors," *Proceedings of SPIE* 6585, 658504, 2007. With permission.)

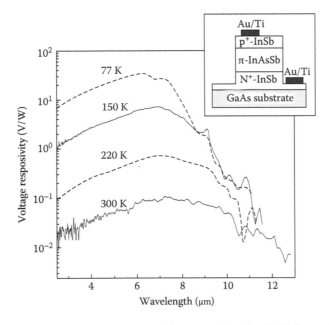

Figure 13.44 Spectral response of the p^+-InSb/π-nAs$_{0.15}$Sb$_{0.85}$/n^+-InSb heterojunction device at various temperatures. Inside a schematic device structure is shown. (From Kim, J. D., and Razeghi, M., *Opto-Electronics Review,* 6, 217–30, 1998. With permission.)

Photoresponse up to 13 μm has been obtained at 300 K in an p^+-InSb/π-nAs$_{0.15}$Sb$_{0.85}$/n^+-InSb heterojunction device. The peak voltage responsivity is 9.13×10^{-2} V/W at 300 K. At 77 K, it is only 2.85×10^1 V/W, which is much lower than the expected value. Possible reasons are the poor inter-face properties due to the lattice mismatch between the absorber and contact layers and high dark current due to the high doping level in the active layer. Introducing the AlInSb buffer layer as the

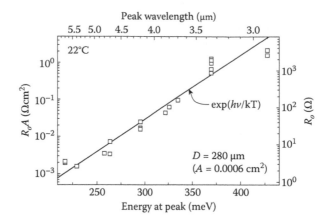

Figure 13.45 R_oA product in series of developed photodiode at room temperature. (From Remennyy, M. A., Matveev, B. A., Zotova, N. V., Karandashev, S. A., Stus, N. M., and Ilinskaya, N. D., "InAs and InAs(Sb)(P) (3–5 µm) Immersion Lens Photodiodes for Potable Optic Sensors," *Proceedings of SPIE* 6585, 658504, 2007. With permission.)

bottom contact layer blocks carriers from the highly dislocated interface resulting in increase of the R_oA product and detectivity [20].

Similar to N-GaInAsSb/n-InAs izotype heterojunction shown in Figure 13.37, also N-GaSb/n-InAsSb rectifying heterostructure has the unique type II broken gap interface [208,209]. A large barrier for electrons is formed in the GaSb side of the interface. Due to differences in electron affinity between two materials, electrons are transferred from the GaSb side to the InAsSb side across the interface. The resultant band banding leads to the formation of a barrier for electrons in the GaSb side and a two-dimensional electron gas in the InAsSb side. The barrier of the N-n interface is comparable to the energy gap of the wider band-gap material (GaSb). Sharabani et al. have shown that N-GaSb/n-In$_{0.91}$As$_{0.09}$Sb heterostructure is a promising material for high operating temperature MWIR detectors [197]. The BLIP temperature was found to be 180 K, and R_oA product of 2.5 and 180 Ωcm² were measured at 300 and 180 K, respectively.

13.4.2 Photodiodes Based on GaSb-Related Ternary and Quaternary Alloys

GaSb-related ternary and quaternary alloys are well established as materials for developing MWIR photodiodes for near-room temperature operation [127–132,159,165,207,210–212]. At present the research efforts are concentrated mainly on double heterostructure devices; their schematic structure is shown in Figure 13.42. This figure also shows different material systems used in the active and cladding layers.

A recently published paper by Remennyy et al. gives a short overview of GaSb-based photodiode performance [132]. Figure 13.45 summarizes experimental data of zero bias resistivity and R_oA product versus photon energy. An exponential dependence of R_oA product, approximated by $\exp(E_g/kT)$, indicates that the diffusion current determines the transport properties of the heterojunctions at room temperature.

Figure 13.46 presents current responsivities and detectivities in dependence on photon energy for the BSI and the coated photodiodes (with immersion lens, IL) [132]. It is shown that the photodiodes developed at the Ioffe Physical-Technical Institute are superior to the others published in literature. The detectivities shown in Figure 13.46b are higher in comparison with that gathered in Table 13.2 for photodiodes operated in similar conditions. It is expected that higher device performance reflects improvements in device design and fabrication: broad mirror contact and radiation collection by inclined mesa walls.

Narrow-gap III-V semiconductors and their alloys are also promising materials for developing high-speed, low-noise APDs. They have found a number of applications in the 2–5 µm spectral range including: laser-diode spectroscopy, mid-IR fiber optics, laser range-finding, free-space optical links for high frequency communications, and so on.

Many articles have discussed the properties of mid-IR APDs and many of the investigations were made at the Ioffe Institute. Recently Mikhailova and Andreev have published a comprehensive review paper devoted to 2–5-µm APDs [165].

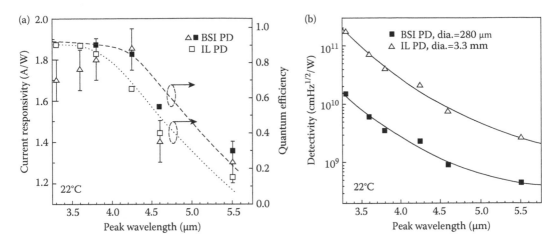

Figure 13.46 (a) Current responsivity and (b) peak detectivity of photodiodes without (back side illuminated, BSI) and with Si lenses (immersion illuminated, IL). (From Remennyy, M. A., Matveev, B. A., Zotova, N. V., Karandashev, S. A., Stus, N. M., and Ilinskaya, N. D., "InAs and InAs(Sb)(P) (3–5 μm) Immersion Lens Photodiodes for Portable Optic Sensors" *Proceedings of SPIE* 6585, 658504, 2007. With permission.)

It is a well-known fact that the excess avalanche noise factor and thus the signal-to-noise ratio of an APD depends on the ratio of electron and hole impact ionization coefficients (α_e and α_h, respectively). To achieve a low noise factor, not only must α_e and α_h be as different as possible, but also the avalanche process must be initiated by the carriers with the higher ionization coefficient. Unlike silicon APDs, it was found that holes dominate the impact ionization process. According to McIntyre's rule, the noise performance of APD can be improved by more than a factor of 10 when the α_h/α_e ionization ratio is increased to 5. For InAs- and GaSb-based alloys, a resonant enhancement of the hole ionization coefficient has been found [165,210,213,214]. This effect is attributed to impact ionization initiated by holes from a split-off valence band: if the spin orbit splitting Δ is equal to the bandgap energy E_g, the threshold energy for hole initiated impact ionization reaches the smallest possible value and the ionization process occurs with zero momentum. This leads to a strong increase of α_h at $\Delta/E_g = 1$.

Figure 13.47 illustrates the electric field dependence of α_e and α_h in the GaInAsSb/GaAlAsSb heterostructure at 230 K [215]. The heterostructure was grown by LPE on < 111 > n-GaSb Te doped substrate with carrier concentration $(5–7) \times 10^{17}$ cm^{-3}. The active region consisted of an unintentionally doped n-type and 2.3 μm thick Ga$_{0.80}$In$_{0.20}$As$_{0.17}$Sb$_{0.83}$ layer ($E_g = 0.54$ eV at 300 K) with carrier density of 2×10^{16} cm^{-3}. The wide-gap "window" layer was 2 μm thick p$^+$-Ga$_{0.66}$Al$_{0.34}$As$_{0.025}$Sb$_{0.975}$ ($E_g = 1.20$ eV at 300 K) and doped with Ge up to $(1–2) \times 10^{18}$ cm^{-3}. The mesa photodiodes were illuminated through the wide-gap GaAlAsSb layer. The relation between the impact ionization coefficients and the excess noise factor was also discussed. It is clear from Figure 13.47 that the hole ionization coefficient was greater than the electron ionization coefficient and their ratio $\alpha_h/\alpha_e \sim 4–5$. The ionization by holes from the spin-orbit splitting valence band predominated in the range of electric fields $E = (1.5–2.3) \times 10^5$ V/cm.

An example of device structure of InGaAsSb APD with separate absorption and multiplication region, (SAM) APD, is shown in Figure 13.48 [215]. This device is sequentially composed of a 2.2 μm thick Te compensated Ga$_{0.78}$In$_{0.22}$As$_{0.18}$Sb$_{0.82}$ layer with electron concentration $(5–7) \times 10^{15}$ cm^{-3}; a 0.3 μm thick n-Ga$_{0.96}$Al$_{0.04}$Sb "resonant" composition layer with electron concentration of 8×10^{16} cm^{-3}; and a 1.5 μm thick Al$_{0.34}$Ga$_{0.66}$As$_{0.014}$Sb$_{0.986}$ window layer with hole concentration of 5×10^{18} cm^{-3}. The location of the p-n junction coincides with the heterointerface between two wide gap materials. The space charge region lies in the n-Ga$_{0.96}$Al$_{0.04}$Sb/p-Al$_{0.34}$Ga$_{0.66}$As$_{0.014}$Sb$_{0.986}$ heterointerface and results in predominant multiplication of holes in the n-Ga$_{0.96}$Al$_{0.04}$Sb multiplication region. The maximum values of the multiplication factor were measured to be $M = 30–40$ at room temperature. The breakdown voltage determined by wide-gap material was about 10–12 V. As a band resonance condition takes place in Ga$_{0.96}$Al$_{0.04}$Sb at 0.76 eV, very high values of α_h/α_e ratio up to 60 are achieved. Thus, an essentially unipolar multiplication by holes is provided that reduces the excess noise problem in these APD.

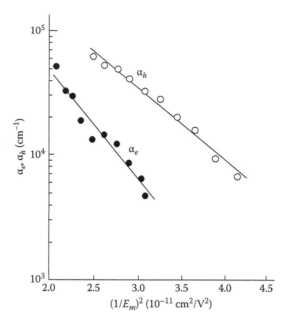

Figure 13.47 Dependence of the hole and electron ionization coefficients on the square of the reciprocal maximum electric field in the $Ga_{0.80}In_{0.20}As_{0.17}Sb_{0.83}$ solid solution at 230 K. (From Andreev, I. A., Mikhailava, M. P., Mel'nikov, S. V., Smorchkova, Yu. P., and Yakovlev, Yu. P., *Soviet Physics-Semiconductor*, 25, 861–65, 1991.)

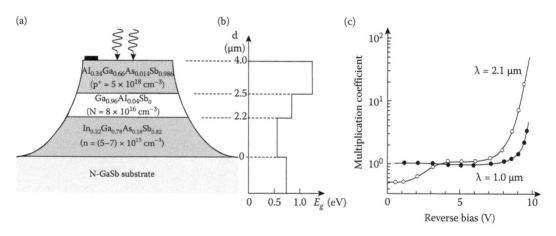

Figure 13.48 SAM APD $Ga_{0.80}In_{0.20}As_{0.17}Sb_{0.83}$ /$Ga_{0.96}Al_{0.04}Sb$ with "resonant" composition in the avalanche region: (a) schematic device structure, (b) band gap structure, and (c) multiplication coefficient versus reverse bias. (From Andreev, I. A., Afrailov, M. A., Baranov, A. N., Marinskaya, N. N., Mirsagatov, M. A., Mikhailova, M. P., Yakovlev, Yu. P., *Soviet Technical Physics Letters*, 15, 692–96, 1989.)

13.5 NOVEL Sв-BASED III-V NARROW GAP PHOTODETECTORS
13.5.1 InTlSb and InTlP

Since an $InAs_{0.35}Sb_{0.65}$ based detector is not sufficient for efficient IR detection operated at lower temperatures in the 8–12 μm range, $In_{1-x}Tl_xSb$ (InTlSb) was proposed as a potential IR material in the LWIR region [217,218]. The TlSb is predicted as a semimetal. By alloying TlSb with InSb, the bandgap of InTlSb could be varied from –1.5 eV to 0.26 eV. Assuming a linear dependence of

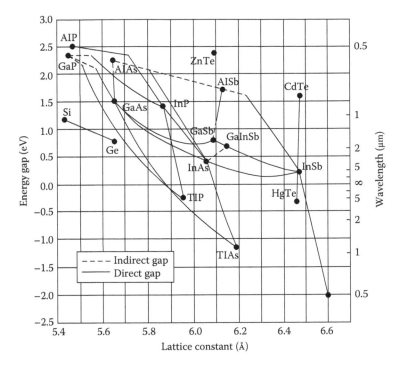

Figure 13.49 Composition and wavelength diagrams of a number of semiconductors with diamond and zinc-blende structure versus their lattice constants. Tl-based III-Vs are also included.

the bandgap on alloy composition, $In_{1-x}Tl_xSb$ can then be expected to reach a bandgap of 0.1 eV at $x = 0.08$, while exhibiting a similar lattice constant as InSb since the radius of Tl atom is very similar to In. At this gap, InTlSb and HgCdTe have very similar band structure. This implies that InTlSb has comparable optical and electrical properties to HgCdTe. In the structural aspect, InTlSb is expected to be more robust due to stronger bonding. The estimated miscibility limit of Tl in zinc-blende InTlSb was estimated to be approximately 15%, which is sufficient to obtain an energy gap down to 0.1 eV. Figure 13.49 shows the expected relationships between bandgap energy and lattice constant for Tl-based III-V zinc-blende alloys [34]. Room temperature operation of InTlSb photodetectors has been demonstrated with a cutoff wavelength of approximately 11 μm [219].

Van Schilfgaarde et al. showed that another ternary alloy, $In_{1-x}Tl_xP$ (InTlP), is a promising material for IR detectors [220]. It was shown that this material can cover the bandgap from 1.42 eV (InP at 0 K) to 0 eV using a small lattice mismatch with InP. Optical measurements verified the reduction of the bandgap by the addition of Tl into the InP [221].

Tl-based III-V alloys will be more widely studied and applied to devices if the difficulties in the crystal growth are overcome.

13.5.2 InSbBi

As another alternative to the HgCdTe material system, $InSb_{1-x}Bi_x$ (InSbBi) has been considered because the incorporation of Bi into InSb produces a rapid reduction in the bandgap of 36 meV/%Bi. Thus, only a few percentage of Bi is required to reduce the bandgap energy.

The growth of an InSbBi epitaxial layer is difficult due to the large solid-phase miscibility gap between InSb and InBi. The successful growth of an InSbBi epitaxial layer on InSb and GaAs (100) substrates with a substantial amount of Bi (~5%) was demonstrated using low-pressure MOCVD [222,223]. The responsivity of the $InSb_{0.95}Bi_{0.05}$ photoconductor at 10.6 μm was 1.9×10^{-3} V/W at room temperature, and the corresponding Johnson noise limited detectivity was 1.2×10^6 cmHz$^{1/2}$/W. The effective carrier lifetime estimated from bias voltage-dependent responsivity was approximately 0.7 ns at 300 K.

13.5.3 InSbN

Dilute nitride alloys of III-V semiconductors have progressed rapidly in recent years following the discovery of strong negative bandgap bowing effects [224,225]. Most papers concentrated on the alloys GaAsN and GaInAsN owing to their technological importance for fiber communications at wavelengths of 1.3 and 1.55 μm.

Initial estimates indicated that the addition of N to InSb would lead to a reduction of its band-gap at a rate similar to that of the wider-gap III-V materials of approximately 100 meV per per-centage of N. Therefore, for some applications the InSbN might provide an alternative that would overcome some of the limitations of the more established materials such as HgCdTe and type II SLs of InAs/GaInSb.

Preliminary estimates of the band structure of $InSb_{1-x}N_x$ were made using a semiempirical **kp** model. The theoretically predicted variation in bandgap structure indicates a decrease in bandgap of 110 meV (fractional change of 63%) at 1% N, which clearly offers potential for long-wavelength applications [226]. These theoretical predictions were experimentally confirmed with measure-ments of response wavelengths of light-emitting diodes. The $InSb_{1-x}N_x$ samples with up to 10% of the N were grown by combining MBE and the N plasma source.

The bandgap reduction has been accompanied by an enhancement in the Auger recombination lifetime of a factor of approximately three in comparison with an equivalent HgCdTe bandgap, due to the higher electron mass and conduction band nonparabolicity [227].

REFERENCES

1. C. Hilsum and A. C. Rose-Innes, *Semiconducting III-V Compounds*, Pergamon Press, Oxford, 1961.

2. O. Madelung, *Physics of III-V Compounds*, Wiley, New York, 1964.

3. T. S. Moss, G. J. Burrel, and B. Ellis, *Semiconductor Optoelectronics*, Butterworths, London, 1973.

4. F. D. Morten and R. E. King, "Photoconductive Indium Antimonide Detectors," *Applied Optics* 4, 659–63, 1965.

5. P. W. Kruse, "Indium Antimonide Photoconductive and Photoelectromagnetic Detectors," in *Semiconductors and Semimetals*, Vol. 5, eds. R. K. Willardson and A. C. Beer, 15–83, Academic Press, New York, 1970.

6. C. T. Elliott and N. T. Gordon, "Infrared Detectors," in *Handbook of Semiconductors*, Vol. 4, ed. C. Hilsum, 841–936, Elsevier, Amsterdam, 1993.

7. K. F. Hulme and J. B. Mullin, "Indium Antimonide: A Review of Its Preparation, Properties and Device Applications," *Solid-State Electronics* 5, 211–47, 1962.

8. K. F. Hulme, "Indium Antimonide," in *Materials Used in Semiconductor Devices*, ed. C. A. Hogarth, 115–62, Wiley-Interscience, New York, 1965.

9. S. Liang, "Preparation of Indium Antimonide," in *Compound Semiconductors*, eds. R. K. Willardson and H. L. Georing, 227–37, Reinhold, New York, 1966.

10. J. B. Mullin, "Melt-Growth of III-V Compounds by the Liquid Encapsulation and Horizontal Growth Techniques," in *III-V Semiconductor Materials and Devices*, ed. R. J. Malik, 1–72, North Holland, Amsterdam, 1989.

11. W. F. M. Micklethwaite and A. J. Johnson, "InSb: Materials and Devices," in *Infrared Detectors and Emitters: Materials and Devices*, eds. P. Capper and C. T. Elliott, 178–204, Kluwer Academic Publishers, Boston, MA, 2001.

12. S. Kasap and P. Capper, eds., *Springer Handbook of Electronic and Photonic Materials*, Springer, Heidelberg, 2006.

13. S. Adachi, *Physical Properties of III-V Semiconducting Compounds: InP, InAs, GaAs, GaP, InGaAs, and InGaAsP*, Wiley-Interscience, New York, 1992; *Properties of Group-IV, III-V and II-VI Semiconductors*, John Wiley & Sons, Ltd., Chichester, 2005.

14. I. Vurgaftman, J. R. Meyer, and L. R. Ram-Mohan, "Band Parameters for III–V Compound Semiconductors and Their Alloys, *Journal of Applied Physics* 89, 5815–75, 2001.

15. M. Ilegems, "In p-based Lattice-matched Heterostructures," in *Properties of Lattice-Matched and Strained Indium Gallium Arsenide*, ed. P. Bhattacharya, 16–25, IEE, London, 1993.

16. G. H. Olsen and V. S. Ban, "InGaAsP: The Next Generation in Photonics Materials," *Solid State Technology*, 99–105, February 1987.

17. A. Rogalski, *New Ternary Alloy Systems for Infrared Detectors*, SPIE Optical Engineering Press, Bellingham, WA, 1994.

18. T. Ashley, A. B. Dean, C. T. Elliott, M. R. Houlton, C. F. McConville, H. A. Tarry, and C. R. Whitehouse, "Multilayer InSb Diodes Grown by Molecular Beam Epitaxy for Near Ambient Temperature Operation," *Proceedings of SPIE* 1361, 238–44, 1990.

19. T. Ashley, A. B. Dean, C. T. Elliott, A. D. Johnson, G. J. Pryce, A. M. White, and C. R. Whitehouse, "A Heterojunction Minority Carrier Barrier for InSb Devices," *Semiconductor Science and Technology* 8, S386–S389, 1993.

20. M. Razeghi, "Overview of Antimonide Based III-V Semiconductor Epitaxial Layers and Their Applications at the Center for Quantum Devices," *European Physical Journal Applied Physics* 23, 149–205, 2003.

21. G. B. Stringfellow and P. R. Greene, "Liquid Phase Epitaxial Growth of $InAs_{1-x}Sb_x$," *Journal of the Electrochemical Society* 118, 805–10, 1971.

22. A. Rogalski, "Infrared Detectors: Status and Trends," *Progress in Quantum Electronics* 27, 59–210, 2003.

23. A. Rogalski, "HgCdTe Infrared Detector Material: History, Status, and Outlook," *Reports on Progress in Physics* 68, 2267–336, 2005.

24. A. Rogalski, "New Trends in Semiconductor Infrared Detectors," *Optical Engineering* 33, 1395–1412, 1994.

25. B. F. Levine, "Quantum Well Infrared Photodetectors," *Journal of Applied Physics* 74, R1–R81, 1993.

26. F. F. Sizov and A. Rogalski, "Semiconductor Superlattices and Quantum Wells for Infrared Optoelectronics," *Progress in Quantum Electronics* 17, 93–164, 1993.

27. S. D. Gunapala and K. M. S. V. Bandara, "Recent Developments in Quantum-Well Infrared Photodetectors," in *Physics of Thin Films*, Vol. 21, eds. M. H. Francombe and J. L. Vossen, 113–237, Academic Press, New York, 1995.

28. S. D. Gunapala and S. V. Bandara, "Quantum Well Infrared Photodetector (QWIP)," in *Handbook of Thin Film Devices*, Vol. 2, edited by M. H. Francombe, 63–99, Academic Press, San Diego, CA, 2000.

29. H. C. Liu, "An Introduction to the Physics of Quantum Well Infrared Photodetectors and Other Related New Devices," in *Handbook of Thin Film Devices*, ed. M. H. Francombe, Vol. 2, pp. 101–34, Academic Press, San Diego, CA, 2000.

30. H. Schneider and H. C. Liu, *Quantum Well Infrared Photodetectors. Physics and Applications*, Springer, Berlin, 2007.

31. S. R. Kurtz, L. R. Dawson, T. E. Zipperian, and R. D. Whaley, "High-detectivity ($> 1 \times 10^{10}$ cmHz$^{1/2}$W^{-1}), InAsSb Strained-Layer Superlattice, Photovoltaic Infrared Detector," *IEEE Electron Device Letters* 11, 54–56, 1990.

32. D. L. Smith and C. Mailhiot, "Proposal for Strained Type II Superlattice Infrared Detectors," *Journal of Applied Physics* 62, 2545–48, 1987.

33. L. Bürkle and F. Fuchs, "InAs/(GaIn)Sb Superlattices: A Promising Material System for Infrared Detection," in *Handbook of Infrared Detection and Technologies*, eds. M. Henini and M. Razeghi, 159–89, Elsevier, Oxford, 2002.

34. H. Asahi, "Tl-Based III-V Alloy Semiconductors," in *Infrared Detectors and Emitters: Materials and Devices*, eds. P. Capper and C. T. Elliott, 233–49, Kluwer Academic Publishers, Boston, MA, 2001.

35. Y. P. Varshni, "Temperature Dependence of the Energy Gap in Semiconductors," *Physica* 34, 149, 1967.

36. H. H. Wieder and A. R. Clawson, "Photo-Electronic Properties of InAs$_{0.07}$Sb$_{0.93}$ Films," *Thin Solid Films* 15, 217–21, 1973.

37. A. Rogalski, K. Adamiec, and J. Rutkowski, *Narrow-Gap Semiconductor Photodiodes*, SPIE Press, Bellingham, WA, 2000.

38. W. Zawadzki, "Electron Transport Phenomena in Small-Gap Semiconductors," *Advances in Physics* 23, 435–522, 1974.

39. F. P. Kesamanli, Yu. V. Malcev, D. N. Nasledov, Yu. I. Uhanov, and A. S. Filipczenko, "Magnetooptical Investigations of InSb Conduction Band," *Fizika Tverdogo Tela* 8, 1176–81, 1966.

40. J. C. Woolley and B. A. Smith, "Solid Solution in III-V Compounds," *Proceedings of the Physical Society* 72, 214–23, 1958.

41. J. C. Woolley and J. Warner, "Preparation of InAs-InSb Alloys," *Journal of the Electrochemical Society* 111, 1142–45, 1964.

42. M. J. Aubin and J. C. Woolley, "Electron Scattering in InAsSb Alloys," *Canadian Journal of Physics* 46, 1191–98, 1968.

43. J. C. Woolley and J. Warner, "Optical Energy-Gap Variation in InAs-InSb Alloys," *Canadian Journal of Physics* 42, 1879–85, 1964.

44. E. H. Van Tongerloo and J. C. Woolley, "Free-Carrier Faraday Rotation in InAs$_{1-x}$Sb$_x$ Alloys," *Canadian Journal of Physics* 46, 1199–1206, 1968.

45. A. Rogalski, "InAsSb Infrared Detectors," *Progress in Quantum Electronics* 13, 191–231, 1989.

46. G. C. Osbourn, "InAsSb Strained-Layer Superlattices for Long Wavelength Detector," *Journal of Vacuum Science and Technology* B2, 176–78, 1984.

47. A. Joullie, F. Jia Hua, F. Karouta, H. Mani, and C. Alibert, "III-V Alloys Based on GaSb for Optical Communications at 2.0–4.5 μm," *Proceedings of SPIE* 587, 46–57, 1985.

48. T. P. Pearsall and M. Papuchon, "The Ga$_{0.47}$In$_{0.53}$As Homojunction Photodiode: A New Avalanche Photodetector in the Near Infrared Between 1.0 and 1.6 μm," *Applied Physics Letters* 33, 640–42, 1978.

49. S. R. Forrest, R. F. Leheny, R. E. Nahory, and M. A. Pollack, "In$_{0.53}$Ga$_{0.47}$As Photodiodes with Dark Current Limited by Generation-Recombination and Tunneling," *Applied Physics Letters* 37, 322–25, 1980.

50. N. Susa, H. Nakagome, O. Mikami, H. Ando, and H. Kanbe, "New InGaAs/InP Avalanche Photodiode Structure for the 1–1.6 μm Wavelength Region," *IEEE Journal of Quantum Electronics* QE-16, 864–70, 1980.

51. S. R. Forrest, R. G. Smith, and O. K. Kim, "Performance of In$_{0.53}$Ga$_{0.47}$As/InP Avalanche Photodiodes," *IEEE Journal of Quantum Electronics* QE-18, 2040–48, 1980.

52. G. E. Stillman, L. W. Cook, G. E. Bulman, N. Tabatabaie, R. Chin, and P. D. Dapkus, "Long-Wavelength (1.3- to 1.6-μm) Detectors for Fiber-Optical Communications," *IEEE Transactions on Electron Devices* ED-29, 1365–71, 1982.

53. G. E. Stillman, L. W. Cook, N. Tabatabaie, G. E. Bulman, and V. M. Robbins, "InGaAsP Photodiodes," *IEEE Transactions Electron Devices* ED-30, 364–81, 1983.

54. T. P. Pearsall and M. A. Pollack, "Compound Semiconductor Photodiodes," in *Semiconductors and Semimetals*, Vol. 22D, ed. W. T. Tang, 173–245, Academic Press, Orlando, FL, 1985.

55. D. E. Ackley, J. Hladky, M. J. Lange, S. Mason, S. R. Forrest, and C. Staller, "Linear Arrays of InGaAs/InP Avalanche Photodiodes for 1.0–1.7 μm," *Proceedings of SPIE* 1308, 261–72, 1990.

56. J. C. Campbell, S. Demiguel, F. Ma, A. Beck, X. Guo, S. Wang, X. Zheng, et al., "Recent Advances in Avalanche Photodiodes," *IEEE Journal of Selected Topics in Quantum Electronics* 10, 777–87, 2004.

57. G. H. Olsen, A. M. Joshi, S. M. Mason, K. M. Woodruff, E. Mykietyn, V. S. Ban, M. J. Lange, J. Hladky, G. C. Erickson, and G. A. Gasparian, "Room-Temperature InGaAs Detector Arrays for 2.5 μm," *Proceedings of SPIE* 1157, 276–82, 1989.

58. G. Olsen, A. Joshi, M. Lange, K. Woodruff, E. Mykietyn, D. Gay, G. Erickson, D. Ackley, V. Ban, and C. Staller, "A 128 × 128 InGaAs Detector Array for 1.0–1.7 Microns," *Proceedings of SPIE* 1341, 432–37, 1990.

59. G. H. Olsen, "InGaAs Fills the Near-IR Detector-Array Vacuum," *Laser Focus World*, A21–A30, March 1991.

60. G. H. Olsen, A. M. Joshi, and V. S. Ban, "Current Status of InGaAs Detector Arrays for 1–3 μm," *Proceedings of SPIE* 1540, 596–605, 1991.

61. A. M. Joshi, V. S. Ban, S. Mason, M. J. Lange, and W. F. Kosonocky, "512 and 1024 Element Linear InGaAs Detector Arrays for Near-Infrared (1–3 μm) Environmental Sensing," *Proceedings of SPIE* 1735, 287–95, 1992.

62. M. J. Cohen and G. H. Osen, "Near-IR Imaging Cameras Operate at Room Temperature," *Laser Focus World*, 109–13, 1993.

63. M. J. Cohen and G. H. Olsen, "Room Temperature InGaAs Camera for NIR Imaging," *Proceedings of SPIE* 1946, 436–43, 1993.

64. L. J. Kozlowski, W. E. Tennant, M. Zandian, J. M. Arias, and J. G. Pasko, "SWIR Staring FPA Performance at Room Temperature," *Proceedings of SPIE* 2746, 93–100, 1996.

65. L. J. Kozlowski, K. Vural, J. M. Arias, W. E. Tennant, and R. E. DeWames, "Performance of HgCdTe, InGaAs and Quantum Well GaAs/AlGaAs Staring Infrared Focal Plane Arrays," *Proceedings of SPIE* 3182, 2–13, 1997.

66. T. Martin, R. Brubaker, P. Dixon, M.-A. Gagliardi, and T. Sudol, "640 × 512 Focal Plane Array Camera for Visible and SWIR Imaging," *Proceedings of SPIE* 5783, 12–20, 2005.

67. A. Hoffman, T. Sessler, J. Rosbeck, D. Acton, and M. Ettenberg, "Megapixel InGaAs for Low Background Applications," *Proceedings of SPIE* 5783, 32–38, 2005.

68. H. Yuan, G. Apgar, J. Kim, J. Laquindanum, V. Nalavade, P. Beer, J. Kimchi, and T. Wong, "FPA Development: From InGaAs, InSb, to HgCdTe," *Proceedings of SPIE* 6940, 69403C, 2008.

69. D. Acton, M. Jack, and T. Sessler, "Large Format Short Wave Infrared (SWIR) Focal Plane Array (FPA) with Extremely Low Noise and High Dynamic Range," *Proceedings of SPIE* 7298, 72983E, 2009.

70. M. J. Cohen and G. H. Olsen, "Near-Infrared Camera Inspects with Clarity," *Laser Focus World*, 269–70, June 1996.

71. I. Gyuro, "MOVPE for InP-based Optoelectronic Device Application," in *Compound Semiconductor Industry Directory*, 58–68, Elsevier Science Ltd., 1996.

72. A. M. Joshi, G. H. Olsen, S. Mason, M. J. Lange, and V. S. Ban, "Near-Infrared (1–3 μm) InGaAs Detectors and Arrays: Crystal Growth, Leakage Current and Reliability," *Proceedings of SPIE* 1715, 585–93, 1992.

73. R. U. Martinelli, T. J. Zamerowski, and P. A. Longeway, "2.6 μm InGaAs Photodiodes," *Applied Physics Letters* 53, 989–91, 1988.

74. G. H. Olsen, M. J. Lange, M. J. Cohen, D. S. Kim, and S. R. Forrest, "Three-Band 1.0–2.5 μm Near-Infrared InGaAs Detector Array," *Proceedings of SPIE* 2235, 151–59, 1994.

75. S. D. Personick, "Receiver Design for Digital Fiber-Optic Communication Systems, Parts I and II," *Bell Systems Technical Journal* 52, 843–86, 1973.

76. R. G. Smith and S. D. Personick, "Receiver Design for Optical Fiber Communications Systems," in *Semiconductor Devices for Optical Communication*, Chapter 4, Springer-Verlag, New York, 1980.

77. S. R. Forrest, "Sensitivity of Avalanche Photodetector Receivers for Highbit-Rate Long-Wavelength Optical Communication Systems," in *Semiconductors and Semimetals*, Vol. 22, Chapter 4, ed. W. T. Tang, Academic Press, Orlando, 1985.

78. B. L. Kasper and J. C. Campbell, "Multigigabit-Per-Second Avalanche Photodiode Lightwave Receivers," *Journal of Lightwave Technology* LT-5, 1351–64, 1987.

79. V. S. Ban, G. H. Olsen, and A. M. Joshi, "High Performance InGaAs Detectors and Arrays for Near-Infrared Spectroscopy," *Spectroscopy* 6(3), 49–52, 1991.

80. M. Gallant, N. Puetz, A. Zemel, and F. R. Shepherd, "Metalorganic Chemical Vapor Deposition InGaAs p-i-n Photodiodes with Extremely Low Dark Current," *Applied Physics Letters* 52, 733–35, 1988.

81. A. Zemel and M. Gallant, "Current-Voltage Characteristics of Metalorganic Chemical Vapor Deposition InP/InGaAs p-i-n Photodiodes: The Influence of Finite Dimensions and Heterointerfaces," *Journal of Applied Physics* 64, 6552–61, 1988.

82. A. Rogalski and R. Ciupa, "Performance Limitation of Short Wavelength Infrared InGaAs and HgCdTe Photodiodes," *Journal of Electronic Materials* 28, 630–36, 1999.

83. H. W. Ruegg, "An Optimized Avalanche Photodiode," *IEEE Transactions on Electron Devices* ED-14, 239–51, 1967.

84. H. Melchior and W. T. Lynch, "Signal and Noise Response of High Speed Germanium Avalanche Photodiodes," *IEEE Transactions on Electron Devices* ED-13, 820–38, 1966.

85. H. Ando, H. Kanbe, T. Kimura, T. Yamaoka, and T. Kaneda, "Characteristics of Germanium Avalanche Photodiodes in the Wavelength Region of 1–1.6 μm," *IEEE Journal of Quantum Electronics* QE-14, 804–9, 1978.

86. T. Mikawa, S. Kagawa, T. Kaneda, Y. Toyama, and O. Mikami, "Crystal Orientation Dependence of Ionization Rates in Germanium," *Applied Physics Letters* 37, 387–89, 1980.

87. J. S. Ng, J. P. R. David, G. J. Rees, and J. Allam, "Avalanche Breakdown Voltage of In$_{0.53}$Ga$_{0.47}$As," *Journal of Applied Physics* 91, 5200–5202, 2002.

88. L. W. Cook, G. E. Bulman, and G. E. Stillman, "Electron and Hole Ionization Coefficients in InP Determined by Photomultiplication Measurements," *Applied Physics Letters* 40, 589–91, 1982.

89. J. C. Campbell, "Recent Advances in Telecommunications Avalanche Photodiodes," *Journal of Lightwave Technology* 25, 109–21, 2007.

90. J. P. R. David and C. H. Tan, "Material Considerations for Avalanche Photodiodes," *IEEE Journal of Selected Topics in Quantum Electronics* 14, 998–1009, 2008.

91. R. W. Zitter, A. J. Strauss, and A. E. Attard, "Recombination Processes in p-Type Indium Antimonide," *Physical Review* 115, 266–73, 1969.

92. M. Y. Pines and O. M. Stafsudd, "Surface Effect in InSb Photoconductors," *Infrared Physics* 19, 559–61, 1979.

93. M. Y. Pines and O. M. Stafsudd, "Characteristics of n-Type InSb," *Infrared Physics* 19, 563–69, 1979.

94. E. Sunde, "Impact of Surface Treatment on the Performance of Cooled Photoconductive Indium Antimonide Detectors," *Physica Scripta* 25, 768–71, 1982.

95. K. M. Van Vliet, "Noise Limitations in Solid State Photodetectors," *Applied Optics* 6, 1145–69, 1967.

96. P. W. Kruse, "Indium Antimonide Photoelectromagnetic Infrared Detector," *Journal of Applied Physics* 30, 770–78, 1959.

97. P. W. Kruse, L. D. McGlauchlin, and R. B. McQuistan, *Elements of Infrared Technology: Generation, Transmission, and Detection,* Wiley, New York, 1962.

98. K. Jóźwikowski, Z. Orman, and A. Rogalski, "On the Performance of Non-Cooled (In,As)Sb Photoelectromagnetic Detectors for 10.6 μm Radiation," *Physica Status Solidi (a)* 91, 745–51, 1985.

99. D. A. Scribner, M. R. Kruer, and J. M. Killiany, "Infrared Focal Plane Array Technology," *Proceedings of IEEE* 79, 66–85, 1991.

100. M. Talley and D. P. Enright, "Photovoltaic Effect in InAs," *Physical Review* 95, 1092–94, 1954.

101. L. L. Chang, "Junction Delineation by Anodic Oxidation in InSb(As,P)," *Solid Sate Electronics* 10, 539–44, 1967.

102. R. L. Mozzi and J. M. Lavine, "Zn-Diffusion Damage in InSb Diodes," *Journal of Applied Physics* 41, 280–85, 1970.

103. C. W. Kim and W. E. Davern, "InAs Charge-Storage Photodiode Infrared Vidicon Targets," *IEEE Transactions on Electron Devices* ED-18, 1062–69, 1971.

104. P. C. Catagnus, C. Polansky, and J. P. Spratt, "Diffusion of Cadmium into InSb," *Solid State Electronics* 16, 633–35, 1973.

105. K. Nishitani, K. Nagahama, and T. Murotani, "Extremely Reproducible Zinc Diffusion into InSb and Its Application to Infrared Detector Array," *Journal of Electronic Materials* 12, 125–41, 1983.

106. J. T. Wimmers and D. S. Smith, "Characteristics of InSb Photovoltaic Detectors at 77 K and Below," *Proceedings of SPIE* 364, 123–31, 1983.

107. R. Adar, Y. Nemirovsky, and I. Kidron, "Bulk Tunneling Contribution to the Reverse Breakdown Characteristics of InSb Gate Controlled Diodes," *Solid-State Electronics* 30, 1289–93, 1987.

108. J. T. Wimmers, R. M. Davis, C. A. Niblack, and D. S. Smith, "Indium Antimonide Detector Technology at Cincinnati Electronics Corporation," *Proceedings of SPIE* 930, 125–38, 1988.

109. T. P. Sun, S. C. Lee, and S. J. Yang, "The Current Leakage Mechanism in InSb p^+-n Diodes," *Journal of Applied Physics* 67, 7092–97, 1990.

110. W. H. Lan, S. L. Tu. Y. T. Cherng, Y. M. Pang, S. J. Yang, and K. F. Huang, "Field-Induced Junction in InSb Gate-Controlled Diodes," *Journal of Applied Physics* 30, L1–L3, 1991.

111. A. G. Foyt, W. T. Lindley, and J. P. Donnelly, "N-p Junction Photodetectors in InSb Fabricated by Proton Bombardment," *Journal of Applied Physics* 16, 335–37, 1970.

112. P. J. McNally, "Ion Implantation in InAs and InSb," *Radiation Effects and Defects in Solids* 6, 149–53, 1970.

113. C. E. Hurwitz and J. P. Donnelly, "Planar InSb Photodiodes Fabricated by Be and Mg Ion Implantation," *Solid State Electronics* 18, 753–56, 1975.

114. R. D. Thom, T. L. Koch, J. D. Langan, and W. L. Parrish, "A Fully Monolithic InSb Infrared CCD Array," *IEEE Transactions on Electron Devices* ED–27, 160–70, 1980.

115. C. Y. Wei, K. L. Wang, E. A. Taft, J. M. Swab, M. D. Gibbons, W. E. Davern, and D. M. Brown, "Technology Developments for InSb Infrared Images," *IEEE Transactions on Electron Devices* ED–27, 170–75, 1980.

116. J. P. Rosbeck, I. Kassi, R. M. Hoendervoog, and T. Lanir, "High Performance Be Implanted InSb Photodiodes," *IEEE IEDM* 81, 161–64, 1981.

117. J. P. Donnelly, "The Electrical Characteristics of Ion Implanted Semiconductors," *Nuclear Instruments & Methods in Physics Research* 182/183, 553–71, 1981.

118. M. Fujisada and T. Sasase, "Effects of Insulated Gate on Ion Implanted InSb p^+-n Junctions," *Journal of Applied Physics* 23, L162–L164, 1984.

119. H. Fujisada and M. Kawada, "Temperature Dependence of Reverse Current in Be Ion Implanted InSb p^+-n Junctions," *Journal of Applied Physics* 24, L76–L78, 1985.

120. S. Shirouzu, T. Tsuji, N. Harada, T. Sado, S. Aihara, R. Tsunoda, and T. Kanno, "64 × 64 InSb Focal Plane Array with Improved Two Layer Structure," *Proceedings of SPIE* 661, 419–25, 1986.

121. J. P. Donnelly, "Ion Implantation in III-V Semiconductors," in *III-V Semiconductor Materials and Devices,* ed. R. J. Malik, 331–428, North-Holland, Amsterdam, 1989.

122. I. Bloom and Y. Nemirovsky, "Quantum Efficiency and Crosstalk of an Improved Backside-Illuminated Indium Antimonide Focal Plane Array," *IEEE Transactions on Electron Devices* 38, 1792–96, 1991.

123. V. P. Astachov, Yu. A. Danilov, V. F. Dutkin, V. P. Lesnikov, G. Yu. Sidorova, L. A. Suslov, I. I. Taukin, and Yu. M. Eskin, "Planar InAs Photodiodes," *Pisma v Zhurnal Tekhnicheskoi Fiziki* 18(3), 1–5, 1992.

124. O. V. Kosogov and L. S. Perevyaskin, "Electrical Properties of Epitaxial p⁺-n Junctions in Indium Antimonide," *Fizika i Tekhnika Poluprovodnikov* 8, 1611–14, 1970.

125. K. Kazaki, A. Yahata, and W. Miyao, "Properties of InSb Photodiodes Fabricated by Liquid Phase Epitaxy," *Journal of Applied Physics* 15, 1329–34, 1976.

126. N. N. Smirnova, S. V. Svobodchikov, and G. N. Talalakin, "Reverse Characteristic and Breakdown of InAs Photodiodes," *Fizika i Tekhnika Poluprovodnikov* 16, 2116–20, 1982.

127. Z. Shellenbarger, M. Mauk, J. Cox, J. South, J. Lesko, P. Sims, M. Jhabvala, and M. K. Fortin, "Recent Progress in GaInAsSb and InAsSbP Photodetectors for Mid-Infrared Wavelengths," *Proceedings of SPIE* 3287, 138–45, 1998.

128. N. D. Stoyanov, M. P. Mikhailova, O. V. Andreichuk, K. D. Moiseev, I. A. Andreev, M. A. Afrailov, and Yu. P. Yakovlev, "Type II Heterojunction Photodiodes in a GaSb/InGaAsSb System for 1.5–4.8 µm Spectral Range," *Fizika i Tekhnika Poluprovodnikov* 35, 467–73, 2001.

129. T. N. Danilova, B. E. Zhurtanov, A. N. Imenkov, and Yu. P. Yuakovlev, "Light Emitting Diodes on GaSb Alloys for Mid-Infrared 1.4–4.4 µm Spectral Range, *Fizika i Tekhnika Poluprovodnikov* 39, 1281–1311, 2005.

130. A. Krier, X. L. Huang, and V. V. Sherstnev, "Mid-Infrared Electroluminescence in LEDs Based on InAs and Related Alloys," in *Mid-Infrared Semiconductor Optoelectronics,* ed. A. Krier, 359–94, Springer-Verlag, London, 2006.

131. B. A. Matveev, "LED-Photodiode Opto-Pairs," in *Mid-Infrared Semiconductor Optoelectronics,* ed. A. Krier, 395–428, Springer-Verlag, London, 2006.

132. M. A. Remennyy, B. A. Matveev, N. V. Zotova, S. A. Karandashev, N. M. Stus, and N. D. Ilinskaya, "InAs and InAs(Sb)(P) (3–5 µm) Immersion Lens Photodiodes for Potable Optic Sensors," *Proceedings of SPIE* 6585, 658504, 2007.

133. T. Ashley, A. B. Dean, C. T. Elliott, C. F. McConville, and C. R. Whitehouse, "Molecular-Beam Growth of Homoepitaxial InSb Photovoltaic Detectors," *Electronics Letters* 24, 1270–72, 1988.

134. G. S. Lee, P. E. Thompson, J. L. Davis, J. P. Omaggio, and W. A. Schmidt, "Characterization of Molecular Beam Epitaxially Grown InSb Layers and Diode Structures," *Solid-State Electronics* 36, 387–89, 1993.

135. C. H. Kuan, R. M. Lin, S. F. Tang, and T. P. Sun, "Analysis of the Dark Current in the Bulk of InAs Diode Detectors," *Journal of Applied Physics* 80, 5454–58, 1996.

136. W. Zhang and M. Razeghi, "Antimony-Based Materials for Electro-Optics," in *Semiconductor Nanostructures for Optoelectronic Applications,* ed. T. Steiner, 229–88, Artech House, Inc., Norwood, MA, 2004.

137. R. M. Biefeld and S. R. Kurtz, "Growth, Properties and Infrared Device Characteristics of Strained InAsSb-Based Materials," in *Infrared Detectors and Emitters: Materials and Devices*, eds. P. Capper and C. T. Elliott, 205–32, Kluwer Academic Publishers, Boston, MA, 2001.

138. W. S. Chan and J. T. Wan, "Auger Analysis of InSb IR Detector Arrays," *Journal of Vacuum Science Technology* 14, 718–22, 1977.

139. P. J. Love, K. J. Ando, R. E. Bornfreund, E. Corrales, R. E. Mills, J. R. Cripe, N. A. Lum, J. P. Rosbeck, and M. S. Smith, "Large-Format Infrared Arrays for Future Space and Ground-Based Astronomy Applications," *Proceedings of SPIE* 4486, 373–84, 2002.

140. T. Ashley, R. A. Ballingall, J. E. P. Beale, I. D. Blenkinsop, T. M. Burke, J. H. Firkins, D. J. Hall, et al., "Large Format MWIR Focal Plane Arrays," *Proceedings of SPIE* 4820, 400–405, 2003.

141. S. R. Jost, V. F. Meikleham, and T. H. Myers, "InSb: A Key Material for IR Detector Applications," *Materials Research Society Symposium Proceedings* 90, 429–35, 1987.

142. M. Zandian, J. D. Garnett, R. E. DeWames, M. Carmody, J. G. Pasko, M. Farris, C. A. Cabelli, et al., "Mid-Wavelength Infrared p-on-on $Hg_{1-x}Cd_xTe$ Heterostructure Detectors: 30–120 Kelvin State-of-the-Art Performance," *Journal of Electronic Materials* 32, 803–9, 2003.

143. I. Bloom and Y. Nemirovsky, "Bulk Lifetime Determination of Etch-Thinned InSb Wafers for Two-Dimensional Infrared Focal Plane Arrays," *IEEE Transactions on Electron Devices* 39, 809–12, 1992.

144. I. Bloom and Y. Nemirovsky, "Surface Passivation of Backside-Illuminated Indium Antimonide Focal Plane Array," *IEEE Transactions on Electron Devices* 40, 309–14, 1993.

145. D. N. B. Hall, R. S. Aikens, R. R. Joyse, and T. W. McCurnin, "Johnson-Noise Limited Operation of Photovoltaic InSb Detectors," *Applied Optics* 14, 450–53, 1975.

146. R. Schoolar and E. Tenescu, "Analysis of InSb Photodiode Low Temperature Characteristics," *Proceedings of SPIE* 686, 2–11, 1986.

147. J. T. Wimmers and D. S. Smith, "Optimization of InSb Detectors for Use at Liquid Helium Temperatures," *Proceedings of SPIE* 510, 21, 1984.

148. T. Ashley and N. T. Gordon, "Epitaxial Structures for Reduced Cooling of High Performance Infrared Detectors," *Proceedings of SPIE* 3287, 236–43, 1998.

149. G. S. Lee, P. E. Thompson, J. L. Davis, J. P. Omaggio, and W. A. Schmidt, "Characterization of Molecular Beam Epitaxially Grown InSb Layers and Diode Structures," *Solid-State Electronics* 36, 387–89, 1993.

150. E. Michel, J. Xu, J. D. Kim, I. Ferguson, and M. Razeghi, "InSb Infrared Photodetectors on Si Substrates Grown by Molecular Beam Epitaxy," *IEEE Photonics Technology Letters* 8, 673–75, 1996.

151. E. Michel and M. Razeghi, "Recent Advances in Sb-Based Materials for Uncooled Infrared Photodetectors," *Opto-Electronics Review* 6, 11–23, 1998.

152. I. Kimukin, N. Biyikli, T. Kartaloglu, O. Aytür, and E. Ozbay, "High-Speed InSb Photodetectors on GaAs for Mid-IR Applications," *IEEE Journal of Selected Topics in Quantum Electronics* 10, 766–70, 2004.

153. E. G. Camargo, K. Ueno, T. Morishita, M. Sato, H. Endo, M. Kurihara, K. Ishibashi, and M. Kuze, "High-Sensitivity Temperature Measurement with Miniaturized InSb Mid-IR Sensor," *IEEE Sensors Journal* 7, 1335–39, 2007.

154. M. Kuze, T. Morishita, E. G. Camargo, K. Ueno, A. Yokoyama, M. Sato, H. Endo, Y. Yanagita, S. Toktuo, and H. Goto, "Development of Uncooled Miniaturized InSb Photovoltaic Infrared Sensors for Temperature Measurements," *Journal of Crystal Growth* 311, 1889–92, 2009.

155. R. M. Lin, S. F. Tang, S. C. Lee, C. H. Kuan, G. S. Chen, T. P. Sun, and J. C. Wu, "Room Temperature Unpassivated InAs p-i-n Photodetectors Grown by Molecular Beam Epitaxy," *IEEE Transactions on Electron Devices* 44, 209–13, 1997.

156. W. Dobbelaere, J. De Boeck, P. Heremens, R. Mertens, and G. Borghs, "InAs p-n Diodes Grown on GaAs and GaAs-Coated Si by Molecular Beam Epitaxy," *Applied Physics Letters* 60, 868–70, 1992.

157. A. P. Davis and A. M. White, "Residual Noise in Auger Suppressed Photodiodes," *Infrared Physics* 31, 73–79, 1991.

158. C. T. Elliott, "Advanced Heterostructures for $In_{1-x}Al_xSb$ and $Hg_{1-x}Cd_xTe$ Detectors and Emiters," *Proceedings of SPIE* 2744, 452–62, 1996.

159. T. Refaat, N. Abedin, V. Bhagwat, I. Bhat, P. Dutta, and U. Singh, "InGaSb Photodetectors Using an InGaSb Substrate for 2-μm Applications," *Applied Physics Letters* 85, 1874–76, 2004.

160. H. Kroemer, "The 6.1 Å family (InAs, GaSb, AlSb) and Its Heterostructures: A Selective Review," *Physica E* 20, 196–203, 2004.

161. H. Sakaki, L. L. Chang, R. Ludeke, C. A. Chang, G. A. Sai-Halasz, and L. Esaki, $In_{1-x}Ga_xAs$-$GaSb_{1-y}As_y$ Heterojunctions by Molecular Beam Epitaxy," *Applied Physics Letters* 31, 211–13, 1977.

162. M. P. Mikhailova, K. D. Moiseev, and Yu. P. Yakovlev, "Interface-Induced Optical and Transport Phenomena in Type II Broken-Gap Single Heterojunctions," *Semiconductor Science and Technology* 19, R109–R128, 2004.

163. J. Benoit, M. Boulou, G. Soulage, A. Joullie, and H. Mani, "Performance Evaluation of GaAlAsSb/GaInAsSb SAM-APDs for High Bit Rate Transmission in the 2.5μm Wavelength Region," *Optics Communication* 9, 55–58, 1988.

164. M. Mikhailova, I. Andreev, A. Baranov, S. Melnikov, Y. Smortchkova, and Y. Yakovlev, "Low-Noise GaInAsSb/GaAlAsSb SAM Avalanche Photodiode in the 1.6–2.5 μm Spectral Range," *Proceedings of SPIE* 1580, 308–12, 1991.

165. M. P. Mikhailova and I. A. Andreev, "High-Speed Avalanche Photodiodes for the 2–5 μm Spectral Range," in *Mid-Infrared Semiconductor Optoelectronics*, ed. A. Krier, 547–92, Springer-Verlag, London, 2006.

166. N. Abedin, T. Refaat, O. Sulima, and U. Singh, "AlGaAsSb/InGaAsSb Heterojunction Phototransistor with High Optical Gain and Wide Dynamic Range," *IEEE Transactions on Electron Devices* 51, 2013–18, 2004.

167. M. R. Reddy, B. S. Naidu, and P. J. Reddy, "Photoresponsive Measurements on $InAs_{0.3}Sb_{0.7}$ Infrared Detector," *Bulletin of Material Science* 8, 373–77, 1986.

168. C. G. Bethea, M. Y. Yen, B. F. Levine, K. K. Choi, and A. Y. Cho, "Long Wavelength $InAs_{1-x}Sb_x$/GaAs Detectors Prepared by Molecular Beam Epitaxy," *Applied Physics Letters* 51, 1431–32, 1987.

169. C. G. Bethea, B. F. Levine, M. Y. Yen, and A. Y. Cho, "Photoconductance Measurements on $InAs_{0.22}Sb_{0.78}$/GaAs Grown Using Molecular Beam Epitaxy," *Applied Physics Letters* 53, 291–92, 1988.

170. W. Dobbelaere, J. De Boeck, M. Van Hove, K. Deneffe, W. De Raedt, R. Martens, and G. Borghs, "Long Wavelength InAs$_{0.2}$Sb$_{0.8}$ Detectors Grown on Patterned Si Substrates by Molecular Beam Epitaxy," *IEDM Technical Digest*, 717–20, 1989.

171. W. Dobbelaere, J. De Boeck, M. Van Hove, K. Deneffe, W. De Raedt, R. Mertens, and G. Borghs, "Long Wavelength Infrared Photoconductive InAsSb Detectors Grown in Si Wells by Molecular Beam Epitaxy," *Electronics Letters* 26, 259–61, 1990.

172. J. De Boeck, W. Dobbelaere, J. Vanhellemont, R. Mertens, and G. Borghs, "Growth and Structural Characterization of Embedded InAsSb on GaAs-Coated Patterned Silicon by Molecular Beam Epitaxy," *Applied Physics Letters* 58, 928–30, 1991.

173. J. Podlecki, L. Gouskov, F. Pascal, F. Pascal-Delannoy, and A. Giani, "Photodetection at 3.65 μm in the Atmospheric Window Using InAs$_{0.91}$Sb$_{0.09}$/GaAs Heteroepitaxy," *Semiconductor Science and Technology* 11, 1127–30, 1996.

174. J. D. Kim, D. Wu, J. Wojkowski, J. Piotrowski, J. Xu, and M. Razeghi, "Long-Wavelength InAsSb Photoconductors Operated at Near Room Temperatures (200–300 K)," *Applied Physics Letters* 68, 99–101, 1996.

175. J. Piotrowski and M. Razeghi, "Improved Performance of IR Photodetectors with 3D Gap Engineering," *Proceedings of SPIE* 2397, 180–192, 1995.

176. D. T. Cheung, A. M. Andrews, E. R. Gertner, G. M. Williams, J. E. Clarke, J. L. Pasko, and J. T. Longo, "Backside-Illuminated InAs$_{1-x}$Sb$_x$-InAs Narrow-Band Photodetectors," *Applied Physics Letters* 30, 587–98, 1977.

177. L. O. Bubulac, E. E. Barrowcliff, W. E. Tennant, J. P. Pasko, G. Williams, A. M. Andrews, D. T. Cheung, and E. R. Gertner, "Be Ion Implantation in InAsSb and GaInSb," *Institute of Physics Conference Series No. 45*, 519–29, 1979.

178. L. O. Bubulac, A. M. Andrews, E. R. Gertner, and D. T. Cheung, "Backside-Illuminated InAsSb/GaSb Broadband Detectors," *Applied Physics Letters* 36, 734–36, 1980.

179. K. Chow, J. P. Rode, D. H. Seib, and J. D. Blackwell, "Hybrid Infrared Focal-Plane Arrays," *IEEE Transactions on Electron Devices* 29, 3–13, 1982.

180. K. Mohammed, F. Capasso, R. A. Logan, J. P. van der Ziel, and A. L. Hutchinson, "High-Detectivity InAs$_{0.85}$Sb$_{0.15}$/InAs Infrared (1.8–4.8 μm) Detectors," *Electronics Letters* 22, 215–16, 1986.

181. J. L. Zyskind, A. K. Srivastava, J. C. De Winter, M. A. Pollack, and J. W. Sulhoff, "Liquid-Phase-Epitaxial InAs$_y$Sb$_{1-y}$ on GaSb Substrates Using GaInAsSb Buffer Layers: Growth, Characterization, and Application to Mid-IR Photodiodes," *Journal of Applied Physics* 61, 2898–903, 1987.

182. W. Dobbelaere, J. De Boeck, P. Heremans, R. Mertens, and G. Borghs, "InAs$_{0.85}$Sb$_{0.15}$ Infrared Photodiodes Grown on GaAs and GaAs-Coated Si by Molecular Beam Epitaxy," *Applied Physics Letters* 60, 3256–58, 1992.

183. J. D. Kim, S. Kim, D. Wu, J. Wojkowski, J. Xu, J. Piotrowski, E. Bigan, and M. Razeghi, "8–13 μm InAsSb Heterojunction Photodiode Operating at Near Room Temperature," *Applied Physics Letters* 67, 2645–47, 1995.

184. M. P. Mikhailova, N. M. Stus, S. V. Slobodchikov, N. V. Zotova, B. A. Matveev, and G. N. Talalakin, "InAs$_{1-x}$Sb Photodiodes for 3–5-μm Spectral Range," *Fizika i Tekhnika Poluprovodnikov* 30, 1613–19, 1996.

185. H. H. Gao, A. Krier, and V. V. Sherstnev, "A Room Temperature $InAs_{0.89}Sb_{0.11}$ Photodetectors for CO Detection at 4.6 µm," *Applied Physics Letters* 77, 872–74, 2000.

186. J. D. Kim and M. Razeghi, "Investigation of InAsSb Infrared Photodetectors for Near-Room Temperature Operation," *Opto-Electronics Review* 6, 217–30, 1998.

187. A. Rakovska, V. Berger, X. Marcadet, B. Vinter, G. Glastre, T. Oksenhendler, and D. Kaplan, "Room Temperature InAsSb Photovoltaic Midinfrared Detector," *Applied Physics Letters* 77, 397–99, 2000.

188. G. Marre, B. Vinter, and V. Berger, "Strategy for the Design of a Non-Cryogenic Quantum Infrared Detector," *Semiconductor Science and Technology* 18, 284–91, 2003.

189. G. Marre, M. Carras, B. Vinter, and V. Berger, "Optimization of a Non-Cryogenic Quantum Infrared Detector," *Physica E* 20, 515–18, 2004.

190. J. L. Reverchon, M. Carras, G. Marre, C. Renard, V. Berger, B. Vinter, and X. Marcadet, "Design and Fabrication of Infrared Detectors Based on Lattice-Matched $InAs_{0.91}Sb_{0.09}$ on GaSb," *Physica E*, 20, 519–22, 2004.

191. H. Ait-Kaci, J. Nieto, J. B. Rodriguez, P. Grech, F. Chevrier, A. Salesse, A. Joullie, and P. Christol, "Optimization of InAsSb Photodetector for Non-Cryogenic Operation in the Mid-Infrared Range," *Physica Status Solidi (a)* 202, 647–51, 2005.

192. M. Carras, J. L. Reverchon, G. Marre, C. Renard, B. Vinter, X. Marcadet, and V. Berger, "Interface Band Gap Engineering in InAsSb Photodiodes," *Applied Physics Letters* 87, 102–3, 2004.

193. H. Shao, W. Li, A. Torfi, D. Moscicka, and W. I. Wang, "Room-Temperature InAsSb Photovoltaic Detectors for Midinfrared Applications," *IEEE Photonics Technology Letters* 18, 1756–58, 2006.

194. A. Krier and W. Suleiman, "Uncooled Photodetectors for the 3–5 µm Spectral Range Based on III-V Heterojunctions," *Applied Physics Letters* 89, 083512, 2006.

195. M. Carras, C. Renard, X. Marcadet, J. L. Reverchon, B. Vinter, and V. Berger, "Generation-Recombination Reduction in InAsSb Photodiodes," *Semiconductor Science and Technology* 21, 1720–23, 2006.

196. A. I. Zakhgeim, N. V. Zotova, N. D. Il'inskaya, S. A. Karandashev, B. A. Matveev, M. A. Remennyi, N. M. Stus, and A. E. Chernyakov, "Room-Temperature Broadband InAsSb Flip-Chip Photodiodes with λ_{cutoff} = 4.5 µm," *Semiconductors* 43, 394–99, 2009.

197. Y. Sharabani, Y. Paltiel, A. Sher, A. Raizman, and A. Zussman, "InAsSb/GaSb Heterostructure Based Mid-Wavelength-Infrared Detector for High Temperature Operation," *Applied Physics Letters* 90, 232106, 2007.

198. W. F. Micklethwaite, R. G. Fines, and D. J. Freschi, "Advances in Infrared Antimonide Technology," *Proceedings of SPIE* 2554, 167–74, 1995.

199. A. Tanaka, J. Shintani, M. Kimura, and T. Sukegawa, "Multi-Step Pulling of GaInSb Bulk Crystal from Ternary Solution," *Journal of Crystal Growth* 209, 625–29, 2000.

200. P. S. Dutta, "III-V Ternary Bulk Substrate Growth Technology: A Review," *Journal of Crystal Growth* 275, 106–12, 2005.

201. W. Dobbelaere, J. De Boeck, W. De Raedt, J. Vanhellemont, G. Zou, M. Van Hove, B. Brijs, R. Martens, and G. Borghs, "InAsSb Photodiodes Grown on InAs, GaAs and Si Substrates by Molecular Beam Epitaxy," *Materials Research Society Proceedings* 216, 181–86, 1991.

202. W. Dobbelaere, J. De Boeck, and G. Borghs, "Growth and Optical Characterization of InAs$_{1-x}$Sb$_x$ ($0 \leq x \leq 1$) on GaAs and on GaAs-Coated Si by Molecular Beam Epitaxy," *Applied Physics Letters* 55, 1856–58, 1989.

203. P. K. Chiang and S. M. Bedair, "p-n Junction Formation in InSb and InAs$_{1-x}$Sb$_x$ by Metalorganic Chemical Vapor Deposition," *Applied Physics Letters* 46, 383–85, 1985.

204. A. M. Litvak and N. A. Charykov, "New Thermodynamic Method for Calculation of the Diagrams of Double and Triple Systems Including In, Ga, As, and Sb," *Zhurnal Inorganic Materials* 27, 225–30, 1990.

205. Yu. P. Yakovlev, I. A. Andreev, S. Kizhayev, E. V. Kunitsyna, and P. Mikhailova, "High-Speed Photothodes for 2.0-4.0 μm Spectral Range," *Proceedings of SPIE* 6636, 66360D, 2007.

206. E. V. Kunitsyna, I. A. Andreev, N. A. Charykov, Yu. V. Soloviev, and Yu. P. Yakovlev, "Growth of Ga$_{1-x}$In$_x$As$_y$Sb$_{1-y}$ Solid Solutions from the Five-Component Ga-In-As-Sb-Pb Melt by Liquid Phase Epitaxy," *Journal of Applied Surface Science* 142, 371–74, 1999.

207. B. A. Matveev, N. V. Zotova, S. A. Karandashev, M. A. Remenniy, N. M. Stus, and G. N. Talalakin, "III-V Optically Pumped Mid-IR LEDs," *Proceedings of SPIE* 4278, 189–96, 2001.

208. A. K. Srivastava, J. L. Zyskind, R. M. Lum, B. V. Dutt, and J. K. Klingert, "Electrical Characteristics of InAsSb/GaSb Heterojunctions," *Applied Physics Letters* 49, 41–43, 1986.

209. M. Mebarki, A. Kadri, and H. Mani, "Electrical Characteristics and Energy-Band Offsets in n-InAs$_{0.89}$Sb$_{0.11}$/n-GaSb Heterojunctions Grown by the Liquid Phase Epitaxy Technique," *Solid State Communication* 72, 795–98, 1989.

210. I. A. Andreev, A. N. Baranov, M. P. Mikhailova, K. D. Moiseev, A. V. Pencov, Yu. P. Smorchkova, V. V. Sherstnev, and Yu. P. Yakovlev, "Non-Cooled InAsSbP and GaInAsSb Photodiodes for 3–5 μm Spectral Range," *Pisma v Zhurnal Tekhnicheskoi Fiziki* 18(17), 50–53, 1992.

211. M. Mebarki, H. Ait-Kaci, J. L. Lazzari, C. Segura-Fouillant, A. Joullie, C. Llinares, and I. Salesse, "High Sensitivity 2.5 μm Photodiodes with Metastable GaInAsSb Absorbing Layer," *Solid-State Electronics* 39, 39–41, 1996.

212. H. Shao, A. Torfi, W. Li, D. Moscicka, and W. I. Wang, "High Detectivity AlGaAsSb/InGaAsSb Photodetectors Grown by Molecular Beam Epitaxy with Cutoff Wavelength Up to 2.6 μm," *Journal of Crystal Growth* 311, 1893–96, 2009.

213. B. A. Matveev, M. P. Mikhailova, S. V. Slobodchikov, N. N. Smirnova, N. M. Stus, and G. N. Talalakin, "Avalanche Multiplication in p-n InAs$_{1-x}$Sb$_x$ Junctions," *Fizika i Tekhnika Poluprovodnikov* 13, 498–503, 1979.

214. O. Hildebrand, W. Kuebart, K. W. Benz, and M.H. Pilkuhn, "Ga$_{1-x}$Al$_x$Sb Avalanche Photodiodes: Resonant Impact Ionization with Very High Ratio of Ionization Coefficients," *IEEE Journal of Quantum Electronics* 17, 284–288, 1981.

215. I. A. Andreev, M. P. Mikhailava, S. V. Mel'nikov, Yu. P. Smorchkova, and Yu. P. Yakovlev, "Avalanche Multiplication and Ionization Coefficients of GaInAsSb," *Soviet Physics-Semiconductor* 25, 861–65, 1991.

216. I. A. Andreev, M. A. Afrailov, A. N. Baranov, N. N. Marinskaya, M. A. Mirsagatov, M. P. Mikhailova, Yu. P. Yakovlev, "Low-Noise Avalanche Photodiodes with Separated Absorption and Multiplication Regions for the Spectral Interval 1.6–2.4 μm," *Soviet Technical Physics Letters* 15, 692–96, 1989.

217. M. Van Schilfgaarde, A. Sher, and A. B. Chen, "InTlSb: An Infrared Detector Material?" *Applied Physics Letters* 62, 1857–59, 1993.

218. A. B. Chen, M. Van Schilfgaarde, and A. Sher, "Comparison of $In_{1-x}Tl_xSb$ and $Hg_{1-x}Cd_xTe$ as Long Wavelength Infrared Materials," *Journal of Electronic Materials* 22, 843–46, 1993.

219. E. Michel and M. Razeghi, "Recent Advances in Sb-Based Materials for Uncooled Infrared Photodetectors," *Opto-Electronics Review* 6, 11–23, 1998.

220. M. Van Schilfgaarde, A. B. Chen, S. Krishnamurthy, and A. Sher, "InTlP: A Proposed Infrared Detector Material," *Applied Physics Letters* 65, 2714–16, 1994.

221. M. Razeghi, "Current Status and Future Trends of Infrared Detectors," *Opto-Electronics Review* 6, 155–94, 1998.

222. J. J. Lee and M. Razeghi, "Novel Sb-Based Materials for Uncooled Infrared Photodetector Applications," *Journal of Crystal Growth* 221, 444–49, 2000.

223. J. L. Lee and M. Razeghi, "Exploration of InSbBi for Uncooled Long-Wavelength Infrared Photodetectors," *Opto-Electronics Review* 6, 25–36, 1998.

224. J. N. Baillargeon, P. J. Pearah, K. Y. Cheng, G. E. Hofler, and K. C. Hsieh, "Growth and Luminescence Properties of GaPN," *Journal of Vacuum Science and Technology B* 10, 829–31, 1992.

225. M. Weyers, M. Sato, and H. Ando, "Red Shift of Photoluminescence and Absorption in Dilute GaAsN Alloy Layers," *Japanese Journal of Applied Physics* 31(7A), L853–L855, 1992.

226. T. Ashley, T. M. Burke, G. J. Pryce, A. R. Adams, A. Andreev, B. N. Murdin, E. P. O'Reilly, and C. R. Pidgeon, "$InSb_{1-x}N_x$ Growth and Devices," *Solid-State Electronics* 47, 387–94, 2003.

227. B. N. Murdin, M. Kamal-Saadi, A. Lindsay, E. P. O'Reilly, A. R. Adams, G. J. Nott, J. G. Crowder, et al., "Auger Recombination in Long-Wavelength Infrared InN_xSb_{1-x} Alloys," *Applied Physics Letters* 78, 1568–70, 2001.

14 HgCdTe Detectors

The 1959 publication by Lawson and coworkers triggered the development of variable bandgap $Hg_{1-x}Cd_xTe$ (HgCdTe) alloys providing an unprecedented degree of freedom in infrared detector design [1]. During the 35th conference in Infrared Technology and Applications held in Orlando, Florida, April 13–17, 2009, a special session was organized to celebrate the 50th anniversary of this first publication [2]. This session brought together most of research centers and industrial companies that have participated in the subsequent development of HgCdTe. Figure 14.1 shows the three Royal Radar Establishment inventors of HgCdTe (W. D. Lawson, S. Nielson, and A. S. Young) that disclosed the compound ternary alloy in a 1957 patent [3]. They were joined by E. H. Putley in the first publication dated in 1959 [1].

HgCdTe is a pseudobinary alloy semiconductor that crystallizes in the zinc blende structure. Because of its bandgap tunability with x, $Hg_{1-x}Cd_xTe$ has evolved to become the most important/ versatile material for detector applications over the entire infrared (IR) range. As the Cd composition increases, the energy gap for $Hg_{1-x}Cd_xTe$ gradually increases from the negative value for HgTe to the positive value for CdTe. The bandgap energy tunability results in IR detector applications that span the short wavelength IR (SWIR: 1–3 μm), middle wavelength (MWIR: 3–5 μm), long wavelength (LWIR: 8–14 μm), and very long wavelength (VLWIR: 14–30 μm) ranges.

HgCdTe technology development was and continues to be primarily for military applications. A negative aspect in support of defense agencies has been the associated secrecy requirements that inhibit meaningful collaborations among research teams on a national and especially on an international level. In addition, the primary focus has been on focal plane array (FPA) demonstration and much less on establishing the knowledge base. Nevertheless, significant progress has been made over five decades. At present HgCdTe is the most widely used variable gap semiconductor for IR photodetectors.

14.1 HgCdTe HISTORICAL PERSPECTIVE

The first paper published by Lawson et al. [1] reported both photoconductive and photovoltaic response at wavelengths extending out to 12 μm, and made the understated observation that this material showed promise for intrinsic infrared detectors. In that time the importance of the 8–12 μm atmospheric transmission window was well known for thermal imaging, which enable night vision by imaging the emitted IR radiation from the scene. Since 1954 Cu-doped Ge extrinsic photoconductive detector was known [4], but its spectral response extended to 30 μm (far longer than required for the 8–12 μm window) and to achieve background-limited performance the Ge:Cu detector was necessary to cool down to liquid helium temperature. In 1962 it was discovered that the Hg acceptor level in Ge has an activation energy of about 0.1 eV [5] and detector arrays were soon made from this material; however, the Ge:Hg detectors were cooled to 30 K to achieve maximum sensitivity. It was also clear from theory that the intrinsic HgCdTe detector (where the optical transitions were direct transitions between the valence band and the conduction band) could achieve the same sensitivity at much higher operating temperature (as high as 77 K). Early recognition of the significance of this fact led to intensive development of HgCdTe detectors in a number of countries including England, France, Germany, Poland, the former Soviet Union, and the United States [6]. However, little has been written about the early development years; for example, the existence of work going on in the United States was classified until the late 1960s.

Figure 14.2 gives approximate dates of significant development efforts for HgCdTe IR detectors; Figure 14.3 gives additional insight in time line of the evolution of detectors and key developments in process technology [7]. Photoconductive devices had been built in the United States as early as 1964 at Texas Instruments after the development of a modified Bridgman crystal growth technique. The first report of a junction intentionally formed to make an HgCdTe photodiode was by Verie and Granger [8], who used Hg in-diffusion into p-type material doped with Hg vacancies. The first important application of HgCdTe photodiodes was as high-speed detectors for CO_2 laser radiation [9]. The French pavilion at the 1967 Montreal Expo illustrated a CO_2 laser system with HgCdTe photodiode. However, the high performance medium wavelength IR (MWIR) and LWIR linear arrays developed and manufactured in the 1970s were n-type photoconductors used in the first generation scanning systems. In 1969 Bartlett et al. [10] reported background-limited performance of photoconductors operated at 77 K in the LWIR spectral region. The advantage in material preparation and detector technology have led to devices approaching theoretical limits of responsivity and detectivity over wide ranges of temperature and background [11].

Figure 14.1 The discoverers of HgCdTe ternary alloy. (From Elliot, T., "Recollections of MCT Work in the UK at Malvern and Southampton," *Proceedings of SPIE* 7298, 72982M, 2009. With permission.)

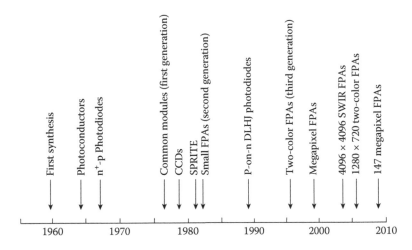

Figure 14.2 History of the development of HgCdTe detectors.

A novel variation of the standard photoconductive device, the SPRITE detector (Signal PRocessing In The Element), was invented in England [12,13]. A family of thermal imaging systems has utilized this device, however its usage has declined. The SPRITE detector provides signal averaging of a scanned image spot that is accomplished by synchronization between the drift velocity of minority carriers along the length of photoconductive bar of material and the scan velocity of the imaging system. Then the image signal builds up a bundle of minority charge that is collected at the end of the photoconductive bar, effectively integrating the signal for a significant length of time and thereby improving the signal-to-noise ratio (SNR).

The scanning system, which does not include multiplexing functions in the focal plane, belongs to the first generation systems. The U.S. common module HgCdTe arrays employ 60, 120, or 180 photoconductive elements depending on the application.

After the invention of charge coupled devices (CCDs) by Boyle and Smith [14], the idea of an all solid-state electronically scanned two-dimensional (2-D) IR detector array caused attention to be directed toward HgCdTe photodiodes. These include p-n junctions, heterojunctions, and MIS photocapacitors. Each of these different types of devices has certain advantages for IR detection, depending on the particular application. More interest has been focused on the first two structures, so further considerations are restricted to p-n junctions and heterostructures. Photodiodes with their very low power dissipation, inherently high impedance, negligible $1/f$ noise, and easy multiplexing on focal plane silicon chip, can be assembled in 2-D arrays containing a very large number of elements, limited only by existing technologies. They can be reverse-biased for even

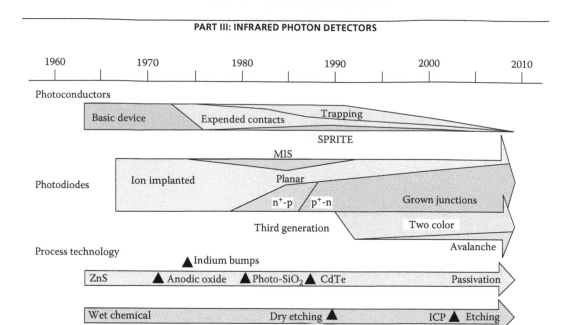

Figure 14.3 A time line of the evolution of HgCdTe IR detectors and key developments in process technology that made them possible. (From Norton, P., *Opto-Electronics Review*, 10, 159–74, 2002. With permission.)

higher impedance, and can therefore match electrically with compact low-noise silicon readout preamplifier circuits. The response of photodiodes remains linear to significantly higher photon flux levels than that of photoconductors (because of higher doping levels in the photodiode absorber layer and because the photogenerated carriers are collected rapidly by the junction). At the end of the 1970s, the emphasis was directed toward large photovoltaic HgCdTe arrays in the MW and LW spectral bands for thermal imaging. Recent efforts have been extended to short wavelengths (e.g., for starlight imaging in the short wavelength (SW) range), as well as to very LWIR (VLWIR) space borne remote sensing beyond 15 µm.

The second-generation systems (full-framing systems) have typically three orders of magnitude more elements (> 10^6) on the focal plane than first generation systems and the detector elements are configured in a 2-D array. These staring arrays are scanned electronically by circuits integrated with the arrays. These readout integrated circuits (ROICs) include, for example, pixel deselecting, antiblooming on each pixel, subframe imaging, output preamplifiers, and other functions. The second generation HgCdTe devices are 2-D arrays of photodiodes. This technology began in the late 1970s and took the next decade to reach volume production. The hybrid architecture was first demonstrated in the mid-1970s [15], indium bump bonding of readout electronics provides for multiplexing the signals from thousands of pixels onto a few output lines, greatly simplifying the interface between the vacuum-enclosed cryogenic sensor and the system electronics. The detector material and multiplexer are optimized independently. Other advantages of the hybrid FPAs are near 100% fill factors and an increased signal-processing area on the multiplexer chip.

The MIS photocapacitor is usually formed on the n-type absorber layer with a thin semitransparent metal film as a gate electrode. The insulator of choice is a thin native oxide. The only motivation for developing HgCdTe MIS detectors was the allure of realizing a monolithic IR CCD, with both the detection and multiplexing taking place in the same material. However, because of the nonequilibrium operation of the MIS detector (usually a bias voltage pulse of several volts is applied across the capacitor to drive the surface into deep depletion), much larger electric fields are set up in the depletion region of the MIS device than in the p-n junction, resulting in a defect-related tunneling current that is orders of magnitude larger than the fundamental dark current. The MIS detector required much higher material quality than photodiode, which still has not been achieved. For this reason, the development of HgCdTe MIS detector was abandoned around 1987 [16,17].

In the last decade of the twentieth century, the third generation of HgCdTe detectors emerged from tremendous impetus in the detector developments (see Chapter 23). This generation of detectors has emerged from technological achievements in the growth of heterostructure devices used in the production of second-generation systems [18].

14.2 HGCDTE: Technology and Properties

High-quality semiconductor material is essential to the production of high performance and affordable IR photodetectors. The material must have a low defect density, a large size of wafers, uniformity, and reproducibility of intrinsic and extrinsic properties. To achieve these characteristics, HgCdTe materials evolved from high-temperature, melt-grown, bulk crystals to low-temperature, liquid and vapor phase epitaxy (VPE). However, the cost and availability of large-area and high-quality $Hg_{1-x}Cd_xTe$ are still the main considerations for producing affordable devices.

14.2.1 Phase Diagrams

A solid understanding of phase diagrams is essential for the proper design of the growth process. The phase diagrams and their implications for $Hg_{1-x}Cd_xTe$ crystal growth have been discussed extensively [19–21]. The ternary Hg-Cd-Te phase diagrams have been established both theoretically and experimentally throughout the Gibbs triangle [22–24]. Brice summarized the works of more than 100 researchers on the Hg-Cd-Te phase diagram as a numerical description [23], which is convenient for the design of growth processes [25].

The generalized associated solution model has proven to be successful in explaining experimental data and to predict the phase diagram of the entire Hg-Cd-Te system. It was assumed that the liquid phase was a mixture of Hg, Te, Cd, HgTe, and CdTe. The gas phase over the material contained Hg, Cd atoms, and Te_2 molecules. The composition of the solid material can be described by a generalized formula $(Hg_{1-x}Cd_x)_{1-y}Te_y$. The familiar $Hg_{1-x}Cd_xTe$ formula corresponds to the pseudobinary CdTe and HgTe alloy ($y = 0.5$) with complete mutual solubilities. At present it is believed that the sphalerite pseudobinary phase region in $Hg_{1-x}Cd_xTe$ is extended in Te-rich material with a width of the order of 1%. The width narrows at lower temperatures. The consequence of such a form of diagram is a tendency for Te precipitation. An excess of Te is due to vacancies in the metal sublattice, which results in p-type conductivity of pure materials. Low-temperature annealing at 200–300 K reduces the native defect (predominantly acceptor) concentration and reveals an uncontrolled (predominantly donor) impurity background. Weak Hg-Te bonding results in low activation energy for defect formation and Hg migration in the matrix. This can cause bulk and surface instabilities.

Most of the problems with crystal growth are due to the marked difference between the solidus and liquidus curves (see Figure 14.4) resulting in the segregation of binaries during crystallization from melts. The segregation coefficient for growth from melts depends on Hg pressure. Serious

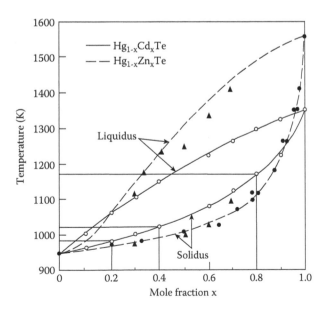

Figure 14.4 Liquidus and solidus lines in the HgTe-CdTe and HgTe-ZnTe pseudobinary systems.

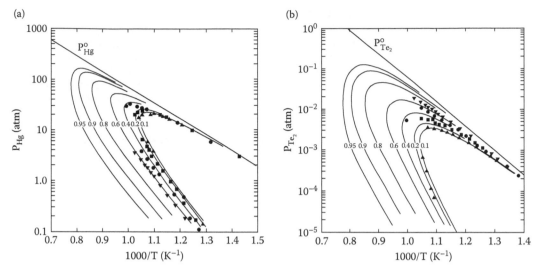

Figure 14.5 Partial pressure of (a) Hg and (b) Te_2 along the three-phase curves for various solid solutions. The labels are the values of x in the formula, $Hg_{1-x}Cd_xTe$. (From Tung, T., Su, C. H., Liao, P. K., and Brebrick, R. F., *Journal of Vacuum Science and Technology*, 21, 117–24, 1982. With permission.)

problems also arise from high Hg pressures over pseudobinary and Hg-rich melts. A full appreciation of the P_{Hg}-T diagram shown in Figure 14.5a is therefore essential [26]. Curves are the partial pressure of Hg along boundaries for solid solutions of composition x where the solid solution is in equilibrium with another condensed phase as well as the vapor phase. We can see that for $x = 0.1$ and $1000/T = 1.3$ K^{-1} HgCdTe exists for Hg pressure of 0.1 (Te-saturated) and 7 (Hg-saturated) atm. The atomic fraction of Te decreases, as the Hg partial pressure increases, over a small but nonzero range near 0.5 atomic fraction. Even at $x = 0.95$ and Te-saturated conditions, Hg is the predominant vapor species and no solid solution contains exactly 0.5 atomic fraction Te. These features are highly significant in controlling the native defect concentrations and hence electrical properties in HgCdTe. From Figure 14.5b results that in comparison with Hg pressure, the partial pressure of Te_2 is several orders of magnitude lower. Figure 14.6 shows the low-temperature liquidus and solid-solution isoconcentration lines in the Hg-rich and Te-rich corners [22]. Figure 14.6a shows that, for example, at 450°C, a solid solution containing 0.90 mole fraction CdTe is in equilibrium with a liquid containing 7×10^{-4} atomic fraction Cd and 0.014 atomic fraction Te. One sees that almost pure CdTe(s) crystallizes from very Hg-rich liquids.

14.2.2 Outlook on Crystal Growth

The time line for the evolution of HgCdTe crystal growth technologies is illustrated in Figure 14.7 [7]. Historically, crystal growth of HgCdTe has been a major problem mainly because a relatively high Hg pressure is present during growth, which makes it difficult to control the stoichiometry and composition of the grown material. The wide separation between the liquidus and solidus, leading to marked segregation between CdTe and HgTe, was instrumental in slowing the development of all the bulk growth techniques to this system. In addition to solidus-liquidus separation, high Hg partial pressures are also influential both during growth and postgrowth heat treatments.

Several historical reviews of the development of bulk HgCdTe have been published [19,27–29]. An excellent review concerning growth of both bulk material and epitaxial layers has been recently given by Capper [30]. Many techniques were tried in the early years (see, for example, Verie and Granger [8] in which Micklethwaite gave comprehensive information on the growth techniques used prior to 1980), but three prime techniques, shown in Figure 14.7, survived: solid state recrystallization (SSR), Bridgman, and traveling heater method (THM).

Early experiments and a significant fraction of early production were undertaken using a quench-anneal or solid-state recrystalization process. In this method the charge of a required composition was synthesized, melted, and quenched. Then, the fine dendritic mass (highly polycrystalline solid) obtained in the process was annealed below the liquidus temperature for a few

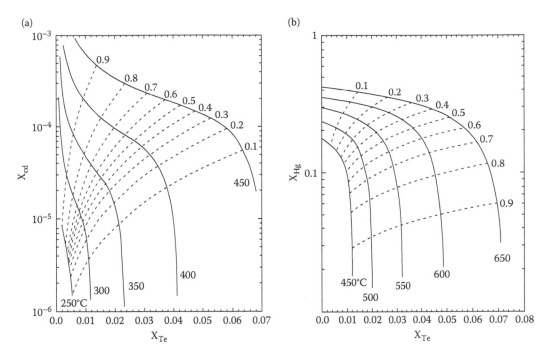

Figure 14.6 Liquidus isotherms for (a) Hg-rich, and (b) Te-rich growth. (From Brebrick, R. F., *Journal of Crystal Growth*, 86, 39–48, 1988. With permission.)

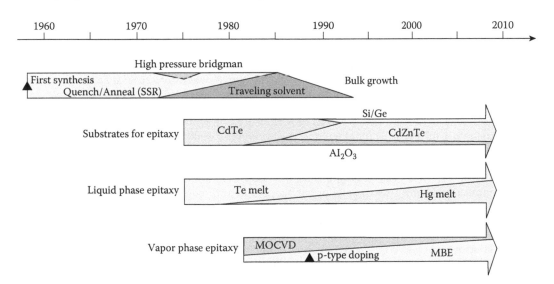

Figure 14.7 Evolution of HgCdTe crystal growth technology from 1958 to present. (From Norton, P., *Opto-Electronics Review*, 10, 159–74, 2002. With permission.)

weeks to recrystallize and homogenize the crystals. Various improvements of the process have been proposed including the temperature gradient annealing and slow cooling down to prevent precipitation of tellurium. The material usually requires low-temperature annealing for adjusting the concentration of native defects. Care must be taken in the quenching stage to avoid pipes/voids, which cannot be removed by the recrystalization step. The material usually requires low-temperature annealing in order to adjust the concentration of native defects. The crystals can also be uniformly doped by the introduction of dopants to the charge.

The SSR method has serious drawbacks. As no segregation occurs, all impurities present in the charge are frozen in the crystal and high-purity starting elements are required. The maximum diameter of the ingots was limited to about 1.5 cm as the cooling rate of large diameter charges is too slow for suppression of the segregation. The crystals contain low grain boundaries.

Alternatives to the basic SSR process have included "slush" growth [31], high pressure growth [32], incremental quenching [33], and horizontal casting [34]. In the slush process the initial homogenous charge is held across the liquidus-solidus gap with the lower end solid and the upper end liquid (temperature gradient of 10 K/cm) [30]. High pressure (30 atm Hg) was used in a attempt to reduce structural defects by improved heat flow control and using intergranual Te as a moving liquid zone during the recrystalization step.

Bridgman growth was attempted for several years near the mid-1970s. Limits on controlling melt mixing in the Bridgman process necessitated a means of stirring melts contained in sealed, pressurized ampoules. The addition of accelerated crucible rotation technique (ACRT) in which the melt is subjected to periodic acceleration/deceleration at rotation rates of up to 60 r.p.m. causes a marked improvement over normal Bridgman growth particularly for high x material: better reproducibility of the process, relatively flat interfaces, and decrease in the number of major grains (typically from 10 to 1 in the $x = 0.2$ region) [35–37]. Crystals were produced up to 20 mm in diameter and with x value up to 0.6 in the tip regions of some crystals.

At the same time, solvent growth methods from Te-rich melts were initiated to reduce the growth temperature. One successful implementation was the THM that resulted in crystals up to 5 cm diameter [38]. The perfect quality of crystals grown by this method is achieved at the cost of a low growth rate [39].

Bulk HgCdTe crystals were initially used for all types of infrared photodetectors. At present they are still used for some infrared applications such as n-type single element photoconductors, SPRITE detectors, and linear arrays. Bulk growth produced thin rods, generally up to 15 mm in diameter, about 20 cm in length, and with a nonuniform distribution of composition. Large two-dimensional arrays could not be realized with bulk crystals. Another drawback to bulk material was the need to thin the bulk wafers, usually cut to about 500 μm down to a final device thickness of about 10 μm. Also, further fabrication steps (polishing the wafers, mounting them to suitable substrates, and polishing to the final device thickness) was very labor intensive.

In comparison with bulk growth techniques, epitaxial techniques offer the possibility of growing large area ($\approx$100 cm^2) epilayers and fabrication of sophisticated device structures with good lateral homogeneity, abrupt and complex composition, and doping profiles that can be configured to improve the performance of photodetectors. The growth is performed at low temperatures (see Figure 14.8) [40], which makes it possible to reduce the native defect density. The properties of HgCdTe grown by the various techniques discussed here are summarized in Table 14.1.

Among the various epitaxial techniques, liquid phase epitaxy (LPE) is the most technologically mature method. The LPE is a single crystal growth process in which growth from a cooling solution occurs onto a substrate. Another technique, vapor phase epitaxial (VPE) growth of HgCdTe is typically carried out by nonequilibrium methods that also apply to metalorganic chemical vapor deposition (MOCVD), molecular beam epitaxy (MBE), and their derivatives. The great potential benefit of MBE and MOCVD over equilibrium methods is the ability to modify the growth conditions dynamically during growth to tailor bandgaps, add and remove dopants, prepare surfaces and interfaces, add passivations, perform anneals, and even grow on selected areas of a substrate. The growth control is exercised with great precision to obtain basic materials properties comparable to those routinely obtained from equilibrium growth.

Epitaxial growth of HgCdTe layers requires a suitable substrate. CdTe was used initially, since it was available from commercial sources in reasonably large sizes. The main drawback to CdTe is that it has a few percentage lattice mismatch with LWIR and MWIR HgCdTe. By the mid-1980s it was demonstrated that the addition of a few percentage of ZnTe to CdTe (typically 4%) could create a lattice-matched substrate. CdTe and closely lattice-matched CdZnTe substrates are typically grown by the modified vertical or horizontal unseeded Bridgman technique. Most commonly the (111) and (100) orientations have been used, although others have been tried. Twinning, which occurs in (111) layers, can be prevented by a suitable misorientation of the substrate. Growth conditions found to be nearly optimal for the (112)B orientation were selected. The limited size, purity problems, Te precipitates, dislocation density (routinely in the low 10^4 cm^{-2} range), nonuniformity of lattice match and high price ($50–500 per cm^2, polished) are remaining problems to be solved. It is believed that these substrates will continue to be important for a long time, particularly for the highest performance devices.

The LPE growth of a thin layer of HgCdTe on CdTe substrates began in the early to mid-1970s. Both Te-solution growth (420–500°C) and Hg-solution growth (360–500°C) have been used with

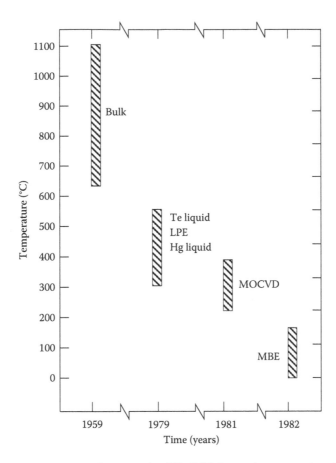

Figure 14.8 Temperature ranges for growth of HgCdTe by various growth techniques versus date of first reported attempt. (From Gertner, E. R., *Annual Review of Materials Science*, 15, 303–28, 1985.)

equal success in a variety of configurations. The pioneering work on the Te corner of the phase diagram of the Hg-Cd-Te system was published in 1980 [41] and the associated LPE growth equipment provided the necessary groundwork that led to several variations of open-tube LPE from Te solutions (see, e.g., [42–45]). Initially, Te solutions with dissolved Cd (Cd has a high solubility in Te) and saturated with Hg vapor were used to efficiently grow HgCdTe in the temperature range 420–600°C. This allowed small volume melts to be used with slider techniques that did not appreciably deplete during the growth run. Experiments with Hg-solvent LPE began in the late 1970s. The Santa Barbara Research Center (SBRC) pioneered the phase diagram study of the Hg corner of the Hg-Cd-Te system [46] and after several years of experimenting a reproducible Hg-solution technology was developed [47]. Because of the limited solubility of Cd in Hg, the volume of the Hg melts had to be much larger then Te melts (typically about 20 kg) in order to minimize melt depletion during layer growth in the temperature range 380–500°C. This precluded the slider growth approach and Hg-melt epitaxy has been developed using large dipping vessels. One major advantage of the Hg-solution technology is its capability to produce layers of excellent surface morphology due to the ease of melt decanting. More recently, two additional unique characteristics have been recognized as essential for the fabrication of high-performance, double-layer heterojunction (DLHJ) photodiodes by LPE; low liquidus temperature (< 400°C) makes the cap-layer growth step feasible and ease of incorporating both p-type and n-type temperature-stable impurities, such as As, Sb, and In, during growth.

In the early 1990s, bulk growth was replaced by LPE and is now very mature for the production of first- and second-generation detectors. However, LPE technology is limited for a variety of advanced HgCdTe structures required for third generation detectors. LPE typically melts off a thin layer of the underlying material each time an additional layer is grown due to the relatively high

Table 14.1: Comparison of the Various Methods Used to Grow HgCdTe

	Bulk			Liquid Phase Epitaxy		Vapor Phase Epitaxy	
		Traveling Heater Method					
	SSR	HCT Melt	Te Melt	Hg Melt	Te Melt	MOCVD	MBE
Temperature (°C)	950	950	500	350–550	400–550	275–400	160–200
Pressure (Torr)	150,000	150,000	760–8000	760–11,400	760–8000	300–760	10^{-3}–10^{-4}
Growth rate (μm/hr)	250	250	80	30–60	5–60	2–10	1–5
Dimensions w (cm)	0.8–1.2 dia	0.8–1.2 dia	2.5 dia	5	5	7.5 dia	7.5 dia
l (cm)	—	—	—	6	5	4	4
t (cm)	15	15	15				
Dislocations (cm^{-2})	$<10^5$	—	$<10^5$	$<10^5$	$<10^5$–10^7	5×10^5–10^7	$<5 \times 10^4$–10^6
Purity (cm^{-3})	$<5 \times 10^{14}$	$<5 \times 10^{14}$	$<5 \times 10^{14}$	$<5 \times 10^{14}$	$<5 \times 10^{14}$	$<1 \times 10^{15}$	$<1 \times 10^{15}$
n-type doping (cm^{-3})	N/A	N/A	N/A	1×10^{14}–1×10^{18}	1×10^{15}–1×10^{16}	5×10^{14}–5×10^{18}	5×10^{14}–1×10^{19}
p-type doping (cm^{-3})	N/A	N/A	N/A	1×10^{15}–1×10^{18}	1×10^{15}–5×10^{16}	3×10^{15}–5×10^{17}	1×10^{16}–5×10^{18}
X-ray rocking curve (arc sec)	—	—	20–60	<20	<20	50–90	20–30
Compositional uniformity (Δx)	<0.002	<0.004	<0.005	<0.002	<0.002	± 0.01–0.0005	± 0.01–0.0006

Source: P. Norton, "HgCdTe Infrared Detectors," *Opto-Electronics Review*, 10, 159–74, 2002. With permission.

growth temperature. Additionally, the gradient in x-value in the base layer of p$^+$-on-n junctions can generate a barrier to carrier transport in certain cases due to interdiffusion. These limitations have provided an opportunity for VPE, especially MBE and MOCVD.

Various VPE methods have been used to grow $Hg_{1-x}Cd_xTe$ layers. One of the oldest is the isothermal vapor phase epitaxy (ISOVPE), which was initially invented in France [48]. ISOVPE is a relatively simple, quasiequilibrium growth technique in which HgTe is transported at a relatively high temperature (400–600°C) from the source (HgTe or $Hg_{1-x}Cd_xTe$) to the substrate by evaporation–condensation mechanisms. An inherent property of the method is in-depth grading because interdiffusion of deposited and substrate materials is involved in layer formation.

The era of MBE and MOCVD growth of HgCdTe began in the early 1980s by adopting both methods that had been well established in the III-V semiconductor materials. Through the following decade, a variety of metalorganic compounds were developed along with a number of reaction-chamber designs [49,50]. The MOCVD growth of HgCdTe can be achieved by one of two alternative processes, direct alloy growth (DAG) and the interdiffused multilayer process (IMP). The DAG is faced with several severe problems, mainly because of the great difference in stability between HgTe and CdTe together with the higher reactivity of Te precursors with the Cd alkyl than with metallic Hg. The IMP technique overcomes these problems [51]. The successive layers of CdTe and HgTe have the combined thickness of two consecutive layers that are approximately 0.1 μm. These layers are deposited and interdiffused during growth or during a short annealing *in situ* at the end of the growth. The activation of As, Sb, and In dopants is easier to achieve with IMP when introduced during the CdTe growth cycle and with a Cd/Te flux ratio above 1 [52]. In comparison with indium, iodine is a more stable donor dopant that can be used to control local doping within a 3×10^{14}–2×10^{18} cm^{-3} range with 100% activation when following a standard stoichiometric anneal [53].

The MBE growth of HgCdTe is carried out with effusive sources that contain Hg, Te$_2$, and CdTe. A specially designed Hg-source oven was successfully designed to overcome the low sticking coefficient of Hg at the growth temperature [54–56]. Surface growth temperature in MBE HgCdTe plays a critical role in the introduction of extended defects. The optimized growth temperature is in the range of 185°C–190°C. At lower temperatures, an excess of Hg is obtained at the surface because the sticking coefficient of Hg increases as the temperature is reduced. The excess Hg produces microtwin defects. These defects are detrimental to the electrical properties of the epilayer and devices. The etch pit density (EPD) values of material grown under these conditions are high (10^6–10^7 cm^{-2} range). If growth temperatures under the same conditions are raised above 190°C, then a deficiency of Hg is obtained at the surface and void defects are formed. Currently, under the best optimized Hg/Te$_2$ flux growth conditions the lowest concentration of void defects that has been observed is around 100 cm^{-2}. Dust particles and/or substrate related surface imperfections may account for this. The EPD values for epilayers grown under these conditions are low (10^4–10^5 cm^{-2}). Significant efforts are being spent on As and Sb-doping to improve incorporation during the MBE process and to reduce the temperature required for activation. The metal-saturation conditions cannot be reached at the temperatures required for high-quality MBE growth. The necessity to activate acceptor dopants at high temperatures eliminates the benefits of low-temperature growth. Recently, near 100% activation was achieved for a 2×10^{18} cm^{-3} As concentration with a 300°C activation anneal followed by a 250°C stoichiometric anneal [57].

At present, MBE is the dominant vapor phase method for the growth of HgCdTe. It offers low temperature growth under an ultrahigh vacuum environment, *in situ* n-type and p-type doping, and control of composition, doping and interfacial profiles. MBE is now the preferred method for growing complex layer structures for multicolor detectors and for avalanche photodiodes. Although the quality of MBE material is not yet on a par with LPE, it has made tremendous progress in the past decade. A key to its success has been the doping ability and the reduction of EPDs to below 10^5 cm^{-2}.

The growth temperature is less than 200°C for MBE, but around 350°C for MOCVD, making it more difficult to control p-type doping in MOCVD due to the formation of Hg vacancies at higher growth temperatures. Arsenic is a preferable dopant for p-type layers, while indium is preferable for n-type layers. Several laboratories consistently reports of undoped MOCVD and MBE grown layers with impurity levels about 10^{14} cm^{-3} indicating that source material purity from commercial vendors now appears adequate, although it could still use some improvement. The remaining problems of the two methods are twin formations, requirement of very good surface preparation prior to growth, uncontrolled doping, dislocation density, and composition inhomogenities.

The lowest reported carrier concentrations and the longest lifetimes in MOCVD and MBE grown layers have been achieved in HgCdTe films grown onto CdZnTe substrates. The substrates are typically grown by the modified vertical and horizontal unseeded Bridgman technique. Near lattice matched CdZnTe substrates have severe drawbacks such as lack of large area, high production cost and, more importantly, a difference in thermal expansion coefficient (TEC) between the CdZnTe substrates and the silicon readout integrated circuit. Furthermore, interest in large area 2-D IR FPAs (1024 × 1024 and larger) have resulted in limited applications of CdZnTe substrates. Currently, readily producible CdZnTe substrates are limited to areas of approximately 50 cm². At this size, the wafers are unable to accommodate more than two 1024 × 1024 FPAs. Not even a single die can be accommodated for very large FPA formats (2048 × 2048 and larger) on substrates of this size.

A viable approach to cheap substrates is the use of hybrid substrates, which consist of laminated structures with wafers of bulk crystal and are covered with buffer lattice-matched layers. Four issues dominate alternative substrates: lattice mismatch, nucleation phenomena, thermal expansion mismatch, and majority species contamination [58,59]. Bulk Si, GaAs, and sapphire are some of the high-quality, low-cost, and readily available crystals that have been shown to be useful substrates for $Hg_{1-x}Cd_xTe$. The buffer layers are a few micrometers thick of CdTe or (Cd,Zn)Te, obtained *in situ* or *ex situ* with a nonequilibrium growth, typically from the vapor phase. The feasibility of growing high-quality $Hg_{1-x}Cd_xTe$ on hybrid substrates was demonstrated first by Rockwell International. This technology is referred to as producible alternative to CdTe for epitaxy (PACE) [60,61]. The substrates are CdTe/sapphire (PACE 1), CdTe/GaAs (PACE 2), and Si/GaAs/CdZnTe (PACE 3).

Sapphire has been widely used as a substrate for HgCdTe epitaxy. In this case a CdTe (CdZnTe) film is deposited on the sapphire prior to the growth of HgCdTe. This substrate has excellent physical properties and can be purchased in large wafer sizes. The large lattice mismatch with HgCdTe is accommodated by a CdTe buffer layer. Sapphire is transparent from the UV to about 6 μm in wavelength and has been used in back side illuminated SWIR and MWIR detectors (it is not acceptable for back side illuminated LWIR arrays because of its opacity beyond 6 μm).

For the 8–12 μm, long wavelength IR band, the CdTe/GaAs (PACE-2) has been developed with detectors fabricated on GaAs substrates [62]. Because GaAs has a TEC comparable to CdZnTe, these PACE-2 FPAs will have the same size limitations as CdZnTe based hybrids unless GaAs readout circuits are used. Moreover, the GaAs initiation layer can result in Ga contamination of the II-VI films and result in undesirable additional cost and process complexity. Recently published papers have indicated that using CdZnTe buffer layers the problem of Ga contamination can be resolved [63].

The use of Si substrates is very attractive in IR FPA technology not only because it is less expensive and available in large area wafers, but also because the coupling of the Si substrates with Si readout circuitry in an FPA structure allows fabrication of very large arrays exhibiting long-term thermal cycle reliability. The 7 × 7 cm² bulk CdZnTe substrate is the largest commercially available, and it is unlikely to increase much larger than its present size. With the cost of 6 inch Si substrates being ≈ $100 versus $10,000 for the 7 × 7 cm² CdZnTe, significant advantages of HgCdTe/Si are evident [64]. Despite the large lattice mismatch (≈19%) between CdTe and Si, MBE has been successfully used for the heteroepitaxial growth of CdTe on Si. Using optimized growth condition for Si(211)B substrates and a CdTe/ZnTe buffer system, epitaxial layers with EPD in the 10^6 cm^{-2} range have been obtained. This value of EPD has little effect on both MWIR and LWIR HgCdTe/Si detectors [64,65]. By comparison, HgCdTe epitaxial layers grown by MBE or LPE on bulk CdZnTe have typical EPD values in the 10^4 to mid-10^5 cm^{-2} range where there is a negligible effect of dislocation density on detector performance. At 77 K, diode performance with cutoff wavelength in LWIR region for HgCdTe on Si is comparable to that on bulk CdZnTe substrates [65].

14.2.3 Defects and Impurities

Native defect properties and impurity incorporation still constitute a field of intensive research. Various aspects of defects in bulk crystals and epilayers, such as electrical activity, segregation, ionization energies, diffusivity, and carrier lifetimes have been summarized in many reviews [66–73].

14.2.3.1 *Native Defects*

The defect structure of undoped and doped $Hg_{1-x}Cd_xTe$ can be explained with the quasichemical approach [74–79]. The dominant native defect in $Hg_{1-x}Cd_xTe$ is a double ionizable acceptor associated with metal lattice vacancies. Some direct measurements show much larger vacancy concentrations than those that follow from Hall measurements, indicating that most vacancies are neutral [80].

In contrast to numerous early findings, now it seems to be established that the native donor defect concentration is negligible. As-grown undoped and pure $Hg_{1-x}Cd_xTe$, including that grown in Hg-rich LPE, always exhibits p-type conductivity with the hole concentration depending on composition, growth temperature, and Hg pressure during growth reflecting correspondence to the concentration of vacancies.

The equilibrium concentration of vacancies and Hg pressures over Te-saturated $Hg_{1-x}Cd_xTe$ are

$$c_V\left[cm^{-3}\right]=\left(5.08\times10^{27}+1.1\times10^{28}x\right)P_{Hg}^{-1}\exp\left(\frac{-\left(1.29+1.36x-1.8x^2+1.375x^3\right)eV}{kT}\right),$$
$$(14.1)$$

$$p_{Hg}\left[atm\right]=1.32\times10^5\exp\left(-\frac{0.635eV}{kT}\right).$$
$$(14.2)$$

The Hg pressure over Hg saturated $Hg_{1-x}Cd_xTe$ is close to saturated Hg pressure

$$p_{Hg}\left[atm\right]=\left(5.0\times10^6+5.0\times10^6x\right)\times\exp\left(\frac{-0.99+0.25x}{kT}eV\right).$$
$$(14.3)$$

Figure 14.9 shows the hole concentration as a function of the partial Hg pressure showing $1/p_{Hg}$ dependence of the native acceptor concentration [68], which is in agreement with the predictions of the quasichemical approach for narrow gap $Hg_{1-x}Cd_xTe$. Annealing in Hg vapors reduces the hole concentration by filling the vacancies. Low-temperature (<300°C) annealing in Hg vapors reveals the background impurity level, causing the p-to-n conversion in some crystals. For example, Figure 14.10 shows the iso-hole concentration plot for $Hg_{0.80}Cd_{0.20}Te$ indicating the possibility of obtaining identical hole concentration for anneals at two different temperatures and partial pressures of Hg [69]. HgCdTe crystals with a vacancy concentration of about 10^{15} cm^{-3} can be obtained either with an anneal at $T = 300°C$ and $p_{Hg} = 7 \times 10^{-2}$ atm or with an anneal at $T = 200°C$ and $p_{Hg} = 6 \times 10^{-5}$ atm, the concentration of Hg interstitial with the lower temperature anneal being

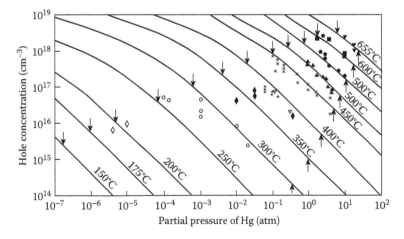

Figure 14.9 The 77 K hole concentration in Hg0.80Cd0.20Te calculated according to the quasichemical approach as a function of the partial pressure of Hg and annealing temperature (150°C to 655°C). Arrows define the material existence region. (From Vydyanath, H. R., *Journal of Vacuum Science and Technology*, B9, 1716–23, 1991. With permission.)

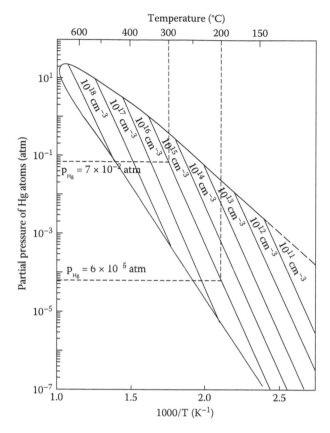

Figure 14.10 Iso-hole concentration plot for $Hg_{0.80}Cd_{0.20}Te$ indicating possibility of obtaining identical hole concentration for anneals at two different temperatures and partial pressures of Hg. (From Vydyanath, H. R., *Journal of Crystal Growth*, 161, 64–72, 1996. With permission.)

lower. If Hg interstitials are the Shockley–Read centers, the minority carrier lifetime in samples prepared at the lower temperature should be higher than that in samples prepared at the highest temperature even though the Hg vacancy concentration in both samples is the same.

Samples with higher residual donor concentration turn n-type at higher temperatures and show higher electron concentration. Unexpected effects may arise from Te precipitates [81]. Hg diffusing into material dissolves precipitates and drives the major impurities ahead of Hg, leaving the core p-type. On further annealing, these impurities may redistribute throughout the slice, turning the whole sample p-type. A variety of effects may cause unexpected n-type behavior contamination, surface layers formed during cool-down, strain, dislocations, twins, grain boundaries, substrate orientation, oxidation, and perhaps other parameters.

Native defects play a dominant role in the diffusion behavior [82]. Vacancies have very high diffusivities even at low temperatures. For example, to form a junction a few micrometers deep in 10^{16} cm^{-3} material requires only about 15 minutes at a temperature of 150°C–200°C. This corresponds to diffusion constants of the order of 10^{-10} cm^{-2}/s. The presence of dislocations can enhance vacation mobility even further, while the presence of Te precipitates may retard the motion of Hg into lattice.

14.2.3.2 Dopants

The electrical behavior of dopants has been extensively reviewed by Capper [70]. Donor behavior is expected for elements from group IIIB on the metal lattice site, and group VIIB elements in the Te site. Indium is most frequently used as a well-controlled dopant for n-type doping due to its high solubility and moderately high diffusion. The experimental data can be explained, assuming that at low (<10^{18} cm^{-3}) concentration In incorporates as a single ionizable donor occupying a metal lattice site. At high In concentration, In incorporates as a neutral complex corresponding to In_2Te_3. The bulk materials are typically doped by direct addition to melts. Indium is frequently introduced during epitaxy and by diffusion; it has been used for many years as a contact material for the n-type photoconductors and the n-type side of photodiodes.

Among the group VIIB elements, only I that was occupying Te sites proved to be a well-behaved donor with concentrations in the 10^{15}–10^{18} cm^{-3} range [72,83]. The electron concentration was found to increase with Hg pressure. Acceptor behavior is expected of elements in the I group (Ag, Cu, and Au) substituting for metal lattice sites, and of elements in the V group (P, As, Sb, Bi) substituting for Te sites.

Ag, Cu, and Au are shallow single acceptors [66,70]. They are very fast diffusers that limit the applications for devices. Significant diffusion of Ag and especially Cu occurs at room temperature [84]. Hole concentrations have been obtained roughly equal to Cu concentration: up to 10^{19} cm^{-3}. But the behavior of Au is more complex. Au seems to be not very useful as a controllable acceptor, though it has proven to be useful for contacts.

The amphoteric behavior of the VB group elements (P, As, Sb) has been established [69,72]. They are acceptors substituting for Te sites and donors at metal sites; therefore, metal-rich conditions are necessary to introduce dopants at Te sites. Arsenic proved to be the most successful p-type dopant to date for formation of stable junctions [85–88]. The main advantages are very low diffusity, stability in lattice, low activation energy, and the possibility of controlling concentration over a wide (10^{15}–10^{18} cm^{-3}) range. Intensive efforts are currently underway to reduce the high temperature (400°C) and high Hg pressures required to activate As as an acceptor.

14.3 FUNDAMENTAL HgCdTe PROPERTIES

HgCdTe ternary alloy is nearly ideal infrared detector material system. Its position is conditioned by three key features [89]:

- tailorable energy bandgap over the 1–30 μm range,

- large optical coefficients that enable high quantum efficiency, and

- favorable inherent recombination mechanisms that lead to high operating temperature (HOT).

These properties are direct consequence of the energy band structure of this zinc-blende semiconductor. Moreover, the specific advantages of HgCdTe are ability to obtain both low and high carrier concentrations, high mobility of electrons, and low dielectric constant. The extremely small change of lattice constant with composition makes it possible to grow high-quality layered and graded gap structures. As a result, HgCdTe can be used for detectors operated at various modes [photoconductor, photodiode, or metal-insulator-semiconductor (MIS) detector].

Table 14.2 summarizes various material properties of $Hg_{1-x}Cd_xTe$ [89]; Table 14.3 compares important parameters of HgCdTe with other narrow gap semiconductors used in IR detector fabrication.

14.3.1 Energy Bandgap

The electrical and optical properties of $Hg_{1-x}Cd_xTe$ are determined by the energy gap structure in the vicinity of the Γ-point of the Brillouin zone, in essentially the same way as InSb. The shape of the electron band and the light-mass hole band are determined by the **k·p** interaction, and hence, by the energy gap and the momentum matrix element. The energy gap of this compound at 4.2 K ranges from −0.300 eV for semimetallic HgTe, goes through zero at about $x = 0.15$, and opens up to 1.648 eV for CdTe.

Figure 14.11 plots the energy bandgap $E_g(x,T)$ for $Hg_{1-x}Cd_xTe$ versus alloy composition parameter x at temperature 77 K and 300 K. Also plotted is the cutoff wavelength $\lambda_c(x,T)$, defined as that wavelength at which the response has dropped to 50% of its peak value.

A number of expressions approximating $E_g(x,T)$ are available at present [73]. The most widely used expression is due to Hansen et al. [90]

$$E_g = -0.302 + 1.93x - 0.81x^2 + 0.832x^3 + 5.35 \times 10^{-4}(1-2x)T,$$

(14.4)

where E_g is in eV and T is in K.

Table 14.2: Summary of the Material Properties for the Hg$_{1-x}$Cd$_x$Te ternary Alloy, Listed for the Binary Components HgTe and CdTe, and for Several Technologically Important Alloy compositions

Property	HgTe	Hg$_{1-x}$Cd$_x$Te						CdTe
x	0	0.194	0.205	0.225	0.31	0.44	0.62	1.0
a (Å)	6.461	6.464	6.464	6.464	6.465	6.468	6.472	6.481
	77 K	77 K	77 K	77 K	140 K	200 K	250 K	300 K
E_g (eV)	−0.261	0.073	0.091	0.123	0.272	0.474	0.749	1.490
λ_c (μm)	—	16.9	13.6	10.1	4.6	2.6	1.7	0.8
n_i (cm^{-3})	—	1.9×10^{14}	5.8×10^{13}	6.3×10^{12}	3.7×10^{12}	7.1×10^{11}	3.1×10^{10}	4.1×10^5
m_c/m_o	—	0.006	0.007	0.010	0.021	0.035	0.053	0.102
g_c	—	−150	−118	−84	−33	−15	−7	−1.2
$\varepsilon_s/\varepsilon_o$	20.0	18.2	18.1	17.9	17.1	15.9	14.2	10.6
$\varepsilon_\infty/\varepsilon_o$	14.4	12.8	12.7	12.5	11.9	10.8	9.3	6.2
n_r	3.79	3.58	3.57	3.54	3.44	3.29	3.06	2.50
μ_e (cm^2/Vs)	—	4.5×10^5	3.0×10^5	1.0×10^5	—	—	—	—
μ_{hh} (cm^2/Vs)	—	450	450	450	—	—	—	—
$b = \mu_e/\mu_\eta$	—	1000	667	222	—	—	—	—
τ_R (μs)	—	16.5	13.9	10.4	11.3	11.2	10.6	2
τ_{A1} (μs)	—	0.45	0.85	1.8	39.6	453	4.75×10^3	
$\tau_{typical}$ (μs)	—	0.4	0.8	1	7	—	—	—
E_p (eV)				19				
Δ (eV)				0.93				
m_{hh}/m_o				0.40–0.53				
ΔE_v (eV)				0.35–0.55				

Source: M. B. Reine, *Encyclopedia of Modern Optics*, Academic Press, London, 2004. With permission.
Note: τ_R and τ_{A1} calculated for n-type HgCdTe with $N_d = 1 \times 10^{15}$ cm^{-3}.
The last four material properties are independent of or relatively insensitive to alloy composition.

The expression that has become the most widely used for intrinsic carrier concentration is that of Hansen and Schmit [91] who used their own $E_g(x,T)$ relationship of Equation 14.4, the **k·p** method and a value of 0.443 m_o for heavy hole effective mass ratio

$$n_i = \left(5.585 - 3.82x + 0.001753T - 0.001364xT\right) \times 10^{14} E_g^{3/4} T^{3/2} \exp\left(-\frac{E_g}{2kT}\right). \tag{14.5}$$

The electron m_e^* and light hole m_{lh}^* effective masses in the narrow gap mercury compounds are close and they can be established according to the Kane band model. Here we used Weiler's expression [92]

$$\frac{m_o}{m_e^*} = 1 + 2F + \frac{E_p}{3}\left(\frac{2}{E_g} + \frac{1}{E_g + \Delta}\right), \tag{14.6}$$

where $E_p = 19$ eV, $\Delta = 1$ eV, and $F = -0.8$. This relationship can be approximated by $m_e^*/m \approx 0.071 E_g$, where E_g is in eV. The effective mass of heavy hole m_{hh}^* is high; the measured values range between 0.3–0.7 m_o. The value of $m_{hh}^* = 0.55 m_o$ is frequently used in modeling of IR detectors.

14.3.2 Mobilities

Due to small effective masses the electron mobilities in HgCdTe are remarkably high, while heavy-hole mobilities are two orders of magnitude lower. A number of scattering mechanisms dominate the electron mobility [93–96]. The x-dependence of the mobility results primary from

Table 14.3: Some Physical Properties of Narrow Gap Semiconductors

Material	E_g (eV)		n_i (cm⁻³)		ε	μ_e(10⁴ cm²/Vs)		μ_h (10⁴ cm²/Vs)	
	77 K	300 K	77 K	300 K		77 K	300 K	77 K	300 K
InAs	0.414	0.359	6.5×10^3	9.3×10^{14}	14.5	8	3	0.07	0.02
InSb	0.228	0.18	2.6×10^9	1.9×10^{16}	17.9	100	8	1	0.08
In$_{0.53}$Ga$_{0.47}$As	0.66	0.75		5.4×10^{11}	14.6	7	1.38		0.05
PbS	0.31	0.42	3×10^7	1.0×10^{15}	172	1.5	0.05	1.5	0.06
PbSe	0.17	0.28	6×10^{11}	2.0×10^{16}	227	3	0.10	3	0.10
PbTe	0.22	0.31	1.5×10^{10}	1.5×10^{16}	428	3	0.17	2	0.08
Pb$_{1-x}$Sn$_x$Te	0.1	0.1	3.0×10^{13}	2.0×10^{16}	400	3	0.12	2	0.08
Hg$_{1-x}$Cd$_x$Te	0.1	0.1	3.2×10^{13}	2.3×10^{16}	18.0	20	1	0.044	0.01
Hg$_{1-x}$Cd$_x$Te	0.25	0.25	7.2×10^8	2.3×10^{15}	16.7	8	0.6	0.044	0.01

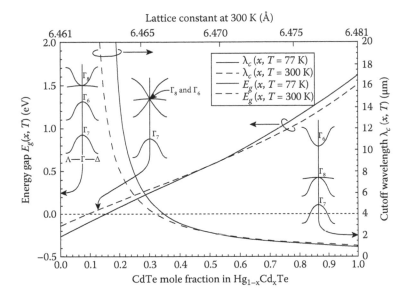

Figure 14.11 The bandgap structure of Hg$_{1-x}$Cd$_x$Te near the Γ-point for three different values of the forbidden energy gap. The energy bandgap is defined at the difference between the Γ_6 and Γ_8 band extrema at $\Gamma = 0$.

the x-dependence of the bandgap, and the temperature dependence primary from the competition among various scattering mechanisms that are temperature dependent.

The electron mobilities in HgCdTe are primarily determined by ionized impurity scattering (CC) in the low-temperature region and by polar longitudinal-optical phonon (LO) scattering above the low-temperature region as shown in Figure 14.12a [97]. Figure 14.12b depicts the composition dependence of mobility in undoped and doped samples at 77 and 300 K. Extremely high values of mobility for high-purity samples are observed near the semiconductor-semimetal transition, where the electron effective mass has its minimum value. It seems that the theory correctly describes the highest mobilities. For Hg$_{0.78}$Cd$_{0.22}$Te LPE layer, the threshold carrier concentration above which the ionized impurity scattering begins to dominate is about 1×10^{16} cm⁻³ for n-type and about 1×10^{17} cm⁻³ for p-type materials [98]. The electron mobility data, when plotted versus temperature, often exhibit a broad peak at $T < 100$ K, particularly for LPE samples. On the other hand, mobility data obtained from high-quality bulk samples do not exhibit these peaks. It is believed that these peaks are associated with scattering by charged centers or are related to the anomalous electrical behavior that are closely associated with sample inhomogeneities [96].

The transport properties of holes are less studied than those of electrons mainly because the contribution of holes to the electrical conduction is relatively small due to their low mobility.

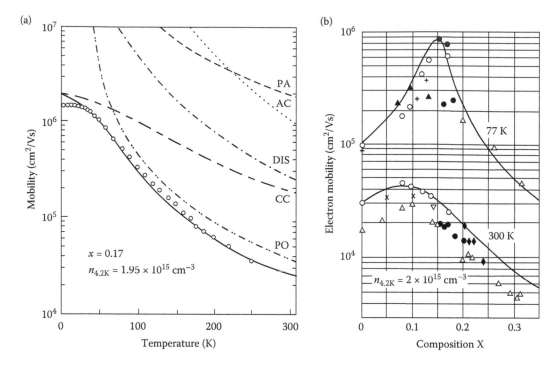

Figure 14.12 Electron mobility in $Hg_{1-x}Cd_xTe$: (a) versus temperature for $Hg_{0.83}Cd_{0.17}Te$; the solid curve is theoretically assumed for mixed scattering modes: by charged center (CC), polar (PO), disorder (DIS), acoustic (AC), and piezoacoustic (PA) scattering modes, (b) versus composition at 77 K and 300 K; curves are calculated with electron concentration of 2×10^{15} cm^{-3} at 4.2 K, experimental points for approximately corresponding concentrations are taken from various works. (From Dubowski, J., Dietl, T., Szymańska, W., and Gałązka, R. R., *Journal of Physics and Chemistry of Solids*, 42, 351–63, 1981. With permission.)

Very often in transport measurements, the electron contribution predominates even in p-type materials unless the electron density is sufficiently low. Comprehensive analysis of different hole scattering mechanisms in $Hg_{1-x}Cd_xTe$ (x = 0.2–0.4) has been carried out by Yadava et al. [99]. They concluded that the heavy hole mobility is largely governed by the ionized impurity scattering, unless the strain field or dislocation scattering below 50 K, or the polar scattering above 200 K, become dominant. The light hole mobility is mainly governed by the acoustic phonon scattering. Figure 14.13 shows the data on hole mobility in $Hg_{1-x}Cd_xTe$ (0.0 ≤ x ≤ 0.4) [100].

The electron mobility in $Hg_{1-x}Cd_xTe$ (expressed in cm^2/Vs), in composition range 0.2 ≤ x ≤ 0.6 and temperature range T > 50 K, can be approximated as [101]:

$$\mu_e = \frac{9 \times 10^8 s}{T^{2r}} \quad \text{where} \quad \begin{aligned} r &= \left(0.2/x\right)^{0.6} \\ s &= \left(0.2/x\right)^{7.5} \end{aligned}. \tag{14.7}$$

Higgins et al. [102] give an empirical formula (valid for 0.18 ≤ x ≤ 0.25) for the variation of μ_e with x at 300 K for the very high-quality melt grown samples they studied:

$$\mu_e = 10^{-4}(8.754x - 1.044)^{-1} \text{ in cm}^2/\text{Vs}. \tag{14.8}$$

The hole mobilities at room temperature range from 40 to 80 cm^2/Vs, and the temperature dependence is relatively weak. A 77 K hole mobility is by one order of magnitude higher. According to Dennis and colleagues [103], the hole mobility measured at 77 K falls as the acceptor

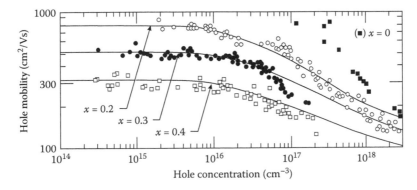

Figure 14.13 Hole mobility as a function of hole concentration for $Hg_{1-x}Cd_xTe$ at 77 K. The solid curves represent calculated data concerning combined lattice and ionized impurity scattering. (From Nelson, D. A., Higgins, W. M., Lancaster, R. A., Roy, R., and Vydyanath, H. R., *Extended Abstracts of U.S. Workshop on the Physics and Chemistry of Mercury Cadmium Telluride*, p. 175, Minneapolis, Minnesota, October 28–30, 1981. With permission.)

concentration is increased and in the composition range 0.20–0.30 yields the following empirical expression:

$$\mu_h = \mu_o \left[1 + \left(\frac{p}{1.8 \times 10^{17}} \right)^2 \right]^{-1/4}, \tag{14.9}$$

where $\mu_o = 440$ cm²/Vs.

For modeling IR photodetectors, the hole mobility is usually calculated assuming that the electron-to-hole mobility ratio $b = \mu_e/\mu_h$ is constant and equal to 100.

The minority carrier mobility is one of the fundamental material properties affecting performance of HgCdTe along with carrier concentration, composition, and minority carrier lifetime. For materials having acceptor concentrations $< 10^{15}$ cm⁻³, literature results give comparable electron mobilities to those found in n-type HgCdTe [$\mu_e(n)$]. As the acceptor concentration increases, the deviation from n-type electron mobilities increases, resulting in lower electron mobilities for p-type material [$\mu_e(p)$]. Typically, for $x = 0.2$ and $N_a = 10^{16}$ cm⁻³, $\mu_e(p)/\mu_e(n) = 0.5$–0.7, while for $x = 0.2$ and $N_a = 10^{17}$ cm⁻³, $\mu_e(p)/\mu_e(n) = 0.25$–0.33. For $x = 0.3$, however, $\mu_e(p)/\mu_e(n)$ ranges from 0.8 for $N_a = 10^{16}$ cm⁻³ to 0.9 for $N_a = 10^{17}$ cm⁻³ [104]. It was found that at temperatures above 200 K there was very little difference between the electron mobility in epitaxial p-type HgCdTe layers and that measured directly in n-type layers

14.3.3 Optical Properties

Optical properties of HgCdTe have been investigated mainly at energies near the bandgap [25,105]. There still appears to be considerable disagreement among the reported results concerning absorption coefficients. This is caused by different concentrations of native defects and impurities, nonuniform composition and doping, thickness inhomogeneities of samples, mechanical strains, and different surface treatments.

In most compound semiconductors, the band structure closely resembles the parabolic energy versus the momentum dispersion relation. The optical absorption coefficient would then have a square-root dependence on energy that follows the electronic density of states, often referred to as the Kane model [106]. The above bandgap absorption coefficient can be calculated for InSb-like band structure semiconductors such as $Hg_{1-x}Cd_xTe$, including the Moss-Burstein shift effect. Corresponding expressions were derived by Anderson [107]. Beattie and White proposed an analytic approximation with a wide range of applicability for band-to-band radiative transition rates in direct, narrow bandgap semiconductors [108].

In high-quality samples the measured absorption in the SW region is in good agreement with the Kane model calculation while the situation appears to be complicated in the long wavelength edge by the appearance of an absorption tail extending at energies lower than the energy gap.

This tail has been attributed to the composition-induced disorder. According to Finkman and Schacham [109], the absorption tail obeys a modified Urbach's rule:

$$\alpha = \alpha_o \exp\left[\frac{\sigma(E - E_o)}{T + T_o}\right] \text{in cm}^{-1}, \tag{14.10}$$

where T is in K, E is in eV, and $\alpha_o = \exp(53.61x - 18.88)$, $E_o = -0.3424 + 1.838x + 0.148x^2$ (in eV), $T_o = 81.9$ (in K), $\sigma = 3.267 \times 10^4(1 + x)$ (in K/eV) are fitting parameters that vary smoothly with composition. The fit was performed with data at $x = 0.215$ and $x = 1$ and for temperatures between 80 and 300 K.

Assuming that the absorber coefficient for large energies can be expressed as

$$\alpha(h\nu) = \beta(h\nu - E_g)^{1/2}, \tag{14.11}$$

many researchers assume that this rule can be applied to HgCdTe. For example, Schacham and Finkman used the following fitting parameter $\beta = 2.109 \times 10^5[(1 + x)/(81.9 + T)]^{1/2}$, which is a function of composition and temperature [110]. The conventional procedure used to locate the energy gap is to use the point inflection; that is, exploit the large change in the slope of $\alpha(h\nu)$ that is expected when the band-to-band transition overtakes the weaker Urbach contribution. To overcome the difficulty in locating the onset of the band-to-band transition, the bandgap was defined as that energy value where $\alpha(h\nu) = 500$ cm^{-1} [109]. Schacham and Finkman analyzed the crossover point and suggested $\alpha = 800$ cm^{-1} was a better choice [110]. Hougen analyzed absorption data of n-type LPE layers and suggested that the best formula was $\alpha = 100 + 5000x$ [111].

Chu et al. [112] have reported similar empirical formulas for absorption coefficient at the Kane and Urbach tail regions. They received the following modified Urbach rule of the form

$$\alpha = \alpha_o \exp\left[\frac{\delta(E - E_o)}{kT}\right], \tag{14.12}$$

where $\ln \alpha_0 = -18.5 + 45.68x$

$$E_o = -0.355 + 1.77x$$

$$\delta/kT = (\ln \alpha_g - \ln \alpha_o)/(E_g - E_o)$$

$$\alpha_g = -65 + 1.88T + (8694 - 10.315T)x$$

$$E_g(x,T) = -0.295 + 1.87x - 0.28x^2 + 10^{-4}(6 - 14x + 3x^2)T + 0.35x^4.$$

The meaning of the parameter α_g is that $\alpha = \alpha_g$ when $E = E_g$, the absorption coefficient at the bandgap energy. When $E < E_g$, $\alpha < \alpha_g$, the absorption coefficient obeys the Urbach rule in Equation 4.12.

Chu et al. [113] have also found an empirical formula for the calculation of the intrinsic optical absorption coefficient at the Kane region

$$\alpha = \alpha_g \exp\left[\beta(E - E_g)\right]^{1/2}, \tag{14.13}$$

where the parameter β depends on the alloy composition and temperature $\beta(x,T) = -1 + 0.083T + (21 - 0.13T)x$. Expanding Equation 14.13 one finds a linear term, $(E - E_g)^{1/2}$, which fits the square-root law between α and E proper for parabolic bands (see Equation 14.11).

Figure 14.14 shows the intrinsic absorption spectrum for Hg$_{1-x}$Cd$_x$Te with $x = 0.170–0.443$ at temperatures 300 and 77 K. The absorption strength generally decreases as the gap becomes smaller due both to the decrease in the conduction band effective mass and to the $\lambda^{-1/2}$ dependence of the absorption coefficient on wavelength λ. It can be seen that the calculated Kane plateaus according to Sharma and colleagues [114] and Equation 14.13 link closely to the calculated Urbach absorption tail from Equation 14.12 at the turning point α_g. Since the tail effect is not included in the Anderson model [107], the curves calculated according to this model fall down sharply at energies adjacent to E_g. At 300 K, the line shapes derived for absorption coefficient above α_g have almost the same tendency; however, the Chu et al. expression (Equation 14.12) shows better agreement with the experimental data. At 77 K, the curves for the expressions of Anderson and Chu et al.

(a)

(b)

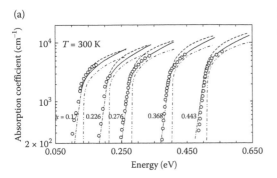

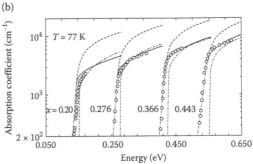

Figure 14.14 Intrinsic absorption spectrum of $Hg_{1-x}Cd_xTe$ samples with $x = 0.170$–0.443 at (a) 300 K, and (b) 77 K. Symbols indicate experimental (Data from Chu, J., Mi, Z., and Tang, D., *Journal of Applied Physics*, 71, 3955–61, 1992; Chu, J., Li, B., Liu, K., and Tang, D., *Journal of Applied Physics*, 75, 1234–35, 1994. With permission), the dash-double-dotted curves are according to Anderson's model; medium dashed curves are (From Sharma, R. K., Verma, D., and Sharma, B. B., *Infrared Physics & Technology*, 35, 673–80, 1994. With permission), solid curves are from Equation 14.13 of Chu et al.; and dash-dotted lines below E_g are from Equation 14.12. (From Li, B., Chu, J. H., Chang, Y., Gui, Y. S., and Tang, D. Y., *Infrared Physics & Technology*, 37, 525–31, 1996. With permission.)

are in agreement with the measurements but discrepancies occur for the empirical parabolic rule of Sharma et al. [114] and these deviations increase with decreasing x. The degree of band nonparabolicity increases as the temperature or x decrease, resulting in increasing discrepancy between the experimental result and the square-root law. In general Chu et al.'s empirical rule and the Anderson model agree well with experimental data for $Hg_{1-x}Cd_xTe$ with x ranging from 0.170 to 0.443 and at temperatures from 4.2 to 300 K, but the Anderson model fails to explain the absorption near E_g [115].

More recently, it has been suggested [116,117] that narrow bandgap semiconductors, such as HgCdTe, more closely resemble a hyperbolic band-structure relationship with an absorption coefficient given by

$$\alpha = \frac{K\sqrt{\left(E - E_g + c\right)^2 - c^2}\left(E - E_g + c\right)}{E}, \tag{14.14}$$

where c is the parameter defining hyperbolic curvature of the band structure and K is the parameter defining the absolute value of absorption coefficient. This theoretical prediction has been recently confirmed by experimental measurements of optical properties of MBE HgCdTe grown samples with uniform compositions [118,119]. The fitting parameters for bandgap tail and hyperbolic regions of the absorption coefficient defined in Equations 4.10 and 14.14 have been extracted by determination of the transition point between both regions. It has been utilized that the fact the derivative of the absorption coefficient has a maximum between the Urbach and hyperbolic regions. Figure 14.15 shows the measured exponential-slope parameter values $\sigma/(T + T_0)$ versus temperature and compares them to values given by Finkman and Schacham [109] for arbitrarily chosen composition of $x = 0.3$. This choice of composition does not have a significant effect on the values obtained, where the values given by Finkman and Schacham have small compositional dependence in the region of interest ($0.2 < x < 0.6$), where the parameter $\sigma/(T + T_0)$ is proportional to $(1 + x)^3$. This parameter shows no clear correlation with composition where there is significant scatter in the data at cryogenic temperatures. The trend of decreasing values with increasing temperature is in agreement with an increase in thermally excited absorption processes, where the values obtained at lower temperatures are more indicative of the quality of the layers grown.

It should be noted that the above discussed expressions do not take into account the influence of doping on the absorption coefficient so they are not very useful in modeling the long wavelength uncooled devices.

$Hg_{1-x}Cd_xTe$ and closely related alloys exhibit significant absorption seen below the absorption edges, which can be related to intraband transitions in both the conduction and valence bands and intervalence band transitions. This absorption does not contribute in optical generation of charge carriers.

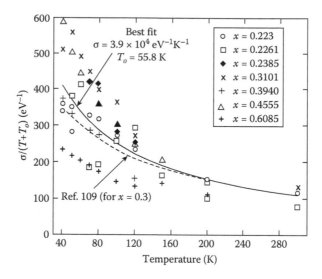

Figure 14.15 Band-tail parameter $\sigma/(T + T_o)$ versus temperature for different compositions and a proposed model based on the best overall fit along with values from (Finkman, E. and Schacham, S. E., *Journal of Applied Physics*, 56, 2896–2900, 1984) for $x = 0.3$. (From Moazzami, K., Phillips, J., Lee, D., Edwall, D., Carmody, M., Piquette, E., Zandian, M., and Arias, J., *Journal of Electronic Materials*, 33, 701–8, 2004. With permission.)

The Kramers and Kronig interrelations are usually used to estimate the dependence of the refractive index on temperature [120–122]. For $Hg_{1-x}Cd_xTe$ with x from 0.276 to 0.540 and temperatures from 4.2 to 300 K the following empirical formula can be used [121]

$$n(\lambda, T)^2 = A + \frac{B}{1 - (C/\lambda)^2} + D\lambda^2,$$ (14.15)

where A, B, C, and D are fitting parameters, which vary with composition x and temperature T. Equation 14.15 can also be used for $Hg_{1-x}Cd_xTe$ with x from 0.205 to 1 at room temperature.

The high frequency dielectric constant, ε_∞, and the static constant, ε_o, are usually derived from reflectivity data in evaluating the real and imaginary parts of ε. The dielectric constants are not a linear function of x and temperature dependence were not observed within the experimental resolution [93]. These dependences can be described by the following relations

$$\varepsilon_\infty = 15.2 - 15.6x + 8.2x^2,$$ (14.16)

$$\varepsilon_o = 20.5 - 15.6x + 5.7x^2.$$ (14.17)

One of the main problems with $Hg_{1-x}Cd_xTe$ detectors is producing a homogeneous material. Recall that the variation of x can be related to the cutoff wavelength by $\lambda_c[\mu m] = 1.24/E_g(x)[eV]$, where E_g is given by Equation 14.4. Substituting and rearranging terms yields

$$\lambda_c[\mu m] = \frac{1}{-0.244 + 1.556x + (4.31 \times 10^{-4})T(1 - 2x) - 0.65x^2 + 0.671x^3}.$$ (14.18)

The differential of Equation 14.17 relates x variation in manufacturing to variation in the cutoff wavelength:

$$d\lambda_c = \lambda_c^2(1.556 - 8.62 \times 10{-}4T - 1.3x + 2.013x^2)dx.$$ (14.19)

Figure 14.16 shows some uncertainty in cutoff wavelength for x variation of 0.1%. This variation in x value is of a very good material. For SW ($\approx$3 μm) and MW ($\approx$5 μm) materials, the variation in cutoff wavelength is not large. However, for the long wavelength materials ($\approx$20 μm) the uncertainty in the cutoff wavelength is large, above 0.5 μm, and cannot be neglected. This response variation causes radiometric-calibration problems in that the radiation is detected over a different spectral region than expected.

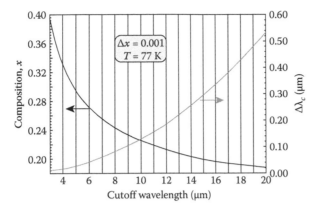

Figure 14.16 The cutoff wavelength variation (right y-axis) of $Hg_{1-x}Cd_xTe$ as a function of cutoff wavelengths (x-axis) with a fixed composition fluctuation of $x = 0.001$ during the growth.

Absorption measurements are possibly the most common routine method to determine and map the composition of bulk crystals and epitaxial layers. Typically, 50% or 1% cut-on wavelengths are used for thick (>0.1 mm) samples [111,123–125]. Various methods have been used for thinner samples. According to Higgins et al. [102] for thick samples,

$$x = \frac{w_n(300K) + 923.3}{10683.98}, \tag{14.20}$$

where w_n is the 1% absolute transmission cut-on wave number. Composition of epitaxial layers is usually determined from a wavelength corresponding to half of the maximum transmission $0.5T_{max}$ [124]. The determination can be complicated by the presence of composition grading.

The UV and visible reflectance measurements are also useful in composition determination, particularly for characterization of the surface (10–30 nm penetration depth) region [125]. Usually, the position of the peak reflectivity at the E_1 bandgap location is measured and the composition is calculated from the experimental expression

$$E_1 = 2.087 + 0.7109x + 0.1421x^2 + 0.3623x^3. \tag{14.21}$$

14.3.4 Thermal Generation–Recombination Processes

The generation processes that compete against the recombination processes directly affect the performance of photodetectors, setting up a steady-state concentration of carriers in semiconductors subjected to the thermal and optical excitation and, frequently, determining the kinetics of photogenerated signals. Generation-recombination processes in semiconductors are widely discussed in literature (see, for example, [73,126,127]). Here, we reproduce only some dependencies directly related to the performance of photodetectors. Assuming bulk processes only, there are three main thermal generation–recombination processes to be considered in the narrow bandgap semiconductors, namely: Shockley–Read (SR), radiative, and Auger.

14.3.4.1 Shockley–Read Processes

The SR mechanism is not an intrinsic and fundamental process as it occurs via levels in the forbidden energy gap. The reported positions of SR centers for both n- and p-type materials range anywhere from near the valence to near the conduction band.

The SR mechanism is probably responsible for lifetimes in lightly doped n- and p-type $Hg_{1-x}Cd_xTe$. The possible factors are SR centers associated with native defects and residual impurities. In n-type material ($x = 0.20$–0.21, 80 K) with carrier concentrations less than 10^{15} cm^{-3}, the lifetimes exhibit a broad range of values (0.4–8 µs) for material prepared by various techniques [127]. Dislocations may also influence the recombination time for dislocation densities $> 5 \times 10^5$ cm^{-2} [128–130].

In p-type HgCdTe SR mechanism is usually blamed for reduction of lifetime with decreasing temperature. The steady-state, low-temperature photoconductive lifetimes are usually much shorter than the transient lifetimes. The low-temperature lifetimes exhibit very different temperature dependencies with a broad range of values over three orders of magnitude, from 1 ns to 1 µs

$(p \approx 10^{16}$ cm^{-3}, $x \approx 0.2$, $T \approx 77$ K, vacancy doping) [127,131]. This is due to many factors, which may affect the measured lifetime including inhomogeneities, inclusions, surface, and contact phenomena. The highest lifetime was measured in high-quality undoped and extrinsically doped materials grown by low temperature epitaxial techniques from Hg-rich LPE [47] and MOCVD [83,132,133]. Typically, Cu or Au doped materials exhibit lifetimes one order of magnitude larger compared to vacancy doped ones of the same hole concentration [131]. It is believed that the increase of lifetime in impurity doped Hg$_{1-x}$Cd$_x$Te arises from a reduction of SR centers. This may be due to the low-temperature growth of doped layers or due to low-temperature annealing of doped samples.

The origin of the SR centers in vacancy-doped, p-type material is not clear at present. These centers seem not to be the vacancies themselves and thus may be removable [134]. Vacancy doped material with the same carrier concentration, but created under different annealing temperatures, may produce different lifetimes. One possible candidate for recombination centers is Hg interstitials [135]. Vacancy doped Hg$_{1-x}$Cd$_x$Te exhibits SR recombination center densities roughly proportional to the vacancy concentration.

Measurements at DSR [136] give lifetime values for extrinsic p-type material

$$\tau_{ext} = 9 \times 10^9 \frac{p_1 + p}{pN_a},$$

(14.22)

where

$$p_1 = N_v \exp\left(\frac{q(E_r - E_g)}{kT}\right),$$

(14.23)

and E_r is the SR center energy relative to the conduction band. Experimentally E_r was found to lie at the intrinsic level for As, Cu, and Au dopants, giving $p_1 = n_i$.

For vacancy doped p-type Hg$_{1-x}$Cd$_x$Te

$$\tau_{vac} = 5 \times 10^9 \frac{n_1}{pN_{vac}},$$

(14.24)

where

$$n_1 = N_c \exp\left(\frac{qE_r}{kT}\right).$$

(14.25)

E_r is ≈ 30 mV from the conduction band ($x = 0.22$–0.30).

As follows from these expressions and Figure 14.17 [137], doping with the foreign impurities (Au, Cu, and As for p-type material) gives lifetimes significantly increased compared to native doping of the same level.

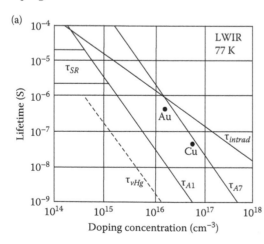
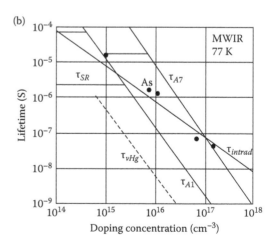

Figure 14.17 Measured lifetimes for n- and p-type (a) LWIR and (b) MWIR at 77 K, compared to theory for Auger 1, Auger 7, Shockley–Read, and internal radiative recombination, as a function of doping concentration. (From Kinch, M. A., Aqariden, F., Chandra, D., Liao, P.-K., Schaake, H. F., and Shih, H. D., *Journal of Electronic Materials*, 34, 880–84, 2005. With permission.)

Although a considerable research effort is still necessary, the SR process does not represent a fundamental limit to the performance of the photodetectors.

14.3.4.2 Radiative Processes

Radiative generation of charge carriers is a result of absorption of internally generated photons. The radiative recombination is an inversed process of annihilation of electron-hole pars with emission of photons. The radiative recombination rates were calculated for conduction-to-heavy-hole-band and conduction-to-light-hole-band transitions using an accurate analytical form [108].

For a long time, internal radiative processes have been considered to be the main fundamental limit to detector performance and the performance of practical devices has been compared to that limit. The role of radiative mechanism in the detection of IR radiation has been critically reexamined [138–140]. Humpreys [139] indicated that most of the photons emitted in photodetectors as a result of radiative decay are immediately reabsorbed, so that the observed radiative lifetime is only a measure of how well photons can escape from the body of the detector. Due to reabsorption the radiative lifetime is highly extended, and dependent on the semiconductor geometry. Therefore, internal combined recombination-generation processes in one detector are essentially noiseless. In contrast, the recombination act with cognate escape of a photon from the detector, or generation of photons by thermal radiation from outside the active body of the detector are noise producing processes. This may readily happen for a case of detector array, where an element may absorb photons emitted by other detectors or a passive part of the structure [140,141]. Deposition of the reflective layers (mirrors) on the back and side of the detector may significantly improve optical insulation preventing noisy emission and absorption of thermal photons.

It should be noted that internal radiative generation could be suppressed in detectors operated under reverse bias where the electron density in the active layer is reduced to well below its equilibrium level [142,143].

As follows from the above considerations, the internal radiative processes, although of fundamental nature, do not limit the ultimate performance of infrared detectors.

14.3.4.3 Auger Processes

Auger mechanisms dominate generation and recombination processes in high-quality narrow gap semiconductors such as $Hg_{1-x}Cd_xTe$ and InSb at near room temperatures [144,145]. The Auger generation is essentially the impact ionization by electrons of holes in the high energy tail of Fermi-Dirac distribution. The band-to-band Auger mechanisms in InSb-like band structure semiconductors are classified in 10 photonless mechanisms. Two of them have the smallest threshold energies ($E_T \approx E_g$) and are denoted as Auger 1 (A1) and Auger 7 (A7; see Figure 14.18). In some wider bandgap materials (e.g., InAs and low x $InAs_{1-x}Sb_x$) in which the split-off band energy Δ is comparable to E_g, and the Auger process involving split-off band (AS process) may also play an important role.

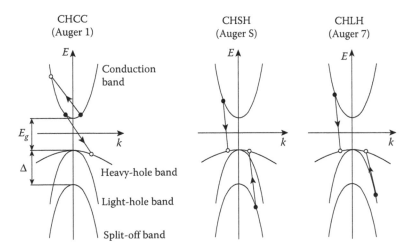

Figure 14.18 The three band-to-band Auger recombination processes. Arrows indicate electron transitions; •, occupied state; ○, unoccupied state.

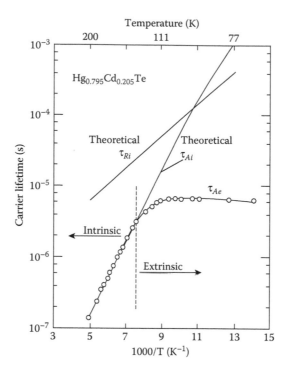

Figure 14.19 Theoretical and experimental lifetime data versus temperature for n-type $Hg_{0.795}Cd_{0.205}Te$ ($n_{77K} = 1.7 \times 10^{14}$ cm^{-3}, $\mu_{77K} = 1.42 \times 10^5$ cm^2/Vs), solid lines represent theoretical values for radiative and Auger 1 recombination, respectively. (From Kinch, M. A., Brau, M. J., and Simmons A., *Journal of Applied Physics*, 44, 1649–63, 1973. With permission.)

The A1 generation is the impact ionization by an electron, generating an electron-hole pair, so this process involves two electrons and one heavy hole. It is well known that the Auger 1 process is an important recombination mechanism in n-type $Hg_{1-x}Cd_xTe$, particularly for x around 0.2 and at higher temperatures [73,127,146,147]. In Figure 14.19 experimental results from Kinch et al. [146] are compared with theoretical data of the intrinsic Auger 1 carrier lifetime τ^i_{A1} and intrinsic radiative carrier lifetime τ^i_R. There is excellent agreement between the experimental and numerical data. Even in the extrinsic range at temperatures below 140 K the measured lifetime seems to be governed by the Auger 1 effect, where the following relation should hold: $\tau_{A1} \approx 2\tau^i_{A1}(n_i/n_o)^2$. An interesting feature is the behavior of Auger 1 generation and recombination with degenerate n-type doping. Due to the low density, the Fermi level moves high into the conduction band with n-type doping, so the concentration of minority holes is strongly reduced and the threshold energy required for the Auger transition increased. This results in suppression of Auger 1 processes in heavy doped n-type material.

Auger 7 generation is the impact generation of electron hole pair by a light hole, involving one heavy hole, one light hole, and one electron [148–150]. This process may dominate in p-type material. Heavy p-type doping has no dramatic effect on the Auger 7 generation and recombination rates due to the much higher density of states. The corresponding Auger recombination mechanisms are inverse processes of electron-hole recombination with energy transferred to electron or hole. Strong temperature and bandgap dependence is expected, since lowered temperature and increased bandgap strongly reduces the probability of these heat-stimulated transitions.

The net generation rate due to the Auger 1 and Auger 7 processes can be described as [151]

$$G_A - R_A = \frac{n_i^2 - np}{2n_i^2}\left[\frac{n}{(1+an)\tau^i_{A1}} + \frac{p}{\tau^i_{A7}}\right], \tag{14.26}$$

where τ^i_{A1} and τ^i_{A7} are the intrinsic Auger 1 and Auger 7 recombination times, and n_i is the intrinsic concentration. The last equation is valid for a wide range of concentrations, including degeneration, which easily occurs in n-type materials. This is expressed by the finite value of a. According to White [151], $a = 5.26 \times 10^{-18}$ cm³. Due to the shape of the valence band, the degeneracy in p-type material occurs only at very high doping levels, which is not achievable in practice.

The Auger 1 intrinsic recombination time is equal [144]

$$\tau_{A1}^i = \frac{h^3 \varepsilon_o^2}{2^{3/2} \pi^{1/2} q^4 m_o} \frac{\varepsilon^2 (1+\mu)^{1/2} (1+2\mu) \exp\left[\left(\frac{1+2\mu}{1+\mu}\right)\frac{E_g}{kT}\right]}{(m_e^* / m)|F_1 F_2|^2 (kT / E_g)^{3/2}}, \tag{14.27}$$

where μ is the ratio of the conduction to the heavy-hole, valence-band effective mass, ε_s is the static-frequency dielectric constant, and $|F_1 F_2|$ are the overlap integrals of the periodic part of the electron wave functions. The overlap integrals cause the biggest uncertainly in the Auger 1 lifetime. Values ranging from 0.1 to 0.3 have been obtained by various authors. In practice it is taken as a constant equal to anywhere between 0.1 and 0.3 leading to changes by almost an order of magnitude in the lifetime.

The ratio of Auger 7 and Auger 1 intrinsic times

$$\gamma = \frac{\tau_{A7}^i}{\tau_{A1}^i}, \tag{14.28}$$

is another term of high uncertainty. According to Casselman et al. [148,149], for $Hg_{1-x}Cd_xTe$ over the range $0.16 \le x \le 0.40$ and 50 K $\le T \le 300$ K, $3 \le \gamma \le 6$. Direct measurements of carrier recombination show the ratio γ larger than expected from previous calculations (≈ 8 for $x \approx 0.2$ at 295 K) [152]. Accurate calculations of the Auger lifetimes have been reported by Beattie and White [153]. The flat valence band model has been used to obtain a simple analytic approximation that requires just two parameters to cover a wide range of temperature and carrier Fermi levels, both degenerated and nondegenerate. More recently published theoretical [154,155] and experimental [137,155] results indicate that this ratio is even higher; the data presented in Figure 14.17 would indicate a value about 60. As γ is higher than unity, higher recombination lifetimes are expected in p-type materials compared to n-type materials of the same doping.

Kinch [136] delivered simplified formula for the Auger 1 intrinsic recombination time

$$\tau_{A1}^i = 8.3 \times 10^{-13} E_g^{1/2} \left(\frac{q}{kT}\right)^{3/2} \exp\left(\frac{qE_g}{kT}\right), \tag{14.29}$$

where E_g is in eV.

As the Equations 14.26 and 14.28 show, the Auger generation and recombination rates are strongly dependent on temperature via dependence of carrier concentration and intrinsic time on temperature. Therefore, cooling is a natural and a very effective way to suppress Auger processes.

Until recently, the n-type Auger 1 lifetime was deemed to be well established. Krishnamurthy et al. [155] have shown that the full band calculations indicate that the radiative and Auger recombination rates are much slower than those predicted by expressions used in the literature (theory of Beattie and Landsberg [144]). It appears that a trap state tracking the conduction band edge with very small activation energy can explain the lifetimes in n-doped MBE samples.

The p-type Auger 7 lifetime has long been subject to controversy. Detailed calculations of the Auger lifetime in p-type HgCdTe reported by Krishnamurthy and Casselman [154] suggest significant deviation from classic $\tau_{A7} \sim p^{-2}$ relation. The decrease of τ_{A7} with doping is much weaker resulting in significantly longer lifetimes in highly doped p-type low x materials (factor of ≈ 20 for $p = 1 \times 10^{17}$ cm⁻³, $x = 0.226$ and $T = 77 \div 300$ K).

14.4 AUGER-DOMINATED PHOTODETECTOR PERFORMANCE

14.4.1 Equilibrium Devices

Let us consider the Auger limited detectivity of photodetectors. At equilibrium the generation and recombination rates are equal. Assuming that both rates contribute to noise (see Equation 3.45),

$$D^* = \frac{\lambda \eta}{2hc(G_A t)^{1/2}} \left(\frac{A_o}{A_e}\right)^{1/2}. \tag{14.30}$$

If nondegenerate statistic is assumed, then

$$G_A = \frac{n}{2\tau_{A1}^1} + \frac{p}{2\tau_{A7}^1} = \frac{1}{2\tau_{A1}^1}\left(n + \frac{p}{\gamma}\right).$$

(14.31)

The resulting Auger generation achieves its minimum in just extrinsic p-type materials with $p = \gamma^{1/2}n_i$. It leads to an important conclusion about optimum doping for the best performance. In practice, the required p-type doping level would be difficult to achieve for LN-cooled and SW devices. Moreover, the p-type devices are more vulnerable to nonfundamental limitations (contacts, surface, SR processes) than the n-type ones. This is the reason why low temperature and the SW photodetectors are typically manufactured from lightly doped n-type materials. In contrast, just p-type doping is clearly advantageous for the near room temperature and long wavelength photodetectors.

The Auger dominated detectivity is

$$D^* = \frac{\lambda}{2^{1/2}hc}\left(\frac{A_o}{A_e}\right)^{1/2}\frac{\eta}{t^{1/2}}\left(\frac{\tau_{A1}^i}{n+p/\gamma}\right)^{1/2}.$$

(14.32)

As follows from Equation 3.52, for the optimum thickness devices,

$$D^* = 0.31k\frac{\lambda\alpha^{1/2}}{hc}\frac{\left(2\tau_{A1}^i\right)^{1/2}}{\left(n+p/\gamma\right)}.$$

(14.33)

This expression can be used for determination of optimum detectivity of Auger 1/Auger 7 as a function of wavelength, material bandgap, and doping.

To estimate the wavelength and temperature dependence of D^* let us assume a constant absorption for photons with energy equal to the bandgap. For extrinsic materials ($p = N_a$ or $n = N_d$)

$$D^* \sim \left(\tau_{A1}^i\right)^{1/2} \sim n_i^{-1} \sim \exp\left(\frac{E_g}{2kT}\right) = \exp\left(\frac{hc/2\lambda_c}{kT}\right).$$

(14.34)

In this case the ultimate detectivity will be inversely proportional to intrinsic concentration. This behavior should be expected at shorter wavelength and lower temperatures when the intrinsic concentration is low.

For intrinsic materials and for materials doped for the minimum thermal generation where $p = \gamma^{1/2}n_i$, $n = n_i/\gamma^{1/2}$ and $n + p = 2\gamma^{-1/2}n_i$, stronger $D^* \sim n_i^{-2}$ dependence can be expected:

$$D^* \sim \frac{\left(\tau_{A1}^i\right)^{1/2}}{n_i} \sim n_i^{-2} \sim \exp\left(\frac{E_g}{kT}\right) = \exp\left(\frac{hc/\lambda_c}{kT}\right).$$

(14.35)

Figure 14.20a shows the calculated detectivity of Auger generation-recombination limited $Hg_{1-x}Cd_xTe$ photodetectors as a function of wavelength and temperature of operation [157]. The calculations have been performed for 10^{14} cm^{-3} doping, which is the lowest donor doping level that at present is achievable in a controllable manner in practice. Values as low as $\approx 1 \times 10^{13}$ cm^{-3} are at present achievable in labs, while values of 3×10^{14} cm^{-3} are more typical in industry. Liquid nitrogen cooling potentially makes it possible to achieve BLIP (300 K) performance over the entire 2–20 μm range. The 200 K cooling, which is achievable with Peltier coolers, would be sufficient for BLIP operation in the middle and SW regions (<5 μm).

The detector performance can be improved further by the use of optical immersion. However, the theoretical limit of detectivity for uncooled detectors remains about one or almost two orders of magnitude below $D^*_{BLIP}(300\ K, 2\pi)$ for ≈ 5 μm and 10 μm wavelength, respectively. Improvement by a factor of ≈ 2 still is possible with the optimum p-type doping.

14.4.2 Nonequilibrium Devices

Auger generation appeared to be a fundamental limitation to the performance of infrared photodetectors. However, British workers [158,159] proposed a new approach to reducing the photodetector cooling requirements, which is based on the nonequilibrium mode of operation. This is one of the most exciting events in the field of IR photodetectors operating without cryogenic cooling.

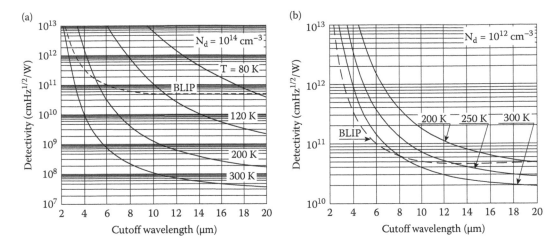

Figure 14.20 Calculated performance of (a) equilibrium and (b) nonequilibrium Auger generation-recombination limited $Hg_{1-x}Cd_xTe$ photodetectors as a function of wavelength and temperature of operation. BLIP detectivity has been calculated for 2π FOV, $T_B = 300$ K, $\eta = 1$.

Their concept relies on dependence of the Auger processes on concentration of free carriers. The suppression of Auger processes may be achieved by decreasing the free carrier concentration below equilibrium values. Nonequilibrium depletion of semiconductors can be applied to reduce concentration of the majority and minority carrier concentration. This can be achieved in some devices based on lightly doped narrow gap semiconductors, for example, in biased low-high (l-h) doped or heterojunction contact structures, MIS structures or using the magnetoconcentration effect. Under strong depletion the concentration of both majority and minority carriers can be reduced below the intrinsic concentration. The majority carrier concentration saturates at the extrinsic level while the concentration of minority carriers is reduced below the extrinsic level. Therefore, the necessary condition for deep depletion is a very light doping of the semiconductor, below the intrinsic concentration.

Let us discuss fundamental limits of performance of Auger suppressed devices. At first, consider detector based on v-type material. At strong depletion, $n = N_d$, and the Auger generation rate is

$$G_A = \frac{N_d}{2\tau_{A1}^i}. \tag{14.36}$$

As follows from Equation 14.36, deep depletion makes the recombination rate negligible compared to the generation rate so it is readily eliminated in as a noise source in depleted materials. Therefore

$$D^* = \frac{\lambda}{hc} \frac{\eta}{t^{1/2}} \left(\frac{\tau_{A1}^i}{N_d}\right)^{1/2}. \tag{14.37}$$

Similarly, for π-type materials,

$$G_A = \frac{N_a}{2\gamma\tau_{A1}^i}, \tag{14.38}$$

and

$$D^* = \frac{\lambda}{hc} \frac{\eta}{t^{1/2}} \left(\frac{\tau_{A1}^i}{N_a/\gamma}\right)^{1/2}. \tag{14.39}$$

Again as in the equilibrium case, the use of π-type material is advantageous, improving the detectivity by a factor of $\gamma^{1/2}$ compared to v-type material of the same doping ($\gamma > 1$).

Comparing the corresponding equations for equilibrium and nonequilibrium modes, we can find that the use of the nonequilibrium mode of operation may reduce the Auger generation rate by a factor n_i/N_d in lightly doped material with corresponding improvement of detectivity by $(2n_i/N_d)^{1/2}$. The additional gain factor of $2^{1/2}$ is due to the negligible recombination rate in the depleted semiconductor. The gain for p-type material is even larger—a factor of $[2(\gamma + 1)n_i/N_a]^{1/2}$, taking into account elimination of Auger 1 and Auger 7 recombination. Additional depletion-related improvement can be also expected from increased absorption due to reduced band-filling effect.

The resulting improvement may be quite large, particularly for LWIR devices operating at near room temperatures as shown in Figure 14.20b for very low doping, 10^{12} cm^{-3}. Potentially, the BLIP performance can be obtained without cooling at all. The BLIP limit can be achieved, by [157]

- using materials with controlled doping at very low levels ($\approx 10^{12}$ cm^{-3}),

- using extremely high quality materials with a very low concentration of SR centers,

- proper design of a device that prevents thermal generation at surfaces, interfaces, and contacts, and

- using a thermal dissipation device whose design makes it possible to achieve a state of strong depletion.

The requirements for the BLIP performance, particularly doping concentration, can be significantly eased by the use of optical immersion.

14.5 PHOTOCONDUCTIVE DETECTORS

The first results on photoconductivity in $Hg_{1-x}Cd_xTe$ were reported by Lawson et al. in 1959 [1]. Ten years later, in 1969, Bartlett et al. reported background-limited performance of photoconductors operated at 77 K in the 8–14 µm LWIR spectral region [10]. The advances in material preparation and detector technology have led to devices approaching theoretical limits of responsivity over wide ranges of temperature and background [11,160–162]. The largest market was for 60-, 120- and 180-element units in the "Common Module" military thermal imaging viewers. Photoconductivity was the most common mode of operation of 3–5 µm and 8–14 µm $Hg_{1-x}Cd_xTe$ n-type photodetectors for many years.

In 1974 Elliott reported a major advance in infrared detectors in which the detection, time delay, and integration functions in serial scan thermal imaging systems are performed within a simple three-lead filament photoconductor, SPRITE (Signal PRocessing In The Element) [126].

The further development of photoconductors is connected with elimination of the deleterious effect of sweep-out [163,164] by the application of accumulated [165–167] or heterojunction [168,169] contacts. Heterojunction passivation has been used to improve stability [170,171]. The operation of 8–14 µm photodetectors has been extended to ambient temperatures [157,172–175]. Means applied to improve the performance of photoconductors operated without cryogenic cooling include the optimized p-type doping, the use of optical immersion, and optical resonant cavities. Elliott and other British scientists introduced Auger suppressed excluded photoconductors [158,159].

The research activity on photoconductors has been significantly reduced in the last two decades on reflecting maturity of the devices. At the same time $Hg_{1-x}Cd_xTe$ photodetectors are still manufactured in large quantities and used in many important applications.

The physics and principle of operation of intrinsic photoconductors are summarized in Section 9.1.1. HgCdTe photoconductive detectors have been reviewed by a number of authors [126,142,157,172,176–178].

14.5.1 Technology

The photoconductors can be prepared either from $Hg_{1-x}Cd_xTe$ bulk crystals or epilayers. Figure 14.21 shows a typical structure of photoconductor. The main part of all structures is a 3–20 µm flake of $Hg_{1-x}Cd_xTe$, supplied with electrodes. The optimum thickness of the active elements (few µm) depends upon the wavelength and temperature of operation and is smaller in uncooled long wavelength devices. The front side surface is usually covered with a passivation layer and AR coating. The back side surface of the device is also being passivated. In contrast, the back side surface of the epitaxial layer grown on the CdZnTe substrate does not require any passivation since increasing bandgap prevents reflects minority carriers. The devices are bonded to heat-conductive substrates.

To increase the absorption of radiation the detectors are frequently supplied with a gold back reflector [172,176], insulated from the photoconductor with a ZnS layer or substrate. The thickness

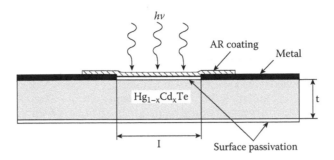

Figure 14.21 Typical structure of HgCdTe photoconductor.

of the semiconductor and two dielectric layers is frequently selected to establish optical resonant cavities with standing waves in the structure with peaks at the front and nodes at the back surfaces. For effective interference the two surfaces must be sufficiently flat.

Numerous fabrication procedures are being used by various manufacturers [172,176,179–182]. Technological methods of modern microelectronics are used in $Hg_{1-x}Cd_xTe$ photoconductor manufacturing, however, extreme care is necessary to prevent any mechanical and thermal damage of the material. Fabrication usually starts from selection of the starting materials, which are $Hg_{1-x}Cd_xTe$ wafers or epilayers. For selection, they are typically characterized by composition, doping, and minority lifetime mapping.

The key processing steps for manufacturing $Hg_{1-x}Cd_xTe$ photoconductors from bulk $Hg_{1-x}Cd_xTe$ include:

- Preparation of the back side surface of the $Hg_{1-x}Cd_xTe$ wafer. The procedure involves careful polishing of one side of the $Hg_{1-x}Cd_xTe$ slab with fine (0.3–1 μm) alumina powder, cleaning in organic solvents followed by etching in 1%–10% bromine in methanol solution for several minutes and washing in methanol. Alternatively, various chemical-mechanical polishing procedures can be used. The back side preparation is completed with passivation, which differs for n- and p-type materials.

- Bonding to a substrate. Sapphire, germanium, Irtran 2, silicon, and alumina ceramics are the most common substrates for bulk-type photoconductors. Epoxy resin is generally used for bonding of $Hg_{1-x}Cd_xTe$ wafers to substrates. The thickness of the epoxy layer should be kept below 1 μm for good heat dissipation.

- Thinning the slab to its final thickness and preparation of the front side surface. This is done by lapping, polishing, and etching, followed by surface passivation and an antireflection coating (usually ZnS). The individual elements are then delineated with wet or dry etching with the use of photolithography. Side walls of active elements are often also passivated.

- Electric contact preparation. Vacuum evaporation, sputtering, electroplating, and galvanic or chemical metallization after further photolithography is used. External contacts are supplied by gold wire ultrasonic bonding, conductive epoxy bonding, or by soldering with indium. Expanded contact pads are sometimes used to prevent damage of the semiconductor.

Preparation of a photoconductor from epilayers is more straightforward as no laborious thinning to a very low thickness and no back side preparation is required. The use of ISOVPE [172,183], LPE [184–187], MOCVD [173,174,188–191], and MBE [192] photoconductors have been reported. The CdZnTe, which is typically used for substrates, has relatively poor thermal conductivity. Therefore, for the best heat dissipation, the substrates must be thinned below 30 μm and the photoconductors must be fixed to a good heat sinking support. The heat dissipation is much easier in small size devices (<50 × 50 μm^2) in which 3-D dissipation is significant and no thinning of substrates is necessary. Another solution is to use epilayers deposited on good heat conducting materials such as sapphire, silicon, and GaAs.

The low temperature epitaxial techniques make it possible to grow complex multilayer photoconductor structures that can be used as multicolor devices or devices with shaped spectral response [173,191].

Passivation is one of the most critical steps in the preparation of photoconductors. The passivation must seal the semiconductor, stabilizing it chemically, and often it also acts as an antireflection coating. An excellent review of $Hg_{1-x}Cd_xTe$ native and deposited insulator layers has been published by Nemirovsky and Bahir [193]. Passivation of n-type materials is commonly performed with the use of anodic oxidation in 0.1 N KOH in a 90% water solution of ethyl glycol [194–196]. Typically, 100 nm thick oxide layers are grown. Good interface properties of the $n-Hg_{1-x}Cd_xTe$-oxide interface are due to accumulation (10^{11}–10^{12} electrons per cm^{-2}) of the semiconductor surface during oxidation. Passivation by pure chemical oxidation in an aqueous solution of $K_3Fe(CN)_6$ and of KOH is also used [197]. Dry methods of native oxide growing have been attempted such as plasma [198] and photochemical oxidation [199]. Passivation can be improved by overcoating with ZnS or SiO_x layers [200]. Another approach to passivation is based on direct accumulation of the surface to repel minority holes by a shallow ion milling [201,202].

Passivation of p-type materials is of strategic importance for near room temperature devices based on p-type absorbers. It should be admitted, the passivation still presents practical difficulties. Oxidation is not useful for p-type $Hg_{1-x}Cd_xTe$ because it causes inversion of the surface. In practice, sputtered or electron beam evaporated ZnS, with an option of a second layer coating [203], is usually used for passivation of p-type materials. Native sulfides [204] and fluorides [205] have also been proposed.

The use of CdTe for passivation is very promising since it has high resistivity, is lattice matched, and is chemically compatible to $Hg_{1-x}Cd_xTe$ [206–208]. Excellent passivation can be obtained with a graded $CdTe-Hg_{1-x}Cd_xTe$ interface [203]. Barriers can be found in both the conduction and valence band. The best heterojunction passivation can be obtained during epitaxial growth [209]. Directly grown *in situ* CdTe layers lead to low fixed interface charge. The indirectly grown CdTe passivation layer are not as good as the directly grown, but acceptable in some applications. Low thickness (10 nm) of CdTe is recommended in some papers to prevent $Hg_{1-x}Cd_xTe$ lattice stress [193].

Contact preparation is another critical step. Evaporated indium has been used for a long time for contact metallization to n-type material [176,179]. Multilayer metallization, Cr-Au, Ti-Au, Mo-Au, is more frequently used at present. Metallization is often preceded by a suitable surface treatment. Ion milling was found very useful to accumulate n-type surfaces and it seems to be the most preferable surface treatment prior to metallization of n-type material. Chemical and dry etch are also used. Preparation of good contacts to p-type material is more difficult. Evaporated, sputtered, or electroless deposited Au and Cr-Au are the most frequently used for contacts to p-type materials.

14.5.2 Performance of Photoconductive Detectors

14.5.2.1 Devices for Operation at 77 K

HgCdTe photoconductive detectors operating at 77 K in the 8–14 µm range are widely used in the first generation thermal imaging systems in linear arrays of up to 200 elements, although custom 2-D arrays up to 10 × 10 have been made for unique applications. The production processes of these devices are well established. The material used is n-type with an extrinsic carrier density of about 1×10^{14} cm^{-3}. The low hole diffusion coefficient makes n-type devices less vulnerable to contact and surface recombination. In addition, n-type materials exhibit a lower concentration of SR centers and there are good methods of surface passivation.

Commercially available HgCdTe photoconductive detectors are typically manufactured in a square configuration with active size from 25 µm to 4 mm. The length of the photoconductors being used in high-resolution thermal imaging systems ($\approx$ 50 µm) is typically less than the minority carrier diffusion and drift length in cooled HgCdTe, resulting in reduction of photoelectric gain due to diffusion and drift of photogenerated carriers to the contact regions, called sweep-out effect [126,163,164,166,176,178,210]. This causes the "saturation" of response with increasing electric field. The behavior of a typical device, showing the saturation in responsivity (at about 10^5 V/W) is shown in Figure 14.22 [126].

The n-type HgCdTe photoconductive detectors (with $E_g \approx 0.1$ eV at 77 K) approaching theoretical limits of performance have been described by Kinch et al. [11,160,161], Borello et al. [162], and Siliquini et al. [170,187,211]. Their generation and recombination carrier mechanisms are clearly dominated by the Auger 1 mechanism. Background radiation has a decisive influence on performance since the concentration of both majority and minority carriers in 77 K 8–14 µm devices and the concentration of minority carriers in 3–5 µm devices are typically determined by background flux. Near-BLIP performance can also be achieved at elevated temperatures, up to about 200 K [161,212]. Figure 14.23 shows the influence of 300 K background photon flux on photoconductor

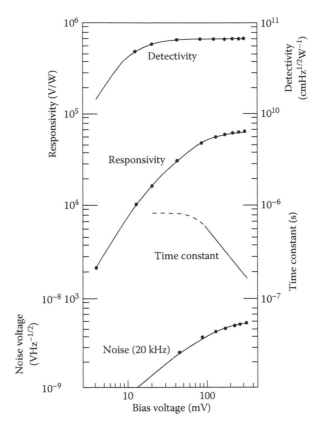

Figure 14.22 Characteristics of a 50 μm HgCdTe photoconductive detector operated at 80 K as a function of voltage. The measurements were made in 30° FOV and the responsivity values refer to the peak wavelength response at 12 μm. (From Elliott, C. T., and Gordon, N. T., *Handbook on Semiconductors*, North-Holland, Amsterdam, Vol. 4, 841–936, 1993. With permission.)

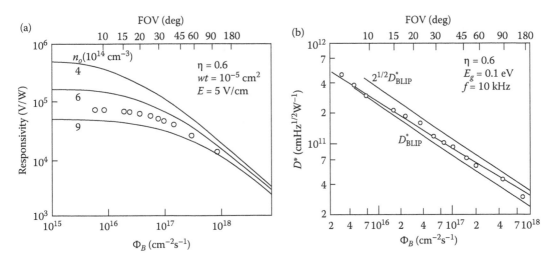

Figure 14.23 Dependence of voltage (a) responsivity and (b) detectivity on background photon flux for a 0.1 eV HgCdTe photoconductive detector. (From Borrello, S. R., Kinch, M., and Lamont, D., *Infrared Physics*, 17, 121–25, 1977. With permission.)

parameters [162]. The density of background-generated holes and, for high fluxes, also electrons may dominate the thermally generated carriers decreasing the recombination time. The effects of background radiation tend to override any nonuniformities that might be present in the bulk material, with regard to element resistance, responsivity, and noise.

Figure 14.24 shows the calculated and measured low-background responsivity and detectivity of a photoconductor as a function of temperature [11]. The generation and recombination rates are clearly dominated by the Auger 1 mechanism. The 77 K detectivity achieves a value of about 10^{12} cmHz$^{1/2}$/W, closely approaching limits predicted by theory.

The dependence of the voltage responsivity on detector length for 8–14 μm HgCdTe photoconductors operated at 77, 200, and 300 K is shown in Figure 14.25 [210]. The marked region inside the figure indicates ranges of voltage responsivity for detector series produced by Judson Infrared, Infrared Associates, and Vigo-System. Improved performance of devices operated at 200 and 300 K are achieved assuming p-type doping materials with hole concentrations $p \approx \gamma^{1/2} n_i$. In the case of ohmic contacts (Rittner model) sweep-out effect significantly reduces responsivity of detector with short active sizes operated in 77 K. This effect is negligible at 300 K. In the range of detector lengths below 100 μm, the experimental results exceed theoretical calculations based on the Rittner model. It may be conditioned by intentional or coincidental processing procedures in which contacts deviate from ohmicity. Application of a high-low doping contact barrier leads to enhancement in responsivity of photoconductive detectors. Ashley and Elliott [166] have shown that by means of a n$^+$-n ion milled contact the responsivity can be enhanced by a factor of five.

The positive feature of sweep-out is an improvement of the high-frequency characteristics. In the high-bias condition the response time is determined by the transit time of the minority carriers between the electrodes rather than the excess-carrier lifetime. Since the recombination processes are partially arranged to take place in the contact region, they do not contribute to recombination noise. As a result, the g-r noise decreases by a factor of $2^{1/2}$ and the sweep-out/g-r-limited devices may exhibit improved detectivity by the same factor. Under high bias D^*_{BLIP} is identical with the photovoltaic case.

The reduced gain may, however, cause the Johnson-Nyquist noise to dominate with deleterious effect to detectivity. Sweep-out has been recognized as a major limitation to the 8–14 μm photoconductor performance, when they are short and operated at low temperatures with low background radiation. The influence of sweep-out is even stronger for shorter wavelength devices. For $\lambda < 5$ μm photoconductors with low generation-recombination rates, these effects become significant at low and elevated temperatures, even for relatively long devices, making the devices Johnson-Nyquist/sweep-out limited.

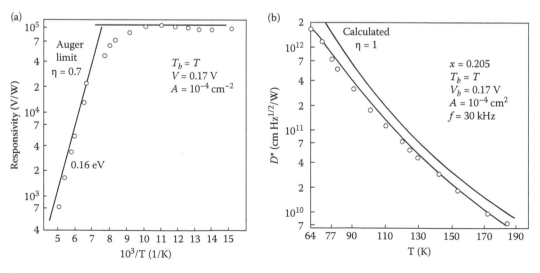

Figure 14.24 Measured and calculated (a) responsivity and (b) detectivity versus temperature for a Hg$_{0.795}$Cd$_{0.205}$Te photoconductor. (From Kinch, M. A., Borrello, S. R., and Simmons, A., *Infrared Physics*, 17, 127–35, 1977. With permission.)

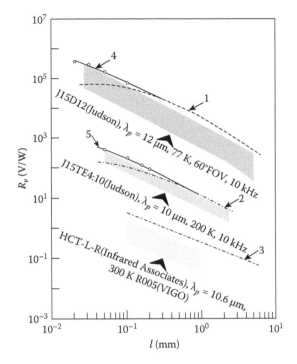

Figure 14.25 Dependence of the voltage responsivity on detector length for 8–14 μm HgCdTe photoconductive detectors operated at 77, 200, and 300 K. The marked regions indicate ranges of voltage responsivity for detector series produced by different manufacturers. The theoretical curves 1–3 are calculated using the Rittner model (see Section 9.1.1); the curves 4 and 5 are calculated assuming a high-low contact regions. (From Jóźwikowska, A., Jóźwikowski, K., and Rogalski, A., *Infrared Physics*, 31, 543–54, 1991. With permission.)

Various approaches to reduce the undesirable sweep-out effects in devices with small sensitive areas have been suggested. Kinch et al. [161] proposed overlap structure geometry for which a device length is greater than the required sensitive length. The end regions are covered with an opaque layer. By this means the effects of carrier diffusion to the contacts, at low bias, or minority-hole sweep-out at the cathode, at high bias, are reduced. The theory of the overlap structure was reported by Smith [213]. According to Shacham-Diamond and Kidron [165], partial blocking of excess carriers occurs at the cathode of n-type photoconductors thus enhancing current responsivity and reducing the time response. The n^+-n contact can be used to isolate regions containing minority holes from high recombination rate regions [166–168]. Due to degeneracy in the n^+ region, the potential barrier to holes is much larger than the simple Boltzmann factor $(kT/q)ln(n/n_o)$. The effective recombination velocity at the blocking contact below 100 cm/s can be achieved. For this, the electron concentration must be degenerate within the distance of 1 μm or less.

The carrier sweep-out can also be virtually eliminated with the heterojunction contact photoconductor proposed by Smith et al. [169,214]. His calculations predict elimination of the "saturation" of responsivity and a very large increase in responsivity. The responsivity increase of an order of magnitude using a heterojunction contact to the HgCdTe with $x = 0.20$ have been obtained [214].

Also combination of different approaches, such as the combined overlap/blocking contact device structure [187] or heterostructure photoconductor with blocking contacts [211], has been used to improve the device performance. The second type of devices are insensitive to the condition of the semiconductor/passivant interface.

The accumulation and heterojunction contacts for reflection of minority carriers are equally applicable to reduce the recombination at the active and back side surface of photoconductors. Excessive accumulation, on other hand, can lead to a large shunt conductance, which can also degrade detectivity [215].

14.5.2.2 Devices for Operation above 77 K

The performance of HgCdTe photoconductors at higher temperatures is reduced. For many applications, however, there are significant advantages in accepting this fact; for example, the input power to cooling engines can be reduced and their life extended, the operation temperature above 180 K can be achieved with thermoelectric coolers.

The carrier lifetimes at higher temperature are short being fundamentally limited by Auger processes, and the g-r noise limited performance is obtained [126,127]. Since $\gamma > 1$ (see Equation 14.32), there is in principle an advantage in using p-type material. In practice, however, p-type photoconductors are difficult to passivate and low $1/f$ noise contacts are difficult to form. For these reasons the majority of devices for the higher temperature operation are n-type. Figure 14.26 shows examples of the detectivity as a function of cutoff wavelength, obtained from 230 μm square n-type devices operated at different temperatures. For comparison theoretical limiting detectivity is shown assuming an extrinsic concentration of 5×10^{14} cm^{-3}, thickness of 7 μm, reflection coefficient at front and back surface of 30%, and $f/1$ optics [126].

The p-type HgCdTe photoconductors are used as laser receivers, where the bandwidth is usually high and $1/f$ noise is unimportant. Intermediate-temperature operation of p-type devices in the LWIR region has been reported by several authors [126,172,182,216–220]. The measured detectivity at 193 K at relatively high modulation frequency of 20 kHz was 7×10^8 cmHz$^{1/2}$W^{-1}.

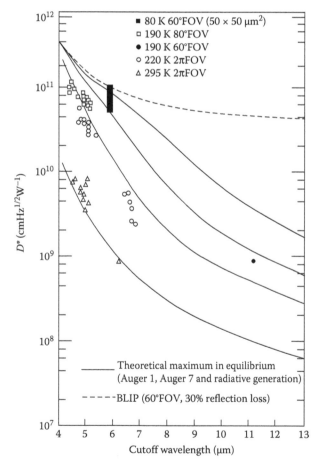

Figure 14.26 Detectivity versus cutoff wavelength for n-type HgCdTe photoconductive detectors. The theoretical curves are calculated including Auger generation and radiative generation only. The experimental points are for 230 μm square n-type detectors, except where indicated. (From Elliott, C. T., and Gordon, N. T., *Handbook on Semiconductors*, North-Holland, Amsterdam, Vol. 4, 841–936, 1993. With permission.)

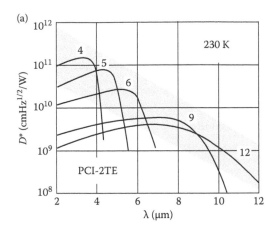
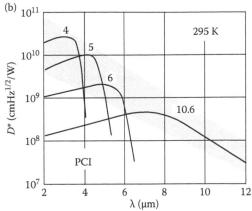

Figure 14.27 Spectral detectivities of optically immersed $Hg_{1-x}Cd_xTe$ photoconductors: (a) uncooled and (b) cooled with 2-stage Peltier coolers. (From Piotrowski, J. and Rogalski, A., *Infrared Physics & Technology*, 46, 115–31, 2004. With permission.)

The minimum value of the NEP for heterodyne detection was observed to be about 1×10^{-19} WHz^{-1} with a bandwidth of 100 MHz [220].

The measured detectivities of optically immersed photoconductors operating at ambient temperature or at temperatures achievable with thermoelectric coolers (200–250 K) are shown in Figure 14.27 [174]. The devices have been fabricated from ISOVPE epilayers, in situ doped with foreign impurities. Recently, due to careful optimization of the compositional and doping profile, the use of metal back reflectors, better surfaces, and contacts processing, the performance of the devices has been significantly improved [157,173,174]. The operation of two-stage cooled photodetectors manufactured at Vigo System has been extended to ≈ 16 μm with detectivities of $\approx 2 \times 10^9$ cmHz$^{1/2}$/W at 12 μm [157].

The optically immersed devices are especially suitable for high frequency operation due to a short lifetime, absorber resistance close to 50 Ω, negligible series resistance and very low capacitance. The measurements using free electron lasers revealed response times of ≈ 0.6 ns and ≈ 4 ns at 300 K and 230 K, respectively, for detectivity-optimized 10.6 μm devices. Shorter response of ≈ 0.3 ns was observed in specially designed devices of small physical size ($\approx 10 \times 10$ μm^2 or less) and more heavy p-type doping [157].

The present high-temperature photoconductors have relatively poor low-frequency properties. Typically, the $1/f$ knee frequencies of 10 kHz have been observed in uncooled 10.6 μm detectors at electrical fields of ≈ 40 V/cm (typical large size Vigo R005 photoconductors). The Hooge's constant of about $\approx 10^{-4}$ has been deduced from the low field measurements. Rapid, nonlinear increase of the Hooge's constant has been observed for stronger electric fields. Therefore, $1/f$ noise is an issue in the near room temperature photoconductors, especially stringent in uncooled devices, which would require a high biasing (≈ 100 V/cm) to approach the generation-recombination limit of performance. Cooling to ≈ 200 K reduces the Hooge's constant by a factor of ≈ 2, which in conjunction with much lower electrical fields requirements makes it possible to achieve the g-r limited operation at frequencies of ≈ 1000 Hz and higher. Proportionality between g-r and $1/f$ noise in $Hg_{0.8}Cd_{0.2}Te$ photoconductors have been observed in a wide temperature range 77–250 K [221].

The reason of the poor low-frequency performance is fully understood at present; inadequate surface passivation and contact technology is usually blamed. No one existing theory can explain qualitatively the observed low frequency noise [222].

An interesting device is a two-lead devices that operates over a wide spectral band with performance improved by a large factor at a given SW ranges [173]. The device consists of several stacked active regions (absorbers) with their outputs connected in parallel so the resulting output signal current is the sum of the signals generated at all active regions. Due to a high photoelectric gain in the wider gap absorbers and low thermal generation and recombination, the devices offer significantly better performance at SWs while the long wavelength response remains essentially unaffected. An example is an uncooled photoconductor operating up to 11 μm, with response at 0.9–4 μm increased by ≈ 3 orders of magnitude in comparison to the response of conventional 11 μm device.

Uncooled two-color photoconductors with individual access to each color region have been also reported [191]. The devices were grown with MOCVD. The $Hg_{1-x}Cd_xTe$ layers are either detectors or IR filters while CdTe layers serve as insulators separating various spectral regions.

It should be noted that the measured detectivity of uncooled ≈ 10 μm photoconductors is many orders of magnitude higher compared to other ambient temperature 10.6 μm detectors with subnanosecond response time such as photon drag detectors, fast thermocouples, bolometers, and pyroelectric ones.

14.5.3 Trapping-Mode Photoconductors

If the minority carrier (hole) is somehow trapped before it can be swept out at the negative electrode, then the electron current keeps flowing for a longer time, which increases the photoconductive gain.

Significant improvements in the gain of photoconductors has been realized by development of trapping-mode HgCdTe detectors in the 1980s [223,224]. The device structure and its energy bandgap profile are shown in Figure 14.28. LPE was used to grow these structures on CdTe substrates. The postgrown low temperature annealing forms n-type lightly doped detector active layer ($n \approx 10^{14}$ cm^{-3}), while retaining p-type layer with the p-type trapping region within a compositionally graded interface between the epitaxial layer and the substrate, underlines the entire n-type layer. Minority-carrier holes are separated from majority-carrier electrons by the junction, producing a reduction in the depletion layer width. Due to low electron concentration, the depletion layer width at the p-n junction is substantial, and tunneling leakage current is minimal. Large photoconductive gain (of the order of 1000 to 2000) results from existing both hole trap region and blocking N-n HgCdTe interface near contacts.

Figure 14.29 compares the responsivity of 12 μm cutoff conventional and trapping-mode HgCdTe photoconductors and 80 K [224]. The presented data are for devices with dimensions of about 50×50 μm². The impedance of all devices is of the order of 100 Ω. Trapping-mode devices have at least two orders of magnitude lower bias lower requirements to achieve 10^5 V/W responsivity—on the order of 0.12 W/cm² compared with 12 W/cm² [225]. This two orders of magnitude lower bias power greatly reduces the bias heat load for large multielement arrays.

A further benefit of trapping-mode devices is significantly lower 1/f noise [226]. In conventional HgCdTe photoconductive detectors the 1/f noise knees are typically 1 kHz, but in their high frequency counterparts at 80 K and for f/2 background flux conditions, is only of the order of a few hundred hertz.

14.5.4 Excluded Photoconductors

Elliott and other English workers [126,142,158,159,178,227–232] proposed and developed a new approach to reducing the photodetector cooling requirements, which is based on the nonequilibrium mode of operation. It appears that the Auger generation process with its associated noise

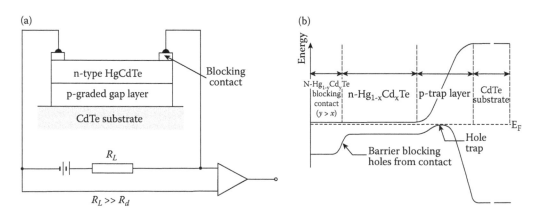

Figure 14.28 Trapping-mode HgCdTe photoconductive detectors: detector structure and its biasing (a), and band diagram (b). (From Norton, P. R., *International Application Published Under The Patent Cooperation Treaty* PCT/US86/002516, International Publication Number WO87/03743, 18 June 1987.)

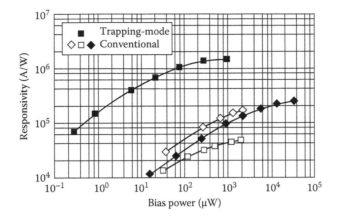

Figure 14.29 Comparison of the bias dependence of responsivity of trapping-mode HgCdTe devices with conventional photoconductive detectors at 80 K. The presented data are for devices with dimensions of about 50×50 μm^2 and 12 μm cutoff wavelengths. The impedance of all devices is of the order of 100 Ω. (From Norton, P. R., *Optical Engineering*, 30, 1649–63, 1991. With permission.)

can be suppressed in devices where the carrier densities are held below their equilibrium values by techniques that do not involve large electric field. The excluded photoconductor was the first demonstrated device of this type.

The principle of operation of the excluding contact photoconductor is shown in Figure 14.30 [159]. The positively biased contact is a highly doped n$^+$ or wide gap material, while the photosensitive area is a near-intrinsic n-type (ν) material. Such a contact does not inject minority carriers but permits the majority electrons to flow out of the device. As a result, the hole concentration in the vicinity of the contact is decreased and the electron concentration also falls (to maintain electroneutrality in the region) to a value close to the extrinsic value $N_d - N_a$. In consequence, the Auger generation and recombination processes become suppressed in the excluded zone. The device must be longer than the exclusion length to avoid the effect of carrier accumulation at the negative bias contact. The length of the excluded region depends on the bias current density, bandgap, temperature, and other factors. Lengths higher than 100 μm have been observed experimentally for MWIR devices [230]. A threshold current is required to counter the back diffusion current between unexcluded and excluded regions. Thereafter, the resistance rises rapidly as the length of the excluded region increases. As a result, the current-voltage characteristic exhibits saturation for above-threshold currents.

The accurate analysis of nonequilibrium devices can be done only by numerical solution of the full continuity equation for electrons and holes. The usual approximations break down in the case of a nonequilibrium mode of operation due to large departures from equilibrium. The properties of the structure can be analyzed taking into account the set of five differential equations, including the phenomenological transport equations for electrons and holes, continuity equations, and the Poisson's equation (see Section 3.4).

Assuming that the Auger generation is reduced below that of other residual processes, and under some simplifications, analytic expressions have been derived for

- the critical field, where exclusion terminates:

$$E_c = \left(\frac{D_e G}{r \mu_e \mu_h N_d} \right)^{1/2},$$

(14.40)

- the length of the exclusion zone:

$$L = \frac{\mu_h J}{\mu_e q G} - \left(\frac{D_e \mu_h N_d}{r \mu_e G} \right)^{1/2}, \text{ and}$$

(14.41)

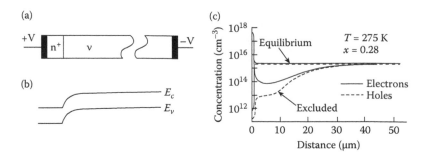

Figure 14.30 (a) Schematic diagram, (b) energy levels, and (c) electron and hole concentrations for an n⁺-v structure with $x = 0.28$, $N_d - N_a = 10^{14}$ cm⁻³, $\tau_{SR} = 4$ μm, $\tau_{Ai} = 2.4$ μs and $J = 48$ A/cm² showing equilibrium and excluding levels. (From Ashley, T., Elliott, T. C., and White, A. M., "Non-Equilibrium Devices for Infrared Detection," *Proceedings of SPIE* 572, 123–32, 1985. With permission.)

■ the threshold bias current:

$$J_o = q\left(\frac{D_e \mu_e G N_d}{r \mu_h}\right)^{1/2}, \qquad (14.42)$$

have been derived by English workers [159,175,227,228]. In the above equations r is some small fraction found by numerical calculations over a wide range of conditions of doping, temperature and bias to be 0.012 ± 0.002; G is constant, appropriate to a constant residual SR process in the exclusion zone.

The exclusion in p⁺-π structures can be modeled in a similar manner. There are two important differences. First, due to low hole mobility, the majority carriers no longer determine the current flow. Second, heavily doped p⁺-contacts exhibit large generation-recombination velocity. This can be solved by using a p-type heterojunction contact with increased bandgap.

The practical realization of nonequilibrium devices depends on several important limitations. The electric field in an excluded region must be sufficiently low to avoid heating a device as a whole and electrons above the lattice temperature. The heating of the structure can be prevented by a proper heat sink design, and it seems not to be a serious limitation—at least for a single element and low-area devices. Electron heating sets a maximum field that has been estimated as 1000 V/cm in materials for uncooled 5 μm devices and a few hundred V/cm in materials for ≈10 μm devices operated at 180 K. Electron heating is not an important constraint in the 3–5 μm band, but it seems to restrict the usefulness of exclusion at 10.6 μm and longer wavelengths in the 8–14 μm band. Very low doping (<10^{14} cm⁻³) is required for effective exclusion; however, the values 3×10^{14} cm⁻³ are typical in industry. Exclusion may be inhibited by any non-Auger generation such as the Shockley–Read or surface generation. A large electrical field may result in flicker noise.

Practical excluded HgCdTe photoconductors have been fabricated from low-concentration, bulk-grown material with the n⁺-regions formed by ion milling [159,175]. In contrast to the equilibrium mode photoconductors that are usually based on extrinsically p-type doped material, the excluded devices are fabricated from very low-concentration, n-type bulk HgCdTe, with the n⁺ regions formed by ion milling or degenerate extrinsic doping. The device is schematically shown in Figure 14.31 [159]. In order to avoid the effect of accumulation at the negative contact the devices are three-lead structures with the sensitive area defined by an opaque mask, and a side-arm potential probe is used for a readout contact. Such device geometry limits the size of the active area to the highly depleted region, which prevents thermal generation in the nondepleted part and at the negative contact from contributing to the noise measured at the readout electrode. ZnS is used for passivation, as the usual native oxide passivation produces an accumulated surface, which shunts the excluded region.

The detector parameters, noise, responsivity, and detectivity are shown for both directions of bias current in Figure 14.32 [126]. The responsivity and noise increase to large values in the excluded direction of bias due to two effects: the increased impedance in the excluded region and the increase in the effective carrier lifetime to the transit time. The improvement of detectivity is more modest due to high flicker noise levels at reverse bias. Reversing the bias direction from direct to reverse has shown improvement of detectivity by a factor of ≈ 3. This can be related to the

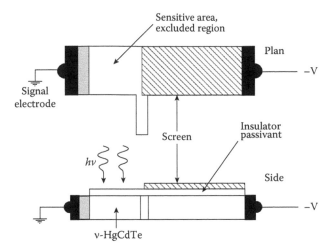

Figure 14.31 Schematic of a three-lead excluding HgCdTe photoconductive detector. (From Ashley, T., Elliott, T. C., and White, A. M., "Non-Equilibrium Devices for Infrared Detection," *Proceedings of SPIE* 572, 123–32, 1985. With permission.)

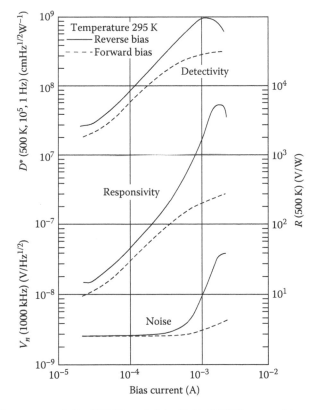

Figure 14.32 Noise, responsivity (500 K) and detectivity (500 K) versus bias current for $Hg_{0.72}Cd_{0.28}Te$ excluding photoconductor, measured at 295 K. (From Elliott, C. T., and Gordon, N. T., *Handbook on Semiconductors*, North-Holland, Amsterdam, Vol. 4, 841–936, 1993. With permission.)

poor HgCdTe-ZnS interface properties and to fluctuations in the injection rate. An uncooled 10 μm by 10 μm photoconductor with 4.2 μm cutoff has exhibited a 500 K blackbody voltage responsivity of 10^6 V/W, detectivity of 1.5×10^9 cm $Hz^{1/2}$/W at a modulation frequency of 20 kHz, and the thermal figure of merit M^* (zero range, 295 K) of 1.5×10^5 $cm^{-1}Hz^{1/2}$ K^{-1}, which exceeds the performance obtained from any other IR detector operated with the same conditions [231].

No practical 8–14 μm excluded photoconductors have been demonstrated to date. This can be attributed to the heating of electrons at the high electric fields required for exclusion.

14.5.5 SPRITE Detectors

The SPRITE detector was originally invented by T. C. Elliott and developed further almost exclusively by British workers [12,13,233–245]. This device has been employed in many imaging systems [243]. Figure 14.33 shows the operating principle of the device [233]. The device is essentially ≈1 mm long, 62.5 μm wide, and 10 μm thick n-type photoconductor with two bias contacts and a readout potential probe. The device is constant current biased with the bias field E set such that the ambipolar drift velocity v_a, which approximates to the minority hole drift velocity v_d, is equal to the image scan velocity v_s along the device. The length of the device L is typically close to or larger than the drift length $v_d\tau$, where τ is the recombination time.

Consider now an element of the image scanned along the device. The excess carrier concentration in the material increases during scan, as illustrated in Figure 14.33. When the illuminated region enters the readout zone, the increased conductivity modulates the output contacts and provides an output signal. Thus, the signal integration that, for a conventional array is done by external delay line and summation circuitry, is done in the SPRITE detector in the element itself.

The integration time approximates the recombination time τ for long devices. It becomes much longer than the dwell time τ_{pixel} on a conventional element in a fast-scanned serial system. Thus, a proportionally larger ($\propto \tau/\tau_{pixel}$) output signal is observed. If Johnson noise or amplifier noise dominates, it leads to a proportional increase in the SNR with respect to a discrete element. In the background-limited detector, the excess carrier concentration due to background also increases by the same factor, but corresponding noise is proportional only to integrated flux. As a result the net gain in the SNR with respect to a discrete element is increased by a factor $(\tau/\tau_{pixel})^{1/2}$.

Elliott et al. derived basic expressions for SPRITE parameters [12]. The voltage responsivity is

$$R_v = \frac{\lambda}{hc} \frac{\eta\tau El}{nw^2 t}\left[1-\exp\left(-\frac{L}{\mu_a E\tau}\right)\right],\tag{14.43}$$

where l is the readout zone length, L is the drift zone length.

The dominant noise is the generation-recombination noise due to the fluctuations in the density of thermal and background-radiation generated carriers. The spectral density of noise at low frequencies is

$$V_n^2 = \frac{4E^2 l\tau}{n^2 wt}\left(p_o + \frac{\eta Q_B\tau}{t}\right)\left(1-\exp\frac{-L}{\mu_a E\tau}\right)\left[1-\frac{\tau}{\tau_a}\left(1-\exp\frac{-\tau_a}{\tau}\right)\right].\tag{14.44}$$

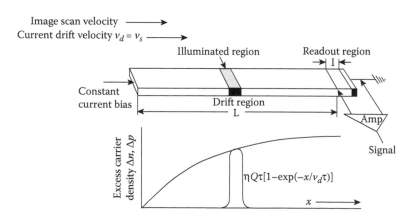

Figure 14.33 The operating principle of a SPRITE detector. The upper part of the figure shows a HgCdTe filament with three ohmic contacts. The lower part shows the buildup of excess carrier density in the device as a point in the image that is scanned along it. (From Elliott, C. T., *Solid State Devices*, Verlag Chemie, Weinheim, 175–201, 1983. With permission.)

For a long, background-limited device in which $L \gg \mu_a E \tau$ and $\eta Q_B \tau / t \gg p_{o}$,

$$D^* = \frac{\lambda \eta^{1/2}}{2hc} \left(\frac{l}{Q_B w} \right)^{1/2} \left[1 - \frac{\tau}{\tau_a} \left(1 - \exp \frac{-\tau_a}{\tau} \right) \right]^{-1/2}, \qquad (14.45)$$

and at sufficiently high speeds such that $\tau_a \ll \tau$, and

$$D^* = (2\eta)^{1/2} D^*_{BLIP} \left(\frac{l}{w} \right)^{1/2} \left(\frac{\tau}{\tau_a} \right)^{1/2}. \qquad (14.46)$$

For a nominal resolution size $w \times w$, the pixel rate is v_a/w and

$$D^* = (2\eta)^{1/2} D^*_{BLIP} (s\tau)^{1/2}. \qquad (14.47)$$

Based on the above considerations, long lifetimes are required to achieve large gains in the SNR. Useful improvement in detectivity relative to a BLIP-limited discrete device can be achieved when the value of $s\tau$ exceeds unity. The performance of the device can be described in terms of the number of BLIP-limited elements in a serial array giving the same SNR,

$$N_{eq}(\text{BLIP}) = 2s\tau. \qquad (14.48)$$

For example, a 60 μm wide element scanned at a speed of 2×10^4 cm/s and with τ equal to 2 μs gives $N_{eq}(\text{BLIP}) = 13$.

Figure 14.34 shows a photograph of an eight-row bifurcated SPRITE detectors [233]. The bifurcated readout allows close-packed arrays without stagger. Readouts are shown at both ends, but only one set is used.

As shown in Table 14.4, to achieve usable performance in the 8–14 μm band the SPRITE devices require LN cooling while 3- or 4-stage Peltier coolers are sufficient for effective operation in the 3–5 μm range. The performance achieved in the 8–12 μm band is illustrated in Figure 14.35 [13]. The detectivity increases with the square root of the bias field except at the higher fields where the

Figure 14.34 Schematic of an eight-row SPRITE array with bifurcated readout zones. (From Elliott, C. T., *Solid State Devices*, Verlag Chemie, Weinheim, 175–201, 1983. With permission.)

Table 14.4: Performance of SPRITE Detectors

Material	Mercury Cadmium Telluride	
Number of elements		8
Filament length (μm)		700
Nominal sensitive area (μm)		62.5 × 62.5
Operating band (μm)	8–14	3–5
Operating temperatures (K)	77	190
Cooling method	Joule-Thompson or heat engine	Thermoelectric
Bias field (V/cm)	30	30
Field of view	$f/2.5$	$f/2.0$
Ambipolar mobility (cm²/Vs)	390	140
Pixel rate per element (pixel/s)	1.8×10^6	7×10^5
Typical element resistance ()	500	4.5×10^3
Power dissipation (mW per element) total	9 <80	1 <10
Mean $D*$ (500 K, 20 kHz, 1Hz) 62.5 × 62.5 μm (10^{10} cmHz$^{1/2}$/W)	>11	4–7
Responsivity (500K), 62.5 × 62.5 μm (10^4 V/W)	6	1

Source: A. Blackburn, M. V. Blackman, D. E., Charlton, W. A. E., Dunn, M. D., Jenner, K. J., Oliver, and J. T. M., Wotherspoon, *Infrared Physics*, 22, 57–64, 1982. With permission.

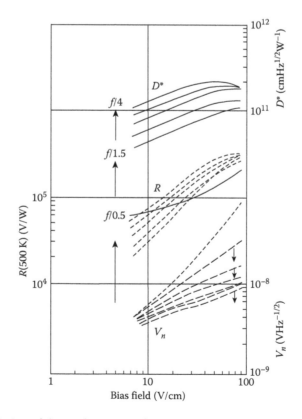

Figure 14.35 Variation of the performance of a SPRITE detector with bias field and field of view: $\lambda_c = 11.5$ μm, $T = 77$ K. (From Blackburn, A., Blackman, M. V., Charlton, D. E., Dunn, W. A. E., Jenner, M. D., Oliver, K. J., and Wotherspoon, J. T. M., *Infrared Physics*, 22, 57–64, 1982. With permission.)

element temperature is raised by Joule heating. In addition, this parameter increases with increasing cold shield effective $f/\#$ to about $f/4$. To avoid reductions of the carrier lifetime resulting from increased carrier density due to the background flux, the efficient cold shielding with $f/\#$ of two or larger is used.

An example of results obtained from a SPRITE operating in the 3–5 μm band is shown in Figure 14.36 [13]. Useful performance in this band can be obtained at temperatures up to about 240 K.

The spatial resolution of the SPRITE detector when the scan velocity and the carrier velocity are matched throughout the device length, is determined by the diffusive spread of the photogenerated carriers and the spatial averaging in the readout zone. This can be expressed through the modulation transfer function (MTF) [233]

$$\text{MTF} = \left(\frac{1}{1 + k_s^2 L_d^2}\right)\left[\frac{2\sin(k_s l/2)}{k_s l}\right],$$
(14.49)

where k_s is spatial frequency, L_d is diffusion length.

The SPRITE detectors are fabricated from lightly doped ($\approx 5 \times 10^{14}$ cm^{-3}) n-type HgCdTe. Both bulk material and epilayers are being used [243]. Single and 2, 4, 8, 16, and 24 element arrays have been demonstrated; the 8-element arrays are the most common at present (Figure 14.34). In order to manufacture the devices in line, it is necessary to reduce the width of the readout zone and corresponding contacts to bring them out parallel to the length of the element within the width of the element as shown in Figure 14.34. Various modifications of the device geometry (Figure 14.37 [240]) have been proposed to improve both the detectivity and spatial resolution. The modifications have included horn geometry of the readout zone to reduce the transit time spread, and slight taper of the drift region to compensate a slight change of drift velocity due to background radiation.

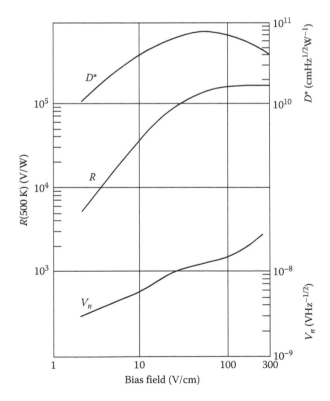

Figure 14.36 Performance of a 3–5 μm SPRITE operating at 190 K. (From Blackburn, A., Blackman, M. V., Charlton, D. E., Dunn, W. A. E., Jenner, M. D., Oliver, K. J., and Wotherspoon, J. T. M., *Infrared Physics*, 22, 57–64, 1982. With permission.)

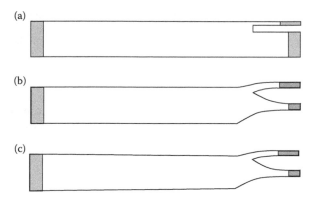

Figure 14.37 The evolution in the geometry of SPRITE devices. (From Elliott, C. T., "SPRITE Detectors and Staring Arrays in $Hg_{1-x}Cd_xTe$," *Proceedings of SPIE* 1038, 2–8, 1988. With permission.)

The response of the SPRITE detector to high spatial frequencies in the image limited fundamentally either by spatial averaging due to limited size of the readout zone or by the diffusive spread of photogenerated carriers in the filament [233]. The response can be further degraded by imperfect matching of carrier drift velocity to the image velocity. The resolution size for 8–14 μm devices is ≈ 55 μm. For 200 K, 3–5 μm devices the spatial resolution is ≈ 140 μm.

One possible method to improve the resolution is to use a short device, where the transit time is less than the lifetime to reduce the diffusive spread. The spatial resolution can also be improved by the use of anamorphic optics, which gives increased magnification of the image in the scan direction [237,240]. The detector length and scan speed are increased in the same proportion as the increase in magnification, but the diffusion length remains constant so that the spatial resolution is improved. Since the SPRITE detectors remain background limited even in low background flux, the SNR remains unaffected.

A number of improvements have been made to the spatial and thermal resolution [242,243]. System thermal sensitivity can be increased by the use of a larger number of elements. SPRITE detectors have been commonly used in 8-row, 16-row, and 24-row arrays and are commercially available in these forms. In addition to parallel arrays, 2-D 8 × 4 parallel/serial arrays with conventional time delay and integration along a row have been demonstrated. Moving from an operating temperature of ≈ 80 K to ≈ 70 K with the cutoff wavelength shift from 12 to 12.5 μm can give both improved signal-to-noise performance and spatial resolution. In the 3–5 μm band, the main improvement has been achieved by the use of more effective 5- and 6-stage Peltier coolers; 500 K blackbody detectivities as high as 5×10^{10} cmHz$^{1/2}$/W have been achieved with 8-element arrays.

Despite remarkable successes, SPRITE detectors have important limitations such as limited size, stringent cooling requirements, and the necessity to use fast mechanical scanning. The ultimate size of SPRITE arrays is limited by the significant heat load imposed by Joule heating. This means that SPRITE detectors are transition-stage devices to the staring 2-D arrays area.

14.6 PHOTOVOLTAIC DETECTORS

The development of HgCdTe photodiodes was stimulated initially by their applications as high-speed detectors, mostly for direct and heterodyne detection of 10.6 μm CO_2 laser radiation [9,246]. Operation of such photodiodes at 77 K in the heterodyne mode, at frequencies up to several GHz is possible because of the low junction capacitance, which in turn is a result of its relatively low static dielectric constant. In the mid-1970s attention turned to the photodiodes for passive IR imaging applications in the two commonly used atmospheric windows at 3–5 μm and 8–14 μm. At that time it was seen that many IR applications in the future would be needed for higher radiometric performance and/or higher spatial resolution than could be achieved with first generation photoconductive detectors. The main limitation of photoconductive detectors is that they cannot easily be multiplexed on the focal plane. In contrast to photoconductors, photodiodes can be assembled in 2-D arrays containing more than megapixel elements, limited only by existing technologies. Systems based upon such FPAs can be smaller, lighter with lower power consumption, and can result in much higher performance than systems based on first generation detectors. Photodiodes can also have less low frequency noise, faster response time, and the potential for a more uniform spatial response across each element. However, the more complex processes needed for photovoltaic detectors have influenced slower development and industrialization of the second-generation systems. Another point is that, unlike photoconductors, there is a large variety of device structures with different passivations, junction-forming techniques, and contact systems.

Initially, the first HgCdTe photodiodes were prepared from bulk materials. Further development was, however, dominated by various epitaxial techniques including ISOVPE, LPE, MBE, and MOCVD. A more detailed historical review of papers published up to the end of the last century is presented in the monograph *Narrow-Gap Semiconductor Photodiodes* [73]. Other important sources of information are included (see, e.g., [182,247–254]), in the publications each year since 1981, workshop-style technical meetings devoted exclusively to the physics and chemistry of HgCdTe and related semiconductor and IR materials (published initially in *Journal of Vacuum Science and Technology* and next in *Journal of Electronic Materials*), and in numerous *Proceedings of SPIE*.

This chapter will concentrate mainly on the present status of HgCdTe photodiode physics and technology important in fabrication of large FPAs. The success in HgCdTe photodiode technology has stimulated programs into third generation IR detector technology being performed in centers around the world (see Chapter 23) [18].

14.6.1 Junction Formation

The p-n HgCdTe junctions have been formed by numerous techniques including Hg in- and out-diffusion, impurity diffusion, ion implantation, electron bombardment, plasma induced type conversion, doping during growth from vapor or liquid phase and other methods [73]. To avoid citing hundreds of related references, please see more recently published monographs and reviews [73,250,251,254].

The low binding energies and ionic bond nature of HgCdTe give rise to two important effects, which are influential in most junctions forming processes. The first is the role of Hg, which is liberated readily by processes such as ion implantation and ion beam milling. This creates a much deeper junction than would be expected from the implantation range. A second effect is the role of dislocations, which may play a part in annihilating vacancies. The role of Hg interstitials, dislocations and ion bombardment in the junction forming process is complex and not well understood in detail. Despite the complex physics involved, manufacturers have received good phenomenological control of the junction depth and dopant profiles with a variety of processes. Recently epitaxial techniques with doping during growth are most often used for preparing p-on-n junctions. MBE and MOCVD have been successfully accomplished with As doping during growth.

14.6.1.1 Hg In-Diffusion

It is relatively straightforward to achieve local type conversion of the material, by neutralizing the vacancies by the in-diffusion of mercury. The n-type conductivity is originated from a background donor impurity. The current knowledge of Hg in-diffusion process has been summarized by Dutton et al. [255]. The Hg diffusion process at 200°C is accompanied by vacancy out-diffusion, which creates a graded distribution of vacancies near the metallurgical junction. To form junctions requires only 10–15 minutes at temperatures of 200°C–250°C in material with vacancy concentrations of low 10^{16} cm^{-3}. This corresponds to diffusion constants of order 10^{-10} cm^2 s^{-1}. The presence of dislocations can enhance vacancy mobility even further, while the presence of Te microprecipitates may retard the motion of Hg into the lattice. Crystal defects such as dislocations and microprecipitates can also harbor background impurities, which will further affect the location and quality of the junction.

Hg in-diffusion into vacancy doped (10^{16}–10^{17} cm^{-3}) HgCdTe, originally proposed by Verie et al. [9,246] at the beginning of the 1970s, has been the most widely used for very fast photodiodes, in which a low concentration (10^{14}–10^{15} cm^{-3}) n-type region is necessary for large depletion width and low junction capacitance. Initially, a mesa configuration was used for this type of device. Spears and Freed [256,257] reported an improvement of the technique using n$^+$-n-p planar structure. A 0.5 μm ZnS was sputtered onto the surface of vacancy doped HgCdTe. After etching openings in the ZnS through a photoresist mask and before removing the photoresist, a ≈ 10 nm In layer was sputtered onto the surface. A low concentration n-type layer about 5 μm deep with thin n$^+$ skin was produced within 30 minutes, 240°C Hg-vapor diffusion in a sealed ampoule. Photoresist lift-off technique was used to define sputtered In-Au bonding pads. A ZnS mask provides passivation around the junction perimeter. Fabrication of similar n$^+$-n-p structures for heterodyne applications was described by Shanley et al. [258].

Further modification in preparing n-p planar junctions was presented by Parat et al. [259]. Following a Hg saturated anneal at 220°C for 25 hour, the MWIR HgCdTe layers grown by MOCVD become n-type with carrier concentrations around 5×10^{14} cm^{-3}. However, the presence of a 0.5–0.8 μm thick CdTe cap layer acts as an effective barrier for Hg diffusion and provides excellent junction passivation. By opening windows in this cap, the underlying HgCdTe layer can be annealed and converted to n-type by a selective manner.

Jenner and Blackman [260] have reported a variation of the Hg in-diffusion method, which uses anodic oxide to act as a source of free mercury. This technique is particularly suitable for high-speed devices in which a low and uniform doping n-type regions are necessary. Anodic oxidation produces a Hg-rich layer at the interface between oxide and semiconductor. During anneal, Hg diffuses into a material background giving rise to n-type characteristics of the material. The anodic oxide layer acts as an out-diffusion mask preventing the loss of Hg into the vacuum on gas ambient. Brogowski and Piotrowski have calibrated this method [261]. The square root dependence of the junction depth on anneal time for short times is a clear indication of a diffusion nature of the p-to-n type conversion. The existence of maximum depth of the junctions shows that the source of free Hg is a finite one. After a prolonged annealing the Hg source becomes exhausted, Hg diffuses deeply into the bulk, resulting in a reconversion of conductivity type at the surface of the material.

14.6.1.2 Ion Milling

Conversion of vacancy doped p-type $Hg_{1-x}Cd_xTe$ to n-type during low-energy ion bombardment became another important technique of junction fabrication [262–268]. Neither donor ions nor post-annealing is required. The ion beam injects a small proportion of the Hg atoms (approximately 0.02% of the gas ions) into the lattice. These then neutralize acceptor-like Hg vacancies, and leave the lattice weakly n-type by background donor atoms.

The ion energy is usually less than 1 keV and the dose normally varies between 10^{16} and 10^{19} cm^{-2}. But even at the lower dose, when the surface is etched very gently, ion milling results in substantial changes of the electrical properties of HgCdTe over a large depth. Blackman et al. have shown that the depth of the p-n junction depends on dose and can extend a few hundred μm from the surface [264]. The electrical properties of ion-etched HgCdTe measured by differential Hall effect have shown that after ion milling a thin n-type degraded layer, of approximately 1 μm in thickness, with low mobility and high concentration of electrons occurs close to the surface [266,269]. Below this damaged region, an n-type doping profile, which decreases exponentially away from the surface, and a low doped n-region of controllable width, and with high electron mobility are created. The n$^+$ grade leads to a very effective reflecting contact for minority carriers, leading to a highly sensitive n-type region, as revealed by electron beam induced current analysis. A deeper junction is created by a higher beam current, longer milling time, lower beam voltage, and higher ion mass. The whole process is carried out at a low temperature preventing the original material and passivation quality, which is an advantage of this technique. The type conversion of p-type material using an ion beam has been used commercially by GEC-Marconi Infrared Ltd. for HgCdTe FPAs since the late 1970s [270].

The diffusion of Hg during ion milling is very quick even in comparison with annealing experiments at 500°C. To explain this phenomena, diffusion mechanisms via fast ways (dislocations, grain boundaries, stacking faults) were discussed. A model of the diffusion of Hg in HgCdTe taking into account recombination of Hg interstitials with Hg vacancies was presented [271,272]. Based on this model it was shown that there is in principle no discrepancy between the high value of the diffusion constant of Hg interstitials and that of the radioactive Hg self-diffusion constant established from equilibrium annealing experiments.

The local damage due to the impinging ions is restricted to a distance of the order of the ion range, while the depth of the converted zone is much larger, and remains roughly proportional to the thickness of the layer removed by ion bombardment.

14.6.1.3 Ion Implantation

The ion implantation in HgCdTe is a well-established approach for fabricating HgCdTe photovoltaic devices with n-on-p type junctions [273,274]. It is a common method of HgCdTe photodiode fabrication since it avoids heating of this metallurgically sensitive material and permits a precise control of junction depth. Many manufacturers obtain the desired p-type level by controlling the density of acceptor-like Hg vacancies within a carrier concentration range of 10^{16} to 10^{17} cm^{-3}. The n$^+$-p structures are produced by Al, Be, In, and B ions implantation into vacancy doped p-type

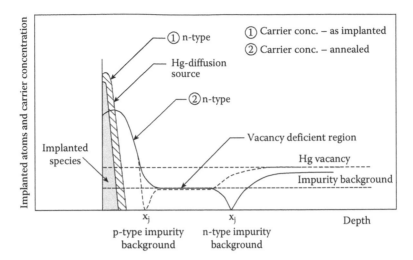

Figure 14.38 Qualitative model of displaced role of Hg on junction formation in ion-implanted HgCdTe. (From Bubulac, L. O., and Tennant, W. E., *Applied Physics Letters,* 51, 355–57, 1987. With permission.)

material, but the technique typically uses ion implantation of light species (usually B and Be) to form n-region. Boron is possibly most frequently used, perhaps due to the fact that boron is also a standard implant for silicon. Regardless of the nature of the implanted species, an n-type electrical activity is associated with implantation damage as observed first by Foyt et al. [275].

A summary of the complex phenomena associated with ion implantation in HgCdTe is shown in Figure 14.38. A fundamental concept developed for light species implantation in HgCdTe is the generation of free Hg atoms by the irradiation process of ion implantation (knockout Hg atoms), and their diffusion from the implanted source [276]. This concept is the basis of n-on-p junction formation in certain conditions of background type and concentration of carriers by the diffusion of Hg from the implant-induced Hg source, as illustrated in Figure 14.38. In this figure, the typical profile implanted species atom concentration and the corresponding carrier concentrations before (curve 1) and after (curve 2) postimplant anneal are sketched. A fraction of Hg interstitials are the diffusing elements primarily responsible for the changes in p-type behavior of the starting material when it is doped by Hg vacancies (i.e., the interstitials annihilate Hg vacancies that are encountered), revealing the net doping due to impurity background that can be either n- or p-type. The junctions form at the end of the annihilated region. Typically, the p-n junctions are located at a depth of 1–3 µm, which is large compared with the submicron range of implanted ions (<300 Å). If the net background doping is p-type, the junctions are of n^+-p type (dashed line). When the net background doping is n-type, the annihilated region is n-type. The junctions thus formed are n^+-n^--p type (full line). This is the most desirable case since the junction is located away from the implant-defect region in a material, which is free of irradiation-induced defects.

The mesa diodes are made by implantation on a passivated surface with subsequent etching for element separation. For planar devices, prior to implantation, the substrates are covered with dielectric layers (photoresists, ZnS, CdTe) with openings acting as a mask for impinging ions and thus defining the junction area. Implantation is typically performed at room temperature; the substrates, if oriented, are inclined to the beam axis to avoid ion channeling. Doses of 10^{12}–10^{15} cm^{-2} and energies of 30–200 keV are applied. No postimplant annealing was found to be necessary to achieve high performance, particularly at lower doses and for SWIR and MWIR devices. Some workers concluded that LWIR photodiode performance can be improved if postimplant annealing is used [277,278]. The postimplant anneals remove the radiation damage with a temperature dependence that varies with implanted species and with implant/anneal conditions. Examples are shown [277,278] for In, and B and Be for which anneals at temperatures of ≥ 300°C annihilated the electrically active defects. Postannealing also modifies the characteristics of the p-type base layer when doped by Hg vacancies. The electrically active ion implantation has been observed by Bahir et al. after rapid thermal annealing (0.3 seconds) of P and B implants with a cw CO_2 laser [279].

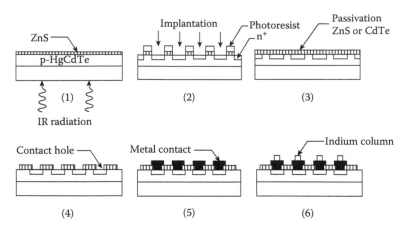

Figure 14.39 PACE-I HgCdTe MWIR processing sequence. (From Bailey, R. B., Kozlowski, L. J., Chen, J., Bui, D. Q., Vural, K., Edwall, D. D., Gil, R. V., Vanderwyck, A. B., Gertner, E. R., and Gubala, M. B., *IEEE Transactions on Electron Devices*, 38, 1104–9, 1991. With permission.)

An example processing sequence of PACE-1 HgCdTe MWIR FPAs is shown in Figure 14.39. At the beginning, a layer of CdTe is grown via MOCVD on the sapphire [280]. HgCdTe is then grown on the CdTe buffer via LPE. The junctions are formed by boron ion implantation and thermal annealing. Planar and mesa junctions are used, depending on the FPA specifications, and are passivated with a ZnS or CdTe film. Metal contacts and indium columns are then deposited and patterned on both detector array and the multiplexing readout, and the hybrid is fabricated by mating the detector array to the readout via indium column interconnects.

Until the 1990s, less attention had been paid to implanted p-on-n photodiodes [273,274]. The junctions were formed by Au, Ag, Cu, P, and As implantation in n-type HgCdTe followed by annealing. Establishing p-on-n junctions in HgCdTe using extrinsic dopants has taken an increasingly critical role in HgCdTe photodiode technology, especially in long wavelength spectral range. In the published results of p-on-n devices, the electrical junctions are controlled by the "tail" components in arsenic diffusion profile, rather than by a classical component which is representative of a volume diffusion mechanism.

The arsenic redistribution from the ion-implanted source as well as from a grown source (As incorporated in a part of the growth cycle) typically exhibits a multicomponent behavior in which the tail component varies from sample to sample. A model that explains the nature of these components has been proposed by Bubulac et al. [281,282]. In this model, the component (1), see Figure 14.40, extends from the surface to its intersection with the deeper Gaussian-like component (2). Arsenic distribution in the near-surface region occurs due to a retarded diffusion mechanism operating in the radiation damage region. It was shown that the depth of component (1) is related to the dislocation density (EPD) in the starting material and on implant-anneal conditions. The diffusion mechanism proposed for deeper component (2) is atomic in nature and vacancy based in which As starts on the Te-sublattice in the surface damage region. Because of a large number of Hg interstitials generated in this region as a result of Hg knockout, the conditions under which As is introduced in the lattice approximates a "Hg-rich" condition, resulting in a larger fraction of As incorporation on the Te-sublattice versus the metal-sublattice. The As diffusion of this component is described by Fick's law with a constant diffusion coefficient of $D = 2.5 \times 10^{-14}$ cm^2/s at 400°C and $D = 2.0 \times 10^{-13}$ cm^2/s at 450°C. It was also determined that the diffusion coefficient in this component (2) is independent of EPD in the material. A linear dependence of the diffusion length of As as a function of the square root of time, confirms the Gaussian redistribution of As in component (2) from the As ion implanted source. The excellent reproducibility of the diffusion coefficient on samples grown by various growth techniques on various substrates (e.g., MOCVD/GaAs/Si, LPE/CdTe, and LPE/sapphire) is evidence that this diffusion mechanism is representative of the bulk diffusion in HgCdTe. The diffusion mechanism in this component is the most desirable mechanism for a controlled junction formation process.

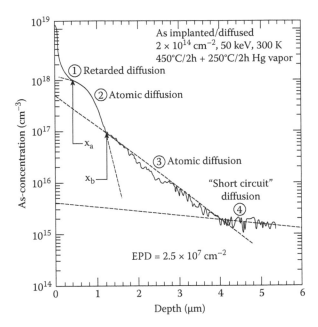

Figure 14.40 Typical example of As redistribution from the ion-implanted source (SIMS) and the proposed model for As diffusion from an ion-implanted source: (1) retarded diffusion; (2) atomic diffusion, As starts on Te sublattice; (3) atomic diffusion, As starts on metal sublattice; and (4) short-circuit diffusion. (From Bubulac, L. O., Edwall, D. D., and Viswanathan, C. R., *Journal of Vacuum Science and Technology*, B9, 1695–1704, 1991. With permission.)

The tailing component (3) is generated by a fraction of As atoms, which is introduced on metal sublattice sites and diffused by an enhanced vacancy-based mechanism. The enhancement is a consequence of departure from point defect equilibrium within the atomic diffusion zone (in the damage region a continuous metal vacancy flux to the diffusion zone is maintained). Since vacancies enhance the substantial diffusion, the atomic diffusion in component (3) is affected by the gradient of vacancy distribution and thus the resulting diffusion coefficient is position dependent. The component (3) of As diffusion controls the location of the electrical junction. This component is electrically complex, consisting of n- and p-type active As, and perhaps also neutral As. A site transfer with the associated change in electrical activity is possible, depending on subsequent thermal treatment and phase equilibria. The principal consideration in minimizing the As tailing component is to reduce the dislocation density of the starting material. An increase of about one order of magnitude in EPD (from 2.6×10^6 to 2.7×10^7 cm^{-2}) causes an increase in surface damage depth by a factor of 9 (from 0.05 to 0.45 μm) and an increase about two orders of magnitude in the arsenic concentration (from 10^{15} to 10^{17} cm^{-3}).

In the presence of nested dislocations in the starting material, a further enhancement of As diffusion was observed (i.e., component, 4). In Si it has been shown that impurities can move in dislocations or in clustered defects by "short circuit" mechanism. The experiments showed that the most effective way to minimize component (4) was to reduce the EPD of the material.

14.6.1.4 *Reactive ion Etching*

Reactive ion etching (RIE) is an effective anisotropic etching technology that is widely used in Si and GaAs semiconductors allowing delineation of a high density of active device elements incorporating small features. This plasma induced technique, as an alternative to ion implantation junction formation technology, has received considerable attention during the past few years [283]. The postimplant annealing, necessary in implantation technology to produce high-quality photodiodes, is not needed in plasma technology.

Figure 14.41 shows the H_2/CH_4 RIE etch depth observed in two vacancy doped p-type HgCdTe samples, as a function of the partial pressure of H_2 and CH_4 [283, 284]. The etch depth increases from zero methane level, reaches a maximum at about (H_2/CH_4) 0.8 and decreases to the zero

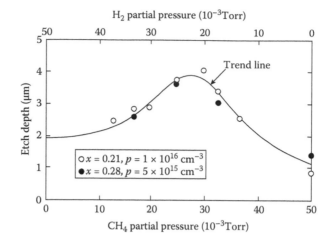

Figure 14.41 Etch depth in $Hg_{1-x}Cd_xTe$ as a function of H_2 and CH_4 partial pressure (H_2/CH_4 RIE): (○) $x = 0.21$, $p = 1 \times 10^{16}$ cm^{-3}; (●) $x = 0.28$, $p = 5 \times 10^{15}$ cm^{-3}. The etch time is 10 minutes, and the rf power is 180 W. (From Agnihorti, O. P., Lee, H.C., and Yang, K., *Semiconductor Science and Technology*, 17, R11–R19, 2002. With permission.)

hydrogen level. The p-n junction depth decreases with the increasing methane fraction in the mixture.

In plasma induced type conversion, the accelerated plasma ions sputter the HgCdTe surface, liberate Hg atoms from ordinary lattice positions, and create a source of Hg interstitials under the etched surface. Some of these atoms diffuse quickly into the material where they decrease the concentration of acceptors by the interaction of point defects, mainly Hg vacancies. Residual or native donor impurities then start to dominate in the conductivity and cause the p-n conversion. In the extrinsically doped material, conversion extends beyond Hg vacancies. White et al. [285] have proposed a p-n conversion model where the junction formation mechanism is thought to be a mixture of RIE induced damage, Hg interstitial formation to which hydrogen forms strong bonds, and hydrogen induced neutralization of acceptors.

Exposure of p-type HgCdTe to CH_4/H_2 RIE plasma has been to fabricate n-p junctions [283,286,287]. The plasma-induced type of conversion may also be used with P-on-n heterostructures to insulate the junction of high-performance photodiodes [288,289].

14.6.1.5 Doping during Growth

Doping during epitaxial growth becomes the preferred technique at present with inherent advantage of integration of the material growth and device processing. High-performance photodiodes can be obtained by successive growth of doped layers using LPE from Te-rich solutions [134,290–293] or Hg-rich solutions [47,294], MBE [54–56,64,295–297], MOCVD [53,83,132,298], and ISOVPE [172]. Recently, Mynbaev and Ivanov-Omski have reviewed available publications concerned with doping of HgCdTe epitaxial layers and heterostructures [299].

Epitaxial techniques make it possible to grow *in-situ* multilayer structures. Stable and readily achievable at temperatures below 300°C dopants to prevent interdiffusion processes are required. Indium and iodine are preferred n-dopants, and arsenic is the preferred p-type dopant for *in-situ* doping.

As was mentioned above, the As dopants must reside on the Te site to accomplish p-doping. This requires either growth or annealing at relatively high temperatures under cation-rich conditions. LPE growth from a Hg melt (at temperatures around 400°C) satisfies this condition automatically, although some postgrowth annealing may be required [291,293]. MOCVD and MBE have been successfully accomplished As doping during growth. Mitra et al. [53] have reviewed progress made in the development of IMP MOCVD for the *in-situ* growth of HgCdTe p-on-n junction devices for FPAs. It is shown that MOCVD-IMP has progressed to the point that sophisticated bandgap-engineered multilayer HgCdTe device structures can be grown *in-situ* on a repeatable and dependable basis. Good run-to-run repeatability and control has been demonstrated for a series of identical grown runs of a MW/LW p-n-N-p double-heterojunction dual-band device structure.

Due to low sticking coefficient of As, *in-situ* p-type doping has been a challenging issue. Incorporation of As into the layer under Te rich conditions (i.e., high concentration of Hg vacancies) facilitates the activation of As [300]. MBE growth under Hg rich conditions results in twin defect formation, and As dopant may incorporate at surface boundaries of twin defects. As a result, As doped samples did not activate even at the highest annealing temperatures. Effective, near 100% activation is achieved for samples annealed under isothermal conditions at temperatures higher than 300°C.

Figure 14.42 illustrates cross-section schematics of two HgCdTe photodiode architectures fabricated using indium and arsenic doping during MBE growth [301]. The two-color detector is an n-p$^+$-n HgCdTe triple-layer heterojunction (TLHJ) back-to-back p$^+$-n photodiode structure grown on a (211) oriented CdZnTe substrate, and the single color detector is a p$^+$-n double layer heterojunction (DLHJ) detector design on a (211) Si substrate with ZnTe and CdTe buffer layers [301,302].

One important advantage of the p-on-n device (see Figure 14.42b) is that the n-type $Hg_{1-x}Cd_xTe$ carrier concentration is easy to control in the 10^{15} cm^{-3} range using extrinsic doping—usually indium or iodine (for the n-on-p device, the p-type carrier concentration at this low level is difficult). The wider bandgap capping $Hg_{1-y}Cd_yTe$ ($y \approx x + 0.04$) is 0.5 to 1 μm thick not intentionally doped. The structures are terminated with a thin (500 Å) CdTe layer for protection of the surface.

The formation of planar p-on-n photodiodes is achieved by first selectively implanting arsenic through windows made on a mask of photoresist/ZnS and then diffusing the arsenic through the cap layer into narrow gap layer [303]. The ion-implanted diffusion source is formed by a shallow implant with As to a dose of about 1×10^{14} cm^{-2}, at energies of 50–350 keV [303,304]. Usually, after implantation the structures are annealed under Hg overpressure. The samples underwent two consecutive annealings; one at high temperatures ($T \geq 350°C$ for a short time; e.g., 20 minutes [305]) and the other at 250°C for 24 hours immediately after. The first annealing is performed to anneal out the radiation-induced damage, and to diffuse and electrically active As by substituting As atoms on the Te sublattice, while the second one is to annihilate Hg vacancies formed in the HgCdTe lattice during growth and diffusion of arsenic to restore the sample background to n-type. The concentration of As dopant in p-type cap layer is at a level of about 10^{18} cm^{-3}.

14.6.1.6 Passivation

The key technology needed to make photodiodes possible was surface passivation. Passivation is a critical step in the HgCdTe photodiode technology that greatly affects surface leakage current and device thermal stability, so it is treated as a proprietary process by most manufacturers. A surface

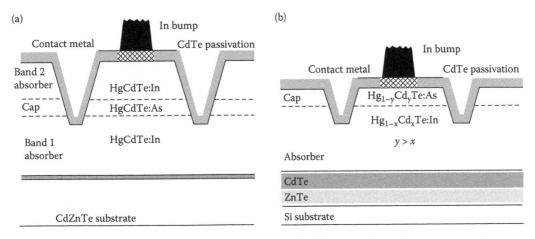

Figure 14.42 Cross-section schematic illustrations of MBE grown HgCdTe detector architectures: (a) two-color triple layer heterojunction (TLHJ) HgCdTe/CdZnTe, and (b) single-color double layer heterojunction (DLHJ) HgCdTe/Si. (From Smith, E. P. G., Bornfreund, R. E., Kasai, I., Pham, L. T., Patten, E. A., Peterson, J. M., Roth, J. A., et al., "Status of Two-Color and Large Format HgCdTe FPA Technology at Raytheon Vision Systems," *Proceedings of SPIE* 6127, 61261F, 2006. With permission.)

potential of less than 100 meV can accumulate, deplete, or invert the surface significantly, thus drastically affecting device performance. In comparison with photoconductive detectors, passivation of photodiodes is more difficult, since the same coating must stabilize simultaneously regions of n- and p-type conductivity. The most difficult is the passivation of p-type material due to its tendency for inversion.

Passivation of HgCdTe has been done by several techniques where a comprehensive review was given by Nemirovsky et al. [193,206,306,307]. Passivation technologies can be classified into three categories: native films (oxides [193,306,307], sulfides [193,204,308], fluorides [205,309]), deposited dielectrics (ZnS [310,311], SiO_x [200,290], Si_3N_4 [312], polymers) and in-situ grown heterostructures where a wider bandgap material is the passivant. A two-layer combination of a thick deposited dielectric film upon a thin native film of heterostructure is often the preferred passivation.

Based on silicon's success, HgCdTe passivation efforts were initially focused mainly on oxides. The native layers of HgCdTe were the most applied and investigated. Two major problems are associated with native films; they are formed by a wet electrochemical process requiring a conducting substrate, and thick native films become porous and do not adhere to the substrate. Hence, the native layers should be considered as a surface treatment and the insulation should be achieved with a deposited dielectric film. Anodic oxide was adequate for n-type photoconductors due to fixed positive charge. Applied to photodiodes, anodic oxide shorted out the devices by inverting the p-type surface. Silicon oxide was employed for photodiode passivation in the early 1980s based upon low-temperature deposition using a photochemical reaction [200]. However, it appears that the excellent surface properties (with low state densities and excellent photodiode properties) could not be maintained when the device were heated in vacuum for extended periods of time, a procedure required for good vacuum packaging integrity [7]. Also surface charge buildup was created when operated in a space-radiation environment.

The recent efforts are concentrated mostly on passivation with CdTe, CdZnTe [307] and heterojunction passivation [313]. These materials have appropriate bandgap, crystal structure, chemical binding, electrical characteristics, adhesion, and infrared transmission. Much pioneering work in this area was initially done by in France at Societe Anonymique de Telecommunicacion (SAT) [314]. It is desirable to use intrinsic CdTe with the layer-stoichiometry, minimal stress, and a low (below 10^{11} cm^{-2}) interface fixed charge density. The layers are sputtered, e-beam evaporated, and grown mainly by MOCVD and MBE.

The CdTe passivation can be obtained during epitaxial growth [209,315,316] or by the post-growth deposition [208,317] or anneal of the grown heterostructures in Hg/Cd vapors [318]. Directly grown in-situ CdTe layers lead to low fixed interface charge, the indirectly grown ones are acceptable, but not as good as the directly grown. Procedures for surface preparation prior to indirect CdTe deposition has been proposed [287,317]. Encouraging results have been achieved in in-situ grown MOCVD and MBE CdTe layers, but additional surface pre- and posttreatments are necessary to achieve near flat band conditions. Annealing at 350°C for one hour exhibited the lowest density of fixed surface states [315]. With low CdTe doping, however, the cap layer will be fully depleted and this may affect the interface charge. Bubulac et al. [319] have evaluated e-beam and MBE grown by SIMS and atomic force microscopy to determine usefulness of the material for passivation of HgCdTe. It has been suggested that CdTe grown by MBE at 90°C is more thermally stable and denser when compared to the e-beam grown material. The wide material is not necessarily single crystalline and in many cases a polycrystalline layer provides the passivation. CdTe passivation is stable during vacuum packaging bake cycles and shows little effect from the radiation found in space applications. Diodes do not shown a variation in R_0A product with diode size, indicating that surface perimeter effects can be neglected.

Bahir et al. [209] have discussed the results of TiAu/ZnS/CdTe/HgCdTe metal insulator semiconductor heterostructures. Samples with indirectly grown CdTe showed hysteresis, which can be attributed to the slow interface traps. A fixed charge density of $(5 \pm 2) \times 10^{10}$ cm^{-2} in p-type $Hg_{0.78}Cd_{0.22}Te$ has been reported. There is also evidence of the presence of slow interface traps in directly grown CdTe. The asymmetry of the CdTe/HgCdTe interface (see Figure 14.43) leads to large band bending and to carrier inversion in p-type HgCdTe. In the presence of the ZnS layer at the top, the band bending in CdTe is eliminated, resulting in flat band conditions in the CdTe/HgCdTe interface. Sarusi et al. [207] have investigated the passivation properties of CdTe/p-$Hg_{0.77}Cd_{0.23}Te$ heterostructure grown by MOCVD. The interface recombination velocity has been determined to be 5000 cm/s, which is a lower value than for ZnS passivated surfaces. The results demonstrated that CdTe is an ideal candidate for passivation of photodiodes based on p-type HgCdTe.

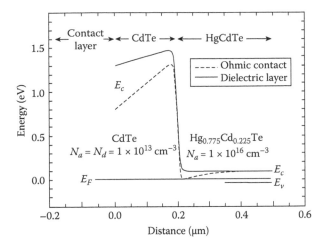

Figure 14.43 Effect of the contact layer on the energy band diagram of CdTe/HgCdTe. Ohmic contact to CdTe causes carrier inversion in p-type HgCdTe while the dielectric layer leads to nearly flat band conditions. (From Bahir, G., Ariel, V., Garber, V., Rosenfeld, D., and Sher, A., *Applied Physics Letters*, 65, 2725–27, 1994. With permission.)

It is well recognized that some degree of interface grading in CdTe/HgCdTe is beneficial for surface passivation [287]. This grading may shift the original HgCdTe defective surface to the wider gap CdTe region, which results in more thermally stable devices and improvements in surface passivation, irrespective of the CdTe growth method.

The passivation with ZnS, CdTe or wider gap $Hg_{1-x}Cd_xTe$ is often overcoated with SiO_x, SiN_x, and ZnS [287]. Thick deposited dielectric films are required to achieve the protection of the interface from environmental conditions, and to insulate the metallization pattern of the contacts and the bonding lads from the substrate. In addition, in front illuminated detectors optical properties of the dielectric films are of great importance: the film must exhibit excellent transmission in the relevant wavelength region and must possess an appropriate index of refraction for achieving a well-matched, antireflection coating at the required wavelength. The most commonly used deposited dielectrics are ZnS (either evaporated or magnetron sputtered), $CVD-SiO_2$ (either thermal or low temperature photo-assisted), and Si_3N_4 (deposited using electron cyclotron resonance or plasma enhanced).

Mestechkin et al. [208] have examined the bake stability for HgCdTe wafers and photodiodes with CdTe surface passivation deposited by thermal evaporation. It was found that the bakeout process (a 10-day vacuum bakeout at 80°C) generated additional defects at the CdTe/HgCdTe interface and degraded photodiode parameters. Annealing at 220°C under a Hg vapor pressure following the CdTe deposition suppressed the interface defect generation process during bakeout and stabilized HgCdTe photodiode performance.

14.6.1.7 Contact Metallization

The issues connected with contacts in photodiodes are contact resistance, contact surface recombination, contact $1/f$ noise, and long-term and thermal stability of devices. The contacts determine the performance and reliability of device [307]. Ideal ohmic contacts are obtained when the work function of the metal is less than the electron affinity of the n-type semiconductor (the barrier is negative) while for a p-type semiconductor it is when the barrier is positive. The various mechanisms proposed for metal/semiconductor barrier formation are reviewed in Section 9.5. Current favored theories are based on the presence of surface or interface states, which when present on a high enough density at a particular energy in the bandgap, can pin the Fermi level producing a barrier height that is approximately metal independent [320,321]. It was found that for $Hg_{1-x}Cd_xTe$ with $x < 0.4$, ohmic contacts are expected on n-type and Schottky barriers are expected on p-type HgCdTe. For $x > 0.4$, Schottky barriers are expected on both n- and p-type material. The common method of preparing ohmic contacts is by highly doping the surface region so that the space charge layer width is substantially reduced and electrons tunnel through yielding low-resistance contacts. In practice, however, ohmic contacts are generally prepared according to empirically derived recipes and the underlying science of the surface is often poorly understood. Practical contacts are often composed of several layers of different metals, which are required in order to promote adhesion and reduce solid state reactions.

Interfacial reactions between HgCdTe and various metal overlayers can be classified in four groups: ultrareactive, reactive, intermediate reactive, and unreactive. It depends on the relative heat of formation of HgTe and the overlayer metal telluride [322], and on the heat of formation of intermetallic compounds Cd and Hg with the overlayer metal [323]. Deposition of an ultrareactive Ti or reactive metal (Al, In, Cr), results in the formation of a metal telluride and induces a loss of Hg from the interfacial regions. Conversely, deposition of an unreactive metal such as Au, results in a stoichiometric interface with little loss of Hg.

Electrical measurements on ohmic contacts have been usually based on surface, which were chemically etched. Electrical properties of various metals to n- and p-type HgCdTe were strongly influenced by interface states [324]. The most popular metal for n-$Hg_{1-x}Cd_xTe$ for many years is indium [17,179,248], which has a low work function. Leech and Reeves examined In/n-HgCdTe contacts that displayed an ohmic nature over the range of composition $x = 0.30$ to 0.68 [325]. Carrier transport in these contacts has been attributed to a process of thermionic field emission. This behavior was attributed to the rapid in-diffusion of In, a substitutional donor in HgCdTe, to form a n$^+$ region below the contact. The value of specific contact resistance ranged from 2.6×10^{-5} Ωcm^2 at $x = 0.68$ through to 2.0×10^{-5} Ωcm^2 at $x = 0.30$, correlating with changes in the sheet resistivity of the HgCdTe.

In general, ohmic contacts to p-type HgCdTe is more difficult to realize since a larger work function of the metal contact is required. Au, Cr/Au, and Ti/Au have been most frequently used for both p-type HgCdTe. Investigations carried out by Beck et al. [326] have shown that Au and Al contacts to p-type $Hg_{0.79}Cd_{0.21}Te$ exhibited ohmic characteristics where the specific contact resistance varies from 9×10^{-4} Ωcm^2 to 3×10^{-3} Ωcm^2 at room temperature. The diameter dependence of $1/f$ noise implied that noise in the Au contacts originated at or near the Au/HgCdTe interface while the noise in the Al contacts originated from a surface conduction layer near the contact.

For lightly doped p-type $Hg_{1-x}Cd_xTe$, no good contacts exist and all metals tend to form Schottky barriers. The problem is especially difficult for high x composition material. This problem could be solved by heavy doping of semiconductor in the region close to the metallization to increase tunnel current, but the required level of doping is difficult to achieve in practice. One practical solution is to use rapid narrowing of bandgap at the $Hg_{1-x}Cd_xTe$-metal interface [327].

14.6.2 Fundamental Limitation to HgCdTe Photodiode Performance

From consideration carried out previously [see Section 14.4], the Auger mechanisms impose fundamental limitations to the HgCdTe photodiode performance. Assuming that the saturation dark current is only due to thermal generation in the base layer and that its thickness is low compared to the diffusion length,

$$J_s = Gtq,$$ (14.50)

where G is the generation rate in the base layer. Then the zero bias resistance-area product is (see Equation 9.83)

$$R_oA = \frac{kT}{q^2Gt}.$$ (14.51)

Taking into account the Auger 7 mechanism in extrinsic p-type region of n$^+$-on-p photodiode, we receive

$$R_oA = \frac{2kT\tau_{A7}^i}{q^2N_at},$$ (14.52)

and the same equation for p-on-n photodiode

$$R_oA = \frac{2kT\tau_{A1}^i}{q^2N_dt},$$ (14.53)

where N_a and N_d are the acceptor and donor concentrations in the base regions, respectively.

As Equations 14.52 and 14.53 show, the R_oA product can be decreased by reduction of the thickness of the base layer. Since $\gamma = \tau^i_{A7}/\tau^i_{A1} > 1$, a higher R_oA value can be achieved in p-type base devices compared to that of n-type devices of the same doping level. Detailed analysis shows that the absolute maximum of R_oA is achievable with base layer doping producing $p = \gamma^{1/2}n_i$, which corresponds to the minimum of thermal generation. The required p-type doping is difficult to achieve in practice for low temperature photodiodes (the control of hole concentration below 5×10^{15} cm^{-3} level is difficult) and the p-type material suffer from some nonfundamental limitations, such as: contacts, surface, and SR processes.

In 1985, Rogalski and Larkowski indicated that, due to the lower minority-carrier diffusion length (lower mobility of holes) in the n-type region of p$^+$-on-n junctions with thick n-type active region, the diffusion-limited R_oA product of such junctions is larger than for n$^+$-on-p ones (see Figure 14.44) [328]. These theoretical predictions were next confirmed by experimental results obtained for p$^+$-on-n HgCdTe junctions.

The thickness of the base region should be optimized for near unity quantum efficiency and a low dark current. This is achieved with a base thickness slightly higher than the inverse absorption coefficient for single pass devices: $t = 1/\alpha$ (which is $\approx$10 µm) or half of the $1/\alpha$ for double pass devices (devices supplied with a retroreflector). Low doping is beneficial for a low thermal generation and high quantum efficiency. Since the diffusion length in the absorbing region is typically longer than its thickness, any carriers generated in the base region can be collected giving rise to the photocurrent.

Different HgCdTe photodiode architectures have been fabricated that are compatible with back side and front side illuminated hybrid FPA technology. The most important eight architectures are included in Table 14.5, which summarizes the applications of HgCdTe photodiode designs by the major FPA manufacturers today.

Figures of configurations II and III show cross sections of the two most important n-on-p HgCdTe junction structures adapted in fabrication of multicolor detectors. The structure III pioneered by SAT has been the most widely developed and used by Sofradir [330]. The second type of

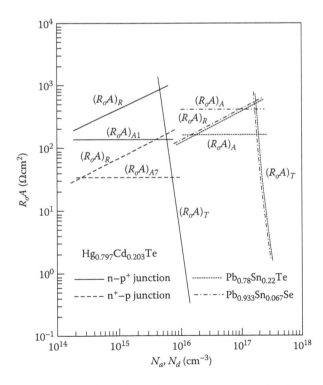

Figure 14.44 The dependence of the R_oA product components on the dopant concentrations for the one-sided abrupt junctions produced in Hg$_{0.797}$Cd$_{0.203}$Te, Pb$_{0.78}$Sn$_{0.22}$Te, and Pb$_{0.933}$Sn$_{0.067}$Se. (From Rogalski, A., and Larkowski, W., *Electron Technology*, 18 (3/4), 55–69, 1985. With permission.)

Table 14.5: HgCdTe Photodiode Architectures Used for Hybrid FPAs

Configuration	Architecture	Junction Formation	Company	References
I	n-on-p VIP		Ion implantation forms n-on-p diode in p-type HgCdTe, grown by Te-solution LPE on CdZnTe and epoxied to silicon ROIC wafer; over the edge contact	DRS Infrared Technologies (formerly Texas Instruments)	329
II	n-p loophole		Ion beam milling forms n-type islands in p-type Hg-vacancy-doped layer grown by Te-solution LPE on CdZnTe, and epoxied onto silicon ROIC wafer; cylindrical lateral collection diodes	GEC-Marconi Infrared (GMIRL)	250,270
III	n$^+$-on-p planar		Ion implant into acceptor-doped p-type LPE film grown by Te-solution slider	Sofradir (Societe Francaise de Detecteurs Infrarouge)	330
IV	n$^+$-n-p planar homojunctions		Boron implant into Hg-vacancy p-type, grown by Hg-solution tipper on 3 inch diameter sapphire with MOCVD CdTe buffer; ZnS passivation	Rockwell/Boeing	331

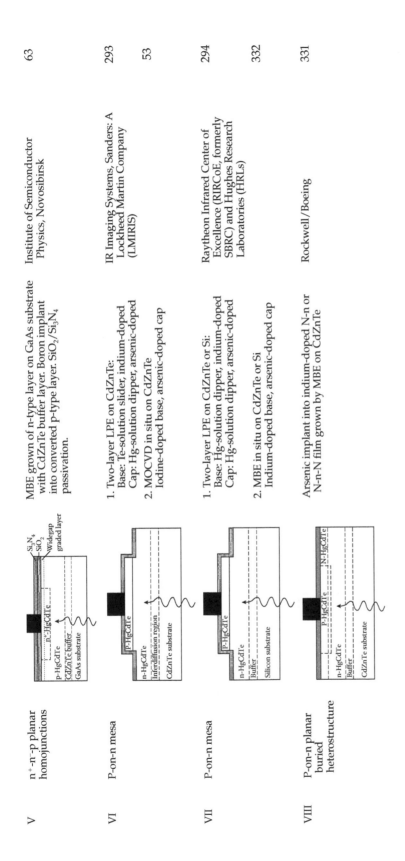

	Structure	Cross-section	Technology	Company	Ref.
V	n⁺-n⁻-p planar homojunctions		MBE grown of n-type layer on GaAs substrate with CdZnTe buffer layer. Boron implant into converted p-type layer. SiO₂/Si₃N₄ passivation.	Institute of Semiconductor Physics, Novosibirsk	63
VI	P-on-n mesa		1. Two-layer LPE on CdZnTe: Base: Te-solution slider, indium-doped Cap: Hg-solution dipper, arsenic-doped	IR Imaging Systems, Sanders: A Lockheed Martin Company (LMIRIS)	293
			2. MOCVD in situ on CdZnTe Iodine-doped base, arsenic-doped cap		53
VII	P-on-n mesa		1. Two-layer LPE on CdZnTe or Si: Base: Hg-solution dipper, indium-doped Cap: Hg-solution dipper, arsenic-doped	Raytheon Infrared Center of Excellence (RIRCoE, formerly SBRC) and Hughes Research Laboratories (HRLs)	294
			2. MBE in situ on CdZnTe or Si Indium-doped base, arsenic-doped cap		332
VIII	P-on-n planar buried heterostructure		Arsenic implant into indium-doped N-n or N-n-N film grown by MBE on CdZnTe	Rockwell/Boeing	331

structure (configuration II) is a vertically integrated photodiode (VIP™) developed by DRS Infrared Technologies [333]. This structure currently referred to as a high-density, vertically integrated photodiode (HDVIP) is similar to the British-developed loophole photodiode [270]. This n^+-n^--p architecture is formed around the via, both by the etching process itself and by a subsequent ion implant step. Low background indium n^--doping levels of 1.5 to 5×10^{14} cm^{-3} are routinely used. The p-type dopant is usually copper (acceptor concentration about 4×10^{16} cm^{-3}). The p$^+$-p noninjection contacts are formed in each cell of the FPA and are joined electrically by a top surface metal grid as shown in Figure 14.45.

The cross sections of the second type of structure used in fabrication of multicolor detectors, heterojunction p-on-n HgCdTe photodiode are illustrated as configuration VI (see also Figure 14.42b). In these so-called double-layer heterojunction (DLHJ) structures, an absorber layer about 10 μm thick is doped with indium at 1×10^{15} cm^{-3} or less, and is sandwiched between the CdZnTe substrate and the highly arsenic-doped, wider-gap region. Contacts are made to the p$^+$-layer in each pixel and to the common n-type layer at the edge of the array (not shown). Infrared flux is incident through the IR-transparent substrate.

The formation of planar p-on-n photodiodes is achieved by selective area arsenic ion implantation through the cap layer into the narrow gap base layer [305]. The dopant activation step is achieved by a two-step thermal anneal under Hg overpressure: the first step, at high temperature, activates the dopant by substituting As atoms on the Te sublattice, and the second, at lower temperature, annihilates the Hg vacancies formed in the HgCdTe lattice during growth and high-temperature annealing step.

Figure 14.46 shows the schematic band profiles of the most commonly used unbiased homo- (n^+-on-p) and heterojunction (P-on-n) photodiodes. To avoid contribution of the tunneling current, the doping concentration in the base region below 10^{16} cm^{-3} is required. In both photodiodes, the lightly doped narrow gap absorbing region ["base" of the photodiode: p(n)-type carrier concentration of about 5×10^{15} cm^{-3} (5×10^{14} cm^{-3})], determines the dark current and photocurrent. The internal electric fields at interfaces are "blocking" for minority carriers and the influence of surface recombination is eliminated. Also suitable passivation prevents the influence of surface recombination. Indium is most frequently used as a well-controlled dopant for n-type doping due to its high solubility and moderately high diffusion. Elements of the VB group are acceptors substituting Te sites. They are very useful for fabrication of stable junctions due to very low diffusivity. Arsenic proved to be the most successful p-type dopant to date. The main advantages are stability in lattice, low activation energy, and the possibility to control concentration over 10^{15}–10^{18} cm^{-3} range. Intensive efforts are currently underway to reduce a high temperature (400°C) required to activate As as an acceptor.

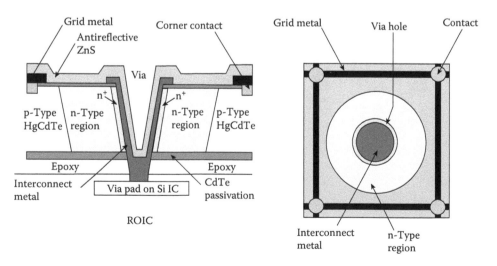

Figure 14.45 DRS's high-density vertically integrated photodiode (HDVIP™) n^+-n^--p HgCdTe photodiode. (From Kinch, M., "HDVIP™ FPA Technology at DRS," *Proceedings of SPIE* 4369, 566–78, 2001. With permission.)

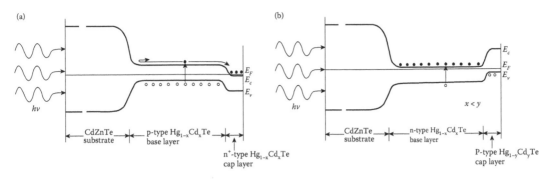

(a)

E_F
E_c
E_v

$h\nu$

CdZnTe substrate | p-type $Hg_{1-x}Cd_xTe$ base layer

n^+-type $Hg_{1-x}Cd_xTe$ cap layer

(b)

E_c
E_F
E_v

$x < y$

$h\nu$

CdZnTe substrate | n-type $Hg_{1-x}Cd_xTe$ base layer

P-type $Hg_{1-y}Cd_yTe$ cap layer

Figure 14.46 Schematic band diagrams of (a) n^+-on-p and (b) p-on-n heterojunction photodiodes.

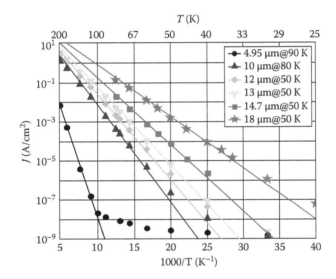

Figure 14.47 Mercury vacancy dark current modeling for n-on-p HgCdTe photodiodes with different cutoff wavelengths. (From Gravrand, O., Chorier, Ph., and Geoffray, H., "Status of Very Long Infrared Wave Focal Plane Array Development at DEFIR," *Proceedings of SPIE* 7298, 7298-75, 2009. With permission.)

The n-on-p junctions are fabricated in two different manners using Hg vacancy doping and extrinsic doping. Hg vacancies (V_{Hg}) provide intrinsic p-type doping in HgCdTe. In this case, the doping level depends on only one annealing temperature. However, the use of Hg vacancy as p-type doping is known to kill the electron lifetime (see Figure 14.17), and the resulting detector exhibits a higher current than in the case of extrinsic doping use. However, for very low doping ($<10^{15}$ cm^{-3}), the hole lifetime becomes SR limited and does not depend on doping anymore [137]. V_{Hg} technology leads to low minority diffusion length of the order of 10–15 μm, depending on doping level. Generally, n-on-p vacancy doped diodes give rather high diffusion currents but lead to a robust technology as its performance weakly depends on doping level and absorbing layer thickness. Simple modeling manages to describe dark current behavior of V_{Hg} doped n-on-p junctions over a range of at least eight orders of magnitude (see Figure 14.47 [334]). In the case of extrinsic doping, Cu, Au, and As are often used. Due to higher minority carrier lifetime, extrinsic doping is used for low dark current (low flux) applications [134]. The extrinsic doping usually leads to larger diffusion length and allows lower diffusion current but might exhibit performance fluctuations, thus affecting yield and uniformity.

In the case of p-on-n configuration, the typical diffusion length is up to 30 μm to 50 μm for low doping levels such as 10^{15} cm^{-3} typically reached with indium doping and the dark currier

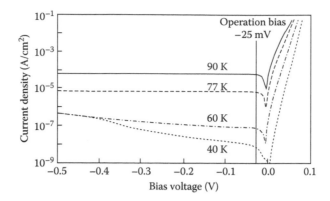

Figure 14.48 I-V characteristics at various temperatures for a 12 µm cutoff p-on-n HgCdTe photodiode. (From Norton, P., "Status of Infrared Detectors," *Proceedings of SPIE* 3379, 102–14, 1998. With permission.)

generation is volume limited by the absorbing layer volume itself [334]. Different behavior of saturation current depending on diffusion length is explained by Equations 9.87 and 9.88. If $t \gg L$, the saturation current is inversely proportional to $N\tau^{1/2}$. On the contrary, if $t \ll L$, the saturation dark current is inversely proportional to $N\tau$.

In ideal photodiodes the diffusion current is dominant, therefore their leakage current is very low and insensitive to the detector bias. Leakage current is the primary contribution of unwanted noise. Figure 14.48 shows typical current-voltage characteristics of an HgCdTe photodiode at temperatures between 40 and 90 K for a 12 µm cutoff detector at 40 K [335]. The leakage current is less than 10^{-5} A/cm² at 77 K. The bias-independent leakage current makes it easier to achieve better FPA uniformity, as well as to reduce the detector bias-control requirements during changes in photocurrent.

The quality of HgCdTe photodiodes has improved steadily over the past 20 years as materials and device processing science evolved and has progressed to the point where there is usually no clear indication of g-r current. The plots of R_oA data versus temperature generally follow a diffusion current dependence at higher temperatures, and transition into a comparatively temperature-independent tunneling-like regime at lower temperatures. An example of such behavior is shown in Figure 14.49 for 10 µm cutoff wavelength boron-implanted n-on-p HgCdTe photodiode [134]. The p-type based layer of photodiode was obtained by LPE deposition from a Te-rich solution on a CdZnTe substrate. Typical sets of reverse current-voltage characteristics as a function of temperatures are given in Figure 14.49a. The continuous curves represent the experimental data, and the dashed curves represent the calculated curves [336]. We can observe an excellent agreement between both types of results in a wide region of temperature down to 60 K, and in the region of reverse bias voltage below 0.1 V where diffusion current dominates. In the region of higher bias voltage, an excess of leakage current is clearly visible. The sharp reverse breakdowns at bias voltages above 0.1 eV that move to progressively lower voltage as temperature is decreased, indicating interband tunneling. In Figure 14.49b, the complete R_oA evolution versus temperature is presented under a 0° field of view for diodes with similar cutoff wavelengths. Both devices show $R_oA \propto n_i^{-2}$ dependence down to $T \approx 50$ K. The R_oA saturation observed at very low temperature ($T < 40$ K) is due to instrumental limitations of impedance measurement ($R > 5 \times 10^{10}\ \Omega$). The n⁻-type region, created during junction fabrication by ion implantation, slightly influences on the R_oA product of photodiode, if the thickness of n⁻-type layer changes from 0 to 5 µm (the R_oA product changes only about 5%) [336]. The same conclusion concerns quantum efficiency.

Measured resistances are also affected by the background-induced shunt resistance effect reported by Rosbeck et al. [101]. Figure 14.50 shows the temperature dependence of the R_oA product of n-on-p $Hg_{0.768}Cd_{0.232}Te$ photodiode for four separate field of view conditions of 180°, 15°, 7°, and 0°. For the 0° FOV condition, only the superposition of current components associated with generation-recombination (g-r) in the depletion region and thermal diffusion current from the bulk material are considered. The larger FOV conditions are analyzed by summing the thermal diffusion, g-r, and background currents. The influence of background radiation was found to create an exceptionally linear reverse I-V characteristics that can be easily distinguished from other current mechanisms. Flux dependent R_oA products have been shown to be a direct result of bias

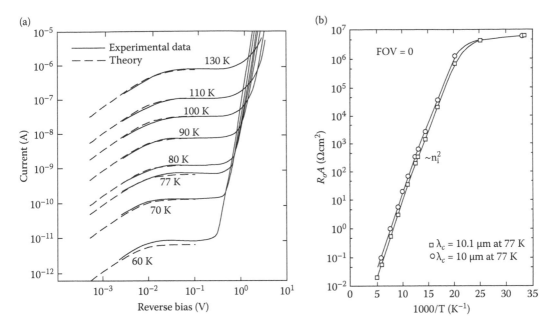

Figure 14.49 Dark characteristics of boron-implanted n⁺-n⁻-p Hg₀.₇₇₆Cd₀.₂₂₄Te photodiode: (a) I-V characteristics versus temperature (continuous curves represent the measured values after Destefanis G. and Chamonal J. P., *Journal of Electronic Materials*, 22, 1027–32, 1993. With permission.) and the dashed lines represent calculated values after (Rogalski, A., and Ciupa, R., *Journal of Applied Physics*, 80, 2483–89, 1996. With permission.) A = 7.2 × 10⁻⁵ cm², λ$_c$ = 10.0 μm at 77 K; (b) R$_o$A product versus 1/T. (From Destefanis, G., and Chamonal, J. P., *Journal of Electronic Materials*, 22, 1027–32, 1993. With permission.)

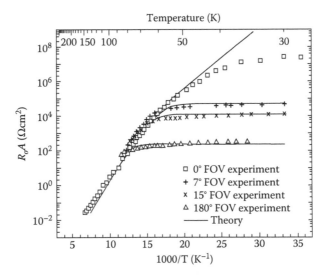

Figure 14.50 R$_o$A product versus temperature for thick base n-side illuminated n-on-p Hg₀.₇₆₈Cd₀.₂₃₂Te photodiode for several background levels. (From Rosbeck, J. P., Star, R. E., Price, S. L., and Riley, K. J., *Journal of Applied Physics*, 53, 6430–40, 1982. With permission.)

dependent quantum efficiency, a mechanism that is much more evident in heterojunction device architectures [337].

Recently, Tennant et al. [338] have developed a simple empirical relationship that describes the dark current behavior with temperature and wavelengths for the better Teledyne HgCdTe diodes and arrays [primarily double layer planar heterojunction (DLPH) structure devices].

It is called Rule 07 and predicts the dark current density within a factor of 2.5 over a 13 order of magnitude range. The formula for Rule 07 is approximately (exact formula given in the reference)

$$J_{dark} = 8367\exp\left(-\frac{1.44212q}{k\lambda_c T}\right) \quad \text{for} \quad \lambda_c \geq 4.635 \, \mu m, \tag{14.54}$$

and

$$J_{dark} = 8367\exp\left\{-\frac{1.44212q}{k\lambda_c T}\left[1 - 0.2008\left(\frac{4.635 - \lambda_c}{4.635\lambda_c}\right)^{0.544}\right]\right\} \quad \text{for} \quad \lambda_c < 4.635 \, \mu m, \tag{14.55}$$

where λ_c is the cutoff wavelength in μm, T is the operating temperature in K, q is the electron charge, and k is the Boltzmann's constant (both of the latter in SI units). Rule 07 was developed for operating temperature cutoff wavelength products between 400 μmK and ~1700 μmK, and for operating temperatures above 77 K. Rule 07 is an excellent tool for a quick comparison of R_oA product of other material systems with HgCdTe. However, caution should be taken when expanding it to other parameters, such as detectivity and lower operating temperatures.

Figure 14.51 shows the fit data, both with Teledyne data and some of the better data from other vendors for comparison including representative data from both InSb (~5.3 μm) and InGaAs (~1.7 μm). InGaAs shows lower dark current density than DLPH HgCdTe but InSb has much higher dark current density. Note that λ_e equals the cutoff wavelength for wavelengths above ~5 μm, but is somewhat longer than the cutoff for wavelengths below 5 μm.

Figure 14.52 is an accumulation of R_oA data with different cutoff wavelength [334], taken from LETI LIR using both V_{Hg} and extrinsic doping n-on-p HgCdTe photodiodes compared with P-on-n data from other laboratories, extracted from various recent literature reports, using various techniques and different diode structures [339–342]. It is clearly shown that extrinsic doping of n-on-p photodiodes leads to higher R_oA product (lower dark current) whereas mercury vacancy doping remains on the bottom part of the plot. P-on-n structures are characterized by the lowest dark current (highest R_oA product); see trend line calculated with Teledyne (formerly Rockwell Scientific) empirical model [338].

An additional insight on dark current achieved by Teledyne gives Figure 14.53 [343], where the dark current scaled for an 18 μm square pixel of MBE grown P-on-n HgCdTe photodiodes is shown. The dark current less than 0.01 electrons per pixel per second for 2.5 and 5 μm cutoff wavelengths can be achieved. It should be marked that the cutoff wavelength is shown with the approximation symbol, since λ_c is a function of temperature and there will be slight variation in cutoff wavelength of a HgCdTe photodiode as it cools.

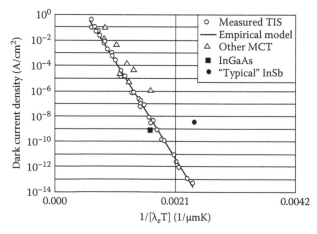

Figure 14.51 The data fit leading to Rule 07. (From Tennant, W. E., Lee, D., Zandian, M., Piquette, E., and Carmody, M., *Journal of Electronic Materials*, 37, 1407–10, 2008. With permission.)

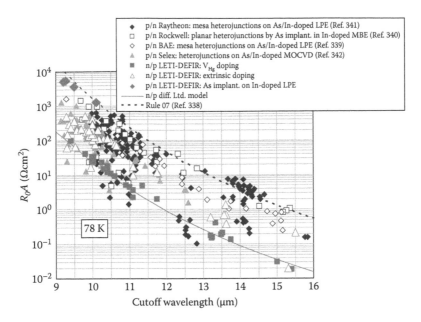

Figure 14.52 R_oA product versus cutoff wavelength at 78 K, summarized with bibliographic data. (From Gravrand, O., Chorier, Ph., and Geoffray, H., "Status of Very Long Infrared Wave Focal Plane Array Development at DEFIR," *Proceedings of SPIE* 7298, 7298-75, 2009.)

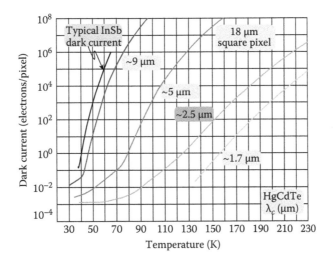

Figure 14.53 The dark current of Teledyne's MBE grown P-on-n HgCdTe photodiodes scaled for an 18 μm square pixel. (From Baletic, J. W., Blank, R., Gulbransen, D., Lee, D., Loose, M., Piquette, E. C., Sprafke, T., Tennant, W. E., Zandian, M., and Zino, J., "Teledyne Imaging Sensors: Infrared Imaging Technologies for Astronomy & Civil Space," *Proceedings of SPIE* 7021, 70210H, 2008. With permission.)

Figure 14.54 illustrates the highest measurable R_oA values of P-on-n HgCdTe photodiodes versus cutoff wavelength at different temperatures [340]. The solid lines are theoretically calculated using a 1-D model that assumes diffusion current from narrower bandgap n-side is dominant, and minority carrier recombination via Auger and radiative process. The R_oA of Teledyne HgCdTe photodiodes exhibit near theoretical performance for various growth material cutoffs at various temperatures. The average value of the R_oA product at 77 K for a 10 μm cutoff HgCdTe photodiode is around 1000 Ωcm^2 and drops to 200 Ωcm^2 at 12 μm. At 40 K, the R_oA product varies between 10^6 and 10^8 Ωcm^2 at 12 μm.

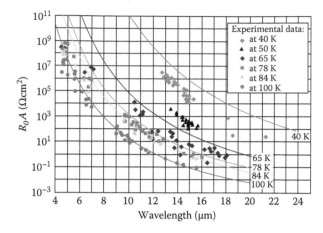

Figure 14.54 R_0A product versus cutoff wavelength for Teledyne's (formerly Rockwell Scientific) P-on-n HgCdTe photodiode data at various temperatures compared to the theoretical 1-D diffusion model. (From Chuh, T., "Recent Developments in Infrared and Visible Imaging for Astronomy, Defense and Homeland Security," *Proceedings of SPIE* 5563, 19–34, 2004. With permission.)

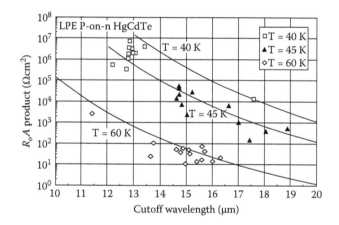

Figure 14.55 Median R_oA data for 20-element arrays of LPE P-on-n HgCdTe photodiodes at 60, 45, and 40 K, compared with calculated R_oA due to n-side diffusion current (solid curves). (From Reine, M. B., *Infrared Detectors and Emitters: Materials and Devices*, Chapman and Hall, London, 313–76, 2000. With permission.)

Several major advantages in development of VLWIR HgCdTe technology have been presented by Reine et al. [17,344]. Continued improvements in defect reduction have enabled R_0A products to follow the diffusion current limit down to temperatures of 40–45 K for cutoff wavelengths out to 19 μm (see Figure 14.55).

The Auger 1 and Auger 7 recombination mechanisms are relatively insignificant in the wide gap HgCdTe alloys needed for SWIR and MWIR applications, thus the only fundamental recombination mechanism to be considered is radiative recombination. This is illustrated in Figure 13.10a for three groups of p-on-n devices fabricated using LPE, MBE, and MOCVD. For photodiodes at temperature 180 K, the R_oA data fall generally about a factor of 10 below the theoretical curves, indicating that a lifetime mechanism other than traditional radiative recombination is lowering the lifetime. According to DeWames et al. [345], a shallow SR recombination center, possibly process-induced, is responsible for the reduced lifetime. It appears that for photodiodes operated at room temperature with cutoff wavelengths less than 3.5 μm, the R_oA product fall short of the limits calculated with traditional radiative recombination equations. However, the highest quality SW photodiodes fabricated with HgCdTe (and also with InGaAs alloys), have performance levels in agreement with the radiative limit.

The performance of conventional p-n junction LW HgCdTe photovoltaic detectors operating at near room temperature is very poor due to a low quantum efficiency (low diffusion length and weak absorption of radiation) and a low dynamic resistance. Even more serious problems are caused by the series resistance and extremely low junction resistance and voltage responsivity. This can be overcome with the development of multiple heterojunction photovoltaic devices in which short elements were connected in series. An example is a device with junction's planes perpendicular to the substrate (Figure 14.56a). The multiheterojunction device consisted of a structure based on back side illuminated n⁺-p-P photodiodes. This device was the first commercially available uncooled and unbiased long wavelength photovoltaic detector introduced in 1995. Such devices are characterized by large voltage responsivity, fast response time, but they suffer from nonuniform response across the active area and dependence of response on polarization of incident radiation.

More promising are the stacked photovoltaic cells monolithically connected in series shown in Figure 14.56b. They are capable of achieving both good quantum efficiency and a large differential resistance. Each cell is composed of p-type doped narrow gap absorber and heavily doped N⁺ and P⁺ heterojunction contacts. The incoming radiation is absorbed only in absorber regions, while the heterojunction contacts collect the photogenerated charge carriers. Such devices are capable of achieving high quantum efficiency, large differential resistance and fast response. The practical problem is the shortage of adjacent N⁺ and P⁺ regions. This can be achieved employing tunnel currents at the N⁺ and P⁺ interface.

At present, the VIGO System offers photovoltaic devices optimized at any wavelength in the LWIR, MWIR, and SWIR range of infrared spectrum [346]. The longest usable wavelength is 11 μm, 13 and 15 μm for uncooled, 2- and 3-stage Peltier coolers, respectively.

Figure 14.57 shows the performance of the HgCdTe devices [347]. Without optical immersion MWIR photovoltaic detectors are sub-BLIP devices with performance close to the g-r limit, but well-designed, optically immersed devices approach BLIP limit when thermoelectrically cooled with 2-stage Peltier coolers. The situation is less favorable for >8 μm LWIR photovoltaic detectors; they show detectivities below the BLIP limit by an order of magnitude. Typically, the devices are used at zero bias. The attempts to use Auger suppressed nonequilibrium devices were not successful due to large 1/f noise extending to ≈100 MHz in extracted photodiode.

The HOT devices are characterized by a very fast response. The uncooled ≈ 10 μm photodetectors show ≈1 ns or less response time. The *RC* time constant of photovoltaic devices can be shortened by the use of optical immersion to reduce the physical area of devices [157,173,174]. The series resistance was minimized to ≈1 Ω using heavily doped N⁺ for base regions of the mesa structures and improved anode contact.

Since HgCdTe ternary alloy is a direct bandgap semiconductor, it is a very efficient absorber of radiation due to high absorption coefficient. The absorption depth defined as the distance over which 63% (1 – 1/e) of the photon flux is absorbed is shown in Figure 14.58 [343]. To receive high quantum efficiency, the thickness of active detector layer should be equal, at least, to the cutoff wavelength. Typical quantum efficiency with antireflection coating is about 90% Substrate removing of the back side illuminated photodiode results in an increase of quantum efficiency in the SW spectral range. This effect is shown in Figure 14.59b after removal of the CdZnTe substrate after the hybridization process [348]. HgCdTe photodiodes are available to cover the spectral range from 1 to 20 μm. Figure 14.59a illustrates representative spectral response from photodiodes. Spectral cutoff can be tailored by adjusting the HgCdTe alloy composition.

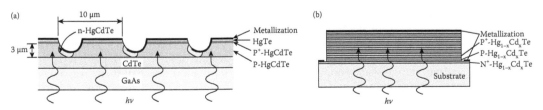

Figure 14.56 Back side illuminated multiple heterojunction devices: (a) junction's planes perpendicular to the surface, and (b) 4-cells stacked multiple detector. (From Piotrowski, J., and Rogalski, A., *High-Operating Temperature Infrared Photodetectors*, SPIE Press, Bellingham, WA, 2007. With permission.)

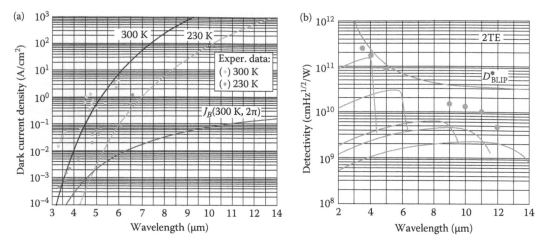

Figure 14.57 Performance of optically immersed HgCdTe devices: (a) dark current density at 300 and 230 K, and (b) typical spectral detectivity of cooled devices with 2-stage coolers; dots indicate the highest experimental data. (From Rogalski, A., "HgTe-Based Photodetectors in Poland," *Proceedings of SPIE* 7298, 72982Q, 2009. With permission.)

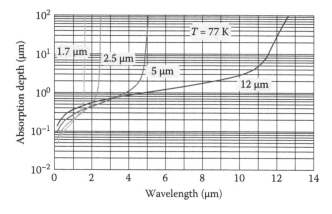

Figure 14.58 Absorption depth of photons in HgCdTe as a function of cutoff wavelength at 77 K. (From Baletic, J. W., Blank, R., Gulbransen, D., Lee, D., Loose, M., Piquette, E. C., Sprafke, T., Tennant, W. E., Zandian, M., and Zino, J., "Teledyne Imaging Sensors: Infrared Imaging Technologies for Astronomy & Civil Space," *Proceedings of SPIE* 7021, 70210H, 2008. With permission.)

14.6.3 Nonfundamental Limitation to HgCdTe Photodiode Performance

Many additional excess mechanisms affect the dark current of HgCdTe photodiodes [73]. They arise from nonfundamental sources located in the base and cap layer, the depletion region, and the surface. As the operating temperature is lowered, the thermal dark current mechanisms become weaker and allow other mechanisms to prevail. In practice, nonfundamental sources dominate the dark current of the present HgCdTe photodiodes, with the exception of specific cases of near room temperature devices and the highest quality 77 K LWIR/VLWIR and 200 K MWIR devices. The main leakage mechanisms of HgCdTe photodiodes are: generation in the depletion region, interband tunneling, trap-assisted tunneling, and impact ionization. Some of them are caused by structural defects in the p-n junction. These mechanisms receive much attention now, particularly because they ultimately determine the array uniformity, yield, and cost for some applications, especially those with lower operating temperatures.

Many authors have successfully modeled the key reverse bias leakage mechanism in LWIR HgCdTe photodiodes at ≤77 K in terms of trap-assisted tunneling [16,349–356]. An alternative model to explain the reverse bias leakage current in LWIR HgCdTe loophole photodiodes has been proposed by Elliott et al. [357] This model is based upon an impact ionization effect

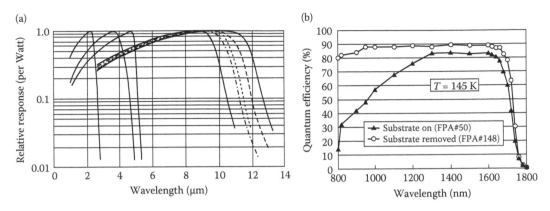

Figure 14.59 Spectral characteristics of HgCdTe photodiodes: (a) representative spectral response data. (From Norton, P. R., *Optical Engineering*, 30, 1649–63, 1991. With permission.) (b) Quantum efficiency comparison for substrate-on versus substrate-removed NIR FPAs. (From Piquette, E. E., Edwall, D. D., Arnold, H., Chen, A., and Auyeung, J., *Journal of Electronic Materials*, 37, 1396–1400, 2008. With permission.)

within the depletion layer and, in fairly low doped devices, this gives a good fit to practical observations. The reverse bias characteristics arising from impact ionization have two main features: firstly, the leakage current is fairly insensitive to temperature over a wide range, and secondly, the current increases much more slowly with reverse bias than currents due to tunneling.

In high quality $Hg_{1-x}Cd_xTe$ photodiodes with $x \approx 0.20$, the diffusion current in the zero-bias and the low-bias region is usually the dominant current down to 40 K [333,337,358,359]. At medium values of reverse bias, the dark current is mostly due to trap-assisted tunneling. Trap-assisted tunneling dominates the dark current also at zero bias and very low temperature (below 30 K). At high values of reverse bias, bulk band-to-band tunneling dominates. At very low temperatures, below 30 K, significant spreads in the R_oA product distributions are typically observed due to the onset of tunneling currents associated with localized defects. Moreover, HgCdTe photodiodes often have an additional surface-related component of the dark current, particularly at low temperatures.

For example, Figure 14.60 shows the measured and modeled data at a temperature 66.7 K for P-on-n HgCdTe photodiode with a cutoff wavelength of 15.6 μm at 78 K, equates with 18.7 μm at 40 K and 20.0 μm at 28 K [358]. The I-V characteristics were found to be diffusion limited by the ideal diffusion currents over much of the temperature and bias ranges analyzed. Band-to-band tunneling (BTB) was found to limit the current at the largest reverse biases (>200 mV) and lowest temperatures. The trap-assisted tunneling (TAT) mechanisms limit the current under low temperatures and moderate bias (50 mV < V < 200 mV).

Figure 14.61 presents, as an example, $I(V)$ curves measured at different temperatures for a LWIR p-on-n and a VLWIR n-on-p photodiodes [334]. It is shown that thermal current strongly decreases with temperature, but the high bias current exhibits the opposite thermal behavior, which is the usual signature of tunnel contributions. In order to minimize tunneling currents, low doping levels are preferable on at least one side of the junction. Note however that the p-on-n photodiode active region is doped to $N_d \leq 10^{15}$ cm^{-3}, whereas the n-on-p doping level is higher, $N_a > 10^{16}$ cm^{-3}. We can also see that the diffusion limited performance extends to larger reverse biases at higher temperatures.

Chen et al. [360] carried out a detailed analysis of the wide distribution of the R_o values of HgCdTe photodiodes operating at 40 K. Figure 14.62 shows the cumulative distribution function, R_o, obtained in devices with a cutoff wavelength between 9.4 and 10.5 μm. It is clear that while some devices exhibit a fair operability with R_o values spanning only two orders of magnitude, other devices show a poor operability with R_o values spanning more than 5–6 orders of magnitude. Lower performance, with R_o values below 7×10^6 Ω at 40 K, is usually due to gross metallurgical defects, such as dislocation clusters and loops, pin holes, striations, Te inclusions, and heavy terracing. However, diodes with R_o values between 7×10^6 Ω and 1×10^9 Ω at 40 K contained no visible defects (Hg interstitials and vacancies).

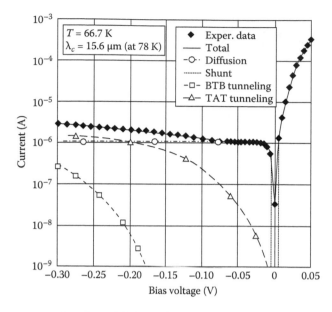

Figure 14.60 The measured and modeled I-V characteristics at a temperature 66.7 K for P-on-n HgCdTe photodiode with a cutoff wavelength of 15.6 μm at 78 K. (From Gilmore, A. S., Bangs, J., and Gerrishi, A., *Journal of Electronic Materials*, 35, 1403–10, 2006. With permission.)

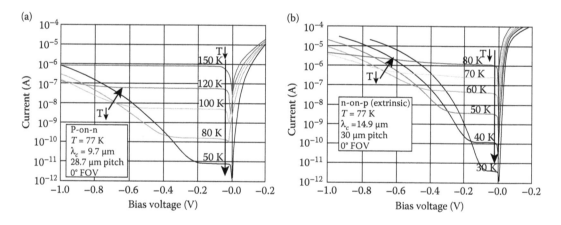

Figure 14.61 Typical dark current-voltage characteristics of (a) a p-on-n and (b) n-on-p HgCdTe photodiodes at different temperatures. (From Gravrand, O., Chorier, Ph., and Geoffray, H., "Status of Very Long Infrared Wave Focal Plane Array Development at DEFIR," *Proceedings of SPIE* 7298, 7298-75, 2009. With permission.)

Dislocations are known to increase the dark current and the $1/f$ noise current. At 77 K the requirement on dislocation density for LWIR material is $< 2 \times 10^5$ cm^{-2}. MWIR, on the other hand, can tolerate higher densities of dislocations at 77 K, but evidence is mounting that this is no longer true at higher operating temperatures [333]. The reverse bias characteristics of HgCdTe diodes depend strongly on the density of dislocations intercepting the junction. Dislocations are a significant source of tunnel current and $1/f$ noise, but are thought to be only a problem if they intersect the depletion region of p-n junction. In the case of vertically integrated geometry, the depletion region presents an extremely small cross section to the threading dislocations relative to planar diode geometries, which is shown in Figure 14.63 [333]. This results in an obvious advantage of vertical structure with regard to potential detector defects.

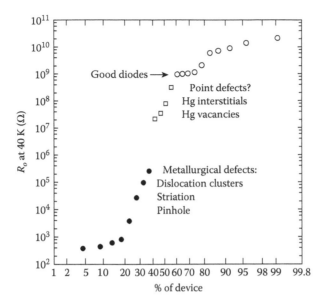

Figure 14.62 Detailed analysis separates the cumulative distribution function of R_oA values of LWIR p-on-n HgCdTe photodiodes (fabricated by LPE) into three regions: good diodes, diodes affected by point defects, and diodes affected by metallurgical defects. (From Chen, M. C., List, R. S., Chandra, D., Bevan, M. J., Colombo, L., and Schaake, H. F., *Journal of Electronic Materials*, 25, 1375–82, 1996. With permission.)

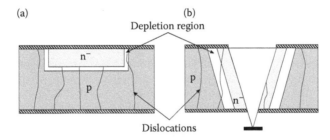

Figure 14.63 Impact of threading dislocations on (a) planar and (b) vertically integrated photodiodes. (From Kinch, M., "HDVIP™ FPA Technology at DRS," *Proceedings of SPIE* 4369, 566–78, 2001. With permission.)

Johnson et al. [129] showed that in the presence of high dislocations densities the R_oA product decreases as the square of the dislocation density; the onset of the square dependence occurs at progressively lower dislocation densities as the temperature decreases, which is shown in Figure 14.64a. At 77 K, R_oA begins to decrease at an EPD of approximately 10^6 cm^{-2}, while at 40 K R_oA is immediately affected by the presence of one or more dislocations in the diode. The scatter in the R_oA data at large EPD may be associated with the presence of an increased number of pairs of "interacting" dislocations in some of those diodes; these pairs are more effective in reducing the R_oA than individual dislocations. To describe the dependence of the R_oA product with dislocation density, a phenomenological model was developed that was based on the conductances of individual and interacting dislocations that shunt the p-n junctions. As Figure 14.64a shows, this model was found to give a reasonable fit to the experimental data.

In general, the $1/f$ noise appears to be associated with the presence of potential barriers at the contacts, interior, or surface of the semiconductor. Reduction of $1/f$ noise to an acceptable level is an art, which depends greatly on the processes employed in preparing the contacts and surfaces. Up until now, no fully satisfactory general theory has been formulated. The two most current

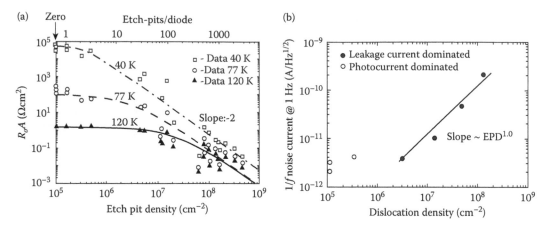

Figure 14.64 Influence of the dislocation density on the parameters of HgCdTe photodiodes: (a) R_oA product versus EPD, showing the fit of model to data, for a 9.5 μm array (at 78 K), measured at 120, 77, and 40 K at zero FOV; and (b) $1/f$ noise current at 1 Hz versus dislocation density measured at 78 K for 10.3 μm HgCdTe photodiode array (f/2 FOV). (From Johnson, S. M., Rhiger, D. R., Rosbeck, J. P., Peterson, J. M., Taylor, S. M., and Boyd, M. E., *Journal of Vacuum Science and Technology*, B10, 1499–1506, 1992. With permission.)

models for the explanation of $1/f$ noise were considered [361]: Hooge's model [362], which assumes fluctuations in the mobility of free charge carriers, and McWhorter's model [363], based on the idea that the free carrier density fluctuates.

Tobin et al. [364] have reported the following relation for $1/f$ noise of implanted n$^+$-p, MWIR HgCdTe photodiodes

$$I_{1/f} = \alpha I_l^\beta f^{-1/2}, \tag{14.56}$$

where α and β are empirical constants with value of 1×10^{-3} and 1 respectively, and I_l is the leakage current. It was found that $1/f$ noise is independent of photocurrent and diffusion current but is linearly related to surface generation current. It was proposed that $1/f$ noise in reverse-biased HgCdTe photodiodes is a result of modulation of the surface generation current by fluctuations in the surface potential. Work by Chung et al. also support this link [365]. Anderson and Hoffman [366] have developed a model involving trap-assisted tunneling across a pinched-off depletion region, and have shown that surface potential fluctuation model can explain the empirical relationship. Bajaj et al. [367] have found a similar relationship with generation-recombination currents originating from junction defects, such as dislocations. The correlation between $1/f$ noise and tunneling process (in particular trap-assisted tunneling) has been supported in other papers [352,356,368,369]. More recently, a relevant model for HgCdTe has been developed by Schiebel [370], which explains the experimental data well including the modulation of the surface generation current by the fluctuations of the surface potential and the influence of trap-assisted tunneling across a pinched-off depletion region.

Johnson et al. [129] presented the effect of dislocations on the $1/f$ noise. Figure 14.64b shows that, at low EPD, the noise current is dominated by the photocurrent, while at higher EPD the noise current varies linearly with EPD. It appears that dislocations are not the direct source of the $1/f$ noise, but rather increase this noise only through their effect on the leakage current. The $1/f$ noise current varies as $I^{0.76}$ (where I is the total diode current); similar to the fit of data taken on undamaged diodes. A similar variation in $1/f$ noise was reported for leakage currents of MWIR PACE-1 HgCdTe photodiodes, where the leakage current varied by changes in temperature, bias voltage, and electron irradiation damage [371].

Measurements of $1/f$ noise taken at DRS for CdTe passivated vertically integrated n$^+$-n$^-$-p HgCdTe photodiodes also indicate a dependence of noise on device dark current density, as shown in Figure 14.65, for material with dislocation density below 2×10^5 cm^{-2} [333]. The noise depends on the absolute value of dark current, regardless of x composition or operating temperature [372].

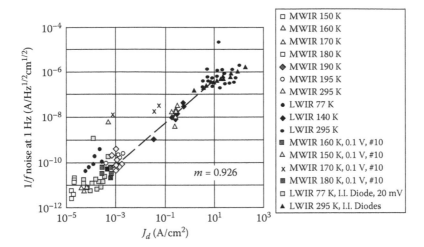

Figure 14.65 $1/f$ noise figure versus dark current density for various HgCdTe photodiodes. (From Kinch, M., "HDVIP™ FPA Technology at DRS," *Proceedings of SPIE* 4369, 566–78, 2001. With permission.)

These observations are in good agreement with Schiebel's theory for realistic surface trap densities in the 10^{12} cm^{-2} range.

14.6.4 Avalanche Photodiodes

HgCdTe, as an attractive material for room temperature avalanche photodiodes (APDs), operates at 1.3–1.6 µm wavelengths for fiber optical communication applications and was recognized in the 1980s [373–375]. The resonant enhancement occurs when the spin-orbit splitting energy in the valence band, Δ, is equal to the fundamental energy gap, E_g. This has the beneficial effect, first pointed out by Verie et al. of making the electron and hole impact ionization rates quite different, which is highly desirable for low-noise APDs. Early work by Alabedra et al. [374] reported on HgCdTe APDs with $x \approx 0.73$ ($E_g = 0.92$ eV). The band structure of HgCdTe gives k-values close to 0—a highly favorable ratio of hole to electron multiplication during avalanche conditions, resulting in very little noise gain. These properties give HgCdTe APDs a figure of merit better than InGaAs APDs, where k is about 0.45. Silicon, in comparison, has an ionization ratio of 0.02, and therefore much lower excess noise, however Si is not sensitive to wavelengths greater than 1.1 µm.

Another APD application is LADAR (laser radar), whereby a pulsed laser system is obtained by 3-D imagery. Three-dimensional imagery has been traditionally obtained with scanned laser systems that operate in the visible and near infrared (NIR) out to 1 µm. However, eye-safe lasers are required if there is a chance humans will be in the scene. The eye-safe range is around 1.55 µm in the NIR region. In addition, the advantages of a single pulse, flood illuminated, laser imaging system caused researchers to consider 2-D arrays of APD sensors. The gated active/passive system promises target detection and identification at longer ranges compared to conventional passive only imaging systems. Because of the desired operating ranges (up to 10 km, typically) and the fact that laser power is limited, there is a need for a NIR solid state detector with a sensitivity approaching single photon [376].

As shown in Figure 14.66, the avalanche properties of $Hg_{1-x}Cd_xTe$ vary dramatically with bandgap. Leveque et al. [378] described two regimes in which the ratio $k = \alpha_h/\alpha_e$ of the hole ionization coefficient to the electron ionization coefficient is either much greater or much less than unity. For cutoff wavelengths shorter than approximately 1.9 µm ($x = 0.65$ at 300 K), they predict $\alpha_h >> \alpha_e$ because of resonant enhancement of the hole ionization coefficient when $E_g \cong \Delta = 0.938$ eV. The situation, when $k = \alpha_h/\alpha_e >> 1$, is favorable for low-noise APDs with hole-initiated avalanche. Using this regime, deLyon et al. [379] demonstrated back-illuminated multilayer separate absorption and multiplication avalanche photodiode (SAM-APD) grown *in situ* by MBE on CdZnTe with cutoff wavelength 1.6 µm and a multiplication cutoff wavelength of 1.3 µm. Avalanche gains in the range of 30–40 were demonstrated at reverse-bias voltages of 80–90 V in 25-element mini arrays.

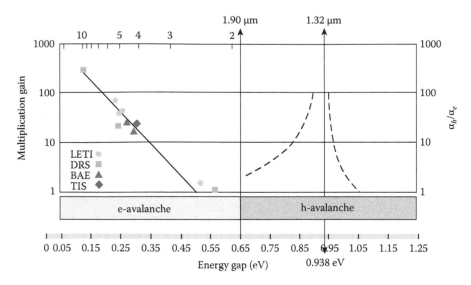

Figure 14.66 The distinct e-APD and h-APD regimes of $Hg_{1-x}Cd_xTe$ cross over at $E_g \approx 0.65$ eV ($\lambda_c \approx 1.9$ μm). At lower bandgaps the e-APD gain increases exponentially (material for four manufacturers shows remarkably consistent results). (From Hall, D. N. B., Rauscher, B. S., Pipher, J. L., Hodapp, K. W., and Luppino, G., *Astro2010: The Astronomy and Astrophysics Decadal Survey*, Technology Development Papers, no. 28.)

The unique crystal lattice properties of HgCdTe allows two types of noise-free linear avalanche in quite distinct modes—pure electron initiated (e-APD) for bandgaps < 0.65 eV (λ_c > 1.9 μm) and pure hole initiated (h-APD) centered on a bandgap of 0.938 eV (λ_c = 1.32 μm) corresponding to a resonance with spin-orbit splitting. Both utilize very similar architectures consisting of a SAM layer graded into a photo-detection layer of a lower bandgap.

Initially, several isolated experimental reports have been published to verify the predicted small values of $k = \alpha_h/\alpha_e$ (≤0.1) in $Hg_{1-x}Cd_xTe$ with λ_c longer than 1.9 μm [380,381]. One of the earliest studies of the electron initiated multiplication on a LWIR HgCdTe (λ_c = 11 μm) showed that reasonable gains could be obtained at low voltages (5.9 at –1.4 V) [357]. However, the clear and compelling advantages of the electron-initiated avalanche process in MWIR lateral-collection n^+-n^--p (with p-type absorber regions) was first reported in 2001 by Beck et al. [382]. Soon thereafter, a theory by Kinch et al. [383] substantiated by Monte Carlo simulations by the University of Texas group [384] has been used to develop an empirical model to fit experimental data obtained at DRS Infrared Technologies. The large inequity between α_e and α_h results from three key features of the HgCdTe energy band structure: (i) the electron effective mass is much smaller than the heavy hole effective mass (electrons have a much higher mobility), (ii) a much lower scattering rate by optical phonons, and (iii) a factor-of-two lower ionization threshold energy (there are no subsidiary minima in the conduction band to which energetic electrons can scatter; and the light holes are not important).

In 1999 DRS researches proposed an APD based on their cylindrical p-around-n HDVIP. This architecture is shown in Figure 14.45 and is used also in production FPAs. It is a front side illuminated photodiode with high quantum efficiency response from visible region to IR cutoff (see Figure 14.67 [385]). The device geometry and operation is illustrated in the next figure (Figure 14.68). If the reverse bias increases from typical 50 mV to several volts, the centralized n-region becomes fully depleted and produces the high field region in which multiplication occurs. The hole-electron pairs are optically generated in the surrounding p-type absorption region and next diffuse to the multiplication region and thus comprise the injection species. Monte Carlo theory modeling predicts the bandwidth of multiplication process typically to above 2 GHz [386]. Large pixels with high bandwidth are obtained by connecting the APDs together in parallel in an N × N configuration with small capacitance due to the cylindrical junction geometry.

Experimental data of HDVIPs reveal almost "ideal" APD; the device is characterized by uniform, exponential gain voltage characteristics that is consistent with a hole to electron ionization

Relative response
Wavelength (μm)
1.2
1.0
0.8
0.6
0.4
0.2
0.0
0
1
2
3
4
5
6

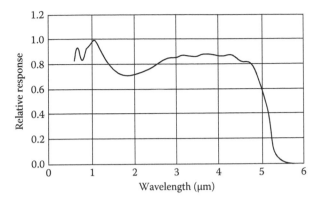

Figure 14.67 Relative spectral response of 5.1 μm HgCdTe HDVIP at 80 K. (From Beck, J., Wan, C., Kinch, M., Robinson, J., Mitra, P., Scritchfield, R., Ma, F., and Campbell, J., *IEEE LEOS Newsletter*, October 8–12, 2006. With permission.)

Top view of pixel unit cell with perimeter absorption region and central cylindrical multiplication region
Absorption region
Multiplication region
n+
hν
Via
Side view:
Electron
Hole
Blowup of multiplication region

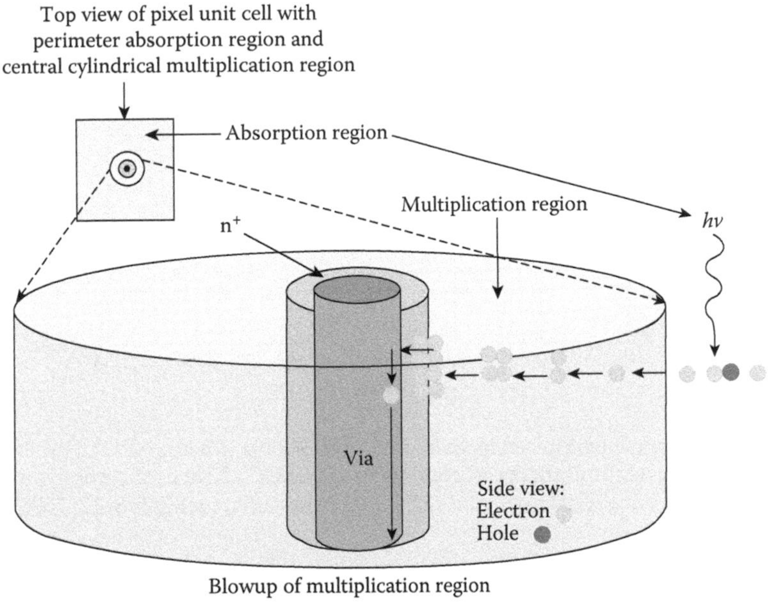

Figure 14.68 Cross-section view of electron avalanche process in HgCdTe HDVIP. (From Beck, J., Wan, C., Kinch, M., Robinson, J., Mitra, P., Scritchfield, R., Ma, F., and Campbell, J., *IEEE LEOS Newsletter*, October 8–12, 2006. With permission.)

coefficient ratio $k = \alpha_h/\alpha_e = 0$. For example, Figure 14.69 shows the measured gain versus bias voltage for two connected diodes in an 8 × 8 array. The mean optical gain at a uniform bias of 13.1 V was 1270 with a σ/mean uniformity of 4.5%. The 8 × 8 array exhibited a median gain normalized dark current of 3.4 nA/cm^2 at a gain of 852. Excess noise data on 4.3 μm cutoff photodiodes indicate a gain independent excess noise factor of 1.3 out to gains of greater than 1000 (see Figure 14.70), suggesting that the electron ionization process is ballistic [384,387].

It appears that the performance of MWIR HgCdTe APDs can be extended to SW and LW photodiodes [386]. Wider bandgap HgCdTe of 0.56 eV (2.2 μm cutoff) also showed similar exponential gain behavior. The operating voltages were, however, higher and the excess noise factor was about 2 suggesting while $k \approx 0$, electron-phonon scattering is present.

Recently, other configurations and architectures have been reported that corroborate the essential features of the electron-initiated avalanche process. Baker et al. [388,389] have reported the

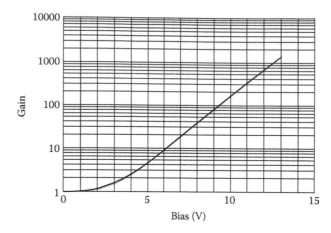

Figure 14.69 Gain versus bias at 80 K for two connected pixels from 8×8, 4.3 μm array. The mean gain at 13.1 V was 1270. (From Beck, J., Wan, C., Kinch, M., Robinson, J., Mitra, P., Scritchfield, R., Ma, F., and Campbell, J., *IEEE LEOS Newsletter*, October 8–12, 2006. With permission.)

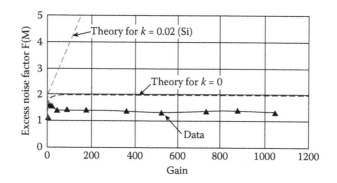

Figure 14.70 Excess noise factor versus gain data at 80 K on a 4.3 μm APD in an 8×8 array compared to McIntyre's original theory. (From Beck, J., Wan, C., Kinch, M., Robinson, J., Mitra, P., Scritchfield, R., Ma, F., and Campbell, J., *IEEE LEOS Newsletter*, October 8–12, 2006. With permission.)

first laser gated imaging operating at 90 K in the eye-safe region at 1.57 μm using a lateral-collection "loophole" HgCdTe e-APD 320 × 256 FPA having 24 × 24 μm unit cells, cutoff wavelengths of 4.2 μm to 4.6 μm, and gains of over 100 at –7 V. Back-illuminated implanted planar n^+-n-p e-APD 1 × 64 arrays grown by MBE on CdZnTe were reported by Vaidyanathan et al. [390] with a gain of 1000 at –10.5 V for a 4.2 μm cutoff at 78 K, and a gain > 100 at –3.5 V for a 10.3 μm cutoff at 78 K. Hall et al. [391] also reported a back-illuminated MWIR n-p-P structure grown *in situ* by MOCVD on GaAs with an $x = 0.36$ multiplication layer and a gain at 78 K of 100 at –12 V.

The HgCdTe APD design, based on front-illuminated lateral-collection "p-around-n" HDVIP architecture developed by DRS, was used to demonstrate the feasibility of a MWIR active/passive 128 × 128 gated imaging system composed of 40 μm-pitch with cutoff wavelengths of 4.2 μm to 5 μm, and custom readout integrated circuit [392,393]. Median gains as high as 946 at –11 V bias and a sensitivity of less than one photon for a 1 μs gate width were measured. Two pixel designs, shown in Figure 14.71, were considered: the single-via and 2 × 2-via pixel designs. The single-via pixel design offers lower risk in terms of achieving operability at the highest possible gain. The second design provides higher bandwidth, however operability at high bias incurs additional risk as the via pitch is reduced.

The highest gain of 5300 at –12.5 V has been recently reported by Perrais et al. [394] for a back-illuminated planar p-i-n 30-μm pitch photodiodes with cutoff wavelength of 5.0 μm, formed in

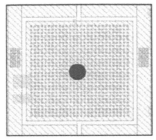 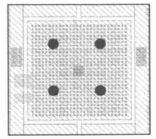

Single, 40 μm pitch via, pixel   2×2, 20 μm pitch via, pixel

Figure 14.71 Forty micron pixel designs. Left: 8 μm diameter via on a 40 μm pitch, single-via pixel. Right: 2×2 of 4 μm diameter vias on 20 μm pitch. (From Beck, J., Woodall, M., Scritchfield, R., Ohlson, M., Wood, L., Mitra, P., and Robinson, J., *Journal of Electronic Materials*, 37, 1334–43, 2008. With permission.)

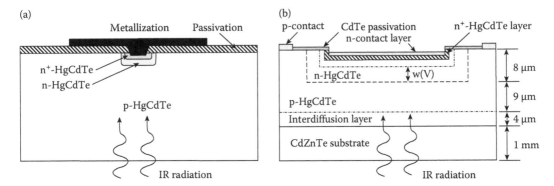

Figure 14.72 Cross section of back side illuminated planar n-on-p HgCdTe e-APD: (a) LETI (After Rothman, J., Perrais, G., Ballet, P., Mollard, L., Gout, S., and Chamonal, J.-P., *Journal of Electronic Materials*, 37, 1303–10, 2008. With permission.) and (b) BAE (After Reine, M. B., Marciniec, J. W., Wong, K. K., Parodos, T., Mullarkey, J. D., Lamarre, P. A., Tobin, S. P., et al., *Journal of Electronic Materials*, 37, 1376–85, 2008. With permission.) architectures.

a HgCdTe layer grown by MBE on a CdZnTe substrate. Similar to shorter wavelength e-APDs behavior, this the highest gain is generally consistent with exponential increasing with both cutoff wavelength and reverse bias voltage.

Figure 14.72a shows schematic illustration of the LETI's p-i-n junction formed by transforming a narrow region close to the surface in a vacancy-doped p-type layer ($N_a = 3 \times 10^{16}$ cm^{-3}) into a n$^+$ region with a doping level of $N_d = 1 \times 10^{18}$ cm^{-3} [395]. During formation of the n$^+$ region, n$^-$ region is generated by suppression of Hg vacancies to the residual doping of the epitaxy typically $N_d = 3 \times 10^{14}$ cm^{-3}. The extension of the n$^-$ layer is correlated to the depth of the n$^+$ layer. The quantum efficiency of the diodes was typically about 50%, due to reduced optical fill factor and use of a nonoptimized antireflecting coating. The width of the n$^-$ layer is of the order of 1–3 μm.

The similar structure elaborated by BAE research group is shown in Figure 14.72b [396]. It is fabricated in a single-layer LPE p-type HgCdTe layer grown on an IR-transparent CdZnTe substrate. The p-side is much more heavily doped than the n-side, so nearly the entire width, w, of the depletion region lies within the n-region.

The APD planar structure has a number of advantages: high fill factor, high quantum efficiency, high bandwidth (photocarriers have a short distance to traverse to the depletion region, and in addition, the favorable compositional grading in the LPE layer with the effective electric field accelerates the photocarriers). Using planar structure the LETI group received the bandwidth value of ≈400 MHz for 1.55 μm radiation, and a gain of ≈2800 at –11 V, corresponding to a gain × bandwidth product of ≈1.1 THz for a device with 5.3 m cutoff wavelength at 80 K [397].

The above described performance of HgCdTe e-APDs has opened the door to new passive/active system capabilities and applications. The combination of the dual-band and avalanche gain functionalities is another technological challenge that will enable many applications, such as dual band detection over a large temperature range [398].

14.6.5 Auger-Suppressed Photodiodes

Practical realization of Auger suppressed photodiodes was impossible for a long time because of the lack of technology to obtain wide gap P contacts to $Hg_{1-x}Cd_xTe$ absorber region. Therefore, the first Auger suppressed devices were III-V heterostructures with InSb absorber [231,232,399,400]. The first reported $Hg_{1-x}Cd_xTe$ Auger extracting diodes were so-called proximity-extracting diode structures (see Figure 14.73), in which additional guard reverse-biased n^+-n^- junctions are placed in the current path between the p^+ and n^+ regions to intercept the electron injected from the p^+ region. Philips Components Ltd. fabricated practical proximity-extracting devices in both a linear and cylindrical geometry [231]. The devices have been fabricated from bulk grown $Hg_{1-x}Cd_xTe$ with a cutoff wavelength of 9.3 µm at 200 K ($x = 0.2$). The acceptor concentration of 8×10^{15} cm^{-3} was due to native doping. The n^+-regions were fabricated by ion milling. The I-V characteristics of such structures are complex and difficult to interpret because of the occurrence of bipolar transistor action and impact ionization, but the general features were predictable when the standard transistor modeling was applied. A current reduction of a factor 48 has been obtained by biasing the guard junction, but the extracted current was much greater than that predicted for an Auger suppressed, SR limited case. This is possibly due to a surface generated current.

The measured 500 K blackbody detectivity of a 320 µm^2 optical area device at a modulation frequency of 20 kHz was 1×10^9 cmHz$^{1/2}$/W, which is the best ever measured value for any photodetector operated at similar conditions.

The use of wide-gap P contact is a straightforward way to eliminate harmful thermal generation in the p-type region. The practical realization of such devices would require a well-established multilayer epitaxial technology capable of growing high-quality heterostructures with complex gap and doping profiles. This technology became available in the early 1990s and three-layer n^+-π-P^+, N^+-π-P^+ heterostructural photodiodes have been demonstrated [401] and gradually improved [400–405]. The arrangement of the photodiode in the optical resonant cavity makes it possible to use a thin extracted zone without loss in quantum efficiency, which is also favorable for reduction of saturation current and results in minimizing noise and bias power dissipation.

Figure 14.74 shows the device structure of N^+-π-P^+ HgCdTe heterostructure intended for operation at temperatures ≥ 145 K and its appropriate energy band diagram [400,403]. Typical parameters for the LW devices would be $x = 0.184$ in the active π region, $x = 0.35$ in the P^+ region and $x = 0.23$ in the N^+ region. The structures were grown on CdZnTe and GaAs substrates using IMP MOCVD. The π and P^+ regions were doped with arsenic to typical levels of 7×10^{15} cm^{-3} and 1×10^{17} cm^{-3}, respectively, and the N^+ region was doped with iodine to a concentration of 3×10^{17} cm^{-3}. Diodes were defined by etching circular trenches to produce 64 element linear arrays with common contact to the P^+ region at each end. These "slotted" mesa devices were passivated with 0.3 µm thick ZnS and metalize with Cr/Au. Finally, the electrical contact to the mesas is achieved by indium bump bonding the array onto a gold lead-out pattern on a sapphire carrier.

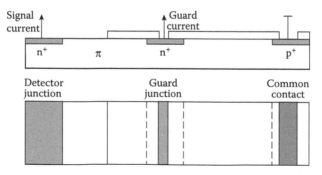

Figure 14.73 Schematic structure of proximity-extracting photodiode. (From Elliott, C. T., *Semiconductor Science and Technology*, 5, S30–S37, 1990. With permission.)

(a)

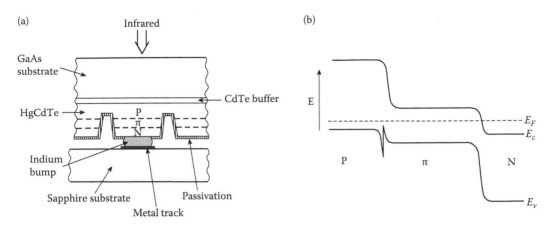

(b)

Figure 14.74 Schematic cross section of (a) the P⁺–π–N⁺ HgCdTe heterostructure photodiode, and (b) its energy band diagram. (From Elliott, C. T., "Advanced Heterostructures for $In_{1-x}Al_xSb$ and $Hg_{1-x}Cd_xTe$ Detectors and Emiters," *Proceedings of SPIE* 2744, 452–62, 1996. With permission.)

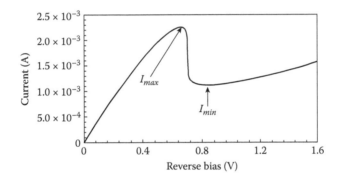

Figure 14.75 An example of current voltage characteristics of a P-p-N heterostructure showing the positions of I_{max} and I_{min}. (From Elliott, C. T., Gordon, N. T., Phillips, T. J., Steen, H., White, A. M., Wilson, D. J., Jones, C. L., Maxey, C. D., and Metcalfe, N. E., *Journal of Electronic Materials*, 25, 1146–50, 1996. With permission.)

For low bias, the device behaves as a linear resistor until the electric field exceeds the critical value for exclusion or extraction (Figure 14.75 [406]). Then the current drops sharply from its maximum value I_{max}, a further increase of voltage gradually decreases the current to minimum I_{min}. At high voltages the current increase became due to diode breakdown. As a result, the dynamic resistance increases to high values in the regions close to the transition range. The devices have shown negative resistances for temperatures above 190 K. The minimum dark current (Figure 14.76) follows the empirical expression

$$I_{min} = 1.3 \times 10^4 \exp\left(-\frac{qE_g}{kT}\right) \quad (E_g \text{ in eV}, \lambda \text{ in micrometers}). \tag{14.57}$$

Comparing the trend line for experimental results with expected dark current, significant Auger suppression is seen for wavelength $\lambda > 6 \ \mu m$. The shot noise detectivity calculated as $R_i(2qJ_{min})^{-1/2}$ is

$$D^* = 1.2 \times 10^7 \eta\lambda \exp\left(\frac{qE_g}{kT}\right) \quad (E_g \text{ in eV}, \lambda \text{ in micrometers}). \tag{14.58}$$

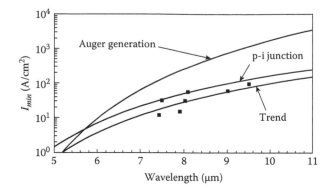

Figure 14.76 I_{min} in P-p-N diodes as a function of cutoff wavelength: trend line (dashed). The results are compared with the leakage current expected for p^+-i junction and also with the current that would be generated in the π-region in the absence of Auger suppression. (From Elliott, C. T., Gordon, N. T., Hall, R. S., Phillips, T. J., White A. M., Jones, C. L., Maxey, C. D., and Metcalfe, N. E., *Journal of Electronic Materials*, 25, 1139–45, 1996. With permission.)

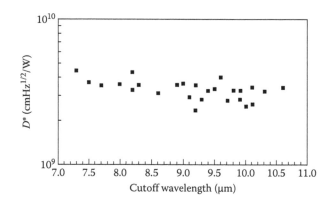

Figure 14.77 Shot-noise detectivity of HgCdTe nonequilibrium detectors at 300 K. (From Elliott, C. T., "Advanced Heterostructures for $In_{1-x}Al_xSb$ and $Hg_{1-x}Cd_xTe$ Detectors and Emiters," *Proceedings of SPIE* 2744, 452–62, 1996. With permission.)

The shot noise limited detectivity as a function of cutoff wavelength is shown in Figure 14.77 [400]. This figure shows a value of 4×10^9 $cmHz^{1/2}/W$ at 7 μm, decreasing only slightly to 3×10^9 $cmHz^{1/2}/W$ at 11 μm. These values are approximately an order of magnitude greater than could be achieved with uncooled thermal detectors. Unfortunately, the shot noise limited D^* cannot be realized in imaging applications because the devices made so far have high levels of $1/f$ noise [407,408]. The shot noise level is only observed experimentally at frequencies in excess of 1 MHz. Applications that will benefit first from these devices, therefore, will be those that can operate at relatively high frequencies; for example, gas detection using an infrared LED as the source. Another application where the $1/f$ noise is not an issue is that of laser heterodyne detection. Elliott et al. [406] have demonstrated a minimally cooled heterodyne receiver for CO_2 laser radiation that has a noise equivalent power (NEP) of 2×10^{-19} WHz^{-1} at 40 MHz (above the $1/f$ knee) at 260 K and 0.3 mW local oscillator power. This NEP is about two orders of magnitude better than any other uncooled device and only a factor of three worse than devices cooled to 80 K.

Auger suppression N-π-P photodiodes grown by MBE using silver as acceptor and indium as donor dopant were also demonstrated [409]. The minimum reverse current density is similar to that obtained in MOVPE-grown material at the same ≈9 μm cutoff wavelength at 300 K. Quantum efficiencies exceeding 100% have been measured and attributed to carrier multiplication due to the relatively high bias across the sample or to mixed conduction effects.

More recently, improved N^+-N^--π-P^--P^+ $Hg_{1-x}Cd_xTe$ heterostructures with refined bandgap and doping profiles were reported [403,408]. The N and P layers were used to minimize generation between regions of different composition. The doping of π-region was $\approx 2 \times 10^{15}$ cm^{-3}. The thickness of π-region was typically 3 μm wide that ensures good quantum efficiency with mesa contact acting as reflector. All the layers have been annealed for 60 hour at 220°C in Hg-rich nitrogen with the same temperature of Hg reservoir. Mesa structures passivated with CdTe and ZnS have been received. Several prepassivation clean procedures have been tried: anodize, HBr/Br etch, and citric acid etch. Some of the CdTe passivated devices were annealed at 220°C for 30 hour in Hg-rich nitrogen with purpose to interdiffuse CdTe/HgCdTe to give a graded interface.

The I-V curves with rapid decrease of dark current from I_{max} to I_{min} have been observed. Some hysteresis effects due to series resistance can be seen. The peak-to-valley ratios up to 35 have been observed with extracted saturation current densities of 10A/cm^2 and shot noise detectivities of $\approx 3 \times 10^9$ cmHz$^{1/2}$/W in uncooled ≈ 10 μm cutoff wavelength diodes, which is by a factor of ≈ 3 improvement compared to the previous results [408]. The residual dark current is a result of residual Auger generation in the depleted zones and at interfaces, SR generation in the whole structure region, and the surface leakage currents.

The present Auger suppressed devices exhibit a high low frequency noise with $1/f$ knee frequencies from 100 to few MHz for ≈ 10 μm devices at room temperature. This reduces their signal-to-noise ratio at frequencies of ≈ 1 kHz to a level below that for equilibrium devices. The $1/f$ noise remains the main obstacle to achieving background limited NETD in 2-D arrays at near room temperatures.

The $1/f$ noise level is much lower in MWIR devices and they look capable of achieving useful D^* values for imaging applications in very high frame rate or high-speed choppers [410]. Typically, the low frequency noise current is proportional to the bias current with $a \approx 2 \times 10^{-4}$.

Recently, the HOTEYE thermal imaging camera based on a 320×256 FPA with cutoff wavelength 4 μm at 210 K has been demonstrated (see Figure 14.78) [410]. The pixel structures were P^+-p-N^+ starting from the GaAs substrate. The histogram peaks at an NEDT of around 60 mK with $f/2$ optics was measured. When the optimum bias was selected for each pixel, an improvement in NEDT histogram was observed. However, this procedure was a problem. A sample of the image of the HOTEYE camera is shown in Figure 14.78b.

The reason for the large $1/f$ noise is not clear [408,411]. Traps in depletion region [412], high field areas [413], hot electrons [414], and background optical generation [415] are among possible reasons.

The attempts to find the source of $1/f$ using the perimeter-area analysis give in part contradictory results [408]. It was found that for ≈ 50 μm diameter diode, perimeter contributes 33% of leakage and 57% of noise power. Some reduction of the $1/f$ noise has been achieved with low temperature annealing. CdTe passivation was found to give better stability than ZnS. The possible

(a)   (b)

Figure 14.78 The HOTEYE camera: (a) housed in a compact encapsulation 12×12×30 cm (including lens); and (b) typical image. (From Bowen, G. J., Blenkinsop, I. D., Catchpole, R., Gordon, N. T., Harper, M. A., Haynes, P. C., Hipwood, L., et al., "HOTEYE: A Novel Thermal Camera Using Higher Operating Temperature Infrared Detectors," *Proceedings of SPIE* 5783, 392–400, 2005. With permission.)

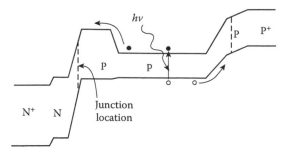

Figure 14.79 Schematic of six-layer extracted photodiode with a wider gap P layer located between absorber and N-N⁺ contact.

routes to reduce the $1/f$ noise are to reduce the dark current or/and the I_n/I ratio with improved material technology and better design of the devices.

For MWIR devices the origin of the noise was suspected to be the depletion region [411], exponentially dependent on bandgap; noise $\propto \exp(E_g/2kT)$.

A modified junction-enhanced semiconductor structure (JEES) with an additional electron barrier near the N⁺ side of the active region (Figure 14.79) has been proposed to reduce $1/f$ noise [411]. Excellent reverse bias characteristics with increased breakdown voltage and high quantum efficiency were observed in this structure, but thus far no significant improvement of low frequency performance has been obtained.

At present, an exotic hetero-junction photo-JFET with a wide bandgap gate and advanced heterostructures with separated extraction and signal contacts is under consideration as a promising HOT device for 2-D arrays, but no clear pathway to eliminating the low frequency problem of this problem has been demonstrated yet [416].

14.6.6 MIS Photodiodes

The interest in HgCdTe MIS structures was mostly connected with the possibility of using them in monolithic FPAs [16,17,417]. These structures enable not only detection of IR radiation but also advanced signal management. For nearly two decades technologists have sought to develop fully monolithic CCD imagers in HgCdTe for detection of IR radiation. Initial works concentrated on p-channel CCDs, due to the maturity of growth and doping control on n-type material [16,417–419]. However, due to the difficulty of forming stable p⁺-n junctions in HgCdTe, readout structures could not be incorporated in the devices. The necessity of utilizing off-chip readout circuitry increased the parasitic capacitance at the sense node and decreased the charge-to-voltage conversion efficiency, resulting in a limited dynamic range. To alleviate these difficulties it became necessary to develop n-channel CCDs. The use of p-type material allowed for stable diode formation by ion implantation and provided a means for the development of HgCdTe MISFETs [420–422]. With the demonstration of MISFET-based amplifiers in HgCdTe [423], the path was cleared for creating a fully monolithic CCD in HgCdTe [424–426]. The MIS structures are also used as a tool for investigation of surface and interface properties of HgCdTe.

The operation and properties of HgCdTe MIS structures have been extensively reviewed by Kinch [417], and general theory of MIS capacitor is described in Section 9.7. The minority carriers generated by absorbed radiation in MIS photodetectors are trapped in the well while the majority carriers are forced into the neutral bulk by the surface potential. Although MIS detectors are essentially capacitors, their dark current can be compared with the dark current in conventional p-n junction photodiodes allowing a reference to the R_oA product. The sources of dark current are essentially the same as in the p-n photodiodes. The dark current limits the maximum storage or integration time at low background and is a source of noise. The source of noise are similar to those in a reverse-biased p-n junction [417,427]. However, this problem is more serious for MIS structures as, in contrast to the p-n junction photodiode, they have to operate in strong depletion to achieve adequate storage capacity. Another reason for the high dark current is the use of weakly doped material for the high breakdown voltage, resulting in high generation-recombination current from the depletion region. Practically, the dark current should be reduced below the background-generated current. This condition sets the maximum operating temperature of the devices. The optimization of the thickness of the base layer can either increase the operating temperature or longer integration time can be achieved.

Trapping the charge in fast interface states at the edge of the CCD well is the primary limitation of the charge transfer efficiency at low frequencies [417]. Since the transfer loss is proportional to the density of the fast states, the latter should be minimized. At high frequencies transfer efficiency sharply decreases due to limited time, which is required for charge transport between adjacent wells. The corner frequency, equal to about 1 MHz for typical p-type channel devices, can be increased by reducing the gate length and increasing the maximum employable gate voltages. A much higher corner frequency would be expected for n-channel devices, as a result of the high electron-to-hole mobility ratio.

For the maximum charge storage capability the MIS structures are operated at maximum effective gate voltages, which are limited by breakdown due to tunnel current. BLIP performance requires that the tunnel current is lower than the current due to incident photon flux, and this criterion is used to define an upper limit on applied electric field, the breakdown field E_{bd}, for the material in question. The breakdown voltage decreases with decreasing gap and increasing doping level. This limitation is especially severe for LWIR devices, which require a large capacity to store background-generated currents. Tactical background flux levels are shown in Figure 14.80a for a typical $f/2$ system, and values for E_{bd} are 3×10^4 V/cm for 5 μm cutoff wavelength, and 8×10^3 V/cm for 11.5 μm [16]. These values are in agreement with experimental data for good quality p-type HgCdTe with doping concentration. E_{bd} values for n-type HgCdTe are typically lower by a factor of 30% due to lower density of states in the conduction band (which is by a factor of 10^2 to 10^3 lower than the density of states in the valence band), and the effects of inversion layer quantization for the p-type case [428].

The dependence of available well capacity on doping concentration is shown in Figure 14.80b, assuming an insulator capacity per area unit of 4×10^{-8} F/cm^2. Integrating MIS FPAs require well capacities in the 10^{-8} to 10^{-7} C/cm^2 range for pixel noise to dominate. Such well capacities are readily available for 5 μm HgCdTe at doping concentrations in the $>10^{15}$ cm^{-3} range. However, the LWIR device operation imposes much more stringent requirements due to the narrow bandgap involved, and the relatively low value of $E_{bd} = 8 \times 10^3$ V/cm. The n-type HgCdTe with low doping concentrations (between 10^{14} cm^{-3} and 10^{15} cm^{-3}) is available, but its performance does not improve at doping concentrations below 8×10^{14} cm^{-3} as predicted by Figure 14.80b, but rather it reaches a plateau at $\approx 2 \times 10^{-8}$ C/cm^2 (equivalent to an applied voltage $V = 0.5$ V). The p-type HgCdTe of this quality is not reproducibility available. An example of the device is the 10 μm bulk n-type HgCdTe based MIS detector reported by Borrello et al. [425], which exhibits at 80 K quantum efficiency above 50% and peak $D^* = 8 \times 10^{10}$ cmHz$^{1/2}$/W at $f/1.3$ shielding.

MIS devices were fabricated using HgCdTe epilayers with thickness lower than the minority carrier diffusion length to reduce the dark diffusion current. For example, the CCD TDI device cross section is shown in Figure 14.81 [426]. The p-type HgCdTe LPE epilayer were grown on (111)

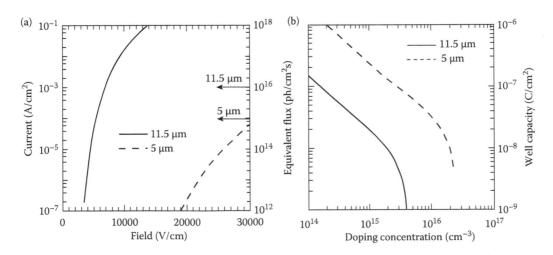

Figure 14.80 (a) Tunnel current versus electric field, and (b) well capacity versus hole concentration, for 5 μm and 11.5 μm p-type HgCdTe. (From Kinch, M. A., in *Properties of Narrow Gap Cadmium-Based Compounds, EMIS Datareviews Series No. 10,* ed. P. Capper, 359–63, IEE, London, 1994. With permission.)

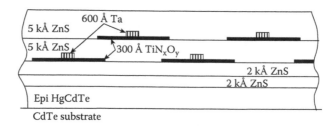

Figure 14.81 CCD TDI HgCdTe array cross-sectional drawing showing the material stack in the TDI section perpendicular to the charge transfer direction. (From Wadsworth, M. V., Borrello, S. R., Dodge, J., Gooh, R., McCardel, W., Nado, G., and Shilhanek, M. D., *IEEE Transactions on Electron Devices*, 42, 244–50, 1995. With permission.)

B CdZnTe substrates. After annealing the dislocation density was typically $(1-4) \times 10^5$ cm^{-2}, and the 77 K carrier concentration was 1.7×10^{15} cm^{-3}. The device channel stops were formed with 100 Å of Ta patterned on top of an initial dielectric layer composed of 2000 Å of ZnS. Similarly MISFET gates were formed from a 600 Å thick layer of Ta also residing on the initial dielectric layer. MISFET source and drain regions were created by ion implantation in the HgCdTe using boron as the dopant species. Ohmic contacts were made to the n$^+$regions and the substrate by etching openings through the overlying dielectric and depositing 1000 Å of Sn into openings. An additional 3000 Å of ZnS was then deposited uniformly across the array. CCD phases 2 and 4 were formed from a 300 Å TiN$_x$O$_y$ film, with a 2 μm wide, 600 Å thick, Ta spine placed down the center of each gate electrode. A second dielectric layer consisting of 5000 Å of ZnS was then deposited across the substrate, covering the first level gate electrodes. CCD gate electrodes were formed similar to that of gates 2 and 4. A 5000 Å coating of ZnS was deposited over the entire device as protective and antireflective coating, providing a 40% improvement to the device quantum efficiency. Finally interconnections were accomplished by bus metal consisting of 3 μm of Pb-In.

Operation of the above described MWIR devices with 5 μm cutoff wavelength at 77 K have produced average detectivity values exceeding 3×10^{13} cmHz$^{1/2}$W^{-1} for a background flux level of 6×10^{12} photon/cm^{2}s [426]. An increase in device operating temperature produced additional dark current, lowering the effective signal storage capacity of the CCD, and lowering the D^* value. For 5.25 μm cutoff CCD, the maximum practical operating temperature is 100 K.

Unlike the photoconductors and p-n junction photodiodes, the MIS device operates under strong nonequilibrium conditions, with large electric fields in the deep depletion regions. This makes the MIS device much more sensitive to material defects than the photoconductors and photodiodes. This sensitivity caused the monolithic approach to be abandoned in favor of various hybrid approaches.

14.6.7 Schottky-Barrier Photodiodes

Photoreceivers based on the Schottky barriers exhibit high bandwidth and simpler fabrication compared to p-n photodiodes. However, the performance of HgCdTe Schottky-barrier photodiodes is not useful for detection of IR radiation.

The traditional models of metal-semiconductor (M-S) interface are described in Section 9.5. However, the physical picture of M-S has been modified and a variety of phenomena have been observed at the microscopic M-S interfaces, which form interfacial regions with new electronic and chemical structures. Several new models have been suggested to explain the Fermi level position in the junction interface, which includes the modified work function model of Freeouf and Woodall [429], metal induced gap states (MIGS) [430], and Spicer's native defects model [431]. According to the MIGS model, when a metal is in intimate contact with a semiconductor surface, the tails of the metal wave functions can tunnel into the bandgap of the semiconductor leading to MIGs that are capable of strong Fermi level pinning.

The Schottky-barrier height and ohmic contacts on Hg$_{1-x}$Cd$_x$Te over the whole composition has been discussed by Spicer et al. [320] in framework of the current theories. Their predictions are summarized in Figure 14.82. The lower limit of the pinning positions due to the various mechanisms are represented by the solid lines across the whole composition range. For the defect based model, two similar lines depending on whether the theoretical calculation of Kobayoshi et al. [432] or Zunger [433] was used. The MIGS derived pinning positions were based on the work of Tersoff

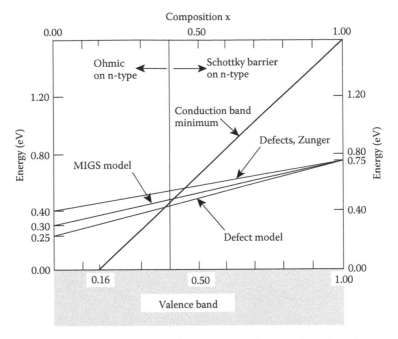

Figure 14.82 Lower limit of the Fermi level pinning position as a function of composition. Two models, MIGS (Tersoff, J. *Physical Review*, 32, 6968–71, 1985. With permission.) and defect (Kobayoshi, A., Shankey, O. F., and Dow, J. D., *Physical Review* 25, 6367–79, 1982; Zunger, A., *Physical Review Letters*, 54, 849–50, 1985. With permission.) were used for the extrapolation. Near $x = 0.4$, the Fermi level moves into the conduction band providing ohmic contacts on n-type material. (From Spicer, W. E., Friedman, D. J., and Carey, G. P., *Journal of Vacuum Science and Technology*, A6, 2746–51, 1988. With permission.)

[430]. It is interesting to note that all models predict a pinning position situated in the conduction band below an x value around 0.4–0.5.

Polla and Sood [434,435] reported Schottky-barrier photodiodes on narrow gap bulk p-type $Hg_{1-x}Cd_xTe$. Barrier metals (Al, Cr, and Mn) were thermally evaporated and ZnS were sputtered as a passivation layer. Schottky diodes on bulk and MOCVD n-type $Hg_{1-x}Cd_xTe$ ($x = 0.6$–0.7) have been reported by Leech and Kibel [436]. All of the metal contacts (Ag, Au, Cu, Pd, Pt, Sb) formed rectifying contacts on the etched surfaces with Schottky-barrier heights in the range 0.7–0.8 eV; only Ti gave near ohmic behavior. The ideality factors for the rectifying contacts were greater than unity, indicating the presence of transport mechanisms other than thermionic emission. The diode characteristics are strongly dependent on surface preparation.

14.7 HG-BASED ALTERNATIVE DETECTORS

Among the small-gap II-VI semiconductors for IR detectors, only Cd, Zn, Mn, and Mg have been shown to open the bandgap of the Hg-based binary semimetals HgTe and HgSe to match the IR wavelength range. It appears that the amount of Mg to introduce in HgTe to match the 10 μm range is insufficient to reinforce the Hg-Te bond [437]. The main obstacles to the technological development of $Hg_{1-x}Cd_xSe$ are the difficulties in obtaining type conversion. From the above alloy systems, $Hg_{1-x}Zn_xTe$ (HgZnTe) and $Hg_{1-x}Mn_xTe$ (HgMnTe) occupy a privileged position.

Neither HgZnTe nor HgMnTe have ever been systematically explored in the device context. There are several reasons for this. First, preliminary investigations of these alloy systems came on the scene after HgCdTe detector development was well on its way. Second, the HgZnTe alloy presents a more serious technological challenge than HgCdTe. In the case of HgMnTe, Mn is not a group II element, and as a result, HgMnTe is not truly a II-VI alloy. This ternary compound was viewed with some suspicion by those not directly familiar with its crystallographic, electrical, and optical behavior. In such a situation, proponents of the parallel development of HgZnTe and HgMnTe for IR detector fabrication encountered considerable difficulty in selling the idea to industry and funding agencies.

In 1985, Sher et al. [438] showed from theoretical consideration that the weak HgTe bond is desta-bilized by alloying it with CdTe, but stabilized by ZnTe. Many groups worldwide have become very interested in this prediction, and more specifically, in the growth and properties of the HgZnTe alloy system as the material for photodetection application in the IR spectral region. But the question of lattice stability in the case of HgMnTe compound is rather ambiguous. According to Wall et al. [439], the Hg-Te bond stability of this alloy is similar to that observed in the binary narrow-gap parent compound. This conclusion contradicts other published results [440]. It has been established that the incorporation of Mn in CdTe destabilizes its lattice because of the Mn 3-D orbitals hybridizing into the tetrahedral bonds [441].

This section reviewed only selected topics on the growth process and physical properties of HgZnTe and HgMnTe ternary alloys. More information can be found in two comprehensive reviews by Rogalski [442,443] and the books cited by Rogalski and colleagues [73,444].

14.7.1 Crystal Growth

The pseudobinary diagram for HgZnTe is responsible for serious problems encountered in crystal growth, including:

- The separation between the liquidus and solidus curves are large and lead to high segregation coefficients.

- The solidus lines that are flat, resulting in a weak variation of the growth temperature that causes a large composition variation.

- The very high Hg pressure over melts that makes the growth of homogeneous bulk crystals quite unfavorable.

For comparison, Figure 14.4 shows HgTe-ZnTe and HgTe-CdTe pseudobinary phase diagrams.

HgTe and MnTe are not completely miscible over the entire range, but the $Hg_{1-x}Mn_xTe$ single-phase region is limited to approximately $x < 0.35$ [445]. As discussed by Becla et al., the solidus-liquidus separation in the pseudobinary HgTe-MnTe system is more than two times narrower than in the corresponding HgTe-CdTe system [446]. This conclusion has been con-firmed by Bodnaruk et al. [447]. Consequently, to meet the same demand for cutoff wavelength homogeneity, the HgMnTe crystals must be much more uniform than similarly grown HgCdTe crystals.

For the growth of bulk HgZnTe and HgMnTe single crystals, three methods are the most popular: Bridgman-Stockbarger, SSR, and the traveling heater method (THM). The best quality HgZnTe crystals have been produced by THM. Using this method, Triboulet et al. [448] produced $Hg_{1-x}Zn_xTe$ crystals ($x \approx 0.15$) with a longitudinal homogeneity of ±0.01 mol and radial homogene-ity of ±0.01 mol. The source material was a cylinder composed of two cylindrical segments—one HgTe, the other ZnTe—the cross section of which was in the ratio corresponding to the desired composition.

To improve the crystalline quality of HgMnTe single crystals, different modified techniques have been used. Gille et al. [449] demonstrated $Hg_{1-x}Mn_xTe$ ($x \approx 0.10$) single crystals grown by THM with standard deviation $\Delta x = ±0.003$ along a 16 mm diameter slice of crystals. Becla et al. [446] decreased the radial macrosegregation and eliminated small-scale compositional undula-tions in the vertical Bridgman-grown material by applying a 30 kG magnetic field. Takeyama and Narita [450] developed an advanced crystal growth method called the modified two-phase mixture method to produce highly homogeneous, large single crystals of ternary and quaternary alloys.

The best performance of modern devices, however, requires more sophisticated structures. These structures are only achieved by using epitaxial growth techniques. Additionally, when compared to bulk growth techniques, epitaxial techniques offer important advantages, includ-ing lower temperatures and Hg vapor pressures, shorter growth times, and reduced precipitation problems that enable the growth of large-area samples with good lateral homogeneity. The above advantages have prompted research in a variety of thin-film growth techniques, such as VPE, LPE, MBE, and MOCVD. The first studies of LPE crystal growth of HgCdZnTe and HgCdMnTe from Hg-rich solutions have demonstrated that the homogeneity of epilayers can be improved by incorporating Zn or Mn during the crystal growth [451]. More recently, considerable progress has been achieved in HgMnTe film fabrication by MOCVD using an IMP [452]. Depending on growth conditions, both n- and p-type layers may be produced with extrinsic electron and hole concentra-tions of the order of 10^{15} cm^{-3} and 10^{14} cm^{-3}, respectively.

All of the epitaxial growth processes depend on the identification of suitable substrates. They require large-area, single-crystal substrates. The large difference in lattice parameters of HgTe and ZnTe induces strong interactions between cations. Vegard's law appears to be obeyed relatively better in HgZnTe than in HgCdTe, and for $Hg_{1-x}Zn_xTe$ at 300 K $a(x) = 6.461-0.361x$ (Å) [444]. In comparison with HgCdTe, the lattice parameter of $Hg_{1-x}Mn_xTe$ $a(x) = 6.461-0.121x$ (Å) varies with x much more rapidly, which is a disadvantage from the point of view of the epitaxial growth of multilayer heterostructures that is required for advanced IR devices. The lattice parameter of the zinc-blende compound $Cd_{1-x}Zn_xTe$ ($Cd_{1-x}Mn_xTe$) indicates a simple matter: to find suitable substrates for epitaxial growth of $Hg_{1-x}Zn_xTe$ ($Hg_{1-x}Mn_xTe$). However, Bridgman-grown CdMnTe crystals are highly twinned and thus unusable as epitaxial substrates.

14.7.2 Physical Properties

The physical properties of both ternary alloys are determined by the energy gap structure near the Γ-point of the Brillouin zone. The shape of the electron band and the light mass hole band is determined by the k·p theory. The bandgap structure of HgZnTe near the Γ-point is similar to that of the HgCdTe ternary illustrated in Figure 14.11. The bandgap energy of HgZnTe varies approximately 1.4 times (two times for HgMnTe) as fast with the composition parameter x as it does for HgCdTe.

Both HgZnTe and HgMnTe exhibit compositional-dependent optical and transport properties similar to HgCdTe materials with the same energy gap. Some physical properties of alternative alloys indicate a structural advantage in comparison with HgCdTe. Introducing ZnTe in HgTe decreases statistically the ionicity of the bond, improving the stability of the alloy. Moreover, because the bond length of ZnTe (2.406 Å) is 14% shorter than that of HgTe (2.797 Å) or CdTe (2.804 Å), the dislocation energy per unit length and the hardness of the HgZnTe alloy are higher than that of HgCdTe. The maximum degree of microhardness for HgZnTe is more than twice that for HgCdTe [453]. HgZnTe is a material that is more resistant to dislocation formation and plastic deformation than HgCdTe.

The as-grown $Hg_{1-x}Zn_xTe$ material is highly p-type in the 10^{17} cm^{-3} range with mobilities in the hundreds of cm^2/Vs. These values indicate that its conduction is dominated by holes arising from Hg vacancies. After a low-temperature anneal ($T \leq 300°$ C) accomplished with an excess of Hg (which annihilates the Hg vacancies), the material is converted to low n-type in the mid 10^{14} cm^{-3} to low 10^{15} cm^{-3} range, which has mobilities ranging from 10^4 to 4×10^5 cm^2/Vs. A study by Rolland et al. [454] showed that the n-type conversion occurs only for crystals with composition $x \leq 0.15$. The Hg diffusion rate is slower in HgZnTe than in HgCdTe. Interdiffusion studies between HgTe and ZnTe indicate that the interdiffusion coefficient is approximately 10 times lower in HgZnTe than in HgCdTe [448].

Berding et al. [455] and Granger et al. [456] gave the theoretical description of the scattering mechanisms in HgZnTe. To obtain a good fit to experimental data for $Hg_{0.866}Zn_{0.134}Te$, they considered phonon dispersion plus ionized impurity scattering plus core dispersion without compensation in their mobility calculations. Theoretical calculations of electron mobilities indicate that the disorder scattering is negligible for HgZnTe alloy [448]. In contrast, the hole mobilities are likely to be limited by alloy scattering, and the predicted alloy hole mobility of HgZnTe is approximately a factor of 2 less than what was found for HgCdTe. Additionally, Abdelhakiem et al. [457] confirmed that the electron mobilities are very close to HgCdTe ones for the same energy gap and the same donor and acceptor concentrations.

The HgMnTe alloy is a semimagnetic narrow-gap semiconductor. The exchange interaction between band electrons and Mn^{2+} electrons modifies their band structure, making it dependent on the magnetic field at very low temperature. In the range of temperatures typical for IR detector operation (≥77 K), the spin-independent properties of HgMnTe are practically identical to the properties of HgCdTe, which are discussed exhaustively in the literature. The studies carried out by Kremer et al. [458] confirmed that the annealing of samples in Hg vapor eliminates the Hg vacancies, with the resulting material being n-type due to some unknown native donor. The diffusion rate of Hg into HgMnTe is the same as into HgCdTe. Measurements of the transport properties of $Hg_{1-x}Mn_xTe$ ($0.095 \leq x \leq 0.15$) indicate deep donor and acceptor levels into the energy gap, which influence not only the temperature dependencies of the Hall coefficient, conductivity, and Hall mobility, but also the minority carrier lifetimes [459,460]. Theoretical considerations of the electron mobilities in HgCdTe and HgMnTe indicate that at room temperature the mobilities are nearly the same. But at 77 K, the electron mobilities are approximately 30% less for HgMnTe when compared with the same concentration of defects [461].

Table 14.6: Standard Relationships for $Hg_{1-x}Zn_xTe$ (0.10 ≤ x ≤ 0.40)

Parameter	Relationship
Lattice constant $a(x)$ (nm) at 300 K	$0.6461-0.0361x$
Density γ (g/cm³) at 300 K	$8.05-2.41x$
Energy gap E_g (eV)	$-0.3 + 0.0324x^{1/2} + 2.731x - 0.629x^2 + 0.533x^3 + 5.3 \times 10^{-4}\, T\,(1 - 0.76x^{1/2} - 1.29x)$
Intrinsic carrier concentration n_i (cm⁻³)	$(3.607 + 11.370x + 6.584 \times 10^{-3}\, T - 3.633 \times 10^{-2}\, xT) \times 10^{14}\, E_g^{3/4}\, T^{3/2} \exp(-5802E_g/T)$
Momentum matrix element P (eVcm)	8.5×10^{-8}
Spin-orbit splitting energy Δ (eV)	1.0
Effective masses: m_e^*/m m_h^*/m	$5.7 \times 10^{-16}E_g/P^2$ E_g in eV; P in eVcm 0.6
Mobilities: μ_e (cm²/Vs) μ_h (cm²/Vs)	$9 \times 10^8 b/T^{2a}$ $a = (0.14/x)^{0.6}$; $b = (0.14/x)^{7.5}$ $\mu_e(x,T)/100$
Static dielectric constant ε_s	$20.206 - 15.153x + 6.5909x^2 - 0.951826x^3$
High-frequency dielectric constant ε_∞	$13.2 + 19.1916x + 19.496x^2 - 6.458x^3$

Source: A. Rogalski, *Progress in Quantum Electronics*, 13, 299–353, 1989. With permission.

Table 14.7: Standard Relationships for $Hg_{1-x}Mn_xTe$ (0.08 ≤ x ≤ 030)

Parameter	Relationship
Lattice constant $a(x)$ (nm) at 300 K	$0.6461-0.0121x$
Density γ (g/cm³) at 300 K	$8.12-3.37x$
Energy gap E_g (eV)	$-0.253 + 3.446x + 4.9 \times 10^{-4}\, xT - 2.55 \times 10^{-3}\, T$
Intrinsic carrier concentration n_i (cm⁻³)	$(4.615 - 1.59x + 2.64 \times 10^{-3}\, T - 1.70 \times 10^{-2}xT + 34.15x^2) \times 10^{14}\, E_g^{3/4}T^{3/2} \exp(-5802E_g/T)$
Momentum matrix element P (eVcm)	$(8.35 - 7.94x) \times 10^{-8}$
Spin-orbit splitting energy Δ (eV)	1.08
Effective masses: m_e^*/m m_h^*/m	$(8.35 - 7.94x) \times 10^{-8}$ E_g in eV; P in eVcm 0.5
Mobilities: μ_e (cm²/Vs) μ_h (cm²/Vs)	$9 \times 10^8 b/T^{2a}$ $a = (0.095/x)^{0.6}$; $b = (0.095/x)^{7.5}$ $\mu_e(x,T)/100$
Static dielectric constant ε_s	$20.5 - 32.6x + 25.1x^2$

Source: A. Rogalski, *Infrared Physics*, 31, 117–66, 1991. With permission.

The measured carrier lifetime in both ternary alloy systems is a sensitive characteristic of semiconductors that depends on material composition, temperature, doping, and defects. The Auger mechanism governs the high-temperature lifetime, and the SR mechanism is mainly responsible for low-temperature lifetimes. The reported positions of SR centers for both n- and p-type materials range anywhere from near the valence to near the conduction band. Comprehensive reviews of generation-recombination mechanisms and the carrier lifetime experimental data for both ternary alloys are given by Rogalski et al. [73,444].

Tables 14.6 and 14.7 contain lists of standard approximate relationships for material properties of HgZnTe and HgMnTe, respectively. Most of these relationships have been taken from Rogalski [442,443]. Some of these parameters, for example, the intrinsic carrier concentration, have since been reexamined. For example, Sha et al. [462] concluded that their improved calculations of intrinsic carrier concentration were approximately 10%–30% higher than those obtained earlier by Jóźwikowski and Rogalski [463]. However, the new calculations also should be treated as approximations since the dependence of the energy gap on composition and temperature [$E_g(x,T)$ is necessary in the calculations of n_i] is still under serious discussion [464].

14.7.3 HgZnTe Photodetectors

The technology for HgZnTe IR detectors has benefited greatly from the HgCdTe device technology base [442,444,465]. In comparison with HgCdTe, HgZnTe detectors are easier to prepare due to their relatively higher hardness. The development of device technology requires reproducible high-quality, electronically stable interfaces with a low interface state density. It was found that the tendency to form surface inversion layers on HgZnTe by anodization is considerably lower than that of HgCdTe [466]. Also, fixed charges at the anodic oxide-HgZnTe interface (2×10^{10} cm^{-2} at 90 K) are lower [467]. Additionally, it has been observed that the anodic oxide-HgZnTe interface is more stable under thermal treatment than the anodic-HgCdTe interface [468].

The first HgZnTe photoconductive detectors were fabricated by Nowak in the early 1970s [469]. Because of their early stage of development, the performance of these devices was inferior to that of HgCdTe. Then Piotrowski et al. demonstrated that p-type $Hg_{0.885}Zn_{0.115}Te$ can be used as a material for high-quality ambient 10.6 μm photoconductors [470,471]. These photoconductors, working at 300 K, can achieve 10^8 $cmHz^{1/2}W^{-1}$ detectivity with optimized composition, doping, and geometry. Aspects of theoretical performance for both photoconductive and photovoltaic detectors are discussed Rogalski and colleagues [442–444,472–475].

A HgZnTe ternary alloy also has been used to fabricate PEM detectors. The first nonoptimal detectors were also fabricated by Nowak in the early 1970s [469]. The theoretical design of HgZnTe PEM detectors was reported by Polish workers [172]. An uncooled 10.6 μm HgZnTe PEM detector made it possible to achieve a detectivity value equal to 1.2×10^8 $cmHz^{1/2}W^{-1}$.

Several different techniques have yielded p-n HgZnTe junctions, including Hg in-diffusion [442,465,476], Au diffusion [442,465,477], ion implantation [442,448,465,478,479], and ion etching [202,442,465,473]. To date, the ion implantation method gives the best-quality n^+-p HgZnTe photodiodes. This technological process, elaborated at SAT, was identical to the one used for HgCdTe

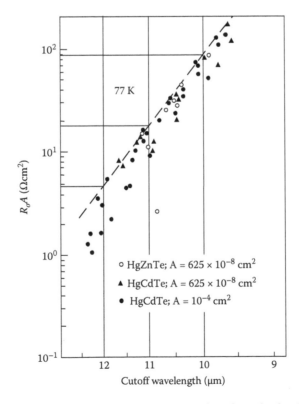

Figure 14.83 R_oA comparison between HgZnTe and HgCdTe photodiodes. The broken line is the upper limit of experimental data. (From Ameurlaine, J., Rousseau, A., Nguyen-Duy, T., and Triboulet, R., "(HgZn)Te Infrared Photovoltaic Detectors," *Proceedings of SPIE* 929, 14–20, 1988. With permission.)

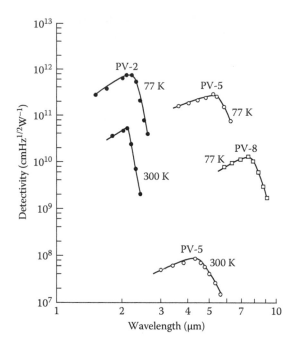

Figure 14.84 Spectral detectivities of three HgCdZnTe photodiodes at 300 and 77 K. (From Kaiser, D. L., and Becla, P., *Materials Research Society Symposium Proceedings*, 90, 397–404, 1987.)

photodiodes in planar structures. The type conversion in p-type material grown by the THM method was achieved by Al implantation [478,479].

The HgZnTe photodiode characteristics at 77 K are similar to those of HgCdTe photodiodes. Comparable values of R_0A for both types of photodiodes have been obtained (see Figure 14.83 [478]). For a staggered, 32-element linear array the following average values were measured at 77 K: $\lambda_c = (10.5 \pm 0.1)$ μm, η = 65%, and detectivity $1.1 \times 10^{11} < D^* < 1.2 \times 10^{11}$ cmHz$^{1/2}$W^{-1} at 30° FOV.

Encouraging results have been achieved using a HgCdZnTe quaternary alloy system. Kaiser and Becla produced high-quality p-n junctions from quaternary $Hg_{1-x-y}Cd_xZn_yTe$ epilayers prepared by ISOVPE on CdZnTe substrates [480]. Typical spectral detectivities of p-n HgCdZnTe junctions are presented in Figure 14.84. The FOV of the diodes was 60°C, with $f = 12$ Hz, and $T_B = 300$ K. Under these conditions, the detectivity of these photodiodes is comparable to those of high-quality HgCdTe photodiodes.

14.7.4 HgMnTe Photodetectors

Among the different types of HgMnTe IR detectors, primarily p-n junction photodiodes have been developed [465,481,482]. Also, the quaternary HgCdMnTe alloy system is an interesting material for IR applications. The presence of Cd, a third cation in this system, makes it possible to use composition to tune not only the bandgap but also other energy levels, in particular the spin-orbit-split band Γ_7 [483,484]. Because of this flexibility, the system appears advantageous, especially for APDs.

Becla produced good-quality p-n HgMnTe and HgCdMnTe junctions by annealing as-grown, p-type samples in Hg-saturated atmospheres [481]. These junctions were made in HgMnTe or HgCdMnTe bulk samples grown by THM, and epitaxial layers grown isothermally on the CdMnTe substrate. Typical spectral detectivities of the HgMnTe and HgCdMnTe photodiodes with 60° FOV are presented in Figure 14.85, which shows that the detectivities in the 3–5 μm and 8–12 μm spectral ranges are close to the background limit. Typical quantum efficiencies were in the 20–40% range without using an AR coating.

More recently, high quality planar and mesa HgMnTe photodiodes have been fabricated by Kosyachenko et al. [485,486] using ion etching in a system generating an argon beam of

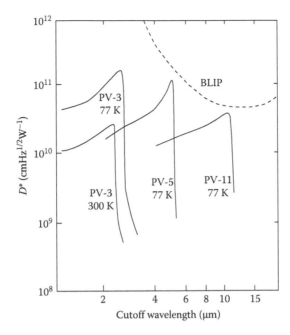

Figure 14.85 Spectral detectivities of HgMnTe and HgCdMnTe photodiodes with 60° FOV. (From Becla, P., *Journal of Vacuum Science and Technology A*, 4, 2014–18, 1986. With permission.)

500–1000 eV energy and 0.5–1 mA/cm² current density. The hole concentration of annealed as-grown Bridgman wafers selected for photodiode preparation was $(2–5) \times 10^{16}$ cm⁻³. For $Hg_{1-x}Mn_xTe$ photodiodes operated at 80 K with cutoff wavelength of 10–11 μm the R_oA product is equal to 20–30 Ωcm², whereas for photodiodes with λ_c of 7–8 μm, $R_oA \approx 500$ Ωcm² has been obtained.

The potential advantage of the HgCdMnTe system is connected with the bandgap spin-orbit-splitting resonance ($E_g = \Delta$) effects in the impact ionization phenomena in APDs. At room temperature, HgCdTe and HgMnTe systems provide an $E_g = \Delta$ resonance at 1.3 μm and 1.8 μm, respectively. To demonstrate the above possibility of obtaining high-performance HgCdMnTe APDs, Shin et al. used the boron-implantation method to fabricate the mesa-type structures for this quaternary alloy [487]. The R_oA product was 2.62×10^2 Ωcm², which is equivalent to a detectivity value of 1.9×10^{11} cmHz$^{1/2}$W⁻¹ at 300 K. The breakdown voltage defined at the dark current 10 μA was over 110 V. The leakage current at –10 V was 3×10^{-5} A/cm². This current density is comparable to that reported by Alabedra et al. for the planar bulk HgCdTe APDs [374]. During the 1980s, the dark current for the 1.46 μm photodiodes was much lower than the dark currents reported for either bulk or LPE HgCdTe APDs.

Becla et al. [488] have developed HgMnTe APDs, which have increased speed and performance compared to its standard line of photodiodes. Several p-n and n-p mesa type structures were fabricated, which permitted the injection of minority carriers from both n- and p-type regions, and led to hole-initiated and electron-initiated avalanche gain. Avalanche gain to 7 μm devices was more than 40, and 10.6 μm detectors showed gains better than 10 (Figure 14.86). Furthermore, it was found that the electron-initiated avalanche multiplication in HgMnTe photodiodes is more desirable for optimization of SNR and gain-bandwidth product.

Other types of HgMnTe detectors also have been fabricated. Becla et al. described $Hg_{0.92}Mn_{0.08}Te$ PEM detectors with an acceptor concentration of approximately 2×10^{17} cm⁻³ [489]. To obtain such concentrations, the wafers cut from bulk crystals grown by THM were annealed in a Hg-saturated atmosphere. The wafers were then thinned to approximately 5 μm and mounted in the narrow slot of a miniature permanent magnet having a field strength of approximately 1.5 T. The best performance of PEM detectors was achieved using $Hg_{1-x}Mn_xTe$ with composition x of approximately 0.08–0.09. At that composition range, the peak detectivity

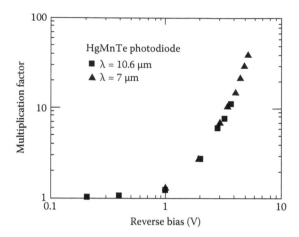

Figure 14.86 Avalanche multiplication factor for 7 and 10.6 μm wavelength HgMnTe photodiodes as a function of reverse bias voltage. (From Becla, P., Motakef, S., and Koehler, T., *Journal of Vacuum Science and Technology*, B10, 1599–1601, 1992. With permission.)

of the detector was in the region of 7–8 μm. Any reduction or increase in the composition of $Hg_{1-x}Mn_xTe$ leads to a performance reduction at 10.6 μm.

REFERENCES

1. W. D. Lawson, S. Nielson, E. H. Putley, and A. S. Young, "Preparation and Properties of HgTe and Mixed Crystals of HgTe-CdTe," *Journal of Physics and Chemistry of Solids* 9, 325–29, 1959.

2. B. J. Andresen, G. F. Fulop, and P. R. Norton, "Infrared Technology and Applications XXXV," *Proceedings of SPIE* 7298, Bellingham, WA, 2009.

3. T. Elliot, "Recollections of MCT Work in the UK at Malvern and Southampton," *Proceedings of SPIE* 7298, 72982M, 2009.

4. E. Burstein, J. W. Davisson, E. E. Bell, W. J. Turner, and H. G. Lipson, "Infrared Photoconductivity Due to Neutral Impurities in Germanium," *Physical Review* 93, 65–68, 1954.

5. S. Borrello and H. Levinstein, "Preparation and Properties of Mercury-Doped Germanium," *Journal of Applied Physics* 33, 2947–50, 1962.

6. D. Long and J. L. Schmit, "Mercury-Cadmium Telluride and Closely Related Alloys," in *Semiconductors and Semimetals*, Vol. 5, eds. R. K. Willardson and A. C. Beer, 175–255, Academic Press, New York, 1970.

7. P. Norton, "HgCdTe Infrared Detectors," *Opto-Electronics Review* 10, 159–74, 2002.

8. C. Verie and R. Granger, "Propriétés de junctions p-n d'alliages $Cd_xHg_{1-x}Te$," *Comptes Rendus de l'Académie des Sciences* 261, 3349–52, 1965.

9. G. C. Verie and M. Sirieix, "Gigahertz Cutoff Frequency Capabilities of CdHgTe Photovoltaic Detectors at 10.6 μm," *IEEE Journal of Quantum Electronics* 8, 180–84, 1972.

10. B. E. Bartlett, D. E. Charlton, W. E. Dunn, P. C. Ellen, M. D. Jenner, and M. H. Jervis, "Background Limited Photoconductive Detectors for Use in the 8–14 Micron Atmospheric Window," *Infrared Physics & Technology* 9, 35–36, 1969.

11. M. A. Kinch, S. R. Borrello, and A. Simmons, "0.1 eV HgCdTe Photoconductive Detector Performance," *Infrared Physics* 17, 127–35, 1977.

12. C. T. Elliott, D. Day, and B. J. Wilson, "An Integrating Detector for Serial Scan Thermal Imaging," *Infrared Physics* 22, 31–42, 1982.

13. A. Blackburn, M. V. Blackman, D. E. Charlton, W. A. E. Dunn, M. D. Jenner, K. J. Oliver, and J. T. M. Wotherspoon, "The Practical Realization and Performance of SPRITE Detectors," *Infrared Physics* 22, 57–64, 1982.

14. W. S. Boyle and G. E. Smith, "Charge-Coupled Semiconductor Devices," *Bell Systems Technology Journal* 49, 587–93, 1970.

15. R. Thom, "High Density Infrared Detector Arrays," U.S. Patent No. 4,039,833, 8/2/77.

16. M. A. Kinch, "MIS Devices in HgCdTe," in *Properties of Narrow Gap Cadmium-Based Compounds, EMIS Datareviews Series No. 10,* ed. P. Capper, 359–63, IEE, London, 1994.

17. M. B. Reine, "Photovoltaic Detectors in HgCdTe," in *Infrared Detectors and Emitters: Materials and Devices,* eds. P. Capper and C. T. Elliott, 313–76, Chapman and Hall, London, 2000.

18. A. Rogalski, J. Antoszewski, and L. Faraone, "Third-Generation Infrared Photodetector Arrays," *Journal of Applied Physics* 105, 091101-44, 2009.

19. W. F. H. Micklethweite, "The Crystal Growth of Mercury Cadmium Telluride," *Semiconductors and Semimetals,* Vol. 18, eds. R. K. Willardson and A. C. Beer, 48–119, Academic Press, New York, 1981.

20. S. Sher, M. A. Berding, M. van Schlifgaarde, and An-Ban Chen, "HgCdTe Status Review with Emphasis on Correlations, Native Defects and Diffusion," *Semiconductor Science and Technology* 6, C59–C70, 1991.

21. H. R. Vydyanath, "Status of Te-Rich and Hg-Rich Liquid Phase Epitaxial Technologies for the Growth of (Hg,Cd)Te Alloys," *Journal of Electronic Materials* 24, 1275–85, 1995.

22. R. F. Brebrick, "Thermodynamic Modeling of the Hg-Cd-Te Systems," *Journal of Crystal Growth* 86, 39–48, 1988.

23. J. C. Brice, "A Numerical Description of the Cd-Hg-Te Phase Diagram," *Progress in Crystal Growth and Characterization of Materials* 13, 39–61, 1986.

24. T. C. Yu and R. F. Brebrick, "Phase Diagrams for HgCdTe," in *Properties of Narrow Gap Cadmium-Based Compounds, EMIS Datareviews Series No. 10,* ed. P. Capper, 55–63, IEE, London, 1994.

25. *Properties of Narrow Gap Cadmium-Based Compounds, EMIS Datareviews Series No. 10,* ed. P. P. Capper, IEE, London, 1994.

26. T. Tung, C. H. Su, P. K. Liao, and R. F. Brebrick, "Measurement and Analysis of the Phase Diagram and Thermodynamic Properties in the Hg-Cd-Te System," *Journal of Vacuum Science and Technology* 21, 117–24, 1982.

27. H. Maier and J. Hesse, "Growth, Properties and Applications of Narrow-Gap Semiconductors," in *Crystal Growth, Properties and Applications,* ed. H. C. Freyhardt, 145–219, Springer Verlag, Berlin, 1980.

28. J. H. Tregilgas, "Developments in Recrystallized Bulk HgCdTe," *Progress in Crystal Growth and Characterization of Materials* 28, 57–83, 1994.

29. P. Capper, "Bulk Growth Techniques," in *Narrow-gap II-VI Compounds for Optoelectronic and Electromagnetic Applications,* ed. P. Capper, 3–29, Chapman & Hall, London, 1997.

30. P. Capper, "Narrow-Bandgap II-VI Semiconductors: Growth," in *Springer Handbook of Electronic and Photonic Materials*, ed. S. Kasap and P. Capper, 303–24, Springer, Heidelberg, 2006.

31. T. C. Harman, "Single Crystal Growth of $Hg_{1-x}Cd_xTe$," *Journal of Electronic Materials* 1, 230–42, 1972.

32. A. W. Vere, B. W. Straughan, D. J. Williams, N. Shaw, A. Royle, J. S. Gough, and J. B. Mullin, "Growth of $Cd_xHg_{1-x}Te$ by a Pressurised Cast-Recrystallise-Anneal Technique," *Journal of Crystal Growth* 59, 121–29, 1982.

33. L. Colombo, A. J. Syllaios, R. W. Perlaky, and M. J. Brau, "Growth of Large Diameter (Hg,Cd) Te Crystals by Incremental Quenching," *Journal of Vacuum Science and Technology* A3, 100–104, 1985.

34. R. K. Sharma, V. K. Singh, N. K. Nayyar, S. R. Gupta, and B. B. Sharma, "Horizontal Casting for the Growth of $Hg_{1-x}Cd_xTe$ by Solid State Recrystalization," *Journal of Crystal Growth* 131, 565–73, 1993.

35. P. Capper, "Bridgman Growth of $Cd_xHg_{1-x}Te$: A Review," *Progress in Crystal Growth and Characterization of Materials* 19, 259–93, 1989.

36. P. Capper, "The Role of Accelerated Crucible Rotation in the Growth of $Hg_{1-x}Cd_xTe$ and CdTe/CdZnTe," *Progress in Crystal Growth and Characterization of Materials* 28, 1–55, 1989.

37. P. Capper, J. Gosney, J. E. Harris, E. O'Keefe, and C. D. Maxey, "Infra-Red Materials Activities at GEC-Marconi Infra-Red Limited: Part I: Bulk Growth Techniques," *GEC Journal of Research* 13, 164–74, 1996.

38. L. Colombo, R. R. Chang, C. J. Chang, and B. A. Baird, "Growth of Hg-Based Alloys by the Travelling Heater Method," *Journal of Vacuum Science and Technology* A6, 2795–99, 1988.

39. R. Triboulet, "The Travelling Heater Method (THM) for $Hg_{1-x}Cd_xTe$ and Related Materials," *Progress in Crystal Growth and Characterization of Materials* 28, 85–114, 1994.

40. E. R. Gertner, "Epitaxial Mercury Cadmium Telluride," *Annual Review of Materials Science* 15, 303–28, 1985.

41. T. C. Harman, "Liquidus Isotherms, Solidus Lines and LPE Growth in the Te-Rich Corner of the Hg-Cd-Te System," *Journal of Electronic Materials* 9, 945–61, 1980.

42. B. Pelliciari, "State of the Art of LPE HgCdTe at LIR," *Journal of Crystal Growth* 86, 146–60, 1988.

43. L. Colombo and G. H. Westphal, "Large Volume Production of HgCdTe by Dipping Liquid Phase Epitaxy," *Proceedings of SPIE* 2228, 66–72, 1994.

44. H. R. Vydyanath, "Status of Te-Rich and Hg-Rich Liquid Phase Epitaxial Technologies for the Growth of (Hg,Cd)Te Alloys," *Journal of Electronic Materials* 24, 1275–85, 1995.

45. P. Capper, T. Tung, and L. Colombo, "Liquid Phase Epitaxy," in *Narrow-gap II-VI Compounds for Optoelectronic and Electromagnetic Applications,* ed. P. Capper, 30–70, Chapman & Hall, London, 1997.

46. P. E. Herning, "Experimental Determination of the Mercury-Rich Corner of the Hg-Cd-Te Phase Diagram," *Journal of Electronic Materials* 13, 1–14, 1984.

47. T. Tung, L. V. DeArmond, R. F. Herald, P. E. Herning, M. H. Kalisher, D. A. Olson, R. F. Risser, A. P. Stevens, and S. J. Tighe, "State of the Art of Hg-melt LPE HgCdTe at Santa Barbara Research Center," *Proceedings of SPIE* 1735, 109–31, 1992.

48. G. Cohen-Solal and Y. Marfaing, "Transport of Photocarriers in $Cd_xHg_{1-x}Te$ Graded-Gap Structures," *Solid State Electronics* 11, 1131–47, 1968.

49. R. F. Hicks, "The Chemistry of the Organometallic Vapor-Phase Epitaxy of Mercury Cadmium Telluride," *Proceedings of IEEE* 80, 1625–40, 1992.

50. S. J. C. Irvine, "Metal-Organic Vapour Phase Epitaxy," *Narrow-gap II-VI Compounds for Optoelectronic and Electromagnetic Applications*, ed. P. Capper, 71–96, Chapman & Hall, London, 1997.

51. J. Tunnicliffe, S. J. C. Irvine, O. D. Dosser, and J. B. Mullin, "A New MOCVD Technique for the Growth of Highly Uniform CMT," *Journal of Crystal Growth* 68, 245–53, 1984.

52. J. C. Irvine, "Recent Development in MOCVD of $Hg_{1-x}Cd_xTe$," *SPIE Proceedings* 1735, 92–99, 1992.

53. P. Mitra, F. C. Case, and M. B. Reine "Progress in MOVPE of HgCdTe for Advanced Infrared Detectors," *Journal of Electronic Materials* 27, 510–20, 1998.

54. J. M. Arias-Cortes, "MBE of HgCdTe for Electro-Optical Infrared Applications," in *II–VI Semiconductor Compounds*, ed. M. Pain, 509–36, World Scientific Publishing, Singapore, 1993.

55. O. K. Wu, T. J. deLyon, R. D. Rajavel, and J. E. Jensen, "Molecular Beam Epitaxy of HgCdTe," in *Narrow-gap II-VI Compounds for Optoelectronic and Electromagnetic Applications*, ed. P. Capper, 97–130, Chapman & Hall, London, 1997.

56. T. J. de Lyon, R. D. Rajavel, J. A. Roth, and J. E. Jensen, "Status of HgCdTe MBE Technology," in *Handbook of Infrared Detection and Technologies*, eds. M. Henini and M. Razeghi, 309–52, Elsevier, Oxford, 2002.

57. T. S. Lee, J. Garland, C. H. Grein, M. Sumstine, A. Jandeska, Y. Selamet, and S. Sivananthan, "Correlation of Arsenic Incorporation and Its Electrical Activation in MBE HgCdTe," *Journal of Electronic Materials* 29, 869–72, 2000.

58. W. E. Tennant, C. A. Cockrum, J. B. Gilpin, M. A. Kinch, M. B. Reine, and R. P. Ruth, "Key Issue in HgCdTe-Based Focal Plane Arrays: An Industry Perspective," *Journal of Vacuum Science and Technology* B10, 1359–69, 1992.

59. R. Triboulet, A. Tromson-Carli, D. Lorans, and T. Nguyen Duy, "Substrate Issues for the Growth of Mercury Cadmium Telluride," *Journal of Electronic Materials* 22, 827–34, 1993.

60. E. R. Gertner, W. E. Tennant, J. D. Blackwell, and J. P. Rode, "HgCdTe on sapphire: A New Approach to Infrared Detector Arrays," *Journal of Crystal Growth* 72, 462–67, 1987.

61. L. J. Kozlowski, S. L. Johnston, W. V. McLevige, A. H. B. Vandervyck, D. E. Cooper, S. A. Cabelli, E. R. Blazejewski, K. Vural, and W. E. Tennant, "128×128 PACE 1 HgCdTe Hybrid FPAs for Thermoelectrically Cooled Applications," *Proceedings of SPIE* 1685, 193–203, 1992.

62. D. D. Edwall, J. S. Chen, J. Bajaj, and E. R. Gertner, "MOCVD HgCdTe/GaA for IR Detectors," *Semiconductor Science and Technology* 5, S221–S224, 1990.

63. V. S. Varavin, V. V. Vasiliev, S. A. Dvoretsky, N. N. Mikhailov, V. N. Ovsyuk, Yu. G. Sidorov, A. O. Suslyakov, M. V. Yakushev, and A. L. Aseev, "HgCdTe Epilayers on GaAs: Growth and Devices," *Opto-Electronics Review* 11, 99–111, 2003.

64. J. M. Peterson, J. A. Franklin, M. Readdy, S. M. Johnson, E. Smith, W. A. Radford, and I. Kasai, "High-Quality Large-Area MBE HgCdTe/Si," *Journal of Electronic Materials* 36, 1283–86, 2006.

65. R. Bornfreund, J. P. Rosbeck, Y. N. Thai, E. P. Smith, D. D. Lofgreen, M. F. Vilela, A. A. Buell, et al., "High-Performance LWIR MBE-Grown HgCdTe/Si Focal Plane Arrays," *Journal of Electronic Materials* 37, 1085–91, 2007.

66. P. Capper, "A Review of Impurity Behavior in Bulk and Epitaxial $Hg_{1-x}Cd_xTe$," *Journal of Vacuum Science and Technology* B9, 1667–86, 1991.

67. S. Sher, M. A. Berding, M. van Schlifgaarde, and An-Ban Chen, "HgCdTe Status Review with Emphasis on Correlations, Native Defects and Diffusion," *Semiconductor Science and Technology* 6, C59–C70, 1991.

68. H. R. Vydyanath, "Mechanisms of Incorporation of Donor and Acceptor Dopants in $Hg_{1-x}Cd_xTe$ Alloys," *Journal of Vacuum Science and Technology* B9, 1716–23, 1991.

69. H. R. Vydyanath, "Incorporation of Dopants and Native Defects in Bulk $Hg_{1-x}Cd_xTe$ Crystals and Epitaxial Layers," *Journal of Crystal Growth* 161, 64–72, 1996.

70. P. Capper, "Intrinsic and Extrinsic Doping," in *Narrow-Gap II-VI Compounds for Optoelectronic and Electromagnetic Applications,* ed. P. Capper, 211–37, Chapman & Hall, London, 1997.

71. Y. Marfaing, "Point Defects in Narrow Gap II-VI Compounds," in *Narrow-Gap II-VI Compounds for Optoelectronic and Electromagnetic Applications,* ed. P. Capper, 238–67, Chapman & Hall, London, 1997.

72. H. R. Vydyanath, V. Nathan, L. S. Becker, and G. Chambers, "Materials and Process Issues in the Fabrication of High-Performance HgCdTe Infrared Detectors," *Proceedings of SPIE* 3629, 81–87, 1999.

73. A. Rogalski, K. Adamiec, and J. Rutkowski, *Narrow-Gap Semiconductor Photodiodes,* SPIE Press, Bellingham, WA, 2000.

74. M. A. Berding, M. van Schilfgaarde, and A. Sher, "$Hg_{0.8}Cd_{0.2}Te$ Native Defect: Densities and Dopant Properties, *Journal of Electronic Materials* 22, 1005–10, 1993.

75. J. L. Melendez and C. R. Helms, "Process Modeling of Point Defect Effects in $Hg_{1-x}Cd_xTe$," *Journal of Electronic Materials* 22, 999–1004, 1993.

76. S. Holander-Gleixner, H. G. Robinson, and C. R. Helms, "Simulation of HgTe/CdTe Interdiffusion Using Fundamental Point Defect Mechanisms," *Journal of Electronic Materials* 27, 672–79, 1998.

77. P. Capper, C. D. Maxey, C. L. Jones, J. E. Gower, E. S. O'Keefe, and D. Shaw, "Low Temperature Thermal Annealing Effects in Bulk and Epitaxial $Cd_xHg_{1-x}Te$," *Journal of Electronic Materials* 28, 637–48, 1999.

78. M. A. Berding "Equilibrium Properties of Indium and Iodine in LWIR HgCdTe," *Journal of Electronic Materials* 29, 664–68, 2000.

79. D. Chandra, H. F. Schaake, J. H. Tregilgas, F. Aqariden, M. A. Kinch, and A. Syllaios, "Vacancies in $Hg_{1-x}Cd_xTe$," *Journal of Electronic Materials* 29, 729–31, 2000.

80. H. Wiedemayer and Y. G. Sha, "The Direct Determination of the Vacancy Concentration and p-T Phase Diagram of $Hg_{0.8}Cd_{0.2}Te$ and $Hg_{0.6}Cd_{0.4}Te$ by Dynamic Mass-Loss Measurements," *Journal of Electronic Materials* 19, 761–71, 1990.

81. H. F. Schaake, J. H. Tregilgas, J. D. Beck, M. A. Kinch, and B. E. Gnade, "The Effect of Low Temperature Annealing on Defects, Impurities and Electrical Properties," *Journal of Vacuum Science and Technology* A3, 143–49, 1985.

82. D. A. Stevenson and M. F. S. Tang, "Diffusion Mechanisms in Mercury Cadmium Telluride," *Journal of Vacuum Science and Technology* B9, 1615–24, 1991.

83. P. Mitra, Y. L. Tyan, F. C. Case, R. Starr, and M. B. Reine, "Improved Arsenic Doping in Metalorganic Chemical Vapor Deposition of HgCdTe and *In Situ* Growth of High Performance Long Wavelength Infrared Photodiodes," *Journal of Electronic Materials* 25, 1328–35, 1996.

84. M. Tanaka, K. Ozaki, H. Nishino, H. Ebe, and Y. Miyamoto, "Electrical Properties of HgCdTe Epilayers Doped with Silver Using an $AgNO_3$ Solution," *Journal of Electronic Materials* 27, 579–82, 1998.

85. S. P. Tobin, G. N. Pultz, E. E. Krueger, M. Kestigian, K. K. Wong, and P. W. Norton, "Hall Effect Characterization of LPE HgCdTe p/n Heterojunctions," *Journal of Electronic Materials* 22, 907–14, 1993.

86. D. Chandra, M. W. Goodwin, M. C. Chen, and L. K. Magel, "Variation of Arsenic Diffusion Coefficients in HgCdTe Alloys with Temperature and Hg Pressure: Tuning of p on n Double Layer Heterojunction Diode Properties," *Journal of Electronic Materials* 24, 599–608, 1995.

87. D. Shaw, "Diffusion in Mercury Cadmium Telluride—An Update," *Journal of Electronic Materials* 24, 587–97, 1995.

88. L. O. Bubulac, D. D. Edwall, S. J. C. Irvine, E. R. Gertner, and S. H. Shin, "p-Type Doping of Double Layer Mercury Cadmium Telluride for Junction Formation," *Journal of Electronic Materials* 24, 617–24, 1995.

89. M. B. Reine, "Fundamental Properties of Mercury Cadmium Telluride," in *Encyclopedia of Modern Optics*, Academic Press, London, 2004.

90. G. L. Hansen, J. L. Schmit, and T. N. Casselman, "Energy Gap Versus Alloy Composition and Temperature in $Hg_{1-x}Cd_xTe$," *Journal of Applied Physics* 53, 7099–7101, 1982.

91. G. L. Hansen and J. L. Schmit, "Calculation of Intrinsic Carrier Concentration in $Hg_{1-x}Cd_xTe$," *Journal of Applied Physics* 54, 1639–40, 1983.

92. M. H. Weiler, "Magnetooptical Properties of $Hg_{1-x}Cd_xTe$ Alloys," in *Semiconductors and Semimetals*, Vol. 16, eds. R. K. Willardson and A. C. Beer, 119–91, Academic Press, New York, 1981.

93. R. Dornhaus, G. Nimtz, and B. Schlicht, *Narrow-Gap Semiconductors*, Springer Verlag, Berlin, 1983.

94. J. R. Meyer, C. A. Hoffman, F. J. Bartoli, D. A. Arnold, S. Sivanathan, and J. P. Faurie, "Methods for Magnetotransport Characterization of IR Detector Materials," *Semiconductor Science and Technology* 8, 805–23, 1991.

95. R. W. Miles, "Electron and Hole Mobilities in HgCdTe," in *Properties of Narrow Gap Cadmium-Based Compounds*, EMIS Datareviews Series No. 10, ed. P. P. Capper, 221–26, IEE, London, 1994.

96. J. S. Kim, J. R. Lowney, and W. R. Thurber, "Transport Properties of Narrow-Gap II–VI Compound Semiconductors," in *Narrow-gap II-VI Compounds for Optoelectronic and Electromagnetic Applications*, ed. P. Capper, 180–210, Chapman & Hall, London, 1997.

97. J. J. Dubowski, T. Dietl, W. Szymańska, and R. R. Gałązka, "Electron Scattering in $Cd_xHg_{1-x}Te$," *Journal of Physics and Chemistry of Solids* 42, 351–63, 1981.

98. M. C. Chen and L. Colombo, "The Majority Carrier Mobility of n-Type and p-Type $Hg_{0.78}Cd_{0.22}Te$ Liquid Phase Epitaxial Films at 77 K," *Journal of Applied Physics* 73, 2916–20, 1993.

99. R. D. S. Yadava, A. K. Gupta, and A. V. R. Warrier, "Hole Scattering Mechanisms in $Hg_{1-x}Cd_xTe$," *Journal of Electronic Materials* 23, 1359–78, 1994.

100. D. A. Nelson, W. M. Higgins, R. A. Lancaster, R. Roy, and H. R. Vydyanath, *Extended Abstracts of U.S. Workshop on the Physics and Chemistry of Mercury Cadmium Telluride*, p. 175, Minneapolis, Minnesota, October 28–30, 1981.

101. J. P. Rosbeck, R. E. Star, S. L. Price, and K. J. Riley, "Background and Temperature Dependent Current-Voltage Characteristics of $Hg_{1-x}Cd_xTe$ Photodiodes," *Journal of Applied Physics* 53, 6430–40, 1982.

102. W. M. Higgins, G. N. Seiler, R. G. Roy, and R. A. Lancaster, "Standard Relationships in the Properties of $Hg_{1-x}Cd_xTe$," *Journal of Vacuum Science and Technology* A7, 271–75, 1989.

103. P. N. J. Dennis, C. T. Elliott, and C. L. Jones, "A Method for Routine Characterization of the Hole Concentration in p-Type Cadmium Mercury Telluride," *Infrared Physics* 22, 167–69, 1982.

104. S. Barton, P. Capper, C. L. Jones, N. Metcalfe, and N. T. Gordon, "Electron Mobility in p-Type Grown $Cd_xHg_{1-x}Te$," *Semiconductor Science and Technology* 10, 56–60, 1995.

105. P. M. Amirtharaj and J. H. Burnett, "Optical Properties of MCT," in *Narrow-gap II-VI Compounds for Optoelectronic and Electromagnetic Applications*, ed. P. Capper, 133–79, Chapman & Hall, London, 1997.

106. E. O. Kane, "Band Structure of InSb," *Journal of Physics and Chemistry of Solids* 1, 249–61, 1957.

107. W. W. Anderson, "Absorption Constant of $Pb_{1-x}Sn_xTe$ and $Hg_{1-x}Cd_xTe$ Alloys," *Infrared Physics* 20, 363–72, 1980.

108. R. Beattie and A. M. White, "An Analytic Approximation with a Wide Range of Applicability for Band-to-Band Radiative Transition Rates in Direct, Narrow-Gap Semiconductors," *Semiconductor Science and Technology* 12, 359–68, 1997.

109. E. Finkman and S. E. Schacham, "The Exponential Optical Absorption Band Tail of $Hg_{1-x}Cd_xTe$," *Journal of Applied Physics* 56, 2896–900, 1984.

110. S. E. Schacham and E. Finkman, "Recombination Mechanisms in p-Type HgCdTe: Freezout and Background Flux Effects," *Journal of Applied Physics* 57, 2001–9, 1985.

111. C. A. Hougen, "Model for Infrared Absorption and Transmission of Liquid-Phase Epitaxy HgCdTe," *Journal of Applied Physics* 66, 3763–66, 1989.

112. J. Chu, Z. Mi, and D. Tang, "Band-to-Band Optical Absorption in Narrow-Gap $Hg_{1-x}Cd_xTe$ Semiconductors," *Journal of Applied Physics* 71, 3955–61, 1992.

113. J. Chu, B. Li, K. Liu, and D. Tang, "Empirical Rule of Intrinsic Absorption Spectroscopy in $Hg_{1-x}Cd_xTe$," *Journal of Applied Physics* 75, 1234–35, 1994.

114. R. K. Sharma, D. Verma, and B. B. Sharma, "Observation of Below Band Gap Photoconductivity in Mercury Cadmium Telluride," *Infrared Physics & Technology* 35, 673–80, 1994.

115. B. Li, J. H. Chu, Y. Chang, Y. S. Gui, and D. Y. Tang, "Optical Absorption Above the Energy Band Gap in $Hg_{1-x}Cd_xTe$," *Infrared Physics & Technology* 37, 525–31, 1996.

116. S. Krishnamurthy, A. B. Chen, and A. Sher, "Near Band Edge Absorption Spectra of Narrow-Gap III-V Semiconductor Alloys," *Journal of Applied Physics* 80, 4045–48, 1996.

117. S. Krishnamurthy, A. B. Chen, and A. Sher, "Electronic Structure, Absorption Coefficient, and Auger Rate in HgCdTe and Thallium-Based Alloys," *Journal of Electronic Materials* 26, 571–77, 1997.

118. K. Moazzami, D. Liao, J. Phillips, D. L. Lee, M. Carmody, M. Zandian, and D. Edwall, "Optical Absorption Properties of HgCdTe Epilayers with Uniform Composition," *Journal of Electronic Materials* 32, 646–50, 2003.

119. K. Moazzami, J. Phillips, D. Lee, D. Edwall, M. Carmody, E. Piquette, M. Zandian, and J. Arias, "Optical-Absorption Model for Molecular-Beam Epitaxy HgCdTe and Application to Infrared Detector Photoresponse," *Journal of Electronic Materials* 33, 701–8, 2004.

120. F. Tong and N. M. Ravindra, "Optical Properties of $Hg_{1-x}Cd_xTe$," *Infrared Physics* 34, 207–12, 1993.

121. K. Liu, J. H. Chu, and D. Y. Tang, "Composition and Temperature Dependence of the Refractive Index in $Hg_{1-x}Cd_xTe$," *Journal of Applied Physics* 75, 4176–79, 1994.

122. S. Rolland, "Dielectric Constant and Refractive Index of HgCdTe," in *Properties of Narrow Gap Cadmium-Based Compounds, EMIS Datareviews Series No. 10,* Chapter 4A, ed. P. P. Capper, 80–85, IEE, London, 1994.

123. J. L. Pautrat and N. Magnea, "Optical Absorption in HgCdTe and Related Heterostructures," in *Properties of Narrow Gap Cadmium-Based Compounds, EMIS Data reviews Series No. 10,* ed. P. Capper, 75–79, IEE, London, 1994.

124. J. Chu and A. Sher, *Physics and Properties of Narrow Gap Semiconductors,* Springer, 2008.

125. S. L. Price and P. R. Boyd, "Overview of Compositional Measurements Techniques for $Hg_{1-x}Cd_xTe$ with Emphasis on IR Transmission, Energy Dispersive X–Ray Analysis and Optical Reflectance," *Semiconductor Science and Technology* 8, 842–59, 1993.

126. C. T. Elliott and N. T. Gordon, "Infrared Detectors," in *Handbook on Semiconductors,* Vol. 4, ed. C. Hilsum, 841–936, North-Holland, Amsterdam, 1993.

127. V. C. Lopes, A. J. Syllaios, and M. C. Chen, "Minority Carrier Lifetime in Mercury Cadmium Telluride," *Semiconductor Science and Technology* 8, 824–41, 1993.

128. Y. Yamamoto, Y. Miyamoto, and K. Tanikawa, "Minority Carrier Lifetime in the Region Close to the Interface Between the Anodic Oxide and CdHgTe," *Journal of Crystal Growth* 72, 270–74, 1985.

129. S. M. Johnson, D. R. Rhiger, J. P. Rosbeck, J. M. Peterson, S. M. Taylor, and M. E. Boyd, "Effect of Dislocations on the Electrical and Optical Properties of Long-Wavelength Infrared HgCdTe Photovoltaic Detectors," *Journal of Vacuum Science and Technology* B10, 1499–1506, 1992.

130. K. Jóźwikowski and A. Rogalski, "Effect of Dislocations on Performance of LWIR HgCdTe Photodiodes," *Journal of Electronic Materials* 29, 736–41, 2000.

131. M. C. Chen, L. Colombo, J. A. Dodge, and J. H. Tregilgas, "The Minority Carrier Lifetime in Doped and Undoped p-Type $Hg_{0.78}Cd_{0.22}Te$ Liquid Phase Epitaxy Films," *Journal of Electronic Materials* 24, 539–44, 1995.

132. P. Mitra, T. R. Schimert, F. C. Case, R. Starr, M. H. Weiler, M. Kestigian, and M. B. Reine, "Metalorganic Chemical Vapor Deposition of HgCdTe for Photodiode Applications," *Journal of Electronic Materials* 24, 661–68, 1995.

133. M. J. Bevan, M. C. Chen, and H. D. Shih, "High-Quality p-Type $Hg_{1-x}Cd_xTe$ Prepared by Metalorganic Chemical Vapor Deposition," *Applied Physics Letters* 67, 3450–52, 1996.

134. G. Destefanis and J. P. Chamonal, "Large Improvement in HgCdTe Photovoltaic Detector Performances at LETI," *Journal of Electronic Materials* 22, 1027–32, 1993.

135. C. L. Littler, E. Maldonado, X. N. Song, Z. You, J. L. Elkind, D. G. Seiler, and J. R. Lowney, "Investigation of Mercury Interstitials in $Hg_{1-x}Cd_xTe$ Alloys Using Resonant Impact-Ionization Spectroscopy," *Journal of Vacuum Science and Technology* B9, 1466–70, 1991.

136. M. A. Kinch, "Fundamental Physics of Infrared Detector Materials," *Journal of Electronic Materials* 29, 809–17, 2000.

137. M. A. Kinch, F. Aqariden, D. Chandra, P.-K. Liao, H. F. Schaake, and H. D. Shih, "Minority Carrier Lifetime in p-HgCdTe," *Journal of Electronic Materials* 34, 880–84, 2005.

138. R. G. Humpreys, "Radiative Lifetime in Semiconductors for Infrared Detectors," *Infrared Physics* 23, 171–75, 1983.

139. R. G. Humpreys, "Radiative Lifetime in Semiconductors for Infrared Detectors," *Infrared Physics* 26, 337–42, 1986.

140. T. Elliott, N. T. Gordon, and A. M. White, "Towards Background-Limited, Room-Temperature, Infrared Photon Detectors in the 3–13 μm Wavelength Range," *Applied Physics Letters* 74, 2881–83, 1999.

141. N. T. Gordon, C. D. Maxey, C. L. Jones, R. Catchpole, and L. Hipwood, "Suppression of Radiatively Generated Currents in Infrared Detectors," *Journal of Applied Physics* 91, 565–68, 2002.

142. C. T. Elliott and C. L. Jones, "Non-Equilibrium Devices in HgCdTe," in *Narrow-Gap II-VI Compounds for Optoelectronic and Electromagnetic Applications,* ed. P. Capper, 474–85, Chapman & Hall, London, 1997.

143. T. Ashley, N. T. Gordon, G. R. Nash, C. L. Jones, C. D. Maxey, and R. A. Catchpole, "Long-Wavelength HgCdTe Negative Luminescent Devices," *Applied Physics Letters* 79, 1136–38, 2001.

144. A. R. Beattie and P. T. Landsberg, "Auger Effect in Semiconductors," *Proceedings of the Royal Society of London* A249, 16–29, 1959.

145. J. S. Blakemore, *Semiconductor Statistics,* Pergamon Press, Oxford, 1962.

146. M. A. Kinch, M. J. Brau, and A. Simmons, "Recombination Mechanisms in 8–14 μm HgCdTe," *Journal of Applied Physics* 44, 1649–63, 1973.

147. G. Nimtz, "Recombination in Narrow-Gap Semiconductors," *Physics Reports* 63, 265–300, 1980.

148. T. N. Casselman and P. E. Petersen, "A Comparison of the Dominant Auger Transitions in p-Type (Hg,Cd)Te," *Solid State Communications* 33, 615–19, 1980.

149. T. N. Casselman, "Calculation of the Auger Lifetime in p-Type $Hg_{1-x}Cd_xTe$," *Journal of Applied Physics* 52, 848–54, 1981.

150. P. E. Petersen, "Auger Recombination in Mercury Cadmium Telluride," in *Semiconductors and Semimetals,* Vol. 18, eds. R. K. Willardson and A. C. Beer, 121–155, Academic Press, New York, 1981.

151. A. M. White, "The Characteristics of Minority-Carrier Exclusion in Narrow Direct Gap Semiconductors," *Infrared Physics* 25, 729–41, 1985.

152. C. M. Ciesla, B. N. Murdin, T. J. Phillips, A. M. White, A. R. Beattie, C. J. G. M. Langerak, C. T. Elliott, C. R. Pidgeon, and S. Sivananthan, "Auger Recombination Dynamics of $Hg_{0.795}Cd_{0.205}Te$ in the High Excitation Regime," *Applied Physics Letters* 71, 491–93, 1997.

153. R. Beattie and A. M. White, "An Analytic Approximation with a Wide Range of Applicability for Electron Initiated Auger Transitions in Narrow-Gap Semiconductors," *Journal of Applied Physics* 79, 802–13, 1996.

154. S. Krishnamurthy and T. N. Casselman, "A Detailed Calculation of the Auger Lifetime in p-Type HgCdTe," *Journal of Electronic Materials* 29, 828–31, 2000.

155. S. Krishnamurthy, M. A. Berding, Z. G. Yu, C. H. Swartz, T. H. Myers, D. D. Edwall, and R. DeWames, "Model of Minority Carrier Lifetimes in Doped HgCdTe," *Journal of Electronic Materials* 34, 873–79, 2005.

156. P. T. Landsberg, *Recombination in Semiconductors*, Cambridge University Press, Cambridge, 2003.

157. J. Piotrowski and A. Rogalski, *High-Operating Temperature Infrared Photodetectors*, SPIE Press, Bellingham, WA, 2007.

158. T. Ashley and C. T. Elliott, "Non-Equilibrium Mode of Operation for Infrared Detection," *Electronics Letters* 21, 451–52, 1985.

159. T. Ashley, T. C. Elliott, and A. M. White, "Non-Equilibrium Devices for Infrared Detection," *Proceedings of SPIE* 572, 123–32, 1985.

160. M. A. Kinch and S. R. Borrello, "0.1 eV HgCdTe Photodetectors," *Infrared Physics* 15, 111–24, 1975.

161. M. A. Kinch, S. R. Borrello, D. H. Breazeale, and A. Simmons, "Geometrical Enhancement of HgCdTe Photoconductive Detectors," *Infrared Physics* 17, 137–45, 1977.

162. S. R. Borrello, M. Kinch, and D. Lamont, "Photoconductive HgCdTe Detector Performance with Background Variations," *Infrared Physics* 17, 121–25, 1977.

163. S. P. Emmons and K. L. Ashley, "Minority-Carrier Sweepout in 0.09-eV HgCdTe," *Applied Physics Letters* 20, 241–42, 1972.

164. M. R. Johnson, "Sweep-Out Effects in $Hg_{1-x}Cd_xTe$ Photoconductors," *Journal of Applied Physics* 43, 3090–93, 1972.

165. Y. J. Shacham-Diamond and I. Kidron, "Contact and Bulk Effects in Intrinsic Photoconductive Infrared Detectors," *Infrared Physics* 21, 105–15, 1981.

166. T. Ashley and C. T. Elliott, "Accumulation Effects at Contacts to n-Type Cadmium-Mercury-Telluride Photoconductors," *Infrared Physics* 22, 367–76, 1982.

167. M. White, "Recombination in a Graded n-n Contact Region in a Narrow-Gap Semiconductor," *Journal of Physics C: Solid State Physics* 17, 4889–96, 1984.

168. D. L. Smith, "Effect of Blocking Contacts on Generation-Recombination Noise and Responsivity in Intrinsic Photoconductors," *Journal of Applied Physics* 56, 1663–69, 1984.

169. D. K. Arch, R. A. Wood, and D. L. Smith, "High Responsivity HgCdTe Heterojunction Photoconductor," *Journal of Applied Physics* 58, 2360–70, 1985.

170. J. F. Siliquini, K. A. Fynn, B. D. Nener, L. Faraone, and R. H. Hartley, "Improved Device Technology for Epitaxial $Hg_{1-x}Cd_xTe$ Infrared Photoconductor Arrays," *Semiconductor Science and Technology* 9, 1515–22, 1994.

171. C. A. Musca, J. F. Siliquini, B. D. Nener, L. Faraone, and R. H. Hartley, "Enhanced Responsivity of HgCdTe Infrared Photoconductors Using MBE Grown Heterostructures," *Infrared Physics & Technology* 38, 163–67, 1997.

172. J. Piotrowski, W. Galus, and M. Grudzień, "Near Room-Temperature IR Photodetectors," *Infrared Physics* 31, 1–48, 1990.

173. J. Piotrowski, "Uncooled Operation of IR Photodetectors," *Opto-Electronics Review* 12, 111–22, 2004.

174. J. Piotrowski and A. Rogalski, "Uncooled Long Wavelength Infrared Photon Detectors," *Infrared Physics & Technology* 46, 115–31, 2004.

175. T. Ashley, C. T. Elliott, and A. M. White, "Infra-Red Detector Using Minority Carrier Exclusion," *Proceedings of SPIE* 588, 62–68, 1985.

176. R. M. Broudy and V. J. Mazurczyck, "(HgCd)Te Photoconductive Detectors," in *Semiconductors and Semimetals*, Vol. 18, ed. R. K. Willardson and A. C. Beer, 157–99, Academic Press, New York, 1981.

177. A. Rogalski, *Infrared Detectors*, Gordon and Breach Science Publishers, Amsterdam, 2000.

178. C. T. Elliott, "Photoconductive and Non-Equilibrium Devices in HgCdTe and Related Alloys," in *Infrared Detectors and Emitters: Materials and Devices*, ed. P. Capper and C. T. Elliott, 279–312, Kluwer Academic Publishers, Boston, MA, 2001.

179. D. Long and J. L. Schmit, "Mercury-Cadmium Telluride and Closely Related Alloys," in *Semiconductors and Semimetals*, Vol. 5, ed. R. K. Willardson and A. C. Beer, 175–255, Academic Press, New York, 1970.

180. D. L. Spears, "Heterodyne and Direct Detection at 10.6 μm with High Temperature p-Type Photoconductors," *Proceedings of IRIS Active Systems*, 1–15, San Diego, 1982.

181. J. F. Siliquini, K. A. Fynn, B. D. Nener, L. Faraone, and R. H. Hartley, "Improved Device Technology for Epitaxial $Hg_{1-x}Cd_xTe$ Infrared Photoconductor Arrays," *Semiconductor Science and Technology* 9, 1515–22, 1994.

182. J. Piotrowski, "$Hg_{1-x}Cd_xTe$ Infrared Photodetectors," in *Infrared Photon Detectors*, ed. A. Rogalski, 391–494, SPIE Optical Engineering Press, Bellingham, WA, 1995.

183. Z. Djuric, "Isothermal Vapour-Phase Epitaxy of Mercury-Cadmium Telluride (Hg,Cd)Te," *Journal of Materials Science: Materials in Electronics* 5, 187–218, 1995.

184. K. Nagahama, R. Ohkata, and T. Murotani, "Preparation of High Quality n-$Hg_{0.8}Cd_{0.2}Te$ Epitaxial Layer and Its Application to Infrared Detector (λ = 8–14 μm)," *Japanese Journal of Applied Physics* 21, L764–L766, 1982.

185. T. Nguyen-Duy and D. Lorans, "Highlights of Recent Results on HgCdTe Thin Film Photoconductors," *Semiconductor Science and Technology* 6, 93–95, 1991.

186. B. Doll, M. Bruder, J. Wendler, J. Ziegler, and H. Maier, "3–5 μm Photoconductive Detectors on Liquid-Phase-Epitaxial-MCT," in *Fourth International Conference on Advanced Infrared Detectors and Systems*, 120–24, IEE, London, 1990.

187. J. F. Siliquini, C. A. Musca, B. D. Nener, and L. Faraone, "Performance of Optimized $Hg_{1-x}Cd_xTe$ Long Wavelength Infrared Photoconductors," *Infrared Physics & Technology* 35, 661–71, 1994.

188. L. T. Specht, W. E. Hoke, S. Oguz, P. J. Lemonias, V. G. Kreismanis, and R. Korenstein, "High Performance HgCdTe Photoconductive Devices Grown by Metalorganic Chemical Vapor Deposition," *Applied Physics Letters* 48, 417–18, 1986.

189. C. G. Bethea, B. F. Levine, P. Y. Lu, L. M. Williams, and M. H. Ross, "Photoconductive $Hg_{1-x}Cd_xTe$ Detectors Grown by Low-Temperature Metalorganic Chemical Vapor Deposition," *Applied Physics Letters* 53, 1629–31, 1988.

190. R. Druilhe, A. Katty, and R. Triboulet, "MOVPE Grown (Hg,Cd)Te Layers for Room Temperature Operating 3–5 μm Photoconductive Detectors," in *Fourth International Conference on Advanced Infrared Detectors and Systems*, 20–24, IEE, London, 1990.

191. M. C. Chen and M. J. Bevan, "Room-Temperature Midwavelength Two-Color Infrared Detectors with HgCdTe Multilayer Structures by Metal-Organic Chemical-Vapor Deposition," *Journal of Applied Physics* 78, 4787–89, 1995.

192. S. Yuan, J. Li He, J. Yu, M. Yu, Y. Qiao, and J. Zhu, "Infrared Photoconductor Fabricated with a Molecular Beam Epitaxially Grown CdTe/HgCdTe Heterostructure," *Applied Physics Letters* 58, 914–16, 1991.

193. Y. Nemirovsky and G. Bahir, "Passivation of Mercury Cadmium Telluride Surfaces," *Journal of Vacuum Science and Technology* A7, 450–59, 1989.

194. P. C. Catagnus and C. T. Baker, "Passivation of Mercury Cadmium Telluride," U.S. Patent 3,977,018, 1976.

195. Y. Nemirovsky and E. Finkman, "Anodic Oxide Films on $Hg_{1-x}Cd_xTe$," *Journal of the Electrochemical Society* 126, 768–70, 1979.

196. E. Bertagnolli, "Improvement of Anodically Grown Native Oxides on n-(Cd,Hg)Te," *Thin Solid Films* 135, 267–75, 1986.

197. A. Gauthier, "Process for Passivation of Photoconductive Detectors Made of HgCdTe," U.S. Patent 4,624,715, 1986.

198. Y. Nemirovsky and R. Goshen, "Plasma Anodization of $Hg_{1-x}Cd_xTe$," *Applied Physics Letters* 37, 813–14, 1980.

199. S. P. Buchner, G. D. Davis, and N. E. Byer, "Summary Abstract: Photochemical Oxidation of (Hg,Cd)Te," *Journal of Vacuum Science and Technology* 21, 446–47, 1982.

200. Y. Shacham-Diamand, T. Chuh, and W. G. Oldham, "The Electrical Properties of Hg-Sensitized 'Photox'-Oxide Layers Deposited at 80°C," *Solid-State Electronics* 30, 227–33, 1987.

201. M. V. Blackman, D. E. Charlton, M. D. Jenner, D. R. Purdy, and J. T. M. Wotherspoon, "Type Conversion in $Hg_{1-x}Cd_xTe$ by Ion Beam Treatment," *Electronics Letters* 23, 978–79, 1987.

202. P. Brogowski, H. Mucha, and J. Piotrowski, "Modification of Mercury Cadmium Telluride, Mercury Manganese Tellurium, and Mercury Zinc Telluride by Ion Etching," *Physica Status Solidi (a)* 114, K37–K40, 1989.

203. P. H. Zimmermann, M. B. Reine, K. Spignese, K. Maschhoff, and J. Schirripa, "Surface Passivation of HgCdTe Photodiodes," *Journal of Vacuum Science and Technology* A8, 1182–84, 1990.

204. Y. Nemirovsky, L. Burstein, and I. Kidron, "Interface of p-type $Hg_{1-x}Cd_xTe$ Passivated with Native Sulfides," *Journal of Applied Physics* 58, 366–73, 1985.

205. E. Weiss and C. R. Helms, "Composition, Growth Mechanism, and Stability of Anodic Fluoride Films on $Hg_{1-x}Cd_xTe$," *Journal of Vacuum Science and Technology* B9, 1879–85, 1991.

206. Y. Nemirovsky, "Passivation with II-VI Compounds," *Journal of Vacuum Science and Technology* A8, 1185–87, 1990.

207. G. Sarusi, G. Cinader, A. Zemel, and D. Eger, "Application of CdTe Epitaxial Layers for Passivation of p-Type $Hg_{0.77}Cd_{0.23}Te$," *Journal of Applied Physics* 71, 5070–76, 1992.

208. A. Mestechkin, D. L. Lee, B. T. Cunningham, and B. D. MacLeod, "Bake Stability of Long-Wavelength Infrared HgCdTe Photodiodes," *Journal of Electronic Materials* 24, 1183–87, 1995.

209. G. Bahir, V. Ariel, V. Garber, D. Rosenfeld, and A. Sher, "Electrical Properties of Epitaxially Grown CdTe Passivation for Long-Wavelength HgCdTe Photodiodes," *Applied Physics Letters* 65, 2725–27, 1994.

210. A. Jóźwikowska, K. Jóźwikowski, and A. Rogalski, "Performance of Mercury Cadmium Telluride Photoconductive Detectors," *Infrared Physics* 31, 543–54, 1991.

211. C. A. Musca, J. F. Siliquini, K. A. Fynn, B. D. Nener, L. Faraone, and S. J. C. Irvine, "MOCVD-Grown Wider-Bandgap Capping Layers in $Hg_{1-x}Cd_xTe$ Long-Wavelength Infrared Photoconductors," *Semiconductor Science & Technology* 11, 1912–22, 1996.

212. J. F. Siliquini, C. A. Musca, B. D. Nener, and L. Faraone, "Temperature Dependence of $Hg_{0.68}Cd_{0.32}Te$ Infrared Photoconductor Performance," *IEEE Transactions on Electron Devices* 42, 1441–48, 1995.

213. D. L. Smith, "Theory of Generation-Recombination Noise and Responsivity in Overlap Structure Photoconductors," *Journal of Applied Physics* 54, 5441–48, 1983.

214. D. L. Smith, D. K. Arch, R. A. Wood, and M. W. Scott, "HgCdTe Heterojunction Contact Photoconductor," *Journal of Applied Physics* 45, 83–85, 1984.

215. J. R. Lowney, D. G. Seiler, W. R. Thurber, Z. Yu, X. N. Song, and C. L. Littler, "Heavily Accumulated Surfaces of Mercury Cadmium Telluride Detectors: Theory and Experiment," *Journal of Electronic Materials* 22, 985–91, 1993.

216. M. C. Wilson and D. J. Dinsdale, "CO_2 Detection with Cadmium Mercury Telluride," in *Third International Conference on Advanced Infrared Detectors and Systems*, 139–45, IEE, London, 1986.

217. D. L. Spears, "IR Detectors: Heterodyne and Direct," in *Optical and Laser Remote Sensing*, eds. D. K. Killinger and A. Mooradian, 278–86, Springer-Verlag, Berlin, 1983.

218. Z. Djuric, Z. Jaksic, Z. Djinovic, M. Matic, and Z. Lazic, "Some Theoretical and Technological Aspects of Uncooled HgCdTe Detectors: A Review," *Microelectronics Journal* 25, 99–114, 1994.

219. F. A. Capocci, A. T. Harker, M. C. Wilson, D. E. Lacklison, and F. E. Wray, "Thermoelectrically-Cooled Cadmium Mercury Telluride Detectors for CO_2 Laser Radiation," in *Second International Conference on Advanced Infrared Detectors and Systems*, 40–44, IEE, London, 1983.

220. D. J. Wilson, R. Foord, and G. D. J. Constant, "Operation of an Intermediate Temperature Detector in a $10.6\,\mu m$ Heterodyne Rangefinder," *Proceedings of SPIE* 663, 155–58, 1986.

221. Y. Li and D. J. Adams, "Experimental Studies on the Performance of $Hg_{0.8}Cd_{0.2}Te$ Photoconductors Operating Near 200 K," *Infrared Physics* 35, 593–95, 1994.

222. Z. Wei-jiann and Z. Xin-Chen, "Experimental Studies on Low Frequency Noise of Photoconductors," *Infrared Physics* 33, 27–31, 1992.

223. P. R. Norton, "Structure and Method of Fabricating a Trapping-Mode Photodetector," *International Application Published Under The Patent Cooperation Treaty* PCT/US86/002516, International Publication Number WO87/03743, 18 June 1987.

224. P. R. Norton, "Infrared Image Sensors," *Optical Engineering* 30, 1649–63, 1991.

225. P. Norton, "HgCdTe for NASA EOS Missions and Detector Uniformity Benchmarks," *Proceedings of the Innovative Long Wavelength Infrared Detector Workshop*, 93–94, Pasadena, Jet Propulsion Laboratory, April 24–26, 1990.

226. D. G. Crove, P. R. Norton, T. Limperis, and J. Mudar, "Detectors," in *The Infrared and Electro-Optical Systems Handbook*, Vol. 3, ed. W. D. Rogatto, 175–283, Infrared Information Analysis Center, Ann Arbor, MI, and SPIE Optical Engineering Press, Bellingham, WA, 1993.

227. T. Ashley, C. T. Elliott, and A. T. Harker, "Non-Equilibrium Mode of Operation for Infrared Detectors," *Infrared Physics* 26, 303–15, 1986.

228. A. M. White, "Generation-Recombination Processes and Auger Suppression in Small-Bandgap Detectors," *Journal of Crystal Growth* 86, 840–48, 1988.

229. A. P. Davis and A. M. White, "Effects of Residual Shockley-Read Traps on Efficiency of Auger-Suppressed IR Detector Diodes," *Semiconductor Science and Technology* 5, S38–S40, 1990.

230. C. T. Elliott, "Future Infrared Detector Technologies," in *Fourth International Conference on Advanced Infrared Detectors and Systems*, 61–66, IEE, London, 1990.

231. C. T. Elliott, "Non-Equilibrium Mode of Operation of Narrow-Gap Semiconductor Devices," *Semiconductor Science and Technology* 5, S30–S37, 1990.

232. T. Ashley and T. C. Elliott, "Operation and Properties of Narrow-Gap Semiconductor Devices Near Room Temperature Using Non-Equilibrium Techniques," *Semiconductor Science and Technology* 8, C99–C105, 1991.

233. C. T. Elliott, "Infrared Detectors with Integrated Signal Processing," in *Solid State Devices*, eds. A. Goetzberger and M. Zerbst, 175–201, Verlag Chemie, Weinheim, 1983.

234. T. Ashley, C. T. Elliott, A. M. White, J. T. M. Wotherspoon, and M. D. Johns, "Optimization of Spatial Resolution in SPRITE Detectors," *Infrared Physics* 24, 25–33, 1984.

235. J. T. M. Wotherspoon, R. J. Dean, M. D. Johns, T. Ashley, C. T. Elliott, and M. A. White, "Developments in SPRITE Infra-Red Detectors," *Proceedings of SPIE* 810, 102–12, 1985.

236. R. F. Leftwich and R. Ward, "Latest Developments in SPRITE Detector Technology," *Proceedings of SPIE* 930, 76–86, 1988.

237. A. Campbell, C. T. Elliott, and A. M. White, "Optimisation of SPRITE Detectors in Anamorphic Imaging Systems," *Infrared Physics* 27, 125–33, 1987.

238. A. B. Dean, P. N. J. Dennis, C. T. Elliott, D. Hibbert, and J. T. M. Wotherspoon, "The Serial Addition of SPRITE Infrared Detectors," *Infrared Physics* 28, 271–78, 1988.

239. C. M. Dyson, "Thermal-Radiation Imaging Devices and Systems," U.K. Patent Application 2,199,986 A, 1988.

240. C. T. Elliott, "SPRITE Detectors and Staring Arrays in $Hg_{1-x}Cd_xTe$," *Proceedings of SPIE* 1038, 2–8, 1988.

241. G. D. Boreman and A. E. Plogstedt, "Modulation Transfer Function and Number of Equivalent Elements for SPRITE Detectors," *Applied Optics* 27, 4331–35, 1988.

242. S. P. Braim, A. Foord, and M. W. Thomas, "System Implementation of a Serial Array of SPRITE Infrared Detectors," *Infrared Physics* 29, 907–14, 1989.

243. J. Severn, D. A. Hibbert, R. Mistry, C. T. Elliott, and A. P. Davis, "The Design and Performance Options for SPRITE Arrays," in *Fourth International Conference on Advanced Infrared Detectors and Systems*, 9–14, IEE, London, 1990.

244. A. P. Davis, "Effect of High Signal Photon Fluxes on the Responsivity of SPRITE Detectors," *Infrared Physics* 33, 301–5, 1992.

245. K. J. Barnard, G. D. Boreman, A. E. Plogstedt, and B. K. Anderson, "Modulation-Transfer Function Measurement of SPRITE Detectors: Sine-Wave Response," *Applied Optics* 31, 144–47, 1992.

246. M. Rodot, C. Verie, Y. Marfaing, J. Besson, and H. Lebloch, "Semiconductor Lasers and Fast Detectors in the Infrared (3 to 15 Microns)," *IEEE Journal of Quantum Electronics* 2, 586–93, 1966.

247. M. B. Reine, A. K. Sood, and T. J. Tredwell, "Photovoltaic Infrared Detectors," in *Semiconductors and Semimetals*, Vol. 18, eds. R. K. Willardson and A. C. Beer, 201–311, Academic Press, New York, 1981.

248. A. Rogalski and J. Piotrowski, "Intrinsic Infrared Detectors," *Progress in Quantum Electronics* 12, 87–289, 1988.

249. A. I. D'Souza, P. S. Wijewarnasuriya, and J. G. Poksheva, "HgCdTe Infrared Detectors," in *Thin Films*, Vol. 28, 193–226, 2001.

250. I. M. Baker, "Photovoltaic IR Detectors," in *Narrow-gap II-VI Compounds for Optoelectronic and Electromagnetic Applications*, ed. P. Capper, 450–73, Chapman & Hall, London, 1997.

251. M. B. Reine, "Photovoltaic Detectors in MCT," in *Infrared Detectors and Emitters: Materials and Devices*, eds. P. Capper and C. T. Elliott, 313–76, Kluwer Academic Publishers, Boston, MA, 2001.

252. I. M. Baker, "HgCdTe 2D Arrays: Technology and Performance Limits," in *Handbook of Infrared Technologies*, eds. M. Henini and M. Razeghi, 269–308, Elsevier, Oxford, 2002.

253. A. Rogalski, "Infrared Detectors: Status and Trends," *Progress in Quantum Electronics* 27, 59–210, 2003.

254. I. M. Baker, "II-VI Narrow-Bandgap Semiconductors for Optoelectronics," in *Springer Handbook of Electronic and Photonic Materials*, eds. S. Kasap and P. Capper, 855–85, Springer, Heidelberg, 2006.

255. D. T. Dutton, E. O'Keefe, P. Capper, C. L. Jones, S. Mugford, and C. Ard, "Type Conversion of $Cd_xHg_{1-x}Te$ Grown by Liquid Phase Epitaxy," *Semiconductor Science & Technology* 8, S266–S269, 1993.

256. D. L. Spears and C. Freed, "HgCdTe Varactor Photodiode Detection of cw CO_2 Laser Beats Beyond 60 GHz," *Applied Physics Letters* 23, 445–47, 1973.

257. D. L. Spears, "Planar HgCdTe Quadrantal Heterodyne Arrays with GHz Response at 10.6 μm," *Infrared Physics* 17, 5–8, 1977.

258. J. F. Shanley, C. T. Flanagan, and M. B. Reine, "Elevated Temperature n^+-p $Hg_{0.8}Cd_{0.2}Te$ Photodiodes for Moderate Bandwidth Infrared Heterodyne Applications," *Proceedings of SPIE* 227, 117–22, 1980.

259. K. K. Parat, H. Ehsani, I. B. Bhat, and S. K. Ghandhi, "Selective Annealing for the Planar Processing of HgCdTe Devices," *Journal of Vacuum Science and Technology* B9, 1625–29, 1991.

260. M. D. Jenner and M. V. Blackman, "Method of Manufacturing an Infrared Detector Device," U.S. Patent 4,318,217, 1982.

261. P. Brogowski and J. Piotrowski, "The p-to-n Conversion of HgCdTe, HgZnTe and HgMnTe by Anodic Oxidation and Subsequent Heat Treatment," *Semiconductor Science and Technology* 5, 530–32, 1990.

262. J. M. T. Wotherspoon, U.S. Patent 4.411.732, 1983.

263. M. A. Lunn and P. S. Dobson, "Ion Beam Milling of $Cd_{0.2}Hg_{0.8}Te$," *Journal of Crystal Growth* 73, 379–84, 1985.

264. M. V. Blackman, D. E. Charlton, M. D. Jenner, D. R. Purdy, and J. T. M. Wotherspoon, "Type Conversion in $Hg_{1-x}Cd_xTe$ by Ion Beam Treatment," *Electronics Letters* 23, 978–79, 1987.

265. J. L. Elkind, "Ion Mill Damage in n–HgCdTe," *Journal of Vacuum Science and Technology* B10, 1460–65, 1992.

266. V. I. Ivanov-Omskii, K. E. Mironov, and K. D. Mynbaev, "$Hg_{1-x}Cd_xTe$ Doping by Ion-Beam Treatment," *Semiconductor Science and Technology* 8, 634–37, 1993.

267. P. Hlidek, E. Belas, J. Franc, and V. Koubele, "Photovoltaic Spectra of HgCdTe Diodes Fabricated by Ion Beam Milling," *Semiconductor Science & Technology* 8, 2069–71, 1993.

268. K. D. Mynbaev and V. I. Ivanov-Omski, "Modification of $Hg_{1-x}Cd_xTe$ Properties by Low-Energy Ions," *Semiconductors* 37, 1127–50, 2003.

269. G. Bahir and E. Finkman, "Ion Beam Milling Effect on Electrical Properties of $Hg_{1-x}Cd_xTe$," *Journal of Vacuum Science and Technology* A7, 348–53, 1989.

270. M. Baker and R. A. Ballingall, "Photovoltaic CdHgTe: Silicon Hybrid Focal Planes," *Proceedings of SPIE* 510, 121–29, 1984.

271. E. Belas, P. Höschl, R. Grill, J. Franc, P. Moravec, K. Lischka, H. Sitter, and A. Toth, "Ultrafast Diffusion of Hg in $Hg_{1-x}Cd_xTe$ ($x \approx 0.21$)," *Journal of Crystal Growth* 138, 940–43, 1994.

272. E. Belas, R. Grill, J. Franc, A. Toth, P. Höschl, H. Sitter, and P. Moravec, "Determination of the Migration Energy of Hg Interstitials in (HgCd)Te from Ion Milling Experiments," *Journal of Crystal Growth* 159, 1117–22, 1996.

273. G. Destefanis, "Electrical Doping of HgCdTe by Ion Implantation and Heat Treatment," *Journal of Crystal Growth* 86, 700–22, 1988.

274. L. O. Bubulac, "Defects, Diffusion and Activation in Ion Implanted HgCdTe," *Journal of Crystal Growth* 86, 723–34, 1988.

275. A. G. Foyt, T. C. Harman, and J. P. Donnelly, "Type Conversion and n–p Junction Formation in $Cd_xHg_{1-x}Te$ Produced by Proton Bombardment," *Applied Physics Letters* 18, 321–23, 1971.

276. L. O. Bubulac and W. E. Tennant, "Role of Hg in Junction Formation in Ion-Implanted HgCdTe," *Applied Physics Letters* 51, 355–57, 1987.

277. P. G. Pitcher, P. L. F. Hemment, and Q. V. Davis, "Formation of Shallow Photodiodes by Implantation of Boron into Mercury Cadmium Telluride," *Electronics Letters* 18, 1090–92, 1982.

278. L. J. Kozlowski, R. B. Bailey, S. C. Cabelli, D. E. Cooper, G. McComas, K. Vural, and W. E. Tennant, "640 × 480 PACE HgCdTe FPA," *Proceedings of SPIE* 1735, 163–73, 1992.

279. G. Bahir, R. Kalish, and Y. Nemirovsky, "Electrical Properties of Donor and Acceptor Implanted $Hg_{1-x}Cd_xTe$ Following CW CO_2 Laser Annealing," *Applied Physics Letters* 41, 1057–59, 1982.

280. R. B. Bailey, L. J. Kozlowski, J. Chen, D. Q. Bui, K. Vural, D. D. Edwall, R. V. Gil, A. B. Vanderwyck, E. R. Gertner, and M. B. Gubala, "256 × 256 Hybrid HgCdTe Infrared Focal Plane Arrays," *IEEE Transactions on Electron Devices* 38, 1104–9, 1991.

281. L. O. Bubulac, D. D. Edwall, and C. R. Viswanathan, "Dynamics of Arsenic Diffusion in Metalorganic Chemical Vapor Deposited HgCdTe on GaAs/Si Substrates," *Journal of Vacuum Science and Technology* B9, 1695–704, 1991.

282. L. O. Bubulac, S. J. Ivine, E. R. Gertner, J. Bajaj, W. P. Lin, and R. Zucca, "As Diffusion in $Hg_{1-x}Cd_xTe$ for Junction Formation," *Semiconductor Science & Technology* 8, S270–S275, 1993.

283. O. P. Agnihorti, H. C. Lee, and K. Yang, "Plasma Induced Type Conversion in Mercury Cadmium Telluride," *Semiconductor Science and Technology* 17, R11–R19, 2002.

284. E. Belas, R. Grill, J. Franc, A. Toth, P. Höschl, H. Sitter, and P. Moravec, "Determination of the Migration Energy of Hg Interstitials in (HgCd)Te from Ion Milling Experiments," *Journal of Crystal Growth* 159, 1117–22, 1996.

285. J. K. White, R. Pal, J. M. Dell, C. A. Musca, J. Antoszewski, L. Faraone, and P. Burke, "p-to-n Type-Conversion Mechanisms for HgCdTe Exposed to H_2/CH_4 Plasma," *Journal of Electronic Materials* 30, 762–67, 2001.

286. J. M. Dell, J. Antoszewski, M. H. Rais, C. Musca, J. K. White, B. D. Nener, and L. Faraone, "HgCdTe Mid-Wavelength IR Photovoltaic Detectors Fabricated Using Plasma Induced Junction Technology," *Journal of Electronic Materials* 29, 841–48, 2000.

287. J. K. White, J. Antoszewski, P. Ravinder, C. A. Musca, J. M. Dell, L. Faraone, and J. Piotrowski, "Passivation Effects on Reactive Ion Etch Formed n-on-p Junctions in HgCdTe," *Journal of Electronic Materials* 31, 743–48, 2002.

288. T. Nguyen, C. A. Musca, J. M. Dell, R. H. Sewell, J. Antoszewski, J. K. White, and L. Faraone, "HgCdTe Long-Wavelength IR Photovoltaic Detectors Fabricated Using Plasma Induced Junction Formation Technology," *Journal of Electronic Materials* 32, 615–21, 2003.

289. E. P. G. Smith, G. M. Venzor, M. D. Newton, M. V. Liguori, J. K. Gleason, R. E. Bornfreund, S. M. Johnson, et al., "Inductively Coupled Plasma Etching Got Large Format HgCdTe Focal Plane Array Fabrication," *Journal of Electronic Materials* 34, 746–53, 2005.

290. M. Lanir and K. J. Riley, "Performance of PV HgCdTe Arrays for 1–14-µm Applications," *IEEE Transactions on Electronic Devices* ED-29, 274–79, 1982.

291. C. C. Wang, "Mercury Cadmium Telluride Junctions Grown by Liquid Phase Epitaxy," *Journal of Vacuum Science and Technology* B9, 1740–45, 1991.

292. A. I. D'Souza, J. Bajaj, R. E. DeWames, D. D. Edwall, P. S. Wijewarnasuriya, and N. Nayar, "MWIR DLPH HgCdTe Photodiode Performance Dependence on Substrate Material," *Journal of Electronic Materials* 27, 727–32, 1998.

293. G. N. Pultz, P. W. Norton, E. E. Krueger, and M. B. Reine, "Growth and Characterization of p-on-n HgCdTe Liquid-Phase Epitaxy Heterojunction Material for 11–18 µm Applications," *Journal of Vacuum Science and Technology* B9, 1724–30, 1991.

294. T. Tung, M. H. Kalisher, A. P. Stevens, and P. E. Herning, "Liquid-Phase Epitaxy of $Hg_{1-x}Cd_xTe$ from Hg Solution: A Route to Infrared Detector Structures," *Materials Research Society Symposium Proceedings* 90, 321–56, 1987.

295. E. P. G. Smith, L. T. Pham, G. M. Venzor, E. M. Norton, M. D. Newton, P. M. Goetz, V. K. Randall, et al., "HgCdTe Focal Plane Arrays for Dual-Color Mid- and Long-Wavelength Infrared Detection," *Journal of Electronic Materials* 33, 509–16, 2004.

296. J. M. Peterson, J. A. Franklin, M. Reddy, S. M. Johnson, E. Smith, W. A. Radford, and I. Kasai, "High-Quality Large-Area MBE HgCdTe/Si," *Journal of Electronic Materials* 35, 1283–86, 2006.

297. G. Destefanis, J. Baylet, P. Ballet, P. Castelein, F. Rothan, O. Gravrand, J. Rothman, J. P. Chamonal, and A. Million, "Status of HgCdTe Bicolor and Dual-Band Infrared Focal Plane Arrays at LETI," *Journal of Electronic Materials* 36, 1031–44, 2007.

298. C. D. Maxey, J. C. Fitzmaurice, H. W. Lau, L. G. Hipwood, C. S. Shaw, C. L. Jones, and P. Capper, "Current Status of Large-Area MOVPE Growth of HgCdTe Device Heterostructures for Infrared Focal Plane Arrays," *Journal of Electronic Materials* 35, 1275–82, 2006.

299. K. D. Mynbaev and V. I. Ivanov-Omski, "Doping of Epitaxial Layers and Heterostructures Based on HgCdTe," *Semiconductors* 40, 1–21, 2006.

300. H. R. Vydyanath, J. A. Ellsworth, and C. M. Devaney, "Electrical Activity, Mode of Incorporation and Distribution Coefficient of Group V Elements in $Hg_{1-x}Cd_xTe$ Grown from Tellurium Rich Liquid Phase Epitaxial Growth Solutions," *Journal of Electronic Materials* 16, 13–25, 1987.

301. E. P. G. Smith, R. E. Bornfreund, I. Kasai, L. T. Pham, E. A. Patten, J. M. Peterson, J. A. Roth, et al., "Status of Two-Color and Large Format HgCdTe FPA Technology at Raytheon Vision Systems," *Proceedings of SPIE* 6127, 61261F, 2006.

302. E. P. G. Smith, G. M. Venzor, Y. Petraitis, M. V. Liguori, A. R. Levy, C. K. Rabkin, J. M. Peterson, M. Reddy, S. M. Johnson, and J. W. Bangs, "Fabrication and Characterization of Small Unit-Cell Molecular Beam Epitaxy Grown HgCdTe-on-Si Mid-Wavelength Infrared Detectors," *Journal of Electronic Materials* 36, 1045–51, 2007.

303. J. M. Arias, J. G. Pasko, M. Zandian, L. J. Kozlowski, and R. E. DeWames, "Molecular Beam Epitaxy HgCdTe Infrared Photovoltaic Detectors," *Optical Engineering* 33, 1422–28, 1994.

304. L. O. Bubulac and C. R. Viswanathan, "Diffusion of As and Sb in HgCdTe," *Journal of Crystal Growth* 123, 555–66, 1992.

305. M. Arias, J. G. Pasko, M. Zandian, S. H. Shin, G. M. Williams, L. O. Bubulac, R. E. DeWames, and W. E. Tennant, "Planar p-on-n HgCdTe Heterostructure Photovoltaic Detectors," *Applied Physics Letters* 62, 976–78, 1993.

306. Y. Nemirovsky, N. Mainzer, and E. Weiss, "Passivation of HgCdTe," in *Properties of Narrow Gap Cadmium-Based Compounds, EMIS Datareviews Series No. 10*, ed. P. P. Capper, 284–90, IEE, London, 1994.

307. Y. Nemirovsky and N. Amir, "Surfaces/Interfaces of Narrow-Gap II–VI Compounds," in *Narrow-gap II-VI Compounds for Optoelectronic and Electromagnetic Applications*, ed. P. Capper, 291–326, Chapman & Hall, London, 1997.

308. Y. Nemirovsky and L. Burstein, "Anodic Sulfide Films on $Hg_{1-x}Cd_xTe$," *Applied Physics Letters* 44, 443–44, 1984.

309. E. Weiss and C. R. Helms, "Composition, Growth Mechanism, and Oxidation of Anodic Fluoride Films on $Hg_{1-x}Cd_xTe$ (x ≈ 0.2)," *Journal of the Electrochemical Society* 138, 993–99, 1991.

310. A. Kolodny and I. Kidron, "Properties of Ion Implanted Junctions in Mercury Cadmium-Telluride," *IEEE Transactions on Electron Devices* ED-27, 37–43, 1980.

311. P. Migliorato, R. F. C. Farrow, A. B. Dean, and G. M. Williams, "CdTe/HgCdTe Indium-Diffused Photodiodes," *Infrared Physics* 22, 331–36, 1982.

312. N. Kajihara, G. Sudo, Y. Miyamoto, and K. Tonikawa, "Silicon Nitride Passivant for HgCdTe n^+–p Diodes," *Journal of the Electrochemical Society* 135, 1252–55, 1988.

313. P. H. Zimmermann, M. B. Reine, K. Spignese, K. Maschhoff, and J. Schirripa, "Surface Passivation of HgCdTe Photodiodes," *Journal of Vacuum Science and Technology* A8, 1182–84, 1990.

314. J. F. Ameurlaire and G. D. Cohen-Solal, U.S. Patent No. 3,845,494, 1974; G. D. Cohen-Solal and A. G. Lussereau, U.S. Patent No. 3,988,774, 1976; J. H. Maille and A. Salaville, U.S. Patent No. 4,132,999, 1979.

315. D. J. Hall, L. Buckle, N. T. Gordon, J. Giess, J. E. Hails, J. W. Cairns, R. M. Lawrence, et al., "High-performance Long-Wavelength HgCdTe Infrared Detectors Grown on Silicon Substrates," *Applied Physics Letters* 84, 2113–15, 2004.

316. Y. Nemirovsky, N. Amir, and L. Djaloshinski, "Metalorganic Chemical Vapor Deposition CdTe Passivation of HgCdTe," *Journal of Electronic Materials* 25, 647–54, 1995.

317. Y. Nemirovsky, N. Amir, D. Goren, G. Asa, N. Mainzer, and E. Weiss, "The Interface of Metalorganic Chemical Vapor Deposition-CdTe/HgCdTe," *Journal of Electronic Materials* 24, 1161–67, 1995.

318. S. Y. An, J. S. Kim, D. W. Seo, and S. H. Suh, "Passivation of HgCdTe p-n Diode Junction by Compositionally Graded HgCdTe Formed by Annealing in a Cd/Hg Atmosphere," *Journal of Electronic Materials* 31, 683–87, 2002.

319. L. O. Bubulac, W. E. Tennant, J. Bajaj, J. Sheng, R. Brigham, A. H. B. Vanderwyck, M. Zandian, and W. V. McLevige, "Characterization of CdTe for HgCdTe Surface Passivation," *Journal of Electronic Materials* 24, 1175–82, 1995.

320. W. E. Spicer, D. J. Friedman, and G. P. Carey, "The Electrical Properties of Metallic Contacts on $Hg_{1-x}Cd_xTe$," *Journal of Vacuum Science and Technology* A6, 2746–51, 1988.

321. S. P. Wilks and R. H. Williams, "Contacts to HgCdTe," in *Properties of Narrow Gap Cadmium-Based Compounds, EMIS Datareviews Series No. 10*, ed. P. P. Capper, 297–99, IEE, London, 1994.

322. G. D. Davis, "Overlayer Interactions with (HgCd)Te," *Journal of Vacuum Science and Technology* A6, 1939–45, 1988.

323. J. F. McGilp and I. T. McGovern, "A Simple Semiquantitative Model for Classifying Metal-Compound Semiconductor Interface Reactivity," *Journal of Vacuum Science and Technology* B3, 1641–44, 1985.

324. J. M. Pawlikowski, "Schematic Energy Band Diagram of Metal $Cd_xHg_{1-x}Te$ Contacts," *Physica Status Solidi (a)* 40, 613–20, 1977.

325. P. W. Leech and G. K. Reeves, "Specific Contact Resistance of Indium Ohmic Contacts to n-Type $Hg_{1-x}Cd_xTe$," *Journal of Vacuum Science and Technology* A10, 105–9, 1992.

326. W. A. Beck, G. D. Davis, and A. C. Goldberg, "Resistance and 1/f Noise of Au, Al, and Ge Contacts to (Hg,Cd)Te," *Journal of Applied Physics* 67, 6340–46, 1990.

327. A. Piotrowski, P. Madejczyk, W. Gawron, K. Kłos, M. Romanis, M. Grudzień, A. Rogalski, and J. Piotrowski, "MOCVD Growth of Hg$_{1-x}$Cd$_x$Te Heterostructures for Uncooled Infrared Photodetectors," *Opto-Electronics Review* 12, 453–58, 2004.

328. A. Rogalski and W. Larkowski, "Comparison of Photodiodes for the 3–5.5 μm and 8–12 μm Spectral Regions," *Electron Technology* 18(3/4), 55–69, 1985.

329. A. Turner, T. Teherani, J. Ehmke, C. Pettitt, P. Conlon, J. Beck, K. McCormack, et al., "Producibility of VIP™ Scanning Focal Plane Arrays," *Proceedings of SPIE* 2228, 237–48, 1994.

330. P. Tribolet, J. P. Chatard, P. Costa, and A. Manissadjian, "Progress in HgCdTe Homojunction Infrared Detectors," *Journal of Crystal Growth* 184/185, 1262–71, 1998.

331. J. Bajaj, "State-of-the-Art HgCdTe Materials and Devices for Infrared Imaging," in *Physics of Semiconductor Devices*, eds. V. Kumar and S. K. Agarwal, 1297–1309, Narosa Publishing House, New Delhi, 1998.

332. T. J. DeLyon, J. E. Jensen, M. D. Gorwitz, C. A. Cockrum, S. M. Johnson, and G. M. Venzor, "MBE Growth of HgCdTe on Silicon Substrates for Large-Area Infrared Focal Plane Arrays: A Review of Recent Progress," *Journal of Electronic Materials* 28, 705–11, 1999.

333. M. Kinch, "HDVIP™ FPA Technology at DRS," *Proceedings of SPIE* 4369, 566–78, 2001.

334. O. Gravrand, Ph. Chorier, and H. Geoffray, "Status of Very Long Infrared Wave Focal Plane Array Development at DEFIR," *Proceedings of SPIE* 7298, 729821, 2009.

335. P. Norton, "Status of Infrared Detectors," *Proceedings of SPIE* 3379, 102–14, 1998.

336. A. Rogalski and R. Ciupa, "Theoretical Modeling of Long Wavelength n⁺-on-p HgCdTe Photodiodes," *Journal of Applied Physics* 80, 2483–89, 1996.

337. G. M. Williams and R. E. DeWames, "Numerical Simulation of HgCdTe Detector Characteristics," *Journal of Electronic Materials* 24, 1239–48, 1995.

338. W. E. Tennant, D. Lee, M. Zandian, E. Piquette, and M. Carmody, "MBE HgCdTe Technology: A Very General Solution to IR Detection, Described by 'Rule 07', A Very Convenient Heuristic," *Journal of Electronic Materials* 37, 1407–10, 2008.

339. J. A. Stobie, S. P. Tobin, P. Norton, M. Hutchins, K.-K. Wong, R. J. Huppi, and R. Huppi, "Update on the Imaging Sensor for GIFTS," *Proceedings of SPIE* 5543, 293–303, 2004.

340. T. Chuh, "Recent Developments in Infrared and Visible Imaging for Astronomy, Defense and Homeland Security," *Proceedings of SPIE* 5563, 19–34, 2004.

341. A. S. Gilmore, J. Bangs, A. Gerrish, A. Stevens, and B. Starr, "Advancements in HgCdTe VLWIR Materials," *Proceedings of SPIE* 5783, 223–30, 2005.

342. C. L. Jones, L. G. Hipwood, C. J. Shaw, J. P. Price, R. A. Catchpole, M. Ordish, C. D. Maxey, et al., "High Performance MW and LW IRFPAs Made from HgCdTe Grown by MOVPE," *Proceedings of SPIE* 6206, 620610, 2006.

343. J. W. Baletic, R. Blank, D. Gulbransen, D. Lee, M. Loose, E. C. Piquette, T. Sprafke, W. E. Tennant, M. Zandian, and J. Zino, "Teledyne Imaging Sensors: Infrared Imaging Technologies for Astronomy & Civil Space," *Proceedings of SPIE* 7021, 70210H, 2008.

344. M. B. Reine, S. P. Tobin, P. W. Norton, and P. LoVecchio, "Very Long Wavelength (>15 μm) HgCdTe Photodiodes by Liquid Phase Epitaxy," *Proceedings of SPIE* 5564, 54–64, 2004.

345. R. E. DeWames, D. D. Edwall, M. Zandian, L. O. Bubulac, J. G. Pasko, W. E. Tennant, J. M. Arias, and A. D'Souza, "Dark Current Generating Mechanisms in Short Wavelength Infrared Photovoltaic Detectors," *Journal of Electronic Materials* 27, 722–26, 1998.

346. J. Piotrowski, "Uncooled IR Detectors Maintain Sensitivity," *Photonics Spectra*, 80–86, May 2004.

347. A. Rogalski, "HgTe-Based Photodetectors in Poland," *Proceedings of SPIE* 7298, 72982Q, 2009.

348. E. C. Piquette, D. D. Edwall, H. Arnold, A. Chen, and J. Auyeung, "Substrate-Removed HgCdTe-Based Focal-Plane Arrays for Short-Wavelength Infrared Astronomy," *Journal of Electronic Materials* 37, 1396–1400, 2008.

349. J. Y. Wong, "Effect of Trap Tunneling on the Performance of Long-Wavelength $Hg_{1-x}Cd_xTe$ Photodiodes," *IEEE Transactions on Electron Devices* ED-27, 48–57, 1980.

350. H. J. Hoffman and W. W. Anderson, "Impurity-to-Band Tunneling in $Hg_{1-x}Cd_xTe$," *Journal of Vacuum Science and Technology* 21, 247–50, 1982.

351. D. K. Blanks, J. D. Beck, M. A. Kinch, and L. Colombo, "Band-to-Band Processes in HgCdTe: Comparison of Experimental and Theoretical Studies," *Journal of Vacuum Science and Technology* A6, 2790–94, 1988.

352. R. E. DeWames, G. M. Williams, J. G. Pasko, and A. H. B. Vanderwyck, "Current Generation Mechanisms in Small Band Gap HgCdTe p-n Junctions Fabricated by Ion Implantation," *Journal of Crystal Growth* 86, 849–58, 1988.

353. Y. Nemirovsky, D. Rosenfeld, R. Adar, and A. Kornfeld, "Tunneling and Dark Currents in HgCdTe Photodiodes," *Journal of Vacuum Science and Technology* A7, 528–35, 1989.

354. Y. Nemirovsky, R. Fastow, M. Meyassed, and A. Unikovsky, "Trapping Effects in HgCdTe," *Journal of Vacuum Science and Technology* B9, 1829–39, 1991.

355. D. Rosenfeld and G. Bahir, "A Model for the Trap-Assisted Tunneling Mechanism in Diffused n-p and Implanted n⁺-p HgCdTe Photodiodes," *IEEE Transactions on Electron Devices* 39, 1638–45, 1992.

356. Y. Nemirovsky and A. Unikovsky, "Tunneling and 1/f Noise Currents in HgCdTe Photodiodes," *Journal of Vacuum Science and Technology* B10, 1602–10, 1992.

357. T. Elliott, N. T. Gordon, and R. S. Hall, "Reverse Breakdown in Long Wavelength Lateral Collection $Cd_xHg_{1-x}Te$ Diodes," *Journal of Vacuum Science and Technology* A8, 1251–53, 1990.

358. A. S. Gilmore, J. Bangs, and A. Gerrishi, "VLWIR HgCdTe Detector Current-Voltage Analysis," *Journal of Electronic Materials* 35, 1403–10, 2006.

359. O. Gravrand, E. De Borniol, S. Bisotto, L. Mollard, and G. Destefanis, "From Long Infrared to Very Long Wavelength Focal Plane Arrays Made with HgCdTe n⁺ n⁻/p Ion Implantation Technology," *Journal of Electronic Materials* 36, 981–87, 2007.

360. M. C. Chen, R. S. List, D. Chandra, M. J. Bevan, L. Colombo, and H. F. Schaake, "Key Performance-Limiting Defects in p-on-n HgCdTe Heterojunction Infrared Photodiodes," *Journal of Electronic Materials* 25, 1375–82, 1996.

361. T. D. Kleinpenning, "1/f Noise in p-n Junction Diodes," *Journal of Vacuum Science and Technology* A3, 176–82, 1985.

362. F. N. Hooge, "1/f Noise is No Surface Effect," *Physics Letters* 29A, 123–40, 1969.

363. A. Van der Ziel, *Fluctuation Phenomena in Semiconductors*, Butterworths, London, 1959.

364. S. P. Tobin, S. Iwasa, and T. J. Tredwell, "1/f Noise in (Hg,Cd)Te Photodiodes," *IEEE Transactions on Electron Devices* ED-27, 43–48, 1980.

365. H. K. Chung, M. A. Rosenberg, and P. H. Zimmermann, "Origin of 1/f Noise Observed in $Hg_{1-x}Cd_xTe$ Variable Area Photodiode Arrays," *Journal of Vacuum Science and Technology* A3, 189–91, 1985.

366. W. W. Anderson and H. J. Hoffman, "Surface-Tunneling-Induced 1/f Noise in $Hg_{1-x}Cd_xTe$ Photodiodes," *Journal of Vacuum Science and Technology* A1, 1730–34, 1983.

367. J. Bajaj, G. M. Williams, N. H. Sheng, M. Hinnrichs, D. T. Cheung, J. P. Rode, and W. E. Tennant, "Excess (1/f) Noise in $Hg_{0.7}Cd_{0.3}Te$ p-n Junctions," *Journal of Vacuum Science and Technology* A3, 192–94, 1985.

368. R. E. DeWames, J. G. Pasko, E. S. Yao, A. H. B. Vanderwyck, and G. M. Williams, "Dark Current Generation Mechanisms and Spectral Noise Current in Long-Wavelength Infrared Photodiodes," *Journal of Vacuum Science and Technology* A6, 2655–63, 1988.

369. Y. Nemirovsky and D. Rosenfeld, "Surface Passivation and 1/f Noise Phenomena in HgCdTe Photodiodes," *Journal of Vacuum Science and Technology* A8, 1159–66, 1990.

370. R. A. Schiebel, "A Model for 1/f Noise in Diffusion Current Based on Surface Recombination Velocity Fluctuations and Insulator Trapping," *IEEE Transactions on Electron Devices* ED-41, 768–78, 1994.

371. J. Bajaj, E. R. Blazejewski, G. M. Williams, R. E. DeWames, and M. Brawn, "Noise (1/f) and Dark Currents in Midwavelength Infrared PACE-I HgCdTe Photodiodes," *Journal of Vacuum Science and Technology* B10, 1617–25, 1992.

372. M. A. Kinch, C.-F. Wan, and J. D. Beck, "Noise in HgCdTe Photodiodes," *Journal of Electronic Materials* 34, 928–32, 2005.

373. C. Verie, F. Raymond, J. Besson, and T. Nquyen Duy, "Bandgap Spin-Orbit Splitting Resonance Effects in $Hg_{1-x}Cd_xTe$ Alloys," *Journal of Crystal Growth* 59, 342–46, 1982.

374. R. Alabedra, B. Orsal, G. Lecoy, G. Pichard, J. Meslage, and P. Fragnon, "An $Hg_{0.3}Cd_{0.7}Te$ Avalanche Photodiode for Optical-Fiber Transmission Systems at $\lambda = 1.3$ µm," *IEEE Transactions on Electron Devices* ED-32, 1302–6, 1985.

375. B. Orsal, R. Alabedra, M. Valenza, G. Lecoy, J. Meslage, and C. Y. Boisrobert, "$Hg_{0.4}Cd_{0.6}Te$ 1.55-µm Avalanche Photodiode Noise Analysis in the Vicinity of Resonant Impact Ionization Connected with the Spin-Orbit Split-Off Band," *IEEE Transactions on Electron Devices* ED-35, 101–7, 1988.

376. J. Vallegra, J. McPhate, L. Dawson, and M. Stapelbroeck, "Mid-IR Couting Array Using HgCdTe APDs and the Medipix 2 ROIC," *Proceedings of SPIE* 6660, 66600O, 2007.

377. D. N. B. Hall, B. S. Rauscher, J. L. Pipher, K. W. Hodapp, and G. Luppino, "HgCdTe Optical and Infrared Focal Plane Array Development in the Next Decade," *Astro2010: The Astronomy and Astrophysics Decadal Survey*, Technology Development Papers, no. 28.

378. G. Leveque, M. Nasser, D. Bertho, B. Orsal, and R. Alabedra, "Ionization Energies in $Cd_xHg_{1-x}Te$ Avalanche Photodiodes," *Semiconductor Science and Technology* 8, 1317–23, 1993.

379. T. J. deLyon, B. A. Baumgratz, G. R. Chapman, E. Gordon, M. D. Gorwitz, A. T. Hunter, M. D. Jack, et al., "Epitaxial Growth of HgCdTe 1.55 µm Avalanche Photodiodes by MBE," *Proceedings of SPIE* 3629, 256–67, 1999.

380. N. Duy, A. Durand, and J. L. Lyot, "Bulk Crystal Growth of $Hg_{1-x}Cd_xTe$ for Avalanche Photodiode Applications," *Materials Research Society Symposium Proceedings* 90, 81–90, 1987.

381. R. E. DeWames, J. G. Pasko, D. L. McConnell, J. S. Chen, J. Bajaj, L. O. Bubulac, E. S. Yao, et al., *Extended Abstracts, 1989 Workshop on the Physics and Chemistry of II-VI Materials,* San Diego, October 3–5, 1989.

382. J. D. Beck, C.-F. Wan, M. A. Kinch, and J. E. Robinson, "MWIR HgCdTe Avalanche Photodiodes," *Proceedings of SPIE* 4454, 188–97, 2001.

383. M. A. Kinch, J. D. Beck, C.-F. Wan, F. Ma, and J. Campbell, "HgCdTe Electron Avalanche Photodiodes," *Journal of Electronic Materials* 33, 630–39, 2004.

384. F. Ma, X. Li, J. C. Campbell, J. D. Beck, C.-F. Wan, and M. A. Kinch, "Monte Carlo Simulations of $Hg_{0.7}Cd_{0.3}Te$ Avalanche Photodiodes and Resonance Phenomenon in the Multiplication Noise," *Applied Physics Letters* 83, 785, 2003.

385. J. Beck, C. Wan, M. Kinch, J. Robinson, P. Mitra, R. Scritchfield, F. Ma, and J. Campbell, "The HgCdTe Electron Avalanche Photodiode," *IEEE LEOS Newsletter*, October 8–12, 2006.

386. J. Beck, C. Wan, M. Kinch, J. Robinson, P. Mitra, R. Scritchfield, F. Ma, and J. Campbell, "The HgCdTe Electron Avalanche Photodiode," *Journal of Electronic Materials* 35, 1166–73, 2006.

387. M. K. Kinch, "A Theoretical Model for the HgCdTe Electron Avalanche Photodiode," *Journal of Electronic Materials* 37, 1453–59, 2008.

388. I. Baker, S. Duncan, and J. Copley, "Low Noise Laser Gated Imaging System for Long Range Target Identification," *Proceedings of SPIE* 5406, 133–44, 2004.

389. I. Baker, P. Thorne, J. Henderson, J. Copley, D. Humphreys, and A. Millar, "Advanced Multifunctional Detectors for Laser-Gated Imaging Applications," *Proceedings of SPIE* 6206, 620608, 2006.

390. M. Vaidyanathan, A. Joshi, S. Xue, B. Hanyaloglu, M. Thomas, M. Zandian, D. Edwall, et al., "High Performance Ladar Focal Plane Arrays for 3D Range Imaging," *2004 IEEE Aerospace Conference Proceedings*, 1776–81, 2004.

391. R. S. Hall, N. T. Gordon, J. Giess, J. E. Hails, A. Graham, D. C. Herbert, D. J. Hall, et al., "Photomultiplication with Low Excess Noise Factor in MWIR to Optical Fiber Compatible Wavelengths in Cooled HgCdTe Mesa Diodes," *Proceedings of SPIE* 5783, 412–423, 2005.

392. J. Beck, M. Woodall, R. Scritchfield, M. Ohlson, L. Wood, P. Mitra, and J. Robinson, "Gated IR Imaging with 128 × 128 HgCdTe Electron Avalanche Photodiode FPA," *Proceedings of SPIE* 6542, 654217, 2007.

393. J. Beck, M. Woodall, R. Scritchfield, M. Ohlson, L. Wood, P. Mitra, and J. Robinson, "Gated IR Imaging with 128 × 128 HgCdTe Electron Avalanche Photodiode FPA," *Journal of Electronic Materials* 37, 1334–43, 2008.

394. G. Perrais, O. Gravrand, J. Baylet, G. Destefanis, and J. Rothman, "Gain and Dark Current Characteristics of Planar HgCdTe Avalanche Photodiodes," *Journal of Electronic Materials* 36, 963–70, 2007.

395. J. Rothman, G. Perrais, P. Ballet, L. Mollard, S. Gout, and J.-P. Chamonal, "Latest Developments of HgCdTe e-APDs at CEA LETI-Minatec," *Journal of Electronic Materials* 37, 1303–10, 2008.

396. M. B. Reine, J. W. Marciniec, K. K. Wong, T. Parodos, J. D. Mullarkey, P. A. Lamarre, S. P. Tobin, et al., "Characterization of HgCdTe MWIR Back-Illuminated Electron-Initiated Avalanche Photodiodes," *Journal of Electronic Materials* 37, 1376–85, 2008.

397. G. Perrais, J. Rothman, G. Destefanis, and J.-P. Chamonal, "Impulse Response Time Measurements in Hg$_{0.7}$Cd$_{0.3}$Te MWIR Avalanche Photodiodes," *Journal of Electronic Materials* 37, 1261–73, 2008.

398. G. Perrais, J. Rothman, G. Destefanis, J. Baylet, P. Castelein, J.-P. Chamonal, and P. Tribolet, "Demonstration of Multifunctional Bi-Colour-Avalanche Gain Detection in HgCdTe FPA," *Proceedings of SPIE* 6395, 63950H, 2006.

399. T. Ashley, A. B. Dean, C. T. Elliott, M. R. Houlton, C. F. McConville, H. A. Tarry, and C. R. Whitehouse, "Multilayer InSb Diodes Grown by Molecular Beam Epitaxy for Near Ambient Temperature Operation," *Proceedings of SPIE* 1361, 238–44, 1990.

400. C. T. Elliott, "Advanced Heterostructures for In$_{1-x}$Al$_x$Sb and Hg$_{1-x}$Cd$_x$Te Detectors and Emiters," *Proceedings of SPIE* 2744, 452–62, 1996.

401. C. T. Elliott, "Non-Equilibrium Devices in HgCdTe," in *Properties of Narrow Gap Cadmium-Based Compounds, EMIS Datareviews Series No. 10*, ed. P. Capper, 339–46, IEE, London, 1994.

402. C. T. Elliott, N. T. Gordon, R. S. Hall, T. J. Phillips, A. M. White, C. L. Jones, C. D. Maxey, and N. E. Metcalfe, "Recent Results on Metalorganic Vapor Phase Epitaxially Grown HgCdTe Heterostructure Devices," *Journal of Electronic Materials* 25, 1139–45, 1996.

403. D. Maxey, C. L. Jones, N. E. Metcalfe, R. Catchpole, M. R. Houlton, A. M. White, N. T. Gordon, and C. T. Elliott, "Growth of Fully Doped Hg$_{1-x}$Cd$_x$Te Heterostructures Using a Novel Iodine Doping Source to Achieve Improved Device Performance at Elevated Temperatures," *Journal of Electronic Materials* 25, 1276–85, 1996.

404. T. Ashley, C. T. Elliott, N. T. Gordon, R. S. Hall, A. D. Johnson, and G. J. Pryce, "Room Temperature Narrow Gap Semiconductor Diodes as Sources and Detectors in the 5–10 μm Wavelength Region," *Journal of Crystal Growth* 159, 1100–03, 1996.

405. M. K. Haigh, G. R. Nash, N. T. Gordon, J. Edwards, A. J. Hydes, D. J. Hall, A. Graham, J. Giess, J. E. Hails, and T. Ashley, "Progress in Negative Luminescent Hg$_{1-x}$Cd$_x$Te Diode Arrays," *Proceedings of SPIE* 5783, 376–83, 2005.

406. C. T. Elliott, N. T. Gordon, T. J. Phillips, H. Steen, A. M. White, D. J. Wilson, C. L. Jones, C. D. Maxey, and N. E. Metcalfe, "Minimally Cooled Heterojunction Laser Heterodyne Detectors in Metalorganic Vapor Phase Epitaxially Grown Hg$_{1-x}$Cd$_x$Te," *Journal of Electronic Materials* 25, 1146–50, 1996.

407. C. T. Elliott, N. T. Gordon, R. S. Hall, T. J. Phillips, C. L. Jones, and A. Best, "1/f Noise Studies in Uncooled Narrow Gap Hg$_{1-x}$Cd$_x$Te Non-Equilibrium Diodes," *Journal of Electronic Materials* 26, 643–48, 1997.

408. C. L. Jones, N. E. Metcalfe, A. Best, R. Catchpole, C. D. Maxey, N. T. Gordon, R. S. Hall, T. Colin, and T. Skauli, "Effect of Device Processing on 1/f Noise in Uncooled Auger-Suppressed CdHgTe Diodes," *Journal of Electronic Materials* 27, 733–39, 1998.

409. T. Skauli, H. Steen, T. Colin, P. Helgesen, S. Lovold, C. T. Elliott, N. T. Gordon, T. J. Phillips, and A. M. White, "Auger Suppression in CdHgTe Heterostructure Diodes Grown by Molecular Beam Epitaxy Using Silver as Acceptor Dopant," *Applied Physics Letters* 68, 1235–37, 1996.

410. G. J. Bowen, I. D. Blenkinsop, R. Catchpole, N. T. Gordon, M. A. Harper, P. C. Haynes, L. Hipwood, et al., "HOTEYE: A Novel Thermal Camera Using Higher Operating Temperature Infrared Detectors," *Proceedings of SPIE* 5783, 392–400, 2005.

411. M. K. Ashby, N. T. Gordon, C. T. Elliott, C. L. Jones, C. D. Maxey, L. Hipwood, and R. Catchpole, "Novel Hg$_{1-x}$Cd$_x$Te Device Structure for Higher Operating Temperature Detectors," *Journal of Electronic Materials* 32, 667–71, 2003.

412. Z. F. Ivasiv, F. F. Sizov, and V. V. Tetyorkin, "Noise Spectra and Dark Current Investigations in n^+-p-type $Hg_{1-x}Cd_xTe$ ($x \approx 0.22$) Photodiodes," *Semiconductor Physics, Quantum Electronics and Optoelectronics*, 2(3), 21–25, 1999.

413. K. Józwikowski, "Numerical Modeling of Fluctuation Phenomena in Semiconductor Devices," *Journal of Applied Physics* 90, 1318–27, 2001.

414. W. Y. Ho, W. K. Fong, C. Surya, K. Y. Tong, L. W. Lu, and W. K. Ge, "Characterization of Hot-Electron Effects on Flicker Noise in III-V Nitride Based Heterojunctions," *MRS Internet Journal of Nitride Semiconductor Research* 4S1, G6.4, 1999.

415. M. K. Ashby, N. T. Gordon, C. T. Elliott, C. L. Jones, C. D. Maxey, L. Hipwood, and R. Catchpole, "Investigation into the Source of 1/f Noise in $Hg_{1-x}Cd_xTe$ Diodes," *Journal of Electronic Materials* 33, 757–65, 2004.

416. S. Horn, D. Lohrman, P. Norton, K. McCormack, and A. Hutchinson, "Reaching for the Sensitivity Limits of Uncooled and Minimally Cooled Thermal and Photon Infrared Detectors," *Proceedings of SPIE* 5783, 401–11, 2005.

417. M. A. Kinch, "Metal-Insulator-Semiconductor Infrared Detectors," in *Semiconductors and Semimetals,* Vol. 18, eds. R. K. Willardson and A. C. Beer, 313–78, Academic Press, New York, 1981.

418. R. A. Chapman, S. R. Borrello, A. Simmons, J. D. Beck, A. J. Lewis, M. A. Kinch, J. Hynecek, and C. G. Roberts, "Monolithic HgCdTe Charge Transfer Device Infrared Imaging Arrays," *IEEE Transactions on Electron Devices* ED-27, 134–46, 1980.

419. A. F. Milton, "Charge Transfer Devices for Infrared Imaging," in *Optical and Infrared Detectors,* ed. R. J. Keyes, 197–228, Springer-Verlag, Berlin, 1980.

420. G. M. Williams and E. R. Gertner, "n-Channel M.I.S.F.E.T. in Epitaxial HgCdTe/CdTe," *Electron Letters* 16, 839, 1980.

421. A. Kolodny, Y. T. Shacham-Diamond, and I. Kidron, "n-Channel MOS Transistors in Mercury-Cadmium Telluride," *IEEE Transactions on Electron Devices* 27, 591–95, 1980.

422. Y. Nemirovsky, S. Margalit, and I. Kidron, "n-Channel Insulated-Gate Field-Effect Transistors in $Hg_{1-x}Cd_xTe$ with $x = 0.215$," *Applied Physics Letters* 36, 466–68, 1980.

423. R. A. Schiebel, "Enhancement Mode HgCdTe MISFETs and Circuits for Focal Plane Applications," *IEDM Technical Digest* 132–35, 1987.

424. T. L. Koch, J. H. De Loo, M. H. Kalisher, and J. D. Phillips, "Monolithic n-Channel HgCdTe Linear Imaging Arrays," *IEEE Transactions on Electron Devices* ED-32, 1592–1607, 1985.

425. S. R. Borrello, C. G. Roberts, M. A. Kinch, C. E. Tew, and J. D. Beck, "HgCdTe MIS Photocapacitor Detectors," in *Fourth International Conference on Advanced Infrared Detectors and Systems,* 41–47, IEE, London, 1990.

426. M. V. Wadsworth, S. R. Borrello, J. Dodge, R. Gooh, W. McCardel, G. Nado, and M. D. Shilhanek, "Monolithic CCD Imagers in HgCdTe," *IEEE Transactions on Electron Devices* 42, 244–50, 1995.

427. J. P. Omaggio, "Analysis of Dark Current in IR Detectors on Thinned p-Type HgCdTe," *IEEE Transactions on Electron Devices* 37, 141–52, 1990.

428. R. A. Chapman, M. A. Kinch, A. Simmons, S. R. Borrello, H. B. Morris, J. S. Wrobel, and D. D. Buss, "$Hg_{0.7}Cd_{0.3}Te$ Charge-Coupled Device Shift Registers," *Applied Physics Letters* 32, 434–36, 1978.

429. J. L. Freeouf and J. M. Woodall, "Schottky Barriers: An Effective Work Function Model," *Applied Physics Letters* 39, 727–29, 1981.

430. J. Tersoff, "Schottky Barriers and Semiconductor Band Structures," *Physical Review* 32, 6968–71, 1985.

431. W. E. Spicer, I. Lindau, P. Skeath, C. Y. Su, and P. Chye, "Unified Mechanism for Schottky-Barrier Formation and III-V Oxide Interface States," *Physical Review Letters* 44, 420–23, 1980.

432. A. Kobayoshi, O. F. Shankey, and J. D. Dow, "Chemical Trends for Defect Energy Levels in $Hg_{1-x}Cd_xTe$," *Physical Review* 25, 6367–79, 1982.

433. A. Zunger, "Composition-Dependence of Deep Impurity Levels in Alloys," *Physical Review Letters* 54, 849–50, 1985.

434. D. L. Polla and A. K. Sood, "Schottky Barrier Photodiodes in $Hg_{1-x}Cd_xTe$," *IEDM Technical Digest*, 419–20, 1978.

435. D. L. Polla and A. K. Sood, "Schottky Barrier Photodiodes in p $Hg_{1-x}Cd_xTe$," *Journal of Applied Physics* 51, 4908–12, 1980.

436. P. W. Leech and M. H. Kibel, "Properties of Schottky Diodes on n-Type $Hg_{1-x}Cd_xTe$," *Journal of Vacuum Science and Technology* B9, 1770–76, 1991.

437. R. Triboulet, "Alternative Small Gap Materials for IR Detection," *Semiconductor Science and Technology* 5, 1073–79, 1990.

438. A. Sher, A. B. Chen, W. E. Spicer, and C. K. Shih, "Effects Influencing the Structural Integrity of Semiconductors and Their Alloys," *Journal of Vacuum Science and Technology* A 3, 105–11, 1985.

439. A. Wall, C. Caprile, A. Franciosi, R. Reifenberger, and U. Debska, "New Ternary Semiconductors for Infrared Applications: $Hg_{1-x}Mn_xTe$," *Journal of Vacuum Science and Technology* A 4, 818–22, 1986.

440. K. Guergouri, R. Troboulet, A. Tromson-Carli, and Y. Marfaing, "Solution Hardening and Dislocation Density Reduction in CdTe Crystals by Zn Addition," *Journal of Crystal Growth* 86, 61–65, 1988.

441. P. Maheswaranathan, R. J. Sladek, and U. Debska, "Elastic Constants and Their Pressure Dependences in $Cd_{1-x}Mn_xTe$ with $0 \leq x \leq 0.52$ and in $Cd_{0.52}Zn_{0.48}Te$," *Physical Review* B31, 5212–16, 1985.

442. A. Rogalski, "$Hg_{1-x}Zn_xTe$ as a Potential Infrared Detector Material," *Progress in Quantum Electronics* 13, 299–353, 1989.

443. A. Rogalski, "$Hg_{1-x}Mn_xTe$ as a New Infrared Detector Material," *Infrared Physics* 31, 117–66, 1991.

444. A. Rogalski, *New Ternary Alloy Systems for Infrared Detectors*, SPIE Press, Bellingham, WA, 1994.

445. R. T. Dalves and B. Lewis, "Zinc Blende Type HgTe-MnTe Solid Solutions: I," *Journal of Physics and Chemistry of Solids* 24, 549–56, 1963.

446. P. Becla, J. C. Han, and S. Matakef, "Application of Strong Vertical Magnetic Fields to Growth of II–VI Pseudo-Binary Alloys: HgMnTe," *Journal of Crystal Growth* 121, 394–98, 1992.

447. O. A. Bodnaruk, I. N. Gorbatiuk, V. I. Kalenik, O. D. Pustylnik, I. M. Rarenko, and B. P. Schafraniuk, "Crystalline Structure and Electro-Physical Parameters of $Hg_{1-x}Mn_xTe$ Crystals," *Nieorganicheskie Materialy* 28, 335–39, 1992 (in Russian).

448. R. Triboulet, "(Hg,Zn)Te: A New Material for IR Detection," *Journal of Crystal Growth* 86, 79–86, 1988.

449. P. Gille, U. Rössner, N. Puhlmann, H. Niebsch, and T. Piotrowski, "Growth of $Hg_{1-x}Mn_xTe$ Crystals by the Travelling Heater Method," *Semiconductor Science and Technology* 10, 353–57, 1995.

450. S. Takeyama and S. Narita, "New Techniques for Growing Highly-Homogeneous Quaternary $Hg_{1-x}Cd_xMn_yTe$ Single Crystals," *Japanese Journal of Applied Physics* 24, 1270–73, 1985.

451. T. Uchino and K. Takita, "Liquid Phase Epitaxial Growth of $Hg_{1-x-y}Cn_xZn_yTe$ and $Hg_{1-x}Cd_xMn_yTe$ from Hg-Rich Solutions," *Journal of Vacuum Science and Technology A* 14, 2871–74, 1996.

452. A. B. Horsfall, S. Oktik, I. Terry, and A. W. Brinkman, "Electrical Measurements of $Hg_{1-x}Mn_xTe$ Films Grown by Metalorganic Vapour Phase Epitaxy," *Journal of Crystal Growth* 159, 1085–89, 1996.

453. R. Triboulet, A. Lasbley, B. Toulouse, and R. Granger, "Growth and Characterization of Bulk HgZnTe Crystals," *Journal of Crystal Growth* 76, 695–700, 1986.

454. S. Rolland, K. Karrari, R. Granger, and R. Triboulet, "P-to-n Conversion in $Hg_{1-x}Zn_xTe$," *Semiconductor Science and Technology* 14, 335–40, 1999.

455. M. A. Berding, S. Krishnamurthy, A. Sher, and A. B. Chen, "Electronic and Transport Properties of HgCdTe and HgZnTe," *Journal of Vacuum Science and Technology A* 5, 3014–18, 1987.

456. R. Granger, A. Lasbley, S. Rolland, C. M. Pelletier, and R. Triboulet, "Carrier Concentration and Transport in $Hg_{1-x}Zn_xTe$ for x Near 0.15," *Journal of Crystal Growth* 86, 682–88, 1988.

457. W. Abdelhakiem, J. D. Patterson, and S. L. Lehoczky, "A Comparison Between Electron Mobility in n-Type $Hg_{1-x}Cd_xTe$ and $Hg_{1-x}Zn_xTe$," *Materials Letters* 11, 47–51, 1991.

458. R. E. Kremer, Y. Tang, and F. G. Moore, "Thermal Annealing of Narrow-Gap HgTe-Based Alloys," *Journal of Crystal Growth* 86, 797–803, 1988.

459. P. I. Baranski, A. E. Bielaiev, O. A. Bodnaruk, I. N. Gorbatiuk, S. M. Kimirenko, I. M. Rarenko, and N. V. Shevchenko, "Transport Properties and Recombination Mechanisms in $Hg_{1-x}Mn_xTe$ Alloys (x ~ 0.1)," *Fizyka i Technika Poluprovodnikov* 24, 1490–93, 1990.

460. M. M. Trifonova, N. S. Baryshev, and M. P. Mezenceva, "Electrical Properties of n-Type $Hg_{1-x}Mn_xTe$ Alloys," *Fizyka i Technika Poluprovodnikov* 25, 1014–17, 1991.

461. W. A. Gobba, J. D. Patterson, and S. L. Lehoczky, "A Comparison Between Electron Mobilities in $Hg_{1-x}Mn_xTe$ and $Hg_{1-x}Cd_xTe$," *Infrared Physics* 34, 311–21, 1993.

462. Y. Sha, C. Su, and S. L. Lehoczky, "Intrinsic Carrier Concentration and Electron Effective Mass in $Hg_{1-x}Zn_xTe$," *Journal of Applied Physics* 81, 2245–49, 1997.

463. K. Jóźwikowski and A. Rogalski, "Intrinsic Carrier Concentrations and Effective Masses in the Potential Infrared Detector Material, $Hg_{1-x}Zn_xTe$," *Infrared Physics* 28, 101–107, 1988.

464. C. Wu, D. Chu, C. Sun, and T. Yang, "Infrared Spectroscopy of $Hg_{1-x}Zn_xTe$ Alloys," *Japanese Journal of Applied Physics* 34, 4687–93, 1995.

465. A. Rogalski, "Hg-Based Alternatives to MCT," in *Infrared Detectors and Emitters: Materials and Devices*, eds. P. Capper and C. T. Elliott, 377–400, Kluwer Academic Publishers, Boston, MA, 2001.

466. D. Eger and A. Zigelman, "Anodic Oxides on HgZnTe," *Proceedings of SPIE* 1484, 48–54, 1991.

467. K. H. Khelland, D. Lemoine, S. Rolland, R. Granger, and R. Triboulet, "Interface Properties of Passivated HgZnTe," *Semiconductor Science and Technology* 8, 56–82, 1993.

468. Yu. V. Medvedev and N. N. Berchenko, "Thermodynamic Properties of the Native Oxide-Hg$_{1-x}$Zn$_x$Te Interface," *Semiconductor Science and Technology* 9, 2253–57, 1994.

469. Z. Nowak, *Doctoral Thesis*, Military University of Technology, Warsaw, 1974 (in Polish).

470. J. Piotrowski, K. Adamiec, A. Maciak, and Z. Nowak, "ZnHgTe as a Material for Ambient Temperature 10.6 μm Photodetectors," *Applied Physics Letters* 54, 143–44, 1989.

471. J. Piotrowski, K. Adamiec, and A. Maciak. "High-Temperature 10,6 μm HgZnTe Photodetectors," *Infrared Physics* 29, 267–70, 1989.

472. J. Piotrowski and T. Niedziela, "Mercury Zinc Telluride Longwavelength High Temperature Photoconductors," *Infrared Physics* 30, 113–19, 1990.

473. A. Rogalski, J. Rutkowski, K. Józwikowski, J. Piotrowski, and Z. Nowak, "The Performance of Hg$_{1-x}$Zn$_x$Te Photodiodes," *Applied Physics A* 50, 379–84, 1990.

474. K. Józwikowski, A. Rogalski, and J. Piotrowski, "On the Performance of Hg$_{1-x}$Zn$_x$Te Photoresistors," *Acta Physica Polonica A* 77, 359–62, 1990.

475. J. Piotrowski, T. Niedziela, and W. Galus, "High-Temperature Long-Wavelength Photoconductors," *Semiconductor Science and Technology* 5, S53–S56, 1990.

476. R. Triboulet, T. Le Floch, and J. Saulnier, "First (Hg,Zn)Te Infrared Detectors," *Proceedings of SPIE* 659, 150–52, 1988.

477. Z. Nowak, J. Piotrowski, and J. Rutkowski, "Growth of HgZnTe by Cast-Recrystallization," *Journal of Crystal Growth* 89, 237–41, 1988.

478. J. Ameurlaine, A. Rousseau, T. Nguyen-Duy, and R. Triboulet, "(HgZn)Te Infrared Photovoltaic Detectors," *Proceedings of SPIE* 929, 14–20, 1988.

479. R. Triboulet, M. Bourdillot, A. Durand, and T. Nguyen Duy, "(Hg,Zn)Te Among the Other Materials for IR Detection," *Proceedings of SPIE* 1106, 40–47, 1989.

480. D. L. Kaiser and P. Becla, "Hg$_{1-x-y}$Cd$_x$Zn$_y$Te: Growth, Properties and Potential for Infrared Detector Applications," *Materials Research Society Symposium Proceedings* 90, 397–404, 1987.

481. P. Becla, "Infrared Photovoltaic Detectors Utilizing Hg$_{1-x}$Mn$_x$Te and Hg$_{1-x-y}$Cd$_x$Mn$_y$Te Alloys," *Journal of Vacuum Science and Technology A* 4, 2014–18, 1986.

482. P. Becla, "Advanced Infrared Photonic Devices Based on HgMnTe," *Proceedings of SPIE* 2021, 22–34, 1993.

483. S. Takeyama and S. Narita, "The Band Structure Parameters Determination of the Quaternary Semimagnetic Semiconductor Alloy Hg$_{1-x-y}$Cd$_x$Mn$_y$Te," *Journal of the Physics Society of Japan* 55, 274–83, 1986.

484. S. Manhas, K. C. Khulbe, D. J. S. Beckett, G. Lamarche, and J. C. Woolley, "Lattice Parameters, Energy Gap, and Magnetic Properties of the Cd$_x$Hg$_y$Mn$_x$Te Alloy System," *Physica Status Solidi (b)* 143, 267–74, 1987.

485. L. A. Kosyachenko, I. M. Rarenko, S. Weiguo, and L. Zheng Xiong, "Charge Transport Mechanisms in HgMnTe Photodiodes with Ion Etched p-n Junctions," *Solid-State Electronics* 44, 1197–1202, 2000.

486. L. A. Kosyachenko, I. M. Rarenko, S. Weiguo, L. Zheng Xiong, and G. Qibing, "Photoelectric Properties of HgMnTe Photodiodes with Ion Etched p-n Junctions," *Opto-Electronics Review* 8, 251–62, 2000.

487. S. H. Shin, J. G. Pasko, D. S. Lo, W. E. Tennant, J. R. Anderson, M. Górska, M. Fotouhi, and C. R. Lu, "$Hg_{1-x-y}Cd_xMn_yTe$ Alloys for 1.3–1.8 μm Photodiode Applications," *Materials Research Society Symposium Proceedings* 89, 267–74, 1987.

488. P. Becla, S. Motakef, and T. Koehler, "Long Wavelength HgMnTe Avalanche Photodiodes," *Journal of Vacuum Science and Technology* B10, 1599–1601, 1992.

489. P. Becla, M. Grudzień, and J. Piotrowski, "Uncooled 10.6 μm Mercury Manganise Telluride Photoelectromagnetic Infrared Detectors," *Journal of Vacuum Science and Technology* B 9, 1777–80, 1991.

15 IV-VI Detectors

Around 1920, Case investigated the thallium sulfide photoconductor—one of the first photoconductors to give a response in the near IR region to approximately 1.1 μm [1]. The next group of materials to be studied was the lead salts (PbS, PbSe, and PbTe), which extended the wavelength response to 7 μm. The PbS photoconductors from natural galena found in Sardinia were originally fabricated by Kutzscher at the University of Berlin in the 1930s [2]. However, for any practical applications it was necessary to develop a technique for producing synthetic crystals. PbS thin-film photoconductors were first produced in Germany, next in the United States at Northwestern University in 1944, and then in England at the Admiralty Research Laboratory in 1945 [3]. During World War II, the Germans produced systems that used PbS detectors to detect hot aircraft engines. Immediately after the war, communications, fire control, and search systems began to stimulate a strong development effort that has extended to the present day. The *Sidewinder* heat-seeking infrared-guided missiles received a great deal of public attention. After 60 years, low-cost, versatile PbS and PbSe polycrystalline thin films remain the photoconductive detectors of choice for many applications in the 1–3 μm and 3–5 μm spectral range. Current development with lead salts is in the focal plane arrays (FPAs) configuration.

The study of the IV-VI semiconductors received fresh impetus in the mid-1960s with a discovery at Lincoln Laboratories [4,5]: PbTe, SnTe, PbSe, and SnSe form solid solutions in which the energy gap varies continuously through zero so that it is possible to obtain any required small energy gap by selecting the appropriate composition. For 10 years, during the late 1960s to mid-1970s, HgCdTe alloy detectors were in serious competition with IV-VI alloy devices (mainly PbSnTe) for developing photodiodes because of the latter's production and storage problems [6,7]. The PbSnTe alloy seemed easier to prepare and appeared more stable. However, development of PbSnTe photodiodes was discontinued because the chalcogenides suffered from two significant drawbacks. The first drawback was a high permittivity that resulted in a high diode capacitance, and therefore a limited frequency response. For scanning systems under development at that time, this was a serious limitation. However, for the staring imaging systems that use 2-D arrays (which are currently under development), this would not be such a significant issue. The second drawback to IV-VI compounds was their very high thermal coefficient of expansion [8] (a factor of 7 higher than Si). This limited their applicability in hybrid configurations with Si multiplexers. Development of ternary lead salt FPAs was nearly entirely stopped in the early 1980s in favor of HgCdTe. Today, with the ability to grow these materials on alternative substrates such as Si, this would not be a fundamental limitation either. Moreover, IV-VI materials remained the only choice to fabricate MWIR laser diodes before the invention of the quantum cascade lasers and continue to be of importance today [9–12].

In this chapter we begin with a survey of fundamental properties of lead salt chalcogenides and go on to describe detailed technology and properties of IV-VI photoconductive and photovoltaic IR detectors.

15.1 MATERIAL PREPARATION AND PROPERTIES

15.1.1 Crystal Growth

The properties of the lead salt binary and ternary alloys have been extensively reviewed [6,7,13–21]. Therefore only some of their most important properties will be mentioned here.

The development of pseudobinary alloy systems, especially $Pb_{1-x}Sn_xTe$ (PbSnTe) and $Pb_{1-x}Sn_xSe$ (PbSnSe), has brought about major advances in the 8–14 μm wavelength region. Their energy gap varies continuously through zero so that it is possible, by selecting the appropriate composition, to obtain any required small energy gap (Figure 15.1). In comparison with $Hg_{1-x}Cd_xTe$ material system, the cutoff wavelength of IV-VI materials is less sensitive to composition.

It should be noted that besides PbSnTe and PbSnSe, a number of other lead salts, such as PbS_xSe_{1-x} (PbSSe) and $PbTe_{1-x}Se_x$, (PbTeSe) are of interest for detection. Moreover, choosing $Pb_{1-x}Y_xZ$ materials with Y = Sr or Eu (Z = Te or Se), wider bandgap compounds can be obtained characterized by lower refractive indices (see Figure 15.2). This allows a high freedom to design more elaborate device structures including epitaxial Bragg mirrors, which can be realized by MBE in different IV-VI optoelectronic devices [22].

The lead chalcogenide semiconductors have the face centered cubic (rock salt) crystal structure, and hence, obtain the name "lead salts." Thus, they have (100) cleavage planes, and tend to grow in the (100) orientation, although they can also be grown in the (111) orientation. Only SnSe possesses

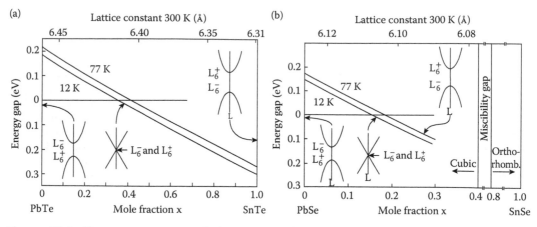

Figure 15.1 Energy gap versus mole fraction x and lattice constant for (a) $Pb_{1-x}Sn_xTe$, and (b) $Pb_{1-x}Sn_xSe$. Schematic representation of the valence and conduction bands.

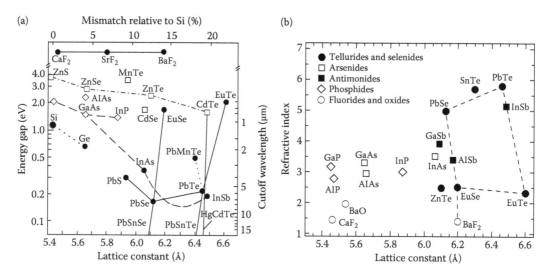

Figure 15.2 (a) Bandgap energy and corresponding emission wavelength (right-hand scale), and (b) bandgap refractive index of various III-V, II-VI, IV-VI, group IV semiconductors and selected fluorides and oxides plotted versus lattice constant. (From Springholz, G., and Bauer, G., *Physica Status Solidi (b)*, 244, 2752–67, 2007. With permission.)

the orthorhombic-B29 structure. Therefore, the ternary compounds PbSnTe and PbSSe exhibit complete solid solubility, while the existence range of PbSnSe with rock-salt structure is restricted to the lead-rich side ($x < 0.4$).

The crystalline properties of ternary alloys (PbSnTe, PbSnSe, and PbSSe) are comparable to those of the binary compounds (PbTe, PbSe, PbS, and SnTe). Despite the less fundamental physical properties relative to HgCdTe PbSnTe and PbSnSe have received a good deal of attention as materials for photodiodes; the main reason being the very much easier materials technology. The separation between the liquidus and solidus curves is much smaller in ternary IV-VI alloys (see Figure 15.3) [23]. As a result, it has been relatively easy to grow PbSnTe and PbSnSe crystals that are homogeneous in composition. A second difference is in the vapor pressure of the elements, which is similar in magnitude for all three elements in ternary IV-VI alloys. Thus, vapor growth techniques have been successfully used in growing lead salts.

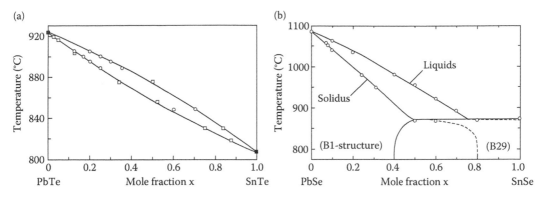

Figure 15.3 The pseudobinary T-x phase diagram of (a) $Pb_{1-x}Sn_xTe$, and (b) $Pb_{1-x}Sn_xSe$. (From Harman, T. C., *Journal of Nonmetals*, 1, 183–94, 1973. With permission.)

Numerous techniques for the preparation of lead salt single crystals and epitaxial layers have been investigated. Several excellent review articles devoted to this topic have been published [15,16,24,25].

Bridgman-type or the Czochralski methods give crystals large in size, of variable composition, and frequently the material show inclusions and rather high dislocation densities. They are mainly used as substrates for the subsequent growth of epilayers. Growth from the solution and the traveling solvent method offer interesting advantages such as higher homogeneity in composition and lower temperatures leading to lower concentrations of lattice defects and impurities. The best results have been achieved with the sublimation growth technique, since lead salts sublime as molecules. A successful growth process requires a very high purity of the source material and a carefully adjusted composition of its metal/chalcogen ratio [16]. For the vapor growth process two different procedures are applied: an unseeded growth and a seeded growth technique. Using the unseeded growth technique, the largest crystals exhibiting 2–3 cm² (100)-facets can be obtained with stoichiometric or slightly metal-rich source material.

A minimal contact with the quartz walls of the ampoule leads to an excellent metallurgical structure. Disadvantage of this growth procedure is the nonreproducible shape of the crystals that implies problems to large-scale device production; but the main advantage is the growth of crystals with controlled carrier concentrations down to the 10^{17} cm⁻³ range without the need for lengthy annealing procedures [16]. Large crystals with excellent homogeneity and structural quality have been grown with seeded growth technique. As a seed, (111)-oriented single crystal lead salt slices or (111)-oriented BaF_2 crystals are used (BaF_2 provides a lattice match and a good thermal expansion match comparable to lead salts). In other methods developed by Tamari and Shtrikman [26], a quartz tip is used and the first deposit of material acts as a nonoriented seed in the growth. Good results have been also obtained by using the method developed by Markov and Davydov [27] for lead salts [28]. In the latter method the crystals do not touch the ampoule wall and the dislocation density is low, typically 10^2–10^3 cm⁻³.

Thin single crystal films of IV-VI compounds have found broad applications in fundamental research and application. The epitaxial layers are usually grown by either VPE or LPE techniques. Recently, the best quality devices have been obtained using MBE [22,25,29].

In the late 1970s, the excellent results were obtained with LPE technique in fabrication PbTe/PbSnTe heterojunctions. A great deal of experimental research has been performed on solid–liquid equilibria in the Pb-Sn-Te system. Szapiro et al. [30] calculated the phase diagram of the Pb-Sn-Te system in the (Pb + Sn)-rich region using the modified model of regular associated solutions and received reasonable agreement with the calculated lines at higher concentrations of Sn in liquid ($x > 0.3$) and poor agreement for $x < 0.2$. Surface morphology of the epilayers was found to be primarily related to the substrate orientation and their surface preparation. Generally, (100)-oriented PbTe or PbSnTe substrates are employed. It should be noticed that the LPE technique leads to some difficulties with other ternary lead salts [31].

Due to the fact that the IV-VI compounds evaporate predominantly in the form of binary molecules, which means that nearly congruent evaporation occurs, in MBE growth the main constituents are supplied from compound effusion sources loaded with PbTe, PbS, PbS, SnTe, SnSe, or GeTe. While the degree of dissociation is only a few percentage, it notably increases for the tin and germanium chalcogenides. By changing the total group IV to group VI flux ratio, the background carrier concentration and type of carriers can be controlled. Excess group IV flux leads to n-type and excess group VI flux to p-type conductivity in the layers.

Concerning the substrate materials, the lattice mismatch of lead salts to common semiconductor substrates such as Si or GaAs is rather large (10% and more, see Figure 15.2a). In addition, the thermal expansion coefficient of the IV-VI compounds of around 20×10^{-6}/K differ strongly from that of Si as well as that of the zinc-blende type III-V or II-VI compounds (typically less than 6×10^{-6}/K). As a consequence, large thermal strains are induced in the epitaxial layers during cooling of the samples to room temperature and below after sample growth. A best compromise in these respects is achieved for BaF_2 substrates, in spite of its different crystal structure (calcium fluoride structure). As shown in Figure 15.2a, BaF_2 shows only a moderate lattice mismatch to PbSe or PbTe (–1.2% and + 4.2%, respectively) and moreover, the thermal expansion coefficient is almost exactly matched to that of the lead salt compounds. Furthermore, BaF_2 is highly insulating and optically transparent in the midinfrared region. Due to its high ionic character, good BaF_2 surfaces can be obtained easily only for the (111) surface orientation. In conclusion, for lead salt MBE growth, BaF_2 has been the most widely used substrate material.

Renewed interest with the growth of IV-VI epitaxial layers started in the mid-1980s with the growth of high-quality MBE layers on Si(111) substrates employing a very thin CaF_2 buffer layer. It appears that the thermal expansion mismatch between the IV-VI layers and Si relaxes through the glide of dislocations on (100) planes, which are inclined with respect to the (111) surface plane. The threading ends of the misfit dislocations glide to remove the mechanical thermal strain built up and therefore increase the structural layer quality [32].

15.1.2 Defects and Impurities

The lead salts can exist with very large deviations from stoichiometry and it is difficult to prepare material with carrier concentrations below about 10^{17} cm^{-3} [16,33–35]. For PbSnTe alloys, the solidus field shifts considerably toward the Te-rich side of the stoichiometric composition with increasing SnTe content, and very high hole concentrations are obtained at the Te-rich solidus lines. The solidus lines for several $Pb_{1-x}Sn_xTe$ alloys shown in Figure 15.4 have been determined by means of an isothermal annealing technique that is also useful for reducing the carrier concentration and converting the carrier type of crystals [16]. Low electron and hole concentrations in the range 10^{15} cm^{-3} have been obtained by isothermal annealing or by LPE at low temperature. In $Pb_{0.80}Sn_{0.20}Te$, which is of particular importance in device applications, the conversion of the carrier type occurs at a temperature of 530°C.

The width of the solidus field is large in IV-VI compounds ($\approx 0.1\%$) making the doping by native defects very efficient. Deviations from stoichiometry create n- or p-type conduction. It is generally accepted that vacancies and interstitials are formed and that they control the conductivity. From the absence of any observable freeze-out it appears unlikely that native defects in lead salts form hydrogen-like states. Native defects associated with excess metal (nonmetal vacancies or possibly metal interstitials) yield acceptor levels, while those which result from excess nonmetal (metal vacancies or possibly nonmetal interstitials) yield donor levels. Parada and Pratt [36,37] first pointed out that the strong perturbations around the defects in PbTe cause valence band states to shift to the conduction band. As a result, a Te vacancy provides two electrons for the conduction band and a Pb vacancy two holes for the valence bands. The Pb interstitial yields a single electron whereas the Te interstitial was found to be neutral. Similar results were obtained by Hemstreet [38] on the basis of scattered-wave cluster calculations for PbS, PbTe, and SnTe. Also Lent et al. [39] have used a simple chemical theory of s- and p-bonded substitutional point defect in PbTe and PbSnTe which corroborates experimental data [40] and the predictions of Parada and Pratt.

In crystals grown from high-purity elements, the effects of foreign impurities are usually negligible when the carrier concentration due to lattice defects is above 10^{17} cm^{-3}. Below this concentration, foreign impurities can play a role by compensating lattice defects and other foreign impurities. A compilation of the results on impurity doping of IV-VI semiconductors is given by Dornhaus, Nimtz, and Richter [41]. Most of the impurities can be assumed to have shallow or even resonant levels [42]. However, deep donor levels of unidentified and identified defects were also found in lead salts [43]. For example, it was found that In has an energy level in $Pb_{1-x}Sn_xTe$, which is resonant with the conduction band for small x values and is situated in

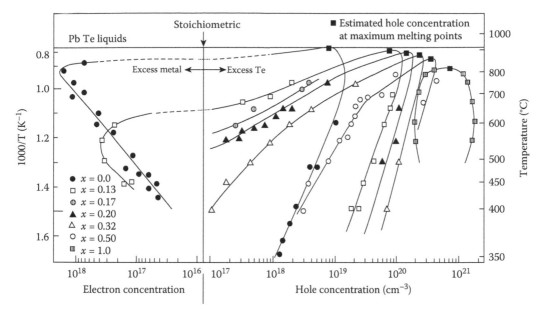

Figure 15.4 Carrier concentrations at 77 K (300 K for SnTe) versus isothermal annealing temperature for several Pb$_{1-x}$Sn$_x$Te compositions. (From Maier, H., and Hesse, J., *Crystal Growth, Properties and Applications*, Springer Verlag, Berlin, 145–219, 1980. With permission.)

the bandgap for higher mole fraction [44]. Such behavior of In level was theoretically explained by Lent et al. [39].

Cadmium is a compensating impurity in Pb$_{1-x}$Sn$_x$Te and has the very important property of reducing carrier concentration and changing conductivity type in as-grown p-type materials. Silberg and Zemel [45] carried out Cd-diffusion in a two temperature zone furnace in which the samples were held at a fixed temperature of 400°C, while the Cd source temperature was varied between 150°C and 310°C. Figure 15.5 shows the dependence of electron concentration of the Cd diffused Pb$_{1-x}$Sn$_x$Te samples measured at 77 K on the Cd concentrations N_{Cd}. The electron concentration in the saturation region decreases appreciably with increasing x values. It appears that after vacancy sites become fully compensated by the Cd atoms most of Cd distributes uniformly in the crystal lattice as electrically inactive impurity.

15.1.3 Some Physical Properties

The lead salts have direct energy gaps, which occur at the Brillouin zone edge at the L point. The effective masses are therefore higher and the mobilities lower than for a zinc blende structure with the same energy gap at the Γ point (the zone center). Due to band inversion in PbSnTe and PbSnSe, the energy gaps approach zero at certain compositions (see Figure 15.1). Therefore, with these ternary compounds, very long wavelength cutoff of infrared detectors can be achieved. The constant energy surfaces are ellipsoids characterized by the longitudinal and transverse effective masses m_l^* and m_t^*, respectively. The anisotropy factor for PbSnTe is of the order of 10 and increases with decreasing bandgap. It is much less, about 2, for PbSnSe and PbSSe.

Table 15.1 contains a list of material parameters for different types of binary and ternary lead salts.

In the vicinity of the energy gap in lead salts a system of three conduction bands and three valence bands has been observed. The temperature and x-dependence of the effective masses at the band edges may be expressed by [15]

$$\frac{1}{m^*(x,T)} = \frac{1}{m_{cv}^*}\frac{E_g(0,0)}{E_g(x,T)} + \frac{1}{m_F^*},$$ (15.1)

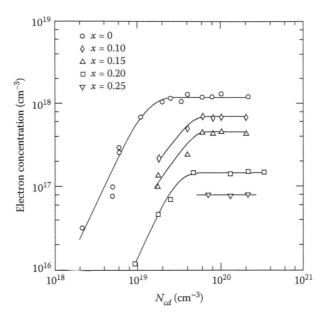

Figure 15.5 Electron concentration as a function of Cd concentration measured at 77 K on Pb$_{1-x}$Sn$_x$Te crystals with different Sn mole fractions. (From Silberg, E., and Zemel, A., *Journal of Electronic Materials*, 8, 99–109, 1979. With permission.)

where m_{cv}^* determines a contribution due to interactions between the nearest extremes of the valence band and conduction band, and m_F^* is the far band contribution. The functions in Equation 15.1 for the four effective masses m_{et}^* (conduction band, transverse), m_{eh}^* (valence band, transverse), m_{el}^* (conduction band, longitudinal), and m_{hl}^* (valence band, longitudinal) are described by Preier [15] and Dornhaus et al. [17].

Exact knowledge of $E_g(x,T)$ dependences for the ternary compounds is required for reliable calculations over a wide range of composition. On the basis of the study of experimental results for the ternary compounds, good agreement with experimental data was obtained using formulas of the Grisar type $E_g(x,T) = E_1 + [E^2 + \alpha(T + \theta)^2]^{1/2}$ [15]:

$$E_g(x,T) = 171.5 - 535x + \sqrt{(12.8)^2 + 0.19(T+20)^2} \quad \text{in meV,} \tag{15.2}$$

$$E_g(x,T) = 125 - 1021x + \sqrt{400 + 0.256T^2} \quad \text{in meV,} \tag{15.3}$$

$$E_g(x,T) = 263 - 138x + \sqrt{400 + 0.265T^2} \quad \text{in meV,} \tag{15.4}$$

for Pb$_{1-x}$Sn$_x$Te, Pb$_{1-x}$Sn$_x$Se, and PbS$_{1-x}$Se$_x$, respectively.

Rogalski and Jóźwikowski [46] have calculated the intrinsic carrier concentrations in Pb$_{1-x}$Sn$_x$Te, Pb$_{1-x}$Sn$_x$Se, and PbS$_{1-x}$Se$_x$ in terms of the six-band **kp** model of Dimmock. By fitting the calculated nonparabolic n_i values to the expression for parabolic bands, the following approximations were obtained:

for Pb$_{1-x}$Sn$_x$Te ($0 \le x \le 0.40$)

$$n_i = \left(8.92 - 34.46x + 2.25 \times 10^{-3}T + 4.12 \times 10^{-2}xT + 97.00x^2\right) \times 10^{14} E_g^{3/4}T^{3/2}\exp\left(-\frac{E_g}{2kT}\right), \tag{15.5}$$

Table 15.1: Physical Properties of Lead Salts

Lattice Structure	T (K)	PbTe	Pb$_{0.8}$Sn$_{0.2}$Te	PbSe	Pb$_{0.93}$Sn$_{0.7}$Se	PbS
		Cubic (NaCl)	Cubic (NaCl)	Cubic (NaCl)	Cubic (NaCl)	Cubic (NaCl)
Lattice constant a (nm)	300	0.6460	0.64321	0.61265	0.6118	0.59356
Thermal expansion coefficient α (10^{-6} K^{-1})	300	19.8	20	19.4		20.3
	77	15.9		16.0		
Heat capacity C_p (Jmol^{-1}K^{-1})	300	50.7		50.3		47.8
Density γ (g/cm^3)	300	8.242	7.91	8.274		7.596
Melting point T_m (K)	300	1197	1168 (sol.) 1178 (liq.)	1354	1325 (sol.) 1340 (liq.)	1400
Bandgap E_g (eV)	300	0.31	0.20	0.28	0.21	0.42
	77	0.22	0.11	0.17	0.10	0.31
	4.2	0.19	0.08	0.15	0.08	0.29
Thermal coefficient of E_g (10^{-4} eVK^{-1})	80–300	4.2	4.2	4.5	4.5	4.5
Effective masses						
$m_{el}^{\star}/m$	4.2	0.022	0.011	0.040	0.037	0.080
$m_{ht}^{\star}/m$		0.025	0.012	0.034	0.021	0.075
$m_{et}^{\star}/m$		0.19	0.11	0.070	0.041	0.105
$m_{hl}^{\star}/m$		0.24	0.13	0.068	0.040	0.105
Mobilities						
μ_e (cm^2/Vs)	77	3×10^4	3×10^4	3×10^4	3×10^4	1.5×10^4
μ_h (cm^2/Vs)		2×10^4	2×10^4	3×10^4	2×10^4	1.5×10^4
Intrinsic carrier concentration n_i (cm^{-3})	77	1.5×10^{10}	3×10^{13}	6×10^{11}	8×10^{13}	3×10^7
Static dielectric constant ε_s	300	380		206		172
	77	428		227		184
High frequency dielectric constant ε_∞	300	32.8	38	22.9	26.0	17.2
	77	36.9	42	25.2	30.9	18.4
Optical phonons						
LO (cm^{-1})	300	114	120	133		212
TO (cm^{-1})	77	32		44		67

Sources: R. Dalven, *Infrared Physics*, 9, 141–84, 1969; H. Preier, *Applied Physics*, 10, 189–206, 1979; H. Maier and J. Hesse, *Crystal Growth, Properties and Applications*, Springer Verlag, Berlin, 145–219, 1980. With permission.

for $Pb_{1-x}Sn_xSe$ ($0 \le x \le 0.12$)

$$n_i = \left(1.73 - 3.68x + 3.77 \times 10^{-4}T + 1.60 \times 10^{-2}xT + 8.92x^2\right) \times 10^{15}E_g^{3/4}T^{3/2}\exp\left(-\frac{E_g}{2kT}\right), \quad (15.6)$$

and for $PbS_{1-x}Se_x$ ($0 \le x \le 1$)

$$n_i = \left(2.14 - 8.85 \times 10^{-1}x + 6.12 \times 10^{-4}T + 6.47 \times 10^{-4}xT + 3.32 \times 10^{-1}x^2\right) \times 10^{15}E_g^{3/4}T^{3/2}\exp\left(-\frac{E_g}{2kT}\right). \quad (15.7)$$

A large number of experimental and theoretical works have been directed toward the elucidation of the dominant scattering mechanisms in lead salts [17,47–50]. As a consequence of the similar valence and conduction bands of lead salts, the electron and hole mobilities are approximately equal for the same temperatures and doping concentrations. Room temperature mobilities in lead salts are 500–2000 cm²/Vs [17]. In many high-quality single crystal samples, the mobility due to lattice scattering varies as $T^{-5/2}$ [14]. This behavior has been ascribed to a combination of polar-optical and acoustical lattice scattering and achieves the limiting values in the range of 10^5–10^6 cm²/Vs due to defect scattering (see Figure 15.6) [51]. The scattering mechanism considerations for the binary alloys are also applicable to the mixed crystals. However, the importance of the acoustical phonon scattering is diminished for the small gap materials, since the mobility due to this type of mechanism is proportional to the reciprocal of the density of states. In these materials scattering by impurity atoms and vacancies is important even at room temperature. Additionally, the lack of chemical order in the mixed crystals results in relatively strong disorder scattering of electrons in samples with comparatively low carrier concentrations at low temperatures. This type of scattering is dominant in PbSnTe with carrier concentrations below approximately 10^{18} cm^{-3} at 4.2 K. Figure 15.7 shows the carrier concentration dependence of the electron mobility for $Pb_{1-x}Sn_xTe$ ($0.17 \le x \le 0.20$) at 77 K [50]. In the concentration range $\ge 10^{18}$ cm^{-3}, the nonelastic scattering on the impurity and vacancy potentials is decisive.

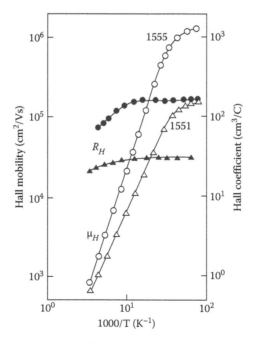

Figure 15.6 Temperature dependence of the Hall mobility and the Hall coefficient for two PbTe epitaxial layers on BaF$_2$ substrates. (From Lopez-Otero, A., *Thin Solid Films*, 49, 3–57, 1978. With permission.)

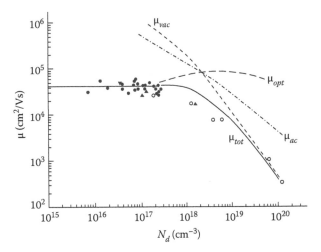

Figure 15.7 Carrier concentration dependence of the electron mobility for $Pb_{1-x}Sn_xTe$ (0.17 ≤ x ≤ 0.20) at 77 K. The curves are calculated for scattering of carriers by the longitudinal optical phonons (μ_{opt}), the acustical phonons (μ_{ac}), and the vacancy potential (μ_{vac}). (From Sizov, F. F., Lashkarev, G. V., Radchenko, M. V., Orletzki, V. B., and Grigorovich, E. T., *Fizika i Tekhnika Poluprovodnikov*, 10, 1801–8, 1976.)

The interband absorption of the lead salts is more complicated as compared with the standard case due to: anisotropic multivalley structure of both conduction and valence bands, nonparabolic Kane-type energy dispersion, and **k**-dependent matrix elements. Analytical expressions for the absorption coefficient for energies near the absorption edge have been given by several workers [52–56]. The temperature and composition dependence of the absorption coefficient near the absorption edge are well described by the two-band model for which the following expression has been obtained [53]:

$$\alpha(z) = \frac{2q^2\left(m_t^{*2}m_l^*\right)^{1/2}}{\varepsilon_o n_r c\pi\hbar^4 E_g^{1/2}} \frac{2P_t^2 + P_l^2}{3} \frac{(z-1)^{1/2}}{z} \frac{(z+1)^{1/2}}{\sqrt{2}} \frac{(1+2z^2)}{3z}(f_v - f_c), \tag{15.8}$$

where $z = h\nu/E_g$, P_t and P_l are the transverse and longitudinal momentum matrix elements, n_r is refractive index and the factor $(f_v - f_c)$ describes the band-filling and nearly equals unity in the nondegenerate case. The formula for the Burstein–Moss factor is given by Anderson [52].

The dielectric properties of the lead salts are characterized by large static and optical dielectric constants and low frequencies of transverse optical phonons. For PbSnTe the observed values of the static dielectric constant have been widely distributed from 400 to 5800, and at the same temperature these values have been scattered in the range up to one order of magnitude [17,57,58]. Recently, Butenko et al. [59] have determined the temperature dependence of the static dielectric constant of PbTe in wide temperature range from 10 K to 300 K according to Barrett's formula [60]

$$\varepsilon_o = \frac{1.356 \times 10^5}{36.14 \coth\left(36.14/T\right) + 49.15}. \tag{15.9}$$

The high frequency dielectric constant as function of x for $Pb_{1-x}Sn_xTe$ can be described by relation [57]:

$$\varepsilon_\infty = 27.4 + 22.0x - 6.4x^2. \tag{15.10}$$

Shani et al. [61] have calculated the refractive index of PbSnTe and good agreements have been obtained between their calculations and experimental results. The similar results for other lead salts have been presented by Jensen and Tarabi [62,63].

The interrelations between the energy bandgap and high-frequency dielectric constant and the refractive index have been summarized by several authors; see, for example, [64,65]. Using a classical oscillator theory, Herve and Vandamme have proposed the following relation between refractive index and energy gap

$$n^2 = 1 + \left(\frac{13.6}{E_g + 3.4}\right)^2. \tag{15.11}$$

This relation is accurate for most of the compounds used in optoelectronics structures and for wide-gap semiconductors, but it cannot properly describe the behavior of the IV-VI group. For lead salts, better agreement between experimental and theoretical results can be obtained using the model of Wemlple and DiDomenico [64].

15.1.4 Generation–Recombination Processes

Although direct one-phonon recombination as well as plasmon recombination have been proposed theoretically [66,67] and identified experimentally [68,69] for very small energy gaps of IV-VI compounds, in samples with gaps of 0.1 eV or more, SRH, radiative, and Auger recombination are dominant [70–73].

Ziep et al. [66] have calculated the lifetime determined by radiative recombination in terms of the Kane-type two-band model case and Boltzmann statistics. However, it appears that in the case of mirror-symmetric band structure of lead salts ($m_e^* \approx m_h^*$), a good approximation for recombination rate is the following equation [6]

$$G_R = \frac{10^{-15} n_r E_g^2 n_i^2}{(kT)^{3/2} K^{1/2} (2+1/K)^{3/2} (m^*/m)^{5/2}} \quad \text{in cm}^3/\text{s}, \tag{15.12}$$

where $K = m_l^*/m_t^*$ is the effective mass anisotropy coefficient. The mass m^* can be determined if we know the longitudinal m_l^* and transverse m_t^* components of the effective mass, since $m^* = [1/3(2/m_t^* + 1/m_l^*)]^{-1}$. In Equation 15.12 the values of kT and E_g should be expressed in electronvolts.

For a long time the Auger process was considered to be a low-efficiency channel for nonradiative recombination in semiconductors of the IV-VI type. The valence band and the conduction band with mirror-reflection symmetry occur at point L of the Brillouin zone (number of valley $w = 4$). In such a case the energy and momentum conservation laws are difficult to fulfill for impact recombination, especially for carriers near the band edges when only single-valley interaction is taken into account. Since pioneering Emtage's paper published in 1976 [74] many theoretical and experimental works have been published [75–83] in which the intervalley interaction of carriers is considered, and it has been found that even at lower temperatures the lifetime of carriers is determined by impact recombination. According to the intervalley carrier interaction model (Figure 15.8) an electron and hole from valley (a), characterized by "heavy" mass m_l^*, and a third carrier from valley (b) with a "light" mass m_t^* (in PbSnTe mass anisotropy coefficient $K > 10$) participate in impact recombination in the given direction. As a result of this interaction the "heavy" electron and hole carriers recombine, and the liberated energy and momentum are transferred to the "light" carrier.

Two cases should be considered:

- all scattered carriers are at a definite point of the Brillouin zone,
- the initial carriers are in different valleys of the band.

In the case of Boltzmann statistics [67]

$$\left(\tau_A^j\right)^{-1} = C_A^j \left(n_o^2 + 2n_o p_o\right), \tag{15.13}$$

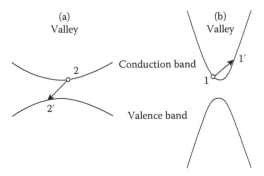

Figure 15.8 Intervalley Auger recombination.

where C_A^j is the Auger coefficient for j-th recombination mechanism. Then

$$\tau_A^{-1} = \left(\tau_A^a\right)^{-1} + \left(\tau_A^b\right)^{-1}. \tag{15.14}$$

Different approximations for C_A, especially for the process (b), can be found in the literature [74,79,80,84–86]. For the intervalley process proposed by Emtage within the parabolic band model [74]

$$C_A = (2\pi)^{5/2} \frac{w-1}{w^2} \frac{q^2}{\left(4\pi\,\varepsilon_0\varepsilon_\infty\right)^2} (kT)^{1/2} E_g^{-7/2} \frac{\hbar^3}{m_l^{*1/3} m_t^{*2/3}} \exp\left(-\frac{E_g K^{-1}}{2kT}\right). \tag{15.15}$$

Nonparabolic bands of the Kane type are expected to reduce the Auger transition rate [78,80]. However, the Emtage expression is a good approximation of more exactly calculated Auger coefficients [79].

Ziep et al. [84] carried out a comprehensive examination of nonradiative and radiative recombination mechanisms in the mixed crystals $Pb_{0.78}Sn_{0.22}Te$ and $Pb_{0.91}Sn_{0.09}Se$ for a temperature interval $20 < T < 400$ K and in a doping range $0 < N_d < 10^{19}$ cm^{-3}. In the calculations, the degeneracy of the carrier gas, the anisotropy, and to some extent the nonparabolicity of the band structure were taken into account. Their calculated data for $Pb_{0.78}Sn_{0.22}Te$ are shown in Figure 15.9. The radiative lifetime as well as the Auger lifetime have a maximum in the transition from the extrinsic to the intrinsic region. At low temperatures in the range of low dopant concentrations, the lifetime is determined by radiative recombination. As the temperature increases, the Auger recombination comes to the fore and determines the carrier lifetimes at room temperature. At comparable energy gaps the radiative lifetimes differ only slightly in PbSnTe and PbSnSe compounds. However, due to the smaller anisotropy of the isoenergetic surface in PbSnSe as compared with PbSnTe, the Auger lifetime is higher in PbSnSe than in PbSnTe. At carrier concentrations above 10^{19} cm^{-3}, the plasmon recombination dominates over radiative and Auger recombinations.

Dmitriev [85] has calculated the Auger carrier lifetime in PbSnTe, taking into account the carrier degeneracy and the exact expressions for the overlap integrals [86]. The calculated lifetimes are greater than those obtained previously.

Experimental investigations of lead salts confirm that the carrier lifetime is determined by band-to-band recombination as well as by any SRH recombination. The results of the first extensive investigation of $Pb_{1-x}Sn_xTe$ ($0.17 \leq x \leq 0.20$) revealed an extremely large variation of the lifetime ($10^{-12} < \tau < 10^{-8}$ s at 77 K) measured from photoconductivity and PEM effects (see Figure 15.10) [72,77,87]. The highest observed lifetime fits quite well to the straight line calculated according to Emtage's theory of Auger recombination [84]. All Cd- and In-doped samples are characterized by considerably lower lifetimes in spite of their lower free-carrier concentration. Donor levels with ionization energies from 12 to 25 meV are presumed to work as recombination centers.

In the case of PbTe epitaxial layers on BaF_2 substrates, a good agreement with the Emtage theory has been obtained by Lischka and Huber [76]. However, these authors observed a second lifetime branch with a much longer lifetime than that for Auger recombination (Figure 15.11). This branch

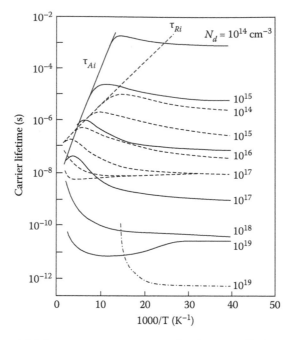

Figure 15.9 Calculated lifetime versus temperature for Auger, radiative, and plasmon recombination in $Pb_{0.78}Sn_{0.22}Te$. (From Ziep, O., Mocker, M., and Genzow, D., *Wissensch. Zeitschr. Humboldt-Univ. Berlin, Math.-Nataruiss. Reihe XXX, 81–97, 1981.*)

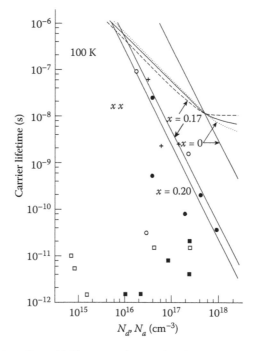

Figure 15.10 PEM lifetime ($T = 100$ K) versus doping level for $Pb_{1-x}Sn_xTe$ ($0.17 \le x \ 0.20$) samples of different origin: Bridgman annealed (●), THM (+), vapor-phase (○), uncompensated (□), and Cd-compensated (■). (From Harrmann, K. H., *Solid-State Electronics*, 21, 1487–91, 1978. With permission.)

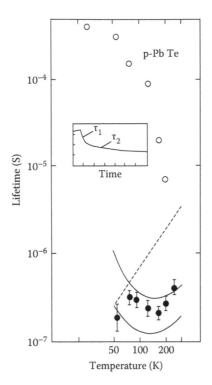

Figure 15.11 Carrier lifetime versus temperature for p-PbTe: experimental values of lifetime τ_1 (•; $p = 9.5 \times 10^{16}$ cm^{-3}), solid curves are calculated for Auger recombination assuming m_l^*/m_t^* (the upper curve) = 10, and $m_l^*/m_t^* = 14$ (the lower curve); experimental data of a second time constant τ_2 occurring in the decay of the photocurrent pulse (O). (From Lischka, K., and Huber, W., *J. Applied Physics*, 48, 2632–33, 1977. With permission.)

was attributed to deep levels acting as minority carrier traps at intermediate temperatures [71]. Zogg et al. [81] observed up to four lifetime branches determined by the transient photoconductivity method in n- and p-type PbSe epitaxial layers. The shortest lifetimes were in agreement with calculated values for direct (Auger and radiative) recombination below ≈ 250 K for samples with carrier concentrations ≤ 2 × 10^{17} cm^{-3}. From the longer observed lifetime branches, three impurity levels (separated between 20 and 50 meV from the nearer band edge), acting principally as minority carrier traps at intermediate temperature, were calculated. Also, Shahar et al. [82] confirmed that the carrier lifetime in undoped Pb$_{0.8}$Sn$_{0.2}$Te layers grown by liquid phase epitaxy is determined by band-to-band radiative and Auger recombination mechanisms in the temperature range 10–110 K, while in In-doped PbTe layers, recombination takes place via nonradiative centers.

Two other groups, Schlicht et al. [88] and Weiser et al. [89], have not obtained agreement between experimental data on PbSnTe epitaxial layers on (111) BaF$_2$ substrates and the calculated Auger lifetime. Measurements of the photoconductive decay time yield values that at 77 K are longer than predicted by the Emtage theory. Also, the dependence of lifetime on carrier density in the extrinsic range of conductivity was not observed. Weiser et al. [89] suggest that the lifetime enhancement by about two orders of magnitude at $T \approx 14$ K is the result of photon recycling and explain the lifetime enhancement by about one order of magnitude at 77 K by overestimation of the recombination rate within the Emtage theory or by the fact that strains in the sample reduce the importance of the multivalley Auger process. Genzow et al. [90] theoretically investigated the influence of uniaxial strain on the Auger and radiative recombinations in PbSnTe and obtained quite good agreement of their theoretical results with the experimental data given by Weiser et al. [89].

We conclude that the mechanism of Auger recombination in small gap IV-VI semiconductors is expected to be important but is not fully clarified. Experimental evidence for this mechanism is not yet unambiguously presented. Some disagreement between the experimental results reported by the various authors may be related to differences in sample preparation as well as to differences in the experimental methods used to measure the lifetime. Also these discrepancies are assumed to be due to the lack of accurate band parameters for lead salts and to the theoretical description of screening in the Auger process [70]. In spite of these differences some conclusions may be drawn:

- With decreasing energy gaps and increasing temperature, impact recombination becomes more important.

- In samples having a large number of lattice defects (impurities, native defects, misfit disloca-tions), the recombination mechanism is attributed to SRH centers.

- In defect-free undoped material, carrier lifetime is determined by direct band-to-band recombination.

The recombination processes in polycrystalline IV-VI films are modified by grain boundary potential barriers. Different theories of photoconductivity in lead chalcogenide polycrystalline layers have been proposed. Their brief review is given by Espevik et al. [91] and Johnson [92]. More recently, Neustroev and Osipov [93] developed a theoretical model based on the knowledge that polycrystalline materials are a two-phase system. Low-ohmic, n-type conductance crystallites are surrounded by oxygen-saturated p-type inversion layers with a large concentration of acceptor-type states. During irradiation, electrons and holes generating in crystallites are separated by surface barriers. Due to spatial division of photocarriers, their lifetime rises abruptly and, conse-quently, so does their photosensitivity.

Similarly, as in single-crystal lead salt samples, the carrier lifetime in polycrystalline materials is determined by three principal recombination processes: SRH, radiative, and Auger. Vaitkus et al. [94] have investigated picosecond photoconductivity in highly excited electrolytically deposited PbS and vacuum-evaporated PbTe polycrystalline films. They described the dependence of the lifetime on nonequilibrium concentrations, taking into account linear, trap-assisted Auger, inter-band radiative, and Auger recombinations [95]

$$\frac{1}{\tau} = \frac{1}{\tau_R} + \left(\gamma_{At} N_t + \gamma_R \right) \Delta N + \gamma_A \Delta N^2, \tag{15.16}$$

where γ_{At} is the trap Auger coefficient, N_t is the trap density, γ_R is the radiative recombination coef-ficient, and γ_A is the interband Auger recombination coefficient. Good agreement of the theoretical curves based on Equation 15.16 with experimental points in all investigated films (freshly made and annealed) has been obtained, assuming $\gamma_{At} N_t + \gamma_R = 2.2 \times 10^{-11}$ cm^3/s and $\gamma_A = 5.3 \times 10^{-29}$ cm^6/s. Interband Auger recombination in the grain bulk is the main mechanism of carriers. The structure and the film preparation process influence only the slower recombination.

15.2 POLYCRYSTALLINE PHOTOCONDUCTIVE DETECTORS

A number of lead salt photoconductive reviews have been published [92,96–106]. One of the best reviews of development efforts in lead salt detectors was published by Johnson [92].

15.2.1 Deposition of Polycrystalline Lead Salts

Although the fabrication methods developed for these photoconductors are not completely under-stood, their properties are well established. Unlike most other semiconductor IR detectors, lead salt photoconductive materials are used in the form of polycrystalline films approximately 1 μm thick and with individual crystallites ranging in size from approximately 0.1–1.0 μm. They are usually prepared by chemical deposition using empirical recipes, which generally yields better uniformity of response and more stable results than the evaporative methods [99–103].

The PbSe and PbS films used in commercial IR detectors are made by chemical bath deposition (CBD); the oldest and most-studied PbSe and PbS thin-film deposition method. It was used to deposit PbS in 1910 [106]. The basis of CBD is a precipitation reaction between a slowly produced anion (S^{2-} or Se^{2-}) and a complexed metal cation. The commonly used precursors are lead salts, $Pb(CH_3COO)_2$ or $Pb(NO_3)_2$, thiourea [$(NH_2)_2CS$] for PbS, and selenourea [$(NH_2)_2CSe$] for PbSe, all in alkaline solutions. Lead may be complexed with citrate, ammonia, triethanolamine, or with selenosulfate itself. Most often, however, the deposition is carried out in a highly alkaline solu-tion where OH$^-$ acts as the complexing agent for Pb^{2+}.

In CBD the film is formed when the product of the concentrations of the free ions is larger than the solubility product of the compound. Thus, CBD demands very strict control over the reaction temperature, pH, and precursor concentrations. In addition, the thickness of the film is limited, the terminal thickness usually being 300–500 nm. Therefore, in order to get a film with a sufficient thickness (approximately 1 μm in IR detectors, for example), several successive depositions must be done. The benefit of CBD compared to gas phase techniques is that CBD is a low-cost temperature method and the substrate may be temperature-sensitive with the various shapes.

As-deposited PbS films exhibit significant photoconductivity. However, a postdeposition baking process is used to achieve final sensitization. In order to obtain high-performance detectors, lead chalcogenide films need to be sensitized by oxidation. The oxidation may be carried out by using additives in the deposition bath, by postdeposition heat treatment in the presence of oxygen, or by chemical oxidation of the film. The effect of the oxidant is to introduce sensitizing centers and additional states into the bandgap and thereby increase the lifetime of the photoexcited holes in the p-type material.

The backing process changes the initial n-type films to p-type films and optimizes performance through the manipulation of resistance. The best material is obtained using a specific level of oxygen and a specific bake time. Only a small percentage (3–9%) of oxygen influences the absorption properties and response of the detector. Temperatures ranging from 100 to 120°C and time periods from a few hours to in excess of 24 hours are commonly employed to achieve final detector performance optimized for a particular application. Other impurities added to the chemical deposition solution for PbS have a considerable effect on the photosensitivity characteristics of the films [103]. The $SbCl_2$, $SbCl_3$, and As_2O_3 prolong the induction period and increase the photosensitivity by up to 10 times that of films prepared without these impurities. The increase is thought to be caused by the increased absorption of CO_2 during the prolonged induction period. This increases $PbCO_3$ formation and thus photosensitivity. Arsine sulfide also changes the oxidation states on the surface. Moreover, it has been found that essentially the same performance characteristics can be achieved by baking in an air or a nitrogen atmosphere. Therefore, all of the constituents necessary for sensitization are contained in the raw PbS films as deposited.

The preparation of PbSe photoconductors is similar to PbS ones. The postdeposition baking process for PbSe detectors operating at 77 K is carried out at a higher temperature (> 400°C) in an oxygen atmosphere. However, for detectors to be used at ambient and/or intermediate temperatures, the oxygen or air bake is immediately followed by baking in a halogen gas atmosphere at temperatures in the range of 300–400°C [92]. According to Torquemada and colleagues [107], iodine plays a key role in sensitization of the PbSe layers obtained by thermal evaporation in a vacuum on thermally oxidized silicon. The halogen behaves as a transport agent during the PbSe recrystallization process, and promotes a fast growth of PbSe microcrystals. Oxygen is trapped in the PbSe lattice during the recrystallization process, as it happens in chemically deposited PbSe films. The introduction of halogens in the PbSe sensitization procedure is a highly efficient technique for the incorporation of oxygen to the semiconductor lattice in electrically active positions. If halogens are not introduced during the PbSe sensitization, the oxygen is incorporated inside the lattice of microcrystals only by diffusion, which is a less efficient way.

A variety of materials can be used as substrates, but the best detector performance is achieved using single-crystal quartz material. PbSe detectors are often matched with Si to obtain higher collection efficiency.

Photoconductors also have been fabricated from epitaxial layers without backing that resulted in devices with uniform sensitivity, uniform response time, and no aging effects. However, these devices do not offset the increased difficulty and cost of fabrication.

15.2.2 Fabrication

The films are deposited, either over or under plated gold electrodes, and on fused quartz, crystal quartz, single crystal sapphire, glass, various ceramics, single crystal strontium titanate, Irtran II (ZnS), Si, and Ge. The most commonly used substrate materials are fused quartz for ambient operation and single crystal sapphire for detectors used at temperatures below 230 K. The very low thermal expansion coefficient of fused quartz relative to PbS films results in poorer detector performance at lower operating temperatures. Different shapes of substrates are used: flat, cylindrical, or spherical. To obtain higher collection efficiency, detectors may be deposited directly by immersion onto optical materials with high indices of refraction (e.g., into strontium titanate). Lead salts cannot be immersed directly; special optical cements must be used between the film and the optical element.

As was mentioned above, in order to obtain high-performance detectors, lead chalcogenide films must be sensitized by oxidation, which may be carried out by using additives in the deposition bath, by postdeposition heat treatment in the presence of oxygen, or by chemical oxidation of the film. Unfortunately, in the older literature the additives are seldom identified and are often referred to only as "an ocidant" [91,100]. A more recent paper deals with the effects of H_2O_2 and $K_2S_2O_8$, both in the deposition bath and in the postdeposition treatment [108]. It was found that both treatments increase the resistivity of PbS films. Although the resistivity usually increases during the oxidation, a different behavior also has been observed [109]. A sensitized PbS film may significantly degrade in air without an overcoating. Possible overcoating materials are As_2S_3, CdTe, ZnSe, Al_2O_3, MgF_2, and SiO_2. Vacuum-deposited As_2S_3 has been found to have the best optical, thermal, and mechanical properties, and it has improved the detector performance. The drawback of the As_2S_3 coating is the toxicity of As and its precursors. Overall, however, the electric properties (resistivity, and in particular, detectivity) of lead salt thin films are rather poorly reported, although there are many papers on PbS and PbSe film growth. The effects of annealing and oxidation treatments on detectivity have not been reported accurately either.

To explain the photoconductivity process in thin film lead salt detectors, three theories have been proposed [102]. The first consists of increases in carrier density during illumination, the oxygen is assumed to introduce a trapping state that inhibits recombination. The second mechanism is based primarily on the increase in the mobility of free carriers in the "barrier model." It is assumed that potential barriers are formed during "sensitization" by heating in oxygen, either between the crystallites of the film or between n- and p-regions in nonuniform films. The third model (commonly referred to as the generalized theory of photoconductivity in semiconducting films in general, and for the lead salts in particular) was proposed by Petritz. In his review paper [102], Bode concluded that the Petritz theory provided a reasonable framework for general use even though the complex mechanisms in lead salts were still unsolved. More recently, Espevik et al. [91] provided significant additional support for the Petritz theory.

The PbS and PbSe materials are peculiar because they have a relatively long response time that affects the significant photoconductive gain. It has been suggested that during the sensitization process the films are oxidized, converting the outer surfaces of exposed PbS and PbSe films to PbO or a mixture of $PbO_xS(Se)_{1-x}$, and forming a heterojunction at the surface (which is shown schematically in Figure 15.12) [110]. Oxide heterointerfaces create conditions for trapping minority carriers or separating majority carriers and thereby extending the lifetime of the material. As was mentioned above, without the sensitization (oxidation) step, lead salt materials have very short lifetimes and a low response.

The basic detector fabrication steps are electrode deposition and delineation, active-layer deposition and delineation, passivation overcoating, mounting (cover/window), and lead wire attachment. Figure 15.13 shows a cross section of a typical PbS detector structure [92]. The entire device is overcoated for environmental protection and sensitivity improvement. Standard packing consists of electrode deposition and placement in a metal case with a window over the top. Wire leads are placed in a groove in a substrate and gold electrodes are usually vacuum-evaporated onto the film using a mask.

Photolithographic delineation methods are used for complex, high-density patterns of small element sizes. The outer electrodes are produced with vacuum deposition of bimetallic films such as TiAu. To passivate and optimize transmission of radiation into the detector, usually a quarter-wavelength thick overcoating of As_2S_3 is used. Normally, detectors are sealed between a cover plate and the substrate with epoxy cement. The cover-plate material is ordinarily quartz, but other

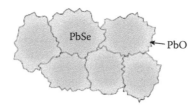

Figure 15.12 PbSe polycrystalline films after the sensitization process. The films are coated with PbO, forming a heterojunction on the surface. (From Horn, S., Lohrmann, D., Norton, P., McCormack, K., and Hutchinson, A., "Reaching for the Sensitivity Limits of Uncooled and Minimaly-Cooled Thermal and Photon Infrared Detectors," *Proceedings of SPIE* 5783, 401–11, 2005. With permission.)

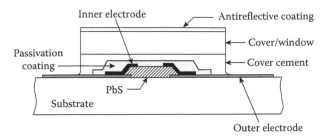

Figure 15.13 Typical structural configuration of PbS detector. (Reprinted from Johnson, T. H., "Lead Salt Detectors and Arrays. PbS and PbSe," *Proceedings of SPIE* 443, 60–94, 1984. With permission.)

materials such as sapphire may be used to transmit longer wavelengths. This technique seals the detector reasonably well against humid environments. The top surface of the cover is normally made antireflective with a material such as MgF_2.

15.2.3 Performance

Standard active area sizes are typically 1, 2, or 3 mm squares. However, most manufacturers offer sizes ranging from 0.08×0.08 to 10×10 mm² for PbSe detectors, and 0.025×0.025 to 10×10 mm² for PbS detectors. Active areas are generally square or rectangular; detectors with more exotic geometries have sometimes not performed up to expectations.

When a lead salt detector is operated below −20°C and exposed to UV radiation, semipermanent changes in responsivity, resistance, and detectivity occur [111]. This is the so-called flash effect. The amount of change and the degree of permanence depend on the intensity of the UV exposure and the length of exposure time. Lead salt detectors should be protected from fluorescent lighting. They are usually stored in a dark enclosure or overcoated with an appropriate UV-opaque material. They are also hermetically sealed so their long-term stability is not compromised by humidity and corrosion.

The spectral distribution of detectivity of lead salt detectors is presented in Figure 15.14. Usually the operating temperatures of detectors are between −196 and 100°C; it is possible to operate at a temperature higher than recommended, but 150°C should never be exceeded. Table 15.2 contains the performance range of detectors fabricated by various manufacturers [106].

Below 230 K, background radiation begins to limit the detectivity of PbS detectors. This effect becomes more pronounced at 77 K, and peak detectivity is no greater than the value obtained at 193 K. Figure 15.15 shows typical peak detectivity values for the PbSe detector [111]. The decrease in detectivity between 150 and 77 K for high background condition occurs because the background flux increases substantially as the spectral cutoff moves to longer wavelengths with cooling. Depending on operating temperature, background flux, and chemical additives, the detector impedance per square can be adjusted in the range of 10^6–10^9 $\Omega/\square$. Quantum efficiency of approximately 30% is limited by incomplete absorption of the incident flux in the relatively thin (1–2 μm) detector material. The responsivity uniformity of PbS and PbSe detectors is generally of the order of 3% to 10%.

The typical frequency response of PbS detectors is shown in Figure 15.16 [112]. An optimum operating frequency arises from the combined effects of the $1/f$ noise at low frequencies and the high-frequency rolloff in responsivity.

15.3 P-N JUNCTION PHOTODIODES

In comparison with photoconductive detectors, lead salt photodiodes have not found wide commercial applications. As was mentioned previously, development of PbSnTe and PbSnSe FPAs ternary was nearly entirely stopped in the early 1980s in favor of HgCdTe. However, through advancements in growth techniques, especially using MBE growth, vast improvements in device performance are currently realizable. One such innovation is microelectromechanical system (MEMS)-based tunable IR detector to deliver voltage-tunable multiband IR FPA [113,114], also called adaptive FPA.

In this section, we begin with a survey of fundamental performance limits of lead salt chalcogenide photovoltaic detectors and go on to describe in more detail the technology and properties of different types of these photodiodes.

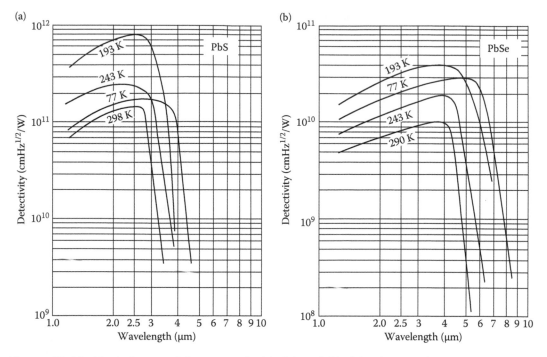

Figure 15.14 Typical spectral detectivity for (a) PbS, and (b) PbSe photoconductors (After New England Photoconductor data sheet, http://www.nepcorp.com.)

Table 15.2: Performance of Lead Salt Detectors (2π FOV, 300 K background)

	T (K)	Spectral Response (μm)	λ_p (μm)	D^* (λ_p, 1000Hz, 1) (cmHz$^{1/2}$W^{-1})	$R/\square$ (MΩ)	τ (μs)
PbS	298	1–3	2.5	$(0.1–1.5) \times 10^{11}$	0.1–10	30–1000
	243	1–3.2	2.7	$(0.3–3) \times 10^{11}$	0.2–35	75–3000
	195	1–4	2.9	$(1–3.5) \times 10^{11}$	0.4–100	100–10000
	77	1–4.5	3.4	$(0.5–2.5) \times 10^{11}$	1–1000	500–50000
PbSe	298	1–4.8	4.3	$(0.05–0.8) \times 10^{10}$	0.05–20	0.5–10
	243	1–5	4.5	$(0.15–3) \times 10^{10}$	0.25–120	5–60
	195	1–5.6	4.7	$(0.8–6) \times 10^{10}$	0.4–150	10–100
	77	1–7	5.2	$(0.7–5) \times 10^{10}$	0.5–200	15–150

Source: R. H. Harris, *Laser Focus/Electro-Optics*, 87–96, December 1983. With permission.

15.3.1 Performance Limit

Considerations carried out in this section are proper for a one-sided abrupt junction model. Surface leakage effects will not be considered since they may be minimized by appropriate surface treatment or the use of a guard ring structure. No distinction needs to be made between n$^+$-on-p and p$^+$-on-n structures owing to the mirror symmetry of valence and conduction bands.

The R_oA product determined by the diffusion current for the p$^+$-n junction is (see Chapter 9, Equation 9.87)

$$\left(R_o A\right)_D = \frac{(kT)^{1/2}}{q^{3/2} n_i^2} N_d \left(\frac{\tau_h}{\mu_h}\right)^{1/2}. \tag{15.17}$$

The R_oA product controlled by depletion layer current is given by Equation 9.116. Relating the width w of the abrupt junction depletion layer to the concentration N_d and assuming that $V_b = E_g/q$ we get

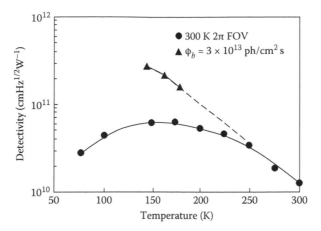

Figure 15.15 Detectivity of PbSe detector at two background flux levels as a function of temperature. (From Norton, P. R., *Optical Engineering*, 30, 1649–63, 1991. With permission.)

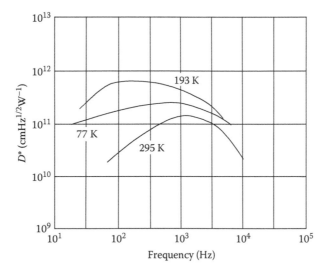

Figure 15.16 Peak spectral detectivity as a function of chopping frequency for a PbS photoconductive detector at 295 K, 193 K, and 77 K operating temperatures. (From Dereniak, E. L., and Boreman, G. D., *Infrared Detectors and Systems*, Wiley, New York, 1996. With permission.)

$$\left(R_o A\right)_{GR} = \frac{E_g^{1/2} \tau_o N_d^{1/2}}{q n_i \left(2\varepsilon_o \varepsilon_s\right)^{1/2}}. \tag{15.18}$$

The ultimate values of the $R_o A$ product for lead salts PbS, PbSe, and PbTe abrupt p-n junctions and ideal Schottky junctions within the temperature range between 77 K and 300 K were calculated by Rogalski and coworkers [115–119]. For example, the dependence of the $R_o A$ product on the concentration of dopands for PbTe photodiodes at 77 K is shown in Figure 15.17 [118]. At 77 K the $R_o A$ product is determined by the generation current of the junction depletion layer. The theoretical estimates yield for the radiative and Auger recombination values of the $R_o A$ product several orders of magnitude larger. Tunneling current produces an abrupt lowering of the $R_o A$ at a concentration of about 10^{18} cm^{-3}.

Figure 15.18 shows the temperature dependence of the $R_o A$ product for PbTe n$^+$-p junctions [119]. For comparison in this figure, experimental data are also included. From the comparison of the

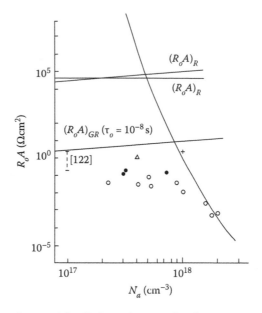

Figure 15.17 The dependence of the R_oA product on the doping concentration for one-sided abrupt PbTe junctions at 77 K. The experimental points are from Refs. 120(●), 118(○), 121(+), 122 and 123(Δ). (From Rogalski, A., Kaszuba, W., and Larkowski, W., *Thin Solid Films*, 103, 343–53, 1983. With permission.)

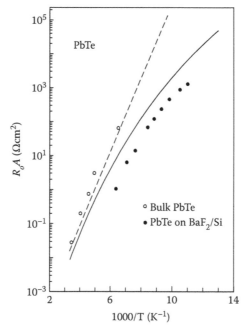

Figure 15.18 The temperature dependence of the R_oA product for n⁺-p PbTe photodiodes. The solid lines are calculated assuming Shockley-Read, radiative, and Auger processes. The dashed lines are calculated regardless Shockley-Read mechanism. The experimental data are taken from Refs. 123(○) and 124(●). (From Rogalski, A., Ciupa, R., and Zogg, H., "Computer Modeling of Carrier Transport in Binary Salt Photodiodes," *Proceedings of SPIE* 2373, 172–81, 1995. With permission.)

theoretical curves with the experimental data it may be seen that a satisfactory consistence has been achieved in a high temperature range of operation, where the dependence of $R_oA(T)$ follows a diffusion-limited behavior as revealed by the approximately $\exp(E_g/kT)$ slope. With lowering temperature the discrepancies increase. These discrepancies seem to be controlled by the state of junction production technology. The dashed line shown in Figure 15.18 is calculated assuming only band-to-band, generation–recombination mechanisms (the SRH mechanism is omitted). It means that for photodiodes with the experimental data situated above the solid curves, influence of Shockley–Read–Hall (SRH) mechanism on the photodiodes performance is considerably diminished. We can see that in the high temperature region of operation, the R_oA product of best quality photodiodes is determined by band-to-band, generation–recombination mechanisms. Some experimental data are situated above dashed lines, which is probably caused by influence of series resistance of photodiodes. It is clearly shown that R_oA product of photodiodes fabricated in PbTe epitaxial layers on BaF_2/Si substrates is lower than that for photodiodes made in bulk material.

The performance of lead salt photodiodes is inferior to HgCdTe photodiodes, and is below theoretical limits. Considerable improvements are possible by improving material quality (to reduce trap concentration) and optimizing the device fabrication technique. Better results should be obtained using buried p-n junctions with a thin wider-bandgap cap layer. This technique is successfully used in fabrication of double-layer heterostructure HgCdTe photodiodes (see Section 14.6.1). The wider bandgap cap layer contributes a negligible amount of thermally generated diffusion current compared with that from the bulk opposite type absorber layer.

During the early 1970s, lead salt ternary alloy (mainly PbSnTe) photodiode technology was advancing rapidly [6,7,19]. The performance of PbSnTe photodiodes was better than HgCdTe ones at that time. The dependence of the R_oA product on the long wavelength cutoff for LWIR PbSnTe photodiodes at 77 K is shown in Figure 15.19 [125]. In this figure a selection of experimental data are also observed. The Auger recombination contribution to R_oA increases with composition x increasing (λ_c increasing) in the base region of photodiode. For the composition range $x > 0.22$ the R_oA product is determined by Auger recombination [131]. A satisfactory agreement between the theoretical curves and the experimental data has been achieved for n⁺-p-p⁺ homojunction

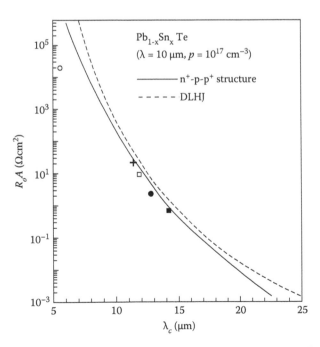

Figure 15.19 Dependence of the R_oA product on the long wavelength cutoff for PbSnTe photodiodes at 77 K. The experimental data are taken from Refs: 123(o), 126(+), 127(•), 128,129(□), and 130(■). The solid line is calculated for n⁺-p-p⁺ PbSnTe homojunction photodiodes, instead the dashed line is calculated for DLHJ PbSnTe photodiode structures. (From Rogalski, A., and Ciupa, R., *Opto-Electronics Review*, 5, 21–29, 1997. With permission.)

structures. In the short wavelength region the discrepancy between the theoretical curve and experimental data increases, which is due to additional currents in the junctions (such as the generation–recombination current of the depletion region or the surface leakage current) that are not considered. The theoretically calculated curve for DLHJ structures (n$^+$-PbSeTe/p-PbSnTe/n-PbSeTe) is situated above experimentally measured values of the R_oA product, which indicates the potential possibilities of constructing higher quality PbSnTe photodiodes. Up until now however, this type of PbSnTe photodiode structure has not been fabricated.

In Figure 15.20, the R_oA product versus temperature is presented under a 0° FOV for Pb$_{0.80}$Sb$_{0.20}$Te photodiode with cutoff wavelength of 11.8 μm at 77 K [125]. Good agreement between experimental data and theoretical calculations (solid line) has been achieved. However, it should be noticed that for more optimized DLHJ structure, theoretically predicted values of the R_oA product are higher (see dashed line). The increase of R_oA product for DLHJ photodiodes will be more emphasized in the case of higher quality p-type base PbSnTe layer, when contribution of Shockley-Read generation will be suppressed (for higher values of τ_{no} and τ_{po}; in our calculations we assumed $\tau_{no} = \tau_{po} = 10^{-8}$ s). It should be noted that due to inherently higher Auger generation rate in PbSnTe in comparison with HgCdTe, the enhancement of R_oA product of PbSnTe photodiodes is more limited in comparison with HgCdTe photodiodes.

The PbSnTe photodiodes were preferred over PbSnSe photodiodes because high-quality single crystals and epitaxial layers could be fabricated more easily from PbSnTe. However, more recently, rapid advancement in the technology of fabricating monolithic PbSnSe Schottky barriers on Si substrates has been achieved by the research group at the Swiss Federal Institute of Technology [124,133–143].

The theoretically limited parameters of PbSnSe photodiodes were determined in several papers [144–146]. Figure 15.21 shows quantitatively the same type of R_oA product for PbSnSe junctions as for PbSnTe junctions, calculated by Rogalski and Kaszuba [145]. The Auger recombination contribution to R_oA increases with composition x increasing (λ_c increasing). For composition $x \approx 0.08$ $(R_oA)_{GR}$ ($\tau_o = 10^{-8}$ s) and $(R_oA)_A$ values are comparable, but for $x > 0.08$ the R_oA product is determined by Auger recombination. In Figure 15.21, experimental values are also shown that were taken from the literature concerning the Schottky diodes. The comparison between results of the calculations and the experimental data shows potential possibilities for constructing higher quality PbSnSe photodiodes.

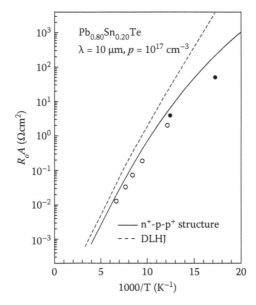

Figure 15.20 Temperature dependence of the R_oA product of Pb$_{0.80}$Sn$_{0.20}$Te photodiode. The experimental data are taken from Refs: 126(o) and 132(•). The solid line is calculated for n$^+$-p-p$^+$ Pb$_{0.80}$Sn$_{0.20}$Te homojunction photodiode, instead the dashed line is calculated for DLHJ Pb$_{0.80}$Sn$_{0.20}$Te photodiode structure. (From Rogalski, A., and Ciupa, R., *Opto-Electronics Review, 5,* 21–29, 1997. With permission.)

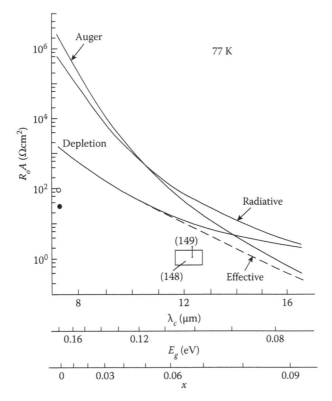

Figure 15.21 The dependence of the R_oA product for the one-sided abrupt $Pb_{1-x}Sn_xSe$ junction (for optimum concentration) on the long wavelength spectral cutoff at 77 K. The experimental data are taken from Refs. 148, 149, 150(o) and 151(•). (From Rogalski, A., and Kaszuba, W., *Infrared Physics*, 21, 251–59, 1981. With permission.)

The Auger recombination in PbSnSe is a less efficient process than in PbSnTe and in this connection $(R_oA)_{APbSnSe} > (R_oA)_{APbSnTe}$ at comparable E_g. It was first noticed by Preier [146] who compared the theoretically determined dependence of R_oA product components on doping concentration in these compounds exhibiting an energy gap of 0.1 eV at 77 K. The results of calculations are plotted in Figure 14.44 and seem to justify efforts to use PbSnSe as a detector material. So far, photodiodes with better performance have been prepared from PbSnTe material.

$PbS_{1-x}Se_x$ ternary alloys have become of some interest as a material for laser diodes with emission wavelengths between 4 and 8 μm [15,29]. This mixed semiconductor also shows promise as a photodetector in the 3–5 μm region at temperatures above 77 K. The R_oA product analysis was carried out by Rogalski and Kaszuba [147]. At 77 K the main contribution to the current flow through the junctions comes from generation–recombination current in the depletion layer. A satisfactory consistency between experimental and theoretical results has been achieved for the effective lifetime in the depletion layer equal to 10^{-10} s. As the temperature increases, the influence of the depletion layer decreases, and instead the influence of Auger recombination increases, especially with increasing composition x (λ_c increasing). At room temperature the Auger process is decisive in a wide range of composition. Only for $x = 0$ are contributions of the diffusion and the depletion currents comparable.

15.3.2 Technology and Properties

A wide variety of techniques have been used to form p-n junctions in lead salts. They have included interdiffusion, diffusion of donors, ion implantation, proton bombardment, and creation of n-type layers on p-type material by vapor epitaxy or liquid phase epitaxy. A summary of works for the fabrication of high-quality p-n junction and Schottky-barrier photodiodes is given in Table 15.3.

Table 15.3: Performance of Lead Salt Photodiodes

Material (Concentration)	Fabrication	T (K)	R_oA (Ωcm^2)	λ_p (μm)	η_p (%)	D^*_λ ($cmHz^{1/2}/W$)	FOV	f (Hz)	Comments	Refs.
n-PbTe/p-Pb$_{0.80}$Sn$_{0.20}$Te	LPE, mesa	82	2.64	≈ 10.5	38	7.3×10^{10}	f/5		Average values for 10-element array	128
n-PbTe/p-Pb$_{0.80}$Sn$_{0.20}$Te	LPE, mesa	80	1.9–4.2	10	15–34	$(1.6–3.2) \times 10^{10}$	2π	1000	Values for 18-element array	127
p-Pb$_{0.80}$Sn$_{0.20}$Te/n-PbTe	LPE, mesa	77	8	≈ 10	≈ 50	6×10^{10}	70	1000	Back side illuminated	159
p-Pb$_{0.76}$Sn$_{0.24}$Te/n-PbTe	LPE, mesa	85	0.75	13		4×10^{10}	2π		Back side illuminated	130
Pb$_{0.79}$Sn$_{0.21}$Te (p $\approx 4 \times 10^{16}$ cm^{-3})	Diffusion, planar	77	1.85	11	40	2.6×10^{10}	2π	1000	Average values for 124-element array	129
Pb$_{0.785}$Sn$_{0.215}$T (p $\approx 1.7 \times 10^{16}$ cm^{-3})	In diffusion, planar	82	2.1	≈ 11		6.3×10^{10}	2π	1000	Average values for 10-element array	128
Pb$_{0.80}$Sn$_{0.20}$Te (p-type)	In diffusion, Mesa	77	21	11	45	2.3×10^{10}	2π	1000	The highest R_oA value	126
Pb$_{0.79}$Sn$_{0.21}$Te (p $\approx 10^{19}$ cm^{-3})	Cd diffusion, mesa	77	5	≈ 11		2.0×10^{10}	2π		n-p$^+$ junction	152
Pb$_{0.80}$Sn$_{0.20}$Te (p $\approx 3 \times 10^{19}$ cm^{-3})	Cd diffusion, mesa	77	2–3	10.6	≈ 30	2.0×10^{10}	2π	800	n-p$^+$ junction	153
PbTe (p $\approx 10^{17}$ cm^{-3})	Sb$^+$ implant, planar	77	2.1×10^4	4.4	40	1.0×10^{12}	0	50		121
PbTe (p $\approx 10^{17}$ cm^{-3})	Sb$^+$ implant, planar	77	1.4×10^3	5.5	55	4.6×10^{11}	0	200	Thin films on BaF$_2$ (MBE)	154
PbTe (p $\approx 10^{17}$ cm^{-3})	M-S(Pb)	80	$(2.5–3.5) \times 10^4$	4.8		$(1.6–1.8) \times 10^{11}$	2π	990	Thin films on BaF$_2$ (MBE), values for 5-element array	155
PbTe (p $\approx 10^{17}$ cm^{-3})	M-S(Pb)	80	$(1.7–25) \times 10^3$	4.6–5.4		$(0.63–1.9) \times 10^{11}$	2π	1000	Values for 6 pinched-off photodiodes	122
PbTe (p $\approx 10^{17}$ cm^{-3})	M-S(Pb)	77	2×10^4	5.5	50	2×10^{12}			MBE epitaxial layers on Si substrate with CaF$_2$-BaF$_2$ bufer layer	124

Material	Structure	T (K)							Comments	Ref.
$PbSe_{0.8}Te_{0.2}$	M-S(Pb)	170	13–100	3.7–4.1	51–85	$(8\text{–}11) \times 10^{10}$	2π	1000	Values for three lateral-collection photodiodes	155
$Pb_{1-x}Sn_xSe$ [$p = (2\text{–}6) \times 10^{17}$ cm^{-3}]	M-S(Pb)	77	0.39–1.6	10.0–11.5	22–53	$(1.5\text{–}5.2) \times 10^{10}$	26–90°	330–1000	$0.062 \leq x \leq 0.70$ thin films on BaF$_2$ (MBE), value for 10 photodiodes	148
$Pb_{1-x}Sn_xSe$ ($p \approx 2 \times 10^{17}$ cm^{-3})	M-S(Pb)	77	1–2	10.6	44	6×10^{10}	20°	510	Thin films on BaF$_2$ (MBE)	149
$PbSe$ ($p = 10^{17}$ cm^{-3})	M-S(Pb)	77	12–88.3	6.1	73	5.7×10^{10}	2π	330	Thin films on BaF$_2$ (MBE)	150
$PbSe$ ($p = 10^{17}$ cm^{-3})	M-S(Pb)		30	6.9	61	2.7×10^{11}	20°	510	Thin films on BaF$_2$ (MBE)	151
$PbS_{0.85}Se_{0.15}$ ($p = 10^{17}$ cm^{-3})	M-S(In)	77	1.5×10^4	4.2	62	9×10^{11}	0	90	Thin films on BaF$_2$ (MBE)	151
$PbS_{0.63}Se_{0.37}$ ($p = 10^{18}$ cm^{-3})	Se$^+$ implant, planar	195	0.7	3.65	30	9×10^9	90°	50		157
		77	5.8–10.3	4.5	36	1.45×10^{11}	90°	50		
PbS ($p \approx 3 \times 10^{18}$ cm^{-3})	Se$^+$ implant, planar	300	0.28	2.55	54	4.8×10^9	90°	100		158
		195	70	2.95	56	1.1×10^{11}	90°	100		
		77	7×10^6	3.40	61	6×10^{11}	90°	100		

PbSnTe photodiodes are the most developed of the lead salt devices, particularly for the 8–14 μm spectral region. Mesa and planar photodiodes are fabricated using standard photolithographic techniques. Performance and stability of the device is especially limited by surface preparation and by the passivation technique that is usually kept proprietary by the producer. It appears that the presence of Pb, Sn, and Te oxides on PbSnTe surfaces almost always produces high leakage currents [160]. The native oxide was found to be an insufficient passivant because the oxidized surfaces contain an unstable TeO_2 [161]. An anodic oxidation is often used for device passivation. The anodic oxide was grown electrolytically from a glycerol-rich solution of water, ethanol, and potassium hydroxide through an anodization/dissolution process [162]. Another type of electrolyte was used by Jimbo et al. [163].

The surrounding atmosphere has a considerable influence on the electrical properties of lead salt materials [164,165]. According to Sun et al. [166] this process is due to the adsorption of oxygen, the diffusion of tin ions from the bulk to the surface, and the oxidation of lead, tin, and tellurium, resulting in the formation of a depletion layer in the n-type samples and an accumulation layer in p-type samples. Photolithographically formed SiO_2 diffusion masks are usually used in the planar PbSnTe photodiodes. About 100 nm of SiO_2 is deposited at a temperature between 340 and 400°C using a silane–oxygen reaction. However, according to Jakobus et al. [167] the PbSnTe surface becomes strongly p-type after coating with pyrolytic SiO_2, therefore rf-sputtered Si_3N_4 was used by them. The junction areas of Schottky-barrier photodiodes are often delineated by windows in a vacuum-deposited layer of BaF_2 [155,168]. The BaF_2 provided a comparable lattice match and the best thermal expansion match to PbSnTe.

Ohmic contacts to n-type regions are usually realized by indium evaporation and to p-type regions by chemical or vacuum deposition of gold. Possible passivating and antireflection surface coatings such as ZnS, Al_2O_3, MgF_2, Al_2S_3, and Al_2Se_3 have been tried [19]. Samples overcoated with As_2S_3 are completely insulated from the effects of oxygen [164]. It was identified that one of the most important factors that limits the performance of small area devices is damage introduced during the bonding of the leads [169]. Therefore, the leads are attached remotely from the sensitive area to obtain high-quality photodiodes.

15.3.2.1 Diffused Photodiodes

Historically, interdiffusion was the first technique used in fabrication of p-n junction in lead salts [5,6], which produced a change in type due to a change in the stoichimetric defects. In PbSnTe, an n-type region was formed in a p-type substrate by diffusion using a metal-rich PbSnTe source at temperatures of 400–500°C for alloy compositions about $x = 0.20$. However, although high-performance devices and detector arrays were produced ($D^* > 10^{10}$ cmHz$^{1/2}$W^{-1} in the 8–14 μm region at 77 K), there were difficulties in obtaining reproducible results. Therefore, detectors have also been made by diffusion of foreign impurities into bulk p-type PbSnTe to form a junction. It was observed that the position or depth of the p-n junction obtained by Al, In, and Cd diffusion varies from slice to slice and tends to drift at temperatures over 100°C [118]. Moreover, the junction diffusion rate increased with decreasing hole concentration in the substrate. Despite the above technological problems, high quality photodiodes formed by diffusing Cd [152,153] or In [126,128,170] into p-type PbSnTe crystals have been fabricated.

According to Wrobel [170], n-p$^+$ junctions should be formed by diffusion of Cd into as-grown $Pb_{1-x}Sn_xTe$ ($x \approx 0.20$) single crystals at a temperature of 400°C for 1.5 hour Cd-diffusion has also been carried out in a two-temperature zone-furnace (specimen at 400°C, source up to 250°C with a 2% Cd in In alloy as a diffusion source) [153,171]. During this process the p$^+$-type material is converted to n-type with a carrier concentration of about 10^{17} cm^{-3}, which is optimum for photodiodes with cutoff wavelengths at about 12 μm. The junction depth is smaller than 10 μm after 1 hour diffusion time. Upon comparing the measured R_oA product with calculated values, approximate values for the lifetimes within depletion layers and for the minority carrier lifetimes within n-type regions have been established. The carrier lifetime at 77 K is determined by Schockley-Read centers and is lower than 10^{-9} s [153]. This value is consistent with carrier lifetimes that were determined from photoconductivity and PEM measurements in homogeneously Cd-doped samples [72,172].

Indium diffusion into p-type PbTe [123] and p-type $Pb_{1-x}Sn_xTe$ ($x \approx 0.20$) [169] has been used to make high-quality photodiodes. Suitable PbSnTe material with an optimum hole concentration of about 10^{16}–10^{17} cm^{-3} can be made only by low temperature growing methods, for instance by LPE

[128], or by annealing in controlled atmospheres [169] in the case of other growing methods. The p-type PbTe material with a hole concentration of 4×10^{17} cm^{-3} can be grown by LPE, if the growth melt is doped with arsenic [123]. A procedure for indium deposition and diffusion is described by Lo Vecchio et al. [169].

In 1975 Chia et al. [128] and DeVaux et al. [173] described results from indium diffused planar arrays of $Pb_{1-x}Sn_xTe$ ($x \approx 0.20$) homojunctions. The diodes of lower base carrier concentrations had a larger R_oA product. From measurements of the forward bias I-V characteristics and from measurements of the R_oA temperature dependence they conclude that bulk diffusion currents limit the R_oA product at temperatures above about 70 K. The measured average R_oA product of a 10-element array equal to 3.8 Ωcm^2 at 78 K is consistent with a minority carrier lifetime of 1.7×10^{-8} s. This array had a $Pb_{0.785}Sn_{0.215}Te$ substrate with 1.7×10^{16} cm^{-3} base hole concentration. At lower temperatures the R_oA product tends to saturate, which was attributed to surface leakage. Figure 15.22 shows the spectral dependence of detectivity of two photodiodes of a 10-element array. Except for some scatter at low wavelength, the curves are indistinguishable. The oscillatory nature of the spectral response is due to the antireflection coating action of the insulator. At 80 K with 10 mV reverse bias, an average D^* of 1.1×10^{11} cmHz$^{1/2}$W^{-1} ($f/5$ FOV) was measured ($\eta = 79\%$).

Wang and Lorenzo [129] have also fabricated high-quality planar devices by impurity diffusion into low-concentration ($p = 4 \times 10^{16}$ cm^{-3} or less) LPE layers. As a diffusion mask they used an oxide layer. At 77 K the average R_oA product was 1.85 Ωcm^2 for an array of 124 elements each with an area 2.5×10^{-5} cm^2. The average quantum efficiency without antireflection coating was 40% and typical peak detectivity of 2.6×10^{10} cmHz$^{1/2}$W^{-1} was measured at 11 μm in a 2π FOV. A striking feature of the arrays was the extreme uniformity in spectral response, the λ_c values varied by less than 1%. The above technique is suitable for the routine fabrication of high-density planar PbSnTe arrays of high quality.

More recently, John and Zogg [174] manufactured p-n PbTe junction photodiode by overgrowth of a Bi doped n$^+$ cap layer onto p-type base layer. Generally, the p-n junction works well with tellurides, while selenides exhibit a too high diffusion in order to obtain reliable devices.

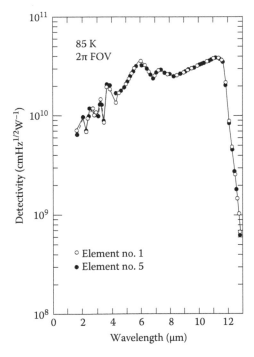

Figure 15.22 Detectivity as a function of wavelength for two photodiodes of a $Pb_{0.785}Sn_{0.215}Te$ array. (From Chia, P. S., Balon, J. R., Lockwood, A. H., Randall, D. M., and Renda, J. J., *Infrared Physics*, 15, 279–85, 1975. With permission.)

15.3.2.2 Ion Implantation

Most of the implant work in IV-VI semiconductors has been devoted to the formation of n-on-p junction photodiodes, mainly covering the 3–5 μm spectral region.

In early works proton bombardment was used to convert p-type PbTe and PbSnTe to n-type [175]. The electron concentration in the n-type layer was about 10^{18} cm^{-3}. Wang et al. [176,177] studied the electrical and annealing properties of proton bombardment of the $Pb_{0.76}Sn_{0.24}Te$ layer. They found that the defects were annealed out at temperatures around 100°C, changing the conductivity back to p-type. Good devices were obtained without knowing details about doping, lattice damage, and annealing. More recently, Donnelly [178] and Palmetshofer [179] have reviewed the investigation and understanding of ion implantation in IV-VI semiconductors.

The best diodes in some of the lead salts were reported by Donnelly et al. [121,157,158]. These diodes were fabricated by Sb$^+$ implantation. It was found that Sb becomes fully activated at an annealing temperature of about 300°C. The Sb$^+$ ion implantation was carried out using a 400 keV beam and a total dose of $(1–2) \times 10^{14}$ cm^{-2}. Following implantation, the photoresist implantation mask was removed and each sample was coated with a 150 nm layer of pyrolitic SiO_2 for 2–5 minutes at a temperature between 340 and 400°C. This annealing step is sufficient to effectively anneal out the radiation damage caused by room temperature implantation. Holes were then opened in the oxide and Au contacts were electroplated on the sample. The β parameter in the forward bias I-V characteristics had a value of about 1.6, while the reverse characteristics showed a soft power-law breakdown due to tunneling. PbTe photodiodes had an R_oA product at 77 K as high as 2.1×10^4 Ωcm^2 and the detectivity limited by background [121]. With a reduced background (77 K shield) at 50 Hz, the measured detectivity was 1.6×10^{12} cmHz$^{1/2}$W^{-1}, which is slightly below the theoretical amplifier plus thermal noise limit. The peak detectivity $\lambda_p = 4$ μm and the cutoff wavelength $\lambda_c = 5.1$ μm are shifted toward a shorter wavelength region, due to band filling in the highly doped p-type substrate ($p > 10^{18}$ cm^{-3}).

Implantation of In seems to be less appropriate for p-n junction fabrication, because of the low electrical activity and a saturation of the carrier concentration at higher doses [179].

Implantation of the constituent elements for the photodiode preparation gave good results in PbSSe diode fabrication [157].

15.3.2.3 Heterojunctions

An alternative technology adopted for the preparation of long wavelength photodiodes is the use of heterojunctions of n-type PbTe (PbSeTe) deposited onto p-type PbSnTe substrates by: LPE [8,128,130,132,159,180–182], VPE [183,184], MBE [185], and HWE [186].

The best results have been obtained using the LPE method. Either type of material with carrier concentrations of 10^{15}–10^{17} cm^{-3} can be grown and used without annealing. Using the LPE method an advanced concept in PbSnTe photodiode design with the fabrication of back side illuminated diodes has been successfully realized (see inset in Figure 15.23 [181]). It permits complete optical utilization of the electrical area of the photodiode, significantly reduces the optical dead area of an array, and increases the optically sensitive area of the diode because of the refractive index mismatch between PbTe and air. Moreover, wider energy gap material on one side of the junction results in reduction of the saturation current. Figure 15.23 shows the spectral response characteristics of the back side illuminated n-PbTe/p-PbSnTe heterojunction at 77 K. Filtering to 6 μm is caused by absorption in the PbTe substrate.

It should be noticed that the spectral response of LPE-grown PbTe/PbSnTe heterojunctions depends on the PbTe growth temperature. Migration of the p-n junction away from the PbTe-PbSnTe interface is possible during epitaxial growth due to the quite large interdiffusion coefficients of the native defects in PbTe and PbSnTe. The p-n junctions shifts into the PbTe layer during PbTe growth at $T > 480$°C [187,188]. A pure PbTe spectral response of n-PbTe/p-PbSnTe heterostructures obtained by HWE technique has been observed elsewhere [189,190]. More recently published papers have indicated that the spectral response of heterojunctions depends on an electric field in the PbSnTe layer [191] and on the relation between the n and p concentrations [192].

Lattice mismatch between PbTe and $Pb_{1-x}Sn_xTe$ (0.4% for $x = 0.2$) introduces strain-relieving misfit dislocations that may be important in determining the performances of photodiodes [193]. Kesemset and Fonstad [194] have found the relatively high interface recombination velocity ($\approx 10^5$ cm/s) in the active region of $Pb_{0.86}Sn_{0.14}Te$/PbTe double-heterostructure laser diodes.

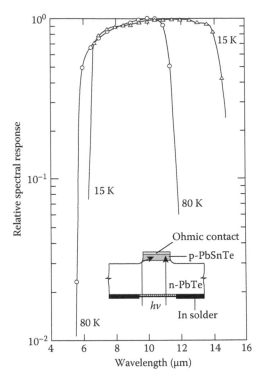

Figure 15.23 Spectral response of n-PbTe/p-PnSnTe back side illuminated mesa photodiode at 77 K. (From Wang, C. C., Kalisher, M. H., Tracy, J. M., Clarke, J. E., and Longo, J. T., *Solid-State Electronics*, 21, 625–32, 1978. With permission.)

With decreasing lattice mismatch the interface recombination also decreases. The lattice-matched PbSnTe/PbTeSe configuration offers a promising solution to the problem of mismatch arising in this system [195]. It was confirmed for n-PbTeSe/p-PbSnTe photodiodes operated at temperatures below 77 K [132].

Wang and Hampton [127] reported high-quality n-PbTe/p-Pb$_{0.80}$Sn$_{0.20}$Te mesa arrays produced by using an LPE wiper technique. The p-type Pb$_{0.80}$Sn$_{0.20}$Te were first grown on the (100) Pb$_{0.80}$Sn$_{0.20}$Te single crystals followed by an n-type PbTe growth. The Pb$_{0.80}$Sn$_{0.20}$Te epilayer was grown at temperatures from 540°C to about 500°C with a cooling rate from 0.1 to 0.25°C/min (typical thickness was 20 μm after about 3 hour of growth). The PbTe layer was started at the temperature where the Pb$_{0.80}$Sn$_{0.20}$Te layer growth stopped. The thickness of the PbTe layer was about 5 μm after 40 min of growth. At 80 K the average R_oA product for an 18-element array was 3.3 Ωcm^2 and the mean D^* at $\lambda_p = 10$ μm was 2.3×10^{10} cmHz$^{1/2}$W^{-1} for 2π FOV and 300 K background.

Studies of the electrical properties of n-Pb$_{0.80}$Sn$_{0.20}$Te/n-PbTe inverted heterostructure diodes have been reported by Wang et al. [181]. Typical forward and reverse I–V characteristics at various temperatures are given in Figure 15.24. In this figure, continuous curves represent the measured values, and the dashed curves represent the theoretical values. Diffusion current is predominant at temperatures above 80 K. The activation energy in the diffusion dominated region is 0.082 eV. As the temperature decreases g-r current becomes important and the activation energy reduces to 0.044 eV. The values of activation energy are roughly equal to E_g and $E_g/2$, respectively, at $T = 15$ K. The g–r current is predominant in the temperature range $30 < T < 80$ K. Below 30 K, the surface related leakage current becomes evident for the small area diodes ($A < 10^{-4}$ cm^2). This current increases as the reverse-bias voltage increases. The excess leakage current of large area diodes is probably conditioned by bulk defects.

The LPE technique can be used in the fabrication of two-color PbTe/PbSnTe detectors. PbTe/PbSnTe heterostructures were proposed that cover both (3–5 μm and 8–14 μm) atmospheric windows [180]. Planar p-n junctions have been made using an indium diffusion technique. The PbTe

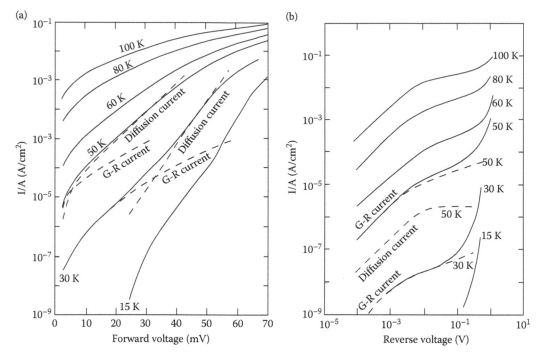

Figure 15.24 (a) Forward and (b) reverse I-V characteristics of p-Pb$_{0.80}$Sn$_{0.20}$Te/n-PbTe inverted heterostructure diode. (From Wang, C. C., Kalisher, M. H., Tracy, J. M., Clarke, J. E., and Longo, J. T., *Solid-State Electronics*, 21, 625–32, 1978. With permission.)

diodes had an average D^* of 10^{11} cmHz$^{1/2}$W^{-1} at $\lambda_p = 4.6$ μm, and the PbSnTe diodes had an average of 2.2×10^{10} cmHz$^{1/2}$W^{-1} at their 9.8 μm peak.

15.4 SCHOTTKY-BARRIER PHOTODIODES

It has been pointed out [120,196,197] that in the case of Schottky-barrier photodiodes with p-type lead salt semiconductors, the effective barrier height ϕ_b is independent of metal work function. In the case of moderate carrier concentrations ($\approx 10^{17}$ cm^{-3}) ϕ_b does not appreciably exceed the energy gap. Assuming $\phi_b = E_g$ the expression (Equation 9.151) takes the form

$$\left(R_oA\right)_{MS} = \frac{h^3}{4\pi q^2 kT}\frac{1}{m^*}\exp\left(\frac{E_g}{kT}\right).$$

(15.19)

15.4.1 Schottky-Barrier Controversial Issue

The experimental values of the R_oA product for the Schottky junctions are consistent with the theoretical curves within the higher temperature range, while at 77 K deviations from the values calculated are very large [117]. This divergence seems to be due to the use of Equation 15.19, which does not take account of additional processes occurring in Schottky junctions with p-type narrow gap semiconductors. Despite the fact that Gupta and colleagues [144], using the above-mentioned relationship, achieved good agreement with experimental data for ternary compounds PbSnTe and PbSnSe, their results seem to be accidental for the energy gap equal to 0.1 eV. The scheme of energy bands for such a junction has been proposed by Walpole and Nill [196] and is shown in Figure 15.25, where three regions may be distinguished: inverted, depleted, and bulk. In the ideal junction model only processes (a); that is, the hole emission from the Fermi level in metal to the valence band for $hv = \phi_b$, are considered. No account is taken of the excitation of hole-electron pairs in the inverted region [processes (b)], or of the band-to-band excitation of hole-electron pairs in the depleted region [processes (c)]. The last excitations are of particular importance, because the depleted region is wide due to a high dielectric constant. According to

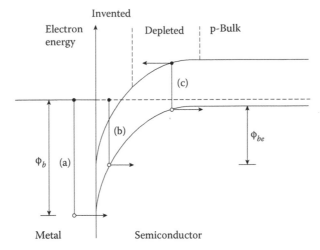

Figure 15.25 Schematic energy band diagram for Schottky barrier with a narrow gap p-type semiconductor. (From Walpole, J. N., and Nill, K. W., *J. Applied Physics*, 42, 5609–17, 1971. With permission.)

Nill and colleagues [156], the barrier height ϕ_b for holes is considerably lowered to take the value ϕ_{be} slightly exceeding the energy gap E_g. Since the hole of kinetic energy slightly exceeds E_g the narrow top of the barrier is transparent due to tunneling effects, the effective barrier ϕ_{be} is for the majority of metals independent of the work function of metal. Also the interaction of metals during the contact formation (even at room temperature) alters the electronic properties and the heights of barriers.

The experimental values of the R_oA product for Schottky junctions with PbTe are comparable with those obtained for p-n junctions at 77 K [117]. Since in the last junctions the R_oA product is determined by generation–recombination processes it should be concluded that for M-S junctions an essential contribution to the R_oA comes from the depletion region. Several researchers [133,155,198,199] observed two saturation current generation mechanisms for Pb barrier lead salt photodiodes: diffusion-limited and depletion-limited behavior. In these papers, a tendency for R_oA saturation at lower temperatures has been explained by a weak temperature-dependent shunt resistance. Instead Maurer [200] explained a temperature dependence of the R_oA product of Pb-PbTe Schottky barriers assuming simple thermionic emission theory and band-to-band tunneling.

Up to the present, knowledge of the metal-semiconductor interface in IV-VI materials is quite fragmentary, which leads to uncertainly about the nature of the p-n junction in metal-barrier photodiodes. However, we cannot exclude the possibility that a shallow diffused n^+-p junction is formed owing to changes in the stoichiometry of the IV-VI layer surface caused by the metal layer. Likewise, Chang et al. [201] and Grishina et al. [202] showed experimentally that evaporated films of indium on PbTe and PbSnTe chemically react at room temperature to give a structure with an intermediate layer between the metal and semiconductor.

Sizov et al. [197] evaluated the activity of the metal with PbTe using the bulk data on standard enthalpies and entropies of the tellurides formation:

Ga, Zn, Mn, Ti, Cd, In, **PbTe**, Mo, Sn, Ge, Cu, Tl, Pt, Ag, Hg, Bi, Sb, As, Au.

The metals to the left of PbTe during the contact formation on the PbTe surface must interact with the semiconductor even at room temperature, forming a new species that must alter the electronic properties and the heights of barriers. The metals to the right of PbTe will not interact with the semiconductor near room temperature.

On account of the above divergence, different methods of Schottky-barrier preparation on IV-VI compounds have been elaborated. Initially specimens were cooled (nominally to 77 K) during Pb deposition [196]. According to a U.S. patent [203] the metal layer (In, Pb) should be evaporated in high vacuum immediately after the epitaxial layer has been cooled to room temperature, without breaking vacuum. A layer of SiO_2 is then deposited by thermal *in situ* evaporation to protect the metal-semiconductor contact from the atmosphere. A contrary method of Schottky-barrier preparation has been described by Buchner et al. [204]. They showed that deposition of Pb onto a $Pb_{0.8}Sn_{0.2}Te$ surface gave a rectifying barrier only if the surface had been previously exposed to the atmosphere. It was interpreted by the presence of oxygen at the surface of PbSnTe that prevents the migration of Sn across the interface. Grishina et al. [205] noticed that improvement of the electrical characteristics of Pb(In)-$Pb_{0.77}Sn_{0.23}Te$ Schottky barriers is caused by the influence of native oxide that prevents interaction between the metal and semiconductor. More recently it has been demonstrated that the usage of a thin chemical oxide as an intermediate layer between metal and semiconductor improves the Schottky-barrier properties while anodic oxide passivation raises thermal and temporal stability as well as lowers the scattering of parameters from element to element in multielement linear arrays [206].

Pb-PbTe rectifying contacts were obtained with both clean and air-exposed surfaces [207]. In contrast to $Pb_{0.8}Sn_{0.2}Te$, the oxidation of PbTe tends to saturate at about monolayer coverage (continuing oxidation of PbSnTe is accompanied by diffusion of Sn from the bulk to the surface). Thermal stability of Pb-PbTe devices has been obtained by washing with pure water and then vacuum baking at 150°C for periods of up to 12 hours [155]. Schoolar et al. [149,151] prepared Schottky-barrier PbSnSe and PbSSe photodiodes using vacuum annealed (at 170°C for 30 min) epitaxial films to desorb a surface oxide layer and cooling the films to 25°C prior to depositing the lead or indium barriers. It was also found that the presence of chlorine in the interface vastly improved I–V characteristics of Schottky junctions [208–210]. Further studies of interfaces are necessary to resolve the present controversial issues concerning Schottky-barrier formation.

A new insight into theory of Schottky-barrier lead salt photodiodes has been presented by Paglino et al. [139]. By using the Schottky-barrier fluctuation model introduced by Werner and Güttler [211], it is assumed that the Schottky-barrier height ϕ_b has a continuous Gaussian distribution σ around a mean value ϕ. Due to exponential dependence of saturation current (see Equation 9.149) on ϕ_b, it follows that the effective barrier responsible for current is given by

$$\phi_b = \phi - \frac{q\sigma^2}{2kT}. \tag{15.20}$$

This barrier ϕ_b is smaller than the mean value ϕ that is derived from capacitance-voltage characteristics. Therefore, since ϕ_b depends on temperature, no straight line is obtained in the Richardson plot. In order to get consistent results with I-V characteristics at $V \neq 0$, Werner and Güttler showed that the changes of the barrier $\Delta\phi$ and square of the barrier fluctuation $\Delta\sigma^2$ with the applied bias voltage V are proportional to this voltage,

$$\Delta\phi_b(V) = \phi_b(V) - \phi_b(0) = \rho_2 V \quad \Delta\sigma^2(V) = \sigma^2(V) - \sigma^2(0) = \rho_3 V, \tag{15.21}$$

where ρ_2 and ρ_3 are negative constants. These settings lead to a temperature dependence of the ideality factors β

$$\frac{1}{\beta} - 1 = -\rho_2 + \rho_3 \frac{q}{2kT}, \tag{15.22}$$

which has been observed with numerous Schottky barriers fabricated with different materials over a large temperature range [211].

A plot of the R_oA product versus temperature for Pb-PbSe Schottky-barrier photodiode on Si substrate is shown in Figure 15.26 [139]. A near perfect fit is obtained over the whole temperature range. The fluctuation σ leads to the saturation of the J_{st} or R_oA product at low temperature. For Pb-PbSnSe Schottky barriers these fluctuations with an assumed Gaussian distribution with a width σ of up to 35 meV. The values depend on the structural quality; higher quality devices show lower σ. The ideality factors β are correctly described with the model even if $\beta > 2$. It is expected that the barrier fluctuation is caused by threading dislocations; lower densities of these dislocations lead to lower σ and higher saturation R_oA values at lower temperatures. The dislocation densities for PbSe were in the 2×10^7 to 5×10^8 cm^{-2} range for the 3–4 µm thick as grown layers.

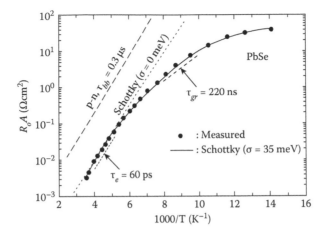

Figure 15.26 Resistance area product R_oA versus inverse temperature for Pb-PbSe Schottky-barrier photodiodes on Si substrates. The values are fitted with the barrier fluctuation model (solid line). For comparison, the values calculated for the ideal Schottky-barrier photodiode are plotted, as well as the values from the p-n theory in the diffusion case for band-to-band recombination limited lifetime $\tau_{bb} = 0.3$ μs as well as for a recombination lifetime $\tau_e = 60$ ps and g-r limited lifetime $\tau_{gr} = 220$ ns. (From Paglino, C., Fach, A., John, J., Müller, P., Zogg, H., and Pescia, D., *Journal of Applied Physics*, 80, 7138–43, 1996. With permission.)

Higher R_oA products are obtainable by lowering these densities by thermal annealing [140], which sweeps the threading ends of the misfit dislocations over appreciable distances (in the cm range) to the edges of the sample [32].

15.4.2 Technology and Properties

In comparison with p-n junction preparation, a considerably simpler technique for Schottky-barrier photodiode fabrication involves evaporation of a thin metal layer onto the semiconductor surface. It is a planar technology with the potential for cheap fabrication of large arrays. This technique has been successfully applied to IV-VI compounds. An excellent review of Schottky-barrier IV-VI photodiodes with emphasis on thin film devices has been given by Holloway [155].

In most cases the Schottky barriers of IV-VI semiconductors are made by evaporation of Pb or In metals. If bulk crystal semiconductor material is used, the photodiodes are front side illuminated through semitransparent electrodes [167,212]. Since 1971, when the first high-performance, thin-film photodiodes were reported by the Ford group [213], the devices have been back side illuminated through BaF$_2$ substrates. Development of these devices is presented by Holloway [155]. The thin-film IV-VI photodiodes are made of epitaxial layers of lead salts using HWE or MBE. These films are p-type with hole concentrations of about 10^{17} cm^{-3}. Development of the MBE technique in the last two decades offers the possibility of growing lead salt epitaxial layers on silicon substrates with an appropriate buffer layer, which is stacked CaF$_2$-BaF$_2$ or CaF$_2$-SrF$_2$-BaF$_2$ of only $\approx$ 200 nm total thickness [124] and of a quality sufficient for infrared device integration. This opens the way to fully monolithic, heteroepitaxial FPAs. Figure 15.27 gives four main configurations in which Schottky-barrier photodiodes are fabricated.

The high-performance, thin-film IV-VI photodiodes were developed in the 1970s by two U.S. research groups at the Ford Motor Company and at the Naval Surface Weapons Center. Pb-(p)PbTe photodiodes prepared in stringent conditions of cleanliness had an R_oA product of 3×10^4 Ωcm^2 at 80–85 K, which together with the quantum efficiency of 90% influenced by interference corresponds to a Johnson-noise-limited $D^* = 10^{13}$ cmHz$^{1/2}$W^{-1} for a peak wavelength near 5 μm [155]. The best photodiodes had the region of contact between the semiconductor and the Pb layer defined by a window in an evaporated layer of BaF$_2$. This configuration of photodiodes has been used in the fabrication of 100-element arrays. The photolithographic methods, included the BaF$_2$ window and delineation techniques for the semiconductor and the Pb and Pt metallization, are briefly described by Holloway [155].

In the last two decades the research group at the Swiss Federal Institute of Technology has continued to pursue thin-film IV-VI photodiode technology and has made significant

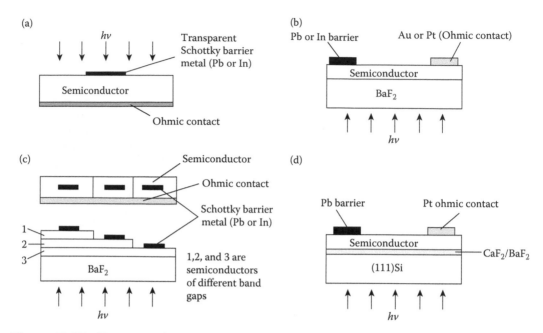

Figure 15.27 Various configurations of Schottky-barrier photodiodes: (a) conventional photodiode front side illuminated, (b) back side illuminated photodiode, (c) multispectral (three-color) detector, and (d) heteroepitaxial photodiode on Si substrate.

progress [9,113,124,138,140,142,143,174]. In 1985 Zogg and Huppi [214] proposed a new monolithic heteroepitaxial PbSe/Si integrated circuit obtained by epitaxial growth of PbSe films onto Si using a (Ca,Ba)F$_2$ buffer layer. The lattice mismatch of $\approx 14\%$ between Si and BaF$_2$ (see Figure 15.2) can be overcome using graded layers consisting of CaF$_2$ at the Si and BaF$_2$ at the IV-VI alloy interface. Thermal expansion coefficient mismatch between the IV-VI alloys and the fluorides (about 20×10^{-6} K^{-1} at 300 K) and silicon (2.6×10^{-6} K^{-1} at 300 K) [215] was not found to be detrimental. It appears that this thermal expansion mismatch relaxes through a glide of dislocations, and the layers survive more than 800 thermal cycles (between 300 and 80 K) without problems [138]. By the development of this technology, considerable improvement in fully monolithic lead salt FPAs has been achieved using different binary and ternary IV-VI alloys (PbS [216,217], PbSe [135,218,219], PbTe [174,199,218], PbEuSe [216], and PbSnSe [133,136,138]). An epitaxial stacked CaF$_2$-BaF$_2$ buffer layer of ≈ 200 nm thickness was usually used.

The fabrication process of PbSnSe photodiodes starts with a layer deposition in the second growth chamber at 350°C–400°C onto a CaF$_2$ buffer on a 3 inch Si(111)-wafer. Typical carrier concentration in p-type layers is $(2–5) \times 10^{17}$ cm^{-3}; thickness of the layers is 3–4 µm. Growth with (111) orientation is preferred, since thermal mismatch strain relaxation occurs by dislocation glide in the main {100} < 110 > glide system with the glide planes inclined to the surface. When grown on (100) oriented substrates, thermal mismatch strain relaxation has to occur via higher glide systems, which normally leads to cracking of the layers as soon their thickness exceeds about 0.5 µm [141,220]. Dislocation densities down to 10^6 cm^{-2} have been obtained in unprocessed layers of a few µm thickness, while for layers used to fabricate detectors, dislocation densities range from the low 10^7 to low 10^8 cm^{-2} range. The further basis steps in the device fabrication are as follows:

- vacuum deposition of ≈ 200 nm Pb *ex situ* or *in situ* in the MBE growth chamber with the sample held at room temperature,

- overgrowth with ≈ 100 nm vacuum-deposited Ti for protection of the Pb,

- vacuum deposition of Pt for better adhesion of the following Au-layer, and delineation of the Pt,

- definition of sensitive areas: etching of Ti with a selective Ti-etchant, etching of Pb with a selective Pb-etchant,

- electroplating of Au for ohmic contacts and on top of the Pb/Ti/Pt blocking contacts,

- etching of the PbSnSe layer,

- spin-on, photolithographic patterning and curve of a polymide for insulation and planarization, and

- vacuum deposition of Al and delineation of the fan-out by wet etching.

The active areas are 30–70 μm in diameter. Illumination is from the back side through the infrared transparent Si-substrate. Figure 15.28 shows a schematic cross section of a device [140]. No surface passivation was needed, surface effects were rather negligible as deduced from the R_oA-values diodes with different sizes.

The spectral response curves of PbTe and $Pb_{0.935}Sn_{0.065}Se$ photodiodes on Si substrates are shown in Figure 15.29 [124]. Typical quantum efficiencies are around 50% without AR coating. Note the pronounced interference effects in the case of $Pb_{0.935}Sn_{0.065}Se$ photodiode, which help to increase considerably the response near the peak wavelength at 50 K but that lead to some decrease near the cutoff at 77 K. In order to optimize the peak response at a specific temperature, one has to optimize the thickness of PbSnSe as well as that of the buffer layer in order to obtain constructive interference in the active region.

Figure 15.30 shows R_oA products at 77 K as a function of cutoff wavelength for different lead salt photodiodes on Si with stacked BaF_2/CaF_2 and CaF_2 buffer layers. Although these values are considerably above the BLIP limit (for 300 K, 2π FOV and $\eta = 50\%$), they are still significantly below theoretical limit given by Auger recombination (see Figure 15.21). The performance of IV-VI photodiodes is inferior to HgCdTe photodiodes; their R_oA products are two orders of magnitude below

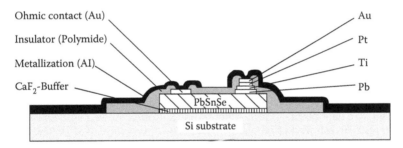

Figure 15.28 Cross section of PbSnSe photovoltaic infrared detector. (From Fach, A., John, J., Müller, P., Paglino, C., and Zogg, H., *Journal of Electronic Materials*, 26, 873–77, 1997. With permission.)

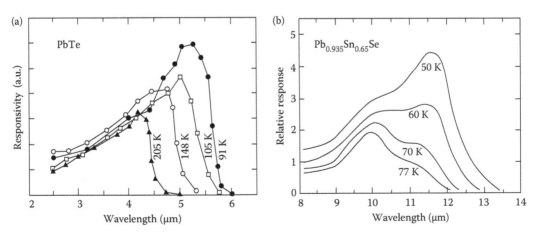

Figure 15.29 Spectral responses of lead salt photodiodes on Si substrates at different temperatures: (a) PbTe photodiode, and (b) $Pb_{0.935}Sn_{0.065}Se$ photodiode. (From Zogg, H., Blunier, S., Hoshino, T., Maissen, C., Masek, J., and Tiwari, A. N., *IEEE Transactions on Electron Devices* 38, 1110–17, 1991. With permission.)

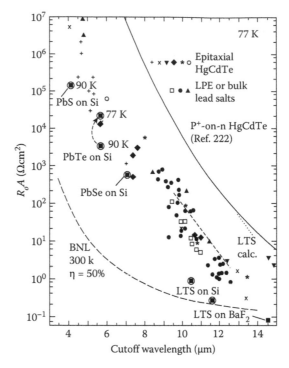

Figure 15.30 Experimental R_oA products at 77 K versus cutoff wavelength for different lead salt photodiodes in comparison with HgCdTe photodiodes. The 300 K background noise limit (BNL) for 180° FOV and 50% quantum efficiency is included. The dotted line labelled LTS is the calculated ultimate value for PbSnSe. (From Rogalski, A., and Piotrowski, J., *Progress in Quantum Electronics,* 12, 87–289, 1988. With permission.) The broken line is the upper limit of numerous experimental HgCdTe data points. (From Ameurlaine, J., Rousseau, A., Nguyen-Duy, T., and Triboulet, R., "(HgZn)Te Infrared Photovoltaic Detectors," *Proceedings of SPIE* 929, 14–20, 1988. With permission.) Solid line represents calculated data for p-on-n HgCdTe photodiodes. (From Rogalski, A. and Ciupa, R., *Journal of Applied Physics* 77, 3505–12, 1995. With permission.) (From Zogg, H., Maissen, C., Masek, J., Hoshino, T., Blunier, S., and Tiwari, A. N., *Semiconductor Science and Technology* 6, C36–C41, 1991. With permission.)

values for p^+-on-n HgCdTe. R_oA products of PbSnSe photodiodes with a cutoff wavelength of 10.5 μm are about 1 Ωcm² at 77 K.

Figure 15.31 shows the temperature dependence of the R_oA product for n^+-p $Pb_{0.935}Sn_{0.065}$Se photodiode [119]. For comparison in this figure, also included are experimental data taken from Zogg and colleagues [133]. For temperatures down to 100 K, the increase in R_oA follows a diffusion-limited behavior as revealed by the approximately $\exp(E_g/kT)$ slope. In the range between 100 and 77 K, the behavior may be attributed to depletion layer generation–recombination mechanism. Theoretical prediction for a back side ohmic contact of the photodiode (dashed curve in Figure 15.31) indicates considerable decreasing of the R_oA product. This structure of photodiode is inferior since the minority carrier diffusion length in base p-type region is larger than the thickness of the photodiode. Results of calculations presented in Figure 15.31 indicate on potential possibilities of constructing higher quality PbSnSe photodiodes. In the region of higher temperature, where the experimental dependence of $R_oA(T)$ follows a diffusion-limited behavior, the dominant generation–recombination mechanism is not fundamental band-to-band mechanism (radiative or Auger processes). It means that diffusion-limited behavior of photodiode is determined by SR generation–recombination mechanism with values of τ_{no} and τ_{po} below 10^{-8} s (which was assumed in calculations). It is probably caused by relatively high trap concentrations, which dominate the excess carrier lifetime in p-type material by the SR generation–recombination mechanism.

It was found that the maximal R_oA products of PbSnSe photodiodes are determined by the dislocation densities. It is shown in Figure 15.32 [143], where the saturation values of the R_oA

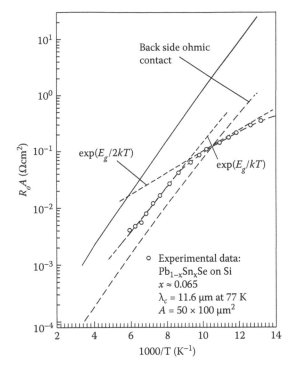

Figure 15.31 The temperature dependence of the R_oA product for n$^+$-p Pb$_{0.935}$Sn$_{0.065}$Se photodiode. The solid line is calculated for back side illuminated photodiode structure with 3 μm thick p-type base layer with carrier concentration 10^{17} cm^{-3} and 0.5 μm thick n$^+$-cap layer with electron concentration 10^{18} cm^{-3}. Instead the dashed line is calculated assuming a back side ohmic contact of the photodiode. The experimental data (o) are taken from Zogg[133]; the diffusion and depletion regimes are indicated by thin dashed lines. (From Rogalski, A., Ciupa, R., and Zogg, H., "Computer Modeling of Carrier Transport in Binary Salt Photodiodes," *Proceedings of SPIE* 2373, 172–81, 1995. With permission.)

products for various PbSnSe photodiodes are plotted versus dislocation density n. We can see that an inverse linear dependence is observed: $R_oA \propto 1/n$. The theoretical ideal leakage resistance according to the Schottky theory without barrier fluctuation is R_{id}. If N dislocations end in the active area A of the device, and if each dislocation gives rise to an additional leakage resistance R_{dis}, the measured differential resistance is [140]

$$\frac{1}{R_{ef}} = \frac{1}{R_{id}} + \frac{N}{R_{dis}},$$

(15.23)

and for $R_{id} \gg R_{dis}/N$

$$R_oA = \frac{R_{disl}}{n},$$

(15.24)

since $N = nA$. Each dislocation therefore causes a shunt resistance. Its value is 1.2 GΩ for a PbSe photodiode at 80 K. The fluctuations σ in the phenomenological model of Werner and Güttler (see Equation 15.20) are therefore explained as due to barrier lowering caused at and in the vicinity of dislocations. Since each dislocation causes a leakage resistance of about 1.2 GΩ, it follows that the dislocation density should be below 2×10^6 cm^{-2} in order that dislocations do not dominate the actual R_oA product of PbSe photodiodes (the theoretical R_oA value of a PbSe Schottky diode at 80 K is about 10^3 cm^2). Dislocation densities in range are obtainable after proper treatment of the layers (after some temperature cycles from room temperature to 77 K and back to room temperature) as was described by Zogg and colleagues [32,143].

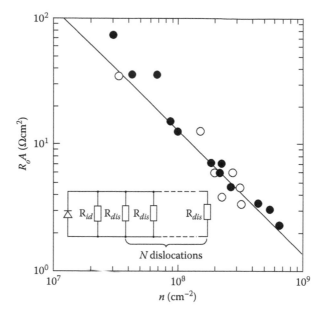

Figure 15.32 Low temperature saturation R_oA values of PbSnSe photodiodes as a function of dislocation density. The influence of the composition is normalized. Model of the influence of dislocations on the leakage resistance in the insert. (From Zogg, H., "Photovoltaic IV–VI on Silicon Infrared Devices for Thermal Imaging Applications," *Proceedings of SPIE* 3629, 52–62, 1999. With permission.)

Much improvement is still possible by improving the material quality and device fabrication technique. Better results should be obtained using buried p-n junctions with a thin, wider bandgap cap layer rather than blocking Pb contacts.

15.5 UNCONVENTIONAL THIN FILM PHOTODIODES

In practice the response speed of a photodiode is determined by the effects of junction capacitance, dynamic resistance and series resistance together with external circuit impedance (see Section 9.2.3). In the case of lead salt photodiodes, the only significant capacitance is the junction space-charge region capacitance C (due to a high dielectric constant) and the only significant resistance is the external resistance R_L; then the upper-frequency limit f_c is [223]

$$f_c = \frac{1}{2\pi R_L C}. \tag{15.25}$$

For the abrupt junction we can take $C = \varepsilon_s \varepsilon_o A/w$ and $w = [2\varepsilon_s \varepsilon_o (V_b - V)/q N_{ef}]^{1/2}$ where $N_{ef} = N_a N_d/(N_a + N_d)$ is the effective doping concentration in the space-charge region, and V_b is the built-in junction voltage.

Holloway and coworkers [122,224–226] developed two unconventional photodiodes that gave reduced junction capacitance (Figure 15.33). One approach uses the pinched-off photodiode (Figure 15.33a), which is operated with its depletion region extended from an n-type region through a semiconductor to an insulating substrate. If the bias is changed, the change in the depletion layer width is confined to the periphery of the diode. Some properties of a typical three-quarter wave PbTe pinched-off photodiode are shown in Figure 15.34. This structure was capable of withstanding 150°C for at least 8 hours without degradation of its 80 K performance and required backbias for pinch-off (quarter-wave structures were unusable). The capacitance of the device decreases from a zero-bias value of 700 pF to about 70 pF after application of backbias greater than about 150 mV. The value of 70 pF is dominated by the lead-out capacitance. The noise current at 1 kHz remains constant for backbiases < 0.3 V and is close to that calculated for the 300 K background; further bias causes significant contribution of $1/f$ noise. Three-quarter wave devices with a larger area (5×10^{-3} cm²) showed almost two orders of

(a)

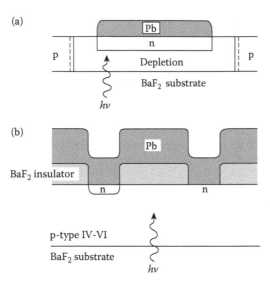

(b)

Figure 15.33 The concept of (a) the pinched-off photodiode, and (b) the lateral-collection photodiode. (From Holloway, H., *Thin Solid Films* 58, 73–78, 1979. With permission.)

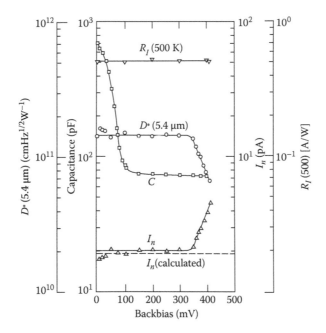

Figure 15.34 Bias-dependent properties of a three-quarter wave pinched-off PbTe photodiode at 80 K and 2π FOV. The blackbody and noise measurements were made at 1 kHz with 10 Hz bandwidth. Thickness of the PbTe film is 0.54 µm. The photodiode area is 6×10^{-4} cm². (From Holloway, H., and Yeung, K. F., *Applied Physics Letters*, 30, 210–12, 1977. With permission.)

magnitude reduction in their junction backbias capacitance. The reduction in capacitance of the pinch-off photodiodes significantly decreases the limitation on operating frequency of IV-VI photodiodes.

The second approach to capacitance reduction is the lateral-collection photodiode (Figure 15.33b), which involves a matrix of small p-n junctions that collect photogenerated minority electrons from the intervening nonjunction p-type region. Detailed analysis carried out by

Holloway [226] shows that useful collection efficiencies can be obtained with collector separations of up to two diffusion lengths. On the other hand, a reduction in junction capacitance can be obtained if the collector diameters are smaller than this dimension. For IV-VI semiconductors the diffusion length is of the order of 10–20 μm.

Thin film lateral-collection photodiodes made of layers of PbTe and PbSeTe on BaF$_2$ substrates have been described by Holloway et al. [224]. With these devices, a capacitance reduction by a factor of 20 and an increase in the Johnson noise-limited D^* by a factor of three relative to conventional photodiodes have been demonstrated.

It should be noticed that the lateral diffusion of photogenerated carriers significantly enhanced photoresponse of bulk p-n junctions with dimensions reduced to values that are comparable to the minority carrier diffusion lengths in the semiconductors of which they are made [227–229].

Thin film IV-VI narrowband photodiodes have been demonstrated by the research group at the Naval Surface Weapons Center [149,151,208]. Epitaxial layers with slightly different energy gaps were deposited (using HWE technique) on either side of a cleaved BaF$_2$ substrate (see insert in Figure 15.35). The Schottky barrier (Pb or In) was made on the layer with the smaller energy gap and the layer with the larger energy gap acted as a cooled absorption filter. The spectral response of narrowband PbSSe photodiodes is shown in Figure 15.35. These photodiodes give half-band-widths of 0.1–0.2 μm and peak quantum efficiencies of 40–50%.

The extension of narrowband detection technique to longer wavelengths with photodiodes prepared from PbSnSe epitaxial films has been described by Schoolar and Jensen [149]. The spectral half-bandwidth of 0.6 μm at 10.6 μm has been achieved for a 6.67 μm thick Pb$_{0.935}$Sn$_{0.065}$Se photodiode with a 4.12 μm thick Pb$_{0.095}$Sn$_{0.05}$Se filter at 77 K (Johnson noise-limited $D^* = 7 \times 10^{10}$ cmHz$^{1/2}$W^{-1} with cold filter and 20° FOV).

An original narrowband device containing a lateral collection photodiode on a BaF$_2$ substrate and a combination of a dielectric stack and a metal reflector has been proposed by Holloway [230].

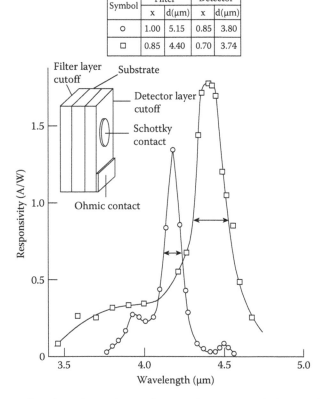

Symbol	Filter		Detector	
	x	d(μm)	x	d(μm)
○	1.00	5.15	0.85	3.80
□	0.85	4.40	0.70	3.74

Figure 15.35 Narrowband configuration and spectral response of PbSSe Schottky-barrier photodiodes. (From Schoolar, R. B., Jensen, J. D., and Black, G. M., *Applied Physics Letters*, 31, 620–22, 1977. With permission.)

The peak position of this device may be tuned by changing the angle of incidence of radiation. This suggests that such a structure may find an application as a simple spectrometer.

Multispectral PbSSe and PbSnSe photovoltaic detectors have been also developed at the Naval Surface Weapons Center Laboratories [208,231]. Two-, three- and four-color detectors were prepared by depositing Pb Schottky-barrier contacts onto stair-stepped structures obtained by subsequent HWE deposition of the lead salt alloy semiconductors on BaF_2 substrates (see Figure 15.27c). The spectral responsivities of these back side illuminated devices cover spectral bands separated by the bandgaps of the underlying layers.

15.5 TUNABLE RESONANT CAVITY ENHANCED DETECTORS

As the IR detectors continue to advance, there is growing interest to multispectral or even hyperspectral sensor arrays for enhanced object discrimination operating in different spectral bands. By use of MEMS fabrication techniques arrays of devices, such as etalons, can be fabricated on an IR detector array that permits tuning of the incident radiation on the detector. If the etalons can be programmed to change distance from the detector surface by the order of IR wavelengths, the detector responds to all wavelengths in a waveband sequentially [232].

Recently, Zogg and coworkers have elaborated tunable resonant cavity enhanced (RCE) IV-VI detectors operated in the mid-wavelength band [233,234]. Figure 15.36 shows a schematic cross section of a tunable resonant cavity PbTe detector on Si substrate [233]. The top mirror is fabricated with piezo-actuators or MEMS technology and hybridized to the detector chip. The Bragg mirror is realized by MBE with IV-VI materials and consist of quarter wavelength pairs with alternating high- and low-refractive index layers. As is shown in Figure 15.2b, their refractive index is large ($n \approx 5$ to 6) for the narrow-band compositions, while, for example, EuSe $n = 2.4$, or BaF_2 $n = 1.4$. A few quarter wavelength pairs is sufficient to obtain near 100% reflectivity across a broad spectral band. It is due to high index contrast between constituents of Bragg mirror.

The detector active layer with an area 7.5×10^{-4} cm^2, shown in Figure 15.36, forms a 0.3 μm thick n$^+$-p structure grown on a 1.5 pair Bragg mirror. In fact, it is heterojunction with a top Bi-doped n$^+$-Pb$_{1-x}$Sr$_x$Te window. The top IR transparent contact is a 100 nm thin Te layer followed by a quarter wavelength TiO$_2$ AR coating. Figure 15.37 illustrates the spectral response at three different cavity lengths [234]. The spectrum exhibits only one peak, which can be tuned by displacing the top mirror.

The temperature dependence of the R_oA product for thin n$^+$-p PbTe junction indicates performance limitation by SR recombination in depletion layer with $\tau_{SR} \approx 10$ ns (see Figure 15.38). This value is higher as compared to PbTe bulk photodiodes fabricated on Si substrates [174,235]. At temperature range between 250 K to 200 K, the experimental points are above the values for an optimized bulk photodiode. It results from limitation of active volume of detector, since the

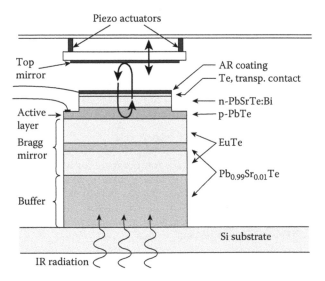

Figure 15.36 Schematic cross section of a PbTe-on-Si resonant cavity enhanced detector. (From Felder, F., Arnold, M., Rahim, M., Ebneter, C., and Zogg, H., *Applied Physics Letters* 91, 101102, 2007. With permission.)

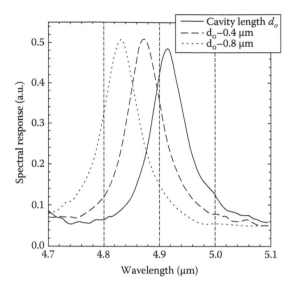

Figure 15.37 Three spectral response with different cavity lengths obtained with the piezo-actuated mirror at 100 K. (From Zogg, H., Arnold, M., Felder, F., Rahim, M., Ebneter, C., Zasavitskiy, I., Quack, N., Blunier, S., and Dual, J., *Journal of Electronic Materials*, 37, 1497–1503, 2008. With permission.)

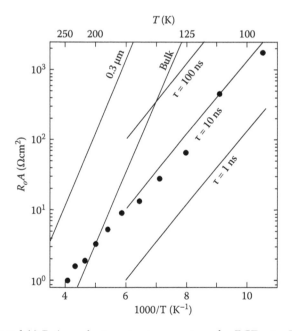

Figure 15.38 Measured (•) R_oA product versus temperature for RCE n+-p PbTe junction on Si substrate. Solid lines are calculated assuming diffusion current and band-to-band recombination limit for thick (bulk) and thin (0.3 μm-thick) photodiodes as well as for generation–recombination in the depletion region for three SR lifetimes (100, 10, and 1 ns). (From Zogg, H., Arnold, M., Felder, F., Rahim, M., Ebneter, C., Zasavitskiy, I., Quack, N., Blunier, S., and Dual, J., *Journal of Electronic Materials*, 37, 1497–1503, 2008. With permission.)

thickness of absorbing layer is smaller than the electron diffusion length (see Equation 9.88). However, the experimental values are still much below the theoretical thin limit since for a 0.3 μm thick absorber layer, the R_oA product should increase 40 times.

Another solution of tunable cavity enhanced detectors is a device with the movable MEMS mirror that consists of gold-coated square mirror plate attached to four symmetrically arranged suspension legs [113]. The counterelectrodes are placed on a glass support wafer at a distance of 10 μm. A movement of more than 3 μm is achieved by applying a voltage of 30 V between the silicon mirror membrane and the electrodes on the glass wafer.

15.6 LEAD SALTS VERSUS HgCdTe

Theoretically, in comparison with HgCdTe, lead salt ternary alloys have

- stronger chemical bonds,
- more resilience to defects (see Figure 15.39, where dependence of the R_oA product on dislocation density is shown for photodiodes with $\lambda_c \approx 10$ μm at 77 K),
- cutoff wavelength is less sensitive to composition,
- less sensitive to passivation requirement,

which should lead to large and highly uniform FPAs operated at higher temperature. The consequence of the above properties is potentially significant, particularly for VLWIR spectral region. Challenges are: soft material, high dielectric constant (low speed), and higher thermal mismatch with Si.

Lead salt ternary alloys (PbSnTe and PbSnSe) seemed easier to prepare and appeared more stable. Development of IV-VI alloy photodiodes was discontinued because the chalcogenides suffered two significant drawbacks: high dielectric constant and very high thermal coefficients of expansion (TEC).

High dielectric constant influences the space-charge region capacitance and the response speed of photodiode. Figure 15.40 shows plots of the cutoff frequency versus applied reverse-bias voltage for one side abrupt n-p$^+$ Pb$_{0.78}$Sn$_{0.22}$Te and Hg$_{0.797}$Cd$_{0.203}$Te photodiodes at 77 K ($\lambda_c \approx 12$ μm) [223]. For reverse-bias voltage above 1 V and doping concentrations in the space-charge region above 10^{15} cm^{-3}, an avalanche breakdown can be observed. The cutoff frequency has been calculated for a load resistance of 50 Ω and for a junction area of 10^{-4} cm^2. Various values of donor concentrations are marked in Figure 15.40. In this figure the space charge region width and the capacitance per unit, C/A, are also shown. We can see that cutoff frequencies of 2 GHz can be realized with HgCdTe photodiodes when n-side doping concentration is not larger than 10^{14} cm^{-3} at reverse-bias voltages. The cutoff frequency for PbSnTe photodiodes is almost an order of magnitude smaller. It should be noted that the cutoff frequency is decreased by any junction series resistance and stray capacitance.

Figure 15.41 shows dependence of the TEC of PbTe, InSb, HgTe and Si on temperature [215]. At room temperature, the TCE HgTe and CdTe is about 5×10^{-6} K^{-1}, while that of PbSnTe is in the

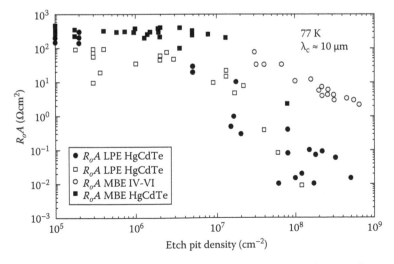

Figure 15.39 Dependence of the R_oA product on dislocation density for HgCdTe and IV-VI photodiodes with cutoff wavelength ≈ 10 μm at 77 K. (From Tidrow, M., private communication, 2007.)

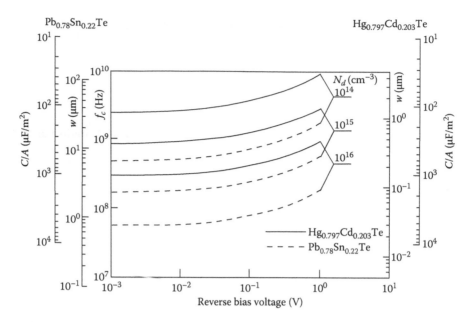

Figure 15.40 Cutoff frequency for one-side abrupt n-p⁺ $Pb_{0.78}Sn_{0.22}Te$ and $Hg_{0.797}Cd_{0.203}Te$ photodiodes ($\lambda_c \approx 12$ μm) at 77 K with an active area of 10^{-4} cm². The additional scales correspond to the depletion width and the junction capacitance per unit area. (From Rogalski, A. and Larkowski, W., *Electron Technology* 18(3/4), 55–69, 1985.)

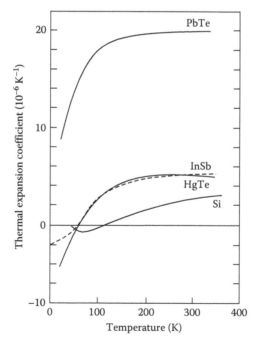

Figure 15.41 Linear TEC of PbTe, InSb, HgTe and Si versus temperature. (From Baars, J., *Physics of Narrow Gap Semiconductors*, eds. E. Gornik., H. Heinrich, and L. Palmetshofer, 280–82, Springer, Berlin, 1982. With permission.)

range of 20×10^{-6} K⁻¹. This results in a much greater TCE mismatch with silicon (TCE about 3×10^{-6} K⁻¹). It should be noted that both Ge and GaAs have TCE values close to HgCdTe, giving detectors on those materials no significant advantage in this respect.

The doping concentrations in the base regions of HgCdTe and PbSnTe photodiodes are different. For both types of photodiodes, the tunneling current (and R_oA product) is critically dependent on doping concentration. To produce high R_oA products for HgCdTe and lead salt photodiodes, the doping concentration of 10^{16} and 10^{17} cm^{-3} (or less) are required, respectively (see Figure 14.44). The maximum available doping levels due to the onset of tunneling are more than an order of magnitude higher with IV-VIs than with HgCdTe photodiodes. This is due to their high permittivities ε_s because the tunneling contribution of the R_oA product contains factors $\exp[const(m^*\varepsilon_s/N)^{1/2}E_g]$ (see Equation 9.120). The maximum allowable concentrations above 10^{17} cm^{-3} are easily controllable in IV-VIs grown by MBE. Tunneling current is not a limiting issue in lead salt photodiodes.

In comparison with HgCdTe detectors, the main performance limit of lead salt is connected with SR centers. It is not clear whether it is due to residual or native defects in material. However, it is obvious that these SR centers exert the dominating influence on the magnitude of the depletion current. Domination of depletion current, unlike the case of HgCdTe with the same bandgap, requires a lower temperature of operation for BLIP performance than for HgCdTe.

REFERENCES

1. T. W. Case, "Notes on the Change of Resistance of Certain Substrates in Light," *Physical Review* 9, 305–10, 1917; "The Thalofide Cell: A New Photoelectric Substance," *Physical Review* 15, 289, 1920.

2. E. W. Kutzscher, "Letter to the Editor," *Electro-Optical Systems Design* 5, 62, June 1973.

3. R. J. Cashman, "Film-Type Infrared Photoconductors," *Proceedings of IRE* 47, 1471–75, 1959.

4. J. O. Dimmock, I. Melngailis, and A. J. Strauss, "Band Structure and Laser Action in $Pb_{1-x}Sn_xTe$," *Physical Review Letters* 16, 1193–96, 1966.

5. A. J. Strauss, "Inversion of Conduction and Valence Bands in $Pb_{1-x}Sn_xSe$ Alloys," *Physical Review* 157, 608–11, 1967.

6. J. Melngailis and T. C. Harman, "Single-Crystal Lead-Tin Chalcogenides," in *Semiconductors and Semimetals*, Vol. 5, eds. R. K. Willardson and A. C. Beer, 111–74, Academic Press, New York, 1970.

7. T. C. Harman and J. Melngailis, "Narrow Gap Semiconductors," in *Applied Solid State Science*, Vol. 4, ed. R. Wolfe, 1–94, Academic Press, New York, 1974.

8. J. T. Longo, D. T. Cheung, A. M. Andrews, C. C. Wang, and J. M. Tracy, "Infrared Focal Planes in Intrinsic Semiconductors," *IEEE Transactions on Electron Devices* ED-25, 213–32, 1978.

9. H. Zogg and A. Ishida, "IV-VI (Lead Chalcogenide) Infrared Sensors and Lasers," in *Infrared Detectors and Emitters: Materials and Devices*, eds. P. Capper and C. T. Elliott, 43–75, Kluwer, Boston, 2001.

10. G. Springholz, "Molecular Beam Epitaxy of IV-VI Heterostructures and Superlattices," in *Lead Chalcogenides: Physics and Applications*, ed. D. Khoklov, 123–207, Taylor & Francis Inc., 2003.

11. H. Zogg, "Lead Chalcogenide Infrared Detectors Grown on Silicon Substrates," in *Lead Chalcogenides: Physics and Applications*, ed. D. Khoklov, 587–615, Taylor & Francis Inc., 2003.

12. G. Springholz, T. Schwarzl, and W. Heiss, "Mid-Infrared Vertical Cavity Surface Emitting Lasers Based on the Lead Salt Compounds," in *Mid-Infrared Semiconductor Optoelectronics*, ed. A. Krier, 265–301, Springer Verlag, 2007.

13. R. Dalven, "A Review of the Semiconductor Properties of PbTe, PbSe, PbS and PbO," *Infrared Physics* 9, 141–84, 1969.

14. Yu. I. Ravich, B. A. Efimova, and I. A. Smirnov, *Semiconducting Lead Chalcogenides*, Plenum Press, New York, 1970.

15. H. Preier, "Recent Advances in Lead-Chalcogenide Diode Lasers," *Applied Physics* 10, 189–206, 1979.

16. H. Maier and J. Hesse, "Growth, Properties and Applications of Narrow-Gap Semiconductors," in *Crystal Growth, Properties and Applications*, ed. H. C. Freyhardt, 145–219, Springer Verlag, Berlin, 1980.

17. R. Dornhaus, G. Nimtz, and B. Schlicht, *Narrow-Gap Semiconductors*, Springer Verlag, Berlin, 1983.

18. A. V. Ljubchenko, E. A. Salkov, and F. F. Sizov, *Physical Fundamentals of Semiconductor Infrared Photoelectronics*, Naukova Dumka, Kiev, 1984 (in Russian).

19. A. Rogalski and J. Piotrowski, "Intrinsic Infrared Detectors," *Progress in Quantum Electronics* 12, 87–289, 1988.

20. A. Rogalski, "IV-VI Detectors," in *Infrared Photon Detectors*, ed. E. Rogalski, 513–59, SPIE Optical Engineering Press, Bellingham, WA, 1995.

21. A. Rogalski, K. Adamiec, and J. Rutkowski, *Narrow-Gap Semiconductor Photodiodes*, SPIE Press, Bellingham, WA, 2000.

22. G. Springholz and G. Bauer, "Molecular Beam Epitaxy of IV-VI Semiconductor Hetero- and Nano-Structures," *Physica Status Solidi (b)* 244, 2752–67, 2007.

23. T. C. Harman, "Control of Imperfections in Crystals of $Pb_{1-x}Sn_xTe$, $Pb_{1-x}Sn_xSe$, and $PbS_{1-x}Se_x$," *Journal of Nonmetals* 1, 183–94, 1973.

24. S. G. Parker and R. E. Johnson, "Preparation and Properties of (Pb,Sn)Te," in *Preparation and Properties of Solid State Materials*, ed. W. R. Wilcox, 1–65, Marcel Dekker, Inc., New York, 1981.

25. D. L. Partin and J. Heremans, "Growth of Narrow Bandgap Semiconductors," in *Handbook on Semiconductors*, Vol. 3, ed. S. Mahajan, 369–450, Elsevier Science B.V., Amsterdam, 1994.

26. N. Tamari and H. Shtrikman, "Growth Study of Large Nonseeded $Pb_{1-x}Sn_xTe$ Single Crystals," *Journal of Electronic Materials* 8, 269–88, 1979.

27. E. V. Markov and A. A. Davydov, "Vapour Phase Growth of Cadmium Sulphide Single Crystals," *Neorganicheskie Materialy* 11, 1755–58, 1975.

28. K. Grasza, "Estimation of the Optimal Conditions for Directional Crystal Growth from the Vapour Phase with No Contact Between Crystal and Ampoule Wall," *Journal of Crystal Growth* 128, 609–12, 1993.

29. D. L. Partin, "Molecular-Beam Epitaxy of IV-VI Compound Heterojunctions and Superlattices," in *Semiconductors and Semimetals*, eds. R. K. Willardson and A. C. Beer, 311–36, Academic Press, Boston, 1991.

30. S. Szapiro, N. Tamari, and H. Shtrikman, "Calculation of the Phase Diagram of the Pb-Sn-Te System in the (Pb + Sn)-Rich Region," *Journal of Electronic Materials* 10, 501–16, 1981.

31. J. Kasai and W. Bassett, "Liquid Phase Epitaxial Growth of $Pb_{1-x}Sn_xSe$," *Journal of Crystal Growth* 27, 215–20, 1974.

32. P. Müller, H. Zogg, A. Fach, J. John, C. Paglino, A. N. Tiwari, M. Krejci, and G. Kostorz, "Reduction of Threading Dislocation Densities in Heavily Lattice Mismatched PbSe on Si(111) by Glide," *Physical Review Letters* 78, 3007–10, 1997.

33. Y. Huand and R. F. Brebrick, "Partial Pressures and Thermodynamic Properties for Lead Telluride," *Journal of the Electrochemical Society* 135, 486–96, 1988.

34. Y. Huand and R. F. Brebrick, "Partial Pressures and Thermodynamic Properties of PbTe-SnTe Solid and Liquid Solutions with 13, 20, and 100 Mole Percent SnTe," *Journal of the Electrochemical Society* 135, 1547–59, 1988.

35. Y. G. Sha and R. F. Brebrick, "Explicit Incorporation of the Energy-Band Structure into an Analysis of the Defect Chemistry of PbTe and SnTe," *Journal of Electronic Materials* 18, 421–43, 1989.

36. N. J. Parada and G. W. Pratt, "New Model for Vacancy States in PbTe," *Physical Review Letters* 22, 180–82, 1969.

37. G. W. Pratt, "Vacancy and Interstitial States in the Lead Salts," *Journal of Nonmetals* 1, 103–9, 1973.

38. L. A. Hemstreet, "Cluster Calculations of the Effects of Lattice Vacancies in PbTe and SnTe," *Physical Review* B12, 1213–16, 1975.

39. C. S. Lent, M. A. Bowen, R. A. Allgaier, J. D. Dow, O. F. Sankey, and E. S. Ho, "Impurity Levels in PbTe and $Pb_{1-x}Sn_xTe$," *Solid State Communications* 61, 83–87, 1987.

40. L. Palmetshofer, "Ion Implantation in IV-VI Semiconductors," *Applied Physics* A34, 139–53, 1984.

41. R. Dornhaus, G. Nimtz, and W. Richter, *Solid-State Physics,* Springer Verlag, Berlin. 1976.

42. K. Lischka, "Bound Defect States in IV-VI Semiconductors," *Applied Physics* A29, 177–89, 1982.

43. V. I. Kaidanov and Yu. I. Ravich, "Deep and Resonant States in $A^{IV}B^{VI}$ Semiconductors," *Uspekhi Fizicheskikh Nauk* 145, 51–86, 1985 (in Russian).

44. B. A. Akimov, L. I. Ryabova, O. B. Yatsenko, and S. M. Chudinov, "Rebuilding of Energy Spectrum in $Pb_{1-x}Sn_xTe$ Alloys Doped with In Under the Pressure and the Variation of Their Composition," *Fizika i Tekhnika Poluprovodnikov* 13, 752–59, 1979; B. A. Akimov, A. V. Dmitriev, D. R. Khokhlov, and L. I. Ryabova, "Carrier Transport and Non-Equillibrium Phenomena in Doped PbTe and Related Materials," *Physica Status Solidi (a)* 137, 9–55, 1993.

45. E. Silberg and A. Zemel, "Cadmium Diffusion Studies of PbTe and $Pb_{1-x}Sn_xTe$ Crystals," *Journal of Electronic Materials* 8, 99–109, 1979.

46. A. Rogalski and K. Jóźwikowski, "The Intrinsic Carrier Concentration in $Pb_{1-x}Sn_xTe$, $Pb_{1-x}Sn_xSe$ and $PbS_{1-x}Se_x$," *Physica Status Solidi (a)* 111, 559–65, 1989.

47. Yu. Ravich, B. A. Efimova, and V. I. Tamarchenko, "Scattering of Current Carriers and Transport Phenomena in Lead Chalcogenides. I. Theory," *Physica Status Solidi (b)* 43, 11–33, 1971.

48. Yu. Ravich, B. A. Efimova, and V. I. Tamarchenko, "Scattering of Current Carriers and Transport Phenomena in Lead Chalcogenides. II. Experiment," *Physica Status Solidi (b)* 43, 453–69, 1971.

49. W. Zawadzki, "Mechanisms of Electron Scattering in Semiconductors," in *Handbook on Semiconductors,* Vol. 1, ed. W. Paul, 423, North Holland, Amsterdam, 1980.

50. F. F. Sizov, G. V. Lashkarev, M. V. Radchenko, V. B. Orletzki, and E. T. Grigorovich, "Peculiarities of Carrier Scattering Mechanisms in Narrow Gap Semiconductors," *Fizika i Tekhnika Poluprovodnikov* 10, 1801–8, 1976.

51. A. Lopez-Otero, "Hot Wall Epitaxy," *Thin Solid Films* 49, 3–57, 1978.

52. W. W. Anderson, "Absorption Constant of $Pb_{1-x}Sn_xTe$ and $Hg_{1-x}Cd_xTe$ Alloys," *Infrared Physics* 20, 363–72, 1980.

53. D. Genzow, K. H. Herrmann, H. Kostial, I. Rechenberg, and A. E. Yunovich, "Interband Absorption Edge in $Pb_{1-x}Sn_xTe$," *Physica Status Solidi (b)* 86, K21–K25, 1978.

54. D. Genzow, A. G. Mironow, and O. Ziep, "On the Interband Absorption in Lead Chalcogenides," *Physica Status Solidi (b)* 90, 535–42, 1978.

55. O. Ziep and D. Genzow, "Calculation of the Interband Absorption in Lead Chalcogenides Using a Multiband Model," *Physica Status Solidi (b)* 96, 359–68, 1979.

56. B. L. Gelmont, T. R. Globus, and A. V. Matveenko, "Optical Absorption and Band Structure of PbTe," *Solid State Communications* 38, 931–34, 1981.

57. N. Suzuki and S. Adachi, "Optical Constants of $Pb_{1-x}Sn_xTe$ Alloys," *Journal of Applied Physics* 79, 2065–69, 1996.

58. S. Adachi, *Properties of Group-IV, III-V and II-VI Semiconductors,* John Wiley & Sons, Ltd, Chichester, 2005.

59. A. V. Butenko, R. Kahatabi, E. Mogilko, R. Strul, V. Sandomirsky, Y. Schlesinger, Z. Dashevsky, V. Kasiyan, and S. Genikhov, "Characterization of High-Temperature PbTe p-n Junctions Prepared by Thermal Diffusion and by Ion Implantation," *Journal of Applied Physics* 103, 024506, 2008.

60. J. H. Barrett, "Dielectric Constant in Perovskite Type Crystals," *Physical Review* 86, 118–20, 1952.

61. Y. Shani, R. Rosman, and A. Katzir, "Calculation of the Refractive Index of Lead-Tin-Telluride for Infrared Devices," *IEEE Journal of Quantum Electronics* QE-20, 1110–14, 1984.

62. B. Jensen and A. Tarabi, "Dispersion of the Refractive Index of Ternary Compound $Pb_{1-x}Sn_xTe$," *Applied Physics Letters* 45, 266–68, 1984.

63. B. Jensen and A. Tarabi, "The Refractive Index of Compounds PbTe, PbSe, and PbS," *IEEE Journal of Quantum Electronics* QE-20, 618–21, 1984.

64. K. H. Harrmann, V. Melzer, and U. Muller, "Interband and Intraband Contributions to Refractive Index and Dispersion in Narrow-Gap Semiconductors," *Infrared Physics* 34, 117–36, 1993.

65. P. Herve and L. K. J. Vandamme, "General Relation Between Refractive Index and Energy Gap in Semiconductors," *Infrared Physics Technology* 35, 609–15, 1994.

66. O. Ziep, D. Genzow, M. Mocker, and K. H. Herrmann, "Nonradiative and Radiative Recombination in Lead Chalcogenides," *Physica Status Solidi (b)* 99, 129–38, 1980.

67. O. Ziep, "Estimation of Second-Order Auger Recombination in Lead Chalcogenides," *Physica Status Solidi (b)* 115, 161–70, 1983.

68. D. A. Cammack, A. V. Nurmikko, G. W. Pratt, and J. R. Lowney, "Enhanced Interband Recombination in $Pb_{1-x}Sn_xTe$," *Journal of Applied Physics* 46, 3965–69, 1975.

69. D. M. Gureev, O. I. Davarashvili, I. I. Zasavitskii, B. N. Matsonashvili, and A. R. Shotov, "Radiative Recombination in Epitaxial $Pb_{1-x}Sn_xSe$ ($0 \le x \le 0.4$) Layers," *Fizika i Tekhnika Poluprovodnikov* 13, 1752–55, 1979.

70. G. Nimtz, "Recombination in Narrow-Gap Semiconductors," *Physics Reports* 63, 265–300, 1980.

71. K. Lischka and W. Huber, "Carrier Recombination and Deep Levels in PbTe," *Solid-State Electronics* 21, 1509–12, 1978.

72. K. H. Harrmann, "Recombination in Small-Gap $Pb_{1-x}Sn_xTe$," *Solid-State Electronics* 21, 1487–91, 1978.

73. D. Khokhlov, ed. Lead Chalcogenides: *Physics and Applications*, Taylor & Francis Inc., 2003.

74. P. R. Emtage, "Auger Recombination and Junction Resistance in Lead-Tin Telluride," *Journal of Applied Physics* 47, 2565–68, 1976.

75. K. Lischka, W. Huber, and H. Heinrich, "Experimental Determination of the Minority Carrier Lifetime in PbTe p–n Junctions," *Solid State Communications* 20, 929–31, 1976.

76. K. Lischka and W. Huber, "Auger Recombination in PbTe," *Journal of Applied Physics* 48, 2632–33, 1977.

77. T. X. Hoai and K. H. Herrmann, "Recombination in $Pb_{0.82}Sn_{0.18}Te$ at High Levels of Optical Excitation," *Physica Status Solidi (b)* 83, 465–70, 1977.

78. O. Ziep, M. Mocker, D. Genzow, and K. H. Herrmann, "Auger Recombination in PbSnTe-Like Semiconductors," *Physica Status Solidi (b)* 90, 197–205, 1978.

79. O. Ziep and M. Mocker, "A New Approach to Auger Recombination," *Physica Status Solidi (b)* 98, 133–42, 1980.

80. R. Rosman and A. Katzir, "Lifetime Calculations for Auger Recombination in Lead-Tin-Telluride," *IEEE Journal of Quantum Electronics* QE18, 814–17, 1982.

81. H. Zogg, W. Vogt, and W. Baumgartner, "Carrier Recombination in Single Crystal PbSe," *Solid-State Electronics* 25, 1147–55, 1982.

82. A. Shahar, M. Oron, and A. Zussman, "Minority-Carrier Diffusion-Length Measurement and Lifetime in $Pb_{0.8}Sn_{0.2}Te$ and Indium-Doped PbTe Liquid Phase Layers," *Journal of Applied Physics* 54, 2477–82, 1983.

83. M. Mocker and O. Ziep, "Intrinsic Recombination in Dependence on Doping Concentration and Excitation Level," *Physica Status Solidi (b)* 115, 415–25, 1983.

84. O. Ziep, M. Mocker, and D. Genzow, "Theorie der Band-Band-Rekombination in Bleichalkogeniden," *Wissensch. Zeitschr. Humboldt-Univ. Berlin, Math.-Nataruiss. Reihe* XXX, 81–97, 1981.

85. A. V. Dmitriev, "Calculation of the Auger Lifetime of Degenerate Carriers in the Many-Valley Narrow Gap Semiconductors," *Solid State Communications* 74, 577–81, 1990.

86. S. D. Beneslavskii and A. V. Dmitriev, "Auger Transitions in Narrow-Gap Semiconductors with Lax Band Structures," *Solid State Communications* 39, 811–14, 1981.

87. P. Berndt, D. Genzow, and K. H. Herrmann, "Recombination analysis in 10 μm $Pb_{1-x}Sn_xTe$," *Physica Status Solidi (a)* 38, 497–503, 1976.

88. A. Schlicht, R. Dornhaus, G. Nimtz, L. D. Haas, and T. Jakobus, "Lifetime Measurements in PbTe and PbSnTe," *Solid-State Electronics* 21, 1481–85, 1981.

89. K. Weiser, E. Ribak, A. Klein, and M. Ainhorn, "Recombination of Photocarriers in Lead-Tin Telluride," *Infrared Physics* 21, 149–54, 1981.

90. A. Genzow, M. Mocker, and E. Normantas, "The Influence of Uniaxial Strain on the Band to Band Recombination in $Pb_{1-x}Sn_xTe$," *Physica Status Solidi (b)* 135, 261–70, 1986.

91. S. Espevik, C. Wu, and R. H. Bube, "Mechanism of Photoconductivity in Chemically Deposited Lead Sulfide Layers," *Journal of Applied Physics* 42, 3513–29, 1971.

92. T. H. Johnson, "Lead Salt Detectors and Arrays. PbS and PbSe," *Proceedings of SPIE* 443, 60–94, 1984.

93. L. N. Neustroev and V. V. Osipov, "Physical Properties of Photosensitive Polycrystalline Lead Chalcogenide Films," *Mikroelektronika* 17, 399–416, 1988.

94. J. Vaitkus, M. Petrauskas, R. Tomasiunas, and R. Masteika, "Photoconductivity of Highly Excited $A^{IV}B^{VI}$ Thin Films," *Applied Physics* A54, 553–55, 1992.

95. J. Vaitkus, R. Tomasiunas, K. Tumkevicius, M. Petrauskas, and R. Masteika, "Picosecond Photoconductivity of Polycrystalline PbTe Films," *Fizika i Tekhnika Poluprovodnikov* 24, 1919–22, 1990.

96. T. S. Moss, "Lead Salt Photoconductors," *Proceedings of IRE* 43, 1869–81, 1955.

97. R. J. Cashman, "Film-Type Infrared Photoconductors," *Proceedings of IRE* 47, 1471–75, 1959.

98. P. W. Kruse, L. D. McGlauchlin, and R. B. McQuistan, *Elements of Infrared Technology*, Wiley, New York, 1962.

99. D. E. Bode, T. H. Johnson, and B. N. McLean, "Lead Selenide Detectors for Intermediate Temperature Operation," *Applied Optics* 4, 327–31, 1965.

100. J. N. Humphrey, "Optimization of Lead Sulfide Infrared Detectors Under Diverse Operating Conditions," *Applied Optics* 4, 665–75, 1965.

101. T. H. Johnson, H. T. Cozine, and B. N. McLean, "Lead Selenide Detectors for Ambient Temperature Operation," *Applied Optics* 4, 693–96, 1965.

102. D. E. Bode, "Lead Salt Detectors," in *Physics of Thin Films*, Vol. 3, eds. G. Hass and R. E. Thun, 275–301, Academic Press, New York, 1966.

103. T. S. Moss, G. J. Burrel, and B. Ellis, *Semiconductor Optoelectronics*, Butterworths, London, 1973.

104. R. E. Harris, "PbS … Mr. Versatility of the Detector World," *Electro-Optical Systems Design* 47–50, December 1976.

105. R. E. Harris, "PbSe … Mr Super Sleuth of the Detector World," *Electro-Optical Systems Design* 42–44, March 1977.

106. R. H. Harris, "Lead-Salt Detectors," *Laser Focus/Electro-Optics*, 87–96, December 1983.

107. M. C. Torquemada, M. T. Rodrigo, G. Vergara, F. J. Sanchez, R. Almazan, M. Verdu, P. Rodrıguez, et al., "Role of Halogens in the Mechanism of Sensitization of Uncooled PbSe Infrared Photodetectors," *Journal of Applied Physics* 93, 1778–84, 2003.

108. C. Nascu, V. Vomir, I. Pop, V. Ionescu, and R. Grecu, "The Study of PbS Films: VI. Influence of Oxidants on the Chemically Deposited PbS Thin Films," *Materials Science and Engineering* B41, 235–240, 1996.

109. I. Grozdanov, M. Najdoski, and S. K. Dey, "A Simple Solution Growth Technique for PbSe Thin Films," *Materials Letters* 38, 28, 1999.

110. S. Horn, D. Lohrmann, P. Norton, K. McCormack, and A. Hutchinson, "Reaching for the Sensitivity Limits of Uncooled and Minimaly-Cooled Thermal and Photon Infrared Detectors," *Proceedings of SPIE* 5783, 401–11, 2005.

111. P. R. Norton, "Infrared Image Sensors," *Optical Engineering* 30, 1649–63, 1991.

112. E. L. Dereniak and G. D. Boreman, *Infrared Detectors and Systems*, Wiley, New York, 1996.

113. N. Quack, S. Blunier, J. Dual, F. Felder, M. Arnold, and H. Zogg, "Mid-Infrared Tunable Resonant Cavity Enhanced Detectors," *Sensors* 8, 5466–78, 2008.

114. A. Rogalski, J. Antoszewski, and L. Faraone, "Third-Generation Infrared Photodetector Arrays," *Journal of Applied Physics* 105, 091101-44, 2009.

115. A. Rogalski, "Detectivity Limits for PbTe Photovoltaic Detectors," *Infrared Physics* 20, 223–29, 1980.

116. A. Rogalski and J. Rutkowski, "R_oA Product for PbS and PbSe Abrupt p-n Junctions," *Optica Applicata* 11, 365–70, 1981.

117. A. Rogalski and J. Rutkowski, "Temperature Dependence of the R_oA Product for Lead Chalcogenides Photovoltaic Detectors," *Infrared Physics* 21, 191–99, 1981.

118. A. Rogalski, W. Kaszuba, and W. Larkowski, "PbTe Photodiodes Prepared by the Hot-Wall Evaporation Technique," *Thin Solid Films* 103, 343–53, 1983.

119. A. Rogalski, R. Ciupa, and H. Zogg, "Computer Modeling of Carrier Transport in Binary Salt Photodiodes," *Proceedings of SPIE* 2373, 172–81, 1995.

120. J. Baars, D. Bassett, and M. Schulz, "Metal-Semiconductor Barrier Studies of PbTe," *Physica Status Solidi (a)* 49, 483–88, 1978.

121. J. P. Donnelly, T. C. Harman, A. G. Foyt, and W. T. Lindley, "PbTe Photodiodes Fabricated by Sb^+ Ion Implantation," *Journal of Nonmetals* 1, 123–28, 1973.

122. H. Holloway and K. F. Yeung, "Low-Capacitance PbTe Photodiodes," *Applied Physics Letters* 30, 210–12, 1977.

123. C. H. Gooch, H. A. Tarry, R. C. Bottomley, and B. Waldock, "Planar Indium-Diffused Lead Telluride Detector Arrays," *Electronics Letters* 14, 209–10, 1978.

124. H. Zogg, S. Blunier, T. Hoshino, C. Maissen, J. Masek, and A. N. Tiwari, "Infrared Sensor Arrays with 3–12 μm Cutoff Wavelengths in Heteroepitaxial Narrow-Gap Semiconductors on Silicon Substrates," *IEEE Transactions on Electron Devices* 38, 1110–17, 1991.

125. A. Rogalski and R. Ciupa, "PbSnTe Photodiodes: Theoretical Predictions and Experimental Data," *Opto-Electronics Review* 5, 21–29, 1997.

126. C. A. Kennedy, K. J. Linden, and D. A. Soderman, "High-Performance 8–14 μm $Pb_{1-x}Sn_xTe$ Photodiodes," *Proceedings of IEEE* 63, 27–32, 1976.

127. C. C. Wang and S. R. Hampton, "Lead Telluride-Lead Tin Telluride Heterojunction Diode Array," *Solid-State Electronics* 18, 121–25, 1975.

128. P. S. Chia, J. R. Balon, A. H. Lockwood, D. M. Randall, and F. J. Renda, "Performance of PbSnTe Diodes at Moderately Backgrounds," *Infrared Physics* 15, 279–85, 1975.

129. C. C. Wang and J. S. Lorenzo, "High-Performance, High-Density, Planar PbSnTe Detector Arrays," *Infrared Physics* 17, 83–88, 1977.

130. C. C. Wang and M. E. Kim, "Long-Wavelength PbSnTe/PbTe Inverted Heterostructure Mosaics," *Journal of Applied Physics* 50, 3733–37, 1979.

131. M. Grudzień and A. Rogalski, "Photovoltaic Detectors $Pb_{1-x}Sn_xTe$ ($0 \leq x \leq 0.25$). Minority Carrier Lifetimes. Resistance-Area Product," *Infrared Physics* 21, 1–8, 1981.

132. V. F. Chishko, V. T. Hryapov, I. L. Kasatkin, V. V. Osipov, and O. V. Smolin, "High Detectivity $Pb_{0.8}Sn_{0.2}Te$–$PbSe_{0.08}Te_{0.92}$ Heterostructure Photodiodes," *Infrared Physics* 33, 275–80, 1992.

133. H. Zogg, C. Maissen, J. Masek, S. Blunier, A. Lambrecht, and M. Tacke, "Heteroepitaxial $Pb_{1-x}Sn_xSe$ on Si Infrared Sensor Array with 12 μm Cutoff Wavelength," *Applied Physics Letters* 55, 969–71, 1989.

134. H. Zogg, C. Maissen, J. Masek, S. Blunier, A. Lambrecht, and M. Tacke, "Epitaxial Lead Chalcogenide IR Sensors on Si for 3–5 and 8–12 μm," *Semiconductor Science and Technology* 5, S49–S52, 1990.

135. H. Zogg, C. Maissen, J. Masek, T. Hoshino, S. Blunier, and A. N. Tiwari, "Photovotaic Infrared Sensor Arrays in Monolithic Lead Chalcogenides on Silicon," *Semiconductor Science and Technology* 6, C36–C41, 1991.

136. T. Hoshino, C. Maissen, H. Zogg, J. Masek, S. Blunier, A. N. Tiwari, S. Teodoropol, and W. J. Bober, "Monolithic $Pb_{1-x}Sn_xSe$ Infrared Sensor Arrays on Si Prepared by Low-Temperature Process," *Infrared Physics* 32, 169–75, 1991.

137. J. Masek, T. Hoshino, C. Maissen, H. Zogg, S. Blunier, J. Vermeiren, and C. Claeys, "Monolithic Lead-Chalcogenide IR-Arrays on Silicon: Fabrication and Use in Thermal Imaging Applications," *Proceedings of SPIE* 1735, 54–61, 1992.

138. H. Zogg, A. Fach, C. Maissen, J. Masek, and S. Blunier, "Photovoltaic Lead-Chalcogenide on Silicon Infrared Sensor Arrays," *Optical Engineering* 33, 1440–49, 1994.

139. C. Paglino, A. Fach, J. John, P. Müller, H. Zogg, and D. Pescia, "Schottky-Barrier Fluctuations in $Pb_{1-x}Sn_xSe$ Infrared Sensors," *Journal of Applied Physics* 80, 7138–43, 1996.

140. A. Fach, J. John, P. Müller, C. Paglino, and H. Zogg, "Material Properties of $Pb_{1-x}Sn_xSe$ Epilayers on Si and Their Correlation with the Performance of Infrared Photodiodes," *Journal of Electronic Materials* 26, 873–77, 1997.

141. H. Zogg, A. Fach, J. John, P. Müller, C. Paglino, and A. N. Tiwari, "PbSnSe-on-Si: Material and IR-Device Properties," *Proceedings of SPIE* 3182, 26–29, 1998.

142. H. Zogg, "Lead Chalcogenide on Silicon Infrared Sensor Arrays," *Opto-Electronics Review* 6, 37–46, 1998.

143. H. Zogg, "Photovoltaic IV–VI on Silicon Infrared Devices for Thermal Imaging Applications," *Proceedings of SPIE* 3629, 52–62, 1999.

144. S. C. Gupta, B. L. Sharma, and V. V. Agashe, "Comparison of Schottky Barrier and Diffused Junction Infrared Detectors," *Infrared Physics* 19, 545–48, 1979.

145. A. Rogalski and W. Kaszuba, "Photovoltaic Detectors $Pb_{1-x}Sn_xSe$ ($0 \leq x \leq 0.12$). Minority Carrier Lifetimes. Resistance-Area Product," *Infrared Physics* 21, 251–59, 1981.

146. H. Preier, "Comparison of the Junction Resistance of (Pb,Sn)Te and (Pb,Sn)Se Infrared Detector Diodes," *Infrared Physics* 18, 43–46, 1979.

147. A. Rogalski and W. Kaszuba, "$PbS_{1-x}Se_x$ ($0 \leq x \leq 1$) Photovoltaic Detectors: Carrier Lifetimes and Resistance-Area Product," *Infrared Physics* 23, 23–32, 1983.

148. D. K. Hohnke, H. Holloway, K. F. Yeung, and M. Hurley, "Thin-Film (Pb,Sn)Se Photodiodes for 8–12 μm Operation," *Applied Physics Letters* 29, 98–100, 1976.

149. R. B. Schoolar and J. D. Jensen, "Narrowband Detection at Long Wavelengths with Epitaxial $Pb_{1-x}Sn_xSe$ Films," *Applied Physics Letters* 31, 536–38, 1977.

150. D. K. Hohnke and H. Holloway, "Epitaxial PbSe Schottky-Barrier Diodes for Infrared Detection," *Applied Physics Letters* 24, 633–35, 1974.

151. R. B. Schoolar, J. D. Jensen, and G. M. Black, "Composition-Tuned PbS_xSe_{1-x} Schottky-Barrier Infrared Detectors," *Applied Physics Letters* 31, 620–22, 1977.

152. M. R. Johnson, R. A. Chapman, and J. S. Wrobel, "Detectivity Limits for Diffused Junction PbSnTe Detectors," *Infrared Physics* 15, 317–29, 1975.

153. J. Rutkowski, A. Rogalski, and W. Larkowski, "Mesa Cd-Diffused $Pb_{0.80}Sn_{0.20}Te$ Photodiodes," *Acta Physica Polonica* A67, 195–98, 1985.

154. J. P. Donnelly and H. Holloway, "Photodiodes Fabricated in Epitaxial PbTe by Sb^+ Ion Implantation," *Applied Physics Letters* 23, 682–83, 1973.

155. H. Holloway, "Thin-Film IV-VI Semiconductor Photodiodes," in *Physics of Thin Films*, Vol. 11, eds. G. Haas, M. H. Francombe, and P. W. Hoffman, 105–203, Academic Press, New York, 1980.

156. K. W. Nill, A. R. Calawa, and T. C. Harman, "Laser Emission from Metal-Semiconductor Barriers on PbTe and $Pb_{0.8}Sn_{0.2}Te$," *Applied Physics Letters* 16, 375–77, 1970.

157. J. P. Donnelly and T. C. Harman, "P-n Junction PbS_xSe_{1-x} Photodiodes Fabricated by Se^+ Ion Implantation," *Solid-State Electronics* 18, 288–90, 1975.

158. J. P. Donnelly, T. C. Harman, A. G. Foyt, and W. T. Lindley, "PbS Photodiodes Fabricated by Sb^+ Ion Implantation," *Solid-State Electronics* 16, 529–34, 1973.

159. A. M. Andrews, J. T. Longo, J. E. Clarke, and E. R. Gertner, "Backside-Illuminated $Pb_{1-x}Sn_xTe$ Heterojunction Photodiode," *Applied Physics Letters* 26, 438–41, 1975.

160. R. W. Grant, J. G. Pasko, J. T. Longo, and A. M. Andrews, "ESCA Surface Studies of $Pb_{1-x}Sn_xTe$ Devices," *Journal of Vacuum Science and Technology* 13, 940–47, 1976.

161. M. Bettini and H. J. Richter, "Oxidation in Air and Thermal Desorption on PbTe, SnTe and $Pb_{0.8}Sn_{0.2}Te$ Surface," *Surface Science* 80, 334–43, 1979.

162. D. L. Partin and C. M. Thrush, "Anodic Oxidation of Lead Telluride and Its Alloys," *Journal of the Electrochemical Society* 133, 1337–40, 1986.

163. T. Jimbo, M. Umeno, H. Shimizu, and Y. Amemiya, "Optical Properties of Native Oxide Film Anodically Grown on PbSnTe," *Surface Science* 86, 389–97, 1979.

164. J. D. Jensen and R. B. Schoolar, "Surface Charge Transport in PbS_xSe_{1-x} and $Pb_{1-x}Sn_xSe$ Epitaxial Films," *Journal of Vacuum Science and Technology* 13, 920–25, 1976.

165. A. Rogalski, "Effect of Air on Electrical Properties of $Pb_{1-x}Sn_xTe$ Layers on a Mica Substrate," *Thin Solid Films* 74, 59–68, 1980.

166. S. Sun, S. P. Buchner, N. E. Byer, and J. M. Chen, "Oxygen Uptake on an Epitaxial PbSnTe (111) Surface," *Journal of Vacuum Science and Technology* 15, 1292–97, 1978.

167. T. Jakobus, W. Rothemund, A. Hurrle, and J. Baars, "$Pb_{0.8}Sn_{0.2}Te$ Infrared Photodiodes by Indium Implantation," *Revue de Physique Applquee* 13, 753–56, 1978.

168. W. Larkowski and A. Rogalski, "High-Performance 8–14 µm PbSnTe Schottky Barrier Photodiodes," *Optica Applicata* 16, 221–29, 1986.

169. P. Lo Vecchio, M. Jasper, J. T. Cox, and M. B. Barber, "Planar $Pb_{0.8}Sn_{0.2}Te$ Photodiode Array Development at the Night Vision Laboratory," *Infrared Physics* 15, 295–301, 1975.

170. J. S. Wrobel, "Method of Forming p-n Junction in PbSnTe and Photovoltaic Infrared Detector Provided Thereby," *Patent USA 3,911,469*, 1975.

171. R. Behrendt and R. Wendlandt, "A Study of Planar Cd-diffused $Pb_{1-x}Sn_xTe$ Photodiodes," *Physica Status Solidi (a)* 61, 373–80, 1980.

172. P. T. Landsberg, *Recombination in Semiconductors*, Cambridge University Press, Cambridge, 2003.

173. L. H. DeVaux, H. Kimura, M. J. Sheets, F. J. Renda, J. R. Balon, P. S. Chia, and A. H. Lockwood, "Thermal Limitations in PbSnTe Detectors," *Infrared Physics* 15, 271–77, 1975.

174. J. John and H. Zogg, "Infrared p-n-Junction Diodes in Epitaxial Narrow Gap PbTe Layers on Si Substrates," *Applied Physics Letters* 85, 3364–66, 1999.

175. E. M. Logothetis, H. Holloway, A. J. Varga, and W. J. Johnson, "N-p Junction IR Detectors Made by Proton Bombardment of Epitaxial PbTe," *Applied Physics Letters* 21, 411–13, 1972.

176. T. F. Tao, C. C. Wang, and J. W. Sunier, "Effect of Proton Bombardment on $Pb_{0.76}Sn_{0.24}Te$," *Applied Physics Letters* 20, 235–37, 1972.

177. C. C. Wang, T. F. Tao, and J. W. Sunier, "Proton Bombardment and Isochronal Annealing of p–type $Pb_{0.76}Sn_{0.24}Te$," *J. Applied Physics* 45, 3981–87, 1974.

178. J. P. Donnelly, "The Electrical Characteristics of Ion Implanted Compound Semiconductors," *Nuclear Instruments and Methods* 182/183, 553–71, 1981.

179. L. Palmetshofer, "Ion Implantation in IV-VI Semiconductors," *Applied Physics* A34, 139–53, 1984.

180. A. H. Lockwood, J. R. Balon, P. S. Chia, and F. J. Renda, "Two-Color Detector Arrays by PbTe/$Pb_{0.8}Sn_{0.2}Te$ Liquid Phase Epitaxy," *Infrared Physics* 16, 509–14, 1976.

181. C. C. Wang, M. H. Kalisher, J. M. Tracy, J. E. Clarke, and J. T. Longo, "Investigation on Leakage Characteristics of PbSnTe/PbTe Inverted Heterostructure Diodes," *Solid-State Electronics* 21, 625–32, 1978.

182. Nugraha, W. Tamura, O. Itoh, K. Suto, and J. Nishizawa, "Te Vapor Pressure Dependence of the p-n Junction Properties of PbTe Liquid Phase Epitaxial Layers," *Journal of Electronic Materials* 27, 438–41, 1999.

183. W. Rolls and D. V. Eddolls, "High Detectivity $Pb_{1-x}Sn_xTe$ Photovoltaic Diodes," *Infrared Physics* 13, 143–47, 1972.

184. R. W. Bicknell, "Electrical and Metallurgical Examination of $Pb_{1-x}Sn_xTe/PbTe$ Heterojunctions," *Journal of Vacuum Science and Technology* 14, 1012–15, 1977.

185. V. V. Tetyorkin, V. B. Alenberg, F. F. Sizov, E. V. Susov, Yu. G. Troyan, A. V. Gusarov, V. Yu. Chopik, and K. S. Medvedev, "Carrier Transport Mechanisms and Photoelectrical Properties of PbSnTe/PbTeSe Heterojunctions," *Infrared Physics* 30, 499–504, 1990.

186. A. Rogalski, "$Pb_{1-x}Sn_xTe$ Photovoltaic Detectors for the Range of Atmospheric Window 8–14 µm Prepared by a Modified Hot-Wall Evaporation Technique," *Electron Technology* 12(4), 99–107, 1978.

187. D. Eger, A. Zemel, S. Rotter, N. Tamari, M. Oron, and A. Zussman, "Junction Migration in PbTe-PbSnTe Heterostructures," *Journal of Applied Physics* 52, 490–95, 1981.

188. D. Yakimchuk, M. S. Davydov, V. F. Chishko, I. J. Tsveibak, V. V. Krapukhin, and I. A. Sokolov, "Current-Voltage Characteristics of $p-Pb_{0.8}Sn_{0.2}Te/n-PbTe_{0.92}Se_{0.08}$ Heterojunctions," *Fizika i Tekhnika Poluprovodnikov* 22, 1474–78, 1988.

189. I. Kasai, D. W. Bassett, and J. Hornung, "PbTe and $Pb_{0.8}Sn_{0.2}Te$ Epitaxial Films on Cleaved BaF_2 Substrates Prepared by a Modified Hot-Wall Technique," *Journal of Applied Physics* 47, 3167–71, 1976.

190. A. Rogalski, "$n-PbTe/p^+-Pb_{1-x}Sn_xTe$ Heterojunctions Prepared by a Modified Hot Wall Technique," *Thin Solid Films* 67, 179–86, 1980.

191. D. Eger, M. Oron, A. Zussman, and A. Zemel, "The Spectral Response of PbTe/PbSnTe Heterostructure Diodes at Low Temperatures," *Infrared Physics* 23, 69–76, 1983.

192. E. Abramof, S. O. Ferreira, C. Boschetti, and I. N. Bendeira, "Influence of Interdiffusion on N-PbTe/P-PbSnTe Heterojunction Diodes," *Infrared Physics* 30, 85–91, 1990.

193. N. Tamari and H. Shtrikman, "Dislocation Etch Pits in LPE-Grown $Pb_{1-x}Sn_xTe$ (LTT) Heterostructures," *Journal of Applied Physics* 50, 5736–42, 1979.

194. D. Kasemset and G. Fonstad, "Reduction of Interface Recombination Velocity with Decreasing Lattice Parameter Mismatch in PbSnTe Heterojunctions," *Journal of Applied Physics* 50, 5028–29, 1979.

195. D. Kasemset, S. Rotter, and C. G. Fonstad, "Liquid Phase Epitaxy of PbTeSe Lattice-Matched to PbSnTe," *Journal of Electronic Materials* 10, 863–78, 1981.

196. J. N. Walpole and K. W. Nill, "Capacitance-Voltage Characteristic of the Metal Barrier on p PbTe and n InAs: Effect of Inversion Layer," *Journal of Applied Physics* 42, 5609–17, 1971.

197. F. F. Sizov, V. V. Tetyorkin, Yu. G. Troyan, and V. Yu. Chopick, "Properties of the Schottky Barriers on Compensated PbTe < Ga > ," *Infrared Physics* 29, 271–77, 1989.

198. W. Vogt, H. Zogg, and H. Melchior, "Preparation and Properties of Epitaxial $PbSe/BaF_2/PbSe$ Structures," *Infrared Physics* 25, 611–14, 1985.

199. C. Maissen, J. Masek, H. Zogg, and S. Blunier, "Photovoltaic Infrared Sensors in Heteroepitaxial PbTe on Si," *Applied Physics Letters* 53, 1608–10, 1988.

200. W. Maurer, "Temperature Dependence of the R_oA Product of PbTe Schottky Diodes," *Infrared Physics* 23, 257–60, 1983.

201. B. Chang, K. E. Singer, and D. C. Northrop, "Indium Contacts to Lead Telluride," *Journal of Physics D: Applied Physics* 13, 715–23, 1980.

202. T. A. Grishina, I. A. Drabkin, Yu. P. Kostikov, A. V. Matveenko, N. G. Protasova, and D. A. Sakseev, "Study of In-Pb$_{1-x}$Sn$_x$Te Interface by Auger Spectroscopy," *Izvestiya Akademii Nauk SSSR. Neorganicheskie Materialy* 18, 1709–13, 1982.

203. K. P. Scharnhorst, R. F. Bis, J. R. Dixon, B. B. Houston, and H. R. Riedl, "Vacuum Deposited Method for Fabricating an Epitaxial PbSnTe Rectifying Metal Semiconductor Contact Photodetector," U.S Patent 3,961,998, 1976.

204. S. Buchner, T. S. Sun, W. A. Beck, N. E. Dyer, and J. M. Chen, "Schottky Barrier Formation on (Pb,Sn)Te," *Journal of Vacuum Science and Technology* 16, 1171–73, 1979.

205. T. A. Grishina, N. N. Berchenko, G. I. Goderdzishvili, I. A. Drabkin, A. V. Matveenko, T. D. Mcheidze, D. A. Sakseev, and E. A. Tremiakova, "Surface-Barrier Pb$_{0.77}$Sn$_{0.23}$Te Structures with Intermediate Layer," *Journal of Technical Physics* 57, 2355–60, 1987 (in Russian).

206. N. N. Berchenko, A. I. Vinnikova, A. Yu. Nikiforov, E. A. Tretyakova, and S. V. Fadyeev, "Growth and Properties of Native Oxides for IV-VI Optoelectronic Devices," *Proceedings of SPIE* 3182, 404–7, 1997.

207. T. S. Sun, N. E. Byer, and J. M. Chen, "Oxygen Uptake on Epitaxial PbTe(111) Surfaces," *Journal of Vacuum Science and Technology* 15, 585–89, 1978.

208. A. C. Bouley, T. K. Chu, and G. M. Black, "Epitaxial Thin Film IV-VI Detectors: Device Performance and Basic Material Properties," *Proceedings of SPIE* 285, 26–32, 1981.

209. A. C. Chu, A. C. Bouley, and G. M. Black, "Preparation of Epitaxial Thin Film Lead Salt Infrared Detectors," *Proceedings of SPIE* 285, 33–35, 1981.

210. M. Drinkwine, J. Rozenbergs, S. Jost, and A. Amith, "The Pb/PbSSe Interface and Performance of Pb/PbSSe Photodiodes," *Proceedings of SPIE* 285, 36–43, 1981.

211. J. H. Werner and H. H. Güttler, "Barrier Inhomogeneities at Schottky Contacts," *Journal of Applied Physics* 69, 1522–33, 1991.

212. D. W. Bellavance and M. R. Johnson, "Open Tube Vapor Transport Growth of Pb$_{1-x}$Sn$_x$Te Epitaxial Films for Infrared Detectors," *Journal of Electronic Materials* 5, 363–80, 1976.

213. E. M. Logothetis, H. Holloway, A. J. Varga, and E. Wilkes, "Infrared Detection by Schottky Barrier in Epitaxial PbTe," *Applied Physics Letters* 19, 318–20, 1971.

214. H. Zogg and M. Huppi, "Growth of High Quality Epitaxial PbSe Onto Si Using a (Ca,Ba)F$_2$ Buffer Layer," *Applied Physics Letters* 47, 133–35, 1985.

215. J. Baars, "New Aspects of the Material and Device Technology of Intrinsic Infrared Photodetectors," in *Physics of Narrow Gap Semiconductors*, eds. E. Gornik, H. Heinrich, and L. Palmetshofer, 280–82, Springer, Berlin, 1982.

216. J. Masek, C. M. Maissen, H. Zogg, S. Blunier, H. Weibel, A. Lambrecht, B. Spanger, H. Bottner, and M. Tacke, "Photovoltaic Infrared Sensor Arrays in Heteroepitaxial Narrow Gap Lead-Chalcogenides on Silicon," *Journal de Physique*, Colloque C4, 587–90, 1988.

217. J. Masek, A. Ishida, H. Zogg, C. Maissen, and S. Blunier, "Monolithic Photovoltaic PbS-on-Si Infrared-Sensor Array," *IEEE Electron Device Letters* 11, 12–14, 1990.

218. H. Zogg and P. Norton, "Heteroepitaxial PbTe-Si and (Pb,Sn)Se-Si Structures for Monolithic 3–5 μm and 8–12 μm Infrared Sensor Arrays," *IEDM Technical Digest* 121–24, 1985.

219. P. Collot, F. Nguyen-Van-Dau, and V. Mathet, "Monolithic Integration of PbSe IR Photodiodes on Si Substrates for Near Ambient Temperature Operation," *Semiconductor Science and Technology* 9, 1133–37, 1994.

220. H. Zogg, A. Fach, J. John, J. Masek, P. Müller, C. Paglino, and W. Buttler, "PbSnSe-on-Si LWIR Sensor Arrays and Thermal Imaging with JFET/CMOS Read-Out," *Journal of Electronic Materials* 25, 1366–70, 1996.

221. J. Ameurlaine, A. Rousseau, T. Nguyen-Duy, and R. Triboulet, "(HgZn)Te Infrared Photovoltaic Detectors," *Proceedings of SPIE* 929, 14–20, 1988

222. A. Rogalski and R. Ciupa, "Long Wavelength HgCdTe Photodiodes: n^+-on-p Versus p-on-n Structures," *Journal of Applied Physics* 77, 3505–12, 1995.

223. A. Rogalski and W. Larkowski, "Comparison of Photodiodes for the 3–5.5 μm and 8–14 μm Spectral Regions," *Electron Technology* 18(3/4), 55–69, 1985.

224. H. Holloway, M. D. Hurley, and E. B. Schermer, "IV-VI Semiconductor Lateral-Collection Photodiodes," *Applied Physics Letters* 32, 65–67, 1978.

225. H. Holloway, "Unconventional Thin Film IV-VI Photodiode Structures," *Thin Solid Films* 58, 73–78, 1979.

226. H. Holloway, "Theory of Lateral-Collection Photodiodes," *Journal of Applied Physics* 49, 4264–69, 1978.

227. H. Holloway and A. D. Brailsford, "Peripheral Photoresponse of a p-n Junction," *Journal of Applied Physics* 54, 4641–56, 1983.

228. H. Holloway and A. D. Brailsford, "Diffusion-Limited Saturation Current of a Finite p-n Junction," *Journal of Applied Physics* 55, 446–53, 1984.

229. H. Holloway, "Peripheral Electron-Beam Induced Current Response of a Shallow p-n Junction," *Journal of Applied Physics* 55, 3669–75, 1984.

230. H. Holloway, "Quantum Efficiencies of Thin-Film IV-VI Semiconductor Photodiodes," *Journal of Applied Physics* 50, 1386–98, 1979.

231. R. B. Schoolar, J. D. Jensen, G. M. Black, S. Foti, and A. C. Bouley, "Multispectral PbS$_x$Se$_{1-x}$ and Pb$_y$Sn$_{1-y}$Se Photovoltaic Infrared Detectors," *Infrared Physics* 20, 271–75, 1980.

232. J. Carrano, J. Brown, P. Perconti, and K. Barnard, "Tuning In to Detection," *SPIE's OEmagazine*, 20–22, April 2004.

233. F. Felder, M. Arnold, M. Rahim, C. Ebneter, and H. Zogg, "Tunable Lead-Chalcogenide on Si Resonant Cavity Enhanced Midinfrared Detector," *Applied Physics Letters* 91, 101102, 2007.

234. H. Zogg, M. Arnold, F. Felder, M. Rahim, C. Ebneter, I. Zasavitskiy, N. Quack, S. Blunier, and J. Dual, "Epitaxial Lead Chalcogenides on Si Got Mid-IR Detectors and Emitters Including Cavities," *Journal of Electronic Materials* 37, 1497–1503, 2008.

235. D. Zimin, K. Alchalabi, and H. Zogg, "Heteroepitaxial PbTe-on-Si pn-Junction IR-Sensors: Correlations Between Material and Device Properties," *Physica E* 13, 1220–23, 2002.

16 Quantum Well Infrared Photodetectors

Since the initial proposal by Esaki and Tsu [1] and the advent of MBE, the interest in semiconductor superlattices (SLs) and quantum well (QW) structures has increased continuously over the years, driven by technological challenges, new physical concepts and phenomena as well as promising applications. A new class of materials and heterojunctions with unique electronic and optical properties has been developed. Here we focus on devices that involve infrared excitation of carriers in low dimensional solids (quantum wells, quantum dots, and superlattices). A distinguishing feature of these infrared detectors is that they can be implemented in chemically stable wide bandgap materials, as a result of the use of intraband processes. On account of this, it is possible to use such material systems as $GaAs/Al_xGa_{1-x}As$ (GaAs/AlGaAs), $In_xGa_{1-x}As/In_xAl_{1-x}As$ (InGaAs/InAlAs), $InSb/InAs_{1-x}Sb_x$ (InSb/InAsSb), $InAs/Ga_{1-x}In_xSb$ (InAs/GaInSb), and $Si_{1-x}Ge_x/Si$ (SiGe/Si), as well as other systems, although most of the experimental works have been carried out with AlGaAs. Some devices are sufficiently advanced that there exists the possibility of their incorporating in high-performance integrated circuits. High uniformity of epitaxial growth over large areas shows promise for the production of large area two-dimensional arrays. In addition, flexibility associated with control over composition during epitaxial growth can be used to tailor the response of quantum well infrared detectors to particular infrared bands or multiple bands.

Among the different types of quantum well infrared photodetectors (QWIPs), technology of the GaAs/AlGaAs multiple quantum well detectors is the most mature. Rapid progress has recently been made in the performance of these detectors [2–13]. Detectivities have improved dramatically and are now high enough so that megapixel focal plane arrays (FPAs) with long wavelength infrared (LWIR) imaging performance comparable to state of the art of HgCdTe are fabricated [14,15]. Despite large research and development efforts, large photovoltaic HgCdTe FPAs remain expensive, primarily because of the low yield of operable arrays. The low yield is due to sensitivity of LWIR HgCdTe devices to defects and surface leakage, which is a consequence of basic material properties. With respect to HgCdTe detectors, GaAs/AlGaAs quantum well devices have a number of potential advantages, including the use of standard manufacturing techniques based on mature GaAs growth and processing technologies, highly uniform and well-controlled molecular beam epitaxy (MBE) growth on greater than 6 inch GaAs wafers, high yield and thus low cost, more thermal stability, and extrinsic radiation hardness.

This chapter is devoted to the properties and applications of quantum well structures for infrared detection. It is difficult to cover such topics since the technology of the above devices is being rapidly developed and new concepts of these devices are currently proposed. It is assumed that the phenomena, materials, and optical and electrical properties of quantum wells and superlattices are well known for the reader. Several introductory textbooks on quantum well physics have been written by Bastard [16], Weisbuch and Vinter [17], Shik [18], Harrison [19], Bimberg, Grundmann, and Ledentsov [20], and Singh [21]. Only elementary properties of quantum wells are described below. Because Chapters 17 and 18 are devoted to superlattice and quantum dot photodetectors, we will concentrate on different types of low dimensional solids in the next section.

16.1 LOW DIMENSIONAL SOLIDS: BACKGROUND

Rapid progress in the development of epitaxial growth techniques has made it possible to grow semiconductor structures at one-monolayer accuracy. The device structure dimensions can be comparable to wavelengths of the relevant electron or hole wave functions, at least in the growth direction. This means that one can do electrical engineering at the quantum mechanical level. The electron confinement within a sufficiently narrow region of semiconductor material can significantly change the carrier energy spectrum and novel physical properties are expected to emerge. These novel properties will give rise to new semiconductor devices as well as to drastically improved device characteristics [22,23]. Most expected improvements in electronic and optoelectronic device performance originate from the change in the density of states.

In addition to quantum well case, where energy barriers for electron motion exist in one direction of propagation, one can also imagine electron confinement in two directions and, as the ultimate case, in all three directions. The structures of these kinds are now known as quantum wires and quantum dots (QDs). Thus, the family of dimensionalities of the device structures involves bulky semiconductor epilayer [three-dimensional (3-D)], thin epitaxial layer of quantum

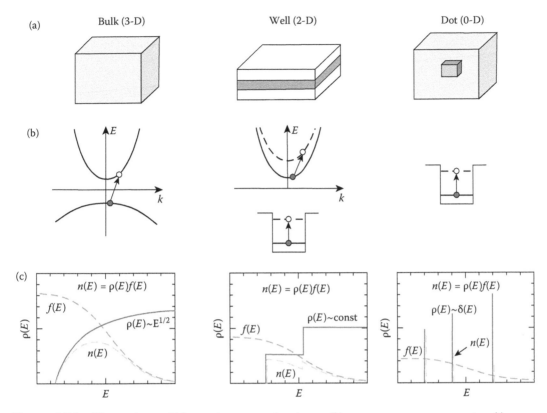

Figure 16.1 Illustrations of (a) quantum nanostructures, (b) energy versus wave vector, (c) density of states and energy spreading (where electrons have the Boltzmann distribution) in bulk, quantum well and quantum dot.

well [two-dimensional (2-D)], elongated tube or quantum wire [one-dimensional (1-D)] and, finally, isolated island of QD [zero-dimensional (0-D)]. These three cases are shown in Figure 16.1.

In a crystalline semiconductor, the electrons and holes that determine the transport and optical properties are considered as "quasi free" with the effective mass $m*$ taking account of the periodic crystal potential. The dimensions of a bulk semiconductor crystal are macroscopic on the scale of the de Broglie wavelength

$$\lambda = \frac{h}{\left(2m*E\right)^{1/2}},$$ (16.1)

and thus so-called size quantization is negligible. The quasi-free electron wave functions are given by the Bloch-functions

$$\Psi_{j\vec{k}}^{3D}\left(r\right) = \frac{1}{V}u_{j\vec{k}}\left(\vec{r}\right)\exp\left(i\vec{k}\vec{r}\right),$$ (16.2)

where V is the macroscopic volume, $k = 2\pi/\lambda$ is the electron wave vector, and j is the band index. The Bloch-functions have the free particle wave function $\exp\left(i\vec{k}\vec{r}\right)$ as the envelope, instead allowed k-values as defined by the crystal boundaries are quasi continuous. The energy dispersion of quasi-free electrons (holes) at the bottom of the conduction band (at the top of the valence band) is given by

$$E^{3D} = \frac{\hbar^2 k}{2m*},$$ (16.3)

where $\hbar$ is Planck's constant. This quadratic dispersion results in a parabolic density of states function

$$\rho^{3D} = \frac{1}{2\pi^2}\left(\frac{2m^*}{\hbar^2}\right)^{3/2} E^{1/2}. \qquad (16.4)$$

While the density of states characterizes the energy distribution of allowed states, Fermi function, f, gives the probability of electron occupation at certain energy even if no allowed state exists at all at that energy. Thus, the product of ρ and Fermi function describe the total concentration of the charge carriers of the given type in a crystal:

$$n = \int \rho(E) f(E)\, dE. \qquad (16.5)$$

The transport and optical properties of semiconductors are basically determined by the uppermost valence band and lowest conduction band, which are separated in energy by the bandgap E_g. The bandgap structure of GaAs consists of a heavy hole, light hole, and split-off valence band and the lowest conduction band. GaAs is a direct gap semiconductor with the maximum of the valence band and minimum of the conduction band at the same position in the Brillouin zone at $k = 0$ (Γ-point). AlAs is an indirect gap semiconductor with the lowest conduction band minimum close to the boundary of the Brillouin zone in (100) direction at the X-point. In addition, the conduction band minimum in AlAs is highly unisotrop with a longitudinal and transverse electron mass = 1.1 m and $m_t^* = 0.2\ m$, respectively.

Confinement of electrons in one or more dimensions modifies the wave functions, dispersion and density of states. The effective potential associated with spatial variation of the conduction and valence band edges is spatially modulated in so called compositional superlattices that consist of alternating layers of two different semiconductors. Figure 16.2 illustrates the electronic states associated with planar quantum wells and superlattices [24]. Widths of minibands in superlattices are related by the uncertainty principle to well-to-well tunneling times. In a typical case, barrier regions would be AlGaAs layers, and wells would be GaAs. Typical distance scales (≈ 50 Å) over which composition is varied is small compared with distances over which ballistic electron propagation has been observed. This supports the notion that wave functions can be coherent throughout structures that have spatially modulated composition. The wave functions just referred to are the envelope wave functions of effective mass theory. Spatial variation of the host composition introduces a number of sublattices, but the basic picture of potentials and wave functions guides much of the effort in the field of quantum devices.

Each quantum well can be considered to be a three-dimensional rectangular potential well. When the thickness of the well is much less than the transverse dimensions ($L_z << L_x, L_y$), and the thickness is comparable to the de Broglie wavelength of the carriers in the well, quantization of the carrier motion in the z-direction must be taken into account in the dispersion carrier dynamics. Motion in the x and y directions is not quantized, so that each state of the system corresponds to a subband. Electrons (or holes) in such a well can be regarded as a two-dimensional electron (or hole) gas. When the well is infinitely deep, the Schrödinger equation energy eigenvalues are

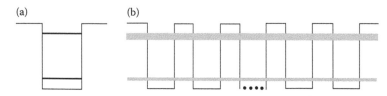

(a)   (b)

Figure 16.2 (a) Electron bound states in a quantum well, and (b) the formation of minibands in a superlattice. The confining potentials are associated with the conduction band edge. (From Coon, D. D., and Bandara, K. M. S. V., *Physics of Thin Films*, Academic Press, Boston, MA, Vol. 15, 219–64, 1991. With permission.)

$$E^{2\text{-D}} = E_{n_z} + \frac{\hbar^2 \left(k_x^2 + k_y^2\right)}{2m^*},\tag{16.6}$$

with the confinement energy

$$E_{n_z} = \frac{\hbar^2}{2m}\left(\frac{n_z\pi}{L_z}\right)^2,\tag{16.7}$$

where k_x and k_y are momentum vectors along the x and y axes, and n_z is the quantum number ($n_z = 1, 2, ...$). The electron wave functions are represented by plane waves in x and y direction and by even or odd harmonic functions in z direction:

$$\psi_{n_z}^{2\text{-D}} = \left(\frac{2}{L}\right)\exp\left(ik_x x\right)\exp\left(ik_y y\right)\left(\frac{2}{L_z}\right)^{1/2}\sin\left(k_{n_z}\right).\tag{16.8}$$

Confinement of electrons by potential wells with finite height (see Figure 16.2a) does not affect the principal features of size quantization as described above; however, it modifies the results in three important respects:

- The confinement energy for a given quantum state characterized by the quantum number n_z is lower for a finite barrier height.

- Only a finite number of quantized states is bound in a well with finite barrier height (for an infinitely high barrier an infinite number of quantized states exists); when the width of a single quantum well is decreased, the first excited state merges from the well and becomes a virtual state (Figure 16.3); for example, Figure 16.4 illustrates the link between bound states and virtual states as the parameters of a single quantum well are varied.

- The electron wave functions do not vanish at the boundary but penetrate into the barrier where the amplitude drops exponentially.

The latter effect actually provides the base for the formation of superlattices. If the overlap, the confined energy levels split into a manifold of levels given by the number of coupled potential wells. For a sufficiently large number of coupled wells these split levels form a quasi-continuous energy band as illustrated in Figure 16.2b.

In general, E_n and ψ_n are eigenvalues and eigenfunctions of a finite one-dimensional potential well. The energy levels in the k_x and k_y directions form a continuum, and for discrete values of E_n (i.e., for each bound state), a two-dimensional in the k_x–k_y plane will form. Each of these two-dimensional energy bands gives rise to a band density of states that is energy independent. The density of states function changes from the smooth parabolic shape to a function

$$\rho^{2\text{-D}} = \frac{m^*}{\pi\hbar^2}\sum_{n_z}\delta\left(E - E_{n_z}\right),\tag{16.9}$$

where $\delta(E)$ is the Heaviside step function with $\delta(E \geq E_{nz}) = 1$ and $\delta(E < E_{nz}) = 0$. The cumulative density of states is steplike in character up to the energy at which the discrete, bound state spectrum gives way to the continuum of free (unbound) states.

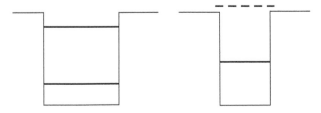

Figure 16.3 Illustration of the formation of a virtual state as the width of the quantum well is decreased. (From Coon, D. D., and Bandara, K. M. S. V., *Physics of Thin Films*, Academic Press, Boston, MA, Vol. 15, 219–64, 1991. With permission.)

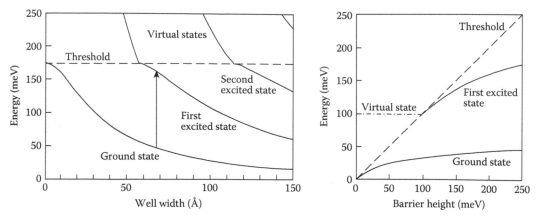

Figure 16.4 Dependence of energy levels on well width in a GaAs/Al$_{0.25}$Ga$_{0.75}$As quantum well. The zero of the energy scale is at the bottom of the well. The barrier height dependence for a 75 Å quantum well is shown on the right. (From Coon, D. D., and Bandara, K. M. S. V., *Physics of Thin Films*, Academic Press, Boston, MA, Vol. 15, 219–64, 1991. With permission.)

Farther confinement in two or finally three dimensions results in size quantization in the corresponding directions and stronger discretization of the energy spectrum and density of state distribution approaching atomic behavior for three dimensional confinement.

An ideal quantum dot, also known as a quantum box, is a structure capable of confining electrons in all three directions, thus allowing zero dimension in their degree of freedom. The energy spectrum is completely discrete, similar to that in an atom. The total energy is the sum of three discrete components:

$$E^{0-D} = E_{nx} + E_{my} + E_{lz} = \frac{\hbar^2 k_{nx}^2}{2m^*} + \frac{\hbar^2 k_{my}^2}{2m^*} + \frac{\hbar^2 k_{lz}^2}{2m^*}, \tag{16.10}$$

where n, m, and l are integers (1, 2, ...) used to index the quantized energy levels and quantized wave numbers, which result from the confinement of the electron motion in the x, y, and z-directions, respectively.

As for the bulk material, the most important characteristic of a QD is its electron density of states in the conduction band given by

$$\rho^{0-D}(E) = 2\sum_{n,m,l} \delta\left[E_{nx} + E_{my} + E_{lz} - E\right]. \tag{16.11}$$

Each QD level can accommodate two electrons with different spin orientation.

The density of states of zero-dimensional electrons consists of Dirac functions, occurring at the discrete energy levels $E(n,m,l)$, as shown in Figure 16.1. The divergences in the density of states shown in Figure 16.1 are for ideal electrons in a QD and are smeared out in reality by a finite electron lifetime ($\Delta E \geq \hbar/\tau$). Since QDs have a discrete, atom-like energy spectrum, they can be visualized and described as "artificial atoms." This discreteness is expected to render the carrier dynamics very different from that in higher-dimensional structures where the density of states is continuous over a range of values of energy.

The energy position of QD (also QW) level essentially depends on geometrical sizes and even one monolayer variation of the size can significantly affect the energy of optical transition. Fluctuations of the geometrical parameters results in corresponding fluctuation of the quantum level over the array of dots. Random fluctuations also affect the density of states on nonuniform array of QDs.

The self-assembling method for fabricating QDs has been recognized as one of the most promising methods for forming QDs that can be practically incorporated into IR photodetectors. In the crystal growth of highly lattice-mismatched materials system, self-assembling formation of nanometer-scale 3-D islands has been reported. The lattice mismatch between a QD and the matrix is the fundamental driving force of self-assembling. In(Ga)As on GaAs is the most commonly used material system because lattice mismatched can be controlled by the In alloy ratio up to about 7%.

Both quantum QW and QD structures are used in fabrication of infrared detectors. In general, quantum dot infrared photodetectors (QDIPs) are similar to QWIPs but with the quantum wells replaced by QDs, which have size confinement in all spatial directions.

Figure 16.5 shows the schematic layers of a QWIP and a QDIP [25]. In both cases, the detection mechanism is based on the intraband photoexcitation of electrons from confined states in the conduction band wells or dots into the continuum. The emitted electrons drift toward the collector in the electric field provided by the applied bias, and photocurrent is created. It is assumed that the potential profile at the conduction band edge along the growth direction for both structure have a similar shape as shown in Figure 16.5b.

The self-assembled QDs in QDIPs are wide in the in-plane direction and narrow in the growth direction. The strong confinement is therefore in the growth direction, while the in-plane confinement is weak, resulting in several levels in the dots (see Figure 16.6). In this situation, the transitions between in-plane confinement levels give rise to the normal incidence response.

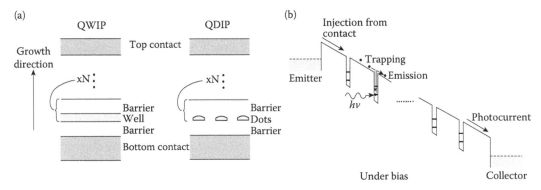

Figure 16.5 (a) Schematic layers of QWIP and QDIP, and (b) potential profile for both structures under bias. For QDIP, influence of wetting layer is neglected. (From Liu, H. C., *Opto-Electronics Review*, 11, 1–5, 2003. With permission.)

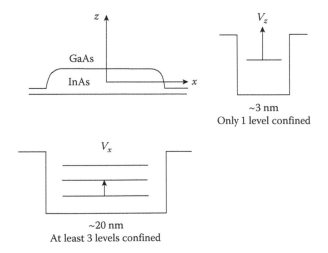

Figure 16.6 Illustration of transitions under polarized light in the growth direction (z) or in the in-plane directions (x or y). The strong confinement in the growth direction is represented by a narrow well; whereas the in-plane wide potential well leads to several states. The upward arrows indicate the strongest transitions for z and x polarized lights. (From Liu, H. C., *Opto-Electronics Review*, 11, 1–5, 2003. With permission.)

16.2 MULTIPLE QUANTUM WELLS AND SUPERLATTICES

All above considerations are devoted to mainly compositional superlattices that consist of alternating layers of two different semiconductors. The second kind of superlattice that was also originated by Esaki and Tsu is doping superlattice [1]. These types of superlattices consist of alternating n- and p-type layers of a single semiconductor. Electrical fields generated by the charged dopants modulate the electronic potential. Superlattices in which both the composition and the doping are modulated have been also considered [26].

16.2.1 Compositional Superlattices

Heteroepitaxy of materials with different bandgap is of fundamental interest for the composition superlattice growth. An important requirement for high quality compositional quantum well and superlattices is the matching of the lattice constants of the constituents with different bandgap, except for the so called pseudomorphic systems, where internal stress or strain due to lattice mismatch is used to tailor the electronic band structure in addition to the effects due to confinement. The status of strained-layer superlattices has been reviewed by Mailhiot and Smith [27].

The condition for lattice match is a severe restriction for the materials that can be considered for heteroepitaxy to form quantum well structures; however, nature still provides enough freedom. Figure 16.7 shows the plot of energy gaps at 4.2 K versus lattice constants for zinc-blende semiconductors together with Si and Ge. Joining lines represents ternary alloys except for Si-Ge, GaAs-Ge and InAs-GaSb. MnSe and MnTe are not shown here because their stable crystal structures are not zinc-blende. The most extensively studied structures are based on the GaAs/AlGaAs material system what is conditioned by almost perfect natural lattice match between GaAs and $Al_xGa_{1-x}As$ for all values of x. From Figure 16.7 results that the energy gap generally decreases with an increase in the lattice constant or the atomic number [28]. It should also be noted that all the binary compounds fall into five distinct columns shown by the shaded areas, suggesting that the lattice constants are alike as long as the mean atomic-numbers of the binary constituents are the same.

The physical properties of the respective quantum well structures are strongly determined by the band discontinuities at the interface (i.e., the band alignment). An abrupt discontinuity in the local band structure is usually associated with a gradual band bending in its neighborhood, which reflects space-charge effects. The conduction- and valence-band discontinuities will determine the character of carrier transport across the interfaces, and so they are the most important quantities, which determine the suitability of present SLs or QWs for IR detector purposes. The presence of an additional SL periodic potential changes the electronic spectrum of a semiconductor in such a manner that the Brillouin zone is divided into a series of minizones giving rise to narrow

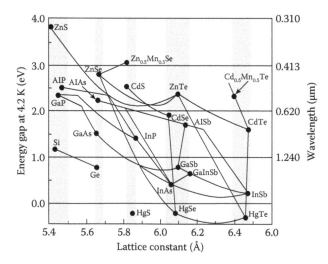

Figure 16.7 A plot of the low temperature energy bandgaps of a number of semiconductors with the diamond and zinc-blende structure versus their lattice constants. The shaded regions highlight several families of semiconductors with similar lattice constants. (From Esaki, L., *IEEE Journal of Quantum Electronics* QE-22, 1611–24, 1986. With permission.)

subbands separated by minigaps (see, e.g., Bastard [16]). Thus, the SLs will possess new properties not exhibited by homogeneous semiconductors. Surprisingly, the corresponding values for the band discontinuities of the conduction band (ΔE_c) and the valence band (ΔE_v) cannot be obtained by simple considerations. Band lineups based on electron affinity do not work in most cases when two semiconductors form a heterostructure. This is because of subtle charge sharing effects that occurs across atoms on the interface. There have been a number of theoretical studies that can predict general trends in how bands live up (e.g., [29–33]). However, the techniques are quite complex and heterostructure design usually depend on experiments to provide line up information [34–36]. It must be taken into account that electrical and optical methods do not measure the band offsets themselves but instead the quantities associated with the electronic structure of the heterostructure. The band offset determination from such experiments requires an appropriate theoretical model. The reported fundamental parameters ΔE_c and ΔE_v are slightly different, even for the most widely studied GaAs/Al$_x$Ga$_{1-x}$As material system. For the latter system values of $\Delta E_c:\Delta E_v$ = 6:4 are widely accepted for composition range $0.1 \leq x \leq 0.4$.

In dependence on the values for the band discontinuities, known heterointerface can be classified into four groups: type I, type II-staggered, type II-misaligned, and type III as shown in Figure 16.8.

Type I occurs for systems such as GaAs/AlAs, GaSb/AlSb, strained layer structure GaAs/GaP, and most II-VI and IV-VI semiconductor structures with a nonzero bandgap. The sum of ΔE_c and ΔE_v is seen to be equal to the bandgap difference $E_{g2} - E_{g1}$ of the two semiconductors. The electrons and holes are confined in one of the semiconductors that are in contact. Such types of SLs and multiple quantum wells (MQWs) are preferentially used as effective injection lasers, in which the threshold currents can be made much lower than those of heterolasers.

Type II structures can be divided into two groups: "staggered" (Figure 16.8b) and "misaligned" (Figure 16.8c) structures. Here it is seen that $\Delta E_c - \Delta E_v$ equals the bandgap difference $E_{g2} - E_{g1}$. The type II staggered structure is found in certain superlattices of ternary and quaternary III-Vs, where the bottom of the conduction band and the top of the valence band of one of the semiconductors are below the corresponding values of the other (e.g., as in the case of InAs$_x$Sb$_{1-x}$/InSb, In$_{1-x}$Ga$_x$As/GaSb$_{1-y}$As$_y$ structures). As a consequence, the bottom of the conduction bands and the top of the valence bands are located in opposite layers of SLs or MQWs, so the spatial separation of confined electrons and holes takes place. Such structures potentially can be used as photodetectors since photoinduced nonequilibrium carriers are spatially separated. The type II misaligned structure is an extension of this in which the conduction band states of semiconductor 1 overlap the valence band states of semiconductor 2. This has been established as occurring, for example, to InAs/GaSb, PbTe/PbS, and PbTe/SnTe systems. Electrons from GaSb valence band enter the InAs conduction band and produce a dipole layer of electron and hole gas shown in Figure 16.8c. With smaller periods of SLs or MQWs it is possible to observe the semimetalic-to-semiconductor transition and to use such systems as photosensitive structures, in which the spectral detectivity range can be changed by the thickness of the components.

Type III structures are formed from one semiconductor with a positive bandgap (e.g., $E_g = E_{\Gamma 6} - E_{\Gamma 8} > 0$, such as CdTe or ZnTe), and one semiconductor with a negative bandgap $E_g = E_{\Gamma 6} - E_{\Gamma 8} < 0$ (e.g., HgTe-type semiconductors). At all temperatures the HgTe-type semiconductors behave like semimetals since there are no activation energies between the light- and heavy-hole states in the Γ_8 band (see Figure 16.8d). This type of superlattice cannot be formed with III-V compounds.

16.2.2 Doping Superlattices

A spatial modulation of the doping in an otherwise homogeneous lattice can produce a superlattice effect; that is, a spatial modulation of the band structure that induces a reduction in the Brillouin zone of electrons and new energy bands in the superlattice direction. The realization of such structures is achieved using periodic n-doped, undoped, p-doped, undoped, n-doped, …, multilayer structure. So far, nearly all of the experimental investigations and most of the theoretical studies on doping superlattices have dealt with GaAs doping superlattice structures not containing intrinsic regions. The term "n-i-p-i crystals," however, became popular for the whole class of doping superlattices. A doping superlattice was first considered in the original proposal of Esaki and Tsu [1] and next pursued especially by Ploog and Döhler [37–39].

The basic concept of the n-i-p-i superlattice is explained with the help of Figure 16.8e. The doping superlattice causes the potential to oscillate between the n and p layers (in the same semiconductor) creating a reduced energy gap E_g^{eff} that separates the electron potential valley in the conduction band from the hole potential valley in the valence band. Charged particles are subject

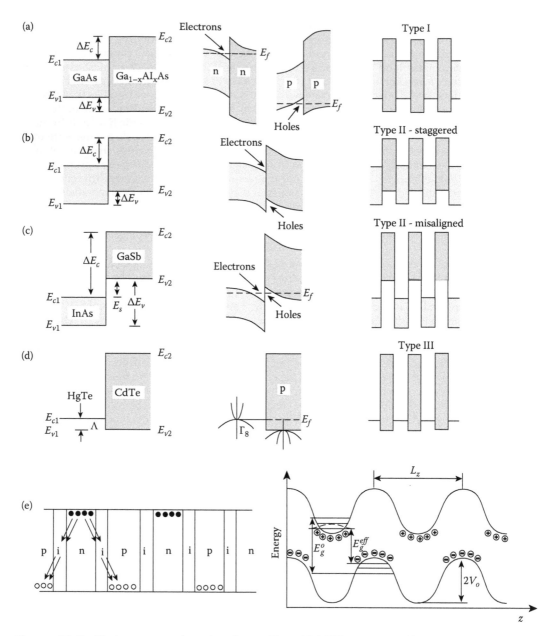

Figure 16.8 Various types of semiconductor SL and MQW structures: (a) type I structure, (b) type II "staggered" structure, (c) type II "misaligned" structure, (d) type III structure, and (e) n-i-p-i-structure. L_z is the period of the structures, $2V_o$ is the modulation potential and E_g^{eff} is the effective bandgap in n-i-p-i-structures.

to a self-consistent potential [40]. The value of the effective potential $2V_o$ and the effective band-gap E_g^{eff} depend on the doping levels N_d, the dielectric constant values ε_r, and the thicknesses of the layers. At equilibrium

$$E_g^{eff} = E_g^o - \frac{q^2 N_d}{\varepsilon_o \varepsilon_r}\left(\frac{a}{4}+b\right), \tag{16.12}$$

where a is the layer thickness of n-type and p-type regions, b is the thickness of the i-type layers, $E_g^o = E_c - E_v$ is the intrinsic band gap, and the donor and acceptor concentrations are assumed equal. For equal uniform doping levels N_d and zero thickness undoped layers, the periodic potential arcs and has an amplitude

$$V_o = \frac{q^2}{8\varepsilon_o\varepsilon_r} N_d a^2.$$ (16.13)

For GaAs with $N_d = 10^{18}$ cm^{-3} and $a = 500$ Å, $V_o = 400$ meV.

Equation 16.12 neglects additional terms due to quantization of energy in the potential valleys of the n- and p-layers when a is very small. The quantized energy levels in the potential wells are approximately the harmonic oscillator levels:

$$E_{e,h} = \hbar \left(\frac{q^2 N_d}{\varepsilon_o \varepsilon_r m_{e,h}^*} \right) \left(n + \frac{1}{2} \right).$$ (16.14)

For electrons in GaAs, for instance, the subband separation is 40.2 meV for the above parameters.

The Equation 16.12 gives the equilibrium value of E_g^{eff} with no applied voltage between the layers. If however, the n- and p-layers can be contacted separately, using methods described by Döhler [38], then the very interesting possibility arises of controlling the energy gap as function of the applied voltage.

The n-i-p-i structures are qualitatively similar to the type-II superlattices in that the separation in free space of electrons and holes reduces the overlap of the electron and hole wave functions and, consequently, the absorption coefficient. This effect is at least partially compensated by the increased carrier lifetime, which is also a consequence of the spatial separation. These structures can also be considered as potential photodetectors due to the spatial separation of photoinduced carriers [41].

16.2.3 Intersubband Optical Transitions

The description of the electron confinement is simplest for a one-dimensional rectangular potential of a well with infinitely high barriers (see Section 16.1). For this model all the results for description of the performance of the optoelectronic devices can be obtained analytically, and though they are not applicable quantitatively for real structures, the insight gained from such a model is transferable to the finite-barrier case.

In the finite-well case the position of the energy levels changes considerably compared to the infinite-well case, even for a parabolic dispersion law. The nonparabolicity of the dispersion law, the many-valley band structure (e.g., the case of n-Si and n-Ge), and the finite heights of the barriers change the description considerably. The electron wave functions already do not vanish at the boundaries of the well but penetrate into the barriers (the amplitudes drop exponentially in the barriers), which is the basis for the formation of SLs. The amplitudes of the envelope wave functions (together with Bloch functions) both in the wells and in the barriers determine the intensity of interband and intersubband (intraband) optical transitions (see Figure 16.9). For more familiar acquaintance with the analysis of optical transitions in SLs and QWs one can see (e.g., [16,21,42]).

To obtain the absorption coefficient values one needs to calculate the dipol matrix element. From theoretical consideration results (see, e.g., [16]) that the allowed dipole optical transitions are split into two classes:

- Interband transitions that take part between QW subbands originating from different band extrema i, j and defined by atomic-like dipole matrix elements.

- Intersubband (intraband, $i = j$) optical transitions that are defined by dipole matrix elements between the envelope functions of the same band.

The optical dipole moment can be expressed as

$$M \sim \int \phi_F(z) \vec{\varepsilon} \cdot \vec{r} \phi_I(z) dr,$$ (16.15)

where ϕ_I and ϕ_F are initial and final envelope wave functions, $\vec{\varepsilon}$ is the polarization vector of the incident photons, and z is the growth direction of the quantum well. These yield a dipole matrix element in the order of the size of the quantum well for intersubband transitions compared to the value of the atomic size for interband transitions. For an infinite well, the value of the dipole matrix element $\langle z \rangle$ between the ground and first excited state is $16L/9\pi^2$ (~0.18 L_w; L_w is the width of the QW). Since the envelope wave functions in Equation 16.15 are orthogonal, M is nonzero due to the component of $\vec{\varepsilon} \cdot \vec{r}$ perpendicular to the QW (along growth direction). Therefore, the optical electric field must also have a component along this direction in order to induce an intersubband transition; thus normal incidence radiation will not be absorbed.

The intensity of intersubband optical transitions is proportional to $\cos^2\phi$, where ϕ is the angle between the plane of the QW and the electromagnetic electric field vector. Levine et al. [43] have shown that the polarization selection rule $\alpha \propto \cos^2\phi$ is experimentally confirmed as shown in Figure 16.10 [2]. The IR intersubband absorption was measured at 8.2 μm in a doped GaAs/AlGaAs

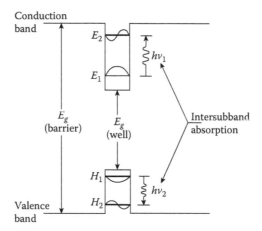

Figure 16.9 Schematic band diagram of a quantum well. Intersubband absorption can take place between the energy levels of a quantum well associated with the conduction band (n-doped) or the valence band (p-doped).

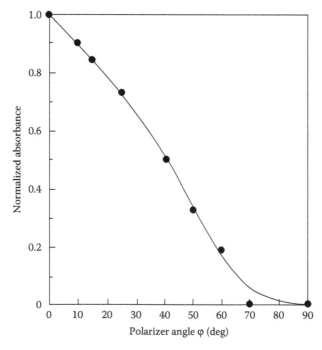

Figure 16.10 Measured intersubband absorbance (normalized to $\phi = 0$) versus polarizer angle at Brewster's angle $\theta_B = 73°$ for a doped 8.2 μm GaAs/AlGaAs quantum well superlattice. The solid line is drawn through the points as a guide to the eye. (From Levine, B. F., *Journal of Applied Physics,* 74, R1–R81, 1993. With permission.)

quantum well superlattice using a multipass waveguide geometry. The multipass waveguide geometry increased the net intersubband absorption by approximately two orders of magnitude, hence allowing accurate measurements of the oscillator strength, the polarization selection rule, and the line shape.

The absorption coefficient $\alpha(h\nu)$ associated with an optical transition of an electron promoted from the ground state E_1 to excited state E_2 by absorbing a photon $h\nu$ can be expressed as [44]

$$\alpha(h\nu) = \frac{\rho^{2\text{-}D}}{L}\frac{\pi q^2 \hbar}{2n_r \varepsilon_o m^* c}\sin^2 \varphi f(E_2)g(E_2),$$
(16.16)

where L is the period length of MQW structure, m^* is the effective mass of the electrons in the well, n_r is the index of refraction, $f(E_2)$ is the oscillator strength, and $g(E_2)$ is the one-dimensional density of the final state. When the scattering effects are neglected, the density of final state $g(E_2)$ in the continuum is given simply by [44]

$$g(E_2) = \frac{L}{2\pi\hbar}\left(\frac{m_b^*}{2}\right)^{1/2}\frac{1}{\sqrt{E_2 - H}},$$
(16.17)

where m_b^* is the effective mass of the electrons in the barrier and H is the barrier height.

According to Equation 16.16, the photon energy at the absorption peak is determined by the product of $f(E_2)$ and $g(E_2)$; however, due to the singularity of $g(E_2)$ at $E_2 = H$, the absorption strongly peaks close to the barrier height. In reality, the density of states in the continuum is strongly modified by the presence of the well and is broadened by impurity scattering and both will tend to smooth out the singularity. As a result, it is safe to assume that local density of states in the continuum varies relatively slowly compared to $f(E_2)$ when E_2 is close to H [44]. Under this assumption, the absorption peak is approximately determined by the energy dependence of the oscillator strength f. Hence, the peak absorption wavelength λ_p can be obtained by computing the value of $E_2 = E_m$ where f is maximized:

$$\lambda_p = \frac{2\pi\hbar c}{E_m - E_1}.$$
(16.18)

Choi [44] has performed calculations on the detector wavelength, the absorption linewidth, and the oscillator strength of a typical GaAs/Al$_x$Ga$_{1-x}$As MQW photodetector with aluminum molar ratio in the barrier ranging from 0.14 to 0.42 and the quantum-well width ranging from 20 to 70 Å. Figure 16.11 shows the absorption peak wavelength λ_p as a function of well width. Within the detector parameters shown, λ_p can be varied from 5 to over 25 μm.

The intersubband absorption has been investigated theoretically and experimentally by a lot of authors as a function of quantum-well width, barrier height, temperature, and doping density in the well [2,5,6]. Bandara et al. [45] have shown that, for high doping ($N_d > 10^{18}$ cm^{-3}), exchange interaction can significantly lower the ground-state subband energy and that the direct Coulomb shift can increase the excited-state subband energy; consequently the peak absorption wavelength shifts to higher energy. In addition to absorption peak shift at high doping densities, the absorption linewidth broadens and the oscillation strength increases linearly with the doping density. Furthermore, there are shifts in the peak absorption wavelength and absorption linewidth with the temperature. The experimentally observed linewidths are within $\Delta\nu = 50$–120 cm$^{-1} \approx 6$–15 meV and are governed by the process of longitudinal optical (LO) phonon scattering, with the intersubband relaxation times between the second excited and the first ground states within $\tau_{21} \approx (0.2$–$0.9)$ ps [4,46]. With temperature lowering, there is a small decrease in the position of the peak absorption wavelength and the absorption linewidth. Hasnain et al. [47] have observed an increase in the peak absorption by approximately 30% in comparison with typical maximum room temperature value of $\alpha \approx 700$ cm^{-1}. Manasreh et al. [48] have explained this temperature shift by including the collective plasma, exciton-like, Coulomb and exchange interactions, nonparabolicity, and the temperature dependence of the band gaps and effective mass.

To demonstrate the difference between the absorption spectra line shapes for different MQW structures, Figure 3.9 shows the normalized absorption spectra at $T = 300$ K. The very large

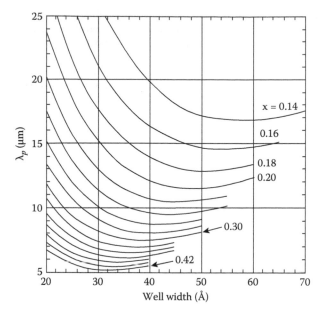

Figure 16.11   The position of the absorption peak λ_p for a given Al molar ratio x as a function of well width. The x changes 0.02 in steps between each curve. (From Choi, K. K., *Journal of Applied Physics*, 73, 5230–36, 1993. With permission.)

difference in spectral width is apparent with the bound excited-state transitions ($\Delta\lambda / \lambda = 9\%$–$11\%$) being three to four times narrower than for the continuum excited-state [49]. Much broader spectra in the case of bound-to-continuum transitions are caused by the broadening of the extended continuum excited states.

The major drawback of using n-type III-V Γ-point extrema MQWs is that intersubband transitions are forbidden for normally incident radiation. Considerable efforts were applied to observe the normal incidence infrared absorption both in n-type and p-type III-V and Si/SiGe MQWs (e.g., see [50–60]).

In the case of n-type structures the normal incidence absorption is possible due to the multivalley band structure when the principal axes of the effective mass tensor in the ellipsoids of equal energy are tilted with respect to the growth direction [50]. In such a situation the normal incidence intersubband absorption may be strong enough and reach in n-Si QWs the values of about 10^3–10^4 cm^{-1} for <110> or <111> growth directions and free carrier concentrations in the ground state of about 10^{19} cm^{-3}. Heavily doped Si is easier to achieve than n-GaAs and corresponding Fermi levels are lower in silicon layers due to the greater density of states.

For p-type Si/SiGe normal incidence MQW photodetectors the great advantage lies in an entirely silicon-based technology. But there exist some metallurgical problems of SiGe layers growth with large Ge content. Also, the p-type MQW devices are restricted in transport and responsivity characteristics due to much lower carrier mobilities compared to n-type III-V compound MQW intersubband devices.

As was earlier shown [51] in p-type QWs the nonorthogonal nature of the hole envelope functions is the reason for allowed hole-intersubband transitions with any polarization of the light since for finite in-plane wave vectors the eigenfunctions of the multiband effective mass Hamiltonian are weighted hole-envelope function linear combinations. In addition to intersubband transitions, transitions between different hole bands were also observed (see Figure 16.12). For these so-called intervalence-subband transitions, the same selection rule is applicable [57–59]. Such transitions occur not only in the 8–12 μm spectral range but also can be extended to the 3–5 μm IR radiation range [58].

The p-type Si/SiGe QWIPs can have broadband photoresponse (8–14 μm), attributed to strain and quantum confinement induced mixing of heavy, light, and split-off hole bands [60].

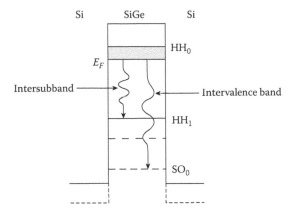

Figure 16.12 Band diagram of the quantum well structures showing possible subband transitions. (From Karunasiri, G., *Japanese Journal of Applied Physics*, 33, 2401–11, 1994. With permission.)

It must be taken into account that the intersubband transitions in SLs or QWs are possible only if their initial states are occupied by carriers. So, the intersubband photodetectors are extrinsic in nature.

16.2.4 Intersubband Relaxation Time

There is a considerable interest in the direct determination of the intersubband relaxation processes as these processes determine the operating frequency of MQW devices, their total quantum efficiency, and the photoconductive gain. The relevant interactions leading to carrier capture are electron-phonon, electron-impurity, and electron-electron scattering. Usually theoretical considerations are restricted to electrons in n-type GaAs/AlGaAs QWIPs, since the band structure is known reasonably well, and since coupling between heavy and light hole states in the valence band introduces additional complications [12].

The most relevant interactions is the Fröhlich interaction between electrons and LO phonons. Because the confined carriers are free to move within the plane, there is no energy gap separating confined from unconfined states, and the density of empty final states satisfying energy-momentum conservation for phonon emission is high. Consequently transitions from extended states ("above" the barriers) to confined states are very fast, resulting in a extremely short excited-carrier lifetime on the order of picoseconds. At large carrier densities (above 10^{18} cm^{-3}), electron-impurity and electron-electron scattering play an important role, whereas radiative relaxation is several orders of magnitude less efficient than electron-phonon interaction and can be neglected in this context.

If LO phonons play a decisive role in the intersubband relaxation processes, then the intersubband relaxation time will depend on whether the energy separation between the subbands is greater or lower than the LO-phonon energy E_{LO} (e.g., in GaAs $E_{LO} = 36.7$ meV). If separation between the ground and excited states in the well is lower than the LO energy then these phonons would not play any role in relaxation processes and the lifetime may be rather long (and so the escape probability of the carriers out of the well under the exterior electric field will be large); if not, then the intersubband relaxation time should be less than 1 ps [61]. The situation discussed was proved in the experiments on picosecond time-resolved Raman spectroscopy. In "broad" GaAs QWs ($L_w = 21.5$ nm), in which the energy separation ΔE_{12} between the ground and excited states is less than the LO-phonon energy ($\Delta E_{12} = 26.8$ meV $< E_{LO} = 36.7$ meV) and thus the carrier scattering via LO-phonon excitation cannot take place, the intersubband relaxation times of several hundred picoseconds were observed [62]. These long lifetimes were explained by longitudinal acoustical phonon scattering of the carriers in the upper excited states of the QWs. In another paper [63], the intersubband relaxation times in excess of 500 ps have been observed at low temperatures. For narrower wells ($L_w = 116$ Å, $\Delta E_{12} = 64.2$ meV $> E_{LO}$), where LO phonons play a major role in relaxation processes, the intersubband relaxation time was too short to be measured with the resolution of about 8 ps of the setup used.

Other experiments on the intersubband relaxation time determination within the conditions $\Delta E_{12} > h\nu_1$ showed that under such conditions the times of intersubband relaxation processes of about $\tau_{12} \approx (1\text{--}10)$ ps were observed in GaAs and related MQWs [4]. The experiments carried out on intersubband relaxation time determination differ substantially on the τ_{12} values. The results obtained depend much on the value of photoexcited carrier densities, the degree of confinement of the particle in the excited state, and so on.

Simple estimations carried out for intersubband energies when $E_2 - E_1 >> E_{LO}$, in the case of bound-to-continuum QWIPs, give [64]

$$\frac{1}{\tau_{LO}} = \frac{q^2 \lambda_c E_{LO} I_1}{4h^2 c L_p} \left(\frac{1}{\varepsilon_\infty} - \frac{1}{\varepsilon_s} \right), \tag{16.19}$$

where λ_c is the cutoff wavelength, L_p is the QWIP period, and $I_1 \approx 2$ a dimensionless integral. This equation gives the capture time about 5 ps for typical QWIP parameters.

The lifetime of excited carriers from the ground to continuum states depends on the energy of the states above the well as the capture probability depends on the energy position of the particle above the well. For AlGaAs MQWs the excited carriers are captured into the well by emission of polar optical phonons and for excited states slightly above the well the lifetime may be less than 20 ps [65]. Thus, taking into account the thin layered structure of intersubband QW photodetectors, high speed operation (> 1 GHz) can be provided for a variety of IR applications such as picosecond CO_2-laser pulse investigations, high frequency heterodyne experiments, new telecommunications requirements with new IR fiber materials, and so forth.

16.3 PHOTOCONDUCTIVE QWIP

The concept of using infrared photoexcitation out of quantum wells as a means of infrared detection was suggested by Smith et al. [66,67]. Coon and Karunasiri [68] made a similar suggestion and pointed out that the optimum response should occur when the first excited state lies near the classical threshold for photoemission from the quantum well. West and Eglash [69] were the first to demonstrate large intersubband absorption between confined states in a 50 GaAs quantum well. In 1987, Levine and coworkers [70] fabricated the first QWIP operating at 10 μm. This detector design was based on a transition between two confined states in the quantum well and subsequent tunneling out of the well by an applied electric field. It appears that transitions between ground state and the first excited state have relatively large oscillator strengths and absorption coefficient. However, this by itself is not useful for detection, since photoexcited carriers may not readily escape from excited bound states. Tunneling from excited bound states is exponentially suppressed. By decreasing the size of a dual-state quantum well, the strong oscillator strength of the excited bound state can be pushed up into the continuum. As long as the virtual state is not far above threshold, then the excited state remains effective in enhancement of photoexcitation [71,72].

Up to the present, several QWIP configurations have been reported based on transitions from bound to extended states, from bound to quasi-continuum states, from bound to quasi-bound states, and from bound to miniband states [12]. All QWIPs are based on bandgap engineering of layered structures of wide bandgap (relative to thermal IR energies) materials. The structure is designed such that the energy separation between two of the states in the structure match the energy of the infrared photons to be detected.

Figure 16.13 shows two detector configurations used in fabrication of multicolor QWIP FPAs. The major advantage of the bound-to-continuum QWIP (Figure 16.13a) is that the photoelectron can escape from the quantum well to the continuum transport states without being required to tunnel through the barrier. As a result, the voltage bias required to efficiently collect the photoelectrons can be reduced dramatically, thereby lowering the dark current. Furthermore, since the photoelectrons are collected without having to tunnel through a barrier, the AlGaAs barriers can be made thicker without reducing the photoelectron collection efficiency. The multilayer structure consists of a periodic array of Si-doped ($N_d \approx 10^{18} \text{cm}^{-3}$) GaAs quantum wells of thickness L_w separated by undoped $Al_x Ga_{1-x} As$ barriers of thickness L_b. The heavy n-type doping in the wells is required to ensure that freezeout does occur at low temperatures and that a sufficient number of electrons are available to absorb the IR radiation. For operation at $\lambda = 7\text{--}11$ μm, typically $L_w = 40$ Å, $L_b = 500$ Å, $x = 0.25\text{--}0.30$, and 50 periods are grown. In order to shift the intersubband absorption to longer wavelength the x value is decreased to $x = 0.15$ and, in addition, in order to maintain the strong optical absorption and reasonably sharp cutoff line shape, the quantum well width is increased from 50 to 60 Å. This optimization allows the same bound state to excited continuum state optical absorption and efficient hot electron transport and collection. It appears that the dark

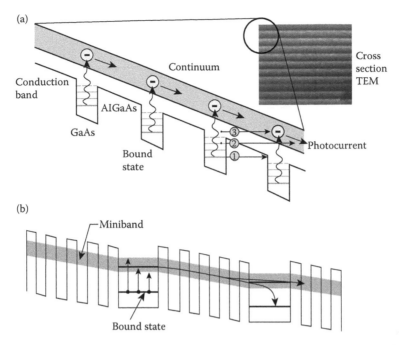

Figure 16.13 Band diagram of demonstrated QWIP structures: (a) bound-to-extended (after Ref. 7) and (b) bound-to-miniband. Three mechanisms creating dark current are also shown in (a): ground-state sequential tunneling (1), intermediate thermally assisted tunneling (2), and thermionic emission (3).

current decreases significantly when the first excited state is decreased in energy from the continuum to the well top in a bound-to-quasi-bound QWIP (Figure 16.14 [71]), without sacrificing responsivity. In comparison with narrow response of bound-to-bound transitions, the bound-to-continuum transitions are characterized by a broader response. The simple QWIP structures shown in Figures 16.5a and 16.13a are based on the photoemission of electrons from the QWs. They are unipolar devices with contacts on both sides, which require typically 50 wells for sufficient absorption (although 10–100 have been used).

A miniband transport QWIP contains two bound states, the higher-energy one being in resonance with the ground state miniband in the superlattice barrier (see Figure 16.13b). In this approach, IR radiation is absorbed in the doped quantum wells, exciting an electron into the miniband that provides the transport mechanism, until it is collected or recaptured into another quantum well. Thus, the operation of this miniband QWIP is analogous to that of a weakly coupled MQW bound-to-continuum QWIP. In this device structure, the continuum states above the barriers are replaced by the miniband of the superlattice barriers. The miniband QWIPs have lower photoconductive gain than bound-to-continuum QWIPs, because the photoexcited electron transport occurs in the miniband where electrons have to pass through many thin heterobarriers, resulting in a lower mobility.

16.3.1 Fabrication

The quantum well AlGaAs/GaAs structures are mainly grown by MBE on semi-insulating GaAs substrate, although excellent superlattice structures can also be grown using MOCVD [73,74]. At present, GaAs substrates are available with a diameter up to 8 inches, however in QWIP processing lines are typically 4 inch substrates. Process technology starts with epitaxial growth of structure with a periodic array of Si-doped ($N_d \approx 10^{18}$ cm^{-3}) GaAs quantum wells and sequential growth of an etch-stop layer (usually AlGaAs) used for substrate removal. The QWIP active region is sandwiched between two n-type GaAs contact layers about 1 μm thick (also heavily doped to $N_d \approx 10^{18}$ cm^{-3}), and an etch stop followed by a sacrificial layer for the grating. For optical coupling, usually 2-D reflective diffraction gratings are fabricated (see Section 16.6). Further process technology includes etching mesas through the superlattice to the bottom contact layer, followed by ohmic contacts to the n^+-doped GaAs contact layers. These steps can be accomplished using wet

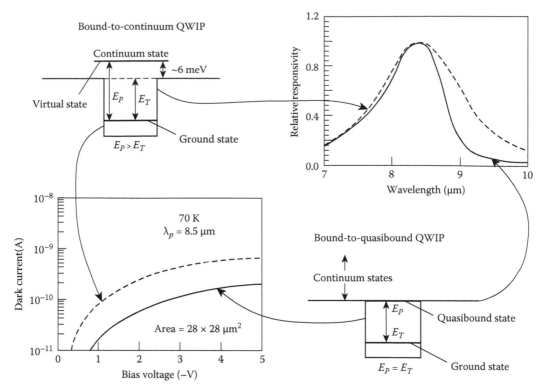

Figure 16.14 In typical photoresponse curves of bound-to-quasibound and bound-to-continuum 8.5 μm QWIPs at a temperature of 77 K the dark current (lower left) decreases significantly when the first excited state is dropped from the continuum to the well top, bound-to-quasibound QWIP, without sacrificing the responsivity (upper right). The first excited state now resonating with barrier top produces sharper absorption and photoresponse. (From Gunapala, S., Sundaram, M., and Bandara, S., *Laser Focus World*, 233–40, June 1996.)

chemical or dray etching techniques. In selective etching usually ion beam etching is practical for pattering the grating coupler into each pixel.

The same processes are used for the fabrication both MWIR and LWIR devices. Difference concerns the mesas definitions. This modification is necessary since the MWIR QWIPs are based on InGaAs/AlGaAs system, whereas the LWIR devices contain In-free GaAs/AlGaAs epilayers. From these reasons, the mesas are defined by reactive ion beam etching for MWIR QWIPs and by chemically assisted ion beam etching for LWIR QWIPs [12].

Ohmic contacts are evaporated (e.g., AuGe/Ni/Au) and alloyed by rapid thermal annealing (e.g., at 425°C for 20 seconds [75]). Usually, the gratings on each pixel is covered by metallization (Au), which is advantageous in comparison with using the ohmic contact metal in order to increase IR absorption in an active detector's region. The surface of the detector array is passivated with silicon nitride and to provide electrical contact to each detector element, openings in the nitride are formed. Finally, in order to facilitate the hybridization to silicon readout, a separate metallization is evaporated. Figure 16.15 shows the cross section of a pixel in a QWIP array.

After dicing the wafers into single chips and hybridization to silicon readout, the GaAs substrate is removed in order to reduce mechanical stress between two chips and to prevent optical crosstalk arising from light propagation between pixels. The process of substrate removal is accomplished using a sequence of mechanical lapping, wet chemical polishing, and a selective wet chemical etching. The last process is stopped at a dedicated etch-stop layer previously deposited.

16.3.2 Dark Current

A good understanding of the dark current is crucial for design and optimization of a QWIP detector because dark current contributes to the detector noise and dictates the operating temperature.

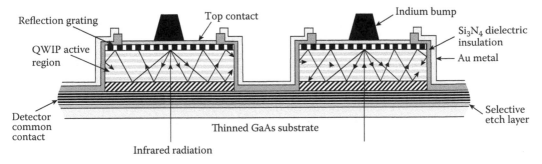

Figure 16.15 Cross section of a detector element in a QWIP array.

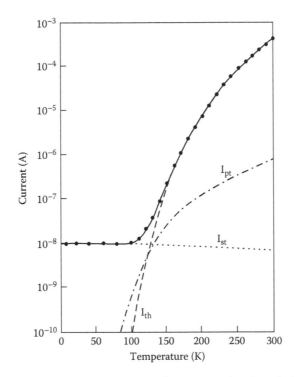

Figure 16.16 Temperature dependence of the dark current at low bias for bound-to-bound multiple quantum well $Al_{0.36}Ga_{0.64}As$/GaAs device (50 wells with 70 Å wells and 140 Å barriers). The subscripts "th," "st," and "pt" refer to thermionic, tunneling, and phonon-assisted tunneling mechanisms, respectively. (From Choi, K. K., Levine, B. F., Bethea, C. G., Walker, J., and Malik, R. J., *Applied Physics Letters*, 50, 1814–16, 1987. With permission.)

Initial efforts with multiple quantum well photoconductive detectors showed appreciable dark currents in which are involved three relevant mechanisms: tunneling, phonon assisted tunneling, and thermionic emission out of quantum wells. In Figure 16.16 [76], individual current contributions for devices with areas $A = 2 \times 10^{-5}$ cm^2 at $V_b = 0.05$ V are shown as a function of temperature. We can see that tunneling is the dominant dark current mechanism in the low-temperature regime, while thermionic emission limits the performance at high operating temperature.

Further rapid progress in multiple quantum well infrared detectors was conditioned by use of thinner quantum wells and pushing the excited state into the continuum. The linear relation between responsivity and the bias voltage for these devices is completely different from the highly nonlinear photoresponse obtained from devices containing two bound states in the quantum wells. In particular, the bound-to-bound state device requires a substantial bias $V_b > 0.5$ V before any photosignal is observed whereas the bound-to-extended detectors generate a photocurrent at a very low bias. The reason for this difference is that the bound-to-bound state detector requires a large electric field to assist the tunneling escape of the photoexcited carriers out of the well, and thus at low bias the excited carriers cannot be collected resulting in a negligible photoresponse. The barrier thickness could be greatly increased (e.g., $L_b \approx 500$ Å) thereby dramatically lowering the undesirable dark current.

In the further considerations we follow after Levine et al. They shown that thermionic-assisted tunneling is a major source of the dark current [2,5,7,77]. To calculate the dark current I_d, we first determine the effective number of electrons n that are thermally excited out of the well into the continuum transport states, as a function of bias voltage V:

$$n = \left(\frac{m^*}{\pi \hbar^2 L_p} \right) \int_{E_o}^{\infty} f(E) T(E, V) \, dE, \tag{16.20}$$

where the first factor containing the effective mass m^* is obtained by dividing the two-dimensional density of states by the superlattice period L_p (to convert it into an average three-dimensional density), and where $f(E)$ is the Fermi factor $f(E) = \{1 + \exp[(E - E_o - E_F)/kT]\}^{-1}$, E_o is the ground state energy, E_F is the two-dimensional Fermi level, and $T(E,V)$ is the bias-dependent tunneling current transmission factor for a single barrier. Equation 16.20 accounts for both thermionic emission above the energy barrier E_b (for $E > E_b$) and thermionically assisted tunneling (for $E < E_b$). The bias-dependent dark current

$$I_d(V) = q n(V) v(V) A, \tag{16.21}$$

where q is the electronic charge, A is the device area, and v is the average transport velocity (drift velocity) given by $v = \mu F[1 + (\mu F/v_s)^2]^{-1/2}$, where μ is the mobility, F is the average field, and v_s is the saturated drift velocity.

A much simpler expression that is a useful low-bias approximation can be obtained by setting $T(E) = 0$ for $E < E_b$ and $T(E) = 1$ for $E > E_b$ (E_b is the barrier energy), resulting in [2,77,78]

$$n = \left(\frac{m^* kT}{\pi \hbar^2 L_p} \right) \exp\left(-\frac{E_c - E_F}{kT} \right), \tag{16.22}$$

where we have set the spectral cutoff energy $E_c = E_b - E_1$. Therefore,

$$\frac{I_d}{T} \propto \exp\left(-\frac{E_c - E_F}{kT} \right), \tag{16.23}$$

where the Fermi energy can be obtained from

$$N_d = \left(\frac{m^* kT}{\pi \hbar^2 L_w} \right) \ln\left[1 + \exp\left(\frac{E_F}{kT} \right) \right]. \tag{16.24}$$

Figure 16.17 compares the experimental (solid curves) and theoretical (dashed) dark I-V curves at various temperatures for a 50-period multiquantum well superlattice [78]. The good agreement between theory and experiment is achieved over a range of eight orders of magnitude in dark current and demonstrates the high quality of the AlGaAs barriers (e.g., no tunneling defects or traps in the barriers).

For AlGaAs/GaAs QWIPs operating at temperatures above 45 K (for 15 μm devices), the thermionic emission dominates the dark current. Dropping the first excited state to the well top (bound-to-quasibound QWIPs; see Figure 16.14) theoretically causes the dark current to drop by a factor of ~6 at temperature 70 K for 9 μm devices [79]. This compares well with the fourfold drop experimentally observed. The bound-to-quasibound QWIP still preserves the photocurrent

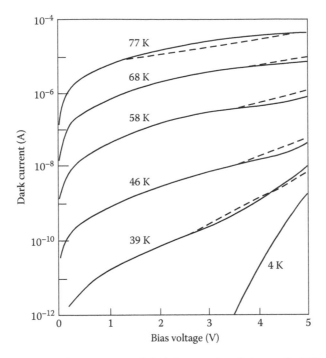

Figure 16.17 Comparison of experimental (solid curves) and theoretical (dashed) dark current-voltage curves at various temperatures for a 50-period, 200 μm diameter mesa $Al_{0.25}Ga_{0.75}As/GaAs$ detector having a doping density of 1.2×10^{18} cm^{-3} ($L_w = 40$ Å, $L_b = 480$ Å, $\lambda_c = 10.7$ μm. (From Gunapala, S. D., Levine, B. F., Pfeifer, L., and West, K., *Journal of Applied Physics*, 69, 6517–20, 1991. With permission.)

[72,79]. The first excited state could be pushed deeper into the well to increase the barrier to thermionic emission, but this would drop the photocurrent to unacceptably low levels. Dark current was also reduced by cutting well-doping density to decrease the ground-state electrons available for thermionic emission and by increasing each barrier thickness in the quantum-well stack.

Using a similar consideration, Kinch and Yariv [80] have presented an investigation of the fundamental physical limitations of individual multiple quantum well infrared detectors as compared to ideal HgCdTe detectors. Figure 16.18 compares the thermal generation current versus temperature for AlGaAs/GaAs multiple quantum well superlattices and HgCdTe alloys at $\lambda_c = 8.3$ μm and 10 μm. Calculations were carried out for a specific set of device parameters ($\tau = 8.5$ ps, $t = 1.7$ μm, $L_w = 40$ Å, $L_p = 340$ μm, and $N_d = 2 \times 10^{18}$ cm^{-3}) chosen to agree with the already published data for detectors with $\lambda_c = 8.3$ μm [81]. For $\lambda_c = 10$ μm the quantum well width is changed to $L_w = 30$ Å and remaining parameters assumed the same. It is apparent from Figure 16.18 that for HgCdTe the thermal generation rate at any specific temperature and cutoff wavelength is approximately five orders of magnitude smaller than for the corresponding AlGaAs/GaAs superlattice. The dominant factor favoring HgCdTe in this comparison is the excess carrier lifetime, which for n-type HgCdTe is above 10^{-6} s at 80 K, compared to 8.5×10^{-12} s for the AlGaAs/GaAs superlattice. Plotted on the right-hand axis of Figure 16.18 is the equivalent minimum temperature of operation in BLIP condition. For example, at a typical system background flux of 10^{16} photons/cm^2 s, the required temperature of operation for the 8.3 μm (10 μm) AlGaAs/GaAs superlattice is below 69 K (58 K) to achieve the BLIP condition.

Even though the model given by Levine et al. [77] has been widely used and is in good agreement with many experimental data, it did not discuss the process of trapping or capture to balance the electron emission or escape. As a result, Equation 16.22 leads to certain misunderstandings. For example, it ignores the implicit dependence of J_d on the photoconductive gain, and it implies an unrealistic proportionality between J_d and $1/L_p$.

The device operation of QWIP is similar to that of extrinsic photodetectors but in contrast with conventional detectors its distinct feature is the discreteness since carriers occupy discrete

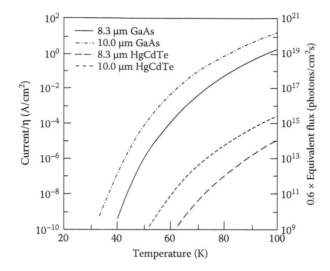

Figure 16.18 Thermal generation current versus temperature for GaAs/AlGaAs MQWs and HgCdTe alloy detectors at $\lambda_c = 8.3$ μm and 10 μm. The assumed effective quantum efficiencies are η = 12.5 and 70% for GaAs/AlGaAs and HgCdTe detectors, respectively. (From Kinch, M. A., and Yariv, A., *Applied Physics Letters*, 55, 2093–95, 1989. With permission.)

quantum wells. Details of the carrier behavior in quantum wells are described in Schneider and Liu's monograph [12]. We further follow after this monograph.

Figure 16.19 (top) shows schematic distribution of the dark current paths. In the barrier region the current flows as a 3-D flux, and the current density, $J_{3\text{-D}}$, is equal dark current, J_d. In the vicinity of each well, the capture (current density J_c) and emission (current density J_e) of electrons from the well must be balanced by the trapping or capture of electrons into the well under a steady state condition, so $J_c = J_e$. If we define a trapping or capture probability, p_c, we must have $J_c = p_c J_{3\text{-D}}$, and the sum of the captured and uncaptured fractions must equal the current in the barrier region [12]:

$$J_{3\text{-D}} = J_c + (1 - p_c) J_{3\text{-D}} = J_e + (1 - p_c) J_{3\text{-D}}. \tag{16.25}$$

The dark current can be determined by calculating either $J_{3\text{-D}}$ directly or by calculating J_e, and in the latter case $J_d = J_e / p_c$.

Liu, in review papers [6,82] and monograph [12], together with critical comments, presented several established physical models of QWIP dark current. Between them we can distinguish:

- carrier drift model,

- emission-capture model, and

- several self-consistent and numerical models.

In the carrier drift model, first presented by Kane et al. [83], only drift carrier contribution is taken into account (diffusion is neglected). The dark current is given by, for example, Equation 16.21, $J_d = q n_{3\text{-D}} v(F)$, where $n_{3\text{-D}}$ is a 3-D electron density on top of the barrier. In this way, the superlattice barriers are treated as a bulk semiconductor, which is justified because the barriers are thick (much thicker than the wells). The only 2-D quantum well effect comes for the evaluation of the Fermi level. Assuming a completely ionization (wells are degenerately doped), the 2-D doping density, N_d, equals the electron density within a given well. Then assuming relation between N_d and the Fermi energy, E_f, as $N_d = (m/\pi\hbar^2) E_f$, a simple calculation yields

$$n_{3\text{-D}} = 2 \left(\frac{m_b k T}{2\pi\hbar^2} \right)^{3/2} \exp\left(-\frac{E_a}{kT} \right), \tag{16.26}$$

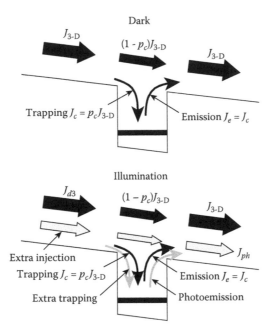

Figure 16.19 Schematic representation of the processes controlling the dark current and photo-current. The top part shows the dark current paths, while the bottom indicates the direct pho-toemission and the extra current injection from the contact to balance the loss of electrons from the well. The dark current paths remain the same under illumination. The collected total photocurrent is the sum of the direct photoexcited and the extra injection contributions. (From Schneider, H., and Liu, H. C., *Quantum Well Infrared Photodetectors*, Springer, Berlin, 2007. With permission.)

if $E_a/kT \gg 1$, which is appropriate for most practical cases. Here, m_b is the barrier effective mass, and E_a is the thermal activation energy equal the energy difference between the top of the barrier and the top of the Fermi level in the well.

The simple carrier drift model compares well with experiments in the regime of a low applied electric field. This is a consequence of the key step in the evaluation of $n_{3\text{-}D}$ by taking the equilibrium value at zero bias with the Fermi level determined by the well doping.

The second emission-capture model, gives adequate results for a large range of applied fields. It was originally presented by Liu et al. [84]. This model calculates first J_e and then dark current as $J_d = J_e/p_c$.

The escape current density can be written as

$$J_e = \frac{q n_{2D}}{\tau_{sc}},$$ (16.27)

where $n_{2\text{-}D}$ is a 2-D electron density electrons on the upper part of the ground state subband and τ_{sc} is the scattering time to transfer these electrons from the 2-D subband to the nonconfined continuum on top of the barrier.

The capture probability is related to the relevant time constants by

$$p_c = \frac{\tau_t}{\tau_c + \tau_t},$$ (16.28)

where τ_c is the capture time (lifetime) for an excited electron back into the well, and τ_t is the transit time for an electron across one QW region. For actual devices at operating electric fields, $p_c \ll 1$ ($\tau_c \gg \tau_t$), and the dark current becomes

$$J_d = \frac{J_e}{p_c} = q \frac{n_{2\text{-}D}}{\tau_{sc}} \frac{1}{p_c} = q \frac{n_{2\text{-}D} \tau_c}{\tau_{sc} \tau_t} = q \frac{n_{2\text{-}D}}{L_p} \frac{\tau_c}{\tau_{sc}} v,$$ (16.29)

where $L_p = L_w + L_b$. $n_{2\text{-}D}/\tau_{sc}$ represents the thermal escape or generation of electrons from the QW, and $1/p_c$, as shown later, is proportional to the photoconductive gain that implies the dependence of dark current on the photoconductive gain.

The final expression for dark current in Liu's model is given by [12,84]

$$J_d = \frac{qv\tau_e}{\tau_{sc}} \int_{E_1}^{\infty} \frac{m}{\pi\hbar^2 L_p} T(E,F) \left[1 + \exp\left(\frac{E - E_F}{kT}\right)\right]^{-1} dE. \tag{16.30}$$

For pure thermionic emission regime, when the transmission coefficient $T(E,F) = 0$ for E below the barrier, the last equation becomes

$$J_d = \frac{qv\tau_c}{\tau_{sc}} \frac{m}{\pi\hbar^2 L_p} kT \exp\left(-\frac{E_a}{kT}\right), \tag{16.31}$$

and closely resembles Equation 16.21 together with Equation 16.26.

Analysis of several established models of dark current in QWIP structures, with varying degrees of complexity, give good agreement between them and convergence with experiments [12]. However, realistic calculations of scattering or trapping rates are extremely complicated and have not performed so far.

The magnitude of QWIP dark current can be modified using different device structures, doping densities, and bias conditions. Figure 16.20 shows the QWIP I-V characteristics for temperatures ranging from 35 to 77 K, measured in a device at the 9.6 µm spectral peak [85]. It shows typical operation at 2 V applied bias in the region where the current varies slowly with bias, between the initial rise in current at low voltage and the later rise at high bias. Typical LWIR QWIP dark current is about 10^{-4} A/cm² at 77 K. Thus, a 9.6 µm QWIP must be cooled to 60 K to have a leakage current comparable to that of a 12 µm HgCdTe photodiode operating at a temperature that is 25° higher.

16.3.3 Photocurrent

The bottom part of Figure 16.19 shows the additional processes that occur in QW as a result of the incident of IR radiation. The direct photoemission of electrons from the well contributes to the observed photocurrent in the collector. All dark current paths remain unchanged.

Photoconductive gain is an important parameter that affects the spectral responsivity and detectivity of detector (see Section 3.2.2). This parameter is defined as the number of electrons flowing through the external circuit for each photon absorbed and is a result of the extra current injection

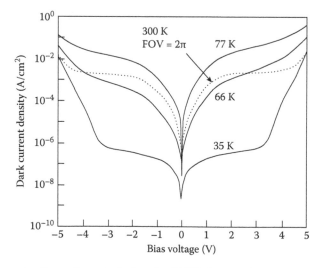

Figure 16.20 Current voltage characteristics of a QWIP detector having a peak response of 9.6 µm at various temperatures, along with the 300 K background window current measured at 30 K with an 180° FOV. (From Tidrow, M. Z., Chiang, J. C., Li, S. S., and Bacher, K., *Applied Physics Letters*, 70, 859–61, 1997. With permission.)

from the contact necessary to balance the loss of electrons from the well due to photoemission. As Figure 16.19 shows, the total photocurrent consists of contributions from the direct photoemission and the extra current injection.

The magnitude of photocurrent is independent of the number of wells if the absorption for each well is the same. Considering two neighboring wells, the processes of photoemission and refilling are identical for both wells. The same arguments can be made for any subsequent wells. This means that the photocurrent is unaffected by adding more wells as long as the magnitude of absorption and hence photoemission from all the wells remains the same.

Assuming n_{ex} as the number of the excited electrons from one well and taking into account a rate equation, we have

$$\frac{dn_{ex}}{dt} = \Phi\eta^{(1)} - \frac{n_{ex}}{\tau_{esc}} - \frac{n_{ex}}{\tau_{relax}}. \tag{16.32}$$

Next, solving Equation 16.32 for n_{ex} with regard to $i_{ph}^{(1)} = qn_{ex}/\tau_{esc}$ for one well under steady state ($dn_{ex}/dt = 0$), gives

$$i_{ph}^{(1)} = q\Phi\eta^{(1)} \frac{\tau_{relax}}{\tau_{relax} + \tau_{esc}} = q\Phi\eta \frac{p_e}{N}, \tag{16.33}$$

where Φ is the incident photon number per unit time, the superscript (1) indicates quantities for one well, τ_{esc} is the escape time, τ_{relax} is the intersubband relaxation time, $\eta = N\eta^{(1)}$ is the total absorption quantum efficiency (we have assumed that the amount of absorption is the same for all the wells), and N is the number of wells. The escape probability for an excited electron from the well is given by

$$p_e = \frac{\tau_{relax}}{\tau_{relax} + \tau_{esc}}. \tag{16.34}$$

The injection current, $i_{ph}^{(1)} = i_{ph}^{(1)}/p_c$, which refills the well to balance the loss due to emission, equals the photocurrent

$$I_{ph} = \frac{i_{ph}^{(1)}}{p_c} = q\Phi\eta \frac{p_e}{Np_c} \tag{16.35}$$

and

$$g_{ph} = \frac{p_e}{Np_c} \tag{16.36}$$

is the photoconductive gain.

We comment on several aspects of Liu's model. In conventional theory of photoconductivity $g_{ph} = \tau_c/\tau_{t,tot}$ (see Equation 9.8 and Rose [86]), where $\tau_{t,tot} = (N + 1)\tau_t$ is the total transit time across the detector active region. Under the approximation $p_e \approx 1$, $p_c \approx \tau_t/\tau_c \ll 1$ and $N \gg 1$, the gain expression given by Equation 16.36 and the conventional theory become the same:

$$g_{ph} \approx \frac{1}{Np_c} \approx \frac{\tau_c}{\tau_{t,tot}} = \frac{\tau_c v}{NL_p}. \tag{16.37}$$

The capture time, also called lifetime of carriers, is associated with scattering electrons (trapping) into the ground state subband. The condition $p_e \approx 1$ is fulfilled for a bound-to-continuum case, while for a bound-to-bound case this is no longer true. If the absorption is proportional to N, the photocurrent is independent of N since g_{ph} is inversely proportional to N. Photocurrent independence of N is equivalent to its independence of device length in the conventional theory. It should also be mentioned that this independence does not mean that the detector performance is independent of the number of wells because of noise considerations.

The dependence of photoconductive gain on the number of wells for different values of p_c and $p_e = 1$ together with some existing experimental data is shown in Figure 16.21 [12,87]. Most of the reported detector samples have 50 QWs and a range of gain values from about 0.25 to 0.80 have been observed.

Concerning estimation of the time scales involved in QWIPs, it can be assumed that τ_c is approximately 5 ps. The transit time is mainly determined by the high-field drift velocity of an excited electron in the barrier region. For typical parameters of $v = 10^7$ cm/s and $L_p = 30$–50 nm, $\tau_t \approx L_p/v$ is estimated to be in the range of 0.3–0.5 ps. One therefore expects a capture probability [$p_c = \tau_t/(\tau_c + \tau_t) \approx \tau_t/\tau_c$] to be in the range of 0.06–0.10 consistent with experiments.

16.3.4 Detector Performance

The conventional theory of photoconductivity (see Section 9.1.1) is normally employed to describe multiple quantum well photoconductors, where the electron recirculates through the superlattice for the time $\tau_{t,tot}$, and thus the hot-electron mean free path can be much larger than the superlattice length. However, because the hot-electron lifetime is very short, so $g_{ph} > 1$ is received only in low-period superlattices (see Figure 16.21).

The current responsivity is given by

$$R_i = \frac{\lambda\eta}{hc}qg_{ph},$$ (16.38)

and depends on both quantum efficiency and photoconductive gain. A high absorption does not necessarily give rise to a large photocurrent. It was shown that the optimum occurs when the excited state is in close resonance with the top of the barrier [88,89]. Then, the photoexcited electrons effectively escape from the wells to give rise to a large photocurrent.

In Section 3.2.1 it is marked that the QWIP absorption spectra, it value and shape, depends on a design of active region. As is shown in Figure 3.9, the spectra of the bound-to-continuum (samples A, B, and C) are much broader than the bound-to-bound or bound-to-quasibound (samples E and F). Table 16.1 gives the absorption values and the corresponding spectral parameters for different n-doped GaAs/AlGaAs QWIP structures. Also included are values of quantum efficiency.

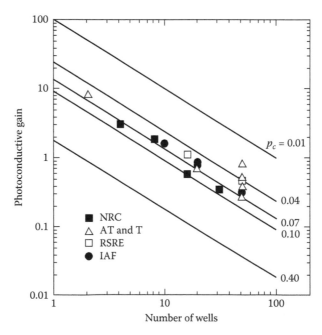

Figure 16.21 Calculated photoconductive gain versus the number of wells for different capture probability. Experimental data are taken from Liu et al. (■), Levine et al. (∆), Kane et al. (□), and Schneider et al. (●). (From Schneider, H., and Liu, H. C., *Quantum Well Infrared Photodetectors*, Springer, Berlin, 2007. With permission.)

Table 16.1: Structure Parameters for Different n-Doped, 50 period Al$_x$Ga$_{1-x}$As QWIP Structures

Sample	Well Width (Å)	Barrier Width (Å)	Composition x	Doping Density (10^{18} cm^{-3})	Intersubband Transition*	λ_p (μm)	λ_c (μm)	$\Delta\lambda$ (μm)	$\Delta\lambda/\lambda$ (%)	α_p(77 K) (cm^{-1})	η(77 K) (%)
A	40	500	0.26	1.0	B-C	9.0	10.3	3.0	33	410	13
B	40	500	0.25	1.6	B-C	9.7	10.9	2.9	30	670	19
C	60	500	0.15	0.5	B-C	13.5	14.5	2.1	16	450	14
E	50	500	0.26	0.42	B-B	8.6	9.0	0.75	9	1820	20
F	45	500	0.30	0.5	B-QB	7.75	8.15	0.85	11	875	14

Source: B. F. Levine, *Journal of Applied Physics*, 74, R1–R81, 1993. With permission.
*Type of transition: bound-to-continuum (B-C), bound-to-bound (B-B), bound-to-quasibound (B-QB).

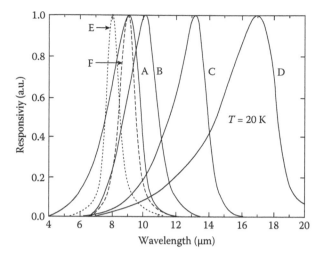

Figure 16.22 Normalized responsivity spectra versus wavelength measured at $T = 20$ K for samples A–F. (From Levine, B. F., *Journal of Applied Physics*, 74, R1–R81, 1993; Gunapala, S. D., and Bandara, S. V., *Handbook of Thin Devices*, Academic Press, San Diego, Vol. 2, 63–99, 2000. With permission.)

Since $\eta \propto N$ and $g_{ph} \propto 1/N$, there is nothing that can be done about the number of wells to improve responsivity. Analysis carried out by Liu [82] indicates that the escape probability must be made close to unity, which is fulfilled for the bound-to-continuum case. If a bound-to-bound design is employed, one must have the excited state close to the top of the barrier. For a typical 10 μm GaAs/AlGaAs QWIP under a typical field of 10 kV/cm, this dictates that the excited state should be lower than about 10 meV below the top of the barrier.

The normalized spectra of responsivity are given in Figure 16.22 for the same samples A–F [2,7]. Again we see that the bound and quasi-bound excited state QWIP samples are much narrower ($\Delta\lambda/\lambda = 10$–12%) than the continuum structure ($\Delta\lambda/\lambda = 19$–28%; $\Delta\lambda$ is the spectral width for which responsivity drops to half value).

The detectivity can be determined using

$$D^* = R_i \frac{\left(A\Delta f\right)^{1/2}}{I_n},$$

(16.39)

where A is the detector area and Δf is the noise bandwidth (taken as $\Delta f = 1$ Hz).

In general, a photoconductive detector has several source of noise. The most important are: Johnson noise, the generation–recombination noise (dark noise), and photon noise (connected with current induced by incident photons). For QWIPs, the $1/f$ noise seldom limits the detector performance. Its nature is complicated and is still an ongoing research topic [12].

In a conventional photoconductor, the noise gain equals the photoconductive gain $g_n = g_{ph}$. However, QWIP detector, g_n is different from g_{ph}, which was explained by Liu [90]. The standard generation–recombination noise can be written as [86]

$$I_n^2 = 4q g_n I_d \Delta f.$$

(16.40)

This equation takes into account influences of both generation and recombination processes. In the case of a QWIP detector, the generation–recombination noise should consist of contributions connected with carrier emission and capture (fluctuations in i_e and i_c). Seeing that

$$I_d = \frac{i_e^{(1)}}{p_c} = \frac{i_e}{N p_c},$$

(16.41)

where $i_e = Ni_e^{(1)}$ is the total emission current from all N wells, and equivalently

$$I_d = \frac{i_c^{(1)}}{p_c} = \frac{i_c}{Np_c},$$

(16.42)

we have

$$I_n^2 = 2q\left(\frac{1}{Np_c}\right)^2 (i_e + i_c)\Delta f = 4q\left(\frac{1}{Np_c}\right)^2 i_e\Delta f = 4q\frac{1}{Np_c}I_d\Delta f.$$

(16.43)

Comparing the last equation with Equation 16.40, the noise gain is defined by

$$g_n = \frac{1}{Np_c},$$

(16.44)

and is different from the photoconductive gain (see Equation 16.36). It appears that this expression is valid for small quantum well capture probabilities (i.e., $p_c \ll 1$). QWIPs satisfy this condition at the usual operating bias (i.e., 2–3 V).

The last equation does not give a correct explanation in the limit of high capture probability $p_c \approx 1$ (or equivalently low noise gain). The appropriate model for this case was first given by Beck [91] using stochastic considerations and taking into account that a high capture probability is not necessarily connected with a low escape probability. A more general expression has a form

$$I_n^2 = 4qg_nI_d\left(1 - \frac{p_c}{2}\right)\Delta f,$$

(16.45)

which can apply even in low-bias conditions where capture probabilities for carriers traversing the wells are high. For the case of $p_c \approx 1$, this expression equals the shot noise expression for N series-connected detectors.

Since $I_d \propto \exp[-(E_c - E_F)/kT]$ (see Equations 16.21 and 16.22) and $D^* \propto (R_i/I_n)$, we have

$$D^* = D_o \exp\left(\frac{E_c}{2kT}\right).$$

(16.46)

Based on this relationship, Levine et al. [2,49] have reported useful empirical D^* values by the best fit $T = 77$ K detectivity for the n-type

$$D^* = 1.1\times10^6 \exp\left(\frac{hc}{2kT\lambda_c}\right)\text{cmHz}^{1/2}\text{W}^{-1}$$

(16.47)

and for p-type GaAs/AlGaAs QWIPs

$$D^* = 2\times10^5 \exp\left(\frac{hc}{2kT\lambda_c}\right)\text{cmHz}^{1/2}\text{W}^{-1}.$$

(16.48)

Figure 16.23 shows the detectivity versus cutoff energy for both n-type and p-type GaAs/AlGaAs QWIPs. It should be noted that the experimental results are for a 45° polished input facet and that optimized gratings and optical cavities can be expected to improve the performance. Note that although Equations 16.47 and 16.48 are fitted to data taken at 77 K, they are expected to be valid over a wide range of temperatures.

Rogalski [92] used simple analytical expressions for detector parameters described by Andersson [64]. Figure 16.24 shows the dependence of detectivity on the long wavelength cutoff for GaAs/AlGaAs QWIPs at different temperatures. The satisfactory agreement with

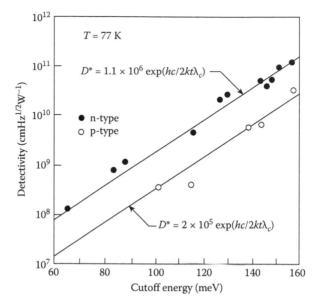

Figure 16.23 Detectivity at 77 K versus cutoff energy for n-doped GaAs/AlGaAs QWIPs (solid circles) and p-doped GaAs/AlGaAs QWIPs (open circles). The straight lines are best fits to the measured data. (From Levine, B. F., Zussman, A., Gunapala, S. D., Asom, M. T., Kuo, J. M., and Hobson, W. S., *Journal of Applied Physics*, 72, 4429–43, 1992. With permission.)

experimental data in a wide range of cutoff wavelength $8 \leq \lambda_c \leq 19$ μm and temperature $35 \leq T \leq 77$ K has been obtained, considering the samples have different doping, different methods of crystal growth (MBE, MOCVD, and gas source MBE), different spectral widths, different excited states (continuum, bound, and quasicontinuum), and even in one case a different materials system (InGaAs). As a matter of fact, Equation 16.47 is in nice agreement with the results presented in Figure 16.24.

16.3.5 QWIP versus HgCdTe

Rogalski [92] has also compared the detectivity of GaAs/AlGaAs QWIPs with the theoretical ultimate performance of n^+-p HgCdTe photodiodes limited by Auger mechanism in the base region. In the range of cutoff wavelength $8 \leq \lambda_c \leq 24$ μm and operating temperature ≤ 77 K, the detectivity of HgCdTe photodiodes is considerably higher. All the QWIP detectivity data for devices with cutoff wavelength near 9 μm is clustered between 10^{10} and 10^{11} cmHz$^{1/2}$/W at an operating temperature close to 77 K. However, the advantage of HgCdTe is less distinct in temperature range below 50 K due to the problems associated with HgCdTe material (p-type doping, Shockley-Read-Hall recombination, trap-assisted tunneling, surface and interface instabilities).

Additional insight into the difference in the temperature dependence of the dark currents is given by Figure 16.25 [96], where the current density versus inverse temperature for a GaAs/AlGaAs QWIP and an HgCdTe photodiode, both with $\lambda_c = 10$ μm, is shown. The current density of both detectors at temperatures lower than 40 K is similar and is limited by tunneling, which is temperature independent. The thermionic emission regime for the QWIP (≥ 40 K) is highly temperature dependent, and "cuts on" very rapidly. At 77 K, the QWIP has a dark current that is approximately two orders of magnitude higher than that of the HgCdTe photodiode.

LWIR QWIPs cannot compete with HgCdTe photodiodes as single devices, especially for higher temperature operation (>70 K) due to fundamental limitations associated with intersubband transitions. In addition, QWIP detectors have relatively low quantum efficiencies, typically less than 10%. Figure 16.26 compares the spectral η of an HgCdTe photodiode to that of a QWIP. A higher bias voltage can be used to boost η in the QWIP. However, an increase in the reverse bias voltage also causes an increase of the leakage current and associated noise, which limits any potential improvement in system performance. HgCdTe has high optical absorption and a wide absorption

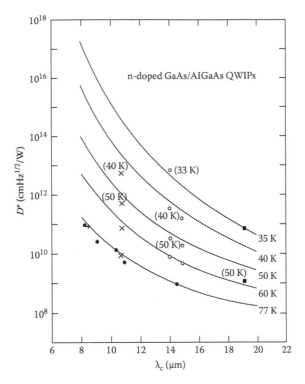

Figure 16.24 Detectivity versus cutoff wavelength for n–doped GaAs/AlGaAs QWIPs at temperatures ≤ 77 K. The solid lines are theoretically calculated. The experimental data are taken from Refs 49 (•), 78 (×), 93 (+), 94 (O), and 95 (■). (From Rogalski, A., *Infrared Physics & Technology*, 38, 295–310, 1997. With permission.)

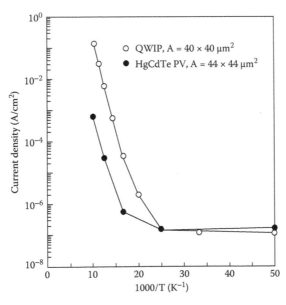

Figure 16.25 Current density versus temperature for a HgCdTe photodiode and a GaAs/AlGaAs QWIP with $\lambda_c = 10$ μm. (From Singh, A., and Manasreh, M. O., "Quantum Well and Superlattice Heterostructures for Space-Based Long Wavelength Infrared Photodetectors," *Proceedings of SPIE* 2397, 193–209, 1995. With permission.)

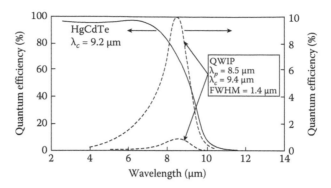

Figure 16.26 Quantum efficiency versus wavelength for a HgCdTe photodiode and GaAs/AlGaAs QWIP detector with similar cutoffs.

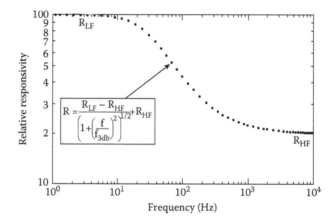

Figure 16.27 Generalized frequency response of QWIP detector. (From Arrington, D. C., Hubbs, J. E., Gramer, M. E., and Dole, G. A., "Nonlinear Response of QWIP Detectors: Summary of Data from Four Manufactures," *Proceedings of SPIE* 4028, 289–99, 2000. With permission.)

band irrespective of the polarization of the radiation, which greatly simplifies the detector array design. The quantum efficiency of HgCdTe photodiodes is routinely around 70% without an antireflection (AR) coating and is in excess of 90% with an AR coating. Moreover, it is independent of the wavelength over the range from less than 1 μm to near the cutoff of the detector. The wide-band spectral sensitivity with a near-perfect η enables greater system collection efficiency and allows a smaller aperture to be used. This makes HgCdTe FPAs useful for imaging, spectral radiometry, and long-range target acquisition. However, it should be noted that because of high photon fluxes, current LWIR staring array performance is mostly limited by the charge handling capacity of the ROIC and the background (warm optics). Thus, the spectral response band of QWIP detectors, with a full-width, half-maximum of about 15%, is not a major drawback at LWIR wavelengths.

At the present stage of technology development, QWIP devices are not suitable for space-based remote sensing applications due to dielectric relaxation effects and flux memory effects. In low irradiance environments and associated low temperature operation, the responsivity of QWIPs depends on frequency and the frequency response depends on the operating conditions (temperature, photon irradiance, bias voltage, and the dynamic resistance of the detector). The typical frequency response is empirically similar to dielectric relaxation effects observed in bulk extrinsic silicon and germanium photoconductors under similar operational conditions. The frequency response has flat regions at both low and at high frequencies and the response rolls off between these two levels at a frequency point that is proportional to the inverse of the dynamic resistance of the detector [97] (see Figure 16.27). The dynamic resistance is set by a combination of detector bias, photon irradiance, and operating temperature. Under typical ambient background conditions, the dynamic resistance is low and the roll-off, which takes place at frequencies in the range of 100 kHz, is not normally evident.

Even though QWIPs are photoconductive devices, several of its properties such as high impedance, fast response time, and low power consumption are well matched with the requirements for large FPA fabrication. The main drawbacks of LWIR QWIP FPA technology are the performance limitations for applications requiring short integration time, and the requirement to operate at a lower temperature than HgCdTe of comparable wavelengths. The main advantages of QWIPs are linked to pixel performance uniformity and to the availability of large size arrays. The large established industrial infrastructure in III-V materials/device growth, processing, and packaging brought about by the application of GaAs-based devices in the telecommunications industry gives QWIPs a potential advantage in producibility and cost, whereas the only major use of HgCdTe, to this date, is for IR detectors.

A more detailed comparison of the two technologies has been given by Tidrow et al. [98] and Rogalski et al. [11,13,99,100].

16.4 PHOTOVOLTAIC QWIP

The standard QWIP structure, pioneered by Levine and his group [2] and discussed in the previous sections, is a photoconductive detector, where the photoexcited carriers are swept out of nominally symmetric quantum wells by an external electric field. A key result of photovoltaic QWIP structures is the application of internal electric fields. These devices in principle can be operated without external bias voltage and it should be expected vanishing of the dark current and suppressing the recombination noise [12]. However, their photocurrent is associated with a much smaller gain by comparison to photoconductive QWIPs. The reduced photocurrent and the reduced noise give rise to detectivities similar to photoconductive devices [101]. In conclusion, photoconductive QWIPs are preferable for applications that require high responsivity (e.g., for sensors operating in the MWIR band), instead photovoltaic QWIPs attractive in camera systems operating in LWIR. The performance of LW FPAs are limited by the storage capacity of the read-out circuit. In this context, the benefits of photovoltaic QWIP arise from two facts: the capacitor is less effective loaded by dark current and the noise associated with the collected photocharge is extremely small [12].

The first experimental work on IR detectors involving the miniband concept was carried out by Kastalsky et al. in 1988 [102]. The spectral response of this GaAs/AlGaAs detector with extremely small quantum efficiency was in the range 3.6–6.3 μm and indicate photovoltaic detection. This detector consists of a bound-to-bound miniband transition (i.e., two minibands below the top of the barrier) and a graded barrier between the superlattice and the collector layer as a blocking barrier for ground miniband tunneling dark current. Electrons excited into the upper miniband traverse the barrier giving rise to a photocurrent without external bias voltage. Further evolution in design of photovoltaic QWIP structures is analyzed by Schneider and Liu in their monograph [12]. Here we concentrate on the final development of photovoltaic structure at Fraunhofer IAF, so called the four-zone QWIP [103–105]. The photovoltaic effect in this structure arises from carrier transfer among an asymmetric set of quantized states rather than asymmetric internal electric fields.

The photoconduction mechanism of the photovoltaic "low-noise" QWIP structure is explained in Figure 16.28 [104]. Because of the period layout, the detector structure has been called a four-zone QWIP [106]. Each period of the active detector region is optimized independently. In the excitation zone (1), carriers are optically excited and emitted into the quasicontinuum of the drift zone (2). The first two zones (1 and 2) are analogous to the barrier and well of a conventional QWIP. Moreover, two additional zones are present in order to control the relaxation of the photoexcited carriers, namely a capture zone (3) and a tunneling zone (4). The tunneling zone has two functions; it blocks the carriers in the quasicontinuum (carriers can be captured efficiently into a capture zone) and transmits the carriers from the ground state of the capture zone into the excitation zone of the subsequent period. This tunneling process has to be fast enough in order to prevent the captured carriers from being reemitted thermoelectrically into the original well. Simultaneously, the tunneling zone provides a large barrier to prevent the photoexcited carriers from being emitted toward the left-hand side of the excitation zone. In this way, the noise associated with the carrier capture is suppressed.

Figure 16.28b depicts several requirements on carrier transport under a finite applied electric field that determines efficient implementation of four-zone structure. As is shown, the tunnel

(a)

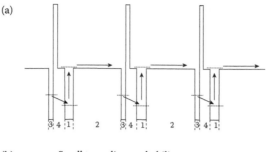

(b)

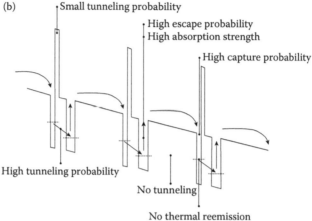

Figure 16.28 Schematic (a) band edge distribution and (b) transport mechanism of the four-zone QWIP. Potential distribution 1: emission zone, 2: drift zone, 3: capture zone, and 4: tunneling zone. (From Schneider, H., Koidl, P., Walther, M., Fleissner, J., Rehm, R., Diwo, E., Schwarz, K., and Weimann, G., *Infrared Physics & Technology*, 42, 283–89, 2001. With permission.)

barrier has to exhibit low probability for tunneling at high energies to receive shorter capture time into the narrow QW than the tunneling escape time. In addition, the time constant for tunneling has to be shorter than that for thermionic reemission from the narrow QW back into the wide QW. An important detail of the tunneling zone is the steplike barrier. The separation between emission zone and the high-energy part of the tunneling zone is required to reduce absorption line to a value comparable to that of conventional QWIP.

The experimentally demonstrated [MBE grown on (100)-oriented semi-insulating GaAs substrate] four-zone structure contains (in the growth direction) an active region with 20 periods of nominally 3.6 nm GaAs (capture zone), 45 nm $Al_{0.24}Ga_{0.76}As$ (drift zone), 4.8 nm GaAs (excitation zone), and a sequence of 3.6 nm $Al_{0.24}Ga_{0.76}As$, 0.6 nm AlAs, 1.8 nm $Al_{0.24}Ga_{0.76}As$, and 0.6 nm AlAs (tunneling zone). The 4.8 nm GaAs wells are n-doped to a sheet concentration of 4×10^{11} cm^{-2} per well. The active region is sandwiched between Si doped (1.0×10^{18} cm^{-3}) n-type contact layers.

Figure 16.29 summarizes the performance of a typical 20-period, low-noise QWIP with a cutoff wavelength of 9.2 µm [105]. The peak responsivity is 11 mV at zero bias (photovoltaic operation) and about 22 mA in the range between –2 and –3 V. Between –1 and –2 V, a gain of about 0.05 is observed. The detectivity has its maximum around –0.8 V and about 70% of this value is obtained at zero bias. Due to the asymmetric nature of the transport process, the detectivity strongly depends on the sign of the bias voltage. This behavior is in strong contrast with a conventional QWIP where the detectivity vanishes at zero bias.

An appropriate noise model for the four-zone QWIP was first given by Beck (see Equation 16.45) and next developed by Schneider [107] in the presence of avalanche multiplication.

Figure 16.30 compares peak detectivities of both conventional and low-noise QWIP structures as a function of the cutoff wavelength [105]. The low-noise QWIPs show similar detectivities as the conventional ones, which are in good agreement with a thermionic emission model.

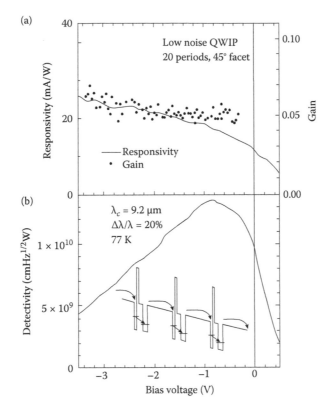

(a)

(b)

$\lambda_c = 9.2\ \mu m$
$\Delta\lambda/\lambda = 20\%$
77 K

Figure 16.29 (a) Peak responsivity, gain, and (b) peak detectivity of a low-noise QWIP versus bias voltage. (From Schneider, H., Walther, M., Schönbein, C., Rehm, R., Fleissner, J., Pletschen, W., Braunstein, J., et al., *Physica E* 7, 101–7, 2000. With permission.)

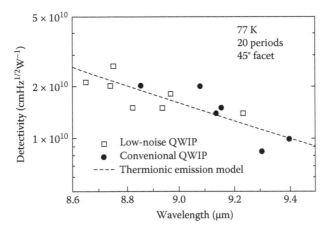

Figure 16.30 Peak detectivities of low-noise QWIPs and photoconductive QWIPs at 77 K versus cutoff wavelength. (From Schneider, H., Walther, M., Schönbein, C., Rehm, R., Fleissner, J., Pletschen, W., Braunstein, J., et al., *Physica E* 7, 101–7, 2000. With permission.)

16.5 SUPERLATTICE MINIBAND QWIPS

In addition to QW, the superlattice is another promising structure for an infrared photodetector but has drawn less attention. The superlattice intersubband photodetector with a graded barrier was fabricated for photovoltaic detection in the range of 3.6–6.3 μm and 8–10.5 μm in 1988 [102] and 1990 [108], respectively.

The superlattices alone were applied in the detection of the wavelength range of 5–10 μm in 1991 [109]. The structures consisted of 100 periods of GaAs quantum wells of either $L_b = 30$ or 45 Å barriers of $Al_{0.28}Ga_{0.72}As$ and $L_w = 40$ Å GaAs wells (doped $N_d = 1 \times 10^{18}$ cm^{-3}) sandwiched between doped GaAs contact layers. The absolute values for the peak absorption coefficients were $\alpha = 3100$ and $\alpha = 1800$ cm^{-1} for the $L_b = 30$ and 45 Å structures, respectively, and detectivities about 2.5×10^9 cmHz$^{1/2}$W^{-1} for $T = 77$ K. Also the superlattices with a blocking layer operated in a low bias region were demonstrated [110,111]. Recently, a modified voltage tunable superlattice infrared photodetector has been implemented to fabricate two-color FPAs [112,113]. The works on the superlattice IR photodetectors (SLIPs) indicate that their advantages include a broader absorption spectrum, lower operation voltage, and more flexible miniband engineering than the conventional QWIP.

The superlattice miniband detectors uses a concept of infrared photoexcitation between minibands (ground state and first excited state) and transport of these photoexcited electrons along the excited state miniband. An energy miniband is formed when the carrier de Broglie wavelength becomes comparable to the barrier thickness of the superlattice. Thus, the wave functions of the individual wells tend to overlap due to tunneling.

Figure 16.31 shows the schematic conduction band diagrams for different miniband structures. Depending on where the upper excited states are located and the barrier layer structure, the intersubband transitions can be based on the bound-to-continuum miniband, bound to miniband, and step bound-to-miniband. Among them, the GaAs/AlGaAs QWIP structures using the bound-to-miniband transitions are the most widely used material systems for the fabrication of large FPAs.

Placing the excited state in the continuum increases the thermionic dark current because of the lower barrier height. This fact is more critical for LWIR detectors because the photoexcitation energy becomes even smaller. To improve the detector performance, a new class of multiple quantum well infrared detectors have received much interest because of their potential for large, uniform FPAs with high sensitivity. Studies by Yu et al. [114–116] revealed that by replacing the bulk AlGaAs barrier in the QWIP with a short-period superlattice barrier layer structure (see Figure 16.31b), a significant improvement of the intersubband absorption and thermionic emission property can be obtained. The physical parameters are chosen so that the first excited state in the enlarged wells is merged and lined up with the ground state of the miniband in the superlattice barrier layer to achieve a large oscillation strength and intersubband absorption. The electron transport in these MQWs is based on the bound-to-miniband transition, superlattice miniband resonant tunneling, and coherent transport mechanism. Thus the operation of this miniband QWIP is analogous to that of a weakly coupled MQW bound-to-continuum QWIP. In this device structure, the continuum states above the barriers are replaced by the miniband of the superlattice barriers. The use of two bound states in the enlarged quantum well removes the requirement

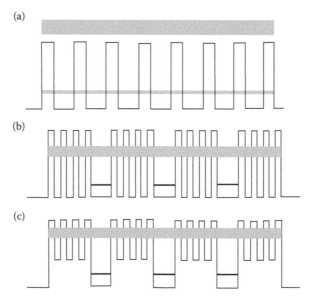

Figure 16.31 Schematic drawing of miniband structures: (a) bound-to-continuum miniband, (b) bound to miniband, and (c) step bound-to-miniband.

imposed by the bound-to-continuum transition design for a unique solution for the well width and barrier height for a given wavelength (i.e., it is possible to obtain the same operating wavelength with a continuous range of well widths and barrier heights). These miniband QWIPs show lower photoconductive gain than bound-to-continuum QWIPs because the photoexcited electron transport occurs in the miniband where electrons have to transport through many thin heterobarriers resulting in a lower mobility.

In the first GaAs/AlGaAs multiple quantum well detector with enlarged wells [114], a 40-period of GaAs quantum wells with a well width of 88 Å and a dopant density 2.0×10^{18} cm^{-3} were used. The barrier layer on each side of the GaAs quantum well consists of five period undoped AlGaAs(58Å)/GaAs(29Å) superlattice layers that were grown alternatively with the GaAs quantum wells. The active structure was sandwiched between a 1 μm thick GaAs buffer layer grown on semi-insulating GaAs and 0.45-μm-thick GaAs cap layer with dopant density 2.0×10^{18} cm^{-3} to facilitate ohmic contacts. To enhance the light coupling efficiency, a planar transmission metal grating coupler (consisted with regularly spaced metal grating fingers) was developed. The dark current in this type of miniband transport QWIPs is dominated by the thermionic-assisted tunneling conduction via miniband for $T \geq 60$ K, whereas resonant tunneling conduction prevails for $T \leq 40$ K. For bias voltage 0.2 V, detectivity $D^* = 1.6 \times 10^{10}$ cmHz$^{1/2}$W^{-1} at $\lambda = 8.9$ μm and $T = 77$ K was found. Beck et al. [117,118] adopted this bound-to-miniband approach and demonstrated excellent IR imagers using FPAs in sizes from 256×256 to 640×480.

In order to further reduce the undesirable electron tunneling from the doped QWs and improve the performance, a step bound-to-miniband QWIP was designed and measured (shown in Figure 16.31c). This QWIP consists of GaAs/AlGaAs superlattice barriers but with a strained quantum well of In$_{0.07}$Ga$_{0.93}$As [116,119].

New ideas of superlattice miniband detectors are still presented; for more details see, for example, Li [120].

16.6 LIGHT COUPLING

A key factor in QWIP FPA performance is the light-coupling scheme. Illumination of the detector at 45° restricts detector geometries to single elements and one-dimensional arrays. The majority of existing gratings are designed for 2-D FPAs. The array illumination is through the substrate backside.

Goossen et al. [121,122] and Hasnain et al. [123] developed a method to couple light efficiency into two-dimensional arrays. They placed gratings on top of the detector that deflect the incoming light away from the direction normal to the surface (see Figure 16.32a). The gratings were made by either depositing fine metal strips on top of the quantum well or etching grooves in a cap layer. These gratings gave a light-coupling efficiency comparable to the 45° illumination scheme, but it still gave a relatively low quantum efficiency of about 10–20% for QWIPs having 50 periods and $N_d = 1 \times 10^{18}$ cm^{-3}. This relatively low quantum efficiency is due to the poor light-coupling efficiency and the fact that only one polarization of the light is absorbed.

The quantum efficiency can be improved by increasing the doping density in the quantum wells, but this leads to a higher dark current. To increase the quantum efficiency without increasing the dark current, Andersson et al. [73,124] and Sarusi et al. [125] developed a two-dimensional grating for QWIPs operating at the 8–10 μm spectral range, which absorbed both polarization components. In this case, the periodicity of the grating is repeated in both directions. The addition of an optical cavity can increase absorption further by making the radiation pass through the MQW structure twice by placing a thin GaAs "mirror" below the QW structure (see Figure 16.32b).

Many more passes of IR radiation and significantly higher absorption can be achieved with a randomly roughened reflecting surface, as shown in Figure 16.32c. Sarusi et al. [126] have demonstrated almost an order of magnitude enhancement in performance compared with the 45° scheme by using carefully designed random reflecting surface above the MQW structure. The randomness prevents the light from being diffracted out of the detector after the second reflection (as happens in Figure 16.32b). Instead the light is scattered at a different random angle after each bounce, and can only escape if it is reflected toward the surface within a critical angle of the normal (which is about 17° for GaAs/air interface). The random surfaces are made from GaAs using standard photolithography and selective dry etching, which allows the feature sizes in the pattern to be controlled accurately and the pixel-to-pixel uniformity needed for high sensitivity imaging arrays to be preserved. To reduce the probability of the light escaping, the surface has three distinct scattering surfaces (see details in [2,5,125]). Experiments show that the maximum response is obtained

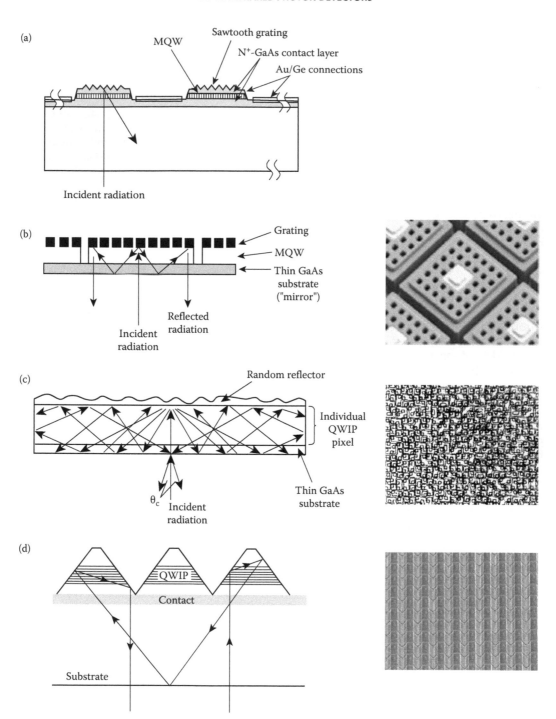

Figure 16.32 Gratings light-coupling mechanisms used in QWIPs: (a) linear or two-dimensional gratings on each detector, (b) gratings with optical cavity, (c) random scatterer reflector, and (d) corrugated quantum wells.

when the unit cell is equal to the wavelength of the QWIP maximum response. If the unit is larger than this, the number of scattering on the detector surface decreases and the light is scattered less efficiently. On the other hand, if the unit cell size is smaller, the scattering surface becomes smoother and the efficiency again decreases. Naturally, thinning down the substrate enables more bounces of light and therefore higher responsivity. This also reduces the amount of light that bounces from a pixel to its neighbor (which is bad for the performance of large detector arrays). The thinning or completely removing of the substrate also allows the GaAs/AlGaAs detector array to stretch and accommodate the thermal expansion mismatch with the Si readout integrated circuit.

It should be noticed that one of the main differences between the effect of the cross grating and the random reflector is the shape of the responsivity curve; unlike the cross grating, the random reflector has little impact on the bandwidth of the response curve since the scattering efficiency of the random reflector is significantly less wavelength dependent than for the regular grating. Therefore, for the QWIPs with random reflectors, the integrated responsivity is enhanced by nearly the same amount as the peak responsivity.

The light coupling, such as diffraction gratings and random gratings achieve high quantum efficiency only when the detector size is large. In addition, because of its wavelength dependence, each grating design is only suitable for a specific wavelength. A size and wavelength independent coupling scheme is much needed.

More recently, Schimert and coworkers [127] reported QWIPs in which the diffraction gratings are etched into the QW stack itself, thereby forming a dielectric grating. This design reduces the conducting detector area (and the leakage current) by a factor of about four while maintaining an quantum efficiency of 15%. The reported peak D^* of 7.7×10^{10} cmHz$^{1/2}$W^{-1} in a QWIP with peak wavelength of 8.5 μm is the highest reported at 77 K.

The gratings are made by etching them into the extra layer grown after the top contact layer and are in the form of either etched pits (see Figure 16.32b) or trenches leaving unetched bumps, and gold metal is then evaporated for near perfect reflection. The grating period should be approximately the wavelength inside the material; that is, $d = \lambda/n_r$, where λ is the wavelength to be detected and n_r is the refractive index. In practice λ is chosen to be close to the cutoff wavelength. The etch depth should be about one fourth of the inside wavelength (i.e., $\lambda/4n_r$). Diffraction gratings for the MWIR with 1.65 μm period were successfully fabricated with contact photolithography and RIE, similar to the LWIR gratings with 2.95 μm period [12].

In order to simplify array production, a new detector structure for normal incidence light coupling, which is referred to as the corrugated QWIP (C-QWIP) has been proposed [128,129]. The device structure is shown in Figure 16.32d. In large FPAs fabricated at present, the entire pixels are occupied by a single corrugation (see top view of pixels in Figure 16.32d) [112,113]. This structure utilizes total internal reflection at the sidewalls of triangular wires to create favorable optical polarization for infrared absorption. These wires are created by chemically etching an array of V grooves through the detector active region along a specific crystallographic direction. During FPA hybridization, an epoxy material is used to connect detector array with silicon readout. This epoxy material that lies on top of the sidewalls can substantially reduce the internal reflection. Therefore, at present C-QWIP contains an MgF$_2$/Au cover layer for sidewall protection (see Figure 16.33 [112]). This cover layer is electrically isolated from the top and bottom contacts of the pixels. The dielectric film MgF$_2$ is chosen for its high dielectric strength, low conducting current, low refractive index, and small excitation coefficient.

In an isotropic optical coupling scheme, 2-D grating is used to eliminate polarization sensitivity. For the polarimetric QWIPs, linear instead of 2-D gratings are used. The use of a microscanner makes it possible for the design of a camera that resolves the polarimetric components of the scene radiation. Such a discriminating imager, added without significant loss of sensitivity or increased

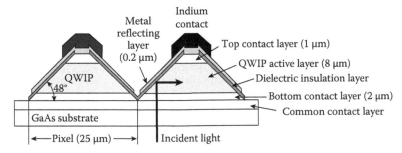

Figure 16.33 The side view of the C-QWIP pixels with 25 μm pitch. (From Choi, K.-K., Monroy, C., Swaminathan, V., Tamir, T., Leung, M., Devitt, J., Forrai, D., and Endres, D., *Infrared Physics & Technology*, 50, 124–35, 2007. With permission.)

Figure 16.34 SEM picture of a polarimetric QWIP array. (From Robo, J. A., Costard, E., Truffer, J. P., Nedelcu, A., Marcadet, X., and Bois, B., "QWIP Focal Plane Arrays Performances from MWIR to VLWIR," *Proceedings of SPIE* 7298, 7298-15, 2009. With permission.)

cost, may be beneficial in locating difficult targets. Thales formed four linear gratings rotated by 45° to each other on a set of four detector elements. This pattern is then replicated across the whole array. The layout and a SEM picture from an actual array are shown in Figure 16.34 [130].

Although gratings are successfully incorporated in commercially available QWIP FPAs, they can be further improved. Development of microfabrication technology and techniques from new fields, such as computer-generated holograms and photonic crystals, could be explored for more efficient optical couplers [12].

16.7 RELATED DEVICES

16.7.1 p-Doped GaAs/AlGaAs QWIPs

Up until now, most of the studies have been centered on the n-type GaAs/AlGaAs QWIPs. However, for n-type QWIPs due to quantum mechanical selection rules, normal incidence absorption is forbidden without use of metal or dielectric grating couplers. The original impetus for the study of p-QWIPs was their ability to absorb light at normal incidence. In p-type QWIPs the normal incidence absorption is allowed due to the mixing between the off zone center ($k \neq 0$) heavy-hole and light-hole states [131]. Because of the larger effective mass (hence lower optical absorption coefficient) and the lower hole mobilities, the performance of p-QWIPs are in general lower than n-QWIPs [2,5,132–135]. However, if the biaxial compressive strain is introduced into the quantum well layers of a p-QWIP, then the effective mass of the heavy holes will be reduced, which in turn can improve the overall device performance [136].

In the type I MQWs the wells for holes, as those for electrons, are in GaAs layers. For moderate levels of doping ($\leq 5 \times 10^{18}$ cm^{-3}) and well thicknesses not exceeding 50 Å, the lowest heavy-hole like subband (HH$_1$) is filled only partly and all other energy subbands (HH$_2$ and light-hole like LH$_1$) are empty at 77 K. Only these three subbands are found to be underneath the Ga$_{1-x}$Al$_x$As ($x \approx 0.3$) barriers. Holes from the HH$_1$ subband can be photoexcited to these subbands or to other energy subbands in the continuum HH$_{ext}$ and LH$_{ext}$ (see Figure 16.35a). The theoretical analysis of the experimental results on the normal incidence absorption and responsivity of p-type GaAs/Ga$_{0.7}$Al$_{0.3}$As QW structures showed the HH$_1 \rightarrow$ LH$_{ext}$ transitions to be the dominant mechanism for IR absorption in the $\lambda \approx 7$ μm region [137]. The HH$_1 \rightarrow$ LH$_1$ transitions are out of observed spectral range.

Levine et al. [54] experimentally demonstrated the first QWIP that uses hole intersubband absorption in the GaAs valence band. The samples were grown on a (100) semi-insulating

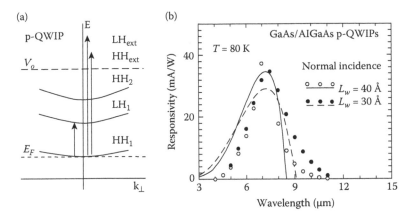

Figure 16.35 p-type GaAs/Ga$_{1-x}$Al$_x$As quantum well IR photodetector: (a) schematic energy-band (the energy of holes is taken to be positive), (b) responsivities of the p-GaAs/Ga$_{0.7}$Al$_{0.3}$As QWIPs. The experimental data (circles) are taken from Levine et al. [54] for structures with L_w = 40 Å and L_w = 30 Å. The solid line and dashed line are calculated. (From Man, P., and Pan, D. S., *Applied Physics Letters*, 61, 2799–2801, 1992. With permission.)

substrate, using gas source MBE, and consisted of 50 periods of L_w = 30 Å (or L_w = 40 Å) quantum wells (doped N_a = 4 × 10^{18} cm^{-3} with Be) separated by L_b = 300 Å barriers of Al$_{0.3}$Ga$_{0.7}$As, and capped by N_a = 4 × 10^{18} cm^{-3} contact layers. The experimental values of the photoconductive gain for these structures are g = 0.024 and g = 0.034, respectively. These values are more than an order of magnitude smaller than for n-type QWIPs. The values of quantum efficiency $\eta \geq$ 15% and escape probability $p_e \geq$ 50% were comparable to those of n-type GaAs/AlGaAs QWIPs in spite of the fact that heavy hole effective mass ($m_{hh} \approx$ 0.5 m_o) is much larger than that of electrons ($m_e \approx$ 0.073 m_o). The calculated and observed responsivities for these structures are shown in Figure 16.35b. Good agreement is obtained in the region of $\lambda < \lambda_c$. The small value of responsivity at $\lambda > \lambda_c$ is due to LH$_1 \rightarrow$ LH$_{ex}$.

The performance of p-type GaAs/GaAlAs normal incidence QWIP intersubband photodetectors is below that of corresponding n-type intersubband detectors for the same wavelength. The detectivities of p-type GaAs/GaAlAs QWIPs are by a factor of 5.5 lower than the detectivities for n-type photodetectors (see Equations 16.47 and 16.48). At present, p-type GaAs/AlGaAs QWIPs are less explored for infrared imaging.

16.7.2 Hot-Electron Transistor Detectors

The essential feature of all designs of QWIPs is that electrons in the lower (ground) state cannot flow in response to the applied electric field, but electrons in the upper (excited) state do flow, thereby yielding photocurrent. The operating temperature of III-V multiple quantum well infrared detectors is still needed to kept lower than 77 K due to the large dark current at high temperatures. To increase the operating temperature, it is desirable to reduce the dark current of the detector while maintaining a high detectivity. To reduce the dark current, Choi et al. [138–140] have proposed a new device structure, the infrared hot-electron transistor (IHET). Its physics was discussed in detail by Choi [141]. In this device, an "energy filter" is added to the QW stack that requires a third terminal, but which preferentially removes leakage current over photocurrent. Under some conditions, the resultant IHET has significantly improved signal-to-noise ratio compared to the standard two terminal QWIP. However, it has not been possible to implement the three-terminal detector into an FPA. The band structure of IHET is shown in Figure 16.36a [140].

The improvement IHET was grown on a (100) semi-insulating substrate [140]. The first layer was a 0.6 µm thick n$^+$-GaAs layer doped to 1.2 × 10^{18} cm^{-3} as the emitter layer. Next, an infrared sensitive 50 period Al$_{0.25}$Ga$_{0.75}$As/GaAs superlattice structure nominally identical to that reported by Levine et al. [77] except that their barrier width was 480 Å instead of 200 Å. On the top of the superlattice structure, thin 300 Å In$_{0.15}$Ga$_{0.85}$As base layer was grown, followed by a 0.2 µm thick Al$_{0.25}$Ga$_{0.75}$As electron energy high pass filter and a 0.1 µm thick n$^+$-GaAs (n = 1.2 × 10^{18} cm^{-3}) as the collector layer. The emitter and collector area of the detector was 7.92 × 10^{-4} cm^2 and 2.25 × 10^{-4} cm^2,

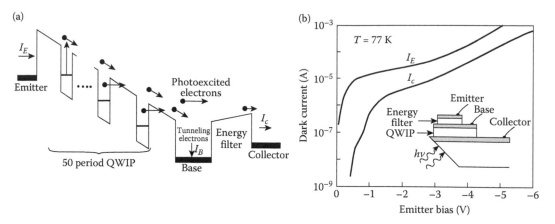

Figure 16.36 Infrared hot-electron transistor: (a) conduction band diagram, and (b) emitter dark current, I_E, and the collector dark current, I_C, as a function of the emitter bias at 77 K. The insert shows the schematic device configuration. (From K. K. Choi, L. Fotiadis, M. Taysing-Lara, W. Chang, and G. J. Iafrate, *Applied Physics Letters*, 59, 3303–5, 1991. With permission.)

respectively. The detector configuration together with emitter, I_E, and collector, I_D, dark currents are shown in Figure 16.36b. Because the thin $In_{0.15}Ga_{0.85}As$ base layer with a large Γ-L valley separation improves the photocurrent transfer ratio, the detectivity of the transistor increase to 1.4×10^{10} cmHz$^{1/2}$W^{-1} at 77 K with a cutoff wavelength of 9.5 μm, two times as large as the companion state of the art GaAs multiple quantum well detector. With further optimization of device parameters a broadband 10 μm IHET with detectivity close to 10^{11} cmHz$^{1/2}$W^{-1} should be achievable at 77 K [140].

Usually, the dark current of IHETs is two to four orders of magnitude lower than that of QWIPs [142]. This fact is especially important in the wavelength range required in many space applications (3–18 μm), where the background photon flux is very low. Also for thermal imaging purposes, it is desirable to incorporate the extended wavelength QWIPs into FPAs. Progress in development of very LWIR IHETs has been achieved [143–145].

Potentially, GaAs based QWIPs can be monolithically integrated with GaAs circuits and a concept of QWIP integration with a high electron mobility transistor was demonstrated [146]. Also other transistor ideas have been proposed to improve the QWIP performance [147] or to achieve monolithic integration [148]. Hitherto however, the GaAs technology has not been developed in the area of readout circuits.

16.7.3 SiGe/Si QWIPs

Depending on the composition, the bandgap of $Si_{1-x}Ge_x$ alloys varies from 1.1 to 0.7 eV and thus, they are suitable for detector operation in 0.5–1.8 μm wavelength region. However, the large lattice mismatch ($\Delta a_o/a_o \approx 4.2\%$ at room temperature) between Ge (lattice constant $a_o = 5.657$ Å) and Si ($a_o = 0.5431$ Å) hamper fabrication of integrated photoelectronic devices on silicon substrates because of large misfit dislocations at the interface that are not desirable for obtaining good device performance. Still, a few years ago it was reported [149] that high quality lattice mismatched structures can be grown pseudomorphically by MBE technique at the low temperature regime ($T \approx 400–500°C$) without misfit dislocations, provided the layers are thinner than the critical thickness h_c. In this case the lattice mismatch is accommodated by a distortion in the layers, giving a built-in coherent strain in them, and photodetectors sensitive in the mid- and long-infrared wavelength regions can be fabricated on the basis of Si/SiGe SLs and MQWs.

The critical thickness of a strained SiGe layer on Si substrate is strongly dependent on the growth parameters, especially the substrate temperature. For a typical growth temperature $T \approx 500°C$ for a $Si_{0.5}Ge_{0.5}$ a critical thickness is about 100 Å. In the case of multiple layer growth, the critical thickness is obtained using the average Ge composition $x_{Ge} = (x_1d_1 + x_2d_2)/(d_1 + d_2)$, where the x's and d's are the Ge content and thickness of each constituent layer, respectively.

The strain not only changes the bandgaps of the constituents, but splits the degeneracy of heavy- and light-hole bands of Si and $Si_{1-x}Ge_x$ layers and also removes the degeneracy of conduction band (many-valley band structure) [150,151]. The changes in the band structure of $Si_{1-x}Ge_x$/Si strained-layer devices, introduced by strain, can be used for several types of photodetector applications based on both p- and n-type conductivity devices. For SiGe/Si QWIPs based on intervalence band absorption, the detector response is strongly dependent on the strain induced splitting of the valence bands [59,152,153].

The major advantage of silicon-based detectors is the fact they are fabricated on Si substrates and thus monolithic integration with Si electronic readout devices makes it feasible to manufacture very large scale arrays. The first observation of intersubband infrared absorption in SiGe/Si MQWs has been described by Karunasiri et al. [154]. The large valence-band offset of SiGe/Si as well as the small hole effective mass, favors the hole intersubband absorption.

The p-type multiple quantum well SiGe/Si structures have been grown in a MBE system on high-resistivity Si(100) wafers kept at about 600°C to improve the epitaxial quality of the layers. Such structures consist of 50 periods of 30 Å thick $Si_{0.85}Ge_{0.15}$ wells (doped $p \approx 10^{19}$ cm^{-3}) and separated by 500 Å thick undoped Si barriers. A quantum well of 30 Å thick can absorb IR energy near 10 μm with the extended state lying above the Si barrier. The entire superlattice is sandwiched between a doped ($p = 1 \times 10^{19}$ cm^{-3}) 1 μm bottom and 0.5 μm top layers for electrical contacts [56].

The photoresponse of the SiGe/Si multiple quantum well (200 μm in diameter mesa structures) with a 45° facet on the edge of the wafer is shown in Figure 16.37. This figure shows the photoresponse at 0° and 90° polarizations at 77 K, with a 2 V bias across the detector. In the 0° polarization case, a peak was found at near 8.6 μm. For the 90° polarization, a peak at 7.2 μm was observed. The responsivities for both cases were the same, 0.3 A/W. The responsivity for an unpolarized beam with peak near 7.5 μm is about 0.6 A/W and is approximately the sum of two polarization cases. The responsivity measured by illuminating the light normally on the back side of the devices is shown by the dashed curve in Figure 16.37. The photoresponse at normal light incidence seemed to be due to the internal photoemission caused by free-carrier absorption. More detailed discussion of absorption mechanisms in SiGe/Si heterostructures is carried out by Park et al. [57] The estimated detectivity for above nonoptimized SiGe/Si multiple quantum well infrared detectors is about 1×10^9 cmHz$^{1/2}$W^{-1} at 9.5 μm and 77 K.

The first intervalence-subband IR detector for the 3–5 μm wavelength region was demonstrated with relatively high detectivity $D^*(3 \; \mu m) = 4 \times 10^{10}$ cmHz$^{1/2}$/W^{-1} [58]. As the Ge composition is increased, the peak photoresponse moves toward a shorter wavelength, in agreement with transmission coefficient data. The photoresponse shows several peaks, not observed in room temperature absorption spectra, which is connected with a different type of transition revealed under the bias, when carriers can tunnel from the excited states through the barrier to reach the contact (except the carriers excited to continuum states).

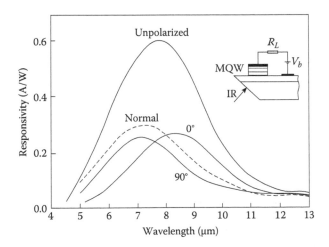

Figure 16.37 Responsivity of $Si_{0.85}Ge_{0.15}$/Si QWIP at 77 K for two polarization angles with a 2-V bias. IR radiation is illuminated on the facet at the normal such that the incidence angle on the MQW structure is 45 deg, as shown in the inset. Dashed curve shows the responsivity for normal backside illumination. (From Park, J. S., Karunasiri, R. P. G., and Wang, K. L., *Applied Physics Letters*, 60, 103–5, 1992. With permission.)

Normal incidence hole intersubband QWIPs are also possible in pseudomorphic GeSi/Si quantum wells. People et al. [155,156] have described the fabrication and performance characteristics of pseudomorphic p-type $Si_{0.75}Ge_{0.25}$/Si QWIP on (001) Si substrate. The 40 Å $Si_{0.75}Ge_{0.25}$ QWs were boron doped by ion implantation of 2 keV and have carrier concentrations $\approx 4 \times 10^{18}$ cm^{-3}. The 300 Å Si barriers layers were undoped. These devices show broadband response (8–14 μm) that is attributed to strain and quantum confinement induced mixing of heavy, light, and split-off hole bands. The 200 μm diameter devices show detectivities 3.3×10^9 cmHz$^{1/2}$W^{-1}, responsivities 0.04 A/W, and differential resistance 10^6 Ω at operating bias –2.4 V and $T = 77$ K with no cold shields (FOV 2π, 300 K).

Strong intersubband IR optical absorption at normal incidence in $Si/Si_{1-x}Ge_x$ strained-layer MQWs is possible not only in p-type structures but also in n-type structures [50], which was first demonstrated for (110) $Si/Si_{1-x}Ge_x$ MQW samples with different Ge compositions, and doping concentrations, with the position of absorption peaks ranging from 4.9 to 5.8 μm [52]. It was earlier shown [50] that for [110] and [111] growth directions the normal intersubband absorption coefficients in Si QWs of the order of 10^4 cm^{-1} can be achieved in Si QWs doped up to 10^{19} cm^{-3}, which is possible in MBE procedure.

The demonstration of the normal incidence of radiation on SiGe/Si superlattices indicates the possible realization of infrared FPAs without the use of the grating couplers normally required for AlGaAs/GaAs intersubband detectors. Also very encouraging are the theoretically predicted performance of SiGe/Si QWIPs [50,157,158]. An important advantage of SiGe/Si QWIP over competing III-V materials in hybrid arrays is that its coefficient of thermal expansion is matched to underlying Si readout circuit; so the array size is not limited by stresses generated during cooling. However, SiGe/Si QWIP detectors have not received detectivities comparable with n-type GaAs/AlGaAs material system; they are about one order lower at 77 K in LWIR region. For this reason, the development of SiGe/Si QWIPs was abandoned in the mid-1990s.

16.7.4 QWIPs With Other Material Systems

The GaAs/AlGaAs multiple quantum well detector response can also be designed to operate in shorter wavelength spectral range. However, the short wavelength limit in the AlGaAs/GaAs materials system imposed by keeping the AlGaAs barriers direct is $\lambda = 5.6$ μm. If the Al concentration x increased beyond $x = 0.45$ the indirect X valley becomes the lowest bandgap. Since Γ–X scattering together with GaAs X-barrier trapping in such structures can result in inefficient carrier collection and thus a poor responsivity, this has been thought to be highly undesirable. The limited conduction band discontinuity of AlGaAs/GaAs system (with acceptable Al mole fractions) makes it impossible to grow epilayer structures that are sensitive in the 3–5 μm MWIR window. For this reason initially, Levine et al. [159] investigated the promise of the AlInAs/InGaAs system for MWIR QWIPs using a 50 well AlInAs-InGaAs epilayer structure with 50 Å InGaAs wells and 100 Å thick AlInAs barriers resulting in a bound-to-bound QWIP with an absorption peak at $\lambda_p = 4.4$ μm and $\Delta\lambda/\lambda_p \approx 7\%$. Latter, Hasnain et al. [160] investigated direct gap system $In_{0.53}Ga_{0.47}As/In_{0.52}Al_{0.48}As$ and demonstrated multiple quantum well infrared detectors operating at $\lambda_p = 4.2$ μm with a detectivity $D^* = 2 \times 10^{10}$ cmHz$^{1/2}$W^{-1} at 77 K and a background limited detectivity 2.3×10^{12} cmHz$^{1/2}$W^{-1} at temperatures up to 120 K.

A larger conduction band discontinuity can be achieved with the AlGaAs/InGaAs material system, which has been the standard material system for MWIR QWIPs in spite of the degrading effects and limitations of lattice-mismatched epitaxy.

An AlInAs/InGaAs lattice-matched structure with a sufficiently large conduction band discontinuity is an alternative to the strained AlGaAs/InGaAs material system for both single-band MWIR and stacked multi-band QWIP focal plane arrays (FPAs). When combined with LWIR InP/InGaAs or InP/InGaAsP quantum well (QW) stacks, this material system offers a completely lattice-matched dual- or multi-band QWIP structure on an InP substrate benefiting from the advantages of InP-based QWIPs, as well as avoiding the limitations of strained layer epitaxy. Therefore, AlInAs/InGaAs is an important material system for QWIP based thermal imaging applications. Recently reported performance of large format 640 × 512 AlInAs/InGaAs FPA [161,162] with a cut-off wavelength of 4.6 μm are comparable to the best results reported for MWIR AlGaAs/InGaAs QWIPs [14].

Intersubband absorption and hot electron transport are not only limited to the $Al_xGa_{1-x}As$/GaAs materials system. The long wavelength superlattice detectors have been demonstrated using InP-based materials system such as lattice matched: $GaAs/Ga_{0.5}In_{0.5}P$ ($\lambda_p = 8$ μm), n-doped (p-doped) 1.3 μm $In_{0.53}Ga_{0.47}As$/InP (7–8 μm and 2.7 μm, respectively), 1.3 μm InGaAsP/InP ($\lambda_c = 13.2$ μm),

1.55 InGaAsP/InP ($\lambda_c = 9.4$ μm) heterosystems. The responsivities of n-doped $In_{0.53}Ga_{0.47}As/$
InP multiple quantum well infrared photoconductors were in fact somewhat larger than those
obtained in equivalent AlGaAs/GaAs ones; detectivity D^* of 9×10^{10} cmHz$^{1/2}$W^{-1} was measured
for a detector operating at 77 K at a wavelength of 7.5 μm [93,163]. The first short wavelength
($\lambda_c = 2.7$ μm) detector in p-type $Ga_{0.47}In_{0.53}As/InP$ material system has been demonstrated by
Gunapala et al. [164]. In the last case, the detector works at normal incidence of infrared radiation. The extensive literature on QWIPs using materials other than GaAs/AlGaAs is reviewed by
Levine [2], Gunapala and Bandara [5], and Li [120].

For the most GaAs-based QWIPs demonstrated thus far, GaAs is the low bandgap well material
and the barriers are lattice matched AlGaAs, GaInP, or AlInP. However, it is interesting to consider
GaAs as the barrier material since the transport in binary GaAs is expected to be superior to that
of a ternary alloy. To achieve this, Gunapala et al. [165,166] have used the lower bandgap nonlattice
matched alloy $In_xGa_{1-x}As$ as well material together with GaAs barriers. It has been demonstrated
that strain-layer heterostructures can be grown for lower In composition ($x < 0.15$), which results
in lower barrier heights. Therefore, this heterobarrier system is very suitable for very long wavelength ($\lambda > 14$ μm) QWIPs. Excellent hot electron transport and high detectivity $D^* = 1.8 \times 10^{10}$
cmHz$^{1/2}$W^{-1} at $\lambda_p = 16.7$ μm were achieved at temperature $T = 40$ K [166]. The large responsivity and
detectivity values are comparable to those achieved with the usual lattice-matched GaAs/AlGaAs
material system [95].

16.7.5 Multicolor Detectors

One of the distinct advantages of the quantum well approach is the ability to easily produce
multicolor (multispectral) detectors, which is desirable for future high-performance IR systems. In
general, a multicolor detector is a device having its spectral response varied with parameters like
applied bias voltage. Three basic approaches to achieving multicolor detection have been proposed: multiple leads, voltage switched, and voltage tuned.

The first two-color GaAs/AlGaAs QWIP has been realized by Kock et al. [167] by stacking on the
same GaAs substrate two series of QWIPs (see Figure 16.38a) with different wavelength selectivity.
This approach involves contacting each intermediate conducting layer separating the one-color
QWIPs. This leads to a separately addressable multicolor QWIP with multiple electrical terminals.
The advantage of this approach is its simplicity in design and its negligible electrical crosstalk
between colors. Moreover, each QWIP detector can be optimized independently for a desired
detection wavelength. The drawback of this approach is the difficulty of fabricating a many color
version. A technical solution of these difficulties is described in many papers, see, for example,

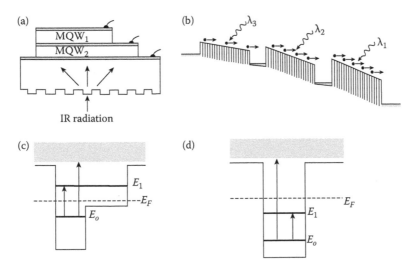

Figure 16.38 Multicolor quantum well infrared detectors based on: (a) intersubband transitions
in two series of QWIPs stacked on the same substrate, (b) three-color voltage tuneable detector at
the highest bias voltage, (c) bound-to-bound and bound-to-extended transitions in asymmetric
step multiquantum well structure, and (d) bound-to-continuum states transition mechanism.

[168–170]. Gunapala et al. [171] demonstrated a four-color imager by separating a large array into four stripes, each responding to a different color (see Section 23.4).

A novel concept of the multicolor detector structure was made by Liu et al. [172] by stacking conventional (one-color) QWIPs, separated by thin, heavily doped layers ($\approx$ 100 nm in test structure). The tunability is achieved by relying on the highly nonlinear and expotential nature of the device dark current voltage characteristics. This implies that an applied voltage across the entire multistack would be distributed among the one-color QWIPs according to their DC resistance values. When the applied voltage is increased from zero, most of the voltage will be dropped across the one-color QWIP with the highest resistance. As the voltage is further increased, an increasing fraction of it will be dropped across the next highest resistance one-color stack of QWs, and so on. The band edge profiles of a three-color version under the highest bias condition is shown schematically in Figure 16.38b. The above structure is similar to that of Grave et al. [173], but with important difference: the voltage division in Grave et al. is accomplished by high-low field domain formation, which is not quantitatively understood for MQWs [174].

In a practically demonstrated three-color version of such a multicolor QWIP concept, the GaAs well widths for the three stacks were 55, 61, and 66 Å, respectively (32 wells in each stack). The $Al_xGa_{1-x}As$ barriers were all 468 Å thick with alloy fractions of 0.26, 022, and 0.19, respectively. The separation between the one-color QWIPs was a 934 Å thick GaAs layer doped with Si to 1.5×10^{18} cm^{-3}. Three well resolved peaks at different biases were observed at 7.0, 8.5, and 9.8 μm. The calculated detectivity for the 8.5 μm response at bias $V_b = -3$ V was 5×10^9 cmHz$^{1/2}$W^{-1} for unpolarized radiation and 45° facet geometry. It can be improved up to $D^* = 3 \times 10^{10}$ cmHz$^{1/2}$W^{-1} for light-coupling geometry with 100% absorption.

The advantage of a voltage tunable approach is simplicity in fabrication (as it requires only two terminals) and implementation of many colors. The drawback is the difficulty to achieve a negligible electrical crosstalk between colors.

Another example of voltage switched two-color detection is schematically shown in Figure 16.39 [112]. In this case the unit cell consists of two superlattices of quantum wells with bound-to-miniband transition mechanism. This idea was suggested for the first time by Wang et al. [175]. One superlattice is tuned to the MWIR band and the other to the LWIR band. Between the superlattices is a graded barrier. Under negative bias, photoelectrons generated in the second superlattice lose energy in the relaxation barrier and are blocked by the first superlattice. The LW photoelectrons generated in the SL2 pass into the highly conducting energy relaxation layer, resulting in no impedance changes. Under positive bias, the reverse situation occurs and only the SL2 LW photoelectrons pass through the graded barrier and produce an impedance change. This design was used for fabrication two-color C-QWIPs [112,113].

Alternative designs of multicolor QWIPs involve special shapes of quantum wells (e.g., a stepped well or an asymmetrically coupled double well). An example of stepped well structure is shown

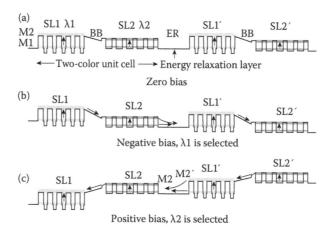

Figure 16.39 Voltage switched two-color detection mechanism in superlattice quantum wells. (From Choi, K.-K., Monroy, C., Swaminathan, V., Tamir, T., Leung, M., Devitt, J., Forrai, D., and Endres, D., *Infrared Physics & Technology*, 50, 124–35, 2007. With permission.)

in Figure 16.38c, where a two-color quantum well photoconductor uses bound-to-bound and the bound-to-extended transitions of similar oscillator strengths in asymmetric step multiquantum well structure [176,177]. An applied electric field excursion of ±40 kV/cm is sufficient to shift the peak responsivity wavelength from 8.5 to 13.5 μm [176]. An asymmetrical energy band bending can promote both photoconductive and photovoltaic mode of operation and using this dual mode of operations, two-color QWIP detectors were demonstrated [178].

Very encouraging results concerning voltage tunable three-color QWIP have been demonstrated by Tidrow et al. [179]. This device uses a couple quantum well units, each containing two coupled quantum wells of different widths separated by a thin barrier (see Figure 16.40a). The device is designed to have two subbands E_1 and E_2 originated from the wide well and one subband E_2' originated from the narrow well. When the wide well is doped, electrons from the first energy state E_1 can be excited by incoming photons to either E_2 or E_2' energy states. Because the parity symmetry is broken in the coupled asymmetric quantum well structure, more than one color can be observed.

The device with 30 periods asymmetric GaAs/AlGaAs coupled double quantum well units was grown on semi-insulating (001) GaAs substrate. The wide well width is 72 Å, the narrow well width is 20 Å, the $Al_{0.31}Ga_{0.69}As$ barrier between the two coupled wells is 40 Å, and the barrier between the coupled well units is 500 Å. The bottom contact layer is 1000 nm doped GaAs and the top contact layer is 500 Å doped $In_{0.08}Ga_{0.92}As$. The doping in the wide quantum wells and the contact layer is n + = 1.0×10^{18} cm^{-3}. The narrow wells are undoped.

The peak detectivity as a function of temperature is shown in Figure 16.40c. We can see that the detectivity is around 10^{10} cmHz$^{1/2}$W^{-1} for the 9.6 and 10.3 μm peaks at 60 K. The responsivity for the 8.4 μm peak is smaller, with the $D^* = 4 \times 10^9$ cmHz$^{1/2}$W^{-1} at 60 K. The detection peak is independently selectable among these three wavelengths by tuning the bias voltage. Generally however, it is difficult to ensure a good QWIP performance for all voltages. To provide a large intersubband transition strength and, at the same time, an easy escape for the excited carriers, the transition final state should be close to the top of the barrier. These two conditions are difficult to fulfill for

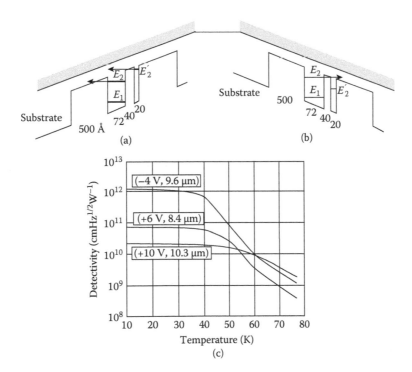

Figure 16.40 Three-color GaAs/AlGaAs QWIP: the energy band structures of the device under (a) positive and (b) negative biases; the peak detectivity of the device at (c) 8.4, 9.6 and 10.3 μm as a function of temperature under bias + 6 V, −4 V, and + 10 V, respectively. (From Tidrow, M. Z., Choi, K. K., Lee, C. Y., Chang, W. H., Towner, F. J., and Ahearn, J. S., *Applied Physics Letters*, 64, 1268–70, 1994. With permission.)

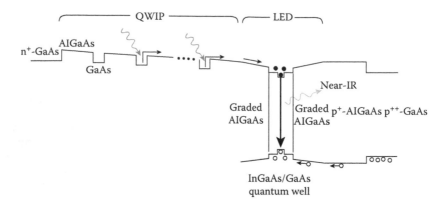

Figure 16.41 Bend edge profile of an integrated QWIP-LED. Under forward bias, the photocurrent generated in QWIP leads to an emission in LED, achieving the up-conversion of QWIP detected IR signal to LED emission near infrared or visible light. (From Liu, H. C., Dupont, E., Byloos, M., Buchanan, M., Song, C.-Y., and Wasilewski, Z. R., *Intersubband Infrared Photodetectors*, World Scientific, Singapore, 299–313, 2003.)

all voltages [12]. Also relatively wide wells in the case of a stepped well, may lead to an enhanced trapping probability, hence a shorter carrier lifetime. Finally, field-induced changes of the subband structure usually require relatively high external fields (large voltages: see Figure 16.40c), which increase dark current and noise.

Also for detectors with symmetric wells, two-color operation is possible [167,180]. This proposal is connected with large filling, when two states in the wells are occupied and optical transitions occur between the different states in the wells (like those shown in Figure 16.38d), or with wells of different thicknesses in which only ground states are occupied by the carriers. Such multiquantum well photodetectors aim to cover both the 8–12 and 3–5 μm spectral regions.

16.7.6 Integrated QWIP-LED

The innovative concept of frequency-up conversion based on the integration of QWIP with light emitting diodes (QWIP-LED) offers an alternative to the standard hybrid technique in making an imaging device. This approach may lead to devices that are difficult to realize by the standard one, such as ultra-large size sensors [181].

The integrated QWIP-LED concept was independently proposed by Liu et al. [182] and by Ryzhii et al. [183] and was first experimentally demonstrated by Liu and colleagues [182]. The basic idea is shown in Figure 16.41. Under a forward bias, photocurrent electrons from the QWIP recombine with injected holes in LED, giving rise to an increase in LED emission. The QWIP is a photoconductor so that under IR light illumination its resistance decreases, which leads to an increase in the voltage drop across the LED and therefore an increase in the amount of emission. This device is therefore an IR converter. Photocarriers in QWIP have a strong lateral locality. The resulting emission around 0.9 μm can be easily imaged using the well-developed Si CCD array.

The electron well charges capacitance in present commercial CCD (typically 4×10^5 electrons) is almost two orders smaller than that of the readout circuit used in LWIR FPAs. Then CCD is usually needed to operate at full well working point mode using QWIP-LED for IR imaging to get a relatively higher thermal image gray level.

The advantage of the integrated QWIP-LED is technologically important since in this scheme, one can make 2-D large format imaging devices without the need of making any circuit readouts. The QWIP-LED device still operates under low temperature. The up-conversion approach can be easily implemented in multicolor imaging devices in a pixelless geometry [184].

The initial demonstration of the pixelless QWIP-LED used a p-type material to simplify fabrication (avoiding gratings). Due to low performance, the next efforts were concentrated on n-type QWIPs and steady improvements were made [185–187]. The latest imaging result is shown in Figure 16.42.

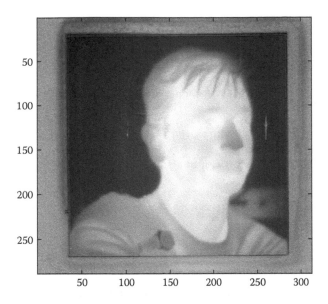

Figure 16.42 Example of thermal imaging with a QWIP-LED. (From Dupont, E., Byloos, M., Oogarah, T., Buchanan, M., and Liu, H. C., *Infrared Physics Technology*, 47, 132–43, 2005. With permission.)

REFERENCES

1. L. Esaki and R. Tsu, "Superlattice and Negative Conductivity in Semiconductors," *IBM Journal of Research and Development* 14, 61–65, 1970.

2. B. F. Levine, "Quantum-Well Infrared Photodetectors," *Journal of Applied Physics* 74, R1–R81, 1993.

3. M. O. Manasreh, ed., *Semiconductor Quantum Wells and Superlattices for Long-Wavelength Infrared Detectors*, Artech House, Norwood, MA, 1993.

4. F. F. Sizov and A. Rogalski, "Semiconductor Superlattices and Quantum Wells for Infrared Optoelectronics," *Progress in Quantum Electronics* 17, 93–164, 1993.

5. S. D. Gunapala and K. M. S. V. Bandara, "Recent Development in Quantum-Well Infrared Photodetectors," in *Thin Films*, Vol. 21, 113–237, Academic Press, New York, 1995.

6. H. C. Liu, "Quantum Well Infrared Photodetector Physics and Novel Devices," in *Semiconductors and Semimetals*, Vol. 62, eds. H. C. Liu and F. Capasso, 129–96, Academic Press, San Diego, 2000.

7. S. D. Gunapala and S. V. Bandara, "Quantum Well Infrared Photodetectors (QWIP)," in *Handbook of Thin Devices*, Vol. 2, ed. M. H. Francombe, 63–99, Academic Press, San Diego, 2000.

8. J. L. Pan and C. G. Fonstad, "Theory, Fabrication and Characterization of Quantum Well Infrared Photodetectors," *Material Science and Engineering* R28, 65–147, 2000.

9. S. D. Gunapala and S. V. Bandara, "GaAs/AlGaAs Based Quantum Well Infrared Photodetector Focal Plane Arrays," in *Handbook of Infrared Detection Technologies*, ed. M. Henini and M. Razeghi, 83–119, Elsevier, Oxford, 2002.

10. V. Ryzhi, ed., *Intersubband Infrared Photodetectors*, World Scientific, New Jersey, 2003.

11. A. Rogalski, "Quantum Well Photoconductors in Infrared Detectors Technology," *Journal of Applied Physics* 93, 4355–91, 2003.

12. H. Schneider and H. C. Liu, *Quantum Well Infrared Photodetectors,* Springer, Berlin, 2007.

13. A. Rogalski, J. Antoszewski, and L. Faraone, "Third-Generation Infrared Photodetector Arrays," *Journal of Applied Physics* 105, 091101-44, 2009.

14. S. D. Gunapala, S. V. Bandara, J. K. Liu, C. J. Hill, B. Rafol, J. M. Mumolo, J. T. Trinh, M. Z. Tidrow, and P. D. LeVan, "1024 × 1024 Pixel Mid-Wavelength and Long-Wavelength Infrared QWIP Focal Plane Arrays for Imaging Applications," *Semiconductor Science and Technology* 20, 473–80, 2005.

15. M. Jhabvala, K. K. Choi, C. Monroy, and A. La, "Development of a 1K × 1K, 8–12 μm QWIP Array," *Infrared Physics & Technology* 50, 234–39, 2007.

16. G. Bastard, *Wave Mechanics Applied to Semiconductor Heterostructures*, Editions de Physique, Les Ulis, 1988.

17. C. Weisbuch and B. Vinter, *Quantum Semiconductor Structures,* Academic Press, New York, 1991.

18. A. Shik, *Quantum Wells,* World Scientific, Singapore, 1997.

19. P. Harrison, *Quantum Wells, Wires and Dots: Theoretical and Computational Physics,* Wiley, New York, 1999.

20. D. Bimberg, M. Grundmann, and N. N. Ledentsov, *Quantum Dot Heterostructures,* Wiley, Chichester, 2001.

21. J. Singh, *Electronic and Optoelectronic Properties of Semiconductor Structures,* Cambridge University Press, Cambridge, 2003.

22. H. Sakaki, "Scattering Suppression and High-Mobility Effect of Size-Quantized Electrons in Ultrafine Semiconductor Wire Structures," *Japanese Journal of Applied Physics* 19, L735–L738, 1980.

23. Y. Arakawa and H. Sakaki, "Multidimensional Quantum-Well Laser and Temperature Dependence of Its Threshold Current," *Applied Physics Letters* 40, 939–41, 1982.

24. D. D. Coon and K. M. S. V. Bandara, "New Quantum Structures," in *Physics of Thin Films,* Vol. 15, eds. M. H. Francombe and J. L. Vossen, 219–64, Academic Press, Boston, MA, 1991.

25. H. C. Liu, "Quantum Dot Infrared Photodetector," *Opto-Electronics Review* 11, 1–5, 2003.

26. D. H. Döhler, "Semiconductor Superlattices: A New Material for Research and Applications," *Physica Scripta* 24, 430, 1981.

27. C. Mailhiot and D. L. Smith, "Strained-Layer Semiconductor Superlattices," *Solid State & Materials Science* 16, 131–60, 1990.

28. L. Esaki, "A Bird's-Eye View on the Evolution of Semiconductor Superlattices and Quantum Wells," *IEEE Journal of Quantum Electronics* QE-22, 1611–24, 1986.

29. H. Krömer, "Barrier Control and Measurements: Abrupt Semiconductor Heterojunctions," *Journal of Vacuum Science and Technology* B2, 433, 1984.

30. J. Tersoff, "Theory of Semiconductor Heterojunctions: The Role of Quantum Dipoles," *Physical Review* B30, 4874, 1984.

31. W. A. Harrison, "Elementary Tight-Binding Theory of Schottky-Barrier and Heterojunction Band Line-Ups," in *Two Dimensional Systems: Physics and New Devices, Springer Series in Solid State Sciences,* Vol. 67, eds. G. Bauer, F. Kuchar, and H. Heinrich, 62, Springer, Berlin, 1986.

32. W. Pollard, "Valence-Band Discontinuities in Semiconductor Heterojunctions," *Journal of Applied Physics* 69, 3154–58, 1991.

33. H. Krömer, "Band Offsets and Chemical Bonding: The Basis for Heterostructure Applications," *Physica Scripta* 68, 10–16, 1996.

34. G. Duggan, "A Critical Review of Semiconductor Heterojunction Band Offsets," *Journal of Vacuum Science and Technology* B3, 1224, 1985.

35. T. W. Hickmott, "Electrical Measurements of Band Discontinuits at Heterostructure Interfaces," in *Two Dimensional Systems: Physics and New Devices, Springer Series in Solid State Sciences,* Vol. 67, eds. G. Bauer, F. Kuchar, and H. Heinrich, 72, Springer, Berlin, 1986.

36. J. Menendez and A. Pinczuk, "Light Scattering Determinations of Band Offsets in Semiconductor Heterojunctions," *IEEE Journal of Quantum Electronics* 24, 1698–711, 1988.

37. K. Ploog and G. H. Döhler, "Compositional and Doping Superlattices in III–V Semiconductors," *Advanced Physics* 32, 285–359, 1983.

38. G. H. Döhler, "Doping Superlattices ("n-i-p-i crystals")," *IEEE Journal of Quantum Electronics* QE-22, 1682–95, 1986.

39. G. H. Döhler, "The Physics and Applications of n-i-p-i Doping Superlattices," *CRC Critical Reviews in Solid State and Materials Sciences* 13, 97–141, 1987.

40. C. Weisbuch, "Fundamental Properties of III–V Semiconductor Two-Dimensional Quantized Structures: The Basis for Optical and Electronic Device Applications," in *Semiconductors and Semimetals,* Vol. 24, eds. R. Willardson and A. C. Beer; *Applications of Multiquantum Wells, Selective Doping, and Superlattices,* ed. R. Dingle, 1–133, Academic Press, New York, 1987.

41. A. Rogalski, *New Ternary Alloy Systems for Infrared Detectors,* SPIE Optical Engineering Press, Bellingham, WA, 1994.

42. D. L. Smith and C. Mailhiot, "Theory of Semiconductor Superlattice Electronic Structure," *Reviews of Modern Physics* 62, 173–234, 1990.

43. B. F. Levine, R. J. Malik, J. Walker, K. K. Choi, C. G. Bethea, D. A. Kleinman, and J. M. Wandenberg, "Strong 8.2 μm Infrared Intersubband Absorption in Doped GaAs/AlAs Quantum Well Waveguides," *Applied Physics Letters* 50, 273–75, 1987.

44. K. K. Choi, "Detection Wavelength of Quantum-Well Infrared Photodetectors," *Journal of Applied Physics* 73, 5230–36, 1993.

45. K. M. S. V. Bandara, D. D. Coon, O. Byungsung, Y. F. Lin, and M. H. Francombe, "Exchange Interactions in Quantum Well Subbands," *Applied Physics Letters* 53, 1931–33, 1988.

46. M. C. Tatham, J. R. Ryan, and C. T. Foxon, "Time-Resolved Raman Scattering Measurement of Electron-Optical Phonon Intersubband Relaxation in GaAs Quantum Wells," *Solid State Electronics* 32, 1497–501, 1989.

47. G. Hasnain, B. F. Levine, C. G. Bethea, R. R. Abbott, and S. J. Hsieh, "Measurement of Intersubband Absorption in Multiquantum Well Structures with Monolithically Integrated Photodetectors," *Journal of Applied Physics* 67, 4361–63, 1990.

48. M. O. Manasreh, F. F. Szmulowicz, D. W. Fischer, K. R. Evans, and C. E. Stutz, "Intersubband Infrared Absorption in a GaAs/Al$_{0.3}$Ga$_{0.7}$As Quantum Well Structure," *Applied Physics Letters* 57, 1790–92, 1990.

49. B. F. Levine, A. Zussman, S. D. Gunapala, M. T. Asom, J. M. Kuo, and W. S. Hobson, "Photoexcited Escape Probability, Optical Gain, and Noise in Quantum Well Infrared Photodetectors," *Journal of Applied Physics* 72, 4429–43, 1992.

50. C. I. Yang and D. S. Pan, "Intersubband Absorption of Silicon-Based Quantum Wells for Infrared Imaging," *Journal of Applied Physics* 64, 1573–75, 1988.

51. Y. C. Chang and R. B. James, "Saturation of Intersubband Transitions in p-Type Semiconductor Quantum Wells," *Physical Review* B39, 12672–81, 1989.

52. Ch. Lee and K. L. Wang, "Intersubband Absorption in Sb δ-Doped Si/Si$_{1-x}$Ge$_x$ Quantum Well Structures Grown on Si (110)," *Applied Physics Letters* 60, 2264–66, 1992.

53. J. Katz, Y. Zhang, and W. I. Wang, "Normal Incidence Infrared Absorption in AlAs/AlGaAs X-Valley Multiquantum Wells," *Applied Physics Letters* 61, 1697–99, 1992.

54. B. F. Levine, S. D. Gunapala, J. M. Kuo, and S. S. Pei, "Normal Incidence Hole Intersubband Absorption Long Wavelength GaAs/Al$_x$Ga$_{1-x}$As Quantum Well Infrared Photodetector," *Applied Physics Letters* 59, 1864–66, 1991.

55. W. S. Hobson, A. Zussman, B. F. Levine, J. de Long, M. Gera, and L. C. Luther, "Carbon Doped GaAs/Al$_x$Ga$_{1-x}$As Quantum Well Infrared Photodetectors Grown by Organometallic Vapor Phase Epitaxy," *Journal of Applied Physics* 71, 3642–44, 1992.

56. J. S. Park, R. P. G. Karunasiri, and K. L. Wang, "Normal Incidence Detector Using p-Type SiGe/Si Multiple Quantum Wells," *Applied Physics Letters* 60, 103–5, 1992.

57. J. S. Park, R. P. G. Karunasiri, and K. L. Wang, "Intervalence-Subband Transition in SiGe/Si Multiple Quantum Wells: Normal Incidence Detection," *Applied Physics Letters* 61, 681–83, 1992.

58. R. P. G. Karunasiri, J. S. Park, and K. L. Wang, "Normal Incidence Infrared Detector Using Intervalence-Subband Transitions in Si$_{1-x}$Ge$_x$/Si Quantum Wells," *Applied Physics Letters* 61, 2434–36, 1992.

59. G. Karunasiri, "Intersubband Transition in Si-Based Quantum Wells and Application for Infrared Photodetectors," *Japanese Journal of Applied Physics* 33, 2401–11, 1994.

60. R. People, J. C. Bean, C. G. Bethea, S. K. Sputz, and L. J. Peticolas, "Broadband (8–14 µm), Normal Incidence, Pseudomorphic Ge$_x$Si$_{1-x}$/Si Strained-Layer Infrared Photodetector Operating between 20 and 77 K," *Applied Physics Letters* 61, 1122–24, 1992.

61. B. K. Ridley, "Electron Scattering by Confined LO Polar Phonons in a Quantum Well," *Physical Review* B39, 5282–86, 1989.

62. D. Y. Oberli, D. R. Wake, M. V. Klein, T. Henderson, and H. Morkoc, "Intersubband Relaxation of Photoexcited Hot Carriers in Quantum Wells," *Solid State Electronics* 31, 413–18, 1988.

63. B. N. Murdin, W. Heiss, C. J. G. M. Langerak, S.-C. Lee, I. Galbraith, G. Strasser, E. Gornik, M. Helm, and C. R. Pidgeon, "Direct Observation of the LO Phonon Bottleneck in Wide GaAs/Al$_x$Ga$_{1-x}$ As Quantum Wells," *Physical Review* B55, 5171–76, 1997.

64. J. Y. Andersson, "Dark Current Mechanisms and Conditions of Background Radiation Limitation of n-Doped AlGaAs/GaAs Quantum-Well Infrared Detectors," *Journal of Applied Physics* 78, 6298–304, 1995.

65. V. D. Shadrin and F. L. Serzhenko, "The Theory of Multiple Quantum-Well GaAs/AlGaAs Infrared Detectors," *Infrared Physics* 33, 345–57, 1992.

66. J. C. Smith, L. C. Chiu, S. Margalit, A. Yariv, and A. Y. Cho, "A New Infrared Detector Using Electron Emission from Multiple Quantum Wells," *Journal of Vacuum Science and Technology* B1, 376–78, 1983.

67. L. C. Chiu, J. S. Smith, S. Margalit, A. Yariv, and A. Y. Cho, "Application of Internal Photoemission from Quantum-Well and Heterojunction Superlattices to Infrared Photodetectors," *Infrared Physics* 23, 93–97, 1983.

68. D. D. Coon and P. G. Karunasiri, "New Mode of IR Detection Using Quantum Wells," *Applied Physics Letters* 45, 649–51, 1984.

69. L. C. West and S. J. Eglash, "First Observation of an Extremely Large-Dipole Infrared Transition within the Conduction Band of a GaAs Quantum Well," *Applied Physics Letters* 46, 1156–58, 1985.

70. B. F. Levine, K. K. Choi, C. G. Bethea, J. Walker, and R. J. Malik, "New 10 μm Infrared Detector Using Intersubband Absorption in Resonant Tunneling GaAlAs Superlattices," *Applied Physics Letters* 50, 1092–94, 1987.

71. S. Gunapala, M. Sundaram, and S. Bandara, "Quantum Wells Stare at Long-Wave IR Scenes," *Laser Focus World*, 233–40, June 1996.

72. S. D. Gunapala, J. K. Liu, J. S. Park, M. Sundaram, C. A. Shott, T. Hoelter, T. L. Lin, et al., "9-μm Cutoff 256 × 256 GaAs/Al$_x$Ga$_{1-x}$As Quantum Well Infrared Photodetector Hand-Held Camera," *IEEE Transactions on Electron Devices* 44, 51–57, 1997.

73. J. Y. Andersson, L. Lundqvist, and Z. F. Paska, "Quantum Efficiency Enhancement of AlGaAs/GaAs Quantum Well Infrared Detectors Using a Waveguide with a Grating Coupler," *Applied Physics Letters* 58, 2264–66, 1991.

74. Z. F. Paska, J. Y. Andersson, L. Lundqvist, and C. O. A. Olsson, "Growth and Characterization of AlGaAs/GaAs Quantum Well Structures for the Fabrication of Long Wavelength Infrared Detectors," *Journal of Crystal Growth* 107, 845–49, 1991.

75. W. Bloss, M. O'Loughlin, and M. Rosenbluth, "Advances in Multiple Quantum Well IR Detectors," *Proceedings of SPIE* 1541, 2–10, 1991.

76. K. K. Choi, B. F. Levine, C. G. Bethea, J. Walker, and R. J. Malik, "Multiple Quantum Well 10 μm GaAs/Al$_x$Ga$_{1-x}$As Infrared Detector with Improved Responsivity," *Applied Physics Letters* 50, 1814–16, 1987.

77. B. F. Levine, C. G. Bethea, G. Hasnain, V. O. Shen, E. Pelve, R. R. Abbot, and S. J. Hsieh, "High Sensitivity Low Dark Current 10 μm GaAs Quantum Well Infrared Photodetectors," *Applied Physics Letters* 56, 851–53, 1990.

78. S. D. Gunapala, B. F. Levine, L. Pfeifer, and K. West, "Dependence of the Performance of GaAs/AlGaAs Quantum Well Infrared Photodetectors on Doping and Bias," *Journal of Applied Physics* 69, 6517–20, 1991.

79. S. D. Gunapala, J. S. Park, G. Sarusi, T. L. Lin, J. K. Liu, P. D. Maker, R. E. Muller, C. A. Shott, and T. Hoelter, "15-μm 128 × 128 GaAs/Al$_x$Ga$_{1-x}$As Quantum Well Infrared Photodetector Focal Plane Array Camera," *IEEE Transactions on Electron Devices* 44, 45–50, 1997.

80. M. A. Kinch and A. Yariv, "Performance Limitations of GaAs/AlGaAs Infrared Superlattices," *Applied Physics Letters* 55, 2093–95, 1989.

81. B. F. Levine, C. G. Bethea, G. Hasnain, J. Walker, and R. J. Malik, "High-Detectivity $D^* = 1.0 \times 10^{10}$ cm $\sqrt{Hz}$ /W GaAs/AlGaAs Multiquantum Well $\lambda = 8.3$ μm Infrared Detector," *Applied Physics Letters* 53, 296–98, 1988.

82. H. C. Liu, "An Introduction to the Physics of Quantum Well Infrared Photodetectors and Other Related Devices," in *Handbook of Thin Film Devices*, Vol. 2, ed. M. H. Francombe, 101–34, Academic Press, San Diego, 2000.

83. M. J. Kane, S. Millidge, M. T. Emeny, D. Lee, D. R. P. Guy, and C. R. Whitehouse, "Performance Trade Offs in the Quantum Well Infra-Red Detector," in *Intersubband Transitions in Quantum Wells*, eds. E. Rosencher, B. Vinter, and B. Levine, 31–42, Plenum Press, New York, 1992.

84. H. C. Liu, A. G. Steele, M. Buchanan, and Z. R. Wasilewski, "Dark Current in Quantum Well Infrared Photodetectors," *Journal of Applied Physics* 73, 2029–31, 1993.

85. M. Z. Tidrow, J. C. Chiang, S. S. Li, and K. Bacher, "A High Strain Two-Stack Two-Color Quantum Well Infrared Photodetector," *Applied Physics Letters* 70, 859–61, 1997.

86. A. Rose, *Concepts in Photoconductivity and Allied Problems*, Interscience, New York, 1963.

87. H. C. Liu, "Photoconductive Gain Mechanism of Quantum-Well Intersubband Infrared Detectors," *Applied Physics Letters* 60, 1507–9, 1992.

88. A. G. Steele, H. C. Liu, M. Buchanan, and Z. R. Wasilewski, "Importance of the Upper State Position in the Performance of Quantum Well Intersubband Infrared Detectors," *Applied Physics Letters* 59, 3625–27, 1991.

89. H. C. Liu, "Dependence of Absorption Spectrum and Responsivity on the Upper State Position in Quantum Well Intersubband Photodetectors," *Applied Physics Letters* 73, 3062–67, 1993.

90. H. C. Liu, "Noise Gain and Operating Temperature of Quantum Well Infrared Photodetectors," *Applied Physics Letters* 61, 2703–5, 1992.

91. W. A. Beck, "Photoconductive Gain and Generation-Recombination Noise in Multiple-Quantum-Well Infrared Detectors," *Applied Physics Letters* 63, 3589–91, 1993.

92. A. Rogalski, "Comparison of the Performance of Quantum Well and Conventional Bulk Infrared Photodetectors," *Infrared Physics & Technology* 38, 295–310, 1997.

93. S. D. Gunapala, B. F. Levine, D. Ritter, R. A. Hamm, and M. B. Panish, "InGaAs/InP Long Wavelength Quantum Well Infrared Photodetectors," *Applied Physics Letters* 58, 2024–26, 1991.

94. A. Zussman, B. F. Levine, J. M. Kuo, and J. de Jong, "Extended Long-Wavelength $\lambda = 11$–15 μm GaAs/Al_xGa_{1-x}As Quantum Well Infrared Photodetectors," *Journal of Applied Physics* 70, 5101–7, 1991.

95. B. F. Levine, A. Zussman, J. M. Kuo, and J. de Jong, "19 μm Cutoff Long-Wavelength GaAs/Al_xGa_{1-x}As Quantum Well Infrared Photodetectors," *Journal of Applied Physics* 71, 5130–35, 1992.

96. A. Singh and M. O. Manasreh, "Quantum Well and Superlattice Heterostructures for Space-Based Long Wavelength Infrared Photodetectors," *Proceedings of SPIE* 2397, 193–209, 1995.

97. D. C. Arrington, J. E. Hubbs, M. E. Gramer, and G. A. Dole, "Nonlinear Response of QWIP Detectors: Summary of Data from Four Manufactures," *Proceedings of SPIE* 4028, 289–99, 2000.

98. M. Z. Tidrow, W. A. Beck, W. W. Clark, H. K. Pollehn, J. W. Little, N. K. Dhar, P. R. Leavitt, et al., "Device Physics and Focal Plane Applications of QWIP and MCT," *Opto-Electronics Review* 7, 283–96, 1999.

99. A. Rogalski, "Third Generation Photon Detectors," *Optical Engineering* 42, 3498–516, 2003.

100. A. Rogalski, "Competitive Technologies of Third Generation Infrared Photon Detectors," *Opto-Electronics Review* 14, 87–101, 2006.

101. C. Schönbein, H. Schneider, R. Rehm, and M. Walther, "Noise Gain and Detectivity of n-Type GaAs/AlAs/AlGaAs Quantum Well Infrared Photodetectors," *Applied Physics Letters* 73, 1251–54, 1998.

102. A. Kastalsky, T. Duffield, S. J. Allen, and J. Harbison, "Photovoltaic Detection of Infrared Light in a GaAs/AlGaAs Superlattice," *Applied Physics Letters* 52, 1320–22, 1988.

103. H. Schneider, M. Walther, J. Fleissner, R. Rehm, E. Diwo, K. Schwarz, P. Koidl, et al., "Low-Noise QWIPs for FPA Sensors with High Thermal Resolution," *Proceedings of SPIE* 4130, 353–62, 2000.

104. H. Schneider, P. Koidl, M. Walther, J. Fleissner, R. Rehm, E. Diwo, K. Schwarz, and G. Weimann, "Ten Years of QWIP Development at Fraunhofer," *Infrared Physics & Technology* 42, 283–89, 2001.

105. H. Schneider, M. Walther, C. Schönbein, R. Rehm, J. Fleissner, W. Pletschen, J. Braunstein, et al., "QWIP FPAs for High-Performance Thermal Imaging," *Physica E* 7, 101–7, 2000.

106. H. Schneider, C. Schönbein, M. Walther, K. Schwarz, J. Fleissner, and P. Koidl, "Photovoltaic Quantum Well Infrared Photodetectors: The Four-Zone Scheme," *Applied Physics Letters* 71, 246–248, 1997.

107. H. Schneider, "Theory of Avalanche Multiplication and Excess Noise in Quantum-Well Infrared Photodetectors," *Applied Physics Letters* 82, 4376–78, 2003.

108. O. Byungsung, J. W. Choe, M. H. Francombe, K. M. S. V. Bandara, D. D. Coon, Y. F. Lin, and W. J. Takei, "Long-Wavelength Infrared Detection in a Kastalsky-Type Superlattice Structure," *Applied Physics Letters* 57, 503–5, 1990.

109. S. D. Gunapala, B. F. Levine, and N. Chand, "Band to Continuum Superlattice Miniband Long Wavelength GaAs/Al$_x$Ga$_{1-x}$As Infrared Detectors," *Journal of Applied Physics* 70, 305–8, 1991.

110. K. M. S. V. Bandara, J. W. Choe, M. H. Francombe, A. G. U. Perera, and Y. F. Lin, "GaAs/AlGaAs Superlattice Miniband Detector with 14.5 μm Peak Response," *Applied Physics Letters* 60, 3022–24, 1992.

111. C. C. Chen, H. C. Chen, C. H. Kuan, S. D. Lin, and C. P. Lee, "Multicolor Infrared Detection Realized with Two Distinct Superlattices Separated by a Blocking Barrier," *Applied Physics Letters* 80, 2251–53, 2002.

112. K-K. Choi, C. Monroy, V. Swaminathan, T. Tamir, M. Leung, J. Devitt, D. Forrai, and D. Endres, "Optimization of Corrungated-QWIP for Large Format, High Quantum Efficiency, and Multi-Color FPAs," *Infrared Physics & Technology* 50, 124–35, 2007.

113. K.-K. Choi, M. D. Jhabvala, and R. J. Peralta, "Voltage-Tunable Two-Color Corrugated-QWIP Focal Plane Arrays," *IEEE Electron Device Letters* 29, 1011–13, 2008.

114. L. S. Yu and S. S. Lu, "A Metal Grating Coupled Bound-to-Miniband Transition GaAs Multiquantum Well/Superlattice Infrared Detector," *Applied Physics Letters* 59, 1332–34, 1991.

115. L. S. Yu, S. S. Li, and P. Ho, "Largely Enhanced Bound-to-Miniband Absorption in an InGaAs Multiple Quantum Well with Short-Period Superlattice InAlAs/InGaAs Barrier," *Applied Physics Letters* 59, 2712–14, 1991.

116. L. S. Yu, Y. H. Wang, S. S. Li, and P. Ho, "Low Dark Current Step-Bound-to-Miniband Transition InGaAs/GaAs/AlGaAs Multiquantum-Well Infrared Detector," *Applied Physics Letters* 60, 992–94, 1992.

117. W. A. Beck, J. W. Little, A. C. Goldberg, and T. S. Faska, "Imaging Performance of LWIR Miniband Transport Multiple Quantum Well Infrared Focal Plane Arrays," in *Quantum Well Intersubband Transition Physics and Devices*, eds. H. C. Liu, B. F. Levine, and J. Y. Anderson, 55–68, Kluwer Academic Publishers, Dordrecht, 1994.

118. W. A. Beck and T. S. Faska, "Current Status of Quantum Well Focal Plane Arrays," *Proceedings of SPIE* 2744, 193–206, 1996.

119. J. Chu and S. S. Li, "The Effect of Compressive Strain on the Performance of p-type Quantum-Well Infrared Photodetectors," *IEEE J. Quantum Electron* 33, 1104–113, 1997.

120. S. S. Li, "Multi-Color, Broadband Quantum Well Infrared Photodetectors for Mid-, Long-, and Very Long-Wavelength Infrared Applications," in *Intersubband Infrared Photodetectors*, ed. V. Ryzhii, 169–209, World Scientific, Singapore, 2003.

121. K. W. Goossen and S. A. Lyon, "Grating Enhanced Quantum Well Detector," *Applied Physics Letters* 47, 1257–1529, 1985.

122. K. W. Goossen, S. A. Lyon, and K. Alavi, "Grating Enhancement of Quantum Well Detector Response," *Applied Physics Letters* 53, 1027–29, 1988.

123. G. Hasnain, B. F. Levine, C. G. Bethea, R. A. Logan, L. Walker, and R. J. Malik, "GaAs/AlGaAs Multiquantum Well Infrared Detector Arrays Using Etched Gratings," *Applied Physics Letters* 54, 2515–17, 1989.

124. J. Y. Andersson, L. Lundqvist, and Z. F. Paska, "Grating-Coupled Quantum-Well Infrared Detectors: Theory and Performance," *Journal of Applied Physics* 71, 3600–3610, 1992.

125. G. Sarusi, B. F. Levine, S. J. Pearton, K. M. S. V. Bandara, and R. E. Leibenguth, "Optimization of Two Dimensional Gratings for Very Long Wavelength Quantum Well Infrared Photodetectors," *Journal of Applied Physics* 76, 4989–94, 1994.

126. G. Sarusi, B. F. Levine, S. J. Pearton, K. M. S. V. Bandara, and R. E. Leibenguth, "Improved Performance of Quantum Well Infrared Photodetectors Using Random Scattering Optical Coupling," *Applied Physics Letters* 64, 960–62, 1994.

127. T. R. Schimert, S. L. Barnes, A. J. Brouns, F. C. Case, P. Mitra, and L. T. Claiborne, "Enhanced Quantum Well Infared Photodetector with Novel Multiple Quantum Well Grating Structure," *Applied Physics Letters* 68, 2846–48, 1996.

128. C. J. Chen, K. K. Choi, M. Z. Tidrow, and D. C. Tsui, "Corrugated Quantum Well Infrared Photodetectors for Normal Incident Light Coupling," *Applied Physics Letters* 68, 1446–48, 1996.

129. C. J. Chen, K. K. Choi, W. H. Chang, and D. C. Tsui, "Performance of Corrugated Quantum Well Infrared Photodetectors," *Applied Physics Letters* 71, 3045–47, 1997.

130. J. A. Robo, E. Costard, J. P. Truffer, A. Nedelcu, X. Marcadet, and P. Bois, "QWIP Focal Plane Arrays Performances from MWIR to VLWIR," *Proceedings of SPIE* 7298, 7298-15, 2009.

131. Y. C. Chang and R. B. James, "Saturation of Intersubband Transitions in p-Type Semiconductor Quantum Wells," *Physical Review* B39, 12672–81, 1989.

132. H. Xie, J. Katz, and W. I. Wang, "Infrared Absorption Enhancement in Light- and Heavy-Hole Inverted $Ga_{1-x}In_xAs/Al_{1-y}In_yAs$ Quantum Wells," *Applied Physics Letters* 59, 3601–3, 1991.

133. H. Xie, J. Katz, W. I. Wang, and Y. C. Chang, "Normal Incidence Infrared Photoabsorption in p-Type $GaSb/Ga_xAl_{1-x}Sb$ Quantum Wells," *Journal of Applied Physics* 71, 2844–47, 1992.

134. F. Szmulowicz, G. J. Brown, H. C. Liu, A. Shen, Z. R. Wasilewski, and M. Buchanan, "GaAs/AlGaAs p-Type Multiple-Quantum Wells for Infrared Detection at Normal Incidence: Model and Experiment," *Opto-Electronics Review* 9, 164–72, 2001.

135. F. Szmulowicz and G. J. Brown, "Whither p-Type GaAs/AlGaAs QWIP?" *Proceedings of SPIE* 4650, 158–66, 2002.

136. K. Hirose, T. Mizutani, and K. Nishi, "Electron and Hole Mobility in Modulation Doped GaInAs-AlInAs Strained Layer Superlattice," *Journal of Crystal Growth* 81, 130–35, 1987.

137. P. Man and D. S. Pan, "Analysis of Normal-Incident Absorption in p-Type Quantum-Well Infrared Photodetectors," *Applied Physics Letters* 61, 2799–801, 1992.

138. K. K. Choi, M. Dutta, P. G. Newman, and M. L. Saunders, "10 µm Infrared Hot-Electron Transistors," *Applied Physics Letters* 57, 1348–50, 1990.

139. K. K. Choi, M. Dutta, R. P. Moekirk, C. H. Kuan, and G. J. Iafrate, "Application of Superlattice Bandpass Filters in 10 µm Infrared Detection," *Applied Physics Letters* 58, 1533–35, 1991.

140. K. K. Choi, L. Fotiadis, M. Taysing-Lara, W. Chang, and G. J. Iafrate, "High Detectivity InGaAs Base Infrared Hot-Electron Transistor," *Applied Physics Letters* 59, 3303–5, 1991.

141. K. K. Choi, *The Physics of Quantum Well Infrared Photodetectors*, World Scientific, Singapore, 1997.

142. K. K. Choi, M. Z. Tidrow, M. Taysing-Lara, W. H. Chang, C. H. Kuan, C. W. Farley, and F. Chang, "Low Dark Current Infrared Hot-Electron Transistor for 77 K Operation," *Applied Physics Letters* 63, 908–10, 1993.

143. C. Y. Lee, M. Z. Tidrow, K. K. Choi, W. H. Chang, and L. F. Eastman, "Long-Wavelength $\lambda_c = 18$ µm Infrared Hot-Electron Transistor," *Journal of Applied Physics* 75, 4731–36, 1994.

144. C. Y. Lee, M. Z. Tidrow, K. K. Choi, W. H. Chang, and L. F. Eastman, "Activation Characteristics of a Long Wavelength Infrared Hot-Electron Transistor," *Applied Physics Letters* 65, 442–44, 1994.

145. S. D. Gunapala, J. S. Park, T. L. Lin, J. K. Liu, K. M. S. V. Bandara, "Very Long-Wavelength $GaAs/Al_xGa_{1-x}As$ Infrared Hot Electron Transistor," *Applied Physics Letters* 64, 3003–5, 1994.

146. D. Mandelik, M. Schniederman, V. Umansky, and I. Bar-Joseph, "Monolithic Integration of a Quantum-Well Infrared Photodetector Array with a Read-Out Circuit," *Applied Physics Letters* 78, 472–74, 2001.

147. V. Ryzhii, "Unipolar Darlington Infrared Phototransistor," *Japanese Journal of Applied Physics* 36, L415–L417, 1997.

148. K. Nanaka, *Semiconductor Devices*, Japan Patent 4-364072, 1992.

149. S. C. Jain, J. R. Willis, and R. Bullough, "A Review of Theoretical and Experimental Work on the Structure of Ge_xSi_{1-x} Strained Layers and Superlattices, with Extensive Bibliography," *Advanced Physics* 39, 127–90, 1990.

150. R. P. G. Karunasiri and K. L. Wang, "Quantum Devices Using SiGe/Si Heterostructures," *Journal of Vacuum Science and Technology* B9, 2064–71, 1991.

151. K. L. Wang and R. P. G. Karunasiri, "SiGe/Si Electronics and Optoelectronics," *Journal of Vacuum Science and Technology* B11, 1159–67, 1993.

152. R. P. G. Karunasiri, J. S. Park, K. L. Wang, and S. K. Chun, "Infrared Photodetectors with SiGe/Si Multiple Quantum Wells," *Optical Engineering* 33, 1468–76, 1994.

153. D. J. Robbins, M. B. Stanaway, W. Y. Leong, J. L. Glasper, and C. Pickering, "$Si_{1-x}Ge_x$/Si Quantum Well Infrared Photodetectors," *Journal of Materials Science: Materials in Electronics* 6, 363–67, 1995.

154. R. P. G. Karunasiri, J. S. Park, Y. J. Mii, and K. L. Wang, "Intersubband Absorption in $Si_{1-x}Ge_x$/Si Multiple Quantum Wells," *Applied Physics Letters* 57, 2585–87, 1990.

155. R. People, J. C. Bean, C. G. Bethea, S. K. Sputz, and L. J. Peticolas, "Broadband (8–14 µm), Normal Incidence Pseudomorphic $Si_{1-x}Ge_x$/Si Strained-Layer Infrared Photodetector Operating between 20 and 77 K," *Applied Physics Letters* 61, 1122–24, 1992.

156. R. People, J. C. Bean, S. K. Sputz, C. G. Bethea, and L. J. Peticolas, "Normal Incidence Hole Intersubband Quantum Well Infrared Photodetectors in Pseudomorphic $Si_{1-x}Ge_x$/Si," *Thin Solid Films* 222, 120–25, 1992.

157. V. D. Shadrin, V. T. Coon, and F. L. Serzhenko, "Photoabsorption in n–Type Si-SiGe Quantum-Well Infrared Photodetectors," *Applied Physics Letters* 62, 2679–81, 1993.

158. V. D. Shadrin, "Background Limited Infrared Performance of n–Type Si–SiGe (111) Quantum Well Infrared Photodetectors," *Applied Physics Letters* 65, 70–72, 1994.

159. B. F. Levine, A. Y. Cho, J. Walker, R. J. Malik, D. A. Kleinmen, and D. L. Sivco, "InGaAs/InAlAs Multiquantum Well Intersubband Absorption at a Wavelength of λ = 4.4 µm," *Applied Physics Letters* 52, 1481–83, 1988.

160. G. Hasnain, B. F. Levine, D. L. Sivco, and A. Y. Cho, "Mid-Infrared Detectors in the 3–5 µm Band Using Bound to Continuum State Absorption in InGaAs/InAlAs Multiquantum Well Structures," *Applied Physics Letters* 56, 770–72, 1990.

161. S. Ozer, U. Tumkaya, and C. Besikci, "Large Format AlInAs-InGaAs Quantum-Well Infrared Photodetector Focal Plane Array for Midwavelength Infrared Thermal Imaging," *IEEE Photonics Technology Letters* 19, 1371–73, 2007.

162. M. Kaldirim, Y. Arslan, S. U. Eker, and C. Besikci, "Lattice-Matched AlInAs-InGaAs Mid-Wavelength Infrared QWIPs: Characteristics and Focal Plane Array Performance," *Semiconductor Science and Technology* 23, 085007, 2008.

163. S. D. Gunapala, B. F. Levine, D. Ritter, R. A. Hamm, and M. B. Panish, "Lattice-Matched InGaAsP/InP Long-Wavelength Quantum Well Infrared Photoconductors," *Applied Physics Letters* 60, 636–38, 1992.

164. S. D. Gunapala, B. F. Levine, D. Ritter, R. Hamm, and M. B. Panish, "InP Based Quantum Well Infrared Photodetectors," *Proceedings of SPIE* 1541, 11–23, 1991.

165. S. D. Gunapala, K. M. S. V. Bandara, B. F. Levine, G. Sarusi, D. L. Sivco, and A. Y. Cho, "Very Long Wavelength $In_xGa_{1-x}As$/GaAs Quantum Well Infrared Photodetectors," *Applied Physics Letters* 64, 2288–90, 1994.

166. S. D. Gunapala, K. M. S. V. Bandara, B. F. Levine, G. Sarusi, J. S. Park, T. L. Lin, W. T. Pike, and J. K. Liu, "High Performance InGaAs/GaAs Quantum Well Infrared Photodetectors," *Applied Physics Letters* 64, 3431–33, 1994.

167. A. Kock, E. Gornik, G. Abstreiter, G. Bohm, M. Walther, and G. Weimann, "Double Wavelength Selective GaAs/AlGaAs Infrared Detector Device," *Applied Physics Letters* 60, 2011–13, 1992.

168. Ph. Bois, E. Costard, J. Y. Duboz, and J. Nagle, "Technology of Multiquantum Well Infrared Detectors," *Proceedings of SPIE* 3061, 764–71, 1997.

169. E. Costard, Ph. Bois, F. Audier, and E. Herniou, "Latest Improvements in QWIP Technology at Thomson-CSF/LCR," *Proceedings of SPIE* 3436, 228–39, 1998.

170. S. D. Gunapala, S. V. Bandara, J. K. Liu, J. M. Mumolo, C. J. Hill, S. B. Rafol, D. Salazar, J. Woollaway, P. D. LeVan, and M. Z. Tidrow, "Towards Dualband Megapixel QWIP Focal Plane Arrays," *Infrared Physics & Technology* 50, 217–26, 2007.

171. S. D. Gunapala, S. V. Bandara, J. K. Liu, S. B. Rafol, J. M. Mumolo, C. A. Shott, R. Jones, et al., "640 × 512 Pixel Narrow-Band, Four-Band, and Broad-Band Quantum Well Infrared Photodetector Focal Plane Arrays," *Infrared Physics & Technology* 44, 411–25, 2003.

172. H. C. Liu, J. Li, J. R. Thompson, Z. R. Wasilewski, M. Buchanan, and J. G. Simmons, "Multicolor Voltage Tunable Quantum Well Infrared Photodetector," *IEEE Electron Device Letters* 14, 566–68, 1993.

173. I. Grave, A. Shakouri, N. Kuze, and A. Yariv, "Voltage-Controlled Tunable GaAs/AlGaAs Multistack Quantum Well Infrared Detector," *Applied Physics Letters* 60, 2362–64, 1992.

174. H. C. Liu, "Recent Progress on GaAs Quantum Well Intersubband Infrared Photodetectors," *Optical Engineering* 33, 1461–67, 1994.

175. Y. H. Wang, S. S. Li, and P. Ho, "Voltage-Tunable Dual-Mode Operation in InAlAs/InGaAs Quantum Well Infared Photodetector for Narrow- and Broadband Detection at 10 μm," *Applied Physics Letters* 62, 621–23, 1993.

176. E. Martinet, F. Luc, E. Rosencher, Ph. Bois, and S. Delaitre, "Electrical Tunability of Infrared Detectors Using Compositionally Asymmetric GaAs/AlGaAs Multiquantum Wells," *Applied Physics Letters* 60, 895–97, 1992.

177. E. Martinet, E. Rosencher, F. Luc, Ph. Bois, E. Costard, and S. Delaitre, "Switchable Bicolor (5.5–9.0 μm) Infrared Detector Using Asymmetric GaAs/AlGaAs Multiquantum Well," *Applied Physics Letters* 61, 246–48, 1992.

178. Y. H. Wang, Sh. S. Li, and P. Ho, "Photovoltaic and Photoconductive Dual-Mode Operation GaAs Quantum Well Infrared Photodetector for Two-Band Detection," *Applied Physics Letters* 62, 93–95, 1993.

179. M. Z. Tidrow, K. K. Choi, C. Y. Lee, W. H. Chang, F. J. Towner, and J. S. Ahearn, "Voltage Tunable Three-Color Quantum Well Infrared Photodetector," *Applied Physics Letters* 64, 1268–70, 1994.

180. K. Kheng, M. Ramsteiner, H. Schneider, J. D. Ralston, F. Fuchs, and P. Koidl, "Two-Color GaAs/AlGaAs Quantum Well Infrared Detector with Voltage-Tunable Spectral Sensitivity at 3–5 and 8–12 μm," *Applied Physics Letters* 61, 666–68, 1992.

181. H. C. Liu, E. Dupont, M. Byloos, M. Buchanan, C.-Y. Song, and Z. R. Wasilewski, "QWIP-LED Pixelless Thermal Imaging Device," in *Intersubband Infrared Photodetectors*, ed. V. Ryzhii, 299–313, World Scientific, Singapore, 2003.

182. H. C. Liu, J. Li, Z. R. Wasilewski, and M. Buchanan, "Integrated Quantum Well Intersub-Band Photodetector and Light Emitting Diode," *Electronics Letters* 31, 832–33, 1995.

183. V. Ryzhii, M. Ershov, M. Ryzhii, and I. Khmyrova, "Quantum Well Infrared Photodetector with Optical Output," *Japanese Journal of Applied Physics* 34, L38–L40, 1995.

184. E. Dupont, M. Gao, Z. R. Wasilewski, and H. C. Liu, "Integration of n-Type and p-Type Quantum-Well Infrared Photodetectors for Sequential Multicolor Operation," *Applied Physics Letters* 78, 2067–69, 2001.

185. E. Dupont, H. C. Liu, M. Buchanan, Z. R. Wasilweski, D. St-Germain, and P. Chevrette, "Pixelless Infrared Imaging Based on the Integration of a n-Type Quantum-Well Infrared Photodetector with a Light Emitting Diode," *Applied Physics Letters* 75, 563–65, 1999.

186. E. Dupont, M. Byloos, M. Gao, M. Buchanan, C.-Y. Song, Z. R. Wasilewski, and H. C. Liu, "Pixel-Less Thermal Imaging with Integrated Quantum-Well Infrared Photo Detector and Light-Emitting Diode," *IEEE Photonics Technology Letters* 14, 182–84, 2002.

187. E. Dupont, M. Byloos, T. Oogarah, M. Buchanan, and H. C. Liu, "Optimization of Quantum-Well Infrared Detectors Integrated with Light-Emitting Diodes," *Infrared Physics Technology* 47, 132–43, 2005.

17 Superlattice Detectors

Many types of optoelectronic devices can be enhanced significantly through the introduction of quantum confinement in reduced-dimensionality heterostructures. This was the main motivation for the study of superlattices (SLs) as an alternative infrared detector materials. The HgTe/CdTe SL system was proposed in 1979, only a few years after the first GaAs/AlGaAs quantum heterostructures were fabricated by MBE. It was anticipated that superlattice infrared materials would have several advantages over bulk HgCdTe (the current industry standard) for this application:

- a higher degree of uniformity, which is importance for detector arrays;

- smaller leakage current due to the suppression of tunneling (larger effective masses) available in superlattices; and

- lower Auger recombination rates due to substantial splitting of the light- and heavy-hole bands and increased electron effective masses.

Early attempts to realize superlattices with properties suitable for infrared detection were unsuccessful, largely because of the difficulties associated with epitaxial deposition of HgTe/CdTe superlattices. More recently, significant interest has been shown in multiple quantum well AlGaAs/GaAs photoconductors. However, these detectors are extrinsic in nature, and have been predicted to be limited to performance inferior to that of intrinsic HgCdTe detectors [1–6]. On account of this—in addition to the use of intersubband absorption [7–14] and absorption in doping superlattice [15]—three additional physical principles are utilized to directly shift bandgaps into the infrared spectral range:

- superlattice quantum confinement without strain: HgTe/HgCdTe,

- superlattice strain-induced bandgap reduction: InAsSb/InSb, and

- superlattice-induced band inversion: InAs/GaInSb.

These types of superlattice rely upon an intrinsic valence to conduction band absorption process.

17.1 HGTE/HGCDTE SUPERLATTICE

The HgTe/CdTe SL system was the first from a new class of quantum-size structures for IR photoelectronics, which was proposed as a promising new alternate structure for the construction of long-wave IR detectors to replace those of HgCdTe alloys [16]. Since that time significant theoretical and experimental attention has been given to the study of this new SL system [17–24]. To date however, attempts to realize HgTe/CdTe SLs with properties suitable for IR detectors with parameters comparable with HgCdTe-alloy photodetectors, have been unsuccessful and in spite of the considerable amount of fundamental research in this field. It seems to be determined by interface instabilities of superlattices due to weak Hg chemical bonding in the material. Significant intermixing even at the very low temperatures (185°C) used in the HgTe/CdTe superlattice preparation have serious implications on device performance. An interdiffusion coefficient of 3.1×10^{-18} cm^2/s has been found at 185°C. Appreciable intermixing of the HgTe and CdTe layers at temperatures as low as 110°C have been observed [25], which prevent the realization of a low dimensional solid system in a stable form. The situation is worse due to certain aspects of device processing, such as impurity activation, native defect reduction, and surface passivation. On account of this, the topic concerning HgTe/CdTe superlattices will be treated shortly and only more recently published data will be included.

17.1.1 Material Properties

The HgTe/CdTe superlattice appears to belong to type III superlattices (see Figure 16.8d). This is due to the inverted band structure (Γ_6 and Γ_8) in zero gap semiconductor HgTe as compared to those of CdTe, which is a normal semiconductor. Thus, the Γ_8 light-hole band in CdTe becomes the conduction band in HgTe. When bulk states made of atomic orbitals of the same symmetry but with effective masses of opposite signs are used, the matching up of bulk states belonging to these bands has, as a consequence, the existence of a quasi-interface state that could contribute significantly to optical and transport properties.

As was shown in many theoretical calculations (see, e.g., [22,24]), the valence band discontinuity between HgTe and CdTe has a crucial influence on the HgTe/CdTe SL band structure. In earlier publications the assumptions of significant advantages of these SLs for LWIR and VLWIR detectors were based on a small valence-band offset [16], $\Delta E_v \geq 40$ meV, which seemed to be in agreement with the common-anion rule of lattice-matched heterojunction interfaces. Of important practical interest were theoretical predictions of a better control of the bandgap in HgTe/CdTe SL compared to HgCdTe alloy, and a sufficient reduction of the tunneling currents due to larger effective masses in the SL growth direction. More recently it has been recognized that numerous aspects of the HgTe/CdTe and related SL properties can only be explained in terms of large valence-band offset $\Delta E_v \geq 350$ meV [24]. According to Becker et al. [26], the valence band offset between HgTe and CdTe

$$\Delta E_v = \Delta E_{v0} + \frac{d(\Delta E_c)}{dT} T, \tag{17.1}$$

where $\Delta E_{v0} = 570$ meV and $d(\Delta E_c)/dT = -0.40$ meV/K [26] and ΔE_v is assumed to vary linearly with x in $Hg_{1-x}Cd_xTe$ [27].

The experimental and calculated optical bandgaps of HgTe/$Hg_{1-x}Cd_x$Te SLs for the (001) and (112)B orientations are shown in Figure 17.1 [28]. If a HgTe thickness, d_w is less than about 6.2 nm, then the band structure is normal and if $d_w > 6.2$ nm then the band structure is inverted. The SL structure is primary determined by that of the quantum well and is influenced to a much lesser degree by the band structure of the barrier ($Hg_{1-x}Cd_x$Te). Only the energy gap changes significantly with alloy composition and temperature. Figure 17.1 shows that experimental and theoretical data are in excellent agreement.

As is mentioned above, the advantages of HgTe/HgCdTe SLs were judged to be insignificant, however, in the very LWIR band they gain sufficient importance to be one of the best detector material. The required precision of the desired band gap or cutoff wavelength is less for a SL in comparison with bulk material. It is demonstrated in Figure 17.2. If $\lambda_c = 17.0 \pm 1.0$ µm at 40 K is wished, the required precision for the alloy is ±1.0% and that of the SL is ±2 and ±8% for the SL with a normal and inverted band structure, respectively.

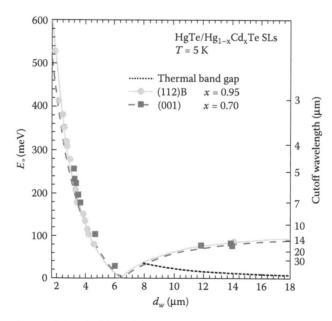

Figure 17.1 Experimental (symbols) and theoretic calculations (solid and dashed lines) band gap: H1→E1 intersubband transition of HgTe based multiple quantum wells, and the thermal bandgap (dotted line) in the inverted band structure regime. (From Becker, C. R., Ortner, K., Zhang, X. C., Oehling, S., Pfeyffer-Jeschke, A., and Latussek, V., *Advanced Infrared Technology and Applications, 2007*, 79–89, Leon, Mexico, 2008.)

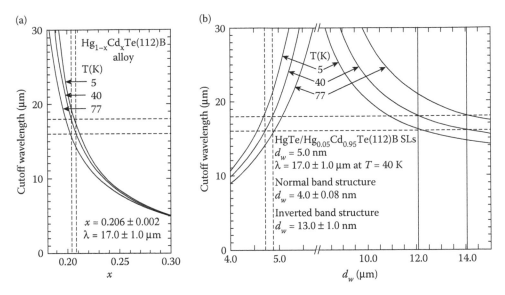

Figure 17.2 (a) The dependence of cutoff wavelength of the $Hg_{1-x}Cd_xTe$ alloy and (b) the HgTe/Hg$_{1-x}$Cd$_x$Te SL with normal and inverted band structure at 5, 40, and 77 K. The precision in x and d_w for these three cases required to produce material with a cutoff wavelength of 17.0 ± 1.0 μm at 40 K is indicated. (From Becker, C. R., Ortner, K., Zhang, X. C., Oehling, S., Pfeyffer-Jeschke, A., and Latussek, V., *Advanced Infrared Technology and Applications 2007*, Leon, Mexico, 79–89, 2008.)

If we take the electron and hole dispersion relations in HgTe/HgCdTe quantum well to have the approximate form given by Equation 16.6, the density of state has the well-known staircase dependence. The j-th steep occurs at the photon energy:

$$h\nu \approx E_g^w + \frac{\hbar^2 j^2 \pi^2}{2 d_w^2}\left(\frac{1}{m_e^w}+\frac{1}{m_h^w}\right),$$ (17.2)

where E_g^w is the bulk energy gap of the well material, m_e^w and m_h^w are the effective masses of electron and hole in the bulk well material before confinement is imposed, and the kinetic energy due to motion in the x,y plane unaffected by the confinement along z. The above equation contrasts with the more gradual $(h\nu - E_g^w)^{1/2}$ dependence for the density of states in a 3-D bulk material such as HgCdTe.

Figure 17.3 shows, for example, the experimental and theoretical absorption coefficient for a SL and an alloy material with similar band gaps near 60 meV ($\lambda_c = 20$ μm). Agreement between experiment and theory is very good. Furthermore, the absorption edge for the SL is much steeper and therefore the absorption of the SL is up to a factor of five larger. The large absorption coefficient in a superlattice represents a distinct advantage over the HgCdTe bulk material, in which α tends to be 1000–2000 cm^{-1} at the same proximity to the band edge. This means that the active detector layers can be significantly thinner in the equivalent alloy, on the order of a few microns. This advantage is more pronounced for LW region, but even for a MW band the absorption is appreciably larger. It appears also that a Burstein–Moss effect produces a negligible shift of the absorption coefficient in comparison with bulk material. This is due to the flatter dispersion in the direction perpendicular to the 2-D plane and the corresponding larger density of states.

One of the main advantages concerning HgTe/HgCdTe SLs for IR photoelectronics lies in a variation of the bandgap not with chemical composition in ternary or quaternary alloys, but in a variation of layer thickness of much more stable binary compounds. As the bandgap, the growth-direction effective mass values of electrons and holes, and thus the carrier mobilities, can be tuned over a wide range by varying the barrier thickness (also to be taken into account is

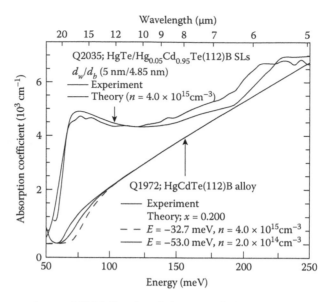

Figure 17.3 The experimental (thick lines) and theoretical (thin lines) absorption coefficient for HgTe/Hg$_{1-x}$Cd$_x$Te SL and a Hg$_{1-x}$Cd$_x$Te alloy with a bandgap near 60 meV ($\lambda_c \approx 20$ μm) at 40 K. The two theoretical spectra for the alloy are for two different electron concentrations and the corresponding Fermi energies. (From Becker, C. R., Ortner, K., Zhang, X. C., Oehling, S., Pfeyffer-Jeschke, A., and Latussek, V., *Advanced Infrared Technology and Applications 2007*, Leon, Mexico, 79–89, 2008.)

a "mass-broadening" effect, which causes the in-plane effective mass to depend strongly on the growth-direction wave vector) [24]. Whereas the mass in the alloy is fixed by its proportionality to the energy gap, in the superlattice the effective mass can be varied independently of E_g simply by adjusting the barrier thickness. A large growth-direction effective mass is desired because the tunneling current scales exponentially with $m^{1/2}$ (see Section 9.2). But to achieve high quantum efficiencies HgTe/HgCdTe SLs should be grown with thin barriers (less than 30 Å [29]) to be beyond the region of hopping mobility from one well to the next one where the miniband conduction model breaks down.

The electronic band of HgTe/HgCdTe SL can be engineered to suppress Auger recombination relative to that in comparable bulk detectors. SRH lifetimes of up to 20 μs have been reported for HgTe/CdTe SLs grown by MBE [30]. Calculated carrier lifetimes at 80 K for hole concentration of 5×10^{15} cm^{-3} are shown in Figure 17.4 as a solid line [31]. The experimental results for photoconductive decay are in reasonable agreement with theory. We can notice the large difference between carrier lifetimes in normal and inverted SL band structure. The extremely fast lifetimes for the inverted SLs are due to the presence of many valence subbands with small energy separations and the corresponding large number of occupied states for efficient Auger recombination [28].

17.1.2 Superlattice Photodiodes

For fabrication of HgTe/HgCdTe SLs mainly the MBE procedure is used. Because Hg has both a high vapor pressure and low sticking coefficient, at the commonly used temperatures of about 180°C or less to minimize interdiffusion processes, special Hg MBE sources are required to let a significant amount of mercury vapor pass through the system. For growth of high quality HgTe/CdTe SLs laser assisted MBE and photoassisted MBE have been used [24].

In spite of HgTe/HgCdTe superlattices being predicted for use in very LW operations, main research efforts were concentrated on MW and LWIR photodiodes. The first reported use of an HgTe/CdTe SL in a MIS detector configuration was given by Goodwin, Kinch, and Koestner [32]. Wroge et al. [33] reported the first photovoltaic device structure based on extrinsic doping using In and Ag for n-type and p-type doping, respectively.

More encouraging results have been obtained by Harris et al. for MWIR photodiodes [34]. It appears that combining the low-temperature technique of photon-assisted MBE with the use of the

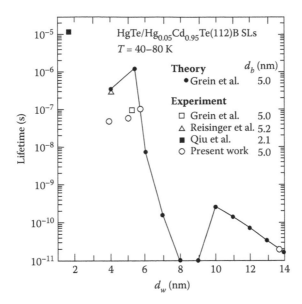

Figure 17.4 The experimental (symbols) and theoretical (solid line) lifetimes for HgTe/Hg$_{1-x}$ Cd$_x$Te SLs at temperatures between 40 and 80 K. (From Becker, C. R., Ortner, K., Zhang, X. C., Oehling, S., Pfeyffer-Jeschke, A., and Latussek, V., *Advanced Infrared Technology and Applications 2007*, Leon, Mexico, 79–89, 2008. With permission.) The theoretical line is calculated for HgTe/ Hg$_{0.05}$Cd$_{0.95}$Te superlattice at 40 K with $d_w = d_b = 5$ nm and an acceptor concentration of 5×10^{15} cm^{-3}. (From Grein, C. H., Jung, H., Singh, R., and Flatte, M. F., *Journal of Electronic Materials*, 34, 905–8, 2005. With permission.)

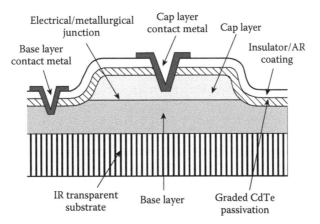

Figure 17.5 Schematic of a processed SL mesa structure. (From Harris, K. A., Myers, T. H., Yanka, R., Mohnkern, L. M., and Otsuka, N., *Journal of Vacuum Science and Technology*, B9, 1752–58, 1991. With permission.)

(211)B orientation has allowed for the growth of multilayers exhibiting a high degree of crystalline perfection and *in-situ* n-type and p-type extrinsic doping. The basic structure consisted of a 3–5 μm n-type layer followed by a 1–2 μm thick p-type cap layer. Figure 17.5 illustrates a schematic cross section of the mesa structure. To ensure p-type and n-type behavior in the layers, As at the 10^{17} cm^{-3} and In at the 10^{16} cm^{-3} level were used, respectively (typical doping levels in an optimized device structure would be about a factor of 10 less than this).

Both MWIR and LWIR SL photodiodes exhibiting high quantum efficiency and uniform response have been fabricated. The MWIR detectors exhibited quantum efficiency as high as 66% (at 140 K; see Figure 17.6) at the peak wavelength and an average over 3–5 μm waveband of 55% [22].

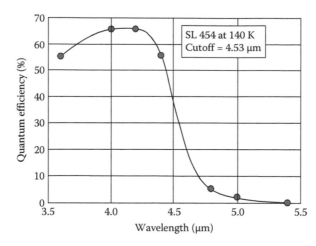

Figure 17.6 Representative spectral response of an HgTe/CdTe SL photodiode with $\lambda_c = 4.53$ µm at 140 K. The peak response corresponds to 66% quantum efficiency. (From Myers, T. H., Meyer, J. R., and Hoffman, C. A., *Semiconductor Quantum Wells and Superlattices for Long-Wavelength Infrared Detectors*, Artech House, Boston, MA, 207–59, 1993.)

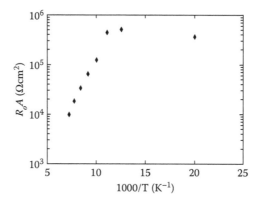

Figure 17.7 Measured $R_o A$ product as a function of temperature for a representative SL photodiode. The low-temperature characteristics are indicative of tunneling processes limiting $R_o A$ (nonoptimal surface passivation). (From Myers, T. H., Meyer, J. R., and Hoffman, C. A., *Semiconductor Quantum Wells and Superlattices for Long-Wavelength Infrared Detectors*, Artech House, Boston, MA, 207–59, 1993.)

Measured quantum efficiency was lower at 78 K, with a peak of 45–50% and cutoff wavelength of 4.9 µm. Figure 17.7 shows measured $R_o A$ product as a function of temperature for a representative SL photodiode. The low-temperature behavior of $R_o A$ product was due to both a bulk tunneling phenomenon as well as surface currents, which was confirmed by measured characteristics of gate-controlled photodiodes. Even with passivation problems, the $R_o A$ values for SL photodiodes fabricated without gate were typically 5×10^5 Ωcm² (see Figure 17.7), comparable to that which were often achieved in the corresponding alloy.

The potential advantages of HgTe/CdTe material system has been also demonstrated in LWIR spectral range. Figure 17.8 shows doping configurations, spectral response, and I-V characteristics for a typical p-on-n LWIR SL photodiode with cutoff wavelength of 9.0 µm [24]. The measured quantum efficiency was 62% and $R_o A$ was 60 Ωcm². Preliminary results confirm that the growth quality of SL photodiodes has become sufficiently advanced that a high-performance SL photodiode technology appears feasible. Figure 17.9 shows values of $R_o A$ product at 80 K as a function of wavelength for four HgTe/CdTe SL photodiodes compared with a number of production n-on-p

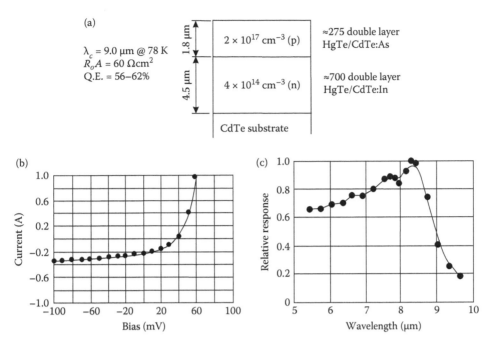

Figure 17.8 (a) Structure, (b) I-V characteristics, and (c) spectral response for (211) HgTe/Hg$_{0.1}$Cd$_{0.9}$Te SL LWIR photodiode. (From Meyer, J. R., Hoffman, C. A., and Bartoli, F. J., *Narrow-Gap II-VI Compounds for Optoelectronic and Electromagnetic Applications*, Chapman & Hall, London, 363–400, 1997. With permission.)

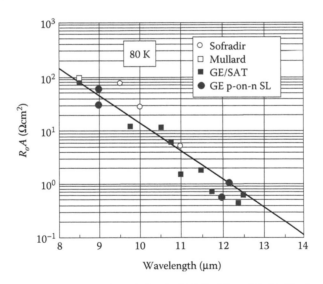

Figure 17.9 R_oA product versus wavelength for (211)HgTe/Hg$_{0.1}$Cd$_{0.9}$Te SL photodiodes (dark circles). Also shown for comparison are analogous results for production n-on-p HgCdTe photodiodes. (From Meyer, J. R., Hoffman, C. A., and Bartoli, F. J., *Narrow-Gap II-VI Compounds for Optoelectronic and Electromagnetic Applications*, Chapman & Hall, London, 363–400, 1997. With permission.)

photodiodes employing HgCdTe bulk materials (Bridgman and traveling heater methods) illustrating comparable performance.

At present, performance of HgTe/HgCdTe SL photodiodes is inferior in comparison with high-quality HgCdTe photodiodes with comparable cutoff wavelength. As a result, a lack of research funding has led to an industry-wide suspension of further efforts to develop HgTe/HgCdTe SL infrared detectors.

17.2 STRAINED LAYER SUPERLATTICES

Infrared optical absorption in mid- and long-wavelength bands in III–V semiconductors for IR detector applications, as was earlier seen, can be effectively used through intersubband transitions. But it is also possible to realize favorable optical properties for long-wavelength detector applications in type II SLs as in this case long-wavelength valence to conduction band optical transitions (between the states in alternate layers due to overlap of the envelope wave functions) may be strong enough to provide appreciable absorption coefficients at normal incidence [35–38]. The bandgaps in such structures occur between electron states localized in one type of layer and hole state, localized in remaining layers. Figure 17.10 presents a schematic band diagram of type II SL and optical transitions between the ground states [39].

The intensity of the band edge optical transitions in type II SLs is determined by exponentially decaying envelope wave function tails. The decay lengths in this case increase with decreasing barrier heights and decreasing effective masses of carriers. Appreciable absorption coefficients are observed for decay wavefunction lengths of barrier thickness; interband absorption coefficient increases at layer thickness diminishing. At the same time the photoexcited carrier relaxation times can be much longer compared to bulk semiconductors with comparable bandgap due to occurrence of excited carrier recombination through the indirect transitions in real space and thus, such type II SLs can possess good performance characteristics.

In the case when the constituent materials are rather closely lattice matched it is possible to design the electronic type II SL or MQW band structure by controlling only the layer thickness and the height of the barriers. But it is also possible to grow high quality III–V type II SLS devices with reduced conduction-valence bandgap for IR detector applications, in which a QW layer can be controlled on the atomic scale too, but with a significantly different lattice constant of the well material compared to the barrier material, which gives additional opportunities to design the electronic band structure by deformation potential effects [37,38], as in the case of SiGe/Si MQWs. A schematic of the typical SLS structure is given in Figure 17.11. The thin SLS layers are alternatively in compression and tension so that the in-plane lattice constants of the individual strained layers are equal. The entire lattice mismatch is accommodated by layer strains without the generation of misfit dislocations if the individual layers are below the critical thickness for dislocation generation. Since misfit defects are not generated in SLS structures, the SLS layer can be of sufficiently high crystalline quality for a variety of scientific and device applications. Strain can change the bandgaps of the constituents and split the degeneracy of heavy- and light-hole bands in such a way that these changes and band splitting can lead not only to energy level reversals in the SL electronic band structure but also to appreciable suppression of recombination rates of photoexcited carriers [40]. In such systems the conduction-valence bandgap can be made much smaller than that in any III-V alloy bulk crystals [41].

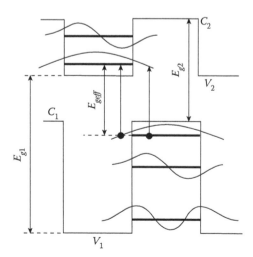

Figure 17.10 Schematic band diagram, envelope wave functions and optical transitions between the ground states in type II SL. E_{geff} is the bandgap of the type II structure. E_{g1} and E_{g2} are the bandgaps of the alternate semiconductor layers, respectively. (From Sizov, F. F., *Infrared Photon Detectors*, SPIE Optical Engineering Press, Bellingham, WA, 561–623, 1995. With permission.)

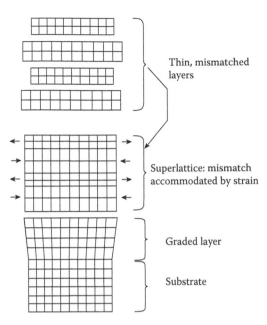

Figure 17.11 Schematic of an SLS.

There exist several systems to realize long wavelength IR absorption in type II SL structures suitable for IR detector applications. In the mid-1970s, investigations of InAs/GaSb SL near lattice-matched systems were performed as alternatives to GaAs/AlAs SLs. The ability of the InAs/GaSb material system to achieve small infrared energy gaps was first realized in 1977 [42]. At once the type II band offset in this system was revealed [43] and the potential use of this system in IR photoelectronics was recognized. However the absorption coefficient data in InAs/GaSb SLs showed a very soft absorption edge and it was concluded that this material couldn't be used as an efficient long-wavelength photodetector. The reason is the large valence band offset in InAs/GaSb heterostructure equal to 0.51 eV resulting in the fact that the InAs conduction band minimum is 0.1 eV below the GaSb valence band maximum. In this case, as a result of the quantum confine-ment effect, the wave function overlap and consequently the optical absorption in SLs, suitable for long IR wavelength region, is too small and for this region these SLs can only be realized in thick layer structures. To avoid these disadvantages, strained-layer type II SLs made of $InAs_{1-x}Sb_x/InSb$ were proposed as a new material system for IR detectors [41]. However, for typical layer thickness of 150 Å, the degree of overlap wavefunctions of valence and conduction electrons is small and thus the optical absorption as well. Only for extremely thin layers does the absorption coefficient become comparable with intrinsic bulk semiconductors. Such thin layer SLs, however, exhibit larger effective optical energy gaps due to the quantum size effect. Therefore, more recently, another material system, namely an $InAs/Ga_{1-x}In_xSb$ one, having significant optical absorption at long wavelength region, was performed [37]. At present, only the last system is considered as a competitor for HgCdTe material system in long IR detector applications.

17.3 InAsSb/InSb STRAINED LAYER SUPERLATTICE PHOTODIODES

The $InAs_{1-x}Sb_x$ ternary alloy has the lowest bandgap of all III-V semiconductors, but this gap is not suitable ($\lambda_c \approx 9$ μm) for operation in 8–14 μm atmospheric window at 77 K (see Section 13.4.1). It was theoretically shown that strain effects in InAsSb SLSs were sufficient to achieve wavelength cutoffs of 12 μm at 77 K independent of the band offset that was unknown at that time [35]. For InSb/InAsSb SLSs it results that InAsSb layers will be in biaxial tension, while InSb layers will be in biaxial compression (at 300 K for InAs lattice constant $a_o = 6.058$ Å and for InSb $a_o = 6.479$ Å). The tensile strain in the small bandgap component of the SLSs, InAsSb is the sum of expansive hydrostatic and compressive uniaxial strain components. The expan-sive hydrostatic strain lowers the energy of the conduction band, and the compressive uniaxial strain splits the degenerate light and heavy hole bands by shifting the light holes to higher energy and lowering the energy of the heavy holes. So, the small band component, InAsSb, is decreased by strain in this SLS, and the InSb-component bandgap is increased. Therefore, from the effects of strain alone, InAsSb SLSs can potentially absorb at longer wavelengths than InAsSb alloys.

Progress in the growth of InAsSb SLSs by both MBE and MOCVD has been observed since Osbourn's proposal. The first decade efforts in development of epitaxial layers are presented in Rogalski's monograph [44]. Difficulties have been encountered in finding the proper growth conditions especially for SLSs in the middle region of composition [45–48]. This ternary alloy tends to be unstable at low temperatures, exhibiting miscibility gaps, and this can generate phase separation or clustering. Control of alloy composition has been problematic especially for MBE. Due to the spontaneous nature of CuPt-orderings, which result in substantial bandgap shrinkage, it is difficult to accurately and reproducibly control the desired bandgap for optoelectronic device applications [49].

The IR absorption spectra at 80 K for high-quality InAsSb SLSs was observed for the first time in 1988 [36]. It was shown that these SLSs absorb at wavelengths longer than the bandgap of corresponding InAsSb alloy. For some SLSs with high As concentration the appreciable absorption was observed in the far IR region up to 20 μm. Small effective masses and low barrier heights of InAsSb/InSb SLS allow the realization of absorption coefficients in rather thick layers of InAsSb/InSb SLSs due to large wave function decay lengths.

The type II SL structures can operate as photodiodes at normal light incidence. In the first InAsSb SLS photodiodes rather low detectivities were observed; below 1×10^{10} cmHz$^{1/2}$/W at 77 K [50,51]. Higher detectivity ($D^* > 1 \times 10^{10}$ cmHz$^{1/2}$W^{-1} at $\lambda \le 10$ μm) InSb/InAsSb SLS photodiodes were fabricated from a p-p⁻-n junction embedded in an InAs$_{0.15}$Sb$_{0.85}$/InSb SLS with equal 150 Å thick layers [52]. The SLS was grown on a thick, composition-graded In$_x$Ga$_{1-x}$Sb ($x = 1.0$–0.9) strain-relief buffer on an InSb substrate. The p- and n-dopants were Be and Se, respectively. The doping level in the i-region represents the background doping level in the MBE system. The photodiodes were mesa-isolated, with an area of 1.2×10^{-3} cm². The InGaSb buffer and n-type substrate are semitransparent at long wavelengths, and with reflecting back contact the optical path length and the quantum efficiency are substantially increased.

The temperature dependence of the R_oA product indicates that performance of the detector is not limited by diffusion or depletion region generation–recombination processes inherent in narrow-gap semiconductors. Noise measurements performed before and after photodiode passivation indicated that a large 1/f noise component was introduced by passivation, and alternative passivation processes must be developed to operate with these detectors at lower modulation frequencies. The zero-bias external current responsivity of an InAs$_{0.15}$Sb$_{0.85}$/InSb SLS diode illuminated at normal incidence is shown in Figure 17.12. The onset of the photoresponse occurs at approximately 119 meV. Both the weak increase in responsivity observed in reverse bias and the magnitudes of responsivity and absorption indicate that the minority diffusion length (electrons), perpendicular to the SLS layers, is about 1–2 μm [52].

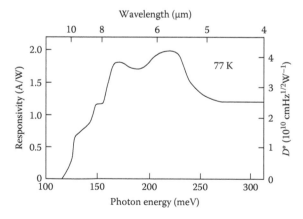

Figure 17.12 External current responsivity of the InAs$_{0.15}$Sb$_{0.85}$/InSb SLS photodiode at zero bias, $T = 77$ K. Detectivity was determined from noise measurements made at 100 kHz, cold-shielded at 77 K. (From Kurtz, S. R., Dawson, L. R., Zipperian, T. E., and Whaley, R. D., *IEEE Electron Device Letters*, 11, 54–56, 1990. With permission.)

17.4 INAS/GAINSB TYPE II STRAINED LAYER SUPERLATTICES

InAs/Ga$_{1-x}$In$_x$Sb (InAs/GaInSb) strained layer superlattices (SLSs) can be considered as an alternative to HgCdTe and GaAs/AlGaAs IR material systems as a candidate for third generation IR detectors. The low quantum efficiency of QWIPs is largely due to fact that the optical transition is forbidden for normal incidence of light. Straylight generated by reflecting gratings is required to achieve reasonable quantum efficiency. In the case of InAs/GaInSb SLS structures the absorption is strong for normal incidence of light. Consequently, the SLS structures provide high responsivity, as already reached with HgCdTe, without any need for gratings. Further advantages are a photovoltaic operation mode, operation at elevated temperatures and well-established III-V process technology.

InAs/GaInSb material system is however in a very early stage of development. Problems exist in material growth, processing, substrate preparation, and device passivation [53]. Optimization of SL growth is a trade-off between interfaces roughness, with smoother interfaces at higher temperature, and residual background carrier concentrations, which are minimized on the low end of this range. The thin nature of InAs and GaInSb layers (<8 nm) necessitate low growth rates for control of each layer thickness to within 1 (or one-half) monolayer (ML). Typical growth rates are less than 1 ML/s for each layer.

17.4.1 Material Properties

InAs and GaInSb form an ideal material system for the growth of semiconductor heterostructure because of their small difference in lattice constant. For example, Ga$_{1-x}$In$_x$Sb with an indium concentration of 15% grows compressively strained on a GaSb substrate with a lattice mismatch of $\Delta a/a = 0.94\%$, while InAs is under tensile with a lattice mismatch of $\Delta a/a = -0.62\%$. In an InAs/Ga$_{1-x}$In$_x$Sb superlattice, the compressively strained Ga$_{1-x}$In$_x$Sb layers can be compensate for the tensile strain in the InAs layers.

InAs/Ga$_{1-x}$In$_x$Sb superlattices were proposed for IR detector applications in the 8–14 μm region in 1987 [37]. It has been suggested that this material system can have some advantages over bulk HgCdTe, including lower leakage currents and greater uniformity. LWIR response in these SLs arises due to a type II band alignment and internal strain that lowers the conduction band minimum of InAs and raises the heavy-hole band in Ga$_{1-x}$In$_x$Sb by the deformation potential effect. However, unlike InAsSb materials, effects due to strain are combined with a substantial valence band offset, which for InAs/GaSb SLs exceeds 500 meV. As can be seen in Figure 17.13, the unstrained conduction band minimum of InAs lies below the unstrained valence band maximum of InSb or GaSb. The GaSb valence band edge lies approximately 150 meV above the InAs conduction band edge at low temperature. LWIR absorption however, can be achieved in InAs/GaSb SLs for an InAs layer thickness greater than approximately 100 Å, resulting in comparatively poor optical absorption due to weak wave function overlap as the barrier heights are large. Substituting a Ga$_{1-x}$In$_x$Sb alloy for GaSb it is possible to reach the important 12 μm IR region for thin SLs to

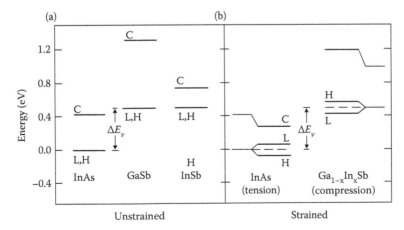

Figure 17.13 (a) Assumed relative energy positions for unstrained InAs, GaSb, and InSb; (b) effect of lattice-mismatch induced internal strain on the energy-band offset for an InAs-Ga$_{1-x}$In$_x$Sb heterointerface.

obtain large optical absorption. A small lattice mismatch (<5%) between the InGaSb and InAs layers causes the tetragonal distortions that shift the bulk energy levels and split the valence band degeneracy's of the light and heavy hole energy levels. The presence of coherent strain shifts the band edges such that the SL energy gap is reduced. In the SLS, the bandgap forms between electron states split upward from the InAs conduction band and heavy hole states split downward from the InGaSb valence band. This reduced bandgap is advantageous because longer cutoff wavelengths can be obtained with reduced layer thickness in the strained SL.

Summarizing, the type-II superlattice has staggered band alignment such that the conduction band of the InAs layer is lower than the valence band of InGaSb layer. This creates a situation in which the energy bandgap of the superlattice can be adjusted to form either a semimetal (for wide InAs and GaInSb layers) or a narrow bandgap (for narrow layers) semiconductor material. In the SL, the electrons are mainly located in the InAs layers, whereas holes are confined to the GaInSb layers, as shown in Figure 17.14 [54]. This suppresses Auger recombination mechanisms and thereby enhances carrier lifetime but optical transitions occur spatially indirectly and, thus, the optical matrix element for such transitions is relatively small. The bandgap of the SL is determined by the energy difference between the electron miniband E_1 and the first heavy hole state HH_1 at the Brillouin zone center and can be varied continuously in a range between 0 and about 250 meV. An example of the wide tunability of the SL is shown in Figure 17.14b.

In comparison with the HgCdTe material system, type II superlattice is mechanically robust and has fairly weak dependence of bandgap on composition. Using InAs/GaInSb SLS we have the ability to fix one component of the material and vary the other to tune the wavelength. As shown in Figure 17.15 [55], by fixing the GaSb layer thickness at 40 Å and varying the thickness of the InAs from 40 to 66 Å, the cutoff wavelength of SLS can be tuned from 5 to 25 μm. There is no material composition change needed, which is a serious problem in the case of LWIR HgCdTe FPA material to fulfill the requirement of high uniformity (see Figure 14.16).

It has been suggested that InAs/Ga$_{1-x}$In$_x$Sb SLSs material system can have some advantages over bulk HgCdTe, including lower leakage currents and greater uniformity [37,39,56]. Electronic properties of SLSs may be superior to those of the HgCdTe alloy [56]. The effective masses are not directly dependent on the bandgap energy, as it is the case in a bulk semiconductor. The electron effective mass of InAs/GaInSb SLS is larger ($m^*/m_o \approx 0.02$–0.03, compared to $m^*/m_o = 0.009$ in HgCdTe alloy with the same bandgap $E_g \approx 0.1$ eV). Thus, diode tunneling currents in the SL can be reduced compared to the HgCdTe alloy [57]. Although in-plane mobilities drop precipitously for thin wells, electron mobilities approaching 10^4 cm^2/Vs have been observed in InAs/GaInSb superlattices with the layers less than 40 Å thick. While mobilities in these SLs are found to be limited by the same interface roughness scattering mechanism, detailed band structure calculations reveal a much weaker dependence on layer thickness, in reasonable agreement with the experiment [58].

A consequence of the type II band alignment of InAs/GaInSb material system is spatial separation of electrons and holes. This is particularly disadvantageous for optical absorption, where

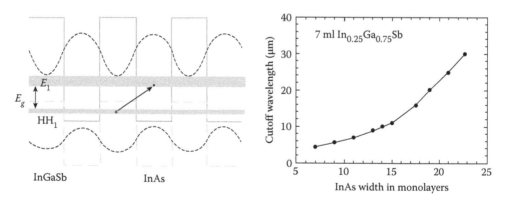

Figure 17.14 InAs/GaInSb strained layer superlattice: (a) band edge diagram illustrating the confined electron and hole minibands that form the energy bandgap; (b) change in cutoff wavelength with change in one superlattice parameter, InAs layer width. (From Brown, G. J., Szmulowicz, F., Mahalingam, K., Houston, S., Wei, Y., Gon, A., and Razeghi, M., "Recent Advances in InAs/GaSb Superlattices for Very Long Wavelength Infrared Detection," *Proceedings of SPIE* 4999, 457–66, 2003. With permission.)

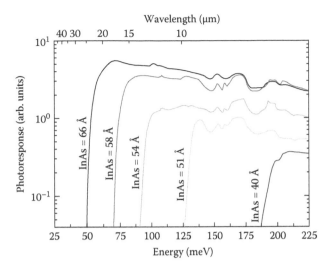

Figure 17.15 Experimental data of type II SLS cutoff wavelengths change with the InAs thickness while GaSb is fixed at 40 Å. (From Wei, Y., and Razeghi, M., "Modeling of Type-II InAs/GaSb Superlattices Using an Empirical Tight-Binding Method and Interface Engineering," *Physical Review* B69, 085316, 2004. With permission.)

a significant overlap of electron and hole wave function is needed. However, a reduction in the electronic confinement can be achieved by growing thinner GaInSb barriers or by introducing more indium into the GaInSb layers leading to optical absorption coefficient comparable to that of HgCdTe.

Theoretical analysis of band-to-band Auger and radiative recombination lifetimes for InAs/GaInSb SLSs showed that in these objects the p-type Auger recombination rates are suppressed by several orders, compared to those of bulk HgCdTe with similar bandgap [40,59], but n-type materials are less advantageous. In p-type superlattice, Auger rates are suppressed due to lattice-mismatch-induced strain that splits the highest two valence bands (the highest light band lies significantly below the heavy hole band and thus limits available phase space for Auger transitions). In n-type superlattice, Auger rates are suppressed by increasing the InGaSb layer widths, thereby flattening the lowest conduction band and thus limiting available phase space for Auger transition.

Comparison of theoretically calculated and experimentally observed lifetimes at 77 K for 10 μm InAs/GaInSb SLS and 10 μm HgCdTe is presented in Figure 17.16. The agreement between theory and experiment for carrier densities above 2×10^{17} cm^{-3} is good. The discrepancy between both types of results for lower carrier densities is due to Shockley-Read recombination processes having a $\tau \approx 6 \times 10^{-9}$ s that has been not taken into account in the calculations. For higher carrier densities, the SL carrier lifetime is two orders of magnitude longer than in HgCdTe, however in low doping region (below 10^{15} cm^{-3}, necessary in fabrication high performance p-on-n HgCdTe photodiodes) experimentally measured carrier lifetime in HgCdTe is more than two orders of magnitude longer than in SL. More recently published upper experimental data [61,62] coincide well with HgCdTe trend-line in the range of lower carrier concentration (see Figure 17.16). In general, however, the SL carrier lifetime is limited by influence of trap centers located in the energy gap (at an energy level of ~1/3 bandgap below the effective conduction band edge [61]). Empirical fitting data gave minority carrier lifetimes from 35 ns to 200 ns, with similar absorption layers but different device structure. There is no clear understanding why the minority carrier lifetime varies within the device structure [63].

InAs/GaInSb SLSs are also employed as the active regions of MWIR lasers operated in the 2.5–6 μm spectral region. Meyer et al. [64] have experimentally determined Auger coefficients for InAs/GaInSb quantum well structures with energy gaps corresponding to 3.1–4.8 μm and have compared their values with that for typical III-V and II-VI type I superlattices. The Auger coefficient is defined by the expression $\gamma_3 \equiv 1/\tau_A n^2$. Figure 17.17 summarizes the Auger coefficients at ≈300 K for different material systems: a wide variety of type I materials (all open points), including bulk and quantum well III-V semiconductors as well as HgCdTe (upside-down

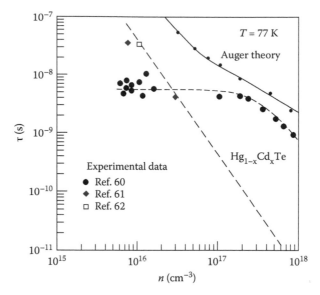

Figure 17.16 Comparison of measured and calculated carrier lifetimes of InAs/GaInSb SLS (about 120 eV energy gap) at 77 K as a function of carrier density. Experimental data are taken from (Youngdale, E. R., Meyer, J. R., Hoffman, C. A., Bartoli, F. J., Grein, C. H., Young, P. M., Ehrenreich, H., Miles, R. H., and Chow, D. H., *Applied Physics Letters*, 64, 3160–62, 1994. With permission.) (●), (Yang, O. K., Pfahler, C., Schmitz, J., Pletschen, W., and Fuchs, F., "Trap Centers and Minority Carrier Lifetimes in InAs/GaInSb Superlattice Long Wavelength Photodetectors," *Proceedings of SPIE* 4999, 448–56, 2003. With permission.) (♦), and (Pellegrini J., and DeWames, R., "Minority Carrier Lifetime Characteristics in Type II InAs/GaSb LWIR Superlattice n⁺πp⁺ Photodiodes," *Proceedings of SPIE* 7298, 7298-67, 2009. With permission.) (□). Theory from (Youngdale, E. R., Meyer, J. R., Hoffman, C. A., Bartoli, F. J., Grein, C. H., Young, P. M., Ehrenreich, H., Miles, R. H., and Chow, D. H., *Applied Physics Letters*, 64, 3160–62, 1994. With permission.)

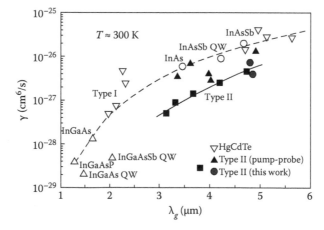

Figure 17.17 Auger coefficient versus gap wavelength for type II InAs/GaInSb quantum wells (filled points) and a variety of type I materials (open points). The filled circles are from the photoconductive response technique and filled boxes from the lasing thresholds, whereas the filled triangles are from pump-probe experiments. The solid and dashed curves are guides to the eye joining data from the present type II and bulk type I experiments, respectively. (From Meyer, J. R., Felix, C. L., Bewley, W. W., Vurgaftman, I., Aifer, E. H., Olafsen, L. J., Lindle, J. R., et al., *Applied Physics Letters*, 73, 2857–59, 1998. With permission.)

triangles). We can see that the room temperature Auger coefficients for seven different InAs/GaInSb SLs are found to be nearly an order of magnitude lower than typical type I results for the same wavelength indicating a significant suppression of Auger losses in the antimonide type II SLs. The data imply that at this temperature, the Auger rate is relatively insensitive to details of the band structure. In contrast to MWIR devices, the promise of Auger suppression has not yet to be observed in practical LWIR type II device materials.

Narrow bandgap materials require the doping to be controlled to at least 1×10^{15} cm^{-3} or below to avoid deleterious high-field tunneling currents across reduced depletion widths at temperatures below 77 K. Lifetimes must be increased to enhance carrier diffusion and reduce related dark currents. At the present stage of development, the residual doping concentration (both n-type as well as p-type) is typically about 5×10^{15} cm^{-3} in superlattices grown at substrate temperature ranging from 360°C to 440°C [53]. Low to mid 10^{15} cm^{-3} residual carrier concentrations are the best that have been achieved so far.

17.4.2 Superlattice Photodiodes

High performance InAs/GaInSb SL photovoltaic detectors are predicted by the theoretical promise of longer intrinsic lifetimes due to the suppression of Auger recombination mechanism.

The first InAs/InGaSb SLS photodiodes with photoresponse out to 10.6 μm, have been presented by Johnson et al. [65]. The detectors consist of double heterojunctions (DH) of the SLS with n-type and p-type GaSb grown on GaSb substrates. The use of heterojunctions in photodiodes offers several advantages over using homojunctions (see discussion in Section 14.6). In 1997 researchers from Franunhofer Institute demonstrated good detectivity (approaching HgCdTe, 8 μm cutoff, 77 K) on individual devices, initiating renewed interest in LWIR detection with type II SLs [66].

At present superlattice photodiodes are typically based on p-i-n double heterostructures with an unintentionally doped, intrinsic region between the heavily doped contact portions of the device. Figure 17.18 shows a cross-section scheme of a completely processed mesa detector and design of 10.5 μm InAs/GaSb SL photodiode. The layers are usually grown by MBE at substrate

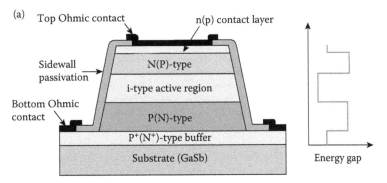

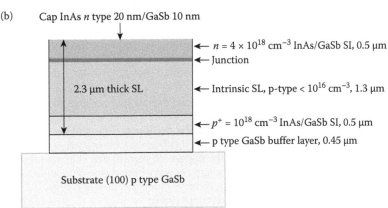

Figure 17.18 (a) Schematic structure of p-i-n double heterojunction photodiode, and (b) design of 10.5 μm photodiode.

temperatures around 400°C on undoped (001) oriented two-inch GaSb substrates. With the addition of cracker cells for the group V sources, the superlattice quality becomes significantly improved. Despite the relatively low absorption coefficients, GaSb substrates require thinning the thickness below 25 μm in order to transmit appreciable IR radiation [67]. Since the GaSb substrates and buffer layers are intrinsically p-type, the p-type contact layer, intentionally doped with beryllium at an acceptor concentration of 1×10^{18} atoms/cm^3, is grown first (see Figure 17.18).

Sensors for the MWIR and LWIR spectral ranges are based on binary InAs/GaSb short-period superlattices [68,69]. The layers needed are already so thin that there is no benefit to using GaInSb alloys. The oscillator strength of the InAs/GaSb SL is weaker than InAs/GaInSb, however, the InAs/GaSb SL, which uses unstrained and minimally strained binary semiconductor layers, may also have material quality advantages over the SLS, which uses a strained ternary semiconductor (GaInSb). For the formation of p-i-n photodiodes, the lower periods of the InAs/GaSb SL are p-doped with 1×10^{17} cm^{-3} Be in the GaSb layers. These acceptor doped SL layers are followed by a 1–2 μm thick, nominally undoped, superlattice region. The width of the intrinsic region does vary in the designs. The width used should be correlated to the carrier diffusion lengths for improved performance. The upper of the SL stack is doped with silicon (1×10^{17}–1×10^{18} cm^{-3}) in the InAs layers and is typically 0.5 μm thick. The top of the SL stack is then capped with an InAs:Si ($n \approx 10^{18}$ cm^{-3}) layer to provide good ohmic contact.

To approach cutoff wavelengths in the 8–12 μm wavelength range, the InAs/GaInSb short period superlattice p-i-n photodiodes, with the indium molar fraction in the ternary GaInSb layers close to 20%, are fabricated [53].

Significant progress has been made on the quality of the material during the past several years. The wafer surface roughness achieved 1–2 Å. The main technological challenge for the fabrication of photodiodes is the growth of thick SLS structures without degrading the materials quality. High-quality SLS materials thick enough to achieve acceptable quantum efficiency is crucial to the success of the technology. Surface passivation is also a serious problem. Besides efficient suppression of surface leakage currents, a passivation layer suitable for production purposes must withstand various treatments occurring during the subsequent processing of the device. The mesa side walls are a source of excess currents. Considerable surface leakage is attributed to the discontinuity in the periodic crystal structure caused by mesa delineation. Several materials and processes have been explored for device passivation. Some of the more prominent thin films studied have been silicon nitride, silicon oxides, ammonium sulfide, aluminum gallium antimonide alloys, and polyimide [68]. Rehm et al. [70] have chosen and demonstrated the good results achieved with lattice matched AlGaAsSb overgrowth by MBE on etched mesas.

It appears that the reproducivity and long-term stability achieved by the SiO$_2$ passivation layer is more critical for photodiodes in the LWIR range. In general, the inversion potentials are bigger for higher bandgap materials, and therefore SiO$_2$ can passivate high bandgap materials (MWIR photodiodes) but not low bandgap materials (LWIR photodiodes). Using this property, a double heterostructure that prevents the inversion of the high bandgap p-type and n-type superlattice contact regions has been proposed [71]. For such structure, the surface leakage channel at the interface between the active region and the p- or n- contacts is considerably decreased (see Figure 17.19). For this structure, an effective passivation is low-temperature, ion-sputtered SiO$_2$ passivation [71]. The best results have been recently obtained using polyimide passivation [72,73] and inductively coupled plasma dry etching [73].

Several additional design modifications that dramatically improve the LWIR photodiode dark current and R_oA product have been described. Since the excess current is due to side walls, one approach is to eliminate them. The very shallow slope of the shallow etched samples demonstrate that it is possible to reduce excess currents [74].

An alternate method of eliminating excess currents due to side walls is shallow-etch mesa isolation with a band-graded junction [75]. The primary effect of the grading is to suppress tunneling and generation–recombination currents in the depletion region at low temperatures. Since both processes depend exponentially on bandgap, it is highly advantageous to substitute a wide gap into depletion region. In this approach, the mesa etch terminates at just past the junction and exposes only a very thin (300 nm), wider bandgap region of the diode. Subsequent passivation is therefore in wider gap material. As a result, it reduces electrical junction area, increases optical fill factor, and eliminates deep trenches within detector array.

The performance of LWIR photodiodes in the high temperature range is limited by a diffusion process. Space charge recombination currents dominate reverse bias at 78 K and taking the dominant recombination centers to be located at the intrinsic Fermi level, which is shown in

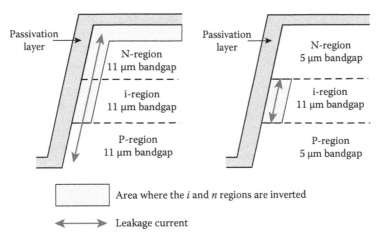

Passivation layer

N-region
11 µm bandgap

i-region
11 µm bandgap

P-region
11 µm bandgap

Passivation layer

N-region
5 µm bandgap

i-region
11 µm bandgap

P-region
5 µm bandgap

☐ Area where the *i* and *n* regions are inverted

◄——► Leakage current

Figure 17.19 Sketch of the technique used to suppress the surface leakage current in a type-II InAs/GaSb double heterostructure when the n-contact region is inverted. (From Delaunay, P.-Y., Hood, A., Nguyen, B.-M., Hoffman, D., Wei, Y., and Razeghi, M., *Applied Physics Letters*, 91, 091112, 2007. With permission.)

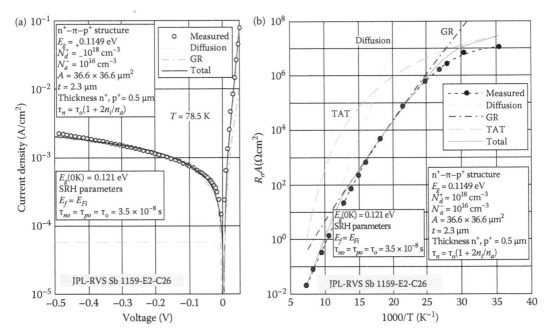

(a)

$n^+-\pi-p^+$ structure
$E_g = 0.1149$ eV
$N_d^+ = 10^{18}$ cm^{-3}
$N_a^- = 10^{16}$ cm^{-3}
$A = 36.6 \times 36.6$ µm^2
$t = 2.3$ µm
Thickness n$^+$, p$^+$ = 0.5 µm
$\tau_n = \tau_o(1 + 2n_i/n_a)$

○ Measured
⋯ Diffusion
─·─ GR
── Total

$T = 78.5$ K

$E_g(0K) = 0.121$ eV
SRH parameters
$E_f = E_{Fi}$
$\tau_{no} = \tau_{po} = \tau_o = 3.5 \times 10^{-8}$ s

JPL-RVS Sb 1159-E2-C26

Current density (A/cm^2)

Voltage (V)

(b)

Diffusion

GR

TAT

─●─ Measured
⋯ Diffusion
─·─ GR
⋯ TAT
── Total

$E_g(0K) = 0.121$ eV
SRH parameters
$E_f = E_{Fi}$
$\tau_{no} = \tau_{po} = \tau_o = 3.5 \times 10^{-8}$ s

$n^+-\pi-p^+$ structure
$E_g = 0.1149$ eV
$N_d^+ = 10^{18}$ cm^{-3}
$N_a^- = 10^{16}$ cm^{-3}
$A = 36.6 \times 36.6$ µm^2
$t = 2.3$ µm
Thickness n$^+$, p$^+$ = 0.5 µm
$\tau_n = \tau_o(1 + 2n_i/n_a)$

JPL-RVS Sb 1159-E2-C26

R_0A (Ωcm^2)

1000/T (K^{-1})

Figure 17.20 Experimental data and theoretical characteristics for InAs/GaInSb photodiode with $\lambda_c = 10.5$ µm at 78.5 K: (a) I-V characteristic at 78 K, (b) the R_0A product as a function of temperature. (From Pellegrini, J., and DeWames, R., "Minority Carrier Lifetime Characteristics in Type II InAs/GaSb LWIR Superlattice n$^+$πp$^+$ Photodiodes," *Proceedings of SPIE 7298*, 7298-67, 2009. With permission.)

Figure 17.20a [62]. At a low temperature, the currents are diffusion limited near zero bias voltage. At larger biases, trap assisted tunneling currents dominate. Figure 17.20b shows the experimental data and theoretical prediction of the R_0A product as a function of temperature for InAs/GaInSb photodiode with 10.5 µm cutoff wavelength at 78 K. The photodiodes are depletion region (generation–recombination) limited in temperature range below 100 K. The trap-assisted tunneling is dominant at $T \leq 40$ K.

Optimization of the SL photodiode architectures is still an open area. Since some of the device design parameters depend on material properties, like carrier lifetime and diffusion lengths, these properties are still being improved. Also additional design modifications dramatically improve the photodiode performance. For example, Aifer at el. [75] have reported W-structured type II superlattice (WSL) LWIR photodiodes with R_0A values comparable for state-of-the-art HgCdTe. These structures initially developed to increase the gain in MWIR lasers, are now showing promise as LWIR and VLWIR photodiode materials. In this design illustrated in Figure 17.21a, the AlSb barriers are replaced with shallower $Al_{0.40}Ga_{0.49}In_{0.11}Sb$ quaternary barrier layers (QBL) that have a much smaller conduction band offset with respect to InAs, resulting in higher electron mobility, with a miniband width of about 35 meV, compared to 20 meV for AlSb barrier layers. The QBL also uses 60% less Al, which improves material quality, since the optimal QBL growth temperature is much closer than that of AlSb (~500°C) to that of the InAs and InGaSb layers around 430°C. In such structure two InAs "electron-wells" are located on either side of an InGaSb "hole-well" and are bound on either side by AlGaInSb "barrier" layers. The barriers confine the electron wavefunctions symmetrically about the hole-well, increasing the electron-hole overlap while nearly localizing the wavefunctions. The resulting quasidimensional densities of states give the WSL its characteristically strong absorption near band edge. However, care is taken to not fully localize the wavefunctions, since an electron miniband is required to allow vertical transport of the photoexcited minority carriers.

The new design W-structured type II SL photodiodes employ a graded bandgap p-i-n design. The grading of the bandgap in the depletion region suppresses tunneling and generation–recombination currents in the depletion region that have resulted in an order of magnitude improvement in dark current performance, with $R_0A = 216$ Ωcm² at 78 K for devices with a 10.5 μm cutoff wavelength. The sidewall resistivity of ≈70 kΩcm for untreated mesas is considerably higher than previously reported for type-II LWIR photodiodes, apparently indicating self-passivation by the graded bandgap [75].

Another type II superlattice photodiode design with the M-structure barrier is shown in Figure 17.21b. This structure significantly reduces the dark current, and on the other hand, does not shown a strong effect on the optical properties of the devices [76]. The AlSb layer in one period of the M structure, having a wider energy gap, blocks the interaction between electrons in two adjacent InAs wells, thus, reducing the tunneling probability and increasing the electron effective mass. The AlSb layer also acts as a barrier for holes in the valence band and converts the GaSb hole-quantum well into a double quantum well. As a result, the effective well width is reduced, and the hole's energy level becomes sensitive to the well dimension. The device with a cutoff wavelength of 10.5 μm exhibits a R_0A product of 200 Ωcm² when a 500 nm thick M structure was used. Recently, using double M-structure heterojunction at the single device level, the R_0A product up to 5300 Ωcm² has been obtained for a 9.3 μm cutoff at 77 K [73].

Figure 17.22 compares the R_0A values of InAs/GaInSb SL and HgCdTe photodiodes in the long wavelength spectral range [77]. The solid line denotes the theoretical diffusion limited

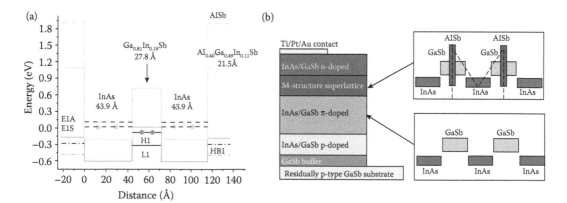

Figure 17.21 Schematic diagrams of modified type-II LWIR photodiodes: (a) band profiles at $k = 0$ of enhanced WSL. (From Aifer, E. H., Tischler, J. G., Warner, J. H., Vurgaftman, I., Bewley, W. W., Meyer, J. R., Canedy, C. L., and Jackson, E. M., *Applied Physics Letters*, 89, 053510, 2006. With permission.) (b) p-π-M-n SL (band alignment of standard and M shape superlattices are shown). (From Nguyen, B.-M., Hoffman, D., Delaunay, P.-Y., and Razeghi, M., *Applied Physics Letters*, 91, 163511, 2007. With permission.)

performance of p-type HgCdTe material. As it can be seen in the figure, the most recent photo-diode results for SL devices rival that of practical HgCdTe devices, indicating substantial improvement has been achieved in SL detector development.

The quantum efficiency of p-i-n photodiode structure shown in Figure 17.18 critically depends on the thickness of i(π)-region. By fitting the quantum efficiencies of a series of photodiodes with i-region thicknesses varying from 1 to 4 µm, Aifer et al. [75] determined that the minority-carrier electron diffusion length in LWIR is 3.5 µm. This value is considerably lower in comparison with a typical one for high quality HgCdTe photodiodes. More recently, an external quantum efficiency of 54% has been obtained for a 12 µm cutoff wavelength photodiodes by extending thickness of the π-region to 6 µm. Figure 17.23a shows the dependence of quantum efficiency on thickness of the π-region but Figure 17.23b presents spectral current responsivity of eight of the structures with different thicknesses of the π-region [78].

Figure 17.24 compares the calculated detectivity of type-II and P-on-n HgCdTe photodiodes as a function of wavelength and temperature of operation with the experimental data of type II detectors operated at 78 K [74]. The solid lines are theoretical thermal limited detectivities for HgCdTe photodiodes, calculated using a 1-D model that assumes diffusion current from narrower bandgap n-side is dominant, and minority carrier recombination via Auger and radiative process. In calculations, typical values for the n-side donor concentration ($N_d = 1 \times 10^{15}$ cm^{-3}), the narrow bandgap active layer thickness (10 µm), and quantum efficiency (60%) have been used. The predicted thermally limited detectivities of the type II SLS are larger than those for HgCdTe [59,79].

From Figure 17.24 results that the measured thermally limited detectivities of type-II SLS photodiodes are as yet inferior to current HgCdTe photodiode performance. Their performance has not achieved theoretical values. This limitation appears to be due to two main factors: relatively high background concentrations (about 5×10^{15} cm^{-3}, although values below 10^{15} cm^{-3} have been reported [77,80]) and a short minority carrier lifetime (typically tens of nanoseconds in lightly doped p-type material). Up until now nonoptimized carrier lifetimes have been observed and desirably low carrier concentrations is limited by Shockley–Read recombination mechanism. The minority carrier diffusion length is in the range of several micrometers. Improving these fundamental parameters is essential to realize the predicted performance of type-II photodiodes.

Type-II based InAs/GaInSb based detectors have made rapid progress over the past few years. The presented results indicate that fundamental material issues of InAs/GaInSb SLs fulfill practical realization of high performance FPAs.

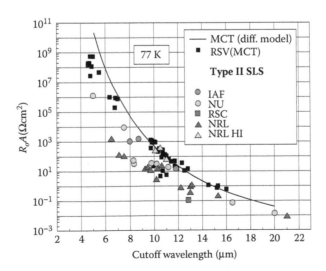

Figure 17.22 Dependence of the R_oA product of InAs/GaInSb SLS photodiodes on cutoff wavelength compared to theoretical and experimental trendlines for comparable HgCdTe photodiodes at 77 K. (From Canedy, C. L., Aifer, H., Vurgaftman, I., Tischler, J. G., Meyer, J. R., Warner, J. H., and Jackson, E. M., *Journal of Electronic Materials*, 36, 852–56, 2007. With permission.)

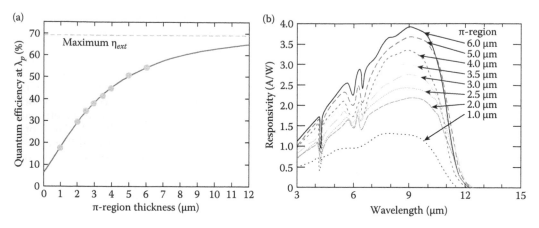

Figure 17.23 Spectral characteristics of InAs/GaInSb SLS photodiodes at 77 K: (a) quantum efficiency as a function of π-region thickness, the dashed line represents the maximum possible quantum efficiency without antireflective coating; (b) measured current responsivity of photodiodes with π-region thickness ranging from 1 to 6 μm, the CO_2 absorption at 4.2 μm is visible as well as the water vapor absorption between 5 and 8 μm. (From Nguyen, B.-M., Hoffman, D., Wei, Y., Delaunay, P.-Y., Hood, A., and Razeghi, M., *Applied Physics Letters*, 90, 231108, 2007. With permission.)

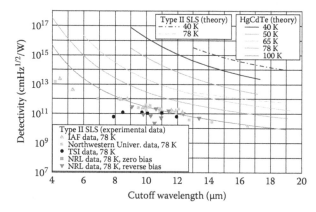

Figure 17.24 The predicted detectivity of type-II and P-on-n HgCdTe photodiodes as a function of wavelength and temperature. The experimental data are taken with several sources. (From Bajaj, J., Sullivan, G., Lee, D., Aifer, E., and Razeghi, M., "Comparison of Type-II Superlattice and HgCdTe Infrared Detector Technologies," *Proceedings of SPIE* 6542, 65420B, 2007. With permission.)

7.4.3 nBn Superlattice Detectors

Type II SLSs are also used in fabrication unipolar barrier photodetectors. Barriers that can block one carrier type (electron or hole) but allow the unimpeded flow of the other can be used to enhance the performance of photodetectors. The nBn detector structure (see Section 9.9) uses a unipolar barrier to suppress dark current without impeding photocurrent flow [81], and to suppress surface leakage current [82]. In practice, unipolar barriers are not always readily available for the desired absorber, as both the absorber and barrier require near lattice matching to available substrates, and the proper band offsets must exist between the absorber and the barrier. These requirements will fulfill the nearly lattice-matched InAs, GaSb, AlSb material systems, which can be epitaxially grown on GaSb or InAs substrates. GaSb/AlSb, InAs/AlSb and InAs/GaInSb give considerable flexibility in forming a variety of alloys and superlattices [83].

Type II InAs/GaInSb SLs can be used as MW [84,85] and LWIR absorber [86] in nBn detectors, however hitherto, efforts are mainly concentrated on MW detectors. Bishop et al. [85] have studied two device structure shown in Figure 17.25–further referred to as structures A and B. Devices

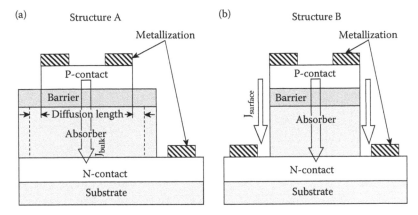

Figure 17.25 Schematic of shallow etched isolation nBn device (a): structure A. The active read is defined by the diffusion length of the minority carriers (holes). Conventionally defined mesa structure is shown in part (b): structure B. (From Bishop, G., Plis, E., Rodriguez, J. B., Sharma, Y. D., Kim, H. S., Dawson, L. R., and Krishna, S., *Journal of Vacuum Science and Technology*, B26, 1145–48, 2008. With permission.)

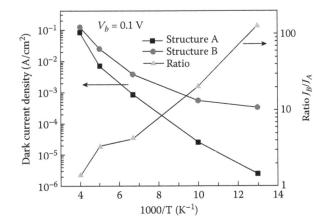

Figure 17.26 Dark current density versus $1000/T$ in structures A and B ($V_b = 0.1$ V). The ratio of dark current densities for both structures measured at the same value of applied bias is also shown. (From Bishop, G., Plis, E., Rodriguez, J. B., Sharma, Y. D., Kim, H. S., Dawson, L. R., and Krishna, S., *Journal of Vacuum Science and Technology*, B26, 1145–48, 2008. With permission.)

consists of 0.25 μm thick n-type bottom contact layer composed of 8 ML InAs:Si ($n = 4 \times 10^{18}$ cm^{-3})/8 ML GaSb followed by the nonintentionally doped 1.4 μm thick 8 ML InAs/8 ML GaSb absorber layer. An Al$_{0.2}$Ga$_{0.8}$Sb barrier layer with thickness of 100 nm was then grown on the top of the absorber layer. The barrier layer was grown to be thick enough to prevent electron tunneling from the top contact SL layer to the SL absorbing layer. The structure was capped by 100 nm thick top contact layer with the same SL composition, thickness, and doping concentration as the bottom contact layer. A detailed description of the growth SL procedure has been reported elsewhere [87]. To ensure ohmic behavior of the contacts, heavily doped contact layers were used. A good solution for ohmic contacts is GeAu-based metallization on the n-type SL [88].

Structures A and B were processed in two distinct ways. An isolation etch was undertaken using H$_3$PO$_4$:H$_2$O$_2$:H$_2$O (1:2:20) solution to the middle of the barrier layer structure A. Then, inductively coupled plasma dry etch to the middle of the bottom contact layer was performed for both structures A and B. For structure A, the active area of the device is defined by the diffusion length of minority carries (holes) rather than by a conventional mesa. Since dangling bonds are expected on the etched mesa side walls of structure B, the surface leakage current is expected to be high. Figure 17.26 shows the dark current densities versus temperature for both structures together with

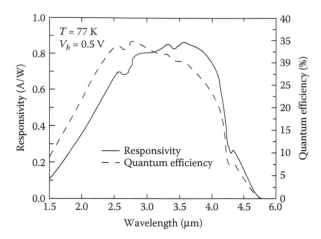

Figure 17.27 Responsivity and quantum efficiency versus wavelength for structure A measured at 77 K with $V_b = 0.5$ V. (From Bishop, G., Plis, E., Rodriguez, J. B., Sharma, Y. D., Kim, H. S., Dawson, L. R., and Krishna, S., *Journal of Vacuum Science and Technology*, B26, 1145–48, 2008. With permission.)

a ratio of dark current densities measured at the same value of applied bias. At high temperature, the influence of the surface current component is not seen as a large mesa (400 × 400 µm²) and the thermally generated carriers dominate. As the temperature decreases and the bulk component of the dark current strongly decreases, the surface current becomes the dominant component. The levels of the dark current in structures A and B are comparable at 250 K but the dark current in structure A is reduced by two orders of magnitude.

Figure 17.27 illustrates the spectral dependences of responsivity and quantum efficiency. The measurements show maximum values of current responsivity and quantum efficiency of 0.74 A/W and 23% at 4.0 µm, respectively.

The lack of stable passivation layer for the etched mesa LWIR SL photodiodes is one of the primary limitations of the type II SL based technology. As is shown above, nBn detectors eliminate the currents associated with SRH centers and mesa lateral surface imperfections, which have resulted in an increase of the operating temperature as compared to the p-i-n design. This prediction has been experimentally confirmed by Khoshakhlagh et al. [86] for LWIR nBn detectors.

Development of an optimized growth procedure with well-ordered constituents and interfacial layers is critical for LWIR SLSs (see discussion by Khoshakhlagh and colleagues [86]). The device design is different from that for MWIR region; the 1.9 µm thick nonintentionally doped active region consisted of 13 MLs InAs/0.75 ML InSb/7 MLs GaSb SL. The structure based on the nBn design showed two orders of magnitude improvement in dark current density as compared to the structure based on the p-i-n design [86].

Recently, the first MWIR FPAs based on type II InAs/GaSb SLSs with nBn detector design have been demonstrated [89,90]. Unlike a p-n junction, the size of the devices is not defined by a mesa etch but confined by the lateral diffusion length of minority carriers. In the first FPA, the depth of the shallow etch was equal to 0.15 µm [89], which corresponds to the middle of the barrier layer. Thus, the active absorber layer underneath is untouched.

REFERENCES

1. M. A. Kinch and A. Yariv, "Performance Limitations of GaAs/AlGaAs Infrared Superlattices," *Applied Physics Letters* 55, 2093–95, 1989.

2. A. Rogalski, "HgCdTe Photodiodes Versus Quantum Well Infrared Photoconductors for Long Wavelength Focal Plane Arrays," *Opto-Electronics Review* 6, 279–94, 1998.

3. M. Z. Tidrow, W. A. Beck, W. W. Clark, H. K. Pollehn, J. W. Little, N. K. Dhar, P. R. Leavitt, et al., "Device Physics and Focal Plane Applications of QWIP and MCT," *Opto-Electronics Review* 7, 283–96, 1999.

4. A. Rogalski, "Quantum Well Photoconductors in Infrared Detectors Technology," *Journal of Applied Physics* 93, 4355–91, 2003.

5. A. Rogalski, "Third Generation Photon Detectors," *Optical Engineering* 42, 3498–516, 2003.

6. A. Rogalski, J. Antoszewski, and L. Faraone, "Third-Generation Infrared Photodetector Arrays," *Journal of Applied Physics* 105, 091101-44, 2009.

7. B. F. Levine, "Quantum-Well Infrared Photodetectors," *Journal of Applied Physics* 74, R1–R81, 1993.

8. S. D. Gunapala and K. M. S. V. Bandara, "Recent Development in Quantum-Well Infrared Photodetectors," in *Thin Films*, Vol. 21, 113–237, Academic Press, New York, 1995.

9. H. C. Liu, "Quantum Well Infrared Photodetector Physics and Novel Devices," in *Semiconductors and Semimetals*, Vol. 62, eds. by H. C. Liu and F. Capasso, 129–96, Academic Press, San Diego, CA, 2000.

10. S. D. Gunapala and S. V. Bandara, "Quantum Well Infrared Photodetectors (QWIP)," in *Handbook of Thin Devices*, Vol. 2, ed. M. H. Francombe, 63–99, Academic Press, San Diego, CA, 2000.

11. J. L. Pan and C. G. Fonstad, "Theory, Fabrication and Characterization of Quantum Well Infrared Photodetectors," *Material Science and Engineering* R28, 65–147, 2000.

12. S. D. Gunapala and S. V. Bandara, "GaAs/AlGaAs Based Quantum Well Infrared Photodetector Focal Plane Arrays," in *Handbook of Infrared Detection Technologies*, eds. M. Henini and M. Razeghi, 83–119, Elsevier, Oxford, 2002.

13. V. Ryzhi, ed., *Intersubband Infrared Photodetectors*, World Scientific, New Jersey, 2003.

14. H. Schneider and H. C. Liu, *Quantum Well Infrared Photodetectors*, Springer, Berlin, 2007.

15. G. H. Döhler, "Doping Superlattices ("n–i–p–i Crystals")," *IEEE Journal of Quantum Electronics* QE-22, 1682–95, 1986.

16. J. N. Schulman and T. C. McGill, "The CdTe/HgTe Superlattice: Proposal for a New Infrared Material," *Applied Physics Letters* 34, 663–65, 1979.

17. D. L. Smith, T. C. McGill, and J. N. Schulman, "Advantages of the HgTe-CdTe Superlattice as an Infrared Detector Material," *Applied Physics Letters* 43, 180–82, 1983.

18. J. P. Faurie, "Growth and Properties of HgTe-CdTe and Other Hg-Based Superlattices," *IEEE Journal of Quantum Electronics* QE-22, 1656–65, 1986.

19. J. M. Berroir, Y. Guldner, and M. Voos, "HgTe-CdTe Superlattices: Magnetooptics and Band Structure," *IEEE Journal of Quantum Electronics* QE-22, 1793–98, 1986.

20. J. R. Meyer, C. A. Hoffman, and R. J. Bartoli, "Narrow-Gap II–VI Superlattices: Correlation of Theory with Experiment," *Semiconductor Science and Technology* 5, S90–S99, 1990.

21. T. H. Myers, J. R. Meyer, C. A. Hoffman, and L. R. Ram-Mohan, "HgTe/CdTe Superlattices for IR Detection Revisited," *Applied Physics Letters* 61, 1814–16, 1992.

22. T. H. Myers, J. R. Meyer, and C. A. Hoffman, "The III Superlattices for Long-Wavelength Infrared Detectors: The HgTe/CdTe System," in *Semiconductor Quantum Wells and Superlattices for Long-Wavelength Infrared Detectors*, ed. M. O. Manasreh, 207–59, Artech House, Boston, MA, 1993.

23. F. F. Sizov and A. Rogalski, "Semiconductor Superlattices and Quantum Wells for Infrared Optoelectronics," *Progress in Quantum Electronics* 17, 93–164, 1993.

24. J. R. Meyer, C. A. Hoffman, and F. J. Bartoli, "Quantum Wells and Superlattices," in *Narrow-Gap II-VI Compounds for Optoelectronic and Electromagnetic Applications,* ed. P. Capper, 363–400, Chapman & Hall, London, 1997.

25. D. K. Arch, J. P. Faurie, J. L. Staudenmann, M. Hibbs-Brenner, and P. Chow, "Interdiffusion in HgTe-CdTe Superlattices," *Journal of Vacuum Science and Technology* A4, 2101–5, 1986.

26. C. R. Becker, V. Latussek, A. Pfeuffer-Jeschke, G. Landwehr, and L. W. Molenkamp, "Band Structure and Its Temperature Dependence for Type-III HgTe/$Hg_{1-x}Cd_xTe$ Superlattices and Their Semimetal Constituent," *Physical Review* B62, 10353–63, 2000.

27. C. K. Shih and W. E. Spicer, "Determination of a Natural Valence-Band Offset: The Case of HgTe-CdTe," *Physical Review Letters* 58, 2594–97, 1987.

28. C. R. Becker, K. Ortner, X. C. Zhang, S. Oehling, A. Pfeyffer-Jeschke, and V. Latussek, "Far Infrared Detectors Based on HgTe/$Hg_{1-x}Cd_xTe$ Superlattices, and In Situ p Type Doping," in *Advanced Infrared Technology and Applications 2007,* ed. M. Strojnik, 79–89, Leon, Mexico, 2008.

29. J. M. Arias, S. H. Shin, D. E. Cooper, M. Zandian, J. G. Pasko, E. R. Gertner, R. E. DeWames, and J. Singh, "P–Type Arsenic Doping of CdTe and HgTe/CdTe Superlattices Grown by Photoassisted and Conventional Molecular-Beam Epitaxy," *Journal of Vacuum Science and Technology* A8, 1025–33, 1990.

30. K. A. Harris, R. W. Yanka, L. M. Mohnkern, A. R. Reisinger, T. H. Myers, Z. Yang, Z. Yu, S. Hwang, and J. F. Schetzina, "Properties of (211)B HgTe–CdTe Superlattices Grown by Photon Assisted Molecular-Beam Epitaxy," *Journal of Vacuum Science and Technology* B10, 1574–81, 1992.

31. C. H. Grein, H. Jung, R. Singh, and M. F. Flatte, "Comparison of Normal and Inverted Band Structure HgTe/CdTe Superlattices for Very Long Wavelength Infrared Detectors," *Journal of Electronic Materials* 34, 905–8, 2005.

32. M. W. Goodwin, M. A. Kinch, and R. J. Koestner, "Metal-Insulator-Semiconductor Properties of HgTe/CdTe Superlattices," *Journal of Vacuum Science and Technology* A6, 2685–92, 1988.

33. M. L. Wroge, D. J. Peterman, B. J. Feldman, B. J. Morris, D. J. Leopold, and J. G. Broerman, "Impurity Doping of HgTe-CdTe Superlattices During Growth by Molecular-Beam Epitaxy," *Journal of Vacuum Science and Technology* A7, 435–39, 1989.

34. K. A. Harris, T. H. Myers, R. W. Yanka, L. M. Mohnkern, and N. Otsuka, "A High Quantum Efficiency *In Situ* Doped Mid-Wavelength Infrared p-on-n Homojunction Superlattice Detector Grown by Photoassisted Molecular-Beam Epitaxy," *Journal of Vacuum Science and Technology* B9, 1752–58, 1991.

35. G. C. Osbourn, "InAsSb Strained-Layer Superlattices for Long Wavelength Detector Applications," *Journal of Vacuum Science and Technology* B2, 176–78, 1984.

36. S. R. Kurtz, G. C. Osbourn, R. M. Biefeld, L. R. Dawson, and H. J. Stein, "Extended Infrared Response of InAsSb Strained-Layer Superlattices," *Applied Physics Letters* 52, 831–33, 1988.

37. D. L. Smith and C. Mailhiot, "Proposal for Strained Type II Superlattice Infrared Detectors," *Journal of Applied Physics* 62, 2545–48, 1987.

38. C. Mailhiot and D. L. Smith, "Long-Wavelength Infrared Detectors Based on Strained InAs–GaInSb Type-II Superlattices," *Journal of Vacuum Science and Technology* A7, 445–49, 1989.

39. F. F. Sizov, "Semiconductor Superlattices and Quantum Well Detectors," in *Infrared Photon Detectors*, ed. A. Rogalski, 561–623, SPIE Optical Engineering Press, Bellingham, WA, 1995.

40. C. H. Grein, P. M. Young, and H. Ehrenreich, "Minority Carrier Lifetimes in Ideal InGaSb/InAs Superlattice," *Applied Physics Letters* 61, 2905–7, 1992.

41. G. C. Osbourn, "Design of III-V Quantum Well Structures for Long-Wavelength Detector Applications," *Semiconductor Science and Technology* 5, S5–S11, 1990.

42. G. A. Sai-Halasz, R. Tsu, and L. Esaki, "A New Semiconductor Superlattice," *Applied Physics Letters* 30, 651–52, 1977.

43. L. L. Chang, N. J. Kawai, G. A. Sai-Halasz, P. Ludeke, and L. Esaki, "Observation of Semiconductor-Semimetal Transition in InAs/GaSb Superlattices," *Applied Physics Letters* 35, 939–42, 1979.

44. A. Rogalski, *New Ternary Alloy Systems for Infrared Detectors*, SPIE Optical Engineering Press, Bellingham, WA, 1994.

45. H. R. Jen, K. Y. Ma, and G. B. Stringfellow, "Long-Range Order in InAsSb," *Applied Physics Letters* 54, 1154–56, 1989.

46. S. R. Kurtz, L. R. Dawson, R. M. Biefeld, D. M. Follstaedt, and B. L. Doyle, "Ordering-Induced Band-Gap Reduction in $InAs_{1-x}Sb_x$ ($x \approx 0.4$) Alloys and Superlattices," *Physical Review* B46, 1909–12, 1992.

47. S. H. Wei and A. Zunger, "InAsSb/InAs: A Type-I or a Type-II Band Alignment," *Physical Review* B52, 12039–44, 1995.

48. R. A. Stradling, S. J. Chung, C. M. Ciesla, C. J. M. Langerak, Y. B. Li, T. A. Malik, B. N. Murdin, et al., "The Evaluation and Control of Quantum Wells and Superlattices of III-V Narrow Gap Semiconductors," *Materials Science and Engineering* B44, 260–65, 1997.

49. Y. H. Zhang, A. Lew, E. Yu, and Y. Chen, "Microstructural Properties of $InAs/InAs_xSb_{1-x}$ Ordered Alloys Grown by Modulated Molecular Beam Epitaxy," *Journal of Crystal Growth* 175/176, 833–37, 1997.

50. S. R. Kurtz, L. R. Dawson, T. E. Zipperian, and S. R. Lee, "Demonstration of an InAsSb Strained-Layer Superlattice Photodiode," *Applied Physics Letters* 52, 1581–83, 1988.

51. S. R. Kurtz, L. R. Dawson, R. M. Biefeld, I. J. Fritz, and T. E. Zipperian, "Long-Wavelength InAsSb Strained-Layer Superlattice Photovoltaic Infrared Detectors," *IEEE Electron Device Letters* 10, 150–52, 1989.

52. S. R. Kurtz, L. R. Dawson, T. E. Zipperian, and R. D. Whaley, "High Detectivity ($>1 \times 10^{10}$ cmHz$^{1/2}$/W), InAsSb Strained-Layer Superlattice, Photovoltaic Infrared Detector," *IEEE Electron Device Letters* 11, 54–56, 1990.

53. L. Bürkle and F. Fuchs, "InAs/(GaIn)Sb Superlattices: A Promising Material System for Infrared Detection," in *Handbook of Infrared Detection and Technologies*, eds. M. Henini and M. Razeghi, 159–89, Elsevier, Oxford, 2002.

54. G. J. Brown, F. Szmulowicz, K. Mahalingam, S. Houston, Y. Wei, A. Gon, and M. Razeghi, "Recent Advances in InAs/GaSb Superlattices for Very Long Wavelength Infrared Detection," *Proceedings of SPIE* 4999, 457–66, 2003.

55. Y. Wei and M. Razeghi, "Modeling of Type-II InAs/GaSb Superlattices Using an Empirical Tight-Binding Method and Interface Engineering," *Physical Review* B69, 085316, 2004.

56. C. Mailhiot, "Far-Infrared Materials Based on InAs/GaInSb Type II, Strained-Layer Superlattices," in *Semiconductor Quantum Wells and Superlattices for Long-Wavelength Infrared Detectors*, ed. M. O. Manasreh, 109–38, Artech House, Boston, MA, 1993.

57. J. P. Omaggio, J. R. Meyer, R. J. Wagner, C. A. Hoffman, M. J. Yang, D. H. Chow, and R. H. Miles, "Determination of Band Gap and Effective Masses in InAs/Ga$_{1-x}$In$_x$Sb Superlattices," *Applied Physics Letters* 61, 207–9, 1992.

58. C. A. Hoffman, J. R. Meyer, E. R. Youngdale, F. J. Bartoli, R. H. Miles, and L. R. Ram-Mohan, "Electron Transport in InAs/Ga$_{1-x}$In$_x$Sb Superlattices," *Solid State Electronics* 37, 1203–6, 1994.

59. C. H. Grein, P. M. Young, M. E. Flatté, and H. Ehrenreich, "Long Wavelength InAs/InGaSb Infrared Detectors: Optimization of Carrier Lifetimes," *Journal of Applied Physics* 78, 7143–52, 1995.

60. E. R. Youngdale, J. R. Meyer, C. A. Hoffman, F. J. Bartoli, C. H. Grein, P. M. Young, H. Ehrenreich, R. H. Miles, and D. H. Chow, "Auger Lifetime Enhancement in InAs-Ga$_{1-x}$In$_x$Sb Superlattices," *Applied Physics Letters* 64, 3160–62, 1994.

61. O. K. Yang, C. Pfahler, J. Schmitz, W. Pletschen, and F. Fuchs, "Trap Centers and Minority Carrier Lifetimes in InAs/GaInSb Superlattice Long Wavelength Photodetectors," *Proceedings of SPIE* 4999, 448–56, 2003.

62. J. Pellegrini and R. DeWames, "Minority Carrier Lifetime Characteristics in Type II InAs/GaSb LWIR Superlattice n$^+$πp$^+$ Photodiodes," *Proceedings of SPIE* 7298, 7298-67, 2009.

63. M. Z. Tidrow, L. Zheng, and H. Barcikowski, "Recent Success on SLS FPAs and MDA's New Direction for Development," *Proceedings of SPIE* 7298, 7298-61, 2009.

64. J. R. Meyer, C. L. Felix, W. W. Bewley, I. Vurgaftman, E. H. Aifer, L. J. Olafsen, J. R. Lindle, et al., "Auger Coefficients in Type-II InAs/Ga$_{1-x}$In$_x$Sb Quantum Wells," *Applied Physics Letters* 73, 2857–59, 1998.

65. J. L. Johnson, L. A. Samoska, A. C. Gossard, J. L. Merz, M. D. Jack, G. H. Chapman, B. A. Baumgratz, K. Kosai, and S. M. Johnson, "Electrical and Optical Properties of Infrared Photodiodes Using the InAs/Ga$_{1-x}$In$_x$Sb Superlattice in Heterojunctions with GaSb," *Journal of Applied Physics* 80, 1116–27, 1996.

66. F. Fuchs, U. Weimer, W. Pletschen, J. Schmitz, E. Ahlswede, M. Walther, J. Wagner, and P. Koidl, "High Performance InAs/Ga$_{1-x}$In$_x$Sb Superlattice Infrared Photodiodes," *Applied Physics Letters* 71, 3251–53, 1997.

67. J. L. Johnson, "The InAs/GaInSb Strained Layer Superlattice as an Infrared Detector Material: An Overview," *Proceedings of SPIE* 3948, 118–32, 2000.

68. G. J. Brown, "Type-II InAs/GaInSb Superlattices for Infrared Detection: An Overview," *Proceedings of SPIE* 5783, 65–77, 2005.

69. M. Razeghi, Y. Wei, A. Gin, A. Hood, V. Yazdanpanah, M. Z. Tidrow, and V. Nathan, "High Performance Type II InAs/GaSb Superlattices for Mid, Long, and Very Long Wavelength Infrared Focal Plane Arrays," *Proceedings of SPIE* 5783, 86–97, 2005.

70. R. Rehm, M. Walther, J. Schmitz, J. Fleißner, F. Fuchs, W. Cabanski, and J. Ziegler, "InAs/(GaIn)Sb Short-Period Superlattices for Focal Plane Arrays," *Proceedings of SPIE* 5783, 123–30, 2005.

71. P.-Y. Delaunay, A. Hood, B.-M. Nguyen, D. Hoffman, Y. Wei, and M. Razeghi, "Passivation of Type-II InAs/GaSb Double Heterostructure," *Applied Physics Letters* 91, 091112, 2007.

72. A. Hood, P.-Y. Delaunay, D. Hoffman, B.-M. Nguyen, Y. Wei, M. Razeghi, and V. Nathan, "Near Bulk-Limited R$_o$A of Long-Wavelength Infrared Type-II InAs/GaSb Superlattice Photodiodes with Polyimide Surface Passivation," *Applied Physics Letters* 90, 233513, 2007.

73. E. K. Huang, D. Hoffman, B.-M. Nguyen, P.-Y. Delaunay, and M. Razeghi, "Surface Leakage Reduction in Narrow Band Gap Type-II Antimonide-Based Superlattice Photodiodes," *Applied Physics Letters* 94, 053506, 2009.

74. J. Bajaj, G. Sullivan, D. Lee, E. Aifer, and M. Razeghi, "Comparison of Type-II Superlattice and HgCdTe Infrared Detector Technologies," *Proceedings of SPIE* 6542, 65420B, 2007.

75. E. H. Aifer, J. G. Tischler, J. H. Warner, I. Vurgaftman, W. W. Bewley, J. R. Meyer, C. L. Canedy, and E. M. Jackson, "W-Structured Type-II Superlattice Long-Wave Infrared Photodiodes with High Quantum Efficiency," *Applied Physics Letters* 89, 053510, 2006.

76. B.-M. Nguyen, D. Hoffman, P-Y. Delaunay, and M. Razeghi, "Dark Current Suppression in Type II InAs/GaSb Superlattice Long Wavelength Infrared Photodiodes with M-Structure Barrier," *Applied Physics Letters* 91, 163511, 2007.

77. C. L. Canedy, H. Aifer, I. Vurgaftman, J. G. Tischler, J. R. Meyer, J. H. Warner, and E. M. Jackson, "Antimonide Type-II W Photodiodes with Long-Wave Infrared R$_o$A Comparable to HgCdTe," *Journal of Electronic Materials* 36, 852–56, 2007.

78. B.-M. Nguyen, D. Hoffman, Y. Wei, P.-Y. Delaunay, A. Hood, and M. Razeghi, "Very High Quantum Efficiency in Type-II InAs/GaSb Superlattice Photodiode with Cutoff of 12 µm," *Applied Physics Letters* 90, 231108, 2007.

79. C. H. Grein, H. Cruz, M. E. Flatte, and H. Ehrenreich, "Theoretical Performance of Very Long Wavelength InAs/In$_x$Ga$_{1-x}$Sb Superlattice Based Infrared Detectors," *Applied Physics Letters* 65, 2530–32, 1994.

80. A. Hood, D. Hoffman, Y. Wei, F. Fuchs, and M. Razeghi, "Capacitance-Voltage Investigation of High-Purity InAs/GaSb Superlattice Photodiodes," *Applied Physics Letters* 88, 052112, 2006.

81. S. Maimon and G. W. Wicks, "nBn Detector, An Infrared Detector with Reduced Dark Current and Higher Operating Temperature," *Applied Physics Letters* 89, 151109, 2006.

82. J. R. Pedrazzani, S. Maimon, and G. W. Wicks, "Use of *nBn* Structure to Suppress Surface Leakage Currents in Unpassivated InAs Infrared Photodetectors," *Electronics Letters* 44, 1487–88, 2008.

83. P. Klipstein, A. Glozman, S. Grossman, E. Harush, O. Klin, J. Oiknine-Schlesinger, E. Weiss, M. Yassen, and B. Yofis, "Barrier Photodetectors for High Sensitivity and High Operating Temperature Infrared Sensors," *Proceedings of SPIE* 6940, 69402U, 2008.

84. J. B. Rodriguez, E. Plis, G. Bishop, Y. D. Sharma, H. Kim, L. R. Dawson, and S. Krishna, "nBn Structure Based on InAs/GaSb Type-II Strained Layer Superlattices," *Applied Physics Letters* 91, 043514, 2007.

85. G. Bishop, E. Plis, J. B. Rodriguez, Y. D. Sharma, H. S. Kim, L. R. Dawson, and S. Krishna, "nBn Detectors Based on InAs/GaSb Type-II Strain Layer Superlattices," *Journal of Vacuum Science and Technology* B26, 1145–48, 2008.

86. A. Khoshakhlagh, H. S. Kim, S. Myers, N. Gautam, S. J. Lee, E. Plis, S. K. Noh, L. R. Dawson, and S. Krishna, "Long Wavelength InAs/GaSb Superlattice Detectors Based on nBn and pin Design," *Proceedings of SPIE* 7298, 72981P, 2009.

87. E. Plis, S. Annamalai, K. T. Posani, and S. Krishna, "Midwave Infrared Type-II InAs/GaSb Superlattice Detectors with Mixed Interfaces," *Journal of Applied Physics* 100, 014510, 2006.

88. H. S. Kim, E. Plis, J. B. Rodriguez, G. Bishop, Y. D. Sharma, and S. Krishna, "N-Type Ohmic Contact on Type-II InAs/GaSb Strained Layer Superlattices," *Electronics Letters* 44, 881–82, 2008.

89. H. S. Kim, E. Pils, J. B. Rodriquez, G. D. Bishop, Y. D. Sharma, L. R. Dawson, S. Krishna, et al., "Mid-IR Focal Plane Array Based on Type-II InAs/GaSb Strain Layer Superlattice Detector with *nBn* Design," *Applied Physics Letters* 92, 183502, 2008.

90. C. J. Hill, A. Soibel, S. A. Keo, J. M. Mumolo, D. Z. Ting, S. D. Gunapala, D. R. Rhiger, R. E. Kvaas, and S. F. Harris, "Demonstration of Mid and Long-Wavelength Infrared Antimonide-Based Focal Plane Arrays," *Proceedings of SPIE* 7298, 729404, 2009.

18 Quantum Dot Infrared Photodetectors

The success of quantum well structures for infrared detection applications has stimulated the development of quantum dot infrared photodetectors (QDIPs). In general, QDIPs are similar to QWIPs but with the quantum wells replaced by quantum dots (QDs), which have size confinement in all spatial directions.

Recent advances in the epitaxial growth of strained heterostructures, such as InGaAs on GaAs, have led to the realization of coherent islands through the process of self-organization. These islands behave electronically as quantum boxes, or QDs. Zero-dimensional quantum confined semiconductor heterostructures have been investigated theoretically and experimentally for some time [1–3]. At present, nearly defect-free quantum dot devices can be fabricated reliably and reproducibly. Also new types of infrared photodetectors taking advantage of the quantum confinement obtained in semiconductor heterostructures have emerged. Like the QWIPs, the QDIPs are based on optical transitions between bound states in the conduction (valence) band in QDs. Also, like the QWIPs, they benefit from a mature technology with large bandgap semiconductors.

First observations of intersublevel transitions in the far infrared were reported in the early 1990s, either in InSb-based electrostatically defined QDs [4] or in structured two-dimensional (2-D) electron gas [5]. The first QDIP was demonstrated in 1998 [6]. Great progress has been made in their development and performance characteristics [7–9] and in their applications to thermal imaging focal plane arrays [10].

Interest in quantum dot research can be traced back to a suggestion by Arakawa and Sakaki in 1982 [1] that the performance of semiconductor lasers could be improved by reducing the dimensionality of the active regions of these devices. Initial efforts at reducing the dimensionality of the active regions focused on using ultrafine lithography coupled with wet or dry chemical etching to form 3-D structures. However, it was soon realized that this approach introduced defects (high density of surface states) that greatly limited the performance of such QDs. Initial efforts were mainly focused on the growth of InGaAs nanometer-sized islands on GaAs substrates. In 1993, the first epitaxial growth of defect-free quantum-dot nanostructures was achieved by using MBE [11]. Most of the practical quantum-dot structures today are synthesized both by MBE and MOCVD.

18.1 QDIP PREPARATION AND PRINCIPLE OF OPERATION

Under certain growth conditions, when the film with the larger lattice constant exceeds a certain critical thickness, the compressive strain within the film is relieved by the formation of coherent island. Figure 18.1 qualitatively shows the changes in the total energy of a mismatched system versus time [12]. The plot can be divided into three sections: period a (2-D deposition), period b (2-D–3-D transition) and period c (ripening of islands). In the beginning of the deposition a 2-D layer by layer mechanism leads to a perfect wetting of the substrate. At the point t_{cw} (the critical wetting layer thickness) the stable 2-D growth enters into an area of the metastable growth. A supercritically thick wetting layer builds up and the epilayer is potentially ready to undergo a transition toward a Stranski–Krastanow morphology [13]. This transition starts around point X in Figure 18.1 and its dynamic depends primarily on the height of the transition barrier E_a. It is presumed that further growth continues without materials supply, simply by consuming the excess material accumulated in the supercritically thick wetting layer. Between the points Y and Z (ripening: period c) the process has lost most of the excess energy; the mobile material is consumed as a result of the potential differences between smaller and larger islands. These islands may be QDs.

Coherent quantum-dot islands are generally formed only when the growth proceeds as Stranski-Krastanow growth model [13]. The onset of the transformation of the growth process from a 2-D layer-by-layer growth mode to a 3-D island growth mode results in a spotty RHEED pattern. This is, in contrast to the conventional streaky pattern, generally observed for the layer-by-layer growth mode. The transition typically occurs after the deposition of a certain number of monolayers. For InAs on GaAs, this transition occurs after about 1.7 monolayers of InAs have been grown; this is the onset of islanding and, hence, quantum-dot formation. Noncoherent islands are typically produced by too high materials supply and contain misfit dislocations at the interface.

The detection mechanism of QDIP is based on the intraband photoexcitation of electrons from confined states in the conduction band wells or dots into the continuum. The emitted electrons drift toward the collector in the electric field provided by the applied bias and photocurrent is created. It is assumed that the potential profile at the conduction band edge along the growth direction have a similar shape as for QWIP shown in Figure 16.5. In practice, since the dots are spontaneously self-assembled during growth, they are not correlated between multilayers in active region.

Two types of QDIP structures have been proposed: conventional structure (vertical, see Figure 18.2) and lateral structure (Figure 18.3). In a vertical QDIP, the photocurrent is collected through the vertical transport of carriers between top and bottom contacts. The device heterostructure comprises repeated InAs QD layers buried between GaAs barriers with top and bottom contact layers at active region boundaries. The mesa height can vary from 1 to 4 μm depending on the device heterostructure. The QDs are directly doped (usually with silicon) in order to provide free

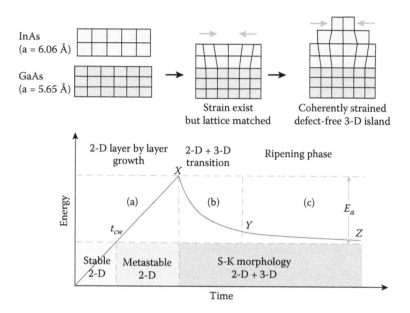

Figure 18.1 Schematics of total energy versus time for the 2-D–3-D morphology transition. t_{cw} is the critical wetting layer thickness, E_a is the barrier for formation of 3-D islands, X is the point where a pure strain-induced transition becomes. Between Y and Z a slow ripening process continues. (From Seifert, W., Carlsson, N., Johansson, J., Pistol, M.-E., and Samuelson, L., *Journal of Crystal Growth*, 170, 39–46, 1997.)

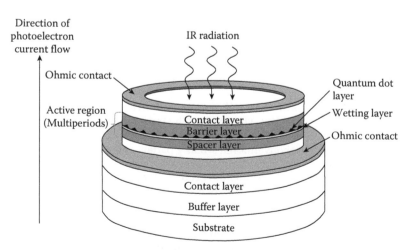

Figure 18.2 Schematic diagram of conventional quantum dot detector structure.

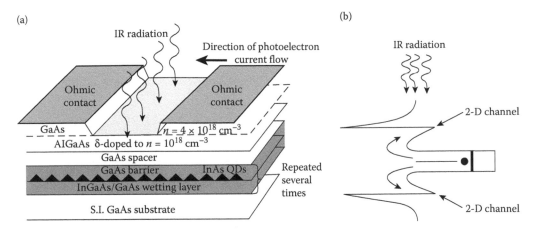

Figure 18.3 Schematic diagram (a) of lateral quantum dot detector structure with a representation (b) of the conduction band profile and photoresponse mechanism.

carriers during photoexcitation, and an AlGaAs barrier can be included in the vertical device heterostructure in order to block dark current created by thermionic emission [14,15].

The lateral QDIP (Figure 18.3a) collects photocurrent through transport of carriers across a high-mobility channel between two top contacts, operating much like a field-effect transistor. As previously, again AlGaAs barriers are present, but instead of blocking the dark current, these barriers are used to both modulation dope the QDs and to provide the high-mobility channel (see Figure 18.3b). The normally incident IR radiation promotes the carriers from the dots to the continuum from where they are quickly transferred to the high mobility 2-D channel on either side, due to the favorable band bending. Lateral QDIPs have demonstrated lower dark currents and higher operating temperatures than vertical QDIPs since the major components of the dark current arise from interdot tunneling and hopping conduction [16]. However, these devices will be difficult to incorporate into a FPA hybrid-bump bonded to a silicon readout circuit. Because of this, more effort is directed to improve the performance of vertical QDIPs, which are more compatible with commercially available readout circuits.

In addition to the standard InAs/GaAs QDIP, several other heterostructure designs have been investigated for use as IR photodetectors [7,8]. An example is InAs QDs embedded in a strain-relieving InGaAs quantum well, which are known as dot-in-a-wall (DWELL) heterostructures (see Figure 18.4) [10,17,18]. This device offers two advantages: challenges in wavelength tuning through dot-size control can be compensated in part by engineering the quantum well sizes, which can be controlled precisely and quantum wells can trap electrons and aid in carrier capture by QDs, thereby facilitating ground state refilling. Figure 18.4b shows DWELL spectral tuning by varying well geometry.

18.2 ANTICIPATED ADVANTAGES OF QDIPs

The quantum-mechanical nature of QDIPs leads to several advantages over QWIPs and other types of IR detectors that are available. As in the HgCdTe, QWIP and type II superlattice technologies, QDIPs provide multiwavelength detection. However, QDs provide many additional parameters for tuning the energy spacing between energy levels, such as QD size and shape, strain, and material composition.

The potential advantages in using QDIPs over quantum wells are as follows:

- Intersubband absorption may be allowed at normal incidence (for n-type material). In QWIPs, only transitions polarized perpendicularly to the growth direction are allowed, due to absorption selection rules. The selection rules in QDIPs are inherently different and normal incidence absorption is observed.

- Thermal generation of electrons is significantly reduced due to the energy quantization in all three dimensions. As a result, the electron relaxation time from excited states increases due to phonon bottleneck. Generation by LO phonons is prohibited unless the gap between the discrete energy levels equals exactly to that of the phonon. This prohibition does not apply to

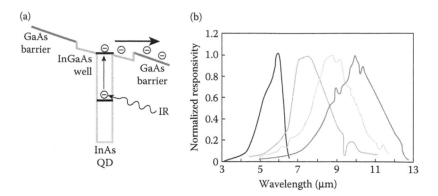

Figure 18.4 DWELL infrared detector: (a) the operation mechanism, and (b) experimentally measured spectral tunability by varying well width from 55 to 100 Å. (From Gunapala, S. D., Bandara, S. V., Hill, C. J., Ting, D. Z., Liu, J. K., Rafol, B., Blazejewski, E. R., et al., *IEEE Journal of Quantum Electronics*, 43, 230–37, 2007. With permission.)

quantum wells, since the levels are quantized only in the growth direction and a continuum exists in the other two directions (hence generation–recombination (g–r) by LO phonons with capture time of a few picoseconds). Thus, it is expected that S/N ratio in QDIPs will be significantly larger than that of QWIPs.

- Lower dark current of QDIPs is expected than of QWIPs due to 3-D quantum confinement of the electron wavefunction.

Both the increased electron lifetime and the reduced dark current indicate that QDIPs should be able to provide high temperature operation. In practice, however, it has been a challenge to meet all of the above expectations.

Carrier relaxation times in QDs are longer than the typical 1–10 ps measured for quantum wells. It is predicted that the carrier relaxation time in QDs is limited by electron-hole scattering [19], rather than phonon scattering. For QDIPs, the lifetime is expected to be even larger, greater than 1 ns, since the QDIPs are majority carrier devices due to absence of holes.

The main disadvantage of the QDIP is the large inhomogeneous linewidth of the quantum-dot ensemble variation of dot size in the Stranski–Krastanow growth mode [20,21]. As a result, the absorption coefficient is reduced, since it is inversely proportional to the ensemble linewidth. Large, inhomogeneously broadened linewidth has a deleterious effect on QDIP performance. Subsequently, the quantum efficiency QD devices tend to be lower than what is predicted theoretically. Vertical coupling of quantum-dot layers also reduces the inhomogeneous linewidth of the quantum-dot ensemble; however, it may also increase the dark current of the device, since carriers can tunnel through adjacent dot layers more easily. As in other types of detectors, nonuniform dopant incorporation adversely affects the performance of the QDIP. Therefore, improving QD uniformity is a key issue in the increasing absorption coefficient and improving the performance. Thus, the growth and design of unique QD heterostructure is one of the most important issues related to achievement of state-of-the art QDIP performance.

18.3 QDIP MODEL

In further considerations a QDIP model developed by Ryzhii et al. is adapted [22,23]. The QDIP consists of a stack of QD layers separated by wide-gap material layers (see Figure 18.5). Each QD layer includes periodically distributed identical QDs with the density Σ_{QD} and sheet density of doping donors equal to Σ_D. In the realistic QDIPs, the lateral size of QDs, a_{QD}, is sufficiently large in comparison with the transverse size, h_{QD}. Consequently, only two energy levels associated with the quantization in the transverse direction exist. Relatively sufficiently large lateral size, l_{QD}, causes a large number of bound states in dots and, consequently, is capable of accepting a large number of electrons. Whereas, the transverse size is small in comparison with the spacing between the QD layers, L. The lateral spacing between QDs is equal to $L_{QD} = \sqrt{\Sigma_{QD}}$. The average number of electrons in a QD belonging to the k-th QD layer, $<N_k>$, can be indicated by a solitary

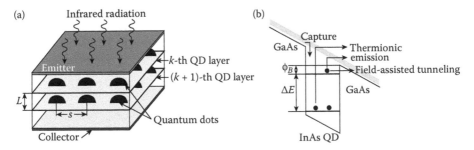

Figure 18.5 Schematic view of (a) the quantum dot structure, and (b) conduction band structure of the dot.

Table 18.1: Typical Parameter Values of QDIP Fabricated From GaAs or InGaAs

a_{QD}	h	Σ_{QD}	Σ_D	L	K	N_{QD}
10–40 nm	4–8 nm	$(1–10)10^{10}$ cm^{-2}	$(0.3–0.6)\Sigma_{QD}$	40–100 nm	10–70	8

QD layer index ($k = 1, 2, ..., K$, where K is the total number of the QD layers). The QDIP active region (the stack of QD arrays) is sandwiched between two heavily doped regions that serve as the emitter and collector contacts.

Due to the discrete nature of QDs, the fill factor F should be included for optical absorption in QDs. This factor can be estimated in a simple way as

$$F = \frac{\sqrt[3]{V}}{s},\tag{18.1}$$

where V is the quantum dot volume.

For self-assembled QDs, a Gaussian distribution has been observed for the electronic and optical spectra. Phillips modeled the absorption spectra for an ensemble of QDs using a Gaussian line shape in the shape [20]

$$\alpha(E) = \alpha_o \frac{n_1}{\delta} \frac{\sigma_{QD}}{\sigma_{ens}} \exp\left[-\frac{(E - E_g)^2}{\sigma_{ens}^2} \right],\tag{18.2}$$

where α_o is the maximum absorption coefficient, n_1 is the areal density of electrons in the quantum dot ground state, δ is the quantum dot density, and $E_g = E_2 - E_1$ is the energy of the optical transition between ground and excited states in the QDs. It should be noticed that Equation 18.2 estimates the absorption coefficient for the necessary presents of electrons in the QD ground state.

The optical absorption between the ground and excited levels is found to have a value [24]

$$\alpha_o \approx \frac{3.5 \times 10^5}{\sigma}, \text{ in cm}^{-1}\tag{18.3}$$

where σ is the linewidth of the transition in meV. Equation 18.3 indicates the trade-off between the absorption coefficient and the absorption linewidth, σ.

The expressions σ_{QD} and σ_{ens} are the standard deviations in the Gaussian line shape for intraband absorption in a single quantum dot and for the distribution in energies for the QD ensemble, respectively. The terms n_1/δ and σ_{QD}/σ_{ens} describe a decrease in absorption due to absence of available electrons in the QD ground state and inhomogeneous broadening, respectively.

Table 18.1 contains the reference values of QD parameters. These values are considered for a QDIP fabricated from GaAs or InGaAs. The self-assembled dots formed by epitaxial growth are typically pyramidal to lens shaped with a base dimension of 10–20 nm and height of 4–8 nm with an areal density determined to be 5×10^{10} cm^{-2} using atomic force microscopy.

Similar with QWIP, the main mechanism producing the dark current in the QDIP device is the thermionic emission of the electrons confined in the QDs. The dark current can be given by

$$J_{dark} = e\upsilon n_{3\text{-}D}, \tag{18.4}$$

where υ is the drift velocity and $n_{3\text{-}D}$ is the three-dimensional density, both for electrons in the barrier [25]. Equation 18.4 neglects the diffusion contribution. The electron density can be estimated by

$$n_{3\text{-}D} = 2\left(\frac{m_b kT}{2\pi\hbar^2}\right)^{3/2} \exp\left(-\frac{E_a}{kT}\right), \tag{18.5}$$

where m_b is the barrier effective mass and E_a is the activation energy, which equals the energy difference between the top of the barrier and the Fermi level in the dot. At higher operating temperature and larger bias voltage, the contribution of field-assisted tunneling through a triangular potential barrier is considerable [26,27].

Figure 18.6 shows, for example, the normalized dark current versus bias for temperature range 20–300 K for QDIP with AlGaAs confinement layers below the QD layer and on top of the GaAs cap layers [8]. In such a case, we have the InAs islands into a quantum wells and AlGaAs blocking layers effectively improve the dark current and detectivity. As it is shown, at low temperature (e.g., 20 K), the dark current increased rapidly as the bias was increased, which is attributed to electron tunneling between the QDs. For higher bias $|0.2| \le V_{bias} \le |1.0|$, the dark current increases slowly. With further increase in bias, the dark current strongly increases, which was largely due to lowering of the potential barriers. Figure 18.6 also shows the photocurrent induced by the room temperature background. It is clear that BLIP temperature varies with bias.

As Equation 9.1 describes, the photocurrent is determined by quantum efficiency and gain, g. The photoconductive gain is defined as the ratio of total collected carriers to total excited carriers,

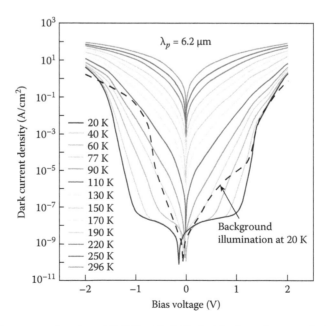

Figure 18.6 Dark current density of QDIP with AlGaAs blocking layer including photocurrent induced by room-temperature background. (From Campbell, J. C., and Madhukar, A., "Quantum-Dot Infrared Photodetectors," *Proceedings of IEEE* 95, 1815–27, 2007. With permission.)

whether these carriers are thermally generated or photogenerated. Usually in photoconductors, the gain is greater than one since the carrier lifetime τ_e, exceeds the carrier transit time τ_t, through the device between contacts

$$g_{ph} = \frac{\tau_e}{\tau_t}. \tag{18.6}$$

In InAs/GaAs QDIPs, the gain has typical values in the 1–5. However, the gain strongly depends on QDIP design and detector polarization. Much higher values, up to several thousands, have been observed [8,21]. The higher gain of the QDIPs in comparison with QWIPs (typically in the range 0.1–50 for similar electric field intensities) is the result of longer carrier lifetimes. The larger photoconductive gain has influence on higher current responsivity (see Equation 3.33).

The photoconductive gain and the noise gain in conventional photoconductive detector are equal to each other. It is not the same in QDIPs since these devices are not homogeneous, nor are they bipolar devices. The photoconductive gain in QWIPs is expressed in terms of the capture probability p_c as [28,29]

$$g_{ph} = \frac{1 - p_c/2}{N p_c}, \tag{18.7}$$

where $p_c \ll 1$ and N is the number of quantum well layers. This equation is approximately correct for QDs after including the fill factor F, in the denominator that takes into account the surface density of discrete dots across the single layer [30]. Then

$$g_{ph} = \frac{1 - p_c/2}{N p_c F}. \tag{18.8}$$

Ye et al. [31] have estimated an average value of F as equal to 0.35. The recently published paper by Lu and colleagues indicate [32] that temperature dependent photoresponsivity is attributable to temperature dependent electron capture probability. The capture probability can be changed in a wide region, from below 0.01 to above 0.1 in dependence on bias voltage and temperature.

The noise current of QDIP contains both g–r noise current and thermal noise (Johnson noise) current

$$I_n^2 = I_{ng-r}^2 + I_{nJ}^2 = 4q g_n I_d \Delta f + \frac{4kT}{R} \Delta f, \tag{18.9}$$

where R is the differential resistance of the QDIP, which can be extracted from the slope of the dark current.

It can be shown that the noise gain is related to the electron capture probability p_c, as

$$g_n = \frac{1}{N p_c F}. \tag{18.10}$$

In typical QDIP, the thermal noise is significant in the very low bias region. For example, Figure 18.7 shows the bias dependence of the noise current at 77, 90, 105, 120, and 150 K and a measurement frequency of 140 Hz for InAs/GaAs QDIP [31]. The calculated thermal noise current is also shown at 77 K. Thermal noise is significant in the very low bias region $|V_{bias}| \leq 0.1$ V. As the bias increases, the detector noise current increases much faster than thermal noise and it is primarily g–r noise.

The larger photoconductive gain has influence on higher current responsivity

$$R_i = \frac{q\lambda}{hc} \eta g_{ph}. \tag{18.11}$$

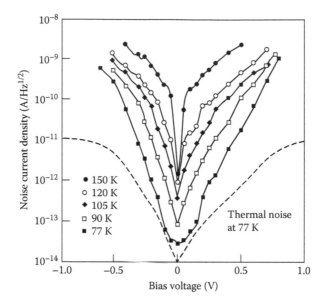

Figure 18.7 Noise current density versus bias voltage at 77, 90, 105, 120, and 150 K. The symbols are measured data. The dashed line is calculated thermal noise current at 77 K. (From Ye, Z., Campbell, J. C., Chen, Z., Kim, E. T., and Madhukar, A., *Applied Physics Letters*, 83, 1234–36, 2003. With permission.)

Detectivity is defined as the rms signal-to-noise ratio in a 1-Hz bandwidth per unit rms incident radiant power per square root of detector area A_d, and can be determined as

$$D^* = \frac{(A_d \Delta f)^{1/2}}{I_n} R_i = \frac{q\lambda}{hc} \frac{\eta g_{ph}}{\left(I_{ng-r}^2 + I_{nJ}^2\right)^{1/2}} (A_d \Delta f)^{1/2}. \tag{18.12}$$

The quantum efficiency often measured in practice is low, typically ≈2%. It should be noticed, however, that rapid progress has recently been made in the performance of QDIP devices, especially at near room temperature. Lim et al. [33] have announced a quantum efficiency of 35% for detectors with peak detection wavelength around 4.1 μm.

Figure 18.8 shows a typical photoconductivity spectrum of an InAs/GaAs vertical QDIP [25]. The wedge coupling geometry is shown in the inset. The S polarization corresponds to an electric field in the layer plane while the electric filed of the P-polarized excitation has a component along the z growth axis and in the layer plane. Clearly the P-polarized response is much stronger than that for S-polarized and the results reported in Figure 18.8 indicate that the photoresponse of a QDIP can be easily measured for an in-plane polarized excitation (i.e., operation at normal incidence is feasible).

The photoconductivity spectra is believed to be due to the fact that self-assembled QDs grown so far for QDIPs are wide in the plane direction (~20 nm) and narrow in the growth direction (~3 nm) [25]. The strong confinement is therefore in the growth direction; while the in-plane confinement is weak, resulting in several levels in the dots (see Figure 16.6). The transitions between in-plane confined levels give rise to the normal incidence response.

Analysis of the fundamental performance limitation of QDIPs carried out by Phillips indicates that the predicted performance of very uniform QDIP [when $\sigma_{ens}/\sigma_{QD} = 1$] rivals with HgCdTe (see Figure 18.9 [20]). However, as was mentioned previously, poor QDIP performance is generally linked to two sources: nonoptimal band structure and nonuniformity in QD size. In the Phillips analysis, two electron energy levels are assumed, where the excited state coincides with the conduction band minimum of the barrier material. If the excited state is below the barrier conduction band, photocurrent is difficult to extract, which is reflected in low responsivity and detectivity. Also usually, QDs contain additional energy levels between the excited and ground state transitions. If these states are similar to the thermal excitations or permit phonon scattering between

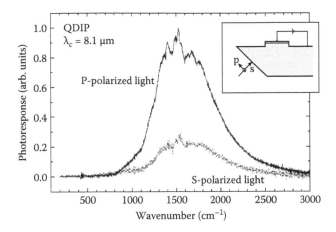

Figure 18.8 P- and S-polarized spectral response curves in the 45° facet detector geometry. The QDIPs have 50 layers of InAs dots separated by 30 nm GaAs barriers. The dot density is about 5×10^9 cm^{-2}. The number of electrons is estimated to be one per dot. (From Liu, H. C., *Opto-Electronics Review*, 11, 1–5, 2003. With permission.)

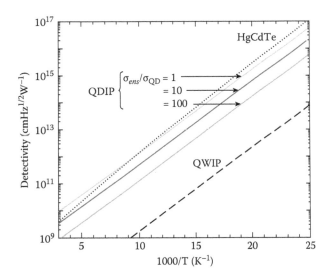

Figure 18.9 Detectivity of HgCdTe, QWIP, and QDIP detectors as a function of temperature with a bandgap energy corresponding to wavelength 10 μm. (From Phillips, J., *Journal of Applied Physics*, 91, 4590–94, 2002. With permission.)

levels, carrier lifetime will be dramatically reduced. In consequence a large increase in dark current and reduction in detectivity are observed. In the case of Stranski-Krastanow growth mode some degradation of self-assembled QDs occurs due to a coupling 2-D "wetting layer."

Fabrication of the QDs also strongly affect absorption properties in the material. If lateral quantum confinement in QDs is small and they more closely resemble QWs, sensitivity to normal incidence radiation is decreased, resulting in an increase in dark current and reduced detectivity. At the current state of QD fabrication technology, the inhomogeneous broadening of QD energy levels modeled by values of σ_{ens}/σ_{QD} is about 100. Moreover, the experimentally measured quantum efficiency is low, several percentages, and has not come to equal that of QWIP. Thus, size uniformity of QDs needs to be improved together with the ability to increase the QD density to improve QDIP's performance.

18.4 PERFORMANCE OF QDIPS

The present status and possible future developments of QDIPs are reviewed by Martyniuk and Rogalski [34]. The performance of QDIPs as compared to other types of infrared photodetectors is also presented in this paper. Our considerations below follow Martyniuk and Rogalski.

18.4.1 R_oA Product

In spite of QDIP being a photoconductor and HgCdTe photodiode, it is interesting to compare their dark currents and incremental resistances. At the present stage of technology development, the dark currents of both MWIR detectors in the region of low bias voltages are comparable [9]. Figure 18.10 displays the dependence of the R_oA product on a wavelength. The QDIP data was determined from dynamic resistance in I-V characteristics at operating bias. Only limited experimental R_oA values for QDIPs marked in Figure 18.10 are available in literature. The highest measured R_oA values for HgCdTe photodiodes operated at 78 K with about 5 μm cutoff wavelength are located between 10^8 and 10^9 Ωcm². The solid line is theoretical R_oA for HgCdTe photodiodes, calculated using a 1-D model that assumes diffusion current from narrower bandgap n-side is dominant, and minority carrier recombination via Auger and radiative process. Theoretical calculations used typical values for the p-side donor concentration ($N_d = 1 \times 10^{15}$ cm^{-3}) and the narrow bandgap active layer thickness (10 μm).

The R_oA product is inherent property of the HgCdTe ternary alloy and depends on cutoff wavelength. The dark current of photodiodes increases with cutoff wavelength, which is an important difference with QDIPs, where dark current is far less sensitive to wavelength and depends on device geometry.

18.4.2 Detectivity at 78 K

A useful figure of merit, for comparing detector performance, is thermally limited detectivity. In the case of photodiodes, this parameter is defined by Equation 3.57. However, for photoconductors the situation is more complicated due to different contribution of thermal noise and g–r noise. As it is discussed above, the noise in QDIP originates from the trapping processes in the QDs and is a more complicated function of detector design and capture probability. As a result, the detectivity depends on several specific quantities, such as the quantum efficiency, photoconductive gain, and contribution of noise current (see Equation 18.12).

Figure 18.11 compares the highest measurable detectivities at 77 K of QDIPs found in literature with the predicted detectivities of P-on-n HgCdTe and type II InAs/GaInSb SLS photodiodes. The solid lines are theoretical thermal limited detectivities for HgCdTe photodiodes, calculated using a 1-D model that assumes diffusion current from narrower bandgap n-side is dominant, and

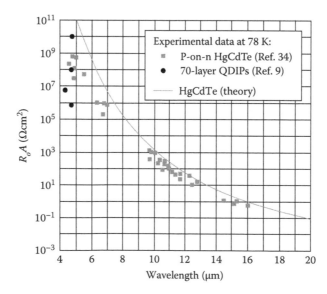

Figure 18.10 R_oA versus wavelength for P-on-n HgCdTe photodiodes and QDIPs at 78 K. Solid line is calculated theoretically assuming 1-D n-side diffusion model.

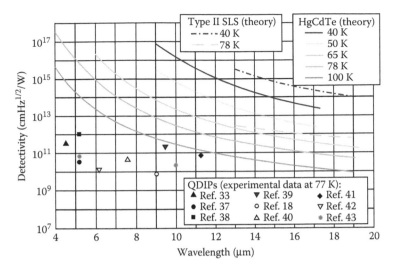

Figure 18.11 The predicted detectivity of P-on-n HgCdTe and type II InAs/GaInSb SLS photo-diodes, compared with measured QDIP detectivities at 77 K.

minority carrier recombination via Auger and radiative process. In calculations of typical values for the p-side donor concentration ($N_d = 1 \times 10^{15}$ cm^{-3}), the narrow bandgap active layer thickness (10 μm), and quantum efficiency (60%) have been used. It should be insisted that for HgCdTe photodiodes theoretically predicted curves for temperature range between 50 and 100 K coincide very well with experimental data (not shown in Figure 18.11). The predicted thermally limited detectivities of the type II SLS are larger than those for HgCdTe [35].

The measured value of QDIPs' detectivities at 77 K gathered in Figure 18.11 indicate that QD device detectivities are as yet considerably inferior to current HgCdTe detector performance. In LWIR region, the upper experimental QDIP data at 77 K coincide with HgCdTe ones at temperature 100 K.

18.4.3 Performance at Higher Temperature

One of the main potential advantages of QDIPs is low dark current. In particular, the lower dark currents enable higher operating temperatures. Up until now, however, most of the QDIP devices reported in the literature have been working in the temperature range of 77–200 K. On account of this fact, it is an interesting insight on achievable QDIP performance in temperature range above 200 K in comparison with other type of detectors.

Most modern infrared FPAs are fabricated from the hybrid devices—a detector array made from compound semiconductor materials and a silicon signal processing chip called a readout integrated circuit (ROIC). To receive high injection efficiency, the input impedance of the MOSFET must be much lower than the internal dynamic resistance of the detector at its operating point, and the condition (19.2) should be fulfilled. Generally, it is not a problem to fulfil the inequality (19.2) for SWIR and MWIR FPAs where the dynamic resistance of detector R_d, is large, but it is very important for LWIR designs where R_d is low. There are more complex injection circuits that effectively reduce the input impedance and allow lower detector resistance to be used.

The above requirement is especially critical for near-room temperature HgCdTe photodetectors operating in LWIR region. Their resistance is very low due to a high thermal generation. In materials with a high electron to hole ration as HgCdTe, the resistance is additionally reduced by ambipolar effects. Small size uncooled 10.6 μm photodiodes (50 × 50 μm²) exhibit less than 1 Ω zero bias junction resistances that are well below the series resistance of a diode [44]. As a result, the performance of conventional devices is very poor, so they are not usable for practical applications. To fulfill inequality (Equation 19.2) to effectively couple the detector with silicon readout, the detector incremental resistance should be $R_d \gg 2\Omega$. The saturation current for 10 μm photodiode achieves 1000 A/cm² and it is by four orders of magnitude larger than the photocurrent due to the 300 K background radiation. The potential advantages of QDIPs is considerably lower dark current and higher R_oA product in comparison with HgCdTe photodiodes (see Figure 18.12).

Figure 18.13 compares the calculated thermal detectivity of HgCdTe photodiodes and QDIPs as a function of wavelength and operating temperature with the experimental data of uncooled

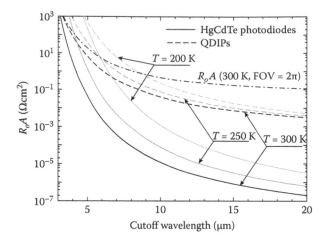

Figure 18.12 R_oA product of HgCdTe photodiodes and QDIPs as a function of wavelength. The calculations for HgCdTe photodiodes have been performed for the optimized doping concentration $p = \gamma^{1/2}n_i$.

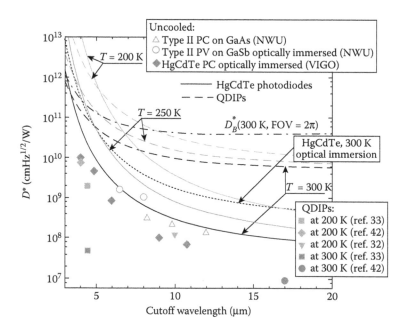

Figure 18.13 Calculated performance of Auger generation-recombination limited HgCdTe photodetectors as a function of wavelength and operating temperature. BLIP detectivity has been calculated for 2π FOV, the background temperature is $T_{BLIP} = 300$ K, and the quantum efficiency $\eta = 1$. The calculations for HgCdTe photodiodes have been performed for the optimized doping concentration $p = \gamma^{1/2}n_i$. The experimental data is taken for commercially available uncooled HgCdTe photoconductors (produced by Vigo System) and uncooled type-II detectors at the Center for Quantum Devices, Northwestern University (Evanston, Illinois). The experimental data for QDIPs are gathered from the marked literature for detectors operated at 200 and 300 K.

HgCdTe and type-II InAs/GaInSb SLS detectors. The Auger mechanism is likely to impose fundamental limitations to the LWIR HgCdTe detector performance. The calculations have been performed for optimized doping concentration $p = \gamma^{1/2}n_i$. The experimental data for QDIPs are gathered from the literature for detectors operated at 200 and 300 K.

Uncooled LWIR HgCdTe photodetectors are commercially available and manufactured in significant quantities, mostly as single-element devices [44]. They have found important applications in

IR systems that require fast response. The results presented in Figure 18.13 confirm that the type-II superlattice is a good candidate for IR detectors operating in the spectral range from the mid-wavelength to the very long-wavelength IR. However, comparison of QDIP performance both with HgCdTe and type II superlattice detectors gives evidence that the QDIP is suitable for high temperature. Especially encouraging results have been achieved for very long-wavelength QDIP devices with a double-barrier resonant tunneling filter with each quantum-dot layer in the absorption region [45]. In this type of device, photoelectrons are selectively collected from the QDs by resonant tunneling, while the same tunnel barriers block electrons of dark current due to their broad energy distribution. For the 17 μm detector, a peak detectivity of $8.5 \times 10^6 \ cmHz^{1/2}/W$ has been measured. Up until now, this novel device demonstrates the highest performance of room-temperature photodetectors. Further improvement in technology and design can result in application of QDIPs in room temperature focal plane arrays with the advantages of larger operating speed (shorter frame time) in comparison with thermal detectors (bolometers and pyroelectric devices).

Thermal detectors seem to be unsuitable for the next generation of IR thermal imaging systems, which are moving toward faster frame rates and multispectral operation. A response time much shorter than that achievable with thermal detectors is required for many nonimaging applications. Improvement in technology and design of QDIP detectors make it possible to achieve both high sensitivity and fast response at room temperature.

The R_0A product is inherent property of the HgCdTe ternary alloy and depends on cutoff wavelength. The dark current of photodiodes increases with cutoff wavelength, which is an important difference with QDIPs, where dark current is far less sensitive to wavelength and depends on device geometry.

REFERENCES

1. Y. Arakawa and H. Sakaki, "Multidimensional Quantum-Well Laser and Temperature Dependence of Its Threshold Current," *Applied Physics Letters* 40, 939–41, 1982.

2. M. Asada, Y. Miyamoto, and Y. Suematsu, "Gain and Threshold of Three Dimensional Quantum-Box Lasers," *IEEE Journal of Quantum Electronics* QE-22, 1915–21, 1986.

3. D. Bimberg, M. Grundmann, and N. N. Ledentsov, *Quantum Dot Heterostructures*, Wiley, Chichester, 1999.

4. Ch. Sikorski and U. Merkt, "Spectroscopy of Electronic States in InSb Quantum Dots," *Physical Review Letters* 62, 2164–67, 1989.

5. T. Demel, D. Heitmann, P. Grambow, and K. Ploog, "Nonlocal Dynamic Response and Level Crossings in Quantum-Dot Structures," *Physical Review Letters* 64, 788–91, 1990.

6. J. Phillips, K. Kamath, and P. Bhattacharya, "Far-Infrared Photoconductivity in Self-Organized InAs Quantum Wells," *Applied Physics Letters* 72, 2020–21, 1998.

7. P. Bhattacharya and Z. Mi, "Quantum-Dot Optoelectronic Devices," *Proceedings of IEEE* 95, 1723–40, 2007.

8. J. C. Campbell and A. Madhukar, "Quantum-Dot Infrared Photodetectors," *Proceedings of IEEE* 95, 1815–27, 2007.

9. P. Bhattacharya, A. D. Stiff-Roberts, and S. Chakrabarti, "Mid-Infrared Quantum Dot Photoconductors," in *Mid-Infrared Semiconductor Optoelectronics*, ed. A. Krier, 487–513, Springer Verlag, Berlin, 2007.

10. S. Krishna, S. D. Gunapala, S. V. Bandara, C. Hill, and D. Z. Ting, "Quantum Dot Based Infrared Focal Plane Arrays," *Proceedings of IEEE* 95, 1838–52, 2007.

11. D. Leonard, M. Krishnamurthy, C. M. Reaves, S. P. Denbaars, and P. M. Petroff, "Direct Formation of Quantum-Sized Dots from Uniform Coherent Islands of InGaAs on GaAs Surface," *Applied Physics Letters* 63, 3203–5, 1993.

12. W. Seifert, N. Carlsson, J. Johansson, M-E. Pistol, and L. Samuelson, "In Situ Growth of Nano-Structures by Metal-Organic Vapour Phase Epitaxy," *Journal of Crystal Growth* 170, 39–46, 1997.

13. I. N. Stranski and L. Krastanow, "Zur theorie der orientierten ausscheidung von Ionenkristallen aufeinander," *Sitzungsberichte d. Akad. d. Wissenschaften in Wein. Abt. IIb,* Vol. 146, 797–810, 1937.

14. S. Y. Wang, S. D. Lin, W. Wu, and C. P. Lee, "Low Dark Current Quantum-Dot Infrared Photodetectors with an AlGaAs Current Blocking Layer," *Applied Physics Letters* 78, 1023–25, 2001.

15. V. Ryzhii, "Physical Model and Analysis of Quantum Dot Infrared Photodetectors with Blocking Layer," *Journal of Applied Physics* 89, 5117–24, 2001.

16. S. W. Lee, K. Hirakawa, and Y. Shimada, "Bound-to-Continuum Intersubband Photoconductivity of Self-Assembled InAs Quantum Dots in Modulation-Doped Heterostructures," *Applied Physics Letters* 75, 1428–30, 1999.

17. S. Krishna, "Quantum Dots-in-a-Well Infrared Photodetectors," *Journal of Physics D: Applied Physics* 38, 2142–50, 2005.

18. S. D. Gunapala, S. V. Bandara, C. J. Hill, D. Z. Ting, J. K. Liu, B. Rafol, E. R. Blazejewski, et al., "640 × 512 Pixels Long-Wavelength Infrared (LWIR) Quantum-Dot Infrared Photoconductor (QDIP) Imaging Focal Plane Array," *IEEE Journal of Quantum Electronics* 43, 230–37, 2007.

19. I. Vurgaftman, Y. Lam, and J. Singh, "Carrier Thermalization in Sub-Three-Dimensional Electronic Systems: Fundamental Limits on Modulation Bandwidth in Semiconductor Lasers," *Physical Review* B50, 14309–26, 1994.

20. J. Phillips, "Evaluation of the Fundamental Properties of Quantum Dot Infrared Detectors," *Journal of Applied Physics* 91, 4590–94, 2002.

21. E. Towe and D. Pan, "Semiconductor Quantum-Dot Nanostructures: Their Application in a New Class of Infrared Photodetectors," *IEEE Journal of Selected Topics in Quantum Electronics* 6, 408–21, 2000.

22. V. Ryzhii, I. Khmyrova, V. Pipa, V. Mitin, and M. Willander, "Device Model for Quantum Dot Infrared Photodetectors and Their Dark-Current Characteristics," *Semiconductor Science and Technology* 16, 331–38, 2001.

23. V. Ryzhii, I. Khmyrova, V. Mitin, M. Stroscio, and M. Willander, "On the Detectivity of Quantum-Dot Infrared Photodetectors," *Applied Physics Letters* 78, 3523–25, 2001.

24. J. Singh, *Electronic and Optoelectronic Properties of Semiconductor Structures*, Cambridge University Press, New York, 2003.

25. H. C. Liu, "Quantum Dot Infrared Photodetector," *Opto-Electronics Review* 11, 1–5, 2003.

26. J.-Y. Duboz, H. C. Liu, Z. R. Wasilewski, M. Byloss, and R. Dudek, "Tunnel Current in Quantum Dot Infrared Photodetectors," *Journal of Applied Physics* 93, 1320–22, 2003.

27. A. D. Stiff-Roberts, X. H. Su, S. Chakrabarti, and P. Bhattacharya, "Contribution of Field-Assisted Tunneling Emission to Dark Current in InAs-GaAs Quantum Dot Infrared Photodetectors," *IEEE Photonics Technology Letters* 16, 867–69, 2004.

28. H. C. Liu, "Noise Gain and Operating Temperature of Quantum Well Infrared Photodetectors," *Applied Physics Letters* 61, 2703–5, 1992.

29. W. A. Beck, "Photoconductive Gain and Generation-Recombination Noise in Multiple-Quantum-Well-Infrared Detectors," *Applied Physics Letters* 63, 3589–91, 1993.

30. J. Phillips, P. Bhattacharya, S. W. Kennerly, D. W. Beekman, and M. Duta, "Self-Assembled InAs-GaAs Quantum-Dot Intersubband Detectors," *IEEE Journal of Quantum Electronics* 35, 936–43, 1999.

31. Z. Ye, J. C. Campbell, Z. Chen, E. T. Kim, and A. Madhukar, "Noise and Photoconductive Gain in InAs Quantum Dot Infrared Photodetectors," *Applied Physics Letters* 83, 1234–36, 2003.

32. X. Lu, J. Vaillancourt, and M. J. Meisner, "Temperature-Dependent Photoresponsivity and High-Temperature (190 K) Operation of a Quantum Dot Infrared Photodetector," *Applied Physics Letters* 91, 051115, 2007.

33. H. Lim, S. Tsao, W. Zhang, and M. Razeghi, "High-Performance InAs Quantum-Dot Infrared Photoconductors Grown on InP Substrate Operating at Room Temperature," *Applied Physics Letters* 90, 131112, 2007.

34. P. Martyniuk and A. Rogalski, "Quantum-Dot Infrared Photodetectors: Status and Outlook," *Progress in Quantum Electronics* 32, 89–120, 2008.

35. T. Chuh, "Recent Developments in Infrared and Visible Imaging for Astronomy, Defense and Homeland Security," *Proceedings of SPIE* 5563, 19–34, 2004.

36. C. H. Grein, H. Cruz, M. E. Flatte, and H. Ehrenreich, "Theoretical Performance of Very Long Wavelength $InAs/In_xGa_{1-x}Sb$ Superlattice Based Infrared Detectors," *Applied Physics Letters* 65, 2530–32, 1994.

37. J. Jiang, S. Tsao, T. O'Sullivan, W. Zhang, H. Lim, T. Sills, K. Mi, M. Razeghi, G. J. Brown, and M. Z. Tidrow, "High Detectivity InGaAs/InGaP Quantum-Dot Infrared Photodetectors Grown by Low Pressure Metalorganic Chemical Vapor Deposition," *Applied Physics Letters* 84, 2166–68, 2004.

38. J. Szafraniec, S. Tsao, W. Zhang, H. Lim, M. Taguchi, A. A. Quivy, B. Movaghar, and M. Razeghi, "High-Detectivity Quantum-Dot Infrared Photodetectors Grown by Metalorganic Chemical-Vapor Deposition," *Applied Physics Letters* 88, 121102, 2006.

39. E.-T. Kim, A. Madhukar, Z. Ye, and J. C. Campbell, "High Detectivity InAs Quantum Dot Infrared Photodetectors," *Applied Physics Letters* 84, 3277–79, 2004.

40. S. Chakrabarti, X. H. Su, P. Bhattacharya, G. Ariyawansa, and A. G. U. Perera, "Characteristics of a Multicolor InGaAs-GaAs Quantum-Dot Infrared Photodetector," *IEEE Photonics Technology Letters* 17, 178180, 2005.

41. R. S. Attaluri, S. Annamalai, K. T. Posani, A. Stintz, and S. Krishna, "Influence of Si Doping on the Performance of Quantum Dots-in-Well Photodetectors," *Journal of Vacuum Science and Technology* B24, 1553–55, 2006.

42. S. Chakrabarti, A. D. Stiff-Roberts, X. H. Su, P. Bhttacharya, G. Ariyawansa, and A. G. U. Perera, "High-Performance Mid-Infrared Quantum Dot Infrared Photodetectors," *Journal of Physics D: Applied Physics* 38, 2135–41, 2005.

43. S. Krishna, D. Forman, S. Annamalai, P. Dowd, P. Varangis, T. Tumolillo, A. Gray, et al., "Two-Color Focal Plane Arrays Based on Self Assembled Quantum Dots in a Well Heterostructure," *Physica Status Solidi (c)* 3, 439–43, 2006.

44. J. Piotrowski and A. Rogalski, *High-Operating Temperature Infrared Photodetectors*, SPIE Press, Bellingham, WA, 2007.

45. X. H. Su, S. Chakrabarti, P. Bhattacharya, A. Ariyawansa, and A. G. U. Perera, "A Resonant Tunneling Quantum-Dot Infrared Photodetector," *IEEE Journal of Quantum Electronics* 41, 974–79, 2005.

PART IV
FOCAL PLANE ARRAYS

19 Overview of Focal Plane Array Architectures

As was mentioned in Chapter 2, the development of IR detectors were connected with thermal detectors at the beginning. The initial spectacular applications of thermal detectors in astronomy are noted in Figure 19.1 [1]. In 1856, Charles Piazzi Smyth [2,3], from the peak of Guajara on Tenerife, detected IR radiation from the Moon using a thermocouple. In the early 1900s, infrared radiation was successfully detected from the planets Jupiter and Saturn and from some bright stars such as Vega and Arcturus. In 1915, William Coblentz [2] at the U.S. National Bureau of Standards developed thermopile detectors, which he uses to measure the infrared radiation from 110 stars. However, the low sensitivity of early infrared instruments prevented the detection of other near-IR sources. Work in infrared astronomy remained at a low level until breakthroughs in the development of new, sensitive infrared detectors were achieved in the late 1950s.

For the first time, the photoconductive effect was observed by Smith in 1873 [4], who noted that the resistance of selenium decreased when exposed to light. However, the photon detectors were mainly developed during the twentieth century. In 1917, Case in the United States was able to produce thallous silfide photoconductive detectors that were sensitive to 1.2 μm [5]. Lead sulphide cells were developed and studied intensively by the Germans beginning before World War II [6]. History of the development in the large family of infrared photodetectors is briefly described in Chapter 2.

IR detector technology development was and continues to be primarily driven by military applications. Many of these advances were transferred to IR astronomy from U.S. Department of Defense research. In the mid-1960s, the first IR survey of the sky was made at the Mount Wilson Observatory, California, using liquid nitrogen cooled PbS photoconductors, which were most sensitive at 2.2 microns. The survey covered approximately 75% of the sky and found about 20,000 infrared sources [2]. Many of these sources were stars that had never been seen before in visible light.

Lately, civilian applications of infrared technology are frequently called "dual technology applications." One should point out the growing utilization of IR technologies in the civilian sphere at the expense of new materials and technologies and also the noticeable price decrease in these high cost technologies. Demands to use these technologies are quickly growing due to their effective applications, for example, in global monitoring of environmental pollution and climate changes, long time prognoses of agriculture crop yield, chemical process monitoring, Fourier transform IR spectroscopy, IR astronomy, car driving, IR imaging in medical diagnostics, and others. Traditionally, IR technologies are connected with controlling functions and night vision problems with earlier applications connected simply with detection of IR radiation, and later by forming IR images form temperature and emissivity differences (systems for recognition and surveillance, tank sight systems, anti-tank missiles, air–air missiles).

In the last five decades, different types of detectors are combined with electronic readouts to make detector arrays. The progress in integrated circuit design and fabrication techniques has resulted in continued rapid growth in the size and performance of these solid state arrays. In the infrared technique, these devices are based on a combination of a readout array connected to an array of detectors.

The term "focal plane array" (FPA) refers to an assemblage of individual detector picture elements ("pixels") located at the focal plane of an imaging system. Although the definition could include one-dimensional ("linear") arrays as well as two-dimensional (2-D) arrays, it is frequently applied to the latter. Usually, the optics part of an optoelectronic images device is limited only to focusing of the image onto the detectors array. These so-called staring arrays are scanned electronically usually using circuits integrated with the arrays. The architecture of detector-readout assemblies has assumed a number of forms that are discussed below [7].

19.1 OVERVIEW

Two families of multielement detectors can be considered; one used for scanning systems and the other used for staring systems. The simplest scanning linear FPA consists of a row of detectors (Figure 19.2a). An image is generated by scanning the scene across the strip using, as a rule, a mechanical scanner. At standard video frame rates, at each pixel (detector) a short integration time has been applied and the total charges are accommodated. A staring array is a 2-D array of detector pixels that are scanned electronically (Figure 19.2b). These types of arrays can provide enhanced sensitivity and gain in camera weight.

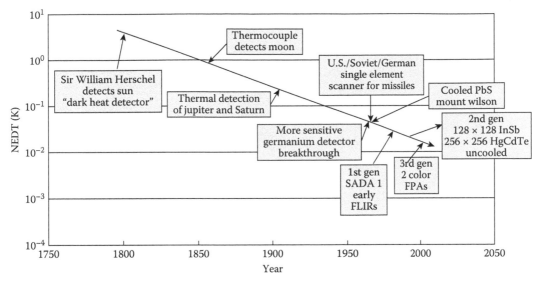

Figure 19.1 Development of infrared detectors: NEDT versus era during. (From Caulfield, J. T., "Next Generation IR Focal Plane Arrays and Applications," *Proceedings of the 32nd Applied Imagery Pattern Recognition Workshop*, 7–10, IEEE, New York, October 2003. With permission.)

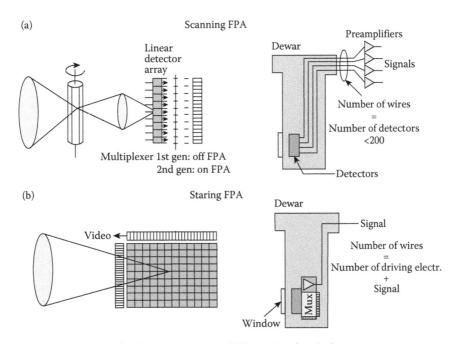

Figure 19.2 (a) Scanning focal plane array, and (b) staring focal plane arrays.

The scanning system, which does not include multiplexing functions in the focal plane, belongs to the first generation systems. A typical example of this kind of detector is a linear photoconductive array (PbS, PbSe, HgCdTe) in which an electrical contact for each element of a multielement array is brought off the cryogenically cooled focal plane to the outside, where there is one electronic channel at ambient temperature for each detector element. The United States common module HgCdTe arrays employ 60, 120, or 180 photoconductive elements depending on the application. An example of the 180 element common module FPA mounted on a dewar stem is shown in Figure 19.3[8].

Figure 19.3 A 180-element common module FPA mounted on a dewar stem. (From Kinch, M. A., "50 Years of HgCdTe at Texas Instruments and Beyond," *Proceedings of SPIE* 7298, 7298-96, 2009. With permission.)

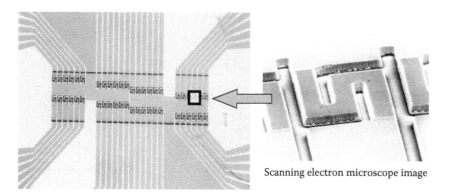

Scanning electron microscope image

Figure 19.4 Photomicrograph of 8 × 6 element photoconductive array of 50 μm square elements using labyrinthed structure for enhanced responsivity. Staggering the elements to solve the connection problems introduces delays between image lines. (From Elliott, T., "Recollections of MCT Work in the UK at Malvern and Southampton," *Proceedings of SPIE* 7298, 7298-89, 2009. With permission.)

The second-generation systems (full-framing systems), which are at present being developed, have at least three orders of magnitude more elements (> 10^6) on the focal plane than first-generation systems and the detectors elements are configured in a 2-D array. These staring arrays are scanned electronically by circuits integrated with the arrays. These readout integrated circuits (ROICs) include, for example, pixel deselecting, antiblooming on each pixel, subframe imaging, output preamplifiers, and some other functions.

Intermediary systems are also fabricated with multiplexed scanned photodetector linear arrays in use and with, as a rule, time delay and integration (TDI) functions. The array illustrated in Figure 19.4 is an 8 × 6 element photoconductive array elaborated in the mid-1970s and intended for use in a serial-parallel scan image [9]. Staggering the elements to solve the connection problems introduces delays between image lines. Typical examples of modern systems are HgCdTe multilinear 288 × 4 arrays fabricated by Sofradir both for 3–5 μm and 8–10.5 μm bands with signal processing in the focal plane (photocurrent integration, skimming, partitioning, TDI function, output preamplification, and some others).

Development in detector FPA technology has revolutionized many kinds of imaging [7]. From γ rays to the infrared and even radio waves, the rate at which images can be acquired has increased by more than a factor of a million in many cases. Figure 19.5 illustrates the trend in

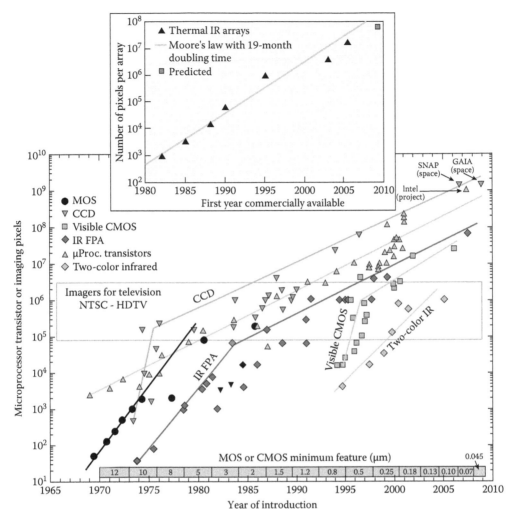

Figure 19.5 Imaging array formats compared with the complexity of microprocessor technology as indicated by transistor count. The timeline design rule of MOS/CMOS features is shown at the bottom. (From Norton, P., *Encyclopedia of Optical Engineering*, Marcel Dekker Inc., New York, 320–48, 2003. With permission.) The number of pixels on an infrared array has been growing exponentially, in accordance with Moore's Law for 25 years with a doubling time of approximately 19 months. (From Hoffman, A., *Laser Focus World*, 81–84, February 2006. With permission.) An 8K × 8K array was predicted for 2009 but is likely it will be at least a year later. (From Beletic, J. W., Blank, R., Gulbransen, D., Lee, D., Loose, M., Piquette, E. C., Sprafke, T., Tennant, W. E., Zandian, M., and Zino, J., "Teledyne Imaging Sensors: Infrared Imaging Technologies for Astronomy & Civil Space," *Proceedings of SPIE* 7021, 70210H, 2008. With permission.)

array size over the past 30 years. Imaging FPAs have developed in proportion to the ability of silicon integrated circuit (ICs) technology to read and process the array signals, and with ability to display the resulting image. The FPAs have nominally the same growth rate as dynamic random access memory (DRAM) ICs, which have had a doubling rate period of approximately 19 months [10]; it is a consequence of Moore's Law, but lag behind in size by about 5–10 years. The graph insert of Figure 19.6 shows the log of the number of pixels per a sensor chip assembly (SCA) as a function of the year first used on astronomy for MWIR SCAs. Arrays exceeded 4K × 4K format—16 million pixels—in 2006, about a year later than the Moore's Law prediction. A subsequent expansion to 8K × 8K with 10 μm pixel array is foreseen for 2010.

The trend of increasing pixel's number is likely to continue in the area of large format arrays. This increasing will be continued using close-butted mosaic of several SCAs. Raytheon manufactured a 4×4 mosaic of $2K \times 2K$ HgCdTe SCAs with 67 million pixels and assisted in assembling it to the final focal-plane configuration (see Figure 19.6) to survey the entire sky in the Southern Hemisphere at four IR wavelengths [10]. Astronomers in particular have eagerly waited for the day when electronic arrays could match the size of photographic film. Development of large format, high sensitivity, mosaic IR sensors for ground-based astronomy is the goal of many observatories around the world (large arrays dramatically multiply the data output of a telescope system). This is somewhat surprising given the comparative budgets of the defense market and the astronomical community.

While the size of individual arrays continues to grow, the very large focal plane arrays (FPAs) required for many space missions by mosaicking a large number of individual arrays. An example of a large mosaic developed by Teledyne Imaging Sensors, is a 147 megapixel FPA that is comprised of 35 arrays, each with 2048×2048 pixels. This is currently the world's largest IR focal plane [11]. Although there are currently limitations to reducing the size of the gaps between active detectors on adjacent SCAs, many of these can be overcome. It is predicted that focal plane of 100 megapixels and larger will be possible, constrained only by budgets but not technology [12].

A number of architectures are used in the development of IR FPAs [13–16]. In general, they may be classified as hybrid and monolithic, but these distinctions are often not as important as proponents and critics state them to be. The central design questions involve performance advantages versus ultimate producibility. Each application may favor a different approach depending on the technical requirements, projected costs, and schedule. Table 19.1 contains a description of representative IR FPAs that are commercially available as standard products and/or catalogue items from major manufacturers.

19.2 MONOLITHIC FPA ARCHITECTURES

In the monolithic approach, both detection of light and signal readout (multiplexing) is done in the detector material rather than in an external readout circuit. The integration of detector and readout onto a single monolithic piece reduces the number of processing steps, increases yields, and reduces costs. Common examples of these FPAs in the visible and near infrared (0.7–1.0 μm) are found in camcorders and digital cameras. Two generic types of silicon technology provide the bulk of devices in these markets: charge coupled devices (CCDs) and complementary metal-oxide-semiconductor (CMOS) imagers. The CCD technology has achieved the highest pixel counts or largest formats with numbers approaching 10^9 (see Figure 19.5). This approach to image acquisition was first proposed in 1970 in a paper written by Bell Lab researchers Boyle and Smith [17]. CMOS imagers are also rapidly moving to large formats and are expected to compete with CCDs for the large format applications within a few years.

Figure 19.7 shows different architectures of monolithic infrared FPAs. The basic element of a monolithic CCD array is a metal-insulator-semiconductor (MIS) structure. Used as part of a charge transfer device, a MIS capacitor detects and integrates the IR-generated photocurrent.

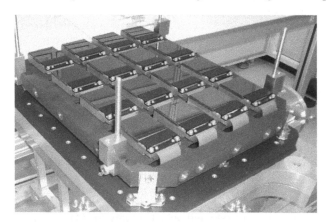

Figure 19.6 Sixteen 2048×2048 HgCdTe SCAs were assembled for the VISTA telescope. The SCA are attached to a precision ground plate that ensures that all pixels are within 12 μm of the desired focus. The detectors are placed in the telescope camera's vacuum chamber and cooled to 72 K. (From Hoffman, A., *Laser Focus World*, 81–84, February 2006.)

Table 19.1: Representative IR FPAs Offered by Some Major Manufacturers

Manufacturer/Web Site	Size/Architecture	Pixel Size (µm)	Detector Material	Spectral Range (µm)	Operational Temperature (K)	NETD (mK)
Raytheon/ www.raytheon.com	256 × 256/H	30 × 30	InSb	1–5.5	10–77	
	620 × 512/H	25 × 25	InSb	1–5.5	10–77	
	2048 × 2048/H	25 × 25	InSb	0.6–5.4	30	
	1024 × 1024/H	30 × 30	Si:As BIB	5–28	8–10	
	128 × 128/H	40 × 40	HgCdTe	9–11	80	
	2048 × 2048/H	20 × 20	HgCdTe	0.85–2.5	70–80	
	320 × 240/M	25 × 25	VO_x(bolometer)	7.5–1.6	−40 to 71°C	<50
	640 × 512/480/M	25 × 25	VO_x(bolometer)	3.5–12.5	−40 to 71°C	<50
Teledyne Imaging Sensors/ www.teledyne-si.com	2048 × 2048/H	18 × 18	HgCdTe	1.65–1.85	140	50
	2048 × 2048/H	18 × 18	HgCdTe	2.45–2.65	77	50
	2048 × 2048/H	18 × 18	HgCdTe	5.3–5.5	40	
Mitsubishi/ www.mitsubishi-imaging.com	320 × 240/M	25 × 25	Si diode bolometer	8–12	300	50
	640 × 480/M	25 × 25	Si diode bolometer	8–12	300	50
BAE Systems/ www.baesystems.com	640 × 480/M	28 × 28	VO_x(bolometer)	8–14	≈300	30–50
	640 × 480/M	17 × 17	VO_x(bolometer)	8–14	≈300	50
Sofradir/ infrared.sofradir.com	320 × 256/H	30 × 30	HgCdTe	7.7–11	70	≤25
	384 × 288/H	25 × 25	HgCdTe	7.7–9.5	77–80	17
	640 × 512/H	15 × 15	HgCdTe	3.7–4.8	<110	≤18
	1000 × 256/H	30 × 30	HgCdTe	0.8–2.5	<200	
	1280 × 1024/H	15 × 15	HgCdTe	3.7–4.8	77–110	18
	640 × 512/H	20 × 20	QWIP	$\lambda_p = 8.5, \Delta\lambda = 1\ \mu m$	70–73	31
Ulis/ www.ulis-ir.com	384 × 288/M	25 × 25	a-Si(bolometer)			<80
	640 × 512/M	25 × 25	a-Si(bolometer)			<80
	1024 × 768/M	17 × 17	a-Si(bolometer)			

Company/Website	Array format/Type	Pixel pitch	Detector material	Spectral range (µm)	Temperature	NETD
L-3/www.L-3com.com	640 × 512/H	28 × 28	InSb	3–5		<20
	1024 × 1024	25 × 25	InSb	3–5		<20
	320 × 240/M	37.5 × 37.5	VO$_x$(bolometer)		≈ 300	50
	640 × 480/M	30 × 30	a-Si(bolometer)		≈ 300	50
	1024 × 768/M	17 × 17	a-Si/a-SiGe(bolometer)		≈ 300	<50
DRS Infrared Technologies/ www.drsinfrared.com	320 × 240/M	25 × 25	VO$_x$(bolometer)	8–12	–20 to 60°C	40–70
	640 × 512/M	25 × 25	VO$_x$(bolometer)	8–12	–20 to 60°C	40–70
Selex/www.selexsgalileo.com	384 × 288/H	20 × 20	HgCdTe	8–10	<90	30
	640 × 512/H	24 × 24	HgCdTe	8–10	<90	24
	640 × 512/H	24 × 24	HgCdTe	3–5	<140	12
	1024 × 768	16 × 16	HgCdTe	3–5	<140	15
	640 × 512/H (dual-band)	24 × 24	HgCdTe	3–5; 8–10	80	13.5/26.6
AIM/www.aim-ir.de	384 × 288/H	24 × 24	HgCdTe	8–9	77	<25
	640 × 512/H	24 × 24	HgCdTe	3–5	77	<15
	384 × 288/H	24 × 24	QWIP	8–10		<20
	640 × 512/H	24 × 24	QWIP	8–10		<20
	384 × 288 × 2/H (dual-band)	24 × 24	QWIP	λ_p = 4.8 and 8.0		<20/<25
	384 × 288/H	24 × 24	Type-II SLS	3–5		<20
	384 × 288 × 2/H (dual-band)	24 × 24	Type-II SLS	3–5		<20/<35
JPL/www.jpl.nasa.gow	128 × 128/H	50 × 50	QWP	15 (λ_c)	45	30
	256 × 256/H	38 × 38	QWIP	9 (λ_c)	70	40
	640 × 486/H	18 × 18	QWIP	9 (λ_c)	70	36
SCD/www.scd.co.il	640 × 512	20 × 20	InSb	3–5		<20
	1280 × 1024	17 × 17	InSb	3–5		20
	384 × 288/M	25 × 25	VO$_x$(bolometer)		≈ 300	<50
	640 × 480/M	25 × 25	VO$_x$(bolometer)		≈ 300	<50
Santa Barbara Focalplane/ www.sbfp.com	640 × 512/H	20 × 20	InSb	1–5.2	77	<20
	1024 × 1024/H	19.5 × 19.5	InSb	1–5.2	77	<20
	1024 × 1024/H	19.5 × 19.5	QWIP	8.5–9.1		<35
Goodrich Corporation/ www.sensorsinc.com	320 × 240/H	25 × 25	InGaAs	0.9–1.7	300	$D^* > 10^{13}$ Jones
	640 × 512/H	25 × 25	InGaAs	0.4–1.7	300	$D^* > 6 \times 10^{12}$ Jones

Note: H: hybrid, M: monolithic

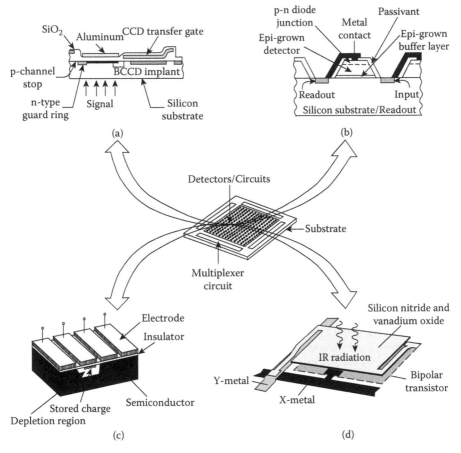

Figure 19.7 Monolithic IR FPAs: (a) all-silicon, (b) heteroepitaxy-on-silicon, (c) nonsilicon (e.g., InSb and HgCdTe CCDs, Ge CMOS), and (d) microbolometer.

CMOS-based imagers for both IR and visible applications use active or passive pixels [14–16,18–22]. Although most IR imaging applications tend to require high charge handling capabilities in the unit cells, an MIS capacitor fabricated in a narrow-gap semiconductor material (e.g., HgCdTe and InSb) has a limited charge capacity because of its low background potential as well as more severe problems involving noise, tunneling effects, and charge trapping when shifting charge through the narrow bandgap CCD to accomplish the readout function. Because of the nonequilibrium operation of the MIS detector, much larger electric fields are set up in the depletion region than in the p-n junction, resulting in defect-related tunneling current that is orders of magnitude larger than the fundamental dark current. The MIS detector required much higher material quality than p-n junction detectors, which still has not been achieved. So, although efforts have been made to develop monolithic FPAs using narrow-gap semiconductors, silicon based FPA technology with Schottky-barrier detectors is the only technology, which has matured to a level of practical use. An example of a fully monolithic silicon unit cell design is shown in Figure 19.7a. Several PtSi Schottky-barrier FPAs with full TV resolution have been commercially available, and arrays with more than one magapixels were reported (see Section 21.3). As the production of Schottky-barrier FPAs is fully compatible with silicon VLSI technology, this technology offers the cost-effective and producible FPA.

19.2.1 CCD Devices

The CCD technology is very mature with respect to fabrication yield and attainment of near-theoretical sensitivity. CCD technique relies on the optoelectronic properties of a well-established semiconductor architecture: the MOS capacitor (see Figure 9.35). A MOS capacitor typically consists of an extrinsic silicon substrate on which is grown an insulating layer of silicon dioxide (SiO_2). When a bias voltage is applied across p-type MOS structure, majority charge carriers (holes) are pushed away from the Si-SiO_2 interface directly below the gate, leaving a region depleted of positive charge and available as a potential energy well for any mobile minority charge carriers (electrons; see Figure 19.8a). Electrons generated in the silicon through absorption (charge generation) will collect in the potential-energy well under the gate (charge collection). Linear or (2-D) arrays of these MOS capacitors can therefore store images in the form of trapped charge carriers beneath the gates. The accumulated charges are transferred from potential well to the next well by using sequentially shifted voltage on each gate (charge transfer). One of the most successful voltage-shifting schemes is called three-phase clocking (Figure 19.8b). Column gates are connected to the separate voltage lines (L_1, L_2, L_3) in contiguous groups of three (G_1, G_2, G_3). The setup enables each gate voltage to be separately controlled.

Figure 19.8c shows the schematic circuit for a typical CCD imager. The photogenerated carriers are first integrated in an electronic well at the pixel and subsequently transferred to slow and fast CCD shift registers. At the end of the CCD register, a charge carrying information on the received signal can be readout and converted into a useful signal (charge measurement).

At present, the following readout techniques are used in CCD devices:

- floating diffusion amplifier in each pixel,
- system with correlated double sampling (CDS), and
- floating gate amplifier.

The floating diffusion amplifier, a typical CCD output preamplifier, can be implemented in each unit cell as shown in the dotted box in Figure 19.9 [24]. The unit cell consists of three transistors and the detector. Photocurrent is integrated onto the stray capacitance, which is the combined capacitance presented by the gate of the source follower $T2$, the interconnection, and the detector capacitance. The capacitance is reset to the voltage level V_R by supplying the reset clock (Φ_R) between successive integration frames. Integration of the signal charge makes the potential of the source follower input node lower. The source follower is active only when the transistor $T3$ is clocked. The drain current of the source follower $T2$ flows through the enable transistor $T3$ and load resistor outside the array.

Figure 19.10 shows a preamplifier, in this example—the source follower per detector (SFD), the output of which is connected to a clamp circuit. The output signal is initially sampled across the clamp capacitor during the onset of photon integration (after the detector is reset). The action of the clamp switch and capacitor subtracts any initial offset voltage from the output waveform. Because the initial sample is made before significant photon charge has been integrated by charging the capacitor, the final integrated photon signal swing is unaltered. However, any offset voltage or drift present at the beginning of integration is, by the action of the circuit subtracted, from the final value. This process of sampling each pixel twice, once at the beginning of the frame and again at the end, and providing the difference is called CDS.

The value of the initial CDS sample represents DC offsets, low-frequency drift and $1/f$ noise, and high-frequency noise; this initial value is subtracted from the final value, which also includes DC offset, low-frequency drift, and high-frequency noise. Since the two samples occur within a short period of time, the DC and lower-frequency drift components of each sample do not change significantly; hence, these terms cancel in the subtraction process. The dominant sources of read noise after CDS include the wideband noise of the output amplifier and excess noise of the video electronics. Both are minimized by minimizing the sense node capacitance and thereby maximizing the conversion gain. To minimize the capacitance various schemes are used (e.g., double stage amplifiers and alternative sense node implementation).

The floating gate amplifier configuration is shown in Figure 19.11. It consists of two MOSFET transistors, the source follower $T2$ and the zeroing transistor $T1$. The floating gate (reading gate) is in the same row as the CCD transfer gates. If a moving charge is under the gate, it causes a change in the gate potential of the transistor of the gate $T2$. At the preamplifier output, a voltage signal appears. This manner of readout does not cause degradation or decay of a moving charge so the charge can be detected at many places. An amplifier in which the same charge is sampled with several floating gates is called a floating diffusion amplifier.

The first CCD imager sensors were developed about 40 years ago primarily for television analog image acquisition, transmission, and display. With increasing demand for digital image data, the

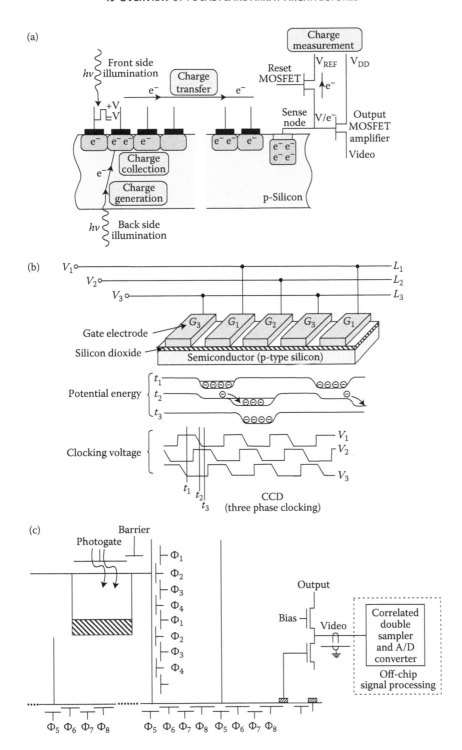

Figure 19.8 CCD image: (a) a cross section of a CCD pixel (charge generation, collection, transfer and measurement are shown). (From Janesick, J., *SPIE's OEmagazine*, 30–33, February 2002. With permission.) (b) Three-phase charge transfer mechanism, and (c) typical readout architecture.

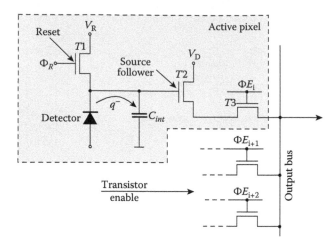

Figure 19.9 Unit cell with floating diffusion amplifier. (From Vampola, J. L., *The Infrared and Electro-Optical Systems Handbook*, SPIE Press, Bellingham, WA, Vol. 3, 285–342, 1993. With permission.)

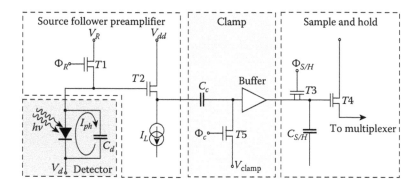

Figure 19.10 Correlated double sampling circuit. (From Vampola, J. L., *The Infrared and Electro-Optical Systems Handbook*, SPIE Press, Bellingham, WA, Vol. 3, 285–342, 1993. With permission.)

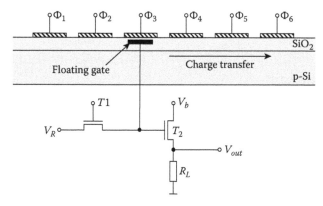

Figure 19.11 Floating gain amplifier circuit.

traditional analog raster scan output of image sensors is of limited use, and there is a strong motivation to fully integrate the control, digital interface, and image sensor on a single chip. In particular, silicon fabrication advances now permit the implementation of CMOS transistor structures that are considerably smaller than the wavelength of visible light and have enabled the practical integration of multiple transistors within a single picture.

19.2.2 CMOS Devices

An attractive alternative to the CCD readout is coordinative addressing with CMOS switches. The configuration of CCD devices requires specialized processing, unlike CMOS imagers that can be built on fabrication lines designed for commercial microprocessors. CMOS have the advantage that existing foundries, intended for application specific integrated circuits (ASICs), can be readily used by adapting their design rules. Design rules of 0.07 μm are currently in production, with preproduction runs of 0.045 μm design rules. As a result of such fine design rules, more functionality has been designed into the unit cells of IR and visible multiplexers with smaller unit cells, leading to large array sizes. Figure 19.5 shows the timelines for minimum circuit features and the resulting CCD, IR FPA, and CMOS visible imager sizes with respect to the number of imaging pixels. Along the horizontal axis is also a scale depicting the general availability of various MOS and CMOS processes. The ongoing migration to even finer lithography will thus enable the rapid development of CMOS-based imagers having even higher resolution, better image quality, higher levels of integration, and lower overall imaging system cost than CCD-based solutions. At present, CMOS with minimum features of ≤ 0.5 μm, makes possible monolithic visible CMOS imagers because the denser photolithography allows for low-noise signal extraction and high performance detection with high optical fill-factor within each pixel. The silicon wafer production infrastructure, which has put high performance personal computers into many homes, makes CMOS-based imaging in consumer products such as video and digital still cameras widely available.

A typical CMOS multiplexer architecture (see Figure 19.12c) consists of fast (column) and slow (row) shift registers at the edges of the active area, and pixels are addressed one by one through the selection of a slow register, while the fast register scans through a column, and so on. Each image sensor is connected in parallel to a storage capacitor located in the unit cell. A column of diodes and storage capacitors is selected one at a time by a digital horizontal scan register and a row bus is selected by the vertical scan register. Therefore, each pixel can be individually addressed.

Monolithic CMOS imagers use active or passive pixels as shown, in simplified form, in Figure 19.12b [22]. In comparison with passive pixel sensors (PPSs), active pixel sensors (APSs) apart from read functions exploit some form of amplification at each pixel. The PPS consists of three transistors (3T): a reset FET, a selective switch, and a source follower (SF) for driving the signal onto the column bus. As a result, circuit overhead is low and the optical collection efficiency [fill factor (FF)] is high even for monolithic devices. A large optical FF of up to 80% maximizes signal selection and minimizes fabrication cost by obviating the need for microlenses. Microlenses, typically used in CCD and CMOS APS imagers for visible application, concentrate the incoming light into the photosensitive region when they are accurately deposited over each pixel (see Figure 19.13 [22]). When the FF is low and microlenses are not used, the light falling elsewhere is either lost or, in some cases, creates artifacts in the imagery by generating electrical currents in the active circuitry.

In the APS three of the metal-oxide-semiconductor field-effect transistors (MOSFETs) have the same function as in PPS. The fourth transistor works as a transfer gate that moves charge from the photodiode to the floating diffusion. Usually, both pixels operate in rolling shutter mode. The APS is capable of performing CDS to eliminate the reset noise (kTC noise) and the pixel offsets. The PPS can only be used with noncorrelated double sampling, which is sufficient to reduce the pixel-to-pixel offsets but does not eliminate the temporal noise (temporal noise can be addressed by other methods like soft reset or tapered reset). Adding these components, however, reduces the FF of monolithic imagers to about 30–50% in 0.5 μm processes at a 5–6 μm pixel pitch or in 0.25 μm processes at a 3.3–4.0 μm pixel pitch [22]. The MOSFETs incorporated in each pixel for readout are optically dead (each pixel requires a minimum of three transistors). CMOS sensors also require several metal layers to interconnect MOSFETs. The busses are stacked and interleaved above the pixel, producing an "optical tunnel" through which incoming photons must pass. In addition, most CMOS imagers are front side illuminated. This limits the visible sensitivity in the red because of a relatively shallow absorption material. For comparison, CCD pixels are constructed so that the entire pixel is sensitive, with a 100% FF.

Figure 19.14 compares the principle of CCDs and CMOS sensors [16]. Both detector technologies use a photosensor to generate and separate the charges in the pixel. Beyond that, however, the two sensor schemes differ significantly. During CCD readout, the collected charge is shifted

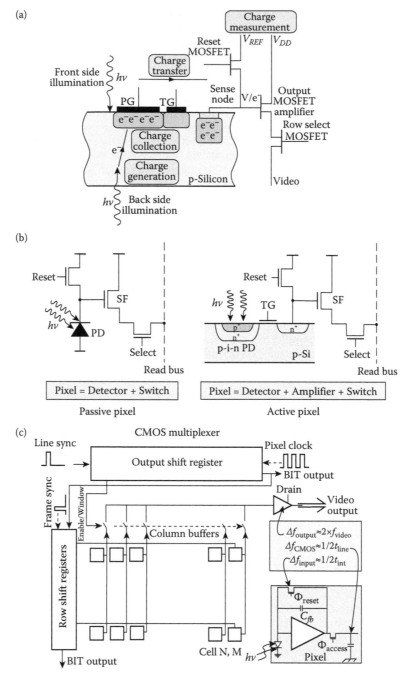

Figure 19.12 CMOS image: (a) a cross section of a CMOS pixel (charge generation, collection, transfer, and measurement are shown). (From Janesick, J., *SPIE's OEmagazine*, 30–33, February 2002. With permission.) (b) Passive and active pixel sensors. (From Kozlowski, L. J., Vural, K., Luo, J., Tomasini, A., Liu, T., and Kleinhans, W. E., *Opto-Electronics Review*, 7, 259–69, 1999. With permission.) (c) Typical readout architecture.

from pixel to pixel all the way to the perimeter. Finally, all charges are sequentially pushed to one common location (floating diffusion), and a single amplifier generates the corresponding output voltages. On the other hand, CMOS detectors have an independent amplifier in each pixel (APS). The amplifier converts the integrated charge into a voltage and thus eliminates the need to transfer charge from pixel to pixel. The voltages are multiplexed onto a common

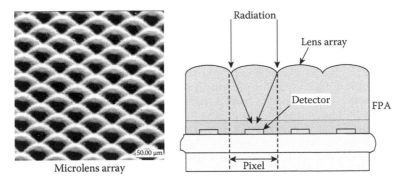

Figure 19.13 Micrograph and cross-sectional drawing of microlensed hybrid FPA. (From Kozlowski, L. J., Vural, K., Luo, J., Tomasini, A., Liu, T., and Kleinhans, W. E., *Opto-Electronics Review*, 7, 259–69, 1999. With permission.)

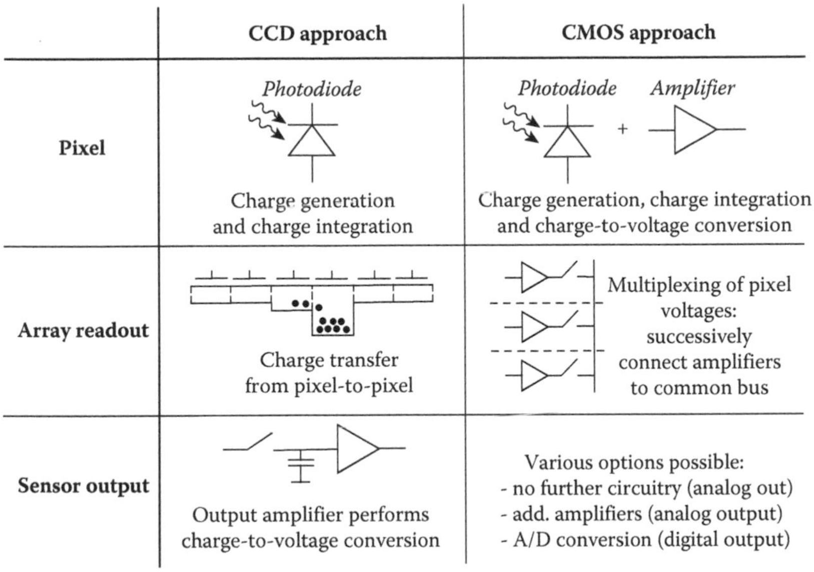

	CCD approach	CMOS approach
Pixel	*Photodiode* Charge generation and charge integration	*Photodiode* *Amplifier* Charge generation, charge integration and charge-to-voltage conversion
Array readout	Charge transfer from pixel-to-pixel	Multiplexing of pixel voltages: successively connect amplifiers to common bus
Sensor output	Output amplifier performs charge-to-voltage conversion	Various options possible: - no further circuitry (analog out) - add. amplifiers (analog output) - A/D conversion (digital output)

Figure 19.14 Comparison between the CCD-based and the CMOS-based image sensor approach. (From Hoffman, A., Loose, M., and Suntharalingam, V., *Experimental Astronomy*, 19, 111–34, 2005.)

bus line using integrated CMOS switches. Analog and digital sensor outputs are possible by implementing either a video output amplifier or an analog-to-digital (A/D) converter on the chip.

The processing technology for CMOS is typically two to three times less complex than standard CCD technology. In comparison with CCDs, the CMOS multiplexers exhibit important advantages due to high circuit density, fewer drive voltages, fewer clocks, much lower voltages (low power consumption) and packing density compatible with many more special functions, lower cost for both digital video and still camera applications. The minimum theoretical read noise of a CCD is limited in large imagers by the output amplifier's thermal noise after CDS is applied in off-chip support circuits. The alternative CMOS paradigm offers lower temporal noise because the relevant noise bandwidth is fundamentally several orders of magnitude smaller and better matches the signal bandwidth. While CCD sensitivity is constrained by the limited design space involving the sense node and the output buffer, CMOS sensitivity is limited only by the desired dynamic range and operating voltage. CMOS-based imagers also offer practical advantages with respect to on-chip integration of camera functions including command and control electronics, digitization and image processing. CMOS is now suitable for TDI-type multiplexers because of the availability from foundries of design rules lower than 1.0 μm, more uniform electrical characteristics and lower noise figures.

19.3 HYBRID FOCAL PLANE ARRAYS

In the case of hybrid technology (see Figure 19.15), we can optimize the detector material and multiplexer independently. Other advantages of hybrid-packaged FPAs are near-100% fill factors and increased signal-processing area on the multiplexer chip. Photodiodes with their very low power dissipation, inherently high impedance, negligible $1/f$ noise, and easy multiplexing via the ROIC, can be assembled in 2-D arrays containing a very large number of pixels, limited only by existing technologies. Photodiodes can be reverse-biased for even higher impedance, and can therefore better match electrically with compact low-noise silicon readout preamplifier circuits. The photo response of photodiodes remains linear for significantly higher photon flux levels than that of photoconductors, primarily because of higher doping levels in the photodiode absorber layer and because the photogenerated carriers are collected rapidly by the junction. Development of hybrid packaging technology began in the late 1970s [25] and took the next decade to reach volume production. In the early 1990s, fully 2-D imaging arrays provided a means for staring sensor systems to enter the production stage. In the hybrid architecture, indium bump bonding with readout electronics provides for multiplexing the signals from thousands or millions of pixels onto a few output lines, greatly simplifying the interface between the vacuum-enclosed cryogenic sensor and the system electronics.

19.3.1 Interconnect Techniques

Two hybridization approaches are in use today. In one approach, indium bumps are formed on both the detector array and the ROIC chip. The array and the ROIC are aligned and force is applied to cause the indium bumps to cold-weld together. In the other approach, indium bumps are formed only on the ROIC; the detector array is brought into alignment and proximity with the ROIC, the temperature is raised to cause the indium to melt, and contact is made by reflow.

The detector array can be illuminated from either the front side (with the photons passing through the transparent silicon multiplexer) or back side (with photons passing through the

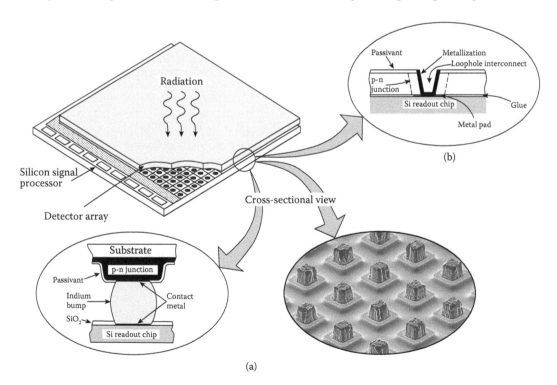

Figure 19.15 Hybrid IR FPA interconnect techniques between a detector array and silicon multiplexer: (a) indium bump technique, and (b) loophole technique.

transparent detector array substrate). In general, the latter approach is most advantageous, as the multiplexer will typically have areas of metallizations and other opaque regions that can reduce the effective optical area of the structure. The epoxy is flowed into the space between the readout and the detectors to increase the bonding strength. In HgCdTe hybrid FPAs, photovoltaic detectors are formed on thin HgCdTe epitaxial layers on transparent CdTe or ZnCdTe substrates. For HgCdTe flip-chip hybrid technology, the maximum chip size is of the order of 20 mm square. To overcome this problem, a producible alternative to CdTe for epitaxy technology is being developed with sapphire or silicon as the substrate of HgCdTe detectors. A SWIR 1024 × 1024 element HgCdTe hybrid FPA was developed using the PACE technology. When using opaque materials, substrates must be thinned to below 10 μm to obtain sufficient quantum efficiencies and reduce crosstalk. In some cases the substrates are completely removed. In the "direct" back side illuminated configuration both the detector array and the silicon ROIC chip are bump mounted side-by-side onto a common circuit board. The "indirect" configuration allows the unit cell area in the silicon ROIC to be larger than the detector area and is usually used for small scanning FPAs, where stray capacitance is not an issue.

Hybrid FPA detectors and multiplexers are also fabricated using loophole interconnection [26,27]. In this case, the detector and the multiplexer chips are glued together to form a single chip before detector fabrication. The photovoltaic detector is formed by ion implantation and loopholes are drilled by ion-milling and electrical interconnection between each detector and its corresponding input circuit is made through a small hole formed in each detector. The junctions are connected down to the silicon circuit by cutting the fine, few μm in diameter holes through the junctions by ion milling, and then backfilling the holes with metallization. The thermal expansion mismatch problem is approached by using about 10 μm thick p-type HgCdTe, bonded rigidly to the silicon so that strain is taken up elastically. This makes the devices mechanically and electrically very robust with contact obscuration typically less than 10%. The disadvantages include the necessity of mechanical thinning of the HgCdTe, which may lead to damage that may affect photodiode performance and the necessity of devising clever low-temperature techniques for junction formation and passivation because of the presence of the epoxy (e.g., the ion milling process to form the n-type regions is done at room temperature). A similar type of hybrid technology called VIP™ (vertically integrated photodiode) was reported by DRS Infrared Technologies (formerly Texas Instruments) [28,29].

Readout circuit wafers are processed in standard commercial foundries and can be constrained in size by the die-size limits of the photolithography step and repeat printers [30]. This limit is currently on the order of 22×22 mm² for submicron lithography. Thus, the array itself can only occupy 18×18 mm² assuming one needs about 2 mm on each side for the peripheral circuitry such as bias supplies, shift registers, column amplifiers, and output drivers. Under these conditions, a 1024×1024 array would need to have pixels no larger than 18 μm.

To build larger sensor arrays, a new photolithographic technique called *stitching* can be used to fabricate detector arrays larger than the reticle field of photolithographic steppers. The large array is divided into smaller subblocks. Later, the complete sensor chips are stitched together from the building blocks in the reticle as shown in Figure 19.16 [31]. Each block can be photocomposed on the wafer by multiple exposures at appropriate locations. Single blocks of the detector array are exposed at one time, as the optical system allows shuttering, or selectively exposing only a desired section of the reticle.

The III-V compound semiconductors are available in large diameter wafers, up to 8 inches. Thus, focal plane technologies such as InSb, QWIP, and type-II SLS are potential candidates for development of large format arrays such as 4096×4096 and larger. Gunapala and coworkers [31] have discussed the possibility of extending the array size up to 16 megapixels.

It should be noted that stitching creates a seamless detector array, as opposed to an assembly of closely butted subarrays. The butting technique is commonly used in the fabrication of very large format HgCdTe hybrid sensor arrays due to the limited size of substrate wafers (usually CdZnTe). For example, Teledyne Scientific & Imaging has developed the world's largest HgCdTe SWIR FPA for astronomy and low background applications. The format of the device is a hybrid 2048×2048 with a unit cell size of 18×18 μm² and with an active size of 37 mm. Sets of four arrays are "tiled" into a 2×2 mosaic configuration giving 4096×4096 pixels [32]. Recently, the first large format MWIR FPAs with pixel dimension of 15 μm have been demonstrated [33,34].

The development of IR FPAs using IC techniques together with development of new material growth techniques and microelectronic innovations began about 30 years ago. The combination of the last two techniques gives many new possibilities for IR systems with increased sensitivity

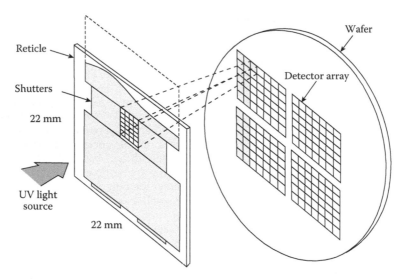

Figure 19.16 The photocomposition of a detector array die using array stitching based on photolithographic stepper. (From Gunapala, S. D., Bandara, S. V., Liu, J. K., Mumolo, J. M., Hill, C. J., Ting, D. Z., Kurth, E., Woolaway, J., LeVan, P. D., and Tidrow, M. Z., "Towards 16 Megapixel Focal Plane Arrays," *Proceedings of SPIE* 6660, 66600E, 2007. With permission.)

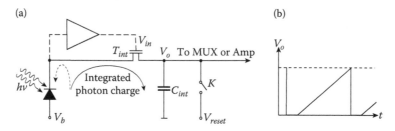

Figure 19.17 Direct injection readout circuit. (From Hewitt, M. J., Vampola, J. L., Black, S. H., and Nielsen, C. J., "Infrared Readout Electronics: A Historical Perspective," *Proceedings of SPIE* 2226, 108–19, 1994. With permission.)

and spatial resolution. Moreover, a number of other important advantages are accrued in terms of simplicity, reliability, and reduced costs. Fifteen years ago, high quality, single element detectors often were priced over $2000, but now some current IR FPA production costs are less than $0.1 per detector, and even greater reductions are expected in the near future. As the commercial market for uncooled imagers expand, the cost of commercial systems will inevitably decrease. At present, the cost of 320×240 bolometer arrays for thermal imagers is below $5000.

19.3.2 Readout Integrated Circuits

The key to the development of ROICs has been the evolution in input preamplifier technology. This evolution has been driven by increased performance requirements and silicon processing technology improvements. A brief discussion of the various circuits is given below.

Hybrid IR FPAs have used CMOS readouts since approximately 1985 for low-noise readout of photo-generated signals. CCD technology is used for not very large scale arrays and their technology is more complicated compared to the CMOS production line.

The direct injection (DI) circuit was one of the first integrated readout preamplifiers and has been used as an input to CCDs and visible imagers for many years. The DI is also commonly used with an input circuit for infrared tactical applications, where the backgrounds are high and detector resistances are moderate [35]. The goal is to fit as large a capacitor as possible into the unit cell, where signal-to-noise ratios can be obtained through longer integration times. Photon current in DI circuits is injected, via the source of the input transistor, onto an integration capacitor (Figure 19.17 [20]). As the photon current charges the capacitor throughout the frame, a simple charge integration takes place (see Figure 19.17b). Next a multiplexer reads out the final value and the capacitor voltage is reset prior to the beginning of the frame. To reduce detector noise, it is important that a uniform, near-zero-voltage bias be maintained across all the detectors.

The operating point of the coupled detector and input DI circuit is found by constructing a load line for the I-V characteristics of the detector and input MOSFET (see Figure 19.18 [36]). The input impedance of a MOSFET is a function of the source-drain current (in this case, the total diode current) and is usually expressed in terms of the transconductance, g_m, given by $qI/(nkT)$ for low injected currents (n is an ideality factor that can vary with temperature and geometry of the transistor and usually is in the range 1–2).

The injection efficiency is approximately given by [36,37]

$$\varepsilon = \frac{IR_d}{IR_d + (nkT/q)},$$

(19.1)

where R_d is the dynamic impedance of the detector and I is the total injected detector current (the sum of the photocurrent and the dark current) equal photocurrent, I_{ph}, in the background-limited case.

To receive high injection efficiency, the input impedance of the MOSFET must be much lower than the internal dynamic resistance of the detector at its operating point, and the following condition should be fulfilled [24]

$$IR_d \gg \frac{nkT}{q}.$$

(19.2)

For most applications, the detector performance depends on operating the detector in a small bias where the dynamic resistance is at a maximum. It is then necessary to minimize extraneous leakage current. The control of these leakage currents and the associated low-frequency noise is therefore of crucial interest. Generally, it is not difficult to fulfill inequality (Equation 19.2) for MWIR staring designs where diode resistance is large (R_oA product in the range of more than 10^6 Ωcm^2), but it may be important for LWIR designs where R_o is small (R_oA product about 10^2 Ωcm^2).

Feedback enhanced direct injection (FEDI) is similar to DI except that inverting amplifier is provided between the detector node and the input MOSFET gate (Figure 19.17, dashed line). The inverting gain provides feedback to yield better control over the detector bias at different photocurrent levels. It can maintain a constant detector bias at medium and high backgrounds. The amplifier reduces the input impedance of the DI and therefore increases the injection efficiency and bandwidth. The minimum operating photon flux range of the FEDI is an order of magnitude below that of the DI, thus the response is linear over a larger range than the DI circuit.

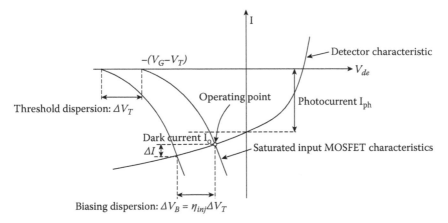

Figure 19.18 Direct injection: DC operating point. Influence of threshold-voltage dispersion. (From Longo, J. T., Cheung, D. T., Andrews, A. M., Wang, C. C., and Tracy, J. M., *IEEE Transactions on Electron Devices*, ED-25, 213–32, 1978. With permission.)

In the background-limited case, Equation 19.1 becomes a simple function of $I_{ph}R_d$. For a very high performance system requiring, the dynamic resistance should be very high, which is a technological challenge for detectors operated in the long wavelength band. There are more complex injection circuits that effectively reduce the input resistance and allow lower detector resistance to be used.

The DI circuit is widely used for simplicity; however, it requires a high impedance detector interface and is not generally used for low backgrounds due to injection efficiency issues. Many times the strategic applications have low backgrounds and require low noise multiplexers interfaced to high resistance detectors. A commonly used input circuit for strategic applications is the capacitative transimpedance amplification (CTIA) input circuit [35,38].

The capacitor-feedback transimpedance amplifier is a reset integrator and addresses a broad range of detector interface and performance requirements across many applications. The CTIA consists of an inverting amplifier with a gain of A, the integration capacitance C_f placed in a feedback loop, and the reset switch K (Figure 19.19). The photoelectron charge causes a slight change in voltage at the inverting input node of an amplifier. The amplifier responds with a sharp reduction in output voltage. As the detector current accumulates over the frame time, uniform illumination results in a linear ramp at the output. At the end of integration, the output voltage is sampled and multiplexed to the output bus. Since the input impedance of the amplifier is low, the integration capacitance can be made extremely small (e.g., 12 fF), yielding low noise performance. The feedback, or integration, capacitor sets the gain. The switch K is cyclically closed to achieve reset. The CTIA provides low input impedance, stable detector bias, high gain, high frequency response, and a high photon current injection efficiency. It has very low noise from low to high backgrounds.

Besides the DI and CTIA inputs mentioned above, we can distinguish other multiplexers; the most important are: source follower per detector (SFD), buffered direct injection (BDI), and gate modulation input (GMI) circuits. These schemes are described in many papers (e.g., [19–22,38–40]). Table 19.2 provides a description of the advantages and disadvantages of DI, CTIA, and SFD circuit [16]. As is mentioned above, the DI circuit is used in higher-flux situations. The CTIA is more complex and higher power but is extremely linear. The SFD is most commonly used in large-format hybrid astronomy arrays as well as commercial monolithic CMOS cameras.

In Table 19.3, gathered specifications of a family of large format ROICs designed by Indigo and Raytheon Vision Systems (RVS). Indigo products provide an off-the-shelf solution for the most demanding applications. The large arrays include a variety of pixel ranging from 25 to 15 microns, for customers with a wide range of optical design, dewar/cooler configurations and resolution requirements. On the contrary, RVS has a rich heritage of developing astronomy FPAs. As the demand for both finer resolution and a larger field of view for astronomy imagery has increased, the RVS sensor chip utilizes various detector materials and silicon readouts with array size up to 4096×4096 pixels.

The combined source follower/detector (SFD) unit cell is shown in Figure 19.20. The unit cell consists of an integration capacitance, a reset transistor ($T1$) operated as a switch, the source-follower transistor ($T2$), and selection transistor ($T3$). The integration capacitance may just be the detector capacitance and transistor $T2$ input capacitance. The integration capacitance is reset to a reference voltage (V_R) by pulsing the reset transistor. The photocurrent is then integrated on the

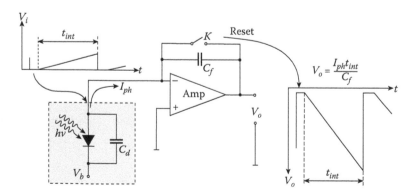

Figure 19.19 Schematic of a capacitive transimpedance amplifier unit cell. (From Felix, P., Moulin, M., Munier, B., Portmann, J., and Reboul, J.-P. *IEEE Trans. Elect. Dev.* ED-27, 1980, fig. 3 (p. 178). With permission.)

Table 19.2: Comparison of Attributes of the Three Most Common Input Circuits

Circuit	Advantages	Disadvantages	Comments
Direct injection (DI)	Large well capacity Gain determined by ROIC design (C_{int}) Detector bias remains constant Low FET glow Low power	Poor performance at low flux	Standard circuit for high flux
Capacitance transimpedance amplifier (CTIA)	Very linear Gain determined by ROIC design (C_f) Detector bias remains constant	More complex circuit FET glow Higher power	Very high gains demonstrated
Source follower per detector (SFD)	Simple Low noise Low FET glow Low power	Gain fixed by detector and ROIC input capacitance Detector bias changes during integration Some nonlinearity	Most common circuit in IR astronomy

Source: Hoffman, A. Loose, M. and Suntharalingam, V. *Experimental Astronomy,* 19, 111–34, 2005. With permission.

capacitance during the integration period. The ramping input voltage of the SFD is buffered by the source follower and then multiplexed, via the $T3$ switch, to a common bus prior to the video output buffer. After the multiplexer read cycle, the input node is reset and the integration cycle begins again. The switch must have very low current leakage characteristics when in the open state, or this will add to the photocurrent signal. The dynamic range of the SFD is limited by the current voltage characteristics of the detector. As the signal is integrated, the detector bias changes with time and incident light level. The SFD has low noise for low bandwidth applications such as astronomy and still has acceptable signal-to-noise at very low backgrounds (e.g., a few photons per pixel per 100 ms). It is nonlinear at medium and high backgrounds, resulting in a limited dynamic range. The gain is set by the detector responsivity and the combined detector plus source-follower-input capacitance. The major noise sources are the kTC noise (resulting from resetting the detector), MOSFET channel thermal, and MOSFET $1/f$ noise.

The resistor load (RL) gate modulation circuit is shown in Figure 19.21a. It was introduced to extend the SFD performance advantages to high irradiance backgrounds and dark currents. This circuit uses the photocurrent to modulate the gate voltage and thereby induce an output current in the MOSFET. The drain current of the MOSFET transistor accumulates onto an integration capacitor. In high background irradiance, this circuit provides a design that can reject much of these background components, as when the background alone is present on the detector, the bias on the detector, or the load resistor can be adjusted to give negligible drain current or integration of a charge. When the signal is then applied, the transistor drain current increases with photon current and thereby allows some level of background flux rejection. The load resistor is designed such that it has low $1/f$ noise, excellent temperature stability, and good cell-to-cell uniformity.

The current mirror (CM) gate modulation (see Figure 19.21b) extends readouts to very high background levels. In this CM preamplifier, the MOSFET replaces the resistor of the RL circuit. The photon current flowing into the drain of the first of two closely matched transistors includes a common gate to source voltage change in both transistors. This results in a similar current in the second transistor. If the source voltage, V_s and V_{ss}, of the two matched transistors are connected, both will have the same gate to source voltages that will induce a current in the output transistor identical to the detector current flowing through the input transistor. In this circuit, the integration current is a linear function of a detector current. This CM interfaces easily to direct access or CCD multiplexers and has low area requirements for the unit cell. The CM circuit requires gain and offset corrections for most applications. The advantages over the RL circuit include its better linearity and absence of a load resistor.

19.4 PERFORMANCE OF FOCAL PLANE ARRAYS

This section discusses concepts associated with the performance of focal plane arrays. For IR imaging systems, the relevant figure of merit for determining the ultimate performance is not the detectivity, D^*, but the noise equivalent difference temperature (NEDT) and the minimum resolvable difference temperature (MRDT). These are considered the primary performance metrics

Table 19.3: Large Format Readout Integrated Circuits

	Indigo						Raytheon Vision Systems				
	ISC9803	ISC002	ISC9901	ISC0402	ISC0403	ISC0404	Aladdin	Orion	Virgo	Phoenix	Aquarius
Format	640×512	640×512	640×512	640×512	640×512	1024×1024	1024×1024	1024×1024, 2048×2048	1024×1024, 2048×2048	1024×1024, 2048×2048	1024×1024
Pixel size (μm)	25	25	20	20	15	18	27	25	20	25	30
ROIC type	DI	CTIA	DI	DI	DI	DI	SFD	SFD	SFD	SFD	SFD
Operating temperature (K)	80–310	80–310	80–310	80	80	80	10–30	30	77	10–30	4–10
Integrated capacity (e⁻)	1.1×10^7	2.5×10^6	7×10^6	1.1×10^7	6.5×10^6	1.2×10^7	2.0×10^5	3.0×10^5	$> 3.5 \times 10^5$	3×10^5	1 or 15×10^6
ROIC noise (e⁻)	≤550	≤360	≤350	≤1279	≤760	≤1026	10–50	<20	<20	6–20	<1000
Full frame rates (Hz)	30	30	30	>30	>30	>30					
Number of outputs	1, 2, or 4	1, 2, or 4	1, 2, or 4	1, 2, or 4	1, 2, or 4	4, 8, or 16	32	64	4 or 16	4	16 or 64
Packaging	LCC*	LCC	LCC	LCC	LCC	LCC	LCC	Module: 2 side buttable	Module: 3 side buttable	LCC	Module: 2 side buttable
Detector	p-on-n	p-on-n	p-on-n	p-on-n	p-on-n	p-on-n					
Compatible detectors	InSb or QWIP	InGaAs or HgCdTe	InSb or QWIP	InSb, InGaAs, HgCdTe, or QWIP	InSb	InSb	InSb, HgCdTe, or IBC	InSb	HgCdTe	InSb or IBC	IBC

*LCC: leadless chip carrier.

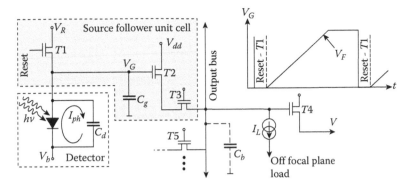

Figure 19.20 Schematic of source-follower per detector unit cell.

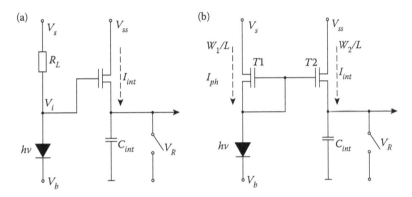

Figure 19.21 (a) Resistor load gate modulation, and (b) current mirror gate modulation.

to thermal imaging systems: thermal sensitivity and spatial resolution. Thermal sensitivity is concerned with the minimum temperature difference that can be discerned above the noise level. The MRDT concerns the spatial resolution and answers on the question: How small an object can be imaged by the system? The general approach of system performance is given by Lloyd in his fundamental monograph [41].

19.4.1 Noise Equivalent Difference Temperature

Noise equivalent difference temperature (NEDT) is a figure of merit for thermal imagers, which is commonly reported. In spite of its widespread use in infrared literature, it is applied to different systems, in different conditions, and with different meanings [42].

NEDT of a detector represents the temperature change that, for incident radiation, gives an output signal equal to the rms noise level. While normally thought of as a system parameter, detector NEDT and system NEDT are the same except for system losses. NEDT is defined

$$\text{NEDT} = \frac{V_n (\partial T / \partial Q)}{(\partial V_s / \partial Q)} = V_n \frac{\Delta T}{\Delta V_s}, \tag{19.3}$$

where V_n is the rms noise, Q is the spectral photon flux density (photons/cm²s) incident on a focal plane, and ΔV_s is the signal measured for the temperature difference ΔT.

To derive the equation estimating NEDT, we consider the configuration of the thermal imaging system shown in Figure 19.22. On the graph shown in this figure, A_s and A_d are, respectively, the surfaces of the object and the detector, r is the distance of the object to the lens (system optics), A_{ap} and D are the surface and the diameter of the lens (aperture, entrance-pupil). The detector is placed in focal plane of system in the distance $\approx f$ to the entrance pupil. The optic's system is opened to F/# (i.e., $F = f/D$ with $A_{ap} = \pi D^2 / 4$).

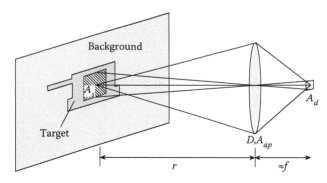

Figure 19.22 Thermal imaging system configuration.

Taking into consideration the methods presented in Section 1.3, we find the flux on the detector, ϕ_d, in terms of the radiance. Within the limitations of small angles ($r >> f$), flux transfer from target to detector is described in terms of the $A\Omega$ product

$$\Phi_d = L A_{ap} \Omega_d. \tag{19.4}$$

Assuming that the solid state of the elementary field of the system is

$$\Omega_d = \frac{A_d}{f^2} = \frac{A_{\text{footprint}}}{r^2}, \tag{19.5}$$

then

$$\Phi_d = L \frac{\pi D_{ap}^2}{4} \frac{A_d}{f^2} = L \frac{\pi}{4} \frac{A_a}{(F/\#)^2}. \tag{19.6}$$

For the geometry of optical system shown in Figure 19.22, we see that the irradiance on the detector, $E_d = \Phi_d/A_d$ is independent of the range r, and depends only on the source radiance and the $F/\#$ of the optics.

To obtain an expression for the signal voltage V_s produced by the detector, the expression (19.6) should be multiplied by detector responsivity and integrating over wavelength of the product of the spectral functions, over the passband of the system:

$$V_s = \frac{\pi}{4} \frac{A_d}{(F/\#)^2} \int_{\lambda_a}^{\lambda_b} R_v(\lambda) L(\lambda) d\lambda. \tag{19.7}$$

Because we are interested in thermal sensitivity of the system,

$$\frac{dV_s}{dT} = \frac{\pi}{4} \frac{A_d}{(F/\#)^2} \int_{\lambda_a}^{\lambda_b} R_v(\lambda) \frac{\partial L(\lambda)}{\partial T} d\lambda. \tag{19.8}$$

Using Equation 2.6, we can rewrite the last equation in the form

$$\frac{dV_s}{V_n} = \frac{\pi}{4} \frac{A_d}{(F/\#)^2} \frac{1}{(A_d \Delta f)^{1/2}} dT \int_{\lambda_a}^{\lambda_b} D^*(\lambda) \frac{\partial L(\lambda)}{\partial T} d\lambda. \tag{19.9}$$

From this equation we can determine the NEDT as the minimal temperature difference ΔT required to get the signal to noise ratio equal 1:

$$\mathrm{NETD} = \frac{4\left(F/\#\right)^2 \Delta f^{1/2}}{\pi A_d^{1/2}} \left[\int_{\lambda_a}^{\lambda_b} \frac{\partial L(\lambda)}{\partial T} D*(\lambda) d\lambda \right]^{-1}. \tag{19.10}$$

Taking into account that $M = \pi L$ (see Equation 1.14) results that

$$\mathrm{NETD} = \frac{4\left(F/\#\right)^2 \Delta f^{1/2}}{A_d^{1/2}} \left[\int_{\lambda_a}^{\lambda_b} \frac{\partial M(\lambda)}{\partial T} D*(\lambda) d\lambda \right]^{-1}. \tag{19.11}$$

In the above consideration, both atmosphere and optics transmissions have been assumed as equal 1.

The NEDT characterizes the thermal sensitivity of an infrared system; that is, the amount of temperature difference required to produce a unity signal-to-noise ratio. A smaller NEDT indicates a better thermal sensitivity.

To receive best sensitivity (lowest NEDT), the spectral integral in Equations 19.10 and 19.11 should be maximized. This can be obtained when the peak of the spectral responsivity and the peak of the exitance contrast coincide. However, the thermal imaging system may not satisfy these conditions because of other constraints such as atmospheric/abscurant transmittance effects or available detector characteristics. Dependence on the square root of bandwidth is intuitive, since the root-mean-square (rms) noise is proportional to $(\Delta f)^{1/2}$. In addition, better NEDT result from lower $F/\#$. A lower $F/\#$ number results in more flux captured by the detector that increases SNR for a given level.

The dependence of NEDT on the detector area is critical. The inverse-square-root dependence of NEDT on detector area results as an effect of two terms: increasing rms noise as the square root of the detector area and proportional increasing of the signal voltage to the area of detector. The net result is that NEDT $\propto 1/(A_d)^{1/2}$. While the thermal sensitivity of an imager is better for larger detectors, the spatial resolution is poorer for larger detectors (pixels). Another parameter, the MRDT, considers both thermal sensitivity and spatial resolution, more appropriate for design.

As Table 19.4 and the last two equations indicate, the best performance of IR imaging devices can be achieved operating in wide spectral range [43]. The spectral range limited to the atmospheric windows 8–14 μm and 3–5.5 μm will reduce the integral value NEDT (see Equations 19.10 and 19.11) to about 33% and about 6% of the 0–∞ range value, respectively. Therefore, IR systems based on unselective detectors, which are optimized for detection of ≈ 300 K objects in the atmosphere must operate in the 8–14 μm region.

Table 19.4: Calculated Radiant Exitance Between λ_a and λ_b at Different Temperatures

λ (μm)		$\int_{\lambda_a}^{\lambda_b} \frac{\partial M(\lambda, T)}{\partial \lambda} d\lambda$ (Wcm^{-2}K^{-1})			
λ_a	λ_b	$T = 280$ K	$T = 290$ K	$T = 300$ K	$T = 310$ K
3	5	1.1×10^{-5}	1.54×10^{-5}	2.1×10^{-5}	2.81×10^{-5}
3	5.5	2.01×10^{-5}	2.73×10^{-5}	3.62×10^{-5}	4.72×10^{-5}
3.5	5	1.06×10^{-5}	1.47×10^{-5}	2.0×10^{-5}	2.65×10^{-5}
3.5	5.5	1.97×10^{-5}	2.66×10^{-5}	3.52×10^{-5}	4.57×10^{-5}
4	5	9.18×10^{-6}	1.26×10^{-5}	1.69×10^{-5}	2.23×10^{-5}
4	5.5	1.83×10^{-5}	2.45×10^{-5}	3.22×10^{-5}	4.14×10^{-5}
8	10	8.47×10^{-5}	9.65×10^{-5}	1.09×10^{-4}	1.21×10^{-4}
8	12	1.54×10^{-4}	1.77×10^{-4}	1.97×10^{-4}	2.17×10^{-4}
8	14	2.15×10^{-4}	2.38×10^{-4}	2.62×10^{-4}	2.86×10^{-4}
10	12	7.34×10^{-5}	8.08×10^{-5}	8.81×10^{-5}	9.55×10^{-5}
10	14	1.3×10^{-4}	1.42×10^{-4}	1.53×10^{-4}	1.65×10^{-4}
12	14	5.67×10^{-5}	6.1×10^{-5}	6.52×10^{-5}	6.92×10^{-5}

Source: G. Gaussorgues, *La Thermographe Infrarouge*, Lavoisier, Paris, 1984.

The previous considerations are valid assuming that the temporal noise of the detector is the main source of noise. However, this assertion is not true to staring arrays, where the nonuniformity of the detectors response is a significant source of noise. This nonuniformity appears as a fixed pattern noise (spatial noise). It is defined in various ways in the literature, however, the most common definition is that it is the dark signal nonuniformity arising from an electronic source (i.e., other than thermal generation of the dark current); for example, clock breakthrough or from offset variations in row, column, or pixel amplifiers/switches. The estimation of IR sensor performance must include a treatment of spatial noise that occurs when FPA nonuniformities cannot be compensated correctly.

Mooney et al. [44] have given a comprehensive discussion of the origin of spatial noise. The total noise of a staring arrays is the composite of the temporal noise and the spatial noise. The spatial noise is the residual nonuniformity u after application of nonuniformity compensation, multiplied by the signal electrons N. Photon noise, equal $N^{1/2}$, is the dominant temporal noise for the high IR background signals for which spatial noise is significant. Then, the total NEDT is

$$\text{NEDT}_{\text{total}} = \frac{\left(N + u^2 N^2\right)^{1/2}}{\partial N / \partial T} = \frac{\left(1/N + u^2\right)^{1/2}}{(1/N)(\partial N / \partial T)}, \tag{19.12}$$

where $\partial N / \partial T$ is the signal change for a 1 K source temperature change. The denominator, $(\partial N / \partial T) N$, is the fractional signal change for a 1 K source temperature change. This is the relative scene contrast.

The dependence of the total NEDT on detectivity for different residual nonuniformity is plotted in Figure 19.23 for a 300 K scene temperature and a set of parameters shown in the figure. When the detectivity is approaching a value above 10^{10} cmHz$^{1/2}$/W, the FPA performance is uniformity limited prior to correction and thus essentially independent of the detectivity. An improvement in nonuniformity from 0.1a to 0.01% after correction could lower the NEDT from 63 to 6.3 mK.

19.4.2 NEDT Limited by Readout Circuit

Usually, the performance of MW and LWIR FPAs is limited by the readout circuits (by storage capacity of the ROIC). In this case [38]

$$\text{NEDT} = \left(\tau C \eta_{\text{BLIP}} \sqrt{N_w}\right)^{-1}, \tag{19.13}$$

where N_w is the number of photogenerated carriers integrated for one integration time, t_{int}

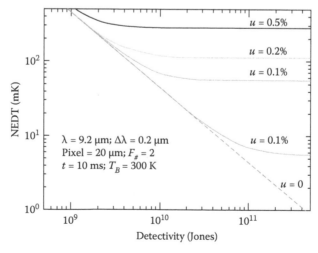

Figure 19.23 NEDT as a function of detectivity. The effects of nonuniformity are included for $u = 0.01\%$, 0.1%, 0.2%, and 0.5%. Note that for $D^* > 10^{10}$ cmHz$^{1/2}$/W, detectivity is not the relevant figure of merit.

$$N_w = \eta A_d t_{int} Q_B. \tag{19.14}$$

Percentage of BLIP, η_{BLIP}, is simply the ratio of photon noise to composite FPA noise

$$\eta_{BLIP} = \left(\frac{N^2_{photon}}{N^2_{photon} + N^2_{FPA}} \right)^{1/2}. \tag{19.15}$$

It results from the above formulas that the charge handling capacity of the readout, the integration time linked to the frame time, and dark current of the sensitive material become the major issues of IR FPAs. The NEDT is inversely proportional to the square root of the integrated charge and therefore the greater the charge, the higher the performance. The well charge capacity is the maximum amount of the charge that can be stored in the storage capacitor of each cell. The size of the unit cell is limited to the dimensions of the detector element in the array.

Figure 19.24 shows the theoretical NEDT versus charge handling capacity for HgCdTe FPAs assuming that the integration capacitor is filled to half the maximum capacity (to preserve dynamic range) under nominal operating conditions in different spectral bandpasses: 3.4–4.2 μm, 4.4–4.8 μm, 3.4–4.8 μm, and 7.8–10 μm [45]. The measured data for TCM2800 at 95 K, other PACE HgCdTe FPAs at 78 K, and representative LWIR FPAs are also shown. We can see that the measured sensitivities agree with the expected values.

In the case of a large LWIR HgCdTe hybrid array, a mismatch between the thermal expansion coefficients of the detector array and the readout, can force the cell pitch to 20 μm or less to minimize lateral displacement. However, the development of heteroepitaxial growth techniques for HgCdTe on Si has opened the possibility of a cost-effective production of significant quantities of large area arrays through the utilization of large diameter Si substrates.

It must be noted the distinction between integration time and FPA's frame time. At high backgrounds it is often impossible to handle the large amount of carriers generated over frame time compatible with standard video rates. Off-FPA frame integration can be used to attain a level of sensor sensitivity that is commensurate with the detector-limited D^* and not the charge-handling limited D^*.

19.4.2.1 Readout Limited NEDT for HgCdTe Photodiode and QWIP

The noise in HgCdTe photodiodes at 77 K is due to two sources; the shot noise from the photocurrent and the Johnson noise from the detector resistance. It can be expressed as [46]

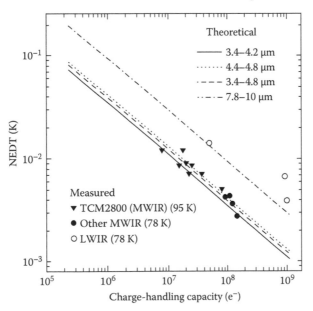

Figure 19.24 NEDT versus charge handling capacity. (From Kozlowski, L. J., "HgCdTe Focal Plane Arrays for High Performance Infrared Cameras," *Proceedings of SPIE* 3179, 200–11, 1997. With permission.)

$$I_n = \sqrt{\left(2qI_{ph} + \frac{4kT_d}{R}\right)\Delta f} \, , \tag{19.16}$$

where k is the Boltzmann constant and R is the dynamic resistance of a photodiode. Assuming that the integration time τ_{int} is such that the readout node capacity is kept half full, we have

$$\Delta f = \frac{1}{2\tau_{int}}, \tag{19.17}$$

and then

$$I_n = \sqrt{\left(2qI_{ph} + \frac{4kT_d}{R}\right)\frac{1}{2\tau_{int}}}. \tag{19.18}$$

At tactical background levels, the Johnson noise is much smaller than the shot noise from the photocurrent. In the case where the number of electrons collected in a frame is limited by the capacity of the ROIC charge well, which is often true, the signal to noise ratio is given by [46]

$$\frac{S}{N} = \frac{qN_w/2\tau_{int}}{\sqrt{2q\left(\dfrac{qN_w}{2\tau}\right)\dfrac{1}{2\tau_{int}}}} = \sqrt{\frac{N_w}{2}}. \tag{19.19}$$

Assuming that the temperature derivative of the background flux Φ can be written to a good approximation as

$$\frac{\partial \Phi}{\partial T} = \frac{hc}{\lambda k T_B^2} Q, \tag{19.20}$$

and using Equation 19.19, the NEDT is equal to

$$\text{NEDT} = \frac{2kT_B^2 \overline{\lambda}}{hc\sqrt{2N_w}}. \tag{19.21}$$

In the last two equations, $\overline{\lambda} = (\lambda_1 + \lambda_2)/2$ is the average wavelength of the spectral band between λ_1 and λ_2.

If one assumes a typical storage capacity of 2×10^7 electrons, $\overline{\lambda} = 10 \ \mu m$, and $T_B = 300$ K, Equation 19.21 yields NEDT of 19.8 mK.

The same estimate can be made for QWIP. In this case the Johnson noise is negligible compared to the generation–recombination noise, therefore

$$I_n = \sqrt{4qg\left(I_{ph} + I_d\right)\frac{1}{2\tau_{int}}}, \tag{19.22}$$

where the dark current may be approximated by

$$I_d = I_0 \exp\left(-\frac{E_a}{kT}\right). \tag{19.23}$$

In the above expressions, I_d is the dark current, I_0 is a constant that depends on the transport properties and the doping level, and E_a is the thermal activation energy, which is usually slightly less than the energy corresponding to the cutoff wavelength of the spectral response. It should be also stressed that g, I_{ph}, and I_0 are bias-dependent.

The signal-to-noise ratio for a storage-capacity-limited QWIP is given by

$$\frac{S}{N} = \frac{qN_w/2\tau_{int}}{\sqrt{4qg\left(\frac{qN_w}{2\tau}\right)\frac{1}{2\tau_{int}}}} = \frac{1}{2}\sqrt{\frac{N_w}{g}},$$ (19.24)

and the NEDT is

$$\text{NEDT} = \frac{2kT_B^2\bar{\lambda}}{hc}\sqrt{\frac{g}{N_w}}.$$ (19.25)

Comparing Equations 19.21 and 19.25 one may notice that the value of NEDT in a charge-limited QWIP detector is better than that of HgCdTe photodiodes by a factor of $(2g)^{1/2}$ since a reasonable value of g is 0.4. Assuming the same operation conditions as for HgCdTe photodiodes, the value of NEDT is 17.7 mK. Thus, a low photoconductive gain actually increases the S/N ratio and a QWIP FPA can have a better NEDT than an HgCdTe FPA with a similar storage capacity.

The performance figures of merit of state-of-the-art QWIP and HgCdTe FPAs are similar because the main limitations come from the readout circuits. Performance is, however, achieved with very different integration times. The very short integration time of LWIR HgCdTe devices (typically below 300 μs) is very useful to freeze a scene with rapidly moving objects. Due to excellent homogeneity and low photoelectrical gain, QWIP devices achieve an even better NEDT; the integration time, however, must be 10–100 times longer, and typically is between 5 and 20 ms. The choice of the best technology is therefore driven by the specific needs of a system.

19.5 MINIMUM RESOLVABLE DIFFERENCE TEMPERATURE

The MRDT is often the preferred figure of merit for imaging sensors. This figure of merit comprises both resolution and sensitivity of the thermal imager. MRDT enables us to estimate probability of detection, recognition, and identification of targets knowing MRDT of the evaluated thermal imager. The MRDT is a subjective parameter that describes ability of the imager-human system for detection of low contrast details of the tested object. It is measured as a minimum temperature difference between the bars of the standard 4-bar target and the background required to resolve the thermal image of the bars by an observer versus spatial frequency of the target [41,42,47]. It can be defined theoretically as:

$$\text{MRDT}(f_s) \approx K(f_s)\frac{\text{NEDT}}{\text{MTF}(f_s)},$$ (19.26)

where f_s is the spatial frequency (in cycles/radian), $\text{MTF}(f_s)$ is the modulation transfer function and $K(f_s)$ is a function containing the response of a human observer to a signal with a modulation given by the MTF embedded in a noisy image characterized by the NEDT [48,49].

Figure 19.25 shows the BLIP (for 70% quantum efficiency) MRDT curves at 300 K background temperature for narrow-field-of-view (high-resolution) sensors in the MWIR and LWIR spectral bands [38]. Two LWIR curves are included to show the impact of matching the diffraction-limited blur to the pixel pitch versus 2 times oversampling of the blur. For comparison, also are included representative curves for first-generation scanning, staring uncooled, staring TE-cooled, and staring PtSi sensors assuming 0.1, 0.1, 0.05, and 0.1 K NEDT, respectively. It can be noticed that theoretically the staring MWIR sensors have order of magnitude better sensitivity while the staring LWIR bands have two orders of magnitude better sensitivity than the first-generation sensors. However in practice, due to charge handling limitations, the LWIR sensor has only slightly better MRDT than the MWIR sensor.

19.6 ADAPTIVE FOCAL PLANE ARRAYS

A number of recent developments in the area of MEMS-based tunable IR detectors have the potential to deliver voltage-tunable, multiband infrared FPAs. These technologies have been developed as part of the DARPA-funded adaptive focal plane array (AFPA) program, and have demonstrated multispectral tunable IR HgCdTe detector structures [50–54]. At present the AFPAs are independently developed by other groups using HgCdTe [55–57] and IV-VI detectors [58].

Figure 19.26 presents a general concept of MEMS-based tunable IR detector. The MEMS filters are individual electrostatically actuated Fabry-Perot tunable filters. In the actual implementation,

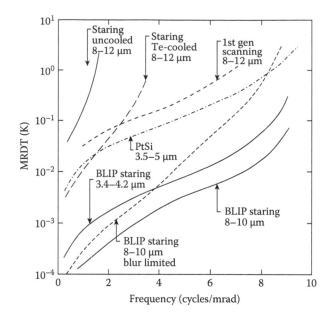

Figure 19.25 BLIP MRDT for staring FPA configurations in the various bands. (From Kozlowski, L. J., and Kosonocky, W. F., *Handbook of Optics,* Chapter 23, McGraw-Hill, Inc., New York, 1995. With permission.)

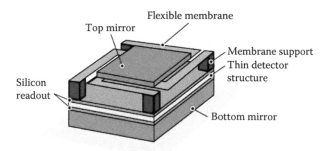

Figure 19.26 General concept of MEMS-based tunable IR detector.

the MEMS filter array is mounted so that the filters are facing toward the detector to minimize spectral crosstalk. By use of MEMS fabrication techniques arrays of devices, such as etalons, can be fabricated on an IR detector array that permits tuning of the incident radiation on the detector. If the etalons can be programmed to change distance from the detector surface by the order of IR wavelengths, the detector responds to all wavelengths in a waveband sequentially.

The integration of various component technologies into an AFPA involves a complex interplay across a broad range of disciplines, involving MEMS device processing, optical coating technology, microlenses, optical system modeling, and FPA devices. The goal of this integration is to produce an image-sensor array in which the wavelength sensitivity of each pixel can be independently tuned. In effect, the device would constitute a large-format array of electronically programmable microspectrometers.

Teledyne Scientific & Imaging has demonstrated simultaneous spectral tuning in the LWIR region while providing broadband imagery in MWIR band using dual-band AFPA (see Figure 19.27 [52]). The filter characteristics, including LWIR passband bandwidth and tuning range, are determined by the integral thin film reflector and antireflection coatings. The nominal dimension of each MEMS filter is between 100 μm and 200 μm on a side and each filter covers a small subarray of the detector pixels. Employing dual-band FPA with 20 μm pixel pitch results in each MEMS filter covering a detector subarray ranging from 5 × 5 to 10 × 10 pixels. The MEMS filter array will then evolve to tunable individual pixels.

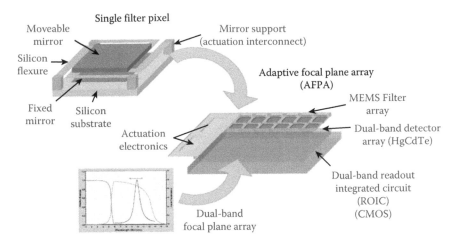

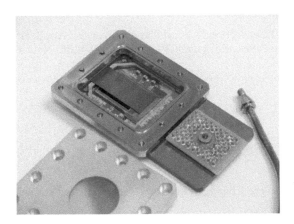

Figure 19.27 Dual-band adaptive focal plane array. (From Gunning, W. I., DeNatale, J., Stupar, P., Borwick, R., Lauxterman, S., Kobrin, P., and Auyeung, J., "Dual Band Adaptive Focal Plane Array. An Example of the Challenge and Potential of Intelligent Integrated Microsystems," *Proceedings of SPIE* 6232, 62320F, 2006. With permission.)

Figure 19.28 Open view of the AFPA integrated test package showing MEMS/MAIC hybrid mounted above dual-band FPA. Connection to the MAIC is made through the connector at the lower right (with pin short connector in place). (From Gunning, W., Lauxtermann, S., Durmas, H., Xu, M., Stupar, P., Borwick, R., Cooper, D., Kobrin, P., Kangas, M., DeNatale, J., and Tennant, W., "MEMS-Based Tunable Filters for Compact IR Spectral Imaging," *Proceedings of SPIE* 7298, 729821, 2009. With permission.)

The device requires a new ROIC to accommodate the additional control functions at each pixel. Therefore MEMS filters, positioned within 100 μm of the backside of the FPA, have been designed to include a separate MEMS Actuation IC (MAIC) that is hybridized to the MEMS chip (see Figure 19.28 [53]). Connection of the MAIC is made via pin block connector on the side of the device.

Figure 19.29 shows the measured spectral response of a typical filter tuned to various wavelengths in the 8–11 μm band [53]. The LWIR passbands have measured bandwidths of about 100 nm. Data at each MEMS filter wavelength are normalized to the peak value to eliminate the dependence on FPA spectral response. The sideband oscillations were found to be caused by interference arising from residual reflectance of the back side of the FPA and the MEMS filter mirror.

The realization of the AFPA concepts offers the potential for dramatic improvements in critical military missions involving reconnaissance, battlefield surveillance, and precision targeting [50].

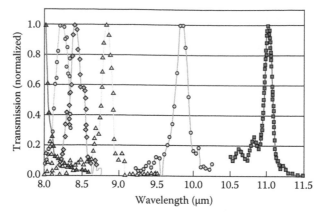

Figure 19.29 Measured normalized transmission of integrated AFPA device showing narrow band spectral response over the 8–11 µm band. For each measurement, a MEMS filter was tuned to a fixed wavelength and the FPA output was recorded as the wavelength of the narrow band incident illumination was scanned. (From Gunning, W., Lauxtermann, S., Durmas, H., Xu, M., Stupar, P., Borwick, R., Cooper, D., Kobrin, P., Kangas, M., DeNatale, J., and Tennant, W., "MEMS-Based Tunable Filters for Compact IR Spectral Imaging," *Proceedings of SPIE* 7298, 729821, 2009. With permission.)

REFERENCES

1. J. T. Caulfield, "Next Generation IR Focal Plane Arrays and Applications," *Proceedings of the 32nd Applied Imagery Pattern Recognition Workshop,* 7–10, IEEE, New York, October 2003.

2. http://coolcosmos.ipac.caltech.edu/cosmic_classroom/timeline/timeline_onepage.html

3. H. J. Walker, "Brief History of Infrared Astronomy," *Astronomy and Geophysics* 41, 10–13, October 2000.

4. W. Smith, "Effect of Light on Selenium During the Passage of an Electric Current," *Nature 7,* 303, 1873.

5. T. W. Case, "Notes on the Change of Resistance of Certain Substrates in Light," *Physical Review* 9, 305–10, 1917.

6. R. J. Cushman, "Film-Type Infrared Photoconductors," *Proceedings of IRE* 47, 1471–75, 1959.

7. P. Norton, "Detector Focal Plane Array Technology," in *Encyclopedia of Optical Engineering,* ed. R. Driggers, 320–48, Marcel Dekker Inc., New York, 2003.

8. M. A. Kinch, "Fifty Years of HgCdTe at Texas Instruments and Beyond," *Proceedings of SPIE* 7298, 72982T, 2009.

9. T. Elliott, "Recollections of MCT Work in the UK at Malvern and Southampton," *Proceedings of SPIE* 7298, 72982M, 2009.

10. A. Hoffman, "Semiconductor Processing Technology Improves Resolution of Infrared Arrays," *Laser Focus World,* 81–84, February 2006.

11. J. W. Beletic, R. Blank, D. Gulbransen, D. Lee, M. Loose, E. C. Piquette, T. Sprafke, W. E. Tennant, M. Zandian, and J. Zino, "Teledyne Imaging Sensors: Infrared Imaging Technologies for Astronomy & Civil Space," *Proceedings of SPIE* 7021, 70210H, 2008.

12. A. W. Hoffman, P. L. Love, and J. P. Rosbeck, "Mega-Pixel Detector Arrays: Visible to 28 µm," *Proceedings of SPIE* 5167, 194–203, 2004.

13. D. A. Scribner, M. R. Kruer, and J. M. Killiany, "Infrared Focal Plane Array Technology," *Proceedings of IEEE* 79, 66–85, 1991.

14. J. Janesick, "Charge Coupled CMOS and Hybrid Detector Arrays," *Proceedings of SPIE* 5167, 1–18, 2003.

15. B. Burke, P. Jorden, and P. Vu, "CCD Technology," *Experimental Astronomy* 19, 69–102, 2005.

16. A. Hoffman, M. Loose, and V. Suntharalingam, "CMOS Detector Technology," *Experimental Astronomy* 19, 111–34, 2005.

17. W. S. Boyle and G. E. Smith, "Charge-Coupled Semiconductor Devices," *Bell Systems Technical Journal* 49, 587–93, 1970.

18. E. R. Fossum, "Active Pixel Sensors: Are CCD's Dinosaurs?" *Proceedings of SPIE* 1900, 2–14, 1993.

19. E. R. Fossum and B. Pain, "Infrared Readout Electronics for Space Science Sensors: State of the Art and Future Directions," *Proceedings of SPIE* 2020, 262–85, 1993.

20. M. J. Hewitt, J. L. Vampola, S. H. Black, and C. J. Nielsen, "Infrared Readout Electronics: A Historical Perspective," *Proceedings of SPIE* 2226, 108–19, 1994.

21. L. J. Kozlowski, J. Montroy, K. Vural, and W. E. Kleinhans, "Ultra-Low Noise Infrared Focal Plane Array Status," *Proceedings of SPIE* 3436, 162–71, 1998.

22. L. J. Kozlowski, K. Vural, J. Luo, A. Tomasini, T. Liu, and W. E. Kleinhans, "Low-Noise Infrared and Visible Focal Plane Arrays," *Opto-Electronics Review* 7, 259–69, 1999.

23. J. Janesick, "Dueling Detectors. CMOS or CCD?" *SPIE's OEmagazine*, 30–33, February 2002.

24. J. L. Vampola, "Readout Electronics for Infrared Sensors," in *The Infrared and Electro-Optical Systems Handbook*, Vol. 3, ed. W. D. Rogatto, 285–342, SPIE Press, Bellingham, WA, 1993.

25. R. Thorn, "High Density Infrared Detector Arrays," U.S. Patent No. 4,039,833, 1977.

26. I. M. Baker and R. A. Ballingall, "Photovoltaic CdHgTe-Silicon Hybrid Focal Planes," *Proceedings of SPIE* 510, 121–29, 1984.

27. I. M. Baker, "Photovoltaic IR Detectors," in *Narrow-Gap II-VI Compounds for Optoelectronic and Electromagnetic Applications*, ed. P. Capper, 450–73, Chapman & Hall, London, 1997.

28. A. Turner, T. Teherani, J. Ehmke, C. Pettitt, P. Conlon, J. Beck, K. McCormack, et al., "Producibility of VIP™ Scanning Focal Plane Arrays," *Proceedings of SPIE* 2228, 237–48, 1994.

29. M. A. Kinch, "HDVIP™ FPA Technology at DRS," *Proceedings of SPIE* 4369, 566–78, 2001.

30. P. Norton, J. Campbell, S. Horn, and D. Reago, "Third-Generation Infrared Imagers," *Proceedings of SPIE* 4130, 226–36, 2000.

31. S. D. Gunapala, S. V. Bandara, J. K. Liu, J. M. Mumolo, C. J. Hill, D. Z. Ting, E. Kurth, J. Woolaway, P. D. LeVan, and M. Z. Tidrow, "Towards 16 Megapixel Focal Plane Arrays," *Proceedings of SPIE* 6660, 66600E, 2007.

32. http://www.teledyne-si.com/infrared_visible_fpas/index.html

33. E. P. G. Smith, G. M. Venzor, Y. Petraitis, M. V. Liguori, A. R. Levy, C. K. Rabkin, J. M. Peterson, M. Reddy, S. M. Johnson, and J. W. Bangs, "Fabrication and Characterization of Small Unit-Cell Molecular Beam Epitaxy Grown HgCdTe-on-Si Mid-Wavelength Infrared Detectors," *Journal of Electronic Materials* 36, 1045–51, 2007.

34. G. Destefanis, J. Baylet, P. Ballet, P. Castelein, F. Rothan, O. Gravrand, J. Rothman, J. P. Chamonal, and Million, "Status of HgCdTe Bicolor and Dual-Band Infrared Focal Plane Arrays at LETI," *Journal of Electronic Materials* 36, 1031–44, 2007.

35. L. J. Kozlowski, S. A. Cabelli, D. E. Cooper, and K. Vural, "Low Background Infrared Hybrid Focal Plane Array Characterization, *Proceedings of SPIE* 1946, 199–213, 1993.

36. J. T. Longo, D. T. Cheung, A. M. Andrews, C. C. Wang, and J. M. Tracy, "Infrared Focal Planes in Intrinsic Semiconductors," *IEEE Transactions on Electron Devices* ED-25, 213–32, 1978.

37. P. Felix, M. Moulin, B. Munier, J. Portmann, and J.-P. Reboul, "CCD Readout of Infrared Hybrid Focal-Plane Arrays," *IEEE Transactions on Electron Devices* ED-27, 175–88, 1980.

38. L. J. Kozlowski and W. F. Kosonocky, "Infrared Detector Arrays," in *Handbook of Optics*, Chapter 23, eds. M. Bass, E. W. Van Stryland, D. R. Williams, and W. L. Wolfe, McGraw-Hill, Inc., New York, 1995.

39. M. Kimata and N. Tubouchi, "Charge Transfer Devices," in *Infrared Photon Detectors*, ed. A. Rogalski, 99–144, SPIE Optical Engineering Press, Bellingham, WA, 1995.

40. J. Bajaj, "State-of-the-Art HgCdTe Materials and Devices for Infrared Imaging," in *Physics of Semiconductor Devices*, eds. V. Kumar and S. K. Agarwal, 1297–1309, Narosa Publishing House, New Delhi, 1998.

41. J. M. Lloyd, *Thermal Imaging Systems*, Plenum Press, New York, 1975.

42. J. M. Lopez-Alonso, "Noise Equivalent Temperature Difference (NETD)," in *Encyclopedia of Optical Engineering*, ed. R. Driggers, 1466–74, Marcel Dekker Inc., New York, 2003.

43. G. Gaussorgues, *La Thermographe Infrarouge*, Lavoisier, Paris, 1984.

44. J. M. Mooney, F. D. Shepherd, W. S. Ewing, and J. Silverman, "Responsivity Nonuniformity Limited Performance of Infrared Staring Cameras," *Optical Engineering* 28, 1151–61, 1989.

45. L. J. Kozlowski, "HgCdTe Focal Plane Arrays for High Performance Infrared Cameras," *Proceedings of SPIE* 3179, 200–11, 1997.

46. A. C. Goldberger, S. W. Kennerly, J. W. Little, H. K. Pollehn, T. A. Shafer, C. L. Mears, H. F. Schaake, M. Winn, M. Taylor, and P. N. Uppal, "Comparison of HgCdTe and QWIP Dual-Band Focal Plane Arrays," *Proceedings of SPIE* 4369, 532–46, 2001.

47. STANAG No. 4349. *Measurement of the Minimum Resolvable Temperature Difference (METD) of Thermal Cameras.*

48. E. Dereniak and G. Boreman, *Infrared Detectors and Systems*, John Wiley and Sons, New York, 1996.

49. K. Krapels, R. Driggers, R. Vollmerhausen, and C. Halford, "Minimum Resolvable Temperature Difference (MRT): Procedure Improvements and Dynamic MRT," *Infrared Physics & Technology* 43, 17–31, 2002.

50. J. Carrano, J. Brown, P. Perconti, and K. Barnard, "Tuning In to Detection," *SPIE's OEmagazine* 20–22, April 2004.

51. W. J. Gunning, J. DeNatale, P. Stupar, R. Borwick, R. Dannenberg, R. Sczupak, and P O Pettersson, "Adaptive Focal Plane Array: An Example of MEMS, Photonics, and Electronics Integration," *Proceedings of SPIE* 5783, 336–75, 2005.

52. W. I. Gunning, J. DeNatale, P. Stupar, R. Borwick, S. Lauxterman, P. Kobrin, and J. Auyeung, "Dual Band Adaptive Focal Plane Array. An Example of the Challenge and Potential of Intelligent Integrated Microsystems," *Proceedings of SPIE* 6232, 62320F, 2006.

53. W. Gunning, S. Lauxtermann, H. Durmas, M. Xu, P. Stupar, R. Borwick, D. Cooper, P. Kobrin, M. Kangas, J. DeNatale, and W. Tennant, "MEMS-Based Tunable Filters for Compact IR Spectral Imaging," *Proceedings of SPIE* 7298, 729821, 2009.

54. C. A. Musca, J. Antoszewski, K. J. Winchester, A. J. Keating, T. Nguyen, K. K. M. B. D. Silva, J. M. Dell, et al., "Monolithic Integration of an Infrared Photon Detector with a MEMS-Based Tunable Filter," *IEEE Electron Device Letters* 26, 888–90, 2005.

55. A. J. Keating, K. K. M. B. D. Silva, J. M. Dell, C. A. Musca, and L. Faraone, "Optical Characteristics of Fabry-Perot MEMS Filters Integrated on Tunable Short-Wave IR Detectors," *IEEE Photonics Technology Letters* 18, 1079–81, 2006.

56. J. Antoszewski, K. J. Winchester, T. Nguyen, A. J. Keating, K. K. M. B. Dilusha Silva, C. A. Musca, J. M. Dell, and L. Faraone, "Materials and Processes for MEMS-Based Infrared Microspectrometer Integrated on HgCdTe Detector," *IEEE Journal of Selected Topics in Quantum Electronics* 14, 1031–41, 2008.

57. L. P. Schuler, J. S. Milne, J. M. Dell, and L. Faraone, "MEMS-Based Microspectrometer Technologies for NIR and MIR Wavelengths," *Journal of Physics D: Applied Physics* 42, 13301, 2009.

58. H. Zogg, M. Arnold, F. Felder, M. Rahim, C. Ebneter, I. Zasavitskiy, N. Quack, S. Blunier, and J. Dual, "Epitaxial Lead Chalcogenides on Si Got Mid-IR Detectors and Emitters Including Cavities," *Journal of Electronic Materials* 37, 1497–1503, 2008.

20 Thermal Detector Focal Plane Arrays

The use of thermal detectors for IR imaging has been the subject of research and development for many decades. Thermal detectors are not useful for high-speed scanning thermal imagers. Only pyroelectric vidicons have found more widespread use. These devices achieved their fundamental limits of performance by about 1970. However, the speed of thermal detectors is quite adequate for nonscanned imagers with 2-D detectors. Figure 20.1 shows the dependence of noise equivalent difference temperature (NEDT) on noise bandwidth for typical detectivities of thermal detectors [1]. The calculations have been carried out assuming $100 \times 100 \ \mu m^2$ pixel size, 8–14 μm spectral range, $f/1$ optics and $t_{op} = 1$ of the IR system. With large arrays of thermal detectors the best values of NEDT below 0.1 K could be reached because effective noise bandwidths less than 100 Hz can be achieved. This compares with a bandwidth of several hundred kilohertz for conventional cooled thermal imagers with a small photon detector array and scanner. Realization of this fact caused a new revolution in thermal imaging, which is underway now. This is due to the development of 2-D electronically scanned arrays, in which moderate sensitivity can be compensated by a large number of elements. Large scale integration combined with micromachining has been used for manufacturing of large 2-D arrays of uncooled IR sensors. This enables fabrication of low cost and high-quality thermal imagers. Although developed for military applications, low-cost IR imagers are used in nonmilitary applications such as: drivers aid, aircraft aid, industrial process monitoring, community services, firefighting, portable mine detection, night vision, border surveillance, law enforcement, search and rescue, and so on.

The typical cost of cryogenically cooled imagers of around $50,000 (U.S.) restricts their installation to critical military applications involving operations in complete darkness. The commercial systems (microbolometer imagers, radiometers, and ferroelectric imagers) are derived from military systems that are too costly for widespread use. Imaging radiometers employ linear thermoelectric arrays operating in the snapshot mode; they are less costly than the TV-rate imaging radiometers that employ microbolometer arrays [2]. As the volume of production increases, the cost of commercial systems will inevitably decrease (see Table 20.1).

The NEDT is a figure of merit for focal plane arrays (FPAs). As is shown in Section 19.4, this parameter takes into account the optics, array, and readout electronics.

The temperature fluctuation noise limit to the performance of FPAs is determined by assuming that all other detector (pixel) and system noise sources are negligible in comparison with temperature fluctuation noise in the detector. By substituting Equation 3.23 into Equation 19.11, the temperature fluctuation noise limited NEDT (i.e., $NEDT_t$) is given by

$$NEDT_t = \frac{8F^2 T_d \left(kG_{th}\right)^{1/2} \Delta f^{1/2}}{\varepsilon t_{op} A_d} \left[\int_{\lambda_a}^{\lambda_b} \frac{dM}{dT} d\lambda \right]^{-1}. \tag{20.1}$$

In a similar way we can determine the background fluctuation noise limit to the NEDT. The $NEDT_b$ is found when radiation exchange is the dominant thermal exchange mechanism. In this case by substituting Equation 3.24 into Equation 19.11, we can obtain

$$NEDT_b = \frac{8F^2 \left[2kG\sigma\left(T_d^5 + T_b^5\right)_{th} \Delta f\right]^{1/2}}{\left(\varepsilon A_d\right)^{1/2} t_{op}} \left[\int_{\lambda_a}^{\lambda_b} \frac{dM}{dT} d\lambda \right]^{-1}. \tag{20.2}$$

The temperature fluctuation noise and background fluctuation noise limited NEDT of FPAs operating at 300 and 85 K against a 300 K background, determined from Equations 20.1 and 20.2, are illustrated in Figure 20.2 [3]. Other parameters used in calculations are listed on the figure. All thermal infrared detectors fall on or above the limits shown in Figure 20.2. Real detectors usually lie above the corresponding lines because of noise greater than temperature fluctuation noise.

The key trade-off with respect to uncooled thermal imaging systems is between sensitivity and response time. The thermal conductance is an extremely important parameter, since NEDT is proportional to $G_{th}^{1/2}$ but the response time of the detector is inversely proportional to G_{th}. Therefore, a change in thermal conductance due to improvements in material processing technique improves sensitivity at the expense of time response. Typical calculations of the trade-off between NEDT and time response carried out by Horn and colleagues [4] are shown in Figure 20.3 [5]

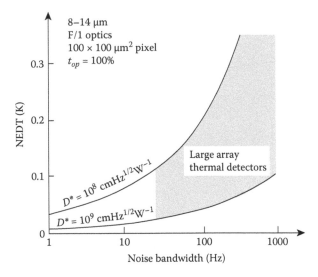

Figure 20.1 The NETD versus equivalent noise bandwidth for typical detectivities of thermal detectors. (From Watton, R., and Mansi, M. V., "Performance of a Thermal Imager Employing a Hybrid Pyroelectric Detector Array with MOSFET Readout," *Proceedings of SPIE* 865, 78–85, 1987. With permission.)

Table 20.1: Approximate Costs of Commercial Uncooled Arrays for Thermal Imagers

Feature	Cost (U.S. $)
640×480 pixel, 25×25 μm bolometer arrays	15,000
384×288 pixel, 35×35 μm or 25×25μm bolometer arrays	4000–5500
320×240 pixel, 50×50 μm bolometer arrays	3500–5000
320×240 pixel, 50×50 μm bolometer arrays for imaging radiometers*	15,000–30,000
120×1 pixel, 50×50 μm thermoelectric arrays for imaging radiometers*	<8000
320×240 pixel, 50×50 μm hybrid ferroelectric bolometer array imagers for driver's vision enhancement	1500–3000
160×120 pixel, 50×50 μm bolometer arrays for thermal imagers	<2000
160×120 pixel, 50×50 μm bolometer arrays for driver's vision enhancement systems	<2000
160×120 pixel, 50×50 μm bolometer arrays for imaging radiometers*	<4000

*The cost of arrays for radiometers is considerably higher and depends on specific performance requirements. Estimations given in table should be treated as approximate.

20.1 THERMOPILE FOCAL PLANE ARRAYS

Thermopiles are widely used, mostly as single-point detectors, in many low-power applications such as radiation temperature sensors and infrared gas detectors. They have very useful characteristics; they are highly linear, require no optical chopper, and have D^* values comparable to resistive bolometers and pyroelectric detectors. They operate over a broad temperature range with little or no temperature stabilization. They have no electrical bias, leading to negligible $1/f$ noise and no voltage pedestal in their output signal. However, array implementations of thermopiles are limited and much less effort has been made in their development. It is mainly owing to the large pixel size required for implementing each thermopile pixel. Pixels size, such as 250×250 μm², especially limits their use for large format detector arrays. Their responsivity (of the order of 5–15 V/W) and noise are orders of magnitude less and thus their applications in thermal imaging systems require very low-noise electronics to realize their potential performance. Thermoelectric detectors found almost no use as matrix arrays in TV frame rate imagers. Instead, they are employed as linear arrays that are mechanically scanned to form an image of stationary or nearly stationary objects. The wide operating-temperature range, lack of temperature stabilization owing to their inherent differential operation between hot and cold junctions, and radiometric accuracy

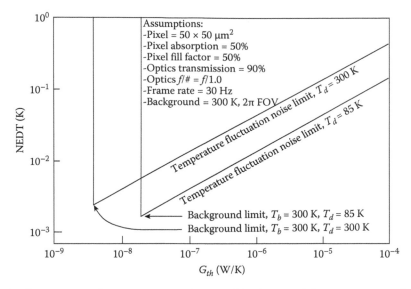

Figure 20.2 Temperature fluctuation noise limit and background fluctuation noise limit to NETD of uncooled and cryogenic thermal detector FPAs as a function of thermal conductance. Other parameters are listed on the figure. (From Kruse, P. W., *Infrared Physics & Technology*, 36, 869–82, 1995. With permission.)

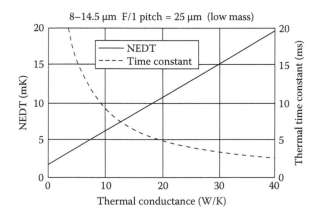

Figure 20.3 Trade-off between sensitivity and response time of uncooled thermal imaging systems. (From Ratches, J. A., *Ferroelectrics*, 342, 183–92, 2006.)

make thermopiles well situated for same space-based scientific imaging applications [6]. However, the temperature gradient in the thermopile array may cause significant offsets. Therefore, careful array design is required to minimize spatial variation in the array temperature. These limitations prevent the use of thermopiles for infrared imagers that require large FPAs, and the attention on uncooled infrared detectors has shifted mainly to microbolometers.

In spite of the above limitations, there are some successful FPA implementations merged with readout electronics. An interesting 128 × 128 array implementation using post-CCD surface micromachining has been described by Kanno et al. [7]. Each thermopile pixel in the array has 32 pairs of p-polySi/n-polySi thermocouples, each 100×100 μm^2, with a fill factor of 67% (see Figure 20.4). Over the CCD, silicon dioxide diaphragms with thickness of 450 nm (for thermal isolation structure) is made using micromachining technology. The polysilicon electrode is 70 nm thick and 0.6 μm wide. The hot junctions are located at the central part of the diaphragm, while the cold junctions are located on the outside edge of the diaphragm, where the heat conductance is very large.

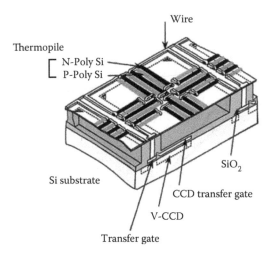

Figure 20.4 Pixel structure for thermopile infrared focal plane array. (From Kanno, T., Saga, M., Matsumoto, S., Uchida, M., Tsukamoto, N., Tanaka, A., Itoh, S., et al., "Uncooled Infrared Focal Plane Array Having 128 × 128 Thermopile Detector Elements," *Proceedings of SPIE* 2269, 450–59, 1994. With permission.)

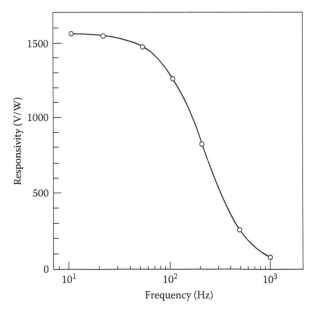

Figure 20.5 Frequency dependence of the thermopile voltage responsivity. (From Kanno, T., Saga, M., Matsumoto, S., Uchida, M., Tsukamoto, N., Tanaka, A., Itoh, S., et al., "Uncooled Infrared Focal Plane Array Having 128 × 128 Thermopile Detector Elements," *Proceedings of SPIE* 2269, 450–59, 1994. With permission.)

The low-frequency voltage responsivity of the 32 pairs thermopile is found to be 1550 V/W (see Figure 20.5). The scanner of 128 × 128 thermopile FPA consists of a vertical and horizontal buried type CCDs. They have overlapping double-layer polysilicon electrodes. The cutoff frequency of 130 Hz was sufficiently large to take a moving object with a 30 or 60 frame/sec frame rate. The reported NEDT is 0.5 K with $f/1$ optics. Although these parameters are very good for a thermopile FPA, requirements of vacuum operation packaging increases its cost. Moreover, CCD technology is today less widespread technology in comparison with CMOS.

Another large-format thermopile FPA suitable for application to various automotive sensor systems has been recently demonstrated by the Nissan Research Center [8,9]. Each detector in a 120 × 90 element array consists of two pairs of p-n polysilicon thermocouples and has external dimensions of 100×100 µm² and internal electrical resistance of 90 kΩ. To thermally isolate the detectors, the front-end bulk etching is used. A precisely patterned Au-black absorbing layer and a lift-off technique utilizing a phosphorus silicate glass sacrificial layer provides high responsivity of detectors—about 3900 V/W. The thermopiles are monolithically integrated with a 0.8 µm CMOS process. Table 20.2 shows device specifications and performance. The measured NEDT is 0.5 K with $f/1$ optics, which is on the same level as the thermoelectric FPA with CCD scanner [7]. An infrared image taken with offset and responsivity compensation is shown in Figure 20.6 [9].

Also researchers at the University of Michigan demonstrated a 32×32 FPA compatible with an in-house 3 µm CMOS process. The pixel size of 375×375 µm² with an active are of 300×300 µm² (fill factor 64%) were implemented with 32 n-p polysilicon thermocouples on dielectric diaphragms [10] using mainly micromachining from the back side of the wafers, but small etch cavities were also placed on the front side of the wafers to achieve heat sink between pixels, to prevent heating of the cold junction and to achieve good thermal isolation between adjacent pixels. The device has a responsivity of 15 V/W, a thermal time constant of 1 ms, and a detectivity of 1.6×10^7 cmHz$^{1/2}$/W.

Table 20.2: Specification of 120×90 Infrared Imaging Thermopile Sensor

Parameter	
Element pitch	100 µm
Fill factor	42%
Thermopile pairs	2
Thermopile width	0.8 µm
Number of beams	2
Beam width	4.4 µm
Responsivity	3900 V/W
Time constant	44 ms
Resistance	90 kΩ
Die size	14×11 mm²
Window	Ge
Package size	44 mm (dia.)

Source: M. Hirota, Y. Nakajima, M. Saito, and M. Uchiyama, *Sensors Actuators,* A135, 146–51, 2007.

Figure 20.6 A infrared image taken with a 120×90 element FPA. (From Hirota, M., Nakajima, Y., Saito, M., and Uchiyama, M., *Sensors Actuators,* A135, 146–51, 2007. With permission.)

Requirements for low cost and manufacturability of two-dimensional thermopile arrays led to the use of polysilicon thermoelectric material, which have relatively low thermoelectric figures of merit. Foote and coworkers [11–13] have improved the performance of thermopile linear arrays by combining Bi-Te and Bi-Sb-Te thermoelectric materials. Compared with most other thermoelectric arrays, their D^* values are highest, which is shown in Figure 5.5. The thermopile linear array technology is described in Ref. [11]. They consist of 0.5 μm thick Si_3N_4 membranes formed back side etching of the underlying silicon substrate. On each membrane there are a number of Bi-Te and Bi-Sb-Te thermocouple running along narrow legs between the substrate and membrane. The detectors are closely spaced with slits through the membrane separating the detectors from each other and defining the detector legs.

The linear thermopile arrays have been bonded to separate CMOS readout electronic chips since Bi-Sb-Te materials are not readily available in a CMOS technology. Next, this technology has been developed to improve the performance of 2-D arrays using a three-level structure with two sacrificial layers. In such a way, it is possible to improve the fill factor and incorporate a large number of thermocouple per pixel. Figure 20.7 shows the thermopile detector structure [13]. The structure allows almost 100% fill factor and the model suggests that optimized detectors will have D^* values over 10^9 cmHz$^{1/2}$/W. Further efforts are continued to fabricate high performance large FPAs.

McManus and Mickelson [14] and Kruse [2] have described a family of imaging radiometers that employ silicon microstructure linear thermoelectric arrays using chromel/constantan thermocouples. One example employs a 120 pixel linear array with a pitch of 50 μm and thermopiles consisting of three series-connected thermocouples. The linear array is mechanically scanned across the focal plane of an f/0.7 germanium lens in 1.44 s. The NEDT is 0.35 K. The specification of thermoelectric linear array employed in the IR SnapShot® imaging radiometer is listed in Table 20.3 [2].

Ann Arbor Sensor Systems, LLC, a small Michigan-based company specializing in noncontact temperature measurement, has teamed up with Malaysia-based MemsTech in developing the first commercial thermal imaging AXT100 camera using thermopile 32 × 32 FPA technology (see Figure 20.8 [15]). The low production cost is achieved through conventional CMOS processing and NO

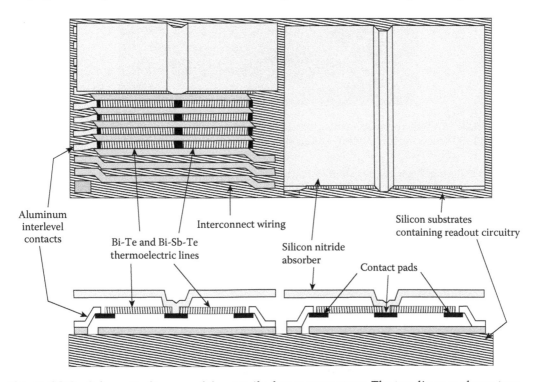

Figure 20.7 Schematic diagram of thermopile detector structure. The top diagram shows two pixels viewed from the top, with part of the left pixel cut away to show the underlying structure. The lower diagram shows a cross-section side view of two pixels. (From Foote, M. C., and Gaalema, S., "Progress Towards High-Performance Thermopile Imaging Arrays," *Proceedings of SPIE* 4369, 350–54, 2001. With permission.)

Table 20.3: Specifications of Thermoelectric Linear Arrays Employed in IR SnapShot® Imaging Radiometers

Parameter	
Number of pixels	128
Number of pixels accessed	120
Pixel size (μm)	50
Number of junctions per pixel	3
Resistance at 300 K (Ω)	2380
Thermal response time (sec)	12
Responsivity (V/W)	265
D^* (cmHz$^{1/2}$/W)	1.7×10^8

Source: P. W. Kruse *Uncooled Thermal Imaging. Arrays, Systems, and Applications,* SPIE Press, Bellingham, WA, 2001. With permission.

Figure 20.8 AXT100 thermopile imaging camera fabricated by Ann Arbor Sensor Systems. (From AXT100 brochure, http://www.aas2.com/products/axt100/index.htm.)

vacuum packaging. Some of the camera's features include image processing that interpolates and smoothes a 32 × 32 image to 128 × 128 resolution. Two manual focus lens options are available: a 29° (f/0.8) or a 22° (f/1.0), with a spectral range of 7–14 μm. The camera provides composite video and S-Video outputs (including NTSC and PAL formats).

20.2 BOLOMETER FOCAL PLANE ARRAYS

Although investment in uncooled thermal imaging goes back to the 1970s, the initial investment was in pyroelectric materials. Today's achievements on uncooled infrared technology were initiated by the U.S. Department of Defense in the 1980s, when it gave large classified contracts to both Honeywell and Texas Instruments to develop two different uncooled infrared technologies [2,16,17]. Texas Instruments concentrated on pyroelectric technology [such as barium strontium titinate (BST)], whereas Honeywell concentrated on microbolometer technology, and they both successfully developed uncooled infrared 320 × 240 format FPAs with sensitivities less than 50 mK. These technologies were unclassified in 1992, and since then many other companies have started working in this technology. Honeywell has licensed this technology to several companies for the development and production of uncooled FPAs for commercial and military systems. The U.S. government allowed American manufactures to sell their devices to foreign countries, but not to divulge manufacturing technologies. Several countries, including the United Kingdom, France, Japan, and Korea have picked up the ball, determined to develop their own uncooled imaging systems. As a result, although the United States has a significant lead, some of the most exciting and promising developments for low-cost uncooled IR systems may come from non-U.S. companies (e.g., microbolometer FPAs with series p-n junction elaborated by Mitsubishi Electric [18]). This approach is unique, based on an all-silicon version of microbolometer. At present the most important manufacturers of uncooled microbolometer FPAs are: Raytheon [19–23], BAE (formerly Honeywell) [24–27], DRS (formerly Boeing) [28–30], Indigo [31,32], InfraredVision Technology [33], and L-3 Communications Infrared Products [34] in the United States; INO in Canada [35–38]; ULIS in France [39–42]; NEC [43–46] and Mitsubishi [18,47–50] in Japan; QuinetiQ [51] in the United Kingdom; XenICs [52] in Belgium; and SCD [53,54] in Israel. Also many research institutions are working on uncooled microbolometer infrared arrays.

Thermal detectors are currently of considerable interest for two-dimensional electronically addressed arrays where the bandwidth is low and the ability of thermal devices to integrate for a frame time is an advantage. The development of the microbridge detector arrays has provided a significant leap forward in sensitivity and array size for uncooled thermal imagers. The sensitivity is not as good as the cooled photon detectors, however it is sufficient for low cost, lightweight, low power IR imagers. Today 1024×768 element arrays with 17 μm pixel size are available with the predicted NEDT less than 50 mK.

As is mentioned in Section 19.4, when the detectors are integrated into arrays, the ability to have high detectivity is important, but the most important figure of merit is the ability to resolve small temperature differences in the field of view. This figure of merit is pressed as the NEDT.

For correct calculation of the NEDT, it is important to analyze noise sources over the proper bandwidths. For pulse biased microbolometer systems (e.g., VO$_x$ bolometers), there are three important bandwidths: the electrical bandwidth, the thermal bandwidth, and the output bandwidth [2,55].

The electrical bandwidth is determined by the integration time of the bias pulse to measure the resistance of the detector. When using pulsed bias, the electrical bandwidth is given by [2]

$$\Delta f = \frac{1}{2\Delta t},$$ (20.3)

where Δt is the bias pulse duration. Assuming typical integration time of 60 μsec, the electrical bandwidth is 8 kHz.

The electrical bandwidth is important for analyzing the system contributions of $1/f$ and Johnson noise. Often, for large FPAs that are readout in a serial manner by pulsed bias, the bandwidth can be sufficiently large that the Johnson noise is much greater than $1/f$ noise over that bandwidth.

The thermal bandwidth is determined by the thermal time constant of the bolometer and is important for analyzing thermal fluctuation noise. Assuming the bolometer as a first order low pass filter, the thermal bandwidth is given by

$$\Delta f = \frac{1}{4\tau_{th}},$$ (20.4)

where τ_{th} is thermal time constant. For typical time constants between 5 and 20 msec, the thermal bandwidth changes between 12 and 50 Hz.

The output bandwidth is the bandwidth at which the bolometer is pulsed. Assuming the frame rate as 30 or 60 Hz, the output bandwidth given by

$$\Delta f = \frac{\text{frame rate}}{2},$$ (20.5)

is 15–30 Hz.

From the above analysis results that the electrical bandwidth is much greater than the thermal or output bandwidth.

It can be shown that contribution of different types of noise to NEDT follows as [55]:

■ Johnson noise

$$NEDT_{Johnson} \propto \frac{G_{th} \left(TR_B \right)^{1/2}}{V_b \alpha},$$ (20.6)

■ thermal fluctuation noise

$$NEDT_{tf} \propto T G_{th}^{1/2},$$ (20.7)

- $1/f$ noise

$$\text{NEDT}_{1/f} \propto \frac{\beta G_{th}}{\alpha}, \tag{20.8}$$

where R_B is the bolometer resistance, β is ratio of the measured $1/f$ noise voltage, $V_{1/f}$, to the detector bias voltage, V_b.

Under the $1/f$ noise model, the noise voltage would be inversely proportional to the square root of the number of carriers, N, [56] and the volume. Therefore

$$\beta = \frac{V_{1/f}}{V_b} \propto \frac{1}{N^{1/2}} \propto \frac{1}{(\text{Volume})^{1/2}} \propto \frac{1}{A^{1/2}}. \tag{20.9}$$

Figure 20.9 shows the $1/f$ noise contribution to noise as a function of the surface area of a VO_x film fabricated by BAE Systems [55]. This dependence follows the relation described by Equation 20.9. Unfortunately, this dependence of $1/f$ noise predicts that smaller pixel bolometers will have higher $1/f$ noise than larger pixels.

Figure 20.10 presents the influence of different detector and electronic noise sources on VO_x BAE Systems microbolometer performance. The key conclusion that can be drawn from this figure

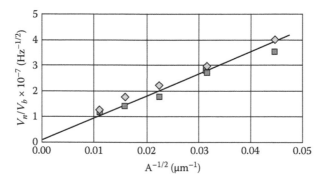

Figure 20.9 Dependence of the V_n/V_b ratio on the square root inversion of the surface area of the VO_x film. (From Kohin, M., and Butler, N., "Performance Limits of Uncooled VO_x Microbolometer Focal-Plane Arrays," *Proceedings of SPIE* 5406, 447–53, 2004. With permission.)

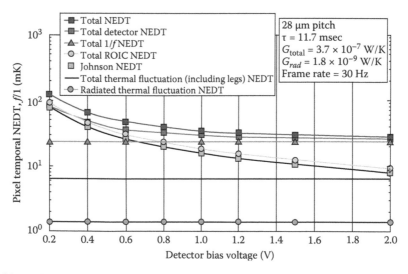

Figure 20.10 Influence of different noise sources on performance of VO_x microbolometer. (From Kohin, M., and Butler, N., "Performance Limits of Uncooled VO_x Microbolometer Focal-Plane Arrays," *Proceedings of SPIE* 5406, 447–53, 2004. With permission.)

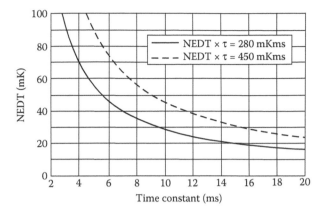

Figure 20.11 Calculated microbolometer NEDT and thermal time constant, τ_{th}, for two NEDT $\times$ τ_{th} products. (From Kohin, M., and Butler, N., "Performance Limits of Uncooled VO$_x$ Microbolometer Focal-Plane Arrays," *Proceedings of SPIE* 5406, 447–53, 2004. With permission.)

is that current microbolometer performance is limited by $1/f$ noise in the VO$_x$ material. Large $1/f$ noise is due to noncrystalline VO$_x$ structure. As Equation 20.6 indicates, the Johnson noise contribution to NEDT is inversely proportional to the bias voltage, V_b. This type of noise does not significantly degrade system performance when the bolometer is highly biased. Both the ROIC and Johnson noise approach the thermal fluctuation noise in a high enough bias region. The thermal fluctuation noise is dominated by thermal conductance of the detector legs, not by radiated conductance.

In order to improve the performance of VO$_x$ microbolometers, the $1/f$ noise must be reduced. From Equation 20.8 results that it can be achieved by:

- reducing the $1/f$ noise of the detector material,

- reducing the thermal conductance of the bolometer legs, and

- increasing the α of the VO$_x$.

It appears that the last approach has the negative effect on the dynamic range requirements for ROIC [55].

If the NEDT is dominated by a noise source that is proportional to G_{th}, which has taken place when Johnson and $1/f$ noises are dominated (see Equations 20.6 and 20.8), and since $\tau_{th} = C_{th}/G_{th}$, then the Figure of Merit given by

$$FOM = NEDT \times \tau_{th} \qquad (20.10)$$

can be introduced [55]. Users are interested not only in the sensitivity, but also in their thermal time constants and the FOM described by Equation 20.10 recognizes the trade-off between thermal time constant and sensitivity. Figure 20.11 shows the dependence of NEDT on thermal time constant for two NEDT $\times$ τ_{th} products.

Various methods for improving microbolometer sensitivity are pointed out in Table 20.4. As is shown, many methods of the sensitivity improving negatively affect the thermal time constant.

20.2.1 Manufacturing Techniques

Most modern microbolometer FPA technology derives from the pioneering efforts of a team under the direction of R. A. Wood at the Honeywell Technology Center that began in 1982 [2]. In 1985, Honeywell Technology Center received military contracts from U.S. Department of Defense, especially Defense Advanced Research Projects Agency (DARPA), and U.S. Army Night Vision and Electronic Sensors Directorate (NVESD). These contracts led to the successful development of an uncooled vanadium oxide 50 μm pixel 240 × 336 arrays operating at the U.S. TV frame rate of 30 Hz.

In the period 1990–1994 Honeywell licensed this technology originally to four companies (Hughes, Amber, Rockwell, and Loral) for the development and production of uncooled FPAs for commercial and military systems. Following acquisitions and merges with defense and aerospace

Table 20.4: Methods for Improving Sensitivity of a Microbolometer System

Design/Process Modification	Impact on NEDT	Impact on Thermal Time Constant	Impact on System Size	Comments
Increase VO_x volume	Reduce	Increase		Pixel resistance must be high enough so VO_x resistance dominates total pixel resistance; increasing length does not negatively affect resistance
Reduce $1/f$ noise inherent in material	Reduce			How?
Increase VO_x TCR	Reduce?			Not known whether higher TCR material will have equivalent or lower $1/f$ noise
Reduce leg thermal conductance	Reduce	Increase		Pixel resistance must be high enough so VO_x resistance dominates total pixel resistance
Reduce bridge heat capacitance	Increase	Reduce		
Increase leg thermal conductance	Increase	Reduce		
Reduce $f/\#$	Reduce		Increase	
Reduce pixel pitch	Increase		Reduce	Essential for smaller, cheaper systems

Source: M. Kohin and N. Butler, "Performance Limits of Uncooled VO_x Microbolometer Focal-Plane Arrays," *Proceedings of SPIE* 5406, 447–53, 2004. With permission.

industries there are now: British Aerospace (the original Honeywell division) and Raytheon. A great deal of activity on VO_x bolometer arrays in different manufacturers modified Honeywell microbolometer support structure to increase fill factor, decrease size of pixels, and improve CMOS readout.

The first 240×336 arrays of VO_x, 50 µm microbolometers were fabricated on industry-standard wafer (4-inch diameter) complete with monolithic readout circuits integrated into underlying silicon (see Section 6.2.4) [57]. To obtain the high thermal isolation of the microbolometer, the ambient gas pressure is typically on the order of 0.01 mbar. Thermal conduction through the bolometer legs can be as low as 3.5×10^{-8} W/K [55]. The bolometers in principle do not need to be thermally stabilized. However, to simplify the problem of pixel nonuniformity correction, the original Honeywell bolometer array incorporated a thermoelectric temperature stabilizer. Another issue was the need to readout the signal by accessing the pixels sequentially. The method chosen was by pulsing the electrical bias to the pixels sequentially. A bipolar input amplifier was normally required, and this was obtained with biCMOS technology. Horizontal and vertical pixel addressing circuitry was integrated with the array but most of the analog readout circuitry was off-chip. The dominant noise was Johnson noise in the sensitive resistor (typically 10–20 kΩ), with some additional contribution from $1/f$ noise and transistor readout noise. In operation, an array consumed about 40 mW [57,58]. An average NEDT of better than 0.05 K was demonstrated with uncooled imager fitted with as $f/1$ optics (see Figure 20.12).

Today, most of the approaches employ CMOS silicon circuitry for which the power dissipation is much less than that of bipolar. Moreover, most of the readout electronics has been moved onto the chip, where it is referred to as the ROIC. Column parallel readout architectures with integrated AD conversion are commonly used in commercial FPAs [55,59].

The surface micromachined bridges on CMOS-processed wafers developed in Honeywell is one of the most widely used monolithic approaches for uncooled imaging. The simplified step processes of monolithic integration are shown in Figure 20.13 [60]. At the beginning, the ROIC is premanufactured and the detector materials are subsequently deposited and patterned on the ROIC wafer. For fabrication of the sacrificial layer, a high-temperature stable polyimide is used typically. Finally, the polyimide layer is removed in an oxygen plasma to obtain free-standing, thermally isolated bolometer membranes. The deposition process for the sensing bolometer material is limited to about 450°C due to risk damaging of the ROIC. Low deposition temperature precludes receiving of monocrystalline materials, which is a potential disadvantage of monolithic integration. This disadvantage is especially serious for poly-SiGe resistive microbolometers developed by

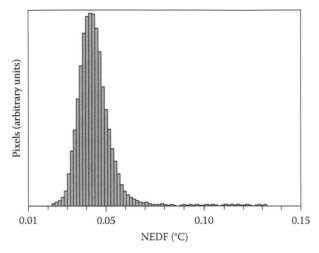

Figure 20.12 Measured pixel NEDT histogram of Honeywell uncooled imager with $f/1$ optics. (From Wood, R. A., "Uncooled Thermal Imaging with Monolithic Silicon Focal Planes," *Proceedings of SPIE* 2020, 322–29, 1993. With permission.)

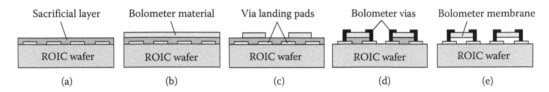

Figure 20.13 Monolithic integration for uncooled infrared bolometer arrays: (a) deposition of sacrificial layer on ROIC wafer, (b) deposition of bolometer materials, (c) patterning of bolometer materials, (d) via formation, and (e) etching of sacrificial layer. (From Niklas, F., Vieider, C., and Jakobsen, H., "MEMS-Based Uncooled Infrared Bolometer Arrays: A Review," *Proceedings of SPIE* 6836, 68360D-1, 2007. With permission.)

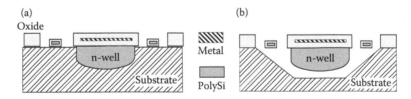

Figure 20.14 Bulk micromachining for uncooled infrared bolometer arrays: (a) formation of the bolometer and the electronics for signal readout (typically side-by-side), and (b) selective etching of the bulk material underneath the bolometer membrane. (From Eminoglu, S., Sabuncuoglu Tezcan, D., Tanrikulu, M. Y., and Akin, T., *Sensors & Actuators A* 109, 102–13, 2003. With permission.)

IMEC, and subsequently transferred to XenICs, Belgium. Poly-SiGe needs to be deposited at high temperatures; therefore, it is not easy to integrate with CMOS [52]. This material exhibits high $1/f$ noise owing to its noncrystalline structure and requires complicated post-CMOS processing to reduce the effects of residual stress.

An alternative to manufactured uncooled bolometers is bulk micromachining shown in Figure 20.14 [61], where the bolometers are formed in the substrate surface of a wafer. Subsequently, the substrate is selectively etched underneath the bolometers to thermally isolate them from the rest of the substrate. The micromachining process is implemented before, in between, or after processing the wafers to implement electronic components. In the bulk micromachining process both electronics and the bolometers can typically be manufactured in a standard

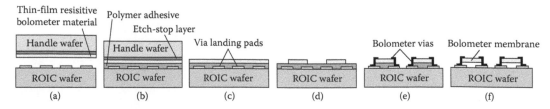

Figure 20.15 Heterogeneous 3-D integration for uncooled infrared bolometer arrays: (a) separate fabrication of ROIC wafer and handle wafer with resistive bolometer material, (b) adhesive wafer bonding, (c) thinning of handle wafer, (d) bolometer definition, (e) via formation, and (f) sacrificial etching of polymer adhesive. (From Niklas, F., Vieider, C., and Jakobsen, H., "MEMS-Based Uncooled Infrared Bolometer Arrays: A Review," *Proceedings of SPIE* 6836, 68360D-1, 2007. With permission.)

CMOS line, which is its advantage. However, a disadvantage of this technology is that the ROIC cannot be placed underneath the bolometer membranes, but has to be placed beside the bolometers that, in consequence, reduces the array fill factor. The bulk micromachining technique is successfully used in commercial fabrication of diode bolometer arrays by Mitsubishi [50].

The third technology of microbolometer array fabrication is the so-called heterogeneous three dimensional (3-D) bolometer integration shown schematically in Figure 20.15 [60]. In this case, the bolometer materials are deposited on a separate handle wafer, and next, the materials are transferred from the handle wafer to the ROIC wafer using low-temperature adhesive wafer bonding. The important advantage of this technology is that it allows the use of high performance monocrystalline sensing bolometer materials on top of standard ROICs. This 3-D bolometer integration process is implemented in Acreo, Sweden [62].

20.2.2 FPA Performance

The microbolometer detectors are now produced in larger volumes than all other IR array technologies together. The demonstrated performance is getting closer to the theoretical limit with the advantages regarding weight, power consumption, and cost. The 240×320 arrays of 50 μm microbolometers are fabricated on industry-standard wafer (4-inch diameter) complete with monolithic readout circuits integrated into underlying silicon. Radford et al. have reported a 240×320 pixel array with 50 μm square vanadium oxide pixels and thermal time constant of about 40 ms, for which the average NEDT ($f/1$ optics) was 8.6 mK [63].

However, there is a strong system need to reduce the pixel size to achieve several potential benefits. The detection range of many uncooled IR imaging systems is limited by pixel resolution rather than sensitivity. The cost of the optics made of Ge, the standard material, depends approximately upon the square of the diameter, so reducing the pixel size causes a reduction in cost of the optics. These reductions in optics size would have a major benefit in reducing the overall size, weight, and cost of manportable IR systems. In addition the reduction in pixel size allows a significantly larger number of FPAs to be fabricated on each wafer. However, the NEDT is inversely proportional to the pixel area, thus, if the pixel size is reduced from 50×50 μm to 17×17 μm, and everything else remained the same, the NEDT would increase by a factor of nine. Improvements in the readout electronics are needed to compensate for this. For future arrays, the $f/1$ NEDT performance of 17 μm pitch microbolometer FPAs is projected to be below 20 mK (see Figure 20.16) [55,64]. The development of highly sensitive 17 μm microbolometer pixels, however, presents significant challenges in both fabrication process improvements and in pixel design. Microbolometer pixels fabricated with conventional single-level micromachining processes suffer severe performance degradation as the unit cell is reduced below 40 μm. This problem can be mitigated to some degree if the microbolometer process capability (design rules) is improved dramatically.

At present, the commercially available bolometer arrays are either made from VO_x, amorphous silicon (a-Si), or silicon diodes. The VO_x is dominated technology apart from its drawback that it is not compatible with a CMOS line and requires a separate line after the CMOS process to prevent contamination of the CMOS line. Figure 20.17 shows scanning electron microscope (SEM) images of commercial bolometers fabricated by different manufacturers [30,55,65].

Conventional single-level bolometer arrays typically have a fill factor between 60 and 70% [30,40]. To increase the fill factor, two-layer bolometer designs have been reported that reach fill

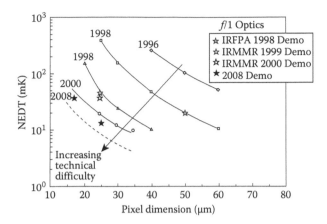

Figure 20.16 VO$_x$ focal plane array development. (From Kohin, M., and Butler, N., "Performance Limits of Uncooled VO$_x$ Microbolometer Focal-Plane Arrays," *Proceedings of SPIE* 5406, 447–53, 2004; Anderson, J., Bradley, D., Chen, D. C., Chin, R., Jurgelewicz, K., Radford, W., Kennedy, A., et al., "Low Cost Microsensors Program," *Proceedings of SPIE* 4369, 559–65, 2001. With permission.)

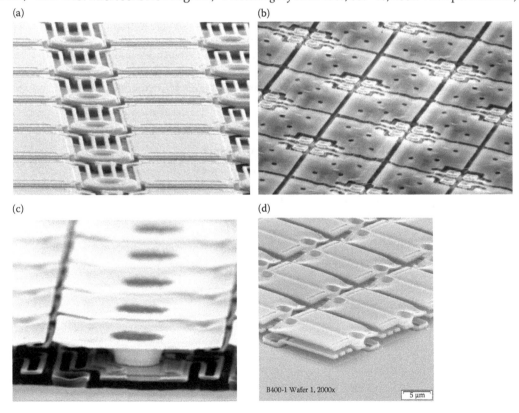

Figure 20.17 Commercial bolometer design: (a) VO$_x$ bolometer with $28 \times 28 \ \mu m^2$ pitch from BAE. (From Kohin, M., and Butler, N., "Performance Limits of Uncooled VO$_x$ Microbolometer Focal-Plane Arrays," *Proceedings of SPIE* 5406, 447–53, 2004. With permission.) (b) a-Si bolometer with $50 \times 50 \ \mu m^2$ pitch from Ulis, (c) VO$_x$ umbrella design bolometer with $17 \times 17 \ \mu m^2$ pitch from DRS. (From Li, C., Skidmore, G. D., Howard, C., Han, C. J., Wood, L., Peysha, D., Williams, E., et al., "Recent Development of Ultra Small Pixel Uncooled Focal Plane Arrays at DRS," *Proceedings of SPIE* 6542, 65421Y, 2007. With permission.) (d) VO$_x$ bolometer with $17 \times 17 \ \mu m^2$ pitch from Raytheon. (From Black, S., Ray, M., Hewitt, C., Wyles, R., Gordon, E., Almada, K., Baur, S., Kuiken, M., Chi, D., and Sessler, T., "RVS Uncooled Sensor Development for Tactical Applications," *Proceedings of SPIE* 6940, 694022, 2008. With permission.)

factor of up to 90% [21,46]. The double deck structures were first presented by research at KAIST in 1998 [66]. Figure 20.17c shows an example of a two-level (umbrella) design. In this case, the bolometer legs, and in some cases the sensing material, are placed underneath the absorbing bolometer membrane [30,46]. The umbrella designs are implemented in arrays with very small pixel sizes up to 17×17 μm^2 [30]. Another way has been chosen by Raytheon, which also implemented an advanced double-layer micromachining fabrication process. In this case, the thermal isolation layer is fabricated on the first level of the structure while the optical absorber level is fabricated on the second level of the structure. This process has been also successfully implemented for 17 μm pitch FPAs (see Figure 20.17d).

Table 20.5 contains an overview of the main suppliers and specifications for existing products and for bolometer arrays that are in the R&D stage, while Table 20.6 summarizes the design and performance parameters for Raytheon's VO_x microbolometers. As we can see, the similar performance has been described by BAE Systems [26,27], DRS [30], Ulis [41,42], L-3 [34], and SCD [54]. Outstanding bolometer performance results in improvements of thermal and spatial resolution and image quality of infrared cameras. For example, Figure 20.18 shows the image of a four bar differential black body ($\Delta T = 20°C$), one with the 640×480 25 μm pixel pitch and the other one with the 1024×768 17 μm pixel pitch [67].

Commercial bolometer arrays that are manufactured using bulk micromachining techniques are series-connected diode microbolometers developed by Mitsubishi using a custom SOI CMOS technology [47–50]. Figure 20.19 shows a schematic of the detector cross section and an SEM view of the recent diode pixels. To lower the thermal conductance without degrading the efficiency of infrared absorption, the most advanced pixels has a three-level structure that has an independent metal reflector for interface infrared absorption between the temperature sensor (bottom level) and the infrared absorbing thin metal film (top level). The MEMS process includes a XeF_2 dry bulk

Table 20.5: Commercial and State-of-the-Art R&D Uncooled Infrared Bolometer Array

Company	Bolometer Type	Array Format	Pixel Pitch (μm)	Detector NEDT (mK) ($f/1$, 20–60 Hz)
FLIR (United States)	VO_x bolometer	160×120–640×480	25	35
L-3 (United States)	VO_x bolometer	320×240	37.5	50
	a-Si bolometer	160×120–640×480	30	50
	a-Si/a-SiGe	320×240–1024×768	R&D:17	30–50
BAE (United States)	VO_x bolometer	320×240–640×480	28	30–50
	VO_x bolometer (standard design)	160×120–640×480	17	50
	VO_x bolometer (standard design)	1024×768	R&D:17	
DRS (United States)	VO_x bolometer (umbrella design)	320×240	25	35
	VO_x bolometer (standard design)	320×240	17	50
	VO_x bolometer (umbrella design)	640×480	R&D:17	
Raytheon (United States)	VO_x bolometer	320×240–640×480	25	30–40
	VO_x bolometer (umbrella design)	320×240–640×480	17	50
	VO_x bolometer (umbrella design)	640×480, 1024×768	R&D:17	
ULIS (France)	a-Si bolometer	160×120, 640×480	25–50	35–80
	a-Si bolometer	1024×768	R&D:17	
Mitsubishi (Japan)	Si diode bolometer	320×240, 640×480	25	50
SCD (Israel)	VO_x bolometer	384×288	25	50
	VO_x bolometer	640×480	25	50
NEC (Japan)	VO_x bolometer	320×240	23.5	75

Table 20.6: Performance Characteristics of Raytheon's VO$_x$ Microbolometers

Performance Parameter	Capability (f/1 and 300 K Scene)		
Array configuration	320×240	320×240	640×480
Pixel size (µm²)	50×50	25×25	20×20
Spectral response (µm)	8–14	8–14	8–14
Signal responsivity (V/W)	>2.5 × 10⁷ V/W or 50 mK/K$_{scene}$	>2.5 × 10⁷ V/W or 20 mK/K$_{scene}$	>2.5 × 10⁷ V/W or 25 mK/K$_{scene}$
NEDT @ f/1 (Mk)	<20	<30	<30
Offset nonuniformity (mV)	<150 p-p	<150 p-p	<150 p-p
Output noise (mV)	1.0 rms	1.0 rms	0.6 rms
Intrascene dynamic range @ f/1 (K)	>40	>100	>100
Pixel operability (%)	>98	>99	>98
Power dissipation (mW)	200	150	390
Nominal operating temperature (°C)	25	25	25

Source: D. Murphy, M. Ray, J. Wyles, C. Hewitt, R. Wyles, E. Gordon, K. Almada, et al., "640 × 512 17 µm Microbolometer FPA and Sensor Development," *Proceedings of SPIE* 6542, 65421Z, 2007; W. Radford, D. Murphy, A. Finch, K. Hay, A. Kennedy, M. Ray, A. Sayed, et al., "Sensitivity Improvements in Uncooled Microbolometer FPAs," *Proceedings of SPIE* 3698, 119–30, 1999; S. Black, M. Ray, C. Hewitt, R. Wyles, E. Gordon, K. Almada, S. Baur, M. Kuiken, D. Chi, and T. Sessler, "RVS Uncooled Sensor Development for Tactical Applications," *Proceedings of SPIE* 6940, 694022, 2008. With permission.

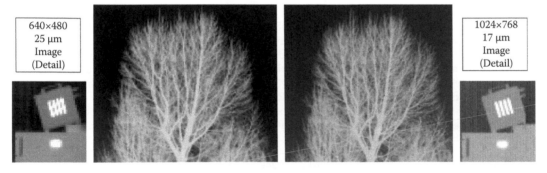

Figure 20.18 Differential black body image (left 640 × 480 25 µm, right 1024 × 768 17 µm) and image quality comparison (details). (From Fieque, B., Robert, P., Minassian, C., Vilain, M., Tissot, J. L., Crastes, A., Legras, O., and Yon, J. J., "Uncooled Amorphous Silicon XGA IRFPA with 17µm Pixel-Pitch for High End Applications," *Proceedings of SPIE* 6940, 69401X, 2008. With permission.)

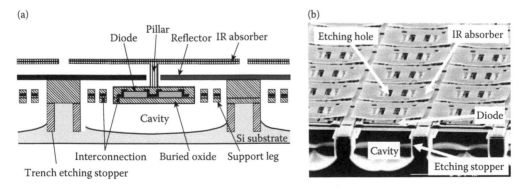

Figure 20.19 SOI diode microbolometer: (a) schematic of the detector cross section, and (b) SEM view of the diode pixels with 40 × 40 µm² pitch. (From Kimata, M., Uenob, M., Takedac, M., and Setod, T., "SOI Diode Uncooled Infrared Focal Plane Arrays," *Proceedings of SPIE* 6127, 61270X, 2006. With permission.)

Table 20.7: Specifications and Performance of SOI Diode Uncooled IR

	320×240	320×240	320×240	640×480
Pixel size (µm)	40×40	28×28	25×25	25×25
Chip size (mm)	17.0×17.0	13.5×13.0	$12.5.0 \times 13.5$	20.0×19.0
Pixel structure	Two-level	Three-level	Three-level	Three-level
Number of diodes	8	6	6	6
Thermal conductance (W/K)	1.1×10^{-7}	4.0×10^{-8}	1.6×10^{-8}	1.6×10^{-8}
Sensitivity (µV/K)	930	801	2 842	2 064
Noise (µV rms)	110	70	102	83
Nonuniformity (%)	1.46	1.25	1.45	0.90
NEDT, $f/1$ (mK)	120	87	36	40

Source: M. Kimata, M. Uenob, M. Takedac, and T. Setod, "SOI Diode Uncooled Infrared Focal Plane Arrays," *Proceedings of SPIE* 6127, 61270X, 2006. With permission.

silicon etching and a double organic sacrificial layer surface micromachining process [50]. Arrays consisting of 640×480 pixels are based on suspended multiple series diodes with 25×25 µm² pixel sizes. The reported NEDT value is 40 mK for $f/1$ optics (see Table 20.7). Although this approach provides very uniform arrays with very good potential for low-cost, high-performance uncooled detectors, its fabrication is based on a dedicated in-house SOI CMOS process. The better approach would be to implement the detector arrays together with readout circuitry fully in standard CMOS process [61].

Encouraging results have also been obtained using semiconducting $YBa_2Cu_3O_{6+x}$ ($0.5 \leq x \leq 1$) thin films on silicon [68–70]. To ensure compatibility and potential integration with CMOS-based processing circuitry, silicon micromachining and ambient temperature processing were employed [68]. Wada et al. have developed 320×240 YBaCuO microbolometer FPA with a pixel pitch of 40 µm, NEDT of 0.08 K with a prototype camera and $f/1.0$ optics [70]. To decrease the resistance of bolometers (which is 10 Ωcm, two orders of magnitude higher than that of conventional VO_x bolometer films), the RF magnetron sputtered films were deposited on silicon with previously prepared SiO_2 isolation layer and platinum comb-shaped electrodes. More recently, efforts to implement YBaCuO detectors in various substrates have been undertaken [71,72].

20.2.3 Packaging

Generally, the bolometer array packages are based on available technologies widely developed for the packaging of mass produced electronic devices. Partially however, the packages are designed by manufacturers in-house. The lead frames enable electronic board integration like it is for standard CMOS devices, insuring high electrical contact reliability in tough and demanding environmental applications such as military operations, fire fighting, automotive applications, process control, or predictable maintenance.

For the best performance, the conventional bolometers operate with vacuum levels below 0.01 mbar [21]. Necessity of vacuum packaging is related to the fact that heat loss due to thermal conduction from bolometers through the gas gap to the substrate underneath causes an increase in the NEDT. The measurements carried out by He et al. have shown that thermal conduction through the gas starts to have an effect from a pressure in the range 0.1 mbar for a device with pixel area 50×50 µm² and an air gap of 20 µm [73].

The most important requirements for the packaging of bolometer arrays follow: good and reliable hermetic seal, integration of IR window material with good IR transmission, and high-yield, low-cost packaging [60]. The packaging may be done at chip level or wafer level. Usually, the bolometer chips are built into a hermetic metal or ceramic package with an IR-transmission lid built into the package cap. Descriptions of different methods for microsystem packaging can be found in a new handbook for MEMS [74].

One of the on-chip hermetic encapsulation methods is schematically shown in Figure 20.20a [75]. The method is based on eutectic solder bonding of two electroplated rims: one is placed on the silicon wafer and surrounds the bolometer array, while the second, with the same geometry, is placed on a wafer made from a material that transmits IR radiation. Usually germanium is used due to its infrared transparency up to 20 µm. After flip-chip bonding, the nanoliter volume

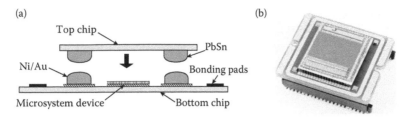

Figure 20.20 Bolometer array packaging: (a) schematic overview of the on-chip hermetic encapsulation of microbolometer array. (From De Moor, P., John, J., Sedky, S., and Van Hoof, C., "Lineal Arrays of Fast Uncooled Poly SiGe Microbolometers for IR Detection," *Proceedings of SPIE* 4028, 27–34, 2000. With permission.) (b) The SB-300 640 × 480 20 μm pitch FPA mounted in the high vacuum, low-mass and low-cost ceramic package fabricated by Raytheon. (From Murphy, D., Ray, M., Kennedy, A., Wyles, J., Hewitt, C., Wyles, R., Gordon, E., et al., "High Sensitivity 640 × 512 (20 μm pitch) Microbolometer FPAs," *Proceedings of SPIE* 6206, 62061A, 2006. With permission.)

between two substrates and two rims is pumped to the required pressure level through a small groove in the rim substrate. Finally, heating two substrates to the eutectic temperature (for PbSn only 240°C) causes the rim material to reflow. For example, Figure 20.20b shows the SB-300 640 × 480 20 μm pitch FPA assembled into ceramic vacuum packages for performance testing and imaging evolution.

20.3 PYROELECTRIC FOCAL PLANE ARRAYS

The use of pyroelectric materials and devices in infrared imaging cameras is now well-established technology. They have been in use for many years in applications ranging from intruder sensing through environmental monitoring to flame and fire detection [76]. In the last two decades there has been a growth of interest in using 2-D arrays of small pyroelectric elements for uncooled thermal imaging [2,77–86].

The imaging systems based on pyroelectric arrays usually need to be operated with optical modulators that chop or defocus the incoming radiation. This may be an important limitation for many applications in which chopperless operation is highly desirable (e.g., guided munitions). A chopper is a less reliable mechanical part, inconvenient and more bulky. However using a chopper blade, the thermal image can be produced by subtracting the field of data output by the detector while viewing the chopper from the field output when viewing the scene. This image difference processing not only removes offset variations between elements in the array but also serves as a temporal high pass filter, eliminating $1/f$ noise components and long-term drifts (low frequency spatial noise).

In recent years, attention has turned to solid state readout of large pyroelectric FPAs and in particular to the possibility of interfacing the pyroelectric arrays directly with a silicon chip. Practical arrays have been demonstrated by several groups including GEC [78], RRSE [79], DERA/BAE Systems [83] in the United Kingdom, Texas Instruments [81,84,86], and others [87,88].

20.3.1 Linear Arrays

Linear arrays are particularly suitable for applications where there is relative motion between the sensor head and the objects being imaged (e.g., intruder alarms and pushbroom linescan) [76,78,89–94]. Arrays from a few tens of elements up to many hundreds are required, but because of the absence of detector scanning along the direction of the array, bandwidths are low. A thin polished wafer of pyroelectric material, about 20 μm thick, is bonded down to a substrate. The manufacture of pyroelectric linear arrays based on $LiTaO_3$ with thicknesses of the self-supporting responsive elements of less than 5 μm has become possible by the development of special thinning techniques (ion-beam etching) [95]. Table 20.8 shows essential properties of different types of $LiTaO_3$ linear arrays.

The pyroelectric layers are reticulated to provide a two-dimensional array of pyroelectric islands or elements separated by grooves. These structures are necessary to inhibit the lateral diffusion of heat in the plane of the arrays. Ion milling enables 2-D reticulation of pyroelectric material with groove widths of 20 μm or less. Without reticulation interelement thermal diffusion becomes severe at modulating frequencies below 100 Hz. Another effect that increases

Table 20.8: Typical Properties of Linear Arrays (Rectangular Chopping With 128 Hz, Array Temperature 25°C)

Number of elements	1×128	1×128	1×128	1×256
Size of elements [μm^2]	90×100	90×100	90×2300	40×50
Pitch [μm]	100	100	100	50
Element thickness	20	5	20	5
Responsivity R_v [V/W]	200 000	500,000	200,000	500,000
Variation R_v [%]	1–2	2–5	1–2	3–6
NEP (nW)	5 (2.5)	2 (1)	6 (3)	2 (1)
NEDT (300K, f/1 optics) [K]	0.8 (0.4)	0.3 (0.15)	0.04 (0.02)	1.4 (0.7)
MTF ($R = 3$ lp/mm)	0.6	0.6	0.6	0.7

Source: V. Norkus, T. Sokoll, G. Gerlach, and G. Hofmann, "Pyroelectric Infrared Arrays and Their Applications," *Proceedings of SPIE* 3122, 409–19, 1997. With permission.

Note: ()values: additional four accumulations (frame rate 32 Hz).

interelement coupling and therefore the crosstalk, is the capacitative coupling due to fringing effects at electrode edges. This effect is also removed by reticulation.

Readout of charge from pyroelectric detectors is by means of conventional FETs. Each detector is connected to its own source follower FET that acts as an impedance buffer. Outputs of these transistors go to a multiplexer that samples the elements in turn at a rate dependent upon the particular applications.

The first research pyroelectric arrays were produced using a CCD design and IC for readout. A rigorous analysis of the interface condition for direct injection mode to optimize the injection efficiency and noise has been given by Watton et al. [93]. It appears however, the large ferroelectric element capacitance, coupled with the CCD sampling within the pixel, led to dominating kTC noise that limited performance. On account of this, attention turned to the use of CMOS ROIC designs.

For a long time there has been an interest in the use of thin pyroelectric films, because of their potential for making low thermal mass elements [96]. Arrays that have been demonstrated include the linear arrays fabricated using bulk micromachining techniques with sputtered $PbTiO_3$ on (100) silicon [97], La-$PbTiO_3$ on MgO [98], PVDF-TrEE on silicon [99], and sputtered $Pb(Zr_{0.15}Ti_{0.85})O_3$ on (100) silicon [100]. In these micromachining techniques silicon from behind the pyroelectric film has been removed, leaving a thin, low thermal mass membrane. Two major solutions have been elaborated for achieving high IR absorption coefficients of the detectors. The first one utilizes black absorption layers; usually porous metal films (black platinum or black gold). The second method uses semitransparent top electrodes (e.g., NiCr) on $\lambda/4$ thick pyroelectric layers. The first method gives larger absorption (about 90%), compared to the second one (60%); however, the total thickness and heat capacity is increased [94]. For example, Figure 20.21 shows the linear array developed for an IR spectroscopy gas sensor. The pyroelectric material is a sol-gel deposited (111)-oriented PZT15/85 thin film [101].

Recently, the integration of a radiation collector with a pyroelectric detectors has been demonstrated successfully. Both pyramidal and hemi-elliptical collector cavities can be made in silicon substrates using wet and dray etching [102,103].

The FPA technologies may be classified into either hybrid or monolithic; hybrid fabrication and micromachining techniques compete at the moment.

20.3.2 Hybrid Architecture

The hybrid approach is based on reticulation of ceramic wafers that are polished down in thickness to 10–15 μm and joined with the readout chip. The interface technology must resolve the conflict between providing electrical connection for signal readout from the element, and on the other hand thermal isolation of the element to avoid loss of signal through thermal loading.

Figure 20.22 shows the structure of an array [87]. A $LiTaO_3$ active volume detector is bound by the common front electrode, the back electrode, and by reticulation cuts. The detector back electrode is connected to the underlying multiplexer by a metalized polymer thermal isolation link that provides connection and support for the detector with controlled thermal conductance. The

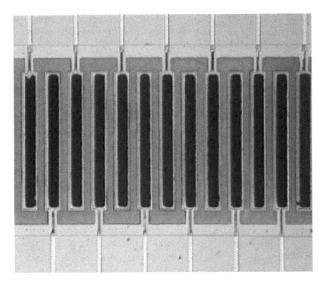

Figure 20.21 Top view of 50 element array with 200 μm period obtained with bulk microma-chining, membrane size: 2 × 11 mm. The black platinum absorbers, the CrAu contact lines, the membrane layers between the elements, and the SiO₂ layer for reduction of parasitic capacitance are visible. (From Willing, B., Kohli, M., Muralt, P., Setter, N., and Oehler, O., *Sensors and Actuators* A66, 109–13, 1998. With permission.)

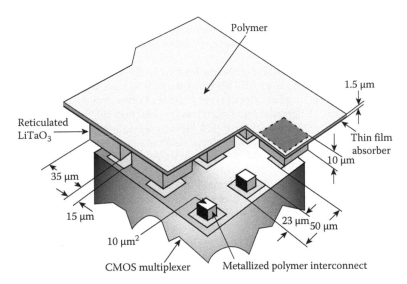

Figure 20.22 Hybrid pyroelectric array structure. (From Butler, N., and Iwasa, S., "Solid State Pyroelectric Imager," *Proceedings of SPIE* 1685, 146–54, 1992. With permission.)

nominal detector size is $35 \times 35 \ \mu m^2$, with 10 μm thickness and a 15 μm gap between detectors for an element pitch of 50 μm. For the detector with $f = 30$ Hz, $R_v \approx 1 \times 10^6$ V/W. To obtain maximum responsivity and minimize thermal fluctuation noise, the thermal conductance from the detector to its surroundings should be minimized. The estimated total thermal conductance is about 3.3×10^{-6} W/K, and is higher than obtained for micromachined silicon bolometers. The thermal time constant is then ≈15 ms. The predicted NEDT of the prototype system with 330×240 elements array has been estimated as 0.07 K with $f/1$ optics, an optic transmission of 0.85 and chopper effi-ciency of 0.85.

As is discussed in Chapter 7, the performance of pyroelectric detectors can be improved with a bias voltage applied to maintain and optimize the pyroelectric effect near the phase transition.

This type of pyroelectric FPAs has been developed by Texas Instruments (TI) [80,81]. The TI detector array comprises 245 × 328 pixels on 48.5 μm centers. Operating near room temperature, ferroelectric BST pixels hybridized with a silicon readout integrated circuit (ROIC) consistently yield devices with system NEDT of 0.047°C with $f/1$ optics.

The fabrication process for forming theses arrays is 95% compatible with standard Si processes. Detector process commonality with a silicon wafer processing format is maintained by the fabricating 100 mm diameter ceramic BST wafers with excellent dielectric properties. Highly dense, sintered ceramic BST offers cost and performance advantages not found in single crystal materials. After edging and polishing, reticulated arrays are formed by writing the array pattern on the wafer surface with a Nd-YAG laser. The laser damage is removed with an acid etchant, after which stoichiometry is restored by annealing in an oxygen ambient. The kerf is approximately 12 μm. Deposition of a parylene backfill, replanarization of the reticulated surface, and application of a common electrode and resonant-cavity IR absorber finishes the processing of the IR sensitive side of the array. The absorber layer is a transparent organic λ/4 layer, sandwiched between a semitransparent metal layer and the common electrode. Information on the composition of the absorber is generally scarce, this part of the process being considered as "technological know-how." The IR absorber provides better than 90% average absorption over the 7.5–13.0 μm spectral band. Dicing the wafer and mounting the individual die to carriers (optical coating side down), prepare the die for thinning and polishing to final thickness approximately 20 μm. Deposition of contact and bond metals followed by removal of parylene by dry etching prepares the array for hybridization. ROICs are prepared for hybridization by forming matching organic mesas with over-the-edge metallization. Mating of the two parts by solder bonding prepares them for packing and final testing. Figure 20.23 shows the details of the completed pyroelectric detector device structure [80].

The CMOS readout unit cell contains a high-pass filter, a gain stage, a tunable low-pass filter, and an address switch. The array output is compatible with standard TV formatting. The array is mounted onto a single-stage thermoelectric cooler for stabilization near the ferroelectric phase transition. The ceramic device interface processor and package is completed by the attachment of the antireflection-coated germanium window that allows IR transmission in the 7.5–13 μm spectral band.

In the mid-1990s, Texas Instruments was clearly leading in the development and production of uncooled, ferroelectric IR systems since NEDTs less than 0.04°C have been measured on systems with $f/1$ optics, without correcting for system-level noise and other losses. The production average was between 70 and 80 mK. A demonstrated sustained production rate in excess of 500 units per month was a small fraction of factory capacity [80]. Hybrid ferroelectric bolometer detector was the first to enter production, and is the most widely used type of thermal detector (in the United States, the Cadillac Division of General Motors pioneered this application, selling thermal imagers to customers for just under $2000) [2].

Large hybrid arrays have been also demonstrated by BAE Systems in the United Kingdom with a pitch of 56 μm and 40 μm. In this case the dielectric bolometers are $Pb(Sc_{0.5}Ta_{0.5})O_3$ biased at 4–5 V/μm (higher than for BST) with F_d levels between 10 to 15×10^{-5} $Pa^{-1/2}$. Conventional unbiased pyroelectrics give values of about 4×10^{-5} $Pa^{-1/2}$. The thin PST wafer is cut and polished from a hot pressed ceramic block and next reticulated by a laser-assisted etching process [104]. Lead/tin solder

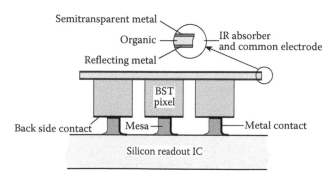

Figure 20.23 BST dielectric bolometer pixel. (From Hanson, C. M., "Uncooled Thermal Imaging at Texas Instruments," *Proceedings of SPIE* 2020, 330–39, 1993. With permission.)

bonds have been used in a liquid phase soldering process. The details of fabrication procedures of these arrays are described by Whatmore and Watton [83].

The performance of hybrid arrays that have been fabricated and tested in the United Kingdom program (BAE Systems and DERA) is listed in Table 20.9.

20.3.3 Monolithic Architecture

Although many applications for this hybrid array technology have been identified, and imagers employing these arrays are in mass production, no hybrid technology advances are foreseen. The reason is that the thermal conductance of the bump bonds is so high that the array NEDT ($f/1$ optics) is limited to about 50 mK. The best NEDT achieved with a hybrid array is about 38 mK, which is consistent with thermal conductance of approximately 4 μW/K. Early BST products suffered from features that poor modulation transfer function (MTF) resulting from thermal conductance between detector pixels and excessive noise resulting from insufficient digital resolution in the system. Pyroelectric array technology therefore is moving toward monolithic silicon microstructure technology. The monolithic process should have fewer steps and shorter cycle time. The detector cost in high volume is limited primarily by detector packing costs, which are not significantly different for hybrid and monolithic arrays. However, the serious problem is loss of ferroelectrics interesting properties as the thickness is reduced.

Thin-film ferroelectric (TFFE) detectors have the performance potential of microbolometers with minimum NEDT below 20 mK [83,84,86]. The properties of the materials and the device structure are sufficient to mach NEDT projections of the bolometer technologies (see Figure 20.24).

The first surface-machining 64×64 PbTiO$_3$ pyroelectric IR imager has been demonstrated by Polla and coworkers [105,106]. Polysilicon microbridges of 1.2 μm thickness have been formed 0.8 μm above the surface of a silicon wafer. The microbridge measures 50×50 μm^2 and forms a low thermal mass support for a 30×30 μm^2 PbTiO$_3$ film with a thickness of 0.36 μm. An NMOS preamplifier cell is located directly beneath each microbridge element. The measured pyroelectric

Table 20.9: Hybrid Arrays Demonstrated in the United Kingdom (DERA/BAE Systems) Program

Array Elements	Pitch (µm)	ROIC Size (mm²)	Package Atmosphere	NEDT (mK)	Array MTF at Nyquist
100×100	100	15.3×13.4	N$_2$	87	65%
256×128	56	17.0×12.4	Xe	90	45%
384×288	40	19.7×19.0	Xe	140	35%

Source: R. W. Whatmore and R. Watton, in *Infrared Detectors and Emitters: Materials and Devices*, 99–147, Kluwer Academic Publishers, Boston, MA, 2000. With permission.

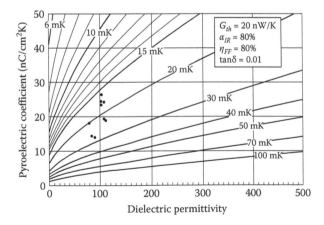

Figure 20.24 Lines of constant NEDT for 25 µm pixels, as functions of dielectric permittivity and pyroelectric coefficient. Data points indicate properties of material samples in test structures. (From Hanson, C. M., Beratan, H. R., and Belcher, J. F., "Uncooled Infrared Imaging Using Thin-Film Ferroelectrics," *Proceedings of SPIE* 4288, 298–303, 2001. With permission.)

coefficient for a single element is 90 nC/cm^{2}K. The measured blackbody voltage responsivity at 30 Hz is 1.2×10^4 V/W, whereas the detectivity is 2×10^8 cmHz$^{1/2}$W^{-1}.

The present TFFE device approach appears remarkably similar to the VO$_x$ microbolometer structure developed by Honeywell. However, there are several key features that distinguish it from that technology [82–86,107,108]. Since the device is a capacitor rather than a resistor (as in a bolometer), the electrodes are located above and below the face of the pixel, are transparent, and do not obscure the active optical area. Usually, the electrical resistance of the leads can be quite large without degrading signal to noise ratio, since the detector capacitance is approximately 3 pF. This enables the use of thin, poorly conducting electrode materials to minimize thermal conductance. A key feature of the design is that the ferroelectric film is self-supporting; there is no underlying membrane necessary to provide mechanical support. In such a way, with the use of transparent oxide electrodes, the ferroelectric material can dominate thermal conductance.

It is well known that absorption of IR radiation is accomplished by means of a resonant optical cavity. In monolithic bridge structure, the cavity is located within the ferroelectric itself or in the space between ferroelectric and the ROIC. This can be realized in two ways (see Figure 20.25) [108,109]:

1. The bottom electrode must be highly reflective, the top electrode must be semitransparent, and the ferroelectric must be approximately 1 μm thick for optimal tuning of the cavity for 10–12 μm radiation.

2. Both electrodes must be semitransparent, a reflective mirror must be present on the ROIC under each pixel, and the pixel must be located approximately 2 μm above the ROIC.

Figure 20.26a shows a cross section of the TFFE pixel designed according to the second way [84], where two top electrodes are connected to one of the posts, which provide electrical connection to

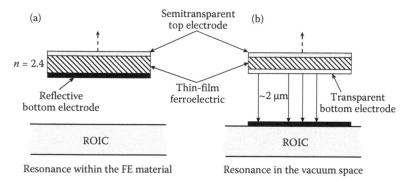

Figure 20.25 Two possible absorption resonant cavities of TFFE detectors. (From Tidrow, M. Z., Clark, W. W., Tipton, W., Hoffman, R., Beck, W., Tidrow, S. C., Robertson, D. N., et al., "Uncooled Infrared Detectors and Focal Plane Arrays," *Proceedings of SPIE* 3553, 178–87, 1997. With permission.)

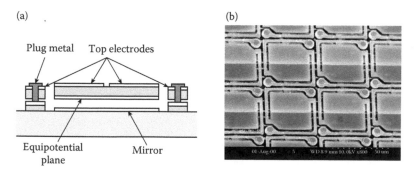

Figure 20.26 TFEE pixels: (a) schematic cross section of pixel element with split top electrode and (b) micrograph of part of a 320 × 240 array with 48.5 μm pixels. (From Hanson, C. M., Beratan, H. R., and Belcher, J. F., "Uncooled Infrared Imaging Using Thin-Film Ferroelectrics," *Proceedings of SPIE* 4288, 298–303, 2001. With permission.)

Figure 20.27 Single-frame $f/1$ 320 × 240 TFFE imagery extracted from a driving video. (From Hanson, C. M., Beratan, H. R., and Belcher, J. F., "Uncooled Infrared Imaging Using Thin-Film Ferroelectrics," *Proceedings of SPIE* 4288, 298–303, 2001. With permission.)

the readout. Thus the pixel capacitance is one-quarter the value of a similar capacitor whose connections are full-face electrodes on top and bottom. In this case the ferroelectric was PCT deposited by metal-organic decomposition of a spun-on solution. Figure 20.26b shows a micrograph of part of a 320 × 240 array with 48.5 μm pixels.

A key factor to the performance of the ceramic thin films is the high temperature processing required achieving the correct ferroelectric crystal phase. The TFFEs of interest are refractory, and require annealing at elevated temperatures to crystallize and develop good pyroelectric properties. Thermal treatments at temperatures that exceed about 450°C may lead to adverse interaction between the silicon and aluminum interconnects. Various techniques for the deposition of thin ferroelectric films have been investigated including spin-on metal-organic decomposition, radio frequency magnetron sputtering, dual ion beam sputtering, sol-gel processing, and laser ablation. Also a number of surface rapid thermal annealing techniques have been investigated to obtain optimum material response while leaving the underlying silicon substrate undamaged [82].

The NEDT of present devices with 48.5 μm pixels is typically about 80–90 mK including all system losses [86]. They are characterized by excellent MTF compared to bulk BST. Figure 20.27 shows an example of the image resolution taken from a driving video using micromachined 320 × 240 TFFE pixels. The low spatial noise is apparent by the uniformity in the extended dark areas of the image. The potential for NEDT improvement exists by reducing thickness and improving thermal isolation and material modification. However, the great challenge is in reducing pixel size. The challenge becomes even greater when we note that several microbolometer manufactures are moving toward 17 μm pixels.

It should be mentioned that the research group from DERA (United Kingdom) has developed integrated and "composite" detector technology [82,110]. In the first technology, the detector material was deposited as a thin film onto free standing microbridge structure defined on the surface of the silicon ROIC. The "composite" technology combines elements of hybrid and integrated technologies (see Figure 20.28). Microbridge pixels are fabricated in a similar fashion to the integrated technology and next are formed onto a high density interconnect silicon wafer. The interconnect wafer uses materials that can withstand the intermediate high temperature processing stage during fabrication of thin ferroelectric films and contains a narrow conducting channel via for every pixel, permitting electrical connection to the underside. Finally, the detector wafer is solder bump bonded to the ROIC as per the established hybrid array process. It is predicted that using PST films and NEDT of 20 mK (50 Hz image rate and $f/1$ optics) is possible to achieve.

20.3.4 Outlook on Commercial Market of Uncooled Focal Plane Arrays

In the beginning of the 1970s in the United States research programmers started to develop uncooled infrared detectors for practical military applications, mainly to bring thermal imaging

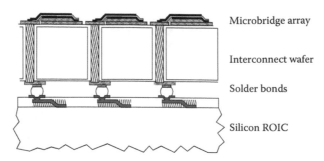

Microbridge array

Interconnect wafer

Solder bonds

Silicon ROIC

Figure 20.28 Schematic cross section of the "composite" detector array design. (From Todd, M. A., Manning, P. A., Donohue, O. D., Brown, A. G., and Watton, R., "Thin Film Ferroelectric Materials for Microbolometer Arrays," *Proceedings of SPIE* 4130, 128–39, 2000. With permission.)

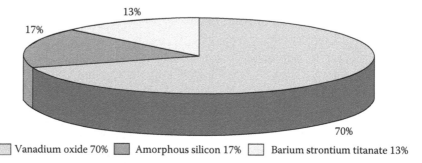

□ Vanadium oxide 70% ■ Amorphous silicon 17% □ Barium strontium titanate 13%

Figure 20.29 Estimated market shares for VO_x, a-Si and BST detectors. (From http://www.flir.com/uploadedFiles/Eurasia/Cores_and_Components/Technical_Notes/uncooled%20detectors%20BST.pdf.)

to every soldier. In order to do so, thermal imaging cameras needed to become more compact, portable, and definitely a lot less expensive than cooled cameras. In 1978, Texas Instruments patented ferroelectric infrared detectors using BST. This technology was demonstrated to the military for the first time in 1979 [111]. At the same time, another technology, micromachining bolometer technology, was developed by Honeywell. The development of both technologies was funded by the U.S. military through the next 15 years. However, about 10 years ago, convinced of the advantages VO_x had over BST, the U.S. military decided not to provide any more funding for research into BST. The loss of government funding for BST meant that research in this technology slowed down drastically. Thus, research efforts supported by DARPA have been moved into development of TFFE detector arrays in Raytheon Commercial Infrared (former Texas Instruments). Further research supported by Missile Defense Agency and L-3 Communication Infrared Products to push technology into marketable products has failed [86]. The research into BST stagnated, while VO_x technology is still continuing to do so with more funding. The size of BST pixels, 50 μm, is still at the same point where it was more than 10 years ago. Moreover, the rotating chopper blocks the detector 50% of the time, which does not benefit the sensitivity of the camera. The mechanical chopper is susceptible to breakdown and sensitive to shock and vibration. As a result, the mean time between failure (MTBF) for a BST camera is shorter than for a microbolometer camera. Also using thermoelectric cooling to stabilize electrical polarization is a disadvantage of BST detectors in comparison to VO_x and a-Si detectors.

In the mid-1990s a third technology, amorphous silicon, was developed. During this time, the big advantage of using a-Si was their fabrication in a silicon foundry. Further, the VO_x technology was controlled by the U.S. military and an export license was required for microbolometer cameras that were sold outside the United States. Today, VO_x bolometers can be produced in a silicon foundry and the above reasons have both disappeared.

At present, VO_x microbolometer arrays is clearly the most used technology for uncooled detectors (see Figure 20.29). The VO_x is winning the battle between the technologies and vanadium oxide detectors are being produced at a much lower cost than either of the two other technologies [111].

20.4 NOVEL UNCOOLED FOCAL PLANE ARRAYS

Development of novel uncooled micromechanical thermal detectors is a result of recent advances in technology of MEMS systems. They show great promise as imaging chips for use in low-cost, high-performance cameras. The high sensitivity of the cantilevers allows the individual pixels in the array to be shrunk from their current size of 50–20 μm while maintaining low-distortion IR imaging [112,113].

Considerable progress to reduce NEDT and time constant has been made through the years by several research groups [112–123]. Multispectral Imaging Inc. has developed a new class of electrically coupled thermal transducers in which bending of the cantilever causes a change in its capacitance. Its first FPA product is a 50 μm pitch microcantilever arrays of 160 × 120 pixel sensors fabricated directly on the array CMOS ROIC wafers using 0.25 μm design rules. Figure 20.30 shows an SEM image of the details of the pixel structure together with the vacuum hybrid metal/ceramic packaging used to house the detector array. The package contains a thermoelectric (TE) cooler to stabilize the array temperature along with a nonevaporable getter to remove residual gas in the package after sealing. Including a thermal compensation technique [112] enables operation of the detector arrays without the TE cooler.

The NEDT values of the early fabricated FPAs were around 1–2 K, much higher than the modeling and sensor structure measurements indicating what these sensors are capable of. Improvements of the camera and control electronics have considerably improved NEDT values that are 3–10 times better. Pixel-to-pixel uniformities of better than ±10% are commonly achieved. The estimated NEDTs for two point correction and gain corrected images is about 300–500 mK (see Figure 20.31) [113].

The measured histograms indicate that as the FPA temperature is reduced, the peak NEDT values also decrease. The best individual pixel NEDTs are in the 1–15 mK range with time constant in the 15 msec range, giving best pixel performance NEDT × τ_{th} in the range 120–200 mKsec [113]. This is comparable to the best published bolometer performance for 50 μm pitch pixel arrays [124].

Recently, several research groups are involved in development of optical-readable imaging arrays. In 2006, Grbovic et al. of the Oak Ridge National Laboratory (Tennessee) reported a high resolution 256 × 256 pixel array [115]. The fabricated arrays had NEDT and response time of below 500 mK and 6 ms, respectively. Image quality can be improved by automatic postprocessing of artifacts arising from noise and nonresponsive pixels, which is shown in Figure 20.32 [116]. The image quality of inpainting method is improved if an accurate mask is determined during the device calibration process. For inpainting methods it is better to work with less true image pixels than to include corrupted pixels in the mask.

Agiltron Inc. produced a 280 × 240 photomechanical IR sensors with an optical readout for both MWIR and LWIR imaging (see Figure 20.33) at a speed of up to 1000 frames per second [117]. Results

(a) (b)

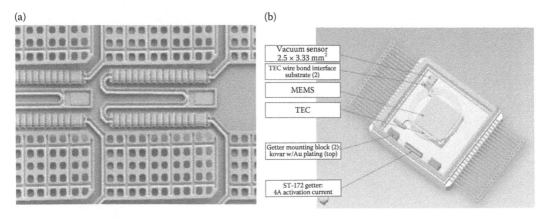

Figure 20.30 50 μm pitch microcantilever 160 × 120 focal plane array: (a) an SEM image of the details of the pixel structure, and (b) schematic of the vacuum sensor package (From Hunter, S. R., Maurer, G., Simelgor, G., Radhakrishnan, S., Gray, J., Bachir, K., Pennell, T., Bauer, M., and Jagadish, U., "Development and Optimization of Microcantilever Based IR Imaging Arrays," *Proceedings of SPIE* 6940, 694013, 2008. With permission.)

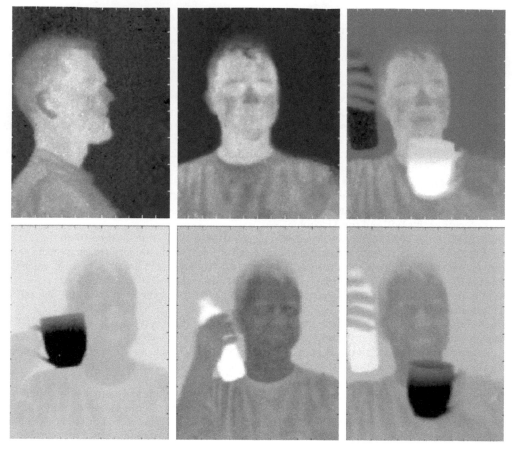

Figure 20.31 Noninverted and inverted IR images showing the result of a two point nonuniformity correction and dead-pixel-substraction. The individual is holding a hot cup of water and a cold water bottle. NEDTs for these images are estimated to be ~300–500 mK. (From Hunter, S. R., Maurer, G., Simelgor, G., Radhakrishnan, S., Gray, J., Bachir, K., Pennell, T., Bauer, M., and Jagadish, U., "Development and Optimization of Microcantilever Based IR Imaging Arrays," *Proceedings of SPIE* 6940, 694013, 2008. With permission.)

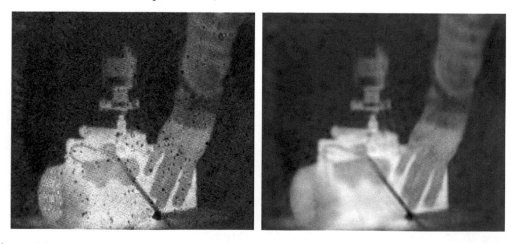

Figure 20.32 Readout of 256 × 256 MEMS IR FPA: (a) after baseline subtraction and with (b) inpainting method. (From Lavrik, N., Archibald, R., Grbovic, D., Rajic, S., and Datskos, P., "Uncooled MEMS IR Imagers with Optical Readout and Image Processing," *Proceedings of SPIE* 6542, 65421E, 2007. With permission.)

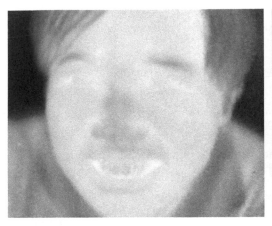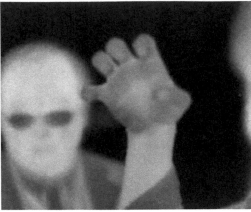

Figure 20.33 Representative LWIR images from 280 × 240 photomechanical camera. (From Salerno, J. P., "High Frame Rate Imaging Using Uncooled Optical Readout Photomechanical IR Sensor," *Proceedings of SPIE* 6542, 65421D, 2007. With permission.)

on the detection of rapid occurrence events, such as gunfire and rocket travel, were reported. At the present stage of development, the imager has a NEDT of approximately 120 mK at $f/1$ optics.

REFERENCES

1. R. Watton and M. V. Mansi, "Performance of a Thermal Imager Employing a Hybrid Pyroelectric Detector Array with MOSFET Readout," *Proceedings of SPIE* 865, 78–85, 1987.

2. P. W. Kruse, *Uncooled Thermal Imaging. Arrays, Systems, and Applications*, SPIE Press, Bellingham, WA, 2001.

3. P. W. Kruse, "A Comparison of the Limits to the Performance of Thermal and Photon Detector Imaging Arrays," *Infrared Physics & Technology* 36, 869–82, 1995.

4. S. Horn, D. Lohrmann, P. Norton, K. McCormack, and A. Hutchinson, "Reaching for the Sensitivity Limits of Uncooled and Minimally-Cooled Thermal and Photon Infrared Detectors," *Proceedings of SPIE* 5783, 401–11, 2005.

5. J. A. Ratches, "Current and Future Trends in Military Night Vision Applications," *Ferroelectrics* 342, 183–92, 2006.

6. A. W. van Herwaarden, F. G. van Herwaarden, S. A. Molenaar, E. J. G. Goudena, M. Laros, P. M. Sarro, C. A. Schot, W. van der Vlist, L. Blarre, and J. P. Krebs, "Design and Fabrication of Infrared Detector Arrays for Satellite Attitude Control," *Sensors Actuators* 83, 101–8, 2000.

7. T. Kanno, M. Saga, S. Matsumoto, M. Uchida, N. Tsukamoto, A. Tanaka, S. Itoh, et al., "Uncooled Infrared Focal Plane Array Having 128 × 128 Thermopile Detector Elements," *Proceedings of SPIE* 2269, 450–59, 1994.

8. M. Hirota, Y. Nakajima, M. Saito, F. Satou, and M. Uchiyama, "Thermoelectric Infrared Imaging Sensors for Automotive Applications," *Proceedings of SPIE* 5359, 111–25, 2004.

9. M. Hirota, Y. Nakajima, M. Saito, and M. Uchiyama, "120 × 90 Element Thermoelectric Infrared Focal Plane Array with Precisely Patterned Au-Black Absorber," *Sensors Actuators* A135, 146–51, 2007.

10. A. D. Oliver and K. D. Wise, "A 1024-Element Bulk-Micromachined Thermopile Infrared Imaging Array," *Sensors Actuators* 73, 222–31, 1999.

11. M. C. Foote, E. W. Jones, and T. Caillat, "Uncooled Thermopile Infrared Detector Linear Arrays with Detectivity Greater than 10^9 cmHz$^{1/2}$/W," *IEEE Transactions on Electron Devices* 45, 1896–1902, 1998.

12. M. C. Foote and E. W. Jones, "High Performance Micromachined Thermopile Linear Arrays," *Proceedings of SPIE* 3379, 192–97, 1998.

13. M. C. Foote and S. Gaalema, "Progress Towards High-Performance Thermopile Imaging Arrays," *Proceedings of SPIE* 4369, 350–54, 2001.

14. T. McManus and S. Mickelson, "Imaging Radiometers Employing Linear Thermoelectric Arrays," *Proceedings of SPIE* 3698, 352–60, 1999.

15. AXT100 brochure, http://www.aas2.com/products/axt100/index.htm

16. R. E. Flannery and J. E. Miller, "Status of Uncooled Infrared Imagers," *Proceedings of SPIE* 1689, 379–95, 1992.

17. R. A. Wood, "Micromachined Bolometer Arrays Achieve Low-Cost Imaging," *Laser Focus World*, 101–6, June 1993.

18. T. Ishikawa, M. Ueno, K. Endo, Y. Nakaki, H. Hata, T. Sone, and M. Kimata, "Low-Cost 320×240 Uncooled IRFPA Using Conventional Silicon IC Process," *Opto-Electronics Review* 7, 297–303, 1999.

19. D. Murphy, M. Ray, R. Wyles, J. Asbrock, N. Lum, J. Wyles, C. Hewitt, A. Kennedy, and D. V. Lue, "High Sensitivity 25 μm Microbolometer FPAs," *Proceedings of SPIE* 4721, 99–110, 2002.

20. D. Murphy, A. Kennedy, M. Ray, R. Wyles, J. Wyles, J. Asbrock, C. Hewitt, D. Van Lue, and T. Sessler, "Resolution and Sensitivity Improvements for VO$_x$ Microbolometer FPAs," *Proceedings of SPIE* 5074, 402–13, 2003.

21. D. Murphy, M. Ray, J. Wyles, J. Asbrock, C. Hewitt, R. Wyles, E. Gordon, et al., "Performance Improvements for VO$_x$ Microbolometer FPAs," *Proceedings of SPIE* 5406, 531–40, 2004.

22. D. Murphy, M. Ray, A. Kennedy, J. Wyles, C. Hewitt, R. Wyles, E. Gordon, et al., "High Sensitivity 640×512 (20 μm pitch) Microbolometer FPAs," *Proceedings of SPIE* 6206, 62061A, 2006.

23. D. Murphy, M. Ray, J. Wyles, C. Hewitt, R. Wyles, E. Gordon, K. Almada, et al., "640×512 17 μm Microbolometer FPA and Sensor Development," *Proceedings of SPIE* 6542, 65421Z, 2007.

24. M. N. Gurnee, M. Kohin, R. Blackwell, N. Butler, J. Whitwam, B. Backer, A. Leary, and T. Nielsen, "Developments in Uncooled IR Technology at BAE Systems," *Proceedings of SPIE* 4369, 287–96, 2001.

25. R. Blackwell, S. Geldart, M. Kohin, A. Leary, and R. Murphy, "Recent Technology Advancements and Applications of advanced uncooled imagers," *Proceedings of SPIE* 5406, 422–27, 2004.

26. R. J. Blackwell, T. Bach, D. O'Donnell, J. Geneczko, and M. Joswick, "17 μm Pixel 640×480 Microbolometer FPA Development at BAE Systems," *Proceedings of SPIE* 6542, 65421U, 2007.

27. R. Blackwell, D. Lacroix, T. Bach, J. Ishii, S. Hyland, J. Geneczko, S. Chan, B. Sujlana, and M. Joswick, "Uncooled VO$_x$ Systems at BAE Systems," *Proceedings of SPIE* 6940, 694021, 2008.

28. P. E. Howard, J. E. Clarke, A. C. Ionescu, and C. Li, "DRS U6000 640×480 VO$_x$ Uncooled IR Focal Plane," *Proceedings of SPIE* 4721, 48–55, 2002.

29. P. E. Howard, J. E. Clarke, A. C. Ionescu, C. Li, and A. Frankenberger, "Advances in Uncooled 1-Mil Pixel Size Focal Plane Products at DRS, *Proceedings of SPIE* 5406, 512–20, 2004.

30. C. Li, G. D. Skidmore, C. Howard, C. J. Han, L. Wood, D. Peysha, E. Williams, et al., "Recent Development of Ultra Small Pixel Uncooled Focal Plane Arrays at DRS," *Proceedings of SPIE* 6542, 65421Y, 2007.

31. W. Parish, J. T. Woolaway, G. Kincaid, J. L. Heath, and J. D. Frank, "Low Cost 160 × 128 Uncooled Infrared Sensor Array," *Proceedings of SPIE* 3360, 111–19, 1998.

32. W. A. Terre, R. F. Cannata, P. Franklin, A. Gonzalez, E. Kurth, W. Parrish, K. Peters, T. Romeo, D. Salazar, and R. Van Ysseldyk, "Microbolometer Production at Indigo Systems," *Proceedings of SPIE* 5406, 557–65, 2004.

33. K. A. Hay, D. Van Deusen, T. Y. Liu, and W. A. Kleinhans, "Uncooled Focal Plane Array Detector Development at InfraredVision Technology Corp.," *Proceedings of SPIE* 5074, 491–99, 2003.

34. T. Schimert, J. Brady, T. Fagan, M. Taylor, W. McCardel, R. Gooch, S. Ajmera, C. Hanson, and A. J. Syllaios, "Amorphous Silicon Based Large Format Uncooled FPA Microbolometer Technology," *Proceedings of SPIE* 6940, 694023, 2008.

35. H. Jerominek, T. D. Pope, C. Alain, R. Zhang, F. Picard, M. Lehoux, F. Cayer, S. Savard, C. Larouche, and C. Grenier, "Miniature VO_2-Based Bolometric Detectors for High-Resolution Uncooled FPAs," *Proceedings of SPIE* 4028, 47–56, 2000.

36. T. D. Pope, H. Jeronimek, C. Alain, F. Cayer, B. Tremblay, C. Grenier, P. Topart, et al., "Commercial and Custom 160 × 120, 256 × 1 and 512 × 3 Pixel Bolometric FPAs, *Proceedings of SPIE* 4721, 64–74, 2002.

37. C. Alain, H. Jerominek, P. A. Topart, T. D. Pope, F. Picard, F. Cayer, C. Larouche, S. Leclair, and B. Tremblay, "Microfabrication Services at INO," *Proceedings of SPIE* 4979, 353–63, 2003.

38. P. Topart, C. Alain, L. LeNoc, S. Leclair, Y. Desroches, B. Tremblay, and H. Jerominek, "Hybrid Micropackaging Technology for Uncooled FPAs," *Proceedings of SPIE* 5783, 544–50, 2005.

39. E. Mottin, J. Martin, J. Ouvrier-Buffet, M. Vilain, A. Bain, J. Yon, J. L. Tissot, and J. P. Chatard, "Enhanced Amorphous Silicon Technology for 320 × 240 Microbolometer Arrays with a Pitch of 35 µm," *Proceedings of SPIE* 4369, 250–56, 2001.

40. E. Mottin, A. Bain, J. Martin, J. Ouvrier-Buffet, S. Bisotto, J. J. Yon, and J. L. Tissot, "Uncooled Amorphous Silicon Technology Enhancement for 25 µm Pixel Pitch Achievement," *Proceedings of SPIE* 4820, 200–07, 2003.

41. J. J. Yon, A. Astier, S. Bisotto, G. Chamming's, A. Durand, J. L. Martin, E. Mottin, J. L. Ouvrier-Buffet, and J. L. Tissot, "First Demonstration of 25µm Pitch Uncooled Amorphous Silicon Microbolometer IRFPA at LETI-LIR," *Proceedings of SPIE* 5783, 432–40, 2005.

42. J. J. Yon, E. Mottin, and J. L. Tissot, "Latest Amorphous Silicon Microbolometer Developments at LETI-LIR," *Proceedings of SPIE* 6940, 69401W, 2008.

43. H. Wada, T. Sone, H. Hata, Y. Nakaki, O. Kaneda, Y. Ohta, M. Ueno, and M. Kimata, "YBaCuO Uncooled Microbolometer IR FPA," *Proceedings of SPIE* 4369, 297–304, 2001.

44. Y. Tanaka, A. Tanaka, K. Iida, T. Sasaki, S. Tohyama, A. Ajisawa, A. Kawahara, et al., "Performance of 320 × 240 Uncooled Bolometer-Type Infrared Focal Plane Arrays, *Proceedings of SPIE* 5074, 414–24, 2003.

45. N. Oda, Y. Tanaka, T. Sasaki, A. Ajisawa, A. Kawahara, and S. Kurashina, "Performance of 320 × 240 Bolometer-Type Uncooled Infrared Detector," *NEC Research & Development* 44, 170–74, 2003.

46. S. Tohyama, M. Miyoshi, S. Kurashina, N. Ito, T. Sasaki, A. Ajisawa, and N. Oda, "New Thermally Isolated Pixel Structure for High-Resolution Uncooled Infrared FPAs," *Proceedings of SPIE* 5406, 428–36, 2004.

47. T. Ishikawa, M. Ueno, Y. Nakaki, K. Endo, Y. Ohta, J. Nakanishi, Y. Kosasayama, H. Yagi, T. Sone, and M. Kimata, "Performance of 320 × 240 Uncooled IRFPA with SOI Diode Detectors," *Proceedings of SPIE* 4130, 1–8, 2000.

48. Y. Kosasayama, T. Sugino, Y. Nakaki, Y. Fujii, H. Inoue, H. Yagi, H. Hata, M. Ueno, M. Takeda, and M. Kimata, "Pixel Scaling for SOI-Diode Uncooled Infrared Focal Plane Arrays," *Proceedings of SPIE* 5406, 504–11, 2004.

49. T. Ishikawa, M. Ueno, K. Endo, Y. Nakaki, H. Hata, T. Sone, and M. Kimata, "640 × 480 Pixel Uncooled Infrared with SOI Diode Detectors," *Proceedings of SPIE* 5783, 566–77, 2005.

50. M. Kimata, M. Uenob, M. Takedac, and T. Setod, "SOI Diode Uncooled Infrared Focal Plane Arrays," *Proceedings of SPIE* 6127, 61270X, 2006.

51. P. A. Manning, J. P. Gillham, N. J. Parkinson, and T. P. Kaushal, "Silicon Foundry Microbolometers: The Route to the Mass-Market Thermal Imager, *Proceedings of SPIE* 5406, 465–72, 2004.

52. V. N. Leonov, Y. Creten, P. De Moor, B. Du Bois, C. Goessens, B. Grietens, P. Merken, et al., "Small Two-Dimensional and Linear Arrays of Polycrystalline SiGe Microbolometers at IMEC-XenICs," *Proceedings of SPIE* 5074, 446–57, 2003.

53. U. Mizrahi, A. Fraenkel, L. Bykov, A. Giladi, A. Adin, E. Ilan, N. Shiloah, et al., "Uncooled Detektor Development Program at SCD," *Proceedings of SPIE* 5783, 551–58, 2005.

54. U. Mizrahi, L. Bikov, A. Giladi, A. Adin, N. Shiloah, E. Malkinson, T. Czyzewski, A. Amsterdam, Y. Sinai, and A. Fraenkel, "New Features and Development Directions in SCD's μ-Bolometer Technology," *Proceedings of SPIE* 6940, 694020, 2008.

55. M. Kohin and N. Butler, "Performance Limits of Uncooled VO_x Microbolometer Focal-Plane Arrays," *Proceedings of SPIE* 5406, 447–53, 2004.

56. A. Van Der Ziel, "Flicker Noise in Electronic Devices," in *Advances in Electronics and Electron Physics*, 49, 225–97, 1979.

57. R. A. Wood, C. J. Han, and P. W. Kruse, "Integrated Uncooled IR Detector Imaging Arrays," *Proceedings of IEEE Solid State Sensor and Actuator Workshop*, 132–35, Hilton Head Island, SC, June 1992.

58. R. A. Wood, "Uncooled Thermal Imaging with Monolithic Silicon Focal Planes," *Proceedings of SPIE* 2020, 322–29, 1993.

59. W. J. Parrish and T. Woolaway, "Improvements in Uncooled Systems Using Bias Equalization", *Proceedings of SPIE* 3698, 748–55, 1999.

60. F. Niklas, C. Vieider, and H. Jakobsen, "MEMS-Based Uncooled Infrared Bolometer Arrays: A Review," *Proceedings of SPIE* 6836, 68360D-1, 2007.

61. S. Eminoglu, D. Sabuncuoglu Tezcan, M. Y. Tanrikulu, and T. Akin, "Low-Cost Uncooled Infrared Detectors in CMOS Process," *Sensors & Actuators A* 109, 102–13, 2003.

62. C. Vieider, S. Wissmar, P. Ericsson, U. Halldin, F. Niklaus, G. Stemme, J.-E. Källhammer, et al., "Low-Cost Far Infrared Bolometer Camera for Automotive Use," *Proceedings of SPIE* 6542, 65421L, 2007.

63. W. Radford, D. Murphy, A. Finch, K. Hay, A. Kennedy, M. Ray, A. Sayed, et al., "Sensitivity Improvements in Uncooled Microbolometer FPAs," *Proceedings of SPIE* 3698, 119–30, 1999.

64. J. Anderson, D. Bradley, D.C. Chen, R. Chin, K. Jurgelewicz, W. Radford, A. Kennedy, et al., "Low Cost Microsensors Program," *Proceedings of SPIE* 4369, 559–65, 2001.

65. S. Black, M. Ray, C. Hewitt, R. Wyles, E. Gordon, K. Almada, S. Baur, M. Kuiken, D. Chi, and T. Sessler, "RVS Uncooled Sensor Development for Tactical Applications," *Proceedings of SPIE* 6940, 694022, 2008.

66. H.-K. Lee, J.-B. Yoon, E. Yoon, S.-B. Ju, Y.-J. Yong, W. Lee, and S.-G. Kim, "A High Fill Factor Infrared Bolometer Using Micromachined Multilevel Electrothermal Structures," *IEEE Transactions on Electron Devices* 46, 1489–91, 1999.

67. B. Fieque, P. Robert, C. Minassian, M. Vilain, J. L. Tissot, A. Crastes, O. Legras, and J. J. Yon, "Uncooled Amorphous Silicon XGA IRFPA with 17μm Pixel-Pitch for High End Applications," *Proceedings of SPIE* 6940, 69401X, 2008.

68. A. Jahanzeb, C. M. Travers, Z. Celik-Butler, D. P. Butler, and S. G. Tan, "A Semiconductor YBaCuO Microbolometer for Room Temperature IR Imaging," *IEEE Transactions on Electron Devices* 44, 1795–1801, 1997.

69. M. Almasri, Z. Celik-Butler, D. P. Butler, A. Yaradanakul, and A. Yildiz, "Semiconducting YBaCuO Microbolometers for Uncooled Broad-Band IR Sensing," *Proceedings of SPIE* 4369, 264–73, 2001.

70. H. Wada, T. Sone, H. Hata, Y. Nakaki, O. Kaneda, Y. Ohta, M. Ueno, and M. Kimata, "YBaCuO Uncooled Microbolometer IRFPA," *Proceedings of SPIE* 4369, 297–304, 2001.

71. A. Yildiz, Z. Celik-Butler, and D. P. Butler, "Microbolometers on a Flexible Substrate for Infrared Detection," *IEEE Sensors Journal* 4, 112–17, 2004.

72. S. A. Dayeh, D. P. Butler, and Z. Celik-Butler, "Micromachined Infrared Bolometers on Flexible Polyimide Substrates," *Sensors & Actuators* A118, 49–56, 2005.

73. X. He, G. Karunasiri, T. Mei, W. J. Zeng, P. Neuzil, and U. Sridhar, "Performance of Microbolometer Focal Plane Arrays Under Varying Pressure," *IEEE Electron Device Letters* 21, 233–35, 2000.

74. V. K. Lindroos, M. Tilli, A. Lehto, and T. Motorka, *Handbook of Silicon Based MEMS Materials and Technologies*, William Andrew Publishing, Norwich, NY, 2008.

75. P. De Moor, J. John, S. Sedky, and C. Van Hoof, "Lineal Arrays of Fast Uncooled Poly SiGe Microbolometers for IR Detection," *Proceedings of SPIE* 4028, 27–34, 2000.

76. R. W. Whatmore, "Pyroelectric Devices and Materials," *Reports on Progress in Physics* 49, 1335–86, 1986.

77. R. Watton, "Ferroelectric Materials and Design in Infrared Detection and Imaging," *Ferroelectrics* 91, 87–108, 1989.

78. R. W. Whatmore, "Pyroelectric Ceramics and Devices for Thermal Infra-Red Detection and Imaging," *Ferroelectrics* 118, 241–59, 1991.

79. R. Watton, "IR Bolometers and Thermal Imaging: The Role of Ferroelectric Materials," *Ferroelectrics* 133, 5–10, 1992.

80. C. M. Hanson, "Uncooled Thermal Imaging at Texas Instruments," *Proceedings of SPIE* 2020, 330–39, 1993.

81. H. Betatan, C. Hanson, and E.DG. Meissner, "Low Cost Uncooled Ferroelectric Detector," *Proceedings of SPIE* 2274, 147–56, 1994.

82. M. A. Todd, P. A. Manning, O. D. Donohue, A. G. Brown, and R. Watton, "Thin Film Ferroelectric Materials for Microbolometer Arrays," *Proceedings of SPIE* 4130, 128–39, 2000.

83. R. W. Whatmore and R. Watton, "Pyroelectric Materials and Devices," in *Infrared Detectors and Emitters: Materials and Devices*, eds. P. Capper and C. T. Elliott, 99–147, Kluwer Academic Publishers, Boston, MA, 2000.

84. C. M. Hanson, H. R. Beratan, and J. F. Belcher, "Uncooled Infrared Imaging Using Thin-Film Ferroelectrics," *Proceedings of SPIE* 4288, 298–303, 2001.

85. R. W. Whatmore, Q. Zhang, C. P. Shaw, R. A. Dorey, and J. R. Alock, "Pyroelectric Ceramics and Thin Films for Applications in Uncooled Infra-Red Sensor Arrays," *Physica Scripta* T 129, 6–11, 2007.

86. C. M. Hanson, H. R. Beratan, and D. L. Arbuthnot, "Uncooled Thermal Imaging with Thin-Film Ferroelectric Detectors," *Proceedings of SPIE* 6940, 694025, 2008.

87. N. Butler and S. Iwasa, "Solid State Pyroelectric Imager," *Proceedings of SPIE* 1685, 146–54, 1992.

88. R. Takayama, Y. Tomita, J. Asayama, K. Nomura, and H. Ogawa, "Pyroelectric Infrared Array Sensors Made of c-Axis-Oriented La-Modified $PbTiO_3$ Thin Films," *Sensors and Actuators* A21–A23, 508–12, 1990.

89. R. Watton, F. Ainger, S. Porter, D. Pedder, and J. Gooding, "Technologies and Performance for Linear and Two Dimensional Pyroelectric Arrays," *Proceedings of SPIE* 510, 139–48, 1984.

90. D. E. Burgess, P. A. Manning, and R. Watton, "The Theoretical and Experimental Performance of a Pyroelectric Array Imager," *Proceedings of SPIE* 572, 2–6, 1985.

91. D. E. Burgess, "Pyroelectric in Harsh Environment," *Proceedings of SPIE* 930, 139–50, 1988.

92. R. Takayama, Y. Tomita, K. Iijima, and I. Ueda, "Pyroelectric Properties and Application to Infrared Sensors of $PbTiO_3$ and $PbZrTiO_3$ Ferroelectric Thin Films," *Ferroelectrics* 118, 325–42, 1991.

93. R. Watton, P. Manning, D. Burgess, and J. Gooding, "The Pyroelectric/CCD Focal Plane Hybrid: Analysis and Design for Direct Charge Injection," *Infrared Physics* 22, 259–75, 1982.

94. P. Muralt, "Micromachined Infrared Detectors Based on Pyroelectric Thin Films," *Reports on Progress in Physics* 64, 1339–88, 2001.

95. V. Norkus, T. Sokoll, G. Gerlach, and G. Hofmann, "Pyroelectric Infrared Arrays and Their Applications," *Proceedings of SPIE* 3122, 409–19, 1997.

96. J. D. Zook and S. T. Liu, "Pyroelectric Effects in Thin Films," *Journal of Applied Physics* 49, 4604–6, 1978.

97. M. Okuyama, H. Seto, M. Kojima, Y. Matsui, and Y. Hamakawa, "Integrated Pyroelectric Infrared Sensor Using $PbTiO_3$ Thin Film," *Japanese Journal of Applied Physics* 22, 465–68, 1983.

98. R. Takayama, Y. Tomita, J. Asayama, K. Nomura, and H. Ogawa, "Pyroelectric Infrared Array Sensors Made of c-Axis-Oriented La-Modified PbTiO₃ Thin Films," *Sensors and Actuators* A21-23, 508–12, 1990.

99. N. Neumann, R. Köhler, and G. Hofmann, "Pyroelectric Thin Film Sensors and Arrays Based on P(VDF/TrFE)," *Integrated Ferroelectrics* 6, 213–30, 1995.

100. M. Kohli, C. Wuethrich, K. Brooks, B. Willing, M. Forster, P. Muralt, N. Setter, and P. Ryser, "Pyroelectric Thin-Film Sensor Array," *Sensors and Actuators* A60, 147–53, 1997.

101. B. Willing, M. Kohli, P. Muralt, N. Setter, and O. Oehler, "Gas Spectrometry Based on Pyroelectric Thin Film Arrays Integrated on Silicon," *Sensors and Actuators* A66, 109–13, 1998.

102. C. Shaw, S. Landi, R. Whatmore, and P. Kirby, "Development Aspects of an Integrated Pyroelectric Array Incorporating and Thin PZT Film and Radiation Collectors," *Integrated Ferroelectrics* 63, 93–97, 2004.

103. R. W. Whatmore, "Uncooled Pyroelectric Detector Arrays Using Ferroelectric Ceramics and Thin Films," *MEMS Sensor Technologies*, 10812, 1–9, 2005.

104. M. A. Todd and R. Watton, "Laser-Assisted Etching of Ferroelectric Ceramics for the Reticulation of IR Detector Arrays," *Proceedings of SPIE* 1320, 95, 1990.

105. D. L. Polla, C. Ye, and T. Tamagawa, "Surface-Micromachined PbTiO₃ Pyroelectric Detectors," *Applied Physics Letters* 59, 3539–41, 1991.

106. L. Pham, W. Tjhen, C. Ye, and D. L. Polla, "Surface-Micromachined Pyroelectric Infrared Imaging Array with Vertically Integrated Signal Processing Circuitry," *IEEE Transactions on Ultrasonics, Ferroelectrics, and Frequency Control* 41, 552–55, 1994.

107. S. G. Porter, R. Watton, and R. K. McEwen, "Ferroelectric Arrays: The Route to Low Cost Uncooled Infrared Imaging," *Proceedings of SPIE* 2552, 573–82, 1995.

108. J. F. Belcher, C. M. Hanson, H. R. Beratan, K. R. Udayakumar, and K. L. Soch, "Uncooled Monolithic Ferroelectric IRFPA Technology," *Proceedings of SPIE* 3436, 611–22, 1998.

109. M. Z. Tidrow, W. W. Clark, W. Tipton, R. Hoffman, W. Beck, S. C. Tidrow, D. N. Robertson, et al., "Uncooled Infrared Detectors and Focal Plane Arrays," *Proceedings of SPIE* 3553, 178–87, 1997.

110. R. K. McEwen and P. A. Manning, "European Uncooled Thermal Imaging Sensors," *Proceedings of SPIE* 3698, 322–37, 1999.

111. http://www.flir.com/uploadedFiles/Eurasia/Cores_and_Components/Technical_Notes/uncooled%20detectors%20BST.pdf

112. S. R. Hunter, G. S. Maurer, G. Simelgor, S. Radhakrishnan, and J. Gray, "High Sensitivity 25µm and 50µm Pitch Microcantilever IR Imaging Arrays," *Proceedings of SPIE* 6542, 65421F, 2007.

113. S. R. Hunter, G. Maurer, G. Simelgor, S. Radhakrishnan, J. Gray, K. Bachir, T. Pennell, M. Bauer, and U. Jagadish, "Development and Optimization of Microcantilever Based IR Imaging Arrays," *Proceedings of SPIE* 6940, 694013, 2008.

114. J. Zhao, "High Sensitivity Photomechanical MW-LWIR Imaging Using an Uncooled MEMS Microcantilever Array and Optical Readout," *Proceedings of SPIE* 5783, 506–13, 2005.

115. D. Grbovic, N. V. Lavrik, and P. G. Datskos, "Uncooled Infrared Imaging Using Bimaterial Microcantilever Arrays," *Applied Physics Letters* 89, 073118, 2006.

116. N. Lavrik, R. Archibald, D. Grbovic, S. Rajic, and P. Datskos, "Uncooled MEMS IR Imagers with Optical Readout and Image Processing," *Proceedings of SPIE* 6542, 65421E, 2007.

117. J. P. Salerno, "High Frame Rate Imaging Using Uncooled Optical Readout Photomechanical IR Sensor," *Proceedings of SPIE* 6542, 65421D, 2007.

118. M. Wagner, E. Ma, J. Heanue, and S. Wu, "Solid State Optical Thermal Imagers," *Proceedings of SPIE* 6542, 65421P, 2007.

119. M. Wagner, "Solid State Optical Thermal Imaging: Performance Update," *Proceedings of SPIE* 6940, 694016, 2008.

120. B. Jiao, C. Li, D. Chen, T. Ye, S. Shi, Y. Qu, L. Dong, et al., "A Novel Opto-Mechanical Uncooled Infrared Detector," *Infrared Physics & Technology* 51, 66–72, 2007.

121. S. Shi, D. Chen, B. Jiao, C. Li, Y. Qu, Y. Jing, T. Ye, et al., "Design of a Novel Substrate-Free Double-Layer-Cantilever FPA Applied for Uncooled Optical-Readable Infrared Imaging System," *IEEE Sensors Journal* 7, 1703–10, 2007.

122. F. Dong, Q. Zhang, D. Chen, Z. Miao, Z. Xiong, Z. Guo, C. Li, B. Jiao, and X. Wu, "Uncooled Infrared Imaging Device Based on Optimized Optomechanical Micro-Cantilever Array," *Ultramicroscopy* 108, 579–88, 2008.

123. X. Yu, Y. Yi, S. Ma, M. Liu, X. Liu, L. Dong, and Y. Zhao, "Design and Fabrication of a High Sensitivity Focal Plane Array for Uncooled IR Imaging," *Journal of Micromechanics and Microengineering* 18, 057001, 2008.

124. P. Kruse, "Can the 300K Radiating Background Noise Limit be Attained by Uncooled Thermal Imagers?" *Proceedings of SPIE* 5406, 437–46, 2004.

21 Photon Detector Focal Plane Arrays

Looking back over the past several hundred years we notice that following the invention and evolution of optical systems (telescopes, microscopes, eyeglasses, cameras, etc.) the optical image was formed on the human retina, photographic plate, or films. The birth of photodetectors can be dated back to 1873 when Smith discovered photoconductivity in selenium. Progress was slow until 1905, when Einstein explained the newly observed photoelectric effect in metals, and Planck solved the blackbody emission puzzle by introducing the quanta hypothesis. Applications and new devices soon flourished, pushed by the dawning technology of vacuum tube sensors developed in the 1920s and 1930s culminating in the advent of television. Zworykin and Morton, the celebrated fathers of videonics, on the last page of their legendary book *Television* (1939) concluded that: *"when rockets will fly to the moon and to other celestial bodies, the first images we will see of them will be those taken by camera tubes, which will open to mankind new horizons."* Their foresight became a reality with the Apollo and Explorer missions. Photolithography enabled the fabrication of silicon monolithic imaging focal planes for the visible spectrum beginning in the early 1960s. Some of these early developments were intended for a picturephone, other efforts were for television cameras, satellite surveillance, and digital imaging. Infrared imaging has been vigorously pursued in parallel with visible imaging because of its utility in military applications. More recently (1997) the CCD camera aboard the Hubble space telescope delivered a deep-space picture, a result of 10 day's integration, featuring galaxies of the 30th magnitude—an unimaginable figure even for astronomers of our generation. Probably, the next effort will be in the big-band age. Thus, photodetectors continue to open to mankind the most amazing new horizons.

Although efforts have been made to develop monolithic structures using a variety of infrared photodetector materials (including narrow-gap semiconductors) over the past 30 years, only a few have matured to a level of practical use. These included Si, PtSi, and more recently PbS, PbTe. Other infrared material systems (InGaAs, InSb, HgCdTe, GaAs/AlGaAs QWIP, and extrinsic silicon) are used in hybrid configurations.

This chapter is a guide over the arrays of photon detectors sensing infrared radiation.

21.1 INTRINSIC SILICON AND GERMANIUM ARRAYS

Visible imaging sensors use three basic architectures:

- Monolithic CCD (both front side and back side illuminated)

- Monolithic CMOS, for which the photodiodes are included within the silicon readout integrated circuit (ROIC)

- Hybrid CMOS that uses a detector layer for detection of light and collection of photocharge into pixels and a CMOS ROIC for signal amplification and readout

At present two monolithic technologies provide the bulk of devices in the markets of camcorders and digital cameras: CCD and CMOS imagers. The fundamental performance parameters common to both CCD and CMOS imagers have been compared by Janesick [1,2]. Compared to CCD, CMOS performance is currently preventing the technology from scientific and high end use [3]. Custom CMOS pixel designs and fabrication process are required to improve performance.

Monolithic imaging FPAs for the visible spectrum began in the early 1960s. Further development is shown in Figure 21.1, where the trend in image sensor migration of CCD and CMOS devices for various market applications is presented [4]. CCD technology was developed for imaging applications, and its fabrication processes were optimized to build an image sensor with the best possible optical properties and image quality. CCDs dominate in high performance, low-volume segments, such as professional digital still cameras, machine vision, medical, and scientific applications. However, due to low power consumption; ability to integrate timing, control, and other signal processing circuitry on chip; and single supply and master clock operation, CMOS emerges the winner in low-cost, high-volume applications, particularly where low power consumption and small system size are key. The majority of growth comes from new products enabled by CMOS imaging technology, such as automotive, computer video, optical mice, imaging phones, toys, biometrics, and a host of hybrid products.

The color film has traditionally been held as the gold standard for photography. It produces rich, warm tones, and incredible color detail that consumers around the world have become accustomed to. Film has achieved this by using three layers of emulsion to capture full color at every point in the image.

Digital CCD silicon image sensors were developed approximately 30 years ago, ushering in the era of digital photography. In visible imaging, CCD arrays integrate the readout and sensor in a combined pixel and integrated red–green–blue (RGB) color filters are included. The most common checkerboard filter pattern dedicates two out of four pixels to green, and one pixel each to red and blue. As a result, the sensor gathers only 50% of the green light, and 25% of the red and blue. Digital post-processing interpolates to fill in the blanks, so more than half of the image is artificially generated and innately imperfect. In this case the fill factor is only a fraction of the pixel area (e.g., about 70%) [5]. Unfortunately, the rich warm tones and detail of color film that the world came to expect suffered over the convenience and immediacy of digital. This was due to the fact that CCD digital image sensors were only capable of recording just one color at each point in the captured image instead of the full range of colors at each location.

Foveon combined the best of what both film and digital have to offer [5]. This is accomplished by the innovative design of Foveon's X3 direct image sensors that have three layers of pixels, just like film has three layers of chemical emulsion (see Figure 21.2) [6]. This silicon sensor is fabricated

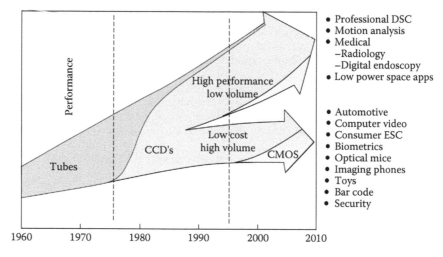

Figure 21.1 Trend in image sensor migration of CCD and CMOS devices for various market applications. (From Titus, H., "Imaging Sensors That Capture Your Attention," *Sensors* 18, February 2001.)

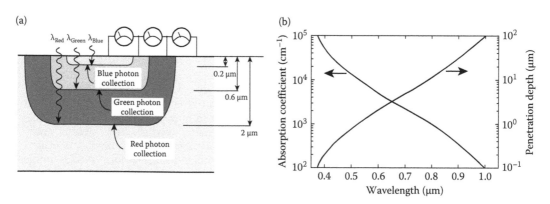

Figure 21.2 Schematic illustration of Foveon's sensor stack: (a) cross section of the sensor, and (b) absorption coefficient and penetration depth in silicon. (From Lyon, R. F., and Hubel, P., *IS&T/ SID Tenth Color Imaging Conference*, 349–55, 2001.)

Figure 21.3 Photograph of 111 million pixel CCD array, with 9 μm pixel size, fabricated by Dalsa. (From http://www.dalsasemi.com)

on a standard CMOS processing line. Foveon's layers are embedded in silicon to take advantage of the fact that red, green, and blue light penetrate silicon to different depths (the photodetectors sensitive to blue light are on top, the green sensitive detectors are in the middle, and the red on the bottom) forming the image sensor that captures full color at every point in the captured image. This is 100% full color with no interpolation. Figure 21.2b plots the absorption coefficient as a function of depth, which is an exponential function of depth for any wavelength. Since the higher energy photons interact more strongly, they have a smaller space constant, and thus the exponential falloff with depth is more rapid.

At present, the largest CCD arrays exceed 100 megapixels. DALSA announces that it has successfully produced a 111 megapixel CCD (see Figure 21.3). The active area measures approximately 4×4 inches and 10560×10560 pixels with 9 μm size [7]. The record-breaking chip is developed for the Astrometry Department of the U.S. Naval Observatory to assist them in the determination of the positions and motions of stars, solar system objects, and the establishment of celestial reference frames. The large CCD sensors are now starting to be produced in large quantities to meet the demands of astronomers. In particular, recent development of buttable CCD arrays could be of considerable interest to the photogrammetric and remote sensing communities. The development of mosaics of area arrays to produce large format (up to gigapixel) frame image is an intriguing idea [8].

For visible light detection, Teledyne Imaging Sensors (TIS) use both monolithic and hybrid CMOS detectors (see Table 21.1) [9,10]. These highest performance silicon-based image arrays are addressed to the astronomical community. Monolithic image sensors as large as 59 million pixels characterized by low noise (2.8 e⁻ readout noise) and low dark current (<10 pA/cm² at 295 K) are being produced with > 99.9% operability.

Teledyne's monolithic CMOS sensors are fully digital system-on-chip, with all bias generation, clocking, and analog-to-digital conversion included with the image array. Some examples of the arrays are listed in Table 21.2 [10].

Until recently, CMOS imagers have been at a disadvantage relative to CCDs for readout noise, since the CMOS's readout circuit is inherently higher noise than a CCD amplifier. One reason for higher noise is the capacitance of the sense node of the three transistor CMOS pixel. For a typical pixel size of 5 μm in a 0.25 μm CMOS image sensor process, the sense node capacitance is about 5 fF, corresponding to a responsivity of 32 μV/electron, which limits the lowest readout noise to about 10 electrons. With the development of the four transistor pixel, monolithic CMOS can achieve the lowest noise levels required by astronomy [10].

Visible hybrids have also been built for special applications to take advantage of the materials flexibility and larger surface area with nearly 100% fill factor [10,11]. Recently, Si p-i-n detector arrays for the astronomy and civil space communities in hybrid configuration with size as large as 4096×4096 have been demonstrated [10]. The prototype arrays demonstrated a high pixel interconnectivity (>99.9%) and high pixel operability (>99.8%). This design is scalable to an array format

Table 21.1: Visible Silicon Imaging Sensors Technologies Produced by Teledyne Imaging Sensors

Sensor Architecture	Monolithic CMOS	Hybrid CMOS
Quantum efficiency	■ Front side: 65% with std. microlenses, 80% with optimized microlenses ■ *Back side: same QE as back side CCD and hybrid silicon p-i-n CMOS*	■ High QE in X-ray, visible, and NIR ■ Detector layer thickness: 50–250 μm ■ Multilayer and graded thickness antireflection coatings available
Dark current (298 K)	■ <10 pA/cm^2 (front side illuminated) ■ *Dark current increase for back side illumination will depend on quality of surface treatment processes*	■ 5–10 nA/cm^2 ■ *Ongoing IR&D effort to reduce dark current to ~1 nA/cm^2*
Readout noise	■ 2.8 e$^-$ demonstrated ■ *<2 e$^-$ readout circuits in development*	■ 7–10 e$^-$ single CDS with source follower, lower with multiple samples ■ *< 4 e$^-$ with specialized high speed CTIA pixel in 900 frames/sec ROIC* ■ 50–100 e$^-$ for high capacity CTIA pixels
Array size	■ 2 megapixel (1936 × 1280) ■ 12 megapixel (3648 × 3375) ■ 59 megapixel (7680 × 7680)	■ 640 × 480 pixels ■ 1024 × 1024 pixels ■ 2048 × 2048 pixels ■ 4096 × 4096 pixels

Source: Y. Bai, J. Bajaj, J. W. Beletic, and M. C. Farris, "Teledyne Imaging Sensors: Silicon CMOS Imaging Technologies for X-Ray, UV, Visible and Near Infrared," *Proceedings of SPIE* 7021, 702102, 2008. With permission.
Note: Future improvements are denoted by italic text.

Table 21.2: Monolithic CMOS Imagers Developed by Teledyne Imaging Sensors

Imager Name	Number of Pixels	Total # of Pixels (millions)	Pixel Pitch (μm)	ADC on Chip	Read-out Mode	Charge Capacity (e$^-$)	Read-out Noise (e$^-$)	QE (%)	Frame Rate (Hz)	Power (mW)
V1M	1280 × 1280	1,6	14			360,000	77	65% peak, lower in blue and red	10	200
V2M	1936 × 1086	2,1	5&10	12-bit	Ripple	150,000	<50		30	300
V12M	3648 × 3375	12,3	5			40,000	<16		20	1250
V59M	7680 × 7680	59	5			45,000	<30		8	3000

Source: Y. Bai, J. Bajaj, J. W. Beletic, and M. C. Farris, "Teledyne Imaging Sensors: Silicon CMOS Imaging Technologies for X-Ray, UV, Visible and Near Infrared," *Proceedings of SPIE* 7021, 702102, 2008. With permission.

up to 16K × 16K. It should be added that due to thin oxides used in fabrication of CMOS ROICs, the FPAs are inherently radiation hard and do not suffer from charge transfer inefficiency that is problematic for CCDs.

During 2007, three types of Teledyne's hybrid silicon p-i-n CMOS sensors, called HyViSi™ (HiRG-19, H2RG-18, H4RG-10; see Figure 21.4) and characterized in Table 21.3, were tested on the 2.1 meter telescope on Kitt Peak (Arizona) [10].

Even though silicon imagers are routinely manufactured with magapixel resolution and excellent reproducibility and uniformity, they are not sensitive to radiation longer than about 1 μm. Today, SiGe alloys are widely used in Si CMOS and bipolar technologies. Recently, this technology is adopted to develop a single-chip SWIR image sensor, which integrates germanium photodetectors on a standard Si process. This innovative technique is illustrated in Figure 10.13. Imaging arrays of 128 × 128 pixels at a 10 μm pitch were fabricated and quick extending of this array to higher resolution is going on [12].

1K × 1K H1RG-18 HyViSi   2K × 2K H2RG-18 HyViSi   4K × 4K H1RG-10 HyViSi

Figure 21.4 Teledyne's hybrid silicon p-i-n CMOS sensors. (From Bai, Y., Bajaj, J., Beletic, J. W., and Farris, M. C., "Teledyne Imaging Sensors: Silicon CMOS Imaging Technologies for X-Ray, UV, Visible and Near Infrared," *Proceedings of SPIE* 7021, 702102, 2008. With permission.)

Table 21.3: ROICs Used for Silicon p-i-n CMOS FPAs

	TCM6604A	TCM8050A	H1RG-18	H2RG-18	H4RG-10
Pixel amplifier	CTIA	DI	SF	SF	SF
Array format (pixels)	640 × 480	1024 × 1024	1024 × 1024	2048 × 2048	4096 × 4096
Pixel pitch (μm)	27	18	18	18	10
Number of outputs	4	4	1, 2, or 16	1, 4, or 32	1,4,16,32, or 64
Pixel rate (MHz)	8	6	0.1 to 5	0.1 to 5	0.1 to 5
Readout mode	Snapshot	Snapshot	Ripple	Ripple	Ripple
Window mode	Programmable	Full, 512, 256	Guide window	Guide window	Guide window
Charge capacity (ke⁻)	700	3,000	100	100	100
Readout noise (e⁻)	<100	<300	<10	<10	<10
Power dissipation (mW)	<70	<100	<1	<4	<14

Source: Y. Bai, J. Bajaj, J. W. Beletic, and M. C. Farris, "Teledyne Imaging Sensors: Silicon CMOS Imaging Technologies for X-Ray, UV, Visible and Near Infrared," *Proceedings of SPIE* 7021, 702102, 2008. With permission.

21.2 EXTRINSIC SILICON AND GERMANIUM ARRAYS

The first extrinsic photoconductive detectors were reported in the early 1950s [13]. They were widely used at wavelengths beyond 10 μm prior to the development of the intrinsic detectors. Since the techniques for controlled impurity introduction became available for germanium at an earlier date, the first high performance extrinsic detectors were based on germanium. The discovery in the early 1960s of Hg-doped germanium led to the first FLIR systems operating in the LWIR spectral window using linear arrays [14]. Because the detection mechanism was based on an extrinsic excitation, it required a two-stage cooler to operate at 25 K. Although doped germanium was the IR detector of choice in the 1960s, doped silicon replaced germanium for most applications during the 1980s.

The first attempt to develop a monolithic extrinsic photodetector array integrating the photosensitive elements and devices for primary signal processing in one crystal was made in the early 1970s. Twenty photosensitive unit cells were arranged on a Si:As plate. Each of them included a photosensitive element, load resistor based on impurity compensation, and MOSFET connected into the circuit as a source follower [15]. This array provided useful performance, but exhibited undesirable electrical and optical crosstalk between detector channels. That was corrected with its improvement.

The first report on the development of monolithic arrays with charge-coupled-device (CCD) multiplexing was published in 1974 [16]. Later, two more monolithic CCD versions were developed [17]. All these devices (Figure 21.5) operate at the temperature of photosensitive substrate corresponding to a low-ionized state of impurity when the concentration of free carriers is low as compared with that of impurity.

A CCD, for the accumulation mode, transfers photogenerated majority charge carriers (holes) that are accumulated at the Si–Si oxide interface. This device has not seen any application because

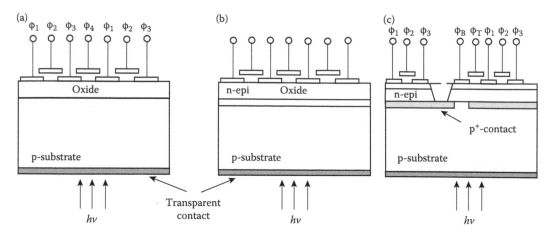

Figure 21.5 Monolithic silicon arrays with CCD multiplexing: (a) accumulation mode, (b) pseudoaccumulation mode, and (c) vias mode. (From Pommerrenig, D. H., "Extrinsic Silicon Focal Plane Arrays," *Proceedings of SPIE* 443, 144–50, 1984. With permission.)

the transfer efficiency was found to be low and the clocking frequency was limited to low frequencies [16]. In the pseudo accumulation mode (PAM), photogenerated holes are injected as minority carriers in the n-epilayer where they can then be clocked out by the CCD. In the vias mode, the photosensitive substrate is equipped with buried individual detector cell contacts. These contacts collect the photogenerated holes that are transferred to the n-epilayer under the action of a CCD transfer gate where they are then clocked out. The PAM has been used with Si:Ga and Si:In to make 32×96 element arrays [18]. It has been operated at background levels of 10^8–10^{14} photons/cm^{2}s.

For the monolithic arrays, biasing and operational temperature were found to be critical in avoiding injection currents from the n-epitaxial, p-substrate contacts. Operating temperatures lower than those dictated by the detector characteristics were required. Further, the detector responsivity obtained for monolithic devices was found to be considerably degraded compared to that obtained on discrete detectors prepared from the same substrate material (lowering by two orders of magnitude). This degradation was supposed to be associated with thermal oxidation, p$^+$, n$^+$ diffusion, epitaxial growths, and polysilicon (gate) depositions. It was found that the loss in responsivity results from an increase in donor concentration to over 10^{14} cm^{-3} occurring with manufacturing a device [19]. It was also found that one order of the magnitude in lifetime and responsivity was reclaimed by a phosphorus gettering process that lowered the net compensation back toward starting values.

To avoid the long shift registers inherent in CCDs, other pseudomonolithic readout mechanisms have been devised. In charge injection device (CID) photodetector arrays, a photosignal charge is assumed to be collected and stored in MOS capacitors of the device cells. Readout of this charge takes place in the same cell where this charge has been collected. The CID photodetector arrays do not include any supplementary elements intended for transfer of the charge. Injection into the semiconductor substrate is used to free the cells of array from the signal charge being accumulated and in some cases it is used as a method of readout.

Originally, CIDs were proposed in which the accumulation of minority charge carriers occurred [20]. In the presence of a negative bias applied to the gate on n-type Si, an electron-depleted layer emerges under this gate. Under the intrinsic irradiation of the substrate, hole-electron pairs are generated and minority charge carriers (holes) are collected and stored in this layer at the Si-Si oxide interface. With removal of the voltage, the accumulated charge carriers are injected into the substrate where they are collected to provide a signal output.

At low temperatures and irradiances when the concentration of free carriers becomes very low, such devices exhibit a capability to collect and store majority charge carriers generated via the photoionization of impurity centers. To realize this, a positive bias should be applied to the gate if the substrate is of n-type conductivity. Then, a pulse of voltage of negative polarity should be applied to the gate for the injection of accumulated charge into the substrate and its readout. A value of this voltage should be great enough to minimize the losses for recombination during the drift through the substrate. Two-dimensional 32×32 and 2×64 Si:Bi CID arrays based on this approach have been constructed [21,22]. However, it was noted that the array well capacity was significantly smaller than predicted. The cutoff frequency of photoresponse appeared to be also considerably lower than the expected value determined by the rate of the dielectric-relaxation processes in the device.

The CCD approach is not compatible with low read noise because of the low operating temperature of the IR detectors. The buried channel CCDs were also used to avoid trapping noise, however, there is no longer sufficient mobile charge to maintain the channel. The devices then operate in surface channel mode, with accompanying high noise. In addition, CCDs suffer due to damage to the devices under extremely high doses of ionizing radiation, which degrades the charge transfer efficiency. Therefore, different architectures were developed in which a readout amplifier is dedicated to each pixel, and transistor switches bring the signal to an output amplifier [23].

The above-mentioned shortcomings of the monolithic arrays with CCD multiplexing have been overcome by the transition to a hybrid fashion of device with CMOS detector array. This makes it possible to use lower temperatures when manufacturing it. With such a design, there also appears an additional Si space for selective input circuits and signal processing. These depend on the application of the array and can include time and add electronics to improve detector performance and gain reduction and DC suppression electronics to increase dynamic range.

The largest extrinsic infrared detector arrays are manufactured for astronomy. Their application began roughly 20 years ago [24] and have doubled about every seven months since then [25]. The speed with which a given region of the sky can be mapped has increased by a factor 10^{18} in 40 years, corresponding to a doubling of speed every 12 months. Sensitivities of individual detectors approach the fundamental limits set by photon noise.

The early detector arrays were small (typically 32×32 pixels) with read noises of more than 1000 electrons. Their basic architecture and processes to produce high-performance arrays came from the military. Further development has followed owing to investments from NASA and the National Science Foundation [26]. At present Raytheon Vision Systems (RVS), DRS Technologies, and TIS (formerly Rockwell Scientific Company) supply the majority of IR arrays used in astronomy between them the most important being BIB detector arrays. The characteristics of the most advanced IR arrays for astronomy from these manufacturers gathered by Rieke are summarized in Table 21.4 [25].

Table 21.4: Si:As BIB Hybrid Arrays

Parameter	DRS Technologies[a] WISE	RVS[b] JWST
Wavelength range (μm)	5–28	5–28
Format	1024×1024	1024×1024
Pixel pitch (μm)	18	25
Operating temperature (K)	7.8	6.7
Read noise (e rms)	42 (Fowler-1; lower noise expected with more reads)	10
Dark current (e/s)	<5	0.1
Well capacity (e)	$>10^5$	2×10^5
Quantum efficiency (%)	>70	>70
Outputs	4	4
Frames/sec	1	0.3

Source: G. H. Rieke, *Annual Review Astronomy and Astrophysics,* 45, 77–115, 2007. With permission.

Note: JWST: James Webb Space Telescope, WISE: Wide-Field Infrared Survey Explorer.

[a] A. K. Mainzer, P. Eisenhardt, E. L. Wright, F.-C. Liu, W. Irace, I. Heinrichsen, R. Cutri, and V. Duval, "Preliminary Design of the Wide-Field Infrared Survey Explorer (WISE)," *Proceedings of SPIE* 5899, 58990R, 2005.

[b] P. J. Love, A. W. Hoffman, N. A. Lum, K. J. Ando, J. Rosbeck, W. D. Ritchie, N. J. Therrien, R. S. Holcombe, and E. Corrales, "1024 × 1024 Si:As IBC Detector Arrays for JWST MIRI," *Proceedings of SPIE* 5902, 590209, 2005.

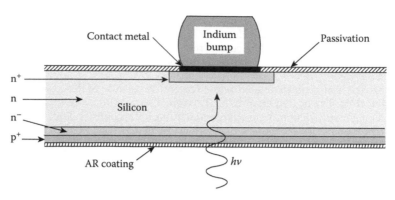

Figure 21.6 Blocked impurity band detector.

A BIB detector structure is displayed in Figure 21.6. This structure, described in detail in Section 11.4, usually has a thin lightly doped n⁻-region (usually grown epitaxially) between the absorbing n-type region and the common back side implant (p⁺-region) in order to block hopping-conduction currents from reaching the p⁺-contact, which are substantial at the doping levels used in the active detection region. Doping of active layer with a thickness value in the 10 µm range is high enough for the onset of an impurity band in order to display high quantum efficiency for impurity ionization. This also prevents the build of large space-charge regions in n-type active regions, which can otherwise result in a dielectric relaxation response that depends on the illumination history.

The blocked impurity band devices, in large staring array formats are now becoming commercially available [27–32]. Impressive progress has been achieved especially in Si:As BIB array technology with formats as large as 1024 × 1024 and pixels as small as 18 µm; operated in spectral band up to 30 µm at about 10 K. The pixel size of 18 µm is smaller than the wavelength at Q band (17–24 µm), however this does not pose a problem since an imager operating at these wavelengths will typically spread the beam out over many pixels to be fully sampled.

The main application of BIB arrays today is for ground- and space-based far-infrared astronomy. The arrays should be operated under the most uniform possible conditions, in the most benign and constant environment possible. Array performance is strongly affected by background levels. The examples described in Table 21.4 have been optimized for low background, where the read noise is minimized by minimizing the detector plus gate integrating capacitance, C. Then, for a given charge, Q, the voltage is maximized since $V = Q/C$. If the detectors are exposed to higher background levels, the readout amplifiers saturate. Counteraction includes simply increasing the integration capacitance or using alternative amplifier architecture with handling larger signals.

To enhance array performance, Fowler and Gatley [33] proposed to reduce the read noise by performing multiple nondestructive passes through the array at the beginning and end of the integration ramp. By resetting each pixel on one pass through the array, and sampling the detector node voltage on subsequent passes through the array, the *true* pedestal is removed from the data. Each time a pixel is selected, a charge is redistributed from the row and column select FETs back onto the detector node capacitance. Fowler claims that the read noise is predominantly due to the kTC noise associated with this charge redistribution. By performing multiple nondestructive passes through the array at the beginning and end of the integration, the read noise is reduced by the square root of the number of passes.

Extrinsic silicon arrays for high background applications are less developed than that for low background applications. Detectors in conventional ground-based systems are operated in thermal backgrounds up to 10^9 photons/second. There are no modern arrays optimized for such backgrounds in MWIR region [25]. The largest available high background Si:As BIB arrays are 256 × 256 or 240 × 320 pixels. Recently, the first 1024 × 1024 high background Si:As BIB detector array has been developed at DRS Technologies [30].

As is mentioned in Section 11.4, the BIB detector active layer should be thick and doped as heavily as possible. This limit is reached at an As concentration of about 10^{18} cm⁻³. The layer thickness is limited by the minority impurity concentration and by incipient avalanching of charge carriers created by bias voltage. According to Love et al. [29], the minority upper limits are 1.44×10^{12} cm⁻³

for a 45 μm thick layer, and 1.85×10^{12} cm^{-3} for a 35 μm thick layer. Therefore, the detectors are designed for arsenic doping of 7×10^{17} cm^{-3} and a thickness of 35 μm.

The readouts for large BIB arrays use a circuit similar to that described in Chapter 19 (Figure 19.13). However, it should be mentioned that silicon-based MOSFETs show a number of operational difficulties conditioned by the very low temperatures required for the readout circuits for these detectors [34]. They are related to freeze-out of thermally generated charge carriers, making the circuits unstable, increasing noise, and causing signal hysteresis. They are described in detail by Glidden et. al [35]. Many of them can be mitigated by growing the circuits that are heavily doped.

Figure 21.7 illustrates 1024 × 1024 Si:As BIB array for the Mid Infrared Instrument on the James Webb Space Telescope [28]. Figure 21.7a shows details of the array mount but 21.7b shows a completed array.

Due to lack of atmospheric transparency in the far-infrared region (above 30 μm), applications are limited to the laboratory, high altitude, and space. Germanium extrinsic detectors have largely been supplanted by silicon detectors for both high and low background applications where comparable spectral response can be obtained, but germanium devices are still of interest for very long wavelengths. For wavelengths longer than 40 μm, there are no appropriate shallow dopants for silicon. Very shallow donors, such as Sb, and acceptors, such as B, In or Ga, provide cutoff wavelengths of extrinsic germanium in the region of 100 μm.

Ge:Ga photoconductors are the best low background photon detectors for the wavelength range from 40 to 120 μm. However, there are a number of problems with the use of germanium. For example, to control dark current the material must be lightly doped and therefore absorption lengths become long (typically 3–5 mm). Because the diffusion lengths are also large (typically 250–300 μm), pixel dimensions of 500–700 μm are required to minimize crosstalk. In space applications, large pixels imply higher hit rates for cosmic radiation. This in turn implies very low readout noise for arrays operated in low background limit, which is difficult to achieve for large pixels with large capacitance and large noise. A solution is using the shortest possible exposure time. Due to small energy bandgap, the germanium detectors must operate well below the silicon "freeze-out" range—typically at liquid helium temperature

Application of uniaxial stress along the (100) axis of Ge:Ga crystals reduces the Ga acceptor binding energy, extending the cutoff wavelength to ≈240 μm. At the same time, the operating temperature must be reduced to less than 2 K. In making practical use of this effect, it is essential to apply and maintain very uniform and controlled pressure to the detector so that the entire detector volume is placed under stress without exceeding its breaking strength at any point. A number of mechanical stress modules have been developed. The stressed Ge:Ga photoconductor systems have found a wide range of astronomical and astrophysical applications [25,36–41].

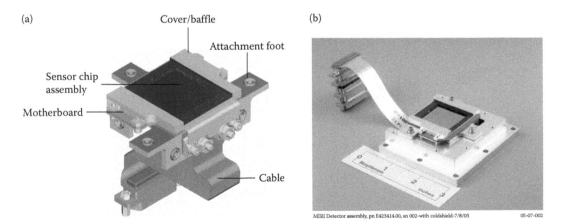

Figure 21.7 1024 × 1024 Si:As BIB array for the Mid Infrared Instrument on the James Webb Space Telescope: (a) details of the array mount, and (b) a completed array. (From Love, P. J., Hoffman, A. W., Lum, N. A., Ando, K. J., Rosbeck, J., Ritchie, W. D., Therrien, N. J., Holcombe, R. S., and Corrales, E., "1024 × 1024 Si:As IBC Detector Arrays for JWST MIRI," *Proceedings of SPIE* 5902, 590209, 2005. With permission.)

The Infrared Astronomical Satellite (IRAS), the Infrared Space Observatory (ISO), and for the far-infrared channels the Spitzer-Space Telescope (Spitzer) have all used bulk germanium photoconductors. In Spitzer mission a 32 × 32 pixel Ge:Ga unstressed array was used for the 70 μm band, while the 160 μm band had a 2 × 20 array of stressed detectors. The detectors are configured in the so-called Z-plane to indicate that the array has substantial size in the third dimension. The poor absorption of the Ge:Ga detector material requires that the detectors in this array are huge—2 mm long. Due to the long absorption path in germanium, the detectors are illuminated edgewise with transverse contacts, and the readouts are hidden behind them, shown in Figure 21.8 [36].

An innovative integral field spectrometer, called the Field Imaging Far-Infrared Line Spectrometer (FIFI-LS), which produces a 5 × 5 pixel image with 16 spectral resolution elements per pixel in each two bands was constructed at the Max Planck Institut für Extraterrestrische Physik under the direction of Albrecht Poglitsch. This array, shown in Figure 21.9 [42], was developed for the Herschel Space Observatory and SOFIA. To accomplish this, the instrument has two 16 × 25 Ge:Ga arrays, unstressed for the 45–110 μm range and stressed for the 110 to 210 μm range. Each detector pixel is stressed in its own subassembly, and a signal wire is routed to preamplifiers housed nearby that obviously limits this type of array to much smaller formats than are available without these constraints.

A modest NASA program to extend BIB detector performance at wavelengths as long as 400 μm using gallium arsenide is rather difficult to achieve due to a problematic requirement for producing the required epitaxial low-doped blocking layer [43,44].

Far-infrared photoconductor arrays suffer from the standard bulk photoconductor photometric issues. Especially, germanium detectors have complicated responses that affect calibration, observing strategies, and data analysis in low background applications. The devices operate at very low bias voltages and even small changes in the operating points of amplifiers can result in unacceptable bias changes on the detectors. More details can be found by Rieke [25].

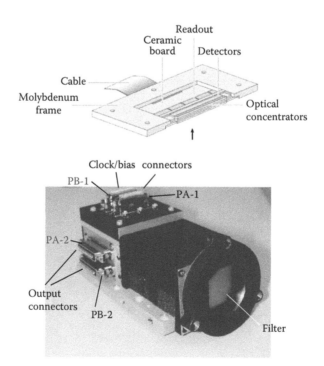

Figure 21.8 The Spitzer Space Telescope 70 μm array design of a 4 × 32 module of the edge-illuminated detectors. Eight such modules are stacked to create the full 32 × 32 array that covers the wavelength range 50–110 μm. (From Young, E. T., Davis, J. T., Thompson, C. L., Rieke, G. H., Rivlis, G., Schnurr, R., Cadien, J., Davidson, L., Winters, G. S., and Kormos, K. A., "Far-Infrared Imaging Array for SIRTF," *Proceedings of SPIE* 3354, 57–65, 1998. With permission.)

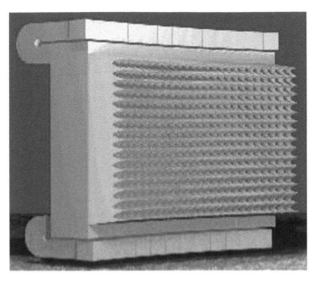

Figure 21.9 Stressed Ge:Ga detector array (FIFI-LS) in a 16 × 25 pixel format. (From http://fifi-ls. mpg-garching.mpg.dr/detector.html)

21.3 PHOTOEMISSIVE ARRAYS

The progress of the Schottky-barrier FPA technology has been constant [45]. The first Schottky-barrier FPA was the 25 × 50 element IR-CCD developed at RCA laboratories under contract to the Rome Air Development Center [46]. At the present time Schottky-barrier FPAs represent the most advanced FPAs technology for medium wavelength applications. Scanning PtSi FPAs with up to 4 × 4096 elements [47] and 2048 × 16 TDI [48] elements were developed for space-borne remote sensing applications. Review of different configuration of staring Schottky-barrier FPAs is given by Kosonocky [49,50] and Kimata et al. [45,51–53], for example. Table 21.5 summarizes the specifications and performance of typical high resolution PtSi Schottky-barrier FPAs that have full TV resolution.

The details of the geometry, and the method of charge transfer differ for different manufacturers. The design of a staring Schottky-barrier FPAs for given pixel size and design rules, involves a trade-off between the charge handling capacity and the fill factor. This trade-off depends also on the choice of the FPA architecture that includes:

- Interline transfer CCD architecture

- Charge sweep device (CSD)

- Line-addressed charge-accumulation (LACA) readout

- Readout by MOS switches

Most of the reported Schottky-barrier FPAs have the interline transfer CCD architecture. Figure 21.10 shows the basic construction and operation of the most popular Schottky-barrier detector in the 3–5 μm spectral range, PtSi/p-Si, integrated with a silicon CCD readout [49]. Radiation is transmitted through the p-type silicon and is absorbed in the metal PtSi, producing hot holes that are then emitted over the potential barrier into the silicon, leaving the silicide charged negatively. Negative charge of silicide is transferred to a CCD by the direct charge injection method. The typical cross-section view of the pixel and its operation in interline transfer CCD architecture is shown in Figure 21.11 [51]. The pixel consists of a Schottky-barrier detector with an optical cavity, a transfer gate, and a stage of vertical CCD. The n-type guard ring on the periphery of the Schottky-barrier diode, reduces the edge electric field and suppresses dark current. The effective detector area is determined by the inner edge of the guard ring. The transfer gate is an enhancement MOS transistor. The connection between detector and the transfer gate is made by an n+ diffusion. A buried-channel CCD is used for the vertical transfer.

During the optical integration time the surface-channel transfer gate is biased into accumulation. The Schottky-barrier detector is isolated from the CCD register in this condition.

Table 21.5: Specifications and Performances of Typical PtSi Schottky-Barrier FPAs

Array Size	Readout	Pixel Size (μm^2)	Fill Factor (%)	Saturation (e^-)	NEDT/ (f/#) (K)	Year	Company
512×512	CSD	26×20	39	1.3×10^6	0.07(1.2)	1987	Mitsubishi
512×488	IL-CCD	31.5×25	36	5.5×10^5	0.07(1.8)	1989	Fairchild
512×512	LACA	30×30	54	4.0×10^5	0.10(1.8)	1989	RADC
640×486	IL-CCD	25×25	54	5.5×10^5	0.10(2.8)	1990	Kodak
640×480	MOS	24×24	38	1.5×10^6	0.06(1.0)	1990	Sarnoff
640×488	IL-CCD	21×21	40	5.0×10^5	0.10(1.0)	1991	NEC
640×480	HB/MOS	20×20	80	7.5×10^5	0.10(2.0)	1991	Hughes
1040×1040	CSD	17×17	53	1.6×10^6	0.10(1.2)	1991	Mitsubishi
512×512	CSD	26×20	71	2.9×10^6	0.03(1.2)	1992	Mitsubishi
656×492	IL-CCD	26.5×26.5	46	8.0×10^5	0.06(1.8)	1993	Fairchild
811×508	IL-CCD	18×21	38	7.5×10^5	0.06(1.2)	1996	Nikon
801×512	CSD	17×20	61	2.1×10^6	0.04(1.2)	1997	Mitsubishi
537×505	IL-CCD	15.2×11.8	32	2.5×10^5	0.13(1.2)	1997	
1968×1968	IL-CCD	30×30	–	–	–	1998	Fairchild

Source: M. Kimata, *Handbook of Infrared Detection Technologies*, 353–92, Elsevier, Oxford, 2002. With permission.
Note: IL-CCD: Interline Transfer CCD, HB: Hybrid, CSD: Charge Sweep Device, MOS: Metal Oxide Semiconductor, LACA: Line-Addressed Charge-Accumulation, CCD: Charge Couple Device.

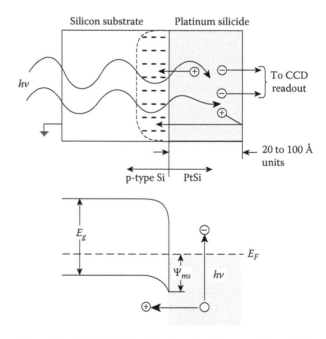

Figure 21.10 Operation of a PtSi/p-Si Schottky-barrier detector. (From Kosonocky, W. F., *Optoelectronics–Devices and Technologies*, 6, 173–203, 1991.)

The IR radiation generates hot holes in the PtSi film and some of the excited hot holes are emitted into the silicon substrate leaving excess electrons in the PtSi electrode. This lowers the electrical potential of the PtSi electrode. At the end of the integration time, the transfer gate is pulsed-on to readout the signal electrons from the detector to the CCD register. At the same time, the electrical potential of the PtSi electrode is reset to the channel level of the transfer gate.

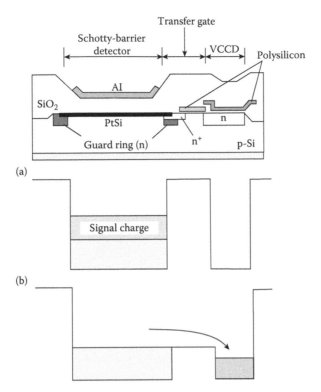

Figure 21.11 Typical construction and operation of PtSi Schottky-barrier IR FPA designed with interline transfer CCD readout architecture. (a) and (b) show the potential diagrams in the integration and readout operations, respectively. (From Kimata, M., and Tsubouchi, N., *Infrared Photon Detectors*, 299–349, SPIE Optical Engineering Press, Bellingham, WA, 1995. With permission.)

A unique feature of the Schottky-barrier IR FPAs is the built-in blooming control (blooming is a form of crosstalk in which a well saturates and the electrons spill over into neighboring pixels). A strong illumination forward biases the detector and no further electrons are accumulated at the detector. The small negative voltage developed at the detector is not sufficient to forward bias the guard ring to the extent that electrons are injected to the CCD register through the silicon region under the transfer gate. Therefore, unless the vertical CCD has an insufficient charge handling capacity, blooming is suppressed perfectly in the Schottky-barrier IR FPA.

The responsivity of the FPAs is proportional to their fill factor, and improvement in the fill factor has been one of the most important issues in the development of imagers. For improving the fill factor a readout architecture called the charge sweep device (CSD) developed by Mitsubishi Corporation is also used. Kimata and coworkers have developed a series of IR image sensors with the CSD readout architecture with array sizes from 256×256 to 1040×1040 elements. Specifications and performance of these devices are summarized in Table 21.6 [45]. The effectiveness of this readout architecture is enhanced as the design rule becomes finer. Using a 1.2 μm CSD technology, a large fill factor of 71% was achieved with a 26×20 μm^2 pixel in the 512×512 monolithic structure [54]. The NETD was estimated at 0.033 K with $f/1.2$ optics at 300 K. At the beginning of the 1990s, the 1040×1040 element CSD FPA had the smallest pixel size (17×17 μm^2) among 2-dimensional IR FPAs [55,56]. The pixel was constructed with 1.5 μm design rules and has 53% fill factor. If the signal charges of 1040×1040 pixels are readout from one output port at the TV compatible frame rate, an unrealistic pixel rate of about 40 MHz was required. Therefore, a 4 output chip design was adopted [57]. The array of 1040×1040 pixels is divided into four blocks of 520×520 pixels. Each block has a horizontal CCD and a floating diffusion amplifier. A 1 million pixel data at a 30 Hz frame rate can be readout by operating each horizontal CCD at a 10 MHz clock frequency.

Table 21.6: Specifications and Performance of 2-D PtSi Schottky-Barrier FPAs with CSD Readout

Array Size	256 × 256	512 × 512	512 × 512	512 × 512	801 × 512	1040 × 1040
Pixel size (μm²)	26 × 26	26 × 20	26 × 20	26 × 20	17 × 20	17 × 17
Fill factor (%)	58	39	58	71	61	53
Chip size (mm²)	9.9 × 8.3	16 × 12	16 × 12	16 × 12	16 × 12	20.6 × 19.4
Pixel capacitor	Normal	Normal	High-C	High-C	High-C	High-C
CSD	4-phase	4-phase	4-phase	4-phase	4-phase	4-phase
HCCD	4-phase	4-phase	4-phase	4-phase	4-phase	4-phase
Number of outputs	1	1	1	1	1	4
Interface	Nonintegration	Field integration	Frame/Field integration	Frame/Field integration	Flexible	Field integration
Number of I/O pins	30	30	30	30	25	40
Process technology	NMOS/CCD 2 poly/2 Al	NMOS/CCD 2 poly/2 Al	NMOS/CCD 2 poly/2 Al	NMOS/CCD 2 poly/2 Al	CMOS/CCD 2 poly/2 Al	NMOS/CCD 2 poly/2 Al
Design rule (μm)	1.5	2	1.5	1.2	1.2	1.5
Thermal response (ke/K)	—	13	—	32	22 Ke/K	9.6 Ke/K
Saturation (e)	0.7×10^6	1.2×10^6	—	2.9×10^6	2.1×10^6	1.6×10^6
NEDT (K)	—	0.07	—	0.033	0.037	0.1

Source: M. Kimata, M. Ueno, H. Yagi, T. Shiraishi, M. Kawai, K. Endo, Y. Kosasayama, T. Sone, T. Ozeki, and N. Tsubouchi, *Opto-Electronics Review,* 6, 1–10, 1998. With permission.

(a)   (b)

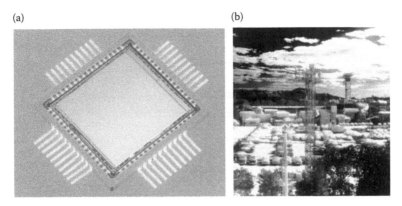

Figure 21.12 1040 × 1040 element PtSi/p-Si Schottky-barrier CSD FPA: (a) photograph of array mounted in a 40-lead ceramic package, and (b) the first mega-pixel infrared image with this array. (From Yutani, N., Yagi, H., Kimata, M., Nakanishi, J., Nagayoshi, S., and Tsubouchi, N., *Technical Digest IEDM*, 175–78, 1991. With permission.)

Vertical charge transfer device (CSD)

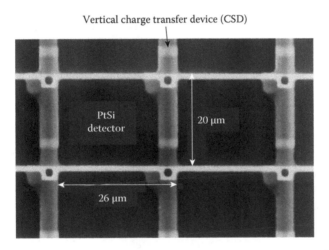

Figure 21.13 Pixel photograph of 512 × 512 PtSi Schottky-barrier CSD FPA taken just before aluminum reflector formation. (From Kimata, M., Ueno, M., Yagi, H., Shiraishi, T., Kawai, M., Endo, K., Kosasayama, Y., Sone, T., Ozeki, T., and Tsubouchi, N., *Opto-Electronics Review*, 6, 1–10, 1998. With permission.)

Figure 21.12 shows a photograph of a 1040 × 1040 array mounted in a 40-lead ceramic package and the first mega-pixel infrared image with this PtSi CSD FPA [55]. The chip size of the device is 20.6 × 19.4 mm². The NEDT of mega-pixel array at 300 K with an $f/1.2$ cold shield and a 30 Hz frame was 0.1 K.

As Table 21.6 shows, all the 512 × 512 FPA fabricated by Mitsubishi Electric Corporation have a pixel size 26 × 20 μm². The earliest array, developed in 1987, was made using design rules of 2 μm and had a relatively small fill factor of 39% [58]. More recently as the design rules have been reduced, a high performance PtSi Schottky-barrier infrared image sensor has been developed with an enhanced CSD readout architecture [45,59,60]. Figure 21.13 shows a photograph of the pixel of 512 × 512 array with 71% fill factor; the array was made with 1.2 μm design rules [45]. The NEDT about 30 mK with $f/1.2$ optics at 300 K has been measured (see Table 21.6). The total power consumption of the 801 × 512 element device was less than 50 mW. Using the finest design rules makes it possible to manufacture a high sensitivity FPA with 78% fill factor [53].

In order to improve the fill factor, other pixel designs have been proposed. One of them is the hybrid structure, which is generally used in compound semiconductor FPA. It attractive advantage

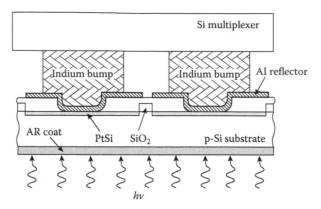

Figure 21.14 Pixel structures of hybrid Schottky-barrier FPA with self-guarded detector. (From Gates, J. L., Connelly, W. G., Franklin, T. D., Mills, R. E., Price, F. W., and Wittwer, T. Y., "488 × 640-Element Hybrid Platinum Silicide Schottky Focal Plane Array," *Proceedings of SPIE* 1540, 262–73, 1991. With permission.)

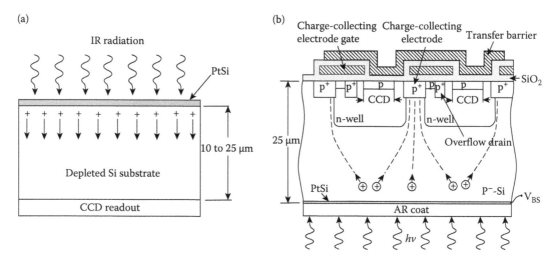

Figure 21.15 Direct Schottky injection (DSI) FPA: (a) concept of the device—holes are injected from the continuous PtSi electrode into the readout CCD fabricated on the opposite surface, and (b) cross section of the IT-CCD DSI FPA across the BCCD channels. (From Kosonocky, W. F., Villani, T. S., Shallcross, F. V., Meray, G. M., and O'Neil, J. J., "A Schottky-Barrier Image Sensor with 100% Fill Factor," *Proceedings of SPIE* 1308, 70–80, 1990. With permission.)

is independent optimization of the detector and multiplexer. As shown in Figure 21.14, individual Schottky electrodes are fabricated so close to each other that their depletion regions merge, and diode isolation is only achieved by a 2 μm oxide gap between the silicide electrodes without the guard ring (the self-guarded detector) [61]. Using this structure, a fill factor of 80% was obtained for a 20 μm pixel array [62].

Kosonocky et al. [63] proposed a novel concept of the Schottky-barrier FPAs called the direct Schottky injection (DSI), which provides a 100% fill factor. This DSI FPA consists of a continuous silicide electrode (DSI surface) formed on one surface of a thinned (10 to 25 μm) silicon substrate with the CCD readout register on the other side as shown in Figure 21.15. During the operation, the silicon substrate is depleted between the DSI surface and charge collecting elements of the readout structure. Injected hot holes from the DSI surface drift along the electric field line toward the collecting elements. The feasibility of the DSI concept was demonstrated at the David Sarnoff Research Center with a 128 × 128 FPA with 50 × 50 μm² pixels and p-channel IT-CCD readout multiplexer [63]. The device has, however, considerable crosstalk to about 20%.

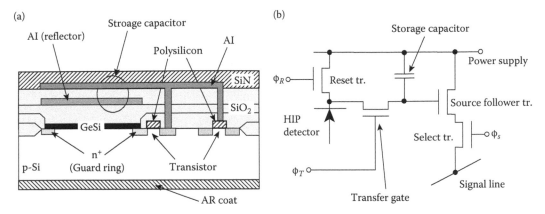

Figure 21.16 Pixel of GeSi HIP FPA: (a) the cross-section structure, and (b) the circuit diagram for a pixel. (From Kimata, M., Yagi, H., Ueno, M., Nakanishi, J., Ishikawa, T., Nakaki, Y., Kawai, M., et al., "Silicon Infrared Focal Plane Arrays," *Proceedings of SPIE* 4288, 286–97, 2001. With permission.)

Current PtSi Schottky-barrier FPAs are mainly manufactured in 300 mm wafer process lines with around 0.15 μm lithography technologies. However, the performance of monolithic PtSi Schottky-barrier FPAs reached a plateau about 10 years ago and further slow progress is expected.

As is described in Chapter 12, besides PtSi, there are other silicides that have been used for Schottky-barrier infrared detectors (Pd$_2$Si, IrSi, Co$_2$Si, and NiSi). However, they have never found wider applications [53]. Also the valence band discontinuity between SiGe and Si is used as an energy barrier for internal photoemission in infrared spectral range. MBE technology gives possibilities to grow high-quality strained SiGe films on Si substrates, which has given another option for extending the cutoff wavelength. Different structures of heterojunction internal photoemission (HIP) detectors based on emission of holes from a highly p-doped SiGe quantum well into an undoped silicon layers have been described by Presting [64].

The first 400 × 400 element GeSi/Si heterojunction array with a CCD readout has been developed by Tsaur et al. [65,66]. They exhibited uncorrected thermal imagers operated at 53 K with cutoff wavelength at 9.3 μm and with minimum resolvable temperature difference of 0.2 K (f/2.35). The responsivity nonuniformity of this array was less than 1%. The performance of Ge$_{1-x}$Si$_x$/Si 320 × 244 and 400 × 400 element arrays with λ_c = 10 μm (pixel sizes 40 × 40 and 28 × 28 μm^2 and fill factors 43% and 40%, respectively) has been described [67]. To improve FPA performance, the monolithic Si microlens arrays have been incorporated. Although these detectors are in an early stage of development, they already outperform IrSi detectors with the same cutoff wavelength with respect to quantum efficiency.

More recently, Wada et al. have developed a high resolution 8–12 μm 512 × 512 element SiGe HIP MOS readout architecture FPA with pixel size 34 × 34 μm^2 and a fill factor of 59% [68]. The array was fabricated with 0.8 μm single-polysilicon and double-aluminum NMOS process technology. Figure 21.16 depicts the pixel design showing the cross section and circuit diagram [69]. The pixel contains a source-follower amplifier and a storage capacitor (four transistors and one capacitor). The storage capacitor is composed of the Al reflector and the other electrode connected with the drain of the transfer gate. A NEDT of 0.08 K (f/2.0) was obtained at 43 K with 300 K background with a very small responsivity dispersion of 2.2% and high pixel yield of 99.998%.

21.4 III-V FOCAL PLANE ARRAYS

21.4.1 InGaAs Arrays

InGaAs ternary alloy is an optimal material choice for SWIR imaging applications due to the ability to operate at room temperature with high quantum efficiency for wavelengths in visible range to about 3 μm. These applications span both commercial and industrial opportunities including semiconductor-wafer inspection, wavefront sensing, astronomy, spectroscopy, machine vision, and military applications (surveillance, active point tracking, and laser radar).

Table 21.7: Pixel and Readout Formats of Linear Arrays Offered by Goodrich Corporation

Multiplexer Type		Pitch (μm)		Wavelength			Number of Pixels	Outputs	Max Pixel Rate	Max Lines Per Second
	Anti-blooming	25	50	1.7 μm	2.2 μm	2.5 μm			Mpixels /s	(24.4 μs ET)
LX			•	50,500	50,250	250	256,512	2,4	10,20	20,661
LD/ LDB	LDB	•		500	250	250	512	2	5	7812
LSB	•		•	500	250	250	256	1	2,5	7812
LE	•	•		25,500	250	250	1024	2	5	4340
LSE	•		•	500	250	250	512	1	2,5	4340

Source: InGaAs Products: Focal Plane Arrays, http://www.sensorsinc.com/arrays.html

Fabrication of traditional InGaAs photodiodes is described in Section 13.2. The arrays of photodiodes are hybridized to CMOS readout integrated circuits (ROICs) and then integrated into cameras for video-rate output to a monitor or for use with a computer for quantitative measurements and machine vision.

Linear array formats of 256, 512, and 1024 elements have been fabricated for operations in three ranges: 0.9 to 1.7 μm, 1.1. to 2.2 μm, or 1.1 to 2.6 μm (see Table 21.7) [70,71]. They are available in various sizes, defined by the detector height, pixel pitch, and the number of pixels, and are packaged with 1-, 2-, or 3-stage thermoelectric cooling, or without a cooler for externally cooled applications. Linear arrays with 1024 square pixels are used for high-resolution imaging of fast moving industrial processes. Arrays with tall pixels are widely used in optical spectrometers [72,73]. The wide range of array pixel and readout formats enable users to make the optimal match to their applications.

The first two-dimensional 128×128 $In_{0.53}Ga_{0.47}As$ hybrid FPA for the 1.0–1.7 μm spectral range was demonstrated by Olsen et al. in 1990 [74]. The 30 μm square pixels had 60 μm spacing and were designed to be compatible with a two-dimensional Reticon multiplexer. Dark current below 100 pA, capacitance near 0.1 pF (–5 V, 300 K), and quantum efficiency above 80% (at 1.3 μm) were measured. During the past 20 years, great strides have been made in the development of these progressing to the large 1280×1024 element arrays now readily available. A 320×240 array operates at room temperature, which allows development of a camera that is smaller than 25 cm^3 in volume, weighs less than 100 g and uses less than 750 mW of power [75].

At present, InGaAs FPAs are fabricated by several manufacturers including Goodrich Corporation (previously Sensors Unlimited) [75–80], Indigo Systems (merged with FLIR Systems) [81–83], Teledyne Judson Technologies [84], Aerious Photonics [85], and XenICs [86].

Table 21.8 lists the measured characteristics of near-IR cameras fabricated by Goodrich [71]. The largest and finest pitched imager in $In_{0.53}Ga_{0.47}As$ material system has been demonstrated recently [79,80,87,88]. The arrays with more than one million detector elements (1280×1024 and 1024×1024 formats) for low background applications have been developed by RVS with detectors provided by Goodrich [87]. The detector elements are as small as 15 μm pitch [80].

Detector dark current and noise are low enough that InGaAs detectors can be considered for astronomy where bandwidth of interest is between 0.9 to 1.7 μm and a high operating temperature for the focal plane is important. In a recently published paper, the low temperature performance of a 1.7 μm cutoff wavelength 1k × 1k InGaAs photodiode against similar 2k × 2k HgCdTe imagers has been compared [89]. The data indicates that InGaAs detector technology is well behaved and comparable to those obtained for state-of-the-art HgCdTe imagers.

Figure 21.17a shows the 1280×1024 1.7 μm InGaAs sensor chip assembly (SCA) for MANTIS (Multispectral Adaptive Networked Tactical Imaging System) program [87]. The detector array is hybridized to an innovative ROIC with unit cell amplifiers designed with a capacitance transimpedance amplifier and a sample/hold circuit. Noise measurements of this SCA at a 30 Hz frame rate are shown in Figure 21.17b. A fit of noise model assuming domination of kTC and the detector g-r noise implies that the detector R_oA product is 8×10^6 Ωcm^2 at 280 K and the ROIC contributes

Table 21.8: Specification of the Near-Infrared InGaAs FPAs Fabricated by Goodrich Corporation

	Configuration		
	Standard (320 × 240)	Standard (640 × 512)	Vis-InGaAs (640 × 512)
Pitch	25 μm	25 μm	25 μm
Optical fill factor	100%	100%	100%
Spectral response	0.9 to 1.7 μm	0.9 to 1.7 μm	0.4 to 1.7 μm
Quantum efficiency (%)	>65% from 1.0 to 1.6 μm	>65% from 1.0 to 1.6 μm	65% from 1.0 to 1.6 μm
Mean detectivity	>1 × 10^{13} cmHz$^{1/2}$/W	>1.5 × 10^{13} cmHz$^{1/2}$/W	>6 × 10^{12} cmHz$^{1/2}$/W
Noise equivalent irradiance	>1.4 × 10^9 photons/cm^{2}s	<9 × 10^8 photons/cm^{2}s	<2.5 × 10^9 photons/cm^{2}s
Noise (rms)	<84 electrons (typical)	<125 electrons (typical)	<300 electrons
Operability	>99%	>99.2%	>99.2%
Full well	>7 × 10^4 electrons	>3.9 × 10^6 electrons	8 × 10^5 electrons
Exposure times	60 μs to 16.57 ms in 16 steps	from 200 μs to 33.19 ms	
Image correction	2-point (offset and gain) pixel by pixel; user selectable	2-point (offset and gain) pixel by pixel; bad pixel replacement	2-point (offset and gain) pixel by pixel; bad pixel replacement
True dynamic range	>800:1	>1000:1	>2500:1
Active area	8 mm × 6.4 mm; 10.2 mm diagonal	16 mm × 12.8 mm; 20.5 mm diagonal	16 mm × 12.8 mm; 20.5 mm diagonal

Source: InGaAs Products: Focal Plane Arrays, http://www.sensorsinc.com/arrays.htm

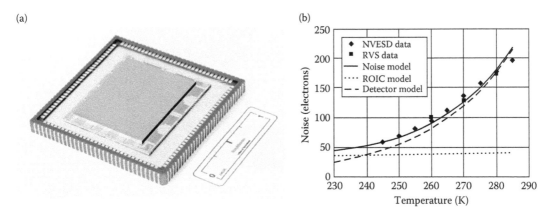

Figure 21.17 1280 × 1024 MANTIS InGaAs: (a) sensor chip assembly, and (b) noise at 30 Hz as a function of temperature. The line is a fit of the data to a noise model that includes detector g-r noise and ROIC kTC noise. (From Hoffman, A., Sessler, T., Rosbeck, J., Acton, D., and Ettenberg, M., "Megapixel InGaAs Arrays for Low Background Applications," *Proceedings of SPIE* 5783, 32–38, 2005. With permission.)

approximately 40 electrons of noise. Above 240 K, the detector noise dominates while below 240 K ROIC noise dominates.

The InGaAs FPAs achieves very high sensitivity in the shortwave infrared bands in addition to the visible response added via substrate removal process post hybridization. The visible InGaAs detector structure (Figure 21.18) is very similar to that shown in Figure 13.8, except that an InGaAs stop layer is added into the epi-wafer structure to allow complete removal of InP substrate [77,78]. The substrate is removed using a combination of mechanical and wet chemical etching techniques. The remaining InP contact layer thickness must be controlled to within 10 nm for consistent

(a)

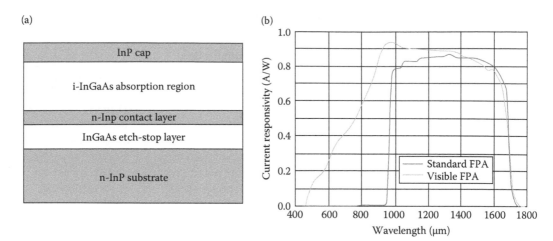

(b)

Figure 21.18 Visible InGaAs detector: (a) epitaxial wafer structure, and (b) quantum efficiency in comparison with the standard detector. (From Martin, T., Brubaker, R., Dixon, P., Gagliardi, M.-A., and Sudol, T., "640 × 512 InGaAs Focal Plane Array Camera for Visible and SWIR Imaging," *Proceedings of SPIE* 5783, 12–20, 2005. With permission.)

Figure 21.19 Image of the development team of an 1280 × 1024 InGaAs visible/SWIR imager taken by their camera. (From Enriguez, M. D., Blessinger, M. A., Groppe, J. V., Sudol, T. M., Battaglia, J., Passe, J., Stern, M., and Onat, B. M., "Performance of High Resolution Visible-InGaAs Imager for Day/Night Vision," *Proceedings of SPIE* 6940, 69400O, 2008. With permission.)

visible quantum efficiency. Thicker InP layers lead to lower quantum efficiency in the visible spectrum, as well as to image retention.

Goodrich has presented a high-resolution 1280 × 1024 InGaAs visible/SWIR imager with 15 μm pixels for day/night imaging. The array with capacitative transimpedance amplification (CTIA) readout unit cells was designated to achieve a noise level of less than 50 electrons, due to its small integration capacitor. The ROIC was readout at 120 frames per second, and had a dynamic range of 3000:1 using rolling, nonsnapshot integration. Total measured noise with the detector was 114 electrons using double sampling. Figure 21.19 shows the image taken with the imager, after nonuniformity correction [80].

It should be also mentioned that greater demand of data transmission in the telecommunications network is solved by transmitting multiple wavelengths of lights down the same fiber known as WDM (Wavelength Division Multiplexing). InGaAs arrays are widely used in spectroscopic monitors of S, C, and L band channels in WDM systems [75].

Also a combination of the InGaAs p-i-n photodiode detector technology on InP substrate with the InP JFET technology for monolithically integrate detector arrays and readout circuits were made [90,91]. The first 4×4 test arrays have been described [92]. The responsivity and NEP of the focal plane array at 1540 nm was found to be 1695 V/W and 45 nW/Hz$^{1/2}$, respectively. However, the stability of the circuit contacts should be improved further prior to applications of these focal plane arrays.

21.4.2 InSb Arrays

InSb photodiodes have been available since the late 1950s. One of the most significant recent advances in infrared technology has been the development of large two-dimensional FPAs for use in the staring arrays. Array formats are available with readouts suitable for both high background $f/2$ operation and for low background astronomy applications. Linear arrays are rarely used.

InSb material is far more mature than HgCdTe and good quality 10 cm diameter bulk substrates are commercially available. Staring arrays of back side illuminated, direct hybrid InSb photodiodes in formats up to $4k \times 4k$ with 15 µm pitch are in fabrication. The $6k \times 6k$ with 10 µm pitch were proposed for 2009 and an 8 µm pitch giga-pixel FPA are set as a goal for realization within the next four years. Figure 21.20 shows the recent results, current efforts, and future directions in development of extremely large format InSb FPAs by L3-Cincinnati Electronics [93].

The earliest arrays fabricated in the mid-1980s were just 58×62 elements in size [94–97] compared to arrays that are up to 4048×4048 today, an increase of more than three orders of magnitude in pixel count. The array noise has improved over this period of time from hundreds electrons to as low as 4 electrons today [98]. Similarly, at the same time, detector dark current has decreased from about 10 electrons/second to as low as 0.004 electrons/second [99,100].

The InSb FPA has been developed with the monolithic architecture as well as hybrid architecture. The monolithic architecture integrates all functions needed for solid-state imaging such as photon detection, charge storage, and multiplexed readout (see Chapter 19). The best performance of an InSb FPA has been obtained using hybrid architecture where the detection and readout parts of the device can be optimized separately.

21.4.2.1 Hybrid InSb Focal Plane Arrays

It is important to keep fabrication process temperatures low in the manufacture of hybrid FPA. This maintains the InSb surface condition to avoid surface leakage. Using anodic oxide and alumina for the surface passivation allows the process temperature to be lower than 100°C after junction formation [101]. Current manufacturing device processes require that detector materials be thinned before or after hybridization to achieve successful back side illumination for reasonable

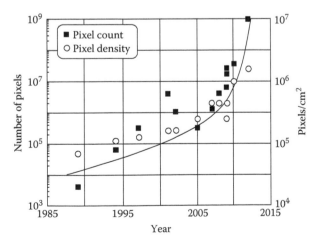

Figure 21.20 Measured and predicted InSb FPA format and pixel density. Notice the gigapixel predicted for 2012. (From Norton, P. R., Andresen, B. F., and Fulom, G. F., "Introduction," *Proceedings of SPIE* 6940, xix–xxxvi, 2008. With permission.)

quantum efficiency and elimination of crosstalk. For p-type mesas rising from an n-type substrate, the substrate must be thinned to 5–15 microns (then photon absorption takes place within a diffusion length of the junction) [102]. It also requires a good quality back surface (to reduce surface recombination), and uniformity in thickness and surface quality (to improve response uniformity). Thus it is difficult to maintain good detector performance while achieving reasonable yields. This accomplishment is a major achievement. In the last decade many manufacturers have made significant process improvements in the fabrication in both InSb detectors and readout electronic chips. In-bumps are deposited on both detector array and silicon ROIC dies, which are then hybridized together using a flip-chip bonder. The gap between bumps is backfilled with epoxy. These improvements in yield and quality have resulted from the implementation of innovative passivation techniques, from specialized antireflection coatings to unique thinning processes.

Until recently, IR FPA sizes have been limited by several constraints. These include lack of availability of large-area, low-cost detector substrates, the inability to fabricate large-area ROICs using submicron processes, and the typically large thermal expansion mismatch characteristics of Si ROICs versus typical IR detector materials. These limitations have been largely overcome by using alternative detector substrates and by using a type of "reticle stitching" (Reticle Image Composition Lithography, or RICL) at silicon foundries to construct ROIC chips much larger than the reticles used in the projection printing photolithography process.

InSb arrays can be scaled to large formats with small pixel sizes because the InSb detectors are thinned after hybridization and are fabricated using a planar, ion-implanted process. Thinning the bulk InSb to ≤10 μm is necessary for operation with back side illumination (to obtain sufficient quantum efficiencies and minimize crosstalk), but it also forces the InSb to match the thermal expansion of the Si ROIC, thereby providing a reliable hybrid structure with respect to repeated thermal cycling.

One of the reasons for selecting InSb for infrared instruments is its broad response, which is shown in Figure 21.21 [100,103]. The internal quantum efficiency is nearly 100% in wide spectral band from 0.4 to 5 μm, which is an advantage of the thinned InSb arrays for this application. The limiting factor in quantum efficiency is the reflection of incident light at the surface, which can be minimized with antireflection coatings.

The first InSb array to exceed one million pixels was the ALADDIN array first produced in 1993 by Santa Barbara Research Center (SBRC) and demonstrated on a telescope by National Optical Astronomy Observations (NOAO), Tucson, Arizona, in 1994 [104–107]. This array had 1024 × 1024 pixels spaced on 27 μm centers and was divided into four independent quadrants, each containing eight output amplifiers. This solution was chosen due to uncertain yield of

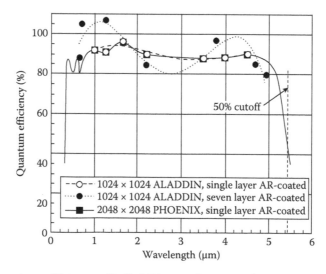

Figure 21.21 Quantum efficiency of InSb SCAs as a function of wavelength for a 1024 × 1024 ALADDIN SCA (single-layer and seven-layer AR-coated) and 2048 × 2048 PHOENIX SCA (single-layer AR-coated).

ALADDIN: 1k × 1k   ORION: 2k × 2k   PHOENIX: 2k × 2k

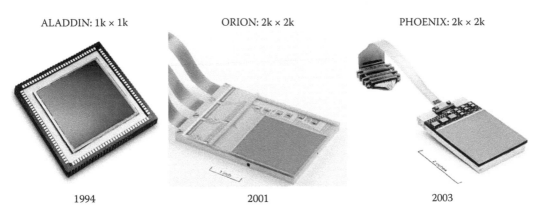

1994   2001   2003

Figure 21.22 Timeline and history development of the InSb RVS astronomy arrays.

Figure 21.23 A demonstration of the 2-sided buttable ORION modules to create a 4k × 4k focal plane. One module contains an InSb SCA while the others have bare readouts. (From Hoffman, A. W., Corrales, E., Love, P. J., Rosbeck, J., Merrill, M., Fowler, A., and McMurtry, C., "2K × 2K InSb for Astronomy," *Proceedings of SPIE* 5499, 59–67, 2004. With permission.)

large arrays at that time. A total of 16 devices were produced in 1996 completing the development program [107].

The ALADDIN has been upgraded with the larger version, the ORION FPA family. A chronological history of the RVS astronomical focal plane arrays is shown in Figure 21.22. The next step in the development of InSb FPAs for astronomy was the 2048 × 2048 ORION SCA (see Figure 21.23) [100]. Four ORION SCAs were deployed as a 4096 × 4096 focal plane in the NOAO near-IR camera, currently in operation at the Mayall 4-meter telescope on Kit Peak [103]. This array has 64 outputs, allowing up to a 10 Hz frame rate. A challenge for large focal planes is maintaining optical focus, and in consequence maintaining the flatness of the detector surface over a large area. The special packaging concept used on the ORION program is described by Fowler and colleagues [108]. Many of the packaging concepts are shared with the 3-side buttable 2k × 2k FPA InSb modules developed by RVS for the James Webb Space Telescope (JWST) mission [109]. The synergy between the ORION and JWST development efforts is summarized elsewhere [100].

InSb photodiodes demonstrate low dark current in large format arrays, which is shown in Figure 21.24 [100]. However, the dark current does not follow the predicted dark current due to generation-recombination mechanisms (see Figure 13.20a). The possibility of surface currents due to nonideal passivation can be investigated pending further funding for development.

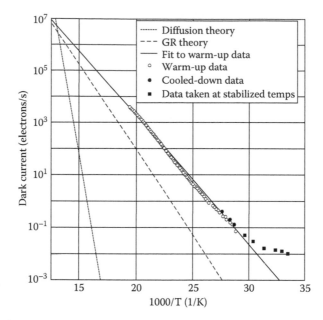

Figure 21.24 Dark current versus inverse temperature for a 2k × 2k InSb array with theoretical plots of diffusion and generation-recombination dark currents for comparison. The measured dark current follows a straight line on this semi-log plot down to 33 K and then flattens out to 0.01 electrons/second at 300 K. Data were obtained for both warming up and cooling down the detectors. (From Hoffman, A. W., Corrales, E., Love, P. J., Rosbeck, J., Merrill, M., Fowler, A., and McMurtry, C., "2K × 2K InSb for Astronomy," *Proceedings of SPIE* 5499, 59–67, 2004. With permission.)

PHOENIX SCA is another 2k × 2k FPA InSb array that has been fabricated and tested. This detector array is identical to ORION (25-μm pixels), however, its readout is optimized for lower frames rates and lower power dissipation. With only four outputs, the full frame read time is typically 10 seconds. The smaller number of outputs allows a smaller module package that is three-side buttable [100].

The characteristics of the most advanced InSb arrays for astronomy are summarized in Table 21.9. The parameters have been taken from a variety of sources.

Different formats of InSb FPAs have found many high background applications including missile systems, interceptor systems, and commercial imaging camera systems. With an increasing need for higher resolution, several manufacturers have developed megapixel detectors. Table 21.10 compares performance of commercially available megapixel InSb FPAs fabricated by L-3 Communications (Cincinnati Electronics), Santa Barbara Focalplane, and SCD. Recently, SCD has developed a new large format InSb detector with 1280 × 1024 elements and a pixel size of 15 μm [109]. A new ROIC was designed for and manufactured with 0.18 μm CMOS technology.

Also other manufacturers are involved in the development of large format 15 μm pitch InSb FPAs. The L-3 Cincinnati Electronics plans further improvements in detector performance through implementation of micro-optics in reticulated small pixel design to achieve 100% fill factor [110].

21.4.2.2 Monolithic InSb Arrays

In the historical evolution of InSb detector technology, the monolithic FPA architecture has been also developed. The main motivation for developing MIS detectors was to integrate all functions needed for solid state imaging (such as photon detection, charge storage, and multiplexed readout; see Chapter 19) together with achievements of high performance imagers. These goals were not achieved due to fundamental limitation of narrow gap semiconductor like InSb.

The charge injection device (CID) was originally fabricated as a MOS silicon device to reduce the number of transfers needed for readout [111]. Shortly thereafter CID devices were fabricated using narrow gap material to form monolithic InSb FPAs [112]. Basic CID mechanisms and

Table 21.9: Megapixel InSb Focal Plane Arrays for High Background Applications

	Configuration		
Parameter	1024×1024	1024×1024 (ALADDIN)	2048×2048 (ORION II)
Architecture	–	Module–2-side buttable	Module–2-side buttable
Pixel pitch (μm)	30	27	25
Operating temperature (K)	50	30	32
Readout structure	SFD	SFD	SFD
Read noise (electrons rms)	10–50	<25	6
Dark current (electrons/sec)	<400	<0.1	0.01
Well capacity (electrons)	2×10^5	3×10^5 (at 1 V)	1.5×10^5
Quantum efficiency (%)	>80%	>80	>80
Outputs	4	32 (8 per quadrant)	64
Frame/sec	1 to 10	20	10
References	www.raytheon.com	106	108

Table 21.10: Performance of Commercially Available Megapixel InSb FPAs

	Configuration		
Parameter	1024×1024 (L-3 Communications)	1024×1024 (Santa Barbara Focalplane)	1280×1024 (SCD)
Pixel pitch (μm)	25	19.5	15
Dynamic range (bits)		14	15
Pixel capacity (electrons)	1.1×10^7	8.1×10^6	6×10^6
Power dissipation (mW)	<100	<150	<120
NEDT (mK)	<20	<20	20
Frame rate (Hz)	1 to 10	120	120
Operability (%)	>99	>99.5	>99.5
References	www.L-3Com.com	www.sbfp.com	www.scd.co.il

readout techniques have been described by Michon and Burke [113]. In a CID, the detection process occurs within the unit cells that are comprised of two MIS structures that are readout in an x–y addressable manner. For readout mechanisms fabricated in narrow gap semiconductors the charge capacity is significantly less than a comparable silicon device. The bulk breakdown voltage V_{bd} of a semiconductor has been empirically related to the semiconductor's bandgap energy E_g as $V_{bd} \propto E_g^{3/2}$ [114]. Therefore, the nominal breakdown voltage of InSb is about 0.1 of silicon. Charge storage is also dependent upon the dielectric constant of the insulator thickness.

An InSb MIS device as a photovoltaic infrared detector was first proposed in 1967 by Phelan and Dimmock [115]. More comprehensive studies have been carried out by Lile and Wieder [116,117].

MIS capacitors were fabricated on both n- and p-type (111)B face InSb wafers with doping density in the low to mid 10^{14} cm^{-3} and an etch pit density of less than 100 cm^{-2}. The wafers were polished in nitric-based acid and rinsed in deionized water. Several dielectric materials have been experimentally evaluated for use as the gate insulator, including SiO_x, Al_2O_3, SiO_xN_y, and In_2O_3, the anodically grown native oxide of InSb, TiO_2, and SiO_2. The SiO_2 deposited by low-temperature CVD has been found to be capable of producing InSb MIS devices with low surface-state density

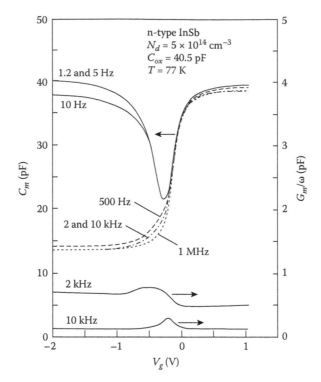

Figure 21.25 Capacitance and conductance curves for an MIS capacitor on n-type InSb at 77 K. (From Wei, C. Y., Wang, K. L., Taft, E. A., Swab, J. M., Gibbons, M. D., Davern, W. E., and Brown, D. M., *IEEE Transactions on Electron Devices*, ED–27, 170–75, 1980. With permission.)

and histeresis, and a near-zero flat-band voltage [118,119]. After the deposition of the 120 nm thickness oxide films, a 15 nm thin film of chromium, followed by 0.5 μm of gold are sputtered onto the wafers [119–121].

The capacitance and conductance characteristics of an n-type InSb MIS capacitor measured at various frequencies are shown in Figure 21.25 for $T = 77$ K [119]. The surface-state density calculated from measured quasistatic and high frequency C-V characteristics has a minimum of 5×10^{10} cm^{-2}eV^{-1} in the upper half of the bandgap and increases to 5×10^{11} cm^{-2}eV^{-1} in the lower half of the bandgap.

Coupling of the InSb MOS technology with that of CID has been presented by Kim [112]. Figure 21.26a shows schematically n-type InSb MIS structure used in CID devices [122]. A fully planar (nonetch-back) process has been employed in InSb CID technology [120,121]. Figure 21.26b shows the cross-sectional and top view of a unit-cell geometry. The wafers were chemically polished and coated with a CVD SiO$_2$ film of about 135 nm thickness at <200°C. Then the column gates and the field plate were patterned using a thin chromium layer. A second layer of 220 nm SiO$_2$ was afterward deposited and the row gates were defined with another thin chromium film. The transmission of the 7.5 nm thick chromium layers was 60–70% at the wavelength of 4 μm when antireflection coating was used. Thicker gold layers were used to form connecting runs, pads, and field shields. Instead of the conventional side-by-side capacitor layout, the improved devices had a concentric design, with one capacitor surrounding the other. The corners of the gates were rounded to minimize the electric fields in their vicinity. Arrays fabricated with that planar process exhibited nonuniformity within the gate oxide thickness of the order of 2%. Injection crosstalk was less than 1%.

The optimization of a CID design strongly depends on the readout scheme by which the array is operated. Although many different readout techniques have been applied to silicon CID [113], only three (i.e., ideal mode, conventional charge-sharing mode, and sequential row injection mode) have been used so far with InSb CID because of its later development [122]. A comparison of the three common readouts has been carried out by Gibbons and colleagues [121,123,124].

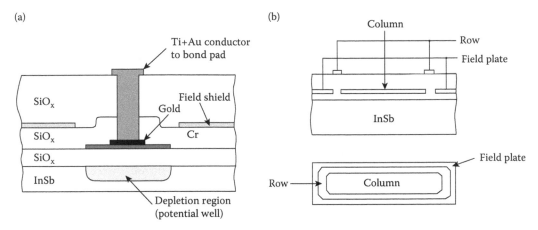

Figure 21.26 The InSb CID device: (a) construction of MIS capacitor. (From Gibbons, M. D., Wang, S. C., Jost, S. R., Meikleham, V. F., Myers, T. H., and Milton, A. F., "Developments in InSb Material and Charge Injection Devices," *Proceedings of SPIE* 865, 52–58, 1987. With permission.) (b) Cross-sectional and top view of a dual-gate CID cell. (From Wang, S. C. H., Wei, c. Y., Woodbury, H. H., and Gibbons, M. D., *IEEE Transactions on Electron Devices*, ED-32, 1599–1607, 1985. With permission.)

The dark current in the InSb CID is proportional to the sum of the depletion generation of carriers and minority carrier diffusion from the bulk material into the depletion layer and may be written as

$$J_{ds} = \frac{q n_i w}{2\tau} + \frac{q n_i^2 L}{N_d \tau}. \tag{21.1}$$

Assuming donor concentration $N_d = 3 \times 10^{14}$ cm^{-3}, diffusion length $L = 25$ μm and operating temperature 80 K, the depletion generation current is nearly three orders of magnitude higher than the diffusion current. The minority carrier lifetime, which is very sensitive to impurities and crystalline defects, has the most significant effect on the dark current. InSb material quality has been improved by growing epitaxial layers of InSb on InSb wafers by LPE [125]. The minority carrier lifetime increases more than two orders of magnitude in comparison with bulk material. Bulk InSb MIS detectors have been typically operated in the 77–90 K range, but LPE devices may allow operating temperatures above 100 K.

Progress in InSb CID technology has led to the development of line arrays with 512 elements [122] and 128 × 128 staring arrays [123,126] with focal-plane Si MOS scanner/preamplifiers. Two-dimensional arrays had a 43 μm center-to-center detector spacing of the 128 × 128 array, the performance efficiency (defined as the number of carriers readout at the preamplifier divided by the number of photons incident on the sensor) measured in the charge-sharing mode was 28% with a standard deviation of 9% [127]. The fill factor was about 75%, which gave a pixel sensing area of 6.2×10^{-5} cm^2. The array was operated with a 2.26 μs element read and an integration time of 0.3 ms. The noise of the array unit (2000 carriers) was contributed by the readout electronics and the NEDT was calculated to be less than 0.25°C. The given parameters indicate that performance of InSb CID devices are considerably inferior in comparison with hybrid FPAs.

The first demonstration of charge-couple devices (CCDs) in semiconductor other than silicon and germanium was achieved with InSb in 1975 [128]. Next Thom et al. have fabricated a fully monolithic 20-element linear p-channel CCD array [118]. This process included planar p+-n junction formation by Be-ion implantation to form the fat-zero input and charge output stages, and an aluminum and SiO$_2$ overlapping CCD gate structure that makes use of low-temperature CVD and plasma etching. The charge transfer efficiency (CTF) of the device was 0.995 and was limited by lateral surface potential variations rather than surface states. The integration time was varied independently of readout rate, and operation in both the multiplexing and time-delay and integration modes was demonstrated. The average detectivity of the array measured in the multiplexing mode for a 5 ms integration time, operating temperature of 65 K, and background flux of 10^{12} photons/ $cm^2 s$ was measured to be 6.4×10^{11} cmHz$^{1/2}$/W.

Fabrication of high performance monolithic arrays in narrow gap semiconductor failed at the end of the 1980s. The detection capability of monolithic InSb devices is limited as a result of specific physical properties. The main problems are: signal handling of the MIS cells especially in conditions of high background operation, high dark current density, and difficulties in achieving high CTE. Especially defect-related tunneling current of the nonequilibrium operated MIS devices is orders of magnitude larger than the fundamental dark current. The MIS capacitor required much higher quality material than photodiode.

21.5 HgCdTe FOCAL PLANE ARRAYS

The main mode of operation of HgCdTe detectors used in FPAs is photovotaic effect. Photodiode offers many key system advantages over photoconductive detectors, especially in LWIR and VLWIR regions: negligible $1/f$ noise, much higher impedance (so that cold preamps or multiplexers are possible), configuration versatility with back side illuminated 2-D arrays of closely spaced elements, better linearity, DC coupling for measuring the total incident photon flux, and a $2^{1/2}$ higher BLIP detectivity limit. However, photoconductive detectors will continue to be the better choice for certain instruments, such as those with relatively small numbers of detectors, or with detection requirements out to extremely long wavelengths. Reine et al. [129,130] have presented an excellent paper that compares the performance of photoconductive and photovoltaic HgCdTe detectors for 15 μm remote sensing applications. Up to the present, photovoltaic HgCdTe FPAs have been mainly based on p-type material.

Higher density detector configuration leads to higher image resolution as well as greater system sensitivity. HgCdTe IR FPAs have been made in linear (240, 288, 480, 960, and 1024), 2-D scanning with time delay and integration (TDI; with common formats of 256×4, 288×4, 480×6), and various 2-D staring formats with size from 64×64 up to 4096×4096 pixels (see Figure 21.27a). Efforts are also underway to develop avalanche photodiode capabilities in the 1.6 μm and at longer wavelength region. Pixel sizes ranging from 15 μm square to over 1 mm have been demonstrated. While the size of individual arrays continues to grow, the very large FPAs required for many space missions by mosaicking a large number of individual arrays. An example of a large mosaic developed by TIS is a 147 megapixel FPA that is comprised of 35 arrays, each with 2048×2048 pixels (see Figure 21.27b). This is currently the world's largest IR focal plane [131].

The 50th anniversary of the first publication devoted to HgCdTe ternary alloy [132] was an occasion to review the historical progress of HgCdTe material and device development in different countries. It was made during a special session of the 35th conference in *Infrared Technology and Applications* held in Orlando, Florida April 13–17, 2009. Several invited papers gathered in *Proceedings of SPIE* Vol. 7298, are excellent sources of historical information about the development of HgCdTe FPAs. For example, Figure 21.28 [133,134] shows the timeline for HgCdTe FPA

(a)   (b)

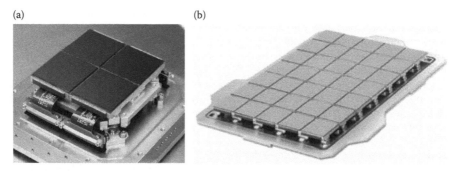

Figure 21.27 Examples of Teledyne Imaging Sensors packaging: (a) a mosaic of four Hawaii-2RGs as is being used for astronomy observations, and (b) a mechanical prototype of a mosaic of 35 Hawaii-2RG arrays as envisioned for the Microlensing Planet Finder. (From Beletic, J. W., Blank, R., Gulbransen, D., Lee, D., Loose, M., Piquette, E. C., Sprafke, T., Tennant, W. E., Zandian, M., and Zino, J., "Teledyne Imaging Sensors: Infrared Imaging Technologies for Astronomy & Civil Space," *Proceedings of SPIE* 7021, 70210H, 2008. With permission.)

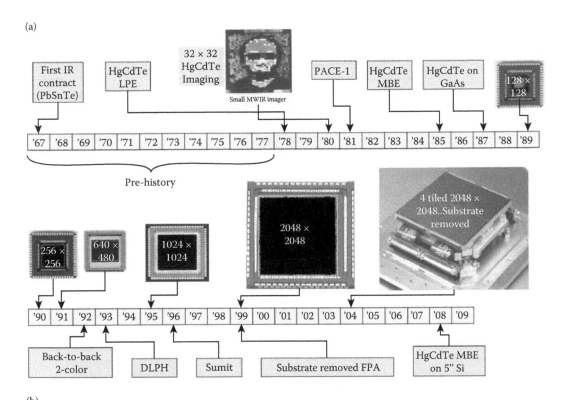

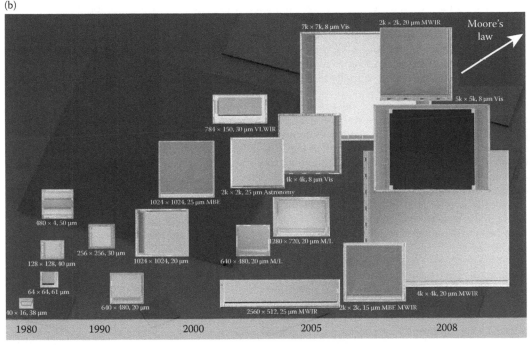

Figure 21.28 Timeline for HgCdTe development (a) at Teledyne Imaging Sensors. (From Tennant, W. E., and Arias, J. M., "HgCdTe at Teledyne," *Proceedings of SPIE* 7298, 72982V, 2009. With permission.) (b) At Raytheon Vision Systems. (From Bratt, P. R., Johnson, S. M., Rhiger, D. R., Tung, T., Kalisher, M. H., Radford, W. A., Garwood, G. A., and Cockrum, C. A., "Historical Perspectives on HgCdTe Material and Device Development at Raytheon Vision Systems," *Proceedings of SPIE* 7298, 72982U, 2009. With permission.)

development at Teledyne Imaging Sensors (TIS, formerly Rockwell Science Center) and Raytheon Vision Systems (RVS, formerly Santa Barbara Research Center, SBRC). Figure 21.28b illustrates the SBRC/RVS growth history in terms of substrate size and corresponding detector array sizes starting from the initial bulk HgCdTe crystal wafers of 3 cm², and progressing through LPE on CdZnTe substrates of 30 cm² up to today's MBE on alternate substrates of 180 cm².

The present LWIR and MWIR arrays are typically operated at liquid nitrogen temperature using Joule–Thompson or engine coolers. Some MWIR and SWIR arrays are thermoelectrically cooled to 190–240 K. Since many FPAs have very high data rates, the housings have coplanar leads to minimize parasitic impedances.

At higher backgrounds, it is impossible to handle the large amounts of carriers generated over frame times compatible with standard video frame rates. The FPAs are often operated at subframe rates much higher than the video update rate. Off-FPA integration of these subframes can be used to attain a level of sensor sensitivity that is commensurate with the detector-limited D^* and not the charge-handling D^*. In this connection, it is noteworthy that the progress on LWIR FPAs was relatively slow during the last decade. While the LWIR band should offer an order of magnitude better sensitivity, staring readout limitations due to charge handling limitations often constrain LWIR camera sensitivity to lower levels than competing MWIR devices due to lower LWIR contrast and similar (or lower) charge-handling capacity.

There are a number of architectures used in the development of IR FPAs that are discussed in Chapter 19. In general, they may be classified as monolithic and hybrid.

21.5.1 Monolithic FPAs

Potentially, monolithic FPAs offer important advantages of minimum focal plane interconnection and no problems with mating different materials for the detector array and readout circuit. Another advantage is the elimination of charge injection losses as the signal detection and integration is performed in the same well. Monolithic HgCdTe MIS charge transfer devices (CTD) were developed for nearly two decades between the mid-1970s to the mid-1990s. The three basic configurations of the HgCdTe CTD have been developed: charge-coupled devices (CCD), charge injection devices (CID), and charge imaging matrices (CIM). However, due to three basic limitations of the monolithic CCD FPAs associated with the use of narrow gap HgCdTe material: limited CTF, high dark currents, and difficulties in achieving the high CTF, these devices did not rival state-of-the-art hybrid diode arrays in midwave and especially in longwave IR bands. From this reason, an approach of monolithic HgCdTe devices is treated marginally in this section. More historical details are given in Rogalski's monograph [23], for example.

Initial work concentrated on p-channel CCDs, due to the maturity of the growth and doping control of n-type HgCdTe material [135–138]. However, due to the difficulty of forming stable p^+-n junctions in HgCdTe, readout structures could not be incorporated in the devices. After the demonstration of MISFET-based amplifiers in HgCdTe [139], Koch et al. [140] reported the development of a monolithic n-channel liner CCD imaging array consisting of two 55-bit CCD multiplexers, each addressing one-half of the 100-element MIS detector array ($x = 0.37$). The device was fabricated on an epitaxial HgCdTe layer grown by ISOVPE with the use of low-temperature photochemically vapor deposited silicon dioxide for the primary gate oxide and subsequent insulator levels in the structure. Using SiO₂/HgCdTe interface, charge transfer efficiencies as high as 0.9995 have been achieved and a value higher than 0.999 was measured for temperatures ranging from 60 K to 140 K.

Fully monolithic 128 × 28 element HgCdTe CCD arrays with 5 μm cutoff for low background applications have been demonstrated by Wadsworth et al. [141]. These arrays incorporate TDI detection, serial readout multiplexing, charge-to-voltage conversion, and buffer amplification in the HgCdTe detector chip. The performance of these arrays (at 77 K the detectivity values exceed 3×10^{13} cmHz$^{1/2}$W^{-1} for a background flux level of 6×10^{12} photon/cm²s) indicated that the monolithic CCD was a promising alternative to diode-based hybrid imaging technology in the mid-1990s.

The low storage capacity of 8–14 μm HgCdTe MIS structures makes them useless even for moderate background flux CCDs as the integration time (limited to ≈ 10 μs) becomes comparable to the transfer time, making the readout of the array impossible. LWIR system applications have been limited to scanning scenarios with relatively short integration times (e.g., 960 × 1 and 480 × 4 elements) [142]. Many device limitations were difficult to eliminate and for these reasons the MIS devices were superseded by the HgCdTe photodiode in hybrid FPAs.

21.5.2 Hybrid FPAs

The combination of existing high-performance HgCdTe arrays and highly developed silicon integrated circuits in the hybrid HgCdTe/silicon planes proved to be useful for thermal imaging systems with the thermal and spatial resolution unmatched by any competing technologies at present.

The linear and small 2-D arrays use the front side illuminated metal strap interconnection approach, with the silicon signal processing chip actually used as a handling substrate for preparation of the HgCdTe array. A HgCdTe wafer is glued to the chip, etched to thickness of about 10 μm and used for subsequent formation of planar or mesa diodes. Alternatively, the thin HgCdTe layer can be prepared separately, passivated, and epoxed to a silicon chip by an evaporated metal track, evaporated over the edge of an island or strip. The major problem in this approach is a reduction of the optically active area by the interconnect-occupied area. The problem becomes acute for small pitched (<50 μm) 2-D arrays, resulting in a low optical fill factor.

Baker et al. [143–146] have developed a unique interconnect technology for front side illuminated detectors, named the loophole technique, shown in Figure 19.16b. This is a lateral collection device with a small central contact. The thermal expansion mismatch problem is approached by using monolith of about 9 μm thick p-type HgCdTe, bonded rigidly to the silicon so that strain is taken up elastically. This makes the devices mechanically and electrically very robust with contact obscuration typically less than 10%. Arrays up to 15 mm in length have been shown to be unaffected by multiple cycling to cryogenic temperatures [147]. The process has two simple masking stages. The first defines a photoresist film with a matrix of holes of, typically 5 μm diameter. Using ion beam milling, the HgCdTe is eroded away in the holes until the aluminum contact pads are exposed. The holes are then backfilled with a conductor, to form the bridge between the HgCdTe and the underlying multiplexer pad. The junction is formed around the hole during the ion milling process. The second masking stage enables the p-type contact to be applied. The junctions are connected down to the silicon circuit by cutting the fine, few μm in diameter holes through the junctions by ion milling, and then backfilling the holes with metallization.

The loophole technology has been applied to both LWIR and MWIR arrays yielding high performance and reliable devices. Current arrays use monolith sizes of over 16×13 mm^2 area for 640×512 2-D arrays. The pixel size as small as 15 μm with 2 μm via holes have been fabricated [145].

A modification of the lateral loophole technology is the vertically MIS approach developed at Texas Instruments [148]. More recently, vertically integrated photodiode, VIP™ technology has been developed at Texas Instruments [149]. In this case, a plasma etching stage is used to cut the via-hole and an ion implantation stage to create a stable HgCdTe junction and damage region near contact. In order to achieve higher lifetimes and lower thermal currents, Cu is introduced at the LPE growth stage. This is swept out during the diode formation and resides selectively in the p-type region, partially neutralizing the S-R centers associated with Hg vacancies. The final effect of this procedure is that dark current approaches this fully doped heterostructures. In the VIP™ process n-on-p photodiode chip is epoxy hybridized directly to the ROICs on large Si wafers by means of vias in the HgCdTe.

A back side illuminated architecture, shown in Figure 19.16a, utilizes separately prepared detector arrays, which are then flipped over and hybridized to a silicon fanout pattern by means of indium bumps [114,124,150–152]. A high optical fill factor is easily achieved with this technique.

Initially the diodes used in this architecture were formed in a single p-type wafer of HgCdTe by ion implantation. After the diode arrays were hybridized, the HgCdTe wafer had to be thinned to about 10 μm to permit optimum absorption of infrared radiation at the junction region and an increased $R_o A$ product by a reduction of the diffusion volume.

Back side illumination is readily achieved by the epitaxial growth of HgCdTe on transparent substrates. No thinning of the material after hybridization is required and the superior quality of epitaxial layers compared to bulk crystals is an additional advantage of this approach. Despite early concern over the stability of the bump interconnections, the devices have exhibited >98% interconnection yield and excellent reliability. At present the operability is typically above 99.5%.

Advances in astronomy have spurred the need for imaging over as large a spectral range as possible, including visible to SWIR and MWIR. Recently a process to remove the visible light blocking substrate has been developed. In addition, this allows the array to accommodate any thermal expansion by eliminating the thermal mismatch between the silicon readout and the detector array and eliminates pixel-to-pixel crosstalk.

Sapphire buffered with CdZnTe became a standard substrate for SWIR and MWIR devices [153]. LWIR devices are typically based on CdZnTe. The progress on GaAs-based substrates has not progressed as fast as had been hoped and the technology returned to laboratory status for further research [154]. However, most of the MOCVD work on silicon has used a GaAs layer to buffer lattice mismatch between silicon and HgCdTe [155], and a variety of device structures are reported using MOVCVD layers grown on 75 mm diameter GaAs on silicon. The next approach to reach production is connected with silicon-based alternative substrates, such as CdZnTe/Si [156].

However, near lattice matched CdZnTe substrates have severe drawbacks such as lack of large area, high production cost and, more importantly, a difference in thermal expansion coefficient (TEC) between the CdZnTe substrates and the silicon readout integrated circuit. Furthermore, interest in large area two-dimensional IR FPAs (1024 × 1024 and larger) have resulted in limited applications of CdZnTe substrates. Currently, readily producible CdZnTe substrates are limited to areas of approximately 50 cm^2. At this size, the wafers are unable to accommodate more than two 1024 × 1024 FPAs. Not even a single die can be accommodated for very large FPA formats (2048 × 2048 and larger) on substrates of this size.

The use of Si substrates is very attractive in IR FPA technology because it is less expensive and available in large area wafers, and because the coupling of the Si substrates with Si readout circuitry in an FPA structure allows fabrication of very large arrays exhibiting long-term thermal cycle reliability [157]. Figure 14.42b shows a schematic cross section of an MBE-grown p-on-n HgCdTe/Si double-layer heterojunction (DLHJ) design. A thin ZnTe buffer, typically 1 μm thick, is used to preserve preferred (211) orientation that can readily twin to form undesirable (552) domains depending on the grown conditions. The CdTe buffer layer is typically 6–9 μm in thickness and helps to reduce the dislocation density by annihilation.

Despite the large lattice mismatch (≈19%) between CdTe and Si, MBE has been successfully used for the heteroepitaxial growth of CdTe on Si. At 77 K, diode performance with cutoff wavelength in LWIR region for HgCdTe on Si is comparable to that on bulk CdZnTe substrates [134,158]. Figure 21.29 shows a 140 K detector median R_oA product versus cutoff trend-line that includes HgCdTe grown by MBE on both bulk CdZnTe and Si and LPE grown on bulk CdZnTe

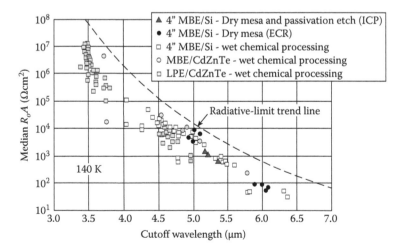

Figure 21.29 40 μm unit cell HgCdTe/Si DLHJ median R_oA detector array trend-line data as a function of measured 140 K cutoff wavelength. Trend-line data include HgCdTe material growth on Si (MBE) and CdZnTe (MBE and LPE) substrates. (From Bratt, P. R., Johnson, S. M., Rhiger, D. R., Tung, T., Kalisher, M. H., Radford, W. A., Garwood, G. A., and Cockrum, C. A., "Historical Perspectives on HgCdTe Material and Device Development at Raytheon Vision Systems," *Proceedings of SPIE* 7298, 72982U, 2009. With permission.)

Figure 21.30 Photograph of a high-performance 2560 × 512 MWIR HgCdTe/Si FPA with 25 μm unit cell. (From Bratt, P. R., Johnson, S. M., Rhiger, D. R., Tung, T., Kalisher, M. H., Radford, W. A., Garwood, G. A., and Cockrum, C. A., "Historical Perspectives on HgCdTe Material and Device Development at Raytheon Vision Systems," *Proceedings of SPIE* 7298, 72982U, 2009. With permission.)

results. A series of high-performance staring SWIR FPAs including 1024 × 1024 and 2048 × 2048 formats along with high-performance staring MWIR FPAs including 640 × 480, 1024 × 1024, 2048 × 2048, and 2560 × 512 formats were demonstrated by RVS with operabilities exceeding 99.4% [134]. For example, Figure 21.30 shows 2560 × 512 format array in comparison with a stick of chewing gum, to illustrate just how large these staring arrays have become.

In the last few years efforts have been devoted to extend the performance of HgCdTe/Si into the LWIR range. This has also been pursued using MOVPE-grown HgCdTe on GaAs and GaAs/Si substrates in the United Kingdom [159–161]. A major challenge has been achieving good I-V characteristics for material having uniformly high dislocation density values in the mid-10^6 cm^{-2} range. The median NEDT of arrays is quite good, but these arrays typically exhibit a noise tail that limits operability, particularly under background flux conditions [134].

Two generic types of silicon addressing circuits have been developed: CCDs and FET switches. The evolution of the indium bumps technology provided the enabling technology required for the ROIC progress achieved. CMOS technology, since 1984, has improved the overall circuit and design of sensor chip assemblies to achieve ROICs with lower noise, higher yields, and higher densities [162]. The choice between CCD and CMOS depends on the application. For example, it seems much more complex to design a TDI linear array using a CMOS processor instead of a CCD processor and on the other hand, a lot of advantages could be found for using a CMOS processor for a staring array. CMOS is now the preferred choice; it operates well at low temperatures.

The CCD structures have been utilized to multiplex information from the detectors in staring systems or/and to perform the TDI function in scanned systems [152,163]. In TDI systems, the charge packets are shifted in synchronization with the scanned image. For fast and low-loss readout, the buried channel CCD is superior to a surface channel CCD. The charge transfer efficiency (CTF) of a silicon CCD achieves a value of 0.99998 at a charge level as low as 1.6×10^5 electrons. The problem in the use of CCD for HgCdTe arrays is the difficulty of efficient charge injection into the source-coupled input of the CCD. To achieve 0.9 injection efficiency, the R_oA value should be about 10 times that for BLIP operation [164]. Even more stringent demands on R_oA are placed by the input gate $1/f$ noise, which is of greater importance in CCD structures. Another problem arises from a limited charge storage capacity ($\approx 10^4$ electrons/μm^2) and clock rate. Saturation of the storage wells by the background photocurrent in a short period makes frame rates impractically high for large arrays, unless additional charge skimming and/or partition circuitry is introduced to remove part of the pedestrial charge. In order to achieve an adequate sensitivity the clocking rate must be high and the background subtraction should be performed within every unit cell of the CCD structure. However, the small unit sizes (less than 50 × 50 μm^2) desired in high-density FPAs constrain the input complexity. Despite higher ultimate performance of the IR image converters operating in LWIR, the stringent demands on R_oA value and the background limitations make SWIR and MWIR FPAs much easier to realize practically, compared to LWIR ones. In CCD technology it is also difficult to envisage ways of disabling defective pixels with additional circuitry, because of the poor packing density of mixed CCD/CMOS processes.

The actual minimum resolvable temperature difference of the system is set by fixed pattern noise. To achieve the potential value of 10 mK, the uniformity of output must be within 0.03%, while the typical standard deviation in present arrays is higher. It clearly illustrates the necessity for correction of the fixed pattern noise. The high injection efficiency of MWIR photodiodes into reedout output and related linear output enables simple two-point correction for nonuniformities. It may be performed by calibrating the FPA at two different uniform background flux levels and storing the calibrated coefficient of each pixel in memory [151]. The DC offset and AC responsivity of all the pixels is then normalized by an addition and multiplication algorithm. Another source of fixed pattern noise is fluctuation of the FPA temperature. Recalibration of the DC offset with one temperature is usually sufficient. CCDs are less suited for long wavelength operation because of the poor storage efficiency.

An attractive alternative to the CCD readout in the SWIR, MWIR, and especially LWIR FPAs is coordinative addressing with CMOS switches (see Figure 19.13). The advantages of CMOS are envisaged in Section 19.2.2. The minimum theoretical read noise of a CCD is limited in large imagers by the output amplifier's thermal noise after correlated double sampling (CDS) is applied in off-chip support circuits. The alternative CMOS paradigm offers lower temporal noise because the relevant noise bandwidth is fundamentally several orders of magnitude smaller and better matches the signal bandwidth. While CCD sensitivity is constrained by the limited design space involving the sense node and the output buffer, CMOS sensitivity is limited only by the desired dynamic range and operating voltage [165].

In hybrid HgCdTe FPAs various detector interface circuits are used to appropriately condition the signal. The readouts used several detector interfaces (see Section 19.2.2). Specifically optimized input circuits are typically required for strategic and tactical applications. For tactical applications, where the backgrounds are high and detector resistances are moderate, direct injection (DI) is a commonly used input circuit. The goal is to fit as large a capacitor as possible into the unit cell, particularly for high tactical applications were signal-to-noise rations can be obtained through longer integration times. This circuit is widely used for simplicity; however, it requires high impedance detector interface and is not generally used for low backgrounds due to injection efficiency issues. The strategic applications many times have low backgrounds and require low noise multiplexers interfaced to high resistance detectors. Commonly used input circuits for strategic applications are the CTIA input circuit. Besides the DI and CTIA inputs, we can distinguish other multiplexers; the most important are: source follower per detector (SFD; see Table 19.3), electronically scanned buffered direct injection (ESBDI), buffered direct injection (BDI), and MOSFET load gate modulation (BGM) input circuits [51,165–169]. Both CTIA and buffered DI give high injection efficiency and also accentuate the $1/f$ noise and the operability, but require higher power to operate.

One of the simplest and most popular readout circuits for IR FPAs is the DI input, where dark current and photocurrent are integrated into a storage capacity. In this case, the bias varies across the array by about ±(5–10) mV due to variations in the transistor thresholds. At 80 K, HgCdTe diodes show very little dependence of leakage current with small changes in reverse-bias near zero volts. The goal is to fit as large a capacitor as possible into the unit cell, particularly for high tactical applications were signal-to-noise rations can be obtained through longer integration times. For high injection efficiency, the resistance of the FET should be small compared to the diode resistance at its operating point (see Section 19.3.2). Generally, it is not a problem to fulfill inequality 19.2 for MWIR HgCdTe staring designs where diode resistance is large (the R_oA product is in the range above 10^6 Ωcm^2), but it can be very important for LWIR designs where diode resistance is small (the R_oA product is several hundred Ωcm^2). In this case of LWIR HgCdTe photodiodes a large bias is desirable, but it strongly depends on the material quality of the array. For very high quality LWIR HgCdTe array, –1 V bias is possible.

As is marked in Section 19.4.2, the performance of MW and LWIR FPAs is limited by the readout circuits and NEDT is estimated by Equation 19.13. High sensitivity can only be achieved if a large number of electrons are integrated and this requires the integration capacitance in each pixel to be fairly high. The charge handling capacity depends on cell pitch. For a 30×30 μm^2 pixel size, the storage capacities are limited to 1 to 5×10^7 electrons (it depends on design feature). For example, for a 5×10^7 electron storage capacity, the total current density of a detector with a 30×30 μm^2 pixel size has to be smaller than 27 $\mu A/cm^2$ with a 33 ms integration time [170]. If the total current density is in 1 mA/cm² range, the integration time has to be reduced to 1 ms. For the LWIR HgCdTe FPAs the integration time is usually below 100 μs. Since the noise power bandwidth $\Delta f = 1/2t_{int}$, a small integration time causes extra noise in integration.

Normally, the capacitance has a thin gate oxide dielectric and capacitance densities as high as 3 fF/μm². The capacitance is restricted to about 1 pF in pixels of around 25 μm square, and the best NEDT that can be expected is about 10 mK per frame.

MWIR and LWIR electronically scanned HgCdTe arrays with CMOS multiplexer are commercially available from several manufactures. Table 19.3 presents the worldwide situation in the industry while Tables 21.11 through 21.14 list typical performance specifications for larger SWIR, MWIR, and LWIR staring arrays fabricated by Raytheon, Sofradir, Teledyne, and Selex. Most manufactures produce their own multiplexer designs because these often have to be tailored to the applications.

Reytheon's SW Virgo-2k 2048 × 2048 pixel array is fabricated for astronomy standard products. This 20 μm pitch array is characterized by high quantum efficiency, low noise, low dark current, and on-chip clocking for ease of operation. Four or 16 outputs can be selected to accommodate a wide range of input flux conditions and readout rates.

Sofradir staring MW and LW snapshot arrays are dedicated to high resolution (TV format) applications (FLIR, IRST, reconnaissance, surveillance, airborne camera, thermography). These FPAs can be offered in different long vacuum-time dewar and cooler configurations in order to meet the different mechanical and cooling needs of the systems. The similar snapshot arrays are offered by Selex.

Teledyne's imaging sensors of Hawaii-2RG™ family are substrate-removed SW and MW HgCdTe arrays with response in the visible spectrum. These arrays built with modularity in mind—four-side-buttable to allow assembly of large mosaics of 2048 × 2048 H2RG modules—are dedicated for visible and IR astronomy in ground-based and space telescope applications. Figure 21.31 shows the typical visible and SWIR spectral response of substrate-removed HgCdTe FPAs with no degradation in detector mechanical and electrical quality and the expected improvements in visible response [9].

One of the most challenging tasks in the development of the next generation of HgCdTe FPAs is the integration of multiple functions into the detection circuit. The efforts are mainly focused on the development of multicolor detectors particularly for target recognition. Avalanche photodiodes are other devices with additional functionalities to the focal plane, in particular in the SW and MW ranges. The extremely low excess noise in the HgCdTe APDs is due to selective electron multiplication for wavelength $\lambda > 2$ μm and a nearly deterministic multiplication processes (see Section 14.7.4) [171].

Table 21.11: Raytheon's Virgo-2k

Parameter	SW	
Array size	2048 × 2048	
Spectral range	0.8–2.5 μm	
Pixel pitch	20 × 20 μm²	
Optical fill factor	>98%	
Architecture	3-side buttable	
Readout structure	SFD unit cell-PMOS	
Detector materials	Double layer heterojunction HgCdTe	
Well capacity	$\geq 3 \times 10^5$ e⁻ at 0.5 V applied bias	
Output performance	≤ 2.5 μs for 0.1% settling per output	
Number of outputs	4 or 16	
Frame time	690 ms per frame	1.43 Hz in 16 output mode
	2.66 seconds per frame	0.376 Hz in 4 output mode
Quantum efficiency	>80%	
Read noise	<20 e⁻/s (fowler 1)	
Dark current	<1 e⁻/s	
Operating temperature	70–80	
Electrical interface	Motherboard with one 51-pin MDM connector	

Table 21.12: HgCdTe Sofradir's Focal Plane Arrays

Parameter	MW (Jupiter)	LW (Venus)
Array size	1280×1024	384×288
Pixel pitch	$15 \times 15\ \mu m^2$	$25 \times 25\ \mu m^2$
Spectral response	3.7–4.8 μm	7.7–9.5 μm
Operating temperature	77–110 K	77–80 K
Max charge capacity	$4.2 \times 10^6\ e^-$	$3.37 \times 10^7\ e^-$
Readout noise	<150 μV (400 e^-)	<130 μV (1460 e^-)
Signal outputs	4 or 8	1 or 4
Pixel output rate	up to 20 MHz	up to 8 MHz
Frame rate	up to 120 Hz full frame rate	up to 300 Hz full frame rate
NEDT	18 mK	17 mK
Operability	>99.5%	>99.5%
Nonuniformity (DC & Resp.)		<5%
Residual fixed pattern noise	<NEDT	<NEDT

Table 21.13: Teledyne Imaging Sensors, Hawaii-2RG™

Parameter	Unit	1.7 μm	2.5 μm	5.4 μm
ROIC			Hawaii-2RG™	
Number of pixels	#		2048×2048	
Pixel size	μm		18	
Outputs			Programmable 1, 4, 32	
Power dissipation	mW		≤0.5	
Detector substrate			CdZnTe, removed	
Cutoff wavelength: 1.7 μm: @ 140 K (50% of peak QE) 2.5 μm: @ 77 K (50% of peak QE) 5.4 μm: @ 40 K (50% of peak QE)	μm	1.65–1.85	2.45–2.65	5.3–5.5
Mean quantum efficiency (QE) 0.4–1.0 μm	%		≥70	
Mean quantum efficiency (QE) 1.7 μm: 1.0–1.6 μm 2.5 μm: 1.0–2.4 μm 5.4 μm: 1.0–5.0 μm	%		≥80	
Mean dark current 1.7 μm: @ 0.25 V bias and 140 K 2.5 μm: @ 0.25 V bias and 77 K 5.4 μm: @ 0.175 V bias and 40 K	e^-/s	≤0.01	≤0.01	≤0.05
Median readout noise (single CDS) at 100 kHz pixel readout rate	e^-	≤25 (goal is 20)	≤20 (goal is 15)	≤16 (goal is 12)
Well capacity at 0.25 V bias (0.175 V) bias for 5.4 μm cutoff	e^-		≥80,000	
Crosstalk	%		≤2	
Operability	%	99	≥99	≥98
Cluster: 50 or more contiguous inoperable pixels within a 2000×2000-pixel area centered on array	#		≤0.5% of array	
SCA flatness	μm		≤30 (goal is 10)	
Planarity	μm		≤50 (goal is 25)	

The HgCdTe e-APDs are applied for gated-active/passive imaging [172–177]. The latter is of particular interest for low flux applications in the MW range, observing in a narrow field of view or spectral range. In addition, the amplification of the photocurrent can improve the linearity of some ROIC designs and a dynamic gain could be used to increase the dynamic range.

Table 21.14: HgCdTe Selex's Focal Plane Arrays

Parameter	MW (Merlin)	LW (Eagle)
Operating waveband	3–5 µm	8–10 µm
Array size	1024×768	640×512
Pixel pitch	16 µm	24 µm
Active area	16.38×12.29 mm²	15.36×12.29 mm²
NEDT	15 mK	24 mK
Operability	>99.5%	>99.5%
Signal uniformity	<5%	<5%
Scan format	Snapshot or rolling readout	Snapshot or rolling readout
Charge capacity	8×10^6 electrons	1.9×10^7 electrons
Number of outputs	8	4
Pixel rate	Up to 10 MHz per output	Up to 10 MHz per output
Technology	CMOS	CMOS
Intrinsic MUX noise	50 µV rms	50 µV rms
Operating temperature	Up to 140 K	Up to 90 K
Power consumption	40 mW	40 mW

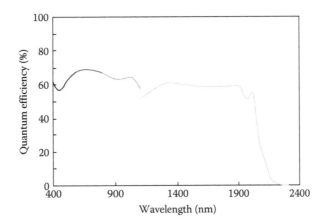

Figure 21.31 Spectral quantum efficiency of substrate-removed 256×256 array. (From Chuh, T., "Recent Developments in Infrared and Visible Imaging for Astronomy, Defense and Homeland Security," *Proceedings of SPIE* 5563, 19–34, 2004. With permission.)

Baker et al. at Selex were the first to demonstrate laser gated imaging in a 320×256 24 µm pitch APD FPA [172]. They reported avalanche gains up to $M = 100$, low excess noise and an input noise equivalent photon noise $NEP_h = 15$ photons rms, for a short integration times $t_{int} = 1$ µs for photodiodes with $\lambda_c = 4.2$ µm. Later [173,174], they described the interest of a dual mode active passive imaging and reported first results on 3-D active imager with a range resolution of 1 m. Next, Beck et al. [175] have specified a 128×128 FPA composed of 40 µm pitch, 4.2 µm to 5 µm cutoff, and the integration time between 50 ns and 500 µs. Median gains as high as 946 at 11 V bias with noise equivalent input as low as 0.4 photon were measured at 80 K. More recently, a first MW 320×256 passive amplified imaging with 30 µm pitch and operability of 99.8% has been demonstrated [176]. Also the electro-optic characteristics of the multifunctional LW-MW-avalanche gain detectors have been reported for 256×256 30 µm pitch arrays hybridized on a bicolor readout circuit, and for direct measurements on 30 µm pitch test arrays [177].

21.6 LEAD SALT ARRAYS

Lead salts were among the first IR detectors used successfully during War World II. After that, important efforts were made in the 1950s and 1960s to develop better and more complex devices

Table 21.15: Typical Performance of PbS and PbSe Linear Arrays with a CMOS Multiplexed Readout

	PbS		PbSe	
Configuration	128	256	128	256
Element dimensions (μm)	91×102 (in line)	38×56 (staggered)	91×102 (in line)	38×56 (staggered)
Center spacing (μm)	101.6	50.8	101.6	50.8
D^* (cmHz$^{1/2}$/W)	3×10^{11}	3×10^{11}	3×10^{10}	3×10^{10}
Responsivity (V/W)	1×10^8	1×10^8	1×10^6	1×10^6
Element time constant (μsec)	≤1000	≤1000	≤20	≤20
Nominal element temperature (K)	220	220	220	220
Operability (%)	≥98	≥98	≥98	≥98
Dynamic range	≤2000:1	≤2000:1	≤2000:1	≤2000:1
Channel uniformity (%)	±10	±10	±10	±10

Source: Northrop/Grumman Electro-Optical Systems data sheet, 2002. With permission.

and arrays and to understand their properties. Their historical prospect is given [23,178–181]. Low-cost, PbS and PbSe polycrystalline thin films remain the photoconductive detectors of choice for many applications in the MWIR spectral range [182,183].

Modern lead salt detector arrays contain more than 1000 elements on a single substrate. Operability exceeding 99% is readily achieved for these arrays. Smaller arrays having 100 or fewer elements have been produced with operability of 100%. Arrays as large as several inches on a side use a linear configuration with a single row and either equal areas equally spaced, or variable sizes. Other configurations include dual rows either in-line or staggered (several staggered rows in staircase fashion, chevron, and double cruciform).

Northrop Grumman EOS coupled 256-pixel PbSe arrays with Si multiplexer readout chips to fabricate assemblies with scanning capabilities [184]. Table 21.15 summarizes the performance of PbS and PbSe arrays in 128- and 256-element configurations [185]. A long-lived thermoelectric element cools the detector/dewar assembly to provide lifetimes greater than 10 years. It should be noted, however, that lead salt photoconductive detectors have a significant $1/f$ noise; for example, for PbSe, a knee frequency is of the order of 300 Hz at 77 K, 750 Hz at 200 K, and 7 kHz at 300 K [186]. This generally limits the use of these materials to scanning imagers.

Figure 21.32 shows the multimode detector/multiplexer/cooler assembly manufactured by Northrop Grumman [182]. This device consists of a linear or bilinear array of 128 or 256 photoconductive PbSe elements integrated with either a single or dual 128-channel multiplexer chip to give the option of an odd/even or a natural sequence pixel analog output. The multiplexed array is thermoelectrically cooled and closed in a long-life evacuated package with an AR-coated sapphire window and mounted on a circuit board, as shown in the photograph. Similar assemblies containing PbS elements are also fabricated.

In FPA fabrication, lead salt chalcogenides are deposited on Si or SiO from wet chemical baths. Such a monolithic solution avoids the use of a thick slab of these materials mated to Si, as is done with typical hybrids. The detector material is deposited from a wet chemical solution to form polycrystalline photoconductive islands on a CMOS multiplexer. Figure 21.33 shows a few of the 30 μm pixels in this detector array format. Northrop Grumman elaborated monolithic PbS FPAs in a 320×240 format (specified in Table 21.16) with a pixel size of 30 μm [182]. Although PbS photoconductors may be operated satisfactorily at ambient temperature, performance is enhanced by utilizing a self-contained thermoelectric cooler.

A first attempt to realize quasimonolithic lead salt detector arrays was described by Barrett, Jhabvala, and Maldari, who elaborated on direct integration of PbS photoconductive detectors with MOS transitions [187,188]. In this process, the PbS films were chemically deposited on the overlaying SiO_2 and metallization. Detectivity at 2.0–2.5 μm of 10^{11} cmHz$^{1/2}$W^{-1} was measured at 300 K on an integrated photoconductive PbS detector-Si MOSFET preamplifier. Elements of 25×25 μm^2 were easily fabricated.

Zogg et al. [189–192] have fabricated a monolithic, staggered linear array with up to 256 PbTe and PbSnSe Schottky-barrier photodiodes (see Figure 15.28) with 30 μm diameter on a 50 μm pitch. The substrates for these arrays contain integrated transistors for each pixel as needed for

Figure 21.32 Multimode PbSe focal plane arrays fabricated by Northrop Grumman. (From Beystrum, T., Himoto, R., Jacksen, N., and Sutton, M., "Low Cost Pb Salt FPAs," *Proceedings of SPIE* 5406, 287–94, 2004. With permission.)

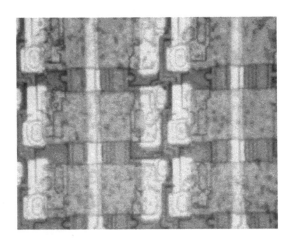

Figure 21.33 Monolithic PbS pixels in a 320 × 240 format. Pixel pitch is 30 μm. (From Beystrum, T., Himoto, R., Jacksen, N., and Sutton, M., "Low Cost Pb Salt FPAs," *Proceedings of SPIE* 5406, 287–94, 2004. With permission.)

the readout. The readout chips were fabricated in a combined CMOS/JFET technology. While CMOS designs require impedances in the MΩ range in order that the amplifier noise does not dominate, JFET input transistors can be designed with negligible noise even for low impedances (down to below 10 KΩ) without demanding high bias currents as, for example, in bipolar designs. For each channel, a charge integrator collects the photogenerated charges over a certain present time. The generated signals are then fed to a common output. Individual offset correction, and multiple correlated sampling to reduce the readout noise was employed. The further processing, background subtraction, and correction of the fixed pattern noise, is performed digitally.

The research group at the Swiss Federal Institute of Technology demonstrated the first realization of monolithic PbTe FPA (96 × 128) on a Si substrate containing the active addressing electronics [193,194]. The monolithic approach used here overcomes the large mismatch in the thermal coefficient of expansion between the group IV-VI materials and Si. Large lattice mismatches between the detector active region and the Si did not impede fabrication of the high quality layers because the easy plastic deformation of the IV-VIs by dislocation glide on their main glide system without causing structural deterioration.

A schematic cross section of a PbTe pixel grown epitaxially by MBE on a Si readout structure is shown in Figure 21.34 [193]. A 2–3 nm thick CaF_2 buffer layer is employed for compatibility

Table 21.16: Specifications for 320 × 240 PbS Focal Plane Arrays

Focal Plane Array Configuration	Monolithic 320 × 240 PbS
Pixel size (μm)	30 × 30
Detectivity (cmHz$^{1/2}$/W)	8×10^{10} (ambient); 3×10^{11} (220 K)
Type of signal processor	CMOS
Time constant (ms)	0.2 (ambient); 1 (220 K)
Integration options	Snapshot
Number of output lines	2
Frame rate (Hz)	60
Integration period	Full frame time
Mux dynamic range (dB)	69
Active heat dissipation (mW)	200 max
Operability (%)	> 99
Mux transimpedance (MΩ)	100
Detector bias (V)	0–6 (user adjustable)

Source: *Northrop/Grumman Electro-Optical Systems* data sheet, 2002. With permission.

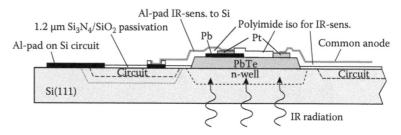

Figure 21.34 Schematic cross section of one pixel showing the PbTe island as a back side illuminated photovoltaic IR detector, the electrical connections to the circuit (access transistor), and a common anode. (From Alchalabi, K., Zimin, D., Zogg, H., and Buttler, W., *IEEE Electron Device Letters*, 22, 110–12, 2001. With permission.)

with the Si substrate. Active layers of 2–3-μm thickness suffice to obtain only a near-reflection-loss limited quantum efficiency. The spectral response curves of PbTe photodiodes are shown in Figure 15.29. Typical quantum efficiencies are around 50% without an AR coating. Metal semiconductor Pb/PbTe detectors are employed. Each Pb-cathode contact is fed to the drain of the access transistor while the anode (sputtered Pt) is common for all pixels. Figure 21.35 shows a complete array [194].

Despite typical dislocation densities in the 10^7 cm^{-2} range in epitaxial lattice and thermal-expansion-mismatched IV-VI on Si(111) layers, useful photodiodes can be fabricated with R_oA products to 200 Ωcm^2 at 95 K (with a 5.5 μm cutoff wavelength). This is due to the high permittivities of the IV-VI materials, which shield the electric field from charged defects over short distances.

Lead salt heterostructure PV detectors also have been fabricated. The heterojunctions were formed directly between the substrate and the IV-VI films. This solution has been employed to fabricate high density monolithic PbS-Si heterojunction arrays [195]. However, the performance of heterostructure arrays is inferior to the Schottky barrier and p-n junction arrays.

Experiments with development of lead salt hybrid array approach during late 1970s and early 1980s failed. A severe problem is the large thermal expansion mismatch of IV-VI materials to silicon. Therefore, the IV-VI detector/silicon hybrid technology lay in small size photodiode arrays with 10^3 elements operating at 77 K in the 8–14 μm region. Such hybrid structures containing 32 × 32 elements of back side illuminated PbTe [196] and PbSnTe [114] photodiodes with solder bump interconnections to silicon chips were demonstrated. Larger arrays require assembly from submodules of $\approx 10^3$ contiguous arrays [114].

Figure 21.35 Part of the completely processed monolithic 96 × 128 PbTe-on-Si infrared FPA for the MWIR with the readout electronics in the Si substrate. Pixel pitch is 75 μm. (From Zogg, H., Alchalabi, K., Zimin, D., and Kellermann, K., *IEEE Transactions on Electron Devices*, 50, 209–14, 2003. With permission.)

Felix et al. [163] described experimental results for a hybrid island structure with front side illuminated $Pb_{0.8}Sn_{0.2}Te$ photodiodes on a silicon CCD. The photodiodes were made by double LPE, PbTe/PbSnTe on a PbTe substrate. This technology is more complex but, because each detector is a separate physical unity, these structures avoid the problems due to thermal expansion mismatch. On the other hand, the island approach suffers from a fill factor loss because of the area contacts. The average R_0A product of photodiodes was close to 0.8 Ωcm², and detectivity was about 2×10^{10} cmHz$^{1/2}$W^{-1} at 77 K for 2π FOV. Felix and colleagues [163] also gave experimental results for the readout of $Pb_{0.8}Sn_{0.2}Te$ photodiodes directly injected to a linear multiplexing silicon CCD. An injection efficiency of 65% has been obtained.

21.7 QWIP ARRAYS

As is marked in Section 16.3.5, the quantum well infrared photoconductors (QWIPs) is an alternative to HgCdTe, hybrid detector for the MW and LW spectral ranges. The advantages of QWIPs are linked to pixel performance uniformity and to the availability of large size arrays, instead their drawbacks are the performance limitations for applications requiring short integration time, and the requirement to operate at a lower temperature than HgCdTe of comparable wavelengths. Moreover, the potential advantages of GaAs/AlGaAs quantum well devices include the use of standard manufacturing techniques based on mature GaAs growth and processing technologies, highly uniform and well-controlled MBE growth on greater than 6 inch GaAs wafers, high yield and thus low cost, more thermal stability, and extrinsic radiation hardness. Figure 21.36 shows evolution of the performance of VLWIR GaAs/AlGaAs QWIP [197]. As can be seen, rapid progress at the beginning of development has been made in detectivity, starting with bound-to-bound QWIPs, which had relatively poor sensitivity, and achieving considerable higher performance of bound-to-quasibound QWIPs with random reflectors.

Quantum well-infrared photoconductors detectors have relatively low quantum efficiencies, typically less than 10%. The spectral response band is also narrow for this detector, with a full-width, half-maximum of about 15%. All the QWIP data with cutoff wavelength about 9 μm is clustered between 10^{10} and 10^{11} cmHz$^{1/2}$/W at about 77 K operating temperature. Investigations of the fundamental physical limitations of HgCdTe photodiodes indicate better performance with this type of detector in comparison with QWIPs operated in the range 40–77 K. However, it has been shown that a low photoconductive gain actually increases the signal-to-noise ratio and a QWIP FPA can have a better temperature resolution than the HgCdTe FPA with similar storage capacity (see Section 19.4.2).

The first long wavelength infrared camera using AlGaAs/GaAs QWIPs was demonstrated by Bethea et al. in 1991 [198]. They used a commercial InSb scanning camera operating in the 3–5 μm spectral region and modified both the optics and electronics to allow operation at

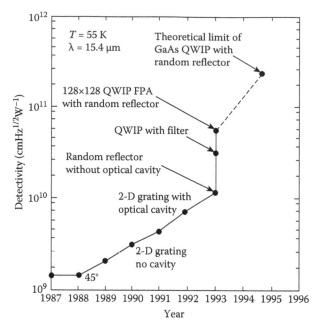

Figure 21.36 Evolution of the performance of very long wavelength GaAs/AlGaAs QWIP. All the data is normalized to wavelength $\lambda = 15.4\ \mu m$ at temperature $T = 55$ K. (From Gunapala, S. D., and Bandara, K. M. S. V., *Thin Films*, Academic Press, New York, Vol. 21, 113–237, 1995. With permission.)

$\lambda = 10\ \mu m$. Also fabrication of GaAs superlattice was changed to obtain geometrical compatibility with the original 10-element linear InSb array. This resulted in a 10 pixel GaAs quantum well array consisting of 200 μm^2 pixels separated by 670 μm. The GaAs substrate was polished at 45° to allow good optical coupling to the quantum wells. Using the above described nonoptimal infrared imaging camera ($\lambda_c = 10.7\ \mu m$) NEDT < 0.1 K has been achieved. Further development of QWIP FPAs is described, for example, by Rogalski [23]. Hitherto, a variety of excellent FPAs have been developed by many groups. These include the first arrays realized at today's Lucent Technologies (Murray Hill, New Jersey) [199], and those of the Jet Propulsion Laboratory (Pasadena, California) [197,200,201], Thales Research and Technology (Palaiseau, France) [202,203], Fraunhofer IAF (Freiburg, Germany) [204,205], Acreo (Kista, Sweden) [206], U.S. Army Research Laboratory (Adelphi, Maryland) [207,208], and BAE Systems (former Lockheed Martin) [209–211], among others. Also several university groups have demonstrated QWIP FPAs, between them, groups at Northwestern University (Evanston, Illinois) [212], Jerusalem College of Technology (Jerusalem, Israel), and Middle East Technical University (Ankara, Turkey) [213,214].

As is described in Section 16.4, the low noise level of photovoltaic QWIP FPAs enable longer integration time and improved thermal resolution of thermal imaging systems as compared to conventional photoconductive QWIPs. On the other side, photoconductive QWIPs are best situated if short integration times (5 ms and below) are required. In this case, thermal resolution is limited by the quantum efficiency of the detector contrary to typical limitation by the storage capacity. Quantum wells of QWIPs with high quantum efficiency are doped to higher electron concentrations (typically $4 \times 10^{11}\ cm^{-2}$, about four times higher than for standard photoconductive QWIPs) [205]. Photoconductive QWIPs with even higher carrier concentration ($2 \times 10^{12}\ cm^{-2}$) are exploited for MWIR with BLIP performance at about 90 K.

Figure 21.37 presents representative NEDT histograms of two types of FPAs with 640×512 pixels—low-noise LWIR and standard MWIR FPAs [215]. For the LW camera system with 24 μm pitch and integration time of 30 ms, a NEDT value as low as 9.6 mK has been observed, which is the best temperature resolutions ever obtained for thermal imagers operating in the 8–12 μm regime. In the case of a typical 640×512 MWIR QWIP FPA (Figure 21.37b), NEDT value of 14.3 mK has been obtained at 88 K.

Properties of QWIP FPAs demonstrated by Fraunhofer IAF are summarized in Table 21.17 [205]. It is interesting to notice that a thermal resolution of 40 mK is possible for only 1.5 ms

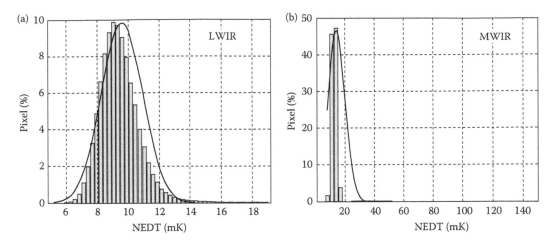

Figure 21.37 NEDT-histogram (a) of a 640×512 LWIR low-noise QWIP FPA for $f/2$ and 30 ms, and (b) of a 640×512 MWIR QWIP FPA for $f/1.5$ and 20 ms. (From Schneider, H., Fleissner, J., Rehm, R., Walther, M., Pletschen, W., Koidl, P., Weimann, G., Ziegler, J., Breiter, R., and Cabanski, W., "High-Resolution QWIP FPAs for the 8–12 μm and 3–5 μm Regimes," *Proceedings of SPIE* 4820, 297–305, 2003. With permission.)

Table 21.17: Properties of QWIP FPAs Demonstrated by Fraunhofer IAF

FPA Type	Array Size	Pitch (μm)	λ (μm)	f/#	τ_{int} (ms)	NEDT (mK)
256×256 PC	256×256	40	8–9.5	$f/2$	16	10
640×512 PC	640×486 512×512	24	8–9.5	$f/2$	16	20
256×256 LN	256×256	40	8–9.5	$f/2$	20 40	7 5
384×288 LN	384×288	24	8–9.5	$f/2$	20	10
640×512 LN	640×486 512×512	24	8–9.5	$f/2$	20	10
384×288 PC-HQE	384×288	24	8–9.5	$f/2$	1.5	40
640×512 PC-HQE	640×486 512×512	24	8–9.5	$f/2$	1.5	40
640×512 PC-MWIR	640×486 512×512	24	4.3–5	$f/1.5$	20	14

Source: H. Schneider and H. C. Liu, *Quantum Well Infrared Photodetectors*, Springer, Berlin, 2007. With permission.

Note: PC: photoconductive, LN: low-noise, HQE: high quantum efficiency, τ_{int}: integration time.

integration time, for arrays with higher doping (4×10^{11} cm^{-2} per QW) and an increased number of periods ($N = 35$).

Various types of MWIR and LWIR high resolution hybrid QWIPs are offered for different applications (FLIR, IRST, reconnaissance, surveillance, airborne camera, etc.). The arrays can be assembled in various long vacuum-life dewar and cooler configuration in order to meet the different mechanical and cooling needs of the systems. The examples are the Catherine XP and Catherine MP cameras manufactured by the Thales Research Technology—the cameras contain Vega and Sirius FPAs hybridized and integrated in sensor chip assembly by Sofradir [216] (see Table 21.18). Also a wide set of QWIP configurations is offered by Lockheed Martin Corporation (see Table 21.19).

Recently, one magapixel hybrid MWIR and LWIR QWIP with 18 μm pixel size has been demonstrated (see Figure 21.38) with excellent imaging performance using transitions from bound to extended states and from bound to miniband states [217–220]. Gunapala et al. [218] have

Table 21.18: QWIP Sofradir's Focal Plane Arrays

Parameter	LW (Sirius)	LW (Vega)
Array size	640×512	384×288
Pixel pitch	$20 \times 20\ \mu m^2$	$25 \times 25\ \mu m^2$
Spectral response	$\lambda_p = 8.5 \pm 0.1\ \mu m$, $\Delta\lambda = 1\ \mu m$ @ 50%	$\lambda_p = 8.5\ \mu m$, $\Delta\lambda = 1\ \mu m$ @ 50%
Operating temperature	70–73 K	73 K
Integration type	Snapshot	Snapshot
Max charge capacity	$1.04 \times 10^7\ e^-$	$1.85 \times 10^7\ e^-$
Readout noise	110 µV for gain 1	950 e^- for gain 1
Signal outputs	1, 2, or 4	1, 2, or 4
Pixel output rate	up to 10 MHz per output	up to 10 MHz per output
Frame rate	up to 120 Hz full frame rate	up to 200 Hz full frame rate
NEDT	31 mK (300 K, $f/2$, 7 ms integration time)	<35 mK (300 K, $f/2$, 7 ms integration time)
Operability	>99.9%	>99.95%
Nonuniformity	<5%	<5%

Table 21.19: QWIP FPAs Assembled in the ImagIR Camera Fabricated by Lockheed Martin Corporation

Parameter	
Spectral range	8.5 µm to 9.1 µm
Resolution/pixel pitch	$1024 \times 1024/19.5$ µm
	$640 \times 512/24$ µm
	$320 \times 256/30$ µm
Integration type	Snapshot
Integration time	<5 µs to full frame time
Dynamic range	14 bits
Data rate	32 Mpixels/sec
Frame rate	$1024 \times 1024 - 114$ Hz
	$640 \times 512 - 94$ Hz
	$320 \times 256 - 366$ Hz
Well capacity	$1024 \times 1024 - 8.1\ Me^-$
	$640 \times 512 - 8.4\ Me^-$
	$320 \times 256 - 20\ Me^-$
NEDT	<35 mK
Operability	>95.5 (>99.95 typ.)
Fixed focal plane	$f/2.3$ (13, 25, 50, 100) mm

demonstrated the MWIR detector arrays with a NEDT of 17 mK at 95 K operating temperature, $f/2.5$ optics and a 300 K background, and the LWIR detector array with a NEDT of 13 mK at 70 K operating temperature, the same optical and background conditions as the MWIR detector array. This technology can be readily extended to a 2K × 2K array. Figure 21.39 shows frames of video images taken with both 5.1 µm and 9 µm cutoff 1024 × 1024 pixel cameras. In addition to the excellent thermal resolution and contrast, both thermal images show high detail that indicates a small optical crosstalk between adjacent pixels and a good modulation transfer function.

Figure 21.40 shows the estimated NEDT as a function temperature for bias voltage of –2 V for MWIR and LWIR 1024 × 1024 QWIP FPAs. The background temperature is 300 K and the area of the pixel is 17.5 × 17.5 µm². The $f/\#$ of the optical system is 2.5, and the frame rates are 10 Hz and 30 Hz for MW and LW arrays, respectively.

Figure 21.38 Picture a 1024 × 1024 pixel QWIP FPA mounted on a 84-pin lead less chip carrier. (From Gunapala, S. D., Bandara, S. V., Liu, J. K., Hill, C. J., Rafol, B., Mumolo, J. M., Trinh, J. T., Tidrow, M. Z., and LeVan, P. D., *Semiconductor Science and Technology,* 20, 473–80, 2005. With permission.)

(a)   (b)

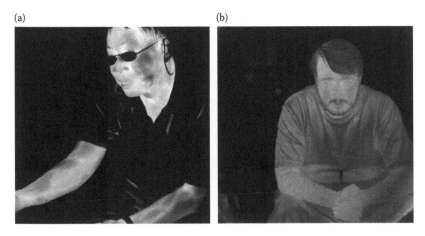

Figure 21.39 One frame of video taken with (a) the 5.1 μm, and (b) 9 μm cutoff 1024 × 1024 pixel QWIP cameras. The MW (LW) video images were taken at a frame rate of 10 Hz (30 Hz) at temperatures 90 K (72 K), using a ROIC capacitor having a charge capacity of 8×10^6 electrons. (From Gunapala, S. D., Bandara, S. V., Liu, J. K., Hill, C. J., Rafol, B., Mumolo, J. M., Trinh, J. T., Tidrow, M. Z., and LeVan, P. D., *Semiconductor Science and Technology,* 20, 473–80, 2005. With permission.)

For a given frame time and maximum storable charge, the way to increase the performance of LW staring systems is to reduce the dark current in order to get photon limited system performance. Bois et al. [221] described their unique approach of using two stacks of identical QWIPs and a new skimmed architecture accommodating the large dark current of the detector. The QWIP at the top stack produced a higher photocurrent than that of the lower stack. Therefore, using a bridge readout arrangement, the dark current of the QWIP at the top stack may be subtracted by the bottom stack without sacrificing much of the photocurrent. With this detector structure, the FPA should be able to operate at much higher temperatures and with a longer integration time. After the selection of the optimum subtraction rates, the expected NEDT is 10 mK at 85 K for $\lambda_c = 9.3$ μm and $f/1$ optics. This performance is usually achieved with standard QWIPs at 65 K (see Figure 21.41).

21.8 InAs/GaInSb SLS ARRAYS

As is marked in Section 17.4, $InAs/Ga_{1-x}In_xSb$ (InAs/GaInSb) strained layer superlattices (SLSs) can be considered as an alternative to HgCdTe and GaAs/AlGaAs IR material systems. The SLS structures provide high responsivity, as already reached with HgCdTe, without any need for gratings necessary in QWIPs. Further advantages are a photovoltaic operation mode, operation at elevated temperatures, and well-established III-V process technology.

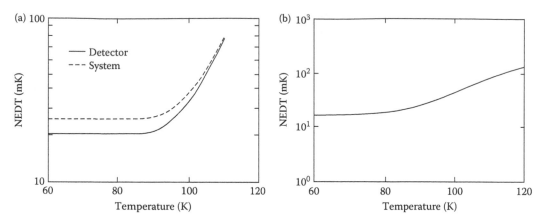

Figure 21.40 NEDT as a function temperature for bias voltage (a) of –2 V for MW, and (b) LW 1024 × 1024 QWIP FPAs. The background temperature is 300 K and the area of the pixel is 17.5 × 17.5 μm². (From Gunapala, S. D., Bandara, S. V., Liu, J. K., Hill, C. J., Rafol, B., Mumolo, J. M., Trinh, J. T., Tidrow, M. Z., and LeVan, P. D., *Semiconductor Science and Technology*, 20, 473–80, 2005. With permission.)

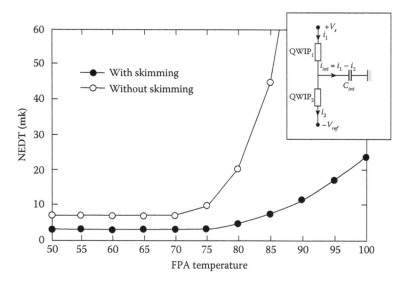

Figure 21.41 NEDT improvement due to the skimming architecture (*f*/1, 25 μm pitch, T_b = 300 K). (From Bois, P., Costard, E., Marcadet, X., and Herniou, E., *Infared Physics & Technology*, 42, 291–300, 2001. With permission.)

Type-II InAs/GaInSb-based detectors have made rapid progress over the past few years [222]. Using double M-structure heterojunction at the single device level, the R_0A product up to 5300 Ωcm² has been obtained for a 9.3 μm cutoff at 77 K and quantum efficiency closed to 80% [223]. The presented results indicate that fundamental material issues of InAs/GaInSb SLs fulfil practical realization of high performance FPAs.

First SL MWIR [224–227] and LWIR [227–229] FPA detectors have been hybridized. The cutoff wavelength of 256 × 256 MWIR detector is 5.3 μm. Excellent NEDT value of approximately 10 mK measured with *f*/2 optics and integration time τ_{int} = 5 ms has been presented (see Figure 21.42) [224]. Tests with reduced time down to 1 ms show that the NEDT scales inversely proportional to the square root of the integration time. It means that even for short integration time the detectors are background limited. A very important feature of InAs/GaInSb FPAs is their high uniformity.

The responsivity spread shows a standard deviation of approximately 3%. It is estimated that the pixel outages are in the order of 1–2% and the pixel are statistically distributed as single pixels without large clusters.

Recently, the demonstration of a high performance type-II FPA with cutoff wavelength of 10 μm has been reported [227–229]. The surface leakage current of photodiodes is suppressed using a double heterostructure design. In the diode structure, a low bandgap primary absorption superlattice layer is sandwiched between n+ (Si-doped) and p+ (Be-doped) high bandgap superlattice layers, finished with Be-doped GaSb p+ capping layer. The R_oA product of diodes passivated with SiO_2 was 23 Ωcm². Using this photodiode design, a 320 × 256 FPA with a pitch of 30 × 30 μm² demonstrated NEDT of 33 mK with an integration time of 0.23 ms (f/2 optics and 300 K background) comparable to HgCdTe. Figure 21.43 shows an image taken with this array at 81 K.

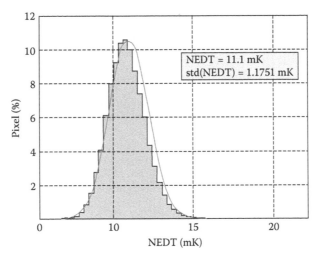

Figure 21.42   NEDT of a 256 × 256 MW type-II superlattice FPAs. f/#, τ_{int} = 5 ms. (From Cabanski, W., Eberhardt, K., Rode, W., Wendler, J., Ziegler, J., Fleißner, J., Fuchs, F., et al.,"3rd Gen Focal Plane Array IR Detection Modules and Applications," *Proceedings of SPIE* 5406, 184–92, 2005. With permission.)

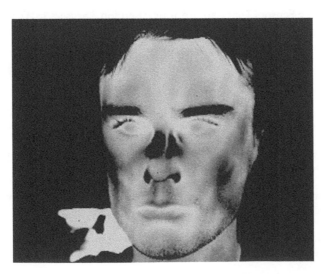

Figure 21.43   Image taken with the LWIR type-II InAs/GaSb FPA at 81 K. (From Delaunay, P.-Y., Nguyen, B. M., Hoffman, D., and Razeghi, M., *IEEE Journal of Quantum Electronics*, 44, 462–67, 2008. With permission.)

(a) (b)

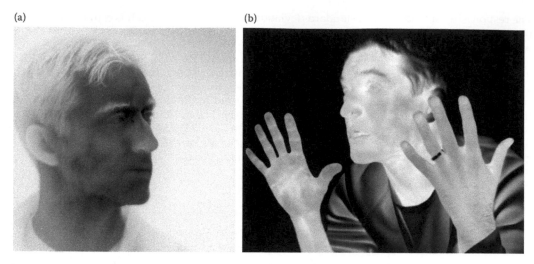

Figure 21.44 Images taken with (a) SL-based megapixel, and (b) nBn 640 × 512 MWIR FPAs. (From Hill, C. J., Soibel, A., Keo, S. A., Mumolo, J. M., Ting, D. Z., Gunapala, S. D., Rhiger, D. R., Kvaas, R. E., and Harris, S. F., "Demonstration of Mid and Long-Wavelength Infrared Antimonide-Based Focal Plane Arrays," *Proceedings of SPIE* 7298, 729404, 2009. With permission.)

Jet Propulsion Laboratory and RVS have demonstrated the first magapixel and the 640 × 512 type-II superlattice barrier infrared photodetector (nBn) MWIR FPAs (see Figure 21.44) [227]. Also the first dual-band LWIR and MWIR FPAs have been demonstrated (see Section 23.5).

REFERENCES

1. J. R. Janesick, *Scientific Charge-Coupled Devices*, SPIE Press, Bellingham, WA, 2001.

2. J. R. Janesick, "Charge-Coupled CMOS and Hybrid Detector Arrays," *Proceedings of SPIE* 5167, 1–18, 2003.

3. "About Image Sensor Solutions," http://www.kodak.com.

4. H. Titus, "Imaging Sensors That Capture Your Attention," *Sensors* 18, February 2001.

5. "X-3: New Single-Chip Colour CCD Technology," *New Technology*, 20–24, March/April 2002.

6. R. F. Lyon and P. Hubel, "Eyeing the Camera: Into the Next Century," *IS&T/SID Tenth Color Imaging Conference*, 349–355, 2001.

7. "DALSA Develops 100 + Megapixel CCD," http://www.dpreview.com/news/0606/06061901dalsa100mp.asp

8. G. Patrie, "Gigapixel Frame Images: Part II. Is the Holy Grail of Airborne Digital Frame Imaging in Sight?" *GeoInformatics*, 24–29, March 2006.

9. T. Chuh, "Recent Developments in Infrared and Visible Imaging for Astronomy, Defense and Homeland Security," *Proceedings of SPIE* 5563, 19–34, 2004.

10. Y. Bai, J. Bajaj, J. W. Beletic, and M. C. Farris, "Teledyne Imaging Sensors: Silicon CMOS Imaging Technologies for X-Ray, UV, Visible and Near Infrared," *Proceedings of SPIE* 7021, 702102, 2008.

11. S. Kilcoyne, N. Malone, M. Harris, J. Vampola, and D. Lindsay, "Silicon p-i-n Focal Plane Arrays at Raytheon," *Proceedings of SPIE* 7082, 70820J, 2008.

12. C. S. Rafferty, C. A. King, B. D. Ackland, I. Aberg, T. S. Sriram, and J. H. O'Neill, "Monolithic Germanium SWIR Imaging Array," *Proceedings of SPIE* 6940-20, 2008.

13. E. Burstein, G. Pines, and N. Sclar, "Optical and Photoconductive Properties of Silicon and Germanium," in *Photoconductivity Conference at Atlantic City*, eds. R. Breckenbridge, B. Russell, and E. Hahn, 353–413, Wiley, New York, 1956.

14. S. Borrello and H. Levinstein, "Preparation and Properties of Mercury Doped Infrared Detectors," *Journal of Applied Physics* 33, 2947–50, 1962.

15. N. Sclar, "Properties of Doped Silicon and Germanium Infrared Detectors," *Progress in Quantum Electronics* 9, 149–257, 1984.

16. R. D. Nelson, "Accumulation-Mode Charge-Coupled Device," *Applied Physics Letters* 25, 568–70, 1974.

17. D. H. Pommerrenig, "Extrinsic Silicon Focal Plane Arrays," *Proceedings of SPIE* 443, 144–50, 1984.

18. R. D. Nelson, "Infrared Charge Transfer Devices: The Silicon Approach," *Optical Engineering* 16, 275–83, 1977.

19. T. T. Braggins, H. M. Hobgood, J. C. Swartz, and R. N. Thomas, "High Infrared Responsivity Indium-Doped Silicon Detector Material Compensated by Neutron Transmutation," *IEEE Transactions on Electron Devices* ED-27, 2–10, 1980.

20. C. J. Michon and H. K. Burke, "Charge Injection Imaging," in *International Solid-State Circuits Conference, Dig. Tech. Papers*, 138–39, 1973.

21. C. M. Parry, "Bismuth-Doped Silicon: An Extrinsic Detector for Long-Wavelength Infrared (LWIR) Applications," *Proceedings of SPIE* 244, 2–8, 1980.

22. M. E. McKelvey, C. R. McCreight, J. H. Goebel, and A. A. Reeves, "Charge-Injection-Device 2×64 Element Infrared Array Performance," *Applied Optics* 24, 2549–57, 1985.

23. A. Rogalski, *Infrared Detectors*, Gordon and Breach Science Publishers, Amsterdam, 2000.

24. W. J. Forrest, A. Moneti, C. E. Woodward, J. L. Pipher, and A. Hoffman, "The New Near-Infrared Array Camera at the University of Rochester," *Publications of the Astronomical Society of the Pacific* 97, 183–98, 1985.

25. G. H. Rieke, "Infrared Detector Arrays for Astronomy," *Annual Review Astronomy and Astrophysics* 45, 77–115, 2007.

26. J. Wu, W. J. Forrest, J. L. Pipher, N. Lum, and A. Hoffman, "Development of Infrared Focal Plane Arrays for Space," *Review of Scientific Instruments* 68, 3566–78, 1997.

27. A. K. Mainzer, P. Eisenhardt, E. L. Wright, F.-C. Liu, W. Irace, I. Heinrichsen, R. Cutri, and V. Duval, "Preliminary Design of the Wide-Field Infrared Survey Explorer (WISE)," *Proceedings of SPIE* 5899, 58990R, 2005.

28. P. J. Love, A. W. Hoffman, N. A. Lum, K. J. Ando, J. Rosbeck, W. D. Ritchie, N. J. Therrien, R. S. Holcombe, and E. Corrales, "1024×1024 Si:As IBC Detector Arrays for JWST MIRI," *Proceedings of SPIE* 5902, 590209, 2005.

29. P. J. Love, K. J. Ando, R. E. Bornfreund, E. Corrales, R. E. Mills, J. R. Cripe, N. A. Lum, J. P. Rosbeck, and M. S. Smith, "Large-Format Infrared Arrays for Future Space and Ground-Based Astronomy Applications," *Proceedings of SPIE* 4486, 373–84, 2002.

30. K. J. Ando, A. W. Hoffman, P. J. Love, A. Toth, C. Anderson, G. Chapman, C. R. McCreight, K. A. Ennico, M. E. McKelvey, and R. E. McMurray, Jr., "Development of Si:As Impurity Band Conduction (IBC) Detectors for Mid-Infrared Applications," *Proceedings of SPIE* 5074, 648–57, 2003.

31. A. K. Mainzer, H. Hogue, M. Stapelbroek, D. Molyneux, J. Hong, M. Werner, M. Ressler, and E. Young, "Characterization of a Megapixel Mid-Infrared Array for High Background Applications," *Proceedings of SPIE* 7021, 70210T, 2008.

32. M. E. Ressler, H. Cho, R. A. M. Lee, K. G. Sukhatme, J. J. Drab, G. Domingo, M. E. McKelvey, R. E. McMurray, Jr., and J. L. Dotson, "Performance of the JWST/MIRI Si:As Detectors," *Proceedings of SPIE* 7021, 70210O, 2008.

33. A. M. Fowler and I. Gatley, "Demonstration of an Algorithm for Read-Noise Reduction in Infrared Arrays," *Astrophysical Journal* 353, L33–L34, 1990.

34. E. T. Young, "Progress on Readout Electronics for Far-Infrared Arrays," *Proceedings of SPIE* 2226, 21–28, 1994.

35. R. M. Glidden, S. C. Lizotte, J. S. Cable, L. W. Mason, and C. Cao, "Optimization of Cryogenic CMOS Processes for Sub-10°K Applications," *Proceedings of SPIE* 1684, 2–39, 1992.

36. E. T. Young, J. T. Davis, C. L. Thompson, G. H. Rieke, G. Rivlis, R. Schnurr, J. Cadien, L. Davidson, G. S. Winters, and K. A. Kormos, "Far-Infrared Imaging Array for SIRTF," *Proceedings of SPIE* 3354, 57–65, 1998.

37. R. Schnurr, C. L. Thompson, J. T. Davis, J. W. Beeman, J. Cadien, E. T. Young, E. E. Haller, and G. H. Rieke, "Design of the Stressed Ge:Ga Far-Infrared Array for SIRTF," *Proceedings of SPIE* 3354, 322–31, 1998.

38. M. Fujiwara, T. Hirao, M. Kawada, H. Shibai, S. Matsuura, H. Kaneda, M. Patrashin, and T. Nakagawa, "Development of a Gallium-Doped Germanium Far-Infrared Photoconductor Direct Hybrid Two-Dimensional Array," *Applied Optics* 42, 2166–73, 2003.

39. M. Shirahata, S. Matsuure, S. Makiuti, M.A. Patrashin, H. Kaneda, T. Nakagawa, M. Fujiwara, et al., "Preflight Performance Measurements of a Monolithic Ge:Ga Array Detector for the Far-Infrared Surveyor Onboard ASTRO-F," *Proceedings of SPIE* 5487, 369–80, 2004.

40. A. Poglitsch, R. O. Katterloher, R. Hoenle, J. W. Beeman, E. E. Haller, H. Richter, U. Grozinger, N. M. Haegel, and A. Krabbe, "Far-Infrared Photoconductors for Herschel and SOFIA," *Proceedings of SPIE* 4855, 115–28, 2003.

41. S. M. Birkmann, K. Eberle, U. Grozinger, D. Lemke, J. Schreiber, L. Barl, R. Katterloher, A. Poglitsch, J. Schubert, and H. Richter, "Characterization of High- and Low-Stressed Ge:Ga Array Cameras for Herschel's PACS Instrument," *Proceedings of SPIE* 5487, 437–47, 2004.

42. E. T. Young, "Germanium Detectors for the Far-Infrared," http://www.stsci.edu/stsci/meetings/space_detectors/pdf/gdfi.pdf

43. N. M. Haegel, "BIB Detector Development for the Far Infrared: From Ge to GaAs," *Proceedings of SPIE* 4999, 182–94, 2003.

44. E. E. Haller and J. W. Beeman, "Far Infrared Photoconductors: Recent Advances and Future Prospects," *Far-IR Sub-MM&MM Detectors Technology Workshop*, 2-06, Monterey, April 1–3, 2002.

45. M. Kimata, M. Ueno, H. Yagi, T. Shiraishi, M. Kawai, K. Endo, Y. Kosasayama, T. Sone, T. Ozeki, and N. Tsubouchi, "PtSi Schottky-Barrier Infrared Focal Plane Arrays," *Opto-Electronics Review* 6, 1–10, 1998.

46. E. S. Kohn, W. F. Kosonocky, and F. V. Shallcross, "Charge-Coupled Scanned IR Imaging Sensors," *Final Report RADC-TR-308*, Rome Air Development Center, 1977.

47. M. Denda, M. Kimata, S. Iwade, N. Yutani, T. Kondo, and N. Tsubouchi, "Schottky-Barrier Infrared Linear Image Sensor with 4-Band × 4096-Element," *IEEE Transactions on Electron Devices* 38, 1145–51, 1991.

48. M. T. Daigle, D. Colvin, E. T. Nelson, S. Brickman, K. Wong, S. Yoshizumi, M. Elzinga, et al., "High Resolution 2048 × 16 TDI PtSi IR Imaging CCD," *Proceedings of SPIE* 1308, 88–98, 1990.

49. W. F. Kosonocky, "Review of Infrared Image Sensors with Schottky-Barrier Detectors," *Optoelectronics–Devices and Technologies* 6, 173–203, 1991.

50. W. F. Kosonocky, "State-of-the-Art in Schottky-Barrier IR Image Sensors," *Proceedings of SPIE* 1682, 2–19, 1992.

51. M. Kimata and N. Tsubouchi, "Schottky Barrier Photoemissive Detectors," in *Infrared Photon Detectors*, ed. A. Rogalski, 299–349, SPIE Optical Engineering Press, Bellingham, WA, 1995.

52. M. Kimata, "Metal Silicide Schottky Infrared Detector Arrays," in *Infrared Detectors and Emitters: Materials and Devices*, eds. P. Capper and C. T. Elliott, 77–98, Kluwer Academic Publishers, Boston, MA, 2000.

53. M. Kimata, "Silicon Infrared Focal Plane Arrays," in *Handbook of Infrared Detection Technologies*, eds. M. Henini and M. Razeghi, 353–92, Elsevier, Oxford, 2002.

54. H. Yagi, N. Yutani, J. Nakanishi, M. Kimata, and M. Nunoshita, "A Monolithic Schottky-Barrier Infrared Image Sensor with 71% Fill Factor," *Optical Engineering* 33, 1454–60, 1994.

55. N. Yutani, H. Yagi, M. Kimata, J. Nakanishi, S. Nagayoshi, and N. Tsubouchi, "1040 × 1040 Element PtSi Schottky-Barrier IR Image Sensor," *Technical Digest IEDM*, 175–78, 1991.

56. M. Kimata, N. Yutani, N. Tsubouchi, and T. Seto, "High Performance 1040 × 1040 Element PtSi Schottky-Barrier Image Sensor," *Proceedings of SPIE* 1762, 350–60, 1992.

57. T. Shiraishi, H. Yagi, K. Endo, M. Kimata, T. Ozeki, K. Kama, and T. Seto, "PtSi FPA with Improved CSD Operation," *Proceedings of SPIE* 2744, 33–43, 1996.

58. M. Kimata, M. Denda, N. Yutani, S. Iwade, and N. Tsubouchi, "512 × 512 Element PtSi Schottky-Barrier Infrared Image Sensor," *IEEE Journal of Solid-State Circuits* SC-22, 1124–29, 1987.

59. M. Inoue, T. Seto, S. Takahashi, S. Itoh, H. Yagi, T. Siraishi, K. Endo, and M. Kimata, "Portable High Performance Camera with 801 × 512 PtSi-SB IRCSD," *Proceedings of SPIE* 3061, 150–58, 1997.

60. M. Kimata, T. Ozeki, M. Nunoshita, and S. Ito, "PtSi Schottky-Barrier Infrared FPAs with CSD Readout," *Proceedings of SPIE* 3179, 212–23, 1997.

61. F. D. Shepherd, "Recent Advances in Platinum Silicide Infrared Focal Plane Arrays," *Technical Digest IEDM*, 370–73, 1984.

62. J. L. Gates, W. G. Connelly, T. D. Franklin, R. E. Mills, F. W. Price, and T. Y. Wittwer, "488 × 640-Element Hybrid Platinum Silicide Schottky Focal Plane Array," *Proceedings of SPIE* 1540, 262–73, 1991.

63. W. F. Kosonocky, T. S. Villani, F. V. Shallcross, G. M. Meray, and J. J. O'Neil, "A Schottky-Barrier Image Sensor with 100% Fill Factor," *Proceedings of SPIE* 1308, 70–80, 1990.

64. H. Presting, "Infarred Silicon/Germanium Detectors," in *Handbook of Infrared Detection Technologies*, eds. M. Henini and M. Razeghi, 393–448, Elsevier, Oxford, 2002.

65. B-Y. Tsaur, C. K. Chen, and S. A. Marino, "Long-Wavelength GeSi/Si Heterojunction Infrared Detectors and 400 × 400-Element Imager Arrays," *IEEE Electron Device Letters* 12, 293–96, 1991.

66. B-Y. Tsaur, C. K. Chen, and S. A. Marino, "Long-Wavelength $Ge_{1-x}Si_x$/Si Heterojunction Infrared Detectors and Focal Plane Arrays," *Proceedings of SPIE* 1540, 580–95, 1991.

67. B-Y. Tsaur, C. K. Chen, and S. A. Marino, "Heterojunction $Ge_{1-x}Si_x$/Si Infrared Detectors and Focal Plane Arrays," *Optical Engineering* 33, 72–78, 1994.

68. H. Wada, M. Nagashima, K. Hayashi, J. Nakanishi, M. Kimata, N. Kumada, and S. Ito, "512 × 512 Element GeSi/Si Heterojunction Infrared Focal Plane Array," *Opto-Electronics Review* 7, 305–11, 1999.

69. M. Kimata, H. Yagi, M. Ueno, J. Nakanishi, T. Ishikawa, Y. Nakaki, M. Kawai, et al., "Silicon Infrared Focal Plane Arrays," *Proceedings of SPIE* 4288, 286–97, 2001.

70. A. M. Joshi, V. S. Ban, S. Mason, M. J. Lange, and W. F. Kosonocky, "512 and 1024 Element Linear InGaAs Detector Arrays for Near-Infrared (1–3 μm) Environmental Sensing," *Proceedings of SPIE* 1735, 287–95, 1992.

71. *InGaAs Products: Focal Plane Arrays*, http://www.sensorsinc.com/arrays.html

72. G. H. Olsen and M. J. Cohen, "Applications of Near-Infrared Imaging," *Proceedings of SPIE* 3379, 300–306, 1998.

73. A. Richards, "Focal-Plane Arrays Open New Near-Infrared Vistas," *Advanced Imaging*, March 2003.

74. G. Olsen, A. Joshi, M. Lange, K. Woodruff, E. Mykietyn, D. Gay, G. Erickson, D. Ackley, V. Ban, and C. Staller, "A 128 × 128 InGaAs Detector Array for 1.0–1.7 Microns," *Proceedings of SPIE* 1341, 432–37, 1990.

75. M. H. Ettenberg, M. J. Cohen, R. M. Brubaker, M. J. Lange, M. T. O'Grady, and G. H. Olsen, "Indium Gallium Arsenide Imaging with Smaller Cameras, Higher Resolution Arrays, and Greater Material Sensitivity," *Proceedings of SPIE* 4721, 26–36, 2002.

76. M. H. Ettenberg, M. J. Lange, M. T. O'Grady, J. S. Vermaak, M. J. Cohen, and G. H. Olsen, "A Room Temperature 640 × 512 Pixel Near-Infrared InGaAs Focal Plane Array," *Proceedings of SPIE* 4028, 201–7, 2000.

77. T. J. Martin, M. J. Cohen, J. C. Dries, and M. J. Lange, "InGaAs/InP Focal Plane Arrays for Visible Light Imaging," *Proceedings of SPIE* 5406, 38–45, 2004.

78. T. Martin, R. Brubaker, P. Dixon, M.-A. Gagliardi, and T. Sudol, "640 × 512 InGaAs Focal Plane Array Camera for Visible and SWIR Imaging," *Proceedings of SPIE* 5783, 12–20, 2005.

79. B. M. Onat, W. Huang, N. Masaun, M. Lange, M. H. Ettenberg, and C. Dries, "Ultra Low Dark Current InGaAs Technology for Focal Plane Arrays for Low-Light Level Visible-Shortwave Infrared Imaging," *Proceedings of SPIE* 6542, 65420L, 2007.

80. M. D. Enriguez, M. A. Blessinger, J. V. Groppe, T. M. Sudol, J. Battaglia, J. Passe, M. Stern, and B. M. Onat, "Performance of High Resolution Visible-InGaAs Imager for Day/Night Vision," *Proceedings of SPIE* 6940, 69400O, 2008.

81. R. F. Cannata, R. J. Hansen, A. N. Costello, and W. J. Parrish, "Very Wide Dynamic Range SWIR Sensors for Very Low Background Applications," *Proceedings of SPIE* 3698, 756–65, 1999.

82. J. Barton, R. Cannata, and S. Petronio, "InGaAs NIR Focal Plane Arrays for Imaging and DWDM Applications," *Proceedings of SPIE* 4721, 37–47, 2002.

83. T. R. Hoelter and J. B. Barton, "Extended Short Wavelength Spectral Response from InGaAs Focal Plane Arrays," *Proceedings of SPIE* 5074, 481–90, 2003.

84. H. Yuan, G. Apgar, J. Kim, J. Laguindanum, V. Nalavade, P. Beer, J. Kimchi, and T. Wong, "FPA Development: From InGaAs, InSb, to HgCdTe," *Proceedings of SPIE* 6940, 69403C, 2008.

85. *InGaAs Focal Plane Arrays*, http://www.aeriousphotonics.com/prod_ingaas.html

86. Near Infrared Cameras, http://www.Xenics.com/en/infrared_camera/visnir-nir_camera_Visual_near_and_near_infrared_cameras_-_ingaas.asp

87. A. Hoffman, T. Sessler, J. Rosbeck, D. Acton, and M. Ettenberg, "Megapixel InGaAs Arrays for Low Background Applications," *Proceedings of SPIE* 5783, 32–38, 2005.

88. J. Getty, E. Hadjiyska, D. Acton, S. Harris, B. Starr, A. Levy, J. Wehner, S. Taylor, and A. Hoffman, "VIS/SWIR Focal Plane and Detector Development at Raytheon Instruments Performance Data and Future Developments at Raytheon," *Proceedings of SPIE* 6660, 66600C, 2007.

89. S. Seshadri, D. M. Cole, B. Hancock, P. Ringold, C. Peay, C. Wrigley, M. Bonati, et al., "Comparison the Low-Temperature Performance of Megapixel NIR InGaAs and HgCdTe Imager Arrays," *Proceedings of SPIE* 6690, 669006, 2007.

90. S. K. Mendis, S. E. Kemeny, R. C. Gee, B. Pain, C. O. Staller, Q. Kim, and E. R. Fossum, "CMOS Active Pixel Sensors for Highly Integrated Image Systems," *IEEE Journal of Solid-State Circuits* 32, 187–97, 1997.

91. Q. Kim, T. J. Cunningham, and B. Pain, "Readout Characteristics of Integrated Monolithic InGaAs Active Pixel Image Sensor Array," *Proceedings of SPIE* 3290, 278–86, 1998.

92. Q. Kim, M. J. Lange, C. J. Wrigley, T. J. Cunningham, and B. Pain, "Two-Dimensional Active Pixel InGaAs Focal Plane Arrays," *Proceedings of SPIE* 4277, 223–29, 2001.

93. P. R. Norton, B. F. Andresen, and G. F. Fulom, "Introduction," *Proceedings of SPIE* 6940, xix–xxxvi, 2008.

94. G. Orias, A. Hoffman, and M. Casselman, "58 × 62 Indium Antimonide Focal Plane Array for Infrared Astronomy," *Proceedings of SPIE* 627, 408–17, 1986.

95. G. C. Baily, C. A. Niblack, and J. T. Wimmers, "Recent Developments on a 128 × 128 Indium Antimonide/FET Switch Hybrid Imager for Low Background Applications," *Proceedings of SPIE* 686, 76–83, 1986.

96. S. Shirouzu, T. Tsuji, N. Harada, T. Sado, S. Aihara, R. Tsunoda, and T. Kanno, "64 × 64 InSb Focal Plane Array with Improved Two Layer Structure," *Proceedings of SPIE* 661, 419–25, 1986.

97. J. T. Wimmers, R. M. Davis, C. A. Niblack, and D. S. Smith, "Indium Antimonide Detector Technology at Cincinnati Electronics Corporation," *Proceedings of SPIE* 930, 125–38, 1988.

98. C. W. McMurtry, W. J. Forrest, J. L. Pipher, and A. C. Moore, "James Webb Space Telescope Characterization of Flight Candidate NIR InSb Array," *Proceedings of SPIE* 5167, 144–58, 2003.

99. G. Finger, R. J. Dorn, A. W. Hoffman, H. Mehrgan, M. Meyer, A. F. M. Moorwood, and J. Stegmeier, "Readout Techniques for Drift and Low Frequency Noise Rejection in Infrared Arrays," in *Scientific Detectors for Astronomy*, eds. P. Amico, J. W. Beletic, and J. E. Beletic, 435–44, Springer, Berlin, 2003.

100. A. W. Hoffman, E. Corrales, P. J. Love, J. Rosbeck, M. Merrill, A. Fowler, and C. McMurtry, "2K × 2K InSb for Astronomy," *Proceedings of SPIE* 5499, 59–67, 2004.

101. H. Fujisada, N. Nakayama, and A. Tanaka, "Compact 128 InSb Focal Plane Assembly for Thermal Imaging," *Proceedings of SPIE* 1341, 80–91, 1990.

102. M. A. Blessinger, R. C. Fischer, C. J. Martin, C. A. Niblack, H. A. Timlin, and G. Finger, "Performance of a InSb Hybrid Focal Plane Array," *Proceedings of SPIE* 1308, 194–201, 1990.

103. E. Beuville, D. Acton, E. Corrales, J. Drab, A. Levy, M. Merrill, R. Peralta, and W. Ritchie, "High Performance Large Infrared and Visible Astronomy Arrays for Low Background Applications: Instruments Performance Data and Future Developments at Raytheon," *Proceedings of SPIE* 6660, 66600B, 2007.

104. A. M. Fowler and J. B. Heynssens, "Evaluation of the SBRC 256 × 256 InSb Focal Plane Array and Preliminary Specifications for the 1024 × 1024 InSb Focal Plane Array," *Proceedings of SPIE* 1946, 25–32, 1993.

105. A. M. Fowler, D. Bass, J. Heynssens, I. Gatley, F. J. Vrba, H. D. Ables, A. Hoffman, M. Smith, and J. Woolaway, "Next Generation in InSb Arrays: ALADDIN, the 1024 × 1024 InSb Focal Plane Array Readout Evaluation Results," *Proceedings of SPIE* 2268, 340–45, 1994.

106. A. M. Fowler, J. B. Heynssens, I. Gatley, F. J. Vrba, H. D. Ables, A. Hoffman, and J. Woolaway, "ALADDIN, the 1024 × 1024 InSb Array: Test Results," *Proceedings of SPIE* 2475, 27–33, 1995.

107. A. M. Fowler, I. Gatley, P. McIntyre, F. J. Vrba, and A. Hoffman, "ALADDIN, the 1024 × 1024 InSb Array: Design, Description, and Results," *Proceedings of SPIE* 2816, 150–60, 1996.

108. A. M. Fowler, K. M. Merrill, W. Ball, A. Henden, F. Vrba, and C. McCreight, "Orion: A 1-5 Micron Focal Plane for the 21st Century," in *Scientific Detectors for Astronomy: The Beginning of a New Era*, ed. P. Amico, 51–58, Kluwer, Dordrecht, 2004.

109. O. Nesher, I. Pivnik, E. Ilan, Z. Calalhorra, A. Koifman, I. Vaserman, J. O. Schlesinger, R. Gazit, and I. Hirsh, "High Resolution 1280 × 1024, 15μm Pitch Compact InSLIR Detector with On-Chip ADC," *Proceeding of SPIE* 7298, 72983K, 2009.

110. R. Rawe, C. Martin, M. Garter, D. Endres, B. Fischer, M. Davis, J. Devitt, and M. Greiner, "Novel High Fill-Factor, Small Pitch, Reticulated InSb IR FPA Design," *Proceedings of SPIE* 5783, 899–906, 2005.

111. H. K. Burke and G. J. Milton, "Charge-Injection Device Imaging; Operating Techniques and Performance Characteristics," *IEEE Transactions on Electron Devices* ED-23, 189–95, 1976.

112. J. C. Kim, "InSb Charge-Injection Device Imaging Array," *IEEE Transactions on Electron Devices* ED-25, 232–46, 1978.

113. G. J. Michon and H. K. Burke, "CID Image Sensing," in *Charge-Coupled Devices*, ed. D. F. Barbe, 5–24, Springer-Verlag, Berlin, 1980.

114. J. T. Longo, D. T. Cheung, A. M. Andrews, C. C. Wang, and J. M. Tracy, "Infrared Focal Plane in Intrinsic Semiconductors," *IEEE Transactions on Electron Devices* ED-25, 213–32, 1978.

115. R. J. Phelan and J. O. Dimmock, "InSb MOS Infrared Detector," *Applied Physics Letters* 10, 55–58, 1967.

116. D. L. Lile and H. H. Wieder, "The Thin Film MIS Surface Photodiode," *Thin Solid Films* 13, 15–20, 1972.

117. D. L. Lile, "Surface Photovoltage and Internal Photoemission at the Anodized InSb Surface," *Surface Science* 34, 337–67, 1973.

118. R. D. Thom, T. L. Koch, J. D. Langan, and W. L. Parrish, "A Fully Monolithic InSb Infrared CCD Array," *IEEE Transactions on Electron Devices* ED–27, 160–70, 1980.

119. C. Y. Wei, K. L. Wang, E. A. Taft, J. M. Swab, M. D. Gibbons, W. E. Davern, and D. M. Brown, "Technology Developments for InSb Infrared Images," *IEEE Transactions on Electron Devices* ED–27, 170–75, 1980.

120. C. Y. Wei and H. H. Woodbury, "Ideal Mode Operation of an InSb Charge Injection Device," *IEEE Transactions on Electron Devices* ED–31, 1773–80, 1984.

121. S. C. H. Wang, C. Y. Wei, H. H. Woodbury, and M. D. Gibbons, "Characteristics and Readout of an InSb CID Two-Dimensional Scanning TDI Array," *IEEE Transactions on Electron Devices* ED-32, 1599–1607, 1985.

122. M. D. Gibbons, S. C. Wang, S. R. Jost, V. F. Meikleham, T. H. Myers, and A. F. Milton, "Developments in InSb Material and Charge Injection Devices," *Proceedings of SPIE* 865, 52–58, 1987.

123. M. D. Gibbons and S. C. Wang, "Status of CID InSb Detector Technology," *Proceedings of SPIE* 443, 151–66, 1984.

124. D. A. Scribner, M. R. Kruer, and J. M. Killiany, "Infrared Focal Plane Array Technology," *Proceedings of IEEE* 79, 66–85, 1991.

125. S. R. Jost, V. F. Meikleham, and T. H. Myers, "InSb: A Key Material for IR Detector Applications," *Materials Research Society Symposium Proceedings* 90, 429–35, 1987.

126. A. Bahraman, C. H. Chen, J. M. Geneczko, M. H. Shelstad, R. N. Ting, and J. G. Vodicka, "Current State of the Art in InSb Infrared Staring Imaging Devices," *Proceedings of SPIE* 750, 27–31, 1987.

127. M. D. Gibbons, J. M. Swab, W. E. Davern, M. L. Winn, T. C. Brusgard, and R. W. Aldrich, "Advances in InSb Charge Injection Device (CID) Focal Planes," *Proceedings of SPIE* 225, 55–63, 1980.

128. R. D. Thom, R. E. Eck, J. D. Philips, and J. B. Scorso, "InSb CCDs and Other MIS Devices for Infrared Applications," in *Proceedings of the 1975 International Conference on the Applications of CCDs*, 31–41, 1975.

129. M. B. Reine, E. E. Krueger, P. O'Dette, C. L. Terzis, B. Denley, J. Hartley, J. Rutter, and D. E. Kleinmann, "Advances in 15 μm HgCdTe Photovoltaic and Photoconductive Detector Technology for Remote Sensing," *Proceedings of SPIE* 2816, 120–37, 1996.

130. M. B. Reine, E. E. Krueger, P. O'Dette, and C. L. Terzic, "Photovoltaic HgCdTe Detectors for Advanced GOES Instruments," *Proceedings of SPIE* 2812, 501–17, 1996.

131. J. W. Beletic, R. Blank, D. Gulbransen, D. Lee, M. Loose, E. C. Piquette, T. Sprafke, W. E. Tennant, M. Zandian, and J. Zino, "Teledyne Imaging Sensors: Infrared Imaging Technologies for Astronomy & Civil Space," *Proceedings of SPIE* 7021, 70210H, 2008.

132. W. D. Lawson, S. Nielson, E. H. Putley, and A. S. Young, "Preparation and Properties of HgTe and Mixed Crystals of HgTe-CdTe," *Journal of Physics and Chemistry of Solids* 9, 325–29, 1959.

133. W. E. Tennant and J. M. Arias, "HgCdTe at Teledyne," *Proceedings of SPIE* 7298, 72982V, 2009.

134. P. R. Bratt, S. M. Johnson, D. R. Rhiger, T. Tung, M. H. Kalisher, W. A. Radford, G. A. Garwood, and C. A. Cockrum, "Historical Perspectives on HgCdTe Material and Device Development at Raytheon Vision Systems," *Proceedings of SPIE* 7298, 72982U, 2009.

135. R. A. Chapman, M. A. Kinch, A. Simmons, S. R. Borrello, H. B. Morris, J. S. Wrobel, and D. D. Buss, "$Hg_{0.7}Cd_{0.3}Te$ Charge-Coupled Device Shift Registers," *Applied Physics Letters* 32, 434–36, 1978.

136. R. A. Chapman, S. R. Borrello, A. Simmons, J. D. Beck, A. J. Lewis, M. A. Kinch, J. Hynecek, and C. G. Roberts, "Monolithic HgCdTe Charge Transfer Device Infrared Imaging Arrays," *IEEE Transactions on Electron Devices* ED-27, 134–46, 1980.

137. A. F. Milton, "Charge Transfer Devices for Infrared Imaging," in *Optical and Infrared Detectors*, ed. R. J. Keyes, 197–228, Springer-Verlag, Berlin, 1980.

138. M. A. Kinch, "Metal-Insulator-Semiconductor Infrared Detectors," in *Semiconductors and Semimetals*, Vol. 18, eds. R. K. Willardson and A. C. Beer, 313–78, Academic Press, New York, 1981.

139. R. A. Schiebel, "Enhancement Mode HgCdTe MISFETs and Circuits for Focal Plane Applications," *IEDM Technical Digest* 132, 1987.

140. T. L. Koch, J. H. De Loo, M. H. Kalisher, and J. D. Phillips, "Monolithic n-Channel HgCdTe Linear Imaging Arrays," *IEEE Transactions on Electron Devices* ED-32, 1592–1607, 1985.

141. M. V. Wadsworth, S. R. Borrello, J. Dodge, R. Gooh, W. McCardel, G. Nado, and M. D. Shilhanek, "Monolithic CCD Imagers in HgCdTe," *IEEE Transactions on Electron Devices* 42, 244–50, 1995.

142. M. A. Kinch, "MIS Devices in HgCdTe," in *Properties of Narrow Gap Cadmium-Based Compounds, EMIS Datareviews Series No. 10*, ed. P. Capper, 359–63, IEE, London, 1994.

143. M. Baker and R. A. Ballingall, "Photovoltaic CdHgTe: Silicon Hybrid Focal Planes," *Proceedings of SPIE* 510, 121–29, 1984.

144. I. M. Baker, "Infrared Radiation Imaging Devices and Methods for Their Manufacture," U.S. Patent 4,521,798, 1985.

145. I. M. Baker, G. J. Crimes, J. E. Parsons, and E. S. O'Keefe, "CdHgTe-CMOS Hybrid Focal Plane Arrays: A Flexible Solution for Advanced Infrared Systems," *Proceedings of SPIE* 2269, 636–47, 1994.

146. I. M. Baker, M. P. Hastings, L. G. Hipwood, C. L. Jones, and P. Knowles, "Infrared Detectors for the Year 2000," *III-Vs Review* 9(2), 50–60, 1996.

147. I. M. Baker, G. J. Crimes, R. A. Lockett, M. E. Marini, and S. Alfuso, "CMOS/HgCdTe, 2D Array Technology for Staring Systems," *Proceedings of SPIE* 2744, 463–72, 1996.

148. R. L. Smythe, "Monolithic HgCdTe Focal Plane Arrays," *Government Microcircuit Applications Conference Proceedings*, 289–92, Fort Monmouth, US Army ERADCOM, New Jersey, 1982.

149. A. Turner, T. Teherani, J. Ehmke, C. Pettitt, P. Conlon, J. Beck, K. McCormack, et al., "Producibility of VIP™ Scanning Focal Plane Arrays," *Proceedings of SPIE* 2228, 237–48, 1994.

150. R. Thorn, "High Density Infrared Detector Arrays," U.S. Patent No. 4,039,833, 1977.

151. K. Chow, J. P. Rode, D. H. Seib, and J. Blackwell, "Hybrid Infrared Focal-Plane Arrays," *IEEE Transactions on Electron Devices* ED-29, 3–13, 1982.

152. J. P. Rode, "HgCdTe Hybrid Focal Plane," *Infrared Physics* 24, 443–53, 1984.

153. W. E. Tennant, "Recent Development in HgCdTe Photovoltaic Device Grown on Alternative Substrates Using Heteroepitaxy," *Technical Digest IEDM*, 704–6, 1983.

154. M. Hewish, "Focal Plane Arrays Revolutionize Thermal Imaging," *International Defense Review* 4, 307–11, 1993.

155. C. D. Maxey, J. P. Camplin, I. T. Guilfoy, J. Gardner, R. A. Lockett, C. L. Jones, P. Capper, M. Houlton, and N. T. Gordon, "Metal-Organic Vapor-Phase Epitaxial Growth of HgCdTe Device Heterostructures on Three-Inch-Diameter Substrates," *Journal of Electronic Materials* 32, 656–60, 2003.

156. T. J. de Lyon, S. M. Johnson, C. A. Cockrum, O. K. Wu, W. J. Hamilton, and G. S. Kamath, "CdZnTe on Si(001) and Si(112): Direct MBE Growth for Large-Area HgCdTe Infrared Focal-Plane Array Applications," *Journal of the Electrochemical Society* 141, 2888–93, 1994.

157. J. M. Peterson, J. A. Franklin, M. Readdy, S. M. Johnson, E. Smith, W. A. Radford, and I. Kasai, "High-Quality Large-Area MBE HgCdTe/Si," *Journal of Electronic Materials* 36, 1283–86, 2006.

158. R. Bornfreund, J. P. Rosbeck, Y. N. Thai, E. P. Smith, D. D. Lofgreen, M. F. Vilela, A. A. Buell, et al., "High-Performance LWIR MBE-Grown HgCdTe/Si Focal Plane Arrays," *Journal of Electronic Materials* 36, 1085–91, 2007.

159. D. J. Hall, L. Buckle, N. T. Gordon, J. Giess, J. E. Hails, J. W. Cairns, R. M. Lawrence, et al., "High-Performance Long-Wavelength HgCdTe Infrared Detectors Grown on Silicon Substrates," *Applied Physics Letters* 85, 2113–15, 2004.

160. C. D. Maxey, J. C. Fitzmaurice, H. W. Lau, L. G. Hipwood, C. S. Shaw, C. L. Jones, and P. Capper. "Current Status of Large-Area MOVPE Growth of HgCdTe Device Structures for Infrared Focal Plane Arrays," *Journal of Electronic Materials* 35, 1275–82, 2006.

161. C. L. Jones, L. G. Hipwood, C. J. Shaw, J. P. Price, R. A. Catchpole, M. Ordish, C. D. Maxey, et al., "High Performance MW and LW IRFPAs Made from HgCdTe Grown by MOVPE," *Proceedings of SPIE* 6206, 620610, 2006.

162. M. J. Hewitt, J. L. Vampola, S. H. Black, and C. J. Nielsen, "Infrared Readout Electronics: A Historical Perspective," *Proceedings of SPIE* 2226, 108–19, 1994.

163. P. Felix, M. Moulin, B. Munier, J. Portmann, and J. P. Reboul, "CCD Readout of Infrared Hybrid Focal-Plane Arrays," *IEEE Transactions on Electron Devices* ED-27, 175–88, 1980.

164. P. Knowles, "Mercury Cadmium Telluride Detectors for Thermal Imaging," *GEC Journal of Research* 2, 141–56, 1984.

165. L. J. Kozlowski, K. Vural, J. Luo, A. Tomasini, T. Liu, and W. E. Kleinhans, "Low-Noise Infrared and Visible Focal Plane Arrays," *Opto-Electronics Review* 7, 259–69, 1999.

166. J. L. Vampola, "Readout Electronics for Infrared Sensors," in *The Infrared and Electro-Optical Systems Handbook*, Vol. 3, ed. W. D. Rogatto, 285–342, SPIE Press, Bellingham, WA, 1993.

167. E. R. Fossum and B. Pain, "Infrared Readout Electronics for Space Science Sensors: State of the Art and Future Directions," *Proceedings of SPIE* 2020, 262–85, 1993.

168. L. J. Kozlowski and W. F. Kosonocky, "Infrared Detector Arrays," in *Handbook of Optics,* Chapter 23, eds. M. Bass, E. W. Van Stryland, D. R. Williams, and W. L. Wolfe, McGraw-Hill, Inc., New York, 1995.

169. A. Rogalski and Z. Bielecki, "Detection of Optical Radiation," in *Handbook of Optoelectronics,* Vol. 1, , eds. J. P. Dakin and R. G. W. Brown, 73–117, Taylor & Francis, New York, 2006.

170. M. Z. Tidrow, W. A. Beck, W. W. Clark, H. K. Pollehn, J. W. Little, N. K. Dhar, R. P. Leavitt, et al., "Device Physics and Focal Plane Array Applications of QWIP and MCT," *Opto-Electronics Review* 7, 283–96, 1999.

171. M. A. Kinch, J. D. Beck, C.-F. Wan, F. Ma, and J. Campbell, "HgCdTe Electron Avalanche Photodiodes," *Journal of Electronic Materials* 33, 630–39, 2004.

172. I. Baker, S. Duncan, and J. Copley, "Low Noise Laser Gated Imaging System for Long Range Target Identification," *Proceedings of SPIE* 5406, 133–44, 2004.

173. I. Baker, P. Thorne, J. Henderson, J. Copley, D. Humphreys, and A. Millar, "Advanced Multifunctional Detectors for Laser-Gated Imaging Applications," *Proceedings of SPIE* 6206, 620608, 2006.

174. I. Baker, D. Owton, K. Trundle, P. Thorne, K. Storie, P. Oakley, and J. Copley, "Advanced Infrared Detectors for Multimode Active and Passive Imaging Applications," *Proceedings of SPIE* 6940, 69402L, 2008.

175. J. Beck, M. Woodall, R. Scritchfield, M. Ohlson, L. Wood, P. Mitra, and J. Robinson, "Gated IR Imaging with 128 × 128 HgCdTe Electron Avalanche Photodiode FPA," *Proceedings of SPIE* 6542, 654217, 2007.

176. J. Rothman, E. de Borniol, P. Ballet, L. Mollard, S. Gout, M. Fournier, J. P. Chamonal, et al., "HgCdTe APD: Focal Plane Array Performance at DEFIR," *Proceedings of SPIE* 7298, 729835, 2009.

177. G. Perrais, J. Rothman, G. Destefanis, J. Baylet, P. Castelein, J.-P. Chamonal, and P. Tribolet, "Demonstration of Multifunctional Bi-Colour-Avalanche Gain Detection in HgCdTe FPA," *Proceedings of SPIE* 6395, 63950H, 2006.

178. R. J. Cashman, "Film-Type Infrared Photoconductors," *Proceedings of IRE* 47, 1471–75, 1959.

179. P. W. Kruse, L. D. McGlauchlin, and R. B. McQuistan, *Elements of Infrared Technology: Generation, Transmission, and Detection,* Wiley, New York, 1962.

180. A. Smith, F. E. Jones, and R. P. Chasmar, *The Detection and Measurement of Infrared Radiation,* Clarendon, Oxford, 1968.

181. A. Rogalski, "IV-VI detectors," in *Infrared Photon Detectors,* ed. E. Rogalski, pp. 513–559, SPIE Optical Engineering Press, Bellingham, WA, 1995.

182. T. Beystrum, R. Himoto, N. Jacksen, and M. Sutton, "Low Cost Pb Salt FPAs," *Proceedings of SPIE* 5406, 287–94, 2004.

183. G. Vergara, M. T. Montojo, M. C. Torquemada, M. T. Rodrigo, F. J. Sanchez, L. J. Gomez, R. M. Almazan, et al., "Polycrystalline Lead Selenide: The Resurgence of an Old Infrared Detector," *Opto-Electronics Review* 15, 110–17, 2007.

184. J. F. Kreider, M. K. Preis, P. C. T. Roberts, L. D. Owen, and W. M. Scott, "Multiplexed Mid-Wavelength IR Long, Linear Photoconductive Focal Plane Arrays," *Proceedings of SPIE* 1488, 376–88, 1991.

185. *Northrop/Grumman Electro-Optical Systems* data sheet, 2002.

186. P. R. Norton, "Infrared Image Sensors," *Optical Engineering* 30, 1649–63, 1991.

187. M. D. Jhabvala and J. R. Barrett, "A Monolithic Lead Sulfide-Silicon MOS Integrated-Circuit Structure," *IEEE Transactions on Electron Devices* ED-29, 1900–1905, 1982.

188. J. R. Barrett, M. D. Jhabvala, and F. S. Maldari, "Monolithic Lead Salt-Silicon Focal Plane Development," *Proceedings of SPIE* 409, 76–88, 1988.

189. H. Zogg, S. Blunier, T. Hoshino, C. Maissen, J. Masek, and A. N. Tiwari, "Infrared Sensor Arrays with 3–12 μm Cutoff Wavelengths in Heteroepitaxial Narrow-Gap Semiconductors on Silicon Substrates," *IEEE Transactions on Electron Devices* 38, 1110–17, 1991.

190. H. Zogg, A. Fach, C. Maissen, J. Masek, and S. Blunier, "Photovoltaic Lead-Chalcogenide on Silicon Infrared Sensor Arrays," *Optical Engineering* 33, 1440–49, 1994.

191. H. Zogg, A. Fach, J. John, P. Müller, C. Paglino, and A. N. Tiwari, "PbSnSe-on-Si: Material and IR-Device Properties," *Proceedings of SPIE* 3182, 26–29, 1998.

192. H. Zogg, "Photovoltaic IV-VI on Silicon Infrared Devices for Thermal Imaging Applications," *Proceedings of SPIE* 3629, 52–62, 1999.

193. K. Alchalabi, D. Zimin, H. Zogg, and W. Buttler, "Monolithic Heteroepiraxial PbTe-on-Si Infrared Focal Plane Array with 96 × 128 Pixels," *IEEE Electron Device Letters* 22, 110–12, 2001.

194. H. Zogg, K. Alchalabi, D. Zimin, and K. Kellermann, "Two-Dimensional Monolithic Lead Chalcogenide Infrared Sensor Arrays on Silicon Read-Out Chips and Noise Mechanisms," *IEEE Transactions on Electron Devices* 50, 209–14, 2003.

195. A. J. Steckl, H. Elabd, K. Y. Tam, S. P. Sheu, and M. E. Motamedi, "The Optical and Detector Properties of the PbS-Si Heterojunction," *IEEE Transactions on Electron Devices* ED-27, 126–33, 1980.

196. D. R. Lamb and N. A. Foss, "The Applications of Charge-Coupled Devices to Infra-Red Image Sensing Systems," *Radio Electronic Engineers* 50, 226–36, 1980.

197. S. D. Gunapala and K. M. S. V. Bandara, "Recent Development in Quantum-Well Infrared Photodetectors," *Thin Films*, Vol. 21, eds. M. M. Francombe and J. L. Vossen, 113–237, Academic Press, New York, 1995.

198. C. C. Bethea, B. F. Levine, V. O. Shen, R. R. Abbott, and S. J. Hseih, "10-μm GaAs/AlGaAs Multiquantum Well Scanned Array Infrared Imaging Camera," *IEEE Transactions on Electron Devices* 38, 1118–23, 1991.

199. B. F. Levine, "Quantum-Well Infrared Photodetectors," *Journal of Applied Physics* 74, R1–R81, 1993.

200. S. D. Gunapala and S. V. Bandara, "Quantum Well Infrared Photodetectors (QWIP)," in *Handbook of Thin Devices*, Vol. 2, ed. M. H. Francombe, 63–99, Academic Press, San Diego, 2000.

201. S. D. Gunapala and S. V. Bandara, "GaAs/AlGaAs Based Quantum Well Infrared Photodetector Focal Plane Arrays," in *Handbook of Infrared Detection Technologies*, eds. M. Henini and M. Razeghi, 83–119, Elsevier, Oxford, 2002.

202. E. Costard, Ph. Bois, F. Audier, and E. Herniou, "Latest Improvements in QWIP Technology at Thomson-CSF/LCR," *Proceedings of SPIE* 3436, 228–39, 1998.

203. J. A. Robo, E. Costard, J. P. Truffer, A. Nedelcu, X. Marcadet, and P. Bois, "QWIP Focal Plane Arrays Performances from MWIR to VLWIR," *Proceedings of SPIE* 7298, 72980F, 2009.

204. H. Schneider, P. Koidl, M. Walther, J. Fleissner, R. Rehm, E. Diwo, K. Schwarz, and G. Weimann, "Ten Years of QWIP Development at Fraunhofer," *Infrared Physics & Technology* 42, 283–89, 2001.

205. H. Schneider and H. C. Liu, *Quantum Well Infrared Photodetectors*, Springer, Berlin, 2007.

206. H. Martijn, S. Smuk, C. Asplund, H. Malm, A. Gromov, J. Alverbro, and H. Bleichner, "Recent Advances of QWIP Development in Sweden," *Proceedings of SPIE* 6542, 6542OV, 2007.

207. K. K. Choi, *The Physics of Quantum Well Infrared Photodetectors*, Word Scientific, Singapore, 1997.

208. K-K. Choi, C. Monroy, V. Swaminathan, T. Tamir, M. Leung, J. Devitt, D. Forrai, and D. Endres, "Optimization of Corrugated-QWIP for Large Format, High Quantum Efficiency, and Multi-Color FPAs," *Infrared Physics & Technology* 50, 124–35, 2007.

209. W. A. Beck and T. S. Faska, "Current Status of Quantum Well Focal Plane Arrays," *Proceedings of SPIE* 2744, 193–206, 1996.

210. T. Whitaker, "Sanders' QWIPs Detect Two Color at Once," *Compound Semiconductors* 5(7), 48–51, 1999.

211. M. Sundaram and S. C. Wang, "2-Color QWIP FPAs," *Proceedings of SPIE* 4028, 311–17, 2000.

212. J. Jiang, S. Tsao, K. Mi, M. Razeghi, G. J. Brown, C. Jelen, and M. Z. Tidrow, "Advanced Monolithic Quantum Well Infrared Photodetector Focal Plane Array Integrated with Silicon Readout Integrated Circuit," *Infared Physics & Technology* 46, 199–207, 2005.

213. S. Ozer, U. Tumkaya, and C. Besikci, "Large Format AlInAs-InGaAs Quantum-Well Infrared Photodetector Focal Plane Array for Midwavelength Infrared Thermal Imaging," *IEEE Photonics Technology Letters* 19, 1371–73, 2007.

214. M. Kaldirim, Y. Arslan, S. U. Eker, and C. Besikci, "Lattice-Matched AlInAs-InGaAs Mid-Wavelength Infrared QWIPs: Characteristics and Focal Plane Array Performance," *Semiconductor Science and Technology* 23, 085007, 2008.

215. H. Schneider, J. Fleissner, R. Rehm, M. Walther, W. Pletschen, P. Koidl, G. Weimann, J. Ziegler, R. Breiter, and W. Cabanski, "High-Resolution QWIP FPAs for the 8–12 μm and 3–5 μm Regimes," *Proceedings of SPIE* 4820, 297–305, 2003.

216. E. Costard and Ph. Bois, "THALES Long Wave QWIP Thermal Imagers," *Infared Physics & Technology* 50, 260–69, 2007.

217. M. Jhabvala, K. Choi, A. Goldberg, A. La, and S. Gunapala, "Development of a 1k × 1k GaAs QWIP Far IR Imaging Array," *Proceedings of SPIE* 5167, 175–85, 2004.

218. S. D. Gunapala, S. V. Bandara, J. K. Liu, C. J. Hill, B. Rafol, J. M. Mumolo, J. T. Trinh, M. Z. Tidrow, and P. D. LeVan, "1024 × 1024 Pixel Mid-Wavelength and Long-Wavelength Infrared QWIP Focal Plane Arrays for Imaging Applications," *Semiconductor Science and Technology* 20, 473–80, 2005.

219. M. Jhabvala, K. K. Choi, C. Monroy, and A. La, "Development of a 1K × 1K, 8–12 μm QWIP Array," *Infared Physics & Technology* 50, 234–39, 2007.

220. S. D. Gunapala, S. V. Bandara, J. K. Liu, J. M. Mumolo, C. J. Hill, S. B. Rafol, D. Salazar, J. Woollaway, P. D. LeVan, and M. Z. Tidrow, "Towards Dualband Megapixel QWIP Focal Plane Arrays," *Infrared Physics & Technology* 50, 217–26, 2007.

221. P. Bois, E. Costard, X. Marcadet, and E. Herniou, "Development of Quantum Well Infrared Photodetectors in France," *Infrared Physics & Technology* 42, 291–300, 2001.

222. M. Z. Tidrow, L. Zheng, and H. Barcikowski, "Recent Success on SLS FPAs and MDA's New Direction for Development," *Proceedings of SPIE* 7298, 7298-61, 2009.

223. E. K. Huang, D. Hoffman, B.-M. Nguyen, P.-Y. Delaunay, and M. Razeghi, "Surface Leakage Reduction in Narrow Band Gap Type-II Antimonide-Based Superlattice Photodiodes," *Applied Physics Letters* 94, 053506, 2009.

224. W. Cabanski, K. Eberhardt, W. Rode, J. Wendler, J. Ziegler, J. Fleißner, F. Fuchs, et al., "3rd Gen Focal Plane Array IR Detection Modules and Applications," *Proceedings of SPIE* 5406, 184–92, 2005.

225. M. Münzberg, R. Breiter, W. Cabanski, H. Lutz, J. Wendler, J. Ziegler, R. Rehm, and M. Walther, "Multi Spectral IR Detection Modules and Applications," *Proceedings of SPIE* 6206, 620627, 2006.

226. F. Rutz, R. Rehm, J. Schmitz, J. Fleissner, M. Walther, R. Scheibner, and J. Ziegler, "InAs/GaSb Superlattice Focal Plane Array Infrared Detectors: Manufacturing Aspects," *Proceedings of SPIE* 7298, 72981R, 2009.

227. C. J. Hill, A. Soibel, S. A. Keo, J. M. Mumolo, D. Z. Ting, S. D. Gunapala, D. R. Rhiger, R. E. Kvaas, and S. F. Harris, "Demonstration of Mid and Long-Wavelength Infrared Antimonide-Based Focal Plane Arrays," *Proceedings of SPIE* 7298, 729404, 2009.

228. P.-Y. Delaunay, B. M. Nguyen, D. Hoffman, and M. Razeghi, "High-Performance Focal Plane Array Based on InAs-GaSb Superlattices with a 10-μm Cutoff Wavelength," *IEEE Journal of Quantum Electronics* 44, 462–67, 2008.

229. M. Razeghi, D. Hoffman, B. M. Nguyen, P.-Y. Delaunay, E. K. Huang, M. Z. Tidrow, and V. Nathan, "Recent Advances in LWIR Type-II InAs/GaSb Superlattice Photodetectors and Focal Plane Arrays at the Center for Quantum Devices," *Proceedings of the IEEE* 97, 1056–66, 2009.

22 Terahertz Detectors and Focal Plane Arrays

The terahertz (THZ) region of electromagnetic spectrum is often described as the final unexplored area of spectrum. First, humans relied on the radiation from the sun. Cave men used torches (approximately 500,000 years ago). Candles appeared around 1000 BC, followed by gas lighting (1772), and incandescent bulbs (Edison, 1897). Radio (1886–1895), X-rays (1895), UV radiation (1901), and radar (1936) were invented at the end of the nineteenth and the beginning of the twentieth centuries. However, THz range of electromagnetic spectrum still presents a challenge for both electronic and photonic technologies.

THz radiation (see Figure 22.1) is frequently treated as the spectral region within frequency range $\nu \approx 1$–10 THz ($\lambda \approx 300$–30 μm) [1–3] and it is partly overlapping with loosely treated sub-millimeter (sub-mm) wavelength band $\nu \approx 0.1$–3 THz ($\lambda \approx 3$ mm – 100 μm) [4]. Even wider region $\nu \approx 0.1$–10 THz [5,6] is treated as THz band overlapping thus with sub-mm wavelength band. As a result, both of these notions are used as equal ones (see, e.g., [7]). Here the THz range is accepted as the range within $\nu \approx 0.1$–10 THz.

The THz region of the electromagnetic spectrum has proven to be one of the most elusive. Being situated between infrared light and microwave radiation, THz radiation is resistant to the techniques commonly employed in these well-established neighboring bands. Historically, the major use of THz spectroscopy has been by chemists and astronomers in the spectral characterization of the rotational and vibrational resonances and thermal-emission lines of simple molecules. Terahertz receivers are also used to study the trace gases in the upper atmosphere, such as ozone and the many gases involved in ozone depletion cycles, such as chloride monoxide. Air efficiently absorbs in wide spectral THz region (except for narrow windows around $\nu \approx 35$ GHz, 96 GHz, 140 GHz, and 220 GHz, and others, see Figure 22.2 [8]). THz and millimeter waves are efficient at detecting the presence of water and thus are efficient to discriminate different objects on human bodies (water content of human body is about 60%) as the clothes are transparent. In the longer wavelength region (cm wavelength region) even persons hiding behind a wall (not very thick), can be visualized. It should be mentioned that about half of the luminosity of the Universe and 98% of all the photons emitted since the Big Bang belong to THz radiation [9]. This relict radiation carries information about the cosmic space, galaxies, stars, and planets formation [10].

The past 20 years have seen a revolution in THz systems, as advanced materials research provided new and higher power sources, and the potential of THz for advanced physics research and commercial applications was demonstrated. Researches evolved with THz technologies are now receiving increasing attention, and devices exploiting this wavelength band are set to become increasingly important in diverse range of human activity applications (e.g., security, biological, drugs and explosion detection, gases fingerprints, imaging, etc.). The interest in the THz range is attracted by the fact that this range is the place where different physical phenomena are revealed that frequently calls for multidisciplinary special knowledge in this research area. Overviews on the various applications of THz technologies can be found (e.g., [1–3,11–18]).

The critical differences between detection at sub-mm wavelengths and IR detection lie in small photon energies (at $\lambda \approx 300$ μm $h\nu \approx 4$ meV, compared to the thermal energy of 26 meV at room temperature. Also the Airy disk diameter (diffraction limit) defined by

$$A_{dif} \approx 2.44 \frac{\lambda f}{D} = 2.44\lambda \left(f/\# \right) \tag{22.1}$$

is large and determines low spatial resolution of THz systems. Here f is the focus length of optical system, D is its input diameter, and $f/\#$ is the f-number of the optical system.

The serious problem that limits an advent of heterodyne sensor arrays in sub-mm (THz) spectral region [e.g., for high-resolution spectroscopy applications ($\nu/\Delta\nu \approx 10^6$) or photometry ($\nu/\Delta\nu \approx 3$–10) and imaging], lies in technology limitations of solid-state local oscillator (LO) power. With the exception of free electron lasers that use relativistic electrons and are capable of reaching kilowatt level THz power [19], other THz sources generate milliwatt or microwatt power levels (see Figure 22.3 [5]). Traditional electronic devices, such as transistors, do not work well much above about 150 GHz. Therefore there are no amplifiers available throughout most of the THz band. Similarly, semiconductor lasers that have long been available in optical and IR bands, are not available in most of the THz band. Although much progress is being made in the area of high frequency transistors [20] and semiconductor lasers [21], it seems clear that the so-called THz gap will remain an important challenge to scientists and engineers for the foreseeable future.

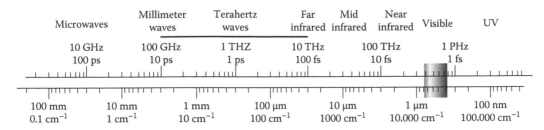

Figure 22.1 The electromagnetic spectrum.

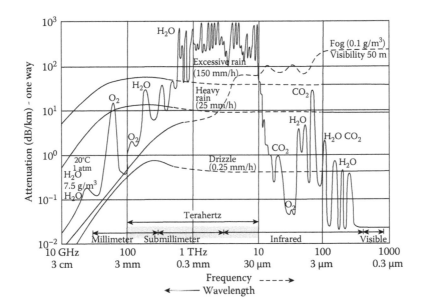

Figure 22.2 Attenuation of the earth's atmosphere from visible to radiofrequency band region. (From Lettington, A. H., Blankson, I. M., Attia, M., and Dunn, D., "Review of Imaging Architecture," *Proceedings of SPIE* 4719, 327–40, 2002. With permission.)

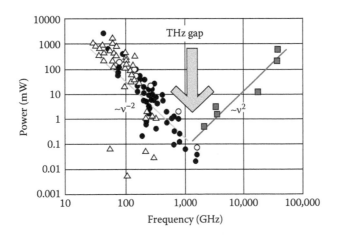

Figure 22.3 THz gap with respect to source technology: (■) quantum cascade lasers (QCL) are progressing downward from high frequencies, the lowest $\nu = 1.2$ THz, $T = 110$ K – CW, $T = 163$ K – pulsed; (●) frequency multipliers dominate other electronic devices (△) above about 150 GHz; cryogenic sources are shown as hollow symbols (○). (From Crowe, T. W., Bishop, W. L., Porterfield, D. W., Hesler, J. L., and Weikle, R. M., *IEEE Journal of Solid-State Circuits*, 40, 2104–10, 2005. With permission.)

22.1 DIRECT AND HETERODYNE TERAHERTZ DETECTION: GENERAL CONSIDERATIONS

The THz detection systems can be divided into two groups: direct (noncoherent) and heterodyne (coherent) systems (see Chapter 4). The first systems measure signal amplitude and, as a rule, are characterized by broadband spectral response. The second systems amplify the incoming photon stream, preserving not only the amplitude of the signal, but also its phase giving additional information about object. Consequently, coherent detectors are subject to a quantum mechanical noise limit. The signals are often downconverted in frequency prior to detection (where extremely low noise electronics can be used), and they have proven particularly useful in applications that require very high sensitivity and spectral resolution (typically $v/\Delta v \approx 10^5$–10^6). The main technical challenges are the need to improve sensitivities at higher frequencies, the need for improved local oscillators, and the desire for large FPAs of receivers [11,22].

The direct detection systems, as a rule, operating in wider spectral range (e.g., when photon background is low), can provide sufficient resolution. They are preferable for moderate spectral resolution $v/\Delta v \approx 10^3$–10^4 or lower [4] and they are also preferable for imaging due to their relative simplicity. Direct detectors can be used in those applications where sensitivity is more important than the spectral resolution.

In comparison to direct detection, heterodyne detection exhibits several additional advantages such as [23]: detection of frequency and phase modulations, sub-BLIP operation (what provides discrimination against background flux, microphonics, etc.), and gain of output signal due to downconversion.

Among the disadvantages of heterodyne detection we can distinguish fulfillment of the following requirements: both signal and LO beams should be coincident and equal in diameter, wavefronts of both beams should have the same radius of curvature and should have the same transverse spatial mode structure, both beams should be polarized in the same direction. In addition, it is very difficult to produce large format arrays for heterodyne systems.

General considerations related to heterodyne detection are given in Chapter 4. Here, we mention additional relations devoted to fundamental performance limits of direct and heterodyne detections.

The merit for THz detectors is noise equivalent power (NEP), which is defined as the value of rms input radiant signal power, P_s, required to produce rms output signal equal to rms noise value (SNR = 1). For BLIP detection it follows [23]

$$\mathrm{NEP}_D^{BLIP} = \left(\frac{2hv}{\eta} P_s \right)^{1/2}, \quad W/Hz^{1/2}. \tag{22.2}$$

The lower NEP means the more sensitive detector.

For heterodyne detection in BLIP regime, it can be shown [23]

$$\mathrm{NEP}_H^{BLIP} = \frac{hv}{\eta}, \quad W/Hz. \tag{22.3}$$

Note that for heterodyne detection, the unit of NEP is W/Hz instead of W/Hz$^{1/2}$, as for direct detection. But frequently NEP still is cited in units of W/Hz$^{1/2}$.

Frequently, the sensitivity of heterodyne detectors is given in terms of mixer noise temperature, T_{mix}, which correlates with the mixer NEP defined as

$$\mathrm{NEP}_{mix} = kT_{mix}. \tag{22.4}$$

For wavelength band $\lambda \approx 3$ mm ($v \approx 100$ GHz), where it is atmospheric transparency window, the value $T_s^{min} = 4.8$ K is the fundamental limit to the noise temperature imposed by the uncertainty principle on any simultaneous measurement of the amplitude and phase of the electromagnetic wave. Frequently, with this fundamental limit are compared the noise temperature limits of heterodyne detectors in mm and sub-mm wavelength bands. Since heterodyne detectors measure both amplitude and phase simultaneously, they are governed by the uncertainty principle, and hence they are quantum-noise limited to an absolute noise-floor of 4.8 K/Hz. For example,

Figure 4.7 compares the noise temperature of Schottky diode mixers, SIS mixers, and HEB mixers operated in THz spectral band.

Beyond 40 μm spectral range, both photon and thermal detectors are used. Different types of cooled semiconductor detectors (hot electron InSb, Si, Ge bolometers, extrinsic Si, and Ge) [24–31] with response time $\tau \approx 10^{-6}$–10^{-8} s and NEP $\approx 10^{-13}$–5×10^{-17} W/Hz$^{1/2}$ for operation temperature $T \leq 4$ K are used. Stressed Ge:Ga extrinsic photoconductors can be sensitive up to wavelength $\lambda \approx 400$ μm [32] and that can be assembled into arrays [33]. A comprehensive review of different kind of detectors has been recently given by Rieke [34].

Among thermal THz detectors we can quote Golay cells [35,36], pyroelectric detectors [37], different types of microbolometers [38–41], which use antennas to couple power to a small thermally absorbing region. The advantage of uncooled thermal detectors, in spite of their relatively low sensitivity, lies in room temperature operability in a wide frequency band. Their NEP is within 10^{-9}–10^{-11} W/Hz$^{1/2}$ [42–47] (see Table 22.1).

The most sensitive direct detectors in sub-mm and mm wavebands are microbolometers cooled to 100–300 mK reaching NEP up to $(0.3-3) \times 10^{-19}$ W/Hz$^{1/2}$ limited by cosmic background radiation fluctuations. The sub-Kelvin superconducting structures are extremely sensitive (NEP $\approx 10^{-20}$ W/Hz$^{1/2}$) at $T \approx 100$–200 mK [7,48]. Because of high sensitivity these detectors are well below the background photon limit.

The detectors used in direct and heterodyne systems are the same but some of them, for example, low-temperature semiconductor hot-electron bolometers (HEBs) are inexpedient to be used in coherent systems because of a relatively "long" response time ($\tau \approx 10^{-7}$ s). The same is valid for most uncooled thermal detectors.

Progress in THz detector sensitivity has been impressive, for example, what is shown in Figure 22.4 in the case of bolometers in a period more than a half century [49]. The NEP value has decreased by a factor of 10^{11} in 50 years, corresponding improvements by a factor of two every two years.

Depending on the specific application, either a passive or active imaging system may be preferred. Most imaging systems used passive direct detection [40,50,51]. Since these systems measure only the energy emitted by the object, usually extremely sensitive detectors are required. In active systems, for operation of what the scene is illuminating, heterodyne detection can also be used in order to increase sensitivity to low radiant levels or to image through the scattering media. In applications where the subjects may be illuminated with modest amounts of THz energy, a source based on diode multipliers can be used to greatly increase the signal strength.

For heterodyne operation, the mixed field must be passed through a nonlinear circuit element or mixer that converts power from the original frequencies to the intermediate (beat) frequency. In the THz region this element is a diode or other nonlinear electrical circuit component. Simple

Table 22.1: Parameters of Some Uncooled THz Detectors

Detector Type	Modulation Frequency (Hz)	Operation Frequency (THz)	Noise Equivalent Power (W/Hz$^{1/2}$)	Refs.
Golay cell	≤ 20	≤ 30	10^{-9}–10^{-10}	—
Piezoelectric	$\leq 10^2$ (decreases with f increasing and depends on dimensions)	≤ 30	$\approx (1-3) \times 10^{-9}$ (decreases with f increasing)	—
Microbolometers	$\leq 10^2$	≤ 30	$\approx 10^{-10}$ (decreases with f increasing)	—
Bi microbolometer	$\leq 10^6$	≤ 3	1.6×10^{-10} (decreases with f increasing)	43
Nb microbolometer	—	≤ 30	5×10^{-11}	44
Schottky diodes	$\leq 10^{10}$	≤ 10	$\leq 10^{-10}$ (decreases several orders with f increasing within 0.1–10 THz)	—
GaAs HEMT	$\leq 2 \times 10^{10}$	≤ 30	$\approx 10^{-10}$ (depends on gate length and gate voltage)	45
Si MOSFET	3×10^4	0.645	3×10^{-10}	46
SHEB	$< 10^8$	≈ 0.03–2	$\approx 4 \times 10^{-10}$ (depends on f)	47

Source: F. Sizov, *Opto-Electronics Review*, 18(1), 10–36, 2010. With permission.

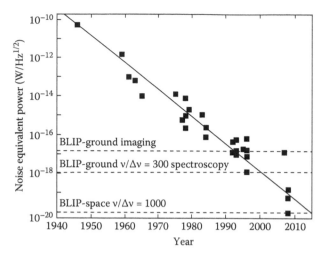

Figure 22.4 Improvement of the bolometer NEP-value for more than half a century. (Adapted from Benford, D. J., "Transition Edge Sensor Bolometers for CMB Polarimetry," http://cmbpol. uchicago.edu/workshops/technology2008/depot/cmbpol_technologies_benford_jcps_4.pdf)

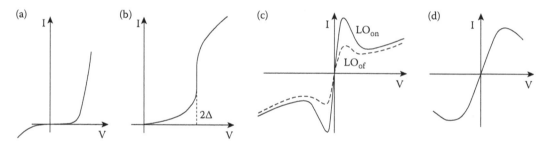

Figure 22.5 Schematic of I-V characteristics of nonlinear elements on which THz heterodyne receivers are based: (a) Schottky diode, (b) SIS, (c) HEB, and (d) SL.

considerations carried out (e.g., in Rieke's monograph [52]) prefer to use "square law" devices as fundamental mixers. In the case of a quadratic I-V curve, $I \propto V^2 \propto P$, an out mixer signal is proportional to power, which is usually what we want to measure.

Figure 22.5 shows the current-voltage characteristics of devices having a strong electric field quadratic nonlinearity. Examples are forward biased Schottky diodes, superconductor-insulator-superconductor (SIS) tunnel junctions, semiconductor and superconducting HEBs, SLs, properties of which are discussed below. Schematic current-voltage characteristics of such devices are shown in Figure 22.5. Simultaneously, with reasonable conversion efficiency and low noise, these nonlinear devices should possess high conversion operation speed for assurance of wide bandpass for consequent signals amplification at much lower frequencies (typically between 1 and 30 GHz). If the time constant of the device is short, the output will simply follow beat frequency; if the time is long, the device will not respond because on average the output signal contains constant value.

22.2 SCHOTTKY-BARRIER STRUCTURES

In spite of achievements of other kind of detectors for THz waveband (mainly SIS and HEBs), the Schottky-barrier diodes (SBDs) are among the basic elements in THz technologies. They are used either in direct detection and as nonlinear elements in heterodyne receiver mixers operating in a temperature range of 4–300 K [1,5,22,42,53]. The cryogenically cooled SBDs were used in mixers preferably in 1980s and early 1990s and then they were replaced by SIS or HEB mixers [11], in which mixing processes are similar to that observed in SBDs, but in SIS structures, for example, the rectification process is based on quantum-mechanical photon-assisted tunneling of quasiparticles (electrons).

Analysis of the state of the art in the development of SBDs and mixers for THz receivers is carried out (e.g., [5,53,54]). In the presence of a THz electric field, one can consider three groups of electrons marked as *a*, *b*, and *c* in Figure 9.32. Only the current generated by the electrons of group *b* are affected by a THz electric field [55]. The electrons of this group suffer from transit time effects. Within half a period of the THz field, they have to transverse the depletion layer in order to cross the barrier. Skin effect, charge inertia, dielectric relaxation, and plasma resonance all lead to degradation of the detector performance. At THz frequencies, the noise due to series resistance and hot electrons dominates over shot noise. However, cooling of the diode lowers this noise.

Historically Schottky-barrier structures were pointed contacts of tapered metal wires (e.g., a tungsten needle) with a semiconductor surface (the so-called crystal detectors). Widely used were, for example, contacts p-Si/W. At operation temperature $T = 300$ K they have NEP $\approx 4 \times 10^{-10}$ W/Hz$^{1/2}$. Also pointed tungsten or beryllium bronze contacts to n-Ge, n-GaAs, n-InSb were used (see, e.g., [56,57]). In the mid-1960s, Young and Irvin [58] developed the first lithographically defined GaAs Schottky diodes for high frequency applications. Their basic diode structure was next replicated by a variety of groups. The basic whiskered diode structure, depicted in Figure 22.6, is similar to the so-called honeycomb chip design and greatly improved the quality of the diode due to inherently low capacity of the wicker contact. This design has been the most important step toward a practical Schottky diode mixer for THz frequency applications [60–64], with several thousand diodes on a single chip and where parasitic losses such as the series resistance and the shunt capacitance are minimized. On highly doped GaAs substrate ($\approx 5 \times 10^{18}$ cm^{-3}) with an ohmic contact on the back side, a thin GaAs epitaxial layer, with thickness of 300 nm to 1 µm, is grown on the top of the substrate. Holes filled with a metal (Pt) in the SiO$_2$ insulating layer on top of the epitaxial layer define the anode area (0.25–1 µm) [62]. In order to couple the signal and the LO radiation to the mixer, a long wire antenna in a 90° corner-cube reflector is used [63,64]. The required LO power ranges from 1 to 10 mW.

The cross-section of whisker SBD with the equivalent circuit of the junction is shown in Figure 22.6b [60]. In heterodyne operation, mixing occurs in the nonlinear junction resistance, R_j. The diode series resistance, R_s, and the voltage-dependent junction capacitance, C_j, are parasitic elements that degrade the performance.

Due to limitation of whisker technology, such as constraints on design and repeatability, starting in the 1980s, the efforts were made to produce planar Schottky diodes. Currently the whisker diodes are almost completely replaced by planar diodes. In the case of the discrete diode chip, the diodes are flip-chip mounted into the circuit with either solder or conductive epoxy. Using advanced technology researched recently, the diodes are integrated with many passive circuit elements (impedance matching, filters, and waveguide probes) onto the same substrate [59]. By improving the mechanical arrangement and reducing loss, the planar technology is pushed well beyond 300 GHz and up to several THz.

(a)

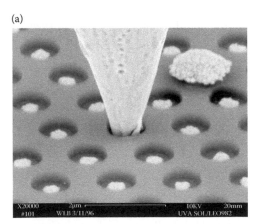

(b)

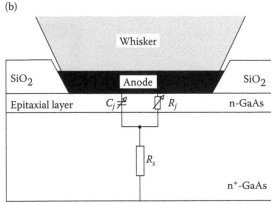

Figure 22.6 GaAs Schottky-barrier whisker contacted diode: (a) SEM image of a contacted chip used up to 5 THz, and (b) cross-section with equivalent circuit of the junction. (After Crowe, T. W., Porterfield, D. P., Hesler, J. L., Bishop, W. L., Kurtz, D. S., and Hui, K., "Terahertz Sources and Detectors," *Proceedings of SPIE* 5790, 271–80, 2005. With permission.)

Schottky diode mixers in a waveguide configuration have an improved coupling efficiency compared to the open structure mount. To overcome a large shunt capacitance caused by the coplanar contact pads, a surface channel diode, shown in Figure 22.7a, was proposed [65]. The planar diode soldered onto a microstrip circuit (e.g., a thin quartz substrate) is mounted into a waveguide mixer block. The required LO power is about 1 mW. For mixers operating at ≈600 GHz, double-sideband (DSB) noise temperature is about 1000 K. To eliminate losses caused by support (influence of surface modes), fabrication of Schottky diodes on a thin membrane was developed [66]. In this approach the diodes are integrated with the matching circuit and most of the GaAs substrate is removed from the chip and the entire circuit is fabricated on the remaining GaAs membrane. Figure 22.7b shows an example of this type of SBD with 3 μm thick membrane in a $600 \times 1400 \ \mu m^2$ GaAs frame.

The current-voltage characteristic of Schottky-barrier junction (see Equation 9.148) for bias voltage values $V > 3 \ kT/q$, can be approximated by

$$J_{MSt} = J_{st} \exp\left(\frac{V}{V_o}\right), \tag{22.5}$$

where device slop parameter $V_o = \beta kT/q$. Differentiating Equation 22.5 we can obtain the junction resistance

$$R_j = \frac{V_o}{I_{MSt}}. \tag{22.6}$$

Parasitic parameters, R_s and C_j, define diode critical frequency called cutoff frequency

$$v_c = \left(2\pi R_s C_j\right)^{-1}, \tag{22.7}$$

which should be notably higher compared to the operation frequency.

The junction space-charge capacitance can be approximated by [67]

$$C_j(V) = \frac{\varepsilon A}{w} + 3\frac{\varepsilon A}{d}. \tag{22.8}$$

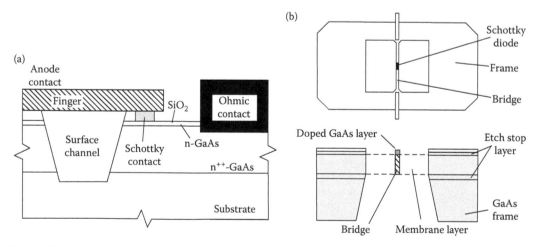

Figure 22.7 Designs of planar Schottky diodes: (a) a surface channel design for frequencies below 1 THz (After Marazita, S. M., Bishop, W. L., Hesler, J. L., Hui, K., Bowen, W. E., and Crowe, T. W., "Integrated GaAs Schottky Mixers by Spin-on-Dielectric Wafer Bonding," *IEEE Transactions on Electron Devices* 47, 1152–56, 2000. With permission.), and (b) a GaAs Schottky diode on a membrane. (From Siegel, P., Smith, R. P., Gaidis, M. C., and Martin, S., *IEEE Transactions on Microwave Theory Technology*, 47, 596–604, 1999. With permission.)

Here ε is semiconductor permittivity, A and d are the anode area and diameter, respectively, and w is the depletion region thickness dependent on the carrier density, diffusion potential, and bias. The second term is the peripheral capacitance. The junction capacitance is voltage dependent as the depletion region depends on the bias applied. The specific capacity as a rule is not less than $C \approx 10^{-7}$ F/cm^2.

To achieve good performance at high frequencies the diode area should be small. When reducing junction area one reduces junction capacities to increase operating frequency. But at the same time one increases the series resistance. The state-of-the-art devices have anode diameters about 0.25 µm and capacitances 0.25 fF. For high-frequency operation, the GaAs layers are doped up to $n \approx (5{-}10) \times 10^{17}$ cm^{-3} [5,60,68].

In the lower frequency range ($\nu < \approx 0.1$ THz) the operation of Schottky-barrier diodes is well understood and could be described by mixer theories taking into account Schottky diode stray parameters (variable-capacitance and series resistance). In the sub-mm (THz) range, however, the design and performance of the devices becomes increasingly complex. At higher frequencies there appears several parasitic mechanisms credited with not only with skin effect, but also with high-frequency processes in semiconductor material such as carrier scattering, carrier transit time through the barrier, dielectric relaxation, and so on, which become important.

Figure 22.8 presents frequency dependent voltage sensitivity characteristic of SBDs at room temperature [53]. The solid line shows the theoretical dependence with allowance for the skin effect, the carrier inertia, the plasma resonance in the epitaxial layer (f_{pe}) and in the substrate (f_{ps}), the phonon absorption (f_t and f_l are the frequencies of the transverse and longitudinal polar optical phonons), and the transit effects. The dashed line shows the same without allowance for the transit effect. The experimental results are related to SBDs with different anode shapes. A satisfactory agreement between the experiment and the calculation takes place on the whole range of frequency. Owing to the further improvement of the antenna, the detector sensitivity presented in the figure was increased by one order of magnitude—to approximately 350 V/W near 1 THz. In the direct detection SBDs reach NEP $\approx 3 \times 10^{-10}{-}10^{-8}$ W/Hz$^{1/2}$ at $\nu = 891$ GHz [53].

The heterodyne SBD receivers are worse compared to cooled HEB receivers and SIS mixers (see Figure 4.7). At the same time, SBD receivers operation without cooling gives opportunities for

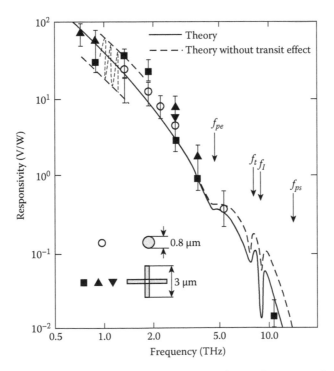

Figure 22.8 Dependence of SBD voltage sensitivity on radiation frequency dependence for diodes with different anode shapes. (From Bozhkov, V. G., *Radiophysics and Quantum Electronics*, 46, 631–56, 2003. With permission.)

using SBD mixers in different mm and sub-mm applications. Their sensitivity is quite suitable for use in mm wave spectrometers with moderate resolution [69,70]. Superconducting mixers typically require microwatt LO powers, which is roughly 3–4 orders of magnitude lower than their SBD predecessors. As a result, a broader range of LO sources can be used. The technologies being used or investigated include diode multipliers, lasers and optoelectronics, and "vacuum tube" oscillators such as klystrons, including novel nano-fabricated versions.

22.3 PAIR-BRAKING PHOTON DETECTORS

The first proposition of the pair-braking detector with superconductor tunnel junction (STJ) was made in the early 1960s [71]. Next, several structures of pair-braking detectors that use different ways to separate quasiparticles from Cooper pairs have been proposed. Among them are: superconductor-insulator-superconductor (SIS), superconductor-insulator-normal metal (SIN) detectors and mixers, radio frequency (RF) kinetic inductance detectors, and superconducting quantum interference device (SQUID) kinetic inductance detectors. Superconducting detectors offer many benefits: outstanding sensitivity, lithographic fabrication, and large array sizes, especially through the development of multiplexing techniques. The basics physics of these devices and recent progress in their developments are described by Zmuidzinas and Richards [11]. Here we concentrate on the most important, SIS detectors.

The SIS detector is a sandwich of two superconductors separated by a thin ($\approx$20 Å) insulating layer, what is schematically shown in Figure 22.9. The Nb and NbTiN are almost exclusively used as superconductors for the electrodes. For a standard junction process, the base electrode is 200 nm sputtered Nb, the tunnel barrier is made using a thin 5 nm sputtered Al layer that is either thermally oxidized (Al_2O_3) or plasma nitridized (AlN). The counterelectrode is 100 nm sputtered Nb or reactively sputtered NbTiN. Typical junction areas are about 1 μm^2. The entire SIS structure is deposited in a single deposited run. The junction is defined by photolithography or electron-beam lithography and reactive ion etching of the counterelectrode. Finally, 200 nm thick thermally evaporated SiO or sputtered SiO_2 are deposited. A SiO_x layer insulates the junction and serves as the dielectric for the wiring and RF tuning circuit on top of the junction.

The SIS operation is based on photoassisted tunneling of quasiparticles through the insulating layer. Although the physics of this effect was demonstrated and theoretically explained in the 1960s [72,73], it took almost two decades to make use of the effect in a mixer [74,75]. Currently, SIS tunnel junctions are mainly used as mixers in heterodyne type mm and sub-mm receivers because of their strong nonlinear I-V characteristic. They can also be used as direct detection detectors [75,76]. The operating temperature of SIS junctions is below 1 K; typically $T \leq 300$ mK.

The SIS operation can be described by using the energy band representation known for semiconductors. The states below the energy gap are considered to be occupied, and those above the gap are empty (Figure 22.10a). The curves indicate the electron density of states. When a bias voltage, V_b, is applied to the junction, there is a relative energy shift of qV_b between the Fermi levels of the two superconductors. If qV_b is lower than the energy gap, 2Δ, no current flows (electrons may only tunnel into unoccupied states at the same energy). However, if the junction is irradiated, photons of energy $h\nu$ may assist the tunneling, which now may occur for $qV_b > 2\Delta - h\nu$.

The current-voltage characteristic of an SIS device is shown in Figure 22.10b. When the bias voltage reaches the gap voltage, a steep increase of the current occurs. At this particular voltage, the divergent densities of states of both superconductor layers cross, and Cooper pairs on one side of the insulating layer break up into two electrons (quasiparticles). Next, these quasiparticles tunnel from one side of the insulator to the other, where they recombine. A sharp onset of normal

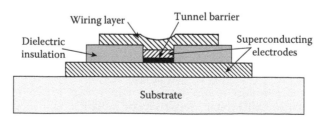

Figure 22.9 A cross section of a typical SIS junction.

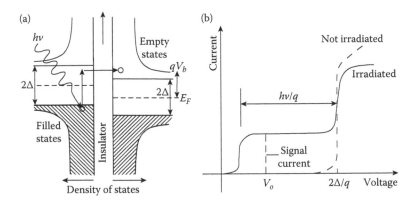

Figure 22.10 SIS junction: (a) energy diagram with applied bias voltage and illustration of photon assisted tunneling, and (b) current-voltage characteristic of a nonirradiated and irradiated barrier. The intensity of the incident radiation is measured as an excess of the current at a certain bias voltage V_o.

tunneling current is arising beyond a DC threshold voltage equal to a superconductor energy gap 2Δ, and this abrupt nonlinearity in the single-particle tunneling is used for mixing. The I-V nonlinearity is small, about a few tenths of a millivolt and might be compared with the nonlinearity of a Schottky diode (in the order of 1 mV). However, the SIS nonlinearity at the gap voltage is substantially less than the energy of THz photons and the classical mixer theory cannot be applied. To explain the possibility of using SIS in THz detection, a detailed description of mixing theory including quantum effects is necessary [76].

Although the IF bandwidth of and the SIS junction itself is very large due to the inherently fast tunneling process, it is limited in practical applications by the electric circuitry. A typical IF band is 4–8 GHz. One of the main challenges for the high-frequency SIS mixer design is dealing with the large parallel-plate capacitance of the SIS junction. Although the area of an SIS junction is similar to the area of a Schottky diode, its parasitic capacitance is much higher (typically 50–100 fF compared to ≈1fF), because the two superconducting electrodes form a parallel plate capacitor. As a consequence, on-chip tuning circuits are needed to compensate for the capacitance, and their proper design is a key issue for any SIS mixer [11,22]. At higher frequencies, especially over 1 THz ($\lambda = 300$ μm), the losses in the tuning circuit become important and cause the mixer performance to deteriorate. Nevertheless, good performance has been obtained up to around 1.5–1.6 THz ($\lambda = 200$–188 μm).

There are two major ways to accomplish proper design issues for SIS mixers: waveguide coupling and quasioptical coupling [11]. Figure 22.11 shows the examples of both configurations. The more traditional approach is waveguide coupling, in which the radiation is first collected by a horn into a single-mode waveguide and then coupled onto a lithographed thin-film transmission line on the SIS chip itself. The chip shown in Figure 22.11b is approximately 2 mm long and 0.24 mm wide and is fabricated on 25 μm thick silicon utilizing silicon-on-insulator bonded wafers. The 1 μm thick gold beam leads extend beyond the edges of substrate are an electrical contact to the metal waveguide probe [11]. Serious complication of waveguide coupling is that the mixer chip must be very narrow and must be fabricated on ultrathin substrate.

In the quasioptical coupling, the intermediate step of radiation collection into a waveguide is omitted, and instead uses a lithographed antenna on the SIS chip itself. Such mixers are substantially simpler to fabricate and may be produced using thick substrates. In the design structure shown in Figure 22.11b, the radial stubs serve as RF short circuits and couple the radiation received by the slots into a Nb/SiO/Nb superconducting microstrip. This chips uses two Nb/Al_2O_3/Nb SIS junctions and the short strip section in between the junctions provides the tuning inductance needed to compensate the junction capacitance.

The SIS mixers are among the most sensitive and low intrinsic noise structures at $\nu \approx 0.3$–0.7 THz. Nb-based SIS mixers with Nb wiring yield almost quantum limited; that is, the noise temperature is below $3h\nu/kT$ (see Figure 4.7) [76,79]. At larger frequencies, $\nu \approx 1.0$–1.3 THz, SIS's intrinsic noise quickly increases due to high frequency losses. Further gain in sensitivity is

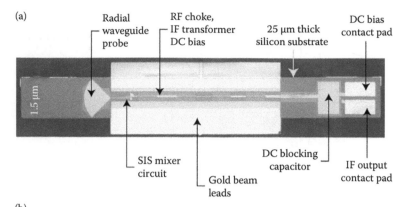

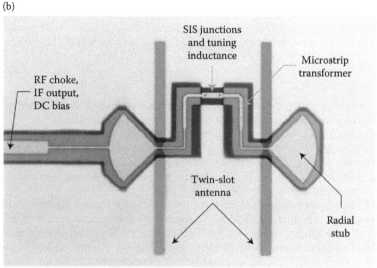

Figure 22.11   Images of (a) a waveguide SIS mixer operated in the 200–300 GHz band and (b) a quasioptical SIS mixer. (From Zmuidzinas, J., and Richards, P. L., *Proceedings of IEEE* 92, 1597–1616, 2004. With permission.)

possible using multielement or matrix arrays. However, up to now SIS detectors are difficult to integrate into large arrays. There is success only in creating a small number of element arrays because of appreciable difficulties in their creation [78]. SIS mixers seem to be the best solution for the ground-based radio-astronomy at mm and sub-mm wavelengths region in frequency range $\nu < 1$ THz [80].

The signal bandwidth of an SIS mixer is 10–30% of its center frequency with the large fractional bandwidth at low frequency band. Up to 1 THz, the mixers are in a waveguide mount, while the 1.2–1.25 THz mixers employ a quasioptical coupling.

Single pixel SIS mixers typically require approximately 40–100 µW of LO pump power, which is appreciably lower compared to LO pump power for single pixel SBD mixers ($P > 1$ mW) [81]. Much lower LO powers require superconductor hot electron bolometer mixers (< 100 nW–1 µW) [82] though they also operate at very low temperature. Unlike Schottky diodes or SIS detectors, the hot electron bolometers are thermal detectors.

22.4 THERMAL DETECTORS

Many types of thermal detectors operated in LWIR and far-IR regions (including Golay cells, bolometers, pyroelectric detectors) are also used in the THz band.

Bolometers, as other thermal devices, traditionally were treated as slow devices. In many applications, their performance is limited by a trade-off between speed and sensitivity. For conventional uncooled microbolometers operated in wavelength range 10–100 µm at room temperature, the typical value of heat capacity is about of 2×10^{-9} J/K (for bolometer with dimensions $50 \times 50 \times 0.5$ μm) and thermal conductance of 10^{-7} W/K (a-Si or VO_x bolometers). Both parameters define time constant, τ, equal to about 20 ms (see Chapter 3). Then, the upper limit of NEP for such a bolometer limited only by radiation exchange with the environment is $NEP_R \approx 2.7 \times 10^{-13}$ W/$Hz^{1/2}$. The upper limit value of $NEP_R \approx 1.7 \times 10^{-19}$ W/$Hz^{1/2}$ can be achieved only at the expense of large τ values ($\tau \approx 3.5 \times 10^4$ s) [42].

If the bolometer is to be used as a THz mixer, it has to be fast enough to follow the IF (i.e., the overall time constant of the processes involved in the mixing has to be a few tens of picoseconds at maximum). In other words, high heat conductivity and small heat capacity are required [38]. These requirements can be fulfilled by such a subsystem as electrons in semiconductor or superconductor interacting with the lattice (phonons). Electron heat capacity is many orders lower compared to the lattice one.

The first bolometer with "hot electrons" (hot electron bolometer, HEB) was low temperature bulk n-InSb [24,83]. InSb HEB has a bandwidth of about 4 MHz. Currently, other semiconductor materials that can be used in fabrication HEBs have been proposed. In spite of the fact that the rate of electron heating is extremely high because of the high rate of photon-electron interaction, the maximum transformation frequency is restricted by the thermal relaxation rate, which in semiconductors is governed by electron–phonon interaction time $\tau \approx 10^{-7}$ s at low temperatures [84]. This response time is relatively short, compared to conventional thermal detectors with lattice heating, but long compared to τ in superconducting HEBs. Thus, for direct detection semiconductor systems the speed of responsivity is quite suitable, but not for the mixers. Their NEP can reach 5×10^{-13} W/$Hz^{1/2}$ at operation temperature about 4 K and below.

Nonlinearity of semiconductor HEB current-voltage characteristic, needed for heterodyne detector operation, is conditioned by the dependence of conductivity on electron mobility, which is a function of applied electric field, thereby the function of electrons temperature. Higher IFs and broader Δf can be obtained in semiconductor HEB frequency converters increasing their temperature to about 80 K (where electron–phonon interaction is much stronger and $\tau \approx 10^{-11}$ s), but in this case the noise level of such frequency converters is increasing appreciably and conversion losses are increasing faster also.

In low-dimensional semiconductor structures the electron–phonon interaction can be substantially increased (τ decreased) and, thus, such kind of structures can be considered as frequency converters with higher IFs and wider bandwidth up to 10^9 Hz [85–87]. Direct measurements of photoresponse relaxation time have shown that τ is about 0.5 ns in the temperature range 4.2–20 K [88]. Thus, the IF can be increased by about three orders compared to bulk semiconductor HEBs.

Historically, HEB mixers using semiconductors were invented in the early 1970s [83] and played an important role in early sub-mm astronomy [89], but were superceded by SIS mixers by the early 1990s. However, the development of superconducting HEB versions led to the most sensitive THz mixers at frequencies beyond the reach of SIS mixers. The main difference between HEB mixers and ordinary bolometers is the speed of their response. HEB mixers are fast enough to allow GHz output IF bandwidths.

In Table 22.2 requirements for typical astronomic instruments containing bolometers have been gathered. To fulfill these requirements, the compromise between response time and NEP is needful.

22.4.1 Semiconductor Bolometers

The classic bolometers contain heavily doped and compensated semiconductors that conduct by a hopping process that yields a resistivity $\alpha \exp(T/T_0)$. The thermistors are made by ion

Table 22.2: Requirements for Typical Astronomic Instruments

Instrument	Wavelength Range (μm)	NEP (W/$Hz^{1/2}$)	τ (ms)	NEP$\tau^{1/2}$ ($\times 10^{-19}$J)	Comments
SCUBA	350–850	1.5×10^{-16}	6	9	High background, needs reasonable τ
SCUBA-2	450–850	7×10^{-17}	1–2	1	Lower background, needs faster τ
BoloCAM	1100–2000	3×10^{-17}	10	3	Lower background, slower device okay
SPIRE	250–500	3×10^{-17}	8	2.4	Space background, slowish device okay
Planck-HFI	350–3000	1×10^{-17}	5	0.5	Lowest background, needs quite fast τ

implantation in Si, or by neutron transmutations doping in Ge [90]. They are typically fabricated by lithography on membranes of Si or SiN. The impedance is selected to a few MΩ to minimize the noise in JFET amplifiers operated at about 100 K. Limitation of this technology is the assertion of thermal mechanical and electrical interface between the bolometers at 100–300 mK and the amplifiers at ≈100 K. There are no practical approaches to multiplexing many such bolometers to one JFET amplifier. Current arrays require one amplifier per pixel and are limited to a few hundred pixels.

In a bolometer the photons are absorbed by metal films that can be continuous or patterned in a mesh. The patterning is designed to select the spectral band, to provide polarization sensitivity, or to control the throughput. Different bolometer architectures are used. In close-packed arrays and spider web, the pop-up structures or two-layer bump bonded structures are fabricated. Agnese et al. have described a different array architecture that is assembled from two wafers by indium bump bonds [91]. Other types of bolometers are integrated in horn-coupled arrays. To minimize low frequency noise an AC bias is used.

The present day technology exists to produce arrays of hundreds of pixels that are operated in spectral range from 40 to 3000 μm in many experiments including NASA pathfinder ground-based instruments and balloon experiments such as BOOMERANG, MAXIMA, and BAM. Figure 22.12 shows a spider web bolometer from the BOOMERANG experiment that is being used on space experiments (Planck and Herschel missions) [90].

An alternative approach is an array for the SHARCII instrument on SOFIA shown in Figure 22.13a [92]. This array construction, with 12×32 pixels, involves a pop-up configuration, where the absorber is deposited on a dielectric film that is subsequently folded. The 12×32 bolometer array, load resistors, and thermally isolated JFETs are housed in a structure approximately $18 \times 17 \times 18$ cm^3 volume, having a total mass of 5 kg, and heat sink to 4 K. Each bolometer is fabricated on a 1 μm silicon membrane and has a collecting area of 1×1 mm^2. The full area is ion implanted with phosphorus and boron, to a depth of ≈0.4 μm, to form a thermistor. Electrical contact between the thermistor and aluminum traces on the silicon frame is accomplished with degenerately doped leads on the edge of the thermistor and running down the bolometer legs. Each of the four thermally isolating legs is 16 μm wide and 420 μm long. Prior to folding, each bolometer is coated with an absorbing ≈200 Å bismuth film and a protective ≈160 Å SiO film. At a base temperature of 0.36 K and in the dark, the bolometers have a peak responsivity of approximately 4×10^8 V/W and a minimum NEP of approximately 6×10^{-17} W/Hz$^{1/2}$ at 10 Hz. Phonon noise is the dominant contributor, followed by bolometer Johnson noise.

With the development of low-noise readouts that can operate near the bolometer temperature, the first true high performance bolometer arrays for the far IR and sub-mm spectral ranges are just becoming available. For example, the Herschel/PACS instrument uses a 2048-pixel array

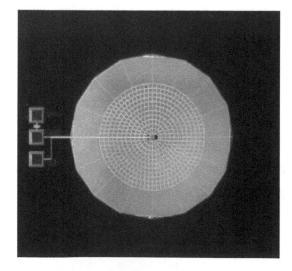

Figure 22.12 BOOMERANG spider web bolometer. (From "Detectors Needs for Long Wavelength Astrophysics," *A Report by the Infrared, Submillimeter, and Millimeter Detector Working Group*, June 2002.)

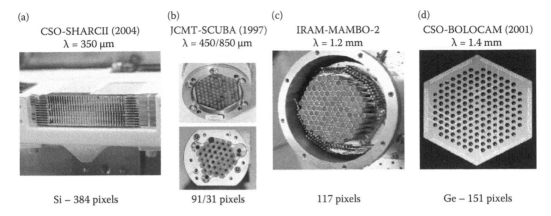

(a)
CSO-SHARCII (2004)
λ = 350 μm

(b)
JCMT-SCUBA (1997)
λ = 450/850 μm

(c)
IRAM-MAMBO-2
λ = 1.2 mm

(d)
CSO-BOLOCAM (2001)
λ = 1.4 mm

Si – 384 pixels

91/31 pixels

117 pixels

Ge – 151 pixels

Figure 22.13 Arrays installed in ground-based telescopes: (a) CSO-SHARCII, (b) JCMT-SCUBA, (c) IRAM-MAMBO-2, and (d) CSO-BOLOCAM.

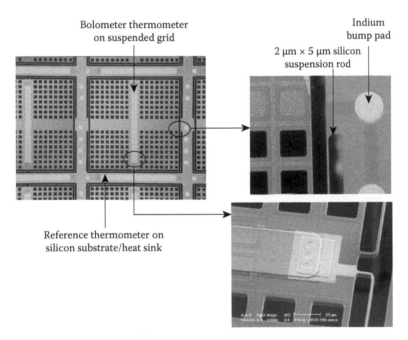

Bolometer thermometer
on suspended grid

Indium
bump pad

2 μm × 5 μm silicon
suspension rod

Reference thermometer on
silicon substrate/heat sink

Figure 22.14 The Herschel/PACS bolometer array with pixel size 750 μm. (From Rieke, G. H., *Annual Review of Astronomy and Astrophysics*, 45, 77–115, 2007.)

of bolometers [93] and is an alternative to JFET amplifiers. The architecture of this array is vaguely similar to the direct hybrid mid-infrared arrays, where one silicon wafer is patterned with bolometers, each in the form of a silicon mesh, as shown in Figure 22.14. As is described in Chapter 6, the development of silicon micromachining has enabled substantial advances in bolometer construction generally and is central to making large-scale arrays. To achieve appropriate response and time constant characteristics, the rods and mesh are designed carefully. The mesh is blackened with a thin layer of titanium nitride with sheet resistance matched to the impedance of free space (377 $\Omega/\square$ section of film) to provide an efficiency of 50% over a broad spectral band. Each bolometer located at the center of the mesh, containing a silicon-based thermometer doped by ion implantation, is characterized by appropriate temperature-sensitive resistance. Their large resistances (>10^{10} Ω) are well adjusted to MOSFET readout amplifiers. In the final step of hybrid array fabrication, the MOSFET-based readouts and silicon bolometer wafer are joined by indium bump bonding. Performance is currently limited by the noise in the MOSFET amplifiers to the NEP $\approx 10^{-16}$ W/Hz$^{1/2}$ regime, but this technology allows the construction of very large arrays suitable for higher background applications. Further details are in Billot et al. [93].

22.4.2 Superconducting Hot-Electron Bolometers

Another approach to thermal detectors is the electron subsystem in superconductor, which weakly interacts with the lattice near superconducting transition or normal metal at low temperatures, when the metal heat capacity is defined by electrons. When a photon is absorbed in a metal, a single electron initially receives the photon energy. This energy is rapidly shared with other electrons, producing a slight increase in the electron temperature. The electron temperature subsequently relaxes to the bath temperature, usually through the emission of phonons. The detailed physics of HEB mixers is not thoroughly understood yet, although significant progress is being made.

A major issue for HEB mixers is achieving a thermal time constant that is fast enough to yield a useful IF output bandwidth of a few GHz. For a fixed thermal relaxation time, the heat capacity sets the required LO power and can be minimized by using a very small volume ($<10^{-2}$ μm^3) of a superconducting film. The required LO power is about one order of magnitude lower (typically between 100 and 500 nW) compared to SIS mixers and much lower (roughly 3–4 orders) in comparison with SBDs. The LO power scales with the volume of the microbridge and decreases with increasing critical temperature [94].

As is described in Section 6.3, two methods are used to obtain useful IF output bandwidth.

- Phonon cooling, using ultrathin NbN or NbTiN films characterized by large electron–phonon interaction

- Diffusion cooling, using submicron Nb, Ta, NbAu, or Al devices coupled to normal-metal cooling "pads" or electrodes

Competitive sensitivities have been demonstrated for both types of devices.

Karasik et al. have demonstrated hot-electron superconducting direct detection Ti nanobolometers fabricated on Si planar bulk substrates with Nb contacts and NEP value of 3×10^{-19} W/Hz$^{1/2}$ at 300 mK [41]. The thermal time constant $\tau_{eph} = 25$ μs at $T = 190$ mK for larger devices has been demonstrated.

Unlike Schottky diodes or SIS mixer, the HEB mixers are thermal detectors, and as such belong to the group of square-law mixers. Figure 22.15a shows an example of the central area of a mixer chip manufactured from a 3.5 nm thick superconducting NbN film on a high resistive Si substrate with an e-beam evaporated MgO buffer layer [95]. Ultrathin NbN films are deposited by reactive magnetron sputtering in the Ar + N_2 gas mixture. The active NbN film area is determined by the dimensions of the 0.2 μm gap between the gold contact pads (typically the length of a superconducting bridge varies between 0.1 and 0.4 μm and the width between 1 and 4 μm). The NbN microstrip is integrated with a planar antenna patterned as log-periodic spiral. The NbN critical temperature depends on the thickness of film deposited on the substrate. An improvement of superconducting properties of NbN films due to the presence of MgO buffer layer on silicon

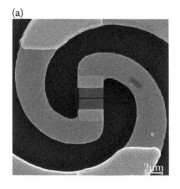
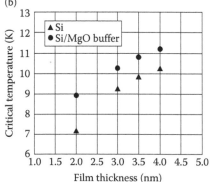

Figure 22.15 NbN HEB mixer chip: (a) SEM micrograph of the central area of mixer, (b) critical temperature versus thickness for NbN films deposited on Si substrates (triangles) and on Si with MgO buffer layer (circles). (From Gol'tsman G. N., Vachtomin, Yu. B., Antipov, S. V., Finkel, M. I., Maslennikov, S. N., Smirnov, K. V., Poluakov, S. L., et al., *Proceedings of SPIE 5727*, 95–106, 2005. With permission.)

substrates is evident (see Figure 22.15b). The superconducting transition temperature is about 9 K and the transition width is ≈0.5 K.

The NbN superconductive HEB mixers are characterized by strong electron–phonon interaction. The response time can achieve a value of 10^{-11} s [96], and because of no principal restrictions for operation at $v > 1$ THz (the absence of their noticeable capacities), these devices can be effectively used for heterodyne detection in wide spectral range up to the visible one, where operation (e.g., of SIS mixers) is hampered. For example, Figure 22.16 shows intermediate frequency dependent output power for the 3.5 nm thick NbN film devices deposited on the plane of MgO substrate (curve A) and on Si substrate with MgO buffer layer (curve B). The frequency limited range to several GHz is due to the influence of free carrier relaxation rate.

The ratio of phonon to electron specific heats, C_p/C_e, controls the energy flow from electrons to phonons and the energy backflow due to reabsorption of nonequilibrium phonons by electrons. This ratio is: 0.85 for Nb, 6.5 for NbN, and 38 for YBaCuO layers [97]. In very thin films the phonons can escape into the substrate before being reabsorbed by the electrons. In thin, below 10 nm thick Nb films deposited on a substrate, $\tau_{phe} > \tau_{eph}$, and the effective escape of phonons to the substrate prevail energy backflow to electrons. As a result, τ_{eph} alone controls the response time, which is approximately equal to about 5 ns. Thus, Nb devices are sensitive in a wide range of spectra, are much faster compared to bulk semiconductor bolometers (operating at $T \approx 4$ K), and can reach NEP $\approx 3 \times 10^{-13}$ W/Hz$^{1/2}$ [98].

The NbN films in comparison to the Nb films have much shorter τ_{eph} and τ_{phe} because of stronger electron–phonon interaction. In ultrathin 3 nm thick NbN films both τ_{eph} and τ_{phe} determine the response time of the detector, which can be about 30 ps near T_c ($\tau_{eph} \approx 10$ ps) [96]. NEP can reach the values of 10^{-12} W/Hz$^{1/2}$ [99].

For YBaCuO detector layers the ratio $C_p/C_e \approx 38$, the layers are mainly phonon-cooled types; the energy backflow from phonons to electrons can be neglected and thermalization time is about an order faster ($\tau_{eph} \approx 1$ ps) compared to NbN layers. In YBaCuO films excited by fs pulses, the non-thermal (hot-electron) and thermal bolometric (phonon) processes are practically decoupled, with the former one dominating the early stage of electron relaxation [96]. To decouple electrons from phonons, nonequilibrium phonons in the film should escape from it (into the substrate) in a short time compared to phonon-electron time, τ_{phe}.

HEB mixers can be made either in a waveguide configuration with a horn antenna or a quasioptical mixer. Above 1 THz, quasioptical hybrid antennas are more common. As shown in Figure 22.15a, the superconducting microbridge is embedded in a planar antenna, for example, a twin-slot or a logarithmic spiral antenna. The substrate with the feed antenna and the microbridge is mounted to the reverse side of a hyperhemispherical or elliptical lens.

HEBs are significantly more sensitive than SBDs but somewhat less sensitive than SIS mixers, see Figure 4.7. This figure shows that DSB noise temperature achieved with HEB mixers range

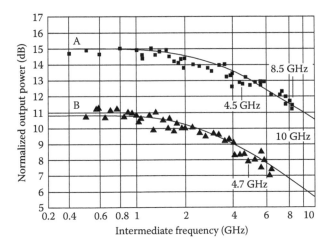

Figure 22.16 Output power as a function of intermediate frequency for the 3.5 nm thick NbN film devices deposited on plane of MgO substrate (curve A) and on Si substrate with MgO buffer layer (curve B). (From Gol'tsman G. N., Vachtomin, Yu. B., Antipov, S. V., Finkel, M. I., Maslennikov, S. N., Smirnov, K. V., Poluakov, S. L., et al., *Proceedings of SPIE* 5727, 95–106, 2005. With permission.)

from 400 K at 600 GHz up to 6800 K at 5.2 THz. In the lower frequency range up to 2.5 THz, the noise temperature follows the $10h\nu/k$ line closely. Above this frequency, the correspondence becomes somewhat worse, which is caused by increasing losses in the optical components, lower efficiency of the antenna, and skin effect contributions in the superconducting bridge. Generally, phonon-cooled HEBs have a lower noise temperature than that of diffusion-cooled devices.

Because of high sensitivity, NbN HEB mixers are currently the first choice for heterodyne spectroscopy above 1 THz. Examples are mixers used in the Herschel mission and SOFIA.

Concerning high temperature superconductor (HTSC) HEBs, it should be noted that no publications devoted to this kind of receiver exists. They have not reached a high state of technological maturity as their complicated composition does not allow fabrication of very thin layers with high critical temperature. HTSC belongs to the phonon-cooled type and electron diffusion mechanism is negligible [97,100]. These receivers are noticeably noisier compared to low temperature devices, as phonon dynamics play an appreciable role due to the relatively high operating temperature and introduce of excess noise. Thus, they do not reach the sensitivity of low temperature HEBs [100], but because of very short electron–phonon relaxation time ($\tau_{eph} \approx 1.1$ ps in YBaCuO [101]), HTSC HEB mixers are the wide bandwidth devices. Lyatti et al. have concluded [102] that HTSC HEB mixers can reach NEP value of 5×10^{-15} W/Hz$^{1/2}$.

22.4.3 Transition Edge Sensor Bolometers

The name of the transmission-edge sensor (TES) bolometer is derived from its thermometer, which is based on thin superconducting films held within transition region, where it can change from the superconducting to the normal state over a temperature range of a few milliKelvin (see Figure 6.15). The film has stable but very steep dependence of resistance on temperature in the transition region. Changes in temperature transition can be set by using a bilayer film consisting of a normal material and a layer of superconductor. Such a design enables diffusion of the Cooper pairs from the superconductor into the normal metal and makes it weakly superconducting—this process is called the proximity effect. As a result, the transition temperature is lowered relative to that for the pure superconducting film. Thus in principle, the transmission-edge sensor (TES) bolometers are quite similar to the HEBs. In the case of HEB, high speed is achieved by allowing the radiation power to be directly absorbed by the electrons in superconductor. In TES bolometers, however, a separate radiation absorber is used that allows the energy to flow to the superconducting TES via phonons, as ordinary bolometers do.

A different type of superconducting metal film pairs (bilayers) can be used including thin Mo/Au, Mo/Cu, Ti/Au, and so on. Two metals behave as a single film with a transition temperature between 800 mK (for Mo) and 0 K (for Au). Transition temperature can be tuned within this temperature range. The lower temperature ($T < 200$ mK) is needed because the energy resolution of these devices scales with temperature.

Traditionally, the superconducting bolometer was biased with a constant current and readout with a voltage amplifier. Then, the bias power $P_b = I^2R$ increased with temperature due to the increase of resistance R near T_c. In consequence, a positive electrothermal feedback leads to instability and even thermal runaway. The new idea of a negative electrothermal feedback proposed by Irwin [103] stabilizes the temperature of the TES at the operating point on the transition. When the TES temperature rises due to power from absorbed photons, their resistance rises, the bias current drops, and the electrical power dissipation in them decreases, partially canceling the effects of the absorbed power and limiting the net thermal excursion. The advantages of the TES with negative feedback include linearity, bandwidth, and immunity of the response to changes in external parameters (e.g., the absorbed optical power and the temperature of the heat sink). Consequently, these devices are suitable for fabrication of a large format horn-coupled and filled array required for many new missions [104–106].

In practice, the bias voltage V_b is chosen so that for small optical power P, the TES will be heated to a steep point on the temperature transition. For intermediate values of P, the electrothermal feedback keeps the total power input $P + V^2/R$ (and, thus, the temperature) constant. The current responsivity is defined as the response of the bolometer current I to a change in the optical power. Then, for a thermal circuit with a single pole response is equal [11,107]

$$R_i = \frac{dI}{dP} = -\frac{1}{V_b}\frac{L}{(L+1)}\frac{1}{(1+i\omega\tau)},$$ (22.9)

where $L = \alpha P/GT$ is the loop gain, $\alpha = (T/R)dR/dT$ is a measure of the steepness of the superconducting transition and is a bolometer figure of merit, $G = dP/dT$ is the differential thermal conductance, and τ is the effective time constant. For a typical loop gain $L \approx 10^2$ [103], the low-frequency responsivity becomes $R_i \approx -1/V_b$, and depends only on the bias voltage and is independent of the signal power and the heat sink temperature. The effective time constant $\tau = \tau_o/(1 + L)$ is much shorter than the time constant without feedback $\tau_o = C/G$ for thermal detectors. Negative electrothermal feedback can make the bolometers operate tens or even hundreds of times faster. The values of α for Mo/Cu proximity-effect layers ($T_c = 190$ mK) consistent with thermal fluctuation noise [108] are within 100–250 [109].

The resistance of a TES is low, so it can deliver significant power only to low-input impedance amplifiers, which rules out JFETs and MOSFETs. Instead, the signals are fed into superconducting quantum interference devices (SQUIDs), which are the basis for a growing family of electronic devices that operate by superconductivity. In this case, the TES is transformer coupled to the SQUID by an input coil. A current-biased shunt resistor is used to provide a constant voltage bias to the TES. When the shunt resistor is operated close to the detector temperature, a negligible Johnson noise from the bias network is given. The SQUID readout has a number of advantages including: it operates near the bolometer temperature, has very low power dissipation and large noise margin, and low sensitivity to microphonic pickup. In addition, the fabrication and lithographic processes used in both SQUID readouts and TES bolometers are similar for help in their integration on the same chip.

Figure 22.17 shows the typical TES circuit wired in series with the coil of a SQUID amplifier. The voltage bias is achieved by current bias of a cold shunt resistor R_{sh} whose 10 mΩ resistance is much smaller than the $R \approx 1$ Ω resistance of the TES. The current through the TES is measured with a SQUID ammeter, and the in-band reactance of the SQUID input is much less than R. When the bias to a SQUID is turned off the whole device goes into a superconducting state where it adds no noise. Thus by switching on or off rows or columns of SQUIDs in an array (one for each pixel) a cold multiplexer may be realized. The biases across the SQUIDs are controlled by the address lines. Each SQUID can be switched from an operational state to a superconducting one if it is biased to carry about 100 μA. The address lines are set so all the SQUIDs in series are superconducting except one, and then only that one contributes to the output voltage. By a suitable series of bias settings, each SQUID amplifier can be readout in turn. To avoid very large numbers of leads leaving the cryostat, lines of 30–50 detectors can be multiplexed before amplification.

In general, SQUID-base multiplexers developed for TES bolometers and microcalorimeters used both time division [111] and frequency division [112] approaches. We have described the time division approach, where the multiplexer uses a SQUID for each bolometer to switch the outputs sequentially through a single SQUID amplifier. In the case of frequency domain case, each TES is biased with a sinusoidally varying voltage and the signals from a number of TESs are encoded in

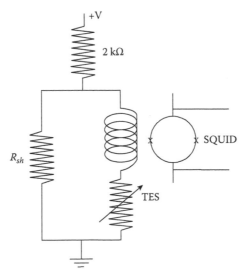

Figure 22.17 Typical TES bias circuit and low noise, low power SQUID.

amplitude-modulated carrier signals by summing them. The signals are then amplified by a single SQUID and recovered with ambient temperature lock-in amplifiers. For more SQUID multiplexers details, see for example, [110,113–116].

The most ambitious example of TES bolometer array is that used in the submillimeter camera SCUBA (Submillimetre Common-User Bolometer Array)-2 with about 10^4 pixels [105,117]. The camera operated at wavelengths of 450 and 850 μm has been mounted on the James Clerk Maxwell Telescope in Hawaii. Each SCUBA-2 array is made of four subarrays, each with 1280 transition-edge sensors. The design is illustrated in Figure 22.18 together with a cross-section of the detector architecture. The detector technology is based on silicon micromachining. Each pixel consists of two silicon wafers bonded together. The upper wafer with its square wells supports the silicon nitride membrane on which TES detectors and the absorbing silicon brick are suspended. The lower wafer is thinned to 1/4 wavelength in silicon (70 μm at 850 μm). The upper surface of this wafer has been previously implanted with phosphorus to match the impedance of free space (377 Ω/□). The detector elements are separated from their heat sinks by a deep etched trench that is bridged by only a thin silicon nitride membrane. The superconducting electronics that read out the bolometers are fabricated on separate wafers. The two components are assembled into an array using indium bump bonding. Further details are in Walton et al. [105] and Woodcraft et al. [118].

In addition to the arrays for SCUBA-2, various forms of TES-based bolometers with SQUID readouts are under active development [49,119]. Large format arrays of antenna-coupled TES bolometers are very attractive candidates for a cosmic microwave background polarization mission. Broadband antennas with RF diplexing and/or interleaved antennas can make efficient use

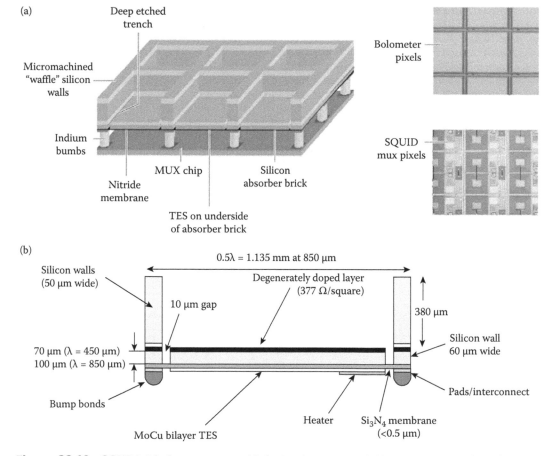

Figure 22.18 SCUBA-2 bolometer array: (a) design features, and (b) cross-section of single pixel. (From Walton, A. J., Parkes, W., Terry, J. G., Dunare, C., Stevenson, J. T. M., Gundlach, A. M., Hilton, G. C., et al., *IEE Proceedings. Science, Measurement and Technology*, 151, 119–20, 2004. With permission.)

(a)   (b)

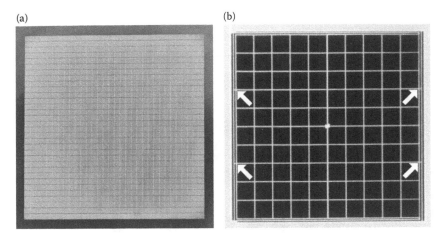

Figure 22.19 Monolithic superconducting bolometer array: (a) silicon nitride structure for a 1024 element array of bolometers (1.5 × 1.5 mm² pixel size), and (b) complete pixel including sensor and metallization. (From Gildemeister, J., Lee, A., and Richards, P., *Applied Physics Letters, 77,* 4040–42, 2000. With permission.)

of the focal plane. Antennas are inherently polarization sensitive, and the excellent gain stability provided by the feedback in TES bolometers facilitates polarization differencing.

The fabrication technologies used for TES bolometers are very flexible and specialized detectors are being developed to meet the needs of specific observations. A bolometer design that allows the production of large monolithic detector arrays with a very high fill factor by standard planar lithography has been described. Figure 22.19 shows the 1024-pixel array structure together with a single pixel [120]. The absorbing element is a square mesh of 1 μm thick low-stress (nonstoichiometric) silicon nitride (LSN), which is metalized with gold to produce an average sheet resistance of 377 $\Omega/\square$. A conducting backshort is located at a distance $\lambda/4$ behind the mesh. This mesh absorber is supported at four points (shown by arrows) by low thermal conductivity beams of LSN. To produce $T_c \approx 400$ mK, a proximity-effect sandwich of Al and Ti is made at the center of the mesh. To connect the thermistors to the edges of the array, fully superconducting leads are used by way of the beams and the dividing strips between pixels. Other designs are using micromachining and folding to bring the leads out of the third dimension [121].

Figure 22.20 shows a close-packed horn-coupled array with radial support legs. The bolometers that are located at the small ends of the horns are then separated sufficiently for easy support and wiring. The array is completely lithographed [11,122].

22.5 FIELD EFFECT TRANSISTOR DETECTORS

Nonlinear properties of plasma wave excitations (the electron density waves) in nanoscale FET channels enable their response at frequencies appreciably higher than the device cutoff frequency, which is due to electron ballistic transport. In the ballistic regime of operation, the momentum relaxation time is longer that the electron transit time. The FETs can be used both for resonant (tuned to a certain wavelength) and nonresonant (broadband) THz detection [123–126] and can be directly tunable by changing the gate voltage.

The transistor receivers operate in wide temperature range up to room temperatures [127]. Different material systems are used in fabrication FET, HEMT, and MOSFET devices including: Si, GaAs/AlGaAs, InGaP/InGaAs/GaAs, and GaN/AlGaN [127–131]. Plasma oscillations can be also observed in a two-dimensional (2-D) electron channel with a reverse-biased Schottky junction [132] and double quantum well FET with a periodic grating gate [133].

The use of FET as detectors of THz radiation was first proposed by Dyakonov and Shur in 1993 [134] on the basis of formal analogy between the equations of the electron transport in a gated two-dimensional transistor channel and those of shallow water, or acoustic waves in music instruments. As a consequence, hydrodynamic-like phenomena should exist also in the carrier dynamics in the channel. Instability of this flow in the form of plasma waves was predicted under certain boundary conditions.

Figure 22.20 Array of 55 TES spider web bolometers and closeup of bolometers. Six wedges of the type shown are assembled to form a 330-element hexagonal horn-coupled array. (From Zmuidzinas, J., and Richards, P. L., "Superconducting Detectors and Mixers for Millimeter and Submillimeter Astrophysics," *Proceedings of IEEE* 92, 1597–1616, 2004. With permission.)

The physical mechanism supporting the development of stable oscillations lies in the refection of plasma waves at the borders of transistor with subsequent amplification of the wave's amplitude. Plasma excitations in FETs with sufficiently high electron mobility can be used for emission as well as detection of THz radiation [135,136].

The plasma waves in FET is characterized by linear dispersion law [134], and in gated region

$$\omega_p = sk = k \left[\frac{q(V_g - V_{th})}{m^*} \right]^{1/2},$$

(22.10)

where the plasma wave velocity $s \approx 10^8$ cm/s in GaAs channel, V_g is the gate voltage, V_{th} is the threshold voltage, k is the wave vector, q is the electron charge, and m^* is the electron effective mass. Figure 22.21 schematically shows the resonant oscillation of plasma waves in the gated region of FET. The dispersion relations in bulk (3-D) and ungated regions of FET differ from Equation 22.10 and are equal

$$\omega_p = \left(\frac{qN}{m^*} \right)^{1/2} \quad \text{and} \quad \omega_p = \left(\frac{q^2 n}{2m^*} k \right)^{1/2},$$

(22.11)

respectively. Here N is the bulk electron concentration for alloyed regions, and n is sheet electron density for channel regions.

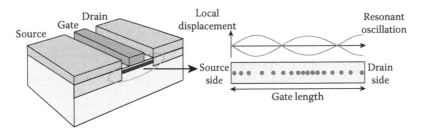

Figure 22.21 Plasma oscillations in a field effect transistor.

The plasma wave velocity in a gated region is typically noticeably larger compared to the electron drift velocity. A short FET channel with length L_g acts as a resonant cavity for these waves with the eigen frequencies $\omega_n = \omega_0(1 + 2n)$ $(n = 1, 2, 3,...)$. The fundamental plasma frequency is

$$\omega_o = \frac{\pi}{2L_g}\left[\frac{q(V_g - V_{th})}{m^*}\right]^{1/2}. \tag{22.12}$$

When $\omega_o\tau \ll 1$, where τ is the momentum relaxation time, the detector response is a smooth function of ω and V_g (broadband detector). When $\omega_o\tau > > 1$, the FET can operate as a resonant detector with tunable by the gate voltage response frequency, and this device can operate in the THz range.

Assuming $m^* \approx 0.1m_o$, where m_o is free electron mass, $L_g \approx 100$ nm, and $V_g - V_{th} \approx 1$ V, the frequency of plasma waves is estimated as $v_o = \omega_o/2\pi \approx 3$ THz. The minimum gate length can approach ≈ 30 nm, and thus, v_o can reach 12–14 THz for FETs with GaAs channels.

Figure 22.22 shows the characteristics of 60 nm gate length InGaAs/InAlAs teransistor [137]. The photoresponse of the device exposed to the radiation of 2.5 THz frequency as a function of the gate voltage measured at various temperatures is shown in Figure 22.22a. At $T > 100$ K only nonresonant detection was observed as a broadband peak. With temperature decreasing below 80 K, the additional peak appears as a shoulder on the temperature-independent background of the nonresonant detection. This behavior can be attributed to the resonant detection of THz radiation by plasma waves. To support this assumption, additional measurements were carried out at 10 K for excitation frequencies of 1.8, 2.5, and at 3.1 THz. The experimental results are displayed in Figure 22.22b. For comparison, the theoretical prediction of plasma frequency as a function of gate voltage according to Equation 22.12 is plotted as a continuous line in Figure 22.22c. One can see that the increasing of excitation frequency from 1.8 to 3.1 THz causes moving of the plasmon resonance with the gate voltage, roughly in agreement with the theory (Equation 22.12).

It appears that driving a transistor into the saturation region enhances the nonresonant detection and can lead to the resonant detection even if the condition $\omega_o\tau \gg 1$ is not satisfied [127,129]. The physical reason is that the effective decay rate for plasma oscillations, in the condition of hot drifting electrons, becomes equal to $1/\tau_{eff} = 1/\tau - 2v/L_g$, where v is the electron drift velocity, and becomes longer when applying current. As $\omega\tau_{eff}$ becomes of the order of unity, the detection becomes resonant.

Tauk et al. have studied Si MOSFETs with 20–300 nm gate lengths at room temperature and frequency $v = 0.7$ THz [128]. It was found that response depends on the gate length and the gate bias. The responsivity value of 200 V/W and NEP $\geq 10^{-10}$ W/Hz$^{1/2}$ were obtained and demonstrates the potential of Si MOSFETs as sensitive detectors of THz radiation. Also a 3×5 Si MOSFET FPA processed by a 0.25 μm CMOS technology was realized [46]. Each pixel of the array consists of a 645-GHz patch antenna coupled to a FET detector and a 43-dB voltage amplifier with a 1.6-MHz bandwidth. The NEP value of 3×10^{-10} W/Hz$^{1/2}$ was achieved that paved the way for the realization of broadband THz detectors and FPAs for high frame-rate imaging on the basis of CMOS technology. The performance of these fast detectors at room temperature is of the order of other uncooled detectors in THz radiation frequency range (see Table 22.1).

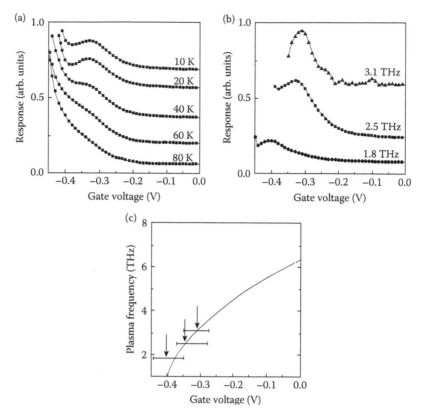

Figure 22.22 Characteristics of 60 nm gate length InGaAs/InAlAs transistor: (a) response at 2.5 THz frequency as a function of the gate voltage at different temperatures (80 K down to 10 K), (b) response at 10 K as a function of the gate voltage at different frequencies (1.8, 2.5, and 3.1 THz), and (c) position (indicated by arrows) of resonance maxima as a function of gate voltage. The calculated plasmon frequency as a function of the gate voltage, using Equation 22.12 for $V_{th} = -0.41$ V, is shown by the solid line. The error bars correspond to the linewidth of the measured plasmon resonance peaks. (From Teppe, F., El Fatimy, A., Boubanga, S., Seliuta, D., Valusis, G., Chenaud, B., and Knap, W., *Acta Physica Polonica A*, 113, 815–20, 2008.)

22.6 CONCLUSIONS

Progress in THz detector technology has achieved the level of performance that many discrete and low pixel arrays (e.g., SISs, HEBs, and TESs) operated at low or sub-Kelvin temperature are close to achieving for ultimate performance at a low background in the whole THz range. But the future sensitivity improvement of instruments will come with the use of large format arrays with readouts in the focal plane to provide the demands of high resolution spectroscopy (with v/ $\Delta v \approx 10^7$ at $v \geq 1$ THz) and vision demands. Superconducting HEB detectors are also promising as single-photon counters in near IR with low dark counts and GHz counting rate. The HTSC HEBs, because of their excess noise, are not expected to reach the sensitivity of low temperature superconducting HEBs, but due to short electron–phonon relaxation time, these materials are candidates for wide-band devices.

Uncooled and cooled heterodyne SBDs can provide relatively high sensitivity appropriate to many applications in mm and sub-mm spectral range, but they can hardly be assembled in the large number pixel arrays, because of the lack of high power compact solid state LO sources with needed power above 1 mW. Today's systems with single pixel coherent SBD detectors and arrays with a modest number of pixels are available but they fail in applications at v > 1 THz because of physical restrictions in their operation.

One of the important components of THz technologies are uncooled or slightly cooled THz sensors, but they require further sensitivity improvements to make the systems less complicated and

798

bulky. Uncooled or slightly cooled sensors based on 2-D electrons plasmon resonance in FETs are promising for use in large format arrays in low-cost systems. Other uncooled THz direct thermal detectors with NEP between 10^{-10} and 10^{-11} W/Hz$^{1/2}$ can be used in many low resolution spectroscopy applications and active vision systems.

REFERENCES

1. P. H. Siegel, "Terahertz Technology," *IEEE Transactions on Microwave Theory Technology* 50, 910–28, 2002.

2. P. H. Siegel and R. J. Dengler, "Terahertz Heterodyne Imaging Part I: Introduction and Techniques," *International Journal of Infrared Millimeter Waves* 27, 465–80, 2006.

3. P. H. Siegel and R. J. Dengler, "Terahertz Heterodyne Imaging Part II: Instruments," *International Journal of Infrared Millimeter Waves* 27, 631–55, 2006.

4. G. Chattopadhyay, "Submillimeter-Wave Coherent and Incoherent Sensors for Space Applications," in *Sensors: Advancements in Modeling, Design Issues, Fabrication and Practical Applications*, eds. S. C. Mukhopadhyay and R. Y. M. Huang, 387–414, Springer, New York, 2008.

5. T. W. Crowe, W. L. Bishop, D. W. Porterfield, J. L. Hesler, and R. M. Weikle, "Opening the Terahertz Window with Integrated Diode Circuits," *IEEE Journal of Solid-State Circuits* 40, 2104–10, 2005.

6. D. Dragoman and M. Dragoman, "Terahertz Fields and Applications," *Progress in Quantum Electronics* 28, 1–66, 2004.

7. J. Wei, D. Olaya, B. S. Karasik, S. V. Pereverzev, A. V. Sergeev, and M. E. Gershenzon, "Ultrasensitive Hot-Electron Nanobolometers for Terahertz Astrophysics," *Nature Nanotechnology* 3, 496–500, 2008.

8. A. H. Lettington, I. M. Blankson, M. Attia, and D. Dunn, "Review of Imaging Architecture," *Proceedings of SPIE* 4719, 327–40, 2002.

9. A. W. Blain, I. Smail, R. J. Ivison, J.-P. Kneib, and D. T. Frayer, "Submillimetre Galaxies," *Physics Reports* 369, 111–76, 2002.

10. D. Leisawitz, W. C. Danchi, M. J. DiPirro, L. D. Feinberg, D. Y. Gezari, M. Hagopian, W. D. Langer, et al., "Scientific Motivation and Technology Requirements for the SPIRIT and SPECS Far-Infrared/Submillimeter Space Interferometers," *Proceedings of SPIE* 4013, 36–46, 2000.

11. J. Zmuidzinas and P. L. Richards, "Superconducting Detectors and Mixers for Millimeter and Submillimeter Astrophysics," *Proceedings of IEEE* 92, 1597–616, 2004.

12. B. Ferguson and X.-C. Zhang, "Materials for Terahertz Science and Technology," *Nature Materials* 1, 26–33, 2002.

13. D. Mittleman, *Sensing with Terahertz Radiation*, Springer-Verlag, Berlin, 2003.

14. E. R. Brown, "Fundamentals of Terrestrial Millimetre-Wave and THz Remote Sensing," *International Journal of High Speed Electronics & Systems* 13, 99–1097, 2003.

15. R. M. Woodward, "Terahertz Technology in Global Homeland Security," *Proceedings of SPIE* 5781, 22–31, 2005.

16. D. L. Woolard, R. Brown, M. Pepper, and M. Kemp, "Terahertz Frequency Sensing and Imaging: A Time of Reckoning Future Applications?" *Proceedings of IEEE* 93, 1722–43, 2005.

17. H. Zhong, A. Redo-Sanchez, and X.-C. Zhang, "Identification and Classification of Chemicals Using Terahertz Reflective Spectroscopic Focal-Plane Imaging System," *Optics Express* 14, 9130–41, 2006.

18. M. Tonouchi, "Cutting-Edge Terahertz Technology," *Nature Photonics* 1, 97–105, 2007.

19. G. L. Carr, M. C. Martin, W. R. McKinney, K. Jordan, G. R. Neil, and G. P. Williams, "High Power Terahertz Radiation from Relativistic Electrons," *Nature* 420, 153–56, 2002.

20. H. Wang, L. Samoska, T. Gaier, A. Peralta, H. Liao, Y. C. Leong, S. Weinreb, Y. C. Chen, M. Nishimoto, and R. Lai, "Power-Amplifier Modules Covering 7–113 GHz Using MMICs," *IEEE Transactions* MTT-49, 9–16, 2001.

21. S. Barbieri, J. Alton, S. S. Dhillon, H. E. Beere, M. Evans, E. H. Linfield, A. G. Davies, et al., "Continuous-Wave Operation of Terahertz Quantum-Cascade Lasers," *IEEE Journal of Quantum Electronics* 39, 586–91, 2003.

22. H.-W. Hübers, "Terahertz Heterodyne Receivers," *IEEE Journal of Selected Topics in Quantum Electronics* 14, 378–91, 2008.

23. N. Kopeika, *A System Engineering Approach to Imaging*, SPIE Optical Engineering Press, Bellingham, WA, 1998.

24. M. A. Kinch and B. V. Rollin, "Detection of Millimetre and Sub-Millimetre Wave Radiation by Free Carrier Absorption in a Semiconductor," *British Journal of Applied Physics* 14, 672–76, 1963.

25. Y. Nakagawa and H. Yoshinaga, "Characteristics of High-Sensitivity Ge Bolometer," *Japanese Journal of Applied Physics* 9, 125–31, 1970.

26. P. R. Bratt, "Impurity Germanium and Silicon Infrared Detectors," in *Semiconductors and Semimetals*, Vol. 12, eds. R. K. Willardson and A. C. Beer, 39–142, Academic Press, New York, 1977.

27. E. E. Haller, M. R. Hueschen, and P. L. Richards, "Ge:Ga Photoconductors in Low Infrared Backgrounds," *Applied Physics Letters* 34, 495–97, 1979.

28. N. Sclar, "Properties of Doped Silicon and Germanium Infrared Detectors," *Progress in Quantum Electronics* 9, 149–257, 1984.

29. R. Padman, G. J. White, R. Barker, D. Bly, N. Johnson, H. Gibson, M. Griffin, et al., "A Dual-Polarization InSb Receiver for 461/492 GHz," *International Journal of Infrared Millimeter Waves* 13, 1487–513, 1992.

30. J. E. Huffman, "Infrared Detectors for 2 to 220 μm Astronomy," *Proceedings of SPIE* 2274, 157–69, 1995.

31. A. G. Kazanskii, P. L. Richards, and E. E. Haller, "Far-Infrared Photoconductivity of Uniaxially Stressed Germanium," *Applied Physics Letters* 31, 496–97, 1977.

32. H.-W. Hübers, S. G. Pavlov, K. Holldack, U. Schade, and G. Wüstefeld, "Long Wavelength Response of Unstressed and Stressed Ge:Ga Detectors," *Proceedings of SPIE* 6275, 627505, 2008.

33. R. Hoenle, J. W. Beeman, E. E. Haller, and U. Groezinger, "Far-Infrared Photoconductors for Herschel and SOFIA," *Proceedings of SPIE* 4855, 115–28, 2003.

34. G. H. Rieke, "Infrared Detector Arrays for Astronomy," *Annual Review of Astronomy and Astrophysics* 45, 77–115, 2007.

35. S. Hargreaves and R. A. Lewis, "Terahertz Imaging: Materials and Methods," *Journal of Materials Science: Materials in Electronics* 18, S299–S303, 2007.

36. N. Karpowicz, H. Zhong, J. Xu, K.-I. Lin, J.-S. Hwang, and X.-C. Zhang, "Nondestructive Sub-THz Imaging," *Proceedings of SPIE* 5727, 132–42, 2005.

37. A. Dobroiu, C. Otani, and K. Kawase, "Terahertz-Wave Sources and Imaging Applications," *Measurement Science and Technology* 17, R161–R174, 2006.

38. P. L. Richards, "Bolometers for Infrared and Millimeter Waves," *Journal of Applied Physics* 76, 1–24, 1994.

39. M. Kenyon, P. K. Day, C. M. Bradford, J. J. Bock, and H. G. Leduc, "Progress on Background-Limited Membrane-Isolated TES Bolometers for Far-IR/Submillimeter Spectroscopy," *Proceedings of SPIE* 6275, 627508, 2006.

40. A. D. Turner, J. J. Bock, J. W. Beeman, J. Glenn, P. C. Hargrave, V. V. Hristov, H. T. Nguyen, F. Rahman, S. Sethuraman, and A. L. Woodcraft, "Silicon Nitride Micromesh Bolometer Array for Submillimeter Astrophysics," *Applied Optics* 40, 4921–32, 2001.

41. B. S. Karasik, D. Olaya, J. Wei, S. Pereverzev, M. E. Gershenson, J. H. Kawamura, W. R. McGrath, and A. V. Sergeev, "Record-Low NEP in Hot-Electron Titanium Nanobolometers," *IEEE Transactions on Applied Superconductivity* 17, 293–97, 2007.

42. F. Sizov, "THz Radiation Sensors," *Opto-Electronics Review* 18, 10–36, 2010.

43. T.-L. Hwang, S. E. Scharz, and D. B. Rutledge, "Microbolometers for Infrared Detection," *Applied Physics Letters* 34, 773–76, 1979.

44. E. N. Grossman and A. J. Miller, "Active Millimeter-Wave Imaging for Concealed Weapons Detection," *Proceedings of SPIE* 5077, 62–70, 2003.

45. W. Knap, J. Łusakowski, F. Teppe, N. Dyakonova, and Y. Meziani, "Terahertz Generation and Detection by Plasma Waves in Nanometer Gate High Electron Mobility Transistors," *Acta Physica Polonica A* 107, 82–90, 2005.

46. A. Lisauskas, D. Glaab, H. G. Roskos, E. U. Oejefors, and R. Pfeiffer, "Terahertz Imaging with Si MOSFET Focal-Plane Arrays," *Proceedings of SPIE* 7215, 72150J, 2009.

47. V. N. Dobrovolsky, F. F. Sizov, Y. E. Kamenev, and A. B. Smirnov, "Ambient Temperature or Moderately Cooled Hot Electron Bolometer for mm and Sub-mm Regions," *Opto-Electronics Review* 16, 172–78, 2008.

48. L. Kuzmin, "Optimal Cold-Electron Bolometer with a Superconductor-Insulator-Normal Tunnel Junction and an Andreev Contact," *17th International Symposium on Space THz Technology*, 183–86, Paris, May 10–12, 2006.

49. D. J. Benford, "Transition Edge Sensor Bolometers for CMB Polarimetry," http://cmbpol.uchicago.edu/workshops/technology2008/depot/cmbpol_technologies_benford_jcps_4.pdf

50. C. A. Allen, M. J. Amato, S. R. Babu, A. E. Bartels, D. J. Benford, R. J. Derro, C. D. Dowell, et al., "Design and Fabrication of Two-Dimensional Semiconducting Bolometer Arrays for HAWC and SHARC-II," *Proceedings of SPIE* 4855, 63–72, 2003.

51. T. J. Ames, T. G. Phillips, and C. Rioux, "Astronomical Demonstration of Superconducting Bolometer Arrays," *Proceedings of SPIE* 4855, 100–107, 2003.

52. G. H. Rieke, *Detection of Light: From the Ultraviolet to the Submillimeter*, Cambridge University Press, Cambridge, 2003.

53. V. G. Bozhkov, "Semiconductor Detectors, Mixers, and Frequency Multipliers for the Terahertz Band," *Radiophysics and Quantum Electronics* 46, 631–56, 2003.

54. T. W. Crowe, R. J. Mattauch, H.-P. Roser, W. L. Bishop, W. C. B. Peatman, and X. Liu, "GaAs Schottky Diodes for THz Mixing Applications," *Proceedings of IEEE* 80, 1827–41, 1992.

55. A. Van Der Ziel, "Infrared Detection and Mixing in Heavily Doped Schottky Barrier Diodes," *Journal of Applied Physics* 47, 2059–68, 1976.

56. H. A. Watson, *Microwave Semiconductor Devices and Their Circuit Applications*, McGraw-Hill, New York, 1969.

57. E. J. Becklake, C. D. Payne, and B. E. Pruer, "Submillimetre Performance of Diode Detectors Using Ge, Si and GaAs," *Journal of Physics D: Applied Physics* 3, 473–81, 1970.

58. D. T. Young and J. C. Irvin, "Millimeter Frequency Conversion Using Au-n-type GaAs Schottky Barrier Epitaxial Diodes with a Novel Contacting Technique," *Proceedings of IEEE* 12, 2130–32, 1965.

59. T. W. Crowe, D. P. Porterfield, J. L. Hesler, W. L. Bishop, D. S. Kurtz, and K. Hui, "Terahertz Sources and Detectors," *Proceedings of SPIE* 5790, 271–80, 2005.

60. H. P. Röser, H.-W. Hübers, E. Bründermann, and M. F. Kimmitt, "Observation of Mesoscopic Effects in Schottky Diodes at 300 K when Used as Mixers at THz Frequencies," *Semiconductor Science and Technology* 11, 1328–32, 1996.

61. L. K. Seidel and T. W. Crowe "Fabrication and Analysis of GaAs Schottky Barrier Diodes Fabricated on Thin Membranes for Terahertz Applications." *International Journal of Infrared and Millimeter Waves* 10, 779–787, 1989.

62. T. W. Crowe, "GaAs Schottky Barrier Mixer Diodes for the Frequency Range 1–10 THz," *International Journal of Infrared Millimeter Waves* 11, 765–77, 1990.

63. H. Kräutle, E. Sauter, and G. V. Schultz, "Antenna Characteristics of Whisker Diodes Used at Submillimeter Receivers," *Infrared Physics* 17, 477–83, 1977.

64. R. Titz, B. Auel, W. Esch, H. P. Röser, and G. W. Schwaab, "Antenna Measurements of Open-Structure Schottky Mixers and Determination of Optical Elements for a Heterodyne System at 184, 214 and 287 μm," *Infrared Physics* 30, 435–41, 1990.

65. S. M. Marazita, W. L. Bishop, J. L. Hesler, K. Hui, W. E. Bowen, and T. W. Crowe, "Integrated GaAs Schottky Mixers by Spin-on-Dielectric Wafer Bonding," *IEEE Transactions on Electron Devices* 47, 1152–56, 2000.

66. P. Siegel, R. P. Smith, M. C. Gaidis, and S. Martin, "2.5-THz GaAs Monolithic Membrane-Diode Mixer," *IEEE Transactions on Microwave Theory Technology* 47, 596–604, 1999.

67. J. A. Copeland, "Diode Edge Effects on Doping Profile Measurements," *IEEE Transactions on Electron Devices* 17, 404–7, 1970.

68. V. I. Piddyachiy, V. M. Shulga, A. M. Korolev, and V. V. Myshenko, "High Doping Density Schottky Diodes in the 3 mm Wavelength Cryogenic Heterodyne Receiver," *International Journal of Infrared Millimeter Waves* 26, 1307–15, 2005.

69. F. Maiwald, F. Lewen, B. Vowinkel, W. Jabs, D. G. Paveljev, M. Winnerwisser, and G. Winnerwisser, "Planar Schottky Diode Frequency Multiplier for Molecular Spectroscopy up to 1.3 THz," *IEEE Microwave Guided Wave Letters* 9, 198–200, 1999.

70. D. H. Martin, *Spectroscopic Techniques for Far-Infrared, Submillimeter and Millimeter Waves*, North-Holland, Amsterdam, 1967.

71. E. Burstein, D. N. Langenberg, and B. N. Taylor, "Superconductors as Quantum Detectors for Microwave and Sub-Millimeter Radiation," *Physical Review Letters* 6, 92–94, 1961.

72. A. H. Dayem and R. J. Martin, "Quantum Interaction of Microwave Radiation with Tunnelling between Superconductors," *Physical Review Letters* 8, 246–48, 1962.

73. P. K. Tien and J. P. Gordon, "Multiphoton Process Observed in the Interaction of Microwave Fields with the Tunnelling between Superconductor Films," *Physical Review* 129, 647–51, 1963.

74. P. L. Richards, T. M. Shen, R. E. Harris, and F. L. Lloyd, "Quasiparticle Heterodyne Mixing in SIS Tunnel Junctions," *Applied Physics Letters* 34, 345–47, 1979.

75. G. J. Dolan, T. G. Phillips, and D. P. Woody, "Low-Noise 115 GHz Mixing in Superconducting Oxide-Barrier Tunnel Junctions," *Applied Physics Letters* 34, 347–49, 1979.

76. J. R. Tucker and M. J. Feldman, "Quantum Detection at Millimeter Wavelength," *Reviews of Modern Physics* 57, 1055–1113, 1985.

77. Ch. Otani, S. Ariyoshi, H. Matsuo, T. Morishima, M. Yamashita, K. Kawase, H. Satoa, and H. M. Shimizu, "Terahertz Direct Detector Using Superconducting Tunnel Junctions," *Proceedings of SPIE* 5354, 86–93, 2004.

78. V. P. Koshelets, S. V. Shitov, L. V. Filippenko, P. N. Dmitriev, A. N. Ermakov, A. S. Sobolev, and M. Yu. Torgashin, "Integrated Superconducting Sub-mm Wave Receivers," *Radiophysics and Quantum Electronics* 46, 618–30, 2003.

79. C. A. Mears, Q. Hu, P. L. Richards, A. H. Worsham, D. E. Prober, and A. V. Raisanen, "Quantum Limited Heterodyne Detection of Millimeter Waves Using Super Conducting Tantalum Tunnel Junctions," *Applied Physics Letters* 57, 2487–89, 1990.

80. A. Karpov, D. Miller, F. Rice, J. A. Stern, B. Bumble, H. G. LeDuc, and J. Zmuidzinas, "Low Noise SIS Mixer for Far Infrared Radio Astronomy," *Proceedings of SPIE* 5498, 616–21, 2004.

81. G. Chattopadhyay, "Future of Heterodyne Receivers at Submillimeter Wavelengths," *Digest IRMMW-THz-2005 Conference*, 461–62, 2005.

82. G. N. Gol'tsman, "Hot Electron Bolometric Mixers: New Terahertz Technology," *Infrared Physics & Technology* 40, 199–206, 1999.

83. T. G. Phillips and K. B. Jefferts, "A Low Temperature Bolometer Heterodyne Receiver for Millimeter Wave Astronomy," *Review of Scientific Instruments* 44, 1009–14, 1973.

84. K. Seeger, *Semiconductor Physics*, Springer, Berlin, 1991.

85. S. M. Smith, M. J. Cronin, R. J. Nicholas, M. A. Brummell, J. J. Harris, and C. T. Foxon, "Millimeter and Submillimeter Detection Using $Ga_{1-x}Al_xAs/GaAs$ Heterostructures," *International Journal of Infrared Millimeter Waves* 8, 793–802, 1987.

86. J.-X. Yang, F. Agahi, D. Dai, C. F. Musante, W. Grammer, K. M. Lau, and K. S. Yngvesson, "Wide-Bandwidth Electron Bolometric Mixers: A 2DEG Prototype and Potential for Low-Noise THz Receivers," *IEEE Transactions on Microwave Theory Technology* 41, 581–89, 1993.

87. G. N. Gol'tsman and K. V. Smirnov, "Electron-Phonon Interaction in a Two-Dimensional Electron Gas of Semiconductor Heterostructures at Low Temperatures," *JETP Letters* 74, 474–79, 2001.

88. A. A. Verevkin, N. G. Ptitsina, K. V. Smirnov, G. N. Gol'tsman, E. M. Gershenzon, and K. S. Ingvesson, "Direct Measurements of Energy Relaxation Times on an AlGaAs/GaAs Heterointerface in the Range 4.2–50 K," *JETP Letters* 64, 404–9, 1996.

89. T. Phillips and D. Woody, "Millimeter-Wave and Submillimeter-Wave Receivers," *Annual Review of Astronomy and Astrophysics* 20, 285–321, 1982.

90. "Detectors Needs for Long Wavelength Astrophysics," *A Report by the Infrared, Submillimeter, and Millimeter Detector Working Group,* June 2002.

91. P. Agnese, C. Buzzi, P. Rey, L. Rodriguez, and J. L. Tissot, "New Technological Development for Far-Infrared Bolometer Arrays," *Proceedings of SPIE* 3698, 284–90, 1999.

92. C. Dowell, C. A. Allen, S. Babu, M. M. Freund, M. B. Gardnera, J. Groseth, M. Jhabvala, et al., "SHARC II: A Caltech Submillimeter Observatory Facility Camera with 384 Pixels," *Proceedings of SPIE* 4855, 73–87, 2003.

93. N. Billot, P. Agnese, J. L. Augueres, A. Beguin, A. Bouere, O. Boulade, C. Cara, et al., "The Herschel/PACS 2560 Bolometers Imaging Camera," *Proceedings of SPIE* 6265, 62650D, 2006.

94. J. J. A. Baselmans, A. Baryshev, S. F. Reker, M. Hajenius, J. Gao, T. Klapwijk, B. Voronov, and G. Gol'tsman, "Influence of the Direct Response on the Heterodyne Sensitivity of Hot Electron Bolometer Mixers," *Journal of Applied Physics* 100, 184103, 2006.

95. G. N. Gol'tsman, Yu. B. Vachtomin, S. V. Antipov, M. I. Finkel, S. N. Maslennikov, K. V. Smirnov, S. L. Poluakov, et al., "NbN Phonon-Cooled Hot-Electron Bolometer Mixer for Terahertz Heterodyne Receivers," *Proceedings of SPIE* 5727, 95–106, 2005.

96. K. S. Il'in, M. Lindgren, M. Currie, A. D. Semenov, G. N. Gol'tsman, R. Sobolewski, S. I. Cherednichenko, and E. M. Gershenzon, "Picosecond Hot-Electron Energy Relaxation in NbN Superconducting Photodetectors," *Applied Physics Letters* 76, 2752–54, 2000.

97. A. D. Semenov, G. N. Gol'tsman, and R. Sobolewski, "Hot-Electron Effect in Semiconductors and Its Applications for Radiation Sensors," *Semiconductor Science and Technology* 15, R1–R16, 2002.

98. E. M. Gershenson, M. E. Gershenson, G. N. Goltsman, B. S. Karasik, A. M. Lyul'kin, and A. D. Semenov, "Ultra-Fast Superconducting Electron Bolometer," *Journal of Technical Physics Letters* 15, 118–19, 1989.

99. Y. Gousev, G. Gol'tsman, A. Semenov, E. Gershenzon, R. Nebosis, M. Heusinger, and K. Renk, "Broad-Band Ultrafast Superconducting NbN Detector for Electromagnetic-Radiation," *Journal of Applied Physics* 75, 3695–97, 1994.

100. A. J. Kreisler and A. Gaugue, "Recent Progress in High-Temperature Superconductor Bolometric Detectors: From the Mid-Infrared to the Far-Infrared (THz) Range," *Semiconductor Science and Technology* 13, 1235–45, 2000.

101. M. Lindgren, M. Currie, C. Williams, T. Y. Hsiang, P. M. Fauchet, R. Sobolewsky, S. H. Moffat, R. A. Hughes, J. S. Preston, and F. A. Hegmann, "Intrinsic Picosecond Response Times of Y-Ba-Cu-O Superconducting Photoresponse, *Applied Physics Letters* 74, 853–55, 1999.

102. M. V. Lyatti, D. A. Tkachev, and Yu. Ya. Divin, "Signal and Noise Characteristics of a Terahertz Frequency-Selective $YBa_2Cu_3O_{7-\delta}$ Josephson Detector," *Technical Physics Letters* 32, 860–62, 2006.

103. K. Irwin, "An Application of Electrothermal Feedback for High-Resolution Cryogenic Particle-Detection," *Applied Physics Letters* 66, 1998–2000, 1995.

104. W. Duncan, W. S. Holland, M. D. Audley, M. Cliffe, T. Hodson, B. D. Kelly, X. Gao, et al., "SCUBA-2: Developing the Detectors," *Proceedings of SPIE* 4855, 19–29, 2003.

105. A. J. Walton, W. Parkes, J. G. Terry, C. Dunare, J. T. M. Stevenson, A. M. Gundlach, G. C. Hilton, et al., "Design and Fabrication of the Detector Technology for SCUBA-2," *IEE Proceedings. Science, Measurement and Technology* 151, 119–20, 2004.

106. A.-D. Brown, D. Chuss, V. Mikula, R. Henry, E. Wollack, Y. Zhao, G. C. Hilton, and J. A. Chervenak, "Auxiliary Components for Kilopixel Transition Edge Sensor Arrays," *Solid State Electronics* 52, 1619–24, 2008.

107. S. Lee, J. Gildemeister, W. Holmes, A. Lee, and P. Richards, "Voltage-Biased Superconducting Transition-Edge Bolometer with Strong Electrothermal Feedback Operated at 370 mK," *Applied Optics* 37, 3391–97, 1998.

108. H. F. C. Hoevers, A. C. Bento, M. P. Bruijn, L. Gottardi, M. A. N. Korevaar, W. A. Mels, and P. A. J. de Korte, "Thermal Fluctuation Noise in a Voltage Biased Superconducting Transition Edge Thermometer," *Applied Physics Letters* 77, 4422–24, 2000.

109. M. D. Audley, D. M. Glowacka, D. J. Goldie, A. N. Lasenby, V. N. Tsaneva, S. Withington, P. K. Grimes, et al., "Tests of Finline-Coupled TES Bolometers for COVER," *Digest IRMMW-THz-2007 Conference*, 180–81, Cardiff, 2007.

110. *The SQUID Handbook, Vol. II: Applications*, eds. J. Clarke and A. I. Braginski, Wiley-VCH, Weinheim, 2006.

111. J. A. Chervenak, K. D. Irwin, E. N. Grossman, J. M. Martinis, C. D. Reintsema, and M. E. Huber, "Superconducting Multiplexer for Arrays of Transition Edge Sensors," *Applied Physics Letters* 74, 4043–45, 1999.

112. P. J. Yoon, J. Clarke, J. M. Gildemeister, A. T. Lee, M. J. Myers, P. L. Richards, and J. T. Skidmore, "Single Superconducting Quantum Interference Device Multiplexer for Arrays of Low-Temperature Sensors," *Applied Physics Letters* 78, 371–73, 2001.

113. K. D. Irvin, "SQUID Multiplexers for Transition-Edge Sensors," *Physica C* 368, 203–10, 2002.

114. K. D. Irwin, M. D. Audley, J. A. Beall, J. Beyer, S. Deiker, W. Doriese, W. D. Duncan, et al., "In-Focal-Plane SQUID Multiplexer," *Nuclear Instruments & Methods in Physics Research* A520, 544–47, 2004.

115. K. D. Irvin and G. C. Hilton, "Transition-Edge Sensors," in *Cryogenic Particle Detection*, ed. C. Enss, 63–149, Springer-Verlag, Berlin, 2005.

116. T. M. Lanting, H. M. Cho, J. Clarke, W. L. Holzapfel, A. T. Lee, M. Lueker, P. L. Richards, M. A. Dobbs, H. Spieler, and A. Smith, "Frequency-Domain Multiplexed Readout of Transition-Edge Sensor Arrays with a Superconducting Quantum Interference Device," *Applied Physics Letters* 86, 112511, 2005.

117. W. S. Holland, W. Duncan, B. D. Kelly, K. D. Irwin, A. J. Walton, P. A. R. Ade, and E. I. Robson, "SCUBA-2: A New Generation Submillimeter Imager for the James Clerk Maxwell Telescope," *Proceedings of SPIE* 4855, 1–18, 2003.

118. A. L. Woodcraft, M. I. Hollister, D. Bintley, M. A. Ellis, X. Gao, W. S. Holland, M. J. MacIntosh, et al., "Characterization of a Prototype SCUBA-2 1280-Pixel Submillimetre Superconducting Bolometer Array," *Proceedings of SPIE* 6275, 62751F, 2006.

119. D. J. Benford, J. G. Steguhn, T. J. Ames, C. A. Allen, J. A. Chervenak, C. R. Kennedy, S. Lefranc, et al., "First Astronomical Images with a Multiplexed Superconducting Bolometer Array," *Proceedings of SPIE* 6275, 62751C, 2006.

120. J. Gildemeister, A. Lee, and P. Richards, "Monolithic Arrays of Absorber-Coupled Voltage-Biased Superconducting Bolometers," *Applied Physics Letters* 77, 4040–42, 2000.

121. D. J. Benford, G. M. Voellmer, J. A. Chervenak, K. D. Irwin, S. H. Moseley, R. A. Shafer, G. J. Stacey, and J. G. Staguhn, "Thousand-Element Multiplexed Superconducting Bolometer Arrays," in *Proceedings on Far-IR, Sub-MM, and MM Detector Workshop*, Vol. NASA/CP-2003-211 408, eds. J. Wolf, J. Farhoomand, and C. R. McCreight, 272–75, 2003.

122. J. Gildemeister, A. Lee, and P. Richards, "A Fully Lithographed Voltage-Biased Superconducting Spiderweb Bolometer," *Applied Physics Letters* 74, 868–70, 1999.

123. A. El Fatimy, F. Teppe, N. Dyakonova, W. Knap, D. Seliuta, G. Valusis, A. Shchepetov, et al., "Resonant and Voltage-Tunable Terahertz Detection in InGaAs/InP Nanometer Transistors, *Applied Physics Letters* 89, 131926, 2006.

124. W. Knap, V. Kachorowskii, Y. Deng, S. Rumyantsev, J.-Q. Lu, R. Gaska, M. S. Shur, et al., "Nonresonant Detection of Terahertz Radiation in Field Effect Transistors," *Journal of Applied Physics* 91, 9346–53, 2002.

125. Y. M. Meziani, J. Lusakowski, N. Dyakonova, W. Knap, D. Seliuta, E. Sirmulis, J. Deverson, G. Valusis, F. Boeuf, and T. Skotnicki, "Non Resonant Response to Terahertz Radiation by Submicron CMOS Transistors," *IEICE Transactions on Electronics* E89-C, 993–98, 2006.

126. G. C. Dyer, J. D. Crossno, G. R. Aizin, J. Mikalopas, E. A. Shaner, M. C. Wanke, J. L. Reno, and S. J. Allen, "A Narrowband Plasmonic Terahertz Detector with a Monolithic Hot Electron Bolometer," *Proceedings of SPIE* 7215, 721503, 2009.

127. F. Teppe, M. Orlov, A. El Fatimy, A. Tiberj, W. Knap, J. Torres, V. Gavrilenko, A. Shchepetov, Y. Roelens, and S. Bollaert, "Room Temperature Tunable Detection of Subterahertz Radiation by Plasma Waves in Nanometer InGaAs Transistors," *Applied Physics Letters* 89, 222109, 2006.

128. R. Tauk, F. Teppe, S. Boubanga, D. Coquillat, W. Knap, Y. M. Meziani, C. Gallon, F. Boeuf, T. Skotnicki, and C. Fenouillet-Beranger, "Plasma Wave Detection of Terahertz Radiation by Silicon Field Effects Transistors: Responsivity and Noise Equivalent Power," *Applied Physics Letters* 89, 253511, 2006.

129. V. I. Gavrilenko, E. V. Demidov, K. V. Marem'yanin, S. V. Morozov, W. Knap, and J. Lusakowski, "Electron Transport and Detection of Terahertz Radiation in a GaN/AlGaN Submicrometer Field-Effect Transistor," *Semiconductors* 41, 232–34, 2007.

130. Y. M. Meziani, M. Hanabe, A. Koizumi, T. Otsuji, and E. Sano, "Self Oscillation of the Plasma Waves in a Dual Grating Gates HEMT Device," *International Conference on Indium Phosphide and Related Materials*, Conference Proceedings, 534–37, Matsue, 2007.

131. A. M. Hashim, S. Kasai, and H. Hasegawa, "Observation of First and Third Harmonic Responses in Two-Dimensional AlGaAs/GaAs HEMT Devices Due to Plasma Wave Interaction," *Superlattice Microstructures* 44, 754–60, 2008.

132. V. Ryzhii, A. Satou, I. Khmyrova, M. Ryzhii, T. Otsuji, V. Mitin, and M. S. Shur, "Plasma Effects in Lateral Schottky Junction Tunneling Transit-Time Terahertz Oscillator," *Journal of Physics: Conference Series* 38, 228–33, 2006.

133. X. G. Peralta, S. J. Allen, M. C. Wanke, N. E. Harff, J. A. Simmons, M. P. Lilly, J. L. Reno, P. J. Burke, and J. P. Eisenstein, "Terahertz Photoconductivity and Plasmon Modes in Double-Quantum-Well Field-Effect Transistors," *Applied Physics Letters* 81, 1627–30, 2002.

134. M. Dyakonov and M. S. Shur, "Shallow Water Analogy for a Ballistic Field Effect Transistor: New Mechanism of Plasma Wave Generation by the DC Current," *Physical Review Letters* 71, 2465–68, 1993.

135. M. Dyakonov and M. Shur, "Plasma Wave Electronics: Novel Terahertz Devices Using Two Dimensional Electron Fluid," *IEEE Transactions on Electron Devices* 43, 1640–46, 1996.

136. M. Shur and V. Ryzhii, "Plasma Wave Electronics," *International Journal of High Speed Electronics and Systems* 13, 575–600, 2003.

137. F. Teppe, A. El Fatimy, S. Boubanga, D. Seliuta, G. Valusis, B. Chenaud, and W. Knap, "Terahertz Resonant Detection by Plasma Waves in Nanometric Transistors," *Acta Physica Polonica A* 113, 815–20, 2008.

23 Third-Generation Infrared Detectors

Multicolor detector capabilities are highly desirable for advanced infrared (IR) imaging systems, since they provide enhanced target discrimination and identification, combined with lower false-alarm rates. Systems that collect data in separate IR spectral bands can discriminate both absolute temperature as well as unique signatures of objects in the scene. By providing this new dimension of contrast, multiband detection also offers advanced color processing algorithms to further improve sensitivity above that of single-color devices. This is extremely important for identifying temperature differences between missile targets, warheads, and decoys. Multispectral IR focal plane arrays (FPAs) are highly beneficial for a variety of applications such as missile warning and guidance, precision strike, airborne surveillance, target detection, recognition, acquisition and tracking, thermal imaging, navigational aids and night vision, and so on. [1,2]. They also play an important role in Earth and planetary remote sensing, astronomy, and so forth. [3].

Military surveillance, target detection, and target tracking can be undertaken using single-color FPAs if the targets are easy to identify. However, in the presence of clutter, or when the target and/or background are uncertain, or in situations where the target and/or background may change during engagement, the single-color system design involves compromises that can degrade overall capability. It is well established that in order to reduce clutter and enhance the desired features/contrast, one will require the use of multispectral FPAs. In such cases, multicolor imaging can greatly improve overall system performance.

Currently, multispectral systems rely on cumbersome imaging techniques that either disperse the optical signal across multiple IR FPAs or use a filter wheel to spectrally discriminate the image focused on a single FPA. These systems include beamsplitters, lenses, and bandpass filters in the optical path to focus the images onto separate FPAs responding to different IR bands. Also, complex optical alignment is required to map the multispectral image pixel-for-pixel. Consequently, these approaches are relatively high cost and place additional burdens on the sensor platform because of their extensive size, complexity, and cooling requirements.

In the future, multispectral imaging systems will include very large sensors feeding an enormous amount of data to the digital mission processing subsystem. The FPAs with the number of pixels above one million are now available. As these imaging arrays grow in detector number for higher resolution, so will the computing requirements for the embedded digital image processing system. One approach to solving this processing bottleneck problem could be to incorporate a certain amount of pixel-level processing within the detector pixel, similar to the technique implemented in biological sensor information processing systems. Currently, several scientific groups in the world have turned to the biological retina for answers as to how to improve man-made sensors [4,5].

In this chapter, we will review the state-of-the-art muticolor detector technologies over a wide IR spectral range. In the wavelength regions of interest such as SWIR, MWIR, and LWIR, four detector technologies that are developing multicolor capability are visited here: HgCdTe, quantum well infrared photodetectors (QWIPs), antimonide based type II superlattices, and quantum dot infrared photodetectors (QDIPs).

Both HgCdTe photodiodes [6–13] and QWIPs [2,14–21] offer multicolor capability in the SWIR, MWIR, and LWIR range. The performance figures of merit of state-of-the-art QWIP and HgCdTe FPAs are similar because the main limitations are related to the readout circuits. A more detailed comparison of both technologies has been given by Tidrow at el. [2] and Rogalski [16,22].

Recently, type II InAs/GaInSb superlattices [18,23–28] and QDIPs [27–34] have emerged as possible candidates for third generation infrared detectors. Table 23.1 compares the essential properties of three types of LWIR devices at 77 K. Whether the low dimensional solid IR photodetectors can outperform the "bulk" narrow gap HgCdTe detectors is one of the most important questions that needs to be addressed for the future of IR photodetectors.

The subsections below describe issues associated with the development and exploitation of materials used in the fabrication of multicolor infrared detectors. Finally, we discuss the on-going detector technology efforts being undertaken to realize third generation FPAs.

23.1 BENEFITS OF MULTICOLOR DETECTION

Infrared multispectral imaging, sometimes referred to in the literature as hyperspectral imaging, is a relatively recent development that combines the information available from spectroscopy with the ability to acquire this information in a spatially resolved manner [35]. Instrumentally, an IR camera is used to record the spatial distribution of infrared radiation in the scene, and the spectral information is gained by scanning a dispersive element to record spectra for each image.

Table 23.1: Essential Properties of LWIR HgCdTe and Type II SL photodiodes and QWIPs at 77 K

Parameter	HgCdTe	QWIP (n-type)	InAs/GaInSb SL
IR absorption	Normal incidence	$E_{optical} \perp$ plane of well required Normal incidence: no absorption	Normal incidence
Quantum efficiency	≥70%	≤10%	≈60%–70%
Spectral sensitivity	Wide-band	Narrow-band (FWHM ≈ 1÷2 μm)	Wide-band
Optical gain	1	0.2 (30–50 wells)	1
Thermal generation lifetime	≈1μs	≈10 ps	≈0.1μs
R_oA product (λ_c = 10 μm)	300 Ωcm²	10^4 Ωcm²	200 Ωcm²
Detectivity (λ_c = 10 μm, FOV = 0)	2×10^{12} cmHz$^{1/2}$W^{-1}	2×10^{10} cmHz$^{1/2}$W^{-1}	1×10^{12} cmHz$^{1/2}$W^{-1}

Single-color FPAs in conjunction with spectral filters, grating spectrometers of Fourier transform spectrometers have been deployed for a variety of NASA spaceborne remote-sensing applications utilizing push-broom scanning to record hyperspectral images of the earth over the visible through very LWIR (VLWIR) spectral range. Dispersive devices based on mechanical scanning (e.g., filter wheels, monochromators) are not desirable because, in addition to their relatively large size, they are prone to vibrations and can be spectrally tuned in a relatively narrow range of a relatively slow speed [36]. Recent advances in material, electronic, and optical technologies have led to the development of novel types of electronically tunable filters, including so called adaptive FPAs [37].

As mentioned previously, simultaneous detection in multiwavelength bands with a single FPA results in reduction or elimination of heavy and complex optical components now required for wavelength differentiation in remote sensors and leads to smaller, lighter, simpler instruments with better performance.

Whenever the target to be detected behaves as a black body, the true temperature inferring from the body is accurate and reliable. On the other hand, when it exhibits a behavior different from a blackbody, emissivity compensation needs to be undertaken. In case of a known emissivity, ε, a single wavelength system can be used whereas, for gray body (unknown but constant emissivity in a narrow bandwidth) a dual-color system is more likely to be utilized.

Let us consider detectors, which are selected so as to cover the spectral range of blackbody emission from the target surface at the desired temperature. Thus, their cutoff wavelength will be shorter for higher temperature objects. Depending on the selected detectors, the two-band technique allows for temperature measurements, for example with either MWIR detectors (higher temperature range) or LWIR detectors (lower temperature range). The detector is aimed toward the scene that is at a temperature T, which initially can be taken as a blackbody given by Planck's law:

$$r(\lambda) = \frac{2h\nu^2}{\lambda^3 \exp\left[(h\nu/kT) - 1\right]},$$

(23.1)

where r is the radiance per unit wavelength and ν is the radiation frequency.

The technique of two-color detection consists of making two measurements of the collected power at two separate wavelengths λ_1 and λ_2. It can be shown that the ratio of the detected signals is equal to

$$R = \left(\frac{\varepsilon_1}{\varepsilon_2}\right)\left(\frac{\lambda_1}{\lambda_2}\right)^5 \left(\frac{\sigma_2 \Delta\lambda_2}{\sigma_1 \Delta\lambda_1}\right)\exp\left[\frac{hc}{kT}\left(\frac{1}{\lambda_2} - \frac{1}{\lambda_1}\right)\right].$$

(23.2)

The signal is now of the type

$$R = C_1 \exp\left(\frac{C_2}{T}\right),$$

(23.3)

where C_1 and C_2 are constants of the instrument. Taking the logarithm of R we have

$$\ln R = \ln C_1 + \frac{C_2}{T} \qquad (23.4)$$

and finally solving for T

$$T = \frac{C_2}{\ln R - \ln C_1} = \frac{(hc/k)[(1/\lambda_2) - (1/\lambda_1)]}{\ln R + \ln(\varepsilon_2/\varepsilon_1) + 5\ln(\lambda_2/\lambda_1) + \ln(\sigma_1 \Delta \lambda_1/\sigma_2 \Delta \lambda_2)}. \qquad (23.5)$$

The two-color detection technique is beneficial because the temperature becomes independent of the object emissivity, providing that emissivity does not vary between λ_1 and λ_2, and is inherently self-calibrated. This method can be extremely useful in missile detection where there is a large difference between the temperature of the surface of the missile and the missile exhaust plume.

23.2 REQUIREMENTS OF THIRD-GENERATION DETECTORS

In the 1990s (see Figure 2.1), third generation IR detectors emerged after the tremendous impetus provided by detector developments. The definition of third generation IR systems is not particularly well established. In the common understanding, third generation IR systems provide enhanced capabilities such as larger number of pixels, higher frame rates, better thermal resolution, as well as multicolor functionality and other on-chip signal processing functions. According to Reago et al. [38], the third generation is defined by the requirement to maintain the current advantage enjoyed by the United States and allied armed forces. This class of devices includes both cooled and uncooled FPAs [1,38]:

- High performance, high resolution cooled imagers having multicolor bands
- Medium- to high-performance uncooled imagers
- Very low cost, expendable uncooled imagers

When developing third generation imagers, the IR community is faced with many challenges. Some of them are shortly considered here:

- Noise equivalent temperature difference (NEDT)
- Pixel and chip size issues
- Uniformity
- Identification and detection ranges

Current readout technology is based upon CMOS circuitry that has benefited from dramatic and continuing progress in miniaturizing circuit dimensions. Second generation imagers provide NEDT of about 20–30 mK with $f/2$ optics. A goal of third generation imagers is to achieve sensitivity improvement corresponding to NEDT of about 1 mK. From Equation 19.13 it can be determined that in a 300 K scene in the LWIR region with thermal contrast of 0.02, the required charge storage capacity is above 10^9 electrons. This high charge-storage density cannot be obtained within the small pixel dimensions using standard CMOS capacitors [1]. Although the reduced oxide thickness of submicrometer CMOS design rules gives large capacitance per unit area, the reduced bias voltage, as illustrated in Figure 23.1, largely cancels any improvement in charge storage density. Ferroelectric capacitors may provide much greater charge storage densities than the oxide-on-silicon capacitors now used. However, such a technology is not yet incorporated into standard CMOS foundries.

To provide an opportunity to significantly increase the charge storage capacity and the dynamic range, the vertically integrated sensor array (VISA) program has been sponsored by DARPA [39–41]. The approach being developed builds on the traditional "hybrid" structure of a detector with a 2-D array of indium-bump interconnects to the silicon readout. The VISA allows additional layers of silicon processing chips to be connected below the readout to provide more complex functionality. It will allow the use of smaller and multicolor detectors without compromising storage capacity. Signal-to-noise ratios will increase for multicolor FPAs. This will permit LWIR focal planes arrays to improve the sensitivity by a factor of 10.

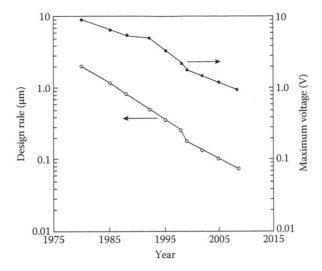

Figure 23.1 Trends for design rule minimum and maximum bias voltage of silicon foundry requirements. (From Norton, P., Campbell, J., Horn, S., and Reago, D., "Third-Generation Infrared Imagers," *Proceedings of SPIE* 4130, 226–36, 2000. With permission.)

Pixel and chip sizes are important issues in association with multicolor imager formats. Small pixels reduce cost by increasing the number of readout and detector dice potentially available from processed wafers. Small pixels also allow smaller, lightweight optics to be used.

The fundamental limit to pixel size is determined by diffraction. The size of the diffraction-limited optical spot or Airy disk is given by

$$d = 2.44\lambda \, f/\#, \tag{23.6}$$

where d is the diameter of the spot, λ is the wavelength, and $f/\#$ is the f-number of the focusing lens. For typical $f/2.0$ optics at 5 μm wavelength, the spot size is 25 μm. Because system users prefer some degree of oversampling, the pixel size may be reduced for MWIR applications to dimensions on the order of 12 μm. Given the track record of human nature, Norton [42] has predicted that MWIR pixel size will eventually be reduced to about 10 μm at some point, just to achieve the smaller pixel record. SWIR pixel size will shrink to correspondingly smaller dimensions for applications seeking maximum spatial resolution. LWIR pixels are not likely to shrink much below 20 μm. However, it is anticipated that LWIR pixels will be made as small as MWIR pixels, since this will allow a single readout design then to be used with both MWIR and LWIR FPAs.

Recently, the first large format MWIR FPAs with pixel dimension of 15 μm has been demonstrated [11,43]. It will be an extreme challenge to deploy a two- or three-color detector structure into a small pixel such as 18×18 μm^2. Current two-color simultaneous mode pixels with two indium bumps per pixel have not been built with pixels smaller than 25 μm on a side.

Figure 14.16 shows the uncertainty in cutoff wavelength of $Hg_{1-x}Cd_xTe$ for x variations of 0.1%. It is shown that the serious changes in cutoff wavelength are observed in VLWIR region. For short wavelength IR ($\approx$3 μm) and MWIR ($\approx$5 μm) materials, the variation in cutoff wavelength is not large. However, the nonuniformity is a serious problem in the case of LWIR HgCdTe detectors. The variation of x across the $Hg_{1-x}Cd_xTe$ wafer causes much larger spectral nonuniformity. At 77 K, a variation of $\Delta x = 0.1\%$ gives a $\Delta\lambda_c$ above 0.5 μm at $\lambda_c = 20$ μm, which cannot be corrected by either two or three point corrections [2]. This cutoff wavelength nonuniformity at the FPA level can be spectrally corrected by using a cold filter, but the dark current variation caused by the variation of cutoff wavelengths will still exist. For applications that require operation in the LWIR band as well as two-color LWIR/VLWIR bands, most probably HgCdTe will not be the optimal solution.

An alternative candidate for third generation IR detectors are the Sb-based III-V material system. These materials are mechanically robust and have fairly weak dependence of bandgap on composition (see Figure 13.49). One advantage of using type II superlattice in LW and VLWIR is the ability to fix one component of the material and vary the other to tune wavelength (see Figure 17.15).

Thermal imaging systems are used first to detect an object and then to identify it. Typically, identification ranges are between two and three times shorter than detection ranges [15]. To increase ranges, better resolution and sensitivity of the infrared systems (and hence the detectors) are required. Third generation cooled imagers are being developed to extend the range of target detection and identification and to ensure that defense forces maintain a technological advantage in night operations over any opposing force.

Identification ranges can be further increased by using multispectral detection to correlate the images at different wavelengths. For example, it appears that in the MWIR spectral range the IR image is washed out to the point that the target and the background cannot be distinguished from each other (see Figure 23.2 [15]). Detectors that cover the entire spectral range will suffer from washout because the background contrast changes from positive to negative. Alternatively, using two band detectors (up to 3.8 μm and from 3.8 up to 5 μm) and summing the inverse of the second band and the output of the first band, will yield a contrast enhancement that is impossible to achieve if an integrated response of the entire spectral range is used.

Figure 23.3 compares the relative detection and identification ranges modeled for third generation imagers using NVESD's (Fort Belvoir, Virginia) NVTherm program. As a range criterion, the standard 70% probability of detection or identification is assumed. Note that the identification range in the MWIR range is almost 70% of the LWIR detection range. For detection, LWIR provides superior range. In the detection mode, the optical system provides a wide field of view (WFOV–f/2.5) since third generation systems will operate as an on-the-move wide area step-scanner with automated target recognition (second generation systems relay on manual target searching) [44]. MWIR offers higher spatial resolution sensing and has an advantage for long-range identification when used with telephoto optics (NFOV–f/6).

23.3 HgCdTe MULTICOLOR DETECTORS

The standard method to detect multiwavelength simultaneously is to use optical components such as lenses, prisms, and gratings to separate the wavelength components before they impinge on the IR detectors. Another simpler method is a stacked arrangement in which the shorter wavelength detector is placed optically ahead of the longer wavelength detector. In such a way, two-color detectors using HgCdTe [45] and InSb/HgCdTe [46] photoconductors have been demonstrated in the early 1970s. At present, however, considerable efforts are directed to fabricating a single FPA with multicolor capability to eliminate the spatial alignment and temporal registration problems that exist whenever separate arrays are used; to simplify optical design; and to reduce size, weight, and power consumption.

The unit cell of integrated multicolor FPAs consists of several collocated detectors, each sensitive to a different spectral band (see Figure 23.4). Radiation is incident on the shorter band detector,

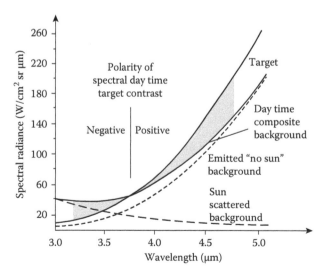

Figure 23.2 Target and background contrast reversal in the MWIR spectral range. (From Sarusi, G., *Infrared Physics & Technology*, 44, 439–44, 2003. With permission.)

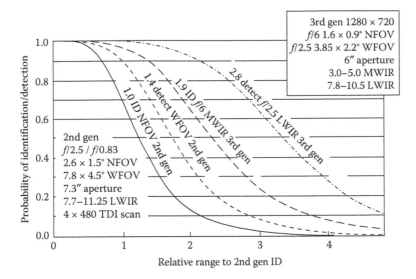

Figure 23.3 Comparison of the detection and identification range between current second generation TDI scanned LWIR imagers and the LWIR and MWIR bands of third generation imager in a 1280 × 720 format with 20 μm pixels. (From Horn, S., Norton, P., Cincotta, T., Stolz, A., Benson, D., Perconti, P., and Campbell, J., "Challenges for Third-Generation Cooled Imagers," *Proceedings of SPIE* 5074, 44–51, 2003. With permission.)

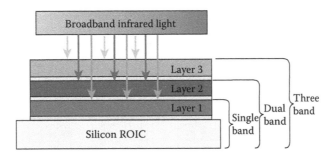

Figure 23.4 Structure of a three-color detector pixel. Infrared flux from the first band is absorbed in Layer 3, while longer wavelength flux is transmitted through the next layers. The thin barriers separate the absorbing bands.

with the longer wave radiation passing through to the next detector. Each layer absorbs radiation up to its cutoff, and hence transparent to the longer wavelengths, which are then collected in subsequent layers. In the case of HgCdTe, this device architecture is realized by placing a longer wavelength HgCdTe photodiode optically behind a shorter wavelength photodiode.

Back-to-back photodiode two-color detectors were first implemented using quaternary III-V alloy ($Ga_xIn_{1-x}As_yP_{1-y}$) absorbing layers in a lattice matched InP structure sensitive to two different SWIR bands [47]. A variation on the original back-to-back concept was implemented using HgCdTe at Rockwell [48] and Santa Barbara Research Center [49]. Following the successful demonstration of multispectral detectors in LPE-grown HgCdTe devices [49], the MBE and MOCVD techniques have been used for the growth of a variety of multispectral detectors at Raytheon [6,7,9,50–53], BAE Systems [54], Leti [10,11,55–58], Selex and QinetiQ [59–62], DRS [12,63–65], Teledyne and NVESD [66,67]. For more than a decade steady progression has been made in a wide variety of pixel sizes (to as small as 20 μm), array formats (up to 1280 × 720) and spectral-band sensitivity (MWIR/MWIR, MWIR/LWIR, and LWIR/LWIR).

23.3.1 Dual-Band HgCdTe Detectors

The unit cell of an integrated two-color FPA consists of two collocated detectors, each sensitive to a different spectral band. In back-illuminated dual-band detectors, the photodiode with the longer cutoff wavelength is grown epitaxially on top of the photodiode with the shorter cutoff wavelength. The shorter cutoff photodiode acts as a long wavelength-pass filter for the longer-cutoff photodiode.

Both sequential mode and simultaneous mode detectors are fabricated from multilayer materials. The simplest two-color HgCdTe detector and the first to be demonstrated was the bias-selectable n-P-N triple-layer heterojunction (TLHJ), back-to-back photodiode shown in Figure 23.5a (capital letter means wider bandgap structure). The n-type base absorbing regions are deliberately doped with indium at a level of about $(1–3) \times 10^{15}$ cm^{-3}. A critical step in device formation is ensuring that the *in situ* p-type As-doped layer (typically 1–2 μm thick) has good structural and electrical properties to prevent internal gain from generating spectral crosstalk. The bandgap engineering effort consists of increasing the CdTe mole fraction and the effective thickness of the p-type layer to keep out-of-band carriers from being collected at the terminal.

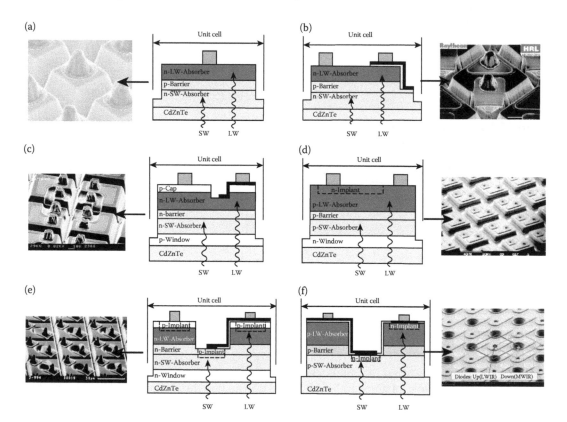

Figure 23.5 Cross-section views of unit cells for various back-illuminated dual-band HgCdTe detector approaches: (a) bias-selectable n-p-n structure reported by Raytheon (Wilson, J. A., Patten, E. A., Chapman, G. R., Kosai, K., Baumgratz, B., Goetz, P., Tighe, S., et al., "Integrated Two-Color Detection for Advanced FPA Applications," *Proceedings of SPIE* 2274, 117–25, 1994. With permission.), (b) simultaneous n-p-n design reported by Raytheon (Rajavel, R. D., Jamba, D. M., Jensen, J. E., Wu, O. K., Wilson, J. A., Johnson, J. L., Patten, E. A., Kasai, K., Goetz, P. M., and Johnson, S. M., *Journal of Electronic Materials*, 27, 747–51, 1998. With permission.), (c) simultaneous p-n-n-p reported by BAE Systems (Reine, M. B., Hairston, A., O'Dette, P., Tobin, S. P., Smith, F. T. J., Musicant, B. L., Mitra, P., and Case, F. C., "Simultaneous MW/LW Dual-Band MOCVD HgCdTe 64 × 64 FPAs," *Proceedings of SPIE* 3379, 200–12, 1998. With permission.), (d) simultaneous n-p-p-p-n design reported by Leti (Zanatta, J. P., Ferret, P., Loyer, R., Petroz, G., Cremer, S., Chamonal, J. P., Bouchut, P., Million, A., and Destefanis, G., "Single and Two Colour Infrared Focal Plane Arrays Made by MBE in HgCdTe," *Proceedings of SPIE* 4130, 441–51, 2000. With permission.), (e) simultaneous structure based on p-on-n junctions reported by Rockwell (Tennant, W. E., Thomas, M., Kozlowski, L. J., McLevige, W. V., Edwall, D. D., Zandian, M., Spariosu, K., et al., *Journal of Electronic Materials*, 30, 590–94, 2001. With permission.), and (f) simultaneous structure based on n-on-p junctions reported by Leti (Destefanis, G., Baylet, J., Ballet, P., Castelein, P., Rothan, F., Gravrand, O., Rothman, J., Chamonal, J. P., and Million, A., *Journal of Electronic Materials*, 36, 1031–44, 2007. With permission.)

The sequential-mode detector has a single indium bump per unit cell that permits sequential bias selectivity of the spectral bands associated with operating back-to-back photodiodes. When the polarity of the bias voltage applied to the bump contact is positive, the top (LW) photodiode is reverse biased and the bottom (SW) photodiode is forward biased. The SW photocurrent is shunted by the low impedance of the forward-biased SW photodiode and the only photocurrent to emerge in the external circuit is the LW photocurrent. When the bias voltage polarity is reversed, the situation reverses; only SW photocurrent is available. Switching times within the detector can be relatively short, on the order of microseconds, so detection of slowly changing targets or images can be achieved by switching rapidly between the MW and LW modes. The problems with the bias-selectable device are the following: its construction does not allow independent selection of the optimum bias voltage for each photodiode, and there can be substantial MW crosstalk in the LW detector.

Multicolor detectors require deep isolation trenches to cut completely through the relatively thick (at least 10 μm) LWIR absorbing layer. The design of small two-color TLHJ detectors of less than 20 μm pitch requires at least 15 μm deep trenches, which are no more than 5 μm wide at the top. Dry etching technology has been used for a number of years to produce two-color detectors. One of the materials technology being developed in order to meet the challenge of shrinking the pixel size to below 20 μm is advanced etching technology. Recently, Raytheon has developed an inductively coupled plasma (ICP) dry mesa etching capability to replace electron cyclotron resonance (ECR) dry mesa etching. The ICP, when compared to ECR, has shown reduced lateral mask erosion during etching, less significant etch-lag effects, and improved etch depth uniformity [66]. For the pseudoplanar devices the etching step is easier to perform because of the lower aspect ratio. Moreover, there is no electrical crosstalk as the pixels are electrically independent.

Many applications require true simultaneous detection in the two spectral bands. This has been achieved in a number of ingenious architectures shown in Figure 23.5b–f. Two different architectures are shown. The first one is the classical n-P-N back-to-back photodiode structure (Figure 23.5b). In the case of the architecture developed at Leti (Figure 23.5d), the two absorption materials are p-type separated by a barrier to prevent any carrier drift between the two n-on-p diodes. Each pixel consists of two standard n-on-p photodiodes, where the p-type layers are usually doped with Hg vacancies. The shorter wavelength diode is realized during epitaxy by simply doping part of the first absorbing layer with In. The longer wavelength junction is obtained by a planar implantation process. It should be noted that the electron mobility is around 100 times greater in n-type material than holes in p-type material and, hence, the n-on-p structures will have a much lower common resistance. This is an important consideration for large area FPAs with detection in the LW range due to the larger incident-photon flux.

The last two architectures shown in Figure 23.5e and f, called pseudoplanar, presents a totally different approach. They are close to the structure proposed by Lockwood et al. [68] in 1976 for PbTe/PbSnTe heterostructure two-color photodiodes. They are based on the concept of two p-on-n (Figure 23.5e) or n-on-p (Figure 23.5f) diodes fabricated by p-type or n-type implantation, respectively, but on two different levels of a three layer heterostructure. The architecture developed by Rockwell is a simultaneous two-color MWIR/LWIR FPA technology based on a double-layer planar heterostructure (DLPH) MBE technology (Figure 23.5e). To prevent the diffusion of carriers between two bands, a wide bandgap 1 μm thick layer separates these two absorbing layers. The diodes are formed by implanting arsenic as a p-type dopant and activating it with an anneal. This results in a unipolar operation for both bands. The implanted area of Band 2 is a concentric ring around the Band 1 dimple. Because the lateral carrier-diffusion length is larger than the pixel pitch in the MWIR material, and the Band 1 junction is shallow, each pixel is isolated by dry-etching a trench around it to reduce carrier crosstalk. The entire structure is capped with a layer of material with a slightly wider bandgap to reduce surface recombination and simplify passivation.

All these simultaneous dual-band detector architectures require an additional electrical contact from an underlying layer in the multijunction structure to both the SW and the LW photodiode. The most important distinction is the requirement of a second readout circuit in each unit cell.

It is expected that with the TLHJ architecture pixel size could decrease to 15 μm and array format could increase to several megapixels. With the pseudoplanar architecture, MWIR/LWIR devices should be produced more easily, with large format arrays having the pixel size around 20 μm.

Having only one bump contact per unit cell, as for single-color hybrid FPAs, is the major advantage of the bias-selectable detector. In addition, it is compatible with existing silicon readout chips. This structure achieves approximately 100% optical fill factor in each band due to total internal

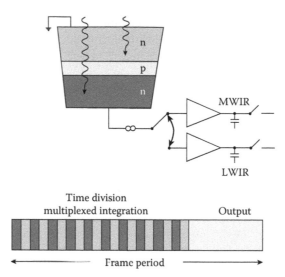

Figure 23.6 Raytheon's two-color FPAs with a time division multiplexed integration scheme in which the detector bias polarity is alternated many times within a single frame period. (From Radford, W. A., Patten, E. A., King, D. F., Pierce, G. K., Vodicka, J., Goetz, P., Venzor, G., et al., "Third Generation FPA Development Status at Raytheon Vision Systems," *Proceedings of SPIE* 5783, 331–39, 2005. With permission.)

reflection of the incident radiation off the mesa sidewalls. Raytheon's approach employs a readout integrated circuit (ROIC) with Time Division Multiplexed Integration (TDMI) [7], see Figure 23.6. As the detector bias is changed the detector current is directed to separate input circuitry and integration capacitors. The bias switching is performed at times much shorter than the frame period. Fast subframe switching of less than 1 ms is typically employed. The MWIR band is integrated by summing the charge collected from the individual subframe integration periods. The LWIR band is integrated by averaging the charge collected from the individual subframe integrations.

Figure 23.7 shows the current-voltage characteristics for a single mesa, single indium bump two-color MWIR1/MWIR2 [69] and MWIR/LWIR [51] TLHJ unit-cell detector design. With appropriate polarity and voltage bias at the pixel contact, the junctions respond to either shorter or longer wavelength infrared radiation. The I-V curves exhibit the serpentine shapes expected for a back-to-back diode structure; the "flatness" displayed in both effective reverse bias regions is a key indicator of high quality two-color diodes. Figure 23.8 illustrates examples of the spectral response from different two-color devices [70]. Note that there is minimal crosstalk between the bands, since the short wavelength detector absorbs nearly 100% of the shorter wavelengths. Test structures indicate that the separate photodiodes in a two-color detector perform exactly like single-color detectors in terms of achievable R_oA variation with wavelength at a given temperature (see Table 23.2) [6].

The best performing bias selectable dual-color FPAs being produced at Reytheon Vision Systems exhibit out-of-band crosstalk below 10%, 99.9% interconnect operability, and 99% response operability that is comparable to state-of-the-art, single-color technology. It is predicted that ongoing development of material growth and fabrication processes will translate to further improvements in dual-color FPA performance.

Recently, Raytheon Vision Systems has developed two-color, large-format infrared FPAs to support the U.S. Army's third generation FLIR systems in both 640 × 480) and "high definition" 1280 × 720 formats with 20 × 20 μm unit cells (see Figure 23.9) [52]. The ROICs share a common chip architecture and incorporate identical unit cell circuit designs and layouts; both FPAs can operate in either dual-band or single-band modes. High-quality MWIR/LWIR 1280 × 720 FPAs with cutoffs ranging out to 11 μm at 78K have demonstrated excellent sensitivity and pixel operabilities exceeding 99.9% in the MW band and greater than 98% in the LW band. Table 23.3 provides a summary of the sensitivity and operability data measured for the three best 1280 × 720 FPAs fabricated to date [9]. Median 300 K NEDT values at f/3.5 of approximately 20 mK for the MW and 25 mK for the LW have been measured for dual-band TDMI operation at 60 Hz frame rate with integration times corresponding to roughly 40% (MW) and 60% (LW) of full well charge capacities. As shown

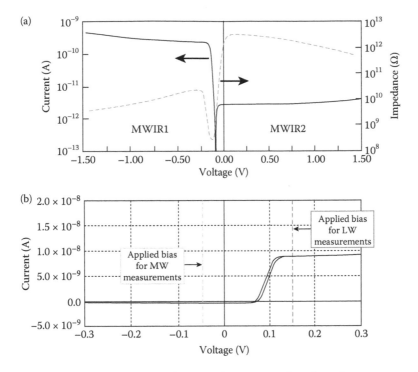

Figure 23.7 Typical I-V characteristics for a single mesa, single indium bump two-color TLHJ unit-cell detector design: (a) MWIR1/MWIR2 25 µm pixel with cutoff wavelength at 3.1 µm and 5.0 µm at 77 K and 30° FOV. (From Baylet, J., Ballet, P., Castelein, P., Rothan, F., Gravrand, O., Fendler, M., Laffosse, E., et al., *Journal of Electronic Materials,* 35, 1153–58, 2006. With permission.) (b) MWIR/LWIR 20 µm pixel with cutoff wavelength at 5.5 µm and 10.5 µm. (From Smith, E. P. G., Patten, E. A., Goetz, P. M., Venzor, G. M., Roth, J. A., Nosho, B. Z., Benson, J. D., et al., *Journal of Electronic Materials,* 35, 1145–52, 2006. With permission.)

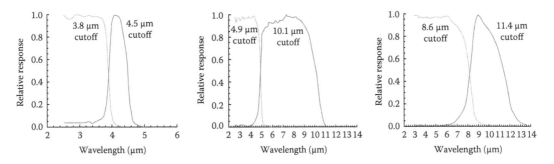

Figure 23.8 Spectral response curves for two-color HgCdTe detectors in various dual-band combinations of MWIR and LWIR spectral bands. (From Norton, P. R., "Status of Infrared Detectors," *Proceedings of SPIE* 3379, 102–14, 1998. With permission.)

in Figure 23.10 [9,71], excellent high resolution IR camera imaging with f/2.8 FOV broadband refractive optics at 60 Hz frame rate has been achieved.

Impressive results have also been demonstrated for other architectures. For example, the NEDT of 128 × 128 simultaneous MWIR1-MWIR2 FPAs (see device architecture in Figure 23.5b) for both bands (2.5–3.9 µm and 3.9–4.6 µm) was below 25 mK (see Figure 23.11) [70], and imagery was acquired at temperatures as high as 180 K with no visible degradation in image quality. The camera used for these measurements had a 50 mm, f/2.3 lens. Also, high performance two-color 128 × 128 FPAs with 40 µm pitch have also been obtained using the pseudoplanar simultaneous architecture shown in Figure 23.5e. Background limited detectivity performance has been

Table 23.2: Typical Measured Performance Parameters for Single- and Dual-Color HgCdTe MWIR and LWIR Detector Configuration for 256 × 256 30 μm unit-cell FPAs

256 × 256 30 μm Unit-Cell Performance Parameters	DLHJ Single Color		TLHJ Sequential Dual Color					
			MWIR/MWIR		MWIR/LWIR		LWIR/LWIR	
	MWIR	LWIR	Band 1	Band 2	Band 1	Band 2	Band 1	Band 2
Spectral band	MWIR	LWIR						
78-K cutoff (μm)	5	10	4	5	5	10	8	10
Operating temperature (K)	78	78	120	120	70	70	70	70
Crosstalk (%)	–	–	<5	<10	<5	<10	<5	<10
Quantum efficiency (%)	>70	>70	>70	>65	>70	>50	>70	>50
R_0A, 0 FOV (Ωcm^2)	>1 × 10⁷	>500						
R_rA^*, 0 FOV (Ωcm^2)	–	–	6×10^5	2×10^5	1×10^6	2×10^2	5×10^4	5×10^2
Interconnect operability (%)	>99.9	>99.9	>99.9	>99.9	>99.9	>99.9	>99.9	>99.9
Response operability (%)	>99	>98	>99	>97	>99	>97	>98	>95

Source: E. P. G. Smith, L. T. Pham, G. M. Venzor, E. M. Norton, M. D. Newton, P. M. Goetz, V. K. Randall, et al., *Journal of Electronic Materials*, 33, 509–16, 2004. With permission.
*Resistance area product at nonzero bias.

Table 23.3: Performance Summary of Three Best 1280 × 720 MW/LW FPAs Fabricated to Date

FPA	W after	MW t_{int} (msec)	MW Median NEDT (mK)	MW Response Operability	LW t_{int} (msec)	LW Median NEDT (mK)	LW Response Operability
7607780	3827	3.14	23.3	99.7%	0.13	30.2	98.5%
7616474	3852	3.40	18.0	99.8%	0.12	27.0	97.0%
7616475	3848	3.40	18.0	99.9%	0.12	26.8	98.7%

Source: D. F. King, W. A. Radford, E. A. Patten, R. W. Graham, T. F. McEwan, J. G. Vodicka, R. F. Bornfreund, P. M. Goetz, G. M. Venzor, and S. M. Johnson, "3rd-Generation 1280 × 720 FPA Development Status at Raytheon Vision Systems," *Proceedings of SPIE* 6206, 62060W, 2006. With permission.

(a)   (b)

Figure 23.9 RVS dual-band MW/LWIR FPAs mounted on dewar platforms: (a) 1280 × 720 format and (b) 640 × 480 format. (From King, D. F., Graham, J. S., Kennedy, A. M., Mullins, R. N., McQuitty, J. C., Radford, W. A., Kostrzewa, T. J., et al., "3rd-Generation MW/LWIR Sensor Engine for Advanced Tactical Systems," *Proceedings of SPIE* 6940, 69402R, 2008. With permission.)

obtained for MWIR (3–5 μm) devices at $T < 130$ K and for LWIR (8 to 10 μm) devices at $T \approx 80$ K (see Figure 23.12) [66]. The FPA also exhibits low NEDT values: 9.3 mK for the MW band and 13.3. mK for the LW band, similar to good quality single color FPAs.

Two-color MWIR/LWIR HgCdTe detectors have been examined theoretically [60,72–74]. It has been shown that it is possible to predict, with relatively good accuracy, the performance of

MWIR

LWIR

Figure 23.10 A still camera image taken at 78 K with f/2.8 FOV and 60 Hz frame rate using two-color 20 μm unit-cell MWIR/LWIR HgCdTe/CdZnTe TLHJ 1280 × 720 FPA hybridized to a 1280 × 720 TDMI ROIC. (From King, D. F., Radford, W. A., Patten, E. A., Graham, R. W., McEwan, T. F., Vodicka, J. G., Bornfreund, R. F., Goetz, P. M., Venzor, G. M., and Johnson, S. M., "3rd-Generation 1280 × 720 FPA Development Status at Raytheon Vision Systems," *Proceedings of SPIE* 6206, 62060W, 2006. With permission.)

complex detectors by using a numerical models. Furthermore, the simulation technique is also useful for understanding the effects of different material parameters and geometrical characteristics on the detector performance.

The HgCdTe HDVIP or loophole concept (see Figure 14.45), developed at DRS and BAE Southampton, represents an alternative approach to IR FPA architecture. It differs from the more entrenched FPA architectures in both its method of diode formation and the manner of its hybridization to the silicon ROIC [64]. The monocolor HDVIP architecture consists of a single HgCdTe epilayer grown on CdZnTe substrate by LPE or MBE [63]. After epitaxial growth, the substrate is removed and the HgCdTe layer is passivated on both surfaces with interdiffused layers of evaporated CdTe (the interdiffusion at 250°C on the Te-rich side of the phase field generates about 10^{16} cm^{-3} metal vacancies). During this process the Cu can also be in-diffused from a doped ZnS source providing an alternative to doping during growth. This single color architecture has been extended to two colors at DRS by gluing two monocolor layers together into a composite, and forming an insulated via through the lower layer in order to readout the upper color, as illustrated in Figure 23.13. Contact to the Si ROIC is obtained by etching holes

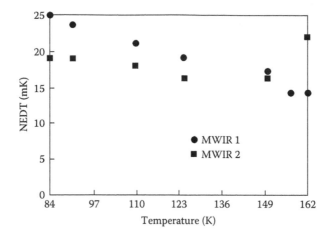

Figure 23.11 NEDT for a two-color camera having 50 mm, $f/2.3$ lens, as a function of the operating temperature. (From Norton, P. R., "Status of Infrared Detectors," *Proceedings of SPIE* 3379, 102–14, 1998. With permission.)

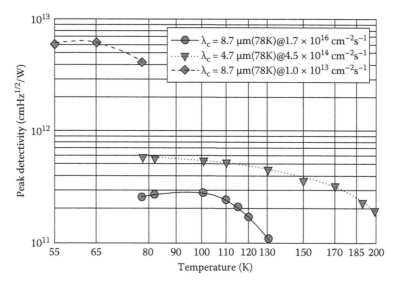

Figure 23.12 Detectivity of two-color pseudoplanar simultaneous MWIR/LWIR 128 × 128 HgCdTe FPA. (From Tennant, W. E., Thomas, M., Kozlowski, L. J., McLevige, W. V., Edwall, D. D., Zandian, M., Spariosu, K., et al., *Journal of Electronic Materials,* 30, 590–94, 2001. With permission.)

(or vias) through the HgCdTe down to contact pads on the Si (see Figure 23.13c). The ROIC used for the dual-band FPA was originally designed for a single-color 640 × 480 array with 25 μm (square) pixels. The even numbered rows of the ROIC have no detectors attached to them, so the chip is operated in a mode that only outputs the odd rows. Odd-numbered columns connect to LWIR detectors, and the MWIR detectors are on the even columns. This approach has been utilized to fabricate both MW-LW and MW-MW 240 × 320 FPAs on a 50 μm pitch. Higher densities are being investigated with dedicated two-color ROIC designs, enabling pitches of <30 μm for two-color FPAs.

Performance data for representative DRS two-color MW-LW and MW-MW 240 × 320 FPAs utilizing $f/3$ optics and a 60-Hz frame rate are shown in Table 23.4 [64]. Composite operabilities >92.5% have been achieved. However, relatively low collection efficiency (the product of quantum efficiency and unit cell fill factor) has been measured on the LWIR layer. A recently published paper describes operabilities in excess of 99% [63].

Table 23.4: Performance Data of Two-Color MW/LW and MW/MW 240 × 320 HgCdTe FPAs: *f*/3 Optics, 60 Hz Frame Rate

	Spectral Bands (μm)	Pitch (μm)	SNR Operability (%)	CE (%)	Response Rniformity (σ/mean)	NEDT (mK)
320 × 240 MW/LW	3.0–5.2/8.0–10.2	50	97.1/96.3	60/35	4.9%/4.2%	9/23
320 × 240 MW/MW	3.0–4.2/4.2–5.2	50	99.4/99.6	58/58	4.3%/3.7%	18.1/8.3

Source: M. A. Kinch, "HDVIP™ FPA Technology at DRS," *Proceedings of SPIE* 4369, 566–78, 2001. With permission.

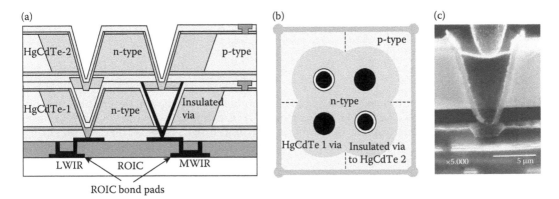

Figure 23.13 Two-color HDVIP architecture is composed of two layers of thinned HgCdTe epoxied to a silicon readout: (a) side view, (b) top view (After Aqariden, F., Dreiske, P. D., Kinch, M. A., Liao, P. K., Murphy, T., Schaake, H. F., Shafer, T. A., Shih, H. D., and Teherant, T. H., *Journal of Electronic Materials*, 36, 900–904, 2007. With permission.), and (c) small hole etched to form junction and to contact the Si readout.

23.3.2 Three-Color HgCdTe Detectors

Some system considerations suggest that three-color FPAs would be more generally useful than two-color ones. The successful development of three-color HgCdTe FPAs requires further improvement in the material quality, adequate processing techniques, and a better understanding of imager operation both in terms of pixel performance and interaction between different pixels in the array.

Recently, the first concept for achieving three-color HgCdTe detectors has been demonstrated by British workers [75]. The concept of a back side illuminated HgCdTe detector is shown schematically in Figure 23.14. The bias dependent cutoff is achieved by using three absorbers in an n-p-n structure in which the first n-layer defines the shorter wavelength (SW) region; the p-type layer, the intermediate wavelength region (IW); and the top layer, the long wavelength (LW) region. Note that the terms SW and LW used here are relative and do not necessarily coincide with the SW and LW infrared bands. The cutoff wavelengths of the SW, IW, and LW regions are respectively marked as λ_{c1}, λ_{c2}, and λ_{c3}. Since the barrier region is low doped, the applied bias mainly falls on this side of the junction. For the device configuration shown in Figure 23.14, the negative bias denotes higher potential of contact A in comparison with contact B.

It is expected that at low biases either the SW or LW response would dominate, depending on which junction is reverse biased. In this case we have the same situation as in the bias selectable two-color detector since the barriers prevent electron flow from the IW layer—both generated photoelectrons as well as direct injected carriers from the forward biased junction. Increasing reverse bias reduces the barrier and electrons photogenerated in the IW layer can cross the junction. As a result, the cutoff wavelength changes from the SW to the IW as the bias is increased. This situation, with negative bias, and corresponding changes in spectral response, is shown in Figure 23.15c. Changing the bias direction to positive shifts the cutoff wavelength to the LW region (see Figure 23.15d). Similarly, increasing positive bias moves the cut-on from being coincident with the IW cutoff to the SW cutoff. It should be noted that the above considerations concern the ideal case

821

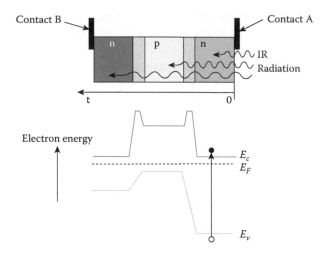

Figure 23.14 Three-color concept and associated zero-bias band diagram. (From Hipwood, L. G., Jones, C. L., Maxey, C. D., Lau, H. W., Fitzmaurice, J., Catchpole, R. A., and Ordish, M., "Three-Color MOVPE MCT Diodes," *Proceedings of SPIE* 6206, 620612, 2006. With permission.)

of the detector structure. The ideal LW response may not be achievable in practice due to the IW absorber being insufficiently thick to absorb all the IW radiation.

The three-color HgCdTe detectors were grown by MOVPE on GaAs substrates oriented off the (100) direction to reduce the size of pyramidal hillock growth defects. Figure 23.16 shows the spectral response with cutoff wavelength 3 (SW), 4 (IW), and 6 µm (LW) [75]. In positive bias mode the LW/IW junction is in reverse bias and a bias independent LW spectrum is obtained above 0.2 V (shown only for + 0.6 V). No barrier lowering at the LW/IW junction at these applied biases is observed due to the chosen doping levels. The response below λ_2 is due to incomplete absorption in the IW layer resulting in carrier generation in the LW layer at these wavelengths (carriers generated in the IW absorber have insufficient energy to surmount the LW barrier). As the positive bias is reduced below 0.2 V, the LW response collapses and a signal from the SW layer appears with the current flowing in the opposite direction. For this bias regime the built-in fields dominate and the largest field is at the SW/IW junction due to the larger bandgap. Further reduction in the bias voltage causes the SW response to grow. The negative voltage puts the SW/IW junction into reverse bias and results in a SW response with cutoff λ_1. Further increase of the negative bias lowers the barrier at this junction and allows a response from the IW layer, thus moving the cutoff to λ_2. The observed increase of the SW signal with increasing negative bias is caused by incomplete absorption in the SW absorption.

Because of the complicated and expensive fabrication process, numerical simulation has become a critical tool for the development of HgCdTe bandgap engineered devices. The numerical simulations can provide valuable guidelines for the design and optimization of the pixel structure and the array geometry. Up until now, only a limited number of theoretical papers have been published that study the performance of three-color detectors [75,76]. Jóźwikowski and Rogalski [76] have shown that the performance of a three-color detector is critically dependent on the barrier doping level and position in relation to the junction. A small shift of the barrier location and doping level causes significant changes in spectral responsivity. This behavior is a serious disadvantage of the considered three-color detector. Therefore, this type of detector structure presents some serious technological challenges.

23.4 MULTIBAND QWIPs

QWIPs are ideal detectors for the fabrication of pixel coregistered simultaneously readable two-color IR FPAs because a QWIP absorbs IR radiation only in a narrow spectral band and is transparent outside of that absorption band. Thus it provides zero spectral crosstalk when two spectral bands are more than a few microns apart. Individual pixels in a multiband QWIP detector array are fabricated using a process similar to that used for their singleband counterparts, except for the via holes that need to be added to electrically connect with the silicon ROIC.

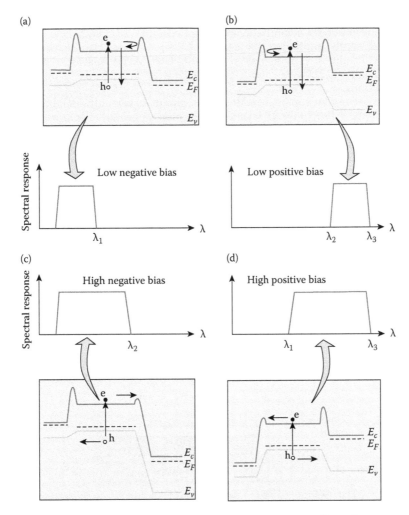

Figure 23.15 Idealized spectral responses of three-color detector. Effect of negative and positive bias voltages on bandgap structure is also shown. (From Hipwood, L. G., Jones, C. L., Maxey, C. D., Lau, H. W., Fitzmaurice, J., Catchpole, R. A., and Ordish, M., "Three-Color MOVPE MCT Diodes," *Proceedings of SPIE* 6206, 620612, 2006. With permission.)

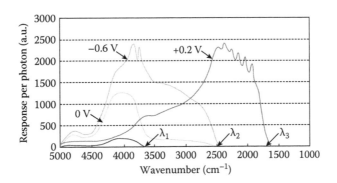

Figure 23.16 Spectral response of a three-color HgCdTe detector with cutoff wavelength 3, 4, and 6 μm at various biases. (From Hipwood, L. G., Jones, C. L., Maxey, C. D., Lau, H. W., Fitzmaurice, J., Catchpole, R. A., and Ordish, M., "Three-Color MOVPE MCT Diodes," *Proceedings of SPIE* 6206, 620612, 2006. With permission.)

Sanders was the first organization to fabricate two-color, 256 × 256 bound-to-miniband QWIP FPAs in each of four important combinations: LWIR/LWIR, MWIR/LWIR, near-IR (NIR)/LWIR, and MWIR/MWIR—with simultaneous integration [77,78]. At present multicolor QWIP detectors are fabricated at Jet Propulsion Laboratory (JPL) [19,20,79–85], Army Research Laboratory [86–88], Goddard [88,89], Thales [17,21,90–93], and AIM [18,94,95] with the majority being based on bound-to-extended transitions.

Devices capable of simultaneously detecting two separate wavelengths can be fabricated by vertical stacking of the different QWIP layers during epitaxial growth. Separate bias voltages can be applied to each QWIP simultaneously via doped contact layers that separate the MQW detector heterostructures. Figure 23.17a shows schematically the structure of a two-color stacked QWIP with contacts to all three ohmic-contact layers [20]. The device epilayers are grown by MBE on up to 6 inches semi-insulating GaAs substrates. An undoped GaAs layer, called an isolator, is grown between two AlGaAs etch stop layers, followed by a 0.5 μm thick doped GaAs layer. Next, the two QWIP heterostructures are grown, separated by another ohmic contact. The long wavelength sensitive stack (red QWIP) is grown above the shorter wavelength sensitive stack (blue QWIP). Typical responsivity spectra at 77 K using a common bias of 1.5 V, recorded simultaneously for two QWIPs at the same pixel are shown in Figure 23.17b. Each QWIP consists of about a 20-period GaAs/$Al_xGa_{1-x}As$ MQW stack in which the thickness of the Si-doped GaAs QWs (with typical electron concentration 5×10^{17} cm^{-3}) and the Al composition of the undoped $Al_xGa_{1-x}As$ barriers (≈550–600 Å thick) is adjusted to yield the desired peak responsivity position and spectral width. The gaps between FPA detectors and the readout multiplexer are backfilled with epoxy. The epoxy backfilling provides the necessary mechanical strength to the detector array and readout hybrid prior to array thinning. The initial GaAs substrate of dual-band FPAs are completely removed leaving

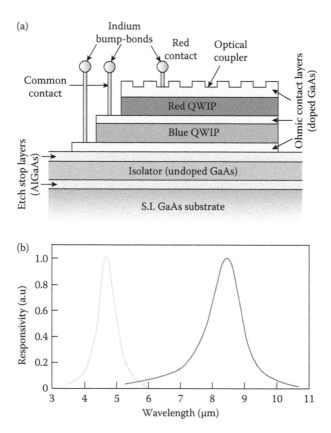

Figure 23.17 Schematic representation of the (a) dual-band QWIP detector structure and (b) typical responsivity spectra at 77 K and a common bias of 1 V, recorded simultaneously for two QWIPs at the same pixel. (From Gunapala, S. D., Bandara, S. V., Liu, J. K., Mumolo, J. M., Hill, C. J., Rafol, S. B., Salazar, D., Woollaway, J., LeVan, P. D., and Tidrow, M. Z., *Infrared Physics & Technology*, 50, 217–26, 2007. With permission.)

only a 50 nm thick GaAs membrane. This allows the array to accommodate any thermal expansion by eliminating the thermal mismatch between the silicon readout and the detector array. It also eliminates pixel-to-pixel crosstalk and, finally, significantly enhances the optical coupling of IR radiation into the QWIP pixels. Using the above described fabrication process, significant progress has been made toward development of a megapixel dual-band QWIP FPA [19,84,85].

Figure 23.18 provides additional insight into dual-band QWIP processing technology developed at JPL [96], based on 4 inch wafers to fabricate 320 × 256 MWIR/LWIR dual-band QWIP devices with pixels collocated and simultaneously readable. As shown in Figure 23.18b, the carriers emitted from each multiquantum well (MQW) region are collected separately using three contacts. The middle contact layer (see Figure 23.18c) is used as the detector common. The electrical connections to the detector common and the LWIR connection are brought to the top of each pixel using via connections. Electrical connections to the common contact and the LWIR pixel connection are brought to the top of each pixel using the gold via connections visible in Figure 23.18. This elaborate processing technology could lead to 2-D imaging arrays that can detect separate bands on a single pixel.

Most QWIP arrays use a 2-D grating, which has the disadvantage of being very wavelength dependent, combined with an efficiency that decreases as the pixel size is reduced. Lockheed Martin has used rectangular and rotated rectangular 2-D gratings for their two-color LW-LW FPAs. Although random reflectors have achieved relatively high quantum efficiencies with large test device structures, it is not possible to achieve comparable quantum efficiencies with random reflectors on small FPA pixels, due to the reduced width-to-height aspect ratios [81]. In addition, it is more difficult to fabricate random reflectors for shorter wavelength detectors, because feature sizes of random reflectors are linearly proportional to the peak wavelength of the detectors. Thus, quantum efficiency becomes a more difficult issue for multicolor QWIP FPAs in comparison to single color arrays. At JPL two different optical coupling techniques have been developed. The first technique uses a dual period Lamar grating structure, and the second is based on multiple diffraction orders (see Fig 23.18c and d) [96].

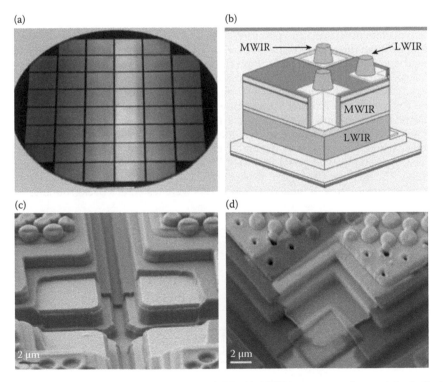

Figure 23.18 Two-color MWIR/LWIR QWIP FPA: (a) 48 FPAs processed on a 4-inch GaAs wafer, (b) 3-D view of pixel structure, (c) electrical connections to the common contact, and (d) the pixel connections are brought to the top of each pixel using the gold via connections. (From Gunapala, S., *Compound Semiconductor*, 10, 25–28, October 2005. With permission.)

Typical operating temperatures for QWIP detectors are in the range of 40–80 K. The bias across each QWIP can be adjusted separately, although it is desirable to apply the same bias to both colors. Results indicate that the complex two-color processing has not compromised the electrical and optical quality of either FPA in the two-color device, since the peak quantum efficiency for each of the 20-period QWIPs was estimated to be ≈10%. For comparison, a normal single-color QWIP with twice the number of periods has a quantum efficiency of around 20%. An accurate design methodology is needed to optimize the detector structure to meet different requirements. In the production process, the fabrication of gratings is still quite an involved process, and the detector quantum efficiency is rather uncertain in small pixels and in pixels with thick material layers.

Development of dual-band QWIP FPAs has been undertaken at JPL over the last decade with the objective of developing 640 × 480 LWIR/VLWIR arrays for moderate background applications [79]. One of the key issues has been the scarcity of appropriate readout multiplexers. To overcome this problem, JPL has chosen to demonstrate initial dual-band concepts with existing multiplexers developed for single-color applications, and use a waveband-interlaced CMOS readout architecture (i.e., odd rows for one color and even rows for the other color). This scheme has the disadvantage that it does not provide a full fill factor for both wavelength bands, resulting in an approximate 50% fill factor for each wavelength band. The device structure, shown in Figure 23.19, consists of a 30-period stack (500-Å AlGaAs barrier and 60-Å GaAs well) of VLWIR structure, and an 18-period stack (500-Å AlGaAs barrier and 40-Å GaAs well) of LWIR structure, separated by a heavily doped 0.5 μm thick intermediate GaAs contact layer. The VLWIR QWIP structure has been designed to have a bound-to-quasibound intersubband absorption peak at 14.5 μm, whereas the LWIR QWIP structure has been designed to have a bound-to-continuum intersubband absorption peak at 8.5 μm, primarily because the photocurrent and dark current of the LWIR device structure are small compared to those of the VLWIR portion.

Figure 23.20 shows a schematic side view of the interlaced dual-band GaAs/AlGaAs FPA [79]. Two different 2-D periodic grating structures were designed to independently couple the 8–9 μm and 14–15 μm radiation into detector pixels in even and odd rows of the FPA, respectively. The top 0.7 μm thick GaAs cap layer was used to fabricate the light-coupling 2-D periodic gratings for 8–9 μm detector pixels, whereas the light-coupling 2-D periodic gratings of the 14–15 μm detector pixels were fabricated through the LWIR MQW layers. Thus, this grating scheme short-circuited all 8–9 μm-sensitive detectors in all odd rows of the FPAs. Next, the LWIR detector pixels were fabricated by dry etching through the photosensitive GaAs/AlGaAs MQW layers into the 0.5 μm thick doped GaAs intermediate contact layer. All VLWIR pixels in the even rows of the FPAs were short circuited. The VLWIR detector pixels were fabricated by dry etching through both MQW stacks into the 0.5 μm thick heavily doped GaAs bottom contact layer. After epoxy backfilling of the gaps between FPA detectors and the readout multiplexer, the substrate was thinned, and finally the remaining GaAs/AlGaAs material contained only the QWIP pixels and a very thin membrane (≈1000 Å).

The 640 × 486 GaAs/AlGaAs array provided images with 99.7% of the LWIR pixels and 98% of VLWIR pixels working, demonstrating the high yield of GaAs technology. The 8–9 μm detectors have shown background-limited performance (BLIP) at 70 K operating temperature, at 300 K background with an f/2 cold stop. The 14–15 μm detectors show BLIP with the same operating conditions at 45 K. The performance of these dual-band FPAs were tested at a background temperature

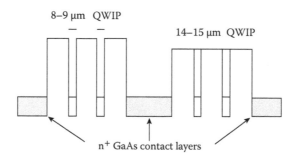

Figure 23.19 Conduction band diagram of a LWIR and a VLWIR two-color detector. (From Gunapala, S. D., Bandara, S. V., Sigh, A., Liu, J. K., Rafol, S. B., Luong, E. M., Mumolo, J. M., et al., "8–9 and 14–15 μm Two-Color 640 × 486 Quantum Well Infrared Photodetector (QWIP) Focal Plane Array Camera," *Proceedings of SPIE* 3698, 687–97, 1999. With permission.)

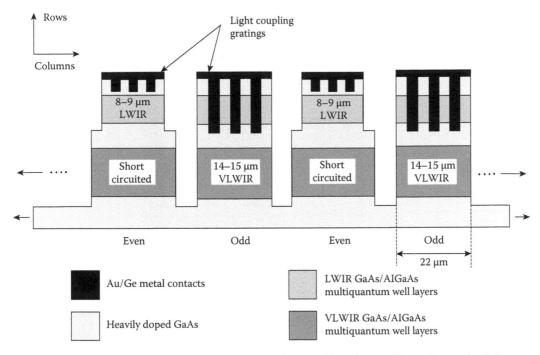

Figure 23.20 Structure cross section of the interlace dual-band FPA. (From Gunapala, S. D., Bandara, S. V., Sigh, A., Liu, J. K., Rafol, S. B., Luong, E. M., Mumolo, J. M., et al., "8–9 and 14–15 μm Two-Color 640 × 486 Quantum Well Infrared Photodetector (QWIP) Focal Plane Array Camera," *Proceedings of SPIE* 3698, 687–97, 1999. With permission.)

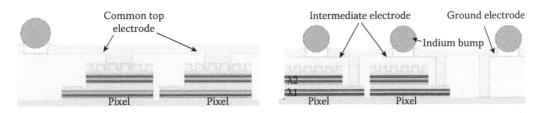

Figure 23.21 Cross section of two different sections of a dual-band QWIP array. (From Robo, J. A., Costard, E., Truffer, J. P., Nedelcu, A., Marcadet, X., and Bois, P., "QWIP Focal Plane Arrays Performances from MWIR to VLWIR," *Proceedings of SPIE* 7298, 72980F, 2009. With permission.)

of 300 K, with *f*/2 cold stop and a 30 Hz frame rate. The estimated NEDTs of LWIR and VLWIR detectors at 40 K were 36 and 44 mK, respectively. The experimentally measured values of the LWIR NEDT, equal to 29 mK, were lower than the estimated ones. This improvement was attributed to the light coupling efficiency of the 2-D periodic grating. However, the experimental VLWIR NEDT value was higher than the estimated value. That was probably a result of inefficient light coupling in the 14–15 μm region, readout multiplexer noise, and noise of the proximity electronics. At 40 K, the performance of detector pixels in both bands was limited by photocurrent noise and readout noise.

Thales selected the ISC0208 Indigo ROIC for demonstration of prototype dual-band QWIP camera. Since this readout is not designed for dual-band applications, the QWIP demonstrator cannot be a temporally coherent MWIR/LWIR array. MWIR/LWIR imagery is based on the two stack quantum structure (see Figure 23.21 [93]) and is described by Castelein and colleagues [91]. The processing is the same as the one developed for *in-situ* skimming FPAs [97]. One band is read when the second one is integrated, hence the two QWIP biases are modulated between two frames. The QWIP wafers were processed with the 25 μm pitch, 384 × 288 format of the ISC0208 ROIC. The SEM picture in Figure 23.22 illustrates details of the processing.

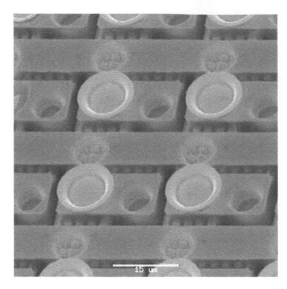

Figure 23.22 Details of a dual-band QWIP array with 25 μm pitch. (From Robo, J. A., Costard, E., Truffer, J. P., Nedelcu, A., Marcadet, X., and Bois, P., "QWIP Focal Plane Arrays Performances from MWIR to VLWIR," *Proceedings of SPIE* 7298, 72980F, 2009. With permission.)

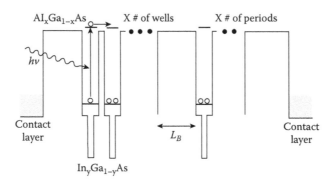

Figure 23.23 Schematic diagram of the conduction band in a bound-to-quasibound QWIP. A couple quantum well structure has been used to broaden the responsivity spectrum. (From Gunapala, S. D., Bandara, S. V., Liu, J. K., Mumolo, J. M., Hill, C. J., Rafol, S. B., Salazar, D., Woollaway, J., LeVan, P. D., and Tidrow, M. Z., *Infrared Physics & Technology*, 50, 217–26, 2007. With permission.)

To cover the MWIR range, a strained-layer InGaAs/AlGaAs material system is used. InGaAs in the MWIR stack produces high in-plane compressive strain, which enhances the responsivity. The MWIR-LWIR FPAs fabricated by the Sanders organization consisted of an 8.6 μm cutoff GaAs/AlGaAs QWIP on top of a 4.7 μm cutoff strained InGaAs/GaAs/AlGaAs heterostructure. The fabrication process allowed for fill factors of 85 and 80% for the MW and LW detectors, respectively. The first FPAs with this configuration had operability in excess of 97%, and NEDT values better than 35 mK with $f/2$ optics.

The first dual-band QWIP FPA with pixel collocation and simultaneous operation in MWIR and LWIR has been described by Goldberg et al. [86]. This 256×256 pixel FPA has achieved a NEDT of 30 mK in the MWIR spectral band and 34 mK in the LWIR spectral band.

More recently, Gunapala et al. [20] have demonstrated a 320×256 MWIR/LWIR pixel collocated and simultaneously readable dual-band QWIP FPA. The device structures of the MWIR and LWIR devices are very similar to the structure shown in Figure 23.23. Each period of the MQW structure consists of coupled QWs of 40 Å containing 10 Å GaAs, 20 Å $In_{0.3}Ga_{0.7}As$, and 10 Å GaAs layers (doped $n = 1 \times 10^{18}$ cm^{-3}) and a 40 Å undoped barrier of $Al_{0.3}Ga_{0.7}As$ between coupled QWs, and a 400 Å thick undoped barrier of $Al_{0.3}Ga_{0.7}As$. It is worth noting that the active MQW region of each

QWIP device is transparent at other wavelengths, which is an important advantage over conventional interband detectors. The experimentally measured NEDT of MWIR and LWIR detectors at 65 K were 28 and 38 mK, respectively.

Another design structure for dual-band MWIR/LWIR QWIPs has been proposed by Schneider et al. [95] This simultaneously integrated 384 × 288 FPA with 40 μm pitch comprises a photovoltaic and a photoconductive QWIP for the LWIR and MWIR, respectively (see Figure 16.29). Excellent NEDT (17 mK) is obtained in the MWIR band (Figure 23.24). Owing to the nonoptimized coupling for LWIR wavelengths, the observed NEDT is higher, but still shows a reasonable value of 43 mK. Due to improvements in the device design, excellent thermal resolution with NEDT < 30 mK ($f/2$ optics and full frame time of 6.8 ms) for both peak wavelengths (4.8 and 8.0 μm) has been demonstrated. Examples of images taken with a dual-band 384 × 288 FPA are shown in Figure 23.25. The features and performance of the dual-band QWIP fabricated by AIM GmbH are summarized in Table 23.5. This table also compares performance of the Eagle camera fabricated by commercial vendor QmagiQ LLC [98].

Recently, the research group from Jet Propulsion Laboratory has implemented a MWIR/LWIR pixel coregistered simultaneously readable 1024 × 1024 dual-band device structure that uses only two indium bumps per pixel (Figure 23.26) compared to three indium bumps per pixel with pixel collocated dual-band devices [85]. In this device structure the detector common (or ground) is shorted to the bottom detector common plane via a metal bridge. Thus, this device structure reduces the number of indium bumps by 30% and has a unique advantage in large format FPAs,

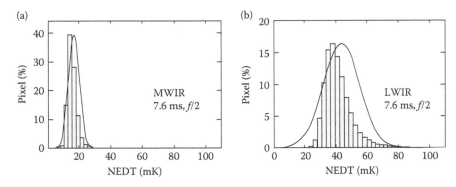

Figure 23.24   NEDT-histogram of the (a) MWIR and (b) LWIR response of a dual-band QWIP FPA. (From Schneider, H., Maier, T., Fleissner, J., Walther, M., Koidl, P., Weimann, G., Cabanski, W., et al., *Infrared Physics & Technology*, 47, 53–58, 2005. With permission.)

Figure 23.25   Images of the scene taken with a dual-band 384 × 288 QWIP demonstration camera with a 100 mm optics under severe weather conditions (cloudy sky, outside temperature below 0°C in winter, 2 p.m.). The church tower is at a distance of 1200 m. The left image shows the scene in the MW, the right image shows in the LW. (From Schneider, H., Maier, T., Fleissner, J., Walther, M., Koidl, P., Weimann, G., Cabanski, W., et al., *Infrared Physics & Technology*, 47, 53–58, 2005. With permission.)

Table 23.5: Specification of the Dual-Band QWIP FPAs

AIM Infrarot-Module GmbH[a]

Technology	QWIP Dual-Band, CMOC MUX
Spectral bands	$\lambda_p = 4.8$ μm; $\lambda_p = 7.8$ μm with temporal coincident integration in both spectral bands
Type	Low noise for LW; photoconductive highly doped for MW
Elements	$388 \times 284 \times 2$; 40 μm pitch
Operability	>99.5%
Biasing	Individually for both bands
NEDT	<30 mK @ $f/2$ and 6.8 ms for both spectral bands
Read put models	Snapshot, stare then scan, temporal signal coincidence in both bands
Subframes	Arbitrary in steps of 8
Data rate digital	80 MHz serial high speed link interface
Full frame rate	50 Hz for $t_{int} = 16.8$ ms; 100 Hz for $t_{int} = 6.8$ ms
IDCA	1.5 W split linear cooler

QmagiQ LLC[b]

Parameter	MW	LW	Conditions
Array format	320×256		
Pixel pitch (μm)	40		
Operating temperature (K)	68		
Optical response (mV/°C)	20 ± 5	20 ± 5	$f/2.3$ cold shield, ROIC gain setting of 1, 1V bias, 300 K scene
Uncorrected response uniformity (%)	5 ± 2	3 ± 2	Aperture shading effects removed
Corrected response uniformity (%)	0.1–0.2	0.1–0.2	30°C scene temperature after a two-point NUC at 20°C and 40°C
Temporal NEDT mean (mK)	35–45	25–35	$f/2.3$, 17 ms integration time, 30-Hz frame rate
Temporal NEDT standard deviation (mK)	3 ± 1	3 ± 1	
Operating temperature (K)	68–70		Dark current and noise increase with operating temperature
Overall operability (%)	> 99.5	> 99.5	Actual value depends on performance specs
ISC0006 power dissipation (MW)	~80		

[a] M. Münzberg, R. Breiter, W. Cabanski, H. Lutz, J. Wendler, J. Ziegler, R. Rehm, and M. Walther, "Multi Spectral IR Detection Modules and Applications," *Proceedings of SPIE* 6206, 620627, 2006. With permission.
[b] http://www.qumagiq.com

since more indium bumps require additional force during the FPA hybridization process. The pitch of the detector array is 30 μm and the actual MWIR and LWIR pixel sizes are 28×28 μm². The estimated NEDT based on single pixel data of MWIR and LWIR detectors at 70 K are 22 and 24 mK, respectively. The experimentally measured NEDT values 27 and 40 mK for MWIR and LWIR, respectively.

The potential of QWIP technology is connected with multicolor detection. A four-band hyper spectral 640×512 QWIP array was successfully developed under a joint Goddard-Jet Propulsion Laboratory-Army Research Laboratory project funded by the Earth Science Technology Office of NASA (see Figure 23.27). The device structure consists of a 15-period stack of 3–5 μm QWIP structure, a 25-period stack of 8.5–10 μm QWIP structure, a 25-period stack of 10–12 μm QWIP structure, and a 30-period stack of 14–15.5 μm QWIP structure [82,83]. The VLWIR QWIP structure has been designed to have bound-to-quasibound intersubband absorption, whereas the other QWIP device structures have been designed to have bound-to-continuum intersubband absorption, since the photocurrent and dark current of these devices are small in comparison to those of the VLWIR device.

The four bands of the QWIP array were fabricated in a manner similar to the two-band system described above (see Figure 23.20). Four separate detector bands were defined by a deep trench

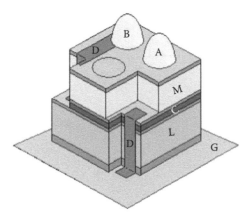

Figure 23.26 3-D view of dual-band QWIP device structure showing via connects for independent access of MWIR and LWIR devices. The color code is as follows: C, isolation layer; L, LWIR QWIP; M, MWIR QWIP; G, contact layer; D, metal bridges between MQW regions; A,B, indium bumps. (From Soibel, A., Gunapala, S. D., Bandara, S. V., Liu, J. K., Mumolo, J. M., Ting, D. Z., Hill, C. J., and Nguyen, J., "Large Format Multicolor QWIP Focal Plane Arrays," *Proceedings of SPIE* 7298, 729806, 2009. With permission.)

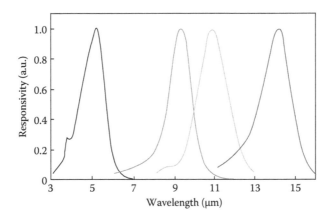

Figure 23.27 Normalized spectral response of the four-band QWIP FPA. (From Gunapala, S. D., Bandara, S. V., Liu, J. K., Rafol, B., and Mumolo, J. M., *IEEE Transactions on Electron Devices,* 50, 2353–60, 2003. With permission.)

etch process and the unwanted spectral bands were eliminated by a detector short-circuiting process using gold-coated reflective 2-D etched gratings as shown in Figure 23.28.

Video images were taken at a frame rate of 30 Hz and at a temperature 45 K, using a ROIC capacitor having a charge capacity of 1.1×10^7 electrons. As shown in Figure 23.29, it is noticeable that the object in the 13–15 μm spectral band is not very clear due to the reduced optical transmission of the germanium lens beyond 14 μm. Figure 23.30 displays the peak detectivities of all spectral bands as a function of operating temperature. From Figure 23.30 it is evident that the BLIP temperatures are 100, 60, 50, and 40 K for the 4–6, 8.5–10, 10–12, and 13–15 μm spectral bands, respectively. The experimentally measured NEDT of 4–6, 8.5–10, 10–12, and 13–15 μm detectors at 40 K are 21, 45, 14, and 44 mK, respectively.

Recently, a novel four-band IR imaging system with simultaneously readable collocated pixels has been proposed [85]. The FPAs divided into 2×2 subpixel areas that function as superpixels marked as Q1, Q2, Q3, and Q4 in Figure 23.31, each sensitive to one of four specific wavelength bands.

The above results indicate that QWIPs have shown significant progress in recent years, especially in their applications to the multiband imaging problem. It is a niche in which they have an intrinsic advantage due to the comparative ease of growing multiband structures by MBE with very low defect density.

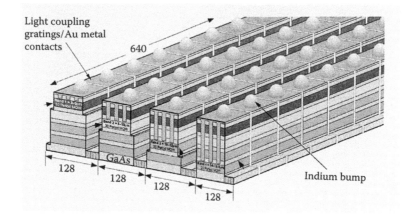

Figure 23.28 Layer diagram of the four-band QWIP device structure and the deep groove 2-D-periodic grating structure. Each pixel represents a 640 × 128 pixel area of the four-band FPA. (From Gunapala, S. D., Bandara, S. V., Liu, J. K., Rafol, B., and Mumolo, J. M., *IEEE Transactions on Electron Devices*, 50, 2353–60, 2003. With permission.)

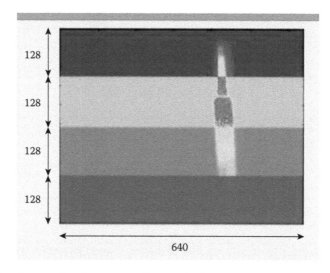

Figure 23.29 One frame of video image taken with the 4–15 μm cutoff four-band 640 × 512 pixel QWIP camera. The image is barely visible in the 13–15 μm spectral band due to the poor optical transmission of the antireflection layer coated germanium lens. (From Gunapala, S. D., Bandara, S. V., Liu, J. K., Rafol, B., and Mumolo, J. M., *IEEE Transactions on Electron Devices*, 50, 2353–60, 2003. With permission.)

23.5 TYPE-II InAs/GaInSb DUAL-BAND DETECTORS

Recently, type II InAs/GaInSb superlattices have emerged as a third candidate for third generation infrared detectors [18,23–28,99–102].

The growth sequence of high quality two-color MWIR type-II SLS FPAs fabricated at the Fraunhofer Institute in Freiburg starts with a 200 nm lattice matched AlGaAsSb buffer layer followed by a 700 nm thick n-type doped GaSb layer. Next, the "blue channel" consisting of 330 periods of p-type of a 7.5 ML InAs/10 ML GaSb is deposited. After the "blue channel" follows a common ground contact layer comprising 500 nm of p-type GaSb follows. The detection of the "red channel" is realized using 150 periods of a 9.5 ML InAs/10 ML GaSb superlattice. Finally, a 20 nm thick InAs terminates the structure. The thickness of the entire vertical pixel structure is only 4.5 μm, which significantly reduces the technological challenge in comparison to dual-band HgCdTe FPAs with a typical total layer thickness around 15 μm. For n-type and p-type doping of the superlattice regions and the contact layers Si, GaTe, and Be are used, respectively.

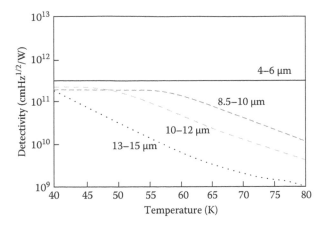

Figure 23.30 Detectivities of each spectral-band of the four-band QWIP FPA as a function of temperature. Detectivities were estimated using the single pixel test detector data taken at $V_b = -1.5$ V and 300 K background with $f/5$ optics. (From Gunapala, S. D., Bandara, S. V., Liu, J. K., Rafol, B., and Mumolo, J. M., *IEEE Transactions on Electron Devices*, 50, 2353–60, 2003. With permission.)

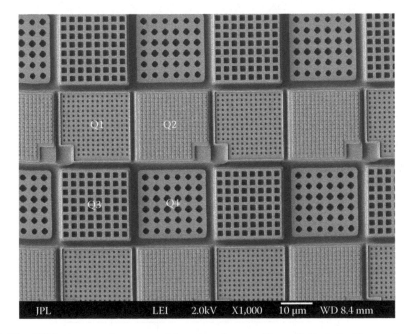

Figure 23.31 SEM picture of processed four-band array. (From Soibel, A., Gunapala, S. D., Bandara, S. V., Liu, J. K., Mumolo, J. M., Ting, D. Z., Hill, C. J., and Nguyen, J., "Large Format Multicolor QWIP Focal Plane Arrays," *Proceedings of SPIE* 7298, 729806, 2009. With permission.)

The first dual-band 288 × 384 MWIR InAs/GaSb camera has already been demonstrated [18]. Figure 23.32 illustrates the device processing. In the first step via holes to the common p-type contact layer and to the n-type contact layer of the lower diode are etched by chlorine-based chemically assisted ion beam etching. Next, another chemical etching is used to fabricate deep trenches for complete electrical isolation of each pixel (see Figure 23.32a). After deposition of the diode passivation, a reactive ion etching is employed to selectively open the passivation to provide access to the contact layers (see Figure 23.32b). Next, the contact metallization is evaporated (see Figure 23.32c). A fully processed dual-color FPA is shown in Figure 23.32d.

In the above approach, simultaneous detection in a 40 μm pixel has been achieved. Solid lines in Figure 23.33 show normalized photocurrent spectra of both channels at 77 K and zero bias. With

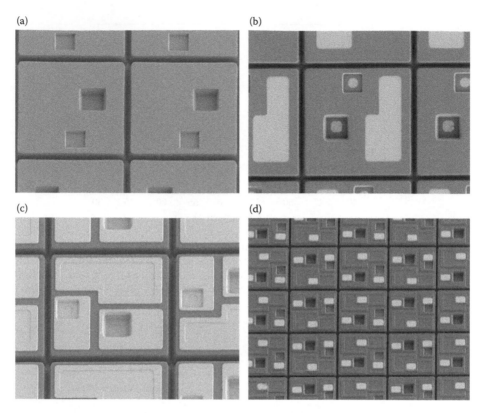

Figure 23.32 SEM images illustrating the processing of 288 × 384 dual color InAs/GaSb SLS FPAs. At a pixel pitch of 40 μm, three contact lands per pixel permit simultaneous and spatially coincident detection of both colors. (From Münzberg, M., Breiter, R., Cabanski, W., Lutz, H., Wendler, J., Ziegler, J., Rehm, R., and Walther, M., "Multi Spectral IR Detection Modules and Applications," *Proceedings of SPIE* 6206, 620627, 2006. With permission.)

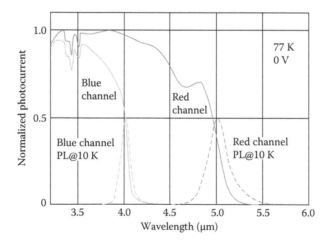

Figure 23.33 Normalized photocurrent at 77 K and the photoluminescence signal at 10 K versus wavelength. (From Münzberg, M., Breiter, R., Cabanski, W., Lutz, H., Wendler, J., Ziegler, J., Rehm, R., and Walther, M., "Multi Spectral IR Detection Modules and Applications," *Proceedings of SPIE* 6206, 620627, 2006. With permission.)

Table 23.6: Key Characteristics of the 384 × 288 Dual Color Superlattice IR-Module

Technology	Antimonide Type II Superlattice, CMOC MUX
Spectral bands	Blue band: 3.4–4.0 μm Red band: 4.0–5.0 μm with temporal coincident integration in both spectral bands
Elements	388 × 284 × 2; 40 μm pitch
Element size	38 μm
Fill factor	>80% for both spectral bands
Biasing	Individually for both bands
Integration capacity for 3–4 μm (in Mio e⁻)	1,2/6 (two gain stages) ± 10%
Integration capacity for 4–5 μm (in Mio e⁻)	7/19 (two gain stages) ± 10%
Operability of elements	>98%
Readout modes	Snapshot, stare then scan, temporal coincidence of signal in both bands
Read put models	Snapshot, stare then scan, temporal signal coincidence in both bands
Subframes	Arbitrary in steps of 8
Outputs	4 analog outputs for each color
Data rate digital	80 MHz serial high speed link interface
Full frame rate	150 Hz @ t_{int} = 2 ms
Environmental temperature range	–54°C to + 71°C
Vibration	MIL-STD-810F
IDCA	1 or 1.5 W linear split Stirling cooler or 0.7 W integral Stirling cooler
Weight of IDCA includes electronics	>2.5 kg for split linear Stirling cooler approximately 1 kg with integral cooler

Source: M. Münzberg, R. Breiter, W. Cabanski, K. Hofmann, H. Lutz, J. Wendler, J. Ziegler, P. Rehm, and M. Walther, "Dual Color IR Detection Modules, Trends and Applications," *Proceedings of SPIE* 6542, 654207, 2007. With permission.

Figure 23.34 Bispectral infrared image of an industrial site taken with a 384 × 288 dual-color InAs/GaSb SL camera. The two color channels 3–4 μm and 4–5 μm are represented by the complementary colors cyan and red, respectively. (From Rutz, F., Rehm, R., Schmitz, J., Fleissner, J., Walther, M., Scheibner, R., and Ziegler, J., "InAs/GaSb Superlattice Focal Plane Array Infrared Detectors: Manufacturing Aspects," *Proceedings of SPIE* 7298, 72981R, 2009. With permission.)

$f/2$ optics, 2.8 ms integration time, and 73 K detector temperature, the superlattice camera achieves an NEDT of 29.5 mK for the blue channel (3.4 μm ≤ λ ≤ 4.1 μm) and 16.5 mK for the red channel (4.1 μm ≤ λ ≤ 5.1 μm). An overview of the figures of merit is given in Table 23.6. As an example, the excellent imagery delivered by the 288 × 384 InAs/GaSb dual-color camera is presented in Figure 23.34. The image is a superposition of the images of the two channels coded in the complimentary colors cyan and red for the detection ranges of 3–4 μm and 4–5 μm, respectively. The red signatures reveal hot CO_2 emissions in the scene, whereas water vapor (e.g., from steam exhausts or in clouds) appear blue due to the frequency dependency of the Rayleigh scattering coefficient.

As one of the first representatives of a third generation system, the dual-color SLS technology is commercialized in a missile approach warning system. A square design of 256 × 256 FPA with a reduced pitch of 30 × 30 μm² is currently in development [100]. The reduction of pixel size can be achieved by using only two indium bumps per pixel. These very promising results confirm that the antimonide SL technology is now a direct competitor to MBE HgCdTe dual color technology.

23.6 MULTIBAND QDIPs

QDIP devices capable of detecting several separate wavelengths can be fabricated by vertical stacking of the different QWIP layers during epitaxial growth. The schematic structure is shown in Figure 23.35. In the case of structure describing by Lu et al. [103] each of QDIP absorption band consists of 10-period of InAs/InGaAs QD layers sandwiched between the top and bottom electrodes. Figure 23.36 shows the simplified band diagram of this structure at different bias levels. The bias voltage selection of detection bands originates from the asymmetric band structure. At low bias voltage, the high energy GaAs barrier blocks the photocurrent generated by LWIR radiation and only responds to the MWIR incidence. On the contrary, as the bias voltage increases, the barrier energy decreases, allowing LWIR signals to be detected at different bias voltage levels.

The first two-color quantum dot FPA demonstration was based on a voltage-tunable InAs/ InGaAs/GaAs DWELL structure [31,32]. As was described in Section 18.1, in this type of structure, InAs QDs are placed in an InGaAs well, which in turn is placed in a GaAs matrix (see Figure 18.4).

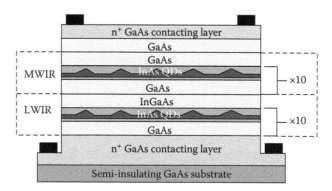

Figure 23.35 Schematic structure of the multispectral QDIP device. (From Lu, X., Vaillancourt, J., and Meisner, M., "A Voltage-Tunable Multiband Quantum Dot Infrared Focal Plane Array with High Photoconductivity," *Proceedings of SPIE* 6542, 65420Q, 2007. With permission.)

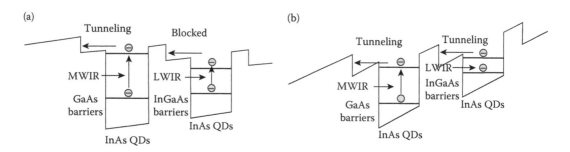

Figure 23.36 Simplified band diagram of the structure shown in Figure 23.35 at different bias levels; (a) low and (b) higher bias voltages. (From Lu, X., Vaillancourt, J., and Meisner, M., "A Voltage-Tunable Multiband Quantum Dot Infrared Focal Plane Array with High Photoconductivity," *Proceedings of SPIE* 6542, 65420Q, 2007. With permission.)

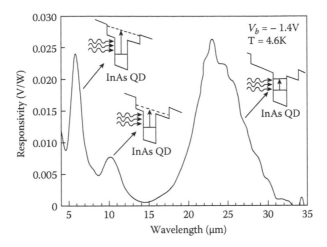

Figure 23.37 Multicolor response from a InAs/In$_{0.15}$Ga$_{0.85}$As/GaAs DWEL detector. The MWIR (LWIR) peak is possibly a transition from a state in the dot to a higher (lower) lying state in the well whereas the VLWIR response is possibly from two quantum-confined levels within the QD. This response is visible to 80 K. (From Krishna, S., *Journal of Physics D: Applied Physics*, 38, 2142–50, 2005. With permission.)

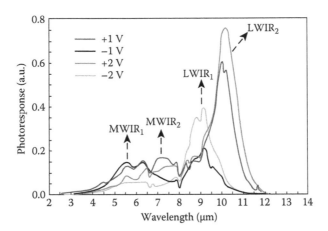

Figure 23.38 Spectral response from a DWELL detector with response at $V_b = +/-1$ V and $+/-2$ V. Note the response in the two MWIR and LWIR bands can be measured using this detector. The relative intensities of the bands can be altered by the applied bias. (From Krishna, S., Gunapala, S. D., Bandara, S. V., Hill, C., and Ting, D. Z., "Quantum Dot Based Infrared Focal Plane Arrays," *Proceedings of IEEE* 95, 1838–52, 2007. With permission.)

Figure 23.37 shows the multicolor response from a DWELL detector [104]. This device has demonstrated multicolor response ranging from the MWIR (3–5 μm) based on a bound-to-continuum transition to the LWIR (8–12 μm), which is based on a bound state in the dot to a bound state in the well. A very long wavelength response (VLWIR) has also been observed and has been attributed to transitions between two bound states in the QDs, since the calculated energy spacing between the dot levels is about 50–60 meV. Moreover, by adjusting the voltage bias of the device, it is possible to modify the ratio of electrons promoted by MWIR, LWIR, and VLWIR absorptions. Typically the MWIR response dominates at low to nominal voltages due to higher escape probability. With increasing voltage, the LWIR and eventually VLWIR responses are enhanced due to the increased tunneling probability of lower states in the DWELL detector (see Figure 23.38) [34]. The bias-dependent shift of the spectral response is observed due to quantum-confined

Stark effect. This voltage-control of spectral response can be exploited to realize spectrally smart sensors whose wavelength and bandwidth can be tuned depending on the desired application [31,104–106].

Typically, the detector structure consists of a 15-stack asymmetric DWELL structure sandwiched between two highly doped n-GaAs contact layers. The DWELL region consists of a 2.2 ML of n-doped InAs QDs in an $In_{0.15}Ga_{0.85}As$ well, itself placed within a GaAs matrix. By varying the width of the bottom InGaAs well from 10 to 60 Å, the operating wavelength of the detector can be changed from 7.2 to 11 μm. The responsivity and detectivity obtained from the test devices at 78 K are shown in Figure 23.39 [32]. The measured detectivities were 2.6×10^{10} cmHz$^{1/2}$/W ($V_b = 2.6$ V) for the LWIR band, and 7.1×10^{10} cmHz$^{1/2}$/W ($V_b = 1$ V) for the MWIR band.

Recently, Varley et al. [33] have demonstrated a two-color, MWIR/LWIR, 320 × 256 FPA based on DWELL detectors. Minimum NEDT values of 55 mK (MWIR) and 70 mK (LWIR) were measured (see Figure 23.40).

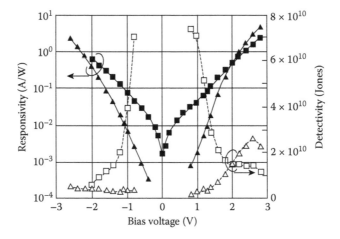

Figure 23.39 Peak responsivity for a 15 stack DWELL detector at 78 K obtained using a calibrated blackbody source. Solid squares: MWIR responsivity; solid triangles: LWIR responsivity; open square: MWIR detectivity; open triangles: LWIR detectivity. (From Krishna, S., Forman, D., Annamalai, S., Dowd, P., Varangis, P., Tumolillo, T., et al., *Physica Status Solidi (c)*, 3, 439–43, 2006. With permission.)

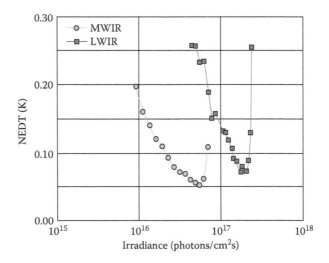

Figure 23.40 NEDT in the MWIR and LWIR bands at 77 K. Irradiance levels for MWIR and LWIR are 3–5 μm (f/2) and 8–12 μm (f/2.3), respectively. (From Varley, E., Lenz, M., Lee, S. J., Brown, J. S., Ramirez, D. A., Stintz, A., and Krishna, S., *Applied Physics Letters* 91, 081120, 2007. With permission.)

REFERENCES

1. P. Norton, J. Campbell, S. Horn, and D. Reago, "Third-Generation Infrared Imagers," *Proceedings of SPIE* 4130, 226–36, 2000.

2. M. Z. Tidrow, W. A. Beck, W. W. Clark, H. K. Pollehn, J. W. Little, N. K. Dhar, P. R. Leavitt, et al., "Device Physics and Focal Plane Applications of QWIP and MCT," *Opto-Electrononics Review* 7, 283–96, 1999.

3. M. N. Abedin, T. F. Refaat, I. Bhat, Y. Xiao, S. Bandara, and S. D. Gunapala, "Progress of Multicolor Single Detector to Detector Array Development for Remote Sensing," *Proceedings of SPIE* 5543, 239–47, 2004.

4. P. McCarley, "Recent Developments in Biologically Inspired Seeker Technology," *Proceedings of SPIE* 4288, 1–12, 2001.

5. J. T. Caulfield, "Next Generation IR Focal Plane Arrays and Applications," *Proceedings of 32nd Applied Imagery Pattern Recognition Workshop*, IEEE, New York, 2003.

6. E. P. G. Smith, L. T. Pham, G. M. Venzor, E. M. Norton, M. D. Newton, P. M. Goetz, V. K. Randall, et al.,"HgCdTe Focal Plane Arrays for Dual-Color Mid- and Long-Wavelength Infrared Detection," *Journal of Electronic Materials* 33, 509–16, 2004.

7. W. A. Radford, E. A. Patten, D. F. King, G. K. Pierce, J. Vodicka, P. Goetz, G. Venzor, et al., "Third Generation FPA Development Status at Raytheon Vision Systems," *Proceedings of SPIE* 5783, 331–39, 2005.

8. A. Rogalski, "HgCdTe Infrared Detector Material: History, Status, and Outlook," *Reports on Progress in Physics* 68, 2267–336, 2005.

9. D. F. King, W. A. Radford, E. A. Patten, R. W. Graham, T. F. McEwan, J. G. Vodicka, R. F. Bornfreund, P. M. Goetz, G. M. Venzor, and S. M. Johnson, "3rd-Generation 1280 × 720 FPA Development Status at Raytheon Vision Systems," *Proceedings of SPIE* 6206, 62060W, 2006.

10. G. Destefanis, P. Ballet, J. Baylet, P. Castelein, O. Gravrand, J. Rothman, F. Rothan, et al., "Bi-Color and Dual-Band HgCdTe Infrared Focal Plane Arrays at DEFIR," *Proceedings of SPIE* 6206, 62060R, 2006.

11. G. Destefanis, J. Baylet, P. Ballet, P. Castelein, F. Rothan, O. Gravrand, J. Rothman, J. P. Chamonal, and A. Million, "Status of HgCdTe Bicolor and Dual-Band Infrared Focal Plane Arrays at LETI," *Journal of Electronic Materials* 36, 1031–44, 2007.

12. P. D. Dreiske, "Development of Two-Color Focal-Plane Arrays Based on HDVIP," *Proceedings of SPIE* 5783, 325–30, 2005.

13. N. T. Gordon, P. Abbott, J. Giess, A. Graham, J. E. Hails, D. J. Hall, L. Hipwood, C. L. Lones, C. D. Maxeh, and J. Price, "Design and Assessment of Metal-Organic Vapour Phase Epitaxy–Grown Dual Wavelength Infrared Detectors," *Journal of Electronic Materials* 36, 931–36, 2007.

14. S. D. Gunapala and S. V. Bandara, "GaAs/AlGaAs Based Quantum Well Infrared Photodetector Focal Plane Arrays," in *Handbook of Infrared Detection Technologies*, eds. M. Henini and M. Razeghi, 83–119, Elsevier, Oxford, 2002.

15. G. Sarusi, "QWIP or Other Alternatives for Third Generation Infrared Systems," *Infrared Physics & Technology* 44, 439–44, 2003.

16. A. Rogalski, "Quantum Well Photoconductors in Infrared Detectors Technology," *Journal of Applied Physics* **93**, 4355–4391 (2003).

17. A. Manissadjian, D. Gohier, E. Costard, and A. Nedelcu, "Single Color and Dual Band QWIP Production Results," *Proceedings of SPIE* 6206, 62060E, 2006.

18. M. Münzberg, R. Breiter, W. Cabanski, H. Lutz, J. Wendler, J. Ziegler, R. Rehm, and M. Walther, "Multi Spectral IR Detection Modules and Applications," *Proceedings of SPIE* 6206, 620627, 2006.

19. S. D. Gunapala, S. V. Bandara, J. K. Liu, J. M. Mumolo, C. J. Hill, D. Z. Ting, E. Kurth, J. Woolaway, P. D. LeVan, and M. Z. Tidrow, "Towards 16 Megapixel Focal Plane Arrays," *Proceedings of SPIE* 6660, 66600E, 2007.

20. S. D. Gunapala, S. V. Bandara, J. K. Liu, J. M. Mumolo, C. J. Hill, S. B. Rafol, D. Salazar, J. Woollaway, P. D. LeVan, and M. Z. Tidrow, "Towards Dualband Megapixel QWIP Focal Plane Arrays," *Infrared Physics & Technology* 50, 217–26, 2007.

21. A. Nedelcu, E. Costard, P. Bois, and X. Marcadet, "Research Topics at Tales Research and Technology: Small Pixels and Third Generation Applications," *Infrared Physics & Technology* 50, 227–33, 2007.

22. A. Rogalski, "Third Generation Photon Detectors," *Optical Engineering* 42, 3498–3516, 2003.

23. R. Rehm, M. Walther, J. Schmitz, J. Fleißner, F. Fuchs, J. Ziegler, and W. Cabanski, "InAs/GaSb Superlattice Focal Plane Arrays for High-Resolution Thermal Imaging," *Opto-Electronics Review* 14, 283–96, 2006.

24. A. Rogalski and P. Martyniuk, "InAs/GaInSb Superlattices as a Promising Material System for Third Generation Infrared Detectors," *Infrared Physics & Technology* 48, 39–52, 2006.

25. R. Rehm, M. Walther, J. Schmitz, J. Fleißner, J. Ziegler, W. Cabanski, and R. Breiter, "Bispectral Thermal Imaging with Quantum-Well Infrared Photodetectors and InAs/GaSb Type II Superlattices," *Proceedings of SPIE* 6292, 629404, 2006.

26. A. Rogalski, "Competitive Technologies of Third Generation Infrared Photon Detectors," *Opto-Electronics Review* 14, 87–101, 2006.

27. A. Rogalski, "New Material Systems for Third Generation Infrared Photodetectors," *Opto-Electronics Review* 16, 458–82, 2008.

28. A. Rogalski, J. Antoszewski, and L. Faraone, "Third-Generation Infrared Photodetector Arrays," *Journal of Applied Physics* 105, 091101-44, 2009.

29. S. M. Kim and J. S. Harris, "Multicolor InGaAs Quantum-Dot Infrared Photodetectors," *IEEE Photonics Technology Letters* 16, 2538–40, 2004.

30. S. Chakrabarti, X. H. Su, P. Bhattacharya, G. Ariyawansa, and A. G. U. Perera, "Characteristics of a Multicolour InGaAs-GaAs Quantum-Dot Infrared Photodetector," *IEEE Photonics Technology Letters* 17, 178–80, 2005.

31. S. Krishna, D. Forman, S. Annamalai, P. Dowd, P. Varangis, T. Tumolillo, A. Gray, et al., "Demonstration of a 320 × 256 Two-Color Focal Plane Array Using InAs/InGaAs Quantum Dots in Well Detectors," *Applied Physics Letters* 86, 193501, 2005.

32. S. Krishna, D. Forman, S. Annamalai, P. Dowd, P. Varangis, T. Tumolillo, et al., "Two-Color Focal Plane Arrays Based on Self Assembled Quantum Dots in a Well Heterostructure," *Physica Status Solidi (c)* 3, 439–43, 2006.

33. E. Varley, M. Lenz, S. J. Lee, J. S. Brown, D. A. Ramirez, A. Stintz, and S. Krishna, "Single Bump, Two-Color Quantum Dot Camera," *Applied Physics Letters* 91, 081120, 2007.

34. S. Krishna, S. D. Gunapala, S. V. Bandara, C. Hill, and D. Z. Ting, "Quantum Dot Based Infrared Focal Plane Arrays," *Proceedings of IEEE* 95, 1838–52, 2007.

35. M. D. Morris, *Microscopic and Spectroscopic Imaging of the Chemical State*, Marcel Dekker, New York, 1993.

36. C. D. Tran, "Principles, Instrumentation, and Applications of Infrared Multispectral Imaging, An Overview," *Analytical Letters* 38, 735–52, 2005.

37. W. J. Gunning, J. DeNatale, P. Stupar, R. Borwick, R. Dannenberg, R. Sczupak, and P. O. Pettersson, "Adaptive Focal Plane Array: An Example of MEMS, Photonics, and Electronics Integration," *Proceedings of SPIE* 5783, 336–75, 2005.

38. D. Reago, S. Horn, J. Campbell, and R. Vollmerhausen, "Third Generation Imaging Sensor System Concepts," *Proceedings of SPIE*, 3701, 108–17, 1999.

39. S. Horn, P. Norton, K. Carson, R. Eden, and R. Clement, "Vertically-Integrated Sensor Arrays – VISA," *Proceedings of SPIE* 5406, 332–40, 2004.

40. R. Balcerak and S. Horn, "Progress in the Development of Vertically-Integrated Sensor Arrays," *Proceedings of SPIE* 5783, 384–91, 2005.

41. P. R. Norton, "Third-Generation Sensors for Night Vision," *Opto-Electronics Review* 14, 283–96, 2006.

42. P. R. Norton, "Infrared Detectors in the Next Millennium," *Proceedings of SPIE* 3698, 652–65, 1999.

43. E. P. G. Smith, G. M. Venzor, Y. Petraitis, M. V. Liguori, A. R. Levy, C. K. Rabkin, J. M. Peterson, M. Reddy, S. M. Johnson, and J. W. Bangs, "Fabrication and Characterization of Small Unit-Cell Molecular Beam Epitaxy Grown HgCdTe-on-Si Mid-Wavelength Infrared Detectors," *Journal of Electronic Materials* 36, 1045–51, 2007.

44. S. Horn, P. Norton, T. Cincotta, A. Stolz, D. Benson, P. Perconti, and J. Campbell, "Challenges for Third-Generation Cooled Imagers," *Proceedings of SPIE* 5074, 44–51, 2003.

45. H. Halpert and B. I. Musicant, "N-Color (Hg,Cd)Te Photodetectors," *Applied Optics* 11, 2157–61, 1972.

46. InSb/HgCdTe Two-Color Detector, http://www.irassociates.com/insbhgcdte.htm

47. J. C. Campbell, A. G. Dentai, T. P. Lee, and C. A. Burrus, "Improved Two-Wavelength Demultiplexing InGaAsP Photodetector," *IEEE Journal of Quantum Electronics* QE-16, 601, 1980.

48. E. R. Blazejewski, J. M. Arias, G. M. Williams, W. McLevige, M. Zandian, and J. Pasko, "Bias-Switchable Dual-Band HgCdTe Infrared Photodetector." *Journal of Vacuum Science and Technology* B10, 1626, 1992.

49. J. A. Wilson, E. A. Patten, G. R. Chapman, K. Kosai, B. Baumgratz, P. Goetz, S. Tighe, et al., "Integrated Two-Color Detection for Advanced FPA Applications," *Proceedings of SPIE* 2274, 117–25, 1994.

50. R. D. Rajavel, D. M. Jamba, J. E. Jensen, O. K. Wu, J. A. Wilson, J. L. Johnson, E. A. Patten, K. Kasai, P. M. Goetz, and S. M. Johnson, "Molecular Beam Epitaxial Growth and Performance of HgCdTe-Based Simultaneous-Mode Two-Color Detectors," *Journal of Electronic Materials* 27, 747–51, 1998.

51. E. P. G. Smith, E. A. Patten, P. M. Goetz, G. M. Venzor, J. A. Roth, B. Z. Nosho, J. D. Benson, et al., "Fabrication and Characterization of Two-Color Midwavelength/Long Wavelength HgCdTe Infrared Detectors," *Journal of Electronic Materials* 35, 1145–52, 2006.

52. D. F. King, J. S. Graham, A. M. Kennedy, R. N. Mullins, J. C. McQuitty, W. A. Radford, T. J. Kostrzewa, et al., "3rd-Generation MW/LWIR Sensor Engine for Advanced Tactical Systems," *Proceedings of SPIE* 6940, 69402R, 2008.

53. E. P. G. Smith, A. M. Gallagher, T. J. Kostrzewa, M. L. Brest, R. W. Graham, C. L. Kuzen, E. T. Hughes, et al., "Large Format HgCdTe Focal Plane Arrays for Dual-Band Long-Wavelength Infrared Detection," *Proceedings of SPIE* 7298, 72981Y, 2009.

54. M. B. Reine, A. Hairston, P. O'Dette, S. P. Tobin, F. T. J. Smith, B. L. Musicant, P. Mitra, and F. C. Case, "Simultaneous MW/LW Dual-Band MOCVD HgCdTe 64×64 FPAs," *Proceedings of SPIE* 3379, 200–12, 1998.

55. J. P. Zanatta, P. Ferret, R. Loyer, G. Petroz, S. Cremer, J. P. Chamonal, P. Bouchut, A. Million, and G. Destefanis, "Single and Two Colour Infrared Focal Plane Arrays Made by MBE in HgCdTe," *Proceedings of SPIE* 4130, 441–51, 2000.

56. P. Tribolet. M. Vuillermet, and G. Destefanis, "The Third Generation Cooled IR Detector Approach in France," *Proceedings of SPIE* 5964, 49–60, 2005.

57. J. P. Zanatta, G. Badano, P. Ballet, C. Largeron, J. Baylet, O. Gravrand, J. Rothman, et al., "Molecular Beam Epitaxy of HgCdTe on Ge for Third-Generation Infrared Detectors," *Journal of Electronic Materials* 35, 1231–36, 2006.

58. P. Tribolet, G. Destefanis, P. Ballet, J. Baylet, O. Gravrand, and J. Rothman, "Advanced HgCdTe Technologies and Dual-Band Developments," *Proceedings of SPIE* 6940, 69402P, 2008.

59. J. Giess, M. A. Glover, N. T. Gordon, A. Graham, M. K. Haigh, J. E. Hails, D. J. Hall, and D. J. Lees, "Dial-Wavelength Infrared Focal Plane Arrays Using MCT Grown by MOVPE on Silicon Substrates," *Proceedings of SPIE* 5783, 316–24, 2005.

60. N. T. Gordon, P. Abbott, J. Giess, A. Graham, J. E. Hails, D. J. Hall, L. Hipwood, C. L. Lones, C. D. Maxeh, and J. Price, "Design and Assessment of Metal-Organic Vapour Phase Epitaxy–Grown Dual Wavelength Infrared Detectors," *Journal of Electronic Materials* 36, 931–36, 2007.

61. C. L. Jones, L. G. Hipwood, J. Price, C. J. Shaw, P. Abbott, C. D. Maxey, H. W. Lau, et al., "Multi-Colour IRFPAs Made from HgCdTe Grown by MOVPE," *Proceedings of SPIE* 6542, 654210, 2007.

62. J. P. G. Price, C. L. Jones, L. G. Hipwood, C. J. Shaw, P. Abbott, C. D. Maxey, H. W. Lau, et al., "Dual-Band MW/LW IRFPAs Made from HgCdTe Grown by MOVPE," *Proceedings of SPIE* 6940, 69402S, 2008.

63. F. Aqariden, P. D. Dreiske, M. A. Kinch, P. K. Liao, T. Murphy, H. F. Schaake, T. A. Shafer, H. D. Shih, and T. H. Teherant, "Development of Molecular Beam Epitaxy Grown $Hg_{1-x}Cd_xTe$ for High-Density Vertically-Integrated Photodiode-Based Focal Plane Arrays," *Journal of Electronic Materials* 36, 900–904, 2007.

64. M. A. Kinch, "HDVIP™ FPA Technology at DRS," *Proceedings of SPIE* 4369, 566–78, 2001.

65. M. A. Kinch, *Fundamentals of Infrared Detector Materials*, SPIE Press, Bellingham, 2007.

66. W.E. Tennant, M. Thomas, L.J. Kozlowski, W.V. McLevige, D.D. Edwall, M. Zandian, K. Spariosu, et al., "A Novel Simultaneous Unipolar Multispectral Integrated Technology Approach for HgCdTe IR Detectors and Focal Plane Arrays," *Journal of Electronic Materials* 30, 590–94, 2001.

67. L. A. Almeida, M. Thomas, W. Larsen, K. Spariosu, D. D. Edwall, J. D. Benson, W. Mason, A. J. Stoltz, and J. H. Dinan, "Development and Fabrication of Two-Color Mid- and Short-Wavelength Infrared Simultaneous Unipolar Multispectral Integrated Technology Focal-Plane Arrays," *Journal of Electronic Materials* 30, 669–76, 2002.

68. A. H. Lockwood, J. R. Balon, P. S. Chia, and F. J. Renda, "Two-Color Detector Arrays by PbTe/Pb$_{0.8}$Sn$_{0.2}$Te Liquid Phase Epitaxy," *Infrared Physics* 16, 509–14, 1976.

69. J. Baylet, P. Ballet, P. Castelein, F. Rothan, O. Gravrand, M. Fendler, E. Laffosse, et al., "TV/4 Dual-Band HgCdTe Infrared Focal Plane Arrays with a 25-μm Pitch and Spatial Coherence," *Journal of Electronic Materials* 35, 1153–58, 2006.

70. P. R. Norton, "Status of Infrared Detectors," *Proceedings of SPIE* 3379, 102–14, 1998.

71. E. P. G. Smith, R. E. Bornfreund, I. Kasai, L. T. Pham, E. A. Patten, J. M. Peterson, J. A. Roth, et al., "Status of Two-Color and Large Format HgCdTe FPA Technology at Reytheon Vision Systems," *Proceedings of SPIE* 6127, 61271F, 2006.

72. K. Jóźwikowski and A. Rogalski, "Computer Modeling of Dual-Band HgCdTe Photovoltaic Detectors," *Journal of Applied Physics* 90, 1286–91, 2001.

73. A. K. Sood, J. E. Egerton, Y. R. Puri, E. Bellotti, D. D'Orsogna, L. Becker, R. Balcerak, K. Freyvogel, and R. Richwine, "Design and Development of Multicolor MWIR/LWIR and LWIR/VLWIR Detector Arrays," *Journal of Electronic Materials* 34, 909–12, 2005.

74. E. Bellotti and D. D'Orsogna, "Numerical Analysis of HgCdTe Simultaneous Two-Color Photovoltaic Infrared Detectors," *IEEE Journal of Quantum Electronics* 42, 418–26, 2006.

75. L. G. Hipwood, C. L. Jones, C. D. Maxey, H. W. Lau, J. Fitzmaurice, R. A. Catchpole, and M. Ordish, "Three-Color MOVPE MCT Diodes," *Proceedings of SPIE* 6206, 620612, 2006.

76. K. Jóźwikowski and A. Rogalski, "Numerical Analysis of Three-Colour HgCdTe Detectors," *Opto-Electronics Review* 15, 215–22, 2007.

77. W. A. Beck and T. S. Faska, "Current Status of Quantum Well Focal Plane Arrays," *Proceedings of SPIE* 2744, 193–206, 1996.

78. M. Sundaram and S. C. Wang, "2-Color QWIP FPAs," *Proceedings of SPIE* 4028, 311–17, 2000.

79. S. D. Gunapala, S. V. Bandara, A. Sigh, J. K. Liu, S. B. Rafol, E. M. Luong, J. M. Mumolo, et al., "8–9 and 14–15 μm Two-Color 640 × 486 Quantum Well Infrared Photodetector (QWIP) Focal Plane Array Camera," *Proceedings of SPIE* 3698, 687–97, 1999.

80. S. D. Gunapala, S. V. Bandara, A. Singh, J. K. Liu, B. Rafol, E. M. Luong, J. M. Mumolo, et al., "640 × 486 Long-Wavelength Two-Color GaAs/AlGaAs Quantum Well Infrared Photodetector (QWIP) Focal Plane Array Camera," *IEEE Transactions on Electron Devices* 47, 963–71, 2000.

81. S. D. Gunapala, S. V. Bandara, J. K. Liu, E. M. Luong, S. B. Rafol, J. M. Mumolo, D. Z. Ting, et al., "Recent Developments and Applications of Quantum Well Infrared Photodetector Focal Plane Arrays," *Opto-Electronics Review* 8, 150–63, 2001.

82. S. D. Gunapala, S. V. Bandara, J. K. Liu, B. Rafol, J. M. Mumolo, C. A. Shott, R. Jones, et al., "640 × 512 Pixel Narrow-Band, Four-Band, and Broad-Band Quantum Well Infrared Photodetector Focal Plane Arrays," *Infrared Physics & Technology* 44, 411–25, 2003.

83. S. D. Gunapala, S. V. Bandara, J. K. Liu, B. Rafol, and J. M. Mumolo, "640 × 512 Pixel Long-Wavelength Infrared Narrowband, Multiband, and Broadband QWIP Focal Plane Arrays," *IEEE Transactions on Electron Devices* 50, 2353–60, 2003.

84. S. D. Gunapala, S. V. Bandara, J. K. Liu, J. M. Mumolo, C. J. Hill, D. Z. Ting, E. Kurth, J. Woolaway, P. D. LeVan, and M. Z. Tidrow, "Development of Megapixel Dual-Band QWIP Focal Plane Array," *Proceedings of SPIE* 6940, 69402T, 2008.

85. A. Soibel, S. D. Gunapala, S. V. Bandara, J. K. Liu, J. M. Mumolo, D. Z. Ting, C. J. Hill, and J. Nguyen, "Large Format Multicolor QWIP Focal Plane Arrays," *Proceedings of SPIE* 7298, 729806, 2009.

86. A. Goldberg, T. Fischer, J. Kennerly, S. Wang, M. Sundaram, P. Uppal, M. Winn, G. Milne, and M. Stevens, "Dual Band QWIP MWIR/LWIR Focal Plane Array Test Results," *Proceedings of SPIE* 4028, 276–87, 2000.

87. A. C. Goldberger, S. W. Kennerly, J. W. Little, H. K. Pollehn, T. A. Shafer, C. L. Mears, H. F. Schaake, M. Winn, M. Taylor, and P. N. Uppal, "Comparison of HgCdTe and QWIP Dual-Band Focal Plane Arrays," *Proceedings of SPIE* 4369, 532–46, 2001.

88. K.-K. Choi, M. D. Jhabvala, and R. J. Peralta, "Voltage-Tunable Two-Color Corrugated-QWIP Focal Plane Arrays," *IEEE Electron Device Letters* 29, 1011–13, 2008.

89. M. Jhabvala, "Applications of GaAs Quantum Well Infrared Photoconductors at the NASA/ Goddard Space Flight Center," *Infrared Physics & Technology* 42, 363–76, 2001.

90. E. Costard, Ph. Bois, X. Marcadet, and A. Nedelcu, "QWIP and Third Generation IR Imagers," *Proceedings of SPIE* 5783, 728–35, 2005.

91. P. Castelein, F. Guellec, F. Rothan, S. Martin, P. Bois, E. Costard, O. Huet, X. Marcadet, and A. Nedelcu, "Demonstration of 256 × 256 Dual-Band QWIP Infrared FPAs," *Proceedings of SPIE* 5783, 804–15, 2005.

92. N. Perrin, E. Belhaire, P. Marquet, V. Besnard, E. Costard, A. Nedelcu, P. Bois, et al., "QWIP Development Status at Thales," *Proceedings of SPIE* 6940, 694008, 2008.

93. J. A. Robo, E. Costard, J. P. Truffer, A. Nedelcu, X. Marcadet, and P. Bois, "QWIP Focal Plane Arrays Performances from MWIR to VLWIR," *Proceedings of SPIE* 7298, 72980F, 2009.

94. W. Cabanski, M. Münzberg, W. Rode, J. Wendler, J. Ziegler, J. Fleißner, F. Fuchs, et al., "Third Generation Focal Plane Array IR Detection Modules and Applications," *Proceedings of SPIE* 5783, 340–49, 2005.

95. H. Schneider, T. Maier, J. Fleissner, M. Walther, P. Koidl, G. Weimann, W. Cabanski, et al., "Dual-Band QWIP Focal Plane Array for the Second and Third Atmospheric Windows," *Infrared Physics & Technology* 47, 53–58, 2005.

96. S. Gunapala, "Megapixel QWIPs Deliver Multi-Color Performance," *Compound Semiconductor*, 10, 25–28, October 2005.

97. P. Bois, E. Costard, X. Marcadet, and E. Herniou, "Development of Quantum Well Infrared Photodetectors in France," *Infrared Physics & Technology* 42, 291–300, 2001.

98. http://www.qumagiq.com

99. E. H. Aifer, J. G. Tischler, J. H. Warner, I. Vurgaftman, and J. R. Meyer, "Dual Band LWIR/ VLWIR Type-II Superlattice Photodiodes," *Proceedings of SPIE* 5783, 112–22, 2005.

100. M. Münzberg, R. Breiter, W. Cabanski, K. Hofmann, H. Lutz, J. Wendler, J. Ziegler, P. Rehm, and M. Walther, "Dual Color IR Detection Modules, Trends and Applications," *Proceedings of SPIE* 6542, 654207, 2007.

101. F. Rutz, R. Rehm, J. Schmitz, J. Fleissner, M. Walther, R. Scheibner, and J. Ziegler, "InAs/GaSb Superlattice Focal Plane Array Infrared Detectors: Manufacturing Aspects," *Proceedings of SPIE* 7298, 72981R, 2009.

102. M. Razeghi, D. Hoffman, B. M. Nguyen, P.-Y. Delaunay, E. K. Huang, M. Z. Tidrow, and V. Nathan, "Recent Advances in LWIR Type-II InAs/GaSb Superlattice Photodetectors and Focal Plane Arrays at the Center for Quantum Devices," *Proceedings of IEEE* 97, 1056–66, 2009.

103. X. Lu, J. Vaillancourt, and M. Meisner, "A Voltage-Tunable Multiband Quantum Dot Infrared Focal Plane Array with High Photoconductivity," *Proceedings of SPIE* 6542, 65420Q, 2007.

104. S. Krishna, "Quantum Dots-in-a-Well Infrared Photodetectors," *Journal of Physics D: Applied Physics* 38, 2142–50, 2005.

105. U. Sakoglu, J. S. Tyo, M. M. Hayat, S. Raghavan, and S. Krishna, "Spectrally Adaptive Infrared Photodetectors with Bias-Tunable Quantum Dots," *Journal of the Optical Society of America B* 21, 7–17, 2004.

106. A. G. U. Perera, "Quantum Structures for Multiband Photon Detection," *Opto-Electronics Review* 14, 99–108, 2006.

Final Remarks

The future applications of IR detector systems require:

- Higher pixel sensitivity
- Further increase in pixel density
- Cost reduction in IR imaging array systems due to less cooling sensor technology combined with integration of detectors and signal processing functions (with much more on-chip signal processing)
- Photon counting in the wide spectral range (e.g., development of e-APD and h-APD HgCdTe photodiodes)
- Improvement in functionality of IR imaging arrays through development of multispectral sensors.

Array sizes will continue to increase but perhaps at a rate that falls below the Moore's Law curve. An increase in array size is already technically feasible. However, the market forces that have demanded larger arrays are not as strong now that the megapixel barrier has been broken.

There are many critical challenges for future civilian and military infrared detector applications. For many systems, such as night-vision goggles, the IR image is viewed by the human eye, which can discern resolution improvements only up to about one megapixel, roughly the same resolution as high-definition television. Most high-volume applications can be completely satisfied with a format of 1280 × 1024. Although wide-area surveillance and astronomy applications could make use of larger formats, funding limits may prevent the exponential growth that was seen in past decades.

Third generation infrared imagers are beginning a challenging road to development. For multiband sensors, boosting the sensitivity in order to maximize identification range is the primary objective. The goal for dual-band MW/LW IR FPAs are 1920 × 1080 pixels, which due to lower cost should be fabricated on silicon wafers. The challenges to attaining those specifications are material uniformity and defects, heterogeneous integration with silicon, and ultra well capacity (on the order of a billion in the LWIR).

It is predicted that HgCdTe technology will continue in the future to expand the envelope of its capabilities because of its excellent properties. Despite serious competition from alternative technologies and slower progress than expected, HgCdTe is unlikely to be seriously challenged for high-performance applications, applications requiring multispectral capability and fast response. However, the nonuniformity is a serious problem in the case of LWIR and VLWIR HgCdTe detectors. For applications that require operation in the LWIR band as well as two-color MWIR/LWIR/VLWIR bands, most probably HgCdTe will not be the optimal solution. Type II InAs/GaInSb superlattice structure is a relatively new alternative IR material system and has great potential for LWIR/VLWIR spectral ranges with performance comparable to HgCdTe with the same cutoff wavelength.

Based on the breakthrough of Sb-based type II SLS technology, it is obvious that this material system is in position to provide high thermal resolution for short integration times comparable to HgCdTe. The fact that Sb-based superlattices are processed close to standard III-V technology raises the potential to be more competitive due to lower costs in series production. The potential low cost compared to HgCdTe is that it can leverage investments in lasers and transistors in the Sb-based industry, and has potential commercial market applications in the future.

Near future high performance uncooled thermal imaging will be dominated by VO_x bolometers. However, their sensitivity limitations and the still significant prices will encourage many research teams to explore other IR sensing techniques with the potential for improved performance with reduced detector costs. Recent advances in MEMS systems have lead to the development of uncooled IR detectors operating as micromechanical thermal detectors. One such attractive approach is optically coupled cantilevers.

Although in an early stage of development, the potential to deliver FPAs that can adapt their spectral response to match the sensor requirements in real time, presents a compelling case for future multispectral IR imaging systems. Such systems have the potential to deliver a much-improved threat and target recognition capabilities for future defence combat systems.

The THz detectors will receive increasing importance in a very diverse range of applications including detection of biological and chemical hazardous agents, explosive detection, building and airport security, radio astronomy and space research, biology and medicine. The future sensitivity improvement of THz instruments will come with the use of large format arrays with readouts in the focal plane to provide the vision demands of high resolution spectroscopy.

Index

A

Absorption coefficients
    for extrinsic semiconductors, 55
    for low-dimensional solids, 55
    for photodetector materials, 54
Absorption edge image converters, 27, 167
Accelerated crucible rotation technique (ACRT), 372
Active pixel sensors (APSs), 657
Adaptive focal plane arrays (AFPA), 673–676
    dual-band, 674–675
    integrated test package view, 675
    measured spectral response of filter, 675–676
    MEMS-based tunable IR detector and, 673–674
    ROIC in, 675
Airy disk, 811
ALADDIN array, 736
Aliane and arsenic acid doped materials
    (ATGSAs), 145
AlInAs/InGaAs, as material system for QWIP, 584
Alkali halides IR transmission, 17
Amplifier in each pixel (APS), 658
Analog-to-digital (A/D) converter, 659
Anisotropic etch process, 100
Antenna-coupled bolometer, 121
Antenna-coupled microbolometers; *see also*
    Cantilevers
    configuration, 168–169
    electrical response, 169
    electromagnetic behavior, 169
    fabrication, 169
    pixels, 169
APD, *see* Avalanche photodiode (APD)
Application specific integrated circuits
    (ASICs), 657
APSs, *see* Active pixel sensors (APSs)
Arsenic doped amorphous selenium, 64
Artificial atoms, *see* Quantum dots (QDs)
ATGSAs, *see* Aliane and arsenic acid doped
    materials (ATGSAs)
Atmosphere, transmission in, 20
Auger-dominated photodetector performance
    equilibrium devices, 391
        BLIP operation in, 392
        calculated performance of, 393
        detectivity, 392
        detector performance, 392
        LN-cooled and SW devices, 392
        nondegenerate statistic, 392
        optimum thickness devices, 392
        wavelength and temperature dependence
            of, 392
    non-equilibrium devices, 392
        BLIP limit, 394
        calculated performance of, 393
        depleted materials, 393
        fundamental limits of, 393
        gain for p-type material, 394
        MIS structures, 393

Avalanche photodiode (APD), 216, 349, 437
    average avalanche gain, 220
    characteristics of, 220
    design of, 221
    electric field profile, 218
    electron-hole pair, 217–218
    electrons drift, 217
    energy band diagram, 218
    gain of device, 219
    ionization coefficients of electrons and holes, 219
    materials selection for, 221
    mean gain, 221
    multiplication process, 217
        representation of, 218
    noise factor and multiplication factor, 220
    n^+-p-π-p^+ Si avalanche photodiode
        characteristics of, 260
        cross section of structure, 260
        distribution of dopants, 260
        gain as function, 261
        parameters, 261
        properties of, 262
    PDF of
        electrons, 220
        ionization path lengths, 221
    per unit bandwidth, 219–220
    shot noise formula, 220
    Si, Ge, and InGaAs APD characteristics, 222
    structure, 217–218
AXT100 camera, 685
    features of, 686

B

Background limited infrared photodetector (BLIP),
    25, 35, 826
Bandgap alloys, 25
Band-to-band tunneling (BTB), 433
Bardeen–Cooper–Schrieffer theory, 117
Bardeen model of n-type semiconductor, 223
Barium strontium titinate (BST), 686
    ceramic, 150
    dielectric bolometer pixel, 700
Beam-splitter, 78
BIB, *see* Blocked impurity band (BIB)
Bimaterial microcantilever IR detector
    operating principle of, 159
Binary III-V detectors
    InAs detectors
        detectivity and wavelength, 333
        diffusion method, 331
        fabrication of, 334
        heterostructure immersion photodiode, 335
        ion implantation, 331
        performance of, 332
        response variation, 332
    InAs gate-controlled photodiode
        dark current, 334
        device structure, 333

849